£12.00

A Specialist Periodical Report

Photochemistry
Volume 2

A Review of the Literature Published between July 1969 and June 1970

Senior Reporter

D. Bryce-Smith, *Department of Chemistry, The University, Reading*

Reporters

A. Gilbert, *The University, Reading*

W. M. Horspool, *University of Dundee*

D. Phillips, *The University, Southampton*

SBN: 85186 015 X

© Copyright 1971

The Chemical Society
Burlington House, London, W1V 0BN

Organic formulae composed by Wright's Symbolset method

Printed in Great Britain by John Wright and Sons Ltd. at The Stonebridge Press, Bristol BS4 5NU

Contents

Introduction and Review of the Year   xi
By D. Bryce-Smith

Part I Physical Aspects of Photochemistry

Chapter 1 Spectroscopic and Theoretical Aspects
By D. Phillips

1 Absorption Spectra and Energy Level Calculations	4
2 Theories of Radiationless Transitions	18
Long Radiative Lifetimes of Small Molecules	20
Statistical Limit	20
Resonance Limit	20
The Dense Intermediate Case	21
The Sparse Intermediate Case	21
Statistical Limit—Radiative Decay of Polyatomic Molecules	25
Interference Effects in the Radiative Decay of Close Spaced Levels	26
3 Theories of Photochemical Reactions	37
Photoisomerizations	37
Predissociation	43
Energy Partitioning in Photochemical Reactions	45
Applications of Unimolecular Reaction-rate Theory to Photochemical Cases	48
Orbital Symmetry Considerations	50
4 Developments in Techniques	58

Chapter 2 Photophysical Processes in Condensed Phases
By D. Phillips

1 Fluorescence and Fluorescence Quenching	82
2 Concentration Quenching, Excimer Emission, and Delayed Fluorescence	109
3 Aspects of the Triplet State	113
Formation	113
Detection	117
Phosphorescence	118
Triplet Decay Times	121
Properties of Triplet States	127
Miscellaneous	132
4 Electronic Energy Transfer	132
Singlet–Singlet Transfer	133
Singlet–Triplet Transfer	134
Triplet–Triplet Transfer	135
5 Physical Aspects of Photochemical Reactions in Solution	141
6 Photochromic Compounds	145
Photochemical Reactions in the Solid State	146

Chapter 3 Gas-phase Photochemistry
By D. Phillips

1 Aromatic Molecules	154
2 Alkanes and Olefins	171
3 Carbonyl Compounds	183
4 Halogen-containing Compounds	199
5 Photochemistry of Nitrogen Compounds	205
6 Sulphur Compounds	215
7 Photochemistry of Oxygen and Ozone; Photo-oxidations	221
8 Atom Sensitizations	226

Part II Inorganic Photochemistry
By D. Phillips

1 Photochemistry of Water, H_2O_2, and Aqueous Anions — 235
2 Photochemistry and Photoluminescence of Transition-metal Co-ordination Complexes — 243
   Cobalt Complexes — 246
   Chromium Complexes — 251
   Iron Complexes — 259
   Plantinum Complexes — 262
   Manganese Compounds — 264
   Copper Complexes and Compounds — 266
   Miscellaneous — 269
   Lanthanide Compounds — 277
   Actinides — 279
3 Gas-phase Studies — 281
4 Solid-phase Luminescence and Photoreactions — 287

Part III Organic Aspects of Photochemistry

Chapter 1 Photolysis of Carbonyl Compounds
By W. M. Horspool

1 Energy Transfer Processes — 300
2 Norrish Type I Reactions — 303
3 Norrish Type II Processes (H-Abstraction Reactions: Cyclobutanol Formation) — 311
4 Rearrangement Reactions — 328
   Cyclopropane Rearrangements — 328
   Cyclobutane and Cyclopentane Rearrangements — 329
   Cyclohexane Rearrangements — 331
   Bicyclic and Spiro Systems — 333
5 Oxetan Formation — 334
6 Alkylation Reactions — 340
7 Ketone Fragmentation Reactions — 341
   Decarbonylations — 343
   Decarboxylations — 346
   Miscellaneous Fragmentations — 350

Contents

**Chapter 2 Enone Rearrangements and Cycloadditions:
Photoreactions of Cyclohexadienones, Tropones,
Quinones, etc.**
By W. M. Horspool

1 Enone Rearrangements — 355
α,β-Unsaturated Carbonyl Compounds — 355
β,γ-Unsaturated Carbonyl Compounds — 367
γ,δ-Unsaturated Carbonyl Compounds — 370
Bridged Unsaturated Carbonyl Compounds — 372
Unsaturated Nitriles, Acids and Esters — 375

2 Addition and Cycloaddition Reactions of Enones — 380
Cyclopentenone-type Cycloadditions — 380
Cyclohexanone-type Cycloadditions — 383

**3 Dimerization, Intramolecular Cycloaddition, and Some
Addition Reactions of Enones** — 389

4 Photoreactions of Thymine, etc. — 394

5 Photochemistry of Dienones — 399
Linearly-conjugated Dienones — 399
Cross-conjugated Cyclohexadienones — 405
Santonin-type Rearrangements — 409
Quinone Methides — 411

6 Quinones — 412
p-Quinones — 412
o-Quinones — 417

7 1,2- and 1,3-Diketones — 418

8 1,4- and Higher Diketones — 422

**Chapter 3 Photochemistry of Olefins, Acetylenes, and
Related Compounds**
By W. M. Horspool

1 Reactions of Alkenes — 427
cis–trans-Isomerization — 427
Rearrangement Reactions — 432
Addition Reactions—Alkenes and Alkynes — 434
Miscellaneous Reactions — 442

Contents

2 Reactions of Dienes 444
    Conjugated Dienes 444
    Non-conjugated Dienes 454
3 Reactions of Trienes and Higher Polyenes 460
4 Dimerizations and Intermolecular Cycloadditions 469
    Dimerizations 469
    Intermolecular Cycloadditions 473
5 Intramolecular Cycloadditions of Dienes, etc. (2+2) Cycloaddition: Formation of Cage Compounds 475
    Valence-bond Isomerization 479
6 Fragmentation Reactions 480
7 Cyclopropyl Compounds 482

Chapter 4 *Photochemistry of Aromatic Compounds*
By A. Gilbert

1 Energy Transfer: Isomerization Reactions 489
2 Addition Reactions 493
3 Substitution Reactions 506
4 Cyclization Reactions 526
5 Dimerization Reactions 541
6 Lateral-nuclear Rearrangements 548
7 Photochemistry of Furan, Thiophen, and other Heterocyclopentadienes 554

Chapter 5 *Photo-oxidation and -reduction Reactions*
By A. Gilbert

1 Conversion of C=O to C—OH 563
2 Reduction of Nitrogen-containing Compounds 586
3 Oxidation of Aromatic Compounds 598

4 Oxidation of Aliphatic Unsaturated Systems	609
5 Oxidation of Nitrogen-containing Compounds	620
6 Miscellaneous Oxidations	627

Chapter 6 Photoreactions of Compounds Containing Heteroatoms other than Oxygen
By A. Gilbert

1 Rearrangement Reactions	630
N-Oxide Rearrangements	630
Reactions of Pyridinium Ylides	637
Ring Contraction and Intramolecular Cyclization of Cyclic Nitrogen-containing Compounds	642
Reactions of Five-membered Heterocyclic Compounds	649
Photochemistry of Oximes	661
Miscellaneous Rearrangements of Nitrogen-containing Compounds	664
2 Synthesis, Substitution, and Addition Reactions of Nitrogen-containing Compounds	675
3 Reactions of Sulphur-containing Compounds	691

Chapter 7 Photoelimination Reactions
By A. Gilbert

1 Elimination of Nitrogen from Azides	697
2 Decomposition of Diazo-compounds	701
3 Decomposition of Azo-compounds	709
4 Decomposition of Other Compounds with N—N Bonds	719
5 Loss of the Elements of Carbon Dioxide	725
6 Fragmentation Reactions of Organosulphur Compounds	730
7 Miscellaneous Decomposition and Elimination Reactions	742
Decomposition into Small Fragments	743
Decomposition into Fragments containing Five or more Atoms	746

Contents

Part IV Polymer Photochemistry
By D. Phillips

1 Photopolymerization	757
2 Optical Properties of Photoexcited Polymers	770
3 Photo-cross-linking and Grafting	776
4 Photodegradation and Stabilization	784
Polypropylene and Vinyl Monomers	784
Polyamides	792
Cellulose	794
Miscellaneous	794
Photostabilizers	795
Carbonyl Oxygen Acceptors	797
Absorbers with Nitrogen Groups	797
Errata	800
Author Index	801

Introduction and Review of the Year

As in Volume 1, this Introduction includes a personal view of some of the more generally significant developments and trends which have appeared during the year 1969–70 under review. The authors have not found it necessary to make any major changes in the organisation of the subject matter in Volume 2, and they regret that it has not proved feasible to introduce a section covering the more biochemical aspects of photochemistry; but even so, Volume 2 is significantly longer than Volume 1, a fact which the authors hope reflects more the increased amount of published work than their verbosity. Minor changes from Volume 1 include the introduction of S.I. units, although not to a comprehensive extent, and increased attention to polymer photochemistry, photochemical aspects of air pollution, and the development of new techniques. It is intended that these last two areas will be covered next year in separate Sections. The Chapter on 'Inorganic Photochemistry' in Volume 1 was written at rather short notice, and the coverage was therefore less complete than we would have wished. It is hoped that this deficiency has been remedied in Volume 2. Photochromism is covered as a separate topic this year within Part 1, Chapter 2, but because of the actual and potential commercial applications, the work actually published is almost certainly only a small fraction of that actually in progress.

One of the trends which seems to have emerged during the year under review lies in the increased attention which is being paid to the photophysical and photochemical importance of charge-transfer phenomena, for example in such diverse areas as fluorescence quenching, the initiation of vinyl polymerization, and studies of acid-catalysis effects. It is not so very long since photochemistry was often treated as if it were a branch of free-radical chemistry, but this attitude has undoubtedly been changed by the recognition of such processes as photonucleophilic and photoelectrophilic substitution, and the quenching of S_1 states by nucleophiles. Yet triplet states in particular often show radical-like behaviour. Thus the subject can be seen to be developing a breadth comparable with that of the more familiar chemistry of molecules in their electronic ground states.

Charge-transfer phenomena have been particularly investigated in the following areas. Firstly, the photoreduction of aromatic ketones by

Introduction

alcohols and amines usually occurs by an initial electron-transfer step, followed by proton transfer: a similar phenomenon appears to be involved in the quenching of excited states of certain quaternary ammonium ions by alcohols. Compounds such as ketosulphides show separate absorption bands corresponding to $n \leftarrow \pi^*$ and charge-transfer excitation, and their photochemical behaviour can depend on the band through which excitation is brought about. Photosensitized electron-transfer between monomer and metal ions (*e.g.* Cu^{2+}) may be involved as the key step in certain metal-ion promoted photopolymerizations. As mentioned above, vinyl polymerization may be initiated by charge transfer, for example, in complexes between amides and maleic anhydride. The importance of polar factors in photochemistry has been given emphasis by the observation that intersystem crossing rates of aryl ketones in solution can decrease markedly with increasing polarity of the solvent, so that singlet lifetimes in polar solvents can be orders of magnitude greater than those normally estimated. The question, almost classical by now, whether diradicals and/or zwitterions are involved as intermediates in the photochemistry of dienones continues to be explored. The balance of evidence certainly appears to favour the intermediacy of zwitterions, but attempts to intercept these have not been particularly successful, and there seems to be a persistent residue of uncertainty about the precise degree of charge separation which may be involved in individual cases. One has the feeling that, particularly for reactions which proceed through the triplet manifold, the zwitterionic structures which are sometimes drawn tend to exaggerate the actual degree of charge separation which may be involved. The classical work of Havinga and his school on photonucleophilic aromatic substitution was particularly important in drawing attention to the importance of polar processes in photochemistry. The reaction has now been investigated with cyanide ion as the nucleophile, and with uncharged nucleophiles such as ammonia, and the great importance of solvent effects in these processes is being recognised. Investigations designed to trap proposed zwitterionic intermediates in certain cycloadditions (*e.g.* of maleic anhydride) to benzene have revealed a number of new acid-catalysed photochemical reactions of benzene. These appear to proceed by protonation of photochemically generated zwitterions, and constitute types of photo-Friedel–Crafts reactions which can be regarded as photoelectrophilic aromatic substitutions. The proposed zwitterionic intermediates are related in structure to the Wheland intermediates familiar in thermal electrophilic aromatic substitution. Acid-catalysis effects have also been noted for the photochemical conversion of *NN*-dialkyl-*o*-nitroanilines into benzimidazole derivatives.

The excellent review by Jortner *et al.* on theoretical treatments of radiationless transitions merits general attention: see Part I, Chapter 1. Jortner's approach seems to be gaining favour in comparison with the various alternatives which have been proposed by other workers.

Introduction

The photochemistry of simple alkanes such as ethane is now beginning to be better understood. Likewise, the photochemistry of simple mono-olefins has hitherto been somewhat neglected because the first optical absorption lies near the transparency limit of quartz. This first optical absorption band of mono-olefins has until recently been generally believed exclusively to involve a $\pi \leftarrow \pi^*$ transition, but there is now evidence that there may be an important contribution from a $\sigma \leftarrow \sigma^*$ component. Baird has criticised those molecular orbital calculations for excited states of ethylene which ignore important changes in C–C bond lengths and distortions from planarity.

The physical photochemistry of simple aromatic molecules such as benzene is continuing to attract much attention. It now seems that benzene is to be regarded as a 'large' molecule in Jortner's classification (rather than 'intermediate' as before), since its excited states can clearly undergo intramolecular non-radiative transitions—a conclusion reached some time ago by the organic chemists. Biacetyl is likewise now regarded as 'large'. Refined pulse and other techniques have established that self-quenching of benzene fluorescence in the vapour phase is small or non-existent, contrary to earlier conclusions based on phenomena which are now believed to be experimental artefacts. At the lowest pressures and at the shorter exciting wavelengths, non-exponential decay can result in emission from the directly populated upper vibrational levels; otherwise emission usually occurs from the vibrationally relaxed S_1 benzene. Little consideration is given in most physical studies to the chemistry of S_1 benzene, although one study with [1,4-^{2}H$_2$]benzene at 2537 Å has rather surprisingly seemed to indicate that the competing non-radiative processes are physical rather than chemical in nature. The quenching of S_1 benzene by oxygen is now thought to produce T_1 benzene. An orbital symmetry analysis of concerted photo-chemical cycloadditions to benzene has indicated that the Woodward–Hoffmann Rules do not apply, although the basic considerations which underlie the Rules are, of course, as equally valid for benzene as for the non-aromatic systems for which they were originally derived. The conclusions from the analysis are in good agreement with experimental observations on the photochemistry of benzene, and have permitted various predictions. There has been particular interest in the higher excited singlet and triplet levels and Rydberg states of benzene. Thus the $S_1 \leftarrow S_4$ transition has been observed, and also the first allowed triplet–triplet transition ($T_1 \leftarrow T_4$). The symmetry of the S_3 state of benzene is confirmed as $^1B_{1u}$. Interesting contrasted effects of dilution by aliphatic solvents on the efficiency of the $S_3 \leftarrow S_1$ conversions in benzene and methylbenzenes have been noted.

Oosterhoff and Van der Lugt have presented a new analysis of potential energy surfaces for the buta-1,3-diene–cyclobutene system which elaborates, and differs in some respects from, the Woodward–Hoffmann treatment.

xiii

Introduction

Much work continues on the mechanism of intermolecular energy-transfer processes. It would be fair to say that the whole area of triplet energy transfer is in a ferment. Until comparatively recently, the Schenck mechanism for photosensitized *cis–trans* isomerization of olefins received comparatively little support outside Mülheim, but opinion now seems to be gravitating towards the involvement of intermediate addition complexes in a number of cases, especially those where the triplet energy level is less than 80 kcal mol^{-1}. There are one or two cases in the distyrylnaphthalene series where *cis–trans* photoisomerization appears to occur *via* formation and dissociation of cyclobutane dimers.

Evidence is accumulating that the triplet states of some cyclic ketones have shorter lifetimes than the excited singlets in the gas phase, and that they are not effectively quenched by triplet quenchers. The previously reported excimer emission from liquid acetone is now attributed to an impurity. There is now evidence that energy transfer from triplet mercury atoms can give rise to *singlet* excited species in the cases of keten and cyclobutanone, contrary to the spin selection rules. The use of cadmium and zinc as triplet photosensitizers has hitherto been limited by the difficulty of obtaining a sufficient concentration of these elements in the gas phase. This difficulty has now been overcome by the use of a new technique in which cadmium and zinc are generated *in situ* by flash photolysis of the corresponding volatile metal alkyl derivatives. Despite complications arising from the presence of the metal alkyls, the technique promises to have valuable applications because the triplet energies of cadmium and zinc lie in a useful range.

It now appears that the T_1 states of *cis-* and *trans-*isomers of 1,3-dienes such as penta-1,3-diene transform into identical biradical-like species, but that the S_1 states retain their chemical individuality. The use of penta-1,3-diene for detection of triplet intermediates must be applied cautiously in view of the reported quenching of fluorescence of various carbonyl compounds at high concentrations of the diene. Likewise, there is slight confusion about the use of triphenylene as an exclusive triplet sensitizer in view of evidence that the S_1 state has a relatively long lifetime, and that the fluorescence of triphenylene can be quenched by certain enones. It is becoming increasingly apparent that parallel singlet and triplet pathways can be involved in the photochemistry of carbonyl compounds and may lead to similar or to chemically distinct products: the former case is one where confusion can easily arise. It is remarkable that discussion still persists about the nature of the primary processes in the photochemistry of acetone. O'Neal and Larson have described an outstandingly thorough new investigation which appears to require certain detailed modifications in the accepted mechanism, in particular that the thermal decomposition of T_1 acetone is unimolecular in the pressure fall-off region.

New results and ideas continue to come forward in the field of photo-oxidation. Dioxetans are being frequently proposed as intermediates in

Introduction

the photo-oxidation of ethylenic bonds, and the dioxetan from tetramethoxyethylene has been prepared and is sufficiently stable to be handled at room temperature. It is known that a great many photo-oxidation reactions are really thermal reactions of photochemically-generated singlet oxygen, but cases have now been described (oxidation of indoles and imines) which seem to involve the reaction of old-fashioned triplet oxygen with photochemically-generated free-radicals. Dye-sensitized photo-initiation of polymerization is an important process having commercial applications. It has been realised for some time that small but finite concentrations of oxygen are necessary for the process to occur. Ingenious use has now been made of Co^{II} triethylenetetramine complexes as reversible oxygen-carriers to stabilise the concentration of oxygen at the desired level. The oxygen-transfer and polymerization processes continue to attract attention, partly because the former provide a model for certain types of enzymic oxidation. And while one is on the topic of similarities between photochemistry and other areas, it is worth noting the accumulating evidence that isomerization and other processes akin to photochemical reactions can occur under mass-spectrometric conditions.

The photochemistry of molecules such as oxygen, ozone, water, hydrogen peroxide, and sulphur dioxide is being intensively studied, partly in connection with photochemical air-pollution problems. The chemically reactive species produced photochemically from sulphur dioxide in the atmosphere has not been positively identified, but it is probably sulphur trioxide. The interesting species C$_2$O appears to be formed in the reaction of acetylene with photochemically-generated ground-state oxygen atoms, and as a primary product in the gas-phase photolysis of ketene and carbon suboxide.

Oxetens have for long been postulated as unstable intermediates in the photoaddition of carbonyl compounds to acetylenes, so it is particularly interesting that tetramethyloxeten has now been prepared from the corresponding $\alpha\beta$-unsaturated ketone, and proves to have reasonable thermal stability. There are indications that allowed $\sigma^2_a + \pi^2_a$ processes are involved in the photoisomerization of certain unsaturated ketones. Work on dienone photochemistry continues unabated, and evidence for the formation of intermediate ketens has been obtained in one or two cases. The cycloaddition of olefins to enones continues to have synthetic utility. In Norrish Type II processes, functionalisation normally involves a γ-hydrogen atom, and more rarely a δ-hydrogen. Breslow and Winnick have now made the important discovery that methylene groups which are much more remote can be involved. They have provided some remarkable examples of remote functionalisation in the steroid field. One wonders whether any corresponding possibilities may exist for the Barton reaction. It is worth noting that remote functionalisation of this type has hitherto only been known to occur in enzyme systems. The formation of 'cage' compounds by photochemical intramolecular cycloaddition reactions of

Introduction

polycyclic compounds is a well-known example of remote functionalisation which has continued to attract interest during the year under review. Some of the bizarre structures accessible in this manner appear to have as much artistic as scientific interest.

Berson and Olin have described a reaction of certain azo-compounds which occurs thermally and photochemically in the same stereochemical sense, in seeming violation of the Woodward–Hoffmann Rules. As noted above for benzene, there is some danger in blind application of these extraordinarily valuable Rules to systems having electronic symmetry characteristics (e.g. orbital degeneracy, $n\pi^*$ and charge-transfer excited states) markedly different from those of aliphatic hydrocarbons. Incidentally, orbital symmetry factors seem to be of only minor importance in reactions of transition-metal complexes (see Part I, Chapter 1, Section 2).

Various previous inconsistencies in the photochemical behaviour of *trans*-azoalkanes have been accounted for in terms of an initial isomerisation to the less-stable *cis*-isomers, followed by *thermal* elimination of nitrogen. Very little has been reported during the year concerning valence-isomerisation of homocyclic benzenoid rings, but pyridine has now been shown to be transformed into 'Dewar-pyridine' on irradiation. There are indications that other aromatic nitrogen heterocycles show analogous behaviour. In fact, the phototransposition of hetero-atoms in aromatic heterocycles is a common and potentially useful reaction which, as with benzene, seems to involve intermediate unstable valence-isomers. It is interesting to note in this connection that the irradiation of thiophens with primary amines has been found to give pyrroles.

Various examples of inter- and intra-molecular 1,3-cycloaddition of ethylenes to aromatic rings have been demonstrated. The processes appear to be stereospecific with respect to the ethylene.

It is becoming increasingly recognised that the pathways of many photoreactions (e.g. those of polyenes) are governed by *ground-state* conformational equilibria. Such reactions often show marked temperature effects.

The mechanism of the photo-Fries rearrangement has been extensively discussed in past years. The possibility that the process is concerted in nature appears to have been ruled out by the demonstration that phenyl acetate does not undergo rearrangement on irradiation in the vapour phase. The reaction definitely seems to be of radical type and to require a solvent cage. The results suggest the need for further examination of the *para-*blocked systems referred to in Volume 1.

Several scattered examples of the photoreduction of conjugated ethylenic bonds by sodium borohydride have been described. This potentially useful reaction merits examination in a wider context.

As far as industrial applications of photochemistry are concerned, one only sees the tip of the iceberg, as it were. Nevertheless, quite a lot of work has appeared on optical brightening agents, the photostabilisation of

Introduction

polymers against degradation in u.v. light, photochemical modification of polymers by cross-linking, *etc.*, and photochromism. Much of this work will also be of interest to the more academic photochemists. Thus ingenious use has been made of energy-transfer phenomena and 'energy sinks' in the photostabilisation of polymers such as nylon and polypropylene. A useful degree of photochemical cross-linking can be achieved in polymers such as poly(vinyl cinnamate) which contain pendant ethylenic bonds through which cyclobutane formation can take place. The incorporation of photochromic units such as spiropyrans into polymer chains has been found to give light-sensitive polymers which are of interest in image-forming processes. Photoconductivity in polymers such as poly(*N*-vinyl-carbazole) has been studied. Photochemical cross-linking and chain-grafting of polymers can produce novel hybrids, for example polystyrene–protein. Energy-transfer from seemingly inert inorganic compounds such as titanium dioxide can play an important part in the photochemical degradation of polymers such as nylon. It seems that radicals and/or peroxides are formed at sites on the titanium dioxide and then attack the polymer, normally by hydrogen-abstraction.

Photochemistry has its sociological aspects, as witnessed by the growing amount of published work on the photodegradation of insecticides and other biologically active compounds which are placed in the environment; and reference has already been made to the large amount of work being carried out on photochemical air-pollution.

Part I, Chapter 1 includes a section on new instrumentation and techniques which we hope will be expanded next year. Porter's rapid development of nanosecond flash photolysis merits particular notice. The molecular modulation technique is also noteworthy because it enables one to obtain u.v. and i.r. spectra of photochemically generated transient species, for example the ClOO· radical formed by irradiation of mixtures of chlorine and oxygen. This technique seems likely to receive greatly increased use in the future.

Finally, the thermal generation of electronically excited states is now clearly established, and makes possible 'photochemistry without light' in certain cases. In particular, the thermal decomposition of certain 1,2-dioxetans gives carbonyl compounds in electronic excited states which may then undergo reactions and energy-transfer processes qualitatively similar to those of the photochemically generated species. At least one case of triplet energy-transfer seems to have been established. This development recalls the work on thermally-generated singlet oxygen which was reported a few years ago.

D. B.-S.

Part I

PHYSICAL ASPECTS OF PHOTOCHEMISTRY

1 Spectroscopic and Theoretical Aspects

As was stated in Volume 1 of this series, the experimental photochemist can learn much from a study of the absorption spectroscopy of molecules under his investigation. Thus, details of the nature of the excited states formed upon absorption, radiative lifetimes of these states, changes in dipole moment and polarizability (and consequently reactivity) may be obtained from a careful study of the electronic absorption spectra. Frequently information obtained in this way is augmented by *ab initio*, or more usually semi-empirical, calculations upon the energy levels and properties of excited states of molecules. Together, theoretical and spectroscopic considerations can often provide a rationale for the photochemists' experimental observations. The volume of material of a spectroscopic and theoretical nature published annually is vast, however, and it would be impossible, and indeed misplaced, to attempt to include a comprehensive survey of this field in this volume. Instead a brief section is included on absorption spectra and energy level calculations which may be of direct interest to the photochemists for whom this volume is intended. References to other theoretical and spectroscopic work will often be found in sections dealing with the photochemistry of specific molecules.

As in Volume 1, a considerable section of this chapter will be devoted to theoretical considerations of radiationless transitions. This is an area in which experimental evidence is building up, and the phenomenon provides a challenge to theoreticians and experimentalists alike to gain a deeper understanding of the complex nature of these processes. It must be stressed that the authors of this volume are primarily experimentalists, and the account given here of the efforts of colleagues engaged upon the difficult task of providing an adequate theory which will quantitatively account for observed rates of internal conversions and intersystem crossings may be coloured by a lack of understanding of the true complexities of the problem.

The extensive data which are available on the rates of decomposition of molecules in the gas phase have been treated in the past in only a semiquantitative manner, but it is encouraging to note that there has been an increasing tendency of late to apply the theories of unimolecular decomposition to excited states of molecules. A short section is devoted to this and other theoretical aspects of photodecomposition, and some further

1 Absorption Spectra and Energy Level Calculations

As stated above, this section is not intended to be comprehensive, but a selection of published papers on molecules which may be of interest to photochemists is given. These molecules will be dealt with in order of increasing complexity. The calculation of Franck–Condon factors is of importance in that these determine the intensities of both radiative and non-radiative transitions. In small molecules, work has continued on investigations into hitherto neglected vibration–rotation interactions in the calculation of Franck–Condon factors. For the hydrogen molecule, the $(B^1\Sigma_u^+ \leftarrow X^1\Sigma_g^+)$, $(I^1\Pi_g \leftarrow B^1\Sigma_u^+)$, $(d^3\Pi_u \leftarrow a^3\Sigma_g^+)$, $(C^1\Pi_u \leftarrow X^1\Sigma_g^+)$, $(D^1\Pi_u \leftarrow X^1\Sigma_g^+)$, systems,[1,2] the $(E^1\Sigma_g^+ \leftarrow B^1\Sigma_u^+)$, $(h^3\Sigma_g^+ \leftarrow c^3\Pi_u)$ systems,[1,2] and $(k^1\Pi_u \leftarrow a^3\Sigma_g^+)$ systems[3] have been extensively studied. In all cases good agreement with experiment is possible if vibration–rotation interactions are considered. Previous computations, for *e.g.* the Lyman bands, have not included the effects of the centrifugal potential, but it has been tacitly assumed that the principal effect of the rotational energy is a shift of the potential curve by a constant energy displacement. This assumption is most severely tested in calculations of Franck–Condon factors for electronic transitions in light molecules, and especially if the transition involves states with appreciably different potential curves, as in the $(B^1\Sigma_u^+ \leftarrow X^1\Sigma_g^+)$ Lyman system. The rotational angular momentum necessarily changes in all transitions (since there is no Q branch, it being a $\Sigma \rightarrow \Sigma$ transition), and this serves to emphasize the vibration–rotation interaction effects. By way of contrast, the $(d^3\Pi_u \leftarrow a^3\Sigma_g^+)$ Fulcher bands arise from states which have similar potential curves, and smaller vibration–rotation interaction effects on intensities would be anticipated.

When the centrifugal potential is explicitly taken into account, the Franck–Condon factor appropriate to the electronic transition $(v', J' \leftarrow v'', J'')$ is given by:

$$q(v', J'; v'', J'') = \left| \int \phi_{v', J'}^{a'}(R) \phi_{v'', J''}^{a''}(R) \, dR \right|^2 \quad (1)$$

[1] R. J. Spindler jun., *J. Quant. Spectroscopy Radiative Transfer*, 1969, **9**, 597.
[2] D. Villarejo, R. Stockbauer, and M. C. Inghram, *J. Chem. Phys.*, 1969, **50**, 1754.
[3] R. J. Spindler jun., *J. Quant. Spectroscopy Radiative Transfer*, 1969, **9**, 1041.

Spectroscopic and Theoretical Aspects

where $\phi_{v',J'}(R)$ and $\phi_{v'',J''}(R)$ are the eigenfunctions of the nuclear motion for the two states. For $^1\Sigma$ states the eigenfunctions are solutions of the Schrödinger equation:

$$-\frac{\hbar^2}{2\mu}\cdot\frac{\mathrm{d}^2\phi_{v,J}}{\mathrm{d}R^2}+\left[V(R)+\frac{J(J+1)\hbar^2}{2\mu R^2}\right]\phi_{v,J}=E_{v,J}\,\phi_{v,J} \qquad (2)$$

Solutions of this equation can be obtained by replacing it with the equivalent finite-difference equation which is solved numerically by a computer procedure. Details will not be given here, but will be found in ref. 2. It is found that vibration–rotation interaction effects for e.g. the Lyman bands are substantial, but small for the Fulcher bands as expected. It should be noted that the calculations of distribution of intensity in these systems have been in the limit of a strictly valid Franck–Condon principle i.e. the molecular electronic dipole moment does not depend on internuclear separation. It is known, however, that this is not strictly true, and in fact there may be an appreciable variation of dipole moment with internuclear distance in some systems. Small variations of dipole moment with internuclear distance tend to be amplified if vibration–rotation effects are appreciable. Thus, if $\mu(R)$ represents the dipole moment function, then expansion of $\mu(R)$ around $R = 0$ leads to

$$\int\phi''_{v'',J''}(R)\,\mu(R)\,\phi_{v',J'}(R)\,\mathrm{d}R$$

$$=\mu_0\int\phi''_{v'',J''}(R)\,\phi_{v',J'}(R)\,\mathrm{d}R + \mu_1\int\phi''_{v'',J''}(R)\,R\phi_{v',J'}(R)\,\mathrm{d}R+\cdots \qquad (3)$$

If the first term, the overlap integral, vanishes, then the other terms may have great significance. It has been shown in the papers discussed briefly above that the overlap integral may vanish because of vibration–rotation interaction, a fact which has not hitherto been appreciated, and clearly any complete theoretical treatment must necessarily involve inclusion of an explicit dipole moment function as well as the effect of perturbations mentioned previously. Franck–Condon factors for the $v = 0$ progression of the N_2 fourth positive system have been derived.[4]

High-resolution studies of the $C\,^2\Pi \leftarrow X\,^2\Pi$ emission bands of NO,[5] the electronic spectrum of cyanogen,[6] and the $(B\,^3\Sigma_u^- \leftarrow X\,^3\Sigma_g^-)$ and $(A\,^3\Sigma_u^- \leftarrow X\,^3\Sigma_g^-)$ bands of SO[7] have been recorded, and forbidden absorption bands of O_2 in the argon continuum region of the spectrum observed.[8] Ethane is the only alkane which exhibits structure in its molecular electronic absorption spectrum. High-resolution photographs reveal that this is diffuse vibrational structure in the case of C_2H_6.[9] However, the

[4] C. R. Herbert and R. W. Nicholls, *J. Phys. (B)*, 1969, **2**, 626.
[5] F. Ackermann and E. Miescher, *J. Mol. Spectroscopy*, 1969, **31**, 400.
[6] S. Bell, C. J. Cartwright, G. B. Fish, D. O. O'Hare, R. K. Ritchie, A. D. Walsh, and P. A. Warsop, *J. Mol. Spectroscopy*, 1969, **30**, 162.
[7] R. Colin, *Canad. J. Phys.*, 1969, **47**, 979.
[8] M. Ogawa and K. R. Yamawaki, *Canad. J. Phys.*, 1969, **47**, 1805.
[9] E. F. Pearson and K. K. Innes, *J. Mol. Spectroscopy*, 1969, **30**, 232.

Photochemistry

0–0 band of C_2D_6 near 1406 Å exhibits a distinctive rotational contour. Computer simulation of this contour shows that the transition moment for the 0–0 band lies perpendicular to the C—C bond. The assignment of the electronic transition is $^1A_g \leftarrow {}^1E_u$. Vibrational analysis shows that the transition effects a large reduction in the C—H stretching frequency, consistent with a 731 cm^{-1} shift of the origin band towards higher frequencies on deuterium substitution, and with the promotion of an electron out of a C—H bond. These results are consistent with the observed photochemistry of this molecule (see Chapter 3). Although the C—C bond energy is lower than that of the C—H bond in the ground state of

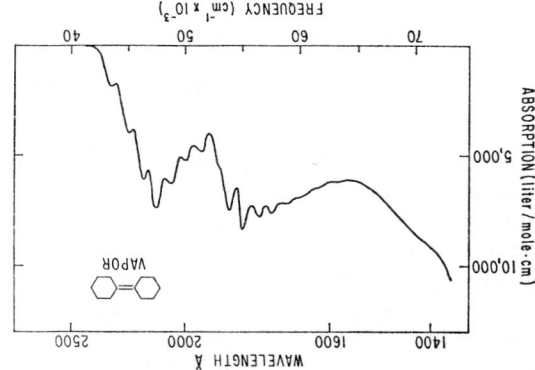

Figure 1 *The absorption spectrum of bicyclohexylidene vapour obtained with a McPherson, double-beam, vacuum spectrophotometer*
Reproduced by permission from *J. Chem. Phys.*, 1970, **52**, 998)

ethane, the photochemistry suggests that C—C cleavage is not the main primary process at *ca.* 1400 Å, indicating either that the ^1E_u state is formed *via* promotion of an electron from the C—H bond, or that excitation of an electron from the C—C bond is followed by a rapid redistribution of energy. The present results favour the former alternative. The linewidths for C_2D_6 are *ca.* 3 cm^{-1} indicating a lifetime of the state of 10^{-11} s. The lifetime of the C_2H_6 excited state is at least ten times shorter.

The absorption spectrum of bicyclohexylidene vapour is shown in Figure 1 to consist of two bands, both associated with the C=C bond.[10] The crystal absorption spectrum is similar, and polarization measurements reveal the very interesting fact that both transitions are polarized in the same direction and correspond to $^1B_{1u}$ excited states. The existence of two strong $^1A_g \leftarrow {}^1B_{1u}$ transitions in the low-energy spectral region of simple olefins is incompatible with the predictions of π-electron theory. There can be only one strong π-electron transition with the observed transition moment direction, the $\pi \rightarrow \pi^*$ $(N \rightarrow V)$ transition. It is thus felt that the

[10] P. A. Snyder and L. B. Clark, *J. Chem. Phys.*, 1970, **52**, 998.

Spectroscopic and Theoretical Aspects

other band must be due to a $\sigma \to \sigma^*$ transition and this recognition of the existence of a strong $\sigma \to \sigma^*$ transition in the energy region previously assumed to be the sole domain of π-electrons is of obvious importance to those concerned with the photochemistry of such molecules. Should other simple olefins also show such bands, the impact of these observations upon photochemistry and indeed π-electron theory could be very great. However, a Rydberg transition should occur in this wavelength region, and although the assignment of the second observed band to this transition has been discounted, further evidence must be presented before the σ–σ^* assignation can be accepted unequivocally.

Singlet–triplet absorptions in mono-olefins are very weak, but the intensity of such bands can be increased by the presence of oxygen at 70 atms pressure. An attempt has been made to seek the absorption spectra of methyl-substituted ethylenes corresponding to the ground state to triplet Rydberg state transition ($T_R \leftarrow N$) under these conditions, but only bands corresponding to the ($T \leftarrow N$) transition and due to contact charge-transfer transitions from the ethylene to O_2 could be observed.[11]

Zero-differential-overlap (ZDO) molecular orbital calculations of the geometry for excited states of unsaturated molecules have been criticized on the grounds that such calculations completely discount one of the major factors influencing bond lengths and bond angles in excited molecules.[12] As an example, ethylene can be considered. For the π-electron energy levels ε_π and ε_{π^*}, according to one-electron LCAO–MO theories which retain the overlap integrals S

$$\varepsilon_\pi = (\alpha + \beta)/(1 + S) \quad (4)$$

$$\varepsilon_{\pi^*} = (\alpha - \beta)/(1 - S) \quad (5)$$

Since the π^* level is more antibonding than the π level is bonding, the total π-electron energy E for a $\pi\pi^*$ electron configuration is net antibonding.

$$E = 2(\alpha - \beta S)/(1 - S^2) \quad (6)$$

In the excited molecule, the energy can be stabilized by decreasing the magnitudes of β and S; this is accomplished by lengthening the C–C bond and/or by twisting the p_π-orbitals away from each other. Any theoretical method which neglects S must predict that the one-electron π-energy of $\pi\pi^*$ configurations is independent of geometry, and thus neglects a major driving force in the determination of the equilibrium geometry.

According to an analysis based on first-order perturbation theory, the net antibonding character of the lowest $^3\pi\pi^*$ state of an acyclic polyene containing $2m$ unsaturated carbon atoms is given approximately by

$$-8S(\beta - \alpha S)/(m+1)^2 \quad (7)$$

The destabilization due to overlap effects decreases rapidly with increasing chain length, and the exothermicity associated with rotation by 90° about a

[11] M. Itoh and R. S. Mulliken, *J. Phys. Chem.*, 1969, **73**, 4332.
[12] N. C. Baird, *Chem. Comm.*, 1970, 199.

7

Allene is a molecule which has excited the interest of photochemists of late, and Gaussian SCF calculations on the ground state of this molecule have generated values for the energy levels of the excited states.[13] These are, however, only in poor agreement with experiment, although they agree qualitatively in that they predict the spectrum of allene to consist of three weak transitions followed by a strong transition at shorter wavelengths, as observed. The absorption and fluorescence spectra of crystalline *trans*-stilbene,[14] and the electronic properties of polyenes and polyphenylacetylenes have been described,[15] the latter from a theoretical point of view.

Benzene is a molecule whose photochemistry has been much studied, and interest in this molecule by theoreticians and spectroscopists is also intense. It is known that the lowest state of benzene is the $^3B_{1u}$ state, but higher triplet levels have been predicted by calculations and have in some cases been observed experimentally. Recently, absorption in the 4300 Å region has been attributed to the first allowed $T-T$ transition in benzene,[16] producing the T_4 state which is of $^3E_{2g}$ symmetry. The absorption decayed with the same lifetime as other well characterized triplet absorptions, and also has been observed in another study.[17] In the latter work, absorption by excited singlet state benzene molecules was also monitored, and the transition attributed to the $S_1 \rightarrow S_t$ states, *i.e.* $^1B_{2u} \rightarrow ^1E_{2g}$. From the positions of the spectral bands, the $^3E_{2g}$ and $^1E_{2g}$ states of benzene can be placed at 52,800 and 58,400 cm^{-1} respectively above the ground state. The figure for the $^3E_{2g}$ level is *ca.* 2 eV below that calculated using the Pariser–Parr–Pople approach. The order of the excited singlet states has also been called into question by a thorough experimental study of higher $\pi-\pi^*$ transitions in solid argon, krypton, xenon, and nitrogen matrices in the spectral region 2800—1700 Å.[18] On the basis of the observed vibrational structure the second excited singlet state of benzene has been confirmed as $^1E_{1u}$ rather than $^1E_{2g}$. Moreover, theoretical calculations of the dynamic electronic vibrational coupling between the $^1B_{1u}$ and the $^1E_{1u}$ states also support the $^1B_{1u}$ assignment of the 2100 Å transition. A phenomenon closely related to electronic relaxation in large molecules is the occurrence of line broadening in the absorption spectra of the higher excited electronic states. In these excited states the Born–Oppenheimer separability conditions for electronic

[13] L. J. Shaad, L. A. Burnelle, and K. P. Dressler, *Theor. Chim. Acta*, 1969, **15**, 91.
[14] M. K. Chaudhuri and S. C. Ganguly, *J. Phys.* (C), 1969, **2**, 1560.
[15] Yu. A. Kruglyak and V. V. Pen'kovskii, *Zhur. strukt. Khim.*, 1969, **10**, 222, 459.
[16] R. Astier and Y. H. Meyer, *Chem. Phys. Letters*, 1969, **3**, 399.
[17] J. B. Birks, *Chem. Phys. Letters*, 1969, **3**, 567.
[18] B. Katz, M. Brith, B. Sharf, and J. Jortner, *J. Chem. Phys.*, 1970, **52**, 88.

6 *Spectroscopic and Theoretical Aspects*

and nuclear motion break down because of vibronic coupling between iso-energetic zero-order Born–Oppenheimer vibronic states which correspond to different electronic configurations. In the so-called 'statistical' limit when the density ρ of vibronic levels is sufficiently high to exceed the reciprocal of the mean vibronic coupling term v between the zero states, inhomogeneous line broadening is expected to be observed. For an isolated resonance, the line shape is expected to be Lorentzian, and the line is given by $\Delta_t = 2\pi v^2 \rho$. In the solid phase medium effects may also cause line broadening by the following mechanisms:

(i) coupling of states with the lattice vibration of the host should lead to temperature-dependent broadening Δ_{ph} of all the vibronic levels for a given electronic state;

(ii) vibrational relaxation effects will lead to the broadening of the higher vibrational components within a given electronic state. These effects should be negligible for the 0–0 band, and a contribution to the line broadening additional to that of the 0–0 band gauges the vibrational relaxation process. This additional broadening due to vibrational relaxation is in the range $\Delta_{vr} = 1$—10 cm^{-1};

(iii) in the statistical limit the radiationless relaxation times should be independent of the medium.

Thus the total linewidth of a large molecule in a solid can be expressed in the form

$$\Delta = \Delta_{ph} + \Delta_{vr} + \Delta_t \qquad (8)$$

In the benzene spectra studied above,[18] the linewidths corresponding to absorption to the $^1B_{1u}$ and $^1E_{1u}$ states are appreciably broadened compared with those of the $^1B_{2u}$ transition, being 350 and 300 cm^{-1} respectively compared with 10—35 cm^{-1} for the first observed transition. From gas-phase data, it is clear that no appreciable non-radiative relaxation to the ground state occurs from the $^1B_{2u}$ state, and thus we can estimate that $\Delta_{ph} + \Delta_{vr} \approx 35$ cm^{-1}. Assuming that the values of Δ_{ph} and Δ_{vr} for different $\pi\pi^*$ states should be roughly the same, it can be estimated that the value of Δ_t for the two upper excited states is of the order of 300 cm^{-1}. Hence electronic relaxation times of these states will be of the order of $h/\Delta_t = 10^{-14}$ s. The present data are not of sufficient precision to allow anything other than the above crude calculation of lifetimes, but improved data would clearly provide much evidence about radiationless transitions from the upper excited states of these and other molecules, and it is hoped that such data will be forthcoming in the future.

The energy separation of the $^1A_{1g}$ and $^3B_{1u}$ states of benzene has been calculated by an *ab initio* AMO approach to be 4.04 eV,[19] which is an improvement over the SCF value of 4.45, and compares quite well with the experimental value of 3.8 eV. Rydberg states of benzene,[20] the elec-

[19] R. R. Gilman and J. DeHeer, *J. Chem. Phys.*, 1970, **52**, 4287.
[20] B. Katz, M. Brith, B. Sharf, and J. Jortner, *J. Chem. Phys.*, 1969, **50**, 5195.

tronic structure of benzene,[21] and the polarizability of benzene in ground and excited states [22] have been discussed. The effect of substitution upon the absorption spectra has been considered theoretically for chloro-[23, 24] and fluoro-[25] substituents, and the influence of fluoromethyl substitutions has also been considered.[26] An analysis of the fluorescence spectra of toluenes and of hydrogenated and deuteriated benzyl radicals has been performed,[27] and the effect of matrix and solute concentration on the u.v. spectra of polymethyl derivatives of benzene at 77 K discussed.[28]

Intramolecular charge transfer states of the general formulas A—Ar, D—Ar, and A—Ar—D where A is an acceptor, D a donor moiety, and Ar an aromatic molecule have been treated in the light of a new theoretical approach.[29] The experimental evidence concerning charge transfer (CT) states can be summarized as follows:

(i) in some molecules D—Ar—A, e.g. 4-nitroaniline, the S_1 singlet state shows a large dipole moment change, whereas the S_2 state is less polar;

(ii) in D—Ar—A molecules where S_1 is of CT type, the energy of higher states S_2 etc. is not much modified from the parent Ar—A molecule;

(iii) in D—Ar and Ar—A molecules the dipole moment changes are generally smaller than in D—Ar—A molecules;

(iv) in a series of D—Ar—A molecules the dipole moment change increases when the energy of the CT state decreases;

(v) there are D—Ar—A molecules which do not show large dipole moment changes;

(vi) CT character is present for o-, m-, and p-isomers D—Ar—A;

(vii) in di(donor)-substituted molecules D,D'—Ar—A there are two independent CT bands when D or D' is in a p-position with respect to A, one CT band of very low energy when neither D nor D' is in a p-position with respect to A;

(viii) if A and Ar are not coplanar, the intensity of the CT band is reduced; if D and Ar are not coplanar, the energy of the CT state is increased.

The results above can be explained on the basis of the following simple model, shown diagrammatically in Figure 2. Here it is assumed that the lowest occupied orbitals π_A and π_{Ar} are of very different energies, whereas the π_A^* and π_{Ar}^* orbitals are almost degenerate. The effect of a donor D would be to increase the energy of the ring orbitals. It has been widely

[21] B. O. Jonsson and E. Lindholm, *Arkiv. Fysik*, 1969, **39**, 65.
[22] A. V. Luzanov, U. B. Malykhanov, and M. M. Mestechkin, *Opt. Spektroskopii*, 1970, **28**, 836.
[23] O. Chalvet and C. Leibovici, *Theor. Chim. Acta*, 1969, **14**, 65.
[24] A. L. Verma and H. D. Bist, *Chem. Phys. Letters*, 1970, **4**, 577.
[25] O. Chalvet and C. Leibovici, *Theor. Chim. Acta*, 1969, **13**, 297.
[26] C. V. Klimuskeva, L. M. Yaguopol'skii, and R. V. Yaremko, *Teor i Eksp. Khim*, 1969, **5**, 392.
[27] L. Wattmann-Crajcar, *J. Chim. Phys.*, 1969, **66**, 1023.
[28] C. Durocher and S. Leach, *J. Chim. Phys.*, 1969, **66**, 628; G. Durnchei, *ibid.*, p. 637.
[29] P. Suppan, *J. Mol. Spectroscopy*, 1969, **30**, 17.

Spectroscopic and Theoretical Aspects

assumed hitherto that the $\pi-\pi^*$ absorption bands are the same as in substituted benzenes as in benzene itself, and merely shifted in wavelength and intensity. The simple orbital interaction model above does not agree with this conclusion since it shows that a simple benzene transition may be split into two new transitions by the interaction of π_{Ar} and π_A^* orbitals, even though the localized excitations $\pi_A \rightarrow \pi_A^*$ of the substituent may be

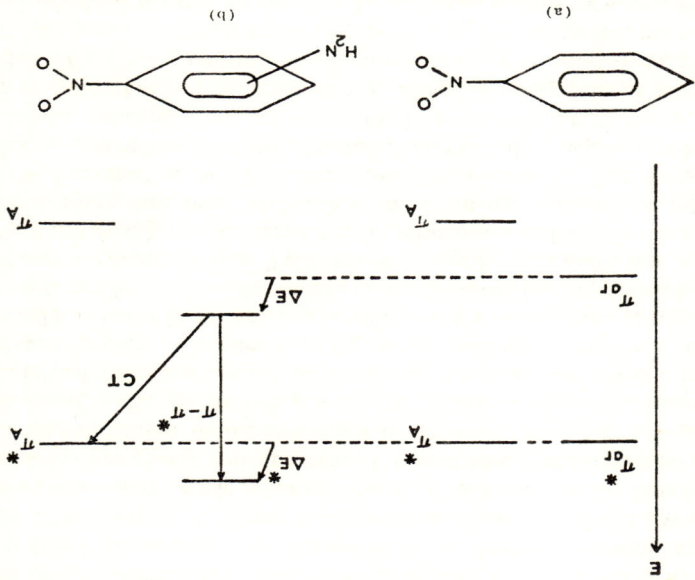

Figure 2 *Theory of charge-transfer spectra. Effect of donor function on nitro-benzene orbitals*
(Reproduced by permission from *J. Mol. Spectroscopy*, 1969, **30**, 17)

at much higher energies than the aromatic transitions. Thus, the 335 nm band of 4-amino-benzophenone in ethanol, and the 245 nm band result from the splitting of the 210 nm benzene band, and both could be labelled ^1L_a. It is, however, the 335 nm band which corresponds to a charge transfer state. In order to clarify the situation arising from the fact that the term 'charge-transfer band' has been used very loosely in the past, it is proposed that the term be used in future to describe states only on the basis of an observed large change in dipole moment. It is clear that $\pi-\pi^*$ states of D—Ar—A molecules derive their CT character from a ring to acceptor substituent $\pi_{Ar} \rightarrow \pi_A^*$ configuration, the donor to ring CT being generally less important and the substituent to substituent CT being negligible.
A general expression for the transition moment of the $^1A \rightarrow {}^3L_a$ transition in linear polyacenes has been derived from the matrix elements of spin—

11

orbit interaction between ($\pi\pi$) and ($\pi\sigma$) states.[30] The results indicate that the oscillator strength of this transition should decrease as the size of the linear polyacene increases although, unfortunately, experimental data are not available to test this conclusion. Theoretical studies on triplet–triplet absorption in 1- and 2-substituted naphthalenes have also been undertaken.[31] With laser irradiation, two-photon absorption in anthracene can be observed.[32] The symmetry-forbidden $^1A_{1g} \rightarrow {}^1B_{2u}$ transition was observed in both crystal and solution, and can be attributed to a two-photon absorption induced vibronically by a non-totally symmetric vibration of frequency 1200 ± 400 cm^{-1}. There is also a crystal state at 3·5—3·6 eV which has no counterpart in the solution spectrum, and this is attributed to a g-symmetry-type charge transfer state. It is also possible that a quadrupole $^1A_{1g} \leftarrow {}^1B_{2u}$ transition to the 0–0 band is observed, since the appearance of this band in the spectrum is unlikely to be due to dipole transitions, as these are forbidden and cannot be vibronically induced (for the 0–0 band). Anthracene presents a challenge in its excited state spectroscopy, since the first singlet is short lived, and also absorbs in the same spectral region as the lowest triplet and as the fluorescence. Nevertheless, by using kinetic spectroscopic methods (see Techniques section) in conjunction with a polarizer the signals due to excited singlet absorption and triplet absorption can be distinguished from each other and from the background fluorescence.[33] The spectra obtained are shown in Figure 3. The long-lived transient is clearly due to the triplet state, and the mildly negative polarization shown indicates that the transition is perpendicular to the $S_1 \leftarrow S_0$ emission (and the $S_1 \rightarrow S_x$ absorption), which is in agreement with the expected $^3B_{1g} \leftarrow {}^3B_{1u}$ nature of this transition. The short-lived singlet absorption is more difficult to identify positively, but, as stated, the transition is polarized parallel to the $S_1 \leftarrow S_0$ emission, indicating a $^1A_{1g}$ assignment for the upper state, S_x. The results obtained here of the $^3B_{1g}$ state at 2·94 eV above $^3B_{1u}$, and the $^1A_{1g}$ 2·93 eV above the $^1B_{2u}$ are in good agreement with calculated values of 2·92 and 3·17 eV respectively.[34] These results are in agreement with observations on the polarization of absorption in phenanthrene, chrysene, and picene, in which it was found that the first three excited singlet states of phenanthrene and picene correspond to ^1L_b, ^1L_a, and ^1B_a states (Platt notation) or $^1A_1$, $^1B_2$, $^1B_2$ (Orloff axial system[35]), with the transition to the $^1A_1$ state characterized by in-plane short-axis polarization and those to the $^1B_2$ states by in-plane long-axis polarization.[36] The lowest triplet state is ^3L_a ($^3B_2$) in nature, and the triplet–triplet absorption which is in-plane long-axis polarized requires that the

[30] V. C. Krishna and W. R. Salzman, J. Chem. Phys., 1969, 50, 3875.
[31] L. Klasinc and U. Sommer, Chem. Phys. Letters, 1969, 3, 107.
[32] I. Webman and J. Jortner, J. Chem. Phys., 1969, 50, 2706.
[33] D. S. Kliger and A. C. Albrecht, J. Chem. Phys., 1969, 50, 4109.
[34] R. Hochstrasser and A. Marchetti, Chem. Phys. Letters, 1968, 1, 597.
[35] M. K. Orloff, J. Chem. Phys., 1967, 47, 235.
[36] J. B. Gallivan and J. S. Brinen, J. Chem. Phys., 1969, 50, 1590.

Spectroscopic and Theoretical Aspects

upper triplet state have A_1 symmetry. In chrysene the lowest three excited singlets have ^1L_b, ^1L_a, and ^1B_b (all ^1B_u states), the lowest triplet is ^3L_a (^3B_u), and the upper triplet must have A_g symmetry. The 3400 Å absorption spectrum of phenanthrene has been discussed with special emphasis on

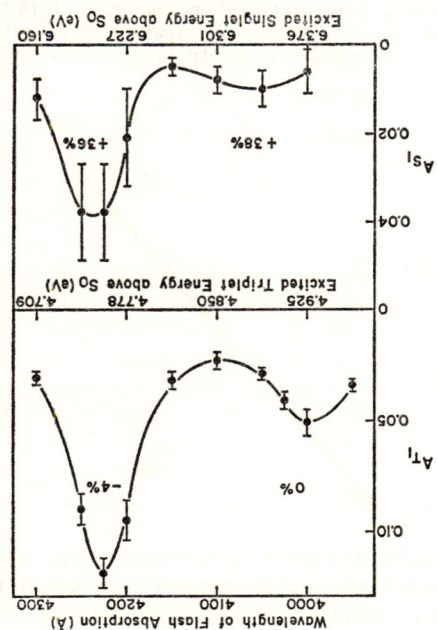

Figure 3 *Polarization and spectra of singlet–singlet and triplet–triplet absorption in anthracene. Top: triplet; bottom: singlet*
(Reproduced by permission from *J. Chem. Phys.*, 1969, **50**, 4109)

vibronic perturbations by totally symmetric vibrations.[37] Such effects can give rise to loss of symmetry between fluorescence and absorption spectra and a radiative lifetime different from that calculated from the integrated absorption curve, and these features can be observed in phenanthrene. The 3000 Å absorption spectrum of phenanthrene in mixed crystals has also been discussed,[38] and some semi-empirical calculations of singlet–triplet and triplet–triplet transitions of conjugated hydrocarbons[39] considered. Triplet–triplet transition energies for alternate hydrocarbons calculated by the Hückel method were found to have a linear correlation between values obtained by an SCF calculation,[40] but such observations are not of much assistance to the experimentalist.

[37] D. P. Craig and C. J. Small, *J. Chem. Phys.*, 1969, **50**, 3827.
[38] C. Fischer, *Mol. Crystals and Liq. Crystals*, 1969, **6**, 105.
[39] J. Pancir and R. Zahradnik, *Theor. Chim. Acta*, 1969, **14**, 426.
[40] D. Lavalette, *Chem. Phys. Letters*, 1969, **3**, 264.

13

14 Photochemistry

Benzophenone is a molecule much used by photochemists, and since equilibrium geometry changes in the excited states of molecules play a significant role in determining their photochemical behaviour, it is of interest to consider such changes on the $n\pi^*$ excited states of benzophenone. Extended Hückel calculations reveal [41] that the ground state of benzophenone has both phenyl rings twisted out of plane to a C_2 geometry by 38° in a conrotatory fashion. The excited state has a considerably steeper potential well for a similar geometry in which the angles of twist are 32°.

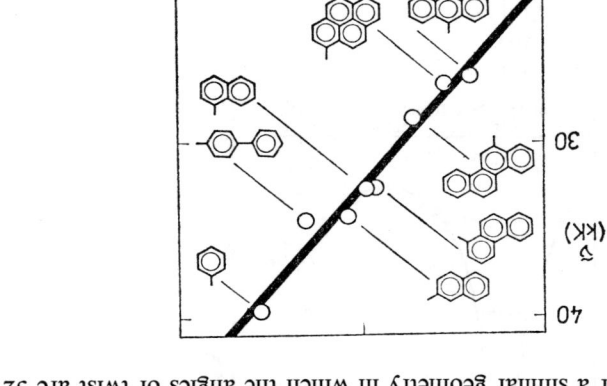

Figure 4 *Plot of experimental* $(\tilde{v})$ *against theoretical LCI* $(S_0 \rightarrow S_1$, *see text) excitation energies for aryl phenyl ketones. Regression line:* $\tilde{v}$ (kK) = 1·170 $S_0 \rightarrow S_1$ (kK) $-2·619$; *correlation coefficient* $r = 0·979$, *number of compounds* $n = 8$. *All values are significant on 1% probability level*
(Reproduced by permission from *J. Phys. Chem.*, 1969, **73**, 1132)

It appears that the carbonyl group remains locally planar in the excited state. Spectral properties of eight aryl phenyl ketones, including benzophenone, have been investigated by Pariser–Parr–Pople type LCI–SCF–MO calculations, and reasonable agreement found between calculated $S_0 \rightarrow S_1$ excitation energies and wavelengths of absorption band maxima of the ketones (see Figure 4).[42] Phosphorescence emission spectra for six of the ketones were obtained, and reasonable agreement found for calculated $S_0 \rightarrow T_1$ excitation energies and the wavelength of the 0–0 emission band for these compounds, for which the lowest triplet has a $\pi\pi^*$ character (Figure 5). In all cases, predicted transition energies were found to be too low (see Figure 4 and 5). Calculated values were found to deviate equally from observed values for both singlet and triplet transitions resulting in accurate predictions of S–T splittings for the lowest $(\pi\pi^*)$ excited states.

[41] R. Hoffmann and J. R. Swenson, *J. Phys. Chem.*, 1970, **74**, 415.
[42] C. Parkanyi, E. J. Baum, J. Wyatt, and J. N. Pitts jun., *J. Phys. Chem.*, 1969, **73**, 1132.

Spectroscopic and Theoretical Aspects

The errors in calculation could be due to choice of calculation parameters for the keto-group, deviations from planarity of the compounds studied, and solvent dependence of the spectral band positions. The positions of the spectral bands for the various ketones and their phosphorescence lifetimes are summarized in Table 1. Photochemical reactivity of aryl

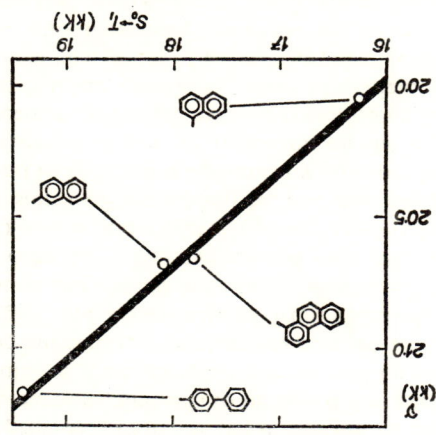

Figure 5 *Plot of experimental ($\tilde{v}$) against theoretical LCI ($S_0 \rightarrow T_1$) excitation energies for aryl phenyl ketones. The position of the 0–0 phosphorescence band is used in the correlation. The point for ketone VIII does not fit the line*
(Reproduced by permission from J. Phys. Chem., 1969, 73, 1132)

Table 1 *Location of lowest singlet ($n\pi^*$) and ($\pi\pi^*$) singlet and triplet states in aryl phenyl ketones*

Compound	Observed $n\pi^*$ singlet energy (10^3 cm^{-1})	Observed $\pi\pi^*$ singlet energy (10^3 cm^{-1})	Observed $n\pi^*$ triplet energy (10^3 cm^{-1})	Observed $\pi\pi^*$ triplet energy (10^3 cm^{-1})	Phosphorescence lifetime (s)
Benzophenone (I)	27.0	39.6	24.6		4.7×10^{-3}
4-Phenylbenzophenone (II)	26.8	34.4		21.2	0.28
1-Naphthylphenyl ketone (III)	26.7	32.6	20.1		0.51
2-Naphthylphenyl ketone (IV)		34.2	23.8(?)	21.3	1.01
1-Phenanthrylphenyl ketone (V)		32.5	20.7		0.92
9-Anthrylphenyl ketone (VI)		26.1			
1-Pyrenylphenyl ketone (VII)		26.3			
6-Chrysenylphenyl ketone (VIII)		28.5	18.9		0.59

2

15

ketones can be correlated with relative ordering of their $n\pi^*$ and $\pi\pi^*$ triplet states, since it is known that $n\pi^*$ states are more reactive than $\pi\pi^*$ states toward hydrogen abstraction reactions. Thus, 4-phenylbenzophenone (II) is less reactive than benzophenone (I) towards intermolecular hydrogen abstraction, and it follows that ketones (III)—(VIII) should be less reactive than benzophenone as well. This kind of thorough spectroscopic study can do much to provide a basis for the discussion of experimental results on rates of photoreactions, and it is to be hoped that such studies on other groups of molecules of photochemical interest will be undertaken.

A configuration interaction procedure has been evolved which is designed to produce energies and wavefunctions from relatively large-scale computations on ground and excited molecular states based on an orthonormal set of molecular orbitals.[43] The method has been used on the singlet and triplet A_2 ($n \to \pi^*$), singlet and triplet A_1 ($n \to \pi^*$, $n \to 3p$), $^1B_1$ ($\sigma \to \pi^*$, $\pi \to 3s$) and $^1B_2$ ($n \to 3s$) states of formaldehyde, and the energies of the states derived were found to be in good agreement with experiment. Of particular interest is the existence of a $n \to 3$ transition on the oxygen atom in the region of the spectrum around 8 eV. $\alpha\beta$-Unsaturated ketones undergo a variety of photochemical reactions, some of which, such as cis–trans isomerizations, must be a result of geometrical changes in an excited state. Generally, enone photochemistry proceeds via triplet states, and information regarding the geometries of such states of enones is useful. Acraldehyde is a convenient model for $\alpha\beta$-unsaturated ketones, and CNDO

$$\underset{\text{Acraldehyde}}{\overset{H}{\underset{H}{>}}C=C\overset{H}{\underset{\,}{-}}C\overset{\,}{\underset{H}{=}}O}$$

calculations on the $n\pi^*$ and $\pi\pi^*$ triplet states of this molecule produce the evidence which is summarized in Figure 6.[44] From this it is apparent that the lowest triplet state of acraldehyde is an $n\pi^*$ type with a planar configuration $^3A''$ symmetry, and the $\pi\pi^*$ triplet is non-planar with the equilibrium angle of the CH_2 group ca. 72°, and the state has $^3A'$ symmetry. The small barrier to rotation of the CH_2 group in the $^3\pi\pi^*$ state is of particular interest. Electronic transitions of o-, m-, and p-anisaldehyde molecules in the vapour phase have been described,[45] and vibronic spin–orbit interactions in 2-naphthaldehyde and 2-naphthyl methyl ketone discussed.[46] The electronic spectrum of fluorenone has been investigated in the vacuum-u.v. region, and two new bands observed attributable to $\pi \to \pi^*$ transitions of the type $^1A_1 \to ^1A_1$ and $^1A_1 \to ^1B_2$.[47] The polarization of $\pi \to \pi^*$ transitions in molecules of the fluorenone type has been discussed.[48]

[43] L. Whitten and M. Hackmeyer, *J. Chem. Phys.*, 1969, **51**, 5584.
[44] J. J. McCullough, H. Ohorodnyk, and D. P. Santry, *Chem. Comm.*, 1969, 570.
[45] C. P. D. Dwivedi, *Indian J. Pure Appl. Phys.*, 1969, **7**, 410.
[46] S. K. Chakrabarti, *Mol. Phys.*, 1969, **16**, 467.
[47] P. A. Mullen and M. K. Orloff, *J. Mol. Spectroscopy*, 1969, **30**, 140.
[48] J. Dehler and K. Fritz, *Tetrahedron Letters*, 1969, 2157.

Spectroscopic and Theoretical Aspects

An analysis of the $X\,^1A_1 \to A\,^1B_1$ electronic transition of pyrimidine and [$^2H_4$]pyrimidine vapours has been carried out.[49]

The $^1A_{1g} \to {}^1B_{2u}$ transition in benzene is symmetry forbidden, and typically in the gas phase the 0–0 band is missing since the transition derives its intensity via vibronic interaction. In solution, however, the

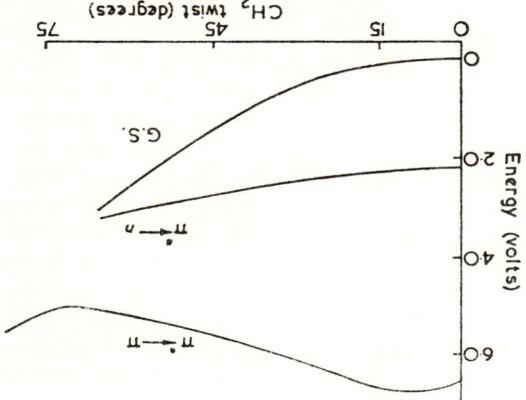

Figure 6 *Potential curves for the ground state, triplet $\pi^* \leftarrow n$ state, and triplet $\pi^* \leftarrow \pi$ state of acraldehyde. The displacements of these curves are only qualitative since each state was calculated at a different level of approximation* (Reproduced by permission from *Chem. Comm.*, 1969, 570)

intensity of the transition can be enhanced, and the 0–0 band may be present. The mechanism by which the solvent interacts with molecules giving rise to such effects has been discussed,[50] and it has been shown that there are two distinct possible interactions between solvent (B) and solute (A): (i) the intensity is enhanced by a borrowing from allowed transitions of A by virtue of the static electric field of the solvent B, (ii) the intensity is increased via borrowing from an intense transition in the solvent B by virtue of terms related to the dispersion interaction between A and B. Second-order perturbation terms are shown to be unimportant. The enhancement in the case of benzene in polar solvents such as water is shown to be due to a borrowing from an allowed transition in the solute (mainly the $^1A_{1g} \to {}^1E_{1u}$), whereas in carbon tetrachloride and other chlorinated solvents the theory supports the mechanism by which the intensity is borrowed from a solvent transition. Matrix effects on the solid phase may affect the position of transitions in molecules. Thus for coronene in polymethylmethacrylate, increasing the pressure between 1 and 30 kbar produces a linear red shift of 22 cm^{-1}kbar^{-1} for the 3414 Å

[49] K. K. Innes, H. D. McSwiney jun., J. D. Simmons, and S. C. Tilford, *J. Mol. Spectroscopy*, 1969, **31**, 76.
[50] N. S. Bayliss, *J. Mol. Spectroscopy*, 1969, **31**, 406.
[51] H. W. Offen, S. A. Balbo, and R. L. Tanquary, *Spectrochim. Acta*, 1969, **25**, A, 1023.

17

($^1L_a \rightarrow {}^1A_1$) transition, and *ca.* 12 cm^{-1}kbar^{-1} for the ($^1L_b \rightarrow {}^1A_1$) weaker 4184 Å transition.[51] The pressure shifts can be explained in terms of the pressure dependence of the refractive index and density of the polymer matrix, and is independent of temperature.

2 Theories of Radiationless Transitions

As in the previous volume, we devote a section of this chapter to a discussion of theoretical treatments of and experimental data on radiationless transitions. The development of a theoretical approach which is sound enough to predict *a priori* rates of radiationless processes, including autoionization, molecular predissociation, internal conversion, intersystem crossing, photodissociation and photoisomerizations, is a problem of such complexity that it has eluded attempts at its solution until recently. Now, however, the basis of an approach has been laid down, largely through the work of Jortner and his co-workers. A very clear exposition of this theoretical treatment has been given in an excellent review article which may become one of the classic papers in the theory of photochemistry.[52] It would be quite out of place for us to attempt to outline in detail this unified approach, and, in any case, the article referred to is written in such an exemplary manner by outstanding theoreticians that it would be impertinent for experimentalists to attempt to emulate it. Nevertheless, the conclusions reached by these authors are of such importance that a summary of them will be given here. The basic difference between this new approach and those previous to it (with the exception of the treatment due to Kasha and Henry[53]) is clearly outlined in ref. 52, and is quoted here directly. 'The key observation in the interpretation of all these relaxation processes is that the system under investigation is in a compound state. Suppose the states of the system under consideration are mentally partitioned into two (or more) sets of zero-order states, and these zero-order states are allowed to interact to generate the exact eigenstates of the system Hamiltonian. Suppose one of the sets of zero-order states (the sparse part) consists of a finite number of discrete energy levels, corresponding to a small subset of the total number of degrees of freedom of the system, while the other set of zero-order states (the dense part) has a continuous spectrum and is associated with an infinite, or effectively infinite, number of degrees of freedom. A relaxation process takes place when a compound state of the system, consisting of some superposition of zero-order discrete states and zero-order continuum states, decays into the continuum.

'It is much more than mere pedanticism to emphasize that the zero-order energy levels of the two subsystems described above cannot be considered to be eigenstates of the Hamiltonian of the total system. The

[52] J. Jortner, S. A. Rice, and R. M. Hochstrasser, *Adv. Photochem.*, 1969, **7**, 149.
[53] B. Henry and M. Kasha, *Ann. Rev. Phys. Chem.*, 1968, **19**, 161.

zero-order levels of the two subsystems are degenerate or quasi-degenerate, and therefore extensive configuration mixing is induced by the (small) interaction which couples the subsystems. As hinted above, the energy levels of the complete system may be represented as a superposition of the zero-order levels of both subsystems. The zero-order states of the system have no real physical significance; such properties of the total system as level widths, absorption coefficients, relaxation times, *etc.* must be described in terms of the properties of compound states.

'It is very important, in the theory of quantum relaxation processes, to understand how an atomic or molecular excited state is prepared, and to know under what circumstances it is meaningful to consider the time development of such a compound state. It is obvious, but nevertheless important to say, that an atomic or molecular system in a stationary state cannot be induced to make transitions to other states by ''small terms in the molecular Hamiltonian''. A stationary state will undergo transition to other stationary states only by coupling with the radiation field, so that all time-dependent transitions between stationary states are radiative in nature. However, if the system is prepared in a nonstationary state of the total Hamiltonian, nonradiative "transitions" will occur. Thus, for example, in the theory of molecular predissociation it is not justified to "prepare" the physical system in a pure Born–Oppenheimer bound state and to force transitions to the manifold of continuum dissociative states. If, on the other hand, the excitation process produces the system in a "mixed" state consisting of a superposition of eigenstates of the total Hamiltonian, a relaxation process will take place. Provided that the absorption line shape is Lorentzian, the relaxation process will follow an exponential decay.'

The review[53] then continues to describe the fundamental quantum mechanics of compound states, surveys the experimental evidence which theoretical treatment must adequately explain, outlines the nature of the Born–Oppenheimer approximation and its breakdown, and develops a simple model of the time evolution of the states of large molecules and an interpretation of the properties of radiationless transitions. The formal theory is shown to yield results in agreement with the simple approach, and a section is devoted to the nature of vibronic coupling. The implications of the analysis of vibronic coupling for the interpretation of molecular relaxation in dense media are surveyed, and a discussion of the radiative decay of molecules and the relationship between radiative decay, radiationless transitions, and the nature of irreversibility is included. Finally, applications of the theory to photochemical reactions, especially *cis–trans* isomerization and photoionization are considered.

Several of these aspects have already been covered in Volume 1 of this series. We thus include below only those features which have not previously been discussed, but the reader seeking a unified comprehensive treatment is strongly recommended to consult the original article.[52]

Long Radiative Lifetimes of Small Molecules.[54]—It has been experimentally demonstrated that the first spin-allowed states of NO_2, SO_2, and CS_2 exhibit anomalously long radiative lifetimes which are considerably longer than those expected on the basis of the integrated oscillator strength. Furthermore it was observed that the absorption spectra of these molecules are very complex, consisting of a large number of lines which could not be classified as corresponding to the usual system of vibrational–rotational manifold of a single electronic state (see references contained in ref. 54). It was suggested[55] that these peculiarities could be due to intramolecular vibronic coupling between a zero-order Born–Oppenheimer state (corresponding to the excited singlet) and a large number of zero-order levels which are quasi-degenerate with this former level, and which do not carry an oscillator strength. A theoretical formulation of this viewpoint has been provided.[54] It is instructive to recall the classification scheme of molecules proposed by these authors.

Statistical Limit. The density of vibronic states is extremely high so that the average vibronic coupling matrix element v appreciably exceeds the mean spacing $\varepsilon = \rho^{-1}$ (where ρ is the density of states) between these levels, so that

$$a\rho \gg 1 \qquad (9)$$

In this limit, inhomogeneous broadening occurs entirely as an intramolecular phenomenon. Intramolecular relaxation occurs on a time scale

$$t \geqslant h\rho \qquad (10)$$

which defines a recurrence time $h\rho$ for the occurrence of the relaxation process. Equation (10) also implies that the radiative bandwidth of the zero-order state which carries oscillator strength exceeds the mean level spacing so that

$$\Gamma \geqslant \varepsilon \qquad (11)$$

Large molecules such as naphthalene, anthracene, and tetracene fall into this category.

Resonance Limit. When the energy levels are coarsely spaced a small number of degenerate or quasi-degenerate zero-order levels may be split by the intrinsic or by external perturbations. Provided that the only decay channel of the coherently excited system (in the absence of external perturbation) will involve a radiative decay process which will exhibit a beat spectrum. Typical examples of this case involve level crossing and level anticrossing in atoms and diatomic molecules (see later). The intramolecular vibronic coupling matrix elements in a diatomic molecule may considerably exceed the radiative widths of the molecular levels. A small molecule (*e.g.* diatomic) molecule may reveal the effects of strong vibronic perturbations between pairs of levels while the interference effects in the radiative

[54] M. Bixon and J. Jortner, *J. Chem. Phys.*, 1969, **50**, 3284.
[55] A. E. Douglas, *J. Chem. Phys.*, 1966, **45**, 1007.

Spectroscopic and Theoretical Aspects

decay are not encountered. In this case, $\Gamma \gg \varepsilon$ while $v\varepsilon \simeq 1$. This can be referred to as coarse level spacing. A typical molecule in this category is CN in which $^2\Sigma-^2\pi$ mixing occurs. The matrix elements connecting the $^2\Sigma$ and $^2\pi$ states are of the order of 1 cm^{-1}, and thus Born–Oppenheimer states which are separated by 1 cm^{-1} will exhibit large perturbations.

Intermediate cases may be separated into two categories.

The Dense Intermediate Case. The eigenstates of the molecule are not sufficiently dense so that equations (9)—(11) are replaced by

$$v a \gtrsim 1 \tag{12}$$

$$t \sim \hbar\rho \tag{13}$$

$$\varepsilon \sim \Gamma \tag{14}$$

Benzene $^1B_{2u}$ has previously been ascribed to this case, but experimental evidence (see Chapter 3) indicates that the decay of the $^1B_{2u}$ state of benzene corresponds to the statistical limit.

The Sparse Intermediate Case. The density of states is rather small, while the vibronic-coupling matrix elements are rather large, due to favourable Franck–Condon vibrational overlap factors. Under these circumstances

$$v a < 1 \tag{15}$$

and the experimental time scale t (for the fluorescence detection) considerably exceeds the recurrence time

$$t \gg \hbar\rho \tag{16}$$

while the coarse spacing of levels exceeds the radiative bandwidth

$$\Gamma \gg \varepsilon \tag{17}$$

The first excited singlet states of SO$_2$, NO$_2$, and CS$_2$ are expected to fall into this category. The density of vibronic states (ground and first triplet) which are quasi-degenerate with the excited singlet is rather low, being of the order of one state per energy interval of 1 cm^{-1}, but these non-linear molecules are characterized by different bond angles in the ground and excited states resulting in large vibrational overlap Franck–Condon factors. Thus, equation (15) is expected to hold so that extensive mixing still occurs.

The theoretical study of the radiative decay of molecules which correspond to the sparse intermediate case is similar to that of the statistical limit, which has been also considered.[56]

The complete Hamiltonian for the system, composed of the molecule and the radiation field, is given by

$$\mathcal{H} = \mathcal{H}_{el} + \mathcal{H}_{rad} + \mathcal{H}_{int} \tag{18}$$

and

$$\mathcal{H}_{el} = \mathcal{H}_{BO} + \mathcal{H}_v \tag{19}$$

[56] M. Bixon and J. Jortner, *J. Chem. Phys.*, 1969, **50**, 4061.

where the molecular Hamiltonian $\mathscr{H}_{el}$ consists of the Born–Oppenheimer Hamiltonian $\mathscr{H}_{BO}$ and an intramolecular perturbation term $\mathscr{H}_v$ which consists of vibronic coupling, spin orbital interactions *etc.* The nuclear kinetic energy operator dominates the coupling although for states of different multiplicity the spin–orbit interaction also has to be included. $\mathscr{H}_{rad}$ is the Hamiltonian corresponding to the free radiation field and $\mathscr{H}_{int}$ is the radiation matter interaction term.

The zero-order states are taken as the eigenstates of the Hamiltonian

$$\mathscr{H}_0 = \mathscr{H}_{BO} + \mathscr{H}_{rad} \qquad (20)$$

They include the vibronic state $\phi_S \equiv |\phi_S; \text{vac}\rangle$ which is the zero-order approximation to the excited state, and the vibronic manifold $\phi_i \equiv |\phi_i; \text{vac}\rangle$, which represent the vibrationally excited states of lower electronic states, while $|\text{vac}\rangle$ corresponds to a zero-photon radiation field. The final zero-order states of the system are $\phi_{E'} = |\phi_0; k, e\rangle$ where ϕ_0 corresponds to the molecular ground state, and $|k, e\rangle$ represents a one-photon state. These final states are normalized per unit energy interval.
The matrix elements of the Hamiltonian between the zero-order states are taken to be

$$\langle \phi_S | \mathscr{H} | \phi_S \rangle = E_S, \qquad \langle \phi_S | \mathscr{H} | \phi_i \rangle = v,$$
$$\langle \phi_i | \mathscr{H} | \phi_j \rangle = E_i \delta_{ij}, \qquad \langle \phi_{E'} | \mathscr{H} | \phi_{E'} \rangle = E\delta(E-E')$$
$$\langle \phi_S | \mathscr{H} | \phi_{E'} \rangle = W, \qquad \langle \phi_i | \mathscr{H} | \phi_{E'} \rangle = 0 \qquad (21)$$

where it is assumed that only the state ϕ_S carried oscillator strength, and that W is independent of E, v is independent of i. A further simplification is that states $\{\phi_i\}$ are uniformly spaced with separation $\varepsilon = \rho^{-1}$.
The diagonalization of the Hamiltonian is carried out in two stages, the first of which involves the diagonalization of the molecular problem which results in the following molecular eigenstates (see ref. 52)

$$\psi_n = \alpha_S{}^n \phi_S + \sum_i \beta_i{}^n \phi_i \qquad (22)$$

with energies E_n. The mixing coefficients are given in the form

$$|\alpha_S{}^n|^2 = v^2[(E_n-E_S)^2 + v^2 + (\pi v^2/\varepsilon)^2]^{-1} \qquad (23)$$

The mixing with the radiation field results in the following (time-independent) eigenfunctions

$$\psi_E = \sum_n a_n(E)\psi_n + \int dE' \, C_{E'}(E)\phi_{E'} \qquad (24)$$

where

$$a_n(E) = [\alpha_S{}^n W/(E-E_n)]\{Z^2(E)/[\pi^2 + z^2(E)]\} \qquad (25)$$

and

$$z(E) = \left\{ \sum_n [|\alpha_S{}^n|^2 W^2/(E-E_n)] \right\}^{-1} \qquad (26)$$

Spectroscopic and Theoretical Aspects   23

The excited state resulting from the optical excitation by a short light pulse is given (only ϕ_s carries oscillator strength) by the expression

$$\psi(t=0) = \int dE \sum_n^u \alpha_s{}^n a_n(E) \psi_E \qquad (27)$$

The time development of this excited state is simply obtained by incorporating the phase development of these eigenstates.

$$\psi(t) = \sum_n^u \alpha_s{}^n \int dE a_n(E) \psi_E \exp\left(-\frac{i}{\hbar} Et\right) \qquad (28)$$

The fluorescence rate is given by

$$\dot{P}(t) = (2\pi/\hbar) |\langle \phi_0 | T | \psi(t) \rangle|^2 \rho_t$$
$$= \Gamma/\hbar |A_s(t)|^2 \qquad (29)$$

where $\Gamma/\hbar$ is the radiative decay probability of the zero-order state ϕ_s; T corresponds to the transition operator, ρ_t is the density of states in the radiation field; and $A_s(t) = \langle \phi_s | \psi(t) \rangle$ is the amplitude of the state ϕ_s in the excited state $\psi(t)$ at time t. This amplitude can be displayed in the form

$$A_s(t) = \frac{2}{\pi \Gamma} \int \frac{\exp(-iE_s t/\hbar) \exp(-|Et|/\hbar) dE}{1 + (16/\Delta^2\Gamma^2) [E^2 + (\Delta^2/4)] [2E/\Delta \coth(\pi\Delta/2) + \coth(\pi E/\varepsilon)]} \qquad (30)$$

where Δ is given by

$$\Delta = 2\pi v^2 \rho \qquad (31)$$

which corresponds to the nonradiative half-width of the distribution of lines.

In the sparse intermediate case, where $\Delta > \varepsilon \gg \Gamma$, two simplifications can be introduced. Firstly $\coth(\pi\Delta/2\varepsilon) \approx 1$ in equation (30), and furthermore, the argument of the Fourier integral (30) is characterized by very small numerical values, except around the points where $E = n$ with $n = 0$, ± 1, ± 2, etc. where the function $\cot(\pi E/\varepsilon)$ diverges. Near these 'critical' points, the cotangent function may be approximated by

$$\cot(\pi E/\varepsilon) \approx (\varepsilon/\pi)(E - n\varepsilon); \quad n = 0, \pm 1, \pm 2, \ldots \qquad (32)$$

The amplitude $A_s(t)$ can be expressed as a sum of integrals whose arguments are functions centred around the points $E = n\varepsilon$:

$$A_s(t) = \frac{2}{\pi \Gamma} \exp\left(-\frac{iE_s t}{\hbar}\right) \sum_n^u \exp\left(-\frac{ien t}{\hbar}\right) \int_{-\infty}^{+\infty} \frac{\exp[-(i/\hbar) Et] dE}{1 + (16/\Delta^2\Gamma^2) [(n\varepsilon)^2 + \Delta^2/4]^2 (\pi E/\varepsilon)^2} \qquad (33)$$

This equation can be rearranged, and taking into account the fact that the integrals obtained are Fourier transforms of Lorentzians, $A_s(t)$ can be expressed as

$$A_s(t) = \left(\frac{\pi\Gamma}{2 \exp[-(i/\hbar) E_s t]}\right) \frac{1}{4}(\varepsilon\Delta\Gamma) \sum_n^u \exp\left(-\frac{i}{\hbar} n\varepsilon t\right) [(n\varepsilon)^2 + \tfrac{1}{4}\Delta^2]^{-1}$$
$$\times \exp\{-(\varepsilon\Delta\Gamma/4\pi\hbar)[(n\varepsilon)^2 + \tfrac{1}{4}\Delta^2]^{-1} t\} \qquad (34)$$

24   *Photochemistry*

The fluorescence rate may then be derived as

$$\dot{P}(t) = (e^2\,\Delta^2\,\Gamma/4\pi^2\hbar) \sum_{n}^{m} (2-\delta_{nm})\cos[(n-m)(\varepsilon t/\hbar)][(n\varepsilon)^2+\tfrac{1}{4}\Delta^2]^{-1}$$

$$\times [(m\varepsilon)^2+\tfrac{1}{4}\Delta^2]\exp\left(-(e^2\Gamma/4\pi\hbar)\{[(n\varepsilon)^2+\tfrac{1}{4}\Delta^2]^{-1}\right.$$

$$\left.+[(m\varepsilon)^2+\tfrac{1}{4}\Delta^2]^{-1}\}\,t\right) \tag{35}$$

Without dwelling upon the complexities of the derivation, it can be seen from equation (35) that the resulting fluorescence rate is characterised by two contributions; exponential decay terms which correspond to the case $n = m$, and interference terms. Thus the fluorescence rate will exhibit a quasi-oscillatory behaviour with periods of the order of the recurrence time h/ε, which in the sparse intermediate case corresponds to $h/\varepsilon \approx 10^{-12}$ s. This recurrence time is appreciably shorter than the experimental time resolution for fluorescence detection, and thus the photon counter will measure an average fluorescence rate where the oscillatory behaviour will be smeared out. Thus the interference terms in (35) can be ignored, and defining the approximate number of states N within the half-width of the inhomogeneously broadened manifold of levels

$$N = \Delta/2\varepsilon = \pi e^2\, \rho_2 \tag{36}$$

the average decay rate of the fluorescence can be displayed in the form

$$\langle \dot{P}(t) \rangle_{\text{Av}} = \frac{\hbar^2\,\pi^2\,e^2\,N}{\Gamma}\sum_{n}^{n}\left[1+\left(\frac{n}{N}\right)^2\right]^{-2}\exp\left\{-\frac{\Gamma_t}{\pi\hbar N[1+(n/N)^2]}\right\} \tag{37}$$

From these results the following conclusions may be reached:

(a) The sparse intermediate case involves a manifold of molecular eigenstates which are well separated relative to their radiative widths. Hence interference effects in the radiative decay of the coarsely spaced levels are not expected.

(b) The fluorescence rate is expressed as a sum of exponentials with decay times τ_e of the order $\tau_e \sim \hbar N/\Gamma$. A non-exponential decay is expected, characterized by a continuous distribution of lifetimes which are approximately in the range $\hbar N/\Gamma$ to $2\hbar N/\Gamma$. A zero-order approximation for the decay law is $\dot{P}(t) \sim (\Gamma/\hbar N)\exp(-\Gamma t/\hbar N)$. The radiative decay time in this limit is characterized by a longer radiative lifetime than that expected on the basis of integrated oscillator strength which is just $\hbar/\Gamma$.

(c) Since equation (37) does not involve interference terms, the same result is expected to hold when the excitation source is not coherent.

(d) The redistribution of the intensity of the zero-order component ϕ_s induces the appearance of many new well-resolved lines which correspond to all the molecular eigenstates ψ_n in the optical spectrum.

(e) In the absence of collisions, the fluorescence quantum yield should be unity.

Spectroscopic and Theoretical Aspects

Thus some of the features of SO_2, NO_2, and CS_2 could be explained on the basis of the treatment above. Photon echoes should also be observable in these molecules.[57]

In the intermediate dense limit, it would be expected that the radiative decay would exhibit an oscillatory behaviour as the manifold of radiatively decaying eigenstates will reveal interference effects and the recurrence time is of the order of h/Γ, although spectroscopic evidence is not available for this case.

Statistical Limit—Radiative Decay of Polyatomic Molecules.—In the statistical limit, where equations (10) and (11) and (30) are obeyed, equation (30) can be evaluated to give a simple expression for the rate of fluorescence emission[56]

$$\dot{P}(t) = (\Gamma/h) \exp\{-[(\Gamma+\Delta)/h]t\} \quad (38)$$

The fluorescence decays exponentially with an apparent lifetime τ_{expt}

$$\tau_{\text{expt}}^{-1} = (\Gamma/h) + (\Delta/h) \quad (39)$$

corresponding to the sum of the inverse radiative lifetime of the state ϕ_s, and the inverse non-radiative lifetime $\tau = h/\Delta$.

The integrated fluorescence yield up to time t is then

$$P(t) = \int_{t_0}^{0} \dot{P}(t')\,dt' = \frac{\Gamma}{\Gamma+\Delta}\left\{1 - \exp\left[-\left(\frac{\Gamma+\Delta}{h}\right)t\right]\right\} \quad (40)$$

Under normalization conditions, the quantum yield for emission $Y = P(\infty)$ so that

$$Y = \Gamma/(\Gamma+\Delta) \quad (41)$$

The results may be summarized as follows:

(a) In the statistical limit, the experimental radiative decay is a pure exponential.

(b) The experimental radiative decay time determined on a time scale which is appreciably shorter than the recurrence time $t \ll h/\rho$ consists of independent contributions of non-radiative and radiative components. In large molecules where the recurrence time is exceedingly long, this establishes the decay law.

(c) The non-radiative decay component Δ/h in the decay time obtained for a manifold of molecular eigenstates in an inhomogeneously broadened line corresponds to a set of indistinguishable levels. The interference effects between a large number of closely spaced indistinguishable levels leads to intramolecular interference effects and to the shortening of the radiative lifetime in the statistical limit.

(d) The quantum yield of fluorescence in the statistical limit is

$$Y = \Gamma/(\Gamma+\Delta)$$

[57] T. Lefebure and J. Savolainen, *Chem. Phys. Letters*, 1969, 3, 449.

26 *Photochemistry*

Interference Effects in the Radiative Decay of Closely Spaced Levels.[58]—Interference effects in the decay of closely spaced atomic or molecular states have been known for some time. As an example, when two levels displaying resonance fluorescence are split in zero magnetic field by fine

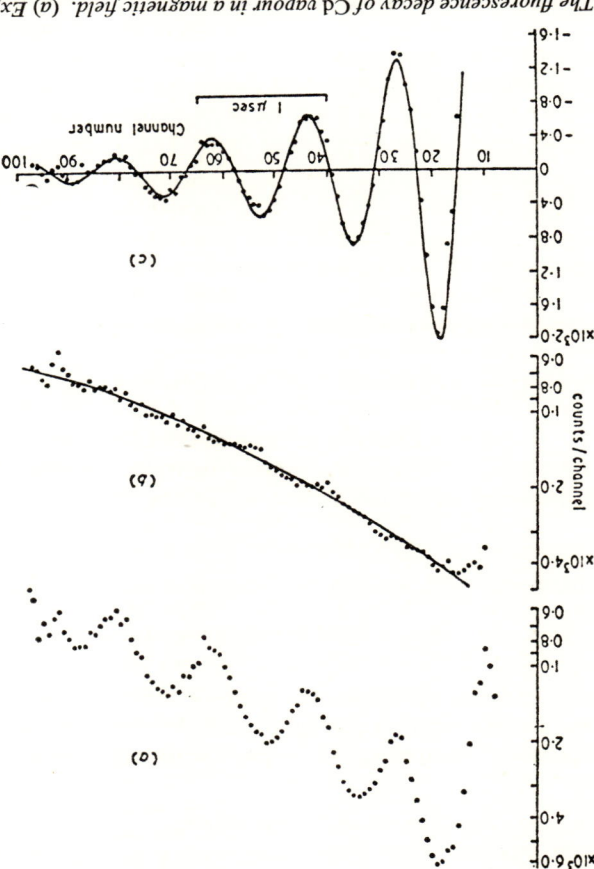

Figure 7 *The fluorescence decay of Cd vapour in a magnetic field.* (a) *Experimental data exhibiting the phenomenon of quantum beats.* (b) *The exponentially decaying component.* (c) *The decaying modulated component*
(Reproduced by permission from *Proc. Phys. Soc.*, 1964, **84**, 176)

and hyperfine interactions, and cross by the application of an external magnetic field, provided that the two levels are closer in energy than their radiative widths, spatial interference in resonant scattering can be observed (again provided that the two levels are connected to the ground state by allowed dipole transitions). A closely related phenomenon is the observa-

[58] M. Bixon, J. Jortner, and Y. Dothan, *Mol. Phys.*, 1969, **17**, 109.

tion of quantum beats in the radiative decay of a set of coherently excited states. Since transitions to the ground state can take place *via* two or more channels, interference effects may be observed (see Figure 7).[58] The experimental conditions required for the observation of quantum beats are that the resolving power of the photon-counting apparatus must be inadequate to resolve the two transitions, because the resolution process involves the separation of the two channels, thus eliminating the possibility of interference. The quantum beat signals reported in the literature (see refs. 58 and 52) involve the interference between two ($m = 1$ and $m = -1$) Zeeman components split by a magnetic field. In this case the two decay channels are characterized by different angular polarizations. Thus, in principle, the two decay channels can be distinguished from one another experimentally. Such levels may be termed 'distinguishable'.[58] On the other hand, the case where a number of decaying levels emitting photons of the same polarization is often encountered. As an example, the degenerate zero-order Born–Oppenheimer singlet and triplet states of a small molecule may be considered. The molecular eigenstates contain equal mixtures of singlet and triplet components and are split by the spin–orbit interaction. The decay channels, which have the same polarization, cannot be separated as the Zeeman levels can, and this is referred to as a system of 'indistinguishable' levels. In a system of distinguishable levels the observation of quantum beats as in Figure 7 require that the photon counting procedure covers only a limited solid angle and interference affects are only observed for partial transition probabilities, since the spatial integration over all angles and polarizations results in the vanishing of the interference terms. On the other hand, for a system of indistinguishable levels the quantum beats also appear in the total integrated emission rate.[58] The theory will not be presented here, but the results [58] can be used to calculate the following quantities for 'distinguishable' and 'indistinguishable' levels:

(*a*) the total differential decay rate $\dot{P}(t)$ which corresponds to the photon flux integrated over all spatial angles and polarizations;

(*b*) the partial differential decay rate $\dot{P}_R{}^n(t)$ determined by setting a photon counter to admit photons in a limited directional and polarization range. Obviously $\dot{P}(t) = \Sigma_R \dot{P}_R(t)$;

(*c*) the integrated photon yield $P(t) = \int_0^t \dot{P}(t')\,dt'$; (*d*) the partial integrated photon yield $P_R(t)$.

The results clearly show that for indistinguishable levels, such as would be encountered in the resonant limit for small molecules, quantum beat signals should be observable in both $\dot{P}_R(t)$ and $\dot{P}(t)$. Experimental confirmation of this would be of great importance.

The theories of the Bixon–Jortner group [52, 59] may be summarized in tabular form (Table 2), in which general features of the decay of molecular

---

[59] J. Jortner and M. Bixon, *Israel J. Chem.*, 1969, **7**, 189.

Table 2 General features of the decay of molecular levels (Reproduced by permission from J. Chem. Phys., 1969, **50**, 3284)

Physical property	Resonance limit	Coarse distribution	Sparse intermediate case	Dense intermediate case	Statistical limit
Number of states $N = v^2 \rho^2$ v^2/ε^2 [a,b]		$N \sim 1$	$N > 1$	$N \sim 1$	$N \gg 1$
Level separation relative to radiative width	$\varepsilon \sim \Gamma$	$\varepsilon \gg \Gamma$	$\rho^{-1} \gg \Gamma$	$\rho^{-1} \sim \Gamma$	$\rho^{-1} \ll \Gamma$
Line shape	Natural radiative + conventional broadening	Natural radiative + conventional broadening	Well-separated lines	No observable effect	Lorentzian line shape $\Delta = \pi v^2 \rho$
Time scale t relative to recurrence time	$t \sim \hbar/\varepsilon$	$t \gg \hbar/\varepsilon$	$t \gg \hbar\rho$	$t \gg \hbar\rho$	$t \ll \hbar\rho$
Mode of decay	Beat spectrum	Normal exponential decay	Sum of (slowly varying) exponentials	Oscillatory (beats) (?)	Exponential
Mean radiative decay time [c] Experimental fluorescence quantum yield [d]	$Y = 1$	$\tau_e \sim \hbar/\Gamma$ $Y = 1$	$\tau_e \sim N\hbar/\Gamma$ $Y = 1$	$\tau_e \sim \hbar/\Gamma$ $Y = 1$	$\tau_e = \hbar/(\Delta + \Gamma)$ $Y = [\Gamma\Delta/(\Gamma + \Delta)] < 1$
Features of relaxation	External	External	External	External	Intramolecular $\tau_{NR} = \hbar/\Delta$
Examples	Level crossing Level anticrossing	CN	SO_2, NO_2, CS_2	?	Naphthalene Anthracene Tetracene

[a] In the resonance limit, we shall just consider two levels with spacing ε. The number of states N is, of course, meaningless in this limit.
[b] In the case of a coarse distribution, we consider a pair of levels with spacing ε, where $v\varepsilon \sim 1$.
[c] In the case of oscillatory decay with exhibits quantum beats, a proper definition of the radiative decay time cannot be provided.
[d] The quantum yield in the statistical limit is measured on a time scale shorter than the recurrence time.

Spectroscopic and Theoretical Aspects

levels for the resonant case, the sparse intermediate case, the dense intermediate case, and the statistical case are considered. The theories outlined above have been carefully reworked [60] and a careful account taken of the difference between distinguishable and indistinguishable levels, and allowance made for variable coupling of the sparse system to the dense system of states. Only certain vibrational modes in the dense manifold have the appropriate symmetries to couple to the sparse manifold and thereby contribute to the radiationless transition. This has been taken into account in the fashion in which the zero-order manifolds of the molecule are classified.[60] External perturbations are also considered,[60] and it is shown under what conditions the external perturbation has either no effect, a small, or a large effect on the radiationless transitions in the statistical, intermediate, and resonance coupling limits respectively. The nature of irreversibility in an isolated molecule has been considered.[61] Recurrence times can be defined for a molecular system in such a way that if the recurrence time is greater than the natural time limit for the experiment, irreversible behaviour will occur, whereas when the recurrence time is shorter than the natural time limits, the molecule will 'resonate' and reversible behaviour will ensue. The intermediate case where some levels are strongly coupled while others are only weakly coupled can provide an instance when the recurrence time for the strongly coupled levels is short compared to the natural time limit, whereas the recurrence time for the weakly coupled states is longer than the natural time limit. Thus, the intermediate case corresponds to both the large and small molecule limits in the same molecule.

The energy gap law for radiationless transitions in large molecules has also been considered.[62] The non-radiative transition rate between the electronic states m and n can be considered to be

$$\Gamma_{mn} = (2\pi/\hbar) \sum_r \sum_s \rho(mr) | v_{mr,ns} |^2 \delta(E_{mr} - E_{ns}) \qquad (42)$$

where $\rho(mr)$ is the Boltzmann factor for the state ψ_{mr} (energy E_{mr}) and $v_{mr,ns}$ is the matrix element of the total molecular Hamiltonian between the mr'th and ns'th (energy E_{ns}) zero-order (Born–Oppenheimer) states. Previous treatments have evaluated expression (42) by factorization into the product of a uniform coupling term and an approximate density of states; a semiempirical approach determined the dependence of Γ_{mn} upon ΔE. In the work considered here [62] the double sum in (42) is evaluated by considering in the limit of zero excitation energy, the general line-shape function encountered in optical absorption in systems with many vibrational degrees of freedom. All molecular vibrations are included, and an instructive analytical expression derived for the rate of non-radiative transitions

---

[60] K. F. Freed and J. Jortner, *J. Chem. Phys.*, 1969, **50**, 2916.
[61] K. F. Freed, *J. Chem. Phys.*, 1970, **52**, 1345.
[62] R. Englmann and J. Jortner, *Mol. Phys.*, 1970, **18**, 145.

without factorization products of coupling terms and densities of states. In the final representation, the new parameters are the reduced displacements in equilibrium configurations of the various normal modes in the electronic states of interest. Two limiting cases are considered.

(a) In the 'weak coupling' limit the horizontal displacement of the potential energy surfaces is small, and it is shown that this situation corresponds to the usual picture of radiationless transitions in large organic molecules.

(b) In the 'strong coupling' limit, the configurational changes are large, possibly corresponding to the important case of intramolecular rearrangement reactions (see next section).

This approach has been extended [63] by a consideration of the consequences of internal rotation in the many-phonon description of non-radiative processes. The treatment applies particularly to *cis–trans* isomerization, and will be considered in the next section.

Similar conclusions to those reached by the above authors have been arrived at independently by another group of workers,[64, 65] and the effect of the bandwidth of the exciting radiation has also been considered. A simplified molecular model is considered in which one Born–Oppenheimer (B–O) state $|i\rangle$ is electric-dipole coupled with the ground state and is bracketed symmetrically by a quasi-continuum of dipole-forbidden B–O states $|f\rangle$ of width $\Delta\alpha$. This model would be applicable to a large isolated model such as naphthalene or anthracene in which $|i\rangle$ represents any vibration–rotation level of the second excited singlet S_2, and $|f\rangle$ represents the manifold of upper vibrational levels of the S_1 state. Assuming that $|i\rangle$ is coupled to each $|f\rangle$ by a constant matrix element V through the non-B–O part of the molecular Hamiltonian $\mathscr{H} - \mathscr{H}_{\text{HB-O}}$, several characteristic times should be considered.

(*i*) Relaxation time for the decay of $|i\rangle$ due to the energy-conserving transition $|i\rangle \to |f\rangle$.

$$\tau_r^{-1}(\text{int}) = \tau(\text{int}) = \frac{2\pi}{\hbar^2}|V|^2 p \qquad (43)$$

where p is density of states $|f\rangle$ and $\tau(\text{int})$ is the broadening of the state $|i\rangle$ due to internal coupling. In the limit of $|f\rangle$ becoming infinite, the line shape becomes Lorentzian (see earlier theories, ref. 52).

(*ii*) The radiative time $\tau_r^{-1} = \tau(\text{rad})$ (natural linewidth).

(*iii*) The correlation time for photon absorption $\tau_c(\text{rad})$ is the effective time of interaction of a photon with the molecule and may be defined as

$$\tau_c(\text{rad}) = \langle 0 | V_r^2 | 0 \rangle^{-1} \int \langle 0 | V_r(t) V_r(0) | 0 \rangle \, \mathrm{d}t \qquad (44)$$

where V_r is the electric dipole coupling between the molecule and field and

---

[63] W. M. Gelbart, K. F. Freed, and S. A. Rice, *J. Chem. Phys.*, 1970, **52**, 2460.
[64] W. Rhodes, B. R. Henry, and M. Kasha, *Proc. Nat. Acad. Sci. U.S.A.*, 1969, **63**, 31.
[65] W. Rhodes, *J. Chem. Phys.*, 1969, **50**, 2885.

Spectroscopic and Theoretical Aspects

averages are taken over the ground state. For the present model this is easily shown to be given by $\tau_c^{-1}(\text{rad}) = \pi^{-1}\Delta\omega$. This is referred to as the excitation correlation time.

(iv) The correlation time for the manifold of transitions $|i\rangle \to |f\rangle$, referred to as the internal correlation time is denoted by $\tau_c(\text{int})$, and is given by $\tau_c^{-1}(\text{int}) = \pi^{-1}\Delta\alpha$.

The relaxation times $\tau_r(\text{rad})$ and $\tau_r(\text{int})$ are the familiar kinetic rate constants for first-order radiative decay and internal conversion respectively. The correlation times are less familiar, but the quantity $c\tau_c(\text{rad}) = c\pi(\Delta\omega)^{-1}$ is the coherence length of the photon, *i.e.* the width of a wavepacket formed by a coherent superposition of waves of equal amplitude over the interval $\Delta\omega$.

The description of photon absorption depends on the relative magnitudes of these times, and several cases of interest arise.

Case (a). $\tau_c(\text{rad}) \ll \tau_r(\text{rad}) \ll \tau_r(\text{int})$ [Figure 8(a)]. This is a trivial case in which the internal broadening of $|i\rangle$ is less than the radiative bandwidth and results from extremely weak internal interactions. Excitations occur to the B–O state $|i\rangle$ which then decays radiatively before internal conversion to $|f\rangle$ can occur.

Case (b). $\tau_c(\text{rad}) \ll \tau_c(\text{int}) \ll \tau_r(\text{int}) \ll \tau_r(\text{rad})$ [Figure 8(b)]. Here internal broadening of $|i\rangle$ is greater than the radiative bandwidth but the radiation field completely covers the spectrum of stationary states containing components of $|i\rangle$. Thus the excitation correlation time is much less than the internal correlation time and photon absorption may be regarded as an excitation to the B–O state followed by a radiationless transition.

Case (c). $\tau_c(\text{int}) \ll \tau_c(\text{rad}) \ll \tau_r(\text{int}) \ll \tau_r(\text{rad})$ [Figure 8(c)]. This is an intermediate case in which the radiation field covers most of the spectrum of stationary states containing components of $|i\rangle$. There are many internal transitions during the excitation correlation time. Strictly, one should consider the transitions as occurring to molecular stationary states, since the transitions $|i\rangle \to |f\rangle$ are part of the radiative process. However, little error is introduced by regarding the process as a radiationless transition following excitation to $|i\rangle$.

Case (d). $\tau_c(\text{int}) \ll \tau_r(\text{int}) \ll \tau_c(\text{rad}) \ll \tau_r(\text{rad})$ [Figure 8(d)]. The radiation field bandwidth is sufficiently narrow to resolve the internally broadened absorption band and covers only a small portion of the stationary states containing components of $|i\rangle$. There are many internal transitions during each photon collision, and it is meaningless to consider photon absorption as a real transition to $|i\rangle$ followed by a radiationless transition. Transitions must be considered as occurring to molecular stationary states. Subsequent emission from these stationary states may be different from the corresponding emission from $|i\rangle$.

For comparison between these results and previous discussion, the reader is referred to Table 2.

The pioneering theoretical approach to radiationless transitions was produced by G. W. Robinson and co-workers, and efforts are still made to use this theory to predict rates of radiationless processes in molecules. It

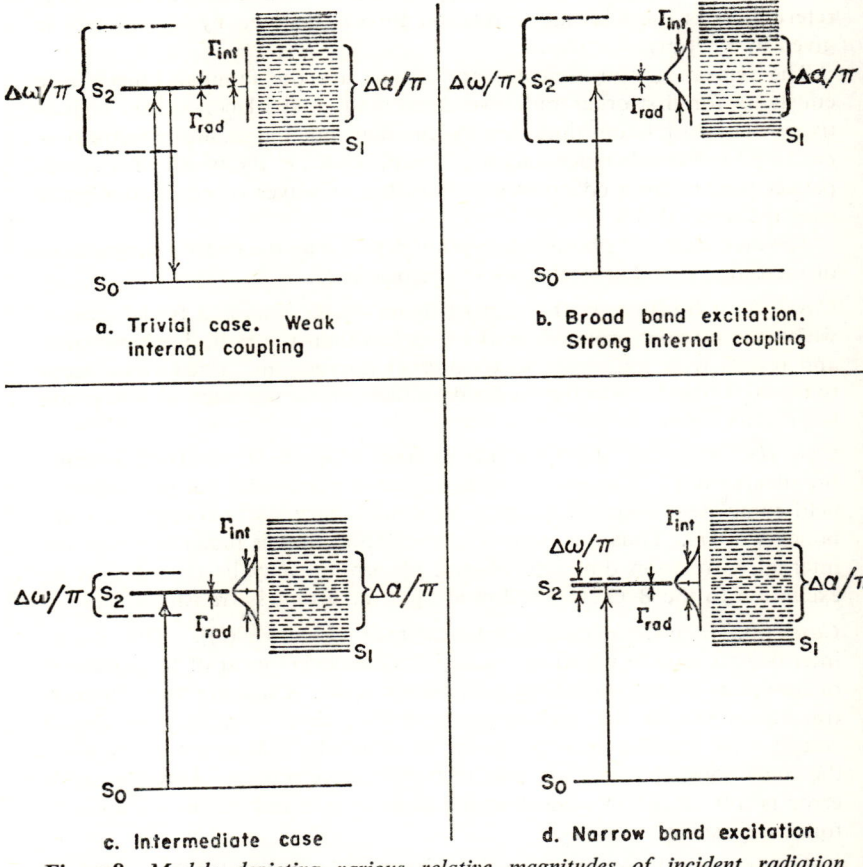

Figure 8 *Models depicting various relative magnitudes of incident radiation bandwidth and internal coupling width.* (*See text for details*)
(Reproduced by permission from *Proc. Nat. Acad. Sci. U.S.A.*, 1969, **63**, 31)

must be noted, however, that not only do previous results conflict badly with experiment (see Chapter 3, sections on benzene and biacetyl), but the whole basis of the Robinson approach is highly questionable, being based on pure Born–Oppenheimer states (see discussion above, ref. 52, and also Volume 1 of this series). Given these criticisms, it would be somewhat surprising if calculations based on this model gave realistic results. Nevertheless, the treatment is useful in that it highlights some of the parameters which may influence rates of radiationless processes. Of paramount importance in the approach is the estimation of Franck–Condon factors for the transition between the upper and lower excited states of concern. For the $^3B_{1u} \to {}^1A_{1g}$ and $^1B_{2u} \to {}^1A_{1g}$ transitions in benzene and perdeuteriobenzene, these have been estimated assuming both harmonic and anharmonic vibrational potentials.[66] The results obtained are shown in Table 3. It can be seen that although there is poor agreement with the

Table 3 *Calculated radiationless transition rates in benzene*

Transition (non-radiative)	Harmonic rate (s^{-1})	Anharmonic rate (s^{-1})	Experimental rate (s^{-1})
$^3B_{1u} \to {}^1A_{1g}$ (C$_6$H$_6$)	9.0×10^{-5}	4.5×10^{-3}	2.4×10^{-2}
$^3B_{1u} \to {}^1A_{1g}$ (C$_6$D$_6$)	1.2×10^{-10}	1.2×10^{-8}	
$^1B_{2u} \to {}^1A_{1g}$ (C$_6$H$_6$)	32.4		

one experimental result available, inclusion of anharmonicity yields results more nearly in agreement with experiment than a purely harmonic model. This observation may go some way to explaining the effect of environment and temperature on triplet lifetimes of molecules, since values have been obtained by experiment for the crystal phase anharmonicities for the C—C and C—H totally symmetric stretching mode, and these have been shown to be different from those in the gas phase (Figure 9). Molecular vibra-

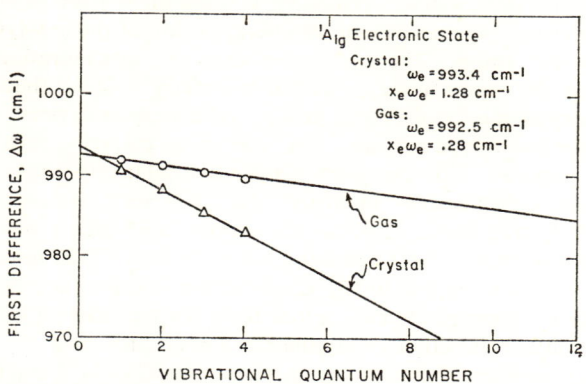

Figure 9 *Anharmonicity plot for the totally symmetric carbon–carbon vibration of benzene in the $^1A_{1g}$ electronic state*
(Reproduced by permission from *J. Chem. Phys.*, 1969, **51**, 4548)

tional overlap integrals have been the subject of further discussion.[67] The average coupling matrix elements of Jortner's theory[52] for intramolecular radiationless transitions have been calculated by the methods of crystal physics, but very poor agreement was obtained, which would necessitate a refinement of the calculations.[68]

[66] D. M. Barland and G. W. Robinson, *J. Chem. Phys.*, 1969, **51**, 4548.
[67] R. J. M. Bennett and W. B. Somerville, *Nature*, 1969, **223**, 489.
[68] R. Englemann, *Israel J. Chem.*, 1969, **7**, 221.

An argument has been advanced that efficient internal conversion in polyatomic molecules occurs when the energy gap between two potential surfaces is sufficiently small so that at finite nuclear velocities the corrections to the Born–Oppenheimer approximation can lead to significant transition probabilities.[69] This will occur particularly near a conical intersection of two surfaces. However, the approach is not well developed, and does not compare favourably with the Jortner treatment. A treatment similar to the latter has been given although different formalism is used.[70] The conclusions reached by this author are in essential agreement with those given in Table 2, and also summarized in Figure 8.

In singlet–triplet transitions spin–orbit interactions are of prime importance. For the aromatic hydrocarbons naphthalene, anthracene, and phenanthrene a formalism has been developed to calculate spin–orbit interactions based on $\sigma\pi$ interactions and which involves singlet and triplet $\sigma\pi^*$ and $\pi\sigma^*$ states and their interactions with both ground and lowest triplet states.[71] C—H and C—C σ-electrons are considered separately, and radiative lifetimes obtained for the above-mentioned molecules are in reasonable agreement with experiment. A very similar approach has extended the treatment to include other aromatic hydrocarbons.[72] Spin-orbit effects have also been considered for aromatic ketones, in particular substituted benzophenones.[73] Aromatic ketones differ from aliphatic ones in that numerous $\pi\pi^*$ states lie in the energy region of the low-lying singlet and triplet $n\pi^*$ states. The $\pi\pi^*$ states may mix configurationally with many of the $n\pi^*$ states if the molecular symmetry is lower than the local C_{2v} symmetry of the C—C(O)—C group. Effectively the system no longer has strictly n and π type orbitals. The $^1\pi\pi^*$ states carry so much oscillator strength compared with the $^1n\pi^*$ states that relatively slight $n\pi^*$–$\pi\pi^*$ mixing can considerably intensify the $A \to {}^1n\pi^*$ transitions. The $(A_1)^1\pi\pi^* \leftarrow A_1$ transitions close to the lowest $^1n\pi^*$ states have $f \sim 10^{-1}$ whereas the $^1n\pi^*(A) \leftarrow {}^1A$ transitions have $f \simeq 10^{-3}$. This indicates that actual $n\pi^*$–$\pi\pi^*$ mixing due to C_{2v}–C_2 distortion is slight, and thus the C_{2v} states make an appropriate conceptual basis for considering excited-state properties. Other quantitative arguments support this view.

(a) The f number of the $^1A_2(n\pi^*) \leftarrow {}^1A_1$ transition in formaldehyde is 2.4×10^{-4}, whereas that for benzophenone is 30×10^{-4}, and thus the additional intensity is ca. 28×10^{-4}, which represents only a few percent of the intensity of the nearby $^1\pi\pi^* \leftarrow {}^1A$ transition with the appropriate polarization.

(b) The $^1n\pi^* \leftarrow {}^1A$ transition of benzophenone polarized parallel to the C—O direction shifts by only 20 cm^{-1} on perdeuteriation, suggesting only slight involvement of the phenyl rings in the transition. That the transition

---

[69] E. Teller, *Israel J. Chem.*, 1969, **7**, 227.
[70] S. E. Webber, *J. Phys. Chem.*, 1970, **74**, 475.
[71] B. R. Henry and W. Siebrand, *J. Chem. Phys.*, 1969, **51**, 2396.
[72] F. J. Adrian, *J. Chem. Phys.*, 1970, **52**, 622.
[73] S. Dym and R. M. Hochstrasser, *J. Chem. Phys.*, 1969, **51**, 2458.

is polarized in the C—O direction in a substituted benzophenone is shown in Figure 10. The presence of relatively low-energy $\pi\pi^*$ states also influences the course of non-radiative transitions. Electronic relaxation from $n\pi^*$ to $\pi\pi^*$ may occur *via* different mechanisms. If the A-type $n\pi^*$ and $\pi\pi^*$

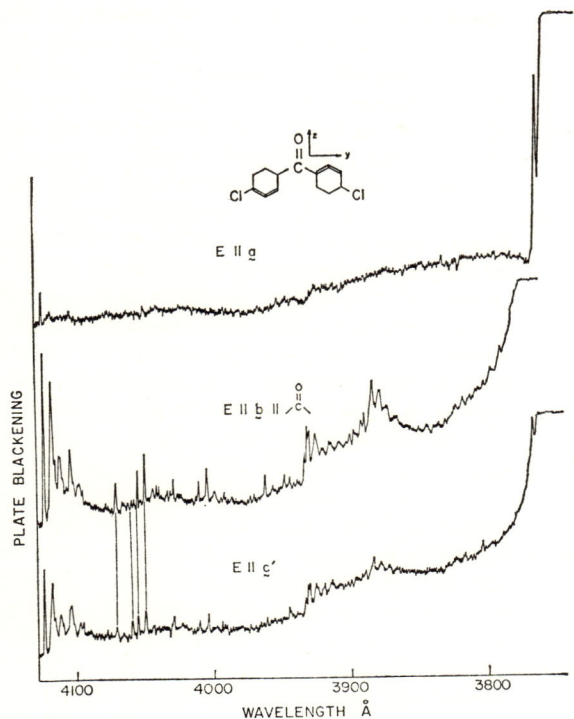

Figure 10 *Singlet–triplet absorption of 4,4'-dichlorobenzophenone crystal at 4·2 K. Note that the singlet origin is polarized exclusively along the* b *axis unlike the triplet. Also notice that the line which appears at 4060 Å is absent from the* b *spectrum; this is a vibronic origin*
(Reproduced by permission from *J. Chem. Phys.*, 1969, **51**, 2458)

states are mixed at some C_2 configuration (distorted from C_{2v}), then the intersystem mixing is controlled by a matrix element of the form

$$\langle F_{S0} | {}^1n\pi^* | \mathcal{H}_{S0}{}^2 | {}^3\pi\pi^* \rangle | F_{TV} \rangle \, C(n\pi^*, \pi\pi^*) \tag{45}$$

where F_{S0} and F_{TV} are vibrational wavefunctions for the configurationally mixed $n\pi^*$ and $\pi\pi^*$ states of singlet (S) and triplet (T) types. The coefficient $C(n\pi^*, \pi\pi^*)$ can be estimated to be *ca.* $(0.1)^{\frac{1}{2}}$ for benzophenone (see ref. 73). If on the other hand singlet and triplet $n\pi^*$ and $\pi\pi^*$ states from C_{2v} are not configurationally mixed, they will be vibronically mixed and the pertinent matrix element will be

$$\langle F_{S0}' | \langle {}^1n\pi^* | \mathcal{H}_{S0}{}^2 | {}^3\pi\pi^* \rangle Q | F_{TV}' \rangle C' \tag{46}$$

where Q is the nuclear displacement that mixes the C_{2v} states and C' is the vibronic interaction coefficient. These two results are essentially identical in principle, although the basic vibrational wavefunctions are different in the two cases. They both represent direct coupling of modified $^1n\pi^*$ and $^3n\pi^*$ states having different geometry, and intersystem crossing is expected to be quite effective even if there are no $^3\pi\pi^*$ states available for direct coupling to $^1n\pi^*$. In the C_2 distorted ketones having nearby but higher-energy $\pi\pi^*$ states the intersystem mixing is expected to be only slightly less than in the case where a $^3\pi\pi^*$ state actually lies below the $^1n\pi^*$ state. The actual relative rates of intersystem crossing will depend upon just how geometrically different are the singlet and triplet states in special cases. The rate of electronic relaxations from $^1n\pi^*$ to the triplet manifold is fixed at $1\cdot5 \times 11\,\mathrm{s}^{-1}$ for benzophenone and 3,3'-dibromobenzophenone. From linewidth measurements on the triplet state, it is concluded that vibrational relaxation in the triplet state of benzophenone occurs at about the same rate. The actual rate of relaxation of $^1n\pi^*$ levels in $4 \times 10^{11}\,\mathrm{s}^{-1}$ seems to be too fast for vibrational relaxation alone, and it is thus proposed that in benzophenones intersystem crossing occurs significantly from vibronic levels of the $^1n\pi^*$ state. It is thus interesting that in aromatic ketones like benzophenone, vibrational relaxation may be the slowest process occurring after light absorption to the singlet states: internal conversion and intersystem crossing are apparently the dominant non-radiative transitions. There is a chance that the increased relaxation rates for the higher vibronic levels of the singlet are caused by the more favourable location of $^3\pi\pi^*$ states with which they may interact directly.

Experimental evidence concerning the rates of radiationless and radiative processes in the condensed phases will be found in Chapter 2, and for the gas phase in Chapter 3. As a conclusion to this section, it would perhaps be advisable to re-emphasize the advantages of the theoretical treatments based on compound states of molecular systems [52, 64, 70] over those based on the use of zero-order states (Robinson, *etc.*, see Volume 1). These are [52]

(*a*) A theory based on the use of compound states automatically focuses attention on the exact molecular eigenstates of the system. It is these states which are needed in the description of absorption line shapes and other spectroscopic properties of the system.

(*b*) An important implication of the breakdown of the B–O approximation in the excited electronic states of medium-sized and large molecules is the prediction of inhomogeneous line broadening. The compound state theory generates a self-consistent scheme for extracting information about electronic relaxation processes from spectroscopic lineshape data.

(*c*) If the widths of two or more inhomogeneously broadened eigenstates exceeds their separation, interference effects may be observed. The rate of

Spectroscopic and Theoretical Aspects

decay and its relationship to the absorption line shape must be analysed by a theory based on molecular eigenstates, since the simpler theories are only capable of evaluating relaxation times corresponding to exponential decay rates and their mated Lorentzian absorption profile.

(*d*) The use of compound molecular states leads to the concept of the irreversibility of electronic relaxation processes without *ad hoc* assumptions.

(*e*) The theory accounts properly for the properties of coupled radiative and radiationless decay processes.

(*f*) For a case when $\rho v \sim 1$, application of the general theory is complex but still formally possible.

(*g*) The approach may be used to describe photochemical reactions.

It must be stressed, however, that as yet reliable numerical calculations are still not possible, owing to the technical difficulties of *a priori* estimates of matrix elements, densities of states, *etc*. Nevertheless, the provision of the theoretical basis for further calculation is a significant step forward, and should induce greater efforts in both theoretically and experimentally oriented workers in the field.

3 Theories of Photochemical Reactions

The initial part of this section will be devoted to an application of the generalized theories outlined in the previous section to photochemical reactions in general, and to *cis–trans* isomerizations in particular. A discussion of predissociation follows, and then a consideration of energy partitioning in photochemical reactions, with emphasis on experimental observations. The application of theories of unimolecular reactions to photochemical cases is next considered, and finally a short discussion on orbital symmetry considerations is included.

Photoisomerizations.—The approach of Jortner and co-workers to photodissociations was included in Volume 1 of this series, and has been reviewed.[52] A similar treatment has recently been applied to the *cis–trans* isomerization of stilbene,[74] a molecule for which there is much experimental data concerning rates of radiative and non-radiative processes (see ref. 74 for summary). The most recent work suggests that the *cis–trans* isomerization may proceed *via* a mechanism summarized in Figure 11. The temperature dependence of the *cis* to *trans* triplet isomerization is assumed to reside in step 3′, *i.e.* the activation energy is interpreted as the thermal energy required to overcome a slight rise in the excited singlet S_1' potential curve as one moves from the *trans*-configuration to the crossing with the second triplet T_2^1. On the *cis* side of the electronic energy plots the potential curves are found to lower upon twisting, consistent with the experimental finding that ϕ_{c-t} is an almost constant function of temperature.

---

[74] W. M. Gelbart and S. A. Rice, *J. Chem. Phys.*, 1969, **50**, 4775.

Because different molecules exhibit different behaviour, it was considered impossible to choose a single theoretical model which might reflect definitively the observed fine details of specific effects. Instead a model was chosen which is faithful to only a given case, that of the unsubstituted stilbene, and then an attempt was made to predict the general properties of such an idealized physical system by a rigorous quantum-mechanical formulation of its time development following preparation in a particular initial state, *e.g.* the lowest singlet reached directly by optical excitation. More specifically the *cis–trans* conversion is understood in terms of the decay of a time-dependent linear superposition of the iso energetic vibronic

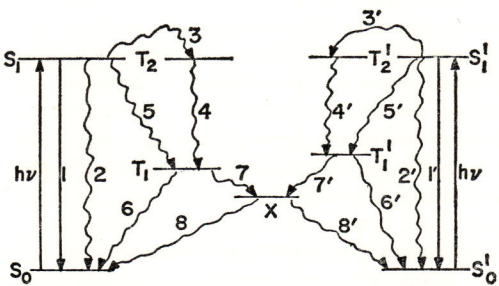

Figure 11 *A schematic plot of electronic energy* vs. *angle of twist* (*about the double bond*) *in unsubstituted stilbene*
(Reproduced by permission from *J. Chem. Phys.*, 1969, **50**, 4775)

states belonging to the various possible isomeric configurations of the molecule. The approach may thus be considered a quantum-mechanical treatment of the view that isomerizations may be an important route for radiationless transitions.[75, 76]

The following prototype of the mixed state representing, say, the excited *trans*-molecule is considered. The zero-order eigenstates are taken to consist of: (*a*) a discrete Born–Oppenheimer state in the lowest excited *trans*-singlet reached directly by optical excitation: in the case of absorption to higher bands, in solution, this initial state is still taken to be the vibrationless S_1 singlet, since internal conversion and vibrational relaxation will be fast; (*b*) a discrete vibronic state in the second *trans*-triplet, T_2^1 which is nearly degenerate with the vibrationless singlet level; (*c*) a dense manifold of vibrational states belonging to the lowest-lying (twisted) triplet *e.g.* T_1^1 which are isoenergetic with the two zero-order states (*a*) and (*b*). Assuming that the likelihoods of the twisted stilbene triplet decaying to *cis* and *trans* ground state are each 0·5, then the probability that the excited *trans* singlet relaxes to the twisted triplet determines the overall rate of isomerization.

[75] D. Phillips, J. Lemaire, C. S. Burton, and W. A. Noyes, jun., *Adv. Photochem.*, 1968, **5**, 329.
[76] A. A. Lamola, C. S. Hammond, and F. B. Mallony, *Photochem. and Photobiol.*, 1965, **4**, 259.

If $|S\rangle$, $|T\rangle$, and $\{|i\rangle\}$ are the eigenfunctions of the zero-order Born-Oppenheimer Hamiltonian $\mathscr{H}_{BO}$ corresponding to the states (a), (b), and (c) respectively, then states $|S\rangle$ and $|T\rangle$ are coupled to each other via the spin-orbit perturbation V_{SO} while $|T\rangle$ is in turn coupled to the quasi-continuum $\{|i\rangle\}$ by the vibronic interaction V_{vib}. Thus we have

$$\langle S | V_{SO} | T \rangle \equiv \omega \tag{47}$$

and

$$\langle T | V_{\text{vib}} | i \rangle = v_i \equiv v \tag{48}$$

for all i. All other matrix elements of the perturbations V_{SO} and V_{vib} between the zero-order states are taken to be zero, that is, the direct coupling of $|S\rangle$ with the manifold of twisted triplets is neglected, and it is assumed that all the zero-order states have been chosen such that there are no non-vanishing diagonal matrix elements of either V_{SO} or V_{vib}, nor any vibronic interactions within the set of closely spaced states $\{|i\rangle\}$.

Since $|S\rangle$ is the only state which carries effective oscillator strength to the ground state, then the model may be thought of rigorously as being prepared in the excited singlet state at time $t = 0$. The problem can be simply stated; it is desired to determine the state function $|\phi(t>0)\rangle$ of the model system in terms of its projections on the complete set of Born-Oppenheimer states, and thereby predict the decay of the prepared singlet. A Green's function formalism is used to achieve this object. The details of this treatment are not given here, but the results will be summarized.

(a) In the statistical limit, simultaneous radiative and non-radiative processes occur independently. The lifetimes and quantum yields characteristic of these several processes are then defined in terms of the relevant densities of states and matrix coupling elements connecting the initial state with the appropriate continua and quasicontinua. The overall decay of the excited state is exponential.

(b) In the formal treatment, the rearrangement kinetics are shown to depend explicitly on the energy difference Δ between the first singlet $|S\rangle$ and the second triplet $|T\rangle$. These results clearly also depend implicitly upon the position of the first triplet in that the density of vibronic levels $\{|i\rangle\}$ in the region of interest is a sharply increasing function of the total vibrational energy of the molecule. It thus appears reasonable to interpret the different quantum yields and varying nature of products in different solvents in terms of the relative shifts in energy of the molecular states involved. Large shifts of this type appear when the singlet state in question is connected to the ground state by a large electric dipole transition moment, or when the static dipole moments of the ground and excited states are very different.

(c) Since the rate processes in the statistical limit were shown to be independent, the formal treatment can be reconciled to a conventional kinetic discussion of isomerization reaction kinetics.

In the model, the intermediate triplet serves directly as a bridge between the initial and final states. The effect of the discrete triplet could be treated in a cruder way by calculating simply the enhancement of the mixing of the singlet and twisted triplet states due to its presence. In the stilbene case, the enhancement would amount to several orders of magnitude, and thus the radiationless processes considered here are not overall rate-determining. Since vibrational relaxation of the twisted triplets will be even faster, the overall rate of the isomerization must be determined by the rate of deactivation of the lowest triplet as it twists to the *cis* and *trans* ground-state isomers.

The formal model presented in the work above [74] should be general enough to describe a wide class of molecular rearrangement reactions, and is not restricted to *cis–trans* isomerizations, although it must be noted that the model chosen is dependent upon experimental evidence. Nonetheless, more thorough experimental work on these and other reactions would render the theoretical treatments useful by providing numerical estimates for the level densities and matrix coupling elements which appear in expressions for the radiative decay of these molecules. The conversion of electronic to vibrational energy in photochemical rearrangements could then be rigorously described in terms of the limiting cases outlined above.

An extension of the considerations above includes contributions from hindered internal rotations.[63] The intramolecular coupling matrix elements are derived in a form which allows the theory of many-phonon radiationless transitions to include torsional motion. It is shown that the strongly coupled degree of freedom (that mode whose equilibrium positions differ greatly in the initial and final electronic states) can be separated out from the remaining vibrations in the general rate expressions. The rate of the non-radiative process in stilbene can be expressed as [63]

$$\Gamma_{mn} = (2\pi/\omega\hbar^2) \, e^{-g} \sum_{nt} [f(n_t)]^2 \, [g(n_t)] \qquad (49)$$

where $f(n_t)$ represents the torsional overlap, and $g(n_t)$ the other vibrational modes. The dependence of these factors upon n_t can be established, and allows the formulation of conclusions concerning the distribution of electronic energy ΔE amongst the internal degrees of freedom. Considering the model in Figure 12 where G is the energy difference between the potential minimum of the upper electronic state and the maximum of the lower state, two general cases can be distinguished.

(*i*) When G is of the order of a few vibrational quanta, then both factors $f(n_t)$ and $g(n_t)$ in equation (49) assume their maxima at values of n_t such that $E_{n_t} \leqslant \Delta E$ (E_{n_t} is the amount of electronic energy which appears as torsional energy in the final state), *i.e.* for this curve geometry most of the electronic energy should go into the torsional mode, rather than the usual vibrational modes which serve as acceptors in purely weak coupling theories.

(*ii*) As G increases, the torsional overlap factors in equation (49) assume their largest values for n_t corresponding to E_{n_t} values which are increasingly

Spectroscopic and Theoretical Aspects

less than ΔE. In this case, the electronic energy can begin to be distributed as well in the stretching modes which are familiarly considered to serve as acceptors.

The above results are of great importance, in that in Case (*i*) a negligible deuterium isotope effect would be expected, in contrast to Case (*ii*). It has been predicted that there might be an onset of a deuterium isotope effect

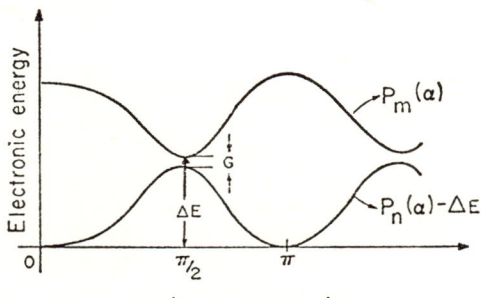

Figure 12 *Model potential curves for the lowest triplet and ground electronic states of stilbenelike molecules, as functions of the angle of twist α about the double-bond axis. The $3n - 7$ vibrational co-ordinates have been fixed at their equilibrium positions. ΔE is the electronic energy gap and $P_n(\alpha)$ and $P_m(\alpha)$ denote the barriers against internal rotation. G is the difference in energy between the maxima and minima of the lower and upper potential curves, respectively, and defines Cases* (i) *and* (ii) *considered in the text*
(Reproduced by permission from *J. Chem. Phys.*, 1970, **52**, 2460)

as one proceeded through a series of substituted stilbenes in which the parameter *G* increased to the order of many vibrational quanta. There is evidence to support the contention that stilbene exhibits Case (*i*) behaviour, in that the lifetime of the lowest triplet state of the stilbene molecule is enhanced when it is formed in glassy solvents of increasing rigidity, and the same effect is found for the substitution ratio Γ_H/Γ_D. These observations could be explained on the basis that increasing the viscosity of the solvent steepens the potential barrier against internal rotation, and therefore confines the torsional motions in the ground state to lower quantum numbers (n_t). Consequently a successively greater fraction of the electronic energy ΔE will be forced into the vibrational degrees of freedom, including the high frequency C—H and C—D modes, which will therefore show an increasing deuterium substitution effect. In the limit of a perfectly rigid solution, the torsional degree of freedom is completely frozen out, and the purely weak coupling situation remains.[62]

The energy barriers involved in the isomerization processes of ethylene in its lowest excited singlet and triplet states, and the lowest doublet and quartet electronic states of its positive ion have been calculated by a semiempirical quantum mechanical method.[77] For the triplet state, two low-energy processes emerge. These are *cis–trans* isomerization, and rearrangement to the ethylidene configuration. These results are in essential agreement with experimental evidence on the sensitized photochemistry of ethylene.

A model similar to that used above [62, 74] has been examined in detail from the point of view of momentum space, and is extended to include orthogonal decay channels.[78] The basis of the model is shown in Figure 13.

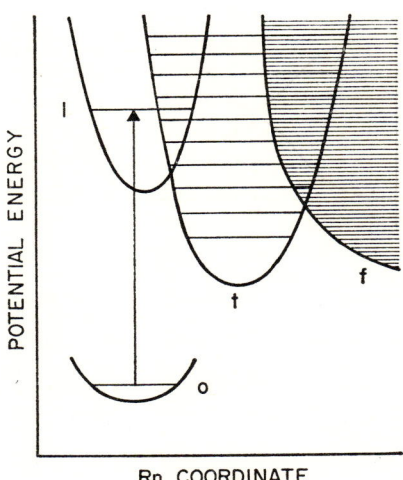

Figure 13 *Schematic representation of the energy levels of a molecule capable of dissociation following photon absorption. Level 1 is the unperturbed state reached by photon absorption, t labels the manifold of discrete transition states in which energy redistribution is possible, and f labels the manifold of continuous fragment levels*
(Reproduced by permission from *J. Chem. Phys.*, 1969, **51**, 4272)

The detailed derivations of rate expressions will not be given here, but the conclusions summarized. The assumption that the manifolds t are discrete means that the initial state $|i\rangle$ cannot decay irreversibly into the transition manifolds, since irreversibility requires continuous final states. It should be noted that Jortner's approach to radiationless transitions in the statistical limit does predict irreversibility, whereas the model considered above does not.[78] Thus the present model is strictly limited to small molecules containing less than ten atoms. The result implies that there should be qualitative differences between the photochemistry of large and small molecules, in that in large molecules simple kinetic analysis will be possible (see discussion on isomerization above) whereas interference effects between

[77] A. J. Lorquet, *J. Phys. Chem.*, 1970, **74**, 895.
[78] M. R. Philpott, *J. Chem. Phys.*, 1969, **51**, 4272.

alternative virtual states should be important in small molecules. The theory shows that a cut-off function in matrix elements between fragments and undissociated molecules is essential for convergent integrals over fragment energies. The approach can be used to consider the case when the energy E_1 of the initial state lies below the manifold t. In this case fragmentation occurs *via* virtual excitations, and the rate is sharply reduced since there is no resonance between $|1\rangle$ and the open fragment channel. (This corresponds to tunnelling.) The model can thus be used to investigate threshold effects, for example the change in decay rate as E_1 approaches the bottom of the transition manifold from below. The literature has many examples of molecules in which new modes of dissociation appear as incident photon energy is increased, and these could be successfully treated with the theory outlined here.

Predissociation.—Predissociation in molecules has been treated using the Jortner approach.[79] Using time-independent perturbation theory and normalization in a closed box, the coupling of a discrete state with a one-dimensional free-particle continuum has been considered. The structure of the exact levels depends on the relative positions of the discrete state and the continuum. When the discrete state is degenerate with a state in the continuum, it is broadened to an asymmetric Lorentzian shape and is slightly shifted to higher energies. The broadening is given by [79]

$$a^{n^2} = 2 |v_{aa}|^2 / [4\pi^2 | V_{aa}|^4 \rho_n^2 + (E^a - E_n)^2] \tag{50}$$

and the shift is approximated by

$$E_n^{\text{shift}} = \Delta E^2 / 2E^a \tag{51}$$

where ΔE is the width given approximately by

$$\Delta E = 2\pi |v_{aa}|^2 \rho a \tag{52}$$

In the above, ρ_n are the density states at energy E_n, v_{aa} is the matrix element which couples the initial state ϕ^a to the state in the continuum degenerate with ϕ^a. A new state is also found below the continuum. When the discrete state lies below the continuum, it is quasishifted towards lower energies. In this case the broadening is not Lorentzian, but is rather a delta function at the exact discrete state and much smeared throughout the continuum. The coupling of two discrete states with a continuum and the coupling of two continua are also discussed.

Predissociation in diatomic spectra, with special reference to the Schumann–Runge bands of O_2, has been considered.[80] From experimental results it is clear that there is a maximum of predissociation at the $v' = 4$ level, and a subsidiary maximum at $v' = 11$, with a minimum at $v' = 9$ (see ref. 80). It is assumed that there is a crossing of potential energy

[79] I. Riess, *J. Chem. Phys.*, 1970, **52**, 871.
[80] J. N. Murrell and J. M. Taylor, *Mol. Phys.*, 1969, **16**, 609.

curves near $v' = 4$ and calculations are performed to evaluate Franck–Condon factors for different types of potential curve shown in Figure 14. The best fit to the experimental data occurs for the single potential crossing at $v' = 4$ by a repulsive curve having the form

$$V(r) = V_\infty + 5 \times 10^5 (R_0/R)^6 \text{ cm}^{-1} \qquad (52)$$

It can thus be seen from Figure 14 that the double maximum in the predissociation of O_2 in the Schumann–Runge bands does not require a

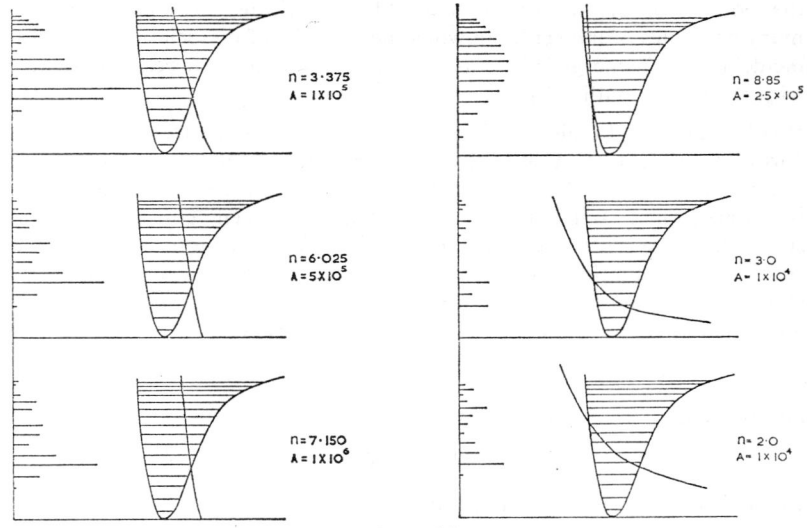

Figure 14

Franck–Condon factors for three $A(R_0/R)^n$ potentials with optimum n

Franck–Condon factors for type (ii) and (iii) crossing $A(R_0/R)^n$ potential

Reproduced by permission from *Mol. Physics*, 1969, **16**, 609.

double crossing of potential surfaces, but is adequately explained on a single crossing basis. Franck–Condon factors have been calculated for the Schumann–Runge bands of oxygen by a different approach,[81] and compared with experiment. Experimental evidence suggests that the $B\,^3\Pi_{0_u}{}^+$ state of I_2 predissociates spontaneously. Calculations have been performed of the relative predissociation rates by means of a model consisting of a repulsive 1_u state which is coupled to the $v'J'$ levels of the B state by rotational and electronic angular momentum terms neglected in the zero-order Hamiltonian.[82] Excellent agreement is found between the calculated and observed relative predissociation rates at regions of

[81] M. Bixon, B. Raz, and J. Jortner, *Mol. Phys.*, 1969, **17**, 593.
[82] A. Chutjian, *J. Chem. Phys.*, 1969, **51**, 5414.

Spectroscopic and Theoretical Aspects

$v' = 14$, 25, and 50 for the B state for one position of the repulsive state lying very near the repulsive wall of the B state over the energy range 17,300 → 20,000 cm^{-1}. The position of this state agrees well with other theoretical predictions.

Energy Partitioning in Photochemical Reactions.—The development of new techniques has allowed for the first time a detailed study of energy partitioning in simple molecules to be quantitatively determined. The technique, variously termed 'translational spectroscopy'[83] or 'photochemical recoil

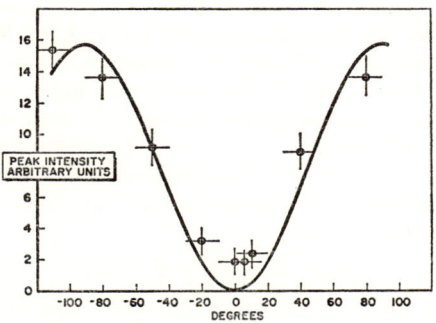

Figure 15 *Observed angular distribution of peak intensity (corrected to centre of mass) of the recoiling chlorine atom as a function of the angle between the Cl_2 bond axis and electric vector of the light. Solid line is a simple $\sin^2 \theta$ function. Above the graph is an oscilloscope trace of the ^{35}Cl intensity from a single 15 ns PRS scan, with time progressing from right to left at 4·9 μs/division*
(Reproduced by permission from *J. Chem. Phys.*, 1969, **50**, 3635)

spectroscopy',[84] essentially consists of a measurement of the velocity distribution of recoiling fragments following photodissociation produced by a pulsed laser beam crossing a molecular beam. A full description of the technique is given in the following section of this chapter on Techniques. Two independent groups of workers have applied the technique to the chlorine molecule.[83, 84]

$$Cl_2 + h\nu \ (3471 \ Å) \longrightarrow Cl + Cl \quad (53)$$

The angular distribution of the recoiling atoms with respect to the direction of the electric vector of the light reveals that the transition dipole moment is parallel or perpendicular to the internuclear axis. A $\cos^2 \theta$ distribution (where θ is measured between the recoil direction and the electric vector) indicates a parallel, and $\sin^2 \theta$ indicates a perpendicular transition. For chlorine, the results clearly indicate a perpendicular transition as shown in Figure 15.[84] The absorption at 3471 Å clearly therefore results from a $^1\Sigma_{0g}{}^+ \rightarrow {}^1\Pi_{1u}$ transition,[83] since the transition to the nearby $^3\Pi_{O_u}{}^+$ state

---

[83] C. E. Busch, R. T. Mahoney, R. I. Morse, and K. R. Wilson, *J. Chem. Phys.*, 1969, **51**, 449.
[84] R. S. Dieson, J. C. Wahr, and S. E. Adler, *J. Chem. Phys.*, 1969, **50**, 3635.

would lead to a parallel angular distribution. The transitional energy distribution peaks at 113·8 kJ mol^{-1}, corresponding to

$$Cl_2(^1\Pi_{1u}) \longrightarrow Cl(^2P_{\frac{3}{2}}) + Cl(^2P_{\frac{3}{2}}) \qquad (54)$$

rather than

$$Cl_2(^3\Pi_{O_u}+) \longrightarrow Cl(^2P_{\frac{3}{2}}) + Cl(^2P_{\frac{1}{2}}) \qquad (54a)$$

Other molecules studied by this means include Br_2,[85] IBr,[86] I_2,[86] ICN,[87] NO_2,[83, 85] $NOCl$,[85] and some alkyl iodides.[88] In the case of bromine, absorption to the $^1\Pi_{1u}$, the $^3\Pi_{O_u}{}^+$, and $^3\Pi_{1u}$ states has been observed. Using the fundamental of the ruby laser (14,405 cm^{-1}) on iodine, only ground-state iodine atoms are produced, and a $\sin^2 \theta$ dependence is observed, leading to an assignation of $^3\Pi_{1u}$ for the upper state. For second-harmonic neodymium light (18,830 cm^{-1}), the results are more complex, since the $B^3\Pi_{O_u}{}^+$ state which is predissociative and the $^1\Pi_{1u}$ state which directly dissociates, are both produced.

For diatomics, it can be seen that the energies of the electronic states of the atoms produced in the dissociation E_{ele} are obtained, even if these states then radiate to lower states.

$$E_{ele} = E_{par} + h\nu - D_0^0 - E_{tm} \qquad (55)$$

E_{par} is the initial thermal energy of the parent molecule, which in an effusive molecular beam is just the thermal distribution at the oven temperature. The laser determines $h\nu$ and commonly the bond dissociation energy D_0^0 is known. E_{tm} is of course measured directly in the experiment.

For triatomics, the distribution of total internal energy of the recoiling fragments from the dissociation of polyatomic molecules may be determined, through energy conservation, from the distribution of fragment recoil energy. In this case the relationship is

$$E_{int} = E_{par} + h\nu - D_0^0 - E_{tm} \qquad (56)$$

where

$$E_{int} \simeq E_{ele} + E_{vib} + E_{rot} \qquad (57)$$

Here the angular distribution of the recoiling fragments is influenced both by the direction of the transition dipole, and by the molecular dynamics of the dissociation process. As an example, the photolysis of ICN may be treated.[87]

$$ICN + h\nu \; (210-300 \text{ nm}) \longrightarrow I + CN \qquad (58)$$

The system is straightforward in that it is linear in the ground state and probably in the upper state, and the absorption spectrum suggests a

---

[85] R. T. Mahoney, R. J. Oldman, R. K. Sander, and K. R. Wilson, *Bull. Amer. Phys. Soc.*, 1969, **14**, 850.
[86] M. L. Blethen, R. T. Mahoney, A. F. Tuck, and K. R. Wilson, *Bull. Amer. Phys. Soc.*, 1969, **14**, 849; C. E. Busch, R. T. Mahoney, R. I. Morse, and K. R. Wilson, *J. Chem. Phys.*, 1969, **51**, 837.
[87] K. E. Holdy, L. C. Klotz, and K. R. Wilson, *J. Chem. Phys.*, 1970, **52**, 4588.
[88] K. R. Wilson, *Discuss. Faraday Soc.*, 1967, **44**, 234.

quasi-diatomic. A model has been proposed which predicts that most of the available energy will appear as translational energy of the recoiling I and CN fragments, with only a minor fraction going into vibrational and rotational excitation of the CN. In the model, an excited-state electronic surface is generated from spectroscopic and thermodynamic data using a quasi-diatomic viewpoint that the photon excitation is localized to the breaking bond. After the molecule arrives on this surface by a 'classical' Franck–Condon leap, the fragments are allowed to scatter from one another in a 'half-collision' and the energy partitioning between translational, vibrational and (crudely) rotational energy is noted when the fragments have separated. The 'half-collision' is treated by using the integration of the classical equations of motion averaged over the phases of the original molecular vibrations to calculate the partitioning between the forms of energy. The partitioning within vibrational states is treated by a classical energy, forced quantum oscillator approximation. There is spectroscopic evidence to suggest that the predictions of the model are in rough agreement.

Energy partitioning in the photolysis of pyrazoline and 4-methyl-1-pyrazoline has been considered in detail in Part I, Chapter 3.[89] In addition to methods which involve direct measurement of the translational energy of photofragments, the rate of chemical reactions of 'hot' fragments, especially atoms with substrates, can be used as a measure of the energy distribution consequent to photodissociation. There is considerable discussion to be found in Chapter 3 concerned with energy partitioning in H_2S and D_2S as a function of incident photon energies. One group of workers claim that compared with excitation at 2130, excitation at 1850 Å produces a strong enhancement of internal excitation in the HS fragment,[90] a result in direct contrast to those of other groups.[91,92] Further work has substantiated that the energy distribution does not change drastically with exciting wavelength,[93] and that nearly all of the excess energy of photodissociation in this system appears as recoil kinetic energy. The difference between the two results appears to arise from substitution of hydrogen for butane as the calibrant standard. The reaction of hot H atoms with C_4D_{10} will not produce any D atoms, whereas every hot H atom reacting with a D_2 molecule will produce a D atom, generally with excess energy. These D atoms so produced may react further, and corrections for these reactions based on thermal rate constants [90] may not be valid, which may thus explain the discrepancy between the two sets of results. Similar experiments have been carried out on the HI system, in which hot hydrogen atoms react with C_4D_{10} and allow an estimate of the initial H atom kinetic

---

[89] P. Cadman, H. M. Meunier, and A. F. Trotman-Dickensen, *J. Amer. Chem. Soc.*, 1969, **91**, 7640; F. H. Dorer, *J. Phys. Chem.*, 1969, **73**, 3109.
[90] G. P. Sturm and J. M. White, *J. Chem. Phys.*, 1969, **50**, 5035.
[91] R. C. Gann and J. Dubrin, *J. Chem. Phys.*, 1967, **47**, 1867.
[92] L. E. Compton, J. L. Gole, and R. M. Martin, *J. Chem. Phys.*, 1969, **73**, 1158.
[93] L. E. Compton and R. M. Martin, *J. Chem. Phys.*, 1970, **52**, 1613.

energy as a function of exciting wavelength.[94] From this knowledge the fraction of excited-state iodine atoms formed in the initial step can be obtained. The results show that the probability of ground-state iodine ($5^2P_{\frac{3}{2}}$) atoms at 2537, 2288, and 1850 Å were 0.93 ± 0.1, 0.81 ± 0.1, and 1.00 ± 0.10 respectively. The transitions responsible for dissociation could be $^1\Sigma^+ \to {}^3\Pi_1$, $^1\Sigma^+ \to {}^3\Pi_0{}^+$, and the $^1\Sigma^+ \to {}^1\Pi$, leading to $I(^2P_{\frac{3}{2}})$, $I(^2P_{\frac{3}{2}})$, and $I(^2P_{\frac{3}{2}})$ respectively. A sharp rise in the absorptivity below 2000 Å may thus be associated with the $^1\Sigma^+ \to {}^1\Pi$ transition, giving rise to a higher fraction of ground-state atom production than at longer wavelengths.

Applications of Unimolecular Reaction-rate Theory to Photochemical Cases.—The Rice–Ramsberger–Kassel–Marcus (RRKM) theory of unimolecular reactions can be applied to several interesting cases arising from photoexcitations. As an example, the photodissociation of azoethane is competitive with collisional deactivation, and the application of the theory to this system allows estimation of the primary photodissociation quantum yields of this molecule.[95] The treatment has been fully described in Part I, Chapter 3 of this volume. The photodissociation of acetone in the gas phase[96] proceeds in part *via* the thermal excitation of triplet state molecules; and this (thermal) reaction has also been fully treated with the RRKM approach, also covered in Chapter 3. An interesting example of a 'thermal' reaction which arises *via* photochemical excitation is the photoisomerization of cycloheptatriene,[97] in which the excited singlet state first formed internally converts to a high vibrational level of the ground state, from which reaction proceeds. In such a system, the reacting state may be prepared with various amounts of vibrational energy dependent upon the photon energy, and reaction is in competition with collisional deactivation, as before. Since all treatments employ the same approach, and several examples have been quoted at length in Chapter 3, only one example will be given here. The example has been chosen since rate constants calculated by the RRKM method allow a (tentative) choice between alternative mechanisms of photochemical dissociation. The photolysis of carbon suboxide in the presence of methyl fluoride yields acetylene and hydrogen fluoride which clearly result from the decomposition of 'hot' $C_2H_3F^*$.[98] This species could be formed in two ways: (*i*) reaction of an electronically excited C_2O species from the primary dissociation of C_3O_2 (59)

$$C_2O + CH_3F \longrightarrow C_2H_3F^* + CO \qquad (59)$$

(*ii*) reaction of free carbon atoms $C(^1D)$ (60)

$$C(^1D) + CH_3F \longrightarrow C_2H_3F^* \qquad (60)$$

[94] L. E. Compton and R. M. Martin, *J. Phys. Chem.*, 1969, **73**, 3474.
[95] P. G. Bowers, *J. Phys. Chem.*, 1970, **74**, 952.
[96] H. E. O'Neal and C. W. Larson, *J. Phys. Chem.*, 1969, **73**, 1011.
[97] R. Atkinson and B. A. Thrush, *Proc. Roy. Soc.*, 1970, **316**, *A*, 123, 131, 143.
[98] E. Tschuikow-Roux and S. Kodoma, *J. Chem. Phys.*, 1969, **50**, 5297.

In order to make a choice between these alternatives the experimental rate constant for the decomposition of the chemically activated vinyl fluoride can be compared with that calculated by RRKM theory.

From thermochemical considerations, the minimum energy content of a vibrationally excited $C_2H_3F^*$ arising via $C(^1D)$ insertion into methyl fluoride is 169 kcal mol^{-1} (709·8 kJ mol^{-1}). The activation energy for HF elimination from $C_2H_3F^*$ is ca. 69 kcal mol^{-1} (289·8 kJ mol^{-1}), and thus the excess energy in $C_2H_3F^*$ arising via (ii) would be 100 kcal (420 kJ mol^{-1}).

According to the Marcus theory of unimolecular reactions, the specific rate constant for decomposition k_E, when neither energized molecule nor activated complex possesses active (internal) rotations, is given by

$$k_E = h^{-1}(\sigma_1/\sigma_1\dagger) \{[\sum_{E_v\dagger E\dagger} P(E_v\dagger)] [N^*(E)]^{-1}\} \quad (61)$$

where $\Sigma P(E_v\dagger)$ is the sum of the energy eigenstates of the active degrees of freedom of the activated complex at energy $E\dagger$; $N^*(E)$ is the density of energy eigenstates of the active degrees of freedom for the molecule at energy $E = E_0 + E\dagger$, and σ_1 and $\sigma_1\dagger$ are respectively the symmetry numbers for the adiabatic (overall) rotations for the molecule and complex. The evaluation of $\Sigma P(E_v\dagger)$ and $N^*(E)$ is simplified by using the 'vibrational-energy-level sums' approximation of Written and Rabinovitch. Thus the expression for k_E becomes

$$k_E = h^{-1}\left(\frac{\sigma_1}{\sigma_1\dagger}\right) \frac{(E\dagger + a\dagger \, E_2\dagger)^{s\dagger}}{(E\dagger + E_0 + aE_2)^{s-1}} [(\prod_i^s \omega_i)(\prod_i^{s\dagger} \omega_i\dagger)^{-1}] \frac{\Gamma(s)}{\Gamma(s\dagger + 1)} \quad (62)$$

where E_0 is the critical energy, $E_2 = \frac{1}{2}\Sigma\omega_i$ is the zero-point energy, S is the total number of vibrational frequencies ω_i, Γ is the gamma function and daggered ($\dagger$) quantities refer to the activated complex. a is a function of the available energy, and represents a correction to the semiclassical approximation. It approaches unity at high energies. The evaluation of k_E requires a knowledge of the vibrational frequencies for molecule and activated complex. These are known (see ref. 98).

The specific rate constants for decomposition were evaluated for three values of the critical energy 60, 70, and 80 kcal mol^{-1} (252, 294, 336 kJ mol^{-1}) and the results are shown in Figure 16. From a plot of stabilization-to-decomposition ratio for the hot vinyl fluoride vs. total pressure, and assuming unit collisional deactivation efficiency and a collision diameter of 4·7 Å, the experimental specific rate constant for decomposition of the $C_2H_3F^*$ at 25 °C was found to be $k_{obs} = 3·84 \times 10^{10}$ s^{-1}. Since the activation energy is of the order of 70 kcal mol^{-1} (294 kJ mol^{-1}), this result may be used to estimate the energy content of the excited $C_2H_3F^*$ molecule, as seen in Figure 16. This is found to be 127 kcal mol^{-1} (5334 kJ mol^{-1}), significantly less than the minimum which could arise via $C(^1D)$ insertion. It is thus concluded that the mechanism (i) involving the C_2O species is the correct one, although it is recognized that the conclusion must be tentative in that the calculation is based upon an uncertain heat of formation of C_2H_3F, and a deduced activation energy for HF elimination. It should further be noted that the simple calculation above presupposes a monoenergetic species, whereas in reality excited vinyl fluoride must be formed with an energy distribution. The exact RRKM calculation of the stabilization-to-decomposition ratio, averaged over all energies, would require

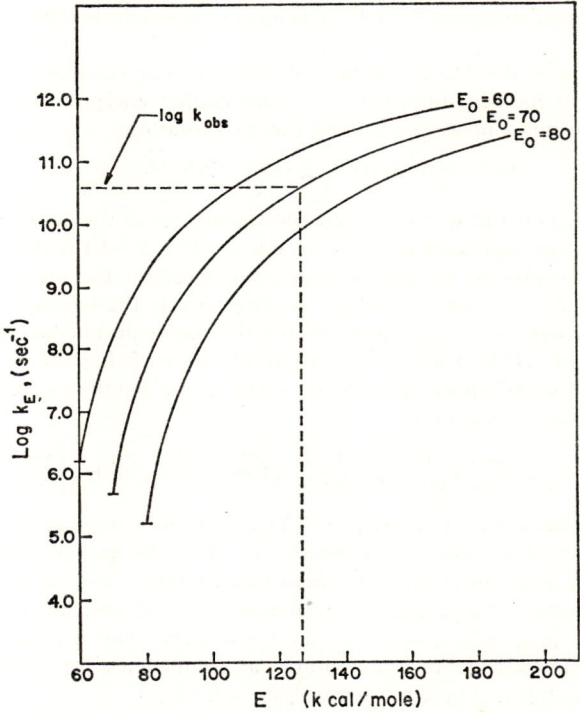

Figure 16 *Specific rate constant for* HF *elimination from vinyl fluoride as a function of the total energy* $E(= E\dagger + E_0)$
(Reproduced by permission from *J. Chem. Phys.*, 1969, **50**, 5297)

knowledge of the non-equilibrium energy distribution function for $C_2H_3F^*$. Such calculations assuming Gaussian and other functions are adequately described in Part I, Chapter 3 in the section dealing with pyrazoline photolysis.

Orbital Symmetry Considerations.—Consideration of the symmetries of molecular orbitals in ground states, intermediates, and products has proved to be of great value in a qualitative understanding of many organic reactions, including photochemical reactions. Nevertheless, difficulties exist, and in Volume 1 of this series a brief communication was mentioned in which the authors demonstrated that the problem of the cyclization of butadiene to cyclobutene was not as simple as had been suggested. The same authors have amplified this view in a recent report.[99] The problem arises because the ground state of cyclobutene lies some 84 kJ mol^{-1} above that of butadiene, and the spectroscopic singlet state of cyclobutene is some 210—252 kJ mol^{-1} higher than that of butadiene. Clearly, therefore, any photochemical formation of cyclobutene from butadiene cannot involve the spectroscopic singlet state of cyclobutene. The authors again demonstrate that a symmetric state of the molecules crosses the antisymmetric spectroscopic state, and provides a pathway for reaction. It is proposed that this phenomenon is general.

In a treatment [100] of cycloadditions to the benzene ring both thermal and photochemical additions have been considered. Possible reactions are shown in Table 4, in which either one or both the reactants is in the ground electronic state. A and F refer to symmetry-allowed and symmetry-forbidden processes respectively. The photochemical reactions listed apply to systems in which the excited state specified (other than the $^1B_{1u}$ state of benzene) is the lowest excited singlet state and photochemical A correlations are with low-lying excited states of products or, in cases 3 and 7, with the ground states of diradicals. Three general conclusions follow from the relationships listed in Table 4. These are that the Woodward–Hoffmann rules for concerted terminal cycloaddition reactions of π-electron systems can be applied successfully to the thermal additions to benzene, but are not applicable to the photoadditions, and that the rules for concerted photoadditions to benzene depend upon the relative energy levels of S_1 benzene and S_1 addend. All the processes should be favoured by charge-transfer interactions. Some of the reactions proposed in Table 4 have not been observed experimentally, but in the cases where comparison is possible there is good agreement between theory and experiment.

Selection rules for the reactions of singlet molecular oxygen ($^1\Sigma$ and $^1\Delta$) and ground state $O_2(^3\Sigma)$ to olefins *via* concerted additions have been generated by the use of molecular orbital and state correlation diagrams.[101] The molecular orbitals for the O_2 free molecule, neglecting electron–electron interaction, are as follows:

$$^3\psi_{xy} = \sqrt{\tfrac{1}{2}}\{ \mid \pi_x^*(1)\, \bar{\pi}_y^*(2) \mid + \mid \bar{\pi}_x^*(1)\, \pi_y^*(2) \mid \} \tag{63}$$

$$^1\psi_{xy} = \sqrt{\tfrac{1}{2}}\{ \mid \pi_x^*(1)\, \bar{\pi}_y^*(2) \mid - \mid \bar{\pi}_x^*(1)\, \pi_y^*(2) \mid \} \tag{64}$$

$$^1\psi_{xx} = \mid \bar{\pi}_x^*(1)\, \pi_x^*(2) \mid \tag{65}$$

$$^1\psi_{yy} = \mid \bar{\pi}_y^*(1)\, \pi_y^*(2) \mid \tag{66}$$

These states are all degenerate but inclusion of electron–electron interaction shifts the states $^3\psi_{xy}$, and $^1\psi_{xx}$ and $^1\psi_{yy}$ are both shifted in energy and

---

[99] W. Th. A. M. van der Lugt and L. G. Oosterhoff, *J. Amer. Chem. Soc.*, 1969, **91**, 6042.
[100] D. Bryce-Smith, *Chem. Comm.*, 1969, 806.
[101] D. R. Kearns, *J. Amer. Chem. Soc.*, 1969, **91**, 6554.

Table 4 Concerted cycloadditions of ethylene and cis- and trans-butadiene to benzene: dimerization of benzene. (Reproduced by permission from *Chem. Comm.*, 1969, 806)

Addend	Addition mode	Product	Excited species			Charge-transfer excitation		Thermal reaction
			Addend (S_1, T_1)*	Benzene B_{2u} (S_1)	Benzene B_{1u} (S_2, T_1)	Donor addend	Donor benzene	
1 Ethylene	cis-ortho/cis-1,2-	(I)	A	F	A^a	A	A	F
2 Ethylene	trans-ortho/cis-1,2-	(II)	A	F	A	A	A	A
3 Ethylene	meta/cis-1,2-	(III) [→ (IIIa)]	A	A	F^b	A	A	F^c
4 Ethylene	para/cis-1,2-	(IV)	A	F	A^d	A	A	A
5 Ethylene	para/trans-1,2	(V)	A	A	A	A	A	F
6 cis-Butadiene	cis-ortho/cis-1,4-	(VI)	A	F^e	F^b	A	A	A
7 cis-Butadiene	meta/cis-1,4-	(VII) [→ (VIIa)]	A	A	A^a	A	A	F^f
8 cis-Butadiene	para/cis-1,4-	(IX)	A	F	A	A	A	F
9 cis-Butadiene	para/trans-1,4-	$(IX)^g$	A	A	A	A	A	A
10 trans-Butadiene	para/cis-1,4-	(X)	A	A	A	A	A	F
11 Benzene	exo-cis-1,2-cis-1',2'-	(XI)	—	F	A	A	A	F
12 Benzene	endo-cis-1,2-cis-1',2'-	(XII)	—	A	A	A	A	F
13 Benzene	cis-1,2-1',4'-	(XIII)	—	F^e	A	A	A	A
14 Benzene	cis-1,4-1',4'-	(XIV)	—	F	A	A	A	F

* Strictly, the relations apply to a T_1 addend only when this is reacting in an effectively planar conformation. T_2 Benzene may be of state $^3E_{1u}$ (S. D. Colson and E. R. Bernstein, *J. Chem. Phys.*, 1965, **43**, 2661). Mixing will occur between states similar in symmetry (*e.g.* S_1 ethylene and S_2 benzene) if the energies are close.

^{a} Correlation noted also by Hoffmann and Woodward.

^{b} Correlation with first excited state of diradical product.

^{c} Correlation with zwitterion (IV).

^{d} The B_{1u} component (XV) which apparently permits concerted *para*-additions of ethylene corresponds to a transition between orbitals ψ_3, ψ_5 which each have zero amplitude at the reaction centres. Nevertheless, the addition should still be possible in principle because the ψ_2, ψ_4 and ψ_3 ψ_5 components are equally part of the B_{1u} state.

^{e} Allowed correlation with S_2 adduct, which might be accessible in practice.

^{f} Correlation with zwitterion (VIII).

^{g} It is possible to envisage the production of adduct (X) by this mode of addition; but Professor R. Hoffmann has suggested convincingly that adduct (IX) should be a preferred product.

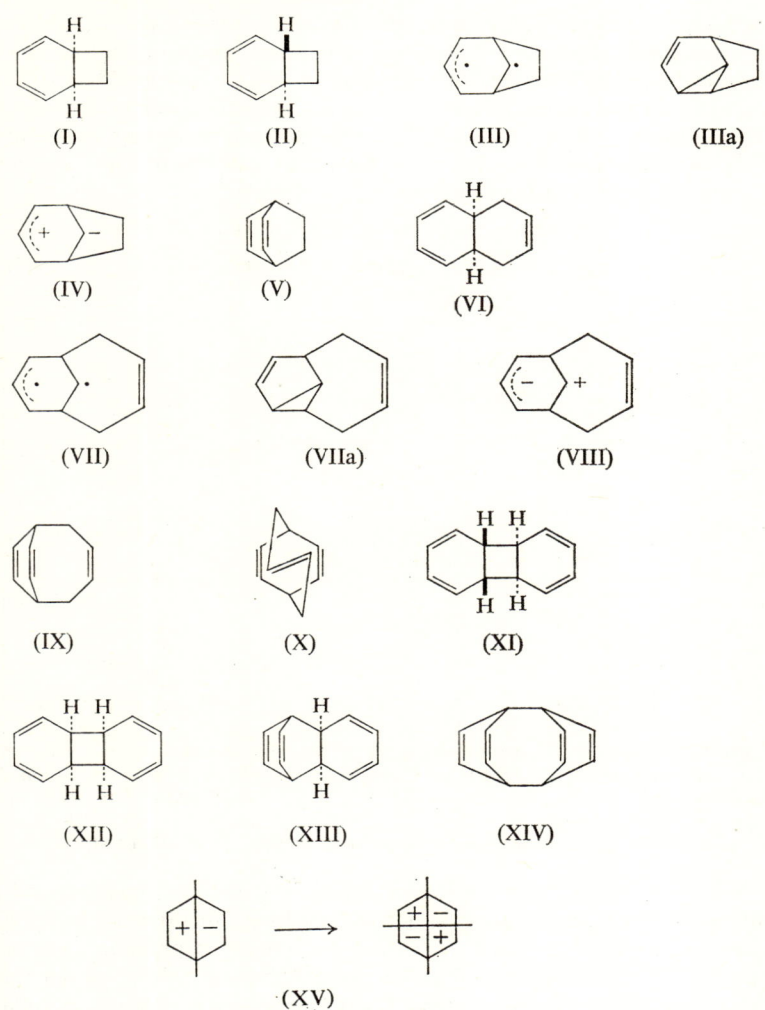

mixed as a result of an exchange interaction

$$K = \langle \pi_x^*(1)\pi_y^*(1) \mid e^2/r_{12} \mid \pi_x^*(2)\pi_y^*(2) \rangle \tag{67}$$

The new functions are then, with their appropriate energies

$$^3\psi_0(^3\Sigma) = {}^3\psi_{xy}, \quad E = 0 \tag{68}$$

$$^1\psi_1(^1\Delta^*) = {}^1\psi_{xy}, \quad E = 22 \text{ kcal } (92\cdot4 \text{ kJ}) \tag{69}$$

$$^1\psi_2(^1\Delta) = \sqrt{\tfrac{1}{2}}({}^1\psi_{xx} - {}^1\psi_{yy}), \quad E = 22 \text{ kcal } (92\cdot4 \text{ kJ}) \tag{70}$$

$$^1\psi_3(^1\Sigma) = \sqrt{\tfrac{1}{2}}({}^1\psi_{xx} + {}^1\psi_{yy}), \quad E = 38 \text{ kcal } (159\cdot6 \text{ kJ}) \tag{71}$$

The above applies to free oxygen molecules where the π_x^* and π_y^* orbitals are degenerate. If oxygen interacts with some appropriate organic acceptor, then the degeneracy between the π_x^* and π_y^* orbitals can be lifted. If it is assumed that the interaction changes the π_x^* orbital energy by an amount ΔE_x and the π_y^* by an amount ΔE_y, but that there is no change in the form of the wavefunctions, we find that the states associated with $^3\psi_{xy}$ and $^1\psi_{xy}$ are unchanged in form, but have new energies

$$^3\psi_0(^3\Sigma) = {}^3\psi_{xy}, \quad E(^3\Sigma) = \Delta E_x + \Delta E_y \qquad (72)$$

$$^1\psi_1(^1\Delta^*) = {}^1\psi_{xy}, \quad E(^1\Delta^*) = 22 + \Delta E_x + \Delta E_y \qquad (73)$$

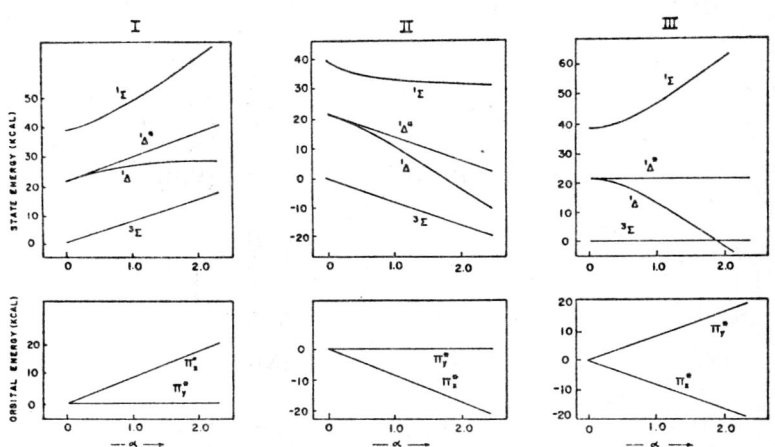

Figure 17 *Behaviour of the orbitals and states of molecular oxygen subjected to a perturbation which lifts the degeneracy of the π_x^* and π_y^* orbitals: case* I, $\Delta E_x = \alpha K, \Delta E_y = 0$; *case* II, $\Delta E_x = -\alpha K, \Delta E_y = 0$; *case* III, $\Delta E_x = -\Delta E_y = -\alpha K$
(Reproduced by permission from *J. Amer. Chem. Soc.*, 1969, **91**, 6554)

When π_x^* and π_y^* are no longer degenerate, the exchange integral K still causes $^1\psi_{xx}$ and $^1\psi_{yy}$ to mix, but the resulting functions are no longer equal mixtures of $^1\psi_{xx}$ and $^1\psi_{yy}$, and the energy expressions are also changed. Without solving for the wave functions, the new energies $E^1(^1\Sigma)$ and $E^1(^1\Delta)$ can be written. These states are those which correlate with the $^1\Sigma$ and $^1\Delta$ states of free oxygen respectively.

$$E^1(^1\Sigma) = 30 + \tfrac{1}{2}\{(\Delta E_x + \Delta E_y) + \sqrt{[4K^2 + (\Delta E_x - \Delta E_y)^2]}\} \qquad (74)$$

$$E^1(^1\Delta) = 30 + \tfrac{1}{2}\{(\Delta E_x + \Delta E_y) - \sqrt{[4K^2 + (\Delta E_x - \Delta E_y)^2]}\} \qquad (75)$$

Figure 17 shows a graphical description of the perturbation on the oxygen states for the following three cases

$$\Delta E_x = +\alpha K, \quad \Delta E_y = 0 \quad \text{Case I} \qquad (76)$$

$$\Delta E_x = -\alpha K, \quad \Delta E_y = 0 \quad \text{Case II} \qquad (77)$$

$$\Delta E_x = \Delta E_y = \alpha K \qquad \text{Case III} \qquad (78)$$

Spectroscopic and Theoretical Aspects

The results in Figure 17 can be summarized as follows: (a) the behaviour of the $^3\Sigma$ (ground) state can be determined immediately by calculating the total orbital energy for an electronic configuration in which one electron is placed in each of the two orbitals which correlate with the π^* antibonding orbitals of free oxygen; (b) the behaviour of the $^1\Delta^*$ state parallels that of the ground state except that it lies 92·4 kJ above it; (c) the $^1\Delta$ state is initially degenerate with the $^1\Delta^*$ state, but as the degeneracy between π_x^* and π_y^* is lifted, it behaves as if its outer two electrons were placed in whichever of the π^* orbitals is lower in energy; (d) the $^1\Sigma$ state behaves as if both of the outer electrons were placed in the higher energy π^* orbital.

In order to follow the reaction of oxygen with an acceptor, a state correlation diagram must be constructed showing how different possible states of the reactants adiabatically correlate with the various states of the product during the course of a reaction. The construction of such diagrams involves the following steps: (a) thermochemical data are used to locate the ground state of the reactants (a weak complex between C_2 and the acceptor) relative to the ground state of the product; (b) spectroscopic data are used to locate the excited states of the reactants relative to their ground state, and similarly for the products; (c) a reasonable geometry for the transition state in the reaction is chosen and an orbital correlation diagram for the reaction is constructed. From the orbital correlation diagram symmetries can be assigned to the various possible reactant and product states, and from a knowledge of how the orbitals correlate, the way in which the states of the reactants and products correlate can be determined.

The approach is best illustrated with an example.[101] The 1,4-addition of singlet oxygen to *cis*-dienes is known to occur. Using cyclopentadiene as the olefin the following reaction occurs.

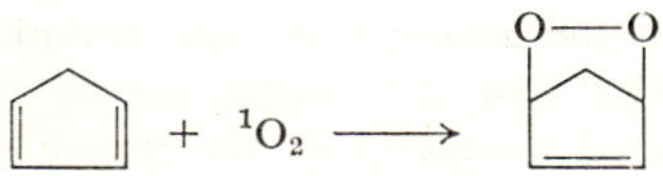

The overall ΔH for this reaction is -29 kcal mol^{-1} ($-121\cdot8$ kJ mol^{-1}) which locates the ground state of the reactants relative to the product. The three lowest-lying excited states of the reactants are all singlets associated with excitation of oxygen to its $^1\Delta$ states or $^1\Sigma$ state. The next highest state is excitation of the diene to its lowest triplet state, at *ca.* 55 kcal mol^{-1} (231 kJ mol^{-1}) above the ground state. Interaction of a triplet diene with ground-state oxygen gives rise to singlet, triplet, and quintet states. Little is known about the product in (79), but it might be expected to have excited singlet and triplet states at between 75 kcal mol^{-1} (315 kJ mol^{-1}) and 85 kcal mol^{-1} (357 kJ mol^{-1}) above its ground state. To construct an orbital correlation diagram it is assumed that there is a plane of symmetry in the transition state, and the resulting diagram is shown in

Figure 18. The parameters used are as follows: $\alpha_O = \alpha_C + 1\cdot 5\beta$; $\alpha_C = 0$; $\beta_{CO}(\pi) = 1\cdot 5\beta$; $\beta_{CC}(\sigma) = 1\cdot 5\beta$; $\beta_{CC}(\pi) = \beta$; $\beta_{CO,CO}(\text{dioxetan}) = 0\cdot 3\beta$; $\beta_{OO}(\pi) = \beta$ $(= 0\cdot 8\beta$ in dioxetan); $\beta_{CO}(\sigma) = \beta$; $\beta_{OO}(\sigma) = 1\cdot 5\beta$ $(= 1\cdot 4\beta$ in oxetan). These parameters are somewhat arbitrary, but the conclusions reached do not depend in any sensitive way upon their choice. The diagram can now be used to assign symmetries to the reactant and product states of interest.

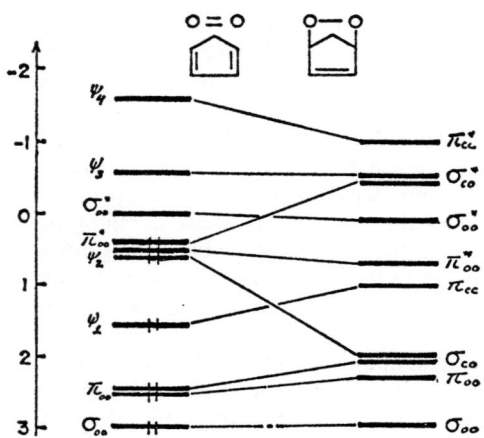

Figure 18 *Orbital correlation diagram for the concerted addition of molecular oxygen to cyclopentadiene. $\psi_1 \cdots \psi_4$ are molecular orbitals of cyclopentadiene, π_{oo}, π_{oo}^*, σ_{oo}, σ_{oo}^* are oxygen orbitals, and σ_{co}'s are linear combinations of two interacting σ_{co} orbitals. Orbital energies are in units of β*
(Reproduced by permission from *J. Amer. Chem. Soc.*, 1969, **91**, 6554)

The ground state of the acceptor–oxygen complex and the three low-lying excited states [$^1O_2$($^1\Delta$ or $^1\Sigma$)+ diene] are all symmetric with respect to the assumed plane of symmetry in the transition state. The ground singlet state of the product and its lower excited singlet and triplet states (configuration $\ldots \pi_{OO}^{*1}\sigma_{OO}^{*1}$) are also symmetric, as is the excited singlet arising from a $\pi_{OO}^* \rightarrow \sigma_{OO}^*$ promotion. With the assignment of spin and symmetry, the non-crossing rule allows a decision on which states in product and reactants correlate.

According to Figure 18, the ground state ($^3\psi_0$) of the [O_2–acceptor] complex has an electronic configuration ($\ldots \psi_1^2\psi_2^2\pi_x^{*1}\pi_y^{*1}$). As molecular oxygen is forced to cycloadd to the diene, this initial state will attempt to correlate with an excited triplet state of the product which has an electronic configuration ($\ldots \sigma_{CO}^2\pi_{CC}^{*2}\pi_{OO}^{*1}\sigma_{CO}^{*1}$). As a result of the crossing of the σ_{OO}^* and π_{OO}^* orbitals, however, the correlation will ultimately be with the lower triplet state of the product ($^3\psi_1$) which has an electronic configuration ($\ldots \sigma_{CO}^2\pi_{CC}^2\pi_{OO}^{*1}\sigma_{OO}^{*1}$), as is shown in Figure

Spectroscopic and Theoretical Aspects

19. The reactant state arising from the interaction with the $^1\Delta^*$ oxygen state behaves like the ground state, and it too passes through a maximum in energy before correlating with the first excited singlet state of the product ($^1\psi_2$). To predict how the $^1\psi_1(\Delta)$ state of the reactant (a complex between $^1\Delta O_2$ and ground state diene) behaves, it is noted that the behaviour of the π_x^* and π_y^* orbitals appears to be similar to Case I shown in Figure 17. Judging from this, it is anticipated that initially the $^1\psi_1(\Delta)$ state of the reactants will be little changed in energy by small displacements along the

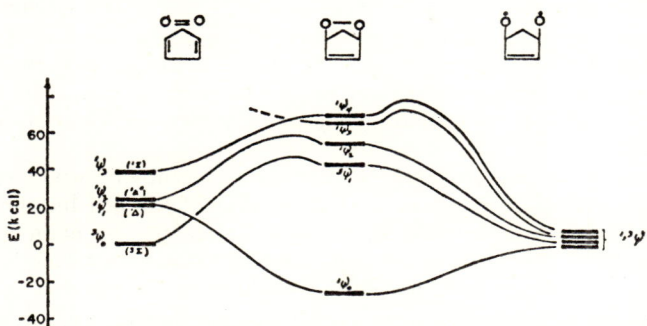

Figure 19 *Left side: State correlation diagram for the concerted addition of molecular oxygen to cyclopentadiene. The states of molecular oxygen in the oxygen–diene complex have been indicated in parentheses. Right side: State correlation diagram depicting the decomposition of the endoperoxide II by cleavage of the O—O bond*
(Reproduced by permission from *J. Amer. Chem. Soc.*, 1969, **91**, 6554)

reaction co-ordinate, but ultimately it will be stabilized, and correlate exothermically with the ground state of the product as shown in Figure 19. The $^1\psi_3(^1\Sigma)$ state of the reactant ($^1\Sigma O_2$ + ground state diene) correlates endothermically with a relatively excited state of the product.

Thus the following predictions can be made for the cycloaddition of oxygen to dienes.

(*a*) The $^3\Sigma$ state of O_2 is not expected to cycloadd to dienes since interaction of $^3\Sigma O_2$ with ground-state diene leads to a triplet-state complex which correlates endothermically with an excited triplet state of the product.

(*b*) $^1\Delta O_2$ is expected to cyloadd since this leads to a complex which correlates exothermically to the ground state of the product.

(*c*) $^1\psi_2(\Delta^*)$ and $^1\psi_3(^1\Sigma)O_2$ are not expected to cycloadd since the complexes formed with ground-state dienes correlate endothermically with high excited states of the product.

These results are in excellent agreement with experimental observations. Further considerations lead to the result that addition of $^1\Delta O_2$ to olefins may be forbidden unless the olefin has a low ionization potential. The lengthy example outlined above has been given to illustrate the point that use of correlation diagrams is not necessarily a simple procedure, but with care an excellent qualitative description of reaction paths can be obtained which will reliably predict actual behaviour.

4 Developments in Techniques

This section is devoted to a discussion of some of the recently developed techniques which will be of assistance to those in the field. The discussion includes details of new light sources, filters, and ancillary equipment for conventional photolysis systems, and new approaches in the treatment of data. However, the major part of the section deals with the development of non-stationary state photolysis systems, mostly using lasers in the nanosecond to picosecond region, and the use of detection methods such as electron paramagnetic resonance, mass spectrometry, *etc*. The recently developed technique of chemically induced nuclear spin polarization is also discussed.

New light sources are always of interest, but recently in two independent investigations has been announced the initiation of photochemical reactions in the absence of an external light source, an example of 'photochemistry without light'.[102, 103] However, the technique is not as revolutionary as might appear at first sight in that a chemiluminescent reaction is used to generate excited species, from which energy transfer to an acceptor molecule capable of undergoing a photochemical change occurs. As an example, oxetans decompose thermally to give excited ketones (and aldehydes).[102] Intermolecular energy transfer from the excited states so

$$\underset{Me}{\overset{Me}{\underset{O-O}{\bigsqcup}}}\underset{H}{\overset{Me}{\bigsqcup}} \xrightarrow{\Delta} \left[\underset{O}{\overset{Me}{\underset{\parallel}{C}}}\overset{Me}{} + \underset{O}{\overset{Me}{\underset{\parallel}{C}}}\overset{H}{} \right]^{*} \quad (80)$$

formed was demonstrated by the addition of anthracene and biacetyl from which fluorescence and phosphorescence were observed respectively. The addition of *cis*-stilbene to the oxetan produced quantities of *trans*-stilbene which presumably arise *via* a triplet-quenching mechanism, and other photoisomerizations, induced by the chemiluminescent reaction, were observed. The technique may not be generally applicable, but has advantages in that problems of simultaneous absorption by more than one species do not arise, since there is no absorption, and the apparatus is extremely inexpensive, since no optical components such as monochromators are required.

Resonance light sources are widely used by photochemists, *e.g.* when a resonance line is absorbed by its own gas, measurement of the absorption can be used to derive a value of the product nf of concentration of absorber and f-value of the transition. The interpretation and success of experiments based on line absorption depend critically on the shape of the line

[102] E. H. White, J. Wiecko, and D. F. Roswell, *J. Amer. Chem. Soc.*, 1969, **91**, 5194.
[103] H. Cuesten and E. F. Ullman, *Chem. Comm.*, 1970, 28.

emitted by the light source, which is strongly influenced by the optical depth in the emitting region, and by the inevitable presence of an absorbing (reversing) layer through which the light must travel on its way out of the lamp. A three-layer model has been devised on the basis of which the optimum conditions for line source operation can be calculated.[104] The model consists of an emitting layer, reversing layer, and absorbing layer, each with its own optical depth, Doppler temperature, and outgoing intensity. The paper will be useful for those designing such resonance lamps in that optimum optical depths (concentrations of atoms) can be calculated for particular systems.

A black-body source of radiation covering a wavelength range from the u.v. to the i.r. has been described.[105] The source consists of an electrically heated graphite tube with an argon atmosphere in which a conically indented graphite plug is located, forming the cavity. The source can attain radiance temperatures of 3000 K and hence provides a valuable background source for temperature measurements by absorptivity–emissivity methods, and also provides a useful standard for radiation calibration from the u.v.—i.r. regions.

A radiation standard for the vacuum-u.v. region of the spectrum has been devised which is also a useful source in this region for photolysis purposes.[106] The source consists of a constricted dc argon arc to which nitrogen or oxygen is added, and is stable and reproducible. The absolute quantum efficiency of sodium salicylate for excitation in the extreme u.v. has been given.[107] Quinine bisulphate has been widely used as a fluorescence standard, but its use has been questioned in recent years (see Volume 1). The conflicting data in the literature have prompted a thorough investigation of the absorptivity, fluorescence, and fluorescence excitation spectra of a number of different commercial samples of quinine sulphate.[108] It is shown that, provided care is taken to use the correct molecular weight for quinine sulphate, taking into account the water of hydration, all samples yield exactly the same spectra and yields. Moreover, for excitations from 240 to 400 nm the spectral distribution of the emission and the fluorescence yield was found to be constant within $\pm 5\%$, with the exception of results obtained at 272 nm (a minimum of the absorption spectrum). It is suggested that discrepancies in the literature may be due to experimental errors associated with particular instrument design, and that quinine sulphate should retain its place as a useful standard.

The optical equipment used to isolate particular wavelength regions from the total distribution emitted by light sources is frequently expensive,

---

[104] W. Braun and T. Carrington, *J. Quant. Spectroscopy and Radiative Transfer*, 1969, **9**, 1133.
[105] K. C. Lapworth, T. J. Quinn, and L. A. Allnutt, *J. Sci. Instr.*, 1970, **3**, 116.
[106] J. C. Morris and R. L. Garrison, *J. Quant. Spectroscopy and Radiative Transfer*, 1969, **9**, 1407.
[107] E. C. Bruner jun., *J. Opt. Soc. Amer.*, 1969, **59**, 204.
[108] A. N. Fletcher, *Photochem. and Photobiol.*, 1969, **9**, 439.

and alternative cheaper methods may be sought. Chemical filters may often be used, but in the wavelength region from 235 to 300 nm, such chemical filters have not been widely used, principally because of the width of their band-pass. However, narrow-band chemical filters for this region have now been developed,[109] based on the systems of Kasha. The composition and characteristics of the filters are given in Table 5, and the spectral

Table 5 *Composition and characteristics of chemical narrow band-pass filters*

Filter number	I	II	III	IV	V	VI
Wavelength of max. transmission in nm	235	240	252	260	277	294
Maximum transmission %	3·0	10·8	16·2	30·2	2·5	1·5
Half-width in nm	17	12	15	8	16	20
Composition:						
9 cm $NiSO_4,7H_2O$, 250 g, $CoSO_4,7H_2O$, 80 g per l of water	+	+	+	+	+	+
2 cm of cyanine perchlorate in water (conc. in mg/l)	100	300	2500	400	50	10
3 cm of acetone (0·5% vol/vol) in ethanol 3 vol, H_2O 1 vol, N_2 saturated	+					
3 cm anisole 170 mg/l in 95% ethanol		+				
3 cm $CHCl_3$ (20% vol/vol) in 95% ethanol			+			
3 cm CCl_4 (10% vol/vol) in cyclohexane				+		
2 mm Pyrex glass						+
3 cm anthracene in cyclohexane, conc in mg/l					100	50
Special filter[a]					+	+
Stability in min 20% variation in flux, for flux shown below	15	20	40	20	90	90
Photon flux in 10^{16} quanta/s	0·18	0·6	1·5	1·9	1·1	2·1

[a] Special filter is made of 2 plates of Corning 9863 glass with 5 mm solution of $N_1SO_4,7H_2O$ (300 g/l) between.

distributions are shown in Figure 20. The only drawback to the use of these filter combinations is their relative instability, but this can be overcome by frequent replacement, or use of continuously flowing solutions.

Recent development of pulse techniques require the use of very sensitive photodetectors. Similar problems of signal-to-noise ratio are met in Raman spectroscopy, and it has been shown that the dark current of an E.M.I. 9558 QA photomultiplier can be drastically reduced by a magnetic defocussing of electrons originating from the inefficient part of the photocathode.[110] The photomultiplier is fitted into a metal tube around which a

[109] B. Muel and C. Malpiece, *Photochem. and Photobiol.*, 1969, **10**, 283.
[110] J. A. Topp, H. W. Schroetter, H. Hacker, and J. Brandmueller *Rev. Sci. Instr.*, 1969, **40**, 1164.

magnetic coil is wound, in a position which can be altered by a screw thread. The position is altered so that the magnetic field between photocathode and the first dynode has optimal divergence. The current in the coil is then varied so as to minimise the dark current. If the detector is cooled also, a great increase in signal-to-noise ratio can be obtained.

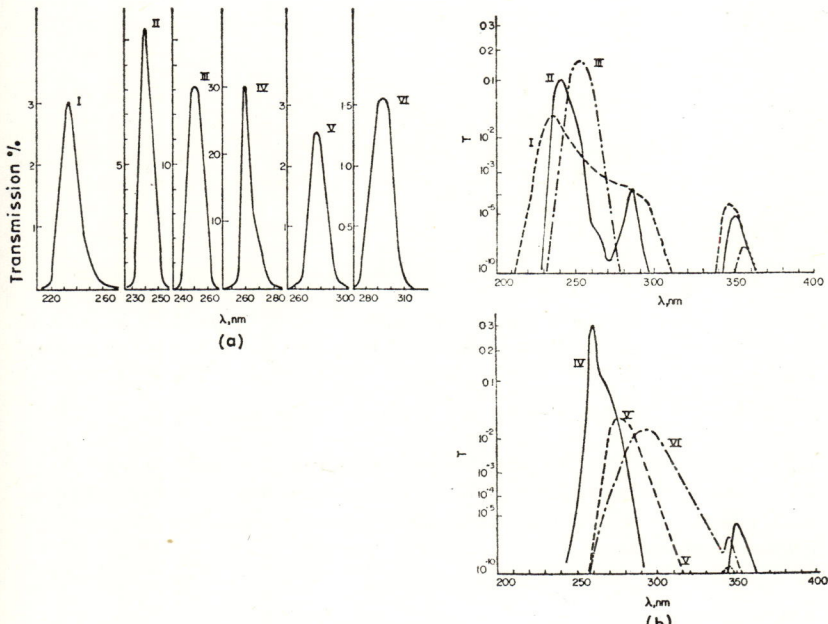

Figure 20 *Transmissions of chemical filter combinations (see Table 5 for details).*
(a) *expanded scale*, (b) *compressed scale*
(Reproduced by permission from *Photochem. and Photobiol.*, 1969, **10**, 283)

A novel method for measuring triplet-state yields based on calorimetry has been developed.[111] The triplet state of organic compounds decaying by radiationless processes to the ground state will produce typically *ca.* 10^{-5}—10^{-4} J of thermal energy. The measurement of this heat is achieved by means of a capacitor microphone transducer which measures volume changes caused by heating. Details of the device will not be given here, but it can be said that changes in volume sensed by an aluminium diaphragm are converted linearly to voltages by the microphone. Details of the experimental reaction cell are shown in Figure 21, together with a block diagram of the complete calorimetric apparatus. Fast radiationless

[111] J. B. Callis, M. Gouterman, and J. D. S. Danielson, *Rev. Sci. Instr.*, 1969, **40**, 1599.

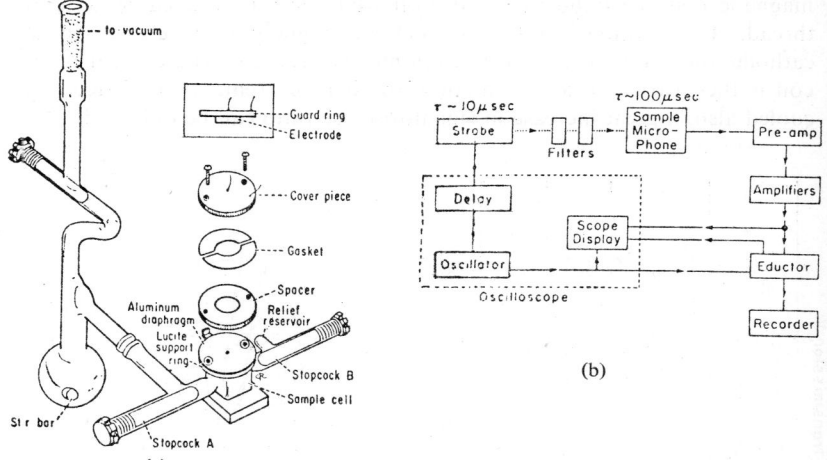

Figure 21 (a) *Calorimetric cell and degassing assembly. Inset shows a side view of the cover piece—the guard ring and stationary electrode.* (b) *Block diagram of the complete calorimetric apparatus. A dotted line indicates light path and arrows show information flow. Response times are indicated*
(Reproduced by permission from *Rev. Sci. Instr.*, 1969, **40**, 1599)

decay processes prior to triplet-state formation will be followed by a slower decay of the triplet state if the system is excited by a short-duration flash lamp. Assuming the triplet state is non-phosphorescent, the total heating is given by

$$Q_{tot} = Q_{fast} + Q_{slow} \quad (81)$$

where

$$Q_{fast} = [h\nu - \Phi_f h\nu_f - \Phi_t E_t] N_{exc} \quad (82)$$

$$Q_{slow} = \Phi_t E_t N_{exc} \quad (83)$$

$h\nu$ is the energy of the incident photon, Φ_f is the fluorescence yield, $h\nu_f$ the fluorescent energy, Φ_t the triplet state yield, E_t the triplet energy and N_{exc} the number of molecules excited initially by the flash. Thus

$$\Phi_t = Q_{slow}(h\nu - \Phi_f h\nu_f)/Q_{tot} E_t \quad (84)$$

Thus Q_{slow}, Q_{tot}, $h\nu$, Φ_f, and $h\nu_f$ are all measurable, and a knowledge of Φ_t allows estimation of E_t, or *vice versa*. Details of the experimental procedure will not be given here, except that it is noted that to improve signal-to-noise ratios, the signal is averaged over 100—500 flashes by means of a waveform eductor, shown in Figure 21. The largest uncertainty in the technique is the estimate of Q_{slow}, partly owing to the necessary extrapolation of long-time behaviour to short time, and partly owing to mechanical acoustical noise. However, results obtained for ϕ_t of anthracene with this method compare very favourably with those obtained using other methods. One of the possibilities of this type of apparatus is that $S_1 \to S_0$ radiationless processes may be measurable directly, a feat which would provide much useful data to resolve inconsistencies in the literature.

Modifications to an Aminco-Bowman spectrophotofluorimeter to allow simultaneous monitoring of different signals has been described,[112] and details of a cryostat for photochemical reactions in the gas phase given.[113] Other miscellaneous items of information of possible interest concern the removal of mercury from systems by gold surfaces, and the prevention of

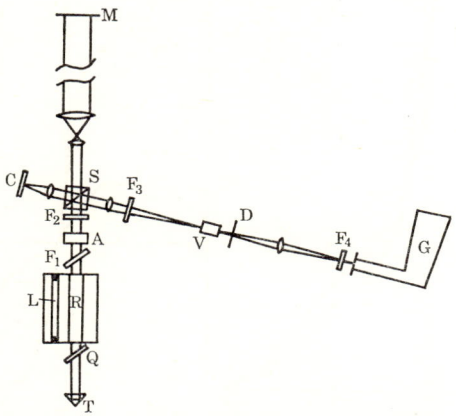

Figure 22 *Nanosecond spectrographic apparatus.* A, ADPh *crystal;* C, *scintillator cell;* D, *aperture stop;* F, *filters;* F_1, *Wratten* 29, *transmits* $\lambda \geqslant 630$ nm; F_2, CuSO$_4$, *attenuates* 694 nm; F_3, *biphenylene, control of intensity of* 347 nm, F_4, *biphenylene, u.v. laser cut-off;* G, *spectrograph;* L, *xenon-filled flash lamp;* M, *mirror;* Q, *passive Q-switch;* S, *beam splitter;* T, *quartz t.i.r. prism;* V, *reaction vessel;* R, *ruby*
(Reproduced by permission from *Proc. Roy. Soc.*, 1970, **315**, *A*, 163)

mercury photosensitization,[114] and the use of monomolecular films as a model to study photosensitized reactions.[115]

In a very short time, nanosecond flash photolysis has become a well-established powerful technique. There are several variations on the basic technique, but that developed using a double flash system will be described.[116] The basic lay-out is shown in Figure 22. Light from a Q-switched ruby laser passes through filters F, and is then frequency-doubled by an ammonium diphosphate crystal. The doubled light then passes through filters F_2 to remove any fundamental, and is split by a beam splitter S.

---

[112] V. A. Mode and R. A. Thomas, *Rev. Sci. Instr.*, 1969, **40**, 1241.
[113] J. H. Sullivan, *Rev. Sci. Instr.*, 1969, **40**, 1347.
[114] S. Lythgoe, P. J. Robinson, and R. D. Sedgwick, *J. Sci. Instr.*, 1970, **3**, 401.
[115] A. Felmeister and R. Schaubman, *J. Soc. Cosmetic Chemists*, 1970, **21**, 155.
[116] G. Porter and M. R. Topp, *Proc. Roy. Soc.*, 1970, **315**, 163; G. Porter, *Israel J. Chem.*, 1969, **7**, 179.

Part of the beam then traverses the reaction vessel, producing excited states. The other fraction of the beam from S traverses some distance to a mirror and returns thereby establishing a time delay in the nanosecond range before striking a scintillator cell C which absorbs the doubled ruby line and emits a broad continuum. This light then passes through the reaction vessel acting as a spectroscopic flash, and enables absorption spectra to be taken on a spectrograph of both singlet and triplet excited states of molecules. The system can be converted for use in a photoelectric

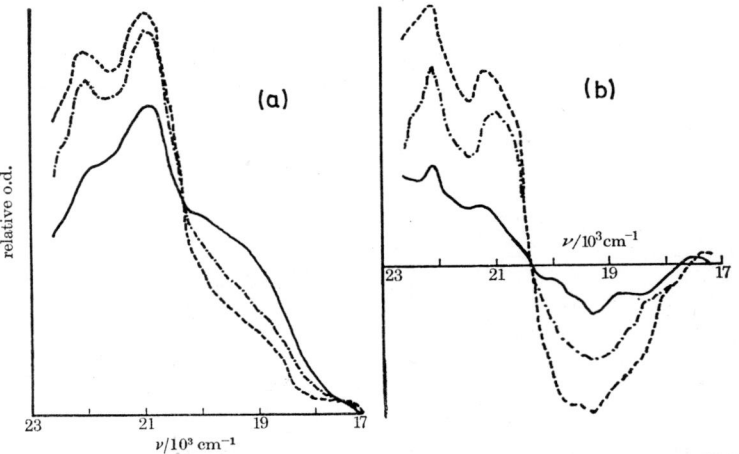

Figure 23 (a) *Time-resolved absorption spectrum of phenanthrene after delays of* ————, 5 nsec, ·—·—·—·—·, 22 nsec, ------, 32 nsec. (b) *Difference spectra at same delays*
(Reproduced by permission from *Proc. Roy. Soc.*, 1970, **315**, *A*, 163)

rather than spectrographic mode. Care must be taken in interpretation of absorption spectra obtained in this way, since frequently the singlet and triplet excited state absorption spectra overlap. Thus the decay of excited singlet 1,2-benzanthracene and naphthalene monitored by fluorescence decay is not matched by growth of triplet absorption, presumably because the absorptions of excited singlet and triplet overlap.[117] This phenomenon has already been commented upon in the case of anthracene,[33] and the use of polarizers in distinguishing between the two absorptions mentioned. Where singlet and triplet absorption spectra do not completely overlap, a time sequence shows decay of one absorption into another. An example of this is shown for phenanthrene in aerated cyclohexane in Figure 23(a). By subtracting the absorption spectrum recorded immediately after the laser-pulse from a spectrum taken at a later time, a difference spectrum may be plotted whose ordinate is a measure of the difference between triplet and singlet extinctions. Such a spectrum for the phenanthrene example is

[117] J. K. Thomas, *J. Chem. Phys.*, 1969, **51**, 770.

shown in Figure 23(b). Results obtained with this method are summarized in Table 6.

The technique can be applied to the gas phase by the employment of a different kind of reaction cell. Several variants on this theme could be quoted, but two examples will suffice. Figure 24 shows the experimental set-up used in a study of the triplet states of aromatic vapours immediately following intersystem crossing, in which the subsequent vibrational relaxation was recorded spectrophotometrically.[118] For anthracene vapour, the vibrationally excited T_1 molecule initially formed in the experiments

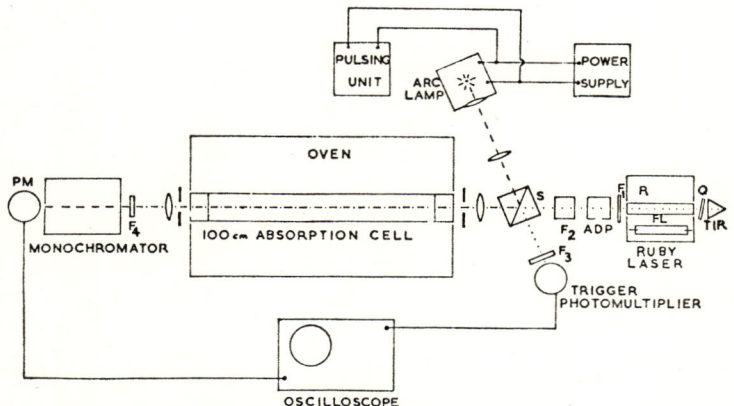

Figure 24 *Arrangement for gas phase nanosecond kinetic spectrophotometry* (Reproduced by permission from *Chem. Phys. Letters*, 1970, **6**, 7)

has 13,200 cm^{-1} excess vibrational energy. The completely relaxed triplet anthracene has an absorption maximum at 402 nm. Inspection of the absorption immediately following the flash revealed that the absorption had a maximum at 406 nm. This is attributed to absorption by the vibrationally excited triplet, and is taken to indicate that the geometry and anharmonicities of the T_1 and T_3 states of anthracene are similar. It must be noted that the singlet state of anthracene absorbs in the same region as T_1, as is revealed by polarization techniques.[33] It is not explicitly stated in ref. 118 that interference from the singlet absorption has been considered, but it is assumed that the decay time of the absorption at 406 nm, which is pressure dependent, is too long to be accounted for in terms of an excited singlet state. The pressure dependency of the growth of the absorption at 402 nm is explicable in terms of collision-induced vibrational relaxation of the excited triplet T_1 molecule. The rate of relaxation corresponds to the removal of *ca.* 800 cm^{-1} vibrational energy per collision (with methane). Argon is a less efficient relaxer, as expected.

[118] S. J. Formosinho, G. Porter, and M. West, *Chem. Phys. Letters*, 1970, **6**, 7.

Table 6 *Higher singlet levels of aromatic molecules and correlation of lifetimes with fluorescence data.* (Reproduced by permission from *Proc. Roy. Soc.*, 1970, **315**, *A*, 163)

Molecule	Solvent	Singlet S_1 absorption (nm)	Level of higher S state (cm^{-1})	Triplet T_1[†] absorption (nm)	Singlet absorption decay time (ns)	Fluorescence decay time (ns)
Phenanthrene (d_{10})	Cx	(545), 515	48 200	520, 510 481, 454	61·1	63·5
Phenanthrene (h_{10})	PMMA	(545), 515	48 200	520, 510, 481 454	65	67·2
Triphenylene	Benzene	500, 465, 433	48 800	428	44·2	43·0
	PMMA	500, 465, 433	48 800	428	45·0	44·0
1,2-Benzanthracene	Cx	550, 495	44 200	540, 490, 461	52·7	49·4
	PMMA	550, 495	44 200	540, 490, 461	51·7	52·5
1,2;3,4-Dibenzanthracene	Cx	540, 500	45 700	445∥	51·2	53·5
	PMMA	540, 500	45 700	445∥	50·3	52·5
Pyrene	Cx	515, (480), 470	48 000	520, 483, 416	296‡	261‡
	PMMA	515, (480), 470	48 000	520, 483, 416	326	319
Coronene	Dx	(600, 570), 530 495, (465)	42 500	525, 480	319	307
	PMMA	(600, 570), 530 495, (465)	42 500	525, 480	390	380
3,4-Benzpyrene	Cx	(590), 535, 510	43 700	480§	49·1	57·5
	PMMA	(590), 535, 510	43 700	480§	39·4	40·5
3,4-Benzphenanthrene	Cx	595, 525	43 200	517	68·5	70
	PMMA	595, 525	43 200	517	76	81
3,4;9,10-Dibenzpyrene	Cx	575, 552	40 400	495∥	140	143
1,2;5,6-Dibenzanthracene	PMMA	(570)	(42 900)	532, 480	(40·0)¶	37·5

† Except where stated, data correlate with those from G. Porter and M. W. Windsor, *Proc. Roy. Soc.*, 1958, **245**, *A*, 238.
‡ 3×10^{-4} mol l^{-1}.
§ D. P. Craig and I. Ross, *J. Chem. Soc.*, 1954, 1589.
∥ Experimentally determined.
¶ Growth of triplet absorption.
Dx, dioxan; PMMA, polymethylmethacrylate; Cx, cyclohexane.

The laser is a perfect tool for spectroscopic purposes operated in a continuous mode. An experimental set-up suitable for measurement of spectral characteristics is shown in Figure 25.[119] The fluorescence cell shown has Brewster angle windows and can be positioned inside the optical resonator. Cells with plane perpendicular windows can be used outside the cavity. This technique was used to study NO_2 fluorescence. A very simple unit for studying transient fluorescent species in the nanosecond

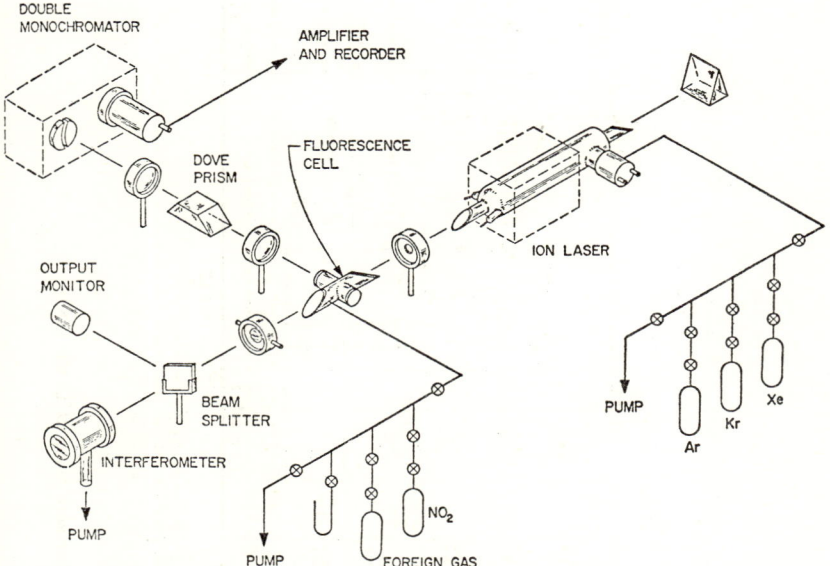

Figure 25 *Experimental arrangement for laser study of fluorescence of* NO_2 (Reproduced by permission from *J. Chem. Phys.*, 1969, **50**, 2404)

range using a pulsed discharge lamp rather than a laser source has been described.[120] In conjunction with adequate computer software simpler units of this type can still prove to be powerful techniques.

Longer duration pulses of monochromatic light are themselves powerful tools for the study of excited states of molecules, and a pulsed 2537 Å mercury source utilizing a microwave pulse has been developed.[121] In this apparatus, a pulser was constructed to operate English Electric magnetrons M561 and M578 on single shot to cover the range 20—800 kW and 0·5—10 μs. The magnetron pulse generator is shown in Figure 26. Microwave power coupled out of the magnetrons into a coaxial line was converted into the TE0*l* mode of rectangular waveguide 10, which con-

[119] K. Sakurai and H. P. Broida, *J. Chem. Phys.*, 1969, **50**, 2404.
[120] B. Selinger and R. Speed, *Chemical Instrumentation*, 1969, **2**, 91.
[121] A. B. Callear, J. Guttridge, and R. E. M. Hedges, *Trans. Faraday Soc.*, 1970, **66**, 1289.

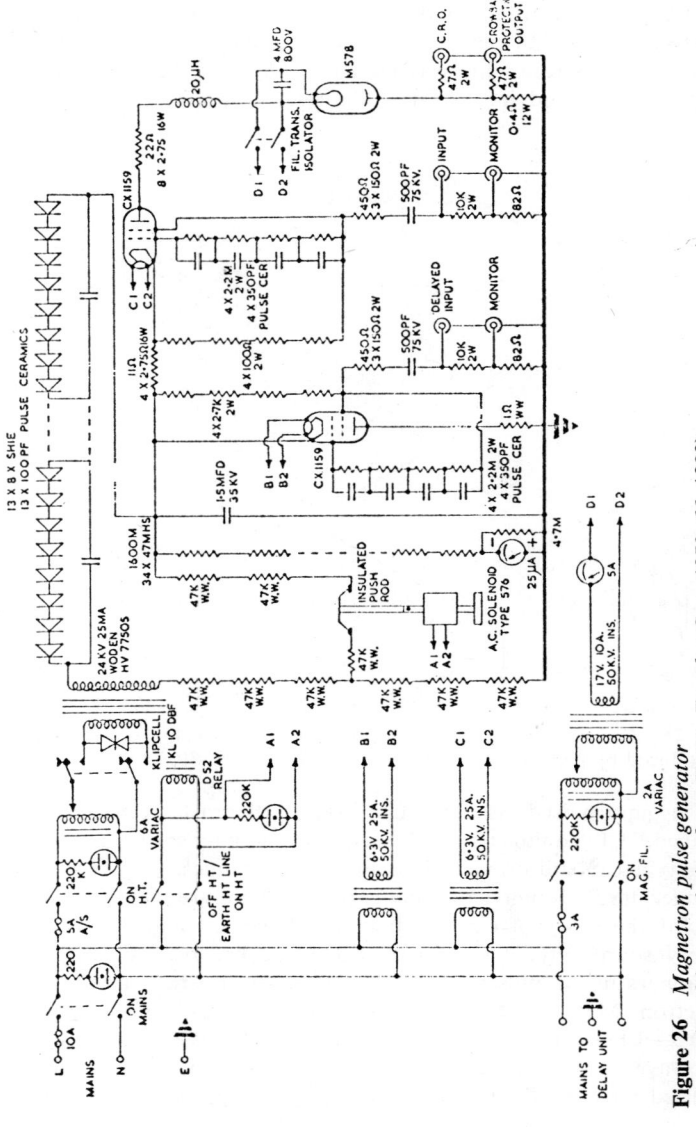

Figure 26 *Magnetron pulse generator*
(Reproduced by permission from *Trans. Faraday Soc.*, 1970, **66**, 1289)

tained a quartz reaction vessel of 5 mm diameter and 0·5 m length (Figure 27). The original system used mixtures of He and CO in the reaction vessel, and the absorption due to C_2 molecules formed was observed ($X\,^3\Pi_u \to A\,^3\Pi_g$) and its relaxation was observed. Relaxation of NO X^2 ($v = 1, 2, 3$) in pulsed NO+He and relaxation of $He_2\,a\,^3\Sigma_u^+$ in pure He were also observed. With mixtures of mercury vapour and nitrogen, an intense flash of 2537 Å radiation was produced.[122] The unit

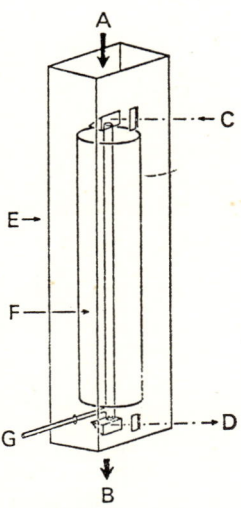

Figure 27 *Open waveguide method of microwave-pulse flash-spectroscopy.* A, *incident microwave power;* B, *microwave exit to matched load;* C, *spectroscopic source;* D, *spectrograph;* E, *S-band waveguide;* F, *refrigerant*
(Reproduced by permission from *Trans. Faraday Soc.*, 1970, **66**, 1289)

provides a high sensitivity for the detection of metastable $Hg(^3P_0)$ atoms, since some 10^{14} quanta cm^{-3} could be absorbed in a 2·5 μs pulse, and yields of metastable atoms were determined in N_2, CO, CO_2, CH_4, C_2H_6, C_3H_8, n-C_4H_{10}, NH_3, and H_2O. Metastable atom yields in H_2D_2, NO, N_2O, O_2, H_2S, C_2H_2, CH_3Br, HBr, HCl, and HCN were reported to be less than 1%.

Processes that have been established spectroscopically include

$$Hg(^3P_1) + H_2S \longrightarrow Hg(^1S_0) + H + SH \qquad (85)$$

$$Hg(^3P_1) + H_2 \longrightarrow HgH + H \qquad (86)$$

$$Hg(^3P_1) + HCl \longrightarrow HgCl + H \qquad (87)$$

$$Hg(^3P_1) + CS_2 \longrightarrow Hg(^1S_0) + CS_2^* \qquad (88)$$

[122] A. B. Callear and R. E. M. Hedges, *Trans. Faraday Soc.*, 1970, **66**, 605, 615.

Quantum yields for reaction (86) are shown in Table 7. Mercury hydride was not detected in the quenching of $Hg(^3P_1)$ by saturated hydrocarbons, HCN, NH_3, H_2O, HCl, HBr, and H_2S. The flash spectroscopic technique employing the cadmium resonance line has also been described.[123]

Table 7 Quantum yields of mercury hydride formation

Reaction	Φ	$\Phi_{^3P_1}/\Phi_{^3P_0}$
$Hg(^3P_1) + H_2$	0.80 ± 0.1	0.87 ± 0.08
$Hg(^3P_1) + D_2$	0.6 ± 0.2	0.9 ± 0.1
$Hg(^3P_0) + H_2$	0.93 ± 0.1	
$Hg(^3P_0) + D_2$	0.65 ± 0.2	
$Hg(^3P_1) + C_2H_2$	0.18 ± 0.07	

Flash techniques must ultimately utilize picosecond pulses, which bring developments to the limits of chemistry. Picosecond pulses have already been produced, and recently these have been used to observe relaxation

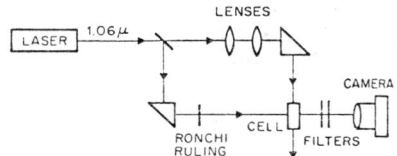

Figure 28 Schematic diagram of the experiment. Collimating lenses focus 90% of the mode-locked laser output to ≈2 mm diameter bleaching light pulse which traverses the rectangular cell along 1 cm path. A beam splitter diverts 10% of the output to form the weak interrogating pulse which traverses the rectangular cell along 2 mm path at right angles to bleaching light pulse. A coarse diffraction grating (500 lines/inch), that is, a Ronchi ruling is placed one metre away from the cell to eliminate the nonuniformities in the interrogating beam. Suitable filters are used so that only the 1·06 μ laser output is recorded on the Polaroid Type 413 film in the camera
(Reproduced by permission from Chem. Phys. Letters, 1969, **3**, 534)

times of polymethine dyes on a picosecond time scale.[124] In this work the output of a mode-locked Neodymium in glass Q-switched (by passive dye) laser is collimated to 2 mm, and forms an intense light pulse which bleaches a solution of a dye (Kodak Q-switch dye No. 9860) (see Figure 28). A fraction of the uncollimated light is split off to form a broad weak beam which, after passage through a Ronchi grating to eliminate irregularities, traverses the dye solution at right angles to the bleaching pulse, forming an interrogating beam. The pulse lengths are of the order of 1 ps, and by measuring the transparency along the path of the bleaching pulse, the relaxation time of the bleachable dye can be established. This beautifully simple set-up, reminiscent of the design in Figure 21, but on a picosecond

---

[123] W. H. Breckenbridge and A. B. Callear, Chem. Phys. Letters, 1970, **5**, 17.
[124] M. M. Malley and P. M. Rentzepis, Chem. Phys. Letters, 1969, **3**, 534.

scale, must surely point the way for future developments in this time régime. The relaxation time for the dye above was 9 ps, and for Kodak Q-switch dye No. 9740 was $\tau = 10$ ps.

A flash photolysis apparatus with a new geometry has been described,[125] and a photomultiplier and amplifier circuit suitable for all forms of kinetic spectrophotometry devised.[126]

The technique of observing relaxation processes after pulsed excitation is widespread, but there are other methods of observing such processes employing continuous illumination which are themselves as elegant and useful as those described above. The development of phase-sensitive detection systems has permitted a rapid spread of the use of modulation techniques, in which light sources of quite moderate output are sufficient to provide high enough concentration of intermediates to be detected on a repetitive basis. The technique is really a sophisticated extension of the rotating sector method of studying free radicals, but is now available for study of intermediates with lifetimes as short as the nanosecond region. As an example of the use of this technique, the apparatus used to detect ClO and ClO_2 radicals in the gas phase will be described.[127] The radical lifetimes in this case will be between 10 and 0·025 s, and consequently the light source generating the free radicals must be modulated at between 0·1 and 40 Hz in order to obtain a reaction period which is of the same order as the lifetime. The low-frequency signal resulting from absorption by these radicals is in the region where electronic noise presents a serious problem and to overcome this, phase-sensitive lock-in amplification techniques are used. Here the concentration oscillations are allowed to modulate a chopped 400 Hz carrier beam of i.r. or u.v. light, which transfers the information to sidebands centred at 400 Hz, and radical lifetimes can be inferred by determination of the phase shift with respect to the photolytic excitation. Both kinetic and spectroscopic information regarding the intermediates of interest is available from this experimental technique. A schematic diagram of the apparatus using i.r. detection of intermediates is shown in Figure 29. In this apparatus, the low-frequency reference generator provides two square waves exactly 90° out of phase, one of which signals triggers the photolytic lamp. The results obtained on the Cl_2–O_2 system have been described in Part I, Chapter 3.

The phase-shift method is especially suitable to lifetime measurements at low fluorescence intensities as it makes efficient use of the available exciting light, and readily allows for long integration periods necessary to overcome shot noise in the optical signal. An apparatus suitable for study of the kinetics of the fluorescence of NO_2 vapour is shown in Figure 30.[128]

[125] J. R. Huber, R. P. Widman, and K. Weiss, *Rev. Sci. Instr.*, 1969, **40**, 1103.
[126] J. P. Keene, E. D. Black, and E. Hayon, *Rev. Sci. Instr.*, 1969, **40**, 1199.
[127] H. S. Johnston, D. E. Morris, and J. van den Bogaer, *J. Amer. Chem. Soc.*, 1969, **91**, 7712.
[128] S. E. Schwartz and H. S. Johnston, *J. Chem. Phys.*, 1969, **51**, 1286.

This system measures the fluorescence phase lag as the ratio of the quadrature component of the optical signal to the in-phase component, and allows integration times to be extended arbitrarily by analogue or digital means. The radiative lifetimes of NO_2 in different vibronic levels of the $^2B_1$ state ranged from 55 to 90 μs, measured using this technique.

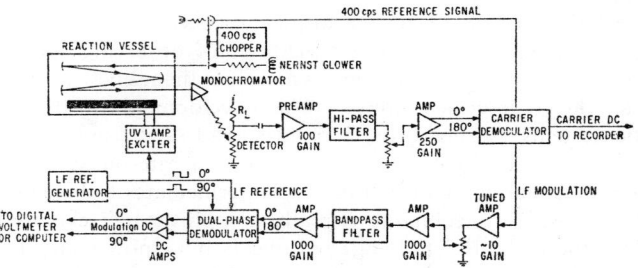

Figure 29 *Apparatus for phase-shift measurements on free radicals*
(Reproduced by permission from *J. Amer. Chem. Soc.*, 1969, **91**, 7712)

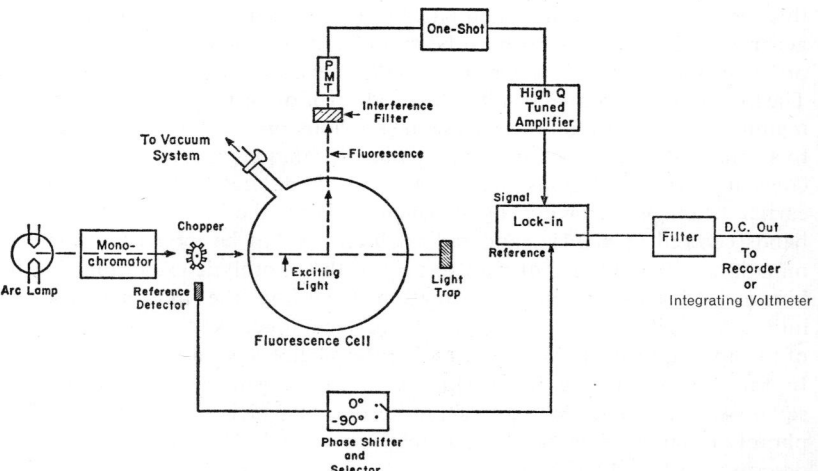

Figure 30 *Block diagram of apparatus for phase-shift measurements at low light intensity*
(Reproduced by permission from *J. Chem. Phys.*, 1969, **51**, 1286)

In general, the theories of electronic relaxation discussed in earlier sections of this chapter are not amenable to experimental testing, simply because in order to produce sufficient signal intensity to measure decay rates of individually excited vibronic levels of a molecule, the conditions of the experiment (condensed phase or high pressure of gas) usually relax the molecule to its ground vibronic level before electronic decay occurs, and thus decay rates for individual vibronic states are not observable.

However, because of the sensitivity of the phase-shift technique, individually excited vibronic states of molecules may be prepared and observed before vibrational relaxation (at low pressures of gas) and thus results obtained in this way used as a basis for discussion of theoretical approaches.

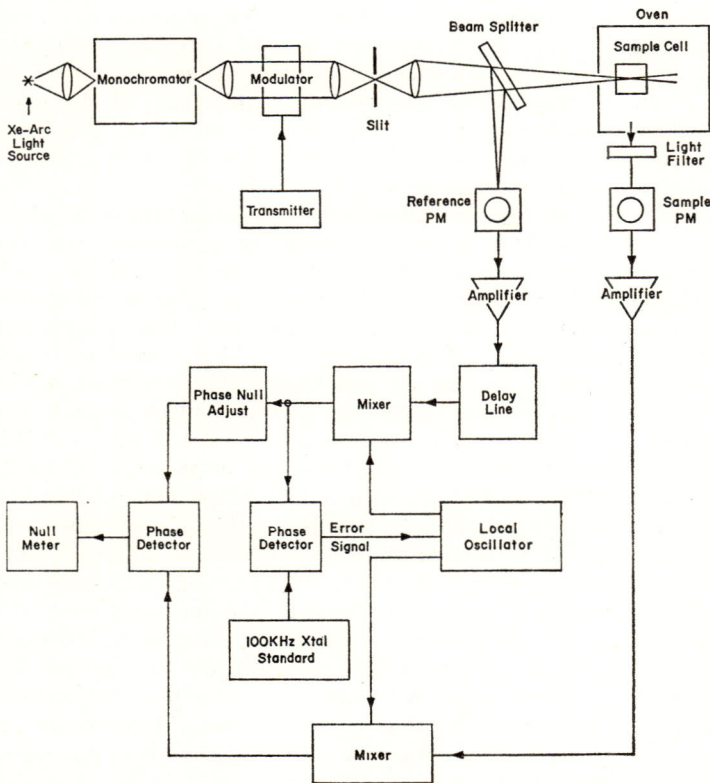

Figure 31 *Kinetic spectrometry using modulated monochromatic radiation* (Reproduced by permission from *J. Chem. Phys.*, 1969, **51**, 2508)

Typically the lifetimes of fluorescing organic molecules in the vapour will be of the order of 10^{-8} s, and in order to carry out phase-shift experiments on such molecules, modulation must be very fast. An apparatus used to observe the decay of β-naphthylamine fluorescence is shown schematically in Figure 31, in which the modulation is of the order of 10 to 30 MHz.[129] A beam splitter focuses a small sample of the modulated radiation on to a photomultiplier which supplies the reference signal. The sample is β-naphthylamine, which radiates with a time delay characteristic

[129] E. W. Schlag, *J. Chem. Phys.*, 1969, **51**, 2508.

of all radiative and non-radiative processes the initial state can undergo. The delay is manifested as a phase delay in the sample signal relative to the phase seen by the reference photomultiplier. The phase angle is measured by use of the phase-sensitive lock-in amplification unit. A null method is used, in which initially zero phase-shift is obtained by by-passing the sample cell with a light pipe which channels the light directly to the sample photomultiplier. A phase adjustment is then performed such that the phase detector operates at a fixed point of 90°. After removal of the light pipe a measurement is made by inserting cable delay in the reference channel until the phase detector is again at a null point. The length of cable required gives the phase delay of the sample signal.

An apparatus has recently been set up which is similar to that above in that it is capable of measuring fluorescence lifetimes of the order of tens of nanoseconds, but which employs a much more powerful light source. This is a low-pressure toroidal 2537 A mercury lamp driven by a 13·5 MHz AM radio frequency transmitter.[130, 131] The lamp responds well to modulation frequencies up to 200 KHz, and is extremely intense. The system can be used in the study of fluorescence decays, by a process similar to that outlined above, but can also be used to measure absorption spectra of intermediates with lifetimes as short as 10 ns. A block diagram is given in Figure 32. The lock-in detection system coupled with a multipath mirror system allows measurement of absorptions as low as 5×10^{-6}, and the technique can also be used in the study of mercury photosensitized reactions. The only drawback of the use of this intense light source is that high pressures of inert gas must be employed to counteract thermal effects, and photochemical effects (*i.e.* build-up of interfering reaction products) can be troublesome. Nevertheless the apparatus, with its light source in a region of the spectrum in which no intense laser radiation is yet available, should prove to be a very powerful tool in the study of the gas-phase photochemistry of molecules which absorb in this region, and in the study of mercury-photosensitized reactions. Reports on results obtained on the naphthalene [130] and benzene [131] systems using this apparatus are given in Part I, Chapter 3.

The use of r.f. signals to drive lamps is also possible in the vacuum-u.v. region of the spectrum, and an experimental arrangement to investigate the vacuum-u.v. photolysis of N_2O is shown in Figure 33.[132] The gases in the lamps were xenon or krypton, and decay times of the order of 2×10^{-4} s can be monitored with this system.

The technique of 'photochemical recoil spectroscopy' or 'translational spectroscopy' has been mentioned earlier in this chapter, and some brief details of the experimental set-up will now be considered. Two such

[130] H. E. Hunziker, *Chem. Phys. Letters*, 1969, **3**, 504.
[131] C. S. Burton and H. E. Hunziker, *J. Chem. Phys.*, 1970, **52**, 3302.
[132] G. Black, T. G. Slanger, G. A. St. John, and R. A. Young, *J. Chem. Phys.*, 1969, **51**, 116.

Spectroscopic and Theoretical Aspects

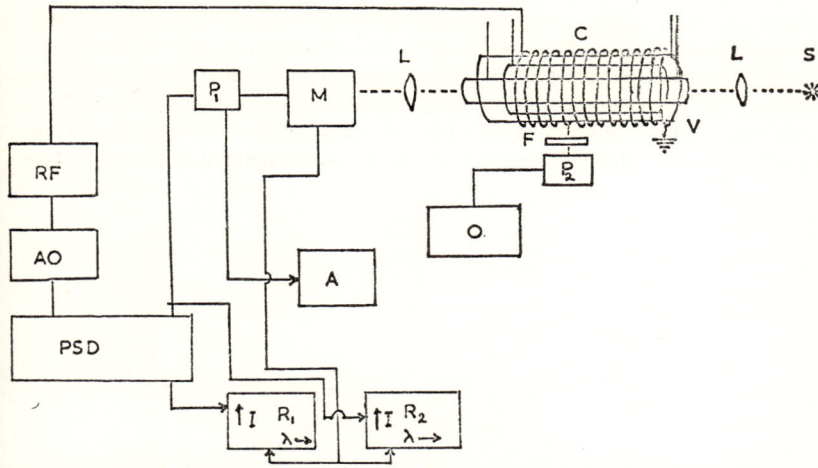

Figure 32 *Block diagram of kinetic spectrometer using Hg 2537 modulated source.*
RF, *radio frequency* AM *generator* (13·5 MHz, 5 kwatt *peak output*) *plus matching system;* AO, *audio oscillator to modulate r.f. frequency;* PSD, *phase sensitive lock in amplification unit;* R_1R_2, *x–y recorders to monitor background radiation and signal from* PSD; M, *grating monochromator, high resolution;* P, *photomultiplier units;* L, *quartz lenses;* V, *multi-jacketed cylindrical quartz, water-cooled reaction vessel with connections to water-cooling, heat exchange unit, vacuum system to fill lamp compartment, vacuum flow system for reactants;* C, *RF coil surrounding lamp;* F, *2537 interference filter;* S, *background source* (500 watt *xenon arc*); A, *pico ammeter to monitor background radiation;* O, *oscilloscope to monitor 2537 fluctuation with time*
(Based on private communication from H. E. Hunziker and C. S. Burton)

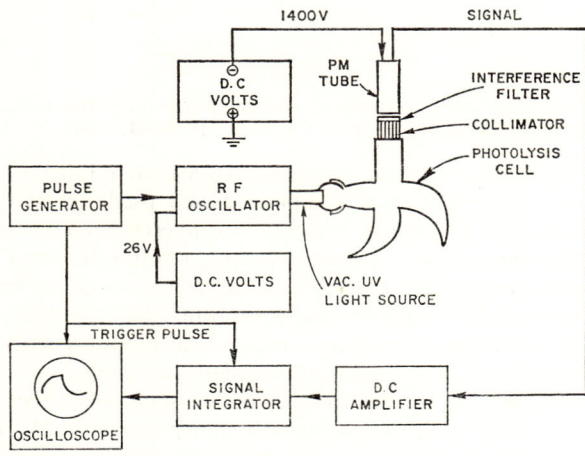

Figure 33 *Vacuum-u.v. kinetic spectrophotometry apparatus*
(Reproduced by permission from *J. Chem. Phys.*, 1969, **51**, 116)

systems are in existence, one of which has been described fully.[133] Essentially the apparatus consists of an ultra-high-vacuum system in which a molecular beam is crossed by pulses of polarized light (from a theta-pinch discharge or from various lasers, including ruby, second harmonic neodymium/glass, and laser pumped tunable dye), and the recoiling photodissociation fragments are detected by a quadrupole mass spectrometer a few centimetres

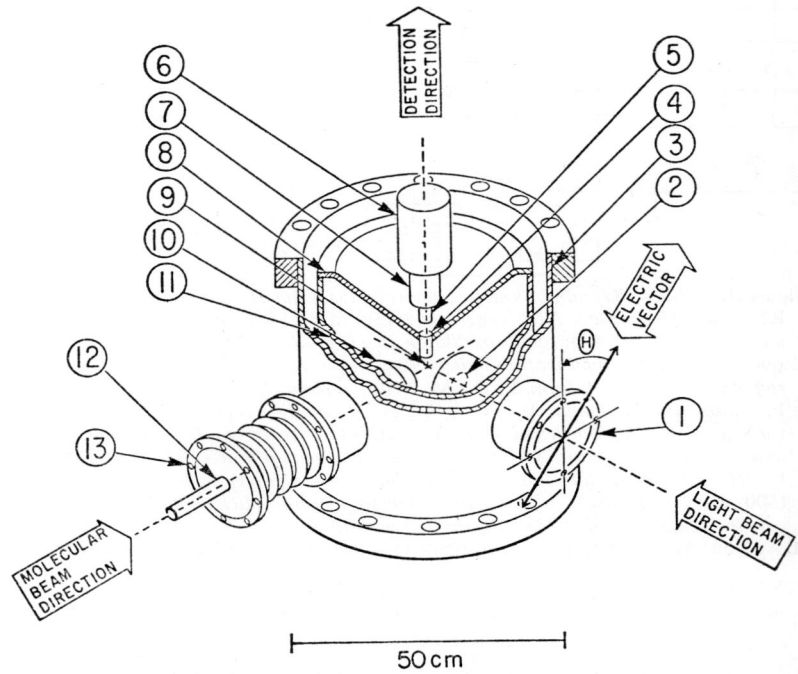

Figure 34 *Photofragment spectrometer. For description of components see text* (Reproduced by permission from *Rev. Sci. Instr.*, 1970, **41**, 1066)

away. The data are collected and analysed by an on-line time-shared computer, and presented on a computer-driven oscilloscope display as plots of number of fragments detected per microsecond *vs.* time after the pulse. Parameters such as photon energy, photon flux, and fragment laboratory recoil angle with respect to the electric vector of the light can be varied to obtain a multi-dimensional spectrum. This spectrum and the probability of photodissociation as a function of photon energy and flux and of fragment mass, centre-of-mass translational energy, and centre-of-mass angle of recoil may be used to study many aspects of molecular states. A cut-away drawing of the spectrometer is shown in Figure 34. The beam

[133] C. E. Busch, J. F. Corneluis, R. T. Mahoney, R. I. Morse, D. W. Schlosser, and K. W. Wilson, *Rev. Sci. Instr.*, 1970, **41**, 1066.

of molecules to be photodissociated enters from the left and is crossed perpendicularly by pulses of polarized light. The photodissociation fragments which recoil upwards are detected by a mass spectrometer as a function of mass, photon energy, photon flux and time t after pulse, and angle H measured from the electric vector of the light. (The H shown in the drawing would be a negative angle of recoil.) The interaction region and the mass spectrometer are in separately pumped chambers connected by a small liquid nitrogen cooled tube, which collimates the fragments. The numbered components are: (1) port for laser beam; (2) lens to match diameter of laser beam to that of molecular beam; (3) outer wall of bakeable ultra-high vacuum chamber; (4) liq. N_2 cooled fragment collimating tube; (5) mass spectrometer electron-bombardment ionizer; (6) mass spectrometer electron multiplier; (7) quadrupole section of mass spectrometer; (8) liq. N_2 cooled partition between interaction and detection chamber; (9) interaction region; (10) liq. N_2 cooled molecular beam collimator and oven shield; (11) liq. N_2 cooled inner wall of interaction chamber; (12) molecular beam oven with capillary slits; (13) molecular beam port.

The operation of the spectrometer as it maps out the multidimensional spectrum is divided into subcycles and cycles. A subcycle contains a single light pulse and a measurement of the resulting output signal $S(t)$ from the mass spectrometer as a function of time t after the light pulse. During a subcycle the parameters of photon energy, photon flux, ion mass, and fragment laboratory angle H are fixed. A cycle is a set of subcycles which measures the variation in $S(t)$ as another parameter is changed, for example H. Usually to achieve sufficient signal-to-noise ratio many identical cycles were run, and the corresponding subcycles averaged. Figure 35 shows the data-handling system. Pulses from the mass spectrometer electron multiplier are amplified, discriminated, and shaped and the conditioned signal is then transmitted over coaxial cable to the computer. Here, noise is removed by a second discrimination, and the data are fed into a high-speed scaler interfaced to the computer's high-speed digital input. When triggered by the laser pulse through a photodiode pickup, the high-speed interface begins counting the mass spectrometer output in 1 μs intervals. Access to the computer is through a high-speed digital input operating at memory cycle speed, and all other computer operations are locked out for the period of data transmission, which lasts for 512 μs out of every 40 second subcycle. Raw data are preserved with temporary storage in core and on magnetic disk, and permanent storage on magnetic tape or cards. The computer communicates with the experimenter through a storage oscilloscope driven by digital-analogue converters.

The results obtained with this very sophisticated system have been discussed earlier in this chapter. It is clear that the size and complexity of such pieces of apparatus preclude the possibility of their use becoming general, but the concept of the experiment is simple, and highly significant and interesting data are being produced on simple diatomic and triatomic molecules, as well as a few polyatomics. It would be as well at this point to stress that highly complex and sophisticated expensive hardware is in itself of no great interest unless the results obtained with such a system are of importance. It is sometimes possible for the experimentalist with the simplest of pieces of apparatus (beakers, lamp, *etc.*) to produce results which are of as much interest as those obtained in a more elaborate fashion.

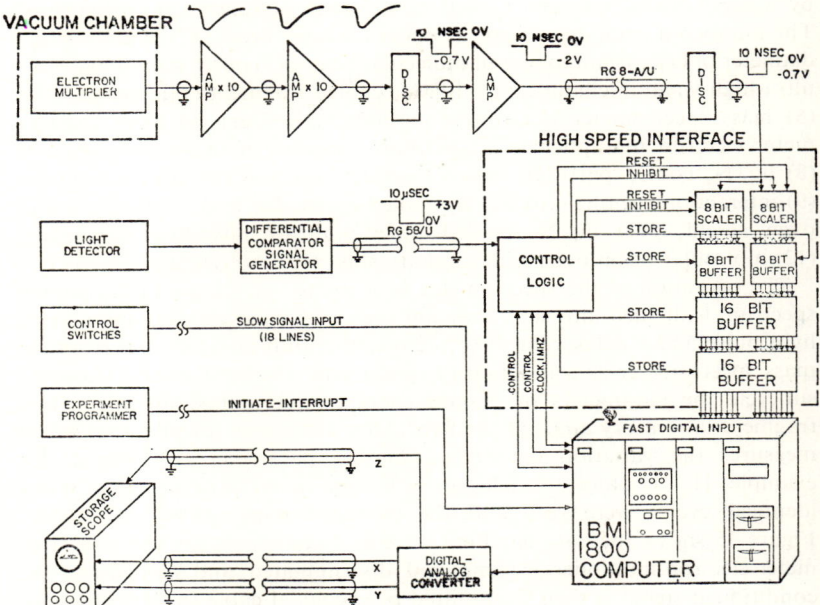

Figure 35 *Data-handling system for photofragment spectrometer*
(Reproduced by permission from *Rev. Sci. Instr.*, 1970, **41**, 1066)

E.s.r. spectroscopy used in conjunction with light sources is an established procedure, and several recent examples are listed. These are the photolysis of formamides studied by e.s.r.,[134] the photolysis of fluoro-*p*-benzoquinone,[135] a study of the stabilities of fluoroalkyl iodides to light using e.s.r. trapping techniques,[136] and the photolytic conversion of free radical systems, again using e.s.r.[137] The most recent development in the coupling of magnetic resonance spectroscopy with photochemistry has been the use of n.m.r. techniques to study the origins of products formed in a photolytic reaction. The phenomenon known as induced dynamic

[134] S. R. Bosco, A. Cirillo, and R. B. Timmons, *J. Amer. Chem. Soc.*, 1969, **91**, 3140.
[135] A. Hudson and J. W. Lewis, *J. Chem. Soc.* (*B*), 1969, 531.
[136] K. J. Klabunde, *J. Amer. Chem. Soc.*, 1970, **92**, 2427.
[137] K. Nishikida and K. Kuwata, *J. Phys. Chem.*, 1969, **73**, 2239.

nuclear spin polarization occurs when products arise *via* free-radical intermediates, the unpaired electrons in which cause polarization of the nuclear spins which is retained in the diamagnetic product and is observed as enhanced n.m.r. absorption or emission. The hyperfine-coupling-induced singlet–triplet mixing in radical pairs can lead to large nuclear polarizations, and as a model, a weakly coupled radical pair (I) generated from a precurser ^mM in a single step is considered,[138] where m denotes the multiplicity of M. A finite fraction of the radical pairs reacts within the cage to give combination or disproportionation products (II) and (III), and the remainder diffuse apart to give free radicals. Assuming only one of the two components of (I) $(HR_1 \cdot)$ has a proton spin strongly coupled to the

$$^mM \longrightarrow \overline{(H)R_1 \cdot \cdot R_2} \xrightarrow{k} (H)R_1 \cdot + \cdot R_2 \qquad (89)$$
$$\text{(I)}$$

$$\searrow (H)R_1\text{-}R_2 + R_1 + R_2H \qquad (90)$$
$$\text{(II)} \quad \text{(III)}$$

electron spin *via* a scalar hyperfine coupling of magnitude A, the spin Hamiltonian describing (I) in a magnetic field H_0 can be written as

$$\mathscr{H} = \mathscr{H}_{EZ} + \mathscr{H}_{NZ} + \mathscr{H}_{SS} + \mathscr{H}_{iS} + \mathscr{H}_{L.S.} + \mathscr{H}_D \qquad (91)$$

where the first two parts are the electron and nuclear Zeeman energies; the third and fourth parts describe the scalar electron exchange and hyperfine couplings, and the last two terms represent spin–orbit coupling and all dipolar interactions respectively. $\mathscr{H}_{NZ}$ is considered to be unimportant, and also neglecting $\mathscr{H}_D$, and combining $\mathscr{H}_{EZ}$ and $\mathscr{H}_{L.S.}$, by introducing g_1 and g_2 as the isotropic g factors of components 1 and 2 of (I) respectively, we get

$$\mathscr{H} = \beta H_0(g_1 S_1 + g_2 S_2) - J(\tfrac{1}{2} + 2S_1 . S_2) + AI . S_1 \qquad (92)$$

where β is the Bohr magneton, S_1 and S_2 and I are the electron and nuclear spin operators, and J is the scalar electron exchange coupling constant. For small values of J, the wave function of I can be written as a mixture of unperturbed singlet function $S = 2^{-\frac{1}{2}}(\alpha\beta - \beta\alpha)$ and the triplet component $T_0 = 2^{-\frac{1}{2}}(\alpha\beta + \beta\alpha)$. This gives by the variation of constants method

$$\psi^+(t) = [C_S^+(t) S + C_{T_0}^+(t) T_0] \alpha_N \qquad (93)$$

$$\psi^-(t) = [C_S^-(t) S + C_{T_0}^-(t) T_0] \beta_N \qquad (94)$$

and from (92) and the time-dependent Schrödinger equation,

$$i\frac{\partial C_S^\pm}{\partial t} = C_S^\pm J + C_{T_0}^\pm(\tfrac{1}{2}\beta H_0 \Delta g \pm \tfrac{1}{4}A) \qquad (95)$$

$$i\frac{\partial C_{T_0}^\pm}{\partial t} = C_S^\pm(\tfrac{1}{2}\beta H_0 \Delta g \pm \tfrac{1}{4}A) - C_{T_0}^\pm J \qquad (96)$$

---

[138] G. L. Closs, *J. Amer. Chem. Soc.*, 1969, **91**, 4552; G. L. Closs and A. D. Trifunac, *ibid.*, 4554; 1970, **92**, 2183, 2185, and 2186.

where $\Delta g = g_1 - g_2$. Integration depends on the state of (I) at the instant of its formation ($t = 0$). If the case of a triplet precurser is considered ($m = 3$), then $C_{T_0}{}^\pm(0) = 1$ and $C_S{}^\pm(0) = 0$

$$C_S(t)^\pm = -i \frac{\frac{1}{2}\beta H_0 \Delta g \pm \frac{1}{4}A}{D^\pm} \sin D^\pm t \qquad (97)$$

$$C_{T_0}{}^\pm(t) = \cos D^\pm t - i \frac{J}{D^\pm} \sin D^\pm t \qquad (98)$$

where

$$D^\pm = [(\tfrac{1}{2}\beta H_0 \Delta g \pm \tfrac{1}{4}A)^2 + J^2]^{\frac{1}{2}} \qquad (99)$$

Since product formation should depend on the degree of singlet character in I, it should be proportional to

$$[C_S{}^\pm(t)]^2 = \frac{(\tfrac{1}{2}\beta H_0 \Delta g \pm \tfrac{1}{4}A)^2}{D^{\pm 2}} \sin^2 D^\pm t \qquad (100)$$

Examination of (100) shows that products with different nuclear spin states will be formed with different rates, attributable to the dependence of the mixing coefficient on the nuclear spin state, $\tfrac{1}{2}\beta H_0 \Delta g + \tfrac{1}{4}A$ for α_N, and $\tfrac{1}{2}\beta H_0 \Delta g - \tfrac{1}{4}A$ for β_N. It also shows that if $\Delta g = 0$, both spin states build up in the product and no polarization occurs. Averaging over the lifetime τ of (I) gives the rates of population increase of the two nuclear spin states (W^+ for α_N and W^- for β_N) in the cage product where K_{SE} is the specific rate constant for cage product formation from the pure singlet state of the radical pair.

$$W^\pm = K_{SE} \frac{2(\tfrac{1}{2}\beta H_0 \Delta g \pm \tfrac{1}{4}A)^2 \tau^2}{1 + 4D^{\pm 2} \tau^2} \qquad (101)$$

If (I) is formed from a singlet precursor, the product formations should be proportional to (102), which leads to the opposite polarization

$$W^\pm = K_{SE}\left[1 - \frac{2(\tfrac{1}{2}\beta H_0 \Delta g \pm \tfrac{1}{4}A)^2 \tau^2}{1 + 4D^{\pm 2} \tau^2}\right] \qquad (102)$$

It is thus possible to distinguish between singlet and triplet precursors of radical intermediates in organic chemical reactions by observation of the transient initial spin polarization in a product molecule. Experimentally the technique consists of u.v. irradiation of solutions of organic molecules in the cavity of a conventional n.m.r. spectrometer, and the technique has been applied to benzophenone,[139] diphenylmethylene,[140] and has also indicated that the photoinduced decomposition of methyldiazoacetate proceeds *via* a radical chain mechanism.[141] The photolysis of benzaldehyde in solution has also been studied using n.m.r. techniques.[142]

---

[139] G. L. Closs and L. E. Closs, *J. Amer. Chem. Soc.*, 1969, **91**, 4550.
[140] G. L. Closs and L. E. Closs, *J. Amer. Chem. Soc.*, 1969, **91**, 4549.
[141] M. Covicera and H. D. Roth, *J. Amer. Chem. Soc.*, 1970, **92**, 2573.
[142] M. Covicera and A. M. Trozzolo, *J. Amer. Chem. Soc.*, 1970, **92**, 1772.

Amongst recent papers concerned with the treatment of data, the use of zero-intensity extrapolations in photochemistry is of interest,[143] and a method of treating photochemical triplet quenching data has been outlined.[144]

D. P.

[143] F. R. Cala, S. L. Chong, H. P. Sperling, and S. Toby, *Canad. J. Chem.*, 1970, **48**, 357.
[144] M. D. Shetlar, *Photochem. and Photobiol.*, 1969, **10**, 407.

2
Photophysical Processes in Condensed Phases

This chapter will be largely concerned with a description of the experimental evidence available on processes which occur in the liquid and solid phases in competition with photochemical reaction. Thus the first section deals with fluorescence and fluorescence quenching, and the next main section deals with intersystem crossing and other aspects of the triplet state. Following this, there is a consideration of electronic energy transfer in the condensed phases. There are also included brief sections dealing with physical aspects of reactions in the condensed phase; reactions in the solid phase, including those on surfaces; and photochromism.

1 Fluorescence and Fluorescence Quenching

A vast amount of data is published annually reporting the fluorescence spectra and variation of fluorescence yield in condensed phases with such parameters as concentration, addition of quenchers, solvent, exciting wavelength, *etc.* In the following section we have dwelt on those reports which may be of exceptional interest, and commented only briefly on many reports.

Of great interest is the observation of fluorescence from saturated hydrocarbons.[1] The fluorescence spectra are shown in Figure 1.

All n-alkanes examined exhibit essentially the same fluorescence spectrum indicated by the single curve in Figure 1(a). The vapour did not fluoresce, when excited at 1470 Å, for any of the compounds studied. However, when excited at wavelengths longer than ~ 1600 Å, a distinct emission was found for sufficiently volatile compounds (*e.g.* pentane, hexane, cyclohexane at $\sim \frac{2}{3}$ vapour pressure at 25 °C). The vapour spectrum, except for a slight blue-shift, is otherwise identical to the liquid fluorescence, suggesting that emission in both cases is of molecular origin with the upper state being strongly predissociated in the vapour phase above ~ 7.7 eV. In the liquid, the fluorescence yield also decreases as the exciting wavelength is reduced, but much less severely than occurs in the vapour. Measurements at exciting wavelengths of 1720, 1660, 1550, and 1470 Å showed no change in fluorescence spectrum. From results of oxygen quenching, the wavelength dependence of Φ_f in the liquid is attributed, at

---

[1] F. Hirayama and S. Lipsky, *J. Chem. Phys.*, 1969, **51**, 3616.

least in part, to the presence of photochemically generated olefins, which quench the alkane emission with increasing importance as the absorption coefficient (and thereby the effective olefin concentration) is increased.

Lifetimes of the excited states in the liquid were estimated using oxygen as quencher, and assuming the quenching reaction to be diffusion-limited. From available viscosity and oxygen solubility data, the following lifetimes

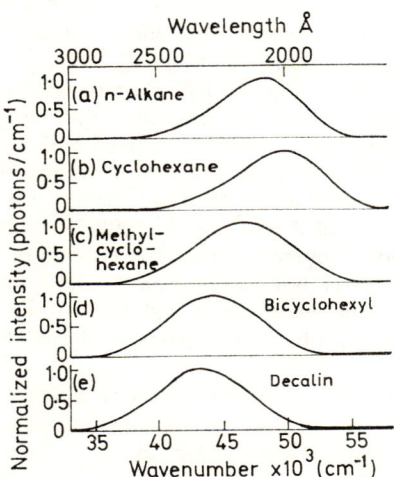

Figure 1 *Corrected fluorescence spectra of some saturated hydrocarbons excited at 1470 Å. All spectra are normalized at the intensity maxima*
(Reproduced by permission from *J. Chem. Phys.*, 1969, **51**, 3616)

were obtained for 1470 Å excitation: pentane (0·3 ns), hexane (2 ns), heptane (6 ns), octane (8 ns), nonane (15 ns), cyclohexane (7 ns). Again, correction for photochemically generated olefins will increase these values, but not alter significantly their trend. The lifetime of excited cyclohexane was also determined using benzene as quencher. A sensitization of benzene fluorescence accompanied the reduction of cyclohexane fluorescence. A value of approximately 6 ns for the cyclohexane lifetime was obtained from analysis of the energy transfer process. Additionally, as predicted for a true sensitization process, oxygen quenching of cyclohexane fluorescence was appropriately reduced in the presence of benzene.

The fluorescence quantum yields of the n-alkanes increase strongly as the number of carbon atoms (N_C) increases. Since the excited-state lifetimes also exhibit approximately this same dependence ($\Phi_f/\tau \sim$ constant $\simeq 10^5$ s^{-1}), it follows that N_C must determine critically some non-radiative transition probability. A possible explanation is that molecular distortions in the excited state are reduced by distribution of the electronic excitation over a larger chain length, thereby also reducing the probability for radiationless transition. Indeed, such a correlation between N_C and a molecular distortion in the excited state may be already apparent from the fact that what is likely the first absorptive transition of the n-alkanes red-shifts with increasing N_C, whereas the position of the emissive transition remains fixed.

Associated with these observations is the observation of excitation energy transfer in alkanes. A model has been proposed to account for the phenomenon based on an exciton model.[2] The exciton model of sigma bonds has been shown to predict that efficient excitation energy transfer will occur in linear and branched saturated hydrocarbon molecules, and that it will have the following particular characteristics:

(i) In linear alkanes, the excitation energy will be essentially divided into two portions, the one which is localized in pairs of C—H bonds attached to a common carbon atom, and the other which is transferred rapidly from one C—C bond to another along the main chain of the molecule.

(ii) The localization of C—H bond excitation will generally lead to scission of one or both members of each C—H bond pair, ejecting H or H_2, but as transfer of C—H bond excitation between different C—H bond pairs is predicted to be small, there will be little or no evidence of an energy-transfer process observable from study of hydrogen-producing reactions.

(iii) The migratory (C—C bond) excitation will be able to transfer to foreign molecules in or outside the alkane molecular chains, as long as these have strong optical absorptions in the 7·5—9·5 eV range. Such diversion of this excitation will not, however, directly affect the C—H bond scission reactions, since migratory C—C bond excitations cannot in general be transformed into C—H bond excitations except in the region of chemical defects (particularly chain ends).

(iv) The migratory excitation will propagate predominantly along the main chain of a particular molecule, in both crystalline and amorphous regions, but will also have a small probability of jumping between chains. A consequence of this is that the efficiency of any observable energy-transfer reaction caused by migratory excitation will be strongly dependent on chain length (below a limiting length which will represent the maximum effective exciton migration distance).

(v) Scission of main-chain C—C bonds (due to excitation) will be much less likely than that of C—H bonds, since the excitation will usually be transferred to another C—C bond before scission can occur. This applies to both linear and branched alkanes.

(vi) Migratory excitation will not cause appreciable scission of any random alkyl side-chains. Such scission as does occur by excitation will be due predominantly to initial deposition of energy directly on to the side-chains themselves. The efficiency of such side-chain scission will be considerably greater if the side-chains are in symmetrical pairs attached to a common carbon atom than if they occur only singly.

[2] R. H. Partridge, *J. Chem. Phys.*, 1970, **52**, 2485, 2491, 2801.

(vii) Some scission of C—C or C—H bonds may be caused by migratory excitons at chain ends.

Experimental evidence is given that such long-range (1500 Å) energy transfer occurs with high efficiency in the radiation chemistry of long-chain alkanes and polymers;[2] the results are clearly of importance in the vacuum-u.v. photochemistry of these molecules also.

The emission observed from benzene solutions may be understood on the basis of the following reaction scheme:

$$M + h\nu_0 \xrightarrow{} M^* \qquad k_1 \qquad (1)$$

$$M^* \xrightarrow{k_2} M + h\nu_1 \qquad (2)$$
$$M^* \xrightarrow{k_3} M \qquad\qquad\quad k_m \qquad (3)$$
$$M^* \xrightarrow{k_4} {}^3M^* \qquad (4)$$

$$M^* + M \xrightarrow{k_5} E^* \qquad k_5[M] \qquad (5)$$

$$E^* \xrightarrow{k_6} M + M^* \qquad k_6 \qquad (6)$$

$$E^* \xrightarrow{k_7} 2M + h\nu_2 \qquad (7)$$
$$E^* \xrightarrow{k_8} 2M \qquad\qquad k_e \qquad (8)$$
$$E^* \xrightarrow{k_9} {}^3E^* \qquad (9)$$

where M^* is monomer aromatic, and E^* is the excimer.

Decay times of neat benzene (C_6H_6 and C_6D_6) fluorescence excited by 250 nm light were respectively measured as 27 and 32 ns at 25 °C.[3] Measurements made over the range of temperature 5—35 °C, both of the pure benzenes and of solutions in cyclohexanes C_6H_{12} and C_6D_{12}, show that the solvent is without effect on the measured rates; that at a particular temperature benzene monomer and excimer have the same rate of de-excitation (corresponding to lifetimes of 28 ns for C_6H_6 at 25 °C and 33 ns for C_6D_6 at 22 °C); and that the activation energy for internal conversion of monomer in both cases (on certain stipulated assumptions) is 0·28 eV, within limits of experimental error. The reduction, by deuterium substitution, in the rate of de-excitation from the monomer is less than the reduction in the rate from the excimer, and the rate for benzene excimer is indicated to be independent of temperature in the range studied.

The efficiency of internal conversion $\beta(\lambda_n)$ from an upper electronic state S_n to the lowest excited state S_1 is defined as the primary quantum yield for production of the S_1 state for excitation at wavelength λ_n. The usual technique for determining $\beta(\lambda_n)$ is to study the dependence on exciting wavelength of the S_1 fluorescence quantum yield $\Phi_f(\lambda_n)$ and to

---

[3] W. P. Helman, *J. Chem. Phys.*, 1969, **51**, 354.

assume that $\Phi_f(\lambda_n)$ is proportional to $\beta(\lambda_n)$. The ratio $\beta_f(\lambda_n) = \Phi_f(\lambda_n)/\Phi_f(\lambda_1)$ is then identified with $\beta(\lambda_n)$.[4]

An equivalent technique for determining the $S_n \to S_1$ conversion efficiency is to measure, for excitation at λ_n, the efficiency of energy transfer $\Phi_t(\lambda_n)$ to a suitable fluorescent solute. The ratio $\beta_t(\lambda_n) = \Phi_t(\lambda_n)/\Phi_t(\lambda_1)$ is identified with $\beta(\lambda_n)$ under the assumptions that (i) $\Phi_t(\lambda_n)$ is proportional to $\beta(\lambda_n)$, and (ii) S_3 decays to S_1 sufficiently rapidly that energy transfer occurs exclusively from S_1.

For $S_3 \leftarrow S_0$ excitation of liquid benzene, toluene, and p-xylene at 1849 Å, $\beta_f(\lambda_3)$ has been reported to increase in the presence of fluorescence quenchers such as O_2 and CCl_4, and to approach at sufficiently high concentrations a limiting value that is independent of the nature of the quencher. Similarly, $\beta_t(\lambda_3)$ for a benzene–PPO (2,5-diphenyloxazole) or benzene–p-terphenyl system has been reported to increase with solute concentration. Both concentration effects have been demonstrated to derive not from an influence of the added quencher on the internal conversion efficiency $\beta(\lambda_3)$ nor from violation of assumption (ii), but rather from the failure of the usual assumption that $\beta_f(\lambda_3) = \beta_t(\lambda_3) = \beta(\lambda_3)$.

The difficulty with this assumption has been traced to the presence of a photochemically generated quencher (fulvene) which is present at substantially higher local concentrations for excitation at λ_3 (1849 Å) than at λ_1 (2537 Å) due to the much smaller penetration depth of λ_3. This conclusion was established for the case of liquid benzene by first showing that, whereas in nitrogenated or degassed solutions $\beta_f(\lambda_3) = 0.22$, at sufficiently high concentrations of added quenchers (C_q), $\beta_f(\lambda_3)$ approached a limiting value of 0.38, independent of the nature of the quencher. Similarly, it was shown that at sufficiently high PPO concentration (C_p), $\beta_t(\lambda_3)$ approached the same limiting value of 0.38. Defining a function $F(C_q)$ such that:

$$\beta_f(\lambda_3) = F(C_q) \lim_{C_q \to \infty} \beta_f(\lambda_3), \qquad (10)$$

it was then demonstrated that the observed effect of quencher on $\beta_f(\lambda_3)$ and $\beta_f(\lambda_1)$ demanded that $F(C_q)$ have the form:

$$F(C_q) = (k + k_q C_q)/(k + k_x + k_q C_q) \qquad (11)$$

where k_q/k is the Stern–Volmer constant for quenching of S_1 by added quencher, and k_x is some positive constant independent of C_q. Substituting equations (10) and (11) into the defining equation for $\beta_f(\lambda_3)$, it simply follows that:

$$\Phi_f(\lambda_3) = [k_f/(k + k_x + k_q C_q)] \; [\lim_{C_p \to \infty} \beta_f(\lambda_3)] \qquad (12)$$

and

$$\Phi_f(\lambda_1) = k_f/(k + k_q C_q) \qquad (13)$$

[4] C. W. Lawson, F. Hirayama, and S. Lipsky, *J. Chem. Phys.*, 1969, **51**, 1590.

where k_f is the S_1 radiative rate-constant. The form of these equations was then shown to be consistent with the identification of $\lim_{C_q \to \infty} \beta_f(\lambda_3)$ with the internal-conversion efficiency $\beta(\lambda_3)$ and k_x with the product of the steady-state concentration of photochemically generated quencher within the illumination zone and the rate constant for its quenching of S_1. Replacing in the above equations $\beta_f(\lambda_3) = \beta_t(\lambda_3)$, $F(C_q) = F(C_p)$, $k_q C_q = k_t C_p$ ($k_t C_p$ is the rate of energy transfer), $\Phi_f(\lambda_3) = \Phi_t(\lambda_3)$, and $k_f = k_t C_p$, it was similarly demonstrated that $\lim_{C_p \to \infty} \beta_t(\lambda_3)$ can be identified with $\beta(\lambda_3)$.

Adequate concentrations of either quencher or fluorescent solute to obtain $F(C) > 0.98$, and therefore $\beta_f(\lambda_3)$ and $\beta_t(\lambda_3)$ within $\sim 2\%$ of $\beta(\lambda_3)$, were found to be $C = C_q = 0.15$ mol l^{-1} CCl$_4$, 0.0070 mol l^{-1} O$_2$ (*i.e.* saturated at 1 atm.), or $C = C_p = 0.01$ mol l^{-1} PPO.

Since penetration depth differences of λ_3 and λ_1 become less significant on dilution of the aromatics with suitable transparent solvents, it would be expected that $F(C_q)$ would increase and, in the absence of any specific solvent effect on the internal-conversion efficiency, such dilution would also increase $\beta_f(\lambda_3)$. An increase in $\beta_f(\lambda_3)$ has indeed been reported for dilution of toluene and *p*-xylene with the oxygen-free solvents cyclohexane and iso-octane. However, benzene has repeatedly been shown to behave anomalously in this respect, with $\beta_f(\lambda_3)$ decreasing considerably on similar dilution, suggesting a possible important effect of solvent perturbation on $\beta(\lambda_3)$. In order to elucidate the internal conversion mechanism ($S_3 \to S_1$), the efficiencies of benzene, toluene, and *p*-xylene [4] were measured in the solvents benzene, methanol, isopropyl alcohol, tetrahydrofuran, ethyl ether, acetonitrile, hexane, cyclohexane, methylcyclohexane, decalin, iso-octane, and perfluorohexane. These results are shown in Table 1, and differences observed in the absorption and emission spectra of benzene vapour and of benzene in different solvents are shown in Table 2.

It is interesting to note in Table 2 how much larger is the solvent shift in emissive transitions as compared to the absorptive transitions. The excited-state lifetime of benzene in these solutions (29 ns in cyclohexane) is apparently sufficiently long to permit the solvent to achieve an energetically more favourable orientation about the benzene before emission occurs.

The enhancement of the 0—0 band has been suggested to require a solvent perturbation which is effective in mixing the dipole-allowed $S_3 (^1E_{1u})$ state of benzene with the dipole-forbidden $S_1 (^1B_{2u})$ state. The results, which here appear to correlate the effect of solvents in enhancing the 0—0 transitions with their effectiveness in increasing $\beta(\lambda_3)$, suggest that the extent of this mixing is also crucial to the $S_3 \to S_1$ internal-conversion efficiency. This view is further substantiated by the results obtained with toluene and *p*-xylene. In toluene, the 0—0 absorptive transition, even in the absence of external perturbation, is much stronger than in benzene, and is stronger still in *p*-xylene, where it becomes the prominent member of the vibrational progression. If one regards the $S_1 \leftarrow S_0$ 0—0 transition in these benzene derivatives as an internally perturbed benzene transition, then the effectiveness of this perturbation in mixing the S_3 and S_1 states clearly increases with increasing methyl substitution. The observed increase in $\beta(\lambda_3)$ from 0·45 (benzene) to 0·76 (toluene) to ~1·0 (p-xylene) emphasizes again, therefore, the importance of such mixing to the $S_3 \to S_1$ internal-conversion efficiency. The effect on $\beta(\lambda_3)$ of dilution with cyclohexane is also consistent with this picture. As the internal perturbation weakens from p-xylene to toluene to benzene, there is observed an increasingly important effect of an external environmental perturbation on the $S_3 \to S_1$ conversion efficiency. As calculated from the

Table 1 $S_3 \to S_1$ *Internal-conversion efficiencies,* $\beta(\lambda_3 = 1849$ Å$)$

Substance	Environment[a]	$\beta(\lambda_3)$
p-Xylene	p-Xylene	1·03
p-Xylene	Cyclohexane	0·95
Toluene	Toluene	0·76
Toluene	Cyclohexane	0·58
Benzene	Benzene	0·45
Benzene	Methanol	0·37
Benzene	Isopropyl alcohol	0·33
Benzene	Tetrahydrofuran	0·33
Benzene	Diethyl ether	0·32
Benzene	Acetonitrile	0·28
Benzene	Hexane	0·25
Benzene	Cyclohexane	0·24
Benzene	Methylcyclohexane	0·24
Benzene	Decalin	0·24
Benzene	Iso-octane	0·22
Benzene	Perfluorohexane	0·04
Benzene	10 Torr, vapour	10^{-4}

[a] For the non-aromatic environments, values are those obtained for 10—20% solutions. For isopropyl alcohol, tetrahydrofuran, and diethyl ether, values of $\beta(\lambda_3)$ are corrected for the fraction of the 1849 Å light directly absorbed by the solvent (5—15%).

Table 2 *Comparison of benzene vapour and solution absorption and emission spectra*

	Absorption ($\nu_{vapour} - \nu_{soln}$)/cm^{-1}			Emission ($\nu_{vapour} - \nu_{soln}$)/cm^{-1}		
Environment	ΔA_0^0	ΔB_0^0	ΔJ_0^0	ΔA_0^0	ΔB_0^0	ΔJ_0^0
Benzene	280	250	210	820	870	830
Methanol	120	130	100	520	480	550
Isopropyl alcohol	190	170	170	630	570	640
Acetonitrile	190	160	160	630	640	600
Hexane	190	200	—	600	570	—
Decalin	210	210	—	780	660	—

$A_0^0 = \nu_{18}'(e_{2g})$ to 0 transition
$B_0^0 = \nu_{18}''(e_{2g})$ to 0 transition
$J_0^0 = 0$—0 transition

data presented in Table 1, the ratio of $\beta(\lambda_3)$ for the pure liquid to that of the 10% cyclohexane solution increases from 1·1 (p-xylene) to 1·3 (toluene) to 1·9 (benzene).

The original observation that $\beta_f(\lambda_3)$ increases on dilution with oxygen-free aliphatic solvents for p-xylene and toluene and decreases for benzene is now very simply explained. The effect of dilution to reduce the concentration of the photochemically generated quencher always tends to increase $\beta_f(\lambda_3)$. However, the solvent perturbation effect in changing from an aromatic to an aliphatic environment operates to oppose this increase. This is so slight an effect for p-xylene and toluene that $\beta_f(\lambda_3)$ remains increased on dilution, but is such a great effect for benzene that the overall dilution effect now causes reduction in $\beta_f(\lambda_3)$.

It is tempting to extend these considerations to the vapour phase and explain similarly the extremely small value of $\beta(\lambda_3)$ observed for vapour benzene as being due to negligible S_3–S_1 mixing, as is evidenced by the complete absence in the vapour of the 0—0 transition in the $S_0 \to S_1$ absorption. Although such extrapolation must be made with caution, it is nevertheless supported by the following results obtained with perfluorohexane as solvent. The absorption spectra of benzene in perfluorohexane and the vapour phase are very similar, and different from that in hexane. Within an uncertainty of 5 cm^{-1}, there is no detectable solvent frequency shift in the $A_0{}^0$ or $B_0{}^0$ transitions either in absorption or emission in perfluorohexane. This is to be contrasted with the behaviour of even the weakly perturbing solvents such as hexane, for which $A_0{}^0 \simeq B_0{}^0 \simeq 600$ cm^{-1} in emission (see Table 2). Additionally, the 0—0 transition could not be detected in perfluorohexane in either emission or absorption. The solvent appears, therefore, to be extremely inert, at least with regard to perturbation of the benzene electronic system and, accordingly, $\beta(\lambda_3)$ is considerably less than that obtained in any other solvent system that has been examined (see Table 1).

Photochemical evidence suggests that, in the vapour phase, the process competing with the benzene $S_3 \to S_1$ internal conversion involves a direct coupling of S_3 to high vibrational levels of S_0. The picture that emerges is that S_3 has two major decay channels, internal conversion to S_1, and internal conversion to S_0, with the solvent playing a very specific role in modifying the efficiency of the $S_3 \to S_1$ process by providing the appropriate perturbation for mixing these states. The possible involvement of (i) the S_2 state, and (ii) chemical reactions from the S_3 state (*e.g.* isomerization) may need further consideration.

The scintillation efficiencies of 1,1'-binaphthyl (1), 2-(4-isopropylphenyl)-5-(4-biphenylyl)-1,3,4-oxadiazole (2), and 2,5-diphenyl-1,3-oxazole (3) have been shown to be strongly solvent dependent.[5] The order of efficiencies is 1,2,4-trimethylbenzene > p-xylene > toluene > benzene. The relative scintil-

---

[5] D. L. Horrocks, *J. Chem. Phys.*, 1970, **52**, 1566.

lation yields seem to follow the same general trend as the efficiency of internal conversion of $S_3 \rightarrow S_1$ of the aromatic solvent. At high solute concentrations (>150 g l^{-1}), the relative scintillation efficiency of these solutions was independent of the particular aromatic solvent. It was assumed that the fluorescence was produced by a process which involved

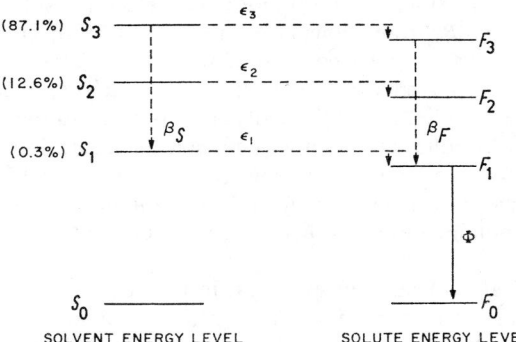

Figure 2 *Effect of solvent on* S_3—S_1 *internal conversion in aromatics* (*see text*)
(Reproduced by permission from *J. Chem. Phys.*, 1969, **51**, 1590)

energy transfer before the $S_3 \rightarrow S_1$ internal conversion. A possible mechanism involves the transfer of energy from the S_3 state of the solvent to an excited state of the solute, F_3. The relative scintillation efficiency of concentrated solutions would then be dependent upon the internal-conversion efficiency for the process $F_3 \rightarrow F_1$. This is illustrated in Figure 2.

A similar correlation to that above [4] has been noted for benzene excited to the first electronic state ($^1B_{2u}$).[6] The u.v.-spectroscopic data for this transition are shown in Table 3, and data for this table shown in Figure 3.

[6] J. W. Eastman and S. J. Rebfeld, *J. Phys. Chem.*, 1970, **74**, 1438.

It is obvious from Figure 3 that a strong correlation exists between solvent-induced enhancement of the radiationless transition from S_1 benzene and the enhancement of the 0—0 transition in absorption. Little temperature dependence (100—300 K) was found in the 0—0 absorption intensity, a

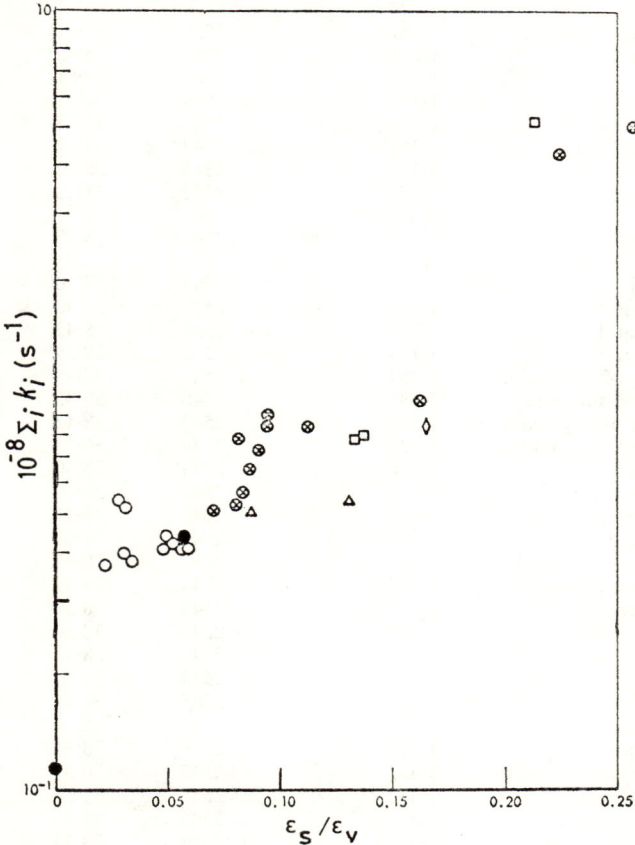

Figure 3 *Correlation of solvent-induced fluorescence quenching at 25 °C with solvent-induced 0—0 u.v. absorption. The total radiationless conversion rate is plotted as a function of the 0—0 absorption intensity ε_s, measured relative to the usual vibronic intensity ε_v. For symbols refer to Table 3*
(Reproduced by permission from *J. Phys. Chem.*, 1970, **74**, 1438)

measure of the average intermolecular force. Thus the thermal quenching of the fluorescence in this same temperature region appears to arise from the thermal population of active vibrational levels above the zero-point level in the fluorescent state of benzene. In any case, only the total rate of

Table 3 Some u.v. spectroscopic properties of benzene as a function of solvent

No.[c]	Solvent	Refractive index, n^b	Symbol[c]	Wavelength, λ_{max} Å ± 1	$10^{-3} \times$ frequency ($10^{-3} \times [\nu_{00} + 522]$), cm^{-1}	Absorptivity,[d] ε_{max} (mol l^{-1} cm)$^{-1}$ ± 4
1a	Vapour	1·000	●	2529	38·61	—
13b	Perfluoro-n-hexane	1·2514		2530	38·61	325
27c	Methanol	1·331	⊗	2543	38·42	268
40d	D$_2$O	1·338	⊗	2537	38·51	179
39e	Water	1·333	⊗	2539	38·49	188
36f	Acetonitrile	1·346	◇	2543	38·43	227
2g	2-Methylbutane	1·351	○	2542	38·42	267
23h	Diethyl ether	1·350	△	2543	38·42	262
4i	n-Pentane	1·357 (15 °C)	○	2542	38·43	258
14j	Tetramethylsilane	1·359 (19 °C)	●	2543	38·40	267
22k	Ethanol	1·359	⊗	2544	38·40	265
5l	n-Hexane	1·375	○	2543	38·40	259
20m	2-Propanol	1·375	⊗	2543	38·42	263
21n	1-Propanol	1·384		2545	38·39	245
16o	2-Methyl-2-propanol	1·384	⊗	2544	38·40	262
35p	Dimethyl sulphate	1·387		2540	38·45	221
3q	2,2,4-Trimethylpentane	1·392	○	2543	38·42	251
18r	2-Methyl-1-propanol	1·398	⊗	2544	38·39	258
6s	n-Octane	1·398	○	2544	38·40	248
10t	Cyclopentane	1·404	○	2544	38·40	261
19u	1-Butanol	1·399	⊗	2545	38·37	256
7v	n-Nonane	1·405	○	2546	38·37	232
15w	3-Methyl-1-butanol	1·408	⊗	2446	38·37	238
8x	Methylcyclohexane	1·424	○	2544	38·39	258
11y	Cyclohexane	1·424	○	2545	38·39	261
33z	2-Methoxyethanol	1·402		2546	38·36	232
31α	2-Ethoxyethanol	1·408		2546	38·37	240
32β	1,4-Dioxan	1·423	△	2544	38·39	241
28γ	2-(2-Ethoxyethoxy)-ethanol			2546	38·36	233
17δ	1-Octanol	1·429		2546	38·36	260
25ε	1,2-Propanediol	1·433	⊗	2547	38·34	206
9ζ	n-Hexadecane	1·432	○	2547	38·36	217
12η	1-Dodecanol			2546	38·36	230
30θ	Dichloromethane	1·425	□	2548	38·34	205
26ι	1,2-Ethanediol	1·432	⊗	2547	38·33	192
29κ	1,2-Dichloroethane	1·448	□	2548	38·33	206
34λ	1,2,3-Propanetriol	1·473	⊗	2549	38·30	197
37μ	Chloroform	1·448	□	2551	38·26	225
24ν	Benzene	1·498	●	2552	38·24	202
38ξ	Carbon tetrachloride	1·604 (15 °C)		2557	38·20	261

[a] Solvents are listed (a—ξ) in order of refractive index, absorption frequency, and line width. The numbers (1—40) give the order of solvent-induced (0—0) absorption and radiationless conversion. [b] Determined at the sodium D-line at $T = 20$ °C. [c] Symbols are the key to Figure 3. [d] Absorption spectra measured with a Cary Model 14 spectrophotometer. [e] Determined as in J. W. Eastman, *Photochem. and Photobiol.*, 1967, **6**, 55. [f] Spontaneous emission rate assumed to be $k_0 = 2·5 \times 10^6$ s^{-1}. [g] F. Almasy and

(Reproduced by permission from *J. Phys. Chem.*, 1970, **74**, 1438)

Line width, $\Delta\nu$ (9/10), cm^{-1}	Oscillator strength,d $10^3 \times f$	$\varepsilon_s/\varepsilon_v$	Fluorescence yield, Φ,[e] at 25 °C	10^{14} $\langle\nu^{-3}\rangle$	$\int(\varepsilon/\nu)\,d\nu$	$10^{-8}\,\Sigma_i\,k_i$,[f] s^{-1}
0	2·0[g]	0	0·18[i]	[h]	—	0·114[f]
27	—	0·049	—	—	—	—
70	1·85	0·095	0·027	2·33	10·7	0·90
114	1·90	0·257	0·0052	[h]	10·8	5·0
120	1·89	0·224	0·0058	2·31	10·8	4·3
89	1·77	0·166	0·029	2·35	10·2	0·84
76	1·83	0·029	0·044	[h]	10·5	0·54
80	1·87	0·088	0·047	[h]	10·8	0·51
79	1·78	0·050	0·054	[h]	10·3	0·44
73	1·88	0·058	0·054	[h]	10·8	0·44
77	1·83	0·091	0·033	2·36	10·6	0·73
73	1·74	0·049	0·058	2·36	10·0	0·41
82	1·80	0·087	0·037	[h]	10·4	0·65
87	1·70	0·097	—	—	—	—
84	1·77	0·082	0·031	[h]	10·2	0·78
89	1·79	0·176	—	—	—	—
89	1·69	0·059	0·057	[h]	9·8	0·41
88	1·81	0·084	0·042	[h]	10·5	0·57
89	1·78	0·058	0·057	[h]	10·3	0·41
80	1·75	0·052	0·056	[h]	10·1	0·42
93	1·89	0·081	0·045	[h]	10·8	0·53
95	1·72	0·031	0·059	[h]	9·9	0·40
81	1·97	0·071	0·045	[h]	11·4	0·53
95	1·82	0·032	0·046	[h]	10·5	0·52
102	1·84	0·035	0·061	2·37	10·6	0·38
111	1·86	0·131	—	—	—	—
108	1·96	0·117	—	—	—	—
119	1·96	0·131	0·044	2·37	11·3	0·54
116	1·94	0·121	—	—	—	—
120	1·78	0·052	—	—	—	—
110	1·81	0·095	0·029	[h]	10·3	0·84
112	1·70	0·023	0·063	[h]	9·7	0·37
130	1·75	0·040	—	—	—	—
133	1·80	0·138	0·031	[h]	10·4	0·78
123	1·65	0·112	0·029	[h]	9·5	0·84
131	1·74	0·134	0·031	[h]	10·1	0·78
133	1·87	0·163	0·025	2·38	10·8	0·98
169	2·27	0·213	0·0049	[h]		5·1[f]
153	1·76	0·091[j]	—	—	—	—
183	—	0·271	—	—	—	—

H. Laemmel, *Helv. Chim. Acta*, 1951, **34**, 462. [h] For undetermined samples, $10^{14}\langle\nu^{-3}\rangle = 2\cdot35$ has been used. [i] W. A. Noyes, jun., W. A. Mulac, and D. A. Harter, *J. Chem. Phys.*, 1966, **44**, 2100; and W. A. Noyes, jun., and D. A. Harter, *ibid.*, 1967, **46**, 674. [j] Benzene crystal gives $\varepsilon_s/\varepsilon_v \approx 0\cdot25$ [V. L. Broude, *Sov. Phys. Usp.* (English Transl.), 1962, **4**, 584].

all non-radiative transitions has been determined. Because the measured thermal quenching may arise from any number of thermally excited vibrations, a single Arrhenius activation energy has only a limited quantitative significance.

The 0—0 absorption depends on the volume fraction of components in several mixed solvents, so it was concluded that no special conformation of benzene with solvent is necessary to explain the solvent-induced radiationless conversion, which correlates with the 0—0 absorption. However, the formation of complexes, which may occur in chloroform, can greatly enhance any effects that the induced-dipole and dispersion interactions exert on the non-radiative transitions. From the correlation between solvent-induced 0—0 absorption and fluorescence quenching (Figure 3), it is possible to estimate the fluorescence quantum yield of monomer benzene in liquid benzene at 25 °C as $\Phi = 0.040$.

The fluorescence spectrum of anthracene crystals has been reported.[7] The extent to which phenyl substitution on naphthalene alters its luminescence properties depends on the number and position of the phenyl groups.[8] In 3-methylpentane at 77 K, successive α-phenyl substitution reduces both singlet- and triplet-state energies, and considerably shortens triplet-state lifetimes. *ortho*-Phenylated naphthalenes examined are characterized by higher excited-state energies and longer phosphorescence lifetimes than less-substituted derivatives. The luminescence features of the 1-(halogenophenyl)naphthalenes are particularly susceptible to the position of the halogen atom. In the *para*-position, the halogen produces effects which one normally associates with the presence of a heavy atom, whereas in the *ortho*-position the heavy-atom effect is partially obscured because of steric factors. Thus the luminescence characteristics of the phenylnaphthalenes vary considerably and are primarily influenced by inter-ring interactions. These interactions are in turn determined by the effective electron density at the substitution site, geometric configurations (including the possibility of planar excited states) and also by steric factors. In the 1-(*o*-halogenophenyl)naphthalenes, the luminescence characteristics are determined primarily by steric factors, while the halogen atoms promote spin–orbit coupling to an extent that is perhaps best described as a 'restricted internal' heavy-atom effect. In 1-(*p*-chlorophenyl)naphthalene, the heavy-atom effect on 1-phenylnaphthalene is operative without the complications due to steric factors. It is apparent that, although steric factors may determine the luminescence characteristics of phenylated aromatic hydrocarbons, it is also clear that ground-state steric considerations need not apply to excited electronic states when additional stabilization can be achieved by the molecule assuming other (*e.g.* planar) geometrical configurations.

---

[7] E. Glockner and H. C. Wolf, *Z. Naturforsch.*, 1969, **24a**, 943.
[8] J. B. Gallivan, *J. Phys. Chem.*, 1969, **73**, 3070.

Radiative and radiationless processes in pyrene and [^{2}H$_{10}$]pyrene have been measured as a function of temperature.[9] The fluorescence yield Φ_f, triplet yield Φ_t, and fluorescence lifetime τ_f were the measured quantities. The following conclusions were reached in this study.

(a) The rate constant for fluorescence for pyrene and [^{2}H$_{10}$]pyrene is independent of temperature between -196 and $+23$ °C, and does not change upon deuteriation.

(b) At -196 °C, for both pyrene and [^{2}H$_{10}$]pyrene, internal conversion to the ground state is unimportant, and may be entirely absent. However, the rate of this process increases with temperature, and at 23 °C it competes significantly with fluorescence emission and intersystem crossing.

(c) At -196 °C, for both pyrene and [^{2}H$_{10}$]pyrene, intersystem crossing is a significant process. Respectively, 22 and 15% of the molecules in S_1 go over to the triplet manifold. This result is in disagreement with previous work which claimed that for pyrene in ethanol, intersystem crossing is negligible at -196 °C. However, whereas for pyrene the intersystem crossing rate is unchanged by an increase in temperature, for [^{2}H$_{10}$]pyrene the rate at 23 °C is more than double that at -196 °C, going from 3.0×10^5 to 7.1×10^5 s^{-1}.

(d) At 23 °C, intersystem crossing and direct internal conversion into the ground state have similar rates. With deuteriation, the latter increases slightly (from 5.9 to 6.5×10^5 s^{-1}) and the former more substantially (from 4.5 to 7.1×10^5 s^{-1}).

(e) The radiative lifetime of the lowest triplet state τ_p is independent of temperature and compound deuteriation, and lies between 40 and 55 s. Radiationless quenching of T_1 shows a five- to seven-fold reduction upon deuteriation and is somewhat less important at lower temperatures. Even for [^{2}H$_{10}$]pyrene at -196 °C, the non-radiative decay of T_1 is 14 times faster than the emission of phosphorescence. Thus the observed triplet lifetime of 3.7 s is controlled almost completely by the non-radiative process. In the absence of such quenching, the triplet lifetime would lie in the range 40—55 s.

These conclusions are in disagreement with those from a recent study [10] of pyrene in poly(methyl methacrylate).

In the latter work, it was found that Φ_t and τ_f varied with temperature for both pyrene and [^{2}H$_{10}$]pyrene. They attributed the entire temperature variation to a change in the rate constant for intersystem crossing. The previous results do not agree with this interpretation, especially for pyrene. Also, values of τ_f at -196 °C of 425 and 434 ns for pyrene and [^{2}H$_{10}$]pyrene,[10] respectively, are considerably less than the values of 515 and 535 ns obtained in ref. 9. This suggests that the chemical environment may effect the emission properties of pyrene. Thus, the differences between

---

[9] J. L. Kropp, W. R. Dawson, and M. W. Windsor, *J. Phys. Chem.*, 1969, **73**, 1747.
[10] P. F. Jones and S. Siegel, *Chem. Phys. Letters*, 1968, **2**, 486.

the two sets of results could be due to different methods of sample preparation.

The quasi-linear fluorescence spectrum of pyrene trapped in methylcyclopentane at very low temperatures [11] and the fluorescence decay time of pyrene in a liquid crystal have been recorded.[12] Values of the rate constants for processes that deactivate the lowest excited singlet state of 1,12-benzperylene have been determined between -196 and 23 °C from measurements of the quantum yields of triplet formation Φ_t, fluorescence Φ_f, and the fluorescence lifetime τ_f.[13] The rate constants for intersystem crossing to the triplet state are nearly the same at -196 and 23 °C. The rate constant for direct radiationless deactivation to the lowest singlet is probably zero at -196 °C, but increases to 6×10^5 s^{-1} at 23 °C. This is similar to results obtained for other aromatic hydrocarbons. However, the fluorescence of 1,12-benzperylene is atypical. An increase in the rate constant of fluorescence from -196 to 23 °C is observed, resulting from an increase in Φ_f with temperature, while τ_f decreases over the same temperature range. The increase in the rate constant for fluorescence with increasing temperature is attributed to a temperature-dependent emission from the second excited singlet state. Since the second excited singlet state of 1,12-benzperylene lies only 1275—1400 cm^{-1} above the ^1L_b state (see Figure 4), there is significant thermal population of the ^1L_a state at 23 °C. The observation of a temperature-dependent anti-Stokes fluorescence supports this interpretation. The effect of solvent upon the energy gap between the ^1L_a and ^1L_b states and the fluorescence properties of 1,12-benzperylene is also discussed.

There is a large difference between the values of Φ_f observed at 23 °C in benzene and ethanol solutions of 1,12-benzperylene. The fluorescence yields of aromatic molecules are usually larger in benzene than in ethanol. Equations used to derive radiation lifetimes show k_f to be proportional to the square of the index of refraction of the solvent. In the case of benzene and ethanol, the ratio of their values of nD^2 is 1·23, so that k_f would be expected to be 23% higher in benzene than in ethanol. However, the value of k_f of 1,12-benzperylene in benzene is 34% higher than the corresponding value of k_f in ethanol. This anomalously large increase in k_f with solvent change can be associated with the effect of solvent change upon the E_p–E_α gap and the resulting difference in the thermal population of the p level. Both the p bands and α bands of aromatic hydrocarbons are displaced toward longer wavelengths upon changing the solvent from ethanol to benzene, but the p band is displaced farther. As a result, $E_\alpha - E_p$ of 1,12-benzperylene is smaller in benzene than in ethanol, and thus the thermal population of the p band is greater in benzene. Consequently, more fluorescence will originate from the p band in benzene than in

[11] A. Pellois and J. Ripoche, *Chem. Phys. Letters*, 1969, **3**, 280.
[12] Y. Tomkiewicz and A. Weinreb, *Chem. Phys. Letters*, 1969, **3**, 229.
[13] W. R. Dawson and J. L. Kropp, *J. Phys. Chem.*, 1969, **73**, 1752.

ethanol, and if Φ_f for the p level is greater than that for the α level, this will account for the large increase in k_f.

The fluorescence spectrum of triphenylmethyl radical trapped in methylcyclopropane at very low temperatures has been recorded.[14] Fission of singlet excitons into pairs of triplet excitons, the reverse of triplet annihilation, has been observed in tetracene crystals.[15] Hexyl azide has been

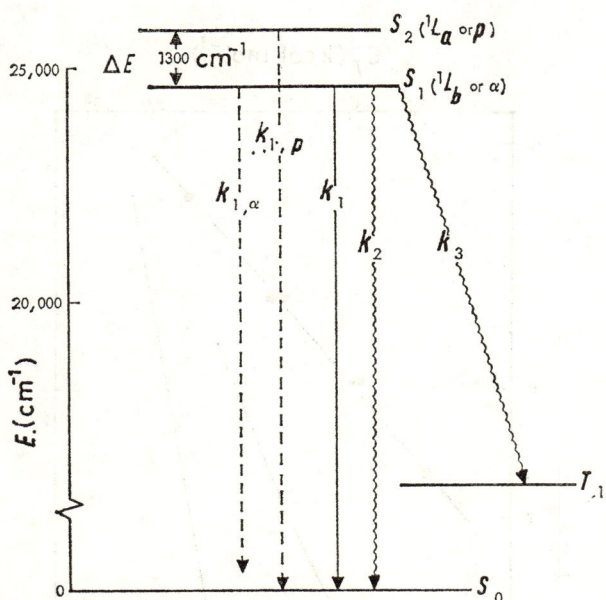

Figure 4 *Energy level diagram of* 1,12-*benzperylene*
(Reproduced by permission from *J. Phys. Chem.*, 1969, **73**, 1752)

found to be an efficient quencher of aromatic hydrocarbon fluorescence.[16] The rate of quenching was related to the singlet energy of the donor, as shown in Figure 5, but energy transfer by donors with low singlet energies is more efficient than that expected for a classical endothermic energy-transfer mechanism. The results are best interpreted by assuming vertical energy transfer to a bent azide ground-state generating a bent excited state. The singlet sensitization leads to decomposition of the azide with an efficiency similar to that in direct photolysis.

The mechanism of the quenching of fluorescence by quadricyclene has been discussed,[17] and the phosphorescence and fluorescence of two-

[14] N. Pellois, J.-C. Navatte, and J. Ripoche, *Compt. rend.*, 1969, **268**, *B*, 1183.
[15] R. E. Merrifield, P. Avakian, and R. P. Groff, *Chem. Phys. Letters*, 1969, **3**, 386.
[16] F. D. Lewis and J. C. Dalton, *J. Amer. Chem. Soc.*, 1969, **91**, 5260.
[17] B. S. Solomon and A. Weller, *Chem. Comm.*, 1969, 927.

component and multicomponent solutions of biphenyl derivatives described.[18] Photoionization in aromatic hydrocarbon solutions[19] and the formation of transient aromatic ions in solution by flash photolysis have been reported.[20]

The luminescence of indole derivatives continues to excite interest. The method of photoselection has been used to determine the polarized fluorescence excitation and emission spectra of indole and several of its

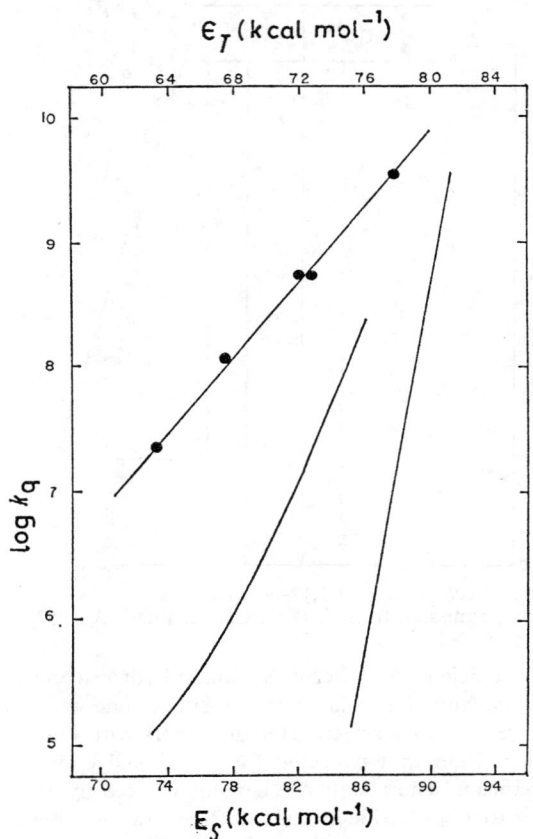

Figure 5 *Plot of log* k_q *against sensitizer energy for energy transfer to hexyl azide. Left to right: singlet transfer, triplet transfer, and classical endothermic transfer*
(Reproduced by permission from *J. Amer. Chem. Soc.*, 1969, **91**, 5260)

[18] A. V. Aristov and E. N. Viktorova, *Optics and Spectroscopy*, 1969, **25**, 283.
[19] S. D. Babenko, V. A. Benderskii, and V. I. Gol'danskii, *Zhur. E.T.F. Letters*, 1969, **10**, 205.
[20] K. Kawai, N. Yamamoto, and H. Tsubomura, *Bull. Chem. Soc. Japan*, 1969, **42**, 369.

derivatives.[21] The results are compared with theoretical predictions. The first and second (π,π^*) states were assigned as ^1L_b and ^1L_a, respectively, in all cases studied. Emission arising from the 0—0 transition from both ^1L_a and ^1L_b states has been confirmed in most of the indoles studied on the basis of sharp changes in the degree of polarization of the fluorescence bands, with the possible exception of indole N-acetic acid. The dual emission seems to occur in both glycerol–methanol (9 : 1, 263 K) and in an EPA rigid glass (77 K). The polarized phosphorescence spectra and mean lifetimes have also been obtained at 77 K in EPA glass. (EPA is a 5 : 5 : 2 mixture of diethyl ether, isopentane, and ethanol.) The triplet–singlet emission in seven indole derivatives has been shown to originate from $^3(\pi,\pi^*)$ states of ^3L_a type. The 0—0 phosphorescence emission bands were found to be negatively polarized, indicating predominant out-of-plane polarizations. Significant vibronic activity along the phosphorescence bands has been revealed, and in particular the degree of polarization increases considerably beyond the 0—0 band of the carboxyl- and formylindoles. The possible contributions of various states (out-of-plane n,π^*, σ,π^*, or π,σ^* and in-plane π,π^*) to the triplet–singlet transition probability via (a) direct spin–orbit, (b) spin–vibronic (first-order), and (c) vibronic–spin–orbit (second-order) couplings have been discussed qualitatively.

The fluorescence solvent shift and excited singlet state pK values of indole derivatives have been reported,[22] and indirect evidence presented for the existence of an excited state of 2H-indole.[23] Anomalous emission from indole derivatives has been observed.[24] The absorption and emission of benzo[e]cyclohept[b]indole and benzo[g]cyclohept[b]indole were taken in dilute cyclohexane solution at ambient temperature. For each isomer, the absorption spectra exhibit a weak, broad, diffuse band at long wavelength, followed by a more intense, narrower band which exhibits more structure. In both cases, the emission spectra appear on the short-wavelength side of the lowest-energy absorption band, indicating that the emitting electronic level lies above the first excited level. This is indicative of a restriction on the normally much faster non-radiative processes in upper singlet levels. The reason for the S_2 emission in this case is not clear.

Quenching of fluorescence from indole derivatives has been observed in several systems. Special derivatives of the basic chromophore 2-phenylindole were prepared [25] (see below), which could not self-quench by normal processes, and these were shown to have decreasing fluorescence yields with increasing concentration in toluene solutions (see Table 4). All componds had a common hydrogen atom on the nitrogen atom of the indole part of the molecule, which if replaced by a methyl group eliminated the self-

---

[21] P. S. Song and W. E. Kuvtin, *J. Amer. Chem. Soc.*, 1969, **91**, 4892.
[22] E. Vander Donckt, *Bull. Soc. chim. belges*, 1969, **78**, 69.
[23] C. M. Chopin and J. H. Wharton, *Chem. Phys. Letters*, 1969, **3**, 552.
[24] J. A. Poole, R. C. Dhingra, and B. Gelernt, *J. Chem. Phys.*, 1970, **52**, 464.
[25] D. L. Horrocks, *J. Chem. Phys.*, 1969, **50**, 4151.

Table 4 Data obtained on fluorescence spectra, self-quenching, excimer, and yields

Compound	Fluorescence spectra characteristics[a] wavelength (nm)			Self-quenching	Excimer	Maximum relative yields Fluorescence[b]
(4)	(350)	365	385	Yes	No	(100)
(5)	(362)	377	(390)	Yes	No	85
(6)	(355)	370	(383)	Yes	No	85
(7)	(360)	375	(387)	No	No	72
(8)	(360)	375	(387)	Yes	No	92
(9)	(360)	377	(390)	No	No	65
(10)		422		No	No	81
(11)		427		No	No	62

[a] Values in parenthesis indicate only a shoulder in the spectra.
[b] Normalized to a value of 100 for (4).

quenching. It is suggested that the phenomenon arises because of hydrogen bonding. There are two possible pathways of self-quenching through hydrogen bonding. One would involve the less-common $\mathrm{>N-H \cdots C <}$ bonding, and the second the $\mathrm{>N-H \cdots N<}$ bonding. The first type of hydrogen bonding would involve the high negative charge on C-3 of the excited molecule acting as a strong attractive force for the weakly acidic hydrogen atom on the nitrogen atom of the like molecule in its ground state:

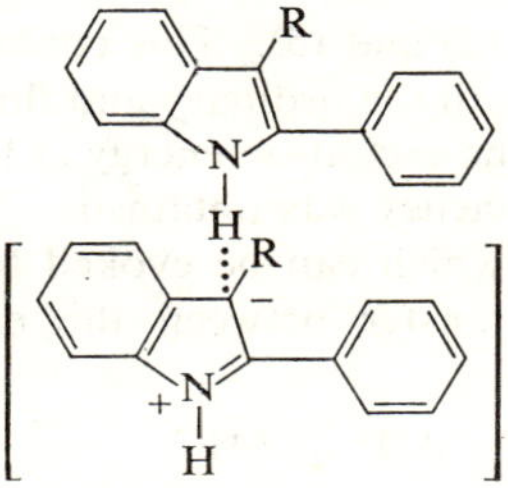

(12)

The second type of hydrogen bonding would require that in the excited state the hydrogen atom on the nitrogen atom became more acidic. This would seem quite logical as the result of the increased positive charge on the nitrogen atom of the excited molecule. The acidic hydrogen could thus form a hydrogen bond with the nitrogen atom of the molecule in the ground state (see formula 13). In both of these proposed mechanisms, increased charges on the C-3 atom and nitrogen atom of the molecule in the excited state have been assumed.

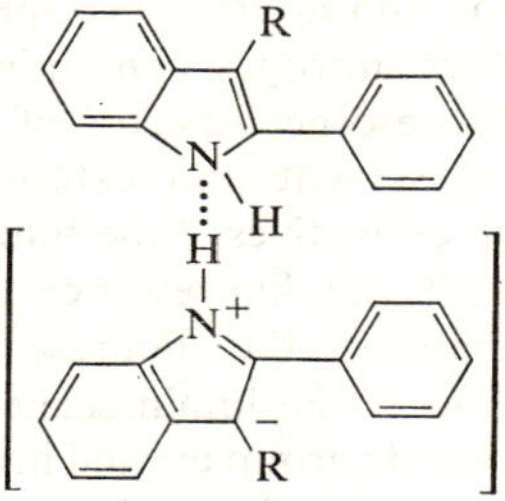

(13)

The compound (10), which has the steric hindrance and a hydrogen atom on the nitrogen atom, did not show self-quenching with increasing concentration. Also, the fluorescence spectra of (10) and (11) were different from those of the other 2-phenylindole compounds. The single-peaked spectra with maximum at 425 nm of (10) and (11) showed that the 3-phenyl group was actually part of the chromophoric system. The phenyl group in the 3-position acts to reduce the charge build-up at that pisition in the excited state. At the same time, this would also reduce the possibility of an increase in acidity by reducing the positive charge increase on the nitrogen atom. Thus the two proposed sources of the hydrogen bonding by the excited molecule have been eliminated in (10).

Also, in all cases of those compounds with N-hydrogen, there was a greater relative fluorescence yield compared to their corresponding N-methyl derivatives. The maximum scintillation yields of the N-hydrogen derivatives were also greater than the corresponding N-methyl derivative except in the compounds (4) and (5). This could be due to the processes of self-quenching of (4), both the ordinary and through hydrogen bonding, effectively competing for the excitation energy at low concentrations before the maximum transfer efficiency was obtained.

An alternative process which can be evoked to explain these results is the process of proton transfer between the excited molecule and an unexcited molecule,

$$AH^* + AH \rightleftharpoons AH^* \ldots AH \longrightarrow A^{*-} + HAH^+, \quad (14)$$

followed by radiationless de-excitation of the excited ion:

$$A^{*-} \longrightarrow A + e^- + \text{energy} \quad (15)$$

A similar explanation has been given to account for the quenching of the excited states of indole derivatives with electron scavengers such as cations (H^+, D^+, Cu^{2+}, Pb^{2+}, Cd^{2+}, or Mn^{2+}), anions (NO_3^-, IO_3^-, or $H_2PO_4^-$), sulphur-containing compounds (CS_2, thiourea, cysteine, or cysteamine), and other amino-acids and peptides.[26] Possible mechanisms of quenching are abstraction of an electron from the excited indole by the scavenger, or attachment of a proton to the excited indole ring. Evidence of intramolecular hydrogen bonding can be obtained from the spectroscopic data on bandwidths of absorption and fluorescence spectra.[27]

The fluorescence of other nitrogen-containing substances has been investigated. Thus, the fluorescence spectra of some benzimidazoles in acid media,[28] the effect of solvent and external heavy atoms on the fluorescence of isoquinoline derivatives,[29] the luminescence of benzonitrile in a solid nitrogen matrix,[30] the fluorescence and phosphorescence of 2- and 4-aminobenzophenone,[31] the fluorescent and phosphorescent behaviour of N-arylcarbazoles,[32] the luminescence of naphthalene derivatives containing a carbonyl amide group excited in the vacuum-u.v. region,[33] and the extinction of luminescence of complex molecules such as 3-amino-

[26] R. F. Steiner and E. P. Kirby, *J. Phys. Chem.*, 1969, **73**, 4130.
[27] I. B. Berlman, *Chem. Phys. Letters*, 1969, **3**, 61.
[28] M. Kondo and H. Kuwano, *Bull. Chem. Soc. Japan*, 1969, **42**, 1433.
[29] T. Okano and H. Matsumoto, *J. Pharm. Soc. Japan*, 1969, **89**, 510.
[30] E. Faure, F. Valadier, and J. Janin, *Compt. rend.*, 1969, **269**, *B*, 431.
[31] E. J. O'Connell jun., *Chem. Comm.*, 1969, 571.
[32] M. Zander, *Z. Naturforsch.*, 1969, **24a**, 870.
[33] N. Y. Dodonova and I. P. Vinogradov, *Opt. i Spektroskopiya*, 1969, **27**, 179.

phthalimide by strong laser beams [34] have been reported. The latter report is interesting in that a Q-switched ruby laser is used to excite the emission from the complex organic molecule. Absorption is a two-quantum process, and for moderate light intensities, the intensity of emission is proportional to the square of the incident intensity, as expected. For high laser output, the emission intensity becomes linear with laser intensity because stimulated emission from the excited state of the organic molecule occurs. The phenomenon is well known in dye molecules, being used in the construction of dye-lasers.

Excimer emission previously reported for acetone in the liquid phase [35] has been reported to be due to an unidentified impurity.[36] The quenching of biacetyl fluorescence by phenols and alkyl- and aryl-amines is reported.[37] Rate constants for this process are given in Table 5. Dienes are frequently

Table 5 *Quenching of biacetyl fluorescence in polar solvents*[a]

$k_q(\times 10^{10} \, \text{l mol}^{-1} \, \text{s}^{-1})$

Quencher	Benzene	Pyridine	Ethanol	Acetonitrile
C_6H_5OH	0·20	0·01	0·01	0·01
Resorcinol	0·65	0·15	0·14	0·04
$C_6H_5NH_2$	1·00	1·00	1·20	0·93
$(C_6H_5)_2NH$	0·74	1·20	1·30	1·20
$(C_6H_5)_3N$	0·49	1·00	1·30	1·30
$C_6H_5N(CH_3)_2$	1·20	1·60	1·60	1·60
$(C_3H_7)_3N$	0·24	0·52	0·12	0·29

[a] Values of k_q are calculated from Stern–Volmer quenching constants, $k_q\tau_f$, and measurement of τ_f by single-photon counting. In benzene, $\tau_f = 10\cdot0$ ns, in pyridine $\tau_f = 5\cdot7$ ns, in ethanol, $\tau_f = 7\cdot7$ ns, and in acetonitrile $\tau_f = 8\cdot2$ ns. Biacetyl concentration $= 0\cdot05 \text{ mol l}^{-1}$. The calculated (P. J. Wagner and I. Kochevar, *J. Amer. Chem. Soc.*, 1968, **90**, 2232) values for diffusion-controlled quenching are: benzene $1\cdot6 \times 10^{10} \text{ l mol}^{-1} \text{s}^{-1}$; pyridine $1\cdot1 \times 10^{10} \text{ l mol}^{-1} \text{s}^{-1}$; ethanol $0\cdot9 \times 10^{10} \text{ l mol}^{-1} \text{s}^{-1}$; acetonitrile $2\cdot9 \times 10^{10} \text{ l mol}^{-1} \text{s}^{-1}$.

used to quench triplet states of molecules in investigations of mechanisms of photoreactions. Recent results show that the excited singlet states of acetone, 2-pentanone, methyl-t-butyl ketone, and norcamphor are all quenched efficiently by penta-1,3-diene with rate constants of 9, 4, 2, and $5 \times 10^7 \text{ l mol}^{-1} \text{ s}^{-1}$ respectively.[38] Clearly, care must be taken in the interpretation of results based on the use of this diene as a triplet quencher. The implications of the lack of effect of change of phase on the fluorescent lifetime of biacetyl [39] on theoretical treatments of radiationless transitions has been discussed in Chapter 3.

[34] M. D. Galanin, B. P. Kirsanov, and Z. A. Chizhikova, *Zhur. E. T. F. Letters*, 1969, **9**, 502.
[35] M. O'Sullivan and A. C. Testa, *J. Amer. Chem. Soc.*, 1968, **90**, 6245.
[36] G. D. Renkes and F. S. Wettack, *J. Amer. Chem. Soc.*, 1969, **91**, 7514.
[37] N. J. Turro and R. Engel, *J. Amer. Chem. Soc.*, 1969, **91**, 7113.
[38] F. S. Wettack, C. D. Renkes, M. G. Rockley, N. J. Turro, and J. C. Dalton, *J. Amer. Chem. Soc.*, 1970, **92**, 1743.
[39] L. G. Anderson and C. S. Parmenter, *J. Chem. Phys.*, 1970, **52**, 466.

The fluorescence of dyestuffs is of great interest, both from an academic viewpoint and industrially. The fluorescence lifetime of acridine has been measured in water–ethanol and water–glycerol mixtures of different composition at 25 °C.[40] Remarkable decrease in lifetime value was found as the amount of organic component increased. A value for the lifetime of 10·3 ns was obtained for a pure aqueous solution. The lifetimes in pure glycerol and in pure ethanol were determined as 3·2 and 0·72 ns, respectively. The observed ratio of the lifetime in the mixture to the lifetime in water, τ/τ_0, has the same functional dependence on solvent composition as the ratio of fluorescence yield, Φ/Φ_0. These quenching phenomena by organic molecules were interpreted as due to the solvent effects upon the radiationless transition rate. The absorption and emission spectra of some Rhodamine dyes,[41] the absorption and fluorescence spectra of 2,5-dipyrazolinylthiophen isomers,[42] the luminescence of actinomycin D in various solvents,[43] the absorption and emission spectra of thiobenzophenone, xanthone, and N-methylthioacridone,[44] the electronic spectra of naphthostyril and its derivatives,[45] have all been recorded. 9-Anthroic acid shows a structureless fluorescence in acidic media and aprotic solvents. This has been ascribed to excimer formation, but recent work [46] has shown that the solvent and pH dependence of the 9-anthroic acid fluorescence can be explained on the basis of an acid–base equilibrium. For the molecular form of 9-anthroic acid, structureless fluorescence, similar to that of the esters of 9-anthroic acid, is observed. The large Stokes shift of the emission is a consequence of an excited-state rotation of the carboxy-group into the plane of the anthracene ring. This rotation, which can result in excited-state six-membered ring formation through intramolecular hydrogen bonding, was shown to be dependent on temperature, solvent matrix, and size of the ester group. For the ionic form of 9-anthroic acid in protonic solvents, rotation is inhibited owing to strong ground-state solvation, and a structured anthracene-like fluorescence is observed. Fluorescence quantum yields and lifetimes measured in benzonitrile and ethanol are consistent with the similarity of the excited-state geometry of the acid and esters. Charge-transfer effects on the fluorescence of anthroic acids have also been considered.

The absorption and fluorescence spectra of 1- and 2-anthroic acids and their anions were investigated.[47] For substitution in the 1-position,

[40] H. Kokuban, *Bull. Chem. Soc. Japan*, 1969, **42**, 919.
[41] L. V. Levshin, Yu. A. Mittsel', and N. Nizamov, *Zhur. priklad. Spektroskopii*, 1969, **11**, 509.
[42] S. V. Tsukerman, Lan Ngok Tkhiem, V. M. Nikitchenko, and V. F. Lavrushin, *Zhur. priklad. Spektroskopii*, 1969, **11**, 529.
[43] V. A. Poltorak, K. A. Vinogradova, and A. B. Silaev, *Vestnik. Moskov Univ.*, 1969, No. 4, 126.
[44] R. N. Nurmukhametov, L. A. Mileshina, D. N. Shigorin, and C. T. Khachaturova, *Russ. J. Phys. Chem.*, 1969, **43**, 24.
[45] I. L. Belaits, R. N. Nurmukhametov, D. N. Shigorin, and C. I. Bystitskii, *Zhur. fiz. Khim.*, 1969, **43**, 937.
[46] T. C. Werner and D. M. Hercules, *J. Phys. Chem.*, 1969, **73**, 2005.
[47] T. C. Werner and D. M. Hercules, *J. Phys. Chem.*, 1970, **74**, 1030.

smearing of the $^1A \to {}^1L_a$ transition by charge-transfer interaction is shown to be greater for carboxyl than for carboxylate substitution. This is attributed to the greater possible resonance interaction between the aromatic ring with the $-CO_2H$ group than with the $-CO_2^-$ group. When the carboxy-group is substituted at the 2-position of anthracene, the $^1A \to {}^1L_b$ transition is enhanced. The fluorescence of 2-anthroic acid originates from the ^1L_b state in polar solvents and from the ^1L_a state in non-polar solvents. Fluorescence lifetime data support these assignments. Comment is made on the increased basicity in the lowest excited singlet state aromatic carboxylic acids and their anions. Greater ΔpK_a is expected for the acid than for the anion due to the greater ability of the CO_2H group to undergo resonance interaction with the ring. Previously published data on the pK_a^* of 9-anthroic acid are found to be too high. An excited-state rotation which occurs for 9-anthroic acid but not for its anion probably invalidates the Forster pK_a equation when fluorescence frequencies are used.

(14)

The quenching of quinine bisulphate fluorescence by H^+ and by halides has been demonstrated.[48] Quinine (14) exists in several forms, depending on the pH of the solution. Absorption maxima occurring at 230 and

$$R^1N-R^2N \xrightarrow{H^+} R^1\overset{+}{N}H-R^2N \xrightarrow{H^+} R^1\overset{+}{N}H-R^2\overset{+}{N}H$$
(I) (II) (III)

$$\longrightarrow R^1\overset{+}{N}H-R^2NH^{2+}$$
(IV)

280 nm belong to the monionic structure (II), whereas maxima at 250 and 345 nm belong to structure (III). In 1N-NaOH structure (I) exists, which is non-fluorescent. The single ion (II) (pH = 4) exhibits violet fluorescence and the double ion (III) shows blue fluorescence. The quenching shown by halides is due to suppression of dissociation of the non-fluorescent form of quinine to the structure (II) by the halide ion. Similar results were obtained with Rhodamine. The flash photolysis of phenosafranine and Neutral Red in aqueous solution has been studied,[49] and the quenching of organic dye fluorescence in laser irradiation described.[50]

Charge-transfer interactions can be important in causing quenching of fluorescence of molecules. Thus, $n-\pi^*$ transitions in dimethyl thio-

---

[48] A. K. Babko and P. Kostyshina, *Ukrain. khim. Zhur.*, 1969, **35**, 544.
[49] A. K. Chibisov, A. V. Karyakin, B. V. Skvortsov, and L. N. Rygalov, *Khim. vysok. Energii*, 1969, **3**, 210.
[50] P. Peretti and P. Ranson *Compt. rend.*, 1970, **270**, *B*, 757.

acetamide–I_2 complexes have been observed.[51] Unusually large Stokes shifts and fluorescence decay times of the charge-transfer fluorescence of electron-donor–acceptor ion pairs have been described,[52] and radiative[53a] and radiationless[53b] processes in some charge-transfer complexes investigated.

The fluorescence spectra and quantum yields of tryptophan (TRP) and several tryptophan derivatives in water and a polar glass have been measured over wide temperature ranges.[54] For TRP, the wavelength of maximum emission shifts from 310 nm at 80 K to 355 nm at room temperature with almost all of the red-shift occurring between 170 and 230 K, which is the temperature range over which the glass softens. The successively more red-shifted spectra have no isoemissive wavelength, which supports the view that reorientation of several solvent molecules in the solvent shell of the excited TRP molecule and not a 1 : 1 exciplex is responsible for the red-shift. The quantum yield remains constant until a temperature is reached at which solvent reorientation is virtually complete. Above that temperature, the quenching of the 355 nm emission can be fitted with a non-radiative de-excitation having an activation energy of 29·4 kJ mol^{-1}. A model for these spectral changes and quenching mechanism is shown in Figure 6. In deuteriated solvents, a large isotope effect of fluorescence yield of TRP has been reported. This effect is clearly not caused by proton transfer in the excited state since it is found to be virtually the same for TRP and 1-Me-TRP. At temperatures below those at which solvent reorientation occurs, the isotope effect vanishes, and above them it approaches an asymptotic value, the quenching activation energy being independent of the isotopic constitution of the solvent. The fluorescence of most proteins and hormones originates in their tryptophan residues.

The polarized fluorescence spectra of some naturally occurring corrins have been recorded,[55] and the absorption and fluorescence spectra of chlorophyll *a* in polar solvents as a function of temperature investigated.[56] The fluorescence spectra and emission quantum yields of porphyrins[57] have also been measured. Measurements on the fluorescence spectra of some retinyl polyenes (15) allowed the following conclusions to be reached.[58]

(i) A comparison of the absorption and emission spectra of the retinyl polyenes enables an unambiguous location of the 0—0 transitions and has shown that the bands are severely Franck–Condon forbidden. This arises from a combination of effects:

[51] A. F. Granduna and M. Tamres, *J. Phys. Chem.*, 1970, **74**, 208.
[52] G. Briegleb, J. Trencséni, and W. Herre, *Chem. Phys. Letters*, 1969, **3**, 146.
[53a] J. Prochorow and R. Siegoczynski, *Chem. Phys. Letters*, 1969, **3**, 635; [b] N. Mataga and Y. Murata, *J. Amer. Chem. Soc.*, 1969, **91**, 3144.
[54] J. Eisinger and G. Navan, *J. Chem. Phys.*, 1969, **50**, 2069.
[55] A. J. Thompson, *J. Amer. Chem. Soc.*, 1969, **91**, 2780.
[56] C. Balny, S. S. Brody, and G. Hui Bon Hoa, *Photochem. and Photobiol.*, 1969, **9**, 445.
[57] P. G. Seybold and M. Gouterman, *J. Mol. Spectroscopy*, 1969, **31**, 1.
[58] A. J. Thomson, *J. Chem. Phys.*, 1969, **51**, 4106.

(a) the existence of bond alternation in the ground state, which leads to a gross change of geometry on excitation;
(b) steric hindrances at various points in the retinyl carbon framework.

(15)

(ii) Varying an electronegative end-group brings about an increase in the intensity of the 0—0 transition, but a breakdown of the mirror-image relationship. This implies a perturbation that is making the minima of the ground- and excited-state potential-energy surfaces occur at similar positions on the configurational co-ordinate, but is making the shapes of the potential-energy surfaces different. Thus the effect is primarily on (ia).

(iii) The various *cis*-isomers of retinal display spectra in which the mirror-image relation is poorly obeyed compared with the spectra of *trans*-retinal. This is an effect arising from (ib).

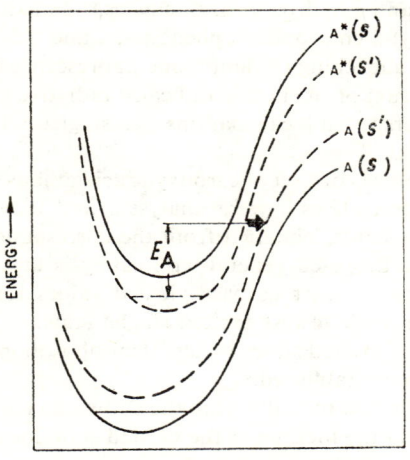

Figure 6 *Schematic diagram illustrating the motion of the potential surfaces of a molecule in its ground and excited states (A and A*) as a result of the reorientation of surrounding polar solvent molecules. s represents the solvent shell in its ground-state configuration, and s′ represents the solvent shell following solvent reorientation in the excited state. The short heavy arrow indicates where tunnelling between the potential surfaces A*(s′) and A(s′) might occur, leading to non-radiative de-excitations characterized by an activation energy E_A which is indicated in the figure*

(Reproduced by permission from *J. Chem. Phys.*, 1969, **50**, 2069)

(iv) Changing the environment in two ways, in polarity or in viscosity, can bring about large changes in the degree of overlap of both the absorption and emission bands; that is, the intensity of the 0—0 transition, and the success of the mirror-image relation. The effect of increasing the viscosity of the environment is to increase the degree of overlap of the absorption and emission bands; that is, to reduce the amount of nuclear relaxation that can occur on excitation. The effect is noted to be particularly pronounced on bands that are Franck–Condon forbidden whether by virtue of (ia), (ib), or both.

(v) The quantum yields of fluorescence of all the retinyl polyenes are very low. The yields change markedly on changing the polarity of the solvent or on increasing the viscosity of the solvent. The yield is shown to be correlated with the extent to which the mirror-image relation is obeyed, rather than with the intensity of the 0—0 band. Thus vibrational relaxation in the excited state is the major process deactivating polyenes in fluid solution, although the efficiency of this process is much reduced on increasing the viscosity of the medium.

The fluorescence quantum yield in retinal and its acetate have also been measured by other workers.[59]

Other topics investigated include fluorescent brightening agents,[60] emission from m- and p-hydroxybenzaldehydes,[61] 2-methylanthraquinone,[62] 2-hydroxy-4-methyl-5-chloroalkanophenones, and 1-hydroxy-2-alkanonaphthones,[63] the quenching of fluorenone fluorescence by triethylamine,[64] the emission spectrum of the benzyl radical,[65] radiative transitions in liquid organic scintillators,[66] and calculations on singlet and triplet states of fluorene.[67]

The quenching effects of a set of carboxylic acid anions on the fluorescence of some phenol derivatives[68] were analysed, and it was found that the quenching rate-constants, obtained from the corresponding Stern–Volmer plots, satisfied the Brønsted general-base catalysis law. Extrapolation of the Brønsted plots to water as base yielded values for the pK_a's of the various phenols in their lowest excited singlet state.

The quenching of excited states by tervalent phosphorus compounds and arylamines has been established.[69]

Jablonski was the first to realize that the Brownian rotations can produce apparent changes in the lifetime of the excited state measured in solutions.

---

[59] D. Lerner, *Compt. rend.*, 1969, **268**, *C*, 1740.
[60] R. D. Desai, *J. Indian Chem. Soc.*, 1969, **46**, 595.
[61] G. D. Baruah, O. N. Singh, and R. S. Singh, *Indian J. Pure Appl. Phys.*, 1969, **7**, 352.
[62] G. D. Baruah, K. P. R. Nair, and D. K. Rai, *Indian J. Pure Appl. Phys.*, 1969, **7**, 520.
[63] B. N. Tripathi and C. L. Garg, *Indian J. Chem.*, 1969, **7**, 778.
[64] R. A. Caldwell, *Tetrahedron Letters*, 1969, 2121.
[65] A. N. Singh and I. S. Singh, *Indian J. Pure Appl. Phys.*, 1969, **7**, 349.
[66] M. N. Gurskii and A. N. Tsay, *Khim. vysok. Energii*, 1969, **3**, 365.
[67] A. Brie and R. Zuravich, *J. Chem. Phys.*, 1969, **51**, 903.
[68] I. Avigail, G. Feitelson, and M. Ottolenghi, *J. Chem. Phys.*, 1969, **50**, 2614.
[69] R. S. Davidson and P. F. Lambeth, *Chem. Comm.*, 1969, 1098.

He showed that the decay time of the fluorescence intensity which is polarized parallel to the exciting, linearly-polarized beam must be expected to be shorter than the overall or true molecular decay time, while the decay time of the component whose polarization is normal to that of the exciting beam is longer than the average. In principle, it is possible to determine the rate of rotation of molecules by measurements of decay of the polarized components. Results obtained indicate that the early theory of Jablonski is only qualitatively correct, and a new treatment has been proposed.[70] The depolarization of the fluorescence of solutions by either Brownian rotations or intermolecular energy transfer may be simply described by a system of first-order linear differential equations containing as sole parameters the rate of fluorescence emission and the rate of transport of the excitation from one orthogonal component of the emission to another. The steady-state solution has the form of Perrin's equation describing the depolarization by Brownian rotations, and the time-dependent depolarization following a unit light impulse is that originally described by Jablonski. The solution for sinusoidal excitation is novel in that: (i) it shows the difference in lifetime between the polarized components of the emission to be a sensitive function of the ratio of the modulation frequency ω to the emission rate λ; for $\omega/\lambda > 1$, the difference between the polarized lifetimes may become many times greater than that observed after a unit light impulse; (ii) it permits the determination of both the rate of transport of the excitation and the limiting polarization of the fluorescence from observations at one fixed temperature and viscosity; and (iii) it allows the definition of conditions under which the true or exponential decay of the fluorescence may be measured. Experimental tests of the theory by phase fluorometry are described. These include observations upon dilute solutions in media of limited viscosity, where Brownian motion is the only cause of depolarization, and observations upon concentrated frozen solutions, where depolarization is due to energy transfer alone.

2 Concentration Quenching, Excimer Emission, and Delayed Fluorescence

Several examples of quenching of fluorescence in solution involving concentration effects have already been alluded to, but we consider here some examples which give rise to excimer emission. *P*-type delayed emission, although involving the triplet states of molecules, is closely related to excimer formation, and so is considered here also. A review of excimers, including the kinetics of their formation, structures, *etc.*, has appeared recently.[71]

Fluorescence spectra and quantum yields have been obtained for methylcyclohexane solutions of benzene, toluene, ethylbenzene, cumene,

---

[70] R. D. Spencer and G. Weber, *J. Chem. Phys.*, 1970, **52**, 1654.
[71] T. Forster, *Angew. Chem.*, 1969, **81**, 364.

p-, m-, o-xylene, and 1,3,5-, 1,2,3-, 1,2,4-trimethylbenzene as a function of aromatic concentration over the temperature range from 25 to $-100\,°C$.[72] At low temperatures, distinct excimer emissions were observed for all compounds studied. The intrinsic emission quantum yields of monomer Φ_m and of excimer Φ_e have been determined by a simple technique which requires no assumptions regarding the details of the monomer–excimer kinetics. With decreasing temperature, Φ_e is observed to decrease, contrary to the behaviour of Φ_m, suggesting that the rate constant for the excimer radiative transition decreases strongly as the temperature is lowered. Such temperature dependence is explained as arising from the existence of a substantial vibronic component in the transition moment that is induced by thermal excitation of upper-state vibrational motions (*e.g.* torsional, tilting, *etc.*) of one monomer with respect to the other. From analysis of the temperature dependence of the fluorescence, lower limits on the excimer binding energies E_b have been determined. The difference between this lower limits and E_b is approximately equal to the activation energy for radiative decay of the excimer. An estimate of this activation energy indicates that, for the case of benzene, $E_b > 0.36$ eV. The probability for association of excited monomer to form excimer and the probability for dissociation of excimer to an excited and unexcited monomer have been determined for benzene at 25 and $-78\,°C$ from an appropriate analysis of the fluorescence quenching effect of CCl_4. Additionally, it has been demonstrated that the observed increase in CCl_4 quenching efficiency at high benzene concentrations is predominantly due to an energy migration process. The probability per encounter of formation of excimer for benzene has been determined to be 1·0 and to decrease with alkyl substitution in a manner consistent with the steric requirements of sandwich-type excimer configurations.

Excimer emission from tetrachloro-p-xylene has also been observed.[73] The formation of excimers of 2-phenyl-5-(4-biphenylyl)-1,3,4-oxadiazole (PBD) was shown by the examination of the fluorescence spectra of concentrated solutions of derivatives of PBD in toluene.[74] A slight increase in the amount of emission in the longer-wavelength region of the spectra was observed with increasing concentration of the PBD derivatives. At high solute concentrations, the amount of the excimer emission was altered by variation of the temperature of the solution. Lower temperatures increased the relative amount of longer-wavelength emission (favoured excimer formation) and higher temperatures decreased the relative amount of longer-wavelength emission (favoured disproportionation of the excimer). It was observed that certain solvents enhanced the relative amount of fluorescence from the excimer of an excited solute molecule. The ratio of intensity of excimer fluorescence I_e to intensity of

---

[72] F. Hirayama and S. Lipsky, *J. Chem. Phys.*, 1969, **51**, 1939.
[73] S. K. Chakrabarti, *Mol. Phys.*, 1969, **16**, 417.
[74] D. L. Horrocks, *J. Chem. Phys.*, 1969, **50**, 4962.

monomer fluorescence I_m for solutions of equal concentration of 2,5-diphenyl-1,3-oxazole (PPO) was approximately four times greater in cyclohexane and ethanol than in the aromatic solvents benzene, toluene, p-xylene, and 1,2,4-trimethylbenzene.[75] Also at room temperature (24 °C) excimer fluorescence of 2-(4-isopropylphenyl)-5-(4-biphenylyl)-1,3,4-oxadiazole (i-p-PBD) was observed only from concentrated solutions in cyclohexane. The effect of temperature on the intensity of excimer fluorescence indicated the role of the solvent in excimer formation and dissociation.

Pyrene continues to be a favourite molecule in which to study excimer formation. Thus the excimers of pyrene in frozen solutions have been reported,[76] but of more interest is the report that the absorption spectrum of the pyrene excimer has been observed.[77]

The fluorescence lifetimes τ_f of a number of aromatic hydrocarbons are shortened by 15—40% when the matrix is compressed by 30 kbar pressure. However, it has been shown that the magnitude of the pressure effect on τ_f does not depend on concentration, deuteriation, or the presence of oxygen for pyrene monomer and excimer fluorescence in poly(methyl methacrylate) and polystyrene matrices.[78] Excimers have been reported to be formed in electrochemiluminescence experiments on anthracene and 9,10-dimethylanthracene.[79]

An attempt has been made to observe triplet excimers by observing the emission from a naphthalene sandwich pair.[80] The photolysis of 1,3-bis(1-naphthyl)propane (16) caused dimerization ($\lambda \sim 280$ nm) to the anthracene-like dimer (17). Further, upon irradiation with a low-pressure mercury lamp, solutions of (17) (77 or 300 K) regenerate (16), identified by

its characteristic u.v. spectrum. The photolytic cleavage of (17) in methylcyclohexane at 77 K generates (16), having a sandwich configuration of the two naphthalene nuclei, an unstable geometry. The luminescence spectra ($\lambda_{exc} = 313$ nm) of the sandwich pair are shown in Figure 7(a). After

---

[75] D. L. Horrocks, *J. Chem. Phys.*, 1969, **51**, 5443.
[76] E. Loewenthal, Y. Tomkiewicz, and A. Weinreb, *Spectrochim. Acta*, 1969, **25**, *A*, 1501.
[77] C. R. Goldschmidt and M. Ottolenghi, *J. Phys. Chem.*, 1970, **74**, 2041.
[78] J. J. Kim, R. A. Beardslee, D. J. Phillips, and H. W. Offen, *J. Chem. Phys.*, 1969, **51**, 2761.
[79] T. C. Werner, J. Chang, and D. M. Hercules, *J. Amer. Chem. Soc.*, 1970, **92**, 763.
[80] E. A. Chandross and C. G. Dempster, *J. Amer. Chem. Soc.*, 1970, **92**, 704.

thawing and refreezing the sample, the spectra in Figure 7(b) are obtained. The u.v. absorption spectrum of the photolysed sample is consistent with a sandwich-pair configuration; after thawing and refreezing it reverts to that of (16). The initial fluorescence of the photodissociated sample is

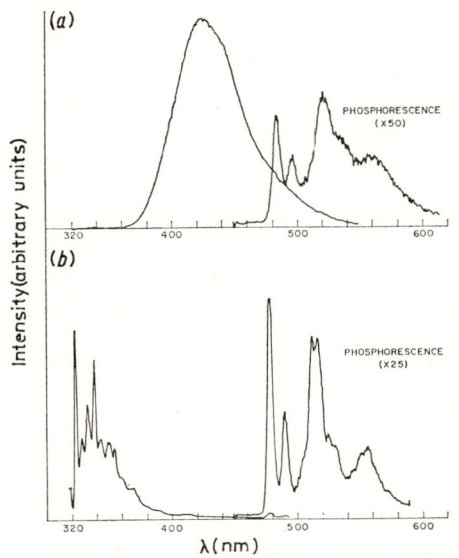

Figure 7 (a) *Luminescence spectra (excited at* 313 nm) *of the sandwich naphthalene pair obtained from* (17) *by photolysis with* 254 nm *light in methylcyclohexane at* 77 K. (b) *Luminescence spectra of the above sample after melting and refreezing. The intensity scale is not corrected for spectrometer response but is the same for both sets of spectra*
(Reproduced by permission from *J. Amer. Chem. Soc.*, 1970, **92**, 704)

similar to the excimer fluorescence of various naphthalenes in fluid solution, including the intramolecular excimer formed by both (16) and the isomeric 2-naphthyl compound. The final fluorescence spectrum is identical with that of (16) at low temperatures, essentially that of 1-methylnaphthalene, and is typical of a simple naphthalene. It is possible to assign the initial phosphorescence spectrum to the sandwich pair, and conclude that a triplet excimer is not formed. The interaction between the naphthalene nuclei is much weaker than in the singlet excimer case. A Hückel MO treatment of excimers as intermediates for photodimers of methoxynaphthalene has been given.[81] Exciplex formation is reported between aromatic hydrocarbons and aliphatic amines.[82]

[81] J. Christie and B. Selinger, *Photochem. and Photobiol.*, 1969, **9**, 471.
[82] M. G. Kuzmin and L. N. Guseva, *Chem. Phys. Letters*, 1969, **3**, 71.

P-Type delayed fluorescence ideally shows a dependence upon the square of the intensity of the exciting radiation. In the crystalline state, deviations from this law for anthracene have been observed [83] which were attributed to trapping and the decay of triplets by bimolecular annihilation. *P*-Type delayed fluorescence has been observed in benzophenone.[84] Delayed luminescence in organic mixed crystals of amino-acids and proteins has been seen,[85] and in DNA–acridine dye complexes in frozen aqueous solution.[86]

Solute re-encounter probabilities have been determined from a measurement of the delayed excimer fluorescence of pyrene in ethanol,[87] and the delayed fluorescence of tetramethylphenylenediamine has been reported.[88] Sensitized delayed fluorescence may be used to measure the $S_1 \rightarrow T_1$ transition probability of acridine dyes.[89]

3 Aspects of the Triplet State

In addition to aspects of the triplet state mentioned briefly so far, an attempt will be made to review the vast amount of data published in the past year upon triplet states of organic molecules. Spectroscopic details are merely mentioned; aspects of more direct use to photochemists will be dealt with in more detail.

Formation.—Triplet states of molecules are populated generally *via* absorption from the ground state to a higher singlet state, followed by intersystem crossing to the triplet manifold, but direct ground-state singlet → excited-state triplet absorption can be seen under special circumstances. The technique of phosphorescence excitation spectroscopy has made observations of singlet–triplet absorption spectra much easier, and this technique has been applied to benzene in the crystalline state,[90] for the $^1A_{1g} \rightarrow {}^3B_{1u}$ transition, using the apparatus shown in Figure 8. The spectrum exhibits very clear evidence of a pseudo-Jahn–Teller distortion of the normally hexagonal benzene molecule upon excitation to the triplet state. Factor-group splitting of the 0—0 and 0—0 + ν_1 exciton bands has also been observed. The position of the mean of the 0—0 exciton band of C_6H_6, when compared to the phosphorescence origin of a C_6H_6 guest in a C_6D_6 crystal, indicates that the 'static' intermolecular interactions between guest and host are different for C_6H_6 and C_6D_6 hosts. This result has important implications for the applicability of mixed-crystal theories to benzene. The method has also been used on acetophenone, *p*-bromoacetophenone, and

---

[83] D. A. Goode and F. R. Lispett, *J. Chem. Phys.*, 1969, **51**, 1222.
[84] J. Saltiel, H. C. Curtis, L. Metts, G. W. Miley, J. Winterle, and M. Wrighton, *J. Amer. Chem. Soc.*, 1970, **92**, 410.
[85] M. E. McCarville and S. P. McGlynn, *Photochem. and Photobiol.*, 1969, **10**, 171.
[86] Y. Kubota, Y. Fujisaki, and M. Miura, *Bull. Chem. Soc. Japan*, 1969, **42**, 853.
[87] B. Stevens, *Chem. Phys. Letters*, 1969, **3**, 233.
[88] M. Ewald and B. Muel, *Compt. rend.*, 1969, **268**, B, 973.
[89] N. Nemoto, H. Kokukun, and M. Koizumi, *Chem. Comm.*, 1969, 1095.
[90] D. McBurland, G. Castro, and G. W. Robinson, *J. Chem. Phys.*, 1970, **52**, 4100.

1-indanone at 77 and 4 K.[91] The studies of 1-indanone demonstrate the following:
(i) the 0—0 band of the $S_0 \to S(n, \pi^*)$ transition is allowed and polarized perpendicular to the plane of the molecule, indicating substantial delocalization of the oxygen lone-pair orbital into the aromatic ring system;

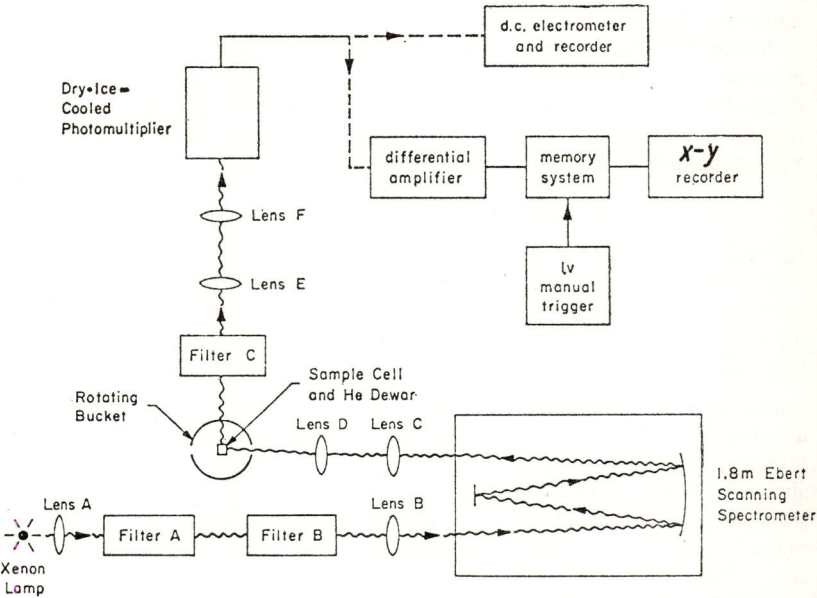

Figure 8 *Phosphorescence excitation spectroscopy. Diagram of experimental apparatus. Lenses A, B, E, and F are fused quartz, lenses C and D Pyrex. Filter A is an* H_2O *filter, B a Corning glass 7–54 filter, and C* p-*dimethylaminobenzaldehyde in methanol*
(Reproduced by permission from *J. Chem. Phys.*, 1970, **52**, 4100)

(ii) most of the intensity (80—90%) in the $S_0 \to S(n, \pi^*)$ transition is vibronically induced;
(iii) a strong carbonyl progression appears to be absent in the $S_0 \to S(n, \pi^*)$ absorption spectrum, and this is attributed to mixing of the carbonyl stretching mode with other vibrational modes in the excited state;
(iv) the $S_0 \to T(n, \pi^*)$ transition has allowed character as a result of spin–orbit interaction between the $T(n, \pi^*)$ state and the $^1L_a(\pi, \pi^*)$ state;
(v) the intensification of the $S_0 \to {}^1L_b$ transition in indanone relative to acetophenone is attributed to mixing of the ^1L_b state with the ^1L_a state induced by *ortho*-substitution. The introduction of ^1L_a character

[91] W. A. Case and D. R. Kearns, *J. Chem. Phys.*, 1970, **52**, 2175.

into the ^1L_b state is also responsible for the intensification of the $S_0 \to T(n, \pi^*)$ transition in indanone relative to acetophenone;
(vi) the 0—0 band of the $S_0 \to T(\pi, \pi^*)$ transition is allowed by spin–orbit coupling with the allowed, out-of-plane-polarized component of the $S_0 \to S(n, \pi^*)$ transition;
(vii) vibronic coupling between the $T(\pi, \pi^*)$ state and a nearby $T(n, \pi^*)$ state introduces significant in-plane-polarized intensity into the $S_0 \to T(\pi, \pi^*)$ transition.

Essentially similar results were obtained with acetophenone and p-bromoacetophenone, with the exception that the strong $S_0 \to T(\pi, \pi^*)$ transition in p-bromoacetophenone was found to be in-plane-polarized parallel to the long-axis-polarized transition to the $^1L_a(\pi, \pi^*)$ state. Diffuseness observed in the 4 K single-crystal $S_0 \to T$ spectra of each of the three compounds is attributed to vibronic interaction of the $T(n, \pi^*)$ with nearly degenerate vibronic levels associated with a lower-lying $T(\pi, \pi^*)$ state. Similar studies have been carried out on santonins.[92] The triplet level of anthracene is non-phosphorescent, and thus the method cannot be used, but a variation using delayed fluorescence monitoring has allowed the recording of the polarized singlet–triplet absorption spectra of normal and deuteriated anthracene crystals.[93]

As said before, the most probable route for triplet state formation is *via* intersystem crossing from an excited singlet state. The determination of the relative importance of the different routes involved in the intersystem crossing process is essential to an understanding of non-radiative processes in polyatomic molecules. This can be accomplished if a method is found by which the rate constants for the intersystem crossing processes involved in the population of the individual zero-field sublevels of the lowest triplet state can be determined for symmetric molecules. A method has been described by which these rate constants can be determined and then applied to [$^2H_4$]pyrazine,[94, 95] for which the decay of the phosphorescence emission from the three individual sublevels of the lowest triplet state has been reported. This is observed because the spin–lattice relaxation (s.l.r.) processes between the three sublevels are slower than the radiative processes from the individual sublevels. Furthermore, in general, the three observed emission lifetimes should be different so that they may be resolved in time. There are a good number of other molecules for which these conditions are satisfied. It thus appears that one can study the radiative and non-radiative properties of the individual sublevels of the lowest triplet state of phosphorescent molecules. Assuming the following system:

(i) A molecule excited to a singlet state loses its energy by internal-conversion processes and finally reaches the lowest singlet state, S_1.

---

[92] G. Marsh, D. R. Kearns, and M. Fisch, *J. Amer. Chem. Soc.*, 1970, **92**, 2252.
[93] G. Durocher and D. F. Williams, *J. Chem. Phys.*, 1969, **51**, 5405.
[94] M. A. El-Sayed and L. Hall, *J. Chem. Phys.*, 1969, **50**, 3113.
[95] M. A. El-Sayed, D. S. Tinti, and D. V. Owens, *Chem. Phys., Letters* 1969, **3**, 339.

(ii) The intersystem crossing process proceeds from S_1 to the different triplet sublevels τ_1, τ_2, and τ_3 of the lowest triplet state, either directly or indirectly *via* the other triplet states located between S_1 and the lowest triplet state. The pumping rates of the τ_1, τ_2, and τ_3 sublevels from S_1 have the rate constants $k_{(IS)1}$, $k_{(IS)2}$, and $k_{(S)3}$, respectively. Each one of these rate constants may or may not be a composite of other rate constants, depending on the exact mechanism involved in the $S_1 \rightsquigarrow \tau_i$ radiationless transitions.

(iii) The system is at temperatures for which the s.l.r. processes between τ_1, τ_2, and τ_3 are much slower than the radiative and non-radiative processes from the three sublevels to the ground state.

Application of the steady-state approximation to the formation and disappearance of any of the multiplets, *e.g.* τ_i, of the above system gives the following relation:

$$k_{(IS)i}[S_1] = k_i[\tau_i], \tag{16}$$

where $k_{(IS)i}$ is the rate constant for the intersystem crossing process from S_1 to the τ_i sublevel, $k_i = k_{(rad)i} + k_{(normal)i}$ is the observed rate constant of the decay of sublevel i, and $[S_1]$, $[\tau_i]$ represent the steady-state concentrations of the lowest singlet state and sublevel i of the lowest triplet state, respectively. The observed steady-state intensity, I_i, of the emission from sublevel τ_i is given by:

$$I_i = k_{(rad)i}[\tau_i] = k_i[\tau_i]\,\Phi_i = k_{(IS)i}[S_1]\,\Phi_i \tag{17}$$

where $\Phi_i = k_{(rad)i}/k_i$ is the phosphorescence quantum yield of sublevel i. Applying equation (17) to the emission from any two of the sublevels, *e.g.* τ_1 and τ_2, one obtains:

$$I_1/I_2 = (k_{(IS)1}/k_{(IS)2}) \times (\Phi_1/\Phi_2). \tag{18}$$

If Φ_1, Φ_2, and Φ_3 are determined at these temperatures, two independent equations can be obtained relating I_1/I_2 and I_1/I_3 to ratios involving $k_{(IS)i}$. If the total $k_{IS} = k_{(IS)1} + k_{(IS)2} + k_{(IS)3}$ is determined in the usual manner, one can solve for $k_{(IS)1}$, $k_{(IS)2}$, and $k_{(IS)3}$ individually.

Since Φ_1, Φ_2, and Φ_3 were not known for any molecule, a molecule with very high phosphorescent quantum yield, [$^2H_4$]pyrazine, was selected. The phosphorescence of this molecule decayed with three components, one short-, one medium-, and one long-lived. No change of the phosphorescence lifetime upon deuteriation was found for the short-lived emission and only small changes for the medium- and long-lived emissions. These results strongly suggest that the observed lifetimes are radiative for a molecule having as high a triplet-state energy as pyrazine. These facts justify making the assumption that $\Phi_1 = \Phi_2 = \Phi_3 = 1$. The initial intensity of the three decay components is determined from the decay curve. From polarization and theoretical considerations, the short-, medium-, and long-lived emissions are assigned to radiation from the τ_y, τ_z, and τ_x sublevels of the $^3B_{3u}$ state respectively. (The N—N and out-of-plane axes are the z and x respectively). Using equation 18, $k_{(IS)z}/k_{(IS)y}$ and $k_{(IS)x}/k_{(IS)y}$ were determined as 3% and 4% respectively. Thus the most probable route is the one which populates the τ_y sublevel of the $^3B_{3u}(\pi, \pi^*)$ state. This is in agreement with predictions on a theoretical basis, which would support the mechanism as:

$$^1B_{3u}(n, \pi^*) \longrightarrow {}^3B_{1u}(\pi\pi^*)\,(\tau_y) \longrightarrow {}^3B_{3u}(n\pi^*)\,(\tau_y) \quad (19)$$

The first step is the only route allowed by direct spin–orbit selection rules. The other sublevels (τ_z and τ_x) can only be populated by spin–orbit–vibronic or spin–vibronic interactions, and it is clear that in pyrazine these are ~30 times less probable. The temperature dependence of intersystem crossing in crystalline anthracene has been investigated.[96]

Detection.—Since triplet states are relatively long-lived, they may be detected readily by spectroscopic methods. Since they are paramagnetic, electron spin resonance may be used (in the condensed phase only for large molecules) in their detection. Thus the triplet states of benzophenone,[97] and other aromatic carbonyl compounds,[98] [$^2H_{10}$]pyrene,[99] phenazine,[100] and of acridine dyes [101] have all been registered using e.s.r. spectrometry. More conveniently, however, optical absorption spectroscopy is used to identify and characterize the triplet excited states of organic molecules. Clearly, for quantitative measurements of concentrations of triplet states, extinction coefficients for absorptions must be known. Extinction coefficients for triplet–triplet absorption have been directly determined for several aromatic hydrocarbons.[102] A high-intensity photolysis lamp was used to populate the triplet state, and the concentration of triplets was measured by monitoring the depletion of the ground state. The values of the extinction coefficients $\varepsilon \times 10^{-4}$ (l mol^{-1} cm^{-1}) and their estimated uncertainties at the most prominent maximum λ (nm) are: naphthalene 4.0 ± 0.6, 414.0; [$^2H_8$]naphthalene 4.0 ± 0.6, 414.0; phenanthrene 3.8 ± 0.6, 492.5; [$^2H_{10}$]phenanthrene 3.1 ± 0.5, 494.5; triphenylene 1.56 ± 0.23, 430.0; and [$^2H_{12}$]triphenylene 1.20 ± 0.18, 431.0. For anthracene, $\varepsilon \geqslant 9 \times 10^4$ l mol^{-1} cm^{-1} at 427.3 nm. Knowledge of the extinction coefficients then allows accurate estimates of the $S_1 \rightarrow T_1$ rates of intersystem crossing to be made. The quantum yields at 77 K in a rigid matrix and their estimated uncertainties were found to be: naphthalene, 0.25 ± 0.05; [$^2H_8$]naphthalene, 0.25 ± 0.05; phenanthrene, 0.35 ± 0.07; [$^2H_{10}$]phenanthrene, 0.45 ± 0.09; triphenylene, 0.54 ± 0.11; and [$^2H_{12}$]triphenylene, 0.88 ± 0.18. No significant deuterium isotope effects on the intersystem-crossing quantum yields were noted. The polarized

[96] W. T. Stacey and C. E. Swenberg, *J. Chem. Phys.*, 1970, **52**, 1962.
[97] M. Sharnoff, *J. Chem. Phys.*, 1969, **51**, 451.
[98] C. H. J. Wells, A. Horsfield, and J. Paxton, *Chem. Comm.*, 1969, 393.
[99] P. H. H. Fischer and A. B. Denison, *Mol. Phys.*, 1969, **17**, 297.
[100] J. P. Grivet and J. M. Lhoste, *Chem. Phys. Letters*, 1969, **3**, 445.
[101] Y. Kobota and M. Miura, *Bull. Chem. Soc. Japan*, 1969, **42**, 2763.
[102] S. C. Hadley and R. A. Keller, *J. Phys. Chem.*, 1969, **73**, 4351, 4356.

triplet–triplet absorption spectrum of pyrene,[103] and other aromatic hydrocarbons,[104] and triplet absorption in duroquinone,[105] chloranil,[106] cyanine dyes,[107] and uracil [108] have been reported. Theoretical calculations on naphthalene predict forbidden triplet–triplet transitions in the range 10 000—20 000 cm^{-1}. Attempts to observe triplet–triplet absorption bands in this region have been as unsuccessful until recently. Weak transient absorptions at 17 500 and 19 000 cm^{-1} in flashed solutions of naphthalene in poly(methyl methacrylate) (PMMA) have now been observed.[109] The triplet spectra of aniline, NN-dimethylaniline, and tetramethyl p-phenylenediamine have been determined in the visible region at 77 K, and polarization measurements have been made.[110] Triplet–triplet absorptions in the i.r. region have been observed in tetracene.[111]

Phosphorescence.—The lowest triplet states of many compounds may, of course, be identified by their characteristic phosphorescence spectra, although low temperatures and condensed phases are necessary to observe this phenomenon in most compounds. The phosphorescence spectrum of naphthalene (showing 94 transitions between 21 224 and 16 880 cm^{-1}) has been obtained.[112] The data help resolve uncertainties in the vibrational assignments, suggest the presence of localized lattice vibrations in isotopic mixed crystals, and may indicate the relative importance of singlet–triplet coupling mechanisms in naphthalene and some halogenated naphthalene derivatives. The data have been used to identify the corresponding absorption and phosphorescence of the pure crystal, and might be used in explaining temperature, environmental, and isotope effects on the radiative or radiationless transition probabilities of the naphthalene molecule.

The phosphorescence and Zeeman spectra of mixed crystals of anthracene were reported.[113] Pyridine has been investigated many times previously, but no phosphorescence has been observed until a recent careful study.[114] From a consideration of the effect of substitution upon the phosphorescence, which is either strong $\pi\pi^*$ (long-lived) or weak $n\pi^*$ (short-lived), it was concluded that the lowest triplet state of pyridine is the $^3A_2(n\pi_5^*)$ state. An explanation of the absence of phosphorescence in pyridine itself is then possible in terms of specific state ordering. The low-lying triplet states of pyridine (see Figure 9) are $^3A_1(\pi_2\pi_4^*$—$\pi_3\pi_5^*)$, $^3B_1(n\pi_4^*)$, and $^3A_2(n\pi_5^*)$. The n-orbital electron after excitation is localized partly on the nitrogen

---

[103] T. G. Pavlopoulos, *J. Chem. Phys.*, 1970, **52**, 3307.
[104] D. Lavalette, *Chem. Phys. Letters*, 1969, **3**, 67.
[105] E. J. Land, *Trans. Faraday Soc.*, 1969, **65**, 2815.
[106] D. R. Kemp and G. Porter, *Chem. Comm.*, 1969, 1029.
[107] R. A. Pierce and R. A. Berg, *J. Chem. Phys.*, 1969, **51**, 1267.
[108] E. Hayon, *J. Amer. Chem. Soc.*, 1969, **91**, 5397.
[109] W. H. Melhuish, *J. Chem. Phys.*, 1969, **50**, 2779.
[110] K. D. Cadogan and A. C. Albrecht, *J. Phys. Chem.*, 1969, **73**, 1868.
[111] R. Astier, A. Bokobza, and Y. H. Meyer, *J. Chem. Phys.*, 1969, **51**, 5174.
[112] D. M. Hanson, *J. Chem. Phys.*, 1969, **51**, 5063.
[113] R. H. Clarke, *J. Chem. Phys.*, 1970, **52**, 2328.
[114] R. G. Hoover and M. Kasha, *J. Amer. Chem. Soc.*, 1969, **91**, 6508.

atom in the $^3B_1(n\pi_4^*)$ state, and on C-2 and C-6 in the $^3A_2(n\pi_5^*)$ state. Methyl substitution at the 2-, 6- positions should blue-shift the A_2 state more than the B_1 state. If the A_2 state were 2500 cm^{-1} (or less) lower in energy than the A_1 state, then 2,6-dimethyl substitution could invert the state ordering (Figure 9) (assuming 500 cm^{-1} blue-shift for the A_2 state

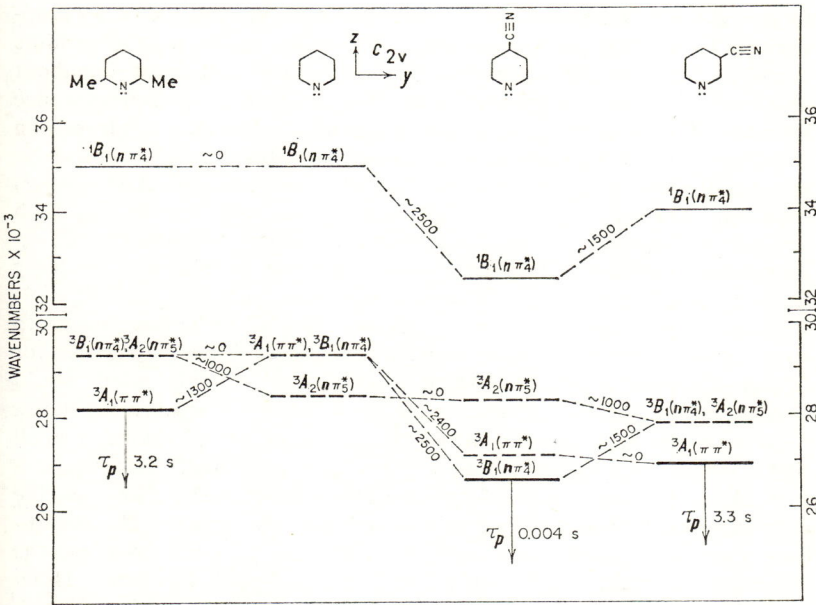

Figure 9 *Lowest excited electronic states of pyridines: solid levels, observed triplet and singlet states; dashed levels, extrapolated estimated state positions*
(Reproduced by permission from *J. Amer. Chem. Soc.*, 1969, **91**, 6508)

and 650 cm^{-1} red-shift for the A_1 state, per methyl group). Experimentally, 2,6-lutidine yields strong phosphorescence (band onset 28 160 ± 200 cm^{-1}, ethanol glass, 77 K) with a lifetime of 3·2 s. Fluorescence (probably $\pi\pi^*$) is observed in alcohol glass but not hydrocarbon glass.

Cyano-substitution red-shifts benzene phosphorescence 2400 cm^{-1}. The pyridine $^1B_1(n\pi^*)$ state is red-shifted 2500 cm^{-1} by 4-cyano-substitution and *ca.* 1000 cm^{-1} for 2- and 3-cyano-substitution. It is assumed that the corresponding triplet state $^3B_1(n\pi_4^*)$ is shifted equivalently. The $^3A_2(n\pi_5^*)$ state should not be shifted significantly by 4-cyano-substitution because of the π_5^* node, while a 3-cyano-group should red-shift this state by about 1000 cm^{-1}. Thus, if it is assumed that the $^3B_1(n\pi_4^*)$ and $^3A_1(\pi_2\pi_4^*$—$\pi_3\pi_5^*)$ states are nearly degenerate (as reasonable values of S–T splits predict) and the $^3A_2(n\pi_5^*)$ state is some 2000 cm^{-1} lower in energy, in 4-cyanopyridine $^3B_1(n\pi_4^*)$ should lie lowest, and in 3-cyanopyridine $^3A_1(\pi\pi^*)$ should be the lowest energy triplet (Figure 9). Experimentally, strong phosphorescence (diffuse) is observed in 4-cyanopyridine with a lifetime of 0·004 s (band onset 26 650 ± 200 cm^{-1}, ethanol glass, 77 K); for 3-cyanopyridine strong phosphorescence (structured) is observed with a lifetime of 3·3 s (0,0 band at 26 946 ± 200 cm^{-1}, ethanol glass, 77 K). The short-lifetime phosphorescence of 4-cyanopyridine is assigned to the $^3B_1(n\pi_4^*)$ state, and the long-lifetime phosphorescence of 3-cyanopyridine to the $^3A_1(\pi_2\pi_4^*$—$\pi_3\pi_5)$ state. The lowest triplet state of pyridine therefore must be the forbidden $^3A_2(n\pi_5^*)$ state, and must be about 28 000 cm^{-1} above the ground state. The absence of phosphorescence from the $^3A_2$ state is attributable to an exceptionally low radiative transition probability, and a high rate of radiationless transition to the ground state. It is easily shown that only ^1L_b and $^1B_b(\pi\pi^*)$ states mix with the $^3A_2(n\pi_5^*)$ state *via* spin–orbit coupling. The out-of-phase coupling of $^3A_2$ with this pair of singlet states results in a nearly vanishing borrowed transition moment for radiation. This is in marked contrast to spin–orbit coupling for the $^3B_1(n\pi_4^*)$ state. Radiationless transition rates for the $^3A_2(n\pi_5^*)$ state to the ground state are, contrariwise, expected to be large owing to large expected molecular distortions, with expected large magnitudes for Franck–Condon factors. The small spacing between triplet states would also be expected to lead to very large vibronic coupling between the states.

Owing to the node through the nitrogen atom in the π_5^* orbital, a small singlet–triplet splitting in the $A_2(n\pi_5^*)$ state is expected. This suggests that the lowest singlet state of pyridine might be $^1A_2(n\pi^*)$. However, the corresponding low-energy absorption band does not appear. The absorption is predicted to be exceedingly weak, and this may be why none was observed in pyridine (at 1·0 mol l^{-1} in 10 cm of 3-methylpentane). However, the extensive charge rearrangement in this state may give a much larger singlet–triplet splitting than expected (*i.e.* large electron-correlation effects). The phosphorescence spectra of phenazine,[115] pyrazine,[116] polycyclic azines and aromatic aldehydes,[117] indan-1-one (which shows anomalous phosphorescence),[118] 2- and 4-aminobenzophenone.[31] acenaphthaquinone,[119] camphorquinone,[120] thymine,[121] naphthalene-*N*-heterocyclics,[122] phenylcarboxylic acids and salts,[123] durene,[124] pyrimidine and 2-chloropyrimidine,[125] carbazole ketones,[126] aminopyridines,[127] matrix-

[115] T. G. Pavlopoulos, *J. Chem. Phys.*, 1969, **51**, 2936.
[116] T. Azumi and Y. Nakano, *J. Chem. Phys.*, 1969, **51**, 2515.
[117] E. C. Lim, R. Li, and Y. H. Li, *J. Chem. Phys.*, 1969, **50**, 4925.
[118] Y. Kanda, J. Stanislaus, and E. C. Lim, *J. Amer. Chem. Soc.*, 1969, **91**, 5085.
[119] S. N. Singh, M. G. Jayswal, and R. S. Singh, *Bull. Chem. Soc. Japan*, 1969, **42**, 2048.
[120] L. Tsai and E. Charney, *J. Phys. Chem.*, 1969, **73**, 2462.
[121] A. C. Szabo, W. D. Riddell, and R. W. Yip, *Canad. J. Chem.*, 1970, **48**, 694.
[122] S. M. Ziegler and M. A. El-Sayed, *J. Chem. Phys.*, 1970, **52**, 3257.
[123] H. G. Maria and S. P. McGlynn, *J. Chem. Phys.*, 1970, **52**, 3399.
[124] D. Owens, M. A. El-Sayed, and S. Ziegler, *J. Chem. Phys.*, 1970, **52**, 4315.
[125] N. Nishi, R. Shimuda, and Y. Kanda, *Bull. Chem. Soc. Japan*, 1970, **43**, 41.
[126] M. Zander, *Z. Naturforsch.*, 1969, **24a**, 1387.
[127] M. R. Padhye and V. V. Bhujle, *Current Sci.*, 1969, **38**, 215.

isolated naphthalene and phenanthrene,[128] and matrix-isolated thiazole[129] have been reported. Calculations have been performed on the vibronic intensity distribution in the phosphorescence of benzene and its deuterio-isomers.[130] The relative intensities of the members of a progression (Franck–Condon factors) can be used to determine changes in molecular geometry upon electronic excitation. A detailed examination of the normal co-ordinates shows that for benzene and similar species only one vibrational mode (v_1, ~990 cm^{-1}) is expected to form progressions, even for deuteriated benzenes having many totally-symmetric vibrations which, according to group theory, might be expected to form progressions. For a D_{2h} distorted benzene, a second co-ordinate (v_8, ~1600 cm^{-1}) is expected to form progressions. The phosphorescence of all isotopically-substituted benzenes has only one main progression frequency (950—990 cm^{-1}), while the dominant progressions for the methylbenzenes involve the 1600 cm^{-1} mode. Quantitative calculations show that, in the lowest triplet state of benzene, the difference between long and short bonds is less than 0·01 Å, while for toluene a value of ~0·07 Å is found.

Triplet Decay Times.—Triplet decay times are dependent upon temperature, environment, intensity, substitution, presence of quenchers, and (in some cases) pressure. As has been mentioned,[95] there has been increased interest in investigating the phosphorescence occurring from the different zero-field multiplet levels of triplet states, each of which decays with a characteristic decay time. Thus SO_2 has been studied in xenon, SF_6, and O_2 matrices at 4 and 20 K.[131] At zero field and 4 K, the decay time for SO_2 is 13·5 ms in SF_6, 5·6 ms in Xe, and 0·50 ms in O_2. In a magnetic field of 26 kG, it is 14·8 ms in SF_6, 5·8 ms in Xe, and 0·56 ms in O_2. In the presence of the magnetic field, the decay time increases by up to 12%. At 20 K in SF_6, increase of the magnetic field causes the lifetime to change from a single component of 11·3 ms at zero field to two components with lifetimes of 10·2 and 15·0 ms at 90 kG. The contribution to the decay of the longer-lived component at 90 kG is about 30%.

A conclusive interpretation of the result is not yet possible because excitation transfer to and from the triplet state of matrix-isolated molecules is not yet understood, and because the Zeeman effect and spin–lattice relaxation (s.l.r.) in solids containing molecules like SO_2 are not yet well studied. However, three factors can be pointed out:

(i) in matrices, molecules can occupy different sites and assume random orientation; the influence of site effects on lifetimes has only recently been discovered, and these effects are not well known;

(ii) the lifetime of $T_3 \rightarrow S_0$ is only a few milliseconds, and at 20 K is not much slower, or comparable in rate, to s.l.r.;

---

[128] J. L. Metzger, B. E. Smith, and B. Meyer, *Spectrochim. Acta*, 1969, **25A**, 1177.
[129] L. Williamson and B. Meyer, *Spectrochim. Acta*, 1970, **26**, 331.
[130] G. C. Nieman, *J. Chem. Phys.*, 1969, **51**, 1660.
[131] J. G. Conway, B. Meyer, J. J. Smith, and L. G. Williamson, *J. Chem. Phys.*, 1969, **51**, 1671.

(iii) the triplet splitting, at zero field, is small compared to kT, which is 14 cm^{-1}. However, at high field it will become comparable with kT. Therefore, the results may be caused by molecules in different sites, by kT imbalance, or s.l.r. dependence on triplet level splitting. The detailed mechanism is expected to be complex, because all three effects can occur simultaneously, and are functions of temperature. Quantitative evaluation is complicated, because the observed lifetimes result from simultaneous depopulation of at least two triplet levels, which are spectroscopically not well resolved. At fields greater than 80 kG, the decay curves show changes, which indicate that a Paschen–Back effect results.

The phosphorescence decay of quinoxaline in durene, pyrazine in cyclohexane and benzene, and other molecules in a variety of hosts is non-exponential and sensitive to an applied magnetic field at very low temperatures (1·6—7 K). It is believed that at these low temperatures the spin–lattice relaxation process between the sublevels of the triplet state is slow relative to the fastest radiative phosphorescent process.

At 77 K, the decay of pyrazine phosphorescence in cyclohexane is exponential, with a lifetime of 18 ms. The decay becomes non-exponential at 1·6 K, and can be resolved into three exponential decays with lifetimes of 6, 100, and 400 ms.[95] Most of the intensity ($\gtrsim 90\%$) decays with the 6 ms lifetime, which is interpreted, on the basis of polarization measurements, as the radiative lifetime of the τ_y spin sublevel of the phosphorescing pyrazine triplet state. The medium and long lifetimes might represent the observed lifetimes for the two processes, τ_x and τ_z phosphorescence (mechanism I),[95] or they could represent the spin–lattice relaxation times from τ_x and τ_z to the emitting spin sublevel τ_y (mechanism II). If mechanism I is correct, or is partially involved in determining the medium and long lifetimes, the polarization of the phosphorescence at short times during the decay would differ from that at medium or long times. On the contrary, mechanism II predicts that the phosphorescence polarization remains constant throughout the decay since the short-, medium-, and long-lived emissions all originate from the τ_y spin sublevel.

Preliminary results have been presented [132] which show that the polarization of the pyrazine phosphorescence in both durene and in 1,4-dichlorobenzene host crystals changes with time during the decay process at 1·6 K (see Figure 10). These results suggest that the radiative phosphorescent process from the τ_z and/or τ_x spin sublevels of the pyrazine triplet state in these hosts determines, at least in part, the observed medium and/or long lifetimes. The time-resolved method employed in this work is the only method presently available by which phosphorescence from the different zero-field multiplets can be observed, since the zero-field splitting is smaller than the phosphorescence linewidth in these crystals by an amount

---

[132] M. A. El-Sayed, W. R. Moomaw, and D. S. Tinti, *J. Chem. Phys.*, 1969, **50**, 1888.

such that it can be resolved energetically. Thus, this method should be extremely useful in elucidating the detailed spin–orbit mechanisms involved in the phosphorescence from the different spin multiplets.

Phosphorescence lifetimes of benzene and methylbenzene derivatives, deuteriated and undeuteriated, have been measured in glassy matrices at

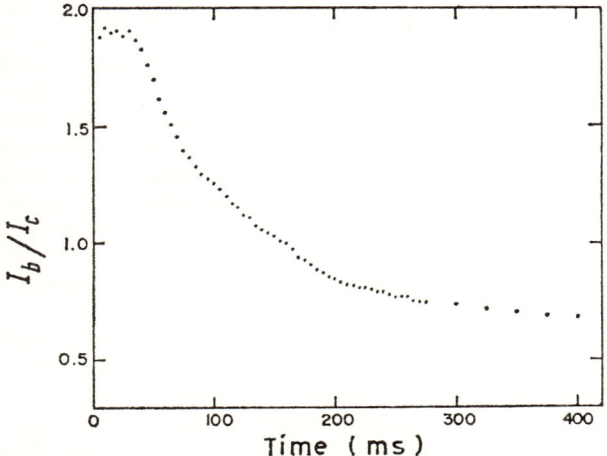

Figure 10 *The time dependence of the I_b/I_c polarization ratio of the pyrazine phosphorescence in 1,4-dichlorobenzene host during its decay at 1·6 K. From refined curves of this kind, it should be possible to elucidate the different mechanisms responsible for the phosphorescence from the different zero-field multiplets of the lowest triplet state of phosphorescent molecules*
(Reproduced by permission from *J. Chem. Phys.*, 1969, **50**, 1888)

4·2 and 77 K.[122] Spin–orbit coupling calculations have been performed for benzene (D_{6h}), distorted benzene (D_{2h}), and toluene; these calculations included spin–own-orbit, spin–other-orbit, and vibronic considerations, and were extended to all the methylbenzenes by a vector-sum semi-empirical method.[133] Franck–Condon calculations were performed for all methylated, deuteriated, and distorted benzene derivatives. The results of these calculations lead to a general understanding of phosphorescence lifetimes, at least insofar as these are affected by either methylation or deuteriation or both. Matrix effects and temperature effects on the phosphorescence lifetimes are largest in the case of benzene and toluene, and decrease with increasing methylation of benzene and with increasing size of the polynuclear hydrocarbon.

The conclusions arising from this work are summarized below.

(i) The lifetime τ_p is not dependent on the time of immersion of the sample in a refrigerant if (*a*) the sample is properly degassed, (*b*) the sample

[133] J. W. Rabalais, H. G. Maria, and S. P. McGlynn, *J. Chem. Phys.*, 1969, **51**, 22 59.

is not subject to irradiation for any considerable length of time, and (c) the thermal contact of sample and refrigerant is good. The third item is most readily satisfied by using small-diameter sample tubes and small sample volumes.

(ii) C—H stretching vibrations of alkyl groups appended to a benzene ring are efficient in modulating non-radiative transition. This quenching ability extends as far as carbon centres which are twice removed from the ring, and not much further.

(iii) The magnitude of Franck–Condon integrals is particularly dependent on the normal-co-ordinate change, denoted R. In benzene, it appears that the vibrations active in quenching are dominantly of skeletal stretching type, and that the effect of deuteriation is to increase this dominance. In polynuclear aromatics, it appears that the dominant quenching of T_1 is caused by C—H stretching modes and that, for constant T_1–S_0 energy gaps, the effect of deuteriation should be correspondingly greater (than in benzene) in increasing τ_p.

(iv) The addition of methyl group(s) to benzene destroys σ, π separability to a significant extent.

(v) The effects of deuteriation should lead to a larger increase of τ_p in polymethylbenzenes than in the less-methylated benzenes. This is not the case experimentally. That it is not is attributable to item (iv) and the effects it exerts on decreasing the radiative lifetimes (τ_p^R) of the polymethylbenzenes. Indeed, the ratio τ_p^N/τ_p^R appears to increase with increasing methylation of benzene, where τ_p^N is the non-radiative lifetime of the state.

(vi) The radiative lifetime of benzene, assumed to be approximately 30 s, is interpretable from two points of view. If the benzene molecule is D_{2h} distorted, a spin–orbit lifetime of 13·3 s results; if it is not distorted, the best calculable lifetime is 24·9 s. It is clear that both assertions provide good agreement with experiment.

(vii) Benzene and the less-methylated derivatives of benzene are exceptional with regard to their sensitivity to both environmental and temperature effects on τ_p. These data are taken as evidence for effects which are related to both the 'nakedness' of benzene and the small size of the space available to its π-electrons. The effect of size is demonstrated in the insensitivity of τ_p of naphthalene and phenanthrene and the influence of 'nakedness' on the τ_p values of the polymethylbenzenes.

(viii) The behaviour of the xylene isomers may be interpreted from the point of view of symmetry, but most probably from the point of view of 'nakedness' specified in item (vii) above.

(ix) The lifetime of all benzene derivatives is always larger in 'harder' solvents at any temperature. A 'hard' solvent is one with high glassification temperature T_g, and of high dipole moment.

(x) The effects of temperature on τ_p are greatest when n, the number of substituent groups, is smallest.

(xi) The effects of 'box' size and 'nakedness', as they relate to gross sensitivity of solvent and temperature, are interpreted from the point of view of solvent-assisted distortions. The activation energies associated with τ_p variations should probably be associated with glass changes and not with any specific property of the T_1 state.

(xii) Benzene is not a prototypical aromatic molecule, and spectroscopists should cease considering it as such.

The effect of various parameters upon phosphorescence decay times are summarized in Figure 11. Changes in phosphorescence lifetime of benzene

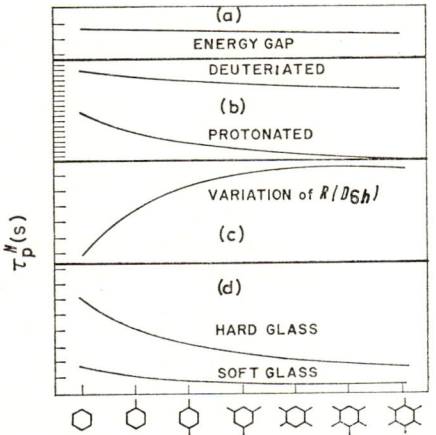

Figure 11 *The effects of the four of the more important factors in the determination of $\tau_p{}^N$ of benzene and its methylated derivatives. Each effect is considered individually. The curves are:* a, *the effect of changes in the T_1–S_0 energy gap;* b, *the contributions of CH and CD modes;* c, *expansion of the benzene ring, D_{6h} symmetry being maintained;* d, *solvent-assisted distortion effects in hard and soft glasses. The ordinate is relative, each mark counts as one second. The curves are a result of semi-empirical calculations*
(Reproduced by permission from *J. Chem. Phys.*, 1969, **51**, 2259)

in rigid solutions at 77 K with exciting wavelength have been attributed to a decrease in the $S_1 \rightarrow T_1$ intersystem crossing rate, implying that some other non-radiative (or chemical) decay process is increasingly important at higher photon energies.[134] Great controversy exists about the magnitude of (ix) and (x) above. The lifetime of benzene phosphorescence was previously found to be temperature dependent, and to change over a period of time which was attributed to glass relaxation, since it occurred especially in glasses which underwent glass transitions near 77 K (see Volume 1 of this series). This viewpoint has been challenged [135] with evidence on benzene and its n-alkyl derivatives ($C_6H_5C_nH_{2n+1}$, where $n = 1$—3), in which the phosphorescence lifetime was apparently independent of temperature below 60 K, depending on the solvent used, and was temperature dependent

[134] M. D. Lumb, C. Lloyd Braga, and L. C. Pereira, *Trans. Faraday Soc.*, 1969, **65**, 1992.

in a relatively small temperature range at higher temperatures.[135] Arrhenius plots indicated that in the temperature region studied there were two or more paths of externally influenced non-radiative decay, and all previous observations can be attributed to this effect. However, a reinvestigation of the original system [136] has produced clear evidence that glass relaxation can be of importance in some systems, especially benzene. There appears to be no way of reconciling the different results except to note that different experimental arrangements and procedures were used, and to say that, as rates of cooling and relaxation in amorphous glasses are difficult to quantify, the techniques may lead to different results. Clearly, more work is required to resolve the deadlock.

The effect of deuteriation on the phosphorescence decay time of naphthalene [137, 138] and phenanthrene [137] has been investigated. The position of the substituent is critical, since 1,4-, 1,5-, and 1,8-α-dideuterionaphthalenes all have similar decay times in 3-methylpentane and ethanol at 77 K of 3·7 s, whereas 2,3- and 2,7-(β)-dideuterionaphthalenes have shorter lifetimes of 3·15 s.[139] Calculations as to the magnitude of this isotope effect have been performed.[140, 141] Benzophenone, a molecule much loved by photochemists, has not been shown previously to phosphoresce in fluid solutions, largely because its triplet state undergoes hydrogen-abstraction reactions which lead to a very short lifetime. In perfluorocarbon solvents, however,[142] the lifetime is longer, and data obtained on related compounds in such solvents are given in Table 6. Benzophenone emission has since been observed in iso-octane solution, as has that of acetophenone and triphenylene.[143] The quantum yields of emission were of the order of 4×10^{-4}, and lifetimes were of the order of 2 ms (compare Table 6). On the basis of these results, it can be concluded that the short lifetime of benzophenone triplets in benzene solution cannot be due to impurity quenching, and thus specific interaction between benzophenone triplets and benzene must be sought. For the aromatic ketones with lowest excited state $^3n\pi^*$ ($^3A_2$), the radiative $T_1 \to S_0$ process is by spin–orbit mixing with an excited $\pi\pi^*$ ($^1A_1$) state, but the radiationless process is by direct spin–orbit coupling between $^3A_2$ and ground ($^1A_1$) states. Quantitative agreement between the experimental rates of radiationless processes with those predicted by theory can be obtained by assuming the Franck–Condon factor to be dominated by the carbonyl stretch.[144]

[135] I. H. Leubner, *J. Phys. Chem.*, 1969, **73**, 2088.
[136] T. E. Martin and A. H. Kalantar, *J. Phys. Chem.*, 1970, **74**, 2030.
[137] T. N. Bolotnikova, T. M. Naumova, F. I. Gurov, and V. G. Kazachkov, *Optics and Spectroscopy*, 1968, **25**, 291.
[138] T. F. Hunter, *Photochem. and Photobiol.*, 1969, **10**, 147.
[139] T. D. Gierke, R. J. Watts, and S. J. Stickler, *J. Chem. Phys.*, 1969, **50**, 5425.
[140] B. R. Henry and W. Siebrand, *Chem. Phys. Letters*, 1969, **3**, 327.
[141] B. R. Henry and W. Siebrand, *Chem. Phys. Letters*, 1969, **3**, 90.
[142] C. A. Parker and T. A. Joyce, *Trans. Faraday Soc.*, 1969, **65**, 2823.
[143] W. D. K. Clark, A. D. Litt, and C. Steel, *J. Amer. Chem. Soc.*, 1969, **91**, 5413.
[144] T. F. Hunter, *Trans. Faraday Soc.*, 1970, **66**, 300.

From a consideration of charge densities on oxygen in the bonding and antibonding π molecular orbitals, benzophenone appears to change shape on $^1A_1 \rightarrow {}^1A_1$ excitation; the $\sim 30°$ angle between the rings in the ground state is lowered following excitation. Discussions on the reactivity of benzophenone excited states in hydrogen-abstraction reactions indicate

Table 6 Triplet data in perfluoromethylcyclohexane at 20 °C

Compound	Φ_p	τ (ms)	τ_p (ms)	type
Acetophenone	0·10	0·39	$3·9 \times 10^{-3}$	n
Xanthone	0·011	<0·1	$<9 \times 10^{-3}$	n
Benzophenone	0·097	0·71	$7·3 \times 10^{-3}$	n
		$(1·0)^a$		
p-Methoxybenzophenone	0·017	0·29	$1·7 \times 10^{-2}$	n
p-Diacetylbenzene	0·0039	0·27	$6·9 \times 10^{-2}$	n (?)
Flavone	$0·00019^b$	$2·3^b$	12^b	π
Methyl 2-naphthyl ketone	$0·00007^b$	$0·6^b$	8^b	π
2-Naphthaldehyde	$0·00009^b$	$0·7^b$	8^b	π
Benzil	0·034	0·47	$1·4 \times 10^{-2}$	n (?)
Phenanthrene	—	1·0	—	π
Naphthalene	—	1·1	—	π
Pyrene	—	15	—	π

^{a} In perfluorodecalin; ^{b} tentative values owing to very low phosphorescence intensity and possible interference by impurity.

that the effective oxygen charge remains largely the same during $\pi \rightarrow \pi^*$ excitation. However, the effective charge is due to all atomic orbitals on oxygen, with small contributions from the $2p_x$ orbital, and thus the result from ref. 144 does not, of necessity, disagree with previous results.

For the substituted aromatic ketones with lowest excited state $^3\pi\pi^*$ ($^3A_1$) only slightly below the $^{1,3}n\pi^*$ states, both radiative and radiationless decay from $^3\pi\pi^*$ are faster than for similar transitions in aromatic hydrocarbons, due to spin–orbit mixing of the $^3\pi\pi^*$ state with $^1n\pi^*$ ($^1A_2$). When the $^3\pi\pi^*$ state in aromatic ketones lies well below the $n\pi^*$ states (fluorenone and 7,8-benzoflavone), the radiative and radiationless decay characteristics become much more similar to those of aromatic hydrocarbons, since $^1A_2$ spin–orbit mixing with $^3A_1$ decreases. The effect of $^1A_2$ mixing on the radiative process is more marked than on the radiationless process. The triplet lifetimes of phenanthrene, chrysene, fluorene, carbazole, and dibenzothiophen in n-heptane at 77 K were measured as 3·18, 2·44, 6·33, 7·16, and 1·28 s respectively.[145] The effect of pressures of up to 30 kbar upon phosphorescence decay times of benzophenone, triphenylene, and other compounds has been described.[146]

Properties of Triplet States.—Misconceptions regarding the degree of biradical character in triplet states have been pointed out.[147] Zero-field

---

[145] D. J. Morantz and T. G. Martin, *Trans. Faraday Soc.*, 1969, **65**, 665.
[146] H. W. Offen and D. E. Hein, *J. Chem. Phys.*, 1969, **50**, 5274.
[147] P. J. Wagner and R. P. Spoerke, *J. Amer. Chem. Soc.*, 1969, **91**, 4437.

splittings in the triplet states of some aromatic hydrocarbons and nitrogen heterocyclics,[148] trimethylenemethane,[149] quinoxaline,[150] phenylalnphthalenes,[151] pyrazine,[152] and deuteriated phenanthrene [153] have been discussed. A theory of photoselection for Zeeman-split phosphorescence has been developed,[154] and an experimental observation of factor-group splitting in the $^3B_{1u}$ state of benzene has been reported.[155] Optical spin-polarization in triplet naphthalene has been observed,[156] and the crystal field effect on triplet states of molecular solid quinoxaline described.[157] Intramolecular triplet exciton transfer in some non-coplanar aromatic systems [158] has been observed, and a multichannel transfer model developed to describe such phenomena.[159] The lifetimes and motion of triplet states in molten aromatic systems have been measured by photoconductivity techniques.[160] Exciton–photon coupling in crystalline anthracene has been described,[161] and an intermolecular charge-transfer triplet state observed.[162] The relationship between light absorption and triplet state concentrations in depopulated systems has been considered.[163] Triplet–triplet annihilation has been observed in toluene [164] and in benz[*f*]indan in crystalline biphenyl and fluorene.[165] Delayed fluorescence referred to previously, of course, arises *via* this phenomenon.

The effect of magnetic fields upon the process in anthracene is of interest.[166] Figure 12 shows the variation of the delayed fluorescence intensity with field strength for two different concentrations of anthracene in *NN*-dimethylformamide (DMF). The intensity in both cases declines with increasing field strength, but it levels off at higher fields. There is no indication of a low-field enhancement of the intensity like that observed with the solid. In addition, the magnitude of the field effect is not as great as that measured even in a randomly oriented polycrystalline sample.

[148] R. P. Mersiner and F. W. Birss, *J. Phys. Chem.*, 1969, **50**, 2085.
[149a] Y. Gondo and A. H. Maki, *J. Chem. Phys.*, 1969, **50**, 3270; [b] Y. Gondo and A. H. Maki, *J. Chem. Phys.*, 1969, **50**, 3638.
[150] J. Schmidt and J. H. van der Waals, *Chem. Phys. Letters*, 1969, **3**, 546.
[151] M. K. Orloff and J. S. Brinen, *Internat. J. Quantum Chem.*, 1969, **3**, 225.
[152] T. Azumi, M. Ito, and S. Nagakura, *Bull. Chem. Soc. Japan*, 1969, **42**, 685.
[153] R. E. Gerkin and A. M. Winer, *J. Chem. Phys.*, 1969, **50**, 3114.
[154] T. Azumi, and S. Nagakura, *Bull. Chem. Soc. Japan*, 1969, **42**, 2203.
[155] D. M. Burland and G. Castro, *J. Chem. Phys.*, 1969, **50**, 4107.
[156] M. Schwoerer and H. Sixl, *Z. Naturforsch.*, 1969, **24a**, 952.
[157] R. H. Clarke, R. M. Hochstrasser, and C. J. Marzzacco, *J. Chem. Phys.*, 1969, **51**, 5015.
[158] A. L. Shain, J. P. Ackerman, and M. W. Teague, *Chem. Phys. Letters*, 1969, **3**, 550.
[159] J. P. Le-Fahler, J. P. Lemaistre, and P. Kottis, *Chem. Phys. Letters*, 1970, **4**, 491.
[160] H. Baessler, G. Loelkes, and G. Vaubel, *J. Chem. Phys.*, 1969, **51**, 3695.
[161] M. R. Philpott, *J. Chem. Phys.*, 1969, **50**, 3925.
[162] G. Briegleb and D. Wolf, *Angew. Chem.*, 1970, **82**, 179.
[163] H. S. Judeikis and S. Siegel, *J. Phys. Chem.*, 1969, **73**, 2036.
[164] I. G. Batekha, M. V. Alfimov, V. I. Gordeev, and U. B. Shekk, *Izvest. Akad. Nauk. S.S.S.R., Ser. fiz.*, 1970, **34**, 675.
[165] T. N. Misra, *J. Chem. Phys.*, 1969, **51**, 2386.
[166] L. R. Faulkner and A. J. Bard, *J. Amer. Chem. Soc.*, 1969, **91**, 6495.

Parker has derived the following equation for the intensity of directly excited delayed fluorescence under steady-state illumination.

$$I_{DF} = (1/2)\, \Phi_f\, k_a (I_a\, \Phi_t \tau_p)^2$$

where Φ_f is the fluorescence efficiency, k_a is the annihilation rate constant, I_a is the rate of light absorption, Φ_t is the triplet formation efficiency, and τ_p is the triplet lifetime. This equation shows five quantities which could be field-dependent, giving rise to the observed effect on delayed fluorescence. Three of the quantities (I_a, Φ_f, and Φ_t) are related to physical properties of a diamagnetic species, and would not be expected to be grossly field-dependent. Indeed, the recent observation that magnetic fields do not

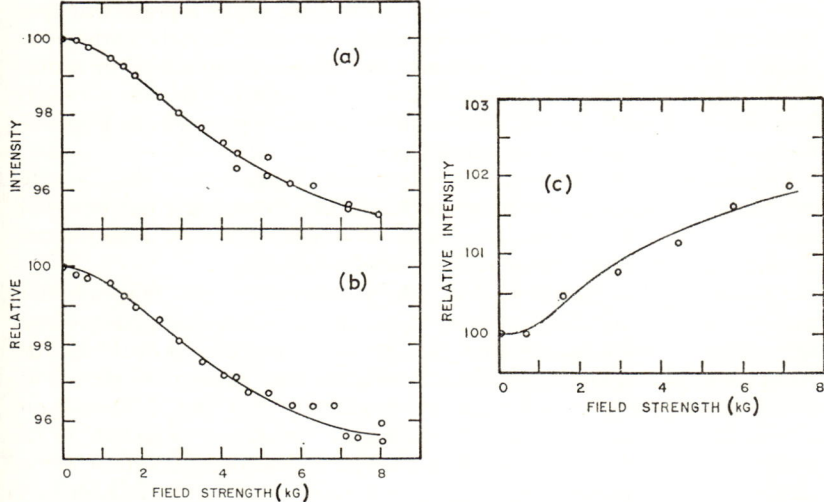

Figure 12 *Magnetic field effect on delayed fluorescence from anthracene solutions in DMF:* (a) 5×10^{-4} mol l^{-1}, *delayed fluorescence lifetime* 2·1 ms; (b) 7×10^{-5} mol l^{-1}, *delayed fluorescence lifetime* 6·3 ms: (c) *from a methylene chloride solution* $1·8 \times 10^{-7}$ mol l^{-1} *in Wurster's blue perchlorate and* 8×10^{-5} mol l^{-1} *in anthracene*
(Reproduced by permission from *J. Amer. Chem. Soc.*, 1969, **91**, 6495, 6497)

affect the intensity of prompt fluorescence from crystalline anthracene indicates that I_a and Φ_f are probably both field-independent. Of the three remaining possibilities, none can be measured accurately enough to be tested directly, and one must rely on indirect evidence to advance further.

Fortunately, the two curves shown in Figure 12 furnish some very useful information concerning possible field effects on τ_p. In solutions having an anthracene concentration greater than about 10^{-4} mol l^{-1}, anthracene triplet lifetimes are drastically shortened by either self- or impurity-quenching. At concentrations below 10^{-4} mol l^{-1} the lifetime is essentially constant. Any field effect on the lifetime would probably be concentration-dependent in the range where the quenching term begins to be important toward determining τ_p. Thus the magnitude of a field effect on delayed fluorescence arising from this source would also be concentration-dependent in the same range. Since experimentally the field variation is independent of concentration in the range where the lifetime begins to change markedly, it seems that the lifetime is not significantly field-dependent.

The results of an experiment dealing with sensitized anthracene delayed fluorescence provided a useful key to the question of a field-influenced Φ_t. In this experiment, a DMF solution 7×10^{-4} mol l^{-1} in phenanthrene and 5×10^{-5} mol l^{-1} in anthracene was illuminated frontally. The field effect on delayed fluorescence of anthracene generated in this way was identical with that shown in Figure 12. Since the anthracene triplet is populated here by energy transfer rather than by intersystem crossing from the anthracene singlet, Φ_t is not a factor governing the delayed fluorescence intensity. Since the field dependence of emission in this experiment was identical with that for directly excited delayed fluorescence, one has strong evidence for a field-independent Φ_t for anthracene.

Eliminating these factors leaves only a field-influenced annihilation rate constant to account for the magnetic-field effect. Thus the effect reported here is apparently the solution-phase analogue to the field-dependent annihilation rate observed in earlier solid-state studies.

This conclusion infers several others concerning the fluid-solution triplet–triplet annihilation process. First, the fact that one observes any field effect at all indicates that the encounter reaction probability for populating the first excited singlet state of anthracene is less than unity, and it is field-dependent. A second implication is that the fluid-solution process differs somewhat from the solid-state interaction, especially in its behaviour toward low fields. Still another interesting aspect of these studies is the observation that there is a solvent effect on the intensity *vs.* field strength curve. Additional studies carried out in methylene chloride and in cyclohexane have shown that the solvent markedly affects both the shape of the curve and the magnitude of the effect. Hitherto, the solvent has not been regarded as an important factor in the annihilation process. However, these results imply that the solvent plays a more significant role than has been previously supposed.

Wurster's blue cation is an effective quencher of anthracene triplet, as is shown by the shortened lifetimes in the presence of the radical. As an example, a methylene chloride solution containing initially 1.8×10^{-7} mol l^{-1} Wurster's blue perchlorate and 8×10^{-5} mol l^{-1} anthracene showed a delayed fluorescence lifetime of 1·4 ms, thus a triplet lifetime of 2·8 ms. This figure is much shorter than the 6·4 ms triplet lifetime recorded for a solution containing the same anthracene concentration without Wurster's blue cation. The results of a study of magnetic field effects on delayed fluorescence from the solution discussed above are shown in Figure 12(*c*).

It should be noted from the Figure that the field enhances delayed fluorescence. When these results are compared to those obtained with methylene chloride solutions containing only anthracene (Figure 12), it can be concluded that the magnetic field inhibits radical quenching of anthracene triplets. Moreover, the longer lifetimes which result from this situation more than compensate for the effect of the field on the annihilation rate; thus there is a positive change in the intensity as the field is applied.

The fluorescence efficiency of tetracene single crystals may be enhanced by as much as 39% in a magnetic field ($H > 2$ kG).[167] The enhancement is anisotropic with respect to the orientation of H in the ab plane. It was shown that a magnetically sensitive coupling of singlet states to a double-triplet-exciton state (T_1T_1) at ~ 2.40 eV is an important channel for radiationless decay in crystalline tetracene above 160 K, with rate constant $\gamma_S = (1.5 \pm 0.07) \times 10^{-12}$ cm^3 molecule^{-1} s^{-1}. The efficiency of this process at room temperature is estimated as 95%, and constitutes an efficient intersystem crossing mechanism. At light intensities $I \gtrsim 10^{15}$ quanta cm^{-2} s^{-1} (366 nm excitation), the triplet densities at 300 K are sufficiently high to produce radiative triplet–triplet annihilation or fusion. As the light intensity is increased, the quantum efficiency of fluorescence (Φ_f) increases, and eventually reaches a constant value (about twice its value in the low-intensity region, where fusion is not important). It is assumed that triplet–triplet fusion gives rise to either an excited singlet [rate constant $\gamma_{TS} = (4.8 \pm 1.2) \times 10^{-10}$ cm^3 molecule^{-1} s^{-1}], or excited triplet [rate constant $\gamma_{TT} = (11 \pm 5) \times 10^{-10}$ cm^3 molecule^{-1} s^{-1}]. The radiative triplet–triplet fusion constant and γ_S both decrease in approximately the same manner when the magnetic field is applied. γ_{TT} is at most slightly dependent on H.

The quenching of triplet β-chloropropiophenone and acetophenone by 1,4-cyclohexadiene without chemical reaction and without involving electronic energy transfer has been reported,[168] and attributed to conversion of the electronic energy to vibrational energy, a collision-induced intersystem crossing. The quenching of triplet 2-pentanone by penta-1,3-diene leads to a rate-constant evaluation of 6.4×10^7 s^{-1} for the Norrish type II reaction from the triplet state in the gas phase, which is different from that obtained in solution, although the latter results were obtained at very high conversions of starting material. Clearly, kinetic analysis under such circumstances is difficult and should be avoided when possible. A re-evaluation of the rate constant for the Norrish type II reaction using low conversions has confirmed the gas-phase result [169] (assuming a diffusion-controlled rate constant for quenching of triplet 2-pentanone by the olefin). This observation is a useful reminder that kinetic analysis should be used

---

[167] M. Pope, N. E. Geactinov, and F. Vogel, *Mol. Crystals and Liq. Crystals*, 1969, **6**, 83.
[168] A. M. Braun, W. B. Hammond, and H. G. Cassidy, *J. Amer. Chem. Soc.*, 1968, **91**, 6196.
[169] C. H. Bibast, M. G. Rockley, and F. S. Wettack, *J. Amer. Chem. Soc.*, 1969, **91**, 2802.

ideally only when reactions are allowed to proceed to low conversions. Further complications in the use of triplet quenchers have been reported for the benzophenone–piperylene system,[170] where quantum yields of *cis–trans* isomerization of the olefins well in excess of unity were obtained at high piperylene concentrations. It was suggested that equation (20) or the diradical addition of triplet piperylene to its ground state [equation (21)] might account for these results. This type of chain reaction may be more general than is supposed at present

$$^3P^* + cis\text{-}P \longrightarrow trans\text{-}P + {}^3P^* \qquad (20)$$

$$^3P^* + cis\text{-}P \longrightarrow [\text{diradical addition}]$$

$$\downarrow$$

$$\alpha\ cis\text{-}P + (2-\alpha)\ trans\text{-}P \qquad (21)$$

where P = piperylene and $\alpha < 1$. The triplet state of tetramethyl *p*-phenylenediamine (10^{-5} molar in aliphatic hydrocarbons) can be made to eject an electron by light of photon energy $\leqslant 3\cdot 86$ eV at room temperature.[171] Electron transfer from an excited triplet-state electron-donor to an acceptor molecule has been described,[172] and the tautomeric and protolytic properties of *o*-aminobenzoic acids in their lowest singlet and triplet states characterized.[173]

Miscellaneous.—The triplet energy levels of some common photosensitizing dyes have been given,[174] and the location and assignment of the lowest triplet state of perylene discussed.[175] Calculations on the triplet states of aromatic hydrocarbons have been performed,[176a] and the triplet states of triphenylamine,[176b] orotic acid and its methyl ester,[177] and dicoumarol[178] have been observed.

4 Electronic Energy Transfer

Electronic energy may be transferred in condensed phases *via* a trivial radiative exchange, *via* long-range resonance interactions, *via* a short-range exchange mechanism, or *via* exciton processes. The spin restrictions upon the first two mechanisms, of which the long-range Forster type is of paramount importance, are rigorous, and in the absence of strong spin–

---

[170] R. Hurley and A. C. Testa, *J. Amer. Chem. Soc.*, 1970, **92**, 211.
[171] N. Houser and R. C. Jarnagin, *J. Chem. Phys.*, 1970, **52**, 1069.
[172] G. Briegleb and H. Schuster, *Angew. Chem.*, 1969, **81**, 790.
[173] A. Tramer, *J. Phys. Chem.*, 1970, **74**, 887.
[174] R. W. Chambers and D. R. Kearns, *Photochem. and Photobiol.*, 1969, **10**, 215.
[175] R. H. Clarke and R. M. Hochstrasser, *J. Mol. Spectroscopy*, 1969, **32**, 309.
[176a] M. J. S. Dewar and N. Trinajstic, *Chem. Comm.*, 1970, 646; [b] G. C. Terry, V. E. Uffindel, and F. W. Willets, *Nature*, 1969, **223**, 1050.
[177] R. W. Yip, W. D. Riddell, and A. G. Szabo, *Canad. J. Chem.*, 1970, **48**, 987.
[178] S. J. Gull, D. R. Graber, and A. Haug, *Photochem. and Photobiol.*, 1969, **10**, 139.

orbit perturbation generally mean that this phenomenon is restricted to singlet–singlet transfer. The exchange (collisional) mechanism is very common, and many examples usually involving triplet–triplet transfer will be considered. Several examples of electronic energy transfer have already been discussed, and remaining examples will be grouped according to type.

Singlet–Singlet Transfer.—In both the trivial mechanism of energy transfer (fluorescence of donor followed by absorption of acceptor) and the Forster long-range mechanism, spectral overlap between donor emission and acceptor absorption is a necessary and rate-determining feature of the phenomenon. This aspect of singlet–singlet energy transfer has been discussed recently.[179] Singlet energy transfer has also been observed in benzene and toluene at 90 K.[180] Excitation energy transfer (exciton model) in alkanes and aliphatic polymers has already been mentioned,[2] as has possible singlet energy transfer from S_3 excited states of aromatics.[5]

It has been reported[181] that singlet oxygen can excite the fluorescence of rubrene (R) in solution, even though it reacts with it. With $E_s \approx 53$ kcal (272·6 kJ), 8 kcal (33·6 kJ) is needed besides the energies of two $^1O_2(^1\Delta_g)$. This 'sensitized delayed fluorescence' is found to be markedly temperature dependent. The results are compatible with the following mechanism:

$$R + {}^1O_2 \longrightarrow {}^3R + {}^3O_2 \qquad (22)$$

$$^3R + {}^1O_2 \longrightarrow {}^1R^* + {}^3O_2 \qquad (23)$$

$$^1R^* \longrightarrow R + h\nu \qquad (24)$$

$$^3R \longrightarrow R \qquad (25)$$

$$R + {}^1O_2 \longrightarrow RO_2 \qquad (26)$$

Under conditions of steady flow, the concentration of $^1O_2 \approx$ constant. By assuming steady states of [$^1R^*$] and [^3R], and that $k_{25} \gg k_{23}[^1O_2]$, one gets:

$$I = k[{}^1O_2]^2 [R] = kI_{635}[R] \qquad (27)$$

where I_{635} = intensity of 635 nm emission band of $^1O_2$, whereas integration of the first-order decay of R during a standard $t = 60$ s run gives:

$$\log \frac{[R]_0}{[R]_f} = k_{26}[{}^1O_2]\frac{t}{2\cdot 3} \propto \sqrt{I_{635}} \qquad (28)$$

Direct excitation of $^1R^*$ by a 'dimol' $(^1O_2)_2$ would yield (27) and (28) also; see Figure 13. The results rule out a triplet–triplet annihilation step $^3R + {}^3R \rightarrow {}^1R^* + R$, which would yield a second-order dependence of I on [R]. In order for steps (22) and (23) to yield singlet excited rubrene,

---

[179] R. P. Haugland, J. Yguerabide, and L. Stryer, *Proc. Nat. Acad. Sci. U.S.A.*, 1969, **63**, 23.
[180] A. Nakahara, M. Koyonagi, Y. Murakami, T. Edamura, and Y. Kanda, *Bull. Chem. Soc. Japan*, 1969, **42**, 3073.
[181] T. Wilson, *J. Amer. Chem. Soc.*, 1969, **91**, 2387.

thermal activation must occur during either of these reactions or else during both. It is not possible to decide whether a two-step excitation of $^1R^*$ occurs or whether a 'dimol' $(^1O_2)_2$ is responsible for the excitation, but the fact that heavy-atom solvents such as iodobenzene perturb the system supports the intermediacy of triplet rubrene. The role of energy transfer in the chemiluminescence seen in peroxide decompositions has been discussed.[182]

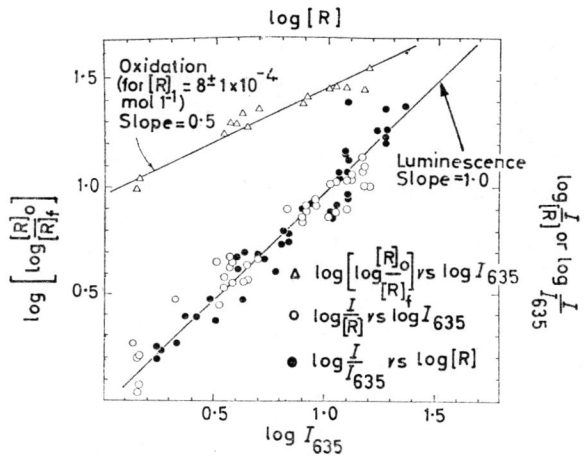

Figure 13 *Effect of $^1O_2(^1\Delta_g)$ and rubrene concentration on the luminescence intensity and rate of oxidation*
(Reproduced by permission from *J. Amer. Chem. Soc.*, 1969, **91**, 2387)

Singlet–Triplet Transfer.—The system by which exchange of electronic energy between donors and acceptors of different spins is described is misleading, since the conventional 'triplet–triplet' transfer involves exchange between excited triplet and ground-state singlet states, with no net change in spin in the system. However, since the term 'triplet–triplet' transfer is now widely used in the sense already explained, the usage will be retained here, and examples which do not fall into this category will be termed singlet–triplet transfer.

Reaction (22) above is an example of energy transfer between two singlet states which produce two triplet states, and could thus be termed a singlet–triplet energy transfer. Other examples involving molecular oxygen have been recorded. Thus, the photoperoxidation of 9,10-dimethylanthracene and of 9,10-dimethyl-1,2-benzanthracene sensitized by azulene, anthracene, and perylene has been investigated as a function of dissolved oxygen concentration in benzene at 25 °C.[183] The analysis of the data presented confirms previous findings that the singlet oxygen molecule, which is

---

[182] S. R. Abbott, S. Ness, and D. M. Hercules, *J. Amer. Chem. Soc.*, 1970, **92**, 1128.
[183] B. Stevens and B. E. Algar, *J. Phys. Chem.*, 1969, **73**, 1711.

almost certainly the intermediate in photosensitized peroxidation, is produced solely by energy transfer from the sensitizer triplet state, even when transfer from the excited singlet state is spin-allowed and exothermic. Other examples have been given.[184-186]

Triplet–Triplet Transfer.—Examples of triplet–triplet transfer are very widespread. Some of the finer details of this process, in particular donor–acceptor pairs, are now being probed. The conservation of spin direction[187] in T–T transfer could be used to prepare triplet molecules of the acceptor in a state of spin alignment or spin polarization (*i.e.* with the populations of the three zero-field levels unequal). Using an intramolecular mechanism involving direct S–S absorption followed by the intersystem crossing process, the state of spin alignment in the triplet state of aromatic molecules has been observed for quinoxaline and a number of other molecules.

The experiment can be done if a donor–acceptor system is found that satisfies the following conditions:

(i) the triplet state of the donor formed by direct absorption is in a known spin-aligned state;
(ii) the absorption of the donor occurs at lower energy than that of the acceptor;
(iii) the experiment is carried out at a temperature at which the electron spin–lattice relaxation (s.l.r.) processes are slower than both the transfer to, and the radiative processes of, the acceptor;
(iv) a convenient method is used to detect the net spin direction of the triplet state of the acceptor.

The system quinoxaline (donor) in naphthalene ($C_{10}H_8$) crystal at 1·6 K is found to satisfy the above conditions. The presence of quinoxaline (10^{-3} mol l^{-1}) in a $C_{10}H_8$ crystal causes the neighbouring naphthalene molecules to 'defect' from the rest of the $C_{10}H_8$ host, and then act as traps for triplet energy. The x-trap emission is observed upon exciting the $C_{10}H_8$ host (3130 Å) or quinoxaline molecules (3600 Å). No quinoxaline emission is observed in $C_{10}H_8$ host due to the T–T energy transfer to the $C_{10}H_8$ x-traps. In $C_{10}D_8$ host, however, quinoxaline shows strong emission due to the fact that its triplet energy is lower than that of the x-traps of the $C_{10}D_8$. The phosphorescence–microwave double-resonance (p.m.d.r.) technique was used to determine the net spin polarization direction for the triplet state of quinoxaline formed by direct excitation in $C_{10}D_8$, and for the triplet state of the acceptor ($C_{10}H_8$ x-traps) sensitized by quinoxaline in $C_{10}H_8$. In this system, the decay of the x-trap emission is found to be non-exponential, with a long-decay-time component. This indicates that the s.l.r. processes are slower than the emission. The saturation of all the

---

[184] V. L. Ermolaev and E. B. Sveshnikova, *Opt. i Spektroskopiya*, 1970, **28**, 601.
[185] V. M. Berenfeld, E. V. Chumaevskii, M. L. Grinev, U. I. Kuriatnikov, E. V. Artemev, and R. V. Dzhagatspanyan, *Izvest. Akad. Nauk S.S.S.R., Ser. fiz.*, 1970, **34**, 678.
[186] R. H. Kummler and M. H. Bortner, *Environ. Sci. Technol.*, 1969, **3**, 944.
[187] M. A. El-Sayed, D. S. Tinti, and E. M. Yee, *J. Chem. Phys.*, 1969, **51**, 5721.

microwave transitions involving the τ_z zero-field levels results in a 10—15% decrease in the intensity of the 0–0 band of the x-trap emission sensitized by exciting the quinoxaline donor. Since the $C_{10}H_8$ emission originates from the τ_z level, these results strongly indicate that the triplet state of the acceptor is formed in a spin-aligned state, with more spins being in the xy plane. The direction of the spin alignment is similar to that of the quinoxaline when formed by direct excitation (in $C_{10}D_8$). This indicates that the donor 'imprints' its initial (non-steady-state) spin polarization onto the triplet state of the acceptor. This observation can be shown to result from two important facts:

(i) In T–T energy transfer, the spin direction of the donor is preserved.

(ii) The probability that either the donor or the acceptor molecule will occupy one of the two different non-equivalent sites of the naphthalene crystal is equal.

The usefulness of this technique in preparing molecules in a spin-polarized state is obvious. The study of the changes in the emission spectrum of molecules in a spin-polarized state upon the application of different types of perturbations is extremely useful for the understanding of the phosphorescence mechanisms of these molecules.

The exchange interaction necessary for the non-radiative triplet–triplet transfer of energy is a short-range interaction sensitive to the overlap of the electronic wave-functions of the participating molecules. Hence, there can be an optimum geometry for transfer between the sensitizer–acceptor pair.

A new technique[188] has been used to investigate this dependence. It involves the excitation of the sensitizer molecules with polarized light and the measurement of the polarization of the triplet–triplet absorption in the acceptor molecules. Thus, a knowledge of the transition directions for the absorbing sensitizer molecules ($S_0 \rightarrow S_1$) and the acceptor molecules ($T_1 \rightarrow T_2$) is required. The $T_1 \rightarrow T_2$ polarization in the acceptor is determined relative to the $S_0 \rightarrow S_1$ in the sensitizer transition. Knowing the molecular directions of these transitions, therefore, yields information about the relative orientation of the sensitizer and acceptor molecules.

In order to completely determine the relative orientation for general pairs of molecules, the polar co-ordinates θ and ϕ between the two molecules must be determined. This may be done by performing measurements involving two orthogonal transitions in the singlet system of the sensitizer relative to a triplet–triplet absorption in the acceptor. An equivalent alternative is the polarization measurements of two orthogonal transitions in the triplet system of the acceptor relative to a singlet transition in the donor. In the work mentioned below, the measurement of the $T_1 \rightarrow T_2$ transition in the acceptor relative to the $S_0 \rightarrow S_1$ sensitizer absorption is supplemented by the polarization measurement of the acceptor's phosphorescence relative to the sensitizer's $S_0 \rightarrow S_1$ transition.

[188] K. B. Eisenthal, *J. Chem. Phys.*, 1969, **50**, 3120.

The systems were studied in EPA glasses at 77 K. The polarization of the triplet–triplet absorption is $(\varepsilon_\| - \varepsilon_\perp)/(\varepsilon_\| + \varepsilon_\perp)$ where $\varepsilon_\|$ and $\varepsilon_\perp$ are the molar absorption coefficients respectively, parallel and perpendicular to the electric field of the excitation light.

The sensitizer–acceptor pairs studied were benzophenone–[^{2}H$_{10}$]phenanthrene and anthrone–[^{2}H$_{10}$]phenanthrene. The sensitizers were selected for the following reasons. First, their triplet excitation is localized in the carbonyl group and, therefore, might be more sensitive to orientation with respect to the phenanthrene acceptor. Secondly, steric effects could require different orientations for the sensitizers since benzophenone is non-planar and anthrone is nearly planar. Thirdly, information about the benzophenone–phenanthrene transfer is available. Lastly, the transition directions for $S_0 \rightarrow S_1$ in the sensitizer molecules and the $T_1 \rightarrow T_2$ transition in phenanthrene are known.

The polarization of the phenanthrene phosphorescence is found to be negatively polarized, $P = -10\% \pm 1\%$ with respect to the $S_0 \rightarrow S_1$ (n, π^*) transition in benzophenone. Since the $S_0 \rightarrow S_1$ (n, π^*) transition in benzophenone is parallel to the C=O axis, and phenanthrene phosphorescence is perpendicular to its own molecular plane, the C=O axis of benzophenone must lie parallel to the molecular plane of phenanthrene. However, the angular orientation of the C=O axis with respect to the short and long axes of phenanthrene cannot be determined from the polarization of the phenanthrene phosphorescence. The angular geometry of the benzophenone and phenanthrene can be determined by measurement of the polarization of the T–T absorption in phenanthrene with regard to the C=O axis, since the T–T transition is in the molecular plane of phenanthrene along its long axis. Thus, if the C=O axis is parallel to the long axis of phenanthrene, the polarization will be positive; if parallel to the short axis, the polarization will be negative. If there is no preferred angular orientation, the T–T polarization will be zero. The measured polarization of the T–T absorption is found to be zero, thus indicating that there is no preferred angular orientation. These experiments have thus determined the configuration of the benzophenone–phenanthrene pair involved in triplet–triplet transfer.

The $S_0 \rightarrow S_1$ (n, π^*) transition in the sensitizer anthrone is primarily perpendicular to the C=O axis, and in the plane of the molecule. The polarization of the phenanthrene phosphorescence with respect to the $S_0 \rightarrow S_1$ (n, π^*) transition in anthrone is $P = -8\% \pm 1\%$. Thus, the molecular planes of anthrone and phenanthrene are parallel. The polarization of the T–T absorption in phenanthrene with respect to the $S_0 \rightarrow S_1$ (n, π^*) transition in anthrone is likewise found to be zero. Thus, the anthrone–phenanthrene molecules involved in triplet–triplet transfer are found to have the following orientation: (i) The molecular planes are parallel; (ii) there is no preferred angular orientation of the molecules in the parallel configuration.

The similar orientations found for the non-planar benzophenone–phenanthrene pair to those of the planar anthrone–phenanthrene pair indicate that only the orientation of the C=O group with respect to the phenanthrene plane is important. Although the sensitizer–acceptor distance must be small for exchange transfer to occur, the molecules are not required to be so close that the planarity of the phenyl groups of the sensitizer plays a significant role in the transfer geometry of the molecular pair.

Orientation effects have been further studied in a molecule containing a tetralin-1,4-dione donor chromophoric group and a fluorene acceptor group held rigidly in the framework of the same molecule [189] (see below).

(18)

Absorption in the tetralindione singlet manifold did not produce any emission in the fluorene singlet system, but strong sensitized phosphorescence in this moiety was observed, indicating that $83\% \pm 5\%$ of triplet energy in the donor species was transferred to the acceptor. Order of magnitude calculations using Dexter's expression gave reasonable agreement with experiment for the probability of energy transfer in this system.

Aromatic carbonyl compounds with low-lying $\pi\pi^*$ triplet states exhibit long triplet lifetimes and low reactivity in photoreduction and Norrish type II cleavage reactions. These properties and the high energies of the triplet states of these compounds make them quite attractive as sensitizers in processes in which chemical reactions of $^3(n, \pi^*)$ sensitizers are undesirable.

A recent report has shown that such sensitizers must be used with great care as mechanistic probes, since the yields of some photosensitized reactions were shown to be strongly dependent upon the concentration of sensitizer.[190] This clearly implies that some form of self-quenching is of importance, and mechanisms previously proposed for reactions based on the use of high concentrations of such sensitizers may be invalid. The sensitizers concerned are p-methoxyacetophenone, m-methoxyacetophenone, 3,4-methylenedioxyacetophenone, and thioxanthone.

[189] N. Filipescu, J. R. DeMember, and F. L. Minn, *J. Amer. Chem. Soc.*, 1969, **91**, 4169.
[190] O. L. Chapman and G. Wampfler, *J. Amer. Chem. Soc.*, 1969, **91**, 5390.

Energy transfer in molecular collisions [191] and *via* virtual phonon processes [192] have been discussed. Energy transfer can be used, as in the gas phase, to measure radiationless transition rates. Thus the intersystem crossing probability, Φ_t^A for fluorescein, eosine, erythrosine, methylene blue, and thionine in ethanol (only for fluorescein, 30 vol % alkaline aqueous ethanol solution) has been determined by comparing the quantum yield of their direct (Φ) and sensitized (Φ^s) photoreduction, on the basis of the following relations,[193]

$$\Phi = \Phi_t^A \beta; \quad \Phi^s = \Phi_t^D \gamma \beta$$

where β is the fraction of triplet dye molecules that are reduced eventually, Φ_t^A and Φ_t^D are respectively for acceptor (dye) and donor (sensitizer) and γ is the energy transfer efficiency, which was assumed as 1. Allyl thiourea was used as a reducing agent. 1,2:5,6-Dibenzanthracene, triphenylene, anthracene, and β-acetonaphthone were used as sensitizers. It has been confirmed that the above relations and assumption hold generally. Different sensitizers usually gave nearly the same Φ_t^A values (with the exception of thionine–anthracene. β-Acetonaphthone gave, in general, somewhat smaller values). The most probable Φ_t^A values obtained are; fluorescein (0·007), eosine 0·43 ± 0·04, erythrosine 1·1 ± 0·06, methylene blue 0·52 ± 0·03, thionine 0·62 ± 0·03. Using the above Φ_t^A values, the ratio between deactivation and genuine reaction between triplet dye and ATU has been evaluated. The ratio increases in the order, fluorescein < eosine < erythrosine, thionine < methylene blue.

Since phosphorescence of thionine has not been observed, the energy of the lowest triplet level of thionine has been determined by triplet–triplet energy transfer experiments using sensitizers of known triplet energy: acridine (E_t = 191 kJ mol⁻¹), eosine (170·6), 9,10-dibromoanthracene (169·3), hematoporphyrin (155·4), chlorophyll *b* (138·6), chlorophyll *a* (120·1).[193] Energy transfer from these sensitizers to thionine has been studied by measuring the bleaching reaction of thionine with allylthiourea in unbuffered methanol. The upper limit for the triplet energy of the basic form of thionine (^{3}TH⁺) is marked by the triplet energy of 9,10-dibromoanthracene (E_t = 169·3 kJ mol⁻¹) from which energy transfer is still obtained, while the lower limit is marked by the triplet energy of hematoporphyrin (E_t = 155·4 kJ mol⁻¹) which quenches the bleaching reaction. The triplet energy of ^{3}TH⁺ is thus rated at 163·8 kJ mol⁻¹ (136 400 ± 550 cm⁻¹). Involvement of reactions other than triplet–triplet energy transfer was excluded on the basis of fluorescence measurements. This result was confirmed using flash experiments.

---

[191] R. D. Levine, *Israel J. Chem.*, 1969, **7**, 237.
[192] R. G. Delosh and W. J. C. Grant, *Phys. Rev. B.*, 1970, **1**, 1754.
[193a] N. Nemoto, H. Kokubun, and M. Koizumi, *Bull. Chem. Soc. Japan*, 1969, **42**, 1223;
  b E. H. A. Kramner, M. Hafner, and M. Zugel, *Z. phys. Chem. (Frankfurt)*, 1969, **65**, 276.

The quenching of biacetyl phosphorescence in solution by diphenylpicrylhydrazyl, bis-galvinoxyl, *trans*-stilbene, ferric acetylacetonate, galvinoxyl, azulene, or ferrocene, proceeds with quenching rate-constants of 5·5, 5·5, 4·0, 3·1, 2·9, 2·4, and 2.4×10^{-9} l mol^{-1} s^{-1} respectively.[194] The rate constants for cyclohexene and biacetyl are orders of magnitude smaller. Of passing interest in the study of diketones is the observation (based on phosphorescence excitation spectra) of an $S_2 \rightarrow T_2$ intersystem crossing in camphorquinone.[195]

Triplet transfer from benzophenone to aniline has been observed[196] and from biacetyl and carbazole to styrylpyridines,[197] and from a variety of donors to all-*trans*-retinal [triplet level 38 kcal (160 kJ) mol^{-1}].[198]

Electronic energy transfer from triplet acridine to paramagnetic ions,[199] from triplet aromatic ketones to rare-earth ions,[200] in rare-earth chelates,[201] and in actinide(III)–β–diketone chelates[202] has been observed. The latter study involves intramolecular energy transfer. Intramolecular singlet energy transfer has been established in 1,4-dimethoxy-5,8-methano-6,7-*exo*[fluorene-9′-spiro-1″-cyclopropane]naphthalene,[203] in contrast to earlier reports that this did not occur.[204]

In the ternary system benzene–2,5-diphenyloxazole (PPO)–2,2-*p*-phenylene-bis-(5-phenyloxazole) (POPOP), both benzene and PPO molecules transfer their energy to POPOP, and the overall transfer efficiency of benzene to POPOP in the presence of PPO depends on the transfer efficiencies for the energy transfer processes in the binary systems benzene–PPO and benzene–POPOP. The effect of PPO–POPOP radiative transfer is discussed.[205]

In crystals, transfer from host (biphenyl) to guest (phenanthrene, naphthalene, and chrysene) has been studied,[206] and the perylene–*N*-isopropylcarbazole system[207] has also been investigated.

Finally, energy transfer from excited gaseous molecules to solids has been reported,[208] the influence of excimer formation on energy transfer

[194] R. B. Cundall, C. B. Evans, and E. J. Lance, *J. Phys. Chem.*, 1969, **73**, 3892.
[195] L. Tsai and E. Charney, *J. Phys. Chem.*, 1969, 73, 2462.
[196] M. Santhanam and V. Ramakrishnan, *Chem. Comm.*, 1970, **344**.
[197] G. Favaro, F. Masetti, and U. Mazzucato, *Z. phys. Chem. (Frankfurt)*, 1969, **66**, 206.
[198] A. V. Guzzo and G. L. Pool, *J. Phys. Chem.*, 1969, **73**, 2512.
[199] D. L. Banfield and D. Husain, *Trans. Faraday Soc.*, 1969, **65**, 1985.
[200] V. L. Ermolaev and V. S. Tachin, *Opt. i Spektroskopiya*, 1969, **27**, 1007.
[201] A. P. Aleksandrov, E. P. Volkova, and V. N. Genkin, *Opt. i Spektroskopiya*, 1969, **27**, 439.
[202] I. J. Nugent, R. D. Baybarz, J. L. Burnett, G. K. Werner, S. P. Tanner, J. R. Tarrant, and O. L. Keller, *J. Phys. Chem.*, 1969, **73**, 1540.
[203] A. A. Lamola, *J. Amer. Chem. Soc.*, 1969, **91**, 4786.
[204] N. Filipescu, J. R. DeMember, and F. L. Morin, *J. Amer. Chem. Soc.*, 1969, **91**, 4169.
[205] J. C. Conte, *Trans. Faraday Soc.*, 1969, **65**, 2382.
[206] K. R. Adam and M. F. O'Dwyer, *Austral. J. Chem.*, 1969, **22**, 2085.
[207] W. Klopffer, *J. Chem. Phys.*, 1969, **50**, 1689.
[208] J. P. Dauchot, J. P. Verhaegen, and J. Van Cakenberghe, *Nature*, 1969, **223**, 824.

rates in the pyrene–perylene system discussed,[209] and triphenylene proposed as a singlet sensitizer.[210]

5 Physical Aspects of Photochemical Reactions in Solution

The demarcation between the physical and organic aspects of photochemistry is somewhat artificial, and it is inevitable in this volume that some overlap between the various parts will occur. Many of the mechanistic considerations, quantum yields, *etc.*, of photochemical reactions of organic molecules in solution will be found in the organic sections of this book, but a few cases will be stressed in this short section where the work is primarily of physical interest.

An account in review form has been given of the primary physical processes which can occur upon irradiation of aromatic compounds in solution.[211] The method of the photosensitized *cis–trans* isomerization of but-2-ene has been used in the liquid phase [212] to study the rates of S_1–T_1 intersystem crossing in liquid benzene and toluene. The triplet yields are 0·57 and 0·45 respectively. Addition of xenon increases both yields to unity, and rate constants for this catalytic process were measured as 2×10^8 l mol^{-1} s^{-1} and 3×10^8 l mol^{-1} s^{-1} respectively. The results imply that the internal conversion efficiencies of benzene and toluene in the liquid phase are 0·41 and 0·48 respectively. The photolyses of the xylenes (mainly *m*-xylene) have been studied at 248 and 275 nm in n-hexane, E.P.A. (a 5:5:2 mixture of diethyl ether, isopentane, and ethanol), and perfluorohexane solution.[213] The quantum yields of isomerization are independent of exciting wavelength, and independent of solvent, but increase with temperature. For the isomerization of *m*-xylene, $\Phi_p =$ 0·0032 and $\Phi_o =$ 0·00075. Both yields have an activation energy of approximately 4·7 kcal mol^{-1}, so that the $p:o$ ratio is independent of all conditions studied. It is not possible to say whether benzvalene- or prismane-type molecules are precursors of the isomerization products. The use of *cis–trans* isomerization of alkenes to measure triplet-state yields, and as a tool diagnostic of triplet states, is very widespread both in the gas phase and in solution. The mechanism involved is thus of great importance. Two mechanisms have been proposed: the Schenck mechanism involving formation of an adduct diradical (19), in which rotation about the central bond is rapid relative to bond breaking [equation (29)], and the triplet mechanism in which olefin triplets are formed by excitation transfer from the sensitizer. The triplet mechanism is favoured when the excitation transfer steps are exothermic. To account for isomerization in cases

---

[209] C. R. Goldschmidt, Y. Romkiewicz, and A. Weinreb, *Spectrochim. Acta*, 1969, **25**, *A*, 1471.
[210] A. B. Smith, tert. and W. C. Agosta, *Chem. Comm.*, 1970, 466.
[211] E. Lippert, *Accounts Chem. Res.*, 1970, **3**, 74.
[212] R. B. Cundall and W. Tippett, *Trans. Faraday Soc.*, 1970, **66**, 350.
[213] D. Anderson, *J. Phys. Chem.*, 1970, **74**, 1686.

$$\underset{(19)}{\overset{H}{\underset{R}{\overset{|}{C}}}\overset{R}{\underset{S^{\cdot}}{\overset{|}{C}}}} \longrightarrow S + \alpha\,{}^0t + (1-\alpha)\,{}^0c \qquad (29)$$

where the sensitizer triplet excitation energy is not sufficient to excite the olefins to planar (spectroscopic) triplet states, non-vertical excitation transfer leading directly to twisted olefin triplet has been suggested. For the case of carbonyl sensitizers with lowest $n\pi^*$ triplet states, formation of the Schenck intermediate has been proposed as a discrete step in the excitation transfer process [equation (30)]. The following observations

$$\underset{(20)}{R^1-\overset{\cdot}{\underset{R^2}{C}}\overset{O-C}{\underset{C^{\cdot}}{\diagdown}}} \longrightarrow R^1R^2C{=}O + {}^3[\text{olefin}] \qquad (30)$$

provide an experimental criterion for choosing between the Schenck and the triplet mechanisms.

The simplest general scheme for sensitized *cis–trans* photoisomerization is given below, where *X represents an unspecified common intermediate,

$$S \xrightarrow{h\nu} {}^1S \longrightarrow {}^3S \qquad (31)$$

$${}^3S \xrightarrow{k_{32}} S \qquad (32)$$

$${}^3S + {}^0t \xrightarrow{k_{33}} {}^*X \qquad (33)$$

$${}^3S + {}^0c \xrightarrow{k_{34}} {}^*X \qquad (34)$$

$${}^*X \xrightarrow{k_{35}} \alpha\,{}^0t + (1-\alpha)\,{}^0c \qquad (35)$$

and other symbols have their usual meanings. Steady-state approximations for ^3S and *X lead to equation (36), which represents the photostationary *trans* : *cis* ratio, and equations (37) and (38) give the dependence

$$([t]/[c])_s = (k_{34}/k_{33})\,[\alpha/(1-\alpha)] \qquad (36)$$

$$\frac{1}{\Phi_{t\to c}} = \frac{1}{1-\alpha}\left(1 + \frac{k_{32}}{k_{33}[t]}\right) \qquad (37)$$

$$\frac{1}{\Phi_{c\to t}} = \frac{1}{\alpha}\left(1 + \frac{k_{32}}{k_{34}[c]}\right) \qquad (38)$$

of *trans* → *cis* and *cis* → *trans* quantum yields on initial concentrations of *cis* and *trans* isomers.

The dependence of *trans* : *cis* photostationary ratios for several olefins on the triplet energies of sensitizers has been attributed entirely to changes in the excitation quotient k_{34}/k_{33}. Direct measurements of rate constants

k_{31} and k_{33} have confirmed this interpretation for the stilbenes and the 1,2-diphenylpropenes. For these olefin pairs, photostationary-state ratios could be predicted using a single decay ratio, $\alpha/(1-\alpha)$, in each case.

Benzene-sensitized photoisomerization of alkenes has been studied in the vapour phase and in solution. Stationary states for several alkene pairs are close to unity. Since triplet excitation transfer should be, in all cases, at least 8—12 kJ mol^{-1} exothermic, k_{31}/k_{33} is expected to be close to unity; hence, for *X = ^{3}[alkene], $\alpha/(1-\alpha) = 1 \cdot 0$.

With acetophenone and acetone as sensitizers of cis–trans isomerization in pent-2-enes, values of $\alpha/(1-\alpha)$ of 1·90 and 1·17 were obtained [214] and in contrast to the stilbenes and the 1,2-diphenylpropenes, a single decay ratio does not account for observations with different sensitizers in the case of the pent-2-enes. The variation of photostationary ratios of oct-2-enes and pent-2-enes with the triplet energy of sensitizers was incorrectly attributed solely to changes in excitation ratios for alkene triplet formation. Since a common decay ratio does not obtain, it is clear that different intermediates are produced with different sensitizers.

The triplet mechanism should be important with sensitizers whose triplet excitation energy is close to that of ethylene. With lower energy sensitizers, the triplet mechanism should diminish in importance except in cases where non-vertical excitation transfer can occur. Thus, for alkenes, deviation of the decay ratio from unity is a measure of the involvement of Schenck intermediates [equation (29)] in the photoisomerization. The small increase of the decay ratio obtained with acetone as sensitizer indicates that at 30 °C the Schenck intermediate is involved to a minor extent, and that triplet excitation transfer represents the major path for the isomerization. On the other hand, the large decay ratio increase obtained with acetophenone as sensitizer suggests that the Schenck mechanism predominates in this case. It may be more than a coincidence that for acetophenone and other $n\pi^*$ sensitizers with even lower triplet-state energies, the photostationary ratios are very close to the thermodynamic ratio. The mechanism accounts for the observations that triphenylene and some carbonyl compounds with lowest $\pi\pi^*$ triplet states are ineffective as sensitizers of the cis–trans isomerization of alkenes. The possibility that the abnormally trans-rich photostationary states obtained for the stilbenes with a few sensitizers were due to competing decay from Schenck intermediates is strong. The ratios $\alpha/(1-\alpha)$ for stilbene with 4-methylbenzophenone, α-(2,4,6-triethylbenzoyl)naphthalene, and β-(2,4,6-triethylbenzoyl)naphthalene were found recently to be 1·36, 1·38, and 1·31 respectively.[215]

cis–trans Isomerizations of this type are of great importance to theoreticians, physical and organic chemists, and it is hoped that a better understanding of the mechanism of such reactions will be developed in the near

---

[214] J. Saltiel, K. R. Neuberger, and M. Wrighton, J. Amer. Chem. Soc., 1969, **91**, 3658.
[215] H. A. Hammond, D. E. DeMayer, and J. L. R. Williams, J. Amer. Chem. Soc., 1969, **91**, 5180.

future. Other relevant studies include further discussion of mechanisms,[216] some stilbene analogues which do not cyclize photochemically to phenanthrene analogues,[217] the *cis–trans* isomerization of arylethylenes,[218] the isomerization of stilbene excited directly to its triplet state, in which the ratio $\Phi_{c \to t}/\Phi_{t \to c}[\alpha/(1-\alpha)]$ was found to be 1·27,[219] the photochemistry of phenylbut-2-ene, in which intramolecular and intermolecular energy transfer have been shown to play a large part,[220] and heavy-atom effects in the *cis–trans* isomerization of *m*-halogenostilbene derivatives.[221]

Excited states and intermediates of the conjugated cyclohexenone 7-keto-13-methyl-5,6,7,9,10,13-hexahydrophenanthrone have been studied flash spectrophotometrically.[222] Two transient species differing in spectral and decay characteristics were observed, the longer-lived being the ketyl radical formed by hydrogen-atom abstraction. The shorter-lived intermediate is presumed to be a triplet state, but is not the triplet precursor to the isomeric rearrangement or reduction reactions; this latter species was not observed directly, but was detected by triplet–triplet energy transfer to naphthalene. Neither directly-observed species was optically active within the limit of detection of the flash spectropolarimeter, although optical activity is retained in the isomeric photoproduct. Flash photolysis of solutions of camphorquinone in benzene, carbon tetrachloride, and isopropyl alcohol, and of solutions of biacetyl in carbon tetrachloride and isopropyl alcohol have also been studied.[223] The irradiations were limited to the visible absorption bands of the two diketones. The absorption spectra of the transients formed were measured at 200—1125 nm. The transient spectra in benzene and carbon tetrachloride solutions contain absorption bands, at ~ 320 and 630—1100 nm, which are unique to the triplet excited molecules of the diketones, and are attributable to (i) the $^3\pi^* \to {}^3\sigma^*$, and (ii) the $^3\pi^* \to {}^3\pi^*$ transitions, respectively. The camphorquinone triplet was photosensitized by benzophenone and quenched by anthracene and oxygen; the anthracene triplet was photosensitized by camphorquinone. Oxygen also quenched the biacetyl triplet. Both the diketones were photoreduced during flash photolysis of their isopropyl alcohol solutions. The spectra of the resulting hydrogen-adduct free radicals were found to be very similar to those of the corresponding triplets in the u.v. region, but the free radicals were much longer lived.

In the Norrish type II fragmentation of 3,4-diphenylbutyrophenone, if the products were formed in excited states, [equation (39)] then the stilbene molecule would be formed in its triplet state, with subsequent relaxation

[216] N. J. Turro, *Photochem. and Photobiol.*, 1969, **9**, 555.
[217] E. V. Blackburn and C. J. Timmons, *J. Chem. Soc. (C).*, 1970, 172.
[218] R. N. Nurmukhametov and G. I. Grishina, *Zhur. fiz. Khim.*, 1969, **43**, 1508.
[219] G. Fisher, K. A. Muszcat, and E. Fisher, *Israel J. Chem.*, 1968, **6**, 965.
[220] C. S. Nakagawa and P. Sigal, *J. Chem. Phys.*, 1970, **52**, 3277.
[221] K. Kruger and E. Lippert, *Z. phys. Chem. (Frankfurt)*, 1969, **66**, 293.
[222] G. Ramme, R. L. Strong, and H. H. Richtol, *J. Amer. Chem. Soc.*, 1969, **91**, 5711.
[223] A. Singh, A. R. Scott, and F. Sopchyshyn, *J. Phys. Chem.*, 1969, **73**, 2633.

to *cis*- and *trans*-isomers. In fact, the *trans*-isomer is preferentially formed [224] and this implies that the products are formed in their ground states from a long-lived diradical.

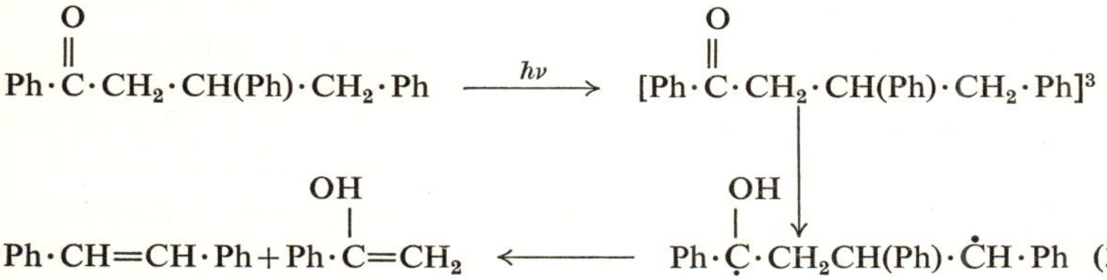

$$Ph \cdot \overset{O}{\underset{\parallel}{C}} \cdot CH_2 \cdot CH(Ph) \cdot CH_2 \cdot Ph \xrightarrow{h\nu} [Ph \cdot \overset{O}{\underset{\parallel}{C}} \cdot CH_2 \cdot CH(Ph) \cdot CH_2 \cdot Ph]^3$$

$$Ph \cdot CH=CH \cdot Ph + Ph \cdot \overset{OH}{\underset{|}{C}}=CH_2 \longleftarrow Ph \cdot \overset{OH}{\underset{|}{C}} \cdot CH_2 CH(Ph) \cdot \overset{\bullet}{C}H \cdot Ph \quad (39)$$

6 Photochromic Compounds

There is considerable industrial interest in the development of photochromic substances. One of the systems exhibiting this behaviour is the spiropyran type of compound. The absorption and emission spectra of indolinobenzospiropyrans (21) have been studied,[225] and transitions shown to occur largely in the chromene half of the molecule. In addition, intramolecular energy transfer between the two halves of the molecule was

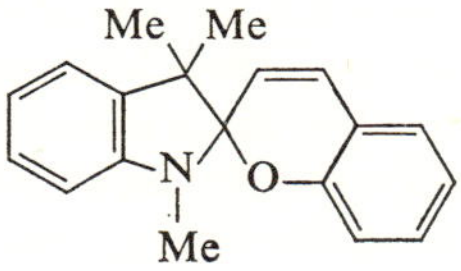

(21)

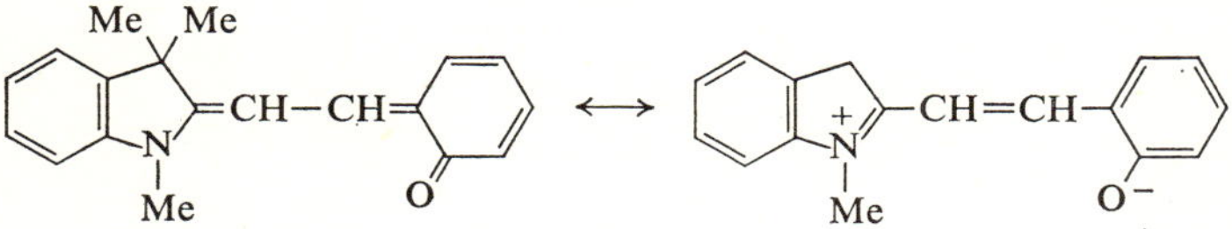

(22) ⟷ (23)

established. The coloured form of the compound produced photochemically is considered to be a resonance hybrid of the forms (22) and (23). Spiro[2,4]heptan-4-one and spiro[2,5]octan-4-one have been studied,[226] and other spiropyrans, including benzothiazole derivatives,[227] have been reported.[228] The possibility of using such substances as actinometers has been discussed.[229]

---

[224] R. A. Caldwell and P. M. Fink, *Tetrahedron Letters*, 1969, 2987.
[225] N. W. Tyer jun., and R. S. Becker, *J. Amer. Chem. Soc.*, 1970, **92**, 1289.
[226] J. K. Crandall and R. J. Seidewand, *J. Org. Chem.*, 1970, **35**, 697.
[227] K. G. Dzhaparidze, Z. M. Elashvili, and L. V. Devadze, *Soobshch Akad. Nauk Gruz. S.S.R.*, 1970, **57**, 77.
[228] K. G. Dzhaparidze, I. Y. Pavlenishvili, M. T. Gugava, and D. P. Maisuradze, *Zhur. fiz. Khim.*, 1970, **44**, 582.
[229] E. G. Akhalkatsi and L. P. Shishkin, *Soobshch Akad. Nauk. Gruz. S.S.R.*, 1969, **55**, 81.

The spectra and photochemistry of the different isomers of some anils of salicylaldehyde and hydroxynaphthaldehyde were investigated over a range of solvents and temperatures. By running flash experiments in fluid solutions, it was shown that both intramolecularly hydrogen-bonded enol (E) and *cis*-keto (Q_A) tautomers yield the *trans*-keto isomer (Q_C) as a common photoproduct.[230] In a rigid paraffin glass, only the E → Q_C, but not the Q_A → Q_C, photoisomerization can be induced. The dark Q_C → E relaxation in low-polarity systems was investigated and was shown to proceed *via* dimer intermediates. The flash photolysis of a keto dimer (Q_B) was also studied. Excitation of Q_B leads to dissociation to enol monomers. An additional path is the formation of a short-lived transient, presumably an excited triplet of the pair. The photochromic transients of bianthrone, 2,2'-dimethylbianthrone, 4,4'-dimethylbianthrone, and 2,2'-dibromobianthrone, as well as their photolysis products and the transients of the photoproducts, have been investigated at 20—55 °C in benzene solvent.[231] The photochromic state is produced *via* the excited singlet state and thermally decays by first-order kinetics (k = 3—11 s^{-1} at 25 °C) with E_a = 58—63 kJ mol^{-1}. The helianthrone and naphthodianthrone photoproducts have triplet transients which decay back to the ground state by second-order diffusion-controlled kinetics (k = 1·3—1·7 × 10^9 l mol^{-1} s^{-1}). 4,4'-Dimethylbianthrone is photochemically stable, and only reversibly cycles through the photochromic state.

Photochromism in 1,2-dihydroquinolines has also been reported,[232] as have the β-tetrachloroketonaphthalene system,[233] and the triphenylimidazolyl dimer system.[234]

Photochemical Reactions in the Solid State.—Photochemical reactions have been observed to proceed in frozen solutions at low temperatures, and photoreactions occurring on surfaces are also mentioned briefly.

The photolysis, using 253·7 nm radiation, of HI in thin films has been investigated at 77 and 85 K. Initially, the quantum yield Φ_{HI} = 2·0.[235] However, as reaction proceeds the rate falls off rapidly and, with films of sufficient thickness, reaction does not go to completion but approaches a photostationary state. This self-inhibition is attributed to the light-filtering action of some product species. A mechanism involving photolysis of polymeric aggregates is proposed. During photolysis, part of the hydrogen which has been formed diffuses out of the solid into the gas phase. The rate of evolution, unlike the overall rate of hydrogen formation, remains almost constant throughout photolysis but, below 90 K, evolution

[230] R. Potashnik and M. Ottenlenghi, *J. Chem. Phys.*, 1969, **51**, 3671.
[231] G. L. Dombrowski, C. L. Groncki, R. L. Strong, and H. H. Richtol, *J. Phys. Chem.*, 1969, **73**, 3481.
[232] J. Kolc and B. S. Becker, *J. Amer. Chem. Soc.*, 1969, **91**, 6513.
[233] E. Inque, H. Kokado, and S. Ohno, *Kogyo Kagaku Zasshi (J. Chem. Soc. Japan, Ind. Chem. Sect.)*, 1970, **73**, 435.
[234] K. Maeda and T. Hayashi, *Bull. Chem. Soc. Japan*, 1969, **42**, 3509.
[235] P. G. Barker, M. P. Halstead, and J. H. Purnell, *Trans. Faraday Soc.*, 1969, **65**, 2389.

stops immediately 253·7 nm irradiation is stopped. Thus, in this temperature region, diffusion is not a simple isothermal process but occurs as a result of local heating of the film by absorbed u.v. radiation. The diffusing species is molecular hydrogen.

If diffusion is slow or non-existent in a photochemical system where a large fraction of the incident light is absorbed and reaction is continued to high percentage conversion, account must be taken of both light intensity and consequent concentration gradients when deriving an expression for the rate of photolysis. This has been done in the above study.

Photolysis of HI in 3-methylpentane (3MP) glass at 77 K produces 'hot' hydrogen atoms, which abstract hydrogen from the matrix to form C_6H_{13} radicals, and thermal hydrogen atoms, which add to olefins present as dilute solutes. The quantum yield of thermal atoms, which is ca. 0·8 in liquid 3MP at 300 K, drops to ca. 0·1 at 77 K, as a result of increasing geminal recombination. Using e.s.r. detection, an investigation has been carried out to determine whether thermal hydrogen atoms can be produced by photolysis of HI in 3MP glass at temperatures below 77 K, and, if so, whether there are conditions under which they can be trapped.[236] Photolysis of HI (0·1 mol %) in [$^2H_{14}$]3MP gave pronounced lines of the ^1H doublet after 60 seconds of photolysis or less, at all temperatures tested between 20 and 50 K. On further photolysis, the two outer lines of the ^2H triplet appeared, the central line being obscured by the C_6D_{13} signal. The ^1H doublet is attributable to the trapping of thermal hydrogen atoms produced from HI photolysis, and the ^2H lines to trapping of thermal deuterium atoms produced by displacement of deuterium atoms from C_6D_{14} by 'hot' hydrogen atoms, or by photolysis of DI formed by disproportionation between an iodine atom and a radical with which it is caged ($C_6D_{13} + I \rightarrow C_6D_{12} + DI$). The radical is formed by attack by the 'hot' hydrogen partner atom split off from the iodine atom by photolysis of HI. Decay kinetics of radicals offer strong evidence for such caging.

At 30 K, the lines of the ^1H doublet grew continuously during a 10 minute photolysis, but at a decreasing rate. Initial decay of the ^1H signal in a sample which received a 30 second photolysis was about 50% in 1 minute, becoming progressively slower thereafter. Initial decay following longer illuminations was slower, indicating a higher fraction of more strongly trapped atoms. When a sample photolysed at 30 K was raised to successively higher temperatures, a relatively rapid initial decay followed by much slower decay occurred at each temperature.

Photolysis of HI in 3MP, under conditions identical with those used with the [$^2H_{14}$]3MP matrix, shows no evidence of the ^1H doublet. This indicates either that 3MP is incapable of trapping ^1H, in contrast to the ability of [$^2H_{14}$]3MP to trap both ^1H and ^2H, or that the protiated matrix reduces the escape of thermal hydrogen atoms formed from the primary photolytic process more than does the deuteriated matrix (by increasing the

[236] D. Timm and G. E. Willard, *J. Amer. Chem. Soc.*, 1969, **91**, 3406.

probability of the 'hot' reaction or of geminate recombination). There is an analogy for such an effect in the slow decay of trapped free radicals in deuteriated, as compared to protiated, matrices at 77 K. Both experimental observations and activation energy considerations exclude the possibility that the latter decay occurs by hydrogen abstraction from the matrix.

The photolysis at 254 nm of liquid allyl chloride at 27 °C and of solid allyl chloride at 77 K has been studied.[237] By the use of aqueous acetone as an actinometer, quantum yields for eighteen products containing six or fewer carbon atoms were determined. The primary process proposed is cleavage of the carbon–chlorine bond to give allyl radical, and a simple mechanism for the formation of the products is given. The trapping of the radical intermediate involved in this primary dissociation has been described.[238]

Ultraviolet and i.r. spectroscopy have been used to study the photochemistry of cyclo-octa-1,3,5-triene and bicyclo[4,2,0]octa-2,4-diene suspended in inert vapour matrices at 20 K.[239] *cis,cis*-Octa-1,3,5,7-tetraene, benzene, and ethylene were found to be common primary photoproducts of the two parent valence isomers. Photolysis of cyclo-octa-1,3,5-triene also yields another primary product. This product reverts to the starting material at room temperature, and has a vibrational spectrum which supports its assignment as a strained cyclic stereoisomer of the starting material.

The photolysis of *m*-disubstituted benzenes at low temperatures,[240] and an e.s.r. study of intermediates arising from the photochemical irradiation of benzene in organic solvents[241] have been carried out. Fluorescence has been observed from radicals generated by the photodissociation of ethyltoluenes enclosed in crystalline matrices,[242] and e.s.r. used to study photoreactions of radicals generated in polyhydric alcohols by γ-irradiation.[243] The photoionization of trapped radicals in the solid phase has also been described.[244]

The yields of alkyl radicals produced from alkyl halide solutes by dissociative electron capture $(RX + e^- \rightarrow R + X^-)$ during photoionization of tetramethyl *p*-phenylenediamine (TMPD) in 3-methylpentane (3MP) glass at 77 K decrease in the order $RCl > RBr > RI$.[245] Phosphorescence yields and lifetimes indicate that the reductions in radical yields from the iodides and bromides in the TMPD system result from the formation of TMPD–RX complexes, in which the lifetime of the TMPD triplet state is reduced.

[237] R. W. Phillips and D. H. Volman, *J. Amer. Chem. Soc.*, 1969, **91**, 3418.
[238] B. B. Jarvis and R. O. Fitch, *Chem. Comm.*, 1970, 408.
[239] P. Datta, T. D. Goldfarb, and R. S. Boikess, *J. Amer. Chem. Soc.*, 1969, **91**, 5429.
[240] A. Egawa, K. Kimura, and H. Tsubomura, *Bull. Chem. Soc. Japan*, 1970, **43**, 944.
[241] T. Tanei and H. Hatano, *Bull. Chem. Soc. Japan*, 1969, **42**, 3369.
[242] A. Pellois, J.-C. Navatte, and J. Ripoche, *Compt rend.*, 1969, **268**, *B*, 1134.
[243] I. E. Makarov and B. G. Ershov, *Izvest. Akad. Nauk. S.S.S.R., Ser. khim.*, 1970, 530.
[244] C. Chachaty, A. Forchioni, and J. Desalos, *Compt. rend.*, 1970, **270**, *C*, 449.
[245] W. G. French and J. E. Willard, *J. Phys. Chem.*, 1970, **74**, 240.

Shortening of the triplet-state lifetime lowers the probability of the two-photon photoionization process. The phosphorescence yields with and without CH_3I present indicate a $[TMPD-CH_3I]/[TMPD][CH_3I]$ ratio in the glass at 77 K of 760. These data help to explain earlier observations on the effect of methyl iodide on the electrical conductivity of solutions of photoionized TMPD in 3MP glass during warm-up from 77 K. Additional observations include: (a) the rate of production of free radicals in the TMPD system is proportional to the second power of the light intensity, as expected; (b) CH_3F undergoes dissociative electron capture in 3MP at 77 K, although such dissociation with thermal electrons is endothermic in the gas phase; and (c) the ratio of yields of CH_3, C_2H_5, and C_3H_7 from dissociative electron capture by the corresponding chlorides, using electrons from the photoionization of TMPD, is 1·00 : 1·25 : 1·75.

The wavelength dependence of the mode of decomposition of some acridans at 93 K by a two-quantum process has been investigated.[246]

Aromatic amino-acids (tryptophan, tyrosine) are ionized by u.v. radiation in frozen aqueous solutions containing various divalent salts.[247] Their triplet → triplet and radical cation absorption spectra have been recorded at 77 K. According to e.s.r. studies, electron photoejection occurs *via* a triplet–triplet absorption process. Photoejected electrons react with M^{2+} ions to give M^+ ions, or with H^+ to give hydrogen atoms. The nature and number of electron traps depend on the salt, as deduced from temperature and light effects. In the presence of $CdCl_2$, photoejected electrons are trapped in many heterogeneous sites, producing a broad e.s.r. signal. Radical ions (Cl_2^-) are also produced when chloride is present. When u.v.-irradiated frozen aqueous solutions of tryptophan containing $CdSO_4$ are further irradiated at longer wavelengths, translocation of trapped electrons is observed. Secondary processes in the two-quantum photoionization of aromatic amines at 77 K have been described.[248] The photochemical production of acetyl manganese carbonyls at 17 K,[249] the photolysis of liquid $CO-C_2H_4$ mixtures at 77 K,[250] and the photolysis of silane isolated in an argon matrix [251] have all been described.

As the temperature of a solution is lowered, the possibility of weak intermolecular interactions leading to association complexes or aggregates becomes increasingly important. Simultaneously, the translational, rotational, and vibrational molecular motions are succesively hindered as the viscosity of the solvent increases to the point where the solute molecules are effectively frozen in the glass. Either effect can lead to large changes in spectroscopic and photochemical behaviour; primary photoprocesses which are dominant at room temperature can be completely suppressed

---

[246] V. Zanker and D. Benicke, *Z. phys. Chem. (Frankfurt)*, 1969, **66**, 34.
[247] R. Santos, A. Helene, C. Helene, and M. Ptak, *J. Phys. Chem.*, 1970, **74**, 530.
[248] V. A. Kondrat'ev and K. S. Bagdasar'yan, *Khim. vysok. Energii*, 1970, **4**, 35.
[249] J. F. Ogilvie, *Chem. Comm.*, 1970, 323.
[250] H. W. Buschmann and W. Groth, *Ber. Bunsengesellschaft Phys. Chem.*, 1969, **73**, 859.
[251] D. E. Milligan and M. E. Jacox, *J. Chem. Phys.*, 1970, **52**, 2544.

when the temperature is reduced, to be replaced by entirely novel photochemical changes. Association may be so important that theoretical interpretation based on the excitation of isolated molecules becomes completely inappropriate, as seen above for HI.

Many experimental studies in different fields have indicated that association occurs between benzene and chloroform in bulk solution, at temperatures ranging from 20 to -70 °C. N.m.r. studies suggest that the chloroform lies with its C—H bond along the six-fold symmetry axis of the benzene ring. Studies of the heats of mixing of chloroform with benzene derivatives suggest that electron-releasing substituents stabilize the complex, while bulky or electron-withdrawing groups destabilize it. In all of these studies, association is described in terms of a transient complex.

When a solution of benzene ($\sim 10^{-3}$ mol l^{-1}) and CHCl$_3$ ($\sim 10^{-1}$ mol l^{-1}) in a 1:1 mixture of isopentane and methyl cyclohexane (IM) was cooled to 77 K, the normal banded u.v. spectrum of the benzene ($\tilde{A}\,^1B_{2n} \leftarrow \tilde{X}\,^1A_{1g}$) at ~ 240—260 nm was progressively replaced by a similar spectrum displaced ~ 1.5 nm to the blue.[252] The change was reversed on warming, and the spectrum returned to its normal profile at > 120 K. The same progressive change could be effected isothermally at 77 K, by altering the [CHCl$_3$]/[C$_6$H$_6$] ratio, or by changing their concentrations by successive dilution with the solvent. The separate spectra were more readily resolved at 77 K than at higher temperature, since the bands broaden as the glassy solution softens. At the highest concentrations of CHCl$_3$, a further blue-shift could be recognized by the presence of a shoulder on the short wavelength edge of each band of the first shifted spectrum.

Similar behaviour was observed when benzene was replaced by its alkyl-substituted derivatives, but (i) two stages could now clearly be distinguished, and (ii) as the inductive effect of the substituents increased, the minimum [CHCl$_3$] required for complete conversion of the original spectrum at 77 K was reduced, and the temperature at which conversion was initiated was increased. In the most favourable cases, the second stage was complete at 77 K when the benzene ring was substituted by electron-withdrawing groups. Thus, addition of CHCl$_3$ had no effect on the near-u.v. absorption bands of monofluorobenzene, benzotrifluoride, or benzotrichloride.

When other polyhalogeno-derivatives were substituted for CHCl$_3$, shifts in the u.v. absorption of benzene in IM at 77 K were promoted by the following: CHCl$_3$, CDCl$_3$, CHCl$_2$CHCl$_2$, CHCl$_2$CCl$_3$, CHCl$_2$CF$_2$Cl, CHCl=CCl$_2$, trans-CHCl=CHCl, and HCl; while CCl$_4$, CH$_2$Cl$_2$, C$_2$Cl$_6$, CH$_3$CCl$_3$, CCl$_2$=CCl$_2$, cis-CHCl=CHCl, CH$_2$=CCl$_2$, C$_2$H$_2$, CH$_3$CN, (CH$_3$)$_2$CO, and (C$_2$H$_5$)$_2$O were ineffective.

All the observations are consistent with a weak hydrogen-bonding interaction between the weakly acidic solute and the π-electron system of

[252] N. C. Perrins and J. P. Simons, *Trans. Faraday Soc.*, 1969, **65**, 390; G. P. Brown and J. P. Simmons, *ibid.*, p. 3245.

the aromatic ring. Although all the effective solutes possessed halogen atoms, there was no interaction with fully substituted solutes, and it is the presence of the hydrogen atom that is crucial. The simplest assumption is that the first shift corresponds to a 1 : 1 complex, and the second with a 2 : 1 complex. The fact that the normal and shifted spectra can be separated and associated with distinct carriers suggests that the complexed molecules must have a specific relative orientation.

The probable structure is represented by (24). By analogy, the same is true of all the other pairs of solutes where association has been detected.

(24)

Since the second spectral shift is promoted by increasing the proportion of the non-aromatic solute at constant temperature, as well as by lowering the temperature, it probably results from the bonding of a second molecule to the ring, rather than an alternative or more rigid orientation of the 1 : 1 complex. Presumably, the second molecule is situated on the opposite face of the ring, perhaps with a similar orientation to the first, but more evidence is needed to establish this. Two possible structures are represented by (25) and (26). Photochemical reactions were also observed in solutions

(25)   (26)

at low temperatures. The formation of octatetraenes from the addition of polysubstituted ethylenes to benzene and its derivatives could follow several possible routes, but the one suggested is addition to isomers of benzene formed photochemically.

All of the reactions are suppressed at low temperature, both by the onset of hydrogen-bonding association and by the increasing rigidity of the solvent. If the hydrogen-bonded complex is oriented with the C—H bond of the olefin lying perpendicular to the plane of the aromatic ring, the π-orbitals of the ring and the olefin will be orthogonal; a rigid solvent will maintain this orientation and so prevent any cycloaddition to the ring. Association may also prevent any alternative photochemical reaction of the aromatic molecule, by rapidly quenching its excited state through energy transfer. Moreover, reaction *via* the route shown in the Scheme will be inhibited, since the isomerization of the benzene will be hindered in a rigid solvent. This would also apply when there is no hydrogen-bonding, as with C_2Cl_4.

Scheme

The chemistry of these electronically excited molecules involves competition between alternative processes. In general, addition of substituted alkenes only occurs in fluid solution, whereas addition of dilute, substituted-alkane solutes, or saturated solvents, to give octatetraenes or hexatrienes, only occurs in a viscous or glassy solution. The contrast implies competition between alternative processes, with the photoaddition of the alkanes taking over at low temperature and high viscosity. A possible scheme is presented in Figure 14.

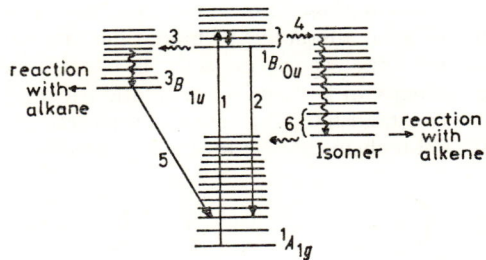

Figure 14 *Possible steps in the photo-reaction of benzene + alkane and benzene + alkene solutions. 1, Excitation; 2, fluorescence; 3, intersystem crossing; 4, radiationless conversion to isomer, viscosity-dependent; 5, phosphorescence; 6, conversion of isomer to benzene*
(Reproduced by permission from *Trans. Faraday Soc.*, 1969, **65**, 390)

It should be stressed that the evidence for the octatetraene product is spectroscopic, and that other reaction products having less characteristic, or weaker, u.v. absorption may have been formed in the above systems. The formation of ionic species in irradiated polyhalogenomethanes at 77 K has also been reported.

Several photochemical reactions on surfaces have been reported. These include the decomposition of isopropyl alcohol sensitized by ZnO,[253] the photochemistry of toluene on a SiO_2–Al_2O_3 surface,[254] photoionization of methylbenzenes on aluminosilicate surfaces,[255] the photodissociation of

---
[253] H. D. Mueller and F. Steinbach, *Nature*, 1970, **225**, 728.
[254] A. A. Pankratov, I. M. Prudnikov, and V. L. Rapoport, *Zhur. fiz. Khim.*, 1969, **43**, 1185.
[255] A. A. Pankratov and E. I. Kotov, *Zhur. fiz. Khim.*, 1969, **43**, 211.

H_2O, D_2O, and H_2O_2 on a MgO surface,[256] effect of visible and u.v. radiation on the oxidation of CO on a Pd catalyst,[257] and the photo-oxidation of paraffins and olefins over anatase at room temperature.[258] Photophysical processes, such as photosorption of O_2 on partially dehydrated Al_2O_3,[259] and photodesorption of the tungsten–CO_2 system at 2537 Å,[260] have also been described.

D. P.

[256] A. A. Pankratov, *Khim. vysok. Energii*, 1970, **4**, 126.
[257] R. F. Baddour and M. Modell, *J. Phys. Chem.*, 1970, **74**, 1392.
[258] M. Formenti, F. Juillet, and S. J. Teichner, *Compt. rend.*, 1970, **270**, *C*, 138.
[259] V. A. Kotel'nikov and I. M. Prudnikov, *Kinetika i Kataliz*, 1969, **10**, 1112.
[260] H. Moesta and N. Trappen, *Naturwiss.*, 1970, **57**, 38.

3
Gas-phase Photochemistry

1 Aromatic Molecules

The basic photochemical processes which occur when simple aromatic molecules are excited to low-lying excited electronic states continue to arouse much interest, particularly since the development of new techniques has allowed properties to be determined quantitatively which were previously not amenable to experimental investigation. Some of the techniques used have been described in Chapter 1 of this Part.

Many aromatic molecules exhibit fluorescence when excited in the first absorption band in the gas phase, but a number of instances have been recorded of emission occurring when such molecules have been excited with light of wavelength such that absorption does not produce the first excited state of the molecule, at least in a one-quantum process. Thus irradiation of benzene derivatives with light in the vacuum-u.v. region causes luminescence,[1] and the competition between photoionization and photodecomposition of aromatic molecules excited in this region has been discussed.[2] Excitation of anthracene and naphthalene by a giant-pulse ruby laser beam results in fluorescence, the spectral distribution of which has been recorded.[3] In this case the first excited singlet state of the aromatics is reached *via* absorption of two quanta of the incident laser radiation. Some special fluorescence properties of the vapour of anthracene derivatives have been briefly discussed.[4] The giant-pulse laser photolysis of benzene and mesitylene[5] has been described, together with the observation of decomposition of benzene upon irradiation by a ruby laser,[6] although this appears not to be photochemical in origin.

The fluorescence emitted when benzene vapour is excited in the first absorption band to produce the $^1B_{2u}$ state has been much studied in the past, but several important new studies have been reported. If a sufficiently high pressure of benzene or added quenching gas is present, vibrational

[1] I. P. Vinogradov and N. Y. Dodonova, *Optika i Spektroskopiya*, 1970, **28**, 170.
[2] H. Tsubomura, *Bull. Chem. Soc. Japan*, 1969, **42**, 3604.
[3] T. S. Jaseja, V. Parkash, and M. K. Dheer, *J. Appl. Phys.*, 1969, **40**, 1882.
[4] A. N. Borisevich and G. B. Tolstorozhev, *Doklady Akad. Nauk S.S.S.R.*, 1969, **188**, 308. (Phys. Sect.)
[5] R. Bonneau, J. Joussot-Dubien, and R. Bensasson, *Chem. Phys. Letters*, 1969, **3**, 353.
[6] O. F. Kulikov, O. V. Bragin, M. V. Gur'ev, M. V. Koz'menko, and G. S. Pashchenko, *Doklady Akad. Nauk S.S.S.R.*, 1969, **187**, 1060. (Chem. Sect.)

relaxation of the excited molecule is complete before emission occurs, and thus a 'normal' fluorescence spectrum results. Under these conditions it is possible to investigate the effect of variation of benzene pressure upon the quantum yield of fluorescence using a steady illumination of the sample, but extreme care must be taken to distinguish between effects which are on the one hand due to a real interaction between ground and excited singlet state benzene molecules, and on the other to an apparent self-quenching which is an experimental artifact due to the geometry of the system viewing the fluorescence. In a careful study using steady illumination and taking such effects into account, it was concluded that the $^1B_{2u}$ state of benzene was self-quenched.[7] However, the development of nanosecond pulse techniques has allowed this point to be tested in an unequivocal manner, since the fluorescence decay time of the aromatic compound should decrease with increase in pressure of that compound for a real interaction, but be independent of pressure if the quenching seen in the static system is due to an experimental artifact. Two studies using pulse techniques have shown the fluorescence decay time to be independent of benzene pressure. In the first, a pulse sampling technique was used,[8] and the lifetime of the vibrationally relaxed $^1B_{2u}$ state of benzene reported as 60 ± 15 ns. The value for toluene vapour was 65 ± 15 ns, and was similarly independent of pressure of the aromatic compound. In the second study, an rf-excited, audio-modulated low-pressure mercury lamp was used to excite 3—10 Torr of benzene in an atmosphere of nitrogen or argon in a flow cell, and from a measurement of the shift in phase between the input and emission signal, the lifetime of the benzene excited state was determined as 80 ± 10 ns,[9] in reasonable agreement with the other measurement. Using the accepted figure for the quantum yield of fluorescence for benzene excited at 2537 Å of 0.18 ± 0.04,[10] the mean radiative lifetime of the excited state can be calculated to be 4.4×10^{-7} s,[9] in fair agreement with the value of 4.07×10^{-7} s obtained by integration of the absorption spectrum.[11] It is evident from these results that self-quenching in benzene vapour must be small or non-existent. This conclusion is in agreement with orbital symmetry arguments which predict 1,2–1,4; 1,2–1,2; or 1,4–1,4 interactions between S_0 and S_1 benzene.[11a] This fact has been confirmed in a study of the effect of pressure upon fluorescence lifetimes using monochromatic excitation.[12] For benzene excited at 2600 Å, which is close to the 0—0 transition, the flourescence follows an exponential decay at all pressures. A plot of reciprocal decay time against benzene pressure follows a good linear

[7] A. Morikawa and R. J. Cvetanovic, *J. Chem. Phys.*, 1968, **49**, 1214.
[8] E. K. C. Lee and G. M. Breuer, *J. Chem. Phys.*, 1969, **51**, 3130.
[9] C. S. Burton and H. E. Hunziker, *J. Chem. Phys.*, 1970, **52**, 3302.
[10] W. A. Noyes jun., W. A. Mulac, and D. A. Harter, *J. Chem. Phys.*, 1966, **44**, 2100.
[11] L. B. Berlmann, 'Handbook of Fluorescence Spectra of Aromatic Molecules', Academic Press, New York, 1965.
[11a] D. Bryce-Smith, *Chem. Comm.*, 1969, 806.
[12] M. Nishikawa and P. K. Ludwig, *J. Chem. Phys.*, 1970, **52**, 107.

relationship with a very small slope over the pressure range 4—75 Torr. Extrapolation of the line to zero pressure yields a value for τ_{2600} of 77 ± 3 nsec, in excellent agreement with the results reported above. Excitation at 2530 Å, however, yields decay curves of two types depending upon the pressure region studied. Above 20 Torr, a single exponential decay is observed which corresponds exactly to that observed with 2600 Å excitation. At lower pressures, however, the initial part of the curves deviates markedly from exponentiality. Evaluation of the initial faster decay of the experimental curves at these lower pressures is complex and depends upon the interpretation outlined below. The results with 2470 Å excitation are similar in that, at high pressures, exponential decay matching that observed at 2600 Å is seen, while at lower pressures the initial part of the curves is non-exponential. However, because of low intensity the results at this wavelength were not subjected to further analysis.

It is clear that for all wavelengths vibrational relaxation is complete before emission occurs for pressures above 20 Torr, which is in agreement with observations in systems under steady illumination. The lifetime measured under these conditions is that of the vibrationally relaxed benzene $^1B_{2u}$ molecule (at 25 °C). At lower pressures and the shorter exciting wavelengths, however, it is evident that collisional loss of vibrational energy cannot compete with emission of radiation from the vibrational levels populated on absorption, leading to non-exponential decay. The results can be explained by a two-level scheme, as outlined below:

$$B + h\nu \longrightarrow {}^1B_n \qquad (1)$$

$$^1B_n \longrightarrow B + h\nu_r \qquad (2)$$

$$^1B_n \longrightarrow {}^3B \text{ and products} \qquad (3)$$

$$^1B_n + B \longrightarrow 2B \qquad (4)$$

$$^1B_n + B \longrightarrow B + {}^1B_0 \qquad (5)$$

$$^1B_0 \longrightarrow B + h\nu_f \qquad (6)$$

$$^1B_0 \longrightarrow {}^3B \text{ and products} \qquad (7)$$

$$^1B_0 + B \longrightarrow 2B \qquad (8)$$

In the above scheme subscript n refers to a vibrationally excited molecule and subscript 0 to a vibrationally relaxed molecule; $h\nu_r$ is resonance fluorescence, and $h\nu_f$ normal fluorescence; B is a ground-state benzene molecule and superscripts refer to multiplicity of excited states. The scheme leads to the following expression for the pressure dependence of the relative steady-state luminescence intensity:

$$\frac{I}{I_0} = \frac{1 + \{k_5 k_6 [B]/k_2(k_6 + k_7 + k_8[B])\}}{1 + \{(k_4 + k_5)/(k_2 + k_3)\}[B]} \qquad (9)$$

For the time dependence of the luminescence intensity, assuming δ-pulse excitation, the expression is obtained:

$$I_\delta(t)/I_0 = \exp(-\lambda' t) + \{k_5 k_6[B]/k_2(\lambda'' - \lambda')\} \times [\exp(-\lambda' t) - \exp(-\lambda'' t)] \quad (10)$$

where I_0 is the intensity at $t = 0$, and $\lambda' = 1/\tau' = k_2 + k_3 + (k_4 + k_5)[B]$, $\lambda'' = 1/\tau'' = k_6 + k_7 + k_8[B]$. (These are the rates of deactivation of the upper and lower states respectively.) However, in the actual experiment a δ-pulse was not used, and because the actual pulse has a finite duration, a convolution integral must be employed:

$$I(t) = \int_0^{t_c} E(\tau) I_\delta(t - \tau) \, d\tau \quad (11)$$

where $E(\tau)$ corresponds to the excitation function defined for $0 < \tau < t_c$, and $I_\delta(t - \tau)$ to the response function of the system which in this case is equation (10): t_c is the cutoff time at which the excitation can be considered negligible.

The excitation pulse has a very fast rise and a decay which can be approximated by an exponential with a decay rate k_0. The time dependence of the total luminescence intensity is then given by:

$$I(t) = \frac{[\exp(-\lambda' t) - \exp(-k_0 t)]}{k_0 - \lambda'}$$

$$+ G\left[\frac{\exp(-\lambda' t) - \exp(-k_0 t)}{k_0 - \lambda'} - \frac{\exp(-\lambda'' t) - \exp(-k_0 t)}{k_0 - \lambda''}\right] \quad (12)$$

For $0 < t < t_c$:

$$I(t) = \frac{[1 - \exp(\lambda' - k_0) t_c] \exp(-\lambda' t)}{k_0 - \lambda'} + G\left\{\frac{[1 - \exp(\tau'' - k_0)t_c] \exp(-\lambda' t)}{k_0 - \lambda'}\right.$$

$$\left. - \frac{[1 - \exp(\lambda'' - k_0) t_c] \exp(-\lambda'' t)}{k_0 - \lambda''}\right\} \quad (13)$$

for $t > t_c$. In the calculations k_0 was taken as $200 \times 10^6 \, \text{s}^{-1}$ and $t_c = 2$ ns. $G = k_5 k_6 P/k_2(\lambda'' - \lambda')$. The first term on the right-hand side of equations (12) and (13) corresponds to the resonance emission and the expression in brackets to the normal fluorescence. Since the lower level is populated from the upper one, the normal emission initially increases and at sufficiently long times decays exponentially. This allows evaluation of λ'' at long times. The value of λ' was estimated by an iterative process via which the second term on the right hand side of equation (13) was evaluated from an estimated value of λ', normalized to the exponential part of the experimental curve, and the process repeated until a satisfactory exponential decay for the faster emission was obtained. The decay of the total luminescence from excitation at 2530 Å at low pressures can thus be quantitatively represented by the sum of two exponential decays, one of which corresponds to emission from the equilibrated level of the benzene singlet state, the other to a faster radiative decay from higher vibrational levels. From a plot of reciprocal decay times against benzene pressure, the lifetime of the state of benzene initially formed upon absorption of 2530 Å radiation can be obtained by extrapolation to zero pressure. Within rather large experimental errors this lifetime is between 40 and 67 ns, shorter than the lifetime of the relaxed level, as expected. By combining the average value of the lifetime of the upper state obtained in this way with the limiting value for the quantum yield of fluorescence of benzene (Φ_0) at low pressures of $\Phi_0 = 0.39$,[13] the sum of the rate constants of all other non-radiative processes depleting the concentration of the upper vibrational level of the excited state is estimated as $\Sigma k = 12.2 \times 10^6 \text{ s}^{-1}$.

The results reported in this study are in essential agreement with an earlier interpretation of results obtained by steady-state illumination of benzene.[14, 15] These authors also considered a two-level mechanism, and obtained an expression for the relative steady-state luminescence intensity of:

$$I/I_0 = (1 + 0.72P)/(1 + 1.23P) \qquad (14)$$

This is of the same form as is obtained in the decay time study, where the expression is:

$$I/I_0 = (1 + 0.153P)/(1 + 0.277P) \qquad (15)$$

The numerical values are, however, different, which may be due in part to the use of the value of Donovan and Duncan for the lifetime of the excited state of benzene in the former equation, a value which is almost certainly an order of magnitude in error. The values of rate constants for various processes occurring from the upper and lower vibrational levels measured by the decay technique are, however, in fair agreement with those calculated on the basis of a two-level model using the steady illumination results, as is shown in Table 1.

Table 1 *Pressure dependence of benzene fluorescence decay curves: summary of rate constants*

Rate constant	Value	Ref.
k_6	$2.6 \times 10^6 \text{ s}^{-1}$	12
k_6	$2.4 \times 10^6 \text{ s}^{-1}$	11
k_7	$10.4 \times 10^6 \text{ s}^{-1}$	12
k_2	$7.8 \times 10^6 \text{ s}^{-1}$	12
k_3	$12.7 \times 10^6 \text{ s}^{-1}$	12
k_8	$8.6 \times 10^{-13} \text{ cc molecule}^{-1} \text{ s}^{-1}$	12
k_4	$1.7 \times 10^{-10} \text{ cc molecule}^{-1} \text{ s}^{-1}$	12
k_2/k_6	3	12
k_2/k_6	3	14
k_7/k_6	3.9	12
k_7/k_6	4	14
k_3/k_6	4.7	12
k_3/k_6	5.8	14
$3.26 \times 10^{16} \, k_4/k_6$	2.1	12
$3.26 \times 10^{16} \, k_4/k_6$	10.8	14

[13] E. M. Anderson and G. B. Kistiakowsky, *J. Chem. Phys.*, 1969, **51**, 182.
[14] S. J. Strickler and R. J. Watts, *J. Chem. Phys.*, 1966, **44**, 426.
[15] G. B. Kistiakowsky and C. S. Parmenter, *J. Chem. Phys.*, 1965, **42**, 2942.

There have been two recent investigations into the quantum yield and intensity distribution of benzene fluorescence in the very-low-pressure region.[13, 16] In the first, [^{1}H$_6$]benzene, [^{2}H$_6$]benzene, and [1,4-^{2}H$_2$]benzene were studied down to 10^{-2} Torr. At pressures below about 0·4 Torr the spectral intensity distribution and quantum yields of the resonance fluorescence of all three benzenes are independent of pressure. The quantum yields of the three benzenes, determined by comparison with that of [^{1}H$_6$]benzene at high pressures, were found to be 0·39, 0·53, and 0·38 respectively. At higher pressures the yields are lower and the spectral intensity distribution is different, corresponding to normal fluorescence. The low-pressure experiments have been extended to the pressure region 0·05—7 × 10^{-5} Torr in a careful study.[16] Here it was shown that at pressures below 0·01 Torr, the intensity of every fluorescence band monitored was directly proportional to benzene pressure, indicating as before that both the quantum yield and the spectrum of benzene fluorescence are invariant in this pressure domain. The constant quantum yield is estimated to be near 0·4, as before, and since these pressures extend by more than two orders of magnitude into the low-pressure region where hard-sphere collisions cannot effect relaxation of $^1B_{2u}$ molecules, significant first-order non-radiative decay must occur. The nature of this non-radiative decay is still the subject of much discussion. At low pressures [1,4-^{2}H$_2$]benzene did not isomerise to the 1,3- isomer when irradiated at 2537 Å.[13] It seems that at this wavelength at least, the competing non-radiative process is not chemical reaction. There remain the possibilities of intersystem crossing to the triplet state, or internal conversion to the ground state.

Since the $^3B_{1u}$ state of benzene is not phosphorescent in the gas phase, any attempt to establish triplet-state formation must involve collisional transfer of the triplet energy from the aromatic molecule to a molecule in which the triplet state can be identified. The two triplet-quenching molecules commonly used are biacetyl and but-2-ene. Attempts have been made in the past to utilise the sensitized *cis–trans* isomerization of but-2-ene in the low-pressure region to establish whether or not intersystem crossing in benzene is a first-order or collisionally-induced process.[17] Apart from the difficulties associated with interpretation of results in which a second-order process is being used to monitor what may or may not be a first-order process, recent work has shown that there may be hitherto neglected experimental difficulties.[13] Thus in the pressure range 0·5—0·1 Torr, the benzene-sensitized isomerization of *cis*-but-2-ene produced data which were consistent with a mechanism in which intersystem crossing was first-order and the triplet aromatic compound either decayed to the ground state *via* a radiationless process, or passed on its energy to the butene to induce isomerization. This is the situation generally held to occur at high pressures. However, at pressures lower than 0·1 Torr, more isomerization

---

[16] C. S. Parmenter and A. H. White, *J. Chem. Phys.*, 1969, **50**, 1631.

occurred than predicted by this mechanism, in qualitative agreement with the earlier results.[17] The deviation, which was previously attributed to a collision-induced intersystem crossing, has in this case been traced to the fact that *cis*-but-2-ene when irradiated alone under these conditions isomerizes to *trans*-but-2-ene at a rate which becomes experimentally significant in the sensitized experiments at benzene pressures lower than 0·1 Torr. The quantum yield of the direct reaction is estimated to be as high as 30, indicating a chain process. It is therefore suggested that the failure to take a direct isomerization into account in the previous interpretation may account for the conclusion of this author that intersystem crossing in benzene is a collision-induced process, and that in fact no data exist at present to suggest that the intersystem crossing is other than an intramolecular first-order step.

The use of two-vibrational level schemes to describe low-pressure data, such as described earlier,[12] has been criticized on the grounds that, while the schemes adequately describe behaviour at the high-pressure and low-pressure extremes, behaviour at intermediate pressures cannot be explained on this basis.[13] These authors prefer a three-level scheme, which, while it does lead to a better fit to experimental data, is in itself only an approximation to the real situation.

In the very-low-pressure study, collisional relaxation of some of the benzene $^1B_{2u}$ vibronic states excited by mercury 2537 Å radiation can be observed at pressures as low as 0·05 Torr. The efficiency of ground-state benzene in effecting these depopulations is estimated to be one in two hard-sphere collisions, and that of carbon monoxide gas about one in four. The rate of vibrational relaxation of the different vibronic levels initially populated was not the same for each level when CO was used as the colliding gas.

These results have considerable impact upon theoretical treatments of radiationless transitions.[18] These theories display such processes as ultimately resulting from the coupling terms that connect the zero-order Born–Oppenheimer singlet and triplet excited states. In benzene, the zero-order singlet appropriate for 2537 Å excitation is a vibronic level of the $^1B_{2u}$ singlet lying about 2000 cm^{-1} above the zero-point energy. The zero-order triplets nearly resonant with this state are the dense array of vibronic levels with about 10,000 cm^{-1} vibrational energy above the zero-point of the $^3B_{1u}$ electronic state. An estimate of the pressure below which $^1B_{2u}$ molecules will decay to the triplet as isolated molecules can be made from consideration of both the inter- and intra-molecular interaction energies in a pure benzene system. Collisions will not significantly affect $^1B_{2u}$ decay if intermolecular perturbations during the $^1B_{2u}$ lifetime are reduced to the order of the average matrix element for intramolecular coupling of the

[17] P. Sigal, *J. Chem. Phys.*, 1965, **42**, 1953; 1967, **46**, 1043.
[18] See this vol., Part I, Chapter 1; also vol. 1, Part I, Chapter 1.

initial and final states. A comparison of the matrix element in benzene for the S_1–T_1 intersystem crossing and collisional perturbation energies has been made by Robinson,[19] who showed that the former is approximately equivalent to the interaction energy of a collision with an impact parameter of 80 Å. On average, an 80 Å collision will occur during the radiative lifetime of $^1B_{2u}$ benzene only when the total gas pressure is greater than 10^{-3} Torr. In the low-pressure experimental work, pressures were used which were an order of magnitude smaller than this, and yet significant non-radiative decay of the benzene singlet was established. The results clearly imply that benzene ought to be considered to be a 'large' molecule in the 'statistical' limit of Jortner,[18] capable of undergoing intramolecular first-order non-radiative transitions, whereas the theoretical treatments have in the past placed benzene in an intermediate category (although such treatments have been acknowledged to be only approximate).

Several experimental and theoretical studies of the quenching of benzene fluorescence by various molecules have been described.[20-22] It has been known for some time that benzene fluorescence is quenched by the addition of molecular oxygen, and the physical processes *via* which this can occur are set out below:

$$^1B_{2u} + O_2\,(^3\Sigma_g^-) \longrightarrow {}^3B_{1u} + O_2\,(^3\Sigma_g^-) \qquad (16)$$

$$^1B_{2u} + O_2\,(^3\Sigma_g^-) \longrightarrow {}^3B_{1u} + O_2\,(^1\Sigma_g^+)\text{ or }{}^1\Delta_g) \qquad (17)$$

$$^1B_{2u} + O_2\,(^3\Sigma_g^-) \longrightarrow {}^1A_{1g} + O_2\,(^3\Sigma_g^-) \qquad (18)$$

$$^1B_{2u} + O_2\,(^3\Sigma_g^-) \longrightarrow {}^1A_{1g} + O_2\,(^1\Sigma_g^+)\text{ or }{}^1\Delta_g) \qquad (19)$$

An attempt has been made to distinguish between these various processes by monitoring the effect of O_2 on the photolysis of mixtures of benzene and *cis*-but-2-ene. In the presence of the olefin the following reactions may be expected to occur:

$$B + h\nu \longrightarrow {}^1B \qquad (20)$$

$$^1B \longrightarrow {}^3B \qquad (21)$$

$$^1B \longrightarrow B + h\nu_f \qquad (22)$$

$$^1B \longrightarrow B \qquad (23)$$

$$^1B + B \longrightarrow B + B \qquad (24)$$

---

[19] G. W. Robinson, *J. Chem. Phys.*, 1967, **47**, 1967.
[20] A. Morikawa and R. J. Cvetanovic, *J. Chem. Phys.*, 1970, **52**, 3237; *Canad. J. Chem.*, 1968, **46**, 1813.
[20a] A. Morikawa, S. Brownstein, and R. J. Cvetanovic, *J. Amer. Chem. Soc.*, 1970, **92**, 1471.
[21] K. E. Wilzbach and L. Kaplan, *J. Amer. Chem. Soc.*, 1966, **8**, 2066; D. Bryce-Smith, A. Gilbert, and B. H. Orgen, *Chem. Comm.*, 1966, 512; D. Bryce-Smith and H. C. Lonquet-Higgins, *Chem. Comm.*, 1966, 593; D. Bryce-Smith, *Pure Appl. Chem.*, 1968, **16**, 47; *Chem. Comm.*, 1969, 806.
[22] E. K. C. Lee, M. W. Schmidt, R. G. Shortridge jun., and G. A. Haninger jun.. *J. Phys. Chem.*, 1969, **73**, 1805.

$$^1B + Bu \longrightarrow B + Bu^* \tag{25}$$

$$^1B + Bu \longrightarrow (adducts) \tag{26}$$

$$^1B + O_2 \longrightarrow ?[\text{one of } (16)\text{—}(19)] $$

$$^3B + B \longrightarrow B + B \tag{27}$$

$$^3B + Bu \longrightarrow B + {}^3Bu \tag{28}$$

$$^3B \longrightarrow B \tag{29}$$

$$^3B + O_2 \longrightarrow ? \tag{30}$$

$$^3Bu \longrightarrow \alpha\ trans\text{-}Bu + (1-\alpha)\ cis\text{-}Bu \tag{31}$$

where B now refers to a benzene molecule, Bu to but-2-ene, and superscripts to multiplicity of excited states, as before. Neglecting equations (25) and (26), and (27) and (29) since a high pressure of *cis*-butene was used initially, the following expressions can be obtained:

$$R_t^0/R_t = (1 + \tau_s k_{16}[O_2])(1 + k_{30}[O_2]/k_{28}[Bu]) \tag{32}$$

$$I_f^0/I_f = (1 + \tau_s k_{16}[O_2]) \tag{33}$$

where R_t^0 and R_t are the rates of isomerization of the butene in the absence and presence of the O_2 respectively, and I_f^0 and I_f are the corresponding intensities of benzene fluorescence: τ_s is the lifetime of the benzene singlet state.

It can be seen that a plot of I_f^0/I_f against oxygen pressure should be a straight line with a unit intercept, and this was found in the experiments. From the slope of the line a value for k_{16} of 2.1×10^{-10} cc molecule^{-1} s^{-1} was obtained which corresponds to almost unit efficiency in collisional quenching. It can be seen from equation (32) that at a constant pressure of oxygen a plot of R_t^0/R_t against butene pressure should also yield a straight line, but that the intercept should in general be greater than unity. However, at all oxygen pressures used, although good straight-line plots were obtained, the intercept was invariably unity. This situation can only arise if the step by which O_2 quenches the benzene singlet state also produces a benzene triplet state. Thus the quenching step can only be either reaction (16) or (17). The authors in the above work [20] favour (17) since $O_2(^1\Delta_g)$ has been observed in the benzene–O_2 system, but it should be noted that similar quenching has been observed in solution in molecules in which the singlet–triplet energy gap is too small to populate the $^1\Delta_g$ state of oxygen upon collision, which would indicate that (16) is very important at least in these systems.[23]

The nature of reaction (26) is of interest. Previous work has established the quenching of the fluorescence of benzene by olefins, and a study has been completed which seeks to relate the quenching parameters found in the earlier work to the rate at which olefins undergo cycloaddition reactions

---

[23] C. S. Parmenter and J. D. Rau, *J. Chem. Phys.*, 1969, **51**, 2242.

to photo-excited benzene.[20a] With *cis*-but-2-ene one major product is found, namely 6,7-dimethyltricyclo[3,3,0,0,[2,8]]oct-3-ene, whereas with *trans*-but-2-ene two stereoisomers of this compound are formed (the *endo,exo* and *exo,endo* forms). Both cycloadditions occur with a retention of the configuration of the olefin, and the rates of adduct formation are proportional to the amount of light absorbed by the system and to the pressure of olefin present. The results suggest that an excited singlet state of benzene is involved in the cycloaddition, but the identification is not unequivocal. The quantum yield of the *cis*-adduct is estimated to be 3×10^{-3}. These results are in essential agreement with earlier work in the condensed phase.[21]

The quenching of benzene and toluene fluorescence by a series of π-bonded molecules in the gas phase has been reported,[22] using an exciting wavelength of 2537 Å. The relative rate-constants and cross-sections are given in Table 2. It can be seen that CS_2, 4-pentenal, 3-methylpentenal,

Table 2 *Relative quenching parameters for benzene and toluene singlet states by π-bonded molecules*

Acceptor	Benzene donor			Toluene donor		
	$P_{\frac{1}{2}}$	$k_{Q_{rel.}}$	$\pi\sigma^2_{rel.}$	$P_{\frac{1}{2}}$	$k_{Q_{rel.}}$	$\pi\sigma^2_{rel.}$
CS_2	1·64	(1·00)	(1·00)	1·81	(1·00)	(1·00)
O_2	2·34	0·70	0·54	—	0·79	0·60
COS	60	0·030	0·026	—	—	—
4-pentenal	1·73	0·95	0·97	—	—	—
3-methylpentenal	—	1·0	1·0	—	—	—
biacetyl	1·96	0·84	0·86	—	—	—
cyclobutanone	2·92	0·56	0·55	3·02	0·60	0·59
cyclopentanone	2·95	0·56	0·57	—	—	—
cyclohexanone	2·92	0·56	0·60	3·06	0·59	0·63
pentan-2-one	—	1·0	1·0	—	—	—
cis-penta-1,3-diene	16·0	0·102	0·100	30·8	0·059	0·056
trans-penta-1,3-diene	17·5	0·094	0·091	33·7	0·054	0·052
buta-1,3-diene	42·0	0·039	0·036	65·0	0·028	0·025
[$^2H_6$]buta-1,3-diene	48·7	0·034	0·032	75·6	0·024	0·022
ethyl vinyl ether	>100	—	—	—	—	—

$P_{\frac{1}{2}}$ = pressure at which aromatic fluorescence is half-quenched.
Rate constants $k_{Q_{rel.}}$ and cross-sections $\pi\sigma^2_{rel.}$ are relative to those for CS_2.

biacetyl, and 2-pentanone are all more efficient singlet quenchers than oxygen, indicating that quenching must occur on every collision between these molecules and the singlet aromatic. Some measure of the efficiencies of these same molecules in quenching triplet-state benzene has been obtained in a study in which different pressures of the gases were added to a mixture of benzene and biacetyl.[24] Having established that the gases do not affect the phosphorescence yield of biacetyl when it is excited directly, any quenching of the sensitized emission from biacetyl must be due to quenching of the aromatic triplet by the added gas. Relative quenching efficiencies are

[24] G. A. Haninger jun. and E. K. C. Lee, *J. Phys. Chem.*, 1969, **73**, 1815.

shown in Table 3. In some cases the molecules added did in fact interact with the biacetyl triplet state, leading to complexity in interpretation. Comparison of Tables 2 and 3 demonstrates the striking difference between the ability of monocarbonyl compounds and the conjugated dienes to accept the electronic excitation energy from singlet and triplet benzene.

Table 3 *Relative quenching efficiencies of π-bonded molecules for benzene triplet state*

Quencher		Biacetyl competition		cis-but-2-ene competition	
	P_Q	k_B/k_Q	k_Q/k_C [a]	k_Q/k_C [b]	k_Q/k_C [c]
cyclopropane	350	1000	0·004	—	—
C_2H_4	5·66	17·8	0·22	0·18	0·22
C_3H_6	2·89	9·10	0·42	0·53	0·47
but-1-ene	2·65	8·34	0·46	0·49	0·50
pent-1-ene	2·80	8·90	0·43	0·49	0·54
cis-but-2-ene	1·223	3·85	(1·00)	(1·00)	(1·00)
trans-but-2-ene	1·185	3·73	1·03	1·09	—
cyclopentene	1·15	3·62	1·06	0·88	—
cyclohexene	0·85	2·7	1·4	—	—
2-methylbut-2-ene	0·65	2·05	1·88	1·70	1·61
tetramethylethylene	0·42	1·32	2·92	3·1	2·67
cyclohexa-1,4-diene	0·83	2·61	1·48	—	—
ethyl vinyl ether	2·4	7·6	0·51	—	—
4-pentenal	0·34	1·07	3·6	3	—
biacetyl	(0·31)	(1·00)	3·8	—	—
buta-1,3-diene	0·58	1·83	21	12	16·2

P_Q = pressure of quencher which competes equally with 0·31 Torr biacetyl.
k_B/k_Q = relative rate-constant ratio for quenching by biacetyl and quencher.
k_Q/k_C = relative rate-constant ratio for quenching by quencher and *cis*-but-2-ene.

[a] Work in ref. 24.
[b] Data from E. K. C. Lee *et al.*, *J. Chem. Phys.*, 1968, **48**, 4547.
[c] Work in ref. 20.

Thus the quenching cross-sections of cyclobutanone and buta-1,3-diene are 1 and 50 Å^2 respectively for triplet benzene, but 36 and 2 Å^2 for the singlet benzene. The reason for this probably lies in the fact that the gap between singlet and triplet levels in dienes is very large whereas in carbonyls it is usually small (3 eV in the former and about 0·5 eV in the latter). As a result, triplet energies of the carbonyls and aromatics and singlet energies of dienes and aromatics are very similar, leading to inefficient transfer.

The existence of a deuterium isotope effect can also be seen from a consideration of Tables 2 and 3. For singlet quenching, the energy transfer cross-section for [^{1}H$_6$]buta-1,3-diene is 1·12 times that for [^{2}H$_6$]buta-1,3-diene, which is a small but real effect. It appears that the effect is manifested in energy transfer studies only when the energetics of the energy transfer process are unfavourable or near the threshold, since no effect was observed for triplet energy transfer from benzene to [^{1}H$_6$]buta-1,3-diene and [^{2}H$_6$]buta-1,3-diene. A much more pronounced deuterium isotope effect has been noted in the intersystem crossing of [²H₆]benzene compared with [¹H₆]benzene. Measurement of the fluorescence decay time of C_6D_6 gave a value of 95 ± 15 nsec,[25] compared with 60 ± 15 nsec reported by these authors for C_6H_6.[8] Thus τ_f^D/τ_f^H is equal to 1.6 ± 0.2. Neglecting internal conversion, it can be shown that

$$k_{\text{ISC}} = (1 - \Phi_f)/\tau_f \tag{34}$$

where k_{ISC} is the rate constant for the intersystem crossing step, Φ_f is the fluorescence yield, and τ_f the fluorescence lifetime. Thus

$$k_{\text{ISC}}^D/k_{\text{ISC}}^H = \{(1 - \Phi_f^D)/(1 - \Phi_f^H)\} \, (\tau_f^H/\tau_f^D) \tag{35}$$

Appropriate substitution gives a value for the rate-constant ratio of 0·5, *i.e.* the intersystem crossing rate of C_6H_6 is twice that of C_6D_6. The magnitude of the effect suggests that intersystem crossing occurs directly from the $^1B_{2u}$ state to the $^3B_{1u}$ state, and not *via* the $^3E_{1u}$ state, for which a much smaller effect would be expected.

It is perhaps of interest to note that although quenching of the excited states of benzene by oxygen has been considered to be a physical process, at least at high temperatures, the photolysis of benzene–oxygen mixtures gives rise to chemical products, among them phenol.[26]

The techniques of the sensitization of phosphorescence of biacetyl and the *cis–trans* isomerization of but-2-ene continue to be used to obtain data about the rate of intersystem crossing in a variety of aromatic molecules, among them substituted benzenes. A series of spectrofluorometric studies using the former technique have provided much data on many of the fluorinated benzenes, which provide an interesting series to study in that all are volatile liquids and can thus be studied in the gas phase. A re-investigation of the triplet state yield of monofluorobenzene [27] has shown that previous results quoted by one of the authors [28] were in error, although the reason for the discrepancy is not stated. In the original work the maximum sensitized phosphorescence yield from biacetyl was quoted as 0·136, leading to an estimate of the triplet-state yield for this molecule of 0·90. However, such an estimation neglected the fact that significant quenching of the singlet state of the aromatic by biacetyl occurs at the pressures used to obtain maximum sensitized emission, and there is an apparent self-quenching of the aromatic compound also. When these factors are taken into account, the apparent triplet-state yields based on the figures given in the early study are alarmingly in excess of unity. The re-investigation has established that in fact the maximum yield of sensitized emission in this system is of the order of 0·06, leading to a crude estimation of the triplet-

---

[25] G. M. Breuer and E. K. C. Lee, *J. Chem. Phys.*, 1969, **51**, 3615.
[26] G. L. Grigoryan, A. A. Mantashyan, and A. B. Nalbandyan, *Armyan. khim. Zhur.*, 1969, **22**, 379.
[27] M. E. MacBeath, G. P. Semeluk, and I. Unger, *J. Phys. Chem.*, 1969, **73**, 995.
[28] I. Unger, *J. Phys. Chem.*, 1965, **69**, 4284.

state yield of 0·38. This result has been confirmed, and if apparent self-quenching and singlet quenching are accounted for, the true triplet-state yield for fluorobenzene excited at 2550 Å is of the order of 0·75.[29] This is rather less than the value obtained with the *cis–trans* isomerization technique,[30] a value which has also been checked and substantially confirmed.[29] It should be noted that quenching cross-sections and rate-constant values quoted in reference 27 are grossly in error, since the authors have taken the rate constant for fluorescence to be equal to the inverse of the actual lifetime of the singlet state, rather than the inverse of the mean radiative lifetime.

In most substituted benzenes studied to date, the effect of wavelength upon the basic photophysical processes of fluorescence and intersystem crossing is drastic. As the energy of the incident photon is increased, both fluorescence yields and usually triplet-state yields are progressively reduced, sometimes to zero. The effect of increase in pressure on such a system under short-wavelength irradiation is generally to enhance vibrational relaxation, causing an increase in fluorescence and triplet-state yields. This effect has been noted for a variety of photochemically inert gas molecules in the excitation of a fluorobenzene.[31] Added gases containing a heavy nucleus have a different effect. It would be expected that a heavy monatomic gas might enhance the rate of intersystem crossing due to spin–orbit coupling, as has been observed in the liquid phase recently.[32] Certainly addition of xenon to fluorobenzene vapour excited at longer wavelengths causes a decrease in the fluorescence yield, but attempts to measure an enhancement in the triplet state using the biacetyl technique were unsuccessful because the xenon apparently also enhanced the non-radiative intersystem crossing of triplet biacetyl to the ground state.[33] (Traces of O_2 impurity in the high pressures of xenon used would have given rise to the same effect.)

It is interesting to consider from an elementary theoretical standpoint what the effect of substitution might be upon the photochemistry of a series of substituted benzenes, assuming that only fluorescence, intersystem crossing to the triplet state, and internal conversion to the ground state were the only important processes. Using an approach of Reiser,[34] we can estimate the magnitude of the transition moment integral for the radiative fluorescence process from a consideration of the group-theoretical symmetry properties of the nominal ground and excited states of the molecules and of the dipole moment operator. Clearly, if the transition moment integral is not totally symmetric, the transition is orbitally forbidden, and a low value of the rate constant for fluorescence would be expected. A similar

---

[29] D. Phillips and Kh. Al-Ani, unpublished results.
[30] D. Phillips, *J. Phys. Chem.*, 1967, **71**, 1839.
[31] S. H. Ng, G. P. Semeluk, and I. Unger, *Ber. Bunsengesellschaft. Phys. Chem.*, 1970, **74**, 29.
[32] R. B. Cundall and W. Tippett, *Trans. Faraday Soc.*, 1970, **66**, 350.
[33] A. Cook, G. P. Semeluk, and I. Unger, *Canad. J. Chem.*, 1969, **47**, 4527.
[34] A. Reiser, personal communication.

treatment can yield a result for the allowedness of the intersystem crossing step, although it is recognized that such an approach is grossly oversimplifying the situation. The results of such a treatment are shown in Table 4, for the different point groups. In the table an orbitally allowed transition is denoted with a $+$, and a forbidden transition with a $-$. There are sufficient data in the literature on the photochemistry of the fluorobenzenes to collate into a table to test the simple approach above. Table 5

Table 4 *Effect of symmetry on rates of photo-processes*

Point group	S_1 / T_1 / S_0	$\hat{M}$ / HS_0	Radiative transition	I.S.C.	I.C.
D_{6h}	B_{2u}	E_u			
	B_{2u}		$-$	$+$	$-$
	A_{1g}	A_{2g}			
C_{2v}	B_1	B_1	$+$	$+$	$-$
	A_1				
	A_1	B_1			
D_{3h}	A_2'	E'	$-$	$+$	$-$
	A_1'				
	A_1'	A_2'			
D_{2h}	B_{3u}	B_{3u}	$+$	$+$	$-$
	B_{2u}				
	A_g	B_{1g}			
C_s	A'	A'	$+$	$+$	$+$
	A'				
	A'	A'			

shows the results of such an exercise. It should be stressed that the data have been selected from the references given such that they satisfy the requirement mentioned above that only three photophysical processes are important. Thus, in general, data obtained at the longest wavelength in each study have been used. The radiative lifetimes quoted in Table 5 have been obtained by integration of the corresponding absorption curve and are not experimental values. Nevertheless, it can be seen that the rate constant data for the fluorescence step in this series of compounds do show a good correlation with the values expected from symmetry considerations. There is little correlation between experimental results and symmetry considerations for the intersystem crossing, but this is not altogether surprising considering the simplicity of the approach, and the fact that symmetry may play only a minor role compared with the effect of substitution upon the relative energy levels of the singlet and triplet states (including intermediate triplets) and the magnitude of spin–orbit coupling. It is of interest to note that increasing substitution of fluorine atoms causes the fluorescence yield of these compounds to fall markedly, and also diminishes the lifetime of the triplet state such that quantitative estimation of the latter

Table 5 *Photochemistry of the fluorobenzenes*

Ref.	Compound	Exciting wavelength (Å)	Radiative lifetime (sec)	Fluorescence yield	Triplet state yield	Deficiency	Point group	k_F	$k_{I.S.C.}$	$k_{I.C.}$
10	benzene	2600	4×10^{-7}	0.18	0.70	~0.12	D_{6h}	2.5×10^6	9.7×10^6	1.6×10^6
30	fluorobenzene	2650	1.15×10^{-7}	0.18	0.86	—	C_{2v}'	8.69×10^6	41.5×10^6	0
35	o-difluorobenzene	2700	1.01×10^{-7}	0.28	0.71	0.03	C_{2v}'	9.90×10^6	25.1×10^6	1.1×10^6
36	m-difluorobenzene	2740	1.17×10^{-7}	0.28	0.77	0	C_{2v}'	8.55×10^6	23.5×10^6	0
36	p-difluorobenzene	2780	0.46×10^{-7}	0.50	0.20	0.30	D_{2h}'	21.7×10^6	8.7×10^6	13.0×10^7
	1,2,3-trifluorobenzene	No data available								
37	1,2,4-trifluorobenzene	2774⎫ 2780⎭	0.56×10^{-7}	0.33 0.28	— 0.64	—	C_{2v}'	—	—	—
37	1,3,5-trifluorobenzene	2680	5.3×10^{-7}	0.035	0.45*	0.08	C_s	17.9×10^6	35.8×10^6	4.5×10^6
38	1,2,3,4-tetrafluorobenzene	2645	—	0.186	0.22	?	D_{3h}''	1.88×10^6	$>24.2 \times 10^6$	?
38	1,2,3,5-tetrafluorobenzene	2680	—	0.03	0.14	0.59	C_{2v}'	—	—	—
38	1,2,4,5-tetrafluorobenzene	2670	0.49×10^{-7}	0.34	0.52	0.83	C_{2v}'	20.4×10^6	30.9×10^6	8.4×10^6
29	pentafluorobenzene	2650	1.35×10^{-7}	0.015	0.10*	0.14	D_{2h}'	7.41×10^6	$>49 \times 10^6$	?
39	hexafluorobenzene	2800⎫ 2700 2650⎭	5×10^{-7}	0.019 0.009 0.005	0.05* 0.02* 0.03*	?	D_{6h}	2×10^6	$>5 \times 10^6$	?

The triplet yields marked * may not be accurate owing to the short lifetime of these states and consequent experimental difficulties.

[35] J. L. Durham, G. P. Semeluk, and I. Unger, *Canad. J. Chem.*, 1968, **46**, 3177.
[36] F. W. Ayer, F. Crein, G. P. Semeluk, and I. Unger, *Ber. Bunsengesellschaft. Phys. Chem*, 1968, **72**, 282.
[37] G. P. Semeluk, R. D. S. Stevens, and I. Unger, *Canad. J. Chem.*, 1969, **47**, 597.
[38] I. Unger, personal communication.
[39] D. Phillips, *J. Chem. Phys.*, 1967, **46**, 4679.

becomes impossible. Other series of substituted benzenes, including methyl and trifluoromethyl, are currently being investigated.[29]

The spectrofluorometric technique has been used [40] to obtain data on toluene in essential agreement with an earlier thorough study,[41] and to investigate the fluorescence of 1,2,3,4-tetrahydronaphthalene excited in the 2850—3100 Å region.[42]

The triplet states of aromatic molecules may be populated directly *via* collision with an excited mercury atom. Quenching data for the mercury (6^3P_1) atom and various aromatic molecules are given in Table 6. It can be

Table 6 *Quenching parameters for quenching of* $Hg(6^3P_1)$ *atoms by aromatics*

Compound	$k_Q \times 10^{-10}$ (l mol^{-1} s^{-1})	σ_Q (Å^2)	σ_V (Å^2)
benzene	24·9	39·4	36·3
toluene	35·3	59·1	41·4
p-xylene	23·6	41·5	45·5
m-xylene	34·8	61·2	45·1
o-xylene	37·9	65·7	44·9
chlorobenzene	39·6	71·0	41·2
fluorobenzene	37·0	62·8	39·0
bromobenzene	28·8	56·9	42·3
ethylbenzene	38·9	68·3	43·9
[^{2}H$_6$]benzene	40·4	65·4	36·3

k_Q = rate constant for quenching, σ_Q = quenching cross-section calculated from rate constant, σ_V = cross-section calculated by assuming the diameter of Hg to be 2·38 Å, and estimating the value of the diameter of the molecule from the '*b*' constant in the van der Waals equation of state.

seen from the table that substituent effects cannot be explained simply in terms of dipole moments or polarizabilities, but a large deuterium isotope effect is indicated.[43]

The remarkable effect of deuteriation on the quenching cross-section for benzene is in accord with the observed large effect upon intersystem crossing in this molecule. A mechanism for the quenching involving such an intersystem crossing can be postulated:

$$Hg + h\nu \longrightarrow Hg^* \tag{36}$$

$$Hg^* \longrightarrow Hg + h\nu \tag{37}$$

$$Hg^* + Ar \longrightarrow (Hg^*-Ar) \tag{38}$$

$$(Hg^*-Ar) \longrightarrow (Hg-Ar^*) \tag{39}$$

$$(Hg^*-Ar) \longrightarrow Hg^* + Ar \tag{40}$$

[40] S. L. Lem, G. P. Semeluk, and I. Unger, *Canad. J. Chem.*, 1969, **47**, 4711.
[41] C. S. Burton and W. A. Noyes jun., *J. Chem. Phys.*, 1968, **49**, 1705.
[42] M. Grossman, G. P. Semeluk, and I. Unger, *J. Phys. Chem.*, 1969, **73**, 1149, 1647.
[43] G. J. Mains and M. Trachtman, *J. Phys. Chem.*, 1970, **74**, 1647.

$$(\text{Hg}-\text{Ar}^*) \longrightarrow \text{Hg} + \text{Ar}^* \quad (41)$$

$$\text{Ar}^* \longrightarrow \text{Ar} \quad (42)$$

where Ar is the aromatic quencher and * represents triplet electronic energy.

Application of the steady-state approximation to the above scheme yields an expression for the quenching rate constant k_Q quoted in Table 6 of:

$$k_Q = k_{38}[k_{39}/(k_{39}+k_{40})] \quad (43)$$

If the formation of a loose complex is assumed, an upper limit of, say, $50 \times 10^{10} < k_{38} < 75 \times 10^{10}$ l mol^{-1} s^{-1} might be placed on the formation of the complex. Variations in k_Q could then be attributed to the relative magnitudes of the rates of intersystem crossing, k_{39}, and decomposition of the complex without energy transfer, k_{40}.

It is also possible that the increase in quenching cross-section on deuteriation is due to more efficient quenching of the Hg ($^3P_1$) atom to the metastable $^3P_0$ level by C_6D_6 than C_6H_6. However, it should be noted that the $^3P_0$ mercury atom is itself capable of being quenched by aromatic compounds to give ground-state mercury and triplet-state aromatic molecule, as a recent study in the case of naphthalene has shown.[44] In this the absorption spectra of both species were monitored using a modulated source, and a rate constant of $2.75 \pm 0.06 \times 10^{10}$ l mol^{-1} s^{-1} measured for the process, which is an order of magnitude less than the values of rate constants for quenching of the $^3P_1$ atoms which are quoted in Table 6.

Intersystem crossing of aromatic molecules from the first excited singlet state to an iso-energetic level of the triplet state is followed in general by a fast collisional vibrational relaxation to the equilibrated level. At low pressures, however, the relaxation becomes the rate-determining step, as has been demonstrated by monitoring the absorption of the equilibrated triplet state of naphthalene, anthracene, and pyrene in the gas phase at low pressures.[45]

Inspection of Table 5 reveals that the majority of excited pentafluorobenzene singlet molecules do not decay *via* fluorescence or triplet-state formation. Some photoisomerization to the two possible 'Dewar' valence isomers of the aromatic molecule has been reported,[46] but no evidence of high quantum yields is available. CF$_3$-, CH$_3$-, and CH$_3$O-substituted benzenes are also reported to photoisomerize to the 'Dewar' forms in the gas phase.[46] In the liquid phase, hexakis(trifluoromethyl)benzene is reported to yield valence isomers photochemically [47,48] and the structure of

[44] H. E. Hunziker, *Chem. Phys. Letters*, 1969, **3**, 504.
[45] C. W. Ashpole, S. J. Formosinho, and G. Porter, *Chem. Comm.*, 1969, 1305.
[46] E. Ratajczak, *Roczniki Chem.*, 1970, **44**, 447.
[47] D. M. Lemal, J. V. Staros, and V. Austel, *J. Amer. Chem. Soc.*, 1969, **91**, 3373.
[48] M. G. Barlow, R. N. Haszeldine, and R. Hubbard, *Chem. Comm.*, 1969, 202.

Gas-phase Photochemistry

hexamethyl-'Dewar' benzene,[49] and some of the cycloaddition reactions of hexafluoro-'Dewar' benzene,[50] have been reported.

2 Alkanes and Olefins

These classes of compounds do not, in general, absorb light in the near-u.v. region of the spectrum, and investigation of their photochemical behaviour requires the use of the specialized techniques appropriate to the vacuum-u.v. spectral range (see Vol. 1 in this series, Part I, Chapter 2). Molecular hydrogen dissociates upon absorption in the vacuum-u.v. region, and a description has been given of this dissociation when the light absorbed is near the threshold energy.[51] The absorption spectrum of hydrogen has also been investigated in the 950—700 Å region, and seventeen Rydberg series identified.[52] Fourteen of these belong to the $X^1\Sigma_g^+ - np\pi^1\Pi_u$ electronic transition, and three to the $X^1\Sigma_g^+ - np^1\Sigma_u^+$ transition. The lowest ionization level of molecular hydrogen was placed spectroscopically at $124,417 \pm 2$ cm^{-1}.

The photodecomposition of methane in the vacuum-u.v. region has been much investigated in the past, but new studies continue to appear. Two theoretical approaches to the problem of the photodissociation of methane have yielded conflicting results. It is known experimentally that at moderate photon energies methane dissociates principally to give methylene and molecular hydrogen (44), with only minor production of methyl radicals and hydrogen atoms (45).

$$CH_4 + h\nu \longrightarrow CH_2 + H_2 \qquad (44)$$

$$CH_4 + h\nu \longrightarrow CH_3\cdot + H\cdot \qquad (45)$$

A valence bond calculation on the ground and six excited states of methane attempts to explain these results,[53] but arrives at the conclusion that (45) should be the predominant mode of decomposition. The method involves distorting the molecule in a C_{2v} fashion to produce incipiently CH_2 and H_2 and in a C_{3v} fashion to obtain $CH_3\cdot$ and $H\cdot$. Potential surfaces are obtained which permit a determination of the shape of the molecule during its dissociation. The crude calculation leads to the following three principal results:

(a) The lowest excited state is a forbidden E state.

(b) In the next highest T_2 state there is a steeper potential gradient in the CH_3 and H direction than in the CH_2 and H_2 direction; the reverse is true for the E state. On these simple grounds (45) would be expected to dominate.

(c) The CH_2 derived from the allowed T_2 state is produced in an electronically excited state, leading to the conclusion that, unless there is a crossover

---

[49] M. J. Cardillo and S. H. Bauer, *J. Amer. Chem. Soc.*, 1970, **92**, 2399.
[50] M. G. Barlow, R. N. Haszeldine, and W. D. Morton, *Chem. Comm.*, 1969, 931.
[51] F. J. Comes and U. Wenning, *Z. Naturforsch.*, 1970, **25a**, 237.
[52] S. Takezawa, *J. Chem. Phys.*, 1970, **52**, 2575.
[53] S. Karplus and R. Bersohn, *J. Chem. Phys.*, 1969, **51**, 2040.

from the T_2 state to the E, the red emission bands of CH_2 should be observable when methane is photolysed.

The above approach has been criticized because it neglects the possibility of ion-pair formation.[54] Thus if in methane one of the two σ-electrons in a particular CH bond is excited to a σ^* orbital, it might be expected that dissociation would cause both electrons to go to the same fragment, forming an ion-pair. This is a very well known phenomenon in diatomic molecules, e.g. the $B^1\Sigma_u^+$ state of H_2 dissociates into an ion-pair, and the same is valid for $^1\Sigma^+$ states of hydrogen halides. The effect in methane will be to make the particular dissociation limit very high, causing a displacement of the state itself to higher energies. It is of special importance when considering the loss of H in photodissociation of CH_4. If the $CH_3\cdot$ radical is presumed to be formed in its ground state, $X\,^2A_2''$, the methane molecule in C_{3v} must be in an excited $^1A_1$ state before the dissociation. However, the $^1A_1$ state of methane is much more likely to dissociate into an ion-pair. Thus $CH_3\cdot$ can only be formed in excited states, and as their energies are comparatively high, at moderate photon energies this mode of decomposition does not occur. Formation of CH_2 and H_2 will therefore be expected to predominate, in agreement with experiment.

The following reactions are possible in the photodissociation of ethane in the vacuum-u.v.

$$C_2H_6 + h\nu \longrightarrow C_2H_6^* \qquad (46)$$

$$C_2H_6^* \longrightarrow C_2H_6^+ + e \qquad (47)$$

$$\longrightarrow C_2H_4 + H_2 \qquad (48)$$

$$\longrightarrow C_2H_5^\cdot + H^\cdot \qquad (49)$$

$$\longrightarrow CH_4 + CH_2 \qquad (50)$$

$$\longrightarrow CH_3^\cdot + CH_3^\cdot \qquad (51)$$

The flash photolysis of ethane using an argon discharge[55] with an output in the range 1600—1040 Å produced mainly H_2, CH_4, and C_2H_4 as products. Moreover, the photolysis of mixtures of C_2H_6 and C_2D_6 under these conditions has shown that most (87%) of the hydrogen formed is produced in a molecular process, presumably via (48). Experimental results on photolysis of CH_3CD_3 reveal that of the hydrogen eliminated in reaction (48), 75% comes from the same carbon atom and about 25% involves the loss of a hydrogen atom from each of the two carbon atoms. Thus process (48) would be better written as

$$C_2H_6^* \longrightarrow CH_3CH + H_2 \qquad (52)$$

The relative importance of reactions (47) to (51) as a function of photon energy has been established in a thorough study using the monochromatic

[54] E. Lindholm, *J. Chem. Phys.*, 1970, **52**, 4921; *Arkiv Fysik*, 1968, **37**, 37, 49.
[55] B. C. Roquitte, *J. Phys. Chem.*, 1970, **74**, 1204.

rare-gas resonance lines.[56] The task of tracing final product molecules back to one of these primary processes is enormously difficult because of secondary reactions. Thus each product molecule may arise from more than one primary step, and in order to interpret the results it is necessary to carry out the photolysis in the presence of added compounds which react with the intermediate species in predictable ways. The available techniques include the following:

(a) Use of free-radical scavengers. Since NO or O_2 will undergo fast reactions with alkyl radicals to form non-hydrocarbon products, any hydrocarbon produced in the presence of these molecules must arise either from a molecular elimination or a fast bimolecular process.

(b) Deuterium labelling. Thus in the photolysis of C_2H_6–C_2D_6–NO mixtures, the methane produced is mainly a mixture of CH_4 and CD_4. This result indicates that it is formed mainly *via* molecular elimination. The acetylene seen is again mainly C_2H_4 and C_2D_4, but the ethylene produced in the presence of NO still contains partially deuteriated products, indicating a fast bimolecular process.

(c) H_2S. If C_2D_6 is photolysed in the presence of H_2S, any deuteriated radicals R_D formed will react with the H_2S to give a monoprotonated product (53).

$$R_D + H_2S \longrightarrow R_DH + HS \qquad (53)$$

Any products which arise *via* unimolecular eliminations or fast bimolecular reactions would be fully deuteriated.

(d) Addition of [$^2H_6$]propylene. This compound scavenges H atoms to form propyl radicals which in turn undergo disproportionation and combination reactions with free radicals to form characteristic products.

Using these techniques and some involved arguments which will not be reproduced here, it has been possible to decide the relative importance of the various primary processes when ethane is photolysed with 11·6—11·8 eV photons (from the argon resonance lamp). The conclusions are as follows:

Reaction (47)—5%

Reaction (48)—26%

Reaction (49)—41%

Reaction (50)—16%

Reaction (51)—15%.

A comparison of these results with those obtained in similar studies using the other rare-gas resonance lines is shown in Figure 1, in which M/N_{ex}. for the various primary processes is plotted as a function of photon energy. (M/N_{ex}. = molecules of product M formed per excited molecule N_{ex}. These values are identical to quantum yields when no ionization occurs.) Dotted lines are used for processes (49) and (48) because the exact yields

---

[56] S. G. Lias, G. J. Collin, R. E. Rebbert, and P. Ausloos, *J. Chem. Phys.*, 1970, **52**, 1841.

of these processes are not well established. It can be seen that C—H bond scission processes increase in importance over C—C bond scission with increasing photon energy. According to theoretical treatments, absorption of an 8·4 eV photon should lead exclusively to excitation in the C—C bond, and yet interestingly enough scission of this bond occurs hardly at all at this energy. Instead the elimination of a hydrogen molecule from one C-atom accounts for 85% of the total dissociation, while the remaining molecules

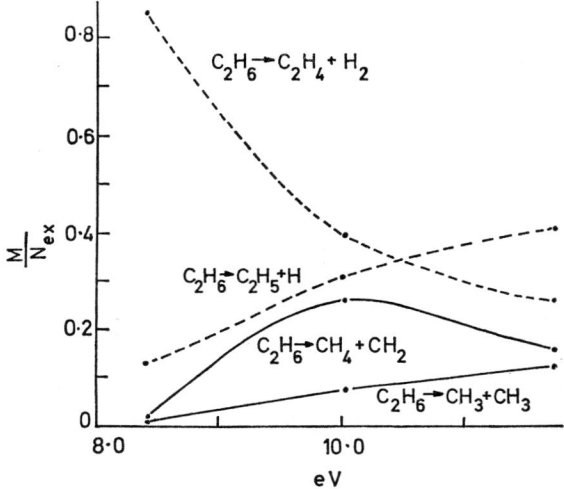

Figure 1 *Yields of primary processes in the photolysis of ethane as a function of energy*
(Reproduced by permission from *J. Chem. Phys.*, 1970, **52**, 1841)

dissociate mainly by C—H bond cleavage (49). For the elimination process to occur, the energy originally absorbed in the C—C bond must be redistributed prior to dissociation. Since fine structure has been seen in the ethane absorption spectrum in this region, the lifetime of the state cannot be as short as has been supposed, and may be long enough to allow the redistribution. Figure 1 clearly shows that as the energy of the incident photon is increased, direct bond cleavage processes (49) and (51) become more important. This result is not surprising since one would expect the lifetime of the excited state to be shorter at higher energies, leading to a diminished probability of those reactions requiring redistribution of energy. The mechanism of the xenon-sensitized photolysis of ethane has also been discussed.[57]

---

[57] G. von-Bunau, K. Nieswandt, D. Henneberg, and G. Schomburg, *Ber. Bunsengesellschaft. Phys. Chem.*, 1969, **73**, 891.

Investigations into the reactions of methylene have usually been carried out in systems in which the methylene has been generated by photolysis of keten or diazomethane, introducing complications into the system because of reactions of the methylene produced with the starting material (see for example the section in this chapter dealing with methylene). The photolysis of propane in the vacuum-u.v. gives rise to methylene and ethane (54), and enables the reactions of methylene with propane to be studied in the absence of a second substrate.[58] These reactions will be discussed later.

$$C_3H_8 + h\nu \longrightarrow C_2H_6 + CH_2 \qquad (54)$$

The techniques described earlier of using nitric oxide, oxygen, and H_2S as radical scavengers have been used to advantage to develop an understanding of the photochemistry of methylcyclopropane.[59] Three primary processes can be postulated:

$$\begin{array}{c}CH_3\\|\\CH\\/\ \backslash\\H_2C-CH_2\end{array} + h\nu \begin{cases}\longrightarrow H_2 + (C_4H_6)^* & (55)\\ \longrightarrow CH_4 + C_3H_4 & (56)\\ \longrightarrow [CH_3-CH(CH_2)_2]\ddagger & (57)\end{cases}$$

where $(C_4H_6)^*$ represents an internally excited species and $[C_4H_8]\ddagger$ a diradical with excess of internal energy.

The occurrence of (55) seems reasonable on the basis of high hydrogen yields obtained in photolysis at 1470 Å, but there are difficulties in the interpretation. From thermochemical data, (55) can be estimated to be about 54 kJ mol^{-1} endothermic; and with 1470 Å photolysis, 810 kJ einstein^{-1} of energy is available. Evidently under these conditions the $(C_4H_6)^*$ species should fragment. However, the likely fragmentation products were not observed, and moreover if the reaction is important, increase in pressure of a chemically inert gas would be expected to promote collisional stabilization of the $(C_4H_6)^*$ species, thereby leading to increased formation of products such as buta-1,3-diene. In fact, increase in inert gas pressure causes a decrease in the rate of formation of this product, indicating that (55) may be of minor importance. Since addition of NO or O_2 does not completely suppress the formation of methane, (56) must occur to some extent, but it cannot account for the high yields of allene or methylacetylene formed in the system. Hence process (57) seems to be the major source of products. There are two structures the biradical might have, and these might be expected to decompose further as follows:

$$\begin{bmatrix}CH_3\\|\\CH\\/\ \backslash\\H_2\dot{C}\quad\dot{C}H_2\end{bmatrix}^{\ddagger} \begin{array}{l}\longrightarrow CH_2 + (C_3H_6)^* \quad (58)\\ \longrightarrow CH_3^{\cdot} + (C_3H_5^{\cdot})^* \quad (59)\end{array}$$

[58] R. D. Koob, *J. Phys. Chem.*, 1969, **73**, 3168.
[59] R. D. Doepker, *J. Phys. Chem.*, 1969, **73**, 3219.

These reactions and subsequent reactions of the 'hot' species formed in reactions (61)—(63) and of the methylene from (61) can explain all of the

$$\left[\begin{array}{c} CH_3 \\ | \\ CH \\ \diagup \cdot \\ H_2C-\dot{C}H_2 \end{array}\right]^{\ddagger} \begin{cases} \longrightarrow 2H\cdot + C_4H_6(1,3) & (60) \\ \longrightarrow CH_2 + (C_3H_6)^* & (61) \\ \longrightarrow CH_3\cdot + (C_3H_5\cdot)^* & (62) \\ \longrightarrow C_2H_4 + (C_2H_4)^* & (63) \end{cases}$$

reaction products found experimentally, and the effects of pressure. Photolysis at 1236 Å increases the yields of products such as buta-1,3-diene which arise *via* elimination of hydrogen atoms from the intermediate biradical. The increased energy of the photon favours this reaction, as might be expected.

Similar experiments with pentane and 2,2,4-trimethylpentane in the liquid phase using 1470 Å radiation have led to the following quantum yields for the primary photodissociative steps.[60]

$$\begin{align}
\text{n-pentane} + h\nu \longrightarrow\ & H\cdot + 1\text{-methylbutyl (57\%)} \\
& + 1\text{-ethylpropyl (33\%)} \\
& + \text{n-pentyl (10\%)} \\
& \Phi = 0.15 & (64)
\end{align}$$

$$\longrightarrow C_2H_5\cdot + \text{n-}C_3H_7\cdot \text{ (79\%)} \\ + \text{i-}C_3H_7\cdot \text{(21\%)} \\ \Phi = 0.07 \quad (65)$$

$$\longrightarrow CH_3\cdot + C_4H_9\cdot \quad \Phi = 0.02 \quad (66)$$

$$\longrightarrow H_2 + C_5H_{10} \quad \Phi = 0.82 \quad (67)$$

$$\longrightarrow C_2H_4 + C_3H_8 \quad \Phi = 0.06 \quad (68)$$

$$\longrightarrow C_2H_6 + C_3H_6 \quad \Phi = 0.05 \quad (69)$$

$$\longrightarrow CH_4 + C_4H_8 \quad \Phi = 0.01 \quad (70)$$

$$\text{2,2,4-trimethylpentane} + h\nu \longrightarrow \text{i-}C_4H_9\cdot \text{ (59\%)} + \text{t-}C_4H_9\cdot \text{ (41\%)} \quad (71) \\ \Phi = 0.046$$

$$\longrightarrow C_3H_7\cdot + \text{neo-}C_5H_{11}\cdot \quad \Phi = 0.007 \quad (72)$$

$$\longrightarrow CH_3\cdot + C_7H_{15}\cdot \quad \Phi = 0.02 \quad (73a)$$

$$\longrightarrow H\cdot + C_8H_{17}\cdot \quad \Phi = 0.02$$

$$\text{or } H\cdot + CH_3\cdot + C_7H_{14} \quad (73b)$$

$$\longrightarrow \text{i-}C_4H_8 + \text{i-}C_4H_{10} \quad \Phi = 0.36 \quad (74)$$

$$\longrightarrow C_3H_6 + \text{neo-}C_5H_{12} \quad \Phi = 0.05 \quad (75)$$

$$\longrightarrow CH_4 + C_7H_{14} \quad \Phi = 0.17 \quad (76)$$

$$\longrightarrow H_2 + C_8H_{16} \quad \Phi = 0.08 \quad (77)$$

[60] R. A. Holroyd, *J. Amer. Chem. Soc.*, 1969, **91**, 2208.

Gas-phase Photochemistry

It can be seen that molecular mechanisms predominate in these decompositions. The correlation with effects expected from the absorption spectra is good in the case of 2,2,4-trimethylpentane in that at 1470 Å excitation involves primarily a $\sigma \to \sigma^*$ transition in the C—C bond, leading to the expectation that cleavage of C—C bonds might be more important than that of C—H bonds. This is borne out by the observation that only 2—3% of the substituted pentane molecules dissociate by cleavage of C—H bonds.

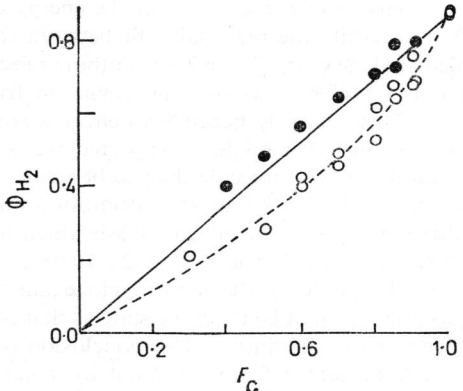

Figure 2 *Quantum yield of hydrogen formation (Φ_{H_2}) as a function of fraction of light absorbed by cyclohexane (F_c) in the photolysis of cyclohexane mixtures at 1470 Å, 95 °C, 750 Torr. Open circles: cyclohexane–benzene mixtures. Filled circles: cyclohexane–N_2O mixtures*
(Reproduced by permission from *J. Phys. Chem.*, 1970, **74**, 1395)

The situation with n-pentane itself is less clear-cut, but the main excitation is again expected to be in the $\sigma \to \sigma^*$ transition of a C—C bond. The main dissociation process with n-pentane is molecular elimination of hydrogen, as was observed in the photolysis of ethane,[56] but C—H bond cleavage is more important than in the case of the substituted pentane since this breakdown mode accounts for about 15% of the total. The vacuum-u.v. photolysis of n-butane at high temperatures has also been described.[61]

It has been stated that cyclohexane excited at 1470 Å in the liquid phase exists long enough to transfer its energy to a benzene molecule in the system, but it has been claimed that this process cannot occur in the gas phase (see Vol. 1, Part I, Chapter 2, page 112, in this series). In order to verify the latter point, mixtures of cyclohexane and benzene, and of cyclohexane and nitrous oxide have been photolysed at pressures up to 750 Torr at 95 °C.[62] The results are shown in Figure 2, in which the quantum yield of formation of molecular hydrogen is plotted against the fraction of light absorbed by the cyclohexane. The results with the C_6H_{12}–N_2O mixtures fall on a reason-

---

[61] J. R. McNesby and R. V. Kelley, *J. Phys. Chem.*, 1969, **73**, 789.
[62] R. R. Hentz and D. B. Peterson, *J. Phys. Chem.*, 1970, **74**, 1395.

able straight line, but those of the C_6H_{12}–C_6H_6 mixtures deviate from this. The deviation could be due either to energy transfer from excited-state cyclohexane molecules, or due to scavenging of hydrogen atoms in the system by benzene. The broken curve is that calculated on the basis of energy transfer using equation (78)

$$\Phi_{H_2} = \Phi^0_{H_2} F_c/(1+Z) \qquad (78)$$

where $\Phi^0_{H_2}$ is the yield of hydrogen in the absence of benzene ($F_c = 1$), and Z the frequency of collision of the C_6H_{12} with the energy acceptor: τ was taken as $4 \cdot 7 \times 10^{-11}$. Despite the reasonable fit between the experimental points and the calculated curve in Figure 2, the authors reject the possibility of energy transfer and prefer to ascribe the deviation from linearity to scavenging effects.[62] This is largely because no effect is seen when N_2O is used as the added gas, and as it might be expected that energy might be transferred more readily to this molecule than to benzene. The conclusion is reached on the basis of Figure 2 that absorption of a photon at 1470 Å produces a cyclohexane molecule in the gas phase which has a lifetime of less than $4 \cdot 7 \times 10^{-11}$ and probably less than 2×10^{-11} s. This being the case, it appears that the excited state of the cyclohexane which produces hydrogen in the gas phase cannot be the same state as that of longer lifetime which produces hydrogen in the liquid. This conclusion is in accord with the observations of fluorescence from saturated hydrocarbons, in which fluorescence from liquid cyclohexane excited at 1470 Å was observed to be quenched by benzene, whereas in the gas phase the same molecule did not exhibit fluorescence at all.[63]

Analysis of the products of the xenon-sensitized photolysis of perfluorocyclobutane indicates that the following primary steps are important:

$$Xe + h\nu \longrightarrow Xe^* \qquad (79)$$
$$Xe^* + c\text{-}C_4F_8 \longrightarrow Xe + c\text{-}C_4F_8^* \qquad (80)$$
$$c\text{-}C_4F_8 \longrightarrow (-CF_2CF_2CF_2CF_2-)^*$$
$$\longrightarrow CF_3CF=CFCF_3 \qquad (81)$$
$$\longrightarrow CF_2 + CF_3CF=CF_2 \qquad (82)$$
$$\longrightarrow 2C_2F_4 \qquad (83)$$
$$\longrightarrow 4CF_2 \qquad (84)$$

Reaction (81) is favoured.[64]

In the direct photolysis of tetrafluoroethylene at 1849 Å, C_2F_6, C_2F_2, cyclo-C_4F_8, cyclo-C_3F_6, C_3F_6, C_3F_8, and a polymer were identified as products. The primary photochemical reactions proposed to account for these products include double-bond fission and fluorine molecule and fluorine atom elimination,[65] processes which do not normally occur in the photolysis of ethylene itself.

[63] F. Hirayama and S. Lipsky, *J. Chem. Phys.*, 1969, **51**, 3616.
[64] J. E. Davenport and G. H. Miller, *J. Phys. Chem.*, 1969, **73**, 809.
[65] J. R. Dacey and J. G. F. Littler, *Canad. J. Chem.*, 1969, **47**, 3871.

Gas-phase Photochemistry

Isobutene has been photolysed using photon energies of 11·6—11·8 eV, 10·0 eV, and 8·4 eV [66] and 6·7 eV.[67] In the 8·4 eV (1470 Å) and 6·7 eV (1849 Å) photolyses, only neutral excited molecules are formed (85), but at the two shorter wavelengths ionization may also occur (86):

$$i\text{-}C_4H_8 + h\nu \longrightarrow i\text{-}C_4H_8^* \qquad (85)$$

$$i\text{-}C_4H_8 + h\nu \longrightarrow [i\text{-}C_4H_8^+]^* + e \qquad (86)$$

The $[i\text{-}C_4H_8^+]^*$ ions can be formed by direct ionization or by autionization of superexcited $C_4H_8^*$. The quantum yields of formation of ions were 0·32 at 1048—1067 Å, and 0·27 at 1236 Å. Although the nature of the products appears to be similar in the study at the three shortest wavelengths to that in the study at 1849 Å, the product distribution is different. Thus throughout the entire energy range in the first investigation the major product reported was allene, followed by propyne, ethane, isobutane, and methane in order of diminishing importance as other major products. Photolysis at 1849 Å apparently leads to the formation of 2-methylbutene as the major product, with isobutane, C_7-compounds, allene, methane, ethane, *trans*-but-2-ene, and propene as other important products. There are other more important differences between the two studies. At the short wavelengths the quantum yields of the products allene, propyne, and isobutane (or the $M/N_{ex.}$ values, where $M/N_{ex.}$ is as defined earlier) are independent of total pressure, whereas at 1849 Å the yield of allene and ethane is strongly quenched by addition of chemically inert gas, as is the amount of isobutene consumed. It appears therefore that at the longest wavelength the excited isobutene molecule exists sufficiently long for it to be quenched by the addition of an inert molecule such as nitrogen (although by an unspecified mechanism), whereas at the higher photon energies this is not the case. If quenching is assumed on every collision, the lifetime of the state produced at 1849 Å must be of the order of 3×10^{-10} s. It is suggested that this state is a Rydberg state of the molecule.[67] Careful studies using deuterium substitution and radical scavengers indicate that in the energy range 8·4—11·8 eV, 80% of the excited isobutene molecules decompose ultimately to give methyl radicals, hydrogen atoms, and allene and propyne.

$$i\text{-}C_4H_8^* \longrightarrow CH_3\cdot + CH_3C{\equiv}CH + H\cdot \qquad (87)$$

$$\longrightarrow CH_3\cdot + CH_2{=}C{=}CH_2 + H\cdot \qquad (88)$$

The initial step must be cleavage of a C—C bond or a C—H bond, and both steps are assigned equal importance at 1849 Å.

$$i\text{-}C_4H_8^* \longrightarrow CH_3\cdot + C_3H_5\cdot \ddagger \qquad (89)$$

$$\longrightarrow H\cdot + C_4H_7\cdot \ddagger \qquad (90)$$

Minor primary fragmentations include the following:

$$i\text{-}C_4H_8^* \longrightarrow C_3H_6 + CH_2 \qquad (91)$$

$$\longrightarrow CH_4 + C_3H_4 \qquad (92)$$

---

[66] J. Herman, K. Herman, and P. Ausloos, *J. Chem. Phys.*, 1970, **52**, 28.
[67] P. Borrel and P. Cashmore, *Trans. Faraday Soc.*, 1969, **65**, 1595.

Interesting reactions are proposed for the ions formed in reaction (86). Initially these undergo a proton transfer reaction with the starting material (93), giving rise to isobutyl radical ions and n-butyl radical ions.

$$\text{i-}C_4H_8^+ + \text{i-}C_4H_8 \longrightarrow C_4H_9\cdot^+ + C_4H_7\cdot \qquad (93)$$

If about 30% of the ions formed in (93) are assumed to be of the iso- form and these subsequently undergo a further hydride ion transfer step (94), the observed yield of isobutane can be explained.

$$\text{i-}C_4H_9^+ + \text{i-}C_4H_8 \longrightarrow \text{i-}C_4H_{10} + C_4H_7^+ \qquad (94)$$

Reaction (94) should be endothermic for ground-state ions, but there should be sufficient energy present in the ions produced by the 10 or 11·6 eV photons to offset this factor. $C_4H_9^+$ ions also disappear by rapid addition to the parent isobutene, eventually producing a polymer. Addition of quenching gas such as argon to the reaction mixture caused a reduction in the rate of formation of isobutene, which could be explained either by a deactivation of the 'hot' isobutene ion, (95), or by a collision-induced isomerization of the isobutyl ion to the more stable tertiary form:

$$[\text{i-}C_4H_8^+]^* + Ar \longrightarrow \text{i-}C_4H_8^+ + Ar \qquad (95)$$

$$(CH_3)_2CHCH_2^+ + Ar \longrightarrow (CH_3)_3C^+ + Ar \qquad (96)$$

When iso-C_4D_8 was irradiated in the presence of alkanes or cycloalkanes AH_2, the interesting H_2^- transfer reaction (97) was observed:

$$\text{i-}C_4D_8^+ + AH_2 \longrightarrow (CD_3)_2CHCD_2H + A^+ \qquad (97)$$

Experiments performed with a partially deuteriated isopentane indicated that process (97) is highly stereospecific.

The photolysis of *cis*-pent-2-ene at 1849 Å is proposed to occur *via* a Rydberg state of the molecule,[68] and eventually leads to the formation of hydrogen, a polymer, and more than twenty hydrocarbons, of which the main ones are ethane, butadiene, but-1-ene, 2- and 3-methylbutene, *trans*-pent-2-ene, methane, and ethylene. A reaction scheme is proposed which can account for the observed products.

Mono-olefins absorb radiation at the high end of the u.v. region extending into the vacuum-u.v. Chemical changes can be induced in such molecules using light of lower photon energy provided that (*a*) a chromophoric group is introduced into the molecule which absorbs the lower-energy radiation, and (*b*) a molecule is introduced into the system which can absorb the radiation and then sensitize the chemical reaction of the olefin. Examples of the latter type have been seen in the preceding section on the photochemistry of aromatics, in which the *cis–trans* isomerization of but-2-ene by aromatics is used to diagnose triplet states. The nature of the intermediates in the liquid-phase sensitized *cis–trans* isomerization of alkenes [69]

---

[68] P. Borrel and P. Cashmore, *Trans. Faraday Soc.*, 1969, **65**, 2412.
[69] J. Saltiel, K. R. Neuberger, and M. Wrighton, *J. Amer. Chem. Soc.*, 1969, **91**, 3658.

and quantum yields for the sensitized photoisomerization of *cis*- and *trans*-stilbene in solution [70] have already been discussed in Chapter 2 of this Part.

As an example of (*a*), the molecular rearrangement of 1-phenylbut-2-ene in the gas phase has been reported using exciting wavelengths in the region 2660—2470 Å.[71] Two isomers were produced, the relative yields of which were strongly dependent upon wavelength. One of these was identified as the *cis* isomer of the (*trans*) starting material, and the other proved to be *trans*-1-phenyl-2-methylcyclopropane. The relative yield of the *cis* isomer to the cyclopropane varied from 1·32 at 2660 Å to 1·02 at 2470 Å. Increase in pressure reduced the yield of the cyclopropane and increased the yield of the *cis* isomer. In addition, the quantum yield of both compounds was drastically reduced with increase in photon energy. This may be due to a radiationless transition associated with the benzene ring which appears to be of increasing importance with increase in energy (see previous section). The thermal and photochemical isomerization [72] of *cis*-3,4-dimethylcyclobutene has been discussed in Chapter 2. An unusual reaction is proposed in the mercury($^3P_1$)-sensitized photolysis of cyclo-octa-1,5-diene vapour involving the molecular elimination of ethylene. The primary processes involved [73] may be written as:

$$Hg + h\nu \longrightarrow Hg^* \qquad (98)$$

$$\text{cyclo-octadiene} + Hg^* \longrightarrow \text{cyclo-octadiene}^* + Hg \qquad (99)$$

$$\text{cyclo-octadiene}^* \longrightarrow \text{benzene}^* + C_2H_4 \quad \Phi = 0\cdot20 \qquad (100)$$

$$\longrightarrow 2\, C_4H_6 \qquad \Phi = 0\cdot13 \qquad (101)$$

Followed by:

$$\text{benzene}^* + M \longrightarrow \text{benzene} \qquad (102)$$

$$\longrightarrow 1,3,5\text{-hexatriene} \qquad (103)$$

$$\longrightarrow \text{benzene} + H_2 \qquad (104)$$

$$\longrightarrow C_2H_4 + 2C_2H_2 \qquad (105)$$

---

[70] H. A. Hammond, D. E. DeMeyer, and J. L. R. Williams, *J. Amer. Chem. Soc.*, 1969, **91**, 5180.
[71] M. Comtet, *J. Amer. Chem. Soc.*, 1969, **91**, 7761.
[72] R. Srinivasan, *J. Amer. Chem. Soc.*, 1969, **91**, 7557.
[73] S. Takamuku and H. Sakurai, *J. Phys. Chem.*, 1969, **73**, 1171.

The cyclohexa-1,3-diene formed in (98) is assumed to be in a vibrationally excited ground-state. The reactions of cyclohexa-1,3-diene photosensitized by benzaldehyde in the gas phase have been reported.[74]

Two investigations of the photochemistry of 1,2-dienes in the gas phase have appeared.[75, 76] In the vacuum-u.v. decomposition of buta-1,2-diene at 1470 Å and 1236 Å, many products were isolated. In the absence of additives these were, in approximate order of importance, C_4H_4, methylacetylene, ethane, acetylene, buta-1,3-diene, and allene, ethylene, methane, and a C_5-compound in roughly equal amounts. Some of the rates of formation of reaction product showed a marked dependence upon pressure, leading to the suggestion that two excited states of the buta-1,2-diene are important in this decomposition. Similar products were observed in photolysis at 2200—2600 Å. Although a reaction mechanism was proposed which can adequately explain the products at all wavelengths, the study reported was brief, and quantum yields were not determined, so a fuller report of this work will be deferred until more experimental evidence becomes available.

The benzene-sensitized photodecomposition of a variety of 1,2-dienes in the vapour phase has established that at least four modes of decomposition are common to these compounds.[76] Thus hexa-1,2-diene gave four products: ethylene ($\Phi = 0.15$); buta-1,3-diene ($\Phi = 0.16$); vinylcyclobutane ($\Phi = 0.07$); and bicyclo[3,1,0]hexane ($\Phi = 0.005$). The vacuum-u.v. photolysis of the same material using light of wavelength 1600—2100 Å yielded many more products, four of which were identified as ethylene, buta-1,3-diene, vinylcyclobutane, and hex-2-ene. Products arising from the benzene-sensitized and vacuum-u.v. photolysis of cyclonona-1,2-diene, hepta-1,2,6-triene, and cyclonona-1,2,6-triene are also reported. A type of reaction which is peculiar to the allenic group was proposed, in which a bond is formed at the central carbon atom of the allenic group, followed by rebonding or rehybridization in the remainder of the molecule. This is not the position at which reaction occurs between free radicals and ground-state allenes, where reaction at a terminal carbon atom seems to be preferred. The reactivity of the central carbon can be explained on the basis of a planar excited state of the allene, of unspecified multiplicity. With such an intermediate, bonding at the central carbon gives an allyl radical directly, whereas with an antiplanar configuration (as in the ground state), reaction at the central carbon must be followed by rotation of the bond by 90° for formation of the allyl radical. The reactions outlined in this work are discussed in detail from the point of view of organic mechanism and in the light of orbital symmetry. Further discussion may be found in the relevant section of the Parts of this volume dealing with Organic Photochemistry.

[74] G. R. Demare, P. Goldfinger, G. Hubrechts, E. Jonas, and M. Toth, *Ber. Bunsengesellschaft. Phys. Chem.*, 1969, **73**, 867.
[75] R. D. Doepker and K. L. Hill, *J. Phys. Chem.*, 1969, **73**, 1313.
[76] H. R. Ward and E. Karafiath, *J. Amer. Chem. Soc.*, 1969, **91**, 7475.

Acetylene which is photolysed at 1236 Å and 1470 Å gives rise to many products, as is typical of molecules irradiated with high-energy photons. In the case of acetylene the products include diacetylene, benzene, ethylene, buta-1,3-diene, and vinylacetylene.[77] The product yields of all molecules except benzene fall with increasing acetylene pressure: the yield of benzene increases with rise in acetylene pressure. As in other studies, a very long complex sequence of reactions is proposed to account for the observed products and their variation with wavelength of excitation, pressure of starting material, and pressure of added inert gases and hydrogen. For reasons of economy of space, readers are referred to the original paper for details of these rather speculative proposals.

3 Carbonyl Compounds

It is astonishing that, since the classic early work on the photochemistry of acetone in the gas phase and the multitude of subsequent publications relating to this molecule, there should still be discussion concerning the primary processes involved. Thus a recent publication gives details of a revised scheme which quantitatively correlates the data from the literature on emission and decomposition yields.[78] Since the new scheme deviates somewhat from the earlier mechanism proposed, it is worth quoting the original and revised schemes from reference 78.

Original Scheme

$$A + h\nu \longrightarrow {}^1A \longrightarrow {}^3A_n^* \quad \text{(sequence of steps)} \tag{106}$$

$$^3A_n^* \longrightarrow CH_3\cdot + CH_3CO\cdot \quad [\text{rate constant} = k'(\varepsilon)^*] \tag{107}$$

$$M + {}^3A_n^* \longrightarrow {}^3A_0 + M^* \quad (\text{rate} = Z, \text{collision frequency}) \tag{108}$$

$$^3A_0 + M \longrightarrow CH_3\cdot + CH_3CO\cdot + M \quad (\text{rate constant} = {}^3k_d) \tag{109}$$

$$^3A_0 \longrightarrow A + h\nu_p \tag{110}$$

With HBr present,

$$HBr + {}^3A \longrightarrow CH_3\dot{C}(OH)CH_3 + Br\cdot \tag{111}$$

where A refers to an acetone molecule. The mechanism above qualitatively satisfies the experimental observations that both a spontaneous and pressure-dependent thermal decomposition of the acetone triplet are involved, but quantitative agreement with experiment is possible only if some rather unsatisfactory assumptions about rates of the processes are made. Thus if the rate constant for reaction (107) is written in the usual Rice–Ramsperger–Kassel way as

$$k'(\varepsilon)^* = A\left(\frac{E - {}^3E_c}{E}\right)^{s-1} \tag{112}$$

and $A = 10^{15.1}\,\text{s}^{-1}$ it is necessary to assume $s = 18$ and that

[77] S. Takita, Y. Mori, and I. Tanaka, *J. Phys. Chem.*, 1969, **73**, 2929.
[78] H. E. O'Neal and C. W. Larson, *J. Phys. Chem.*, 1969, **73**, 1011.

$^3E_c = 71\cdot4$ kJ mol^{-1} in order to get a fit with the experimental data. Since the maximum number of degrees of freedom is only 24, 18 is too large a number to have to invoke at the temperatures used. Moreover, the experimental value for the activation energy at 200 Torr was about 42 kJ mol^{-1}, and thus it was necessary to postulate that the observed decomposition was a unimolecular process in its low-pressure region.

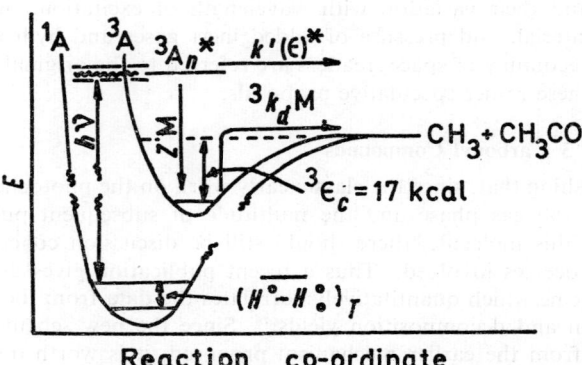

Figure 3 *Original primary process mechanism*
(Reproduced by permission from *J. Phys. Chem.*, 1969, **73**, 1011)

However, this assumption seems unreasonable in view of the fact that the spontaneous decomposition was competitive with, but slower than, collisional deactivation at a total pressure of 75 Torr, implying that the thermal decomposition under similar conditions would be in the *fall-off region* and not near the low-pressure limit. A pictorial representation of the processes involved in this mechanism is given in Figure 3.

The revised mechanism obviates this difficulty by assuming that the thermal decomposition of the acetone triplet is a unimolecular reaction in its pressure fall-off region, as the experimental data would imply.

Revised Mechanism

$^3A_m^* \xrightarrow{k(\varepsilon)^*} CH_3\cdot + CH_3CO\cdot$ (spontaneous decomposition) (113)

$M + {}^3A_m^* \longrightarrow {}^3A_0 + M^*$ (collisional deactivation) (114)

$^3A + M \underset{Z}{\overset{k_0}{\rightleftarrows}} {}^3A_n + M$ ⎫
$\phantom{^3A + M \xrightarrow{k_0} {}^3A_n + M}$ ⎬ (thermal decomposition) (115)
$^3A_n \xrightarrow{k(\varepsilon)} CH_3\cdot + CH_3CO\cdot$ ⎭

$^3A_0 \longrightarrow A + h\nu_p$ (phosphorescence) (116)

$^3A_0 \longrightarrow A$ (intersystem crossing) (117)

$^3A + HBr \longrightarrow$ quenching (118)

Gas-phase Photochemistry

The processes are represented pictorially in Figure 4, and a good fit with experimental data is obtained with the following parameters:

$$k(\varepsilon)^* = A_\infty^* \left(\frac{E-E_c}{E}\right)^{s-1}, \text{ where } A_\infty^* = 10^{11.1}\text{s}^{-1}, E_c = 38\cdot2\text{---}43\cdot6 \text{ kJ mol}^{-1}, \text{ and}$$

$$s = 5\cdot7\text{---}7\cdot2.$$

$$^3k_d = \frac{k_0 k(\varepsilon)\text{M}}{Z\text{M} + k(\varepsilon)}, \text{ where } k_0 = 10^{9\cdot81-5\cdot95/\theta} \text{ l mol}^{-1}\text{s}^{-1} \text{ and } k = \frac{k_0 k(\varepsilon)}{Z}$$

$$= 10^{10\cdot14-9\cdot6/\theta} \text{ s}^{-1}.$$

$$k_{117} = 10^{4\cdot96-1\cdot84/\theta} \text{ s}^{-1} \quad \text{and} \quad k_{118} = 10^{9\cdot6-1\cdot1/\theta} \text{ l mol}^{-1}\text{s}^{-1}.$$

In the equations above, θ refers to the absolute temperature, and all data discussed here refer to excitation at 3130 Å.

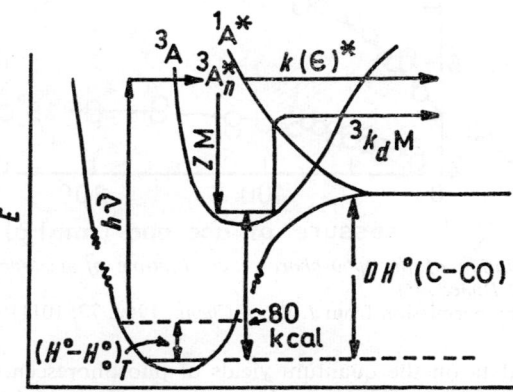

Figure 4 *Reaction co-ordinate representation of the primary processes in the acetone photochemical system*
(Reproduced by permission from *J. Phys. Chem.*, 1969, **73**, 1011)

Data from the literature can be used to test the validity of the revised mechanism. Phosphorescence lifetimes of acetone have been found to be strongly pressure dependent,[79] as shown in Figure 5. Usual Stern–Volmer behaviour was not observed, and the lifetime of phosphorescence was best represented by an equation of the type:

$$\tau^{-1} = k_1 + k_2 P/(1 + k_3 P) \tag{119}$$

The curves in Figure 5 show a marked similarity to unimolecular rate-constant fall-off curves, and use of the steady-state approximation on the revised mechanism above gives the following relations:

$$^3k_d = \frac{k_0 \text{M}}{1 + [Z\text{M}/k(\varepsilon)]} \tag{120}$$

$$^3\tau^{-1} = k_{116} + k_{117} + {}^3k_d \tag{121}$$

[79] W. E. Kaskan and A. B. F. Duncan, *J. Chem. Phys.*, 1950, **18**, 427.

Therefore

$$^3\tau^{-1} = \frac{(k_{116}+k_{117}) + \{(k_{116}+k_{117})Z/k(\varepsilon) + k_0\}M}{1 + [ZM/k(\varepsilon)]} \quad (122)$$

This is of the same form as (119), and thus the curvature in the plots in Figure 5 may well be due to the pressure-dependence of decomposition of the acetone triplet molecules. The possibility that the reaction product biacetyl is a cause of the curvature in these experiments is discounted on good experimental grounds.

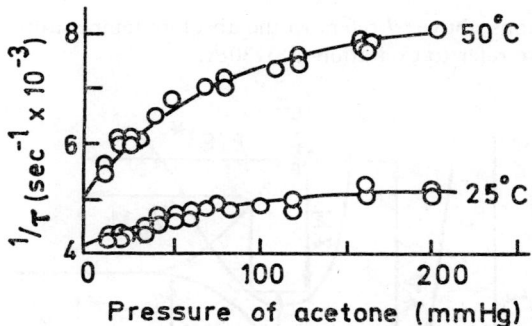

Figure 5 *Variation of the 'phosphorescence' lifetime of acetone with pressure* (*Kaskan and Duncan* [79])
(Reproduced by permission from *J. Phys. Chem.*, 1969, **73**, 1011)

Available data on the quantum yields of phosphorescence can also be re-evaluated on the basis of the scheme above. From it the value of the phosphorescence yield (Φ_p) is given by:

$$\Phi_p = \left(\frac{ZM}{k(\varepsilon)^* + ZM}\right) \times \left(\frac{k_{116}}{k_{116} + k_{117} + {}^3k_d}\right) \quad (123)$$

This can be rearranged to

$$\Phi_p^{-1} F_m^{-1} = \left(\frac{k_{117}+k_{116}}{k_{116}}\right) + \frac{k_0}{k_{116}}\left(\frac{M}{X_m}\right) \quad (124)$$

where

$$F_m = 1 + \frac{k(\varepsilon)^*}{ZM} \quad \text{and} \quad X_m = 1 + \frac{M}{M(\tfrac{1}{2})}$$

$$M(\tfrac{1}{2}) = k(\varepsilon)/Z = k_\infty/k_0$$

From values of F_m and X_m obtained from decomposition work the literature values of phosphorescence yields can be used to evaluate the rate-constant ratios in equation (124). The agreement between widely different sets of data treated in this manner is shown in Table 7. The data relating phosphorescence yields to pressure and exciting wavelength [82] are worthy of further consideration.

Table 7 Rate parameters from the photodecomposition of acetone

Thermal decomposition of triplet acetone

Ref.	Log A_0	E_0 (kJ mol^{-1})	Log A	E (kJ mol^{-1})	s
79	10·05	25·00	9·81	38·1	5·7
80	—	25·84	—	—	—
81	—	—	10·46	42·63	7·5
82	—	23·10	—	—	—

Spontaneous decomposition of acetone triplets

Ref.	λ(Å)	T(°C)	$(E-E_c)$	$[k(\varepsilon)^*/Z]$	$[k'(\varepsilon)^*/Z]$	s
78	3140	44	10·92	5·5	—	6·7
78	3140	96	13·44	15·4	—	6·7
78	3140	126	13·02	24·4	—	6·7
78	3140	150	16·80	35·3	—	6·7
82	3130	40	11·76	8·3	—	6·6
82	3020	40	25·84	37·5	—	8·1
82	2970	40	32·76	44	18	9·1
82	2890	40	43·68	110	35	9·8
82	2800	40	57·12	320	30	9·7

Units of A_0 are l mol^{-1} s^{-1}, A_∞ s^{-1}: s = number of effective oscillators.
For the data at 3140 Å, s has been assumed constant.
$k'(\varepsilon)^*/Z$ represents decomposition: stabilization ratios assuming a two-step collisional deactivation.
$k(\varepsilon)^* = A_\infty^*(E-E_c)^{s-1}$; ε_{00} = 338·1 kJ mol^{-1}; $A_\infty^* = 10^{11\cdot1}$ s^{-1}.

Rearrangement of (124) gives

$$\Phi_p F_m = k_{116}/(k_{117} + k_{116} + {}^3k_0) \tag{125}$$

At any constant pressure, the right-hand side of (125) is constant, and thus phosphorescence data at different wavelengths can be compared with those at 3130 Å since

$$(\Phi_p F_m)_\lambda = (\Phi_p F_m)_{3130} = S, \quad \text{say.} \tag{126}$$

Now the value of F_m at 3130 Å is known from thermal data, so S is also known. Rearranging (126) we obtain

$$(S)/\Phi_p(\lambda) = 1 + [k(\varepsilon)^*_\lambda Z] \, (\text{M}^{-1}) \tag{127}$$

Thus plots of $(S)/\Phi_p$ against M^{-1} should be straight lines with unit intercepts and slopes of $[k(\varepsilon)^*/Z]$. Figure 6 shows that this is the case for the longest wavelength, 3021 Å, but that at other wavelengths there is pronounced curvature in the plots. These departures from linearity can be explained. The model used has assumed that 'hot' triplets are deactivated in a single collision, *i.e.* a strong-collision model has been assumed. However, at short wavelengths, where the vibrational energy content of the triplet acetone may be considerable, such a model may be inadequate.

[80] G. W. Luckey and W. A. Noyes jun., *J. Chem. Phys.*, 1951, **19**, 227.
[81] H. J. Groh, G. W. Luckey, and W. A. Noyes jun., *J. Chem. Phys.*, 1953, **21**, 115.
[82] J. Heicklen, *J. Amer. Chem. Soc.*, 1959, **81**, 3863.

Assuming for the shorter wavelengths that a two-collision deactivation is necessary, it can be shown that (127) should be revised to

$$\frac{(\Phi_p F_m)_{3130}}{\Phi_p(\lambda)} = 1 + \frac{k(\varepsilon)^* + k'(\varepsilon)^*}{ZM} + \frac{k(\varepsilon)^* k'(\varepsilon)^*}{Z^2 M^2} \quad (128)$$

where $k'(\varepsilon)^*$ is the rate constant for the spontaneous decomposition of the triplet at the average energy state produced by the first collision. The data in Figure 6 allow an approximate evaluation of the rate parameters in

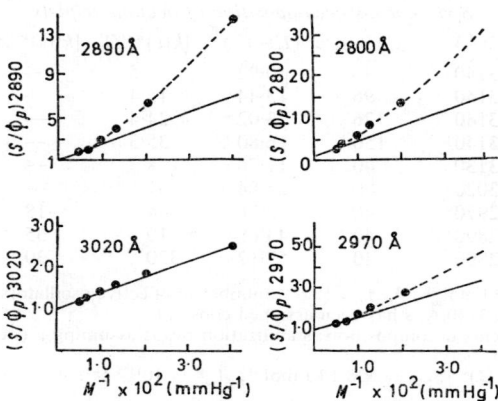

Figure 6 *Test of equation* (127) *for acetone triplets*
(Reproduced by permission from *J. Phys. Chem.*, 1969, **73**, 1011)

equation (128), which are given in Table 7. The numbers correspond to removal of about 28·6, 17·6, and 13·9 kJ mol⁻¹ of vibrational energy in the first collision at 2800, 2890, and 2970 Å respectively. The number of effective oscillators contributing to the thermal decomposition of a unimolecular reaction can be estimated from the limiting values of the activation energy at high and low pressures using the relation

$$E - E_0 = (s - 3/2) RT \quad (129)$$

The values obtained from different data are given in Table 7. The Rice-Ramsperger–Kassel treatment also allows an estimate to be made of the zero-point energy of the acetone triplet state above the ground state, and the value obtained was 336 kJ mol⁻¹.

The work described above is a good example of the parameters which can be measured in a photochemical study provided all the data pertinent to particular excited states, including physical as well as chemical processes, are collected. It is hoped that many more such thorough investigations of photochemical systems will be undertaken in the future.

Methyl radicals formed in the photolysis of acetone vapour will, among other reactions, abstract a hydrogen atom from the starting material to form methane and an acetonyl radical. Hitherto this reaction has been supposed to occur in the gas phase only, but it has been shown that below a pressure of about 8 Torr, reaction (130) contributes significantly to the amount of methane produced.[83] The rate of reaction (130) varies markedly with the nature of the surface. On a Pyrex surface the reaction has an activation energy of 27.7 ± 4.2 kJ mol^{-1}, and the rate increases with increase in acetone pressure, and is always faster on a 'per-collision' basis than the corresponding reaction in the gas phase.

$$CH_3\cdot + (CH_3COCH_3)_{adsorbed} \longrightarrow CH_4 + CH_3COCH_2\cdot \quad (130)$$

Acetone has been used to sensitize the decomposition of 1,4-dichlorobutane and other chlorobutanes.[84] The singlet state of the ketone is said to be responsible.

Hexafluoroacetone photolysed at 1470 Å yields only carbon monoxide and hexafluoroethane, as in irradiation at longer wavelengths.[85] If the quantum yield of formation of N_2 from the photolysis of N_2O at the same wavelength is taken as 1.44, then the quantum yield of formation of CO from the ketone, Φ_{CO}, is 0.97 ± 0.05 at 25 °C. Since carbon monoxide is the only product not condensable at 77 K, hexafluoroacetone is a convenient actinometer to use in this wavelength region. The relative yield of hexafluoroethane to CO was 0.9, similar to that in the near-u.v. experiments, a fact which points to some loss of trifluoromethyl radicals. Perfluoropropylene quenches the excited state of hexafluoroacetone, giving an oxetan.[86] Addition of small amounts of triplet quenchers greatly reduced the yield of oxetan, indicating that the triplet state of the ketone was involved in the oxetan formation, in contrast with the acetone–perfluoropropylene system in which the singlet state of the ketone is the reacting species. The differences between the two systems may be due to the acetone having a higher triplet level than the olefin or the hexafluoroacetone, so that triplet energy transfer to the olefin dominates in the case of acetone, whereas if the triplet level of the hexafluoroacetone lies lower than that of the olefin no transfer can occur, and oxetan formation becomes a possibility. One wonders why this proposal was not verified experimentally. Other substituted acetones have been photolysed. Thus chloropentafluoroacetone at 3130 Å decomposes predominantly to give carbon monoxide and two radicals (131a), but at shorter wavelengths a primary mode of decomposition involving the loss of chlorine atom becomes important (131b):

$$CF_3COCF_2Cl + h\nu \longrightarrow CF_3 + CO + CF_2Cl \quad (131a)$$
$$CF_3COCF_2Cl + h\nu \longrightarrow CF_3COCF_2 + Cl \quad (131b)$$

The reactions of the radicals produced in these primary steps are discussed, but not the states responsible for the breakdown.[87] This Report is not

---

[83] J. Konstantatos and C. P. Quinn, *Trans. Faraday Soc.*, 1969, **65**, 2693.
[84] M. A. Golub, *J. Amer. Chem. Soc.*, 1969, **91**, 4925; 1970, **92**, 2615.
[85] J. J. Magenheimer and R. B. Timmons, *J. Chem. Phys.*, 1970, **52**, 2790.
[86] R. H. Knipe, A. S. Gordon, and W. R. Ware, *J. Chem. Phys.*, 1969, **51**, 840.
[87] J. R. Majer, C. Olavesen, and J. C. Robb, *J. Chem. Soc. (A)*, 1969, 893.

primarily concerned with reactions of free radicals produced photochemically, and these will not be discussed here. However, those interested will find discussion of the reaction of $CH_3\cdot$ and $CH_3CO\cdot$ radicals with nitric oxide,[88] $CH_3\cdot$ and $CF_3\cdot$ with sulphur hexafluoride,[89] $CF_3\cdot$ with NH_3,[90] $CF_3\cdot$ with silanes,[91, 92] $CF_3\cdot$ with aromatic aldehydes,[93] and hydrogen-abstraction reactions of $\cdot CF_2Cl$ radicals,[94] addition of $\cdot CF_2Cl$ radicals to olefins,[95] and the absolute rate of combination of $\cdot CF_2Cl$ radicals,[96] and their heat of formation,[97] in the references given.

The absolute quantum yield of fluorescence of pentan-2-one is dependent upon exciting wavelength,[98] being $2\cdot 67 \times 10^{-3}$ at 3130 Å and falling to $1\cdot 49 \times 10^{-3}$ at 2654 Å. This decrease by a factor of two is accompanied by a doubling of the quantum yield of the Norrish Type II reaction, indicating that the reaction and fluorescence are competitive at all wavelengths. No quenching of the singlet state of the ketone by buta-1,3-diene was observed, which is interpreted as meaning that the singlet state of the diolefin must lie above that of the ketone. Since efficient quenching of aromatic molecules by the olefin occurs, it seems likely that its singlet level must lie between 2650 Å and 2800 Å. The values of quenching parameters for pentan-2-one and olefins in the literature are different for the gas phase and the liquid, and this has been shown to be due to absorption by the product acetone at high conversions in the liquid phase.[99] Clearly comparisons between the phases should only be made with data obtained where product concentrations are not allowed to reach levels at which significant absorption of light can occur.

Biacetyl is one of the few molecules of which change of phase does not drastically alter luminescence yields. Thus the fluorescence lifetime of biacetyl vapour at 25 Torr and 300 K was measured as $10\cdot 6 \pm 1\cdot 3$ ns, and that of a $0\cdot 05$ mol l^{-1} solution in benzene as $12\cdot 3 \pm 1\cdot 8$ ns.[100] Since decay of biacetyl singlets is predominantly *via* intersystem crossing to the triplet state, it is clear that this radiationless process is not subject to medium effects. It would appear, therefore, that intersystem crossing in singlet biacetyl to the lowest triplet state is an intramolecular phenomenon, and that biacetyl can therefore be classed as a 'large' molecule. Any theoretical

---

[88] H. E. Avery, D. M. Hayes, and L. Phillips, *J. Phys. Chem.*, 1969, **73**, 3498.
[89] H. F. Le Fèvre, J. D. Kale, and R. B. Timmons, *J. Phys. Chem.*, 1969, **73**, 1614.
[90] H. F. Le Fèvre and R. B. Timmons, *J. Phys. Chem.*, 1969, **73**, 3854.
[91] E. R. Morris and J. C. J. Thynne, *Trans. Faraday Soc.*, 1970, **66**, 183.
[92] E. Jakubowski, H. S. Sandhu, H. E. Gunning, and O. P. Strausz, *J. Chem. Phys.*, 1970, **52**, 4242.
[93] J. R. Majer, S.-A. M. A. Naman, and J. C. Robb, *Trans. Faraday Soc.*, 1969, **65**, 3295.
[94] L. M. Leyland, J. R. Majer, and J. C. Robb, *Trans. Faraday Soc.*, 1970, **66**, 901.
[95] L. M. Leyland, J. R. Majer, and J. C. Robb, *Trans. Faraday Soc.*, 1970, **66**, 904.
[96] J. R. Majer, C. Olavesen, and J. C. Robb, *Trans. Faraday Soc.*, 1969, **65**, 2988.
[97] L. M. Leyland, J. R. Majer, and J. C. Robb, *Trans. Faraday Soc.*, 1970, **66**, 898.
[98] F. S. Wettack, *J. Phys. Chem.*, 1969, **73**, 1167.
[99] C. H. Bibart, M. G. Rockley, and F. S. Wettack, *J. Amer. Chem. Soc.*, 1969, **91**, 2802.
[100] L. G. Anderson and C. S. Parmenter, *J. Chem. Phys.*, 1970, **52**, 466.

Gas-phase Photochemistry

description of the transition must be based only on those eigenstates of the system accessible to the isolated molecule, and must not include states of the surrounding medium. This point has been further developed in an investigation of the fluorescence, phosphorescence, and triplet formation of biacetyl at low pressures.[101] Both singlet and triplet yields are independent of pressure down to 0·1 Torr, well below the pressure at which collisions can be important. Since theoretical approaches would place biacetyl outside of the 'large' molecule limit, implying that radiationless processes should only occur for this molecule when an external perturbation occurs, these important experimental observations suggest that the relevant theories must be regarded as very approximate until developed further.

The mechanism of the photochemical decomposition of cyclopentanone in the gas phase has been further examined by obtaining quantum yields of formation of the products CO, C_2H_4, cyclobutane, and pent-1-en-5-al as a function of pressure, temperature, and of wavelength. At 3130 Å the yield of the pentenal decreases with increase in temperature and increases with increase in pressure, indicating that it is formed predominantly from lower vibrational levels of an excited state. The ratio of ethylene to cyclobutane is apparently independent of conditions. Since a triplet quencher, penta-1,3-diene, apparently exerts no influence on the decomposition other than might be expected for a vibrational relaxer, there is no necessity to invoke the intermediacy of an excited triplet state. However, a step must be included which relaxes the excited molecule *via* a first-order process which does not lead to products. The following simple scheme suffices to describe the decomposition.

$$C + h\nu \longrightarrow {}^1C^* \quad (132)$$

$$ {}^1C^* \longrightarrow 2C_2H_4 + CO \quad (133)$$

$$ {}^1C^* \longrightarrow \text{cyclo-}C_4H_8 + CO \quad (134)$$

$$ {}^1C^* \longrightarrow \text{pent-1-en-5-al} \quad (135)$$

$$ {}^1C^* \longrightarrow C \quad (136)$$

$$ {}^1C^* + M \longrightarrow {}^1C + M \quad (137)$$

$$ {}^1C \longrightarrow \text{pent-1-en-5-al} \quad (138)$$

$$ {}^1C \longrightarrow C \quad (139)$$

where C refers to a cyclopentanone molecule, and * to excess of vibrational energy. Rate constant ratios have been determined from this scheme.[102] The conclusions regarding the triplet state in the study above are in disagreement with results obtained in the photolysis of *trans*-2,3-dimethylcyclopentanone,[103] and with earlier studies on the direct and benzene-

---

[101] C. S. Parmenter and H. M. Poland, *J. Chem. Phys.*, 1969, **51**, 1551.
[102] C. Y. Mok, *J. Phys. Chem.*, 1970, **74**, 1432.
[103] H. M. Frey and D. H. Lister, *J. Chem. Soc. (A)*, 1970, 627.

photosensitized decompositions of cyclopentanone and [2-^{3}H$_1$]cyclopentanone [104] in which powerful arguments are developed for the triplet state of the ketones being the precursor of the product aldehyde. The differences can be reconciled if it is assumed that the triplet state of the ketone is formed, but that it has an extremely short lifetime, so that triplet quenchers have no effect. It should be noted that the maximum pressure of quencher used in the first study [102] was only 80 Torr, which would not be sufficient to show an effect on a short-lived triplet. The absence of stereospecificity in the hydrocarbon products in the photolysis of the substituted cyclopentanone is further proof that short-lived biradical intermediates are present. The products observed in this study were *cis*-2,3-dimethylcyclopentanone, *cis*- and *trans*-4-methylhex-4-enal, 2,3-dimethylpent-4-enal, *cis*- and *trans*-1,2-dimethylcyclobutane, *cis*- and *trans*-but-2-ene, ethylene, and carbon monoxide. Several mechanisms are discussed, but the simple scheme outlined below is preferred, although there are insufficient data to support it as the unique explanation of the results:

$$trans\text{-}2,3\text{-DMCP} + h\nu \longrightarrow {}^1C^* \qquad (140)$$

$$^1C^* + M \longrightarrow {}^1C + M \qquad (141)$$

$$^1C^* \longrightarrow {}^3C^* \qquad (142)$$

$$^1C \longrightarrow {}^3C \qquad (143)$$

$$^3C^* \longrightarrow \text{hydrocarbons} + CO \qquad (144)$$

$$^3C \longrightarrow \text{aldehydes} + \text{ketones} \qquad (145)$$

where C now refers to the substituted cyclopentanone, superscripts to multiplicity, and * to a vibrationally 'hot' molecule. If it is supposed that the singlet states of the ketone are relatively long-lived whereas the triplets are very short-lived, agreement with experimental observations can be obtained with this scheme. Moreover, if a first-order deactivation step is added to the latter scheme, it becomes similar to that discussed earlier,[102] except that the products are formed *via* an undetectable triplet state rather than the initially formed singlet.

A scheme very similar to the one above has been proposed to account for the direct and benzene-sensitized photolysis of cyclohexanone.[105] The only major difference is that hydrocarbon products are assumed to be produced directly from the singlet, and not *via* the 'hot' triplet [*cf.* (133) and (134) rather than (142) and (144)]. Cyclohexanone has a quenching cross-section for benzene singlet molecules which is about fifty times that for triplets. The effects of pressure, wavelength, and foreign gases upon product distribution are similar to those reported for cyclopentanone.

The photolysis of keten and subsequent reactions of the singlet and triplet methylene products still attract much attention. Attempts to observe

[104] E. K. C. Lee, *J. Phys. Chem.*, 1967, **71**, 2804.
[105] R. G. Shortridge and E. K. C. Lee, *J. Amer. Chem. Soc.*, 1970, **92**, 2228.

sensitized emission from keten–biacetyl mixtures failed to produce any emission other than that which occurs *via* direct excitation of the biacetyl.[106] Since the excited triplet state of keten lies some 25·2 kJ mol^{-1} above that of biacetyl, the authors conclude that the absence of any sensitized emission points to the lack of formation of triplet keten in the system. This may be the case, but as experiments on cyclic ketones have shown (see above), triplet states of ketones may be very short-lived, and not affected by triplet quenchers. At the highest pressure of quencher used in the keten system, the detection of a triplet keten species with a lifetime shorter than 10^{-8} sec would be impossible, and the failure to observe emission from biacetyl may be due to this cause rather than absence of a triplet keten in the system. It is generally held that singlet and triplet methylenes result from the predissociation of singlet and triplet ketens respectively, the latter arising *via* intersystem crossing from the initially formed singlet keten. If collision-induced intersystem crossing of the singlet keten to the triplet can be ignored, which appears to be reasonable at least at short wavelengths where the singlet lifetime will be short, a study of the rate of formation of ethylene in the presence of triplet quenching molecules such as O_2 can yield information about the collisional deactivation of singlet methylene to the ground triplet state. The simple scheme below suffices to describe the system.[107]

$$CH_2CO + h\nu \longrightarrow {}^1CH_2 + CO \quad (146)$$
$$\longrightarrow {}^3CH_2 + CO \quad (147)$$
$$CH_2CO + {}^1CH_2 \longrightarrow C_2H_4 + CO \quad (148)$$
$$^1CH_2 + M \longrightarrow {}^3CH_2 + M \quad (149)$$

With a high pressure of O_2 or CO, the only reaction of the $^3CH_2$ will be:

$$^3CH_2 + O_2 \longrightarrow CO, CO_2, H_2 \text{ etc.} \quad (150)$$
$$^3CH_2 + CO \longrightarrow CH_2CO \quad (151)$$

Applying the steady-state hypothesis to the scheme, the following expression may be obtained for the quantum yields, Φ_0 and Φ respectively, of ethylene formation in the absence and the presence of the additive M.

$$\Phi_0(\Phi)^{-1} - 1 = k_{149}[M]/k_{148}[CH_2CO] \quad (152)$$

The data for keten–oxygen mixtures photolysed at 2490 Å are shown in Figure 7. Data with carbon monoxide as added gas can be treated similarly, and the rate-constant ratios obtained are given in Table 8. It can be seen that polyatomic molecules quench efficiently, but of the monatomic gases xenon is by far the most efficient, a fact which may mean that the spin inversion in the methylene occurs *via* a complex with the noble gas atom, or during a collision with it, in which the latter contributes to the spin–orbit

---

[106] M. Grossman, G. P. Semeluk, and I. Unger, *Canad. J. Chem.*, 1969, **47**, 3079.
[107] R. A. Cox and K. F. Preston, *Canad. J. Chem.*, 1969, **47**, 3345.

Table 8 *Rate-constant ratios for deactivation of singlet methylene to reaction in photolysis of CH_2CO-O_2-M mixtures and $CH_2CO-CO-M$ mixtures*[107]

Quencher, M	Exciting wavelength, λ (Å)	Scavenger	Φ_s	$k_{149}/k_{148} \times 10^2$
He	2490	O_2	—	1·83 ± 0·04
	2800	CO	—	1·73 ± 0·46
Ar	2800	O_2	0·60	1·4 ± 0·2
Kr	2490	O_2	—	3·3 ± 0·2
CF_4	2490	O_2	—	4·7 ± 0·1
N_2	2490	O_2	—	5·6 ± 0·1
	2800	O_2	0·68	5·96 ± 0·2
	2800	CO	—	4·6 ± 0·1
SF_6	2490	O_2	—	4·53 ± 0·03
Xe	2490	O_2	—	7·3 ± 0·3
	2800	O_2	0·77	9·6 ± 4·0
	2800	CO	—	7·8 ± 0·5
C_2F_6	2800	O_2	0·64	6·3 ± 4·6
	2800	CO	—	11·0 ± 1·0
N_2O	2490	O_2	—	10·0 ± 0·3
	2800	CO	—	17 ± 2
				$k_{CO}/k_{148} \times 10^2$
CO	2800	CO	0·60	12·1 ± 0·5

Φ_s = quantum yield of formation of singlet CH_2, k_{CO} = rate constant of reaction of $^1CH_2$ with CO to form keten.

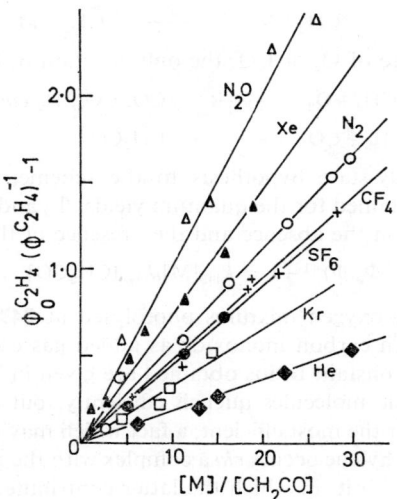

Figure 7 *Plot of $\Phi_0^{C_2H_4}(\Phi^{C_2H_4})^{-1} - 1$ against concentration ratio $[M]/[CH_2CO]$ for the 2490 Å photolysis of mixtures of CH_2CO, O_2, and various added gases M*
(Reproduced by permission from *Canad. J. Chem.*, 1969, **47**, 3345)

perturbation necessary for the mixing of the singlet and triplet states. The rate-constant ratios in Table 8 do not afford evidence about the absolute rate of reaction (149), but there is good evidence that singlet methylene persists in systems even when the pressure of added gas is very high.[108] A new primary process is suggested in the photolysis of keten at photon energies from 3·96 eV (3130 Å) to 8·43 eV (1470 Å).[109] The production of monoisotopic molecular hydrogen and deuterium from the photolysis of CH_2CO–CD_2CO mixtures leads to the conclusion that reaction (153) must be a primary process.

$$CH_2CO + h\nu \longrightarrow H_2 + C_2O \qquad (153)$$

The yield of D_2 relative to that of CO in the photolysis of CD_2CO as a function of photon energy is shown in Figure 8. Alternative reactions

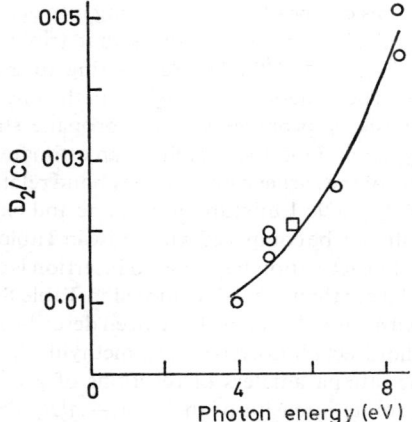

Figure 8 D_2/CO *ratio from photolysis of* CD_2CO *as a function of energy:* ○, *this work;* □, *ref. 109a*
(Reproduced by permission from *J. Phys. Chem.*, 1969, **73**, 959)

which could lead to molecular hydrogen formation can be discounted on energetic grounds or on the basis of isotopic content. The fate of the C_2O species thought to be formed in process (153) is not determined. Although there is good evidence that the relative amounts of singlet and triplet methylene produced photochemically from keten in the near-u.v. region are strongly dependent upon excitation wavelength, it seems that in the case of the photolysis of diazomethane the relative amounts of the singlet and triplet carbene species are insensitive to wavelength changes.[110] This

---

[108] T. W. Eder, R. W. Carr jun., and J. W. Koenst, *Chem. Phys. Letters*, 1969, **3**, 520.
[109] A. H. Laufer, *J. Phys. Chem.*, 1969, **73**, 959.
[109a] T. A. Walter, Ph.D. thesis, Harvard University, 1967.
[110] G. W. Taylor and J. W. Simons, *Canad. J. Chem.*, 1970, **48**, 1016.

observation would suggest that the energy difference between the singlet and triplet states of diazomethane must be small, and that in all probability both are repulsive or weakly bonding with a low-lying intersystem crossing point on the potential-energy surfaces.

The reactions of singlet and triplet methylene are still the subject of much attention and debate. It has been demonstrated convincingly that the reaction of triplet methylene with *cis*- and *trans*-but-2-ene leads to quite different ratios of the stereoisomers of the two major products, dimethylcyclopropane and pent-2-ene. In the case of addition to the *trans* olefin, the ratio of *trans* products to *cis* is of the order of 3·5, whereas addition to the *cis* olefin produces a *trans/cis* ratio of about 1·6 for the cyclopropane product, and of the order of unity for the pentene product.[111] Triplet methylene has been shown to be very unreactive towards alkanes, since reaction products arising from interaction of these methylenes with alkyl radicals in these systems can be observed. Taking into account the short lifetime of the alkyl radicals, the implication is that triplet methylenes must survive $> 10^9$ collisions with CD_4 before reacting to give ethylene and a hydrogen atom.[112] The reaction of singlet methylene produced in the vacuum-u.v. photolysis of propane with the propane starting material in the presence of oxygen and inert gases allows an estimate to be made of the efficiency of insertion of the carbene into a C—H bond relative to deactivation to triplet carbene.[58] For a 9 : 1 mixture of propane and oxygen the two rates are equal. This result may be compared with those in Table 8. In the presence of argon, the ratio of deactivation by argon to insertion is equal to $2·4 \times 10^{-2}$, which is somewhat larger than the value quoted in Table 8. The reactions of singlet methylene with trimethylsilane have been described,[113] and these may be compared with the direct photolysis of hexamethyldisilane.[114] Finally, data are available on the rate parameters of reactions of ground-state difluorocarbene and the absolute intensity of the X^1A_1—A^1B_1 absorption bands.[115]

The photolysis of *trans*-crotonaldehyde in the vapour phase at wavelengths in the region of 2550 Å depends strongly upon temperature.[116] At 25 °C, the only observable reaction was polymerization, but at higher temperatures carbon monoxide and propylene were formed in significant amounts, with ethylene, allene, methylacetylene, cyclopropane, ethylketen, and enol-crotonaldehyde,[117] hydrogen, and methane [116] being reported as minor products. There are significant differences between these two studies which probably arise because, in the former, light was used which caused excitation in both the π—π^* band of the olefin moiety and the n—π^* transition of the carbonyl. In the latter study, 95% of the light absorbed was

---

[111] T. W. Eder and R. W. Carr jun., *J. Phys. Chem.*, 1969, **73**, 2074.
[112] P. S. T. Lee, R. L. Russell, and F. S. Rowland, *Chem. Comm.*, 1970, 18.
[113] W. L. Hase and J. W. Simons, *J. Chem. Phys.*, 1970, **52**, 4004.
[114] S. P. Narula, *J. Indian Chem. Soc.*, 1969, **46**, 1067.
[115] W. J. R. Tyerman, *Trans. Faraday Soc.*, 1969, **65**, 1188.
[116] E. R. Allen and J. N. Pitts jun., *J. Amer. Chem. Soc.*, 1969, **91**, 3135.
[117] J. W. Coomber and J. N. Pitts jun., *J. Amer. Chem. Soc.*, 1969, **91**, 4955.

by the carbonyl group. In this region the chemistry is best explained in terms of decomposition from, and multistage deactivation of, a vibrationally excited upper singlet state, some intersystem crossing to an unstable triplet state, and some internal conversion to the ground state *via* unstable isomeric intermediates. It is worthwhile quoting the complete scheme in view of its interesting features.

$$A + h\nu \longrightarrow {}^1A_n \tag{154}$$

$${}^1A_n \longrightarrow C_3H_6 + CO \tag{155}$$

$${}^1A_n \longrightarrow C_3H_5 + HCO \tag{156}$$

$${}^1A_n \longrightarrow \text{ethylketen} \tag{157}$$

$${}^1A_n + A \longrightarrow {}^1A_{n-1} + A \tag{158}$$

$${}^1A_{n-1} \longrightarrow C_3H_6 + CO \tag{159}$$

$${}^1A_{n-1} \longrightarrow C_3H_5 + HCO \tag{160}$$

$${}^1A_{n-1} \longrightarrow \text{ethylketen} \tag{161}$$

$${}^1A_{n-1} + A \longrightarrow {}^1A_{n-2} \tag{162}$$

etc.

$${}^1A_{m-1} + A \longrightarrow {}^1A_m + A \tag{163}$$

$${}^1A_m + A \longrightarrow {}^1A_{m-1} + A \tag{164}$$

etc.

$${}^1A_1 + A \longrightarrow {}^1A_0 + A \tag{165}$$

$${}^1A_0 \longrightarrow A \tag{166}$$

$${}^1A_0 \longrightarrow {}^3A_0 \tag{167}$$

$${}^1A_0 \longrightarrow \text{enol-crotonaldehyde} \tag{168}$$

$${}^3A_0 \longrightarrow C_3H_3 + CO \tag{169}$$

$${}^3A_0 \longrightarrow C_3H_5 \cdot + HCO \cdot \tag{170}$$

$${}^3A_0 \longrightarrow A \tag{171}$$

$$\text{ethylketen} \longrightarrow A \tag{172}$$

$$\text{enol-crotonaldehyde} \longrightarrow A \tag{173}$$

where A refers to the crotonaldehyde (but-2-en-1-al) molecule, and subscripts vibrational energy content. $^1A_{m+1}$ is taken to be the lowest level at which dissociation occurs. One of the tests for multistep deactivation proposed by Kutschke [118] involves the function f, where

$$f = (1-\Phi)[A]^{-1} \tag{174}$$

(Φ being the quantum yield of some product). Here

$$f = (1-\Phi_\infty)/([A]+a_i) \quad \text{if} \quad n = m+1 \tag{175}$$

[118] A. N. Strachan, R. K. Boyd, and K. O. Kutschke, *Canad. J. Chem.*, 1964, **42**, 1345.

and

$$f = (1-\Phi_\infty)[A]^{-1} \Big/ \prod_{i=m+1}^{n} (1+a_i/[A]) \quad \text{for} \quad n > m+1 \qquad (176)$$

Here

$$a_i = k_{155}/k_{158}, k_{159}/k_{162}, \text{ etc.},$$

and

$$\Phi_\infty = k_{167}(k_{169}+k_{170})/(k_{166}+k_{167}+k_{168})(k_{169}+k_{170}+k_{171})$$

Thus if $n = m+1$, plots of f against [A] start at $f = (1-\Phi_\infty)/ai$ for [A] = 0, and decrease smoothly to zero as [A] $\to \infty$; the slope is always negative and continuously decreases in magnitude as [A] increases. If,

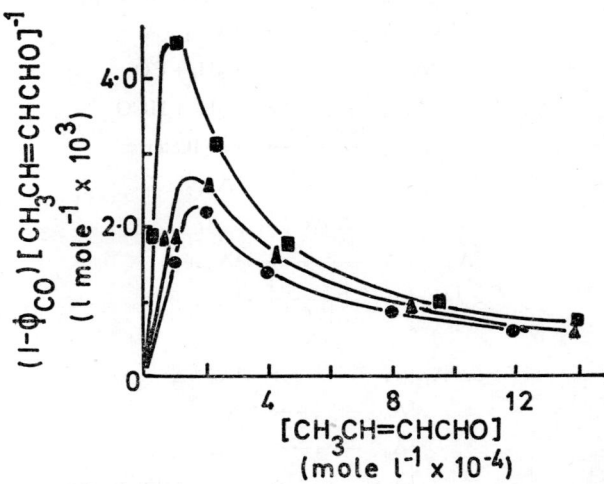

Figure 9 *The function* $f_1 = (1-\Phi_{CO})[CH_3CH=CHCHO]^{-1}$ *plotted against trans-crotonaldehyde concentration at 70, 100, and 130 °C:* ■, 70 °C; ▲, 100 °C; ●, 130 °C
(Reproduced by permission from *J. Amer. Chem. Soc.*, 1969, **91**, 4955)

however, $n > m+1$, then $f = 0$ when [A] = 0, and the function increases with [A] when [A] is small and passes through a maximum when

$$\sum_{i=m+1}^{n} a_i/(a_i+[A]) = 1 \qquad (177)$$

and then decreases to zero as [A] $\to \infty$. Such plots for quantum yields of carbon monoxide formation are shown in Figure 9, and show clearly that for crotonaldehyde at these wavelengths, $n > m+1$. Similar expressions may be derived for propylene formation. There is other similar good evidence for the multistep cascade in this molecule. Reactions (172) and (173) are of interest in that they correspond to a radiationless transition to a ground state *via* intermediate unstable isomers, as was proposed for aromatic compounds in the past.[119] There is no good evidence for aromatic systems that isolable or detectable isomeric intermediates may participate in relaxations to the ground state: indeed it has been proposed that some of these intermediates may be very short-lived diradicals. Thus it is of great interest to see that isolable intermediates have been proposed in the crotonaldehyde system.

The irradiation of benzaldehyde and pentafluorobenzaldehyde using unfiltered light from a medium-pressure mercury lamp gives polymers plus benzene and CO, and pentafluorobenzene and CO respectively. A mechanism has been proposed.[120] The mercury-photosensitized decomposition of γ-crotonolactone[121] and cyclic anhydrides,[122] and the vapour-phase photolysis of phenyl acetate[123] are of principal interest from a mechanistic point of view, and are discussed in other sections of this Volume. An α-hydroxy aliphatic ketone, 3-hydroxybutan-2-one (acetoin), yields the same major product (acetaldehyde) in the gas phase and in solution at 3130 Å, and with the same quantum yield ($\Phi = 0.3$). Temperature-dependent yields of CO and CH_4 were also obtained in the gas phase.[124] Two primary dissociative steps were proposed, one an intramolecular rearrangement of an excited acetoin molecule to give two molecules of acetaldehyde, and the other forming free radicals which undergo secondary reactions. The products arising from 1-hydroxy-2-methylbutan-3-one were more complex.[124] Both ketones appeared to have lowest-lying $n\pi^*$ singlet and triplet states, but in the β-hydroxy compound the triplet state appears not to be of importance.

The photolysis of tetrahydrofuran,[125] methylfurans,[126] and dimethyl carbonate vapour[127] has been described.

4 Halogen-containing Compounds

Apart from a report on the photochemical reaction of fluorine with carbon dioxide,[128] the halogens considered here are chlorine, bromine, and iodine in that order.

The importance of 'hot' deuterium atoms in photolyses of $DCl-Cl_2$ mixtures at 1849 Å has been established and the kinetics of the system described.[129] The decomposition of the DCl was followed by mass-

---

[119] W. A. Noyes jun., D. Phillips, J. Lemaire, and C. S. Burton, *Adv. Photochem.*, 1968, **5**, 329; D. Bryce-Smith and H. C. Longuet-Higgins, *Chem. Comm.*, 1966, 593.
[120] J. R. Majer, S.-A. M. A. Naman, and J. C. Robb, *Trans. Faraday Soc.*, 1969, **65**, 1846.
[121] I. S. Krull and D. R. Arnold, *Tetrahedron Letters*, 1969, 1247.
[122] I. S. Krull and D. R. Arnold, *Tetrahedron Letters*, 1969, 4349.
[123] J. W. Meyer and G. S. Hammond, *J. Amer. Chem. Soc.*, 1970, **92**, 2187.
[124] E. J. Baum, L. D. Hess, J. R. Wyatt, and J. N. Pitts jun., *J. Amer. Chem. Soc.*, 1969, **91**, 2461.
[125] B. C. Roquitte, *J. Amer. Chem. Soc.*, 1969, **91**, 7664.
[126] H. Hiraoka, *J. Phys. Chem.*, 1970, **74**, 574.
[127] T. Ibuki and Y. Takezaki, *Bull. Inst. Chem. Res., Kyoto Univ.*, 1969, **47**, 239.
[128] A. H. Jubert, J. E. Sicre, and H. J. Schumaker, *Z. phys. Chem. (Frankfurt)*, 1969, **67**, 138.
[129] G. O. Wood and J. M. White, *J. Chem. Phys.*, 1970, **52**, 2613.

spectrometric analysis of the D_2 formed after successive intervals of time, and the observed decreases in the initial rate of D_2 formation upon the addition of Cl_2 and the inert gases CO_2, Xe, and CF_4 may be readily interpreted in terms of a 'hot-atom' mechanism, but cannot be made compatible with a completely thermal mechanism. The 'hot-atom' reactions are

$$D^* + DCl \longrightarrow D_2 + Cl\cdot \qquad (178)$$

$$D^* + M \longrightarrow D\cdot + M \qquad (179)$$

$$D^* + Cl_2 \longrightarrow DCl + Cl\cdot \qquad (180)$$

The following rate-constant ratios were found: $k_{180}/k_{178} = 6\cdot5$; $k_{179}^{DCl}/k_{178} = 0\cdot65$; $k_{179}^{CO_2}/k_{178} = 1\cdot15$; $k_{179}^{CF_4}/k_{178} = 1\cdot15$; and $k_{179}^{Xe}/k_{178} = 0\cdot1$. The ratio k_{180}/k_{178} increased from 6·5 in the absence of an inert gas to 9·2 when $[CO_2]/[DCl] = 1\cdot87$. This ratio increases to about 300 for a thermal atom distribution at 300 K. Comparisons of thermalizing efficiencies for several gases reveal that the efficiency increases noticeably with the number of internal degrees of freedom, and it is thus apparent that for 'hot' D atoms possessing 2·1 eV energy, inelastic collisions are largely responsible for their deactivation. This conclusion is in contrast with results on 0·8 and 2·0 eV hydrogen atoms in which deactivation can be reasonably explained on the basis of simple elastic collisions.[130] Other 'hot' H atom reactions are discussed in the section of this Chapter dealing with the photochemistry of sulphur-containing compounds.

The photolysis of chloroform and carbon tetrachloride wavelengths at shorter than 2500 Å can be represented by the familiar general mechanism:

$$RCl + h\nu \longrightarrow R\cdot + Cl\cdot \qquad (181)$$

$$Cl\cdot + R'H \longrightarrow R'\cdot + HCl \qquad (182)$$

The reaction products observed in each case can be explained in terms of subsequent reactions of the free radicals formed in processes (181) and (182).[131] The direct photolysis of vinyl chloride [132] leads to fragmentation of a chlorine atom and the molecular elimination of HCl.

$$C_2H_3Cl + h\nu \longrightarrow C_2H_3Cl^* \qquad (183)$$

$$C_2H_3Cl^* \longrightarrow C_2H_3\cdot + Cl\cdot \qquad (184)$$

$$C_2H_3Cl^* \longrightarrow C_2H_2 + HCl \qquad (185)$$

The ratio k_{184}/k_{185} was found to be invariant with exciting wavelength, pressure, and temperature, and equal to $1\cdot4 \pm 0\cdot1$. The lifetime of the excited vinyl chloride molecule is estimated to be less than $3\cdot5 \times 10^{-11}$ s. Reaction (184) could provide a valuable experimental source of free vinyl radicals.

---

[130] R. D. Penzhorn and B. de B. Darwent, *J. Phys. Chem.*, 1968, **72**, 1639.
[131] W. H. S. Yu and M. H. J. Wijnen, *J. Chem. Phys.*, 1970, **52**, 2736, 4166.
[132] T. Fujimoto, A. M. Rennert, and M. J. H. Wijnen, *Ber. Bunsengesellschaft. Phys. Chem.*, 1970, **74**, 282.

In the flash photolysis of five 1,1-dichloroethylenes, there is evidence of C=C bond scission to give CCl_2 species which then undergo secondary photolysis to produce a longer-lived CCl transient, the absorption spectrum of which has been recorded.[133] The proposal that the CCl is formed *via* an intermediate carbene is supported on the basis of a strong inverse correlation between the strength of the C=C bond in the parent ethylene and the intensity of the CCl absorption observed. The primary dissociation (186) probably occurs as a result of absorption in the π—π^* transition. The olefins can be constructed from two bent carbenes which have a_1 and b_1 orbitals as the highest occupied and lowest vacant orbitals respectively, for C_{2v} symmetry. The a_1 orbitals become σ and σ^* in the olefin, and the b_1 become the π and π^*. The states arising from these configurations are shown in Table 9. In a symmetrical olefin such as C_2F_4, the σ^* orbital lies close to the π^*, so optically metastable $^1B_{2g}$ molecules can be formed after absorption in the π—π^* system (X^1A_{1g}—A^1B_{1u}). As can be seen from Table 9, radiationless transition from the initially formed state to the $\sigma^2\sigma^{*2}$

Table 9 *Electronic configurations and symmetry species for 1,1-dichloro-ethylenes* [133]

Configuration of olefin	Point group		
	D_{2h}	C_{2v}	C_s
$(\sigma)^2 (\sigma^*)^2$	^1A_g	$^1A_1$	$^1A'$
$(\sigma)^2 (\pi) (\sigma^*)$	$^1B_{2g}$	$^1B_1$	$^1A''$
$(\sigma)^2 (\pi) (\pi^*)$	A^1B_{1u}	$^1A_1$	$^1A'$
$(\sigma)^2 (\pi)^2$	X^1A_g	$^1A_1$	$^1A'$

configuration from which dissociation to two ground-state carbenes is possible is forbidden for the D_{2h} species, but is allowed for the C_{2v} and C_s groups. Since carbene formation is implied even in the D_{2h} species (CCl is formed from C_2Cl_4), another route must be available. If the CCl_2 group were rotated through 90° in C_2Cl_4, an ^1E state would be produced in point group D_{2d} from which the transition to the dissociative state would be allowed. A similar explanation has been proposed in the past for ethylene.

$$CXY = CCl_2 + h\nu \longrightarrow CXY + CCl_2 \qquad (186)$$

The fate of ClO and the postulated ClO_2 radicals in the Cl_2–O_2 photochemical system have long been the subject of debate, and a further postulated mechanism for the recombination of ClO radicals has appeared.[134] However, much of the speculation about this system must now surely be reconsidered in the light of a penetrating study performed using the molecular modulation technique described in Chapter 1.[135] Using this method, the u.v. and i.r. spectra of the ClO_2 peroxyl radical in the gas

---

[133] W. J. R. Tyerman, *Trans. Faraday Soc.*, 1969, **65**, 2948.
[134] Yu. A. Kiryushin and V. A. Poluektov, *Khim. vysok. Energii*, 1969, **3**, 316.
[135] H. S. Johnston, E. D. Morris jun., and J. Van den Bogaerde, *J. Amer. Chem. Soc.*, 1969, **91**, 7712.

phase have been recorded, establishing beyond doubt the intermediacy of the species in this photochemical system. The u.v. spectrum of the ClO radical was also observed, and the direct monitoring of these intermediates allowed a complete kinetic study to be carried out in which radical half-lives were inferred from phase shifts between the square-wave photolysing light and the radical concentration. Previous apparently conflicting data [136–138] (see references contained in ref. 135) can be completely reconciled on the basis of the following mechanism:

$$Cl_2 + h\nu \xrightarrow{a} Cl\cdot + Cl\cdot \quad \text{Rate} = aI_0[Cl_2] \quad (187)$$

$$Cl\cdot + O_2 + M \underset{c}{\overset{b}{\rightleftarrows}} ClOO\cdot + M \quad K_1 = \frac{[ClOO]}{[Cl][O_2]} = \frac{b}{c} \quad (188)$$

$$Cl\cdot + ClOO\cdot \underset{e}{\overset{d}{\rightleftarrows}} ClO\cdot + ClO\cdot \quad K_2 = \frac{[ClO]^2}{[Cl][ClOO]} = \frac{d}{e} \quad (189)$$

$$Cl\cdot + ClOO\cdot \xrightarrow{f} Cl_2 + O_2 \quad (190)$$

$$ClO\cdot + ClO\cdot + M \underset{h}{\overset{g}{\rightleftarrows}} Cl_2O_2 + M \quad K_3 = \frac{[Cl_2O_2]}{[ClO]^2} = \frac{g}{h} \quad (191)$$

$$Cl_2O_2 + M \xrightarrow{i} Cl_2 + O_2 + M \quad (192)$$

$$Cl\cdot + Cl\cdot + M \xrightarrow{j} Cl_2 + M \quad (193)$$

The analysis of the data which allows estimates of the rate constants for the scheme above is tedious, and will not be given here. It suffices to say that non-stationary-state kinetics must be employed, and the values obtained are given in Table 10. The previous disagreements between

Table 10 Rate constants from photolysis of $Cl-O_2$ mixtures

Quantity†	Units	Source	Value
a	cm^2	Ref. 135, u.v. data	9.35×10^{-20}
		Ref. 135, i.r. data	9.35×10^{-20}
I_0	photons cm^{-2} s^{-1}	Ref. 135, u.v. data	4.2×10^{16}
		Ref. 135, i.r. data	0.85×10^{16}
aI_0	photons s^{-1}	Ref. 135, u.v. data	3.92×10^{-3}
		Ref. 135, i.r. data	0.80×10^{-3}
K_1	ml molecule^{-1}	Thermodynamic	3.62×10^{-21}
b	ml^2 molecule^{-2} s^{-1}	Ref. 136	1.7×10^{-33}
c	ml^2 molecule^{-2} s^{-1}	b/K_1	4.7×10^{-13}
K_2		Thermodynamic	227
d	ml molecule^{-1} s^{-1}	Ref. 135, u.v. data	1.44×10^{-12}
e	ml molecule^{-1} s^{-1}	d/K_2	6.3×10^{-15}
f	ml molecule^{-1} s^{-1}	Ref. 135, u.v. data	1.56×10^{-10}
ig/h	ml^2 molecule^{-2} s^{-1}	Ref. 135, u.v. data	5.0×10^{-32} (O$_2$)
			3.3×10^{-32} (Ar)
$K_1(d+f)$	ml^2 molecule^{-2} s^{-1}	Ref. 135, u.v. data	5.7×10^{-31}
		Ref. 135, i.r. data	4.8×10^{-31}
		Ref. 137	5.4×10^{-31}
j	ml^2 molecule^{-2} s^{-1}	Ref. 138	1.17×10^{-32}

† For the meaning of the parameters in the first column, see text.

second-order rate constants for the recombination of ClO· radicals to give Cl_2 and O_2 can be largely explained by the foreign-gas catalysis of this process (191) and (192), which has been confirmed in this study. Photochemical chlorinations of methylene chloride and 1,1-dichloroethane in solution have been described.[139]

The recombination of bromine atoms produced in the flash photolysis of molecular bromine in the temperature range from 300 to 1275 K can be assumed to occur *via* the formation of a radical–atom or radical–molecule complex (194)

$$Br + M \rightleftarrows BrM \quad (194)$$

$$Br + BrM \longrightarrow Br_2 + M \quad (195)$$

Values of the relevant rate constants for many different species M have been measured,[140] and calculations performed on models of the BrM complex in an attempt to predict experimental behaviour. It was shown that complexes in bound, metastable, and BrM quasi-dimers all contribute to the recombination, and that interaction potentials between Br and M are several times larger than for Kr and M.

Br_2–N_2O and Br_2–NO_2 mixtures both yield BrO radicals under flash illumination.[141] The mechanism has not been well established, but probably involves primary dissociation of the nitrogen oxides to give oxygen atoms followed by reaction of these with molecular bromine to form the BrO radical. The kinetics of the photoaddition of HBr to propylene,[142] allyl halides,[143] and the acetone-sensitized addition of HBr to ethylene [144] have been described.

Equivalent-width measurements on rotational lines in six vibrational bands of the $X^1\Sigma_{0_g^+}$—$B^3\Pi_{0_u^+}$ transition in molecular iodine enable an estimation of quantum yields of emission back to the ground state and of predissociation.[145] In the absence of collisions, predissociation to ground-state iodine atoms accounts for 66% of the initially excited molecules, the remainder undergoing emission. At higher pressures (0·238 Torr) about 20% spontaneously predissociate, 10% radiate, and the remainder decay *via* a collision-induced non-radiative process, which is largely an induced predissociation. The interference of resonance emission with laser-Raman measurements in halogen molecules has been discussed.[146] The photolysis of cyanogen halides and hydrogen cyanide in the vacuum-u.v. region [147]

[136] J. E. Nicholas and R. G. W. Norrish, *Proc. Roy. Soc.*, 1968, **A307**, 391.
[137] G. Porter and F. J. Wright, *Discuss. Faraday Soc.*, 1953, No. 14, 23.
[138] E. Hatton and M. Wright, *Trans. Faraday Soc.*, 1965, **61**, 78.
[139] A. S. Bratolyubov and L. I. Vasil'kova, *Zhur. priklad. Khim.*, 1970, **43**, 199; A. S. Bratolyubov and G. F. Aleshina, *ibid.*, p. 651.
[140] J. K. K. Ip and G. Burns, *J. Chem. Phys.*, 1969, **51**, 3425, 3414.
[141] R. E. Tomalesky and J. E. Sturm, *J. Chem. Phys.*, 1970, **52**, 472.
[142] R. J. Field and P. I. Abell, *J. Amer. Chem. Soc.*, 1969, **91**, 7226.
[143] P. I. Abell and P. K. Adolf, *Internat. J. Chem. Kinetics*, 1969, **1**, 499.
[144] K. T. Wong and D. A. Armstrong, *Canad. J. Chem.*, 1969, **47**, 4183.
[145] A. Chutjian and T. C. James, *J. Chem. Phys.*, 1969, **51**, 1242.
[146] W. Holzer, W. F. Murphy, and H. J. Bernstein, *J. Chem. Phys.*, 1970, **52**, 399.
[147] A. Mele and H. Okabe, *J. Chem. Phys.*, 1969, **51**, 4798.

produces the CN $B^2\Pi$ species, in which only 20% of the available energy appears as vibrational excitation (up to levels $v = 3$), and a further 10—20% as rotational energy (in ICN and ClCN). The remaining energy must appear as kinetic energy in the CN species and in the atoms (presumably as electronic and kinetic). The internal energy in the $B^2\Pi$ CN radical is much less than would be expected from the equipartition of energy among all the degrees of freedom in the molecule.

The flash photolysis of methyl iodide produces 'hot' methyl radicals which at low intensities react largely to give methane (196).

$$CH_3^* \cdot + CH_3I \longrightarrow CH_4 + \cdot CH_2I \quad (196)$$

At high absorbed intensities, however, results indicate that the activation energy necessary for this reaction is not carried by the methyl radicals, and a sequence involving energy transfer to a methyl iodide molecule *via* iodine atom abstraction is invoked:

$$CH_3^* \cdot + CH_3I \longrightarrow CH_3I^* + CH_3 \cdot \quad (197)$$

The excited CH_3I can then react with thermal methyl radicals to give the product methane. A number of other speculations are advanced to account for the other observed products.[148]

CF_3I flashed in the u.v. and vacuum-u.v. is a much 'cleaner' system, and allows a study of the reactions of the electronically excited iodine atoms $I(5^2P_{\frac{1}{2}})$ initially produced. The electronic quenching of these atoms by added olefins appears to involve a collisional spin–orbit relaxation to ground-state iodine atoms rather than a chemical interaction leading to product formation.[149] The values of the second-order rate constant for the deactivation of the excited iodine atoms are shown in Figure 10. Two other values not shown in Figure 10 were obtained with tetrafluoroethylene and trifluoroethylene, *viz.* 3.7×10^{-15} and 5.3×10^{-14} ml molecule^{-1} sec^{-1}, respectively. Figure 10 shows that there is a good correlation between the rate constants obtained for the hydrocarbon olefins and the ionization potential of each olefin. Similar correlations have been found between log (quenching rate-constant) and the excitation energy of the π—π^* transition in the olefin, and with heats of formation of the olefin (see ref. 149). Such correlations are to be expected for reactions involving electrophilic attack on a double bond and may be represented in terms of a charge-transfer interaction. The wave-function of the charge-transfer complex (π-complex) may be envisaged to include contributions from structures of the form (olefin)$^+(I^1S_0)^-$ which will then facilitate a breakdown of the spin–orbit coupling in $I(5^2P_{\frac{1}{2}})$ causing relaxation from $J = \frac{1}{2}$ to $J = \frac{3}{2}$ on separation into an olefin and an iodine atom. It should be stressed that this mechanism is not general and does not apply to all quenchers.

---

[148] G. J. Mains and D. Lewis, *J. Phys. Chem.*, 1970, **74**, 1649.
[149] R. J. Stevenson, D. Husain, and C. D. Stevenson, *Trans. Faraday Soc.*, 1969, **65**, 2941.

The laser-induced photolysis of iodoform in solution,[150] and the photolysis of HI [151] and C_2H_5I [152] at 77 K have been described in Chapter 2 of this Part.

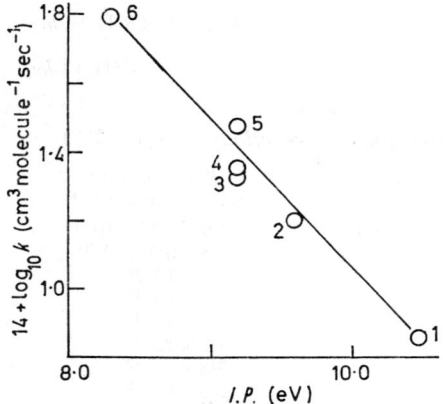

Figure 10 *Correlation between second-order rate constant* k *for the collisional quenching of* I $(5P_\frac{3}{2})$ *and the ionization potential* (I.P.) *of the olefin.* 1, *ethylene;* 2, *propylene and but-1-ene;* 3, cis-*but-2-ene;* 4, trans-*but-2-ene;* 5, *isobutene;* 6, *tetramethylethylene*
(Reproduced by permission from *Trans. Faraday Soc.*, 1969, **65**, 2941)

5 Photochemistry of Nitrogen Compounds

A number of reactions of nitrogen itself in the $A^3\Sigma_u^+$ state have been reported. These include energy transfer to atomic mercury as in equation (198), and nitric oxide,[153] excitation of the auroral green line of atomic oxygen (199),[154] and a triplet–triplet annihilation process (200).[155] The rate constants measured for these processes are shown in Table 11.

$$N_2\,(A^3\Sigma_u^+) + \text{Hg}\,(6^1S_0) \longrightarrow N_2\,(X^1\Sigma_g^+) + \text{Hg}\,(^3P_1) \qquad (198)$$

$$N_2\,(A^3\Sigma_u^+) + \text{O}\,(^3P) \longrightarrow N_2\,(X^1\Sigma_g^+) + \text{O}\,(^1S) \qquad (199)$$

$$N_2\,(A^3\Sigma_u^+) + N_2\,(A^3\Sigma_u^+) \longrightarrow N_2\,(X^1\Sigma_g^+) + N_2\,(C^3\Pi_u) \qquad (200)$$

The metastable $E^3\Sigma_g^+$ state of molecular nitrogen which lies 11·87 eV above the ground state can be produced by electron impact.[156] The state

---

[150] S. Speiser and S. Kimel, *J. Chem. Phys.*, 1969, **51**, 5614.
[151] P. G. Barker, M. P. Halstead, and J. H. Purnell, *Trans. Faraday Soc.*, 1969, **65**, 2389.
[152] P. G. Barker and J. H. Purnell, *Trans. Faraday Soc.*, 1970, **66**, 163.
[153] A. B. Callear and P. M. Wood, *Chem. Phys. Letters*, 1970, **5**, 128.
[154] J. A. Meyer, D. W. Setser, and D. H. Stedman, *The Astrophysical Journal*, 1969, **157**, 1023.
[155] D. H. Stedman and D. W. Setser, *J. Chem. Phys.*, 1969, **50**, 2256.
[156] R. S. Freund, *J. Chem. Phys.*, 1969, **50**, 3734.

has a radiative lifetime of 270 ± 100 μs, and decays *via* the three forbidden transitions:

$$N_2 (E^3\Sigma_g^+) \longrightarrow N_2 (A^3\Sigma_u^+) + h\nu_1 \quad (201)$$

$$\longrightarrow N_2 (B^3\Pi_g) + h\nu_2 \quad (202)$$

$$\longrightarrow N_2 (C^3\Pi_u) + h\nu_3 \quad (203)$$

Table 11 Rate constants for reactions of $N_2 (A^3\Sigma_u^+)$

Reacting gas	Rate constant (ml molecule^{-1} s^{-1})	Ref.
Hg (reaction 198)	$2.9 \pm 0.15 \times 10^{-10}$	153
NO	$5.45 \pm 0.3 \times 10^{-11}$	153
NO	3×10^{-11}	a
NO	2.3×10^{-11}	154
NO	8×10^{-11}	158
O_2	1.2×10^{-12}	154
O_2	3.8×10^{-12}	b
CO	2.5×10^{-12}	158
N_2O	1×10^{-11}	158
CH_4	3×10^{-15}	158
CO_2	5×10^{-14}	158
O (^3P)	3.2×10^{-12}	154
(reaction 199)	(10.1×10^{-12})*	154
$N_2 (A^3\Sigma_u^+)$	8.3×10^{-13} (i)	155
(reaction 200)	2.1×10^{-11} (ii)	155

* This is the value if the value for quenching by O_2 in ref. (*b*) is taken as correct.
(i) is based on a radiative lifetime for $N_2(A^3\Sigma_u^+)$ of 10 s.
(ii) is based on a radiative lifetime for $N_2(A^3\Sigma_u^+)$ of 2.0 s.
a R. A. Young and G. A. St. John, *J. Chem. Phys.*, 1968, **48**, 898.
b R. A. Young, G. Black, and T. G. Slanger, *J. Chem. Phys.*, 1969, **50**, 303.

Absorption of light in the region from 1000 to 450 Å does not produce detectable emission from any stable or metastable states of molecular nitrogen, but emission from highly excited states of atomic nitrogen is observed.[157] From the atoms identified, it is apparent that the photoexcited nitrogen dissociates as shown:

$$N_2 (X^1\Sigma_g^+) + h\nu \longrightarrow N (3S^4P) + N (^4S^0) \quad (204)$$

$$\longrightarrow N (3S^2P) + N (^2D^0) \quad (205)$$

The photolysis of N_2O at 1470 Å produces an emission which is identified as the $B^2\Pi - X^2\Pi$ emission of NO. There are several possible ways in which the excited NO can be formed (for a resumé of the species formed

[157] K. D. Beyer and K. H. Welge, *J. Chem. Phys.*, 1969, **51**, 5323.
[158] G. Black, T. G. Slanger, G. A. St. John, and R. A. Young, *J. Chem. Phys.*, 1969, **51**, 116.

from this photolysis see Volume 1, Part I, Chapter 2) but the most likely are as follows:

$$N(^2D) + N_2O \longrightarrow N_2(X^1\Sigma_g^+) + NO\ (B^2\Pi) \qquad (206)$$

$$O(^1S) + N_2O \longrightarrow NO\ (X^2\Pi) + NO\ (B^2\Pi) \qquad (207)$$

It has been demonstrated that process (206) is the major source of the excited NO species,[158] and rate-constant data for the removal of the $N(^2D)$ atoms by other molecules based on competition with N_2O are presented. Rate constants for the reactions of $N_2\ (A^3\Sigma_u^+)$, which is also present in this system, are given in Table 11.

Nitrosyl fluoride irradiated with the full arc of a medium-pressure mercury lamp decomposes in the expected manner to give nitric oxide and fluorine atoms.[159] The latter, in the presence of ethylene and other hydrocarbons and inert gases, react to give 'hot' species which decomposes at low pressures, but which can be collisionally deactivated at higher pressures. Products arising from the industrially important photochemical nitrosation of cyclohexane [160] and chlorocyclohexane have been described.[161]

The 2491 Å absorption spectrum of NO_2 reveals features which can best be explained by the existence of a double-minimum potential in the antisymmetrical co-ordinate.[162] The fluorescence intensity and lifetime of NO_2 in the $(^2B_1)$ state have been measured as a function of excitation wavelength, fluorescence wavelength, and pressure, using the phase-shift technique described earlier.[163] The Stern–Volmer analysis of the fluorescence kinetics can be generalized to a multilevel system, and from this values can be obtained for a parameter which is the product of energy-transfer rate-constants and efficiencies. The analysis indicates that for excited-state NO_2 molecules containing excess of vibrational energy from 12,500 to 25,000 cm^{-1}, at least one quantum of this energy (average value 1230 cm^{-1}) is lost per gas-kinetic collision. These results may be compared with those obtained by excitation of NO_2 fluorescence with light from the argon and krypton ion lasers.[164] The fluorescence spectrum from such a narrow excitation source consists of sharp lines superimposed on a continuum. If the assumption is made that the individual quantum states populated by the line excitation have radiative lifetimes similar to those measured with the phase-shift apparatus, data on the quenching of the discrete fluorescence can be considered to yield depopulation cross-sections which are approximately gas-kinetic. The decrease in the observed Stern–Volmer constant may be explained by the step-wise deactivation model proposed in ref. 163.

---

[159] A. L. Flores and B. de B. Darwent, *J. Phys. Chem.*, 1969, **73**, 2203.
[160] H. Miyama, K. Fukuzawa, N. Harumiya, Y. Ito, and S. Wakamatsu, *J. Phys. Chem.*, 1969, **73**, 4345.
[161] G. N. Semina, L. G. Zelenskaya, L. A. Levashova, K. E. Kuznetsova, and A. A. Strel'tsova, *Neftekhimiya*, 1970, **10**, 103.
[162] J. B. Coon, F. A. Cesani, and F. P. Huberman, *J. Chem. Phys.*, 1970, **52**, 1647.
[163] S. E. Schwartz and H. S. Johnston, *J. Chem. Phys.*, 1969, **51**, 1286.
[164] K. Sakurai and H. P. Broida, *J. Chem. Phys.*, 1969, **50**, 2404.

The continuum radiation is quenched much less strongly than the discrete,[164] and this continuous emission is considered to arise *via* collisional energy transfer from the initially populated discrete states.

NO_2 photolysed at 3660 Å yields NO and O atoms, and in the presence of propylene, the latter add rapidly to the olefin to give intermediates which are either stabilized by collision or dissociate to give free radicals.[165] Nitrogen dioxide reacts rapidly with the free radicals formed to produce stable products. The photolysis of NO_2 at high pressures has also been investigated.[166] The species discussed in this section so far are all of importance in the atmosphere and upper atmosphere. Discussion of other molecules which are of importance from an atmospheric standpoint, including atmospheric pollutants, will be found in subsequent sections on sulphur compounds, including SO_2, and in the section on oxygen.

The photolytic decomposition of azoalkanes alone and in the presence of hydrocarbons and other reactive molecules provides systems in which the modern theories of unimolecular decomposition can be applied to the excited molecule, to primary products, and to intermediates formed with substrates. As an example of the first case, the absolute quantum yield of dissociation of azoethane has been calculated as a function of pressure, temperature, and exciting wavelength, and the results compared with experimental values from the literature.[167] Using the simple reaction scheme:

$$A + h\nu \longrightarrow {}^1A^* \qquad (208)$$

$$^1A^* \longrightarrow N_2 + 2C_2H_5 \cdot \qquad (209)$$

$$^1A^* + A \longrightarrow \text{deactivation} \qquad (210)$$

where A refers to an azoethane molecule; the dissociation quantum yield is given by

$$\Phi^{-1} = 1 + k_{210}[A]/k_{209} \qquad (211)$$

For the purpose of the calculation we consider that the dissociating molecules have a vibrational distribution function $g(E^*)$ which incorporates both the thermal energy distribution factor $f(E)$ of the ground state, and the energy profile of the absorbed light. Instead of (211), the yield Φ is given by

$$\Phi = \int_{E_{\min}}^{\infty} \frac{[k_{209}(E^*)][g(E^*)]}{k_{209}(E^*) + k_{210}[A]} dE^* \qquad (212)$$

The Marcus expression for $k_{209}(E^*)$ is

$$k_{209}(E^*) = \frac{1}{h}\left(\frac{Z^+}{Z^*}\right)\left(\frac{S^+(E^+)}{N^*(E^*)}\right) \qquad (213)$$

where $E^* = E_\lambda - E_{00} + E$; and $E^+ = E^* - E_{\min}$. (see Figure 11). The functions S^+, N^*, and $g(E^*)$ can be calculated from vibrational frequency

[165] S. Jaffe and R. C. S. Grant, *J. Chem. Phys.*, 1969, **50**, 3477.
[166] J. Troe, *Ber. Bunsengesellschaft. Phys. Chem.*, 1969, **73**, 906.
[167] P. G. Bowers, *J. Phys. Chem.*, 1970, **74**, 952.

assignments for the species involved (see ref. 167), and since *a priori* values of E_{00} and $E_{\min}$ were not available, calculations were carried out over a range of values of these parameters. The agreement with experiment is shown in Figure 12, in which the values for $E_{00} = 5260$ Å, and $E_{\min.} = 51.7$ kJ mol^{-1} have been selected since they give the best fit.

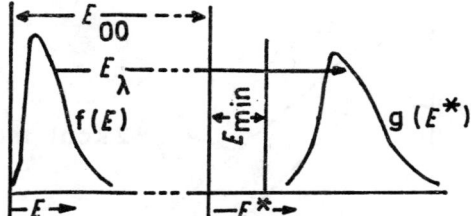

Figure 11 *The Boltzmann distribution $f(E)$, of total vibrational energy in ground-state azoethane, and the non-equilibrium distribution, $g(E^*)$ in the excited $^1(n, \pi^*)$ state*
(Reproduced by permission from *J. Phys. Chem.*, 1970, **74**, 952)

Calculations on the temperature dependence are summarized in Figure 13, where the same energy parameters have been selected. Agreement is moderately good, and in particular it should be noted that both calculated and experimental lines are curved. The mean 'activation' energy derived

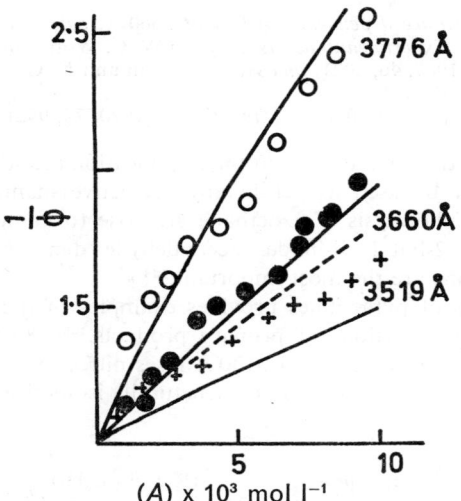

Figure 12 *Calculated wavelength dependence of the primary dissociation yield in azoethane at 27 °C. Data points are experimental results from W. C. Worsham and O. K. Rice, J. Chem. Phys., 1967, 46, 2021. Lines are calculated as in text. Dotted line includes correction for spectral width of 3660 Å radiation*
Reproduced by permission from *J. Phys. Chem.* 1970, **74**, 952)

from the average slope in Figure 13 is about six times the value of E_{min} assumed here, and it is thus clear that serious misinterpretation of Arrhenius data could occur for any excited molecule or radical reaction in which the reacting species has not had time to thermalize its vibrational energy. Although calculations of this type are useful at present in demonstrating

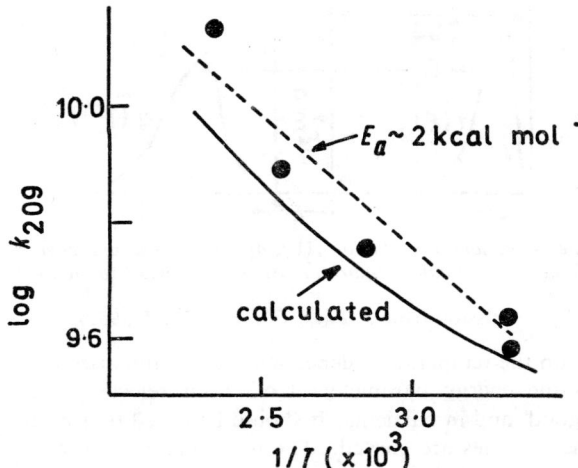

Figure 13 *Temperature dependence of* k_{209} *at* 3660 Å, *calculated from equations* (212) *and* (213). *Experimental points are from* W. C. Worsham and O. K. Rice, *J. Chem. Phys.*, 1967, **46**, 2021; *and* H. Cerfontain and K. O. Kutschke, *Canad. J. Chem.*, 1958, **36**, 344
(Reproduced by permission from *J. Phys. Chem.*, 1970, **74**, 952)

general features, the uncertainties in their application are still considerable, and they cannot be used as yet in any predictive manner. Secondary reactions in the photolysis of azoethane give rise to numerous products, of which ethyl 2-butyl di-imide, acetaldehyde diethylhydrazone, and tetraethylhydrazine are the most important.[168]

The photolysis of pyrazolines provides examples of the application of RRKM theory to reactions of primary products.[169, 170] A 4-substituted pyrazoline (I) upon irradiation at 3130 Å loses nitrogen to give an excited cyclic hydrocarbon, which will either structurally isomerize or be collisionally stabilized.

$$\underset{(I)}{\overset{N=}{\underset{N-}{\big\rangle}}\!\!-R} + h\nu \longrightarrow (R\text{-cyclo-}C_3H_5)^* \qquad (214)$$

---

[168] O. P. Strausz, R. E. Berkley, and H. E. Gunning, *Canad. J. Chem.*, 1969, **47**, 3470.
[169] P. Cadman, H. M. Meunier, and A. F. Trotman-Dickenson, *J. Amer. Chem. Soc.*, 1969, **91**, 7640.
[170] F. H. Dorer, *J. Phys. Chem,*, 1969, **73**, 3109.

Gas-phase Photochemistry

$$(\text{R-cyclo-}C_3H_5)^* \longrightarrow R-CH_2-CH=CH_2 \quad etc. \quad (215)$$

$$(\text{R-cyclo-}C_3H_5)^* + M \longrightarrow \text{R-cyclo-}C_3H_5 + M \quad (216)$$

The multiplicities of the excited states through which these reactions proceed are unimportant for the purpose of the present discussion. The ratio of total isomeric products to cyclopropane derivative is given by the expression

isomers/cyclopropane $= k_E/\omega$, where $k_E = k_{215}$ and $\omega = k_{216}M$

The ratio is thus given by the following expression when the substitution is made for k_E

$$k_E = \int_{E_{\min}}^{E_T} \frac{k_E f(E)}{\omega + k_E} \Big/ \int_{E_{\min}}^{E_T} \frac{\omega f(E)}{\omega + k_E} \quad (217)$$

The value of k_E can be evaluated by making certain assumptions about the parameters in equation (217). In the case of pyrazoline itself there is only one isomer formed, namely propylene.[169] The situation is slightly complicated by the observation that propylene is still formed at very high pressures, indicating a second source of this compound, but after due allowance is made for this, one can obtain experimental values for the rate of formation of propylene in (215) relative to that of cyclopropane as a function of pressure. In order to compute values, it is necessary to know $E_{\min}$ and E_T, and to make some assumption about the distribution function $f(E)$. Now $E_{\min}$ is known from thermal studies, and E_T is given by

$$E_T = -\Delta H_R + h\nu + E_{\text{thermal}} \quad (218)$$

where $-\Delta H_R$ is the heat of reaction (214) which can be estimated to be -168 kJ mol^{-1}, and E_{thermal} is the thermal energy of cyclopropane. Thus for 3130 Å photons, $E_T = 546$—588 kJ mol^{-1}. In order to get a best fit with the experimental data, several trial functions for $f(E)$ were used. The first of these was a triangular function which has no theoretical significance. The fraction of molecules $f(E)$ at energy E was calculated by dividing the triangular function into intervals of 21 kJ, and finding the relative height at each point. The sum of the heights and hence fractions $f(E)$ were then normalized to unity. In order to get a fit with the experimental value of the propylene/cyclopropane yields (0·8), the only reasonable triangular distribution was that with a peak at 294 kJ and a width of ±168 kJ. Using this function in equation (217), the variation of propylene/cyclopropane could be calculated, and the results are shown in Figure 14. It can be seen that there is good agreement over the critical pressure range from 0—10 Torr.

Another function tried was a Gaussian one (219)

$$f(E) = \frac{\exp\left[-\frac{1}{2}\left(\frac{E_{\text{mp}} - E_i}{\sigma}\right)^2\right]}{\sum_i \exp\left[-\frac{1}{2}\left(\frac{E_{\text{mp}} - E_i}{\sigma}\right)^2\right]} \quad (219)$$

where E_{mp} = most probable energy and is related to the width of the function. The results obtained were almost identical with those shown in Figure 14 for the triangular function. The last function tried was a statistical distribution, in which the fraction of cyclopropane molecules

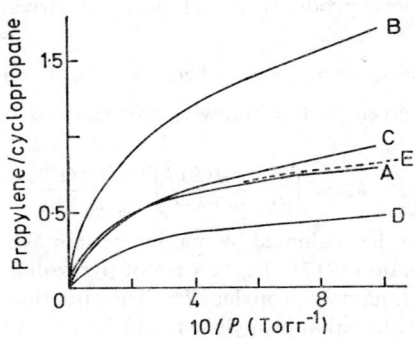

Figure 14 *Plot of propylene/cyclopropane versus p^{-1} for various calculated distributions.* (A) *best triangular function*, (B) *statistical distribution with* $E_T = 462$ kJ mol^{-1}, (C) *statistical distribution with* $E_T = 441$ kJ mol^{-1}, (D) *statistical distribution with* $E_T = 420$ kJ mol^{-1}, (E) *experimental result*
(Reproduced by permission from *J. Amer. Chem. Soc.*, 1969, **91**, 7640)

formed at any energy is assumed proportional to the energy-level density at that energy.

$$f(E) = \frac{N\nabla(E_i)N_R(E_T - E_i)}{\sum_{E_i} N\nabla(E_i)N_R(E_T - E_i)} \tag{219a}$$

where $N\nabla$ = energy-level density of the cyclopropane part of the pyrazoline, N_R = energy level density of the remaining degrees of freedom of the pyrazoline, and E_T = total available energy, as before. $N\nabla$ and N_R can be calculated from vibrational frequencies (see ref. 169), and the results for different assumed values of E_T are shown in Figure 14. It can be seen that if the value for E_T obtained from thermochemical data of 546 kjoule mol^{-1} is used, there is no agreement with the experimental results. A match is obtained if it is assumed that not all of the excess energy is available to the cyclopropane, *i.e.* that $E_T = 441$ kJ mol^{-1}. Very similar results were obtained for the 4-methylpyrazoline system.[170] Here only a Gaussian function was used and it was discovered that the most probable energy of the methylcyclopropane formed was only 62% of the total energy available for distribution between this species and the nitrogen molecule. On a purely statistical model for partitioning of all the excess of energy between the products, it would be expected that about 77% of the energy would reside in the methylcyclopropane. Part of the reason for this failure of statistical considerations to describe the energy partitioning in such systems may be due to the fact that the N=N bond distance in pyrazoline is much greater (1·25 Å) than in molecular nitrogen (1·09 Å). Thus formation of the nitrogen must be accompanied by a considerable shortening of the bond, with the consequence that the nitrogen could be formed with considerably more vibrational energy than expected from purely statistical considerations. However, from a Morse potential function it can be shown that only 117·6 kJ mol^{-1} is necessary to extend the N—N bond length in nitrogen from 1·09 to 1·25 Å, which still leaves 92·4 kJ mol^{-1} unaccounted for. This presumably appears as translational and rotational energy in the products.

Similar treatments to the ones outlined above have been applied to the diazomethane–*cis*-but-2-ene–O_2 photochemical system in which the singlet methylene which is initially produced reacts with the olefin to give 'hot' pent-2-ene and 'hot' dimethylcyclopropane, both of which then undergo unimolecular reactions in competition with collisional deactivation. RRKM treatments for the former [171] and the latter [172] 'hot' molecules have led to a satisfactory correlation between theory and experiment, and similar results were obtained with the diazomethane–propane and diazomethane–isobutane photochemical systems.[173] The photochemistry of diphenyl-diazidomethane [174] and the photoisomerization of azoalkanes [175] have been discussed.

Methylamine decomposes in a variety of ways in the vacuum-u.v., the main process being decomposition to give hydrogen atoms.[176] In contrast with results from studies at longer wavelengths, it was found by deuterium substitution that the majority of these hydrogen atoms fragment from the carbon atom in methylamine rather than the nitrogen atom. Direct molecular elimination of hydrogen occurs to a small extent at 1470 Å. The quantum yields determined for the various primary processes are shown in the scheme below:

$$CD_3NH_2 + h\nu \longrightarrow CD_3NH\cdot + H\cdot \quad \Phi = 0\cdot21 \quad (220)$$

$$\longrightarrow \cdot CD_2NH_2 + D\cdot \quad \Phi = 0\cdot47 \quad (221)$$

$$\longrightarrow CD_3N + H_2 \quad \Phi = 0\cdot05 \quad (222)$$

$$\longrightarrow CDNH_2 + D_2 \quad \Phi = 0\cdot05 \quad (223)$$

$$\longrightarrow CD_2NH + HD \quad \Phi = 0\cdot09 \quad (224)$$

$$\longrightarrow \cdot CD_3 + \cdot NH_2 \quad \Phi = 0\cdot13 \quad (225)$$

[171] G. W. Taylor and J. W. Simons, *J. Phys. Chem.*, 1970, **74**, 464.
[172] G. W. Taylor and J. W. Simons, *J. Phys. Chem.*, 1969, **73**, 1274.
[173] R. L. Johnson, W. L. Hase, and J. W. Simons, *J. Chem. Phys.*, 1970, **52**, 3911.
[174] G. Ege and G. Jooss, *Chem. Z.T.G. Chem. App.*, 1970, **94**, 215.
[175] T. Mill and R. S. Stringham, *Tetrahedron Letters*, 1969, 1853.
[176] J. J. Magenheimer, R. E. Varnerin, and R. B. Timmons, *J. Phys. Chem.*, 1969, **73**, 3904.

The photodissociation of NCN_3 in the gas phase and vacuum-u.v. region of the spectrum yields excited CN and NCN species as in equations (226) and (227).[177]

$$NCN_3 + h\nu \longrightarrow CN(B^2\Sigma) + N_3 \quad (226)$$

$$\longrightarrow NCN(X^3\Sigma) + N_2(A^3\Sigma) \quad (227)$$

A plot of the emission intensity of the CN ($B^2\Sigma$) species against exciting wavelength shows pronounced structure, indicating that process (226) is predissociative. The threshold wavelength for the production of excited CN radicals is 1685 Å. The reaction leading to excited NCN is due to the $N_2(A^3\Sigma)$ produced in (227)

$$NCN_3 + N_2(A^3\Sigma) \longrightarrow NCN(A^3\Pi) + 2N_2 \quad (228)$$

The bond dissociation energies in NCN_3 can be estimated from threshold energies, and are $D(NC-N_3) = 403 \pm 8$ kJ mol^{-1} and $D(NCN-N_2) = 29.4 \pm 8.4$ kJ mol^{-1}.

Examples of a new Type II photoelimination reaction in acetyl and ethyl isocyanates have been described,[178] and an e.s.r. technique has been used to study the photochemical production of free radicals from acetonitrile.[179] Flash photolysis of formaldoxime and acetaldoxime under isothermal and adiabatic conditions yields reactive hydroxyl radicals and a more stable

$$R^1R^2CNOH + h\nu \longrightarrow R^1R^2CN + \cdot OH \quad (229)$$

species.[180] The rate constant for the hydrogen abstraction reaction of the hydroxyl radicals with the parent oxime was determined as 3.8×10^8 l mol^{-1} sec^{-1} for the formaldoxime, and 1.39×10^9 l mol^{-1} s^{-1} for the acetaldoxime. The flash photolysis of formaldazine, acetaldazine, and dimethyl ketazine allowed the recording of the absorption spectra of three new transients, viz. CH_2N, CH_3CHN, and $(CH_3)_2CN$.[181]

Formamide vapour excited at 2062 Å decomposes in three ways, as shown below.[182]

$$NH_2CHO + h\nu \longrightarrow \cdot NH_2 + CO + H\cdot \quad \Phi = 0.35 \quad (230)$$

$$\longrightarrow \cdot H + \cdot NHCHO \quad \Phi = 0.22 \quad (231)$$

$$\longrightarrow NH_3 + CO \quad \Phi = 0.45 \quad (232)$$

Hydrogen-abstraction reactions by the H atoms and NH_2 species accounted for the other reaction products. It was found that hydrogen abstraction occurred almost exclusively from the formyl position. Irradiation of acetanilide in the liquid phase leads to rearrangement reactions, but in the

---

[177] H. Okabe and A. Mele, *J. Chem. Phys.*, 1969, **51**, 2100.
[178] N. J. Friswell and R. A. Back, *Canad. J. Chem.*, 1969, **47**, 4169.
[179] P. Svejda and D. H. Volman, *J. Phys. Chem.*, 1970, **74**, 1872.
[180] D. G. Horne and R. G. W. Norrish, *Proc. Roy. Soc.*, 1970, **A315**, 287.
[181] D. G. Horne and R. G. W. Norrish, *Proc. Roy. Soc.*, 1970, **A315**, 301.
[182] J. C. Boden and R. A. Back, *Trans. Faraday Soc.*, 1970, **66**, 175.

gas phase only decomposition is observed.[183] At 2537 Å and 120 °C the decomposition is adequately described by the following mechanism:

$$S_0 + h\nu \longrightarrow S_2(\pi\pi^*) \quad (233)$$
$$S_2(\pi\pi^*) \longrightarrow S_1(\pi\pi^*) \quad (234)$$
$$S_1(\pi\pi^*) \longrightarrow S_0 \quad (235)$$
$$S_1(\pi\pi^*) \longrightarrow S_0 + h\nu \quad (236)$$
$$S_1(\pi\pi^*) \longrightarrow T_1(\pi\pi^*) \quad (237)$$
$$S_1(\pi\pi^*) \longrightarrow C_6H_5-NH\cdot + \cdot COCH_3 \quad (238)$$
$$\cdot COCH_3 \longrightarrow CO + \cdot CH_3 \quad (239)$$
$$2CH_3\cdot \longrightarrow C_2H_6 \quad (240)$$
$$CH_3\cdot + RH \longrightarrow CH_4 + R\cdot \quad (241)$$

where S and T have the usual meaning, referring here to the acetanilide molecule. The yield of ethane was approximately half that of the carbon monoxide, as would be expected from the above mechanism. The quantum yield of process (238) was found to be 0·28. The dissociation is assumed to occur *via* an intersystem crossing from the $S_1(\pi\pi^*)$ state to a $^3\sigma_0(N-C)$ state, which is dissociative. The absence of fragmentation products in the liquid phase may be accounted for by the strong solvent cage surrounding the excited molecule which ensures that the back-reaction of the species formed in (238) is always predominant. Tetramethylurea in the gas phase gives dimethylamino radicals and carbon monoxide when irradiated with a medium-pressure mercury arc.[184] These radicals initiate polymerization and undergo further reactions to form stable products. The fluorescence lifetime of some substituted phthalimide vapours at elevated temperatures has been measured as a function of exciting wavelength and concentration of added gases, including helium.[185] The lifetimes measured decreased markedly with increase in photon energy, the greatest effect being noted in the absence of chemically inert gas. Increase in pressure of inert gas also caused a significant decrease in the lifetime of the fluorescence of these compounds. A mechanism is discussed.

6 Sulphur Compounds

The flash photolysis of molecular sulphur vapour (S_8) gave rise to transient S_2 molecules which could be identified by their absorption in the u.v. region corresponding to the $^3\Sigma_g^- - {^3\Sigma_u^-}$ transition.[186] The decay of the S_2 molecules was studied as a function of pressure and temperature.

The photolysis of H_2S in the u.v. produces 'hot' H atoms and HS radicals, as in (242). In the presence of a sufficiently high concentration of

[183] H. Shizuka and I. Tanaka, *Bull. Chem. Soc. Japan*, 1969, **42**, 909.
[184] J. R. Majer, S.-A. M. A. Naman, and J. C. Robb, *J. Chem. Soc. (B)*, 1970, 93.
[185] V. A. Yakovenko, L. G. Pikulik, and M. Ya. Kostko, *Zhur. priklad. Spektroskopii*, 1969, **10**, 933.
[186] M. Elbanowski, *Roczniki Chem.*, 1969, **43**, 1883.

an inert gas, the 'hot' atoms can be thermalized before they undergo further reaction, and a technique has been developed to measure the relative rates of addition of the thermal H atoms to olefins.[187] This involves simultaneous irradiation of two reaction cells, one of which contains only H_2S and an atmosphere of carbon dioxide, and the other contains the same pressure of H_2S and varying pressures of olefin, together with CO_2. The reaction sequence is shown below.

$$H_2S + h\nu \; (2490 \text{ Å}) \longrightarrow \cdot H^* + HS \cdot \quad (242)$$

$$\cdot H^* + M \longrightarrow H \cdot + M \quad (243)$$

$$\cdot H + H_2S \longrightarrow H_2 + HS \cdot \quad (244)$$

$$\cdot H + Ol \longrightarrow OlH \quad (245)$$

$$H + Ol \longrightarrow H_2 + R \quad (246)$$

where Ol = olefin. The rate-constant ratios k_{244}/k_{245} and k_{246}/k_{245} are given by the slopes and intercepts of plots of $f(H_2)$ where

$$f(H_2) = \frac{R_{H_2(Ol)}}{R_{H_2(H_2S)} - R_{H_2(Ol)}} = \frac{k_{244}[H_2S]}{k_{245}[Ol]} + \frac{k_{246}}{k_{245}} \quad (247)$$

$R_{H_2(Ol)}$ is the rate of production of hydrogen in the presence of the olefin, and R_{H_2} in its absence. Values of rate-constant ratios obtained in this manner are shown in Table 12, compared with values from the literature.

Table 12 *Rate constant data for reactions of hydrogen atoms with olefins*

Olefin	k_{244}/k_{245}	k_{246}/k_{245}	k_{246}/k_{245}		Relative value of k_{245}			
	Ref. 187	Ref. 187	Ref. (a)	Ref. 187	Ref. (a)	Ref. 188	Ref. (b)	Ref. (c)
ethylene	0.828	0.067	—	0.67	0.56	0.64	0.66	4.17
propylene	0.565	0.082	0.045	1.00	1.00	1.00	1.00	1.00
isobutene	0.221	0.032	0.020	2.56	2.46	2.55	9.52	3.17
but-1-ene	0.571	0.093	0.081	0.99	1.08	—	—	—
buta-1,3-diene	0.114	0.015	0.032	4.96	4.84	5.3	13.0	—

^a K. R. Jennings and R. J. Cvetanovic, *J. Chem. Phys.*, 1961, **35**, 1233.
^b K. Yang, *J. Amer. Chem. Soc.*, 1962, **84**, 3795.
^c P. E. M. Allen, H. W. Melville, and J. C. Robb, *Proc. Roy. Soc*, 1953, **A218**, 311.

The technique has also been used to measure the rate constants of H atom addition and abstraction from olefins relative to propylene.[188] In this case the H atoms were generated by the mercury-sensitized photolysis of H_2–olefin mixtures. It can be seen from Table 12 that the rate of hydrogen abstraction by thermal H atoms is relatively small in comparison with the rate of their addition to olefins. 'Hot' H atoms abstract much more readily however, as the last row of figures in Table 12 illustrates. The H atoms in this study were generated by the photolysis of H_2S at 2537 Å [189] without the

[187] G. R. Woolley and R. J. Cvetanovic, *J. Chem. Phys.*, 1969, **50**, 4697.
[188] R. J. Cvetanovic and L. C. Doyle, *J. Chem. Phys.*, 1969, **50**, 4705.
[189] B. G. Dzantiev, A. K. Lubimova, and A. V. Shishkov, *Khim. vysok. Energii*, 1969, **3**, 478.

Gas-phase Photochemistry

large excess of moderating gas used in the studies described earlier. The average initial kinetic energies of the H atoms produced from the photolysis of a source HY (Y = Br, HS, SCH_3) can be determined by a competition method using a deuteriated scavenger such as D_2 or C_4D_{10}.[190] The variation of the function $2[H_2]/[HD]$ with the reactant ratio $[HY]/[D_2]$ or $[HY]/[C_4D_{10}]$ gives straight line plots whose intercept I_0 can be used to estimate the fraction of hydrogen atoms which react while 'hot' in the limit of infinite dilution in deuteriated scavenger. This fraction is given by $(I_0+1)^{-1}$. By calibration with HBr it is possible to determine the amount of translational energy E_0 possessed by the H atoms compared with the total amount of energy, E_{max}, which is available from the photon absorbed: (for HBr, $E_0 = E_{max}$.). At 2537 Å, 2288 Å, and 2138 Å, studies using D_2[191] and C_4D_{10}[192] are in agreement that the majority of excess energy from the photolysis of H_2S and CH_3SH appears as kinetic energy in the H atom, but the two systems give conflicting results at 1849 Å. E_{max} at this wavelength is 2·72 eV for H_2S and 2·83 eV for CH_3SH. The limiting value for $(I_0+1)^{-1}$ was determined as 0·43 for the H_2S–D_2 system,[191, 192] but was 0·18 for H_2S–C_4D_{10}. The first value leads to the conclusion that about 75% of the available excess energy appears in the H atom, whereas use of the second figure results in a situation in which only about half of the available energy is found in the H atom, the remainder appearing as internal energy in the SH or CH_3S fragments. Since the two studies are in essential agreement in the experimental values for $(I_0+1)^{-1}$, the different conclusions reached must be due to differences in the calibrations with HBr. If the higher figure is taken as correct, the result at 1849 Å is in essential agreement with the results at longer wavelengths, but the lower value can be rationalized by assuming that additional electronic states of the H_2S and CH_3SH are important in this wavelength region, and/or by the involvement of additional primary processes. These studies on the partitioning of excess energy among the products of photochemical reactions are of great interest, but it is clear in this case that further experiments may be necessary to resolve the differences between the two studies. 'Hot' atom reactions produced in the photolysis of DCl at 1849 Å have been discussed previously. Such reactions arising in the photolysis of water vapour will be discussed in Part II of this Volume. The reaction of O (^3P) atoms with H_2S has been investigated using an e.s.r. technique,[193] and the reactions of thiyl radicals (produced photochemically from methanethiol) with acetylene and buta-1,3-diene have been described.[194]

Thiophen vapour absorbs light at 2139 and 2288 Å in the π—π^* band to produce an excited thiophen molecule which can be collisionally

---

[190] R. G. Gann and J. Dubrin, *J. Chem. Phys.*, 1967, **47**, 1867.
[191] G. P. Sturm jun. and J. M. White, *J. Chem. Phys.*, 1969, **50**, 5035.
[192] L. E. Compton, J. L. Cole, and R. M. Martin, *J. Phys. Chem.*, 1969, **73**, 1158.
[193] G. A. Hollinden, M. J. Kurylo, and R. B. Timmons, *J. Phys. Chem.*, 1970, **74**, 988.
[194] D. M. Graham and J. F. Soltys, *Canad. J. Chem.*, 1969, **47**, 2529, 2719.

quenched or can dissociate to give acetylene, allene, methylacetylene, carbon disulphide, vinylacetylene, and a polymer.[195] In the presence of oxygen, CO_2, COS, SO_2, and CO were also produced. The hydrocarbons were all produced in primary processes, and arguments were advanced to support a complex reaction mechanism which was proposed to explain the results.

The flash photolysis of mercaptans in aqueous solution gives rise to a transient absorption spectrum which has been attributed to the $\overline{RSSR}^{\cdot}$ radical-ion.[196] This species arises *via* a fast reaction between an RS^- anion and the RS radical produced on photolysis. Cystine has been photolysed in solution in the presence of benzyl chloride.[197]

The u.v. absorption spectrum of carbonyl sulphide in the spectral region from 2000—2650 Å shows band structure superimposed upon a continuum.[198] 'Hot' bands are also visible, and the transition is shown to produce a bent upper state of the COS, either $(^1\Delta)^1 A'$ or $(^1\Sigma^-)^1 A''$. Absorption in this region causes dissociation, yielding ground-state ^3P sulphur atoms. These have been used to initiate the *cis–trans* isomerization of but-2-ene, by a chain reaction having an approximate chain length of 2×10^3. The photostationary state ratio of *trans* to *cis* was found to be 2.83 ± 0.06, and the initial isomerization-rate ratio R_{c-t}/R_{t-c} was 2.8 ± 0.2: these figures are in good agreement.[199] The primary process in this photolysis probably produces $S\,(^1D)$ atoms, which may collisionally relax before initiating the chain reaction. The sequence is thus:

$$COS\,(^1\Sigma^+) + h\nu \longrightarrow COS\,(^1\Delta \text{ or } ^1\Sigma^-) \quad (248)$$

$$COS\,(^1\Delta \text{ or } ^1\Sigma^-) \longrightarrow COS\,(^1\Pi) \quad (249)$$

$$COS\,(^1\Pi) \longrightarrow CO\,(^1\Sigma^+) + S\,(^1D) \quad (250)$$

The $S\,(^1D)$ atoms produced have been observed to undergo insertion reactions with alkanes in solution,[200] but not in other solvents. In the vacuum-u.v. region, a more highly excited sulphur atom is produced, *viz.* the $S\,(3^1 S)$ atom.[201] This atom is apparently less reactive towards H_2 than the $S\,(^1D)$ excited atom. The rate of the reaction of $S\,(3^1 D)$ atoms with COS is now known.[202] The photolysis of thionyl chloride apparently proceeds as follows:

$$SOCl_2 + h\nu \longrightarrow \cdot SOCl^* + \cdot Cl \quad (251)$$

$$\cdot SOCl^* \longrightarrow SO + \cdot Cl \quad (252)$$

$$\cdot SOCl^* + M \longrightarrow \cdot SOCl + M \quad (253)$$

[195] J. Heicklen, *Canad. J. Chem.*, 1969, **47**, 2965.
[196] G. Caspari and A. Granzow, *J. Phys. Chem.*, 1970, **74**, 836.
[197] C. J. Dixon and D. W. Grant, *J. Phys. Chem.*, 1970, **74**, 941.
[198] W. H. Breckenridge and H. Taube, *J. Chem. Phys.*, 1970, **52**, 1713.
[199] M. W. Schmidt and E. K. C. Lee, *J. Chem. Phys.*, 1969, **51**, 2024.
[200] K. Gollnik and E. Leppin, *J. Amer. Chem. Soc.*, 1970, **92**, 2217; E. Leppin and K. Gollnik, *ibid.*, p. 2221.
[201] R. J. Donovan, *Trans. Faraday Soc.*, 1969, **65**, 1419.
[202] R. J. Donovan, L. J. Kirsch, and D. Husain, *Nature*, 1969, **222**, 1164.

Two new absorption spectra of the SO species were observed in the vacuum-u.v. region [203] corresponding to the transitions SO ($X^3\Sigma^-$)—SO ($D^3\Pi$) and SO ($X^3\Sigma^-$)—SO ($E^3\Pi$). New bands in the S_2 and CS species have also been seen in the vacuum-u.v.[204] The radiative lifetimes of CS_2^+, SO_2, C_2N_2, and N_2O^+ have been measured using a phase-shift technique and electron-beam excitation.[205] There are many reports in the literature on reactions of small molecules, such as SO, and reactions of atoms with this and many other gaseous substrates, but for reasons of space it is only possible to consider species produced photochemically, despite the relevance of the other studies to the above discussion.

Sulphur dioxide is a small molecule whose photochemistry has been studied exhaustively in recent years, undoubtedly because of its notoriety as an atmospheric pollutant. The photochemistry can be explained on the basis of the mechanism below.

$$SO_2 + h\nu \longrightarrow {}^1SO_2 \quad (254)$$

$$^1SO_2 + SO_2 \longrightarrow (2SO_2) \longrightarrow 2SO_2 \quad (255)$$

$$\longrightarrow {}^3SO_2 + SO_2 \quad (256)$$

$$^1SO_2 \longrightarrow SO_2 + h\nu_f \quad (257)$$

$$\longrightarrow SO_2 \quad (258)$$

$$\longrightarrow {}^3SO_2 \quad (259)$$

$$^3SO_2 \longrightarrow SO_2 + h\nu_p \quad (260)$$

$$\longrightarrow SO_2 \quad (261)$$

$$^3SO_2 + {}^3SO_2 \longrightarrow (2SO_2) \longrightarrow SO_3 + SO \quad (262)$$

($^1SO_2$ is the first excited singlet state.)

Controversy exists over some features of this mechanism. In an earlier study it was concluded that at limiting low pressures reaction (257) was the only fate of singlet excited SO_2.[206] However, other results showed that processes (258) and (259) were more important than (257),[207] and further work has been undertaken to decide this point. A reinvestigation of the fluorescence yields and phosphorescence yields as a function of pressure for SO_2 excited at several different wavelengths shows that whereas there is reasonable agreement between the results of the earlier work at longer wavelengths, serious disagreement exists over the behaviour at the shorter wavelengths.[208] The relevant data are summarized in Table 13. The new data clearly support the contention that at limiting low pressures of SO_2

---

[203] R. J. Donovan, D. Husain, and P. T. Jackson, *Trans. Faraday Soc.*, 1969, **65**, 2930.
[204] R. J. Donovan, D. Husain, and C. D. Stevenson, *Trans. Faraday Soc.*, 1970, **66**, 1.
[205] W. H. Smith, *J. Chem. Phys.*, 1969, **51**, 3140.
[206] H. D. Mettee, *J. Chem. Phys.*, 1968, **49**, 1784.
[207] T. N. Rao, S. S. Collier, and J. G. Calvert, *J. Amer. Chem. Soc.*, 1969, **91**, 1609, 1616.
[208] T. N. Rao and J. G. Calvert, *J. Phys. Chem.*, 1970, **74**, 681.

significant non-radiative decay persists. Experiments with added biacetyl allowed an estimation of $k_{256} : (k_{256} + k_{255})$ at the two wavelengths, 2963 Å and 3020 Å, and the values obtained were 0·09 and 0·10 respectively, in good agreement with the previously reported value of 0·08 at 2875 Å.[207] The ratio Φ_P/Φ_F varies with wavelength of excitation, and since in theory

Table 13 *Emission of sulphur dioxide vapour*

Wavelength (Å)	Φ_P/Φ_F	$(k_{255}+k_{257}):k_{257} \times 10^6$	$\dfrac{(k_{255}+k_{256})^*}{(k_{255}+k_{256}+k_{257})}$	Φ_F at $[SO_2] = 0$	Ref.
2650	0·08	5·3	—	0·55	208
2650	0·08	1·9	—	1·4	206
2750	0·09	7·0	—	0·13	208
2750	0·10	3·4	—	1·2	206
2850—2894	0·14	6·6	2·7	0·14	208
2850—2894	—	5·6	—	1·0	206
2960—2967	0·26	13·8	3·4	0·15	208
2960—2967	0·26	10·2	—	0·86	206
3020—3021	0·35	15·9	3·9	0·061	208
	0·39	10·7	—	0·97	206
3130—3132	—	20·1	7·2	0·67	208

* Data from S. J. Strickler and D. B. Howell, *J. Chem. Phys.*, 1968, **49**, 1947.

the ratio should be given by a rate-constant ratio at high SO_2 pressures as in equation (263), it is of interest to speculate about the reason for the observed variation.

$$\Phi_P/\Phi_F = k_{260}\, k_{256}/k_{262}\, k_{257} \qquad (263)$$

Previous studies suggest that emission occurs from the vibrationally relaxed triplet SO_2, and thus k_{260} and k_{262} are unlikely to vary with wavelength. Similarly it can be seen above that the $k_{256} : (k_{256} + k_{255})$ ratio does not vary significantly with wavelength, so it appears that the change in the ratio is principally due to a variation of k_{257} with wavelength of excitation. Confirmation of the fact that vibrational relaxation of the $^3SO_2$ is complete before reaction or emission is complete comes from studies on the lifetime of this species when excited in two different ways. In the first set of experiments the triplet state of the SO_2 was produced directly by absorption of 3828·8 Å radiation from a frequency-doubled, Raman-shifted, ruby laser (see Techniques section, Chapter 1), and in the second the triplet level was populated *via* absorption to the excited singlet state and subsequent intersystem crossing. Despite the different conditions under which the triplet state was populated, the values obtained in the two experiments for the rate constant k_{262} were very similar, being $3·8 \pm 0·1 \times 10^8$ l mol^{-1} s^{-1} and $3·9 \pm 0·7 \times 10^8$ l mol^{-1} s^{-1} respectively.[209] The lifetime of the triplet SO_2 at zero pressure was estimated to be $7·9 \pm 1·7 \times 10^{-4}$ s, in agreement with other values in the literature: see ref. 209. The interest in the SO_2 molecule from the viewpoint of air pollution is that it photochemically produces an

[209] S. S. Collier, A. Morikawa, D. H. Slater, J. G. Calvert, G. Reinhardt, and E. Damon, *J. Amer. Chem. Soc.*, 1970, **92**, 217.

active species. This species has never been identified positively, but is probably SO_3. The photodissociation of SO_2 is impossible at wavelength longer than 2180 Å, so it has been suggested that a likely route for formation of SO_3 is interaction of an electronically excited SO_2 with a ground-state molecule (264).

$$SO_2^* + SO_2 \longrightarrow SO_3 + SO \qquad (264)$$

However, reactions of SO_2^* with oxygen, or of SO_2 with ozone, seem more likely to be important at the concentrations which occur in the atmosphere. The state responsible for formation of SO_3 has never been established, but there is now strong evidence that it is the triplet.[210] In the presence of biacetyl, which preferentially quenches the triplet state of SO_2, the rate of formation of SO_3 was drastically reduced. The quantum yield of formation of SO_3 at 3126—2537 Å was 0·08. The quenching of the excited singlet and triplet states of sulphur dioxide by fifteen collision partners has been investigated, and correlation sought between efficiencies of quenching and polarizability of quencher, product of reduced mass and polarizability, and energy differences. It was found that the best correlation was between polarizability and quenching rate, and it was further found that the singlet SO_2 was quenched efficiently, requiring from one to ten gas-kinetic collisions, whereas the triplet state of the SO_2 requires of the order of thousands of collisions before quenching occurs.[211]

7 Photochemistry of Oxygen and Ozone: Photo-oxidations

The oxygen molecule plays a large part in atmospheric studies, both because of its high concentration and because of the number of available electronic states which can be populated upon absorption of u.v. and vacuum-u.v. radiation. The latter transitions are especially important in studies of the upper atmosphere. The absorption coefficient of molecular oxygen near 1215 Å has been measured.[212] When excited in the region from 1000—450 Å, molecular oxygen dissociates to give highly excited oxygen atoms.[157] Three distinct processes have been identified:

$$O_2 + h\nu \longrightarrow O(3s^3S^0) + O(^3P) \qquad (265)$$

$$\longrightarrow O(3p^3P) + O(^3P) \qquad (266)$$

$$\longrightarrow O(3s^1D^0) + O(^3P) \qquad (267)$$

The excited oxygen atoms were identified by their emission spectra in the vacuum-u.v. region. Thus the atom formed in (265) emits at 1305 Å, and that in (266) decays via a two-step mechanism.

$$O(3p^3P) \longrightarrow O(3s^3S^0) + h\nu \,(= 8447 \text{ Å})$$

$$\longrightarrow O(^3P) + h\nu \,(= 1305 \text{ Å}) \qquad (268)$$

[210] S. Okuda, T. N. Rao, D. H. Slater, and J. G. Calvert, *J. Phys. Chem.*, 1969, **73**, 4412.
[211] H. D. Mettee, *J. Phys. Chem.*, 1969, **73**, 1071.
[212] T. D. Gaily, *J. Opt. Soc. Amer.*, 1969, **59**, 536.

The ^1D oxygen atoms formed in (267) emit at 6300 Å and they are also produced in the photolysis of oxygen at 1470 Å.[213] The production of O(^1S) atoms is also spin-allowed, and this species has been observed in the vacuum-u.v. photolysis of O_2.[214] The relaxation of the O(^1D) atoms produced in reaction (267) and in other systems may be influenced by the presence of added gases. The interaction with ground-state O_2 has been shown to yield O_2 ($^1\Sigma$) with high efficiency,[213] although this conclusion is in contradiction with earlier reports. The relative rates of reaction of O (^1D) with a number of inert and reactive atoms and molecules have been determined,[215] and similar values for argon, krypton, and neon have been obtained in an independent study.[216] The reader is referred to the profuse data listed in references 213 and 215, and to the additional information contained in reference 216.

The reactions of ground-state oxygen atoms are also of interest in atmospheric studies, and in general are different from those of excited atoms. The interaction of the former with acetylene produces carbon monoxide, C_3H_4, H_2, and a polymer.[217] The two primary steps appear to be

$$O + C_2H_2 \longrightarrow CH_2 + CO \qquad (269)$$
$$\longrightarrow H_2 + C_2O \qquad (270)$$

It may be recalled that the C_2O species has also been proposed as a primary product of the photolysis of keten [109] and carbon suboxide. Much of the oxygen initially present appears in the polymer. This is also the case with oxygen-atom attack on cyclopropane, in which about 44% of the oxygen produces formaldehyde (which polymerizes), and the remainder produces water and carbon monoxide.[218] The mechanism of addition of ground-state oxygen atoms to olefins has been the subject of fierce debate. One author prefers the concept of a 'triplet biradical' (a term to which purists will object) in which the atom attacks perpendicular to the plane of the molecule,[219] whereas another group prefer the notion of oxygen-atom attack in the plane of the ring to give 'transition states'.[220] Experimental evidence can be rationalized in terms of either scheme, but the former allows a simpler interpretation and there is evidence to show that the latter on its own is insufficient to account for results observed with cyclopropane.[218] It is possible that perpendicular attack to produce the diradical is followed by interaction of the oxygen atom with both the carbon atom and a hydrogen atom as suggested by the second group,[220] and that a combination of the two approaches is necessary. The reaction of oxygen

[213] J. F. Noxon, *J. Chem. Phys.*, 1970, **52**, 1852.
[214] S. V. Filseth and K. H. Welge, *J. Chem. Phys.*, 1969, **51**, 839.
[215] G. Paraskevopolous and R. J. Cvetanovic, *J. Amer. Chem. Soc.*, 1969, **91**, 7572.
[216] M. Clerc, A. Reiffsteck, and B. Lesigne, *J. Chem. Phys.*, 1969, **50**, 3721.
[217] D. G. Williamson and K. D. Bayes, *J. Phys. Chem.*, 1969, **73**, 1232.
[218] A. A. Scala and W.-T. Wu, *J. Phys. Chem.*, 1970, **74**, 1852.
[219] R. J. Cvetanovic, *J. Phys. Chem.*, 1970, **74**, 2730.
[220] M. D. Scheer and R. Klein, *J. Phys. Chem.*, 1970, **74**, 2732.

atoms with H_2S has been mentioned in the previous section, and the reactivity of oxygen atoms in aqueous solution has also been the subject of a publication.[221]

The u.v. photolysis of ozone is probably the major source of electronically excited atomic and molecular oxygen in the atmosphere, and at 2537 Å this reaction has been well-established.[222–226] At longer wavelengths excited singlet molecular oxygen is still apparently formed,[222, 226] but the atom formed under these circumstances is in the ground state, in violation of the spin-conservation rules.

The flash photolysis of ozone–cyanogen mixtures gives rise to two unstable products, both of molecular formula C_2N_2O.[227] One of these arises from the insertion of $O(^1D)$ atoms into the C—C bond in the cyanogen, and the other results from the terminal addition of $O(^3P)$. Photolysis of ozone in liquid carbon dioxide at $-45\,°C$ apparently produces the CO_3 intermediate by the sequence of reactions outlined below.[228]

$$O_3 + h\nu\ (2537\ \text{Å}) \longrightarrow O_2 + O(^1D) \qquad (271)$$

$$O(^1D) + CO_2 + M \longrightarrow CO_3 + M \qquad (272)$$

$$CO_3 + CO_3 \longrightarrow 2CO_2 + O_2 \qquad (273)$$

The same intermediate species has been observed in the gas-phase photolysis of CO_2 at 1470 Å, and its lifetime determined.[229]

Undoubtedly $O_2(^1\Delta_g)$–'singlet oxygen'–is the major species responsible for photo-oxidations in the gas and condensed phases, and it has also been recognized as playing a major role in polluted atmospheres of the Los Angeles type. For this reason it is essential to know details of the deactivation of this species in the gas phase, and also its rate of reaction with organic molecules and other species present in urban atmospheres. Many studies have been undertaken recently to supply information on $O_2(^1\Delta_g)$, and rate-constant data for reaction and deactivation are collated in Table 14. Included are data on deactivation by ground-state O_2[230, 231] and N_2,[230, 231] atomic nitrogen and atomic oxygen,[232, 233] and other species.[234] Rate constants have been measured for reactions with ozone[235] and

---

[221] O. Amichai and A. Treinin, *Chem. Phys. Letters*, 1969, **3**, 611.
[222] I. T. N. Jones and R. P. Wayne, *J. Chem. Phys.*, 1969, **51**, 3617.
[223] R. E. Huffman, J. C. Larrabee, and V. C. Baisley, *J. Chem. Phys.*, 1969, **50**, 4594.
[224] I. T. N. Jones, U. B. Kaczmar, and R. P. Wayne, *Proc. Roy. Soc.*, 1970, **A316**, 431.
[225] M. Gauthier and D. R. Snelling, *Chem. Phys. Letters*, 1970, **5**, 93.
[226] E. Castellano and H. J. Schumaker, *Z. phys. Chem. (Frankfurt)*, 1969, **65**, 62.
[227] C. W. Hand and R. M. Hexter, *J. Amer. Chem. Soc.*, 1970, **92**, 1828.
[228] W. B. DeMore and C. W. Jacobsen, *J. Phys. Chem.*, 1969, **73**, 2935.
[229] M. Arvis, *J. Chim. phys.*, 1969, **66**, 517.
[230] I. D. Clark and R. P. Wayne, *Chem. Phys. Letters*, 1969, **3**, 93.
[231] R. P. Steer, R. A. Ackerman, and J. N. Pitts jun., *J. Chem. Phys.*, 1969, **51**, 843.
[232] I. D. Clark and R. P. Wayne, *Chem. Phys. Letters*, 1969, **3**, 405.
[233] I. D. Clark and R. P. Wayne, *Proc. Roy. Soc.*, 1970, **A316**, 539.
[234] F. D. Findlay, C. J. Fortin, and D. R. Snelling, *Chem. Phys. Letters*, 1969, **3**, 204.
[235] R. P. Wayne and J. N. Pitts jun., *J. Chem. Phys.*, 1969, **50**, 3644.

olefins.[236, 237] The quenching by ozone produces ground-state molecular oxygen and oxygen atoms (274).[235]

$$O_2\,(^1\Delta_g) + O_3 \longrightarrow 2O_2 + O \qquad (274)$$

The quenching of $O_2\,(^1\Delta_g)$ by tetramethylethylene was largely chemical in nature, producing 2,3-dimethyl-3-hydroperoxybut-1-ene, whereas significant

Table 14 Reactions and deactivation of $O_2\,(^1\Delta_g)$

Reactant	Rate constant (l mol^{-1} s^{-1})	Ref.
O_3	$1.7 \pm 0.5 \times 10^6$	235
O_3	$2.1 \pm 0.6 \times 10^6$	a
O_3	$1.5 \pm 0.3 \times 10^6$	b
O_2 (ground)	$1.32 \pm 0.3 \times 10^3$	230
O_2 (ground)	1.22×10^3	231
O_2 (ground)	1.2×10^3	234
N_2	1.2×10^2	230
N_2	25.2	231
O	$\leqslant 7.8 \times 10^4$	232
N	$1.7 \pm 1.2 \times 10^6$	232, 233
methyl chloride	$9.6 \pm 1.7 \times 10^2$	236
ethylene	$1.1 \pm 0.19 \times 10^3$	236
propylene	$1.3 \pm 0.11 \times 10^3$	236
but-1-ene	$1.4 \pm 0.03 \times 10^3$	236
pent-1-ene	$1.9 \pm 0.34 \times 10^3$	236
2,3-dimethylbut-2-ene	$3.0 \pm 2 \times 10^5$	236
tetramethylethylene	$1.0 \pm 0.5 \times 10^5$	234
2,5-dimethylfuran	$3.7 \pm 1.4 \times 10^5$	237
CO_2	$2.3 \pm 0.5 \times 10^3$	232, 233
H_2O	$9 \pm 3 \times 10^3$	232, 233
Ar	$\leqslant 1.3 \times 10^2$	232, 233

[a] I. D. Clark and R. P. Wayne, unpublished results.
[b] R. J. McNeal and G. R. Cook, *J. Chem. Phys.*, 1967, **47**, 585.

physical quenching was observed with other molecules, including 2,5-dimethylfuran; but the latter also gave 2,3,7-trioxa-1,4-dimethylbicyclo-[2,2,1]hept-5-ene as a product. The rate constants in Table 14 represent the sum of rate constants for chemical and physical quenching. It can be seen from Table 14 that in atmospheres not containing significant concentrations of highly reactive pollutants, the lifetime of $O_2\,(^1\Delta_g)$ will be determined largely by the concentration of ground-state O_2.

The photolysis of ozone is undoubtedly one source of atmospheric $O_2\,(^1\Delta_g)$, but there is recent evidence that the species may also be formed *via* energy transfer in the gas phase.[238] Other studies on air pollution include a review,[239] a description of atmospheric photochemical reactions

[236] R. A. Ackerman, J. N. Pitts jun., and R. P. Steer, *J. Chem. Phys.*, 1970, **52**, 1603.
[237] W. S. Gleason, A. D. Broadbent, E. Whittle, and J. N. Pitts jun., *J. Amer. Chem. Soc.*, 1970, **92**, 2068.
[238] R. P. Steer, J. L. Sprung, and J. N. Pitts jun., *Environ. Sci. Technol.*, 1969, **3**, 946.
[239] R. D. Cadle and E. R. Allen, *Science*, 1970, **167**, 243.

Gas-phase Photochemistry 225

in a tube flow reactor,[240] and other reactions,[241] the inhibition of atmospheric photo-oxidation of hydrocarbons by nitric oxide,[242] and a proposed dynamic model of photochemical smog.[243] In future volumes of this series it is hoped to include a comprehensive section on the rapidly growing field of photochemical air pollution and related areas.

As stated above, O_2 ($^1\Delta_g$) is of great importance in photo-oxidations in the condensed phases also. Discussion of these reactions will be found in the appropriate sections of this volume, but for convenience a number of reactions involving singlet molecular oxygen will be listed here, *viz.* the photo-oxidation of anthracene,[244] the addition of singlet oxygen to monocyclic aromatic rings,[245] the abstraction of hydrogen from a phenol,[246] and the reaction with a simple thiophen:[247] and the mechanism of the reaction with olefins has been discussed.[248] Brief reviews on sensitized photo-oxygenations [249, 250] have appeared, and the selection rules pertinent to singlet oxygen reactions have been treated using molecular orbital and state correlation diagrams.[251] The treatment has been described fully in Chapter 1 of this section. Dye-sensitized photo-oxygenations involving leucofluorescein,[252] xanthine,[253] and Methylene Blue [254] have been reported, as has the photoperoxidation of unsaturated organic molecules in solution.[255] The fluorescence or rubrene can be excited by electronic energy transfer from singlet oxygen.[256] The last two studies are described in detail in the previous chapter of this Part.

Finally in this section is included a brief description of the photolysis of carbon suboxide (C_3O_2) since this compound has not been included in any of the previous categories.

Photolysis at 1470 Å in the presence of methyl fluoride produces acetylene, vinyl fluoride, CO, and CO_2 by the following mechanism.[257]

$$C_3O_2\,(^1\Sigma) + h\nu \longrightarrow C_3O_2^*\,(^1\ ) \qquad (275)$$

$$C_3O_2^*\,(^1\ ) \longrightarrow C_2O\,(^1\ ) + CO\,(^1\Sigma) \qquad (276)$$

[240] E. R. Stephens and M. A. Price, *Atmos. Environ.*, 1969, **3**, 573.
[241] M. Katz, *Canad. J. Chem. Eng.*, 1970, **48**, 3.
[242] W. A. Glasson and C. S. Tuesday, *Environ. Sci. Technol.*, 1970, **4**, 1.
[243] S. K. Friedlander and J. H. Seinfeld, *Environ. Sci. Technol.*, 1969, **3**, 1175.
[244] N. Sugiyama, M. Iwata, M. Yoshioka, K. Yamada, and H. Aoyama, *Bull. Chem. Soc. Japan*, 1969, **42**, 1377.
[245] I. Saito, S. Kato, and T. Matsuura, *Tetrahedron Letters*, 1970, 239.
[246] T. Matsuura, A. Nishinaga, N. Yoshimura, T. Arai, K. Omura, H. Matsushima, S. Kato, and I. Saito, *Tetrahedron Letters*, 1970, 1669, 1673.
[247] C. N. Skold and R. H. Schlessinger, *Tetrahedron Letters*, 1970, 791.
[248] W. Fenical, D. R. Kearns, and P. Radlick, *J. Amer. Chem. Soc.*, 1969, **91**, 3396, 7771.
[249] I. Moritani and T. Sato, *J. Synthetic Org. Chem. Japan*, 1969, **27**, 1165.
[250] D. R. Kearns and A. U. Khan, *Photochem. and Photobiol.*, 1969, **10**, 193.
[251] D. R. Kearns, *J. Amer. Chem. Soc.*, 1969, **91**, 6554.
[252] Y. Usui, C. Iwanga, and M. Koizumi, *Bull. Chem. Soc. Japan*, 1969, **42**, 1231.
[253] Y. Le-Roux, J. C. Ginisty, and C. Nofre, *Compt. rend.*, 1969, **269**, C, 744.
[254] J. Stauff and H. Fuhr, *Ber. Bunsengesellschaft. Phys. Chem.*, 1969, **73**, 245.
[255] B. Stevens and B. E. Algar, *J. Phys. Chem.*, 1969, **73**, 1711.
[256] T. Wilson, *J. Amer. Chem. Soc.*, 1969, **91**, 2387.
[257] E. Tschuikow-Roux and S. Kodama, *J. Chem. Phys.*, 1969, **50**, 5297.

$$C_2O\,(^1\ ) + CH_3F \longrightarrow C_2H_3F^* + CO\,(^1\Sigma) \quad (277)$$

$$C_2H_3F^* \longrightarrow C_2H_2 + HF \quad (278)$$

$$C_2H_3F^* + M \longrightarrow C_2H_3F + M' \quad (279)$$

$$C_2H_3F^* \longrightarrow C_2H_3F^{**} \quad (280)$$

$$M + C_2H_3F^{**} \longrightarrow C_2H_3F + M'' \quad (281)$$

$$C_2O\,(^1\ ) + C_3O_2\,(^1\Sigma) \longrightarrow CO_2 + C_4O\,(?) \quad (282)$$

$$C_2O\,(^1\ ) + C_3O_2\,(^1\Sigma) \longrightarrow CO\,(^1\Sigma) + C_4O_2\,(?) \quad (283)$$

The production of vibrationally excited vinyl fluoride in the system allows a RRKM treatment to be applied. Details of this will be found in the original paper, but the results indicate that the singlet C_2O species formed initially must have about 3·1 eV of energy above its ground state. This is considerably in excess of the energy levels of the $^1\Delta$, $^1\Sigma$, and $^3\Pi$ states reported for C_2O, and for this reason the state involved in reactions (277), (282), and (283) is left unspecified, as is the singlet state of the parent molecule produced upon absorption of a photon at 1470 Å. At 2537 Å, in the presence of ethylene oxide, carbon suboxide produces ethylene and traces of acetylene.[258] With trimethylene oxide (oxetan) the products are ethylene, acetylene, propylene, and cyclopropane, the latter being formed with 344 kJ mol^{-1} excess energy, causing it to isomerize to the propylene unless collisionally deactivated. The flash photolysis of carbon suboxide has also been investigated.[259]

8 Atom Sensitizations

Mercury atoms continue to be the most widely used atoms for sensitization of reactions in the gas phase. There has been further discussion of the oscillator strength of the Hg$(6^1S_0$—$6^1P_1)$ transition at 1849 Å [260] and of low-lying excited levels of the ^{200}Hg isotope.[261] Quenching cross-sections for the $^1P_1$ atom with substrates varying in complexity from helium atoms to cyclohexane show little marked variation,[262] in sharp contrast with the situation with $^3P_1$ atoms and the same quenchers: see Table 15. The results suggest that there may be a common quenching mechanism for all gases with the singlet atom. The triplet atom is quenched by the noble gases also,[263] and this phenomenon has been ascribed in the past to the formation of a loosely bonded diatomic van der Waals molecule which can cross to the ground state without emission of radiation. This explanation has now been shown to be in error, since the complex emits intensely in the form of

[258] R. T. K. Baker and J. A. Kerr, *J. Chem. Soc.* (*A*), 1969, 390.
[259] W. Braun, A. M. Bass, D. D. Davis, and J. N. Simmons, *Proc. Roy. Soc.*, 1969, **A312**, 417.
[260] A. Skeberle and E. N. Lassette, *J. Chem. Phys.*, 1970, **52**, 2708.
[261] M. Sakai, *J. Phys. Soc. Japan*, 1969, **26**, 879.
[262] A. Granzow, M. Z. Hoffman, and N. N. Lichtin, *J. Phys. Chem.*, 1969, **73**, 4289.
[263] H. E. Gunning, S. Penzes, H. S. Sandhu, and O. P. Strausz, *J. Amer. Chem. Soc.*, 1969, **91**, 7684.

bands extending to 2780 Å, with a quantum yield of unity. The mechanism is thus:

$$Hg + h\nu \longrightarrow Hg^* \qquad (284)$$

$$Hg^* \longrightarrow Hg + h\nu \ (2537 \text{ Å}) \qquad (285)$$

$$Hg + Ar \longrightarrow [Hg-Ar]^* \longrightarrow [Hg-Ar] + h\nu_b \qquad (286)$$

From the quenching of the 2537 Å radiation, rate constants for process (286) have been determined, and these are shown in Table 16. The

Table 15 *Relative cross-sections for the quenching of* $Hg\ (^1P_1)$ *and* $Hg\ (^3P_1)$ *atoms*

Quencher	Mass, M_Q	Cross-section for $^1P_1$ ($\sigma_{N_2} = 1.00$)	Cross-section for $^3P_1$ ($\sigma_{N_2} = 1.00$)*
H_2	2	0.2	31
He	4	0.0	0.0
CH_4	16	0.3	0.3
NH_3	17	0.8—1.6	15
H_2O	18	0.9	5.2
Ne	20	0.3	0.0
CO	28	1.0	21
C_2H_4	28	1.0—2.0	135
Ar	40	1.1	0.0
CO_2	44	0.9	13
N_2O	44	1.3	66
cyclo-C_6H_{12}	78	3.0	75
cyclo-C_4F_8	200	2.8	—

* Data from J. G. Calvert and J. N. Pitts jun., 'Photochemistry'. John Wiley and Sons, New York, 1966.

untenability of the crossing theory is further demonstrated by simple Lennard-Jones potential-energy calculations on the Hg–Xe, Hg–Kr, and Hg–Ar systems. As can be seen in Figure 15, the potential energy curves for the excited-state complexes do not cross the ground-state curves at any reasonable energy, indicating that energy dissipation must occur radiatively. The relative cross-sections for the quenching of $Hg\ (6^3P_1)$ atoms by aromatic compounds have already been discussed [43] (see Table 6). A theoretical study of the $^3P_1$-sensitized decomposition of molecular hydro-

Table 16 *Rate constants and cross-section values for quenching of* $Hg\ (6^3P_1)$ *atoms by rare gas atoms* [263]

Rare gas	Rate constant (l mol^{-1} s^{-1})	Cross-section (Å^2)
He	5.9	0.20
Ne	3.9	0.27
Ar	5.8	0.32
Kr	5.5	0.34
Xe	3.1	0.12

gen [264] indicated that the yields of H atoms, HgH, and undissociated H_2 should be 0.52, 0.16, and 0.58 respectively. This prediction has prompted a reinvestigation of this simple system using the ethylene scavenger technique: the previous experimental values of Φ_H = unity were confirmed.[265]

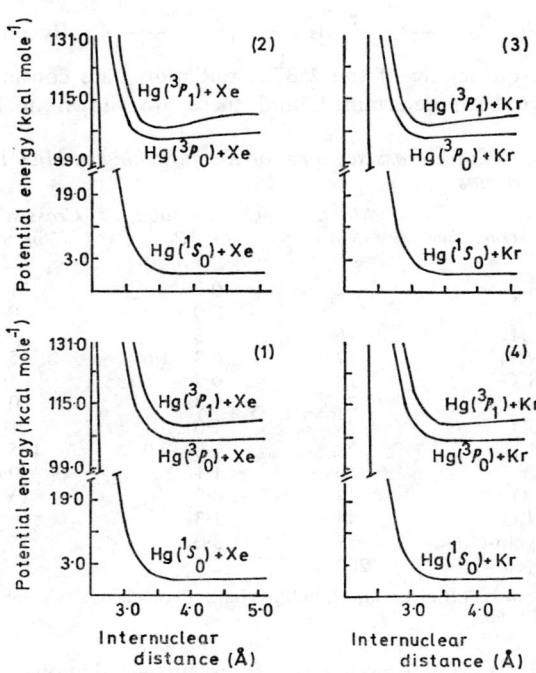

Figure 15 *Calculated Lennard-Jones potentials for* Hg–Xe, *and* Hg–Kr *complexes, in ground and excited states. Curves* 1 *and* 4 *assume* $\sigma(ex.) = \sigma(ground)$, *and* 2 *and* 3 *assume* $\sigma(ex.) = 0.9\ \sigma(ground)$
(Reproduced by permission from *J. Amer. Chem. Soc.*, 1969, **91**, 7684)

Hydrogen atoms produced in this reaction have been used to investigate the relative importance of reactions (287) and (288)

$$H\cdot + C_3H_8 \longrightarrow i\text{-}C_3H_7\cdot + H_2 \quad (287)$$
$$\longrightarrow n\text{-}C_3H_7\cdot + H_2 \quad (288)$$

From the results, k_{287}/k_{288} was given by:

$$k_{287}/k_{288} = 1\cdot 29 \exp(-8400/RT) \quad (289)$$

where R is in J mol^{-1} deg^{-1}.

[264] K. Yang, J. D. Paden, and C. L. Hassell, *J. Chem. Phys.*, 1967, **47**, 3824.
[265] T. L. Pollock, E. Jakubowski, H. E. Gunning, and O. P. Strausz, *Canad. J. Chem.*, 1969, **47**, 3474; J. M. Campbell, O. P. Strausz, and H. E. Gunning, *ibid.*, p. 3759.

The corresponding ratio for the direct attack of Hg $^3P_1$ atoms on propane, leading to isopropyl and n-propyl radicals, is as follows:[266]

$$\log k_n/k_i = -0.206 \pm 0.041 - (4502 \pm 260)/(2.303RT) \quad (290)$$

Mercury-photosensitized decomposition of ethyl fluoride, 1,1-difluoroethane, and 1,1,1-trifluoroethane gave a complex mixture of reaction products, some of which arise *via* the elimination of HF from vibrationally excited species.[267] The addition of hydrogen atoms to vinyl fluoride, 1,1-difluoroethylene, and trifluoroethylene has also been discussed.[268] The sensitized decomposition of CH_2ClF produces CH_2F radicals, which can be used to generate 'hot' 1,2-$C_2H_4F_2$.[269] A RRKM treatment of the data for HF elimination from the 'hot' species revealed that the critical energy for the process was 260 kJ mol^{-1}.

Other photosensitized decompositions reported during the year include those of methyl ethyl ether,[270] methyl vinyl ether,[271] 2,5-dimethylfuran,[272] 2,5-dihydrofuran,[273] and cyclobutanone.[274] In the latter study it was concluded that some *singlet* excited cyclobutanone molecules are formed by energy transfer from the *triplet* mercury atom, in apparent violation of the spin selection rules. A similar phenomenon was previously noted in the sensitization of keten.

Cadmium atoms have probably been used as a sensitizers more than any metal atoms other than mercury. The metal suffers from the drawback that the vapour pressure at room temperature is very low, and in order to see significant photosensitization, high temperatures must be employed in the reaction vessel, with consequent loss in flexibility of experimental conditions, plus the added complications of thermally induced reactions. A brief study of acetone under such conditions has been reported.[275] A technique is available, however, which can increase the concentration of cadmium atoms at room temperature by a factor of eight orders of magnitude for a sufficient length of time to induce photosensitization.[276] The technique involves flash photolysis of the vapours of volatile metal alkyls, such as dimethylcadmium, which can produce several hundred microns pressure, of free atoms, persisting for several milliseconds. Because of pressure broadening of the absorption line, these atoms can be used to initiate photosensitization reactions. Under these conditions the $^3P_1$ atoms are produced *via* the $^1P_1$ since absorption of 2288 Å radiation by ground-state cadmium atoms has

---

[266] J. M. Campbell, H. E. Gunning, and O. P. Strausz, *Canad. J. Chem.*, 1969, **47**, 3763.
[267] P. M. Scott and K. R. Jennings, *J. Phys. Chem.*, 1969, **73**, 1513.
[268] P. M. Scott and K. R. Jennings, *J. Phys. Chem.*, 1969, **73**, 1521.
[269] H. W. Chang and D. W. Setser, *J. Amer. Chem. Soc.*, 1969, **91**, 7648.
[270] S. V. Filseth, *J. Phys. Chem.*, 1969, **73**, 793.
[271] R. V. Morris and S. V. Filseth, *Canad. J. Chem.*, 1970, **48**, 924.
[272] S. Boue and R. Srinivasan, *J. Amer. Chem. Soc.*, 1970, **92**, 1824.
[273] B. Francis and A. G. Sherwood, *Canad. J. Chem.*, 1970, **48**, 25.
[274] D. C. Montague and F. S. Rowland, *J. Amer. Chem. Soc.*, 1969, **91**, 7230.
[275] B. L. Kalra and A. R. Knight, *Canad. J. Chem.*, 1970, **48**, 1333.
[276] P. J. Young, G. Greig, and O. P. Strausz, *J. Amer. Chem. Soc.*, 1970, **92**, 413.

an extinction coefficient several orders of magnitude greater than that at 3261 Å. Similar results were obtained with dimethylzinc. The technique promises to be useful in that reactions may now be studied at room temperature, and the energies of the cadmium and zinc triplet atoms are in a very useful range, but of course the presence of the starting material and products of the primary reaction will cause complications. Nevertheless, further experiments with this system are awaited with interest. The cadmium photosensitization of but-2-ene–benzene mixtures at 270 °C indicates that the triplet benzene so formed cannot transfer its energy to the but-2-ene with unit efficiency, as is the case at ambient temperatures.[277] The cause may well be associated with the high temperatures used, which could promote chemical transformations such as the transient production of 'Dewar'-benzene.[119]

There are a few reports in the literature of the quenching of the fluorescence of electronically excited alkali-metal atoms by rare gases,[278] nitrogen,[279] and hydrocarbons,[280, 281] but there is no report of the reactions, if any, which result from such interactions.

Rare-gas atoms are capable of initiating reactions in substrates, and argon ($^3P_{2,0}$) atoms can produce electronically excited OH and OD radicals following collision with H_2O, HCO_2H, and D_2O.[282] Xenon sensitizes the $CO(^1\Sigma^-)$—$CO(^1\Sigma^+)$ emission, even though the production of the singlet CO by collision with an excited triplet Xe atom is spin forbidden.[283] However, in view of the known effects of xenon in other photochemical systems, this breakdown in spin conservation is not surprising. The xenon-sensitized photolysis of carbon dioxide can be explained on the basis of the mechanism below, involving the CO_3 intermediate [284] which has been proposed in other studies (see section on oxygen photochemistry).

$$Xe + h\nu\ (1470\ \text{Å}) \longrightarrow Xe^* \qquad (291)$$

$$Xe^* + CO_2 \longrightarrow Xe + CO + O(^1D) \qquad (292)$$

$$O(^1D) + CO_2 \longrightarrow CO_3^* \qquad (293)$$

$$CO_3^* \longrightarrow CO + O_2 \qquad (294)$$

$$CO_3^* + M \longrightarrow O(^3P)\ \text{or}\ O(^1D) + M \qquad (295)$$

At low pressures of CO_2, $\Phi_{CO} = 2$, but the value decreases towards unity if gases are added: the relative quenching efficiencies were $CO_2 = 1\cdot00$, $Xe = 0\cdot90$, $He = 0\cdot14$, $Kr = 0\cdot06$, and $C_2F_6 = 0\cdot0$.

---

[277] S. Tsunashima, S. Satoh, and S. Sato, *Bull. Chem. Soc. Japan*, 1969, **42**, 1531.
[278] M. G. Edwards, *J. Phys. (B)*, 1969, **2**, 719.
[279] E. Bauer, E. R. Fisher, and F. R. Gilmore, *J. Chem. Phys.*, 1969, **51**, 4173.
[280] N. Fronimos, J. Fricke, and J. Haas, *Z. Naturforsch.*, 1969, **24a**, 1030.
[281] M. Stupavsky and L. Krause, *Canad. J. Phys.*, 1969, **47**, 1249.
[282] M. A. A. Clyne, J. A. Coxon, D. W. Setser, and D. H. Stedman, *Trans. Faraday Soc.*, 1969, **65**, 1177.
[283] T. G. Slanger and G. Black, *J. Chem. Phys.*, 1969, **51**, 4534.
[284] L. J. Stief, V. J. De-Carlo, and W. A. Payne, *J. Chem. Phys.*, 1969, **51**, 3336.

Other miscellaneous atom reactions include luminescent reactions of $H(^2P)$,[285] the quenching of the hydrogen Lyman α fluorescence,[286] the collisional deactivation of S (3^1D_2) and S (3^1S_0) atoms by the noble gases,[287] and the quenching of the resonance fluorescence of Cu atoms by H_2, N_2, CO_2, and Ar.[288]

D. P.

[285] T. S. Wauchop and L. F. Phillips, *J. Chem. Phys.*, 1969, **51**, 1167.
[286] T. S. Wauchop, M. J. McEwan, and L. F. Phillips, *J. Chem. Phys.*, 1969, **51**, 4227.
[287] R. J. Donovan, L. J. Kirsch, and D. Husain, *Trans. Faraday Soc.*, 1970, **66**, 774.
[288] R. Bleekrode and W. Van Benthem, *J. Chem. Phys.*, 1969, **51**, 2757.

Part II

INORGANIC PHOTOCHEMISTRY

The approach in this part of Volume 2 is similar to that in Volume 1. The initial section deals with the photochemistry of water, hydrogen peroxide, and aqueous anions, and the second, major, section is concerned with the luminescence and photochemistry of co-ordination complexes of transition metals, generally in solution. The final brief sections report on examples of the effect of light on inorganic materials in the gas and solid phases, including a discussion of phosphors.

1 Photolysis of Water, H_2O_2, and Aqueous Anions

The position in the spectral region of the absorption spectrum of water is dependent upon the phase in which it is measured. The complete spectrum

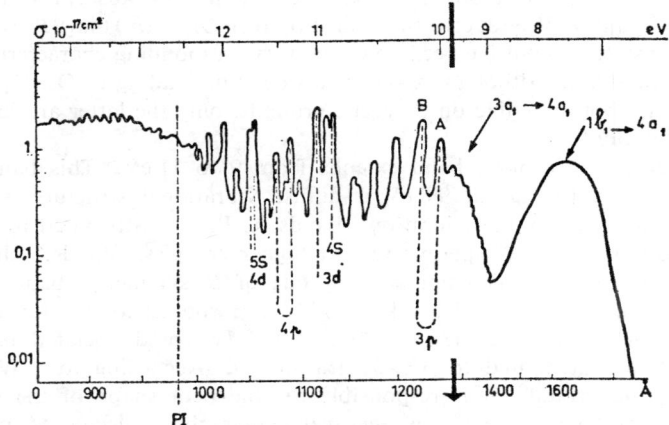

Figure 1 *U.v. absorption spectrum of* H_2O
(Reproduced by permission from *Israel. J. Chem.*, 1969, **7**, 275)

in the vapour phase under low resolution in the region from 900 to 2000 Å is shown in Figure 1.[1] It can be seen that the spectrum is entirely diffuse,

[1] F. Fiquet-Fayard, *Israel J. Chem.*, 1969, **7**, 275.

and thus the photoexcitation of water always leads to dissociation. The energy thresholds of the various processes of fragmentation are shown in reactions (1)—(7).[1]

$$
\begin{align}
H_2O + h\nu \quad &\longrightarrow \quad H_2(^1\Sigma_g^+) + O(^3P) \quad &5.02 \text{ eV} \quad &(1)\\
&\longrightarrow \quad H(^2S) + OH(^2\Pi) \quad &5.14 \text{ eV} \quad &(2)\\
&\longrightarrow \quad H_2(^1\Sigma_g^+) + O(^1D) \quad &6.98 \text{ eV} \quad &(3)\\
&\longrightarrow \quad H_2(^1\Sigma_g^+) + O(^1S) \quad &9.20 \text{ eV} \quad &(4)\\
&\longrightarrow \quad H(^2S) + OH(^2\Sigma^+) \quad &9.19 \text{ eV} \quad &(5)\\
&\longrightarrow \quad H^* (n=2) + OH(^2\Pi) \quad &15.33 \text{ eV} \quad &(6)\\
&\longrightarrow \quad H_2(^1\Sigma_u^+) + O(^3P) \quad &16.39 \text{ eV} \quad &(7)
\end{align}
$$

Fragments in excited states can be identified by emission measurements. Thus, when excited between 600 and 800 Å, strong emission is seen between 8.2 and 11.8 eV, which corresponds to emission from the H^* ($n = 2$) and $H_2^*(^1\Sigma_u^+)$ fragments. When excited at longer wavelengths, u.v. emission is seen at 4.05 eV, which corresponds to emission from the excited $OH(^2\Sigma^+)$ radical, reaction (8). Rotational analysis reveals that the excited

$$OH(^2\Sigma^+) \quad \longrightarrow \quad OH(^2\Pi) + h\nu \quad (8)$$

OH radical is highly rotationally excited also, implying that the excited H_2O molecule must dissociate by the departure of a hydrogen atom perpendicularly to the OH bond.[1] Absorption in the first absorption band between 7 and 9 eV is due to the transition $(b_1)^2\, ^1A_1 \to (b_1)(4a_1)^1 B_1$ where the upper state is repulsive because of the very antibonding characteristics of the orbital $4a_1$. Although modes of dissociation leading to $O + H_2$ and $H + OH$ are both possible on energetic grounds, only the latter are found experimentally.[1]

The second absorption band extends from 9 to 11 eV. This band is continuous, exhibiting till 9.8 eV a diffuse vibrational structure which corresponds to a bending vibration. Bands A, B, etc., attributed to predissociative states, are superimposed at higher energies. Band A shows rotational structure which points to a state of B_1 symmetry; band B is entirely diffuse. The broad band probably corresponds to the transition $\ldots (3a_1)^2 (b_1)^2 A_1 \to \ldots (3a_1)(b_1)^2 (4a_1)^1 A_1$. The predissociated bands A and B are attributed to $b_1 \to 3p$ transitions. According to a Walsh diagram, the orbital $3a_1$ is responsible for the bent shape of the H_2O molecule, and consequently its removal causes the molecule to move towards linearity, thus exciting a bending vibration, and also, if in the act of dissociating, causing the H atom to leave perpendicularly to the OH bond. In this energy region, O, $OH(^2\Sigma^+)$ and $OH(^2\Pi)$ are all formed simultaneously, with a relative abundance of 25%, 5% and 70% respectively.

The third absorption region from 11 eV beyond the ionization potential encompasses several Rydberg states, all predissociated.

Inorganic Photochemistry

The results can be understood on the basis of a correlation diagram, shown in Figure 2. Along a vertical line drawn above the C_{2v} group are the states observed in absorption experiments and, to the left, are the same states in a linear configuration. The energy of these states is not known, except that their position relative to the C_{2v} conformation is known

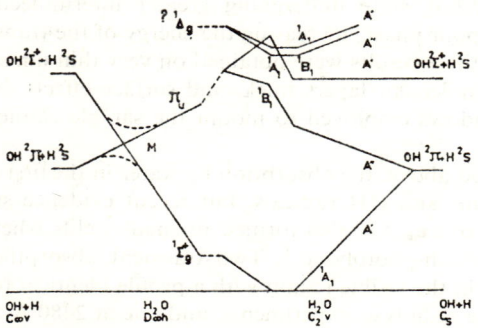

Figure 2 *Correlation diagram of H_2O neutral. On the left part, linear conformations, on the right, bent conformations. There is a conical intersection in M*
(Reproduced by permission from *Israel. J. Chem.*, 1969, **7**, 275)

qualitatively. These two states are correlated to the fragments obtained, to the right for the bent structure and to the left for the linear one. It can be seen from Figure 2 that the correlations differ. Water excited to the first $^1B_1$ level will immediately dissociate *via* reaction (2). When excited to the second level 'A', it could either dissociate *via* reaction (5), or begin to widen the H–O–H angle to adopt a linear conformation. It has been observed that all of the excess energy appears as rotational energy in the fragments and furthermore, the weak vibrational structure seen in the absorption extends beyond the limit for reaction (5), and it thus appears that the principal process is a movement toward the linear conformation. During this process, the molecule vibrates as if alternately linear and bent, and a representative point moves in the region where the surface offers a conical intersection, point M. In this region several processes are in competition:
(a) direct dissociation into $OH(^2\Sigma^+) + H(^2S)$ when crossing the corresponding saddle point;
(b) direct dissociation into $OH(^2\Pi) + H(^2S)$ when crossing the lower part of the conical surface;
(c) direct dissociation into $H_2(^1\Sigma_g^+) + O(^1D)$ when going through the saddle point, since $^1B_1$ is directly correlated to these fragments;
(d) several predissociations *via* triplets or from the $^1B_1$ state.

Experimental evidence suggests all of these possibilities occur. As can be seen from the reaction scheme above, photolysis of water vapour (and D_2O vapour) in the far-u.v. can give rise to hot hydrogen (and deuterium) atoms. Some of the reactions of these have been described,[2] and may be compared with other hot atom reactions discussed in Part I, Chapter 3 of this volume.

The liquid-phase spectrum of the first absorption band of water is shifted to the blue by some 7000 cm^{-1} compared with that in the vapour phase.[3] The reason for the shift is uncertain, but it seems likely that it is due to the upper antibonding state undergoing greater intermolecular interactions than in the vapour phase, increasing the energy of the transition. It should be noted that these results were obtained on very thin films of water, of the order of 900 molecular layers thick, and surface effects due to the highly polar LiF windows employed to mount the sample cannot be discounted completely.

As was stated above, the absorption by water in the first continuum gives rise to H atoms and OH radicals, but recent evidence suggests that the hydrated electron e_{aq}^- is also formed in small yields when pure, oxygen-free water is flash-photolysed.[4] Two transient absorption spectra were obtained, one in the visible region with a profile identical to that attributed to e_{aq}^- in pulse radiolysis experiments, and one at 2480 Å attributed to the OH radical. The shape of the visible spectrum due to e_{aq}^- was further confirmed by flash-photolysing a solution of OH$^-$ saturated with H$_2$ where e_{aq}^- is produced directly, and also from H atoms *via* the sequence shown in reactions (9) and (10).

$$H_2 + OH \longrightarrow H + H_2O \qquad (9)$$

$$nH_2O + H + OH^- \longrightarrow e_{aq}^- + H_2O \qquad (10)$$

No absorption in the visible was obtained when sufficient concentration of electron scavengers, such as O$_2$, HF, N$_2$O were present. The rate constant for the reaction of two e_{aq}^- species was estimated as $k = 0.58 \pm 0.02 \times 10^{10}$ l mol^{-1} s^{-1} at 24 °C.

The minimum amount of energy required to form the hydrated electron was found to be 6.5 ± 0.1 eV (1920 Å), and the estimated quantum yield for its formation is 4×10^{-3}. The mechanism of its formation is probably:

$$2H_2O(aq) + h\nu \longrightarrow e_{aq}^- + OH(aq) + H_3O^+(aq) \qquad (11)$$

and this may require that the excited state of the water molecule be produced near an existing electron trap (a configuration of suitably oriented water dipoles) or is actually part of the trap. As the electron is trapped, either an H atom is abstracted from, or a proton is shifted to, a neighbouring hydrogen-bonded water molecule to complete reaction (11). Possibly a greater fraction of initial excitations leads to formation of e_{aq}^-, and H atoms are then produced by geminate recombination of e_{aq}^- with H$_3$O$^+$(aq).

[2] J. Mansanet and C. Vermeil, *J. Chim. phys.*, 1969, **66**, 1248.
[3] R. E. Verral and W. A. Senior, *J. Chem. Phys.*, 1969, **50**, 2746.
[4] J. W. Boyle, J. A. Ghormley, C. J. Hochanadel, and J. F. Riley, *J. Phys. Chem.*, 1969, **73**, 2886.

Inorganic Photochemistry

The formation of hydroxyl radicals during the sensitized photo-oxidation of water has been discussed.[5] The dissociation of ammonia and water in excited states[6] has been described.

The photodissociation of H_2O, D_2O, and H_2O_2 adsorbed on magnesium oxide surfaces has been studied by e.p.r. techniques.[7] Four types of radical were observed, including that of the hydrogen atom. Two of the other intermediates were identified as adsorbed OH radicals and HO_2 radicals.

Flash photolysis of hydrogen peroxide in aqueous alkaline solutions gives rise to an absorption of 4300 Å due to a transient identified as the ozonide ion, O_3^-.[8] The decay of this transient was found to be first-order, but the rate constant was dependent upon the concentrations of H_2O_2 base, and dissolved molecular oxygen. A longer-lived species was formed as the ozonide ion decayed, and this was shown to be ozone itself.

The results can be explained by reactions (12)—(14).

$$O_3^- \longrightarrow O_2 + O^- \qquad (12)$$

$$O^- + O_2 \longrightarrow O_3^- \qquad (13)$$

$$O^- + HO_2^- \longrightarrow O_2^- + HO^- \qquad (14)$$

The dependence of O_3^- on oxygen and hydrogen peroxide concentrations is consistent with competition by reactions (13) and (14) for O^- in the photolysis of HO_2^-. The further decay, reaction (15), accounts for the

$$O_3^- + O_2^- \xrightarrow{H_2O} O_3 + HO_2^- + OH^- \qquad (15)$$

appearance of O_3 as a product. k_{12} was found to be $2·81 \pm 0·14 \times 10^3$ s^{-1}, and $k_{13}/k_{14} = 3·6 \, (\pm 0·3)$. The marked increase in the stability of O_3^- at high concentrations which was observed can be explained on the basis of the ion pair:

$$Na^+ + O_3^- \rightleftharpoons NaO_3 \qquad (16)$$

the equilibrium constant for which is $2·24 \pm 0·11$ l mol^{-1}.

The photolysis of aqueous H_2O_2 in the presence of CO and O_2 can be discussed in terms of reactions (17)—(21):[9]

$$H_2O_2 + h\nu \longrightarrow 2OH \qquad (17)$$

$$OH + CO \longrightarrow COOH \qquad (18)$$

$$COOH + O_2 \longrightarrow CO_2 + HO_2 \qquad (19)$$

$$OH + H_2O_2 \longrightarrow H_2O + HO_2 \qquad (20)$$

$$2HO_2 \longrightarrow O_2 + H_2O_2 \qquad (21)$$

---

[5] A. K. Chibisov, V. A. Kuz'min, and A. P. Vinogradov, *Doklady Akad. Nauk S.S.S.R.*, 1969, **187**, 142.
[6] J. A. Horsley and F. Flouquet, *Chem. Phys. Letters*, 1970, **5**, 165.
[7] M. K. Dmitriev and N. K. Kitroskii, *Doklady. Acad. Nauk. S.S.S.R.*, 1969, **189**, 1286.
[8] V. R. Landi and L. J. Heidt, *J. Phys. Chem.*, 1969, **73**, 2361.
[9] F. P. Laming, G. Buxton, and W. K. Wilmarth, *J. Phys. Chem.*, 1969, **73**, 867.

In the absence of CO and other scavengers, the photolysis of H_2O_2 is a non-chain reaction under the conditions employed. In the presence of sufficient CO, reactions (17), (18), and (21) should cause the rate of destruction of H_2O_2 to become zero, and this has been observed. The rate-constant ratio k_{18}/k_{20} was evaluated as 72 ± 3.6.

The photolysis of hydrogen peroxide in the presence of acrylonitrile and methanol gives rise to transient free radicals which have been studied by e.p.r. techniques.[10] The formation of OH radicals in the phototransfer reaction of an electron from OH^- to an anthraquinone derivative has also been described.[11]

The flash photolysis of BrO_3^-, BrO_2^-, and BrO^- in strongly alkaline solutions produces the ozonide ion O_3^-, and thus the initial step in the photolysis is probably [12] reaction (22), followed by reaction (23).

$$BrO_n^- + h\nu \longrightarrow BrO_{n-1} + O^- \qquad (22)$$

$$O^- + O_2 \rightleftharpoons O_3^- \qquad (23)$$

The decay rate of O_3^- is enhanced by the presence of the oxybromine anions, and this can be ascribed to reactions (24) and (25).

$$O^- + BrO_n^- \xrightarrow{H_2O} BrO_n + 2OH^- \qquad (24)$$

$$OH + BrO_n^- \longrightarrow BrO_n + OH^- \qquad (25)$$

The production of BrO_2 from BrO_3^- in an amount equal to that of O^- was demonstrated, in accordance with the scheme proposed. BrO_2 is also produced when BrO_2^- solutions are flashed, presumably *via* the secondary reaction (26).

$$BrO + BrO_2^- \longrightarrow BrO_2 + BrO^- \qquad (26)$$

XO_2 radicals have also been established as products in the flash photolysis of XO_3^- ions, where X is Cl, Br, and I.[13] The authors attribute the formation of the XO_2 species to reaction (27), rather than a sequence such as (22)

$$XO_3^-, H_2O \xrightarrow{h\nu} (XO_3^-, H_2O)^* \longrightarrow XO_2 + OH + OH^- \qquad (27)$$

to (25), but it is clearly difficult to distinguish between the steps unless a transient is observed. The fact that O_3^- is clearly seen in the case of the experiments with BrO_3^- suggests that a sequential formation of BrO_2 is correct. There was no evidence of the formation of the hydrated electron.[13] Photochemical cage effects in aqueous solutions of halide ions have been discussed.[14]

[10] H. Hefler and H. Fischer, *Ber. Bunsengesellschaft Phys. Chem.*, 1969, **73**, 633.
[11] S. I. Sholina, G. V. Fomin, and L. A. Blyumenfeld, *Zhur. fiz. Khim.*, 1969, **43**, 447.
[12] O. Amichai, G. Czapskii, and A. Treinin, *Israel J. Chem.*, 1969, **7**, 357.
[13] F. Barat, J. Gilles, B. Hickel, and J. Sutton, *Chem. Comm.*, 1969, 1485.
[14] G. Czapskii, J. Ogdan, and M. Ottolenghi, *Chem. Phys. Letters*, 1969, **3**, 383.

A new absorption system of the NO_2^- ion has been observed in large single crystals of sodium nitrite,[15,16] belonging to the $^1A_1 \to {}^3B_1$ transition. The principal source of the intensity of this singlet–triplet transition is intensity stealing from the $^1A_1 \to {}^1B_2$ transition at 2100 Å. The $^1A_1 \to {}^3B_1$ transition occurs in the visible with a maximum at 4400 Å.

The photochemistry of aqueous nitrate solutions excited in the high-energy 195 nm ($\pi\pi^*$) band of the ion has been investigated, and the formation of hydroxyl radicals demonstrated by observing the absorption of transients formed by the reaction of $\cdot$OH (or O$^-$) with CO_3^{2-}, CNS$^-$, and O_2.[17] Four primary independent processes appear to take place from this state:

(a) the formation of pernitrite. No satisfactory explanation of this is offered, but the process might involve equilibria between ion pairs, reactions (28) and (29).

$$(NO_2 + \cdot OH) \rightleftharpoons ONO_2^- + H^+ \qquad (28)$$

$$(NO^+, HO_2^-) \rightleftharpoons ONO_2 + H^+ \qquad (29)$$

(b) the generation of hydroxyl radicals, reaction (30).

$$NO_3^{-*} \longrightarrow NO_2 + O^- \xrightarrow{H_2O} OH^- + \cdot OH \qquad (30)$$

(c) formation of nitrite (NO_2^-). An internal recombination process is assumed to account for this, leading directly to the formation of O_2.

(d) formation of an intermediate leading to O_2 evolution via reaction with NO_3^-. This problem is little understood, but probably involves a metastable isomer of NO_3^-, such as: $\left(\begin{smallmatrix} O \\ | \\ O \end{smallmatrix} \!\!> N - O \right)^-$ since the assignment of oxygen atoms as the intermediate has been shown to be in error.

The steady state photolysis of ammonium nitrate has also been investigated with excitation in the near-u.v.[18] The decomposition leads to the formation of nitrite, which in turn decomposes to nitrogen and water, and also leads to the formation of unidentified oxidising transients.

The photochemistry of the N_3^- ion in aqueous solution at 2288 Å and 2139 Å has been investigated.[19] The behaviour at the two wavelengths is different, reflecting the different states formed upon absorption. At 2288 Å the $n\pi^*$ state of N_3^- is populated on absorption, whereas at 2139 Å the absorption is due to a charge-transfer-to-solvent (CTTS) transition. At the longer wavelength, the reaction is essentially expressed by reaction (31). At the shorter wavelength, NH_3 becomes a significant product, and together

$$N_3^- + 2H_2O \xrightarrow{h\nu} NH_2OH + H_2 + OH^- \qquad (31)$$

[15] W. C. Allen and R. N. Dixon, *Trans. Faraday Soc.*, 1969, **65**, 1168.
[16] R. M. Hochstrasser and A. P. Marchetti, *J. Chem. Phys.*, 1969, **50**, 1727.
[17] U. Shuali, M. Ottolenghi, J. Rabani, and Z. Yelin, *J. Phys. Chem.*, 1969, **73**, 3445.
[18] G. L. Petriconi and H. M. Papee, *Ricerca sci.*, 1969, **39**, 235, 242.
[19] I. Burak, D. Shapira, and A. Treinin, *J. Phys. Chem.*, 1970, **74**, 568.

with N_2 and NH_2OH, provides a nitrogen balance. The quantum yield for N_3^- depletion is 0·31, and some process occurs which consumes NH_2OH and produces NH_3. The initiation of this process is due to a reaction of solvated electrons and the N_3 radical with hydroxylamine. When this product reaches its limiting yield (see Figure 3), both e_{aq}^- and N_3 radicals

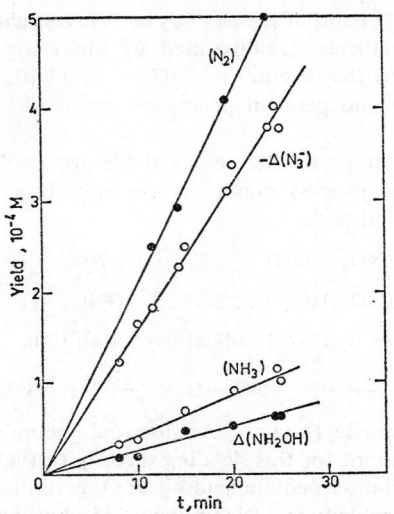

Figure 3 *The 2139 Å photolysis of* 10^{-3} M N_3^- *in air-free aqueous solutions, with* 3×10^{-4} M NH_2OH *initially added*
(Reproduced by permission from *J. Phys. Chem.*, 1970, **74**, 568)

should be totally scavenged in the system, and the mechanism can be represented by reactions (32)—(39). Reactions (33) and (35) are identical

$$N_3^- \xrightarrow{h\nu} N_3^{-*} \quad (32)$$

$$N_3^{-*} + H_2O \longrightarrow NH + OH^- + N_2 \quad (33)$$

$$N_3^{-*} \longrightarrow N_3 + e_{aq}^- \quad (34)$$

$$NH + H_2O \longrightarrow NH_2OH \quad (35)$$

$$e_{aq}^- + NH_2OH \longrightarrow NH_2 + OH^- \quad (36)$$

$$NH_2 + NH_2OH \longrightarrow NH_3 + NHOH \quad (37)$$

$$N_3 + NH_2OH \longrightarrow N_3^- + H^+ + NHOH \quad (38)$$

$$2NHOH \longrightarrow N_2 + 2H_2O \quad (39)$$

with the overall reaction proposed for the $n\pi^*$ state, and probably also involve this state at 2139 Å (the CTTS and $n\pi^*$ absorptions overlap).

Scavengers which completely remove e_{aq}^- and convert N_3 to N_3^- should lead to the overall conversion of N_3^- to NH_2OH and N_2, and this is apparently the case for acetone and ethanol scavengers.

2 Photochemistry and Photoluminescence of Transition-metal Co-ordination Compounds

Inorganic photochemistry is a subject which has grown rapidly of late, and the majority of research effort has been directed to studies of the effects of visible and u.v. light on the chemical and physical properties of co-ordination compounds of transition metals. Both the photochemistry of such complexes [20] and their luminescence [21] have been the subject of recent reviews. In these reviews compounds are classified according to the

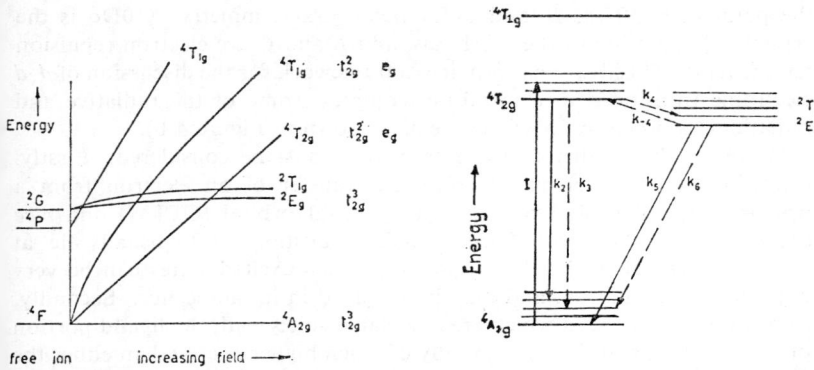

Figure 4(a) *Schematic correlation diagram (lowest levels) for octahedral* d^3 *metal complexes*
(Reproduced by permission from *Chem. Rev.*, 1970, **70**, 199)

Figure 4(b) *Level diagram for a typical* d^3 *complex in an* O_h *field, showing some processes which can occur after light absorption*
(Reproduced by permission from *Chem. Rev.*, 1970, **70**, 199)

number of d-electrons in the metal, but in this chapter, since the examples published in the last year are relatively few, classification will be by metals. It is important to discuss the treatment by which energy levels of various states of transition-metal complexes may be described, following ref. 21.

Considering the metal ion first, in the absence of any field due to ligands, it is evident that the d-electrons can couple their orbital and spin angular momenta to produce a series of different states, designated by different term symbols. The ordering for a d^3 metal is shown on the left side of Figure 4(a). Hund's rules generally determine this ordering. On the right side of Figure 4(a) the situation is shown for the same d^3 metal ion placed in an octahedral arrangement of ligands which produce a very strong field. The designations here refer to the placement of electrons in one or other sets of one-electron d-orbitals which result from the splitting produced by the ligand field. In O_h symmetry the three-fold degenerate t_{2g} orbitals are

[20] A. W. Adamson, *Pure. Appl. Chem.*, 1969, **20**, 25.
[21] P. D. Fleischauer and P. Fleischauer, *Chem. Rev.*, 1970, **70**, 199.

lower in energy than the two-fold e_g orbitals by an amount 10Dq. The two extreme cases of free atom and infinitely strong ligand field may then be correlated by one of three different approximations, weak-, medium-, and strong-field calculations. A common practice when discussing the various energy levels of a co-ordination complex is to choose a point along the abscissa of a diagram such as Figure 4(a) which represents the appropriate value of the term separation for the particular complex involved and construct a new figure (Figure 4b) depicting the lower states, showing their energy separation, the magnitude of which is expressed in terms of the parameters 10Dq, B, and C for octahedral symmetry. (10Dq is the crystal field splitting of the d-orbitals, and B and C are electron repulsion parameters.) This treatment provides a framework for the discussion of d–d electronic transitions in a metal-ion complex, some of the radiative and radiationless processes of which are also shown in Figure 4(b).

However, three other types of transition must be considered. Firstly, transitions arise which result from the transfer of an electron from a primarily ligand orbital to a primarily metal orbital (CTTM) and *vice versa* (CTTL), and these charge-transfer transitions (CT) usually lie at higher energies than the d–d transitions. Such excited states can be very important in excitation, and may be involved in luminescence. Secondly, transitions may be seen which are associated solely with the ligand portion of the complex, and these are usually of much higher energy than either the d–d or CT excited states. Transitions in these ligand states in a complex molecule often differ little from the corresponding transitions in the free ligand, but there are differences which arise occasionally, especially in ligands with low-lying $n\pi^*$ and $\pi\pi^*$ states. In the free ligand, frequently the $n\pi^*$ state is lowest in energy. If in the complex the n-electrons are used to form the co-ordinate bond, then in the complex the transition corresponding to $n \rightarrow \pi^*$ excitation is absent, thus leaving the $\pi \rightarrow \pi^*$ excitation as the lowest in energy. The last type of excited state to be covered results from the promotion of an electron from a metal or ligand orbital into a conduction or otherwise delocalised band (*i.e.* electron hole transitions). These transitions give rise to luminescence in the solid state and are also found in simple metal salts.

The above classification of transitions applies strictly to transition-metal complexes. The luminescence and photoreactions of lanthanide complexes are different because the nature of the metal transitions (involving 4f-electrons) is different, but similar classification to that above is possible. Finally, before considering individual metals, some discussion on the application of orbital symmetry arguments to the reactions of transition-metal complexes is in order.[22] The method of Woodward and Hoffmann involves the construction of orbital correlation diagrams for the one-electron energy levels of starting material and product, the correlations

[22] T. H. Whitesides, *J. Amer. Chem. Soc.*, 1969, **91**, 2395.

being made according to the symmetries of the activated complexes of two separate pathways. If these correlations predict substantial electronic destabilisation of one transition state relative to the other, then the corresponding mechanism is disallowed; otherwise it is allowed. Merely correlating the orbitals of the starting material and the product without regard to the symmetry of the activated complex does not guarantee that this type of information will be obtained. Thus, in the isomerisation reactions of four-co-ordinate d^8 species, lack of consideration of the intermediate activated complex symmetry leads to the conclusions that square-planar to tetrahedral reactions are allowed, but that square-planar cis–trans isomerisations are disallowed. These two statements are contradictory since attainment of tetrahedral geometry assures an allowed pathway for cis–trans isomerisation. The reason for the contradiction lies in part with the fact that the activated complex for isomerisation (tetrahedral) is of higher symmetry than either the starting material or the product. It is thus more informative to look at orbital correlation between the square plane and the tetrahedron (Figure 5). Consideration of this diagram leads to several conclusions:

(a) isomerisation of a triplet tetrahedral complex to a square-planar complex is 'disallowed';
(b) the reverse reaction is 'disallowed';
(c) excitation of a square-planar complex to the ^3E_g level makes the reaction 'allowed'.

Thus, photochemical cis–trans isomerisation is allowed, as observed experimentally, whereas thermal reactions should not occur. Similar treatment of axial–equatorial interchange of ligands of a trigonal-bipyramidal complex (through an intermediate of square-pyramidal geometry) shows that this reaction is thermally allowed, as is the substitution of square-planar complexes. Tetrahedral substitution is disallowed. For a d^3 complex, racemisation is allowed for weak ligand fields but disallowed for strong fields.

Experimentally, however, there is little correlation between the predictions above and observed reaction rates. This observation is hardly surprising since the theory explicitly ignores changes in energy in the bonding levels of the complex. The Woodward–Hoffmann treatment depends on the general assumption that if a reaction A → B can occur by two pathways, then, other factors being equal, the allowed pathway will have the lower activation energy. Implicit in this statement is the necessity for a comparison between two modes of reaction. For systems containing only carbon atoms, the energy difference between the two pathways appears, at least in one case, to be quite large, but this empirical observation is not guaranteed by the qualitative arguments of the theory. Unfortunately, in the reactions of transition-metal complexes, distinguishable pathways for the same reaction are usually not available, and the 'allowedness' or 'disallowedness' of any single reaction is not experimentally defined. In many of the reactions considered here (and particularly in substitutions), the rate will be determined by changes in binding energies, in overall ligand field stabilization, and in the strengths of bonds being made or broken during the reaction. The contribution of orbital symmetry conservation might be expected to be much smaller.

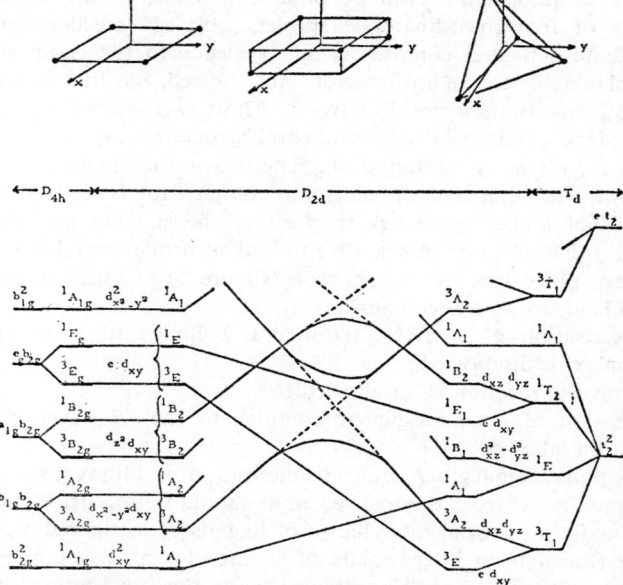

Figure 5 *State correlation diagram for the isomerisation of a square-planar to a tetrahedral d^8 complex. The symbol e stands for the degenerate set of orbitals (d_{xz}, d_{yz}). The orbital labels show the distribution of holes among the d-orbitals for each state*
(Reproduced by permission from *J. Amer. Chem. Soc.*, 1969, **91**, 2395)

Thus, orbital symmetry arguments concerning reactions of transition elements must be viewed with caution. The qualitative theory described above may be useful in rationalizing differences in thermal and photochemical behaviour of a complex with regard to the same reaction, or in the comparison of two distinguishable paths from the same starting material to slightly different products, if appropriate experiments can be devised.

Some of the aspects of the photochemistry and luminescence of compounds of various metals will now be considered.

Cobalt Complexes.—Cobalt(II) complexes are of the octahedral type, but a simple weak crystal field treatment has been applied to the interpretation of the polarised optical spectra of pseudotetrahedral cobalt(II) and nickel(II) complexes of the type NLX_3 where L = N-ethyl-1,4,-diazabicyclo[2,2,2,]octonium cation.[23] Assignments of the electronic transitions are shown in Table 1, together with polarisations. Similar assignments of

Table 1 Assignments of observed absorption bands

Transition	Polarisation	Absorption maxima, kK	Relative direction of polarisation
	$[Co(L_N^+)Br_3]^a$		
$^4A_2(F)-^4A_1(T_2, F)$	Forbidden x, y, z	Not observed	
$^4A_2(F)-^4E(T_2, F)$	Allowed x, y	Not observed	
$^4A_2(F)-^4A_2(T_1, F)$	Allowed z	4·42	b
$^4A_2(F)-^4E(T_1, F)$	Allowed x, y	7·27	a
$^4A_2(F)-^4A_2(T_1, P)$	Allowed z	14·99⎫ 15·70⎬ Av. 15·6 16·10⎭	a
$^4A_2(F)-^4E(T_1, P)$	Allowed x, y	14·62⎫ 14·81⎬ Av. 14·9 15·38⎭	b
	$[Ni(L_N^+)Br_3]^a$		
$^3E(T_1, F)-^3A_2(T_1, F)$	Allowed x, y	Not observed	
$^3E(T_1, F)-^3E(T_2, F)$	Allowed x, y, z	5·10	a, b
$^3E(T_1, F)-^3A_1(T_2, F)$	Allowed x, y	6·41	b
$^3E(T_1, F)-^3A_2(F)$	Allowed x, y	8·70	b
$^3E(T_1, F)-^3E(T_1, P)$	Allowed x, y, z	15·0⎫ 15·2⎬ Av. 15·4 16·1⎭	a
$^3E(T_1, F)-^3A_2(T_1, P)$	Allowed	16·2	b

^{a} $L_N^+ = $ N-ethyl-1,4-diazabicyclo[2,2,2]octonium cation.

unpolarised spectra of several other cobalt(II) complexes with similar geometry indicate that the trigonal splittings behave in a regular manner, consistent with the ligand spectrochemical series.

The photochemical reduction of thionine by cobalt(II)–edta complexes under anaerobic conditions yields cobalt(III) and leucothionine.[24] The back reaction can largely be prevented by extraction of the leuco-dye into ether during irradiation. In this manner 54% separation of the products of the light reaction can be achieved at pH 7·2. When the delayed back reaction is allowed to proceed in a galvanic cell, a potential difference of 0·3 V is observed. On complete separation of the products of the light reaction, this potential rises to 0·52 V. This system represents a significant step towards efficient chemical storage of the energy of visible light.

Cobalt(III) complexes with d_6 electronic configuration are widely studied. Most of the photochemistry of such complexes arises from the irradiation of absorption bands of CT character and result in oxidation–reduction reactions, and to a lesser extent ligand exchange reactions. It has been

[23] B. B. Garrett, V. L. Goedken, and J. V. Quagliano, *J. Amer. Chem. Soc.*, 1970, **92**, 489.
[24] V. Srinivasan and E. Rabinowitch, *J. Chem. Phys.*, 1970, **52**, 1165.

difficult to distinguish between excited-state and radical pair mechanisms which have been proposed to account for the photochemistry of cobalt(III) complexes. Non-absorbing reactive chemical scavengers provide one tool which has not been hitherto widely used.[25] Thus, in the presence of methanol and ethanol, it was established that absorption of 2537 Å radiation by the $Co(NH_3)_5O_2CMe^{2+}$ ion leads to two chemically indistinguishable intermediates: X* which is not scavengeable and leads to gaseous products, and Y* which normally returns to the ground state and is reduced by the alcohols. The scheme is set out in reaction (40). It seems likely that the

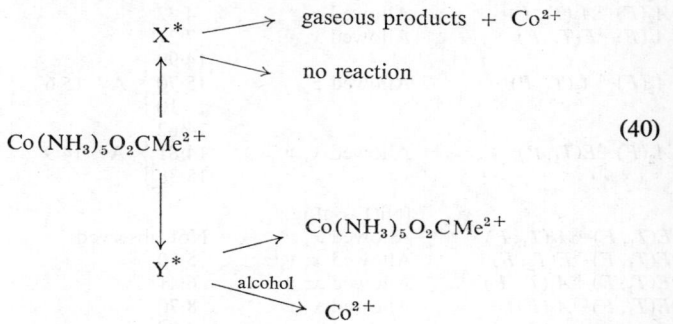

(40)

intermediate X* is a radical pair, and X* and Y* probably arise from different excited states.

The aqueous ion $Co(NH_3)_5(tsc)^{2+}$, where tsc denotes *trans*-4-stilbenecarboxylate, appears in most respects to be a normal member of the carboxylatopenta-amminecobalt(III) family.[26] The second ligand field band is masked, however, by a strong absorption at 320 nm essentially identical with that for the free tsc molecule. The first singlet–singlet transition of the tsc moiety thus appears virtually to be associated with an isolated chromophoric group of the complex. Irradiation of the 320 nm band of the free tsc ligand leads to *trans* to *cis* isomerisation; however, irradiation of this same band in the complex produces cobalt(II) in a quantum yield of 0·16 and both some free (unisomerised) and some oxidised tsc ligand. Other members of the carboxylatopenta-amminecobalt(III) series are nearly inert to this wavelength of irradiation. Free tsc⁻ ion shows a peak fluorescent emission at 400 nm, which is partially quenched in the presence of $Co(NH_3)_5(H_2O)^{3+}$ and essentially completely so when the ligand is co-ordinated. It is concluded that an intramolecular excitation energy transfer occurs, probably to produce a charge-transfer triplet excited state of the complex, which then undergoes redox decomposition. The suggested process is one of intersystem crossing from the first singlet excited state of the tsc ligand.

[25] E. R. Kantrowitz, J. F. Endicott, and M. Z. Hoffmann, *J. Amer. Chem. Soc.*, 1970, **92** 1776.
[26] A. W. Adamson, A. Vogler, and I. Lantzke, *J. Phys. Chem.*, 1969, **73**, 4183. 4250,

When bis(acetylacetonato)aminenitro-cobalt complexes are photolysed at 2310, 2800, 3130, and 3650 Å, photolinkage isomerisation occurs.[27] However, irradiation with a broad band at wavelengths longer than 4500 Å did not lead to appreciable isomerisation. The results suggest that the d–d excited states of Co^{III} nitro-complexes are not highly active toward photolinkage isomerisation. It seems likely that the excited state responsible is a CTTM state.

Three mixed oxalatoamminecobalt(III) complexes $[Co(NH_3)_5C_2O_4H]Cl_2$, $[Co(NH_3)_4C_2O_4]Cl$, and $K[Co(NH_3)_2(C_2O_4)_2],H_2O$ have been found to undergo photoredox decomposition when irradiated in the near-u.v. or in the visible spectral regions.[28] The products were Co^{II}, NH_3, CO_2, and $C_2O_4^{2-}$.

The mechanism is shown in reactions (41)—(47).

$$Co^{III}(NH_3)_4C_2O_4^+ + h\nu \longrightarrow Co^{III}(NH_3)_4C_2O_4^{+*} \quad (41)$$

$$Co^{III}(NH_3)_4C_2O_4^{+*} \longrightarrow Co^{III}(NH_3)_4C_2O_4^+ + \text{heat} \quad (42)$$

$$Co^{III}(NH_3)_4C_2O_4^{+*} \longrightarrow Co^{II}(NH_3)_4C_2O_4^+ \quad (43)$$

$$Co^{II}(NH_3)_4C_2O_4^+ \longrightarrow Co^{III}(NH_3)_4C_2O_4^+ \quad (44)$$

$$Co^{III}(NH_3)_4C_2O_4^{+*} \longrightarrow Co^{2+} + 4NH_3 + C_2O_4^- \quad (45)$$

$$C_2O_4^- + Co^{III}(NH_3)_4C_2O_4^+ \longrightarrow Co^{2+} + 2CO_2 + C_2O_4^{2-} \quad (46)$$

$$2C_2O_4^- \longrightarrow 2CO_2 + C_2O_4^{2-} \quad (47)$$

The quantum yield for the reaction is wavelength dependent, varying from 10^{-3} at 5000 Å to 0·38 at 2537 Å for the $Co(NH_3)_5C_2O_4H^{2+}$ complex.

The principal products of the 2537 Å irradiation of $Co(NH_3)_5N_3^{2+}$ are Co^{2+}, N_2, and probably $Co(NH_3)_4OH_2N_3^{2+}$.[29] The yields of Co^{2+} and N_2 have been determined directly under a variety of conditions and exhibit a peculiar dependence on the intensity of absorbed radiation. This dependence of product yields on I_a as well as the spectral features of irradiated solutions imply the formation and subsequent photolysis of a different azido-ammine complex of cobalt(III). The source of product N_2 has been shown to be co-ordinated azide, and experiments with $Co(^{15}NH_3)_5N_3^{2+}$ show that $Co(NH_3)_5OH_2^{3+}$ cannot be an important product of photolysis. $Co(NH_3)_4OH_2N_3^{2+}$ exhibits kinetic behaviour very similar to that of $Co(NH_3)_5N_3^{2+}$. The photochemistry of $Co(tetraen)N_3^{2+}$, cis-$Co(NH_3)_4$-$(N_3)_2^+$, and trans-$Co(NH_3)_4(N_3)_2^+$ appears to be much more straightforward and may not involve photo-labilisation of an ammine ligand.

Various aspects of the photochemistry of Co^{III} complexes in solution have been discussed, including the influence of the acidic nature of the ligand on the rate of photoreduction of the central ion,[30] and the nature of

---

[27] D. A. Johnson and J. E. Martin, *Inorg. Chem.*, 1969, **8**, 2509.
[28] H. Way and N. Filipescu, *Inorg. Chem.*, 1969, **8**, 1609.
[29] J. F. Endicott, M. Z. Hoffman, and L. S. Beres, *J. Phys. Chem.*, 1970, **74**, 1021.
[30] G. A. Shagisultanova and R. M. Orisheva, *Khim. vysok. Energii*, 1969, **3**, 265, 266.

the amino-ligand on the character of the photochemical redox reaction.[20] Triethylenediamine complexes of Co^{III}, Pt^{IV}, and Cr^{III} have also been discussed from the point of view of photochemical reactions.[31]

Irradiation of the first ligand field band of trans-$Co(en)_2(NCS)Cl^+$ in aqueous solution a 22—20 °C leads to photo-aquation in very low quantum yield, with a ratio of thiocyanate to chloride aquation of 1·6.[32] Photoredox decomposition is negligible at this wavelength. Irradiation of the first CT band leads to reaction of total quantum yield 0·013, of which 66% comprises Co^{2+} production and the remainder, aquation. The ratio of the two aquation modes is now 6·3. The thermal reaction chemistry of the complex is entirely one of chloride aquation. The results conform to qualitative photolysis rules for Co^{III} ammines, and those for irradiation of the first CT band are discussed further in terms of a previously proposed homolytic bond fission mechanism. A general conclusion is that the ligand field and CT excited states exhibit distinctive chemistries.

Photo-aquation quantum yields for aqueous $Co(CN)_6^{3-}$ have been reported for 340—380 nm, at various temperatures and ionic strengths.[33] In the presence of added sodium azide, thiocyanate, and iodide, both $Co(CN)_6^{3-}$ and $Co(CN)_5(H_2O)^{2-}$ exhibit direct photoaquation, again for this wavelength region and under various medium conditions. The results can be interpreted in terms of a photoproduced intermediate, $Co(CN)_5^{2-}$, which is scavenged either by solvent or, if present, by a co-ordinating anion. The reaction rates of the intermediate with azide and thiocyanate ions, relative to iodide ion, are similar to the literature values for the analogous thermal reaction systems. Photochemically produced $Co(CN)_5^{2-}$ reacts much faster with solvent water, however, than does that produced thermally. It is suggested that the photochemically nascent $Co(CN)_5^{2-}$ has a non-equilibrium structure and must escape cage recombination reactions before participating in ordinary scavenging competitions. Analogies with the photochemistry of the isoelectronic $Cr(CO)_6$ are noted. Photoaquation yields are also reported for aqueous $Co(CN)_5I^{3-}$, for 380 and 500 nm, again for various temperatures and ionic strengths. The yields are found to decrease with increasing electrolyte concentration, interpreted in terms of ion pairing which increases the cage recombination efficiency of the primary photoproducts. Photo-exchange with labelled iodide ion was found to be small. Data on the temperature and ionic strength dependence of the thermal aquation rate of $Co(CN)_5I^{3-}$ and of the thermal aquation rate of $Co(CN)_5(H_2O)^{2-}$ by iodide ion are also reported.

In aqueous acid solutions, the decomposition of potassium hexacyanatocobaltate(III), $K_3Co(CN)_6$, occurs cleanly with a quantum yield of 0·31 for the production of the mono-aquated ion independent of excitation wavelength and nearly of temperature. The lowest triplet state of the complex

[31] G. A. Shagisultanova and R. M. Orisheva, *Khim. vysok. Energii*, 1969, **3**, 459.
[32] A. Vogler and A. Adamson, *J. Phys. Chem.*, 1970, **74**, 67.
[33] A. W. Adamson, A. Chiang, and E. Zinato, *J. Amer. Chem. Soc.*, 1969, **91**, 5467.

has been identified as the $^3T_{1g}$ state, with its origin at 600 nm. Biacetyl should thus be able to sensitize the decomposition of the state, and this has been shown to occur,[34] with a rate constant of $2\cdot6 \times 10^7$ l mol^{-1} s^{-1}, considerably slower than the diffusion-controlled rate constant. The yield of the photosensitized aquation reaction was found to be $0\cdot23 \pm 0\cdot04$. A kinetic study of the thermal and photochemical decomposition of the tris-malonatocobalt(III) ion has been carried out.[35]

Tris(acetylacetonato)cobalt(III), tris(benzoylacetonato)cobalt(III), and tris(dibenzoylmethanato)cobalt(III) have been photolysed in different organic solvents.[36] The complexed Co^{III} ion is reduced photochemically to Co^{II} with simultaneous oxidative fragmentation of one β-ketoenolate group. The reaction was partly reversible in the presence of oxygen. The quantum yields were found to be wavelength dependent and ranged from around 10^{-3} for excitation in the first d–d band to $0\cdot6$ in the u.v. region. A reaction mechanism (see Scheme) consistent with the identified photoproducts is proposed.

The thermal and photochemical reactions in the solid state of [Co(NH$_3$)$_5$OH$_2$]X$_3$ where X = Cl$^-$, Br$^-$, and I$^-$ have been studied.[37] The reactions proceed like the analogous aqueous reactions except that tetrahedral Co^{II} species are produced in the solid state. The steps in the thermal reaction were studied in some detail and the photochemical reaction is shown to follow the same path. In addition, the photodecomposition of K$_3$[Co(C$_2$O$_4$)$_3$],xH$_2$O in the solid state is shown to involve a radical intermediate.

A study of the electronic spectrum of the hexafluorocobaltate(IV) anion in Cs$_2$CoF$_6$ by diffuse reflection, between 4 and 45 kK, indicates the presence of a low-spin $^2T_{2g}(t_{2g}^5)$ ground state.[38] Two very weak bands at $6\cdot4$ and $10\cdot3$ kK are assigned to transitions to $^4T_{1g}$ and $^4T_{2g}$ states and the complex absorption between 15 and 26 kK is resolved into four bands at $17\cdot3$, $18\cdot9$, $21\cdot4$, and $24\cdot4$ kK. These are assigned, respectively, to transitions to $^2A_{2g}$ and $^2T_{1g}$, $^2T_{2g}$, ^2E_g, and $^2T_{1g}(t_{2g}^4 e_g)$ states, and the intense broad bands at $28\cdot3$ and $37\cdot1$ kK are assigned to charge-transfer excitations. The bands at $18\cdot9$ and $21\cdot4$ kK both show vibrational structure which is attributed to a symmetric α_{1g} progression. Least-squares fitting of the d–d bands yields Dq = 2030 cm^{-1}, B = 635 cm^{-1}, and $\beta = 0\cdot54$, indicating a degree of covalency which is high for a fluoride complex, although not as great as for the NiF$_6^{2-}$ ion.

Chromium Complexes.—There have been several additions recently to the already vast amount of data in the literature relating to this metal and its mainly octahedral complexes. Several descriptions of the absorption

[34] G. P. Parker, *J. Amer. Chem. Soc.*, 1969, **91**, 3980.
[35] R. Van-Eldik and J. A. Van-den-Berg, *J. S. African Chem. Inst.*, 1969, **22**, 175.
[36] N. Filipescu and H. Way, *Inorg. Chem.*, 1969, **8**, 1863.
[37] S. T. Spees and P. Z. Petrak, *J. Inorg. Nuclear Chem.*, 1970, **32**, 1229.
[38] G. C. Allen and K. D. Warren, *Inorg. Chem.*, 1969, **8**, 1902.

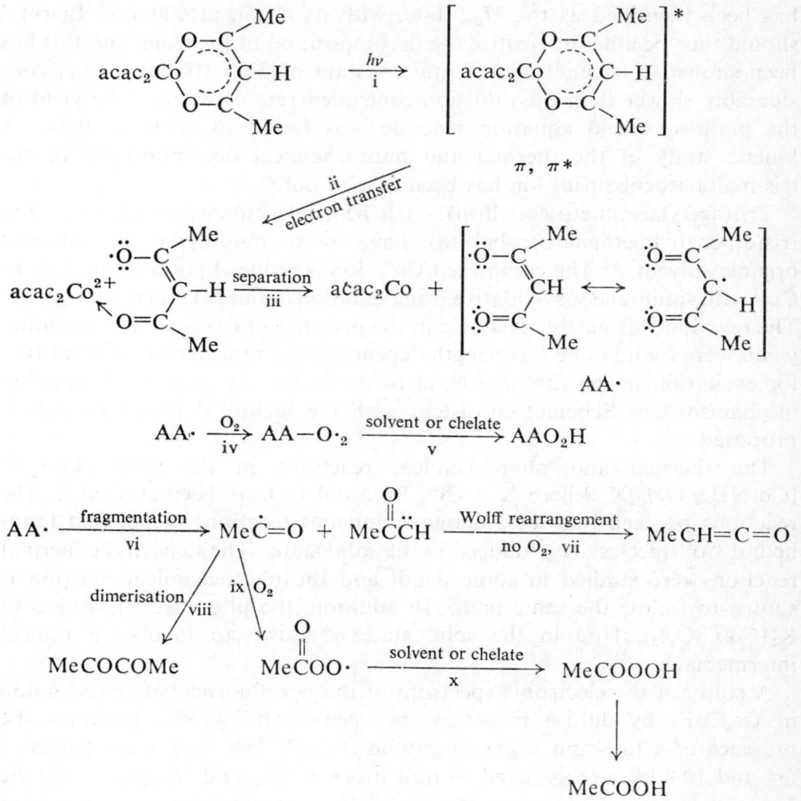

Scheme

$acac_2Co\begin{pmatrix}O-C(Me)\\ CH\\O-C(Me)\end{pmatrix} \xrightarrow[i]{h\nu} \left[acac_2Co\begin{pmatrix}O-C(Me)\\ C-H\\O-C(Me)\end{pmatrix}\right]^* \quad \pi,\pi^*$

↙ ii electron transfer

$acac_2Co^{2+} \begin{pmatrix}\cdot\ddot{O}-C(Me)\\ C-H\\ O=C(Me)\end{pmatrix} \xrightarrow[iii]{separation} acac_2Co + \left[\begin{pmatrix}\cdot\ddot{O}-C-Me\\ CH\\ \ddot{O}=C-Me\end{pmatrix} \longleftrightarrow \begin{pmatrix}\ddot{O}=C-Me\\ \dot{C}-H\\ \ddot{O}=C-Me\end{pmatrix}\right]$

AA·

$AA\cdot \xrightarrow[iv]{O_2} AA-O\cdot_2 \xrightarrow[v]{solvent\ or\ chelate} AAO_2H$

$AA\cdot \xrightarrow[vi]{fragmentation} Me\overset{\cdot}{C}=O + Me\overset{O}{\overset{\parallel}{C}}\overset{\cdot\cdot}{C}H \xrightarrow[no\ O_2,\ vii]{Wolff\ rearrangement} MeCH=C=O$

↙ dimerisation viii   ↓ ix O_2

MeCOCOMe $Me\overset{O}{\overset{\parallel}{C}}OO\cdot \xrightarrow[x]{solvent\ or\ chelate} MeCOOOH$

↓

MeCOOH

spectra [39-42] have been given, and a consideration of the excited states and their significance for photochemistry and luminescence in these complexes has appeared.[43] One of the studies involved a flash photolysis technique.[41]

At $-196\ °C$ transient species with strong absorption bands in visible and near-u.v. regions were observed to decay at the same rate as their phosphorescence for $Cr^{III}\ acac_3$, $K_3[Cr^{III}(NCS)_6]$, $NH_4[Cr^{III}(NCS)_4(NH_3)_2]$, and $Cr^{III}\ exan_3$. They are therefore ^2E states (phosphorescent states). A strong absorption band for $Cr^{III}\ acac_3$ is assigned to a charge-transfer transition from ^2E state based on the intensity and energy of the transition. A transition to localised excited states in the ligand with a small extinction

---

[39] L. S. Forster, *Transition Metal Chem.*, 1969, **5**, 1.
[40] K. Mizutani, K. Sone, and T. Sakaki, *Z. anorg. Chem.*, 1969, **365**, 217.
[41] T. Ohno and S. Kato, *Bull. Chem. Soc. Japan*, 1970, **43**, 8.
[42] F. D. Camassei and L. S. Forster, *J. Mol. Spectroscopy*, 1969, **31**, 129.
[43] H. L. Schläfer, *Z. Chem.*, 1970, **10**, 9.

coefficent was observed in the near-i.r. region for Cr^{III} $acac_3$. [*N.B.* acac = acetylacetonate ion; exan = ethylxanthogenate ion.] Figure 6 shows the energy levels of Cr^{III} $acac_3$.

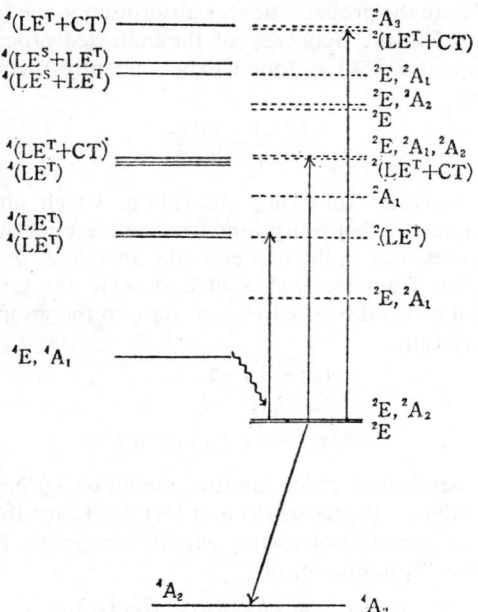

Figure 6 *Energy levels of Cr^{III} $acac_3$. Solid line shows energy level of the quartet state, broken line a calculated energy level of the doublet state, and solid arrow the transition observed. LE^T: triplet state on localised excitation in ligand; LE^S: singlet state on localised excitation in ligand; CT: charge transfer excited state*
(Reproduced by permission from *Bull. Chem. Soc. Japan*, 1970, **43**, 8)

Calculations have been performed on the energy levels of chromium(III) compounds which provide a basis for the understanding of intersystem crossing in such complexes.[44] In particular, these crystal field calculations provide a scheme for predicting the ratio of fluorescence to total luminescence for effectively octahedral chromium(III) systems. The ratio of fluorescent to phosphorescent intensity for a given compound is given by equation (48), where P is the probability of emission per molecule from the

$$\frac{I(F)}{I(P)} = \frac{N(^4T_2)P(F)}{N(^2E)P(P)} \quad (48)$$

lowest vibrational state and N the number of molecules in that state. In

---

[44] J. C. Hempel and F. A. Matsen, *J. Phys. Chem.*, 1969, **73**, 2502.

equation (49),

$$\frac{P(F)}{P(P)} = \frac{P(^4T_2)}{P(^2E)} \frac{\nu(F)^3}{\nu(P)^2 \nu(^4T_2)} \qquad (49)$$

$P(^2E)$ and $P(^4T_2)$ are the probabilities for absorption to the ^2E and $^4T_2$ states, respectively, and ν is the frequency of the indicated process.[10] The ratio expressed in equation (50) is found experimentally to be always much greater than one.

$$\frac{I(^4T_2)}{I(^2E)} = \frac{P(^4T_2)}{P(^2E)} \qquad (50)$$

Among the processes following absorption, which affect $N(^2E)$ and $N(^4T_2)$, are (*i*) radiationless transitions from one excited state to another; (*ii*) intersystem crossover in the first excited state (state $|u'; I\rangle$); (*iii*) radiationless transitions from the first excited state to the ground state; (*iv*) radiative transitions from the first excited state to the ground state.

For phosphorescence

$$\nu(P) = \varepsilon^0(^2E) \qquad (51)$$

and

$$\varepsilon^0(^4T_2) = [\nu(^4T_2)+\nu(F)]/2 \qquad (52)$$

For effectively octahedral chromium(III) complexes, process (*i*) is much faster than all others. Processes (*ii*) and (*iii*) are faster than process (*iv*). Assuming that a pseudo-Boltzmann equilibrium exists between the ^2E minimum and the $^4T_2$ minimum of

$$\frac{N(^4T_2)}{N(^2E)} = \frac{W(^4T_2) \exp[-\varepsilon^0(^4T_2)/kT]}{W(^2E) \exp[-\varepsilon^0(^2E)/kT]} \qquad (53)$$

where W represents the statistical weight of the $^m\alpha$ state. On substituting equations (51) and (52) into (53) and taking the natural log of both sides we obtain equation (54).

$$\ln\left(\frac{N(^4T_2)}{N(^2E)}\right) = \ln(3) + \frac{2\nu(P)-\nu(^4T_2)-\nu(F)}{2kT} \qquad (54)$$

Taking the quantum yields Φ_F and Φ_P to be directly proportional to the intensity of the observed emission, the natural log of both sides of equation (48) yields equation (55).

$$\ln\left(\frac{\Phi_F}{\Phi_P}\right) = A + Z/T \qquad (55)$$

Calculated and experimental values of Φ_F/Φ_P are shown in Table 2. A is expected to vary somewhat from system to system but to remain a positive number since it is an ln function, and the experimentally observed intensity of related absorption curves indicates the argument will be greater than unity. A quantitative study of Φ_F/Φ_P vs. $1/T$ for a given chromium(III) compound will yield both A and the quantity Z. It can be predicted that

Table 2 Ratios of quantum yields ($\Phi_F : \Phi_P$) at T = 130 K

Compound[a]	Z/T	Calculated[b]	Experimental[c]
$(CrCl_6)^{3-}$	23·7	>1000	F
$(CrF_6)^{3-}$	20·3	>1000	F
$(NH_4)_3(CrF_6)$	19·1	>1000	F
$[CrF(H_2O)_5]SiF_6$	14·3	>1000	F
$K_2[CrF_5(H_2O)]$	9·7	>1000	F
$[CrF_3(H_2O)_3]H_2O$	5·2	652	F
$[CrF_3(H_2O)_3]\tfrac{1}{2}H_2O$	5·2	652	F
$[CrF_3(H_2O)_3]$	5·2	652	F
$[CrF_3(H_2O)_3]2H_2O$	4·2	240	F
$(Cr\ antip_6)^{3+}$	2·4	39·5	F+P
$(Cr\ urea_6)^{3+}$	0·0	3·61	3·61
$[Cr(H_2O)_6]F_3$	−0·4	2·46	F+P
$[Cr(D_2O)_6]^{3+}$	—	—	0·10
$NH_4(CrF_4\ en)$	−4·2	0·054	F+P

[a] See ref. 44, Table II, for spectral data and references.
[b] This is the ratio predicted with an A determined for $(Cr\ urea_6)^{3+}$ in a water–glycerol glass.
[c] Observed luminescence indicated by F and P (see ref. 44, Table II).

Experimental quantum yields given in K. K. Chatterjee and L. S. Forster, *Spectrochim. Acta*, 1964, **20**, 1603.

(*i*) lowering the temperature enhances fluorescence for compounds with a positive Z; (*ii*) lowering the temperature enhances phosphorescence for compounds with a negative Z.

In addition, and independently of the Boltzmann approximation, chromium(III) complexes may be divided into three categories using absorption spectra data.

(*a*) When

$$R \equiv \nu(^4T_2)/\nu(^2E) \tag{56}$$

is less than 1, $\varepsilon^0(^4T_2) \ll T^0(^2E)$ and only fluorescence will be observed;
(*b*) when R is greater than 1 but less than 2, $\varepsilon^0(^4T_2) \simeq \varepsilon^0(^2E)$ so that there is a possibility of observing both phosphorescence and fluorescence;
(*c*) for ratios about 2 or greater, $\varepsilon^0(^2E) \ll \varepsilon^0(^4T_2)$ and only phosphorescence will be observed.

The photochemical aquation of $Cr(NH_3)_5Br^{2+}$ in acid aqueous solution (10^{-3} M-HClO$_4$, $\mu = 0.5$) has been studied using light corresponding to absorption in the ligand-field and the charge-transfer bands (580—250 nm). Quantum yields for NH_3 and Br^- production have been determined.[45] Irradiation in the *d–d* band region yields the aquobromotetra-ammine ion, mainly in the *cis*-configuration, and $NH_3(\Phi = 0.34$—$0.37)$. Simultaneously, to a much smaller extent, the monoaquopenta-ammine ion and Br^- are photochemically produced [$\Phi(Br^-) = 0.009$—0.011]. Excitation in the charge-transfer band gives a decrease in $\Phi(NH_3)$ and an increase in $\Phi(Br^-)$ compared to the values in the *d–d* band region [$\Phi(NH_3) = 0.17$—0.20, $\Phi(Br^-) = 0.18$—0.26].

[45] P. Riccieri and H. L. Schläfer, *Inorg. Chem.*, 1970, **9**, 727.

The photo-aquation of $[Cr(CN)_6]^{3-}$ is shown to occur in the following steps:[46]

$$H_2O + [Cr(CN)_6]^{3-} + h\nu \longrightarrow [Cr(CN)_5H_2O]^{2-} + CN^- \quad (57)$$

$$[Cr(CN)_5H_2O]^{2-} + H_2O \longrightarrow [Cr(CN)_4(H_2O)_2]^- + CN^- \quad (58)$$

$$[Cr(CN)(H_2O)_5]^{2+} + H_2O \longrightarrow [Cr(H_2O)_6]^{3+} + CN^- \quad (59)$$

The quantum yield of this reaction was independent of wavelength of excitation, but showed a temperature dependence, with an activation energy of 12·18 kJ mol^{-1}. Photo-aquation of Cr^{III} complexes can also be photosensitized by organic compounds known to have relatively stable triplet excited states.[47] Some of the results obtained with different complexes and donors are summarised in Table 3.

Table 3 *Photosensitization experiments with Cr^{III} complexes*

Complex	Sensitizer	Solvent[a]	Irradiating wavelength (nm)	Result[b]
$[Cr(NH_3)_5(NCS)](ClO_4)_2$	Biacetyl	Water[c]	410	NH_3[d]
	Biacetyl	Water–acetone[e]	410	NH_3
	Acridine	Water–acetone[e]	410	NH_3, NCS^-
	Michler's ketone	Water–acetone[e]	410	NH_3, (NCS^-?)
$(NH_4)[Cr(NH_3)_2(NCS)_4]$	Biacetyl	Water–acetone[e]	410	NCS^-
	Cr urea$_6^{3+}$	Water	676	$\Phi_{NCS} < 10^{-4}$
	Methylene blue	Water–acetone–dioxan	674	NCS^-
$K_3Cr(NCS)_6$	Biacetyl	Water–acetone[e]	410	NCS^-

[a] Containing dissolved air.
[b] Sensitized aquation of indicated ligand observed.
[c] 0·1N sulphuric acid, and either in the presence or absence of dissolved oxygen.
[d] Ratio of ammonia to thiocyanate ion produced greater than 100.
[e] 0·05N in sulphuric acid.

The photolysis of the tris(ethylenediamine)chromium(III) complex ion in acid aqueous solution has also been described.[48]

Energy transfer in Cr^{III} complex systems has already been shown to be of importance, since sensitized photoreactions occur. This phenomenon has also been demonstrated with the triplet excited states of aromatic molecules

[46] H. F. Wasgestian, *Z. phys. Chem. (Frankfurt)*, 1969, **67**, 39.
[47] A. W. Adamson, J. E. Martin, and F. D. Camessei, *J. Amer. Chem. Soc.*, 1969, **91** 7530.
[48] W. Geis and H. L. Schläfer, *Z. phys. Chem. (Frankfurt)*, 1969, **65**, 107.

as donors.[49] Internal energy transfer from chromium in the cation to chromium in the anion in some crystalline compounds of the type (CrA$_6$) [Cr(CN)$_6$],xH$_2$O has been described,[50] but this result has been challenged.[51] The evidence presented supporting the phenomenon was three-fold.

Irradiation at 77 K with light of a wavelength absorbed predominantly by the cation apparently gave no emission from the cation. A high yield of anion emission was seen and, in some of the most favourable cases, this anion emission was obtained with an apparent quantum yield some 10 to 20 times that obtained by direct irradiation of the anion in the potassium salt. It was recognized in this work that such an energy transfer might occur through the intermediacy of either the $^4T_{2g}$ or the ^2E_g states of the donor, and that the existence of the interesting and impressive 'amplification factor' might depend either on differences in the intersystem crossing rates in the donor and acceptor ions or on changes in their luminescence properties when residing in different lattices. However, for complexes in which A = antipyrine (atp), results are provided which do not support the energy transfer proposal.[51]

The temperature dependencies of ^2E_g[Cr(CN)$_6$]$^{3-}$ emission intensity from K$_3$[Cr(CN)$_6$] and [Cr atp$_6$][Cr(CN)$_6$] and of $^4T_{2g}$[Cr atp$_6$]$^{3+}$ emission intensity from [Cr atp$_6$][Cr(CN)$_6$] and [Cr atp$_6$](ClO$_4$)$_3$ were compared for excitation into the [Cr atp$_6$]$^{3+}$ and [Cr(CN)$_6$]$^{3-}$ absorption regions. In addition, the emission intensity differences were measured. Lifetime measurements at 77 K on the ^2E_g anion and $^4T_{2g}$ cation emissions from these compounds show the former to be 1000 times longer lived. Studies of the excitation spectra and of the emission spectrum of the double salt as a function of excitation wavelength were also described.

These new extensive data on the antipyrine system suggest that energy transfer from cation to anion is not the major mechanism of ^2E_g emission from the [Cr(CN)$_6$]$^{3-}$ ion. It appears that the effects observed previously and confirmed by this work all arise from a marked change in the luminescence properties of [Cr(CN)$_6$]$^{3-}$, i.e. quantum yield of phosphorescence and its temperature dependence, upon placing this ion in the lattice of the double salt. Since the case chosen for study was one where the effects were particularly large and apparently unambiguous, it seems rather likely that energy transfer does not occur significantly in any of the compounds so far studied.

Energy transfer between two large complex ions has been described.[52] Tris(oxalato)chromate(III) [Cr ox$_3$]$^{3-}$ in a crystal or in a rigid solution at low temperature emits a weak phosphorescence ($^2E_g \rightarrow {}^4A_{2g}$) with a characteristic fine structure in the region of 13 118—14 398 cm^{-1} when it is

[49] T. Ohno and S. Kato, *Bull. Chem. Soc. Japan*, 1969, **42**, 3385.
[50] H. L. Schläfer, H. Gausmann, and C. H. Mobius, *Inorg. Chem.*, 1969, **8**, 1137.
[51] A. D. Kirk and H. L. Schläfer, *J. Chem. Phys.*, 1970, **52**, 2411.
[52] I. Fujita and H. Kobayashi, *J. Chem. Phys.*, 1970, **52**, 4904.

excited by a spin-allowed ligand-field band ($^4T_{2g} \leftarrow {}^4A_{2g}$, $\nu_{max} = 17\,600$ cm^{-1} or $^4T_{1g} \leftarrow {}^4A_{2g}$, $\nu_{max} = 23\,900$ cm^{-1}) or by a 'ligand-to-metal' charge-transfer band ($\nu_{max} = 37\,400$ cm^{-1}). Tris(2,2'-bipyridyl)ruthenium(II) complex [Ru bipy$_3$]$^{2+}$ in a crystal or in a solution even at room temperature emits a band of strong phosphorescence in the region of $13\,300$—$17,250$ cm^{-1} when it is excited by a 'metal-to-ligand' charge-transfer band in the visible

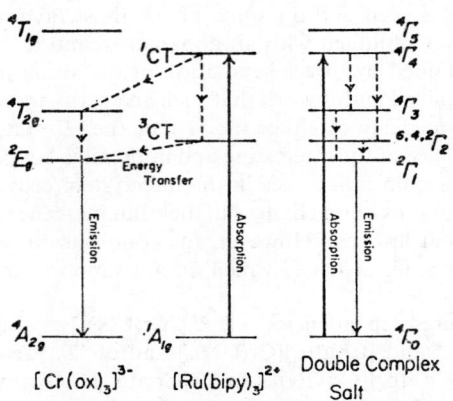

Figure 7 *Energy diagram of the double complex salt of* [Cr ox$_3$]$^{3-}$ *and* [Ru bipy$_3$]$^{2+}$. ← : *radiative process;* —←— : *non-radiative process. A strong phosphorescence of* [Cr ox$_3$]$^{3-}$ *is observed upon the irradiation in the allowed charge-transfer band of* [Ru bipy$_3$]$^{2+}$. *The strong emission of* [Ru bipy$_3$]$^{2+}$ *is wholly quenched in the double complex salt.*
(Reproduced by permission from *J. Chem. Phys.*, 1970, **52**, 4904)

region ($\nu_{max} = 22\,100$ cm^{-1}) or by a ligand band ($\nu_{max} = 34\,900$ cm^{-1} or $\nu_{max} = 41,000$ cm^{-1}). The emission shows the characteristic vibrational structure of bipyridyl and has been assigned to a triplet–singlet 'ligand-to-metal' charge-transfer transition.

From a 2:3 mixture of the equimolar solutions of [Cr ox$_3$]$^{3-}$ and [Ru bipy$_3$]$^{2+}$, a double complex salt crystal was obtained, the solid-state absorption spectrum of which was observed very close to that of [Ru bipy$_3$]Cl$_2$,6H$_2$O, indicating no existence of strong electronic interaction such as charge transfer between the complex ions in the double complex salt. Since the spectral intensity of [Ru bipy$_3$]$^{2+}$ in the visible region is about 14 000, while that of [Cr ox$_3$]$^{3-}$ is about 80, their superposed absorption spectra in the double complex salt in the solid state should be very similar to that of [Ru bipy$_3$]Cl$_2$,6H$_2$O.

At 77 K the double complex salt showed a strong phosphorescence of [Cr ox$_3$]$^{3-}$ instead of that of [Ru bipy$_3$]$^{2+}$ upon the irradiation at $21\,000$ cm^{-1}, which could excite only [Ru bipy$_3$]$^{2+}$ by the allowed charge-transfer transition but not [Cr ox$_3$]$^{3-}$, since it is in the midway region of two ligand-field bands of [Cr ox$_3$]$^{3-}$. The phosphorescence of [Cr ox$_3$]$^{3-}$ in the absence of [Ru bipy$_3$]$^{2+}$ was very weak even when it was excited at the absorption maxima of [Cr ox$_3$]$^{3-}$. The strong phosphorescence of [Ru bipy$_3$]$^{2+}$, which was so strong that it could be observed even from the undiluted solid sample at room temperature, was wholly quenched in the presence of [Cr ox$_3$]$^{3-}$ in the double complex salt. These observations provide evidence for electronic excitation energy transfer from the ruthenium complex ion to the chromium in the [Cr ox$_3$]$^{3-}$ (see Figure 7). However, an enhancement of the population of the ^2E_g state in the chromium(III) ion should be ascribed not only to an effective energy transfer but also to a stronger energy absorption in [Ru bipy$_3$]$^{2+}$. Further analytical studies along the lines described [51] must be carried out before this evidence can be accepted unequivocally.

The half-sandwich type complexes LCr(CO)$_3$, where L = Et$_3$B$_3$N$_3$Me$_3$ Me$_3$B$_3$N$_3$Et$_3$, C$_6$H$_6$, sym-(C$_6$H$_3$Me$_3$), have been prepared by the photochemical reaction of Cr(CO)$_6$ with the appropriate ligand L at highly reduced pressures.[53]

The transition state and hydrogen isotope effect in the photochemical oxidation of Cr^{II} in solution have been discussed.[54]

Iron Complexes.—Charge-transfer spectra of bisnioximebis(substituted pyridine)iron(II) complexes have been described.[55a] The quantum yields of production of iron(II) and of cobalt(II) in the photolysis of solid potassium trisoxalatoferrate(III) trihydrate and potassium trisoxalatocobaltate(III) trihydrate have been measured.[55b]

The values of quantum yields of production of Fe^{II}, $\Phi_{Fe^{II}}$, are about an order of magnitude lower in the microcrystalline solid compared with solution values, ranging between about 0·07 and 0·17 and increasing with increasing intensity and with decreasing wavelength. Values of $\Phi_{Co^{II}}$ in the microcrystalline solid vary between about 0·07 and 0·55, are somewhat less than values for solutions, and are independent of intensity, but decrease with increasing exposure at the shorter wavelengths. It is suggested that hole–electron pairs are generated within the crystals; these subsequently generate radical ions at the surface.

The photosensitized autoxidation of iron(III),[56] and the formation and reactions of hydrated electrons in u.v.-irradiated hexacyanoferrate(III) solutions [57] have been the subjects of recent reports. The indirect photochemical preparation of monosubstituted metal hexacarbonyls and aromatic metal tricarbonyls was described,[58] and the photolytic formation of a ferrocenyl radical from iodoferrocene was established.[59] Temperature

[53] K. Deckelmann and H. Wemer, *Helv. Chim. Acta*, 1970, **53**, 139.
[54] H. Hartmann and P. Filss, *Z. phys. Chem. (Frankfurt)*, 1969, **66**, 48, 60.
[55a] N. Sanders and P. Day, *J. Chem. Soc. (A)*, 1969, 2303.
[55b] H. E. Spencer, *J. Phys. Chem.*, 1969, **73**, 2316.
[56] S. Bagger and U. Ulstrup, *Chem. Comm.*, 1969, 1190.
[57] S. Ohno, *J. Chem. Soc. Japan*, 1970, **91**, 91.
[58] W. Strohmeier and F. J. Mueller, *Chem. Ber.*, 1969, **102**, 3608.
[59] T. Sato, S. Shimada, and K. Hata, *Bull. Chem. Soc. Japan*, 1969, **42**, 2731.

effects on charge-transfer luminescence in d^6 metal-ion chelate compounds have been observed to be large in some cases.[60] The compounds studied are shown in Table 4, and the effect of temperature is shown in Table 5.

Table 4 *Luminescence properties of some imine chelates of* Fe^{II}, Ru^{II}, *and* Ir^{III}

Chelates luminescent in absolute EtOH soln. at room temp. and at ~80 K	Chelates nonluminescent in absolute EtOH soln. at room temp. but luminescent in EtOH glasses at ~80 K	Chelates nonluminescent in absolute EtOH soln. at room temp. and at ~80 K
$[Ir^{III} bipy]^{3+}$	$[Ir^{III}(2\text{-Me-phen})]^{3+}$	$[Fe^{II} bipy_3]^{2+}$
$[Ir^{III} phen]^{3+}$	$[Ru^{II}(2\text{-Me-phen})_3]^{2+}$	$[Fe^{II} phen_3]^{2+}$
$[Ir^{III} ter]^{3+}$	$[Ru^{II} ter_2]^{2+}$	$[Fe^{II} ter_2]^{2+}$
$[Ru^{II} bipy_3]^{2+}$		$[Fe^{II}(2\text{-Me-phen})_2]^{2+}$
$[Ru^{II} phen_3]^{2+}$		
$[Ru^{II}(5\text{-Me-phen})_3]^{2+}$		
$[Ru^{II}(4,7\text{-diMe-phen})_3]^{2+}$		

Table 5 *Temperature effects on luminescence intensity*

Species	Wavelength, nm Excitation	Wavelength, nm Emission	Average % increase in luminescence intensity 30—20 °C	20—10 °C	10—0 °C	Factor increase in I_L; room temp. to ~80K
$[Ir^{III} ter]^{3+}$	365	520	33	35	40	×27
$[Ir^{III} phen]^{3+}$	365	530	39	56	55	×50
$[Ru^{II}(5\text{-Me-phen})_3]^{2+}$	436	580	66	62	56	×50
$[Ru^{II} phen_3]^{2+}$	436	580	33	36	46	×10
$[Ru bipy_3]^{2+}$	365	590	57	40	26	×5
$[In^{III} Q_3]^a$	365	530	11	13	19	×8
$QSO_4{}^b$	365	450	2·5	3·9	—	—

[a] Tris(8-quinolinolato)indium(III) in absolute ethanol.
[b] 1 ppm of quinine sulphate in 0·1-HN_2SO_4(aq).

Intersystem crossing rates have very important effects on luminescence characteristics of excited species. An increase in the rate of intersystem crossing results in decreased fluorescence intensity and perhaps increased phosphorescence intensity if the sample is examined in a rigid matrix. However, when the intersystem crossing rate becomes very large, the phosphorescence may be quenched also. The intersystem crossing rate constant is large for diamagnetic species containing heavy atoms (due to the strong spin–orbit coupling) and is very large for species that are paramagnetic. Of the complexes studied in the work above, all of the lower members are high-spin and paramagnetic, but the highest complexes [tris species of bipy, phen, and 5-Me-phen, and the bis species of ter (see Table 4)] are diamagnetic.

[60] D. W. Fink and W. E. Ohnesorge, *J. Amer. Chem. Soc.*, 1969, **91**, 4995.

Considerable interest has recently been directed to studying the temperature dependence of magnetic properties and the identification of magnetic crossover points for complexes that are capable of existing both in high-spin and in low-spin configurations. The low-spin state is favoured at lower temperatures. For octahedral complexes of Fe^{II} and Co^{II} the energy difference between these states may be rather small compared to the energy supplied to the complex upon excitation by absorption of u.v.–visible radiation.

The magnetic moment of $[Fe^{II}(2\text{-Me-phen})_3]^{2+}$ at room temperature has the value expected for a high-spin d^6 species (*ca.* 5 BM). However, upon cooling, the magnetic moment becomes smaller, approaching 3 BM at 100 K, and behaves as expected if an equilibrium between approximately equienergetic high-spin (t_{2g}^4, e_g^2) and low-spin (t_{2g}^6) configurations was shifting to favour increased population of the low-spin state as the temperature is lowered: $^1A_1 \rightleftharpoons {}^5T_2$. Accompanying the change in spin states will be a decreased metal ion–ligand bond distance and an increase in the ligand field strength. An additional observation consistent with these explanations is a change in the colour of solutions containing the $(2\text{-Me-phen})_3Fe^{II}$ ion upon cooling; this is attributed to the enhancement of the metal-to-ligand charge-transfer absorption transition probability with decreased ligand-to-metal ion bond distances in the low-spin complex.

The Ru^{II} and the Ir^{III} imines are probably in the low-spin configuration in the ground state at all temperatures used in this work. Chelates of these metal ions studied in the present work showed the characteristic metal-to-ligand charge-transfer absorption bands near 450 nm at room temperature. Furthermore, no noticeable colour changes were observed when absolute ethanol solutions containing these chelates were converted to glasses by cooling to liquid nitrogen temperatures. Upon excitation to a charge-transfer excited state (π^*, d) there will be considerable redistribution of the electron density in the chelate because of transfer of the excited electron from an orbital centred at the metal ion to one largely delocalised on the ligands. Thus, it is anticipated that there are large and important increases in the metal ion–ligand bond distances when these chelates are excited to charge-transfer states; in consequence the ligand-field splitting will be lower in this excited state of the chelate.

If the ligand-field splitting is lowered sufficiently upon excitation and is then near the magnetic crossover point at room temperature, an excited-state equilibrium will be established between the diamagnetic (t_{2g}^5, π^*) and the paramagnetic (t_{2g}^3, e_g^2, π^*) configurations. The luminescence intensity will be decreased accordingly, approximately in proportion to the extent of crossover to the high-spin state. In this case not only is the intersystem crossing rate-constant greatly enhanced by the formation of the high-spin paramagnetic species, but a group of closely spaced energy levels, *e.g.* ligand-field states of the high-spin species, is also created; these are available to the excited species and their presence would enhance further the probability of non-radiative decay of the excitation energy.

The extent of crossover will be governed by a Boltzmann-like distribution, and a very marked temperature dependence is anticipated. The smaller excited-state ligand fields, Dq*, that exist with the Fe^{II}-imines and with the most hindered Ru^{II} and Ir^{III} chelates, could result in nearly complete crossover to the high-spin forms at room temperature. Thus, luminescence would not be observed. Reducing the sample solution temperature to form a rigid glass might permit observation of the characteristic charge-transfer luminescence. The experimental facts are consistent with these explanations: the imine chelates of Fe^{II}, the metal ion which would show the smallest ligand-field splittings, show no luminescence even at the lowest temperatures. The most hindered ligands, 2-Me-phen and ter, form chelates with Ru^{II} that do not luminesce at room temperature but do emit the characteristic red-orange luminescence at the temperatures of liquid nitrogen. With Ir^{III}, the metal ion that would be subject to the largest ligand-field splittings, only the most hindered ligand, 2-Me-phen, forms a chelate that fails to show charge-transfer luminescence in fluid solution although this chelate too emits in rigid media.

These considerations can account for the observed phenomena under discussion regardless of the kind of luminescence (fluorescence or phosphorescence) and the multiplicity (singlet or triplet) of the emitting excited state in the low-spin cases. Two alternatives should also be considered, particularly for the Fe^{II}-imines. One is the existence of triplet ligand-field excited state(s) ($^3T_{1g}$, $^3T_{2g}$) at energies below the lowest energy charge-transfer excited state (emission from such states might be in the near-i.r. and not detectable in the present work). Distortion of the excited chelate species from octahedral microsymmetry might cause a triplet ($^3T_{1g}$) to assume lowest energy among the ligand field states; paramagnetic quenching of the luminescence would probably result in this event.

Platinum Complexes.—The photochemical behaviour of Pt dien Br$^+$ in aqueous solutions containing various amounts of NO_2^- and Br^- has been investigated.[61] The excitation was performed with 313 nm radiation, corresponding to a ligand field band of the complex. In neutral solutions, the photosubstitution of Br^- by NO_2^- was observed. The quantum yield for the photosubstitution was in the range 0·007—0·04, and it increased with increasing [NO_2^-] and decreased with increasing [Br^-]. When the irradiation was carried out in basic solution, only Pt dien OH$^+$ was obtained and its quantum yield of formation ($\Phi = 0·24$) was independent of the amounts of NO_2^- and Br^- present in the solution. The results obtained are interpreted on the basis of a photochemical Pt–Br heterolytic bond fission, leading to an ion-pair intermediate [presumably, Pt dien H_2O^{2+},Br^- or Pt(dien)$^{2+}$,Br^-]. In neutral solutions, the intermediate can (a) react with NO_2^- to give the product, Pt dien NO_2^+;

[61] C. Bartocci, F. Scandola, and V. Balzani, *J. Amer. Chem. Soc.*, 1969, **91**, 6948.

(b) react with external Br^- to give Pt dien Br^+; and (c) undergo geminate recombination. The photochemical behaviour in alkaline solution is accounted for by an efficient capture of the intermediate by the OH^- ions.

The behaviour can be explained on the basis of the mechanism shown in reactions (60)—(65), where dien = diethylenetriamine.

$$Pt \text{ dien } Br^+ \xrightarrow[\text{photoexcitation}]{h\nu} {}^*Pt \text{ dien } Br^+ \qquad (60)$$

$$*Pt \text{ dien } Br^+ \xrightarrow[\text{deactivation}]{} Pt \text{ dien } Br^+ \qquad (61)$$

$$*Pt \text{ dien } Br^+ + H_2O \xrightarrow[\text{primary photoreaction}]{} Pt \text{ dien } H_2O^{2+}, Br^- \qquad (62)$$

$$Pt \text{ dien } H_2O^{2+}, Br^- + NO_2^- \xrightarrow[\text{thermal } NO_2^- \text{ substitution}]{}$$
$$Pt \text{ dien } NO_2^+ + Br^- + H_2O \qquad (63)$$

$$Pt \text{ dien } H_2O^{2+}, Br^- + Br^- \xrightarrow[\text{thermal } Br^- \text{ substitution}]{} Pt \text{ dien } Br^+ + Br^- + H_2O \qquad (64)$$

$$Pt \text{ dien } H_2O^{2+}, Br^- \xrightarrow[\text{geminate recombination}]{} Pt \text{ dien } Br^+ + H_2O \qquad (65)$$

The expression of the quantum yield of the NO_2^- photosubstitution is:

$$\Phi = \left\{\frac{k_{62}}{k_{61}+k_{62}}\right\}\left\{\frac{k_{63}[NO_2^-]}{k_{63}[NO_2^-]+k_{64}[Br^-]+k_{65}}\right\} = \Phi' \frac{k_{63}[NO_2^-]}{k_{63}[NO_2^-]+k_{64}[Br^-]+k_{65}} \qquad (66)$$

where Φ' is the primary quantum yield.

Rearrangement of equation (66) gives the following expression:

$$\frac{1}{\Phi} = \frac{1}{\Phi'} + \frac{k_{65}}{k_{63}\Phi'[NO_2^-]} + \frac{k_{64}[Br^-]}{k_{63}\Phi'[NO_2^-]} \qquad (67)$$

It can be verified that all the experimental results obtained in neutral solutions are consistent with equation (66) and (67).

Biacetyl has been used in the photo-aquation of the tetrachloroplatinate(II) system.[62] Quantitative measurements of the quantum yield for the photo-aquation of $PtCl_4^{2-}$ both in the presence and in the absence of biacetyl as sensitizer have been made. With irradiation at 404 nm and 0·034M-$PtCl_4^{2-}$ and 0·001M-$HClO_4$, a quantum yield of 0·16 ± 0·02 was obtained, which is in good agreement with the value of 0·14 reported in the literature. A solution 0·5M in biacetyl, 0·005M in $PtCl_4^{2-}$, and 0·001M in $HClO_4$ gave a value of $\Phi = 0·26 \pm 0·02$ with 404 nm radiation, using reineckate and ferrioxalate actinometry. It is possible to interpret the constancy of the quantum yield of 0·14 at 313, 404, and 472 nm as due to a complete deactivation of the ligand field singlets (313, 304 nm) and the triplet (472 nm) to the lowest triplet (555 nm) state, which could be the

---

[62] V. S. Sastri and C. H. Langford, *J. Amer. Chem. Soc.*, 1969, **91**, 7533.

common precursor of the photo-aquation having a quantum yield of 0·14. On the basis of results on quantum yields the following scheme is proposed:

$$(\text{biacetyl})_{S_0} \xrightarrow{h\nu} (\text{biacetyl})_{S_1} \quad (68)$$

$$(\text{biacetyl})_{S_1} \longrightarrow (\text{biacetyl})_T \quad (69)$$

$$(\text{biacetyl})_T \longrightarrow (\text{biacetyl})_{S_0} + h\nu_P(\text{phosphorescence}) \quad (70)$$

$$(\text{biacetyl})_T + (\text{complex})_{S_0} \longrightarrow (\text{complex})_T(^1A_{1g} \longrightarrow {}^3E_g) \quad (71)$$

$$(\text{complex})_T \longrightarrow \text{product} \quad (72)$$

The higher yield of the sensitized reaction suggests the important result that the unsensitized reaction does not proceed by complete conversion of all higher excited states to the lowest triplet. In fact, it is true that the actual sensitized yield is larger than 0·26, and it may well be true that the quantum yield from the lowest triplet of the Pt complex is quite close to unity.

The excitation of azide to platinum charge-transfer transitions in the azidodiethylenetriamineplatinum(II) complex ion causes the formation of N_3 radicals.[63] The *trans*-isomers of dihalogenobis(triphenylphosphine)-platinum(II) complexes, difficult to obtain by other methods, are readily synthesised by photochemical isomerisation of the *cis*-complexes,[64] and the method can be applied to other platinum(II) complexes, reaction (73). Luminescent emission from $PtCl_4\,py_2$ compounds has been described.[65]

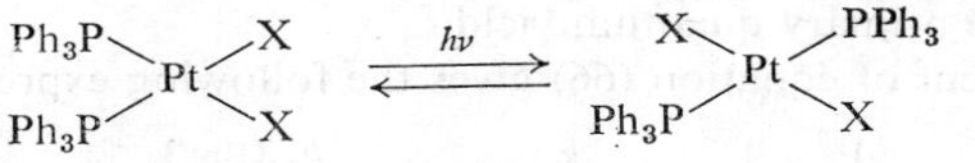

$$\begin{array}{c} Ph_3P \diagdown \diagup X \\ Pt \\ Ph_3P \diagup \diagdown X \end{array} \xrightleftharpoons{h\nu} \begin{array}{c} X \diagdown \diagup PPh_3 \\ Pt \\ Ph_3P \diagup \diagdown X \end{array} \quad (73)$$

Manganese Compounds.—Absorption spectra of several $3d$ transition-metal ions in molten fluoride solution have been measured.[66] Fluorescence has been observed in molten manganese salts also.[67] Carefully purified samples of $(Bu_4A)_2MnX_4^{2-}$, where A = N or P and X = Cl, Br, or I, exhibited the pronounced yellow-green fluorescence commonly seen in tetrahedral Mn^{II}. However, it was noted that the fluorescence of all these compounds persisted into the liquid state to more than 50 °C above the melting points. The fluorescence of tetrahedral Mn^{II} in liquid systems has not been observed previously, presumably because the relatively long natural lifetimes associated with spin-forbidden transitions make the ions quite susceptible to quenching. The fluorescence spectra of these solutions confirm the expectation that the transition responsible for the emission is the $T_1(^4G) \to$ ground state (^6S). The long lifetime is a consequence of the fact that the

[63] C. Bartocci and F. Scandola, *Chem. Comm.*, 1970, 531.
[64] S. H. Mastin and P. Haake, *Chem. Comm.*, 1970, 202.
[65] L. A. Rosiello, *J. Chem. Phys.*, 1969, **51**, 5191.
[66] J. P. Young, *Inorg. Chem.*, 1969, **8**, 825.
[67] B. Howard and B. R. Sundheim, *J. Chem. Phys.*, 1969, **50**, 5035.

transition is spin forbidden. The absence of sextet excited states obviates intersystem crossing as a mode of quenching. The ability of this species to withstand radiationless transition to the ground state is probably a consequence of the near perfection of the tetrahedral symmetry maintained in the fused salt system by the large cation-to-anion ratio. The small number of vibrations and their low frequency may also be significant. The relative absence of admixture of excited-state configurations in the ground state is also responsible for the fact that well-resolved hyperfine structure can be observed in the e.p.r. spectra of these molten salts.

A re-investigation of the electronic spectrum of potassium hexafluoromanganate(IV) has been carried out,[68] and the photochemical reduction of Mn^{III} tetrasulphophthalocyanine described.[69] A polarographic investigation of the photolysis of $[Mn(CN)_5NO]^{3-}$ ion has been carried out.[70] The reaction proceeds as in (74).

$$[(CN)_5Mn^I{=}N{=}O]^{3-} + h\nu \longrightarrow {}^*[(CN)_5Mn^{II}{-}\dot{N}{=}O]^{3-}$$

$$\downarrow$$

$$Mn^{II} + 5CN^- + NO \qquad (74)$$

In the presence of excess $(CN)^-$ the reaction can be suppressed:

$$\{[(CN)_5MnNO]^{3-}\}^* + CN^- \longrightarrow [Mn(CN)_6]^{4-} + NO \qquad (75)$$

In the presence of molecular oxygen, other reactions occur:

$$2\,{}^*[(CN)_5Mn^{II}{-}\dot{N}{=}O]^{3-} + O_2 \qquad (76)$$

$$\downarrow$$

$$[(CN)_5Mn^{II}{-}\underset{O}{N}{-}O{-}O{-}\underset{O}{N}{-}Mn^{II}(CN)_5]^{6-}$$

$$\downarrow$$

$$2[(CN)_5Mn^{III}NO_2]^{3-} \longrightarrow 2[Mn^{III}(CN)_6]^{3-} + 2NO_2^- \qquad (77)$$

---

[68] G. C. Allen, G. A. M. El-Sharkarwy, and K. D. Warren, *Inorg. Nuclear Chem. Letters*, 1969, **5**, 725.

[69] L. N. Zavgorodnyaya and T. S. Glikman, *Zhur. obshchei. Khim.*, 1969, **39**, 1443.

[70] W. Jakob, T. Senkowski, J. Czaja, and D. Rublowsku, *Roczniki. Chem.*, 1969, **43**, 253.

The fluorescence of some manganese chloride compounds with alkylpyridine hydrochloride bases,[71] and the photoinduced reaction of cyclopentadienyl manganese tricarbonyl with cyclopentadiene have been reported.[72]

Copper Complexes and Compounds.—The electronic absorption spectra of isothiocyanato- and thiocyanato-polyaminecopper(II) complexes have been reported.[73] The polarised single-crystal electronic spectra and e.s.r. spectra of CuH_2 edta (H_2O) and Cu dien$_2$Br$_2$,H_2O have been measured; both have been interpreted in D_2 symmetry and the interpretation of the latter in C_{2v} symmetry discussed.[74] The polarisation data from CuH_2 edta (H_2O) yields the tentative one-electron orbital sequence $d_{xy} > d_{z^2} > d_{xz} > d_{x^2} > d_{y^2} > d_{yz}$ and that from Cu dien$_2$Br$_2$,H_2O the tentative sequence $d_{x^2-y^2} > d_{z^2} > d_{xy} > d_{xz} > d_{yz}$. The ground states correspond with the x- and y-axes bisecting the angles between the equatorial bonds in the former complex, but lying along the bonds in the latter.

The combined orbital and spin–orbit reduction parameters (k) were calculated for CuH_2 edta (H_2O). The energy levels for both complexes are discussed in terms of the restricted tetragonal distortion imposed by the presence of ligands which are involved in chelation out of the equatorial plane. A corresponding interpretation of the electronic spectra of Cu en$_3$SO$_4$ and Cu bipy$_3$Br$_2$,6H_2O is suggested. The effect of a $d_{x^2-y^2}$ and a d_{xy} ground state on the relative intensities of the x-, y-, and z-polarised spectra are discussed.

The electronic, e.s.r., and i.r. spectra of a series of [Cu bipy$_2$X]Y complexes, where X = ONO$^-$, HCO$_2^-$, MeCO$_2^-$, PhCO$_2^-$, NCS$^-$, or $\frac{1}{2}$C$_2$O$_4^{2-}$, and Y = ClO$_4^-$; where X = Y = ClO$_4^-$, NO$_3^-$, or BF$_4^-$; and [Cu bipy$_2$NH$_3$]X$_2$, where X = ClO$_4^-$ or BF$_4^-$, are reported.[75] By comparing these electronic properties with those of the trigonal-bipyramidal (Cu bipy$_3$) and the *cis*-distorted octahedral [Cu bipy$_2$(ONO)]NO$_3$ complexes, possible molecular structures are suggested. The complexes with X = ONO$^-$ or a carboxylate anion have a *cis*-distorted octahedral stereochemistry; with X = NCS$^-$ or NH$_3$ an approximately trigonal-bipyramidal stereochemistry is suggested. Possible stereochemistries for the complexes Cu bipy$_2$X$_2$, where X = ClO$_4^-$, BF$_4^-$, or NO$_3^-$, are discussed. The molecular structures of the corresponding [Cu phen$_2$X]Y complexes are considered to be analogous.

Photochemical reduction of bis-(2,9-dimethyl-1,10-phenanthroline)copper(II) undergoes photoreduction in acidic aqueous media.[76] The quantum

---

[71] I. Burich, K. Nikolich, and K. Velashevich, *Zhur. priklad Spektroskopii*, 1969, **11**, 304.
[72] W. Bathelt, M. Herberhold, and E. O. Fischer, *J. Organometallic Chem.*, 1970, **21**, 395.
[73] R. Barbucci, G. Cialdi, G. Ponticelli, and P. Paoletti, *J. Chem. Soc. (A)*, 1969, 1775.
[74] B. J. Hathaway, I. M. Procter, R. C. Slade, and A. A. G. Tomlinson, *J. Chem. Soc. (A)*, 1969, 2219.
[75] B. J. Hathaway, M. J. Bew, D. E. Billing, R. J. Dudley, and P. Nicholls, *J. Chem. Soc. (A)*, 1969, 2312.
[76] S. Sundararajan and E. L. Wehry, *Chem. Comm.*, 1970, 267.

yield for appearance of Cu dmp^{2+} has been determined as a function of incident frequency, using incident power values of $(1.05 \pm 0.10) \times 10^{14}$ photon s^{-1}. The results are indicated in Figure 8, along with the electronic

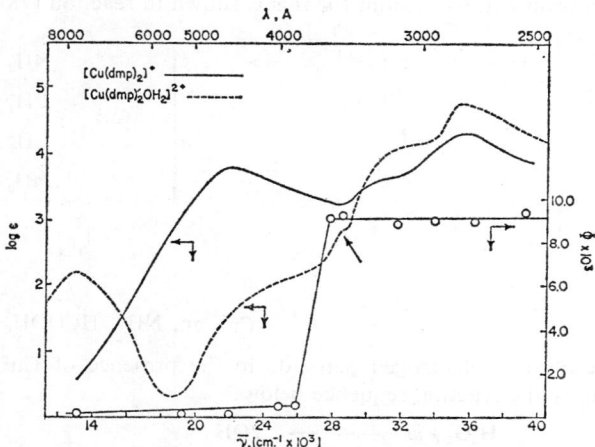

Figure 8 *Variation of Φ for Cudmp$_2^+$ formation with incident frequency (right ordinate); electronic spectra of Cu dmp$_2$OH$_2^{2+}$ and Cu dmp$_2^+$ (left ordinate). Charge-transfer band in Cu dmp$_2$OH$_2^{2+}$ indicated by arrow*
(Reproduced by permission from *Chem. Comm.*, 1970, 267)

absorption spectra of Cu dmp$_2$OH$_2^{2+}$ and Cu dmp$_2^+$. A very sharp decrease of Φ is noted at approximately 27 700 cm^{-1}. In an assignment of the electronic absorption spectrum of bis-(2,9-dimethyl-1,10-phenanthroline) chelates of Cu^{2+}, it was postulated that there existed a $\pi \to d$ (ligand $\to$ metal) charge-transfer band at 28 600 cm^{-1}. Hence, there is an extremely close correlation between photoreduction activity of Cu dmp$_2$OH$_2^{2+}$ and the assigned position of the lowest spin-allowed charge-transfer excited state. That Φ is essentially constant at all frequencies greater than 27 700 cm^{-1} may indicate that the charge-transfer state is indeed reactive, with internal conversion thereto from higher $\pi\pi^*$ intraligand states being highly efficient. It is also noteworthy that excitation in the $d \to d$ absorption region (12 000—17 000 cm^{-1}) does not effect detectable photoreduction of Cu dmp$_2$OH$_2^{2+}$. This result strongly implies that internal conversion from the lowest spin-allowed charge-transfer state into lower-lying doublet ligand-field excited states is not rapid with respect to the photoreduction process.

Cu dmp$_2^+$ is resistant to photochemical oxidation under all experimental conditions. The Cu–(dmp) system is thus analogous to Fe phen$_3^{3+}$–Fe phen$_3^{2+}$, wherein the former undergoes photochemical reduction [5] but the latter is resistant to photo-oxidation.

The photolysis of some Cu^{II} complexes in solution, including $Cu\,en_2^{2+}$, $Cu\,en_2(H_2O)_2SO_4$, $Cu\,dien_2Cl_2$, $Cu\,dien_2Br_2$, and $Cu\,dien\,Cl_2$, where en = ethylendiamine and dien = diethylenetriamine, gave rise to NH_3, CH_3NH_2, CO_2, H_2CO, and $HCOOH$ as reaction products.[77] The mechanism proposed to account for this is shown in reaction (78).

$$(Cu^{II}en_2)^{2+} + h\nu \longrightarrow (Cu\,en_2)^{2+\,*} \longrightarrow \left[enCu^{I} \begin{array}{c} NH_2 \\ | \\ CH_2 \\ | \\ CH_2 \\ | \\ NH_2 \end{array} \right]^{+} \quad (78)$$

$$\downarrow H_2O$$

$$Cu^{I},\ en,\ NH_3,\ HCOOH,\ HCHO,$$

The photolysis of hydrogen peroxide in the presence of Cu^{2+} ions is accelerated by the reaction sequence below:[78]

$$H_2O_2 + h\nu \longrightarrow 2OH \quad (79)$$

$$OH + H_2O_2 \longrightarrow HO_2 + H_2O \quad (80)$$

$$Cu^{2+} + HO_2 \longrightarrow Cu^{+} + O_2 + H^{+} \quad (81)$$

$$Cu^{+} + H_2O_2 \longrightarrow Cu^{2+} + OH + OH^{-} \quad (82)$$

$$Cu^{+} + HO_2 \longrightarrow Cu^{2+} + HO_2^{-} \quad (83)$$

Isopropyl alcohol–H_2O_2–metallic copper mixtures show photochromic behaviour in that the mixture turns from colourless to light blue on exposure to light.[79] After irradiation the inner walls of the reaction vessel were covered with a metallic red-purple thin layer, and a fine black suspension was seen in the solution. Cu_2O, CuO, copper acetate, acetone, acetic acid, and pinacone were all detected in the bulk. Upon shutting off the light the system reverted to colourless in 7—10 minutes. The suggested mechanism is shown in reactions (84) and (85).

$$(MeCOO)_2Cu + Cu \rightleftharpoons 2MeCOOCu \quad (84)$$

$$2(MeCOO)_2Cu + Me_2CHOH \xrightarrow{h\nu} 2MeCOOCu + MeCOMe + 2MeCOOH \quad (85)$$

The photolysis of frozen alcoholic solutions of cupric chloride has also been described.[80]

[77] L. A. Il'yikevich and G. A. Shagisultanova, *Khim. vysok. Energii*, 1969, **3**, 207.
[78] V. M. Berdnikov, Yu. N. Kozlov, and A. P. Purmal, *Khim. vysok. Energii*, 1969, **3**, 321, 370.
[79] S. Paszyc, *Roczniki Chem.*, 1969, **43**, 1783.
[80] A. L. Poznyak, *Khim. vysok. Energii*, 1969, **3**, 380.

Miscellaneous.—The electronic spectra of the hexafluoronickelate(III), hexafluorocuprate(III), and hexafluoroargentate(III) anions have been observed,[81] and that of the hexafluoronickelate(IV) ion also seen.[82]

There is considerable current interest in the nitrogenammineruthenium(II) compounds as possible models for biological nitrogen reduction. During the course of an investigation of the chemistry of this ion [83] it was observed that the characteristic absorption peak at 221 nm of anaerobic solutions of the ion stored in quartz cells decreased significantly when kept in the light. On exposure of a helium-deoxygenated solution of $Ru(NH_3)_5N_2^{2+}$ to u.v. light from a mercury lamp it was found that the 221 nm absorption disappeared completely and the absorption characteristic of an Ru^{III} species appeared after a short period of irradiation (15 min—3 hr, depending on concentration and light intensity). Further studies revealed that the photochemical oxidation proceeds equally well in acidic, neutral, or basic solutions. In HCl, $Ru(NH_3)_5Cl^{2+}$, identified by its absorption maximum at 330 nm, is the product, while in basic solution containing excess NH_3, $Ru(NH_3)_6^{3+}$, $\lambda_{max} = 280$ nm, is formed. In neutral solutions, $Ru(NH_3)_5OH^{2+}$, absorbing at 290 nm, is formed. When the photolysed solutions are treated with amalgamated zinc in dilute H_2SO_4, the Ru^{III} species is reduced, as indicated by the disappearance of its characteristic absorbance. By passing N_2 through the reduced solution, both $Ru(NH_3)_5N_2^{2+}$ and the dimer, $[Ru(NH_3)_5]_2N_2^{4+}$, are formed, again identified by their absorbance at 221 and 262 nm respectively. The identification of the reduction product is clearly of the greatest importance. Since it is known that some Ru^{II} species reduce H_2O, it seemed probable that H_2 is the product. Repeated experiments under widely different conditions, however, showed only the presence of N_2 and a small amount of O_2 in the gas above the solution, as determined by mass spectrometric analysis. As H_2 is absent, it is quite possible that the bound N_2 is reduced, since no other reducible species is present. Qualitative tests of the photolysed solution for $NH_2 \cdot NH_2$ and NH_2OH were negative.

The oxidation of ruthenium(II) is reported above [83] as the principal photochemical reaction pathway for the Ru^{II}–molecular nitrogen complexes, $[Ru(NH_3)_5N_2]^{2+}$ and $[Ru(NH_3)_5]_2N_2^{2+}$. In a study [84] of the photochemical reactions of several related penta-ammineruthenium(II) complexes, $Ru(NH_3)_5py^{2+}$, $Ru(NH_3)_5(CH_3CN)^{2+}$, $Ru(NH_3)_5H_2O^{2+}$, and $Ru(NH_3)_6^{2+}$, it was found that aquation of the complex as well as oxidation of Ru^{II} was a major photochemical reaction pathway in each case. Not surprisingly, preliminary studies show that the rôle played by the various pathways is dependent on the wavelength of the excitation radiation.

---

[81] G. C. Allen and K. D. Warren, *Inorg. Chem.*, 1969, **8**, 1895.
[82] G. C. Allen and K. D. Warren, *Inorg. Chem.*, 1969, **8**, 753.
[83] C. Sigwart and J. T. Spence, *J. Amer. Chem. Soc.*, 1969, **91**, 3991.
[84] P. C. Ford, D. H. Stuermer, and D. P. McDonald, *J. Amer. Chem. Soc.*, 1969, **91**, 6209.

The spectrum of the pyridine complex, $Ru(NH_3)_5py^{2+}$, displays an intense metal-to-ligand charge-transfer (CT) band at 408 nm ($\varepsilon = 7800$) and a ligand $\pi-\pi^*$ band at 244 nm ($\varepsilon = 4570$) as the only discernible features. Irradiation of anaerobic aqueous $Ru(NH_3)_5py^{2+}$ (0·001M-HCl) with 406 nm light (half-band width ~ 100 Å) leads to photo-aquation of this relatively substitution-inert complex [reaction (86)]. The reactions proceed with an overall quantum yield of 0.35 ± 0.1, determined by ferric

$$Ru(NH_3)_5\,py^{2+} \xrightarrow[25\,°C]{h\nu} Ru(NH_3)_5H_2O^{2+} + cis\text{-}Ru(NH_3)_4(H_2O)\,py^{2+} \\ \phantom{Ru(NH_3)_5\,py^{2+} \xrightarrow[25\,°C]{h\nu}} 19\% \qquad\qquad\qquad 45\% \\ + trans\text{-}Ru(NH_3)_4(H_2O)\,py^{2+} \\ \phantom{Ru(NH_3)_5\,py^{2+}}14\% \\ + \text{unidentified products} \qquad (86)$$

oxalate actinometry. Since product ratios are somewhat dependent upon the extent of total reaction, the product yields in reaction (86) and the quantum yield are reported for $\sim 30\%$ disappearance of starting material to minimise complications derived from secondary photolysis.

Photo-oxidation of Ru^{II} necessitates the corresponding reduction of some solution species. It was suggested[83] that the reduction was of co-ordinated N_2 in the photolysis of Ru^{II}–N_2 complexes, since mass spectrometer experiments did not reveal H_2 as a reaction product. However, in photolysis of argon-deaerated $Ru(NH_3)_6^{2+}$ and $Ru(NH_3)_5H_2O^{2+}$ solutions, H_3O^+ or H_2O appear to be most likely oxidants. It has been demonstrated in non-quantitative mass spectrometer experiments that the full-beam photolysis of $Ru(NH_3)_5\,py^{2+}$ ($\sim 4 \times 10^{-3}$ mol l^{-1}, BF_4^- salt) does produce H_2 in easily detectable quantities. Whether the formation of H_2 is a minor or major pathway in the photolysis of this ion and of the other Ru^{II} ammines remains to be determined.

An investigation of the photolytic production of hydrated electrons from aqueous solutions containing transition-metal cyanide complexes and also $IrCl_6^{3-}$ is reported.[85] A transient identified as due to e_{aq}^- has been observed in flash photolysis experiments with aqueous $Fe(CN)_6^{4-}$, $Mo(CN)_8^{4-}$, $W(CN)_8^{4-}$, and $Ru(CN)_6^{4-}$. In addition, steady illumination experiments have been made using N_2O as electron scavenger, to obtain the quantum yields $\Phi(N_2)\,[=\Phi(e_{aq}^-)]$ for the above complexes and for $IrCl_6^{3-}$. Values ranging from 0·031 to 0·66 are reported, using 254 nm radiation. In the case of $Mo(CN)_8^{4-}$, $\Phi(N_2)$ drops to essentially zero with irradiation at wavelengths greater than 300 nm.

The results are summarised in Table 6. The nominal reaction for photoelectron production is (87), and in the presence of N_2O, the overall reaction will be (88).

$$ML_n^{x+} \longrightarrow ML_n^{(x+1)+} + e_{aq}^- \qquad (87)$$

$$2ML_n^{x+} + N_2O + H_2O \longrightarrow 2ML_n^{(x+1)+} + 2OH^- + N_2 \qquad (88)$$

[85] W. L. Waltz and A. W. Adamson, *J. Phys. Chem.*, 1969, **73**, 4250.

Table 6 Photoelectron production from N_2O scavenging

Complex (absorption max. or shoulder nm: extinction coefficient)	$k_e{}^a$ l mol^{-1} s^{-1}	Concentrations Complex mol l^{-1} × 10^{-3}	N_2O mol l^{-1} × 10^{-2}	% photo-lysis^b	Wavelength of irradiation nm	ΦN_2	$\varepsilon^0 v^c$
$Fe(CN)_6{}^{4-}$ (422: 4·73; 322·5: 302·0; 270sh: ~10^3; 218: 24,200; 200: 23,700)^d	<10^{5e}	1·5	2·2	12	254	0·66 0·46^f	−0·356^g
$Ru(CN)_6{}^{4-}$ (323: —; 206: 9, 550; 192: —)^d	<10^{7h}	3·0	2·2	3	254	0·36	−0·86^i
$Mo(CN)_8{}^{4-}$ (510sh: 2·7; 431sh: 69; 367·6: 170; 308·2sh: 262; 267·4sh: 1350; 240·0: 15,540)^j	4·0 × 10$^{8\,k}$	1·5 1·5 1·0	2·2 1·6 2·2	11 10 100	254 254 >300	0·28 0·26 <·01	−0·726^g
$W(CN)_8{}^{4-}$ (625·0sh: 1·9; 502·4sh: 4·8; 434·8: —111·0; 370·3: 251·0; 303·1sh: 520; 273·75sh: 3000; 249·0: 25,060)^j	2·0 × 10$^{8\,k}$	0·75	2·2	19	254	0·34	−0·457^l
$IrCl_6{}^{3-}$ (400sh: 480; ~365: 6820; 206: 28,000)^m	9·4 × 10$^{9\,n}$	1·0	2·2	6	254	0·031	−1·02°

^{a} Rate constants for reaction with $e_{aq}{}^-$. ^{b} Per cent photolysis calculated from the nitrogen yield on the basis that hydroxyl radical oxidised a second molecule of complex. ^{c} Standard half-cell potential for oxidation. ^{d} Taken in part from J. J. Alexander and H. B. Gray, *Co-ordination Chem. Rev.*, 1967, **2**, 29. ^{e} M. Anbar, *Chem. Comm.*, 1966, 416. ^{f} A minimum value as small misalignment of the lamp and reaction cell was noticed after the experiment. ^{g} I. M. Kolthoff and W. J. Tomsicek, *J. Phys. Chem.*, 1935, **39**, 945; *ibid.*, 1936, **40**, 247. ^{h} M. Anbar, E. M. Fielden, and E. J. Hart, unpublished results cited in M. Anbar and P. Neta, *J. Appl. Radiat. Isotopes*, 1967, **18**, 493. ^{i} D. D. DeFord and A. W. Davidson, *J. Amer. Chem. Soc.*, 1951, **73**, 1469. ^{j} J. R. Perumareddi, A. D. Liehr, and A. W. Adamson, *ibid.*, 1963, **85**, 249. ^{k} ref. 85. ^{l} H. Baadsgaard and W. D. Treadwell, *Helv. Chim. Acta*, 1955, **38**, 1669. ^{m} Taken in part from C. K. Jørgensen, *Mol. Phys.*, 1959, **2**, 309. ^{n} M. Anbar and E. J. Hart, unpublished results cited in M. Anbar and P. Neta, *J. Appl. Radiat. Isotopes*, 1967, **18**, 493. ^{o} W. M. Latimer, 'Oxidation Potentials,' 2nd ed, Prentice-Hall, Inc., New York, N.Y., 1952, p. 217.

Photoelectron production in transition-metal co-ordination compounds has been a rare phenomenon; the only previously reported case being that of $Fe(CN)_6^{4-}$. The present combined flash photolysis and N_2O scavenging results not only certify the similar behaviour of other cyano-complexes, but also, with the example of $IrCl_6^{3-}$, show that the process is not unique to complex cyanides. It may, in fact, eventually prove to be quite common among transition-metal complexes.

It was noted that in selecting systems for study certain criteria were important. These are (a) that the irradiation should be in the wavelength region of a CT band of the complex; (b) that a reasonably stable complex of the same stoicheiometry and one higher valence state exists; and (c) that a stable valence state of the same stoicheiometry and one lower oxidation number should not exist.

Considering the reasons in reverse order, (c) applies primarily to flash photolysis studies. In order to observe the e_{aq}^- transient, it is essential that scavenging be minimised, and the existence of an easily attainable lower valence state of the complex would ensure that excessive scavenging would occur. Criterion (b) follows from energetic considerations. If the next higher valence state is not known, at least to the extent of being observable as an electrode reaction, it is doubtful if light quanta corresponding to wavelengths accessible in aqueous media would have sufficient energy to bring about reaction (87); also implied in this reasoning is that the one higher oxidation state should be stable in the same stoicheiometry as the parent complex, as otherwise the activation energy for ejection of an electron might be excessively large relative to those of other possible photochemical reaction modes. The first criterion followed from a general conclusion that the photochemistry of co-ordination compounds derives usually from specific excited-state chemistries rather than from hot ground-state molecules produced during de-excitation. Thus, photoelectron production is associated with an excited state in which electron density has moved either towards the periphery of the complex or into the adjacent solvent. Absorptions of this type have been termed 'charge transfer to ligand' (CTTL), or 'charge transfer to solvent' (CTTS). Conversely, if photoreduction of the central metal ion occurs, as with many Co^{III} complex ammines, the band involved would by this criterion be considered to be 'charge transfer to metal' (CTTM) in type. Where ligand-field or d–d bands are irradiated, the general observation so far has been that substitution predominates. Thus, it can be assumed that the electronic nature of an excited state is qualitatively deducible from that of the chemical reaction to which the state is precursor; criterion (a) above is simply an application of this assumption.

It does seem correct, therefore, that photoelectron production is confined to CT-type excited states. Longer wavelength irradiation of $Fe(CN)_6^{4-}$ and of $Mo(CN)_8^{4-}$ leads to aquation rather than e_{aq}^- production. The present observation in the case of $Mo(CN)_8^{4-}$ is that $\Phi(N_2)$ drops to the vanishing point if the irradiation is at wavelengths above 300 nm, in keeping with the proposed mechanism.

The known photochemistry of metal cations other than those of the transition elements, lanthanides, or actinides is very sparse since only a single well-defined oxidation state of main-group elements usually exists in aqueous solution. However, in perchloric acid media the $+1$ and $+3$ oxidation states of thallium can co-exist and in the u.v. spectra both of these oxidation states exhibit high-intensity bands which may be assigned to charge-transfer transitions.[86] At 2537 Å, the principal absorbing species in 1M-perchloric acid is the hydrolysed thallium(III) species TlOH^{2+} and a minor absorption is due to the aquated species Tl$_{aq}^{3+}$. Absorption by Tl$_{aq}^+$ only becomes significant at wavelengths shorter than 2537 Å. Absorption of 2537 Å quanta by thallium(III) solutions would be expected to induce reactions (89) and (90).

$$\text{TlOH}^{2+} + h\nu \longrightarrow \text{Tl}^{2+} + {}^\bullet\text{OH} \tag{89}$$

$$\text{Tl}_{aq}^{3+} + h\nu \longrightarrow \text{Tl}^{2+} + {}^\bullet\text{OH} + \text{H}^+ \tag{90}$$

In the presence of added Tl$_{aq}^+$, scavenging of all hydroxyl radicals which escape primary and secondary recombination would occur.[6]

$$\text{Tl}_{aq}^+ + {}^\bullet\text{OH} \longrightarrow \text{Tl}^{2+} + \text{OH}^- \tag{91}$$

The 2537 Å photolysis of a thallium(III)–thallium(I) mixture should therefore result in the overall initiation outlined in reaction (92), with a

$$\text{TlOH}^{2+} + \text{Tl}_{aq}^+ + h\nu \longrightarrow 2\text{Tl}^{2+} + \text{OH}^- \tag{92}$$

quantum yield Φ_1, together with a minor contribution from reaction (93).

$$\text{Tl}_{aq}^{3+} + \text{Tl}_{aq}^+ \longrightarrow 2\text{Tl}_{aq}^{2+} \tag{93}$$

A study has sought evidence for the photolytic formation of the unstable intermediate oxidation state thallium(II) by investigating the photocatalysis of the thallium(I)–thallium(III) homonuclear electron-exchange reaction:[86]

$$*\text{Tl}^{III} + \text{Tl}^{I} \longrightarrow *\text{Tl}^{I} + \text{Tl}^{III}$$

detected by isotopic labelling with ^{204}Tl. If reaction (92) accurately represents the initiation induced by 2537 Å photolysis of a thallium(I)–thallium(III) mixture, then this could be followed by chain-propagation involving a Tl^{2+} chain carrier

$$*\text{Tl}^{3+} + \text{Tl}^{2+} \longrightarrow *\text{Tl}^{2+} + \text{Tl}^{3+} \tag{94}$$

$$*\text{Tl}^{2+} + \text{Tl}^{+} \longrightarrow *\text{Tl}^{+} + \text{Tl}^{2+} \tag{95}$$

and a quadratic chain-termination

$$\text{Tl}^{2+} + \text{Tl}^{2+} \longrightarrow \text{Tl}^{+} + \text{Tl}^{3+} \tag{96}$$

---

[86] D. R. Stranks and J. K. Yandell, *J. Phys. Chem.*, 1969, 73, 840.

A generalised treatment of photoinduced chain reactions suggests that, for this anticipated mechanism, the quantum yield for the overall electron-exchange reaction is:

$$\Phi_{ex} = \Phi_1 \left\{ \frac{k_{95}[Tl^I] \, k_{94}[Tl^{III}]}{k_{95}[Tl^I] + k_{94}[Tl^{III}] + 2(k_{96} \, I_{abs} \, \Phi_1)^{\frac{1}{2}}} \left(\frac{1}{k_{95} \, I_{abs} \, \Phi_1}\right)^{\frac{1}{2}} \right.$$
$$\left. + \frac{k_{95}[Tl^I] + k_{94}[Tl^{III}] + (k_{96} \, I_{abs} \, \Phi_1)^{\frac{1}{2}}}{k_{95}[Tl^I] + k_{94}[Tl^{III}] + 2(k_{96} \, I_{abs} \, \Phi_1)^{\frac{1}{2}}} \right\} \quad (97)$$

where Φ_{ex} is the observed rate of electron exchange per absorbed light intensity, i.e. R_{ex}/I_{abs}. The second term of equation (97) represents the contribution to electron exchange from the termination in the absence of any propagation. Alternatively, if the Tl^{2+} chain carrier were to undergo linear termination by reactions with a trace impurity or at the vessel walls, both proceeding at the same rate, as in

$$Tl^{2+} + S^1 \longrightarrow Tl^I \quad \text{and} \quad Tl^{2+} + S^2 \longrightarrow Tl^{III} \quad (98)$$

then Φ_{ex} would be independent of light intensity, giving equation (99).

$$\Phi_{ex} = \Phi_1 \left\{ \frac{k_{95}[Tl^I] \, k_{94}[Tl^{III}]}{k_{95}[Tl^I] + k_{94}[Tl^{III}] + 2k_{98}} \left(\frac{1}{k_{98}}\right) + \frac{k_{95}[Tl^I] + k_{94}[Tl^{III}] + k_{98}}{k_{95}[Tl^I] + k_{94}[Tl^{III}] + 2k_{98}} \right\} \quad (99)$$

Experiments have shown that the behaviour of this system is consistent with equation (97).

The consistent observation of quantum yields well in excess of unity is compelling evidence for the photo-generation of a thallium(II) species. The quantitative treatment of possible exchange mechanisms requires that the chain carrier be capable of one-electron exchange with both thallium(III) and thallium(I). A thallium(II) species seems to be the only conceivable possibility.

The initiation step of the mechanism arises from photochemical reactions of thallium(III) species. Light absorption by Tl_{aq}^+ does not lead to significant initiation as judged by measurements at 2400 and 2300 Å (where Tl_{aq}^+ is a major absorber) and by plateau quantum yields at Tl^+ concentrations ranging from 0.05 to 0.15 mol l^{-1}. The strong absorption due to Tl_{aq}^+ below 2500 Å could be interpreted as involving charge transfer to solvent:

$$Tl_{aq}^+ + h\nu \longrightarrow (Tl^{2+}, e_{aq}^-) \longrightarrow Tl^{2+} + H \quad (100)$$

but an alternative interpretation is that this absorption is a d–s transition on the metal cation. This may account for the relative ineffectiveness of the absorption.

Initiation by a thallium(III) absorbing species is consistent with the proposed charge-transfer assignment. Thus, the first excited state for $TlOH^{2+}$ is expected to have a charge distribution shifted from that represented by $Tl^{3+} \cdot OH^-$ toward $Tl^{2+} \cdot OH$. The value of $\Phi_1 = 0.5$, for the generation of geminate radicals which have escaped primary and secondary recombination, is significantly less than 1. This situation may be represented as the generation of an excited species in a solvent cage followed by de-excitation and diffusion apart of geminate radicals, thus

$$Tl^{3+} \cdot OH^- + h\nu \longrightarrow (Tl^{2+} \cdot OH) \overset{\Phi}{\rightleftharpoons} Tl_{aq}^{2+} + OH \quad (101)$$

and likewise for Tl_{aq}^{3+} absorption:

$$Tl_{aq}^{3+} + h\nu \longrightarrow (Tl_{aq}^{2+} OH_2^+) \overset{\Phi}{\longrightarrow} Tl_{aq}^{2+} + OH + H^+ \quad (102)$$

It is interesting that within the wavelength range 2700—2300 Å there is no significant evidence for a rise in quantum yields, even though an additional 18·5 kcal mol^{-1} (77·7 kJ mol^{-1}) is delivered with the 2300 Å quanta. The simplified Noyes treatment of diffusive recombination of geminate radical pairs predicts a strong wavelength dependence. Although the scavenging reaction (103) proceeds at almost diffusion-controlled

$$OH + Tl^+ \longrightarrow OH^- + Tl^{2+} \quad (103)$$

rates ($k_{103} = 9 \times 10^9$ l mol^{-1} s^{-1}) and, by the Noyes theory, would give an enhanced scavenging efficiency of up to 0·2, the recombination of two Tl^{2+}, as discussed below, also has a rate constant of a similar magnitude. Thus the effect of the kinetic energy of geminate radicals may well be removed by this unusual kinetic aspect of the scavenging reactions. Moreover there is unlikely to be any significant effect upon primary quantum yields due to possible competition of Tl^+ for secondary recombination of Tl^{2+} and OH.

The effectiveness of halogeno- and sulphato-thallium(III) species would suggest, in analogy to $TlOH^{2+}$, net reactions:

$$TlX^{2+} + h\nu \longrightarrow Tl^{2+} + X \cdot \quad (104)$$

followed by scavenging of the radical X· by added Tl^+:

$$Tl^+ + X \cdot \longrightarrow Tl^{2+} + X \quad (105)$$

The one-electron propagation reactions of thallium(II) with thallium(III) probably involve both Tl_{aq}^{3+} and $TlOH^{2+}$. For the thermal thallium(I)–thallium(III) electron exchange, the rate constants at 30 °C are

$$k(Tl_{aq}^{3+} - Tl_{aq}^+) = 1·25 \times 10^{-2} \text{ l mol}^{-1} \text{ s}^{-1} \quad (106)$$

and

$$k(TlOH^{2+} - Tl_{aq}^+) = 5·83 \times 10^{-3} \text{ l mol}^{-1} \text{ s}^{-1} \quad (107)$$

The lack of any marked dependence of quantum yields on acid concentration is consistent with propagation proceeding through reactions of Tl_{aq}^{3+} and $TlOH^{2+}$ with very similar rate-constants.

The rotating-sector measurements suggests that the quadratic termination proceeds with a rate-constant characteristic of diffusion-controlled reactions. The disproportionation of Tl^{II}, generated in substantial concentrations by pulse radiolysis, has been shown to exhibit a rate-constant of $2 \cdot 3 \pm 0 \cdot 8 \times 10^9$ l mol^{-1} s^{-1} and this is consistent with the photochemical evidence. The less important linear terminations probably arise from the same impurities responsible for the small induction periods.

The effect of u.v. irradiation on the luminescent properties of alcoholic solutions of thallium and lead salts has been discussed,[87] and a method of photochemical analysis employing photonometric titration of the thallic ion by uranyl photo-eduction has been developed.[88] A physicochemical study of the centres of luminescence in solutions of Tl$^+$ and Pb^{2+} has been

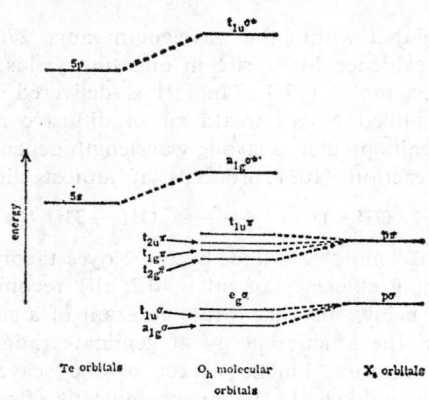

Figure 9 *Relative orbital energies estimated for* TeX$_6^{2-}$ *complexes* (Reproduced by permission from *J. Amer. Chem. Soc.*, 1970, **92**, 307)

reported,[89] as has the e.s.r. spectrum of the CF$_3$ radical produced by the photolysis of lead(IV) trifluoroacetate.[90] The absorption spectrum of InO and InO$^+$ has been investigated and progressions for vibrational bands given.[91] The absorption and fluorescence spectra of trivalent molybdenum in aluminoborophosphate glasses have been described,[92] and the characterisation and photolysis of octacyanomolybdic(IV) acid and the isolation of a red photoproduct described.[93] The product is the anion [Mo(CN)$_7$H$_2$O]$^{3-}$.

The absorption spectra of octahedral hexahalogenotellurate(IV) complexes have been reported and assigned.[94]

The $s \rightarrow p$ transition $a_{1g}\sigma^* \rightarrow t_{1u}\sigma^*$ gives rise to three spin-allowed bands in TeCl$_6^{2-}$ (32 880, 34 480, 35 680 cm^{-1}) and TeBr$_6^{2-}$ (28 470, 29 730

[87] M. V. Belyi and V. V. Bursevitch, *Izvest. Akad. Nauk S.S.S.R., Ser. fiz.*, 1970, **34**, 667
[88] Y. Yokoyama and S. Ikeda, *Bull. Chem. Soc. Japan*, 1969, **42**, 2254.
[89] R. I. Gindina, *Zhur. priklad. Spekroskopii*, 1969, **10**, 287.
[90] H. Loeliger, *Helv. Chim. Acta*, 1969, **52**, 1516.
[91] V. F. Shevel'kov, D. I. Kataev, and A. A. Mal'tsev, *Vestnik Moskov. Univ.*, 1969, **4**, 108.
[92] S. Parke and A. I. Watson, *Phys. and Chem. Glasses*, 1969, **10**, 37.
[93] R. P. Mitra, B. K. Sharma, and H. Mohan, *Canad. J. Chem.*, 1969, **47**, 2317.
[94] D. A. Couch, C. J. Wilkins, G. R. Rossman, and H. B. Gray, *J. Amer. Chem. Soc.*, 1970, **92**, 307.

30 850 cm^{-1}). These data are taken as evidence that the excited singlet states are distorted to lower symmetry, probably C_{2v}. The lowest allowed halide → Te^{IV} charge transfer transitions are at 44 170 cm^{-1} and somewhat higher than 52 000 cm^{-1} in TeCl$_6^{2-}$; analogous transitions appear at 37 000 and 42 600 cm^{-1} in TeBr$_6^{2-}$. The observed spectral data indicate that the 'inert' $5s^2$Te^{IV} electrons, which reside in the $a_{1g}\sigma^*$ level, are partially delocalised to the halide ligands. The relative orbital energies of TeX$_6^{2-}$ complexes are shown in Figure 9.

Lanthanide Compounds.—Intramolecular energy transfer in rare-earth chelates under excitation in the near-u.v. has been much studied. The mechanism by which the excitation energy is transferred from the ligand to the rare-earth (RE) ion has not been well established, and three possibilities have been suggested. These are:
(a) energy transfer from S_1 (ligand) via T_1 (ligand);
(b) direct transfer from S_1 (ligand);
(c) transfer from S_1 (ligand) to upper level of rare-earth, back transfer to T_1 (ligand) and further transfer to rare-earth lower level.

The ligand triplet-state energies were obtained from the phosphorescence spectra of the Gd chelates.[95] These data are presented in Figure 10 along with several electronic states of Eu^{3+} and Tb^{3+}.

The triplet states of all the Eu chelates except Eu BZDTF$_4$ lie above the $^5D_1$ state of Eu^{3+}. In these cases the time-resolved emission spectra can be accounted for as follows: the population of the $^5D_1$ state occurs in a time of the order of ∼ 10^{-8} s; then the depopulation of the $^5D_1$ state occurs in about 2 μs through radiative and non-radiative transitions to the ^7F manifold as well as through simultaneous non-radiative transition to the $^5D_0$ state. On the other hand, the population of $^5D_0$ occurs in about 2 μs, which is nearly the same as the decay time of $^5D_1$; finally, radiative and non-radiative transitions from $^5D_0$ to the ^7F manifold occur in about 500 μs. However, in the case of Eu BZDTF$_4$, where the lowest ligand triplet is situated between $^5D_1$ and $^5D_0$, no emission from the $^5D_1$ level is observed. The population of $^5D_0$ occurs in a time of the order of ∼ 10^{-8} s, and this is followed by radiative and non-radiative transitions to the ^7F manifold in about 500 μs. In the cases of the Tb chelates, Tb BFA$_4$, Tb BFA$_3$,2H$_2$O, and Tb TTA$_3$,2H$_2$O, the population of $^5D_4$ occurs in a time of the order of ∼ 10^{-8} s and decays in about 500 μs.

The experimental luminescence-rise time of the $^5D_0 \to {}^7F_2$ transition (613 nm) of Eu BZDTF$_4$ and the rise times of the $^5D_1 \to {}^7F_1$ transition (538 nm) in other Eu chelates and of the $^5D_4 \to {}^7F_5$ transition (545 nm) in Tb chelates, all are of the order of 10^{-8} s.

From the results stated above, the following conclusions can be reached. First, since the luminescence from RE levels which are higher than T_1 is not

[95] M. Tanaka, G. Yamaguchi, J. Shiokawa, and C. Yamanaka, *Bull. Chem. Soc. Japan*, 1970, **43**, 549.

observed, and since the time for energy transfer is too long to account for the direct transfer from S_1, there is no possibility of the mechanism (b). Second, the mechanism (c) can also be rejected, because the rise times of the RE-ion luminescence in several RE chelates investigated are nearly the same, i.e. 10^{-8} s, in spite of the fact that the energy differences between

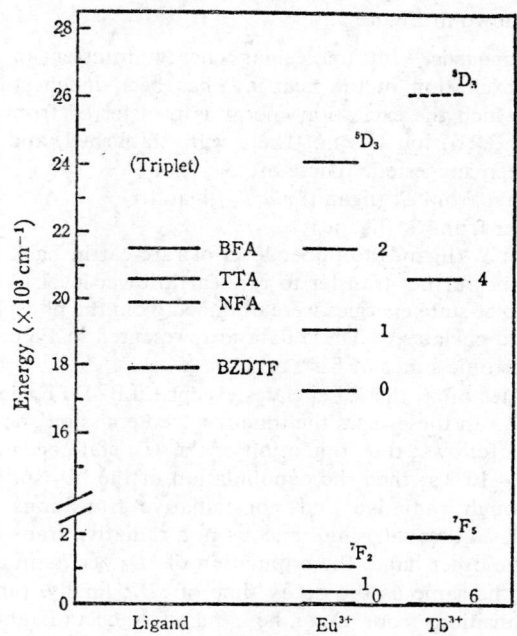

Figure 10 *The triplet states of organic ligands and the f-electronic states of* Eu^{3+} *and* Tb^{3+} *ions.* NFA = *napthhoyl trifluoroacetone;* TTA = *thenoyl trifluoroacetone;* BFA = *benzoyl trifluoroacetone;* BZDTF = *benzylidene trifluoroacetylacetone*
(Reproduced by permission from *Bull. Chem. Soc. Japan*, 1970, **43**, 549)

the intermediate RE level and T_1 are not similar. Therefore, it can be concluded that the energy transfer occurs by means of the mechanism (a).

New fluorescence emission bands of Eu^{III} solutions in light and heavy water and in methyl cyanide have been reported,[96] and an analysis of the fluorescence spectra of Eu^{III}-bis-(1,10-phenanthroline)tris(carboxylate) complexes given.[97] The luminescence of tris(salicylato)bis(1,10-phenanthroline)samarium and tris(salicaldehydato)-(1,10-phenanthroline)samar-

---

[93] Y. Haas and G. Stein, *Chem. Phys. Letters*, 1969, **3**, 313.
[97] S. P. Sinha and E. Buller, *Mol. Phys.*, 1969, **16**, 285.

ium complexes has also been studied,[98] and the integrated intensities of the yellow-green system of lanthanum monoxide spectral bands given.[99]

Actinides.—The electronic spectra of the U^{3+} ion have been discussed.[100] The uranyl oxalate actinometer was widely used until superseded, except for some applications, by the more convenient ferrioxalate system. The influence of pH upon the photolysis of this actinometer system has been investigated.[101]

The study was made at 25 °C of the influence of pH upon the relative quantum yields for the consumption of oxalate and the production of carbon dioxide, carbon monoxide, and uranous ion by light of 254 nm absorbed by the uranyl oxalate actinometer system. The uranyl ion was at 0·01 F, the oxalate at 0·06 F, and initial pH was at 0·6. About 12% of the oxalate was decomposed. The quantum yields for the consumption of oxalate were found to be independent of pH between pH 1 and 5 but to decrease outside this range. The reaction at all values of pH was found to consume acid. The molar ratio of carbon dioxide produced (in all forms: CO_2, H_2CO_3, HCO_3^-, CO_3^{2-}, and the uranyl carbonate complexes) to oxalate consumed very nearly equalled unity at pH 0 to 5. The molar ratio of carbon monoxide produced to oxalate consumed was always less than unity; it abruptly decreased with increasing pH between pH 1 and 2, and became negligible above pH 3. The molar ratio of uranous ion produced to oxalate consumed increased slowly from about 0·03 between pH 1·5 and 4 to about 0·08 at pH 6.

Insight into the mechanism of the uranyl oxalate actinometer system has been obscured by several problems. The first is the nature of the photo-sensitive complex. Almost all possible forms of the complex have been proposed, namely, $UO_2C_2O_4$, $UO_2HC_2O_4^+$, $UO_2(HC_2O_4)_2$, $UO_2H_2C_2O_4^{2+}$, and $UO_2(C_2O_4)_2^{2-}$. The protonated species have been assumed in the analysis of experimental results but have not been shown to offer a unique solution. Conductance measurements of uranyl oxalate solutions, pH measurements of solutions containing uranyl ion with excess oxalic acid, and spectroscopic measurements indicate strongly that the predominant uranyl species in solutions at pH 1 to 7 are $UO_2C_2O_4$ and $UO_2(C_2O_4)_2^{2-}$.

The second problem is the nature of the excited uranyl oxalate complex which is produced by the absorption of a quanta of light at wavelengths of 436 nm and less. The absorption of light by the complexes could result in an energy transfer to the oxalate or in a charge transfer from the oxalate to the uranium. The products do not indicate which process is occurring but some analogy can be drawn with similar processes. The photolysis of uranyl ion with an excess of formic or acetic acid produces mainly U^{IV}.

---

[98] K. K. Rohatgi and S. K. Sengupta, *J. Indian Chem. Soc.*, 1969, **46**, 579.
[99] N. S. Murthy and B. Narasimhamurthy, *Nature*, 1969, **223**, 181.
[100] B. Jezowska-Trzebiatowska, J. Drozdzynski, and K. Butietynska, *Bull. Acad. polon. Sci., Sér. Sci. chim.*, 1969, **17**, 295.
[101] L. J. Heidt, G. W. Tregay, and F. A. Middleton, *J. Phys. Chem.*, 1970, **74**, 1876.

The photolysis of cobalt, iron, and manganese oxalate complexes also results in a reduction of the metal ion even at wavelengths where the incident light is absorbed in a transition which is localised on the metal atom. Thus, it seems likely that when uranyl oxalate absorbs light of wavelength even as long as 436 nm, then a charge transfer from the oxalate to the uranyl ion results.

The reaction sequence may thus be:

$$O_2U^{VI} C_2O_4 + light \rightarrow O_2U^V + OCO + CO_2$$

$$O_2U^V + OCO^- + H_2O \rightarrow (O_2U^{VI} OOCH^+)^* + OH^-$$

$$(O_2U^{VI} OOCH^+)^* + H^+ \rightarrow O_2U^{VI} OH^+ + HCO^+$$

followed by dissociation of HCO^+ to H^+ and CO.

Several photochemical reactions of the uranyl ion have been reported. These include the reaction with propionaldehyde,[102] and the quenching of fluorescence of uranyl ions by organic substrates.[103] The structure and luminescent properties of complexes of uranyl nitrate with bipyridyl [104] and the temperature quenching of the luminescence of organic solutions of uranyl compounds [105] have also been described.

Photochemical oxidation of plutonium is dependent upon the acid solution used.[106] In nitric acid, plutonium(II) can be oxidised to plutonium(VI), whereas in HCl, $HClO_4$, or H_2SO_4 only Pu^{IV} is formed.

The absorption spectra of magenta-coloured $CsNpF_6$, both as a mull of the crystalline solid and as a solution in CsF,2HF, have been obtained under various conditions.[107] The room-temperature spectra were unique to the electronic energy levels of the seven Russell–Saunders states for the $5f^2$ configuration, split by spin–orbit coupling into 13 free-ion levels; this confirms the chemical evidence for the presence of pentavalent neptunium. At liquid-nitrogen temperatures the spectra of $CsNpF_6$ mulls were interpreted in terms of the free-ion $f-f$ transitions of Np^{5+} split by an O_h crystal field. Term assignments to the following experimental energy levels obtained at room temperature were made on the basis of the observed Stark splitting in the low-temperature spectra, and agreement with calculated energy levels (in cm^{-1}): $^3F_2 = 4337$, $^3H_5 = 8425$, $^3F_3 = 8797$, $^3F_4 = 9701$, $^3H_6 = 13 738$, $^3P_0 = 16 129$, $^1D_2 = 16 736$, $^1G_4 = 18 265$, $^3P_1 = 19 531$, $^1I_6 = 20 619$, $^3P_2 = 26 667$. All levels above the $^3H_4$ ground state were observed except $^1S_0$. A least-squares fit of these experimental energy levels was made to the Racah Coulomb interaction parameters, $E^1 = 2807$ cm^{-1}, $E^2 = 19.98$ cm^{-1}, $E^3 = 284.1$ cm^{-1}; the spin–orbit coupling parameter $\xi =$

---

[102] R. M. Haas and K. H. Gayer, *Z. anorg. Chem.*, 1969, **367**, 102.
[103] K. Venkaturao and M. Santuppa, *Z. phys. Chem.* (*Frankfurt*), 1969, **66**, 308.
[104] I. M. Kopashova, D. S. Umreiko, and R. I. Shamanovskaya, *Zhur. priklad. Spektroskopii*, 1969, **10**, 675.
[105] L. V. Volob'ko and E. A. Turetskaya, *Zhur. priklad. Spektroskopii*, 1969, **10**, 294.
[106] P. N. Patei, A. A. Nemodruk, and E. V. Bezrogova, *Radiokhimiya* 1969, **11**, 300.
[107] L. P. Varga, L. B. Asprey, T. K. Keenan, and R. A. Penneman, *J. Chem. Phys.*, 1970 **52**, 1664.

2316 cm^{-1}; and the configuration interaction parameters, $\alpha = 2\cdot7$ cm^{-1} and $\beta = 52\cdot9$ cm^{-1}. The position of the $5f^2$ levels of Np^{5+} on the intermediate coupling diagram, compared to the f^2 levels of the isoelectronic series Pr^{3+}, Nd^{4+}, Th^{2+}, U^{4+}, and Pu^{6+} (in PuF$_6$), was intermediate between U^{4+} and Pu^{6+} as expected.

Survey experiments have been performed to determine whether u.v.-excited sharp-line sensitized luminescence (SLSL) could be detected and studied in some α-active actinide (An^{3+}) β-diketones, namely, the hexafluoroacetylacetonate chelates CsAn hfa$_4$,nH$_2$O with the An^{3+} ions ^{241}Am^{3+}, ^{243}Am^{3+}, ^{244}Cm^{3+}, ^{249}Bk^{3+}, ^{249}Cf^{3+}, and ^{253}Es^{3+}. The relatively heavy hfa$^-$ ligand was chosen as the chelating agent so the work could be performed effectively at room temperature. Measurements were made on the pure crystals CsAn hfa$_4$,H$_2$O only in the Am^{3+} and Cm^{3+} cases, on the chelates in anhydrous ethanol solutions in all cases, and with An^{3+} as a dopant in the CsGd hfa$_4$ crystal matrix in all cases. SLSL was detected only in the case of Cm^{3+}; it is highly efficient in all three media, resembling Eu^{3+} in red SLSL colour and in high quantum efficiency. The results indicate that laser emission could be demonstrated in certain Cm^{3+}-chelate solutions. The absence of SLSL over the 4000—10 200 Å experimental range in the case of Es^{3+} suggests that the first excited state of this $5f^{10}$ electronic configuration is below 9800 cm^{-1}. Measurements of self-excited luminescence of the crystalline Cs^{244}Cm hfa$_4$,H$_2$O show that it is essentially the same as the u.v.-excited luminescence, and that the radiolytic decomposition is linear with a total decomposition time of 6 hr. This is approximately the time required for the time-integrated molar radioheat to equal the sum of the molar bond energies and crystal binding energy. Evidently, the individual ligands are excited by the α-activity and the energy is transferred *via* the ligand singlet and/or triplet state to the Cm^{3+}, as it is with u.v. excitation. Estimates are made of the radiolysis lifetime for each sample studied, assuming the radiolytic decomposition is in each case linear in time and that the decomposition rate is proportional to the specific radioheat of the isotope under consideration. These results indicate that radiolytic decomposition over the time required to do the experiments is not responsible for the absence of SLSL in the Am^{3+}, Bk^{3+}, Cf^{3+}, and Es^{3+} cases.

3 Gas-phase Studies

This brief section is devoted to the relatively few examples of the study of inorganic compounds in the gas phase. Spectral characteristics of several metals and simple metal compounds are mentioned, and a few examples of the gas-phase photolysis of organometallic compounds are included. Other studies on organometallic compounds will be found in the parts of this volume concerned with organic photochemistry.

The time dependence of Na(3^2P) fluorescence produced by pulsed, far-u.v. photodissociation of NaI vapour at 600 °C has been directly observed *via* the method of single-photon counting.[108] Sodium iodide molecules photodissociate in a time much less than 1 ns, and the fluorescence decay is exponential with $\tau = 16.0 \pm 0.2$ ns. No evidence of collision-induced fluorescence is seen, and quenching of the fluorescence by NaI vapour itself is not observed. Quenching by iodine vapour is studied as a function of pressure, and I_2 molecules are shown to be at least 30 times more effective as quenchers than I atoms. The velocity dependence of the I_2 cross-section in the range ~ 12—25×10^4 cm s^{-1} was studied by varying the photodissociation pulse wavelength. The fluorescence of $Na(3^2P)$ following photodissociation of NaBr, and of $Tl(7^2S)$ following photodissociation of TlI are also observed. An average cross-section for the quenching of $Na(3^2P)$ by Br_2 is reported. Both TlI and I_2 are observed to quench $Tl(7^2S)$ quite strongly, with I_2 having a cross-section of 490 Å^2 at 4×10^4 cm s^{-1}. The quenching process is discussed in terms of the 'harpooning' mechanism and comparison is made with analogous molecular beam experiments.

A semiquantitative calculation has been made of the cross-section for the quenching of $Na(3^2P)$ by molecular nitrogen as a function of initial kinetic energy and of final vibrational quantum number, v_f of the nitrogen molecule.[109] The large observed cross-section, which is of gas-kinetic order, can be explained in terms of an intermediate ionic state, involving Na^+ and N_2^-. This state is unstable at infinite separation of Na and N_2, but because of the Coulomb attraction it becomes stable at collision distances below about 3 Å. As a result of the vibrational structure of both the intermediate and final states, the reaction can be treated in terms of a diffusion of the probability flux through a two-dimensional network of potential energy curves parametrised by both the electronic state and also the vibrational quantum numbers v^- and v_f. At each potential-energy curve crossing, the transition matrix element for insertion into a Landau–Zener type of transition probability can be computed. Results are presented on the quenching of $Na(4^2P)$ by N_2, and on the quenching of $Na(3^2P)$ by CO. All the results have the same general character: the total cross-section is of gas-kinetic order and depends only weakly on kinetic energy. The partial cross-sections for excitation of the different final vibrational levels ν show a rather broad distribution, with somewhat more than half the energy of electronic excitation ending up as vibrational excitation.

The resonance fluorescence of Cu atoms in the gas phase quenched by H_2, N_2, CO_2, and Ar has been investigated.[110] Cross-sections in Å^2 for the $^2P^0_{\frac{1}{2}}$ state were 22, 14, 50, and 0.2 respectively, and for the $^2P^0_{\frac{3}{2}}$ state 23, 19, 36, and 0.8. The shifts produced by argon quenching in the emission lines of some strong lines of Al, Ca, Cu, Mn, Mo, and Na have been

[108] L. E. Brus, *J. Chem. Phys.*, 1970, **52**, 1716.
[109] E. Bauer, E. R. Fischer, and F. R. Gilmore, *J. Chem. Phys.*, 1969, **51**, 4173.
[110] R. Bleekrode and W. van Bentham, *J. Chem. Phys.*, 1969, **51**, 2757.

described.[111] Atomic and molecular fluorescence from laser-excited diatomic caesium and rubidium has been observed.[112]

The measurement of the absorption and emission spectra of tin iodide (SnI_2) vapour in the u.v. and visible regions has allowed a tentative identification of the energy levels of states responsible. These are summarised in Figure 11.[113] The band spectrum of tin mono-iodide has also

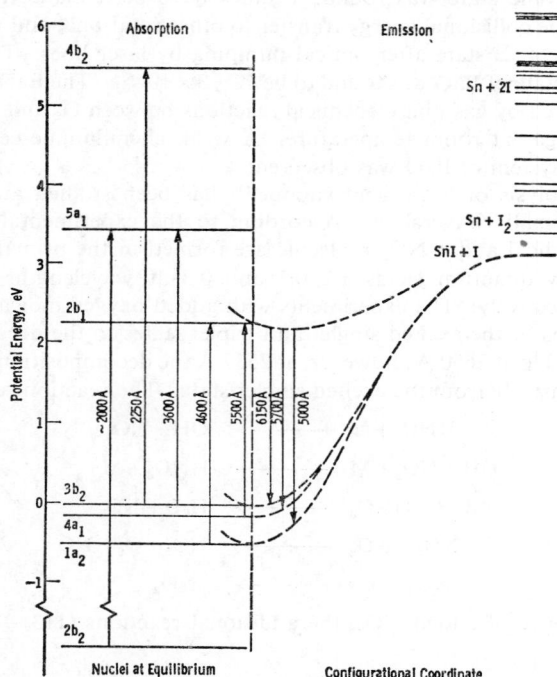

Figure 11 *Interpretation of SnI_2 spectra*
(Reproduced by permission from *J. Chem. Phys.*, 1969, **50**, 3281)

been described,[114] and the visible bands of SnBr have been assigned to the $X^2\Pi \rightarrow A^2\Sigma^+$ transition of the molecule.[115] The laser-induced fluorescence of BaO in the gas phase has been studied.[116] Fluorescence spectra of the $A\,^1\Sigma - X\,^1\Sigma$ system of BaO excited by seven visible Ar ion laser lines have

---

[111] Q. A. Holmes, M. Takeo, and S. Y. Ch'en, *J. Quant. Spectroscopy Radiative Transfer*, 1969, **9**, 769.
[112] M. McClintock and L. C. Balling, *J. Quant. Spectroscopy Radiative Transfer*, 1969, **9**, 1209.
[113] R. J. Zollweg and L. S. Frost, *J. Chem. Phys.*, 1969, **50**, 3281.
[114] A. A. Murthy and P. B. V. Haranath, *Current Sci.*, 1969, **38**, 211.
[115] A. Chatalic, P. Deschamps, and G. Pannetier, *Compt. rend.*, 1969, **269 C**, 584.
[116] K. Sakurai, S. E. Johnson, and H. P. Broida, *J. Chem. Phys.*, 1970, **52**, 1625.

been observed over a wavelength region from the exciting lines to beyond 1000 nm. The strongest fluorescence was observed from $v' = 8$, $J' = 49$ excited by 488·0 nm and $v' = 7$, $J' = 3$ and 7 excited by 496·5 nm. The rotational and vibrational assignments of the various excitation transitions have been made, and rotational and vibrational constants for the lower electronic state were obtained. A rotational perturbation in $v' = 7$ of the upper electronic state was found. Studies have been made from 0·4 to 40 Torr of the collisional energy transfer to other rotational and vibrational levels of the $A\ ^1\Sigma$ state after optical pumping by laser lines. The lifetime of the $A\ ^1\Sigma$ state of BaO was found to be $12 \pm 3 \times 10^{-6}$ s. The BaO molecules were produced by gas-phase chemical reactions between O_2 and Ba vapour in an inert gas at room temperature. A weak chemiluminescence of the $A\ ^1\Sigma - X\ ^1\Sigma$ system of BaO was observed.

The photolysis of nitric acid vapour [117] has been studied at 2650 and 2537 Å at small conversions. According to the experimental results, a hydroxyl radical and a NO_2 molecule are formed in the primary process. The primary quantum yields are 0·1 and 0·3 at wavelengths 2650 and 2537 Å respectively. The experiments with added oxygen indicate that the contributions of the excited singlet and triplet states to the decomposition are comparable at 2650 Å; however, at 2537 Å the decomposition originates almost exclusively from the excited singlet state. The reaction sequence is:

$$HNO_3 + h\nu \longrightarrow OH + NO_2 \qquad (108)$$

$$OH + NO_2 + M \longrightarrow HNO_3 + M \qquad (109)$$

$$OH + HNO_3 \longrightarrow H_2O + NO_3 \qquad (110)$$

$$NO_3 + NO_2 \longrightarrow NO_2 + O_2 + NO \qquad (111)$$

$$NO_3 + NO \longrightarrow 2NO_2 \qquad (112)$$

In the presence of added NO_2, the additional reactions (113)—(119) were considered.

$$NO_2 + h\nu \longrightarrow NO + O \qquad (113)$$

$$O + HNO_3 \longrightarrow OH + NO_3 \qquad (114)$$

$$O + NO_2 \longrightarrow NO + O_2 \qquad (115)$$

$$O + NO_2 \longrightarrow NO_3^* \qquad (116)$$

$$NO_3^* \longrightarrow NO_2 + O \qquad (117)$$

$$NO_3^* + M \longrightarrow NO_3 + M \qquad (118)$$

$$NO_3^* + NO_2 \longrightarrow NO_2 + O_2 + NO \qquad (119)$$

Upon the addition of CO and H_2 several other processes occur.

$$OH + H_2 \longrightarrow H_2O + H \qquad (120a)$$

$$OH + CO \longrightarrow CO_2 + H \qquad (120b)$$

[117] T. Berces and S. Foevgeteg, *Trans. Faraday Soc.*, 1970, **66**, 633, 640, 648.

$$H + NO_2 \longrightarrow OH + NO \qquad (121)$$

$$H + HNO_3 \longrightarrow H_2O + NO_2 \qquad (122)$$

$$H + HNO_3 \longrightarrow H_2 + NO_3 \qquad (123)$$

$$H + HNO_3 \longrightarrow OH + HNO_2 \qquad (124)$$

$$HNO_2 + HNO_3 \longrightarrow H_2O + 2NO_2 \qquad (125)$$

The flash photolysis of dimethylmercury produces excited methyl radicals whose relaxation has been studied.[118]

The photolysis of ethyl-lithium with mercury resonance radiation apparently proceeds by two competing photolytic mechanisms: a lithium hydride elimination which yields ethylene, and a homolytic process which yields lithium metal, ethane, and ethylene.[119a] The absence of butane and deuteriated ethane (when the photolysis is carried out in C_6D_6 or C_6D_{12}) indicates that the homolytic process occurs *via* an intra-aggregate disproportionation. Photolysis in the solid state yields ethane, ethylene, and butane as well as a polymeric material. Solution photolysis in the presence of a mercury pool yields only ethane (and no LiH). [In comparison, the thermolysis of alkyl-lithium compounds appears to involve only hydride elimination.[119b]]

The adsorption of tetraethyl-lead on the surface of the quartz crystal of an electronic microbalance and the deposition rate on the crystal as a function of light intensity and pressure during the photolysis of tetraethyl-lead (TEL) have been measured.[120]

The primary reaction in the gas-phase photolysis of TEL is:

$$PbEt_4 + h\nu \longrightarrow PbEt_3 \cdot + Et \cdot \qquad (126)$$

and subsequent reactions may take place by the sequential removal of ethyl radicals followed by combination and disproportionation of these radicals. Evidence for this reaction sequence was provided by the complete oxidation of the gaseous products and the conservation of the overall rate of reaction when oxygen was added. Ethyl radicals spontaneously react with oxygen and a similar process may also occur with unstable organometallic intermediates. The removal of these radicals would affect secondary chain processes if they occurred and the maintenance of the overall reaction rate implied that photon absorption was the rate-determining reaction in the gaseous photolysis of TEL. Kinetic measurements showed that the deposition rates on the quartz crystal microbalance were controlled by the gas-phase reaction, but a secondary surface reaction controlled the delineation of the deposit. Oxygen reduced the amount of TEL adsorbed and increased the photodeposition rate by a secondary surface process.

---

[118] A. B. Callear and H. E. Van-den-Bergh, *Chem. Phys. Letters*, 1970, **5**, 23.
[119] a W. H. Glaze and T. L. Brewer, *J. Amer. Chem. Soc.*, 1969, **91**, 4490. b D. Bryce-Smith, *J. Chem. Soc.*, 1955, 1712.
[120] L. J. Rigby, *Trans. Faraday Soc.*, 1969, **65**, 2421.

Carbon was detected in the photodeposit from pure TEL, but a photodeposit obtained in oxygen did not contain either carbon or oxygen.

A free-radical trapping technique has been described which can be used to detect and identify reactive short-lived free radicals in reacting systems by obtaining the e.s.r. spectrum of the radical addition product (a 'spin adduct') of phenyl N-t-butylnitrone (a 'spin trap').[121] The nitrogen and β-hydrogen hyperfine coupling constants of the spin adduct (an α-substituted benzyl t-butylnitroxide) are a unique set of parameters for each reactive radical trapped. Verification for the structure of the spin adduct can be readily obtained for alkyl and aryl spin adducts by an addition reaction between the organolithium or Grignard compound of corresponding structure and phenyl t-butylnitrone followed by oxygen oxidation. It has been shown that in the photolysis of organo-lead, -tin, and -mercury compounds the order of cleavage is phenyl > alkyl ≫ acetate or halide. Phenyl- or n-butyl-carbon–tin bond cleavage occurs more readily than tin–tin bond cleavage. Stable radicals, e.g. triphenylmethyl, are not trapped. Very short-lived radicals, e.g. acetoxyl, are trapped without difficulty.

Photolysis of tetra-alkyl (methyl, ethyl, and n-propyl) tin compounds produces free radicals according to reactions (127)—(129).

$$R_4Sn + h\nu \longrightarrow R_3Sn\cdot + R \qquad (127)$$

$$2R_3Sn\cdot \longrightarrow R_3SnSnR_3 \qquad (128)$$

$$R_3SnSnR_3 + h\nu \longrightarrow R_3SnSnR_2 + R\cdot \qquad (129)$$

The photolysis of tetraphenyl cyclobutadiene nickel bromide has been described.[122] Other miscellaneous reactions reported, not necessarily in the gas phase, include the photochemical exchange of borazine with deuterium,[123] the photoreaction between difluoroazine and sulphur dioxide,[124] the synthesis of trifluoroamine oxide by photochemical fluorination of nitrosyl fluoride,[125] and the u.v. spectrum and photolysis of diphosphorus tetra-iodide and phosphorus.[126]

A description of singlet–triplet transitions in cyclic molecules has been presented such that the selection rules take full advantage of the cyclic symmetry.[127] It is found that unlike singlet–singlet transitions, all component triplet states of a cyclic tetramer have allowed transitions to the ground state in electric dipole radiation. The relative intensities of transitions to the various exciton-like states is determined by the geometrical transformation involved in switching the axis of spin quantisation

[121] E. O. Janzen and B. J. Blackburn, *J. Amer. Chem. Soc.*, 1969, **91**, 4481.
[122] T. Hosokawa and I. Moritani, *Bull. Chem. Soc. Japan*, 1970, **43**, 959.
[123] M. P. Nadler and R. F. Porter, *Inorg. Chem.*, 1969, **8**, 599.
[124] M. Lustig, *Inorg. Nuclear Chem. Letters*, 1969, **5**, 723.
[125] W. B. Fox, J. S. Mackenzie, and R. Vitek, *Inorg. Nuclear Chem. Letters*, 1970, **6**, 177.
[126] T. Kennedy and R. S. Sinclair, *J. Inorg. Nuclear Chem.*, 1970, **32**, 1125.
[127] R. M. Hochstrasser and A. P. Marchetti, *J. Chem. Phys.*, 1970, **52**, 1360.

from the monomer to the cyclic tetramer basis. The theoretical predictions are convincingly exemplified by the $T_1 \leftarrow S_0$ low-temperature absorption spectra of crystals of the tetraphenyl Group IV (C, Si, Ge, Sn, and Pb) compounds, all of which have S_4 symmetry. The absorption spectra of the singlet states of neat crystals of these materials have also been presented and these also demonstrate the cyclic exciton selection rules, but in this case only two of the four states are seen as predicted. The tetraphenyl-X triplets all show four exciton-like states—^3A, ^3B, and ^3E (doubly degenerate). The total splittings are: X = Si, $^3E-^3B = 1\cdot1$ cm^{-1}; X = Ge, $^3E-^3B = 2\cdot8$ cm^{-1}; X = Sn, $^3E-^3A = 7\cdot4$ cm^{-1}; X = Pb, $^3E-^3A = 6\cdot6$ cm^{-1}. The ^3E state is always at highest energy. In the singlet states only the E and B states are seen and the splittings are: X = Ge, $^3B-^3E = 11$ cm^{-1}; X = Sn, $^3B-^3E = 10$ cm^{-1}; and X = Si, $^3B-^3E = 6$ cm^{-1}. The spectra in high magnetic fields (Zeeman effect) confirm the detailed assignments and the theoretical model. The spin–orbit interaction between different monomer-based triplet states was also observed.

4 Solid-phase Luminescence and Photoreactions

Possibly the solid-phase photochemical reaction of greatest interest commercially is the effect of light on the silver halides. A review of progress in the use of this photosensitive material has appeared.[128] The effect of photolytic silver on the silver halide lattice has been discussed,[129] and the photolysis of silver halides formed by the decomposition of their ammoniates described.[130] The phototarnishing of silver by alkyl iodides has also been reported.[131]

The photolysis of tetracarbonylnickel [Ni(CO)$_4$] in rare-gas matrices at 15 K has been shown to produce Ni(CO)$_3$.[132]

Absorption and laser-excited fluorescence spectra of CuO trapped in various matrices have been observed.[133] The matrix data and previously observed gas-phase data are consistent with the assignment in Table 7 for the lowest states of CuO.

Table 7

	ν_{00} (matrix) (cm^{-1})	ν_{00} (gas) (cm^{-1})	ν (matrix) (cm^{-1})	(cm^{-1})
C		23 550		
$B(^2\Sigma$?)	20 490	20 953	624	
$A(^2\Pi$?)	3900	4460	605	($^2\Pi_{\frac{1}{2}}-^2\Pi_{\frac{3}{2}} = 275$)
$X(^2\Pi)$	0	0	665	($^2\Pi_{\frac{1}{2}}-^2\Pi_{\frac{3}{2}} \simeq 200$)

[128] F. W. H. Mueller, *Photograph. Sci. Eng.*, 1970, **14**, 157.
[129] F. Orban, *Bull. Soc. roy. Sci. Liège*, 1969, **38**, 481.
[130] F. Orban, *Bull. Soc. roy. Sci. Liège*, 1969, **38**, 476.
[131] R. F. Cross, P. T. McTique, and D. J. Young, *Trans. Faraday Soc.*, 1969, **65**, 3355.
[132] A. J. Rest and J. J. Turner, *Chem. Comm.*, 1969, 1026.
[133] J. S. Shirk and A. M. Bass, *J. Chem. Phys.*, 1970, **52**, 1894.

It was shown that the matrix-isolation technique can be used to 'tune' an absorption into coincidence with a laser line in order to observe fluorescence. The vibrational relaxation of CuO in a solid matrix requires on the order of 10^5 vibrations.

The electronic structure of luminescence centres of ZnS phosphors activated with impurity ions of S^2 configuration has been studied extensively. The impurity ions studied were Bi, and Pb in both zinc-blende and wurtzite forms.[134]

The excitation, reflection, and fluorescence spectra and glow curves of various ZnS : Bi phosphors were measured accurately. A, B, and C bands with peaks at 6200, 4350, and 4060 Å (room-temperature values), respectively, were found in the visible region of excitation or reflection spectra. These bands were analysed by using molecular-orbital theory. The positions of A, B, and C bands and the calculated ratio of the dipole strength of C band to that of A band are consistent with each other. Thus, it is possible to conclude unambiguously that the resonance transitions between the ground $[A_1(^1S_0)]$ and excited states $[T_2(^1P_1), ET_1(^3P_2),$ and $T_2(^3P_1)]$ of the localised Bi^{3+} centre produce the A, B, and C bands or the observed red (7600 Å), orange (6200 Å), and green (5100 Å) emission bands.

A, B, and C bands with peaks at about 4910, 4220, and 3990 Å (mean values for split sub-bands at 77 K), respectively, were found in the excitation or reflection spectra of visible region of ZnS : Pb (zinc-blende) phosphors. These bands were also analysed by means of molecular-orbital theory and identified unambiguously as those due to the resonance transitions from the ground state $(A_1, {}^1S_0)$ to the excited states $(T_2, {}^3P_1; T_1$ and $E, {}^3P_2;$ and $T_2, {}^1P_1$, respectively) in the localised Pb^{2+} centre at lattice site. The observed orange (6400 Å), green (5000 Å), and yellow (5850 Å) emission bands were also unambiguously correlated to the inverse transitions to the A, B, and C absorption transitions, respectively. The observed singlet, doublet, and triplet structure of C, B, and A excitation bands, respectively, or the triplet structure of the green emission band were interpreted in detail in terms of dynamic Jahn–Teller effect.

In the accurate measurement of excitation and reflection spectra of Pb-activated wurtzite phosphors, $A(A_1, A_2, A_3)$, B, and C absorption or excitation bands with peaks at about 5070 (5360, 5050, 4800), 4240, and 4060 Å (room-temperature values), respectively, were found in the visible region of the spectra. Quantitative analysis of these bands justifies again the ligand-field model established for impurity-activated ZnS phosphors. A number of bands or lines observed in the u.v. region of excitation spectra were also analysed satisfactorily in terms of charge-transfer and exciton- or phonon-coupled exciton transitions. Unambiguous assignment for the observed green, yellow, orange, and red emission bands was given.

The energy levels and electronic structure of the three phosphors are shown in Figure 12. Planar electron traps and i.r. stimulation in zinc

[134] Y. Uehara, *J. Chem. Phys.*, 1969, **50**, 961; **51**, 4385, 4401.

sulphide phosphors,[135] and luminescence damage in ZnS : Cu phosphors [136] have been discussed. Phosphorescence emission in sulphoselenides Zn $(S_{100-n}Se_n)$: Cu, I has been observed.[137]

Single crystals of CdF_2 doped with Eu^{3+} have been shown by fluorescence and absorption spectra to contain no detectable Eu^{2+}, irrespective of the

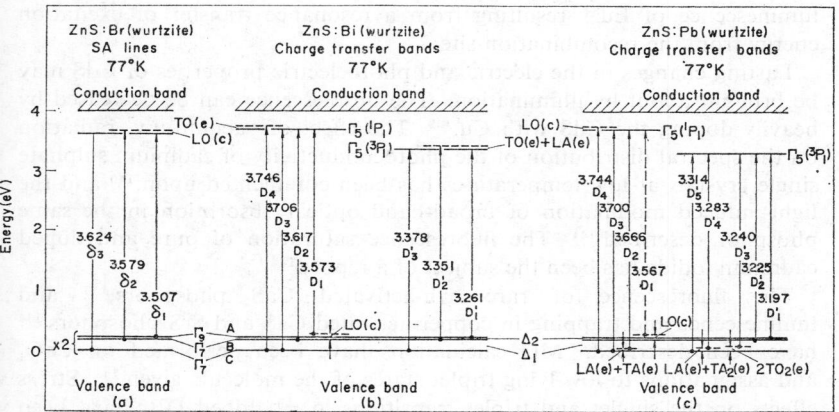

Comparison of electronic structure and mode of transitions for charge-transfer bands or lines in ZnS:Br, ZnS:Bi, and ZnS:Pb phosphors

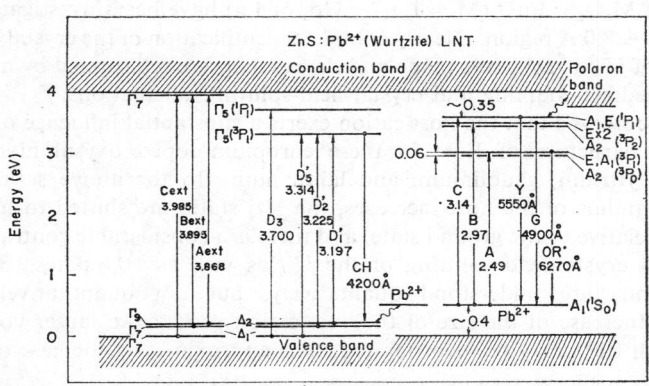

Schematic energy-level diagram of ZnS:Pb phosphors (at 77 K). Levels split by the Jahn-Teller effect or trigonal field and also phonon-coupled transitions are omitted for the sake of simplicity

Figure 12
(Reproduced by permission from *J. Chem. Phys.*, 1969, **51**, 4385)

---

[135] N. Riehl, G. Baur, and L. Mader, *Z. Naturforsch.*, 1969, **24a**, 1296.
[136] R. Grasser and A. Scharmann, *Z. Naturforsch.*, 1969, **24a**, 937.
[137] A. G. Goldman, B. N. Korolko, and S. F. Lysenko, *Optics and Spectroscopy*, 1968, **25**, 324.

presence of Na^+ or Y^{3+} as added impurities.[138] U.v. irradiation at 77 K produced a new broad absorption band near 3400 Å, which disappeared on warming, with associated thermoluminescence. A complex e.s.r. signal associated with this band is attributed to Eu^{2+}, formed during the irradiation. A mechanism proposed for the luminescence processes involves formation of Eu^{2+} in the excitation and trapping steps, with the luminescence of Eu^{3+} resulting from a resonance transfer of excitation energy from the recombination site.

Lasting changes in the electric and photoelectric properties of CdS may be brought about by illumination. The obsolescence can be decreased by heavily doping the CdS with Cu.[139] The effect of vacuum-u.v. radiation on the spectral distribution of the photoconductivity of cadmium sulphate single crystals at low temperatures has been commented upon,[140] and the light-induced modulation of broad-band optical absorption in the same phosphor described.[141] The fluorescence saturation of pure and doped cadmium iodide has been the subject of a report.[142]

The fluorescence of rare-earth-activated CaS phosphors,[143] and luminescence and trapping in copper-activated CaS and SrS phosphors[144] have been described. MO calculations have been performed on CaO, and assignments to low-lying triplet states of the molecule given.[145] Stress effects on the singlet and triplet transitions in π^+ doped DBr have been reported.[146]

Phosphors based on europium ions are widespread. Fluorescence spectra of $M_2O_2S:Eu^{3+}$ (M = Lu, Y, Gd, or La) have been investigated in the 4000—9000 Å region. Nearly complete identification of the crystal components of $^7F_{0-6}$ $^5D_{0-3}$ was made and the results were interpreted by means of intermediate-coupling and crystal-field-splitting calculations.[147]

It was observed that the host cation exerts a substantial influence on the electronic structure of Eu^{3+} in these europium-doped oxysulphides of lutetium, yttrium, gadolinium, and lanthanum. In the above series, as the ionic radius of the host increases, the ^5D_J states are shifted to higher energies relative to the ground state, and there is a considerable contraction of overall crystal-field splitting of the ^7F_J as well as ^5D_J states. These observations are understood qualitatively, but not quantitatively at present. Increase of the size of the host cation produces a larger volume into which Eu^{3+} may be incorporated, thus approximating the case of the

[138] E. Banks and R. W. Schwartz, *J. Chem. Phys.*, 1969, **51**, 1956.
[139] S. Kanev, V. Stoganov, and M. Lakova, *Doklady Bolg. Akad. Nauk.*, 1969, **22**, 863.
[140] R. V. Grigor'ev, B. V. Novikov, and A. E. Cherednichenko, *Vestnik. Leningrad Univ. Fiz. Khim.* 1970, **70**, 75.
[141] E. J. Conway, *J. Appl. Phys.*, 1970, **41**, 1689.
[142] A. T'Kint de Roodenbeke, M. F. Coillot-Demay, and A. Lashley, *J. Chim. phys.*, 1969, **66**, 1073.
[143] B. R. Malhotra and D. R. Bhawalkar, *Indian J. Pure Appl. Phys.*, 1969, **7**, 573.
[144] D. Sharma and A. Singh, *Indian J. Pure Appl. Phys.*, 1969, **7**, 310.
[145] K. D. Carlson, K. Kaiser, C. Moser, and A. C. Wahl, *J. Chem. Phys.*, 1970, **52**, 4678.
[146] W. Dultz and W. Gebhardt, *Solid State Comm.*, 1969, **7**, 1153.
[147] O. J. Sovers and T. Yoshioka, *J. Chem. Phys.*, 1970, **51**, 5330.

free ion, in which the ^5D_J levels are considerably elevated and the crystal-field splitting non-existent.

Elevation of the barycentres of $^6P_{\frac{7}{2}}$ and $^6P_{\frac{5}{2}}$ in Gd^{3+}, as well as a contraction of their crystal-field splitting with increasing host cation radius, was observed in gadolinium-doped fluorides of Cd, Ca, Sr, and Ba. In some incomplete work on europium-doped vanadates and oxides, the same qualitative variation has been observed; this leads to the hypothesis that the effect may hold quite generally, independent of host anion as well as of the activator ion. If further accumulation of experimental data supports this hypothesis, then it will be quite useful in semiquantitative systematisation of energy levels of lanthanide-doped solids.

The status of quantitative explanation of these observations is considerably less satisfactory. Separate theoretical calculations of the free-ion and crystal-field split europium energy-level structures yield agreement with experiment to within 10 cm^{-1} or better. Despite this agreement, the parametrised intermediate coupling calculations employing a spin–orbit plus electrostatic repulsion Hamiltonian do not yield parameters for these two interactions that vary in a reasonable fashion along the series of four europium-doped oxysulphides. Apparently the quantitative key to the phenomenon is to be sought in some other intra-atomic or covalent effect or in the modification of intra-atomic interactions by the crystal field. The main requirements of a theory which would explain the above observations are at present unknown. They are: (i) information concerning the geometry of the crystal in the vicinity of the doping ion, and (ii) adequate wavefunctions for the atomic orbitals of the lanthanide ion. As such information becomes available, it will be quite useful in formulating a detailed theory of rare-earth-doped solids.

The luminescence of europium in a fluoroberyllate glass,[148] and in EuSc[149] has been reported. Evidence has been presented for Eu^{3+} emission from two symmetry sites in Y$_2$O Eu^{3+}.[150] Luminescence properties of rare-earth tellurates have also been described.[151]

The luminescence of the unactivated and Bi^{3+}- and Eu^{3+}-activated compounds ABO$_4$ (A is Sc or In; B is P, V, or Nb)[152] has been reported, and compared with the luminescence of the corresponding rare-earth and yttrium compounds. ScNbO$_4$ shows a weak, blue emission, ScVO$_4$ a blue-green, and InVO$_4$ a green emission. The other host lattices show no or only a very weak emission at room temperature. The colour of the vanadate emission depends strongly on the choice of A. This was studied by replacing part of the A ions by other ions or a combination of ions. The shift of the vanadate emission is ascribed to two effects, viz. metal–

---

[148] I. V. Kovaleva, V. P. Kolobkov, G. T. Petrovskii, and G. A. Tsurikova, *Zhur. priklad. Spektroskopii*, 1969, **10**, 805.
[149] P. Streit and P. Wachler, *Helv. Phys. Acta*, 1969, **42**, 606.
[150] H. Forest and G. Ban, *J. Electrochem. Soc.*, 1969, **116**, 474.
[151] S. Natansohn, *J. Electrochem. Soc.*, 1969, **116**, 1250.
[152] G. Blasse and A. Bril, *J. Chem. Phys.*, 1969, **50**, 2974.

metal interaction (*e.g.* Bi^{3+}–VO_4) and the effect of the size of the metal ions at the A sites (*e.g.* Sc^{3+}). The Eu^{3+}-activated phosphors show some interesting features. The intensity of the forced electric dipole emission from the Eu^{3+} ions seem to depend strongly on the position of the lowest strong absorption band. In the zircon structure the $^5D_0$–$^7F_2$ emission is hypersensitive to the surroundings, although no linear crystal field component can be present. Both (In, Eu)PO_4 and (In, Eu)VO_4 show an orange ($^5D_0$–$^7F_1$) emission due to the presence of inversion symmetry at the In^{3+} site.

Excitation spectra of the $^5D_4 \to {}^7F$ emission of Tb^{3+} have been measured in Sc(or In)BO_3, Y(or Gd)BO_3, $Y_3Ga_5O_{12}$, Sc(or Y)PO_4, and Sc_2(or Y_2)O_2 at room temperature.[153] The $4f$–$5d$ excitation band was distinguished from the host lattice excitation band by inspection of the concentration dependences of the excitation spectra. The energy and the shape of the $4f$–$5d$ excitation band are almost the same in the lattices with the same structure. The band splittings are related to the symmetry of the crystal field acting on Tb^{3+}. The $4f$–$5d$ excitation band is strongly enhanced by lowering the temperature to 77 K.

In the series of the trivalent lanthanides, Ce^{3+} has the smallest separation of the lowest levels of the $4f^n$ and the $4f^{n-1}5d$ configurations. Then the lowest $4f$–$5d$ excitation energy increases with increasing n until $n = 7$ (Gd^{3+}), and then it switches down to the lower value for $n = 8$ (Tb^{3+}) and again increases with n until $n = 14$ (Lu^{3+}). The charge-transfer band has the opposite trend for the variation of n. The Tb^{3+} has the low $4f$–$5d$ excitation energy and the high charge-transfer one in solids, and the latter band is usually hidden by host lattice absorption.

The energy levels of the $4f^7(^8S_{\frac{7}{2}})5d$ configuration of a free Gd^{2+} ion which is isoelectronic with Tb^{3+} are described in terms of LS coupling rather than jj coupling, and the resulting energy levels are ^7D and ^9D, the latter lying lower. The energy separation between ^7D and ^9D is about 8250 cm^{-1} in Gd^{2+}. When the Tb^{3+} ion is embedded in a lattice, the crystal electric field acting on $5d$-electrons is so strong that it exceeds the Coulomb interaction of the $5d$-electron with the $4f^7$ core and the interaction of $4f$-electrons with the crystal electric field. The crystal-field splitting of the $5d$-orbital level is of the order of 10^4 cm^{-1}. Therefore, in order to determine the energy levels of $4f^7\,5d$, the secular determinant of the Hamiltonian for the crystal-field potential of the $5d$-electron and the Coulomb interaction between the $5d$-electron and the $4f^n$ core must be determined. The calculation is, however, not easy to perform and instead it can merely be said that the lower group of the $4f$–$5d$ excitation band may be involved with the lower levels of the crystal splitting of the $5d$-electron level as is discussed below, and may have predominantly the character of the spin-forbidden transition of $4f^8\,{}^7F \to 4f^7(^8S_{\frac{7}{2}})5d\,^9D$, for the case of Gd^{2+} the ^9D level lies lower than D. The restriction of this

---
[153] T. Hoshina, *J. Chem. Phys.*, 1969, **50**, 5158.

spin-forbidden transition is partly removed by admixing of the ^7D character to ^9D by the spin–orbit interaction. A weak absorption band has been observed at 271·5 nm in addition to the strong one at 233·8 nm in $TbCl_6^{3-}$, and there is also a weak band at 278 nm in $TbBr_6^{3-}$. These weak bands are attributed to the spin-forbidden transitions. The lower energy group of the $4f$–$5d$ excitation band lies close to the same energy region as these weak absorption bands.

The d-orbital level is split into the t_2 and e levels by the crystal field of cubic symmetry. In an octahedral ligand co-ordination around Tb^{3+} as in Sc(or In, Y, Gd)BO_3, the t_2 level lies lower. The orbital triplet t_2 is split further into three by the field of lower symmetry of S_6($ScBO_3$, $InBO_3$) or by the field of D_3(YBO_3 and $GdBO_3$) and the spin–orbit interaction. The three peaks seen in the lower group of the $4f$–$5d$ band in these crystals may correspond to these crystal-field splittings of the t_2 level.

In Sc(or Y)PO_4, four nearest-neighbour oxygen ions are co-ordinated tetrahedrally, and in $Y_3Ga_5O_{12}$ eight nearest-neighbour oxygen ions are situated at the vertices of a (skewed) cube. In these cases the lower level is e, and it is split into two by the field of lower symmetry. The observed lower band was a simple broad one. The splittings may be hidden in the overall band shapes.

In Sc_2(or Y_2)O_3, the different cation sites correspond to the two ways of removing oxygen ions from the cubicly co-ordinated eight oxygen ions. In these cases, the e level lies lower and is split into two by the field of lower symmetry of C_2 or S_6. The splittings were not observed by the band broadening and the poor separation of the bands for the two different sites.

The lowest $4f$–$5d$ excitation energy of Tb^{3+} is always smaller than the highest $4f$–$4f$ excitation energy. Such low energy of the $4f^7\,5d$ configuration of Tb^{3+} results in the high degree of the $4f$–$5d$ mixing by odd-parity crystal-field terms and, therefore, the probability of the electric dipole transition can become large in Tb^{3+} in solids; while in Eu^{3+}, which has the higher $4f$–$5d$ excitation energy, the degree of the $4f$–$5d$ mixing is so small that in the lattices studied, except for Y_2O_3 and Sc_2O_2, the intensity of $^5D_0 \rightarrow {}^7F_1$ (magnetic dipole transition) is stronger than that of $^5D_0 \rightarrow {}^7F_2$ (electric dipole transition).

The luminescence of Tb^{3+}-activated CeF_3, $CePO_4$, $CeBO_3$, and $CeAlO_3$ has been investigated.[154] Energy transfer from the Ce^{3+} host lattice to the Tb^{3+} ion occurs by Ce^{3+}–Ce^{3+} and Ce^{3+}–Tb^{3+} transfer. From a study of the Ce^{3+} luminescence in the analogous La compounds it was concluded that the probability of the former transfer is different for these host lattices. An analogy with Eu^{3+}-activated host lattices was pointed out.

The optical absorption, luminescence, and excitation spectra of $LiNbO_3$ and $LiTaO_3$ doped with chromium impurities have been examined

[154] G. Blasse and A. Bril, *J. Chem. Phys.*, 1969, **51**, 3252.

as a function of temperature.[155] Several interesting differences between these spectra and those of other chromium-doped oxides were noted. The most important differences are the large half-width of the Cr^{3+} ion 'R' lines (~ 50 cm^{-1} at 4·2 K) and the absence of any 'R' line luminescence down to 4·2 K. Instead a broad-band emission peaking at about 1 μm is observed. It was shown that these differences arise from the low value of the crystal-field parameter Dq at the Cr^{3+} ion site in LiNbO$_3$ and LiTaO$_3$, so that the zero-phonon $^4T_2$ state of the Cr^{3+} ion lies at a lower energy than the ^2E state. Point-charge calculations of Dq for several oxides, including LiNbO$_3$ and LiTaO$_3$, and other considerations, suggest that the most probable sites of the Cr^{3+} ion impurities are the Nb/Ta sites and not the Li sites.

Luminescence from Bi^{3+}-activated rare-earth orthovanadates,[156] erbium-activated Group II—VI compounds with alkali-metal compensators,[157] gallates,[158] LiF/U^{6+} single crystals,[159] SrTiO$_3$,[160] Ce^{3+}-activated phosphors,[161] rutile[162] and phenacite,[163] and the absorption spectrum of the Tm^{3+} ion in a lanthanum aluminate matrix[164] have been reported.

The effects of temperature on the fluorescence lines of Sm^{2+} in CaF$_2$, and BaClF have been studied.[165] The results for the first two samples were obtained in the region 10—100 K. For CaF$_2$ a strong dependence of the linewidth on temperature was found; this width is $\sim 1·7$ cm^{-1} at 15 K and ~ 40 cm^{-1} at 95 K. For SrF$_2$, the linewidth is constant and equal to $\sim 2·15$ cm^{-1} up to ~ 77 K. This difference in behaviour is explained qualitatively as due to the fact that in CaF$_2$ the metastable state belongs to the $4f^5 5d$ configuration and, therefore, the electron charge is strongly affected by the thermal perturbations of the environment. The widths and positions of several fluorescence lines of BaClF : Sm^{2+} in the 30—600 K region were also measured. The thermal broadening of most of these lines is explained as due to microscopic strains and Raman scattering of phonons: the thermal shifts as due to emission, and absorption of virtual phonons.

A model has been proposed to explain qualitatively the dependence of the quenching temperature of a given fluorescent centre on the choice of the host lattice.[166] It is well known that the quenching temperature of a given luminescent centre with characteristic emission depends strongly on

[155] A. M. Glass, *J. Chem. Phys.*, 1969, **50**, 1501.
[156] G. Boulon, C. Pedrini, F. Gaume-Mahn, and J. Loners, *Compt. rend.*, 1969, **269 B**, 133.
[157] S. Larach, R. E. Shrader, and P. N. Yocom, *J. Electrochem. Soc.*, 1969, **116**, 471.
[158] W. L. Wanmaker and J. W. ter Vrugt, *J. Electrochem. Soc.*, 1969, **116**, 871.
[159] H. Hartmann and A. Sharmann, *Z. Naturforsch.*, 1969, **24a**, 1117.
[160] L. Grabner, *Phys. Rev.*, 1969, **177**, 1315.
[161] A. H. Gomes de Mesquita and A. Bril, *J. Electrochem. Soc.*, 1969, **116**, 871.
[162] A. K. Ghosh, F. G. Wakim, and R. R. Addiss, *Phys. Rev.*, 1969, **184**, 979.
[163] F. Lozykowskii and F. Holuj, *J. Chem. Phys.*, 1969, **51**, 2315.
[164] F. Martin-Brunetiere and J. Fuerxer, *Compt. rend.*, 1969, **268 B**, 1264.
[165] B. Birang, A. S. M. Mahbub'ul Alam, and R. D. Bartolo, *J. Chem. Phys.*, 1969, **50**, 2750.
[166] G. Blasse, *J. Chem. Phys.*, 1969, **51**, 3529.

the chemical composition of the host lattice even in an isomorphous series.

It has been argued that Δr, the difference between the equilibrium distance of the ground and excited state will determine the quenching temperature of the fluorescence to a first and rough approximation. The sign of Δr depends on the nature of the luminescent centre. If an electron of a cation is excited, the electronic charge distribution extends further away from the nucleus. The cation is effectively made more positive and has greater attraction for the neighbouring anions. This means Δr is negative. A well-known example is KCl–Tl$^+$. Upon excitation of the Tl$^+$ ion the luminescent centre shrinks. In the case of anion excitation Δr is predicted to be positive. This is found, for example, upon excitation into charge-transfer absorption bands (tungstate, Eu^{3+}). There are two properties of the host lattice that influence Δr: (a) the size of the host lattice cation for which the activator is substituted, (b) the size and the charge of the cations in the neighbourhood of the activator. If the activator ion is larger than the host lattice ion [e.g. Eu^{3+} ($r = 0.98$ Å) or Ce^{3+} (1.07 Å) in a Lu^{3+} (0.85 Å) host lattice] the surroundings of the activator will be forced to expand. If, upon excitation, Δr is positive, this means a further expansion which will be opposed by the lattice, so that Δr will be relatively small and the quenching temperature T_q high. If, however, Δr is negative, excitation brings about a shrinkage. Since the lattice was expanded in the ground state, it will shrink easily, so that Δr is relatively large and T_q low. The reverse is true if the activator is smaller than the host lattice ions. The influence of the neighbouring cations can be considered. It is obvious to assume that Δr will be relatively small if the luminescent centre is surrounded by a rigid lattice, i.e. a lattice containing small cations with high charge. In such lattices we may, therefore, expect a high T_q.

Table 8 summarises these considerations. For $\Delta r > 0$ the value of T_q will be high if the lattice site of activator is relatively small, and the other

Table 8 *Summary of proposed model for temperature quenching of phosphors*

Δr	Negative e.g. Tl$^+$, Bi^{3+}, Tb^{3+}, Ce^{3+}, Eu^{2+}	Positive e.g. Eu^{3+}, WO$_4$, VO$_4$
Activator > host lattice ion	T_q low	T_q high
Activator < host lattice ion	T_q high	T_q low
Lattice with small ions with high charge	T_q high	T_q high

cations in the lattice are also small. These conditions can be realised so that it should be possible to find relations between T_q and composition. If, however, Δr is negative, a high T_q of the activator emission requires a large lattice site for the activator and at the same time small cations in the lattice. Large lattice sites can only be realised in compounds with large cations. The value of T_q is determined, therefore, by two phenomena that counteract each other, so that for $\Delta r < 0$ no clear relations can be expected. The model has been applied to several phosphors known to exhibit temperature quenching, with qualitative success.

Optical absorption and photoconductivity in thin films of cuprous and lead azides have been studied.[167] On the basis of the spectral evidence, the energy levels shown in Figure 13 were suggested for these two compounds.

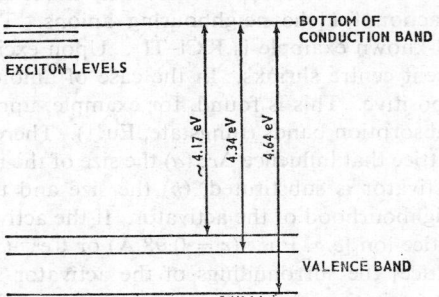

Suggested energy level diagram for CuN_3

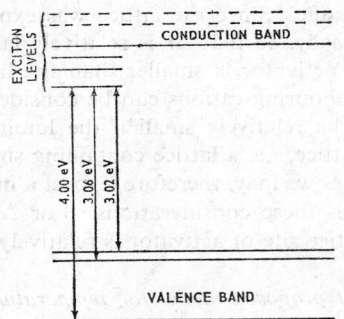

Figure 13 *Proposed energy level diagram for optical transitions in PbN_6*
(Reproduced by permission from *Trans. Faraday Soc.*, 1969, **65**, 3074, 3187)

D. P.

[167] S. K. Deb, *Trans. Faraday Soc.*, 1969, **65**, 3074, 3187.

Part III

ORGANIC ASPECTS OF PHOTOCHEMISTRY

1
Photolysis of Carbonyl Compounds

The long-standing interest in the photochemistry of simple carbonyl compounds has continued throughout the past year. There has been no appreciable shift of effort, and studies of the Norrish Type I and II photoprocesses are still well to the fore as the most popular area for examination.

Numerous review articles have been published. Tezuka [1] has published articles on photoreactions arising from $n \to \pi^*$ excitation of carbonyl compounds. Unfortunately for most readers, these papers are in Japanese, but presumably the work described is an extension of that [2] outlined last year. Reviews dealing with the use of photochemistry in the synthesis of natural products,[3] and the action of light on dyes [4] and cosmetic preparations [5] have appeared.

Interesting publications have dealt with some of the problems associated with the actual irradiation of materials. Some of the problems of the distribution of light in a cylindrical flow reactor have been treated mathematically.[6] It has been calculated that rate constants can be measured to within 5% regardless of the fact that the light in a radial reactor is partially diffuse. Muel and Malpiece [7] have described a system of filters which is suitable for the isolation of narrow wavelength bands between 235 and 300 nm. The interesting report of a modified n.m.r. spectrometer involving the use of an aluminised mirror to direct u.v. light from a broad spectrum lamp vertically down the sample tube has appeared.[8] The use of the mirror means that the sample homogeneity, and consequently the resolution of the spectrometer, is not altered. The technique will be of use for the detection of photochemically produced intermediates.

[1] T. Tezuka, *J. Synthetic Org. Chem. Japan*, 1969, **27**, 309, 430.
[2] T. Tezuka, cited by W. M. Horspool, in 'Photochemistry,' ed. D. Bryce-Smith (Specialist Periodical Reports), The Chemical Society, London, 1970, Vol. I, p. 206.
[3] P. G. Sammes, *Quart. Rev.*, 1970, **24**, 37.
[4] G. S. Egerton and A. G. Morgan, *J. Soc. Dyers and Colourists*, 1970, **86**, 79.
[5] N. Lewin, *J. Soc. Cosmetic Chem.*, 1969, **20**, 761.
[6] T. Matsuura and J. M. Smith, *Amer. Inst. Chem. Engineers J.*, 1970, **16**, 321.
[7] B. Muel and C. Malpiece, *Photochem. and Photobiol.*, 1969, **10**, 283.
[8] T. F. Page, *Chem. and Ind.*, 1969, 1462.

1 Energy Transfer Processes

A review describing the many ways in which an excited state can be deactivated to the ground state has been published.[9] This article should prove to be an extremely useful reference source as well as an authoritative dissertation on the subject. Associated with the problems of deactivation and quenching is the cautionary note [10] concerning the use of penta-1,3-diene, the commonly used quencher for the detection of triplet states in photochemical reactions. At high diene concentrations it is usually assumed that the reduction in the quantum yield is attributable to the triplet process, while the unquenched portion of the reaction arises from the singlet state. However, in the new study [10] it has been found that the fluorescence of pentan-2-one, norcamphor, t-butylmethyl ketone, and acetone is quenched by the same diene. Thus it is necessary to treat results for triplet reactions with a certain amount of caution since at high diene concentrations there may be some quenching of the singlet state, and a correction must be made for such an occurrence. It is suggested that this singlet quenching explains why Stern–Volmer plots of the quenching of Norrish Type I and II reactions have non-zero slopes in the high diene concentration region.

Smith and Agosta [11] report studies on the use of triphenylene as a sensitizer. It has been found that triphenylene ($E_T = 280 \cdot 1$ kJ mol^{-1}; 67 kcal mol^{-1}), often used as a triplet sensitizer, can behave as a singlet sensitizer. The singlet state of this molecule ($E_S = 342 \cdot 8$ kJ mol^{-1}; 82 kcal mol^{-1}) is relatively long lived ($\tau = 36 \cdot 3$ ns) and could allow for

Table 1 *Rate constants (k_q) for quenching of triphenylene fluorescence*[11] *($k_q \times 10^9$, l mol^{-1} s^{-1})*

Quencher ($E_S = 313 \cdot 5$ kJ mol^{-1}; 75 kcal mol^{-1})	Methanol	t-butanol
Cyclohexenone	3·1	1·8
5,5-dimethylcyclopent-2-enone	2·9	3·1
$k_{\text{diffusion}}$	19·0	3·0

energy transfer from this excited state. They [11] have found that energy transfer can take place rapidly (Table 1) as measured by the quenching of triphenylene fluorescence. The authors suggest that considerable caution be used in the use of this compound as a triplet sensitizer.*

---
[9] L. M. Stephenson and G. S. Hammond, *Angew. Chem.*, 1969, **81**, 279 (*Internat. Edn.*, 1969, **8**, 261).
[10] F. S. Wettack, G. D. Renkes, M. G. Rockley, N. J. Turro, and J. C. Dalton, *J. Amer. Chem. Soc.*, 1970, **92**, 1793.
[11] A. B. Smith and W. C. Agosta, *Chem. Comm.*, 1970, 466.

* A recent authoritative text suggests the use of triphenylene as a triplet sensitizer. ('Energy Transfer and Organic Photochemistry,' by A. A. Lamola and N. J. Turro, Wiley, New York, 1969, p. 94; and T. R. Evans, *ibid.*, p. 344.)

The use of ketones as triplet energy donors continues to be of interest. The calculations of Hoffmann and Swenson [12] on the equilibrium geometries for benzophenone in the ground and the $n\pi^*$ states are pertinent to this point. It appears that in the ground state the phenyl rings are twisted out of plane by 38°, while in the excited state the angles of twist are 32°. The carbonyl group remains planar in the excited state. Benzaldehyde has been calculated to have a planar configuration.

Studies by Chapman and Wampfler on energy transfer from benzophenone and acetophenone have shown that triplet energy transfer from aromatic ketones which have long-lived triplets arising from a $\pi \to \pi^*$ state is concentration dependent.[13] The test system used for the study was the rearrangement of 4,4-dimethylcyclohexenone to 6,6-dimethylbicyclo[3,1,0]hexan-2-one and 3-isopropyl-2-cyclopentenone which could be effectively sensitized by acetophenone and benzophenone or quenched by di-t-butylnitroxide. The efficiency of the sensitizers used with relation to the concentration is shown in Table 2. The reason for the dependence is not fully understood, but it is suggested that excimer formation or self-quenching could play a major part in the process.

Table 2 *Concentration dependence of the efficiency of triplet energy transfer*[13]

Sensitizer	Conc. (mol l^{-1})	Φ_0/Φ_{sens}	Φ_{redn}	τ_{rad} (s)
p-Methoxyacetophenone	0.10	1.03	0.04^a	0.38^a
	0.30	1.08		
	0.65	1.05		
	1.00	1.29		
m-Methoxyacetophenone	0.02	1.04		
	0.10	1.14		
	0.20	1.39	0.006^a	0.71^a
	0.60	2.25		
	1.00	3.09		
3,4-Methylenedioxyacetophenone	0.01	1.40		
	0.05	1.58	0.002^a	1.20^a
	0.10	2.31		
	0.20	3.02		
Thioxanthone	0.005	1.65		
	0.01	2.59		

^{a} Taken from N. C. Yang, D. S. McClure, S. L. Murov, J. J. Houser, and R. L. Dusenbury, *J. Amer. Chem. Soc.*, 1967, **89**, 5466.
$\Phi_0 = 0.014$ at 43 °C.

Considerable interest has been shown in the generation of electronically excited states of ketones by the thermal decomposition of 1,2-dioxetans. This approach to 'photochemical' reactions without the agency of light has been developed by White, Wiecko, and Roswell.[14] Unlike the other

[12] R. Hoffmann and J. R. Swenson, *J. Phys. Chem.*, 1970, **74**, 415.
[13] O. L. Chapman and G. Wampfler, *J. Amer. Chem. Soc.*, 1969, **91**, 5390.
[14] E. H. White, J. Wiecko, and D. F. Roswell, *J. Amer. Chem. Soc.*, 1969, **91**, 5194.

reports of reactions of this class,[15] the isomerisation reported [the well-known conversion of (1) into (2)] involves excitation to the triplet state and is not the consequence of a ground-state reaction. The reaction involves thermal decomposition of a cyclic peroxide (3). The decomposition of such dioxetans, which are recognised chemiluminescent compounds, has been shown to generate electronically excited states.[16] The present researches establish that the electronic energy of the generated excited states can be transferred to acceptors, e.g. the isomerisation of *trans*- to *cis*-stilbene, the dimerisation of acenaphthylene to produce the *anti*-dimer, or the previously mentioned isomerisation [(1) → (2)]. Evidence for an intramolecular energy transfer has also been obtained as the result of the oxidation of the *trans*-hydrazide (4). The products from the reaction were shown to contain 3% of the *cis*-styrylphthalic acid as well as the *trans*-acid. Further studies[17] have shown that the dioxetan (3a) is more effective in bringing about photoreactions of acenaphthylene, stilbene, and 4,4-diphenylcyclohexa-2,5-dienone than is the phenyl analogue (3b). A further example involves the decomposition of the high energy dioxetandione (5), formed by the oxidation of certain oxalates (6), in the presence of fluorescers.[18] The transfer of energy has been demonstrated by the cyclisation of the diketone (7) to the indanone (8) with a chemical sensitization efficiency (Φ_C) of 0·006, considerably less than the quantum yield

[15] H. E. Zimmerman, D. Döpp, and P. S. Huyffer, *J. Amer. Chem. Soc.*, 1969, **91**, 434.
[16] K. R. Kopecky and C. Mumford, *Canad. J. Chem.*, 1969, **47**, 709, and reference therein.
[17] E. H. White, J. Wiecko, and C. C. Wei, *J. Amer. Chem. Soc.*, 1970, **92**, 2167.
[18] G. Güsten and E. F. Ullman, *Chem. Comm.*, 1970, 28.

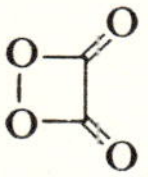

(5) [cyclic dione structure]

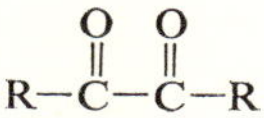

(6) R-C(=O)-C(=O)-R
(a) R = 2,4-(NO$_2$)$_2$-C$_6$H$_3$O
(b) R = 2,4,6-Cl$_3$-C$_6$H$_2$O
(c) R = N-2-pyridone

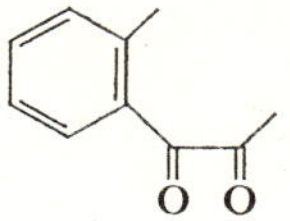

(7) [o-methylphenyl glyoxal structure]

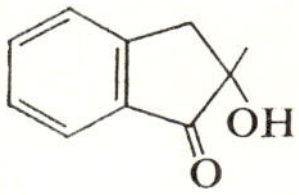

(8) [indanol structure]

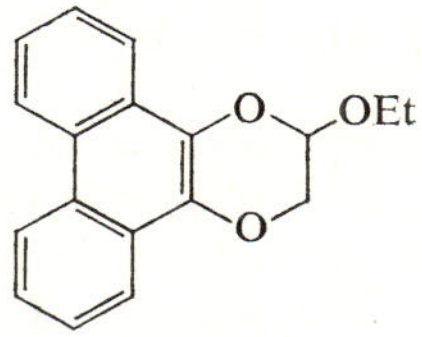

(9) [phenanthrene-dioxine-OEt adduct structure]

of the direct photolysis experiments. A higher chemical sensitization efficiency (0·06) was obtained for the formation of the adduct (9) from phenanthraquinone and ethoxyethylene. The highest efficiencies were obtained in the cis–trans-isomerisation studies with a stilbene (Table 3).

Table 3 Chemical sensitization studies for trans–cis-*isomerisation of* trans-4-methoxy-4′-nitrostilbene [18]

	Oxalate (mol l^{-1})	trans-*stilbene* (mol l^{-1})	cis-*stilbene* (% yield)	Φ_C
(6a)	5 × 10^{-3}	1 × 10^{-3}	6	0·03
(6a)	7·5 × 10^{-3}	2 × 10^{-3}	6	0·04
(6a)	2 × 10^{-2}	4 × 10^{-3}	9	0·045
(6b)	4 × 10^{-2}	8 × 10^{-3}	2	0·01
(6c)	1·5 × 10^{-2}	2 × 10^{-3}	6	0·02

The authors [18] point out that although the yields are lower than those reported by White *et al.*,[14] the chemical efficiencies are significantly higher. It is of passing interest that the highly unstable dione (5) is a new oxide of carbon, C$_2$O$_4$, the dimer of carbon dioxide.

2 Norrish Type I Reactions

Chapman *et al.*[19] have continued to report uses of their low-temperature i.r. technique for the identification of reactive intermediates produced by photochemical means. The latest results [19] pertain to the spectroscopic identification of the keten (10) as the intermediate in the photochemical solvolysis (at −190 °C) of dihydrocoumarin (11) in methanol. No evidence was seen for the presence of the dienone (12) proposed by Gutsche and Oude-Alink [20] as the reactive intermediate. Thus this evidence is fairly

---

[19] O. L. Chapman and C. L. McIntosh, *J. Amer. Chem. Soc.*, 1969, **91**, 4309.
[20] C. D. Gutsche and B. A. M. Oude-Alink, *J. Amer. Chem. Soc.*, 1968, **90**, 5855.

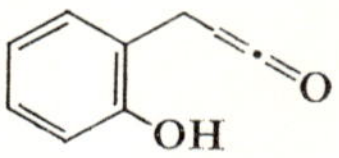

(10) 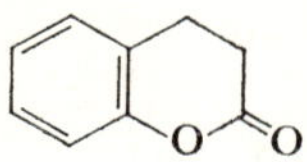   (11) 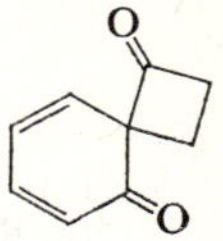   (12)

conclusive that the keten[21] mechanism is the more acceptable for the reaction.

A study of the photochemical reactions of pivalophenones (13a—d) in two solvents has shown that two major processes predominate.[22] In benzene as solvent the main reaction is α-fission giving rise to aldehydes (13e—h). The eliminated t-butyl substituent forms isobutylene and isobutane. In propan-2-ol as solvent, two processes are operative, *viz.* the α-fission described above, and reduction to give the products (14). The use of piperylene has shown that both processes can be quenched, thereby suggesting that these reactions may be occurring from the triplet state; but see also ref. 10 for a possible complication. In one example (13d)

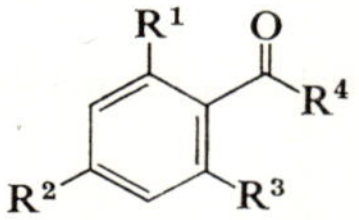

(13)

(a) $R^1 = R^2 = R^3 = H$, $R^4 = Bu^t$
(b) $R^1 = R^3 = H$, $R^2 = R^4 = Bu^t$
(c) $R^1 = R^2 = R^3 = Me$, $R^4 = Bu^t$
(d) $R^1 = H$, $R^2 = R^3 = Me$, $R^4 = Bu^t$
(e) $R^1 = R^2 = R^3 = R^4 = H$
(f) $R^1 = R^3 = R^4 = H$, $R^2 = Bu^t$
(g) $R^1 = R^2 = R^3 = Me$, $R^4 = H$
(h) $R^1 = R^4 = H$, $R^2 = R^3 = Me$

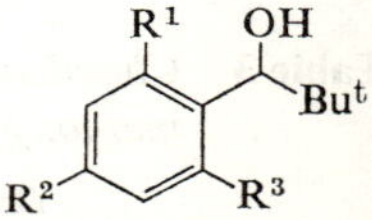

(14) (a),(b),(c)

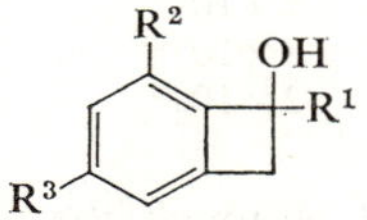

(a) $R^2 = H$, $R^1 = Bu^t$, $R^3 = Me$
(b) $R^2 = R^3 = Me$, $R^1 = H$

(15)

intramolecular hydrogen abstraction and ring-closure affords the cyclobutenol (15a). A cyclobutenol (15b) is also formed during the photolysis of the aldehyde (13g) in propan-2-ol. Photolysis of the pivalophenones (13a—d) in the presence of oxygen affords oxidation products (16a—d) which presumably arise from the photo-oxidation of the aldehydes (13e—h) formed by the Norrish Type I reaction. This interpretation, however, is difficult to reconcile with the observation that the α-fission process can be quenched by piperylene since triplet reactions which might be essential for the α-fission should be susceptible to quenching by oxygen.

[21] D. A. Plank, Ph.D. Thesis, Purdue University, 1966.
[22] H. G. Heine, *Annalen*, 1970, **732**, 165.

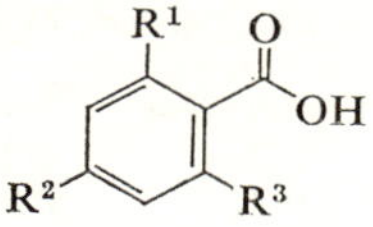

(a) $R^1 = R^2 = R^3 = H$
(b) $R^1 = R^3 = H$, $R^2 = Bu^t$
(c) $R^1 = R^2 = R^3 = Me$
(d) $R^1 = H$, $R^2 = R^3 = Me$

(16)

The interest in the ring-opening reactions of cyclohexanones by a Norrish Type I process has been maintained. Strong evidence for a diradical mechanism is put forward by Barltrop and Coyle.[23] These authors partially separated the *cis*- and *trans*-isomers of 2,3-dimethylcyclohexanone (17) and irradiated the enriched isomers. From both mixtures of isomers the same ratio (1·0 : 2·1) of stereoisomers of the product unsaturated aldehyde (18)

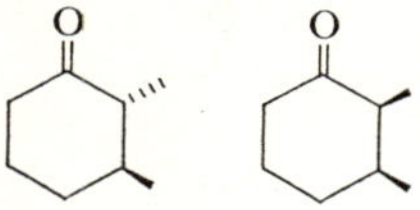

(17) 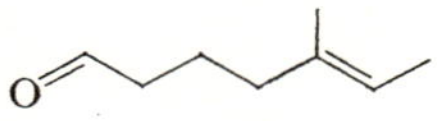   (18a)    (18b)

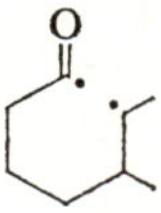

(19)

was obtained. Quenching experiments showed that the reaction took place from the triplet excited state and from these results it was proposed that one-third of the biradicals (19) ring-close to the ground-state ketones with a preference for closure to the *trans*-isomer. Closely related to this work is the report of a study on the photorearrangement of a series of cyclic ketones to ω-alkenals.[24] The quantum yields for the reaction and triplet decay rates are shown in Table 4. The results show that the triplet decay rates vary with ring size and with the degree of substitution at the two and three positions in the ring. This is entirely in agreement with the proposal [23] that excitation brings about cleavage to a diradical (20). Thus the more strained the ring the faster the bond breaks. While the quantum yields do not approach unity, it is worthwhile mentioning that α-substituents increase the quantum yield by a substantial amount. On the other hand, the influence of β-substituents, which decrease the yield considerably, is difficult to

[23] J. A. Barltrop and J. D. Coyle, *Chem. Comm.*, 1969, 1081.
[24] P. J. Wagner and R. W. Spoerke, *J. Amer. Chem. Soc.*, 1969, **91**, 4437.

Table 4 *Quantum yield values for Norrish Type I ring cleavage* [24]

Ketone	$\Phi_{\text{diss of ketone}}$	Φ_{alkenal}	$k_q\tau$ (l mol^{-1})	$1/\tau$ (10^8 s^{-1})
Cyclopentanone	0·28	0·24	47	1·1
Cyclohexanone	0·20	0·09	152	0·33
3-Methylcyclohexanone	0·083	0·033	209	0·25
3,5-Dimethylcyclohexanone	0·033	0·005	206	0·24
3,3,5-Trimethylcyclohexanone	0·024	0·002	200	0·25
2-Methylcyclohexanone	0·50	0·42	10·6	4·7
2-Phenylcyclohexanone	0·51	0·04	15·2	3·3
2,6-Dimethylcyclohexanone	0·55	0·40	5·4	9·3
2,2-Dimethylcyclohexanone	0·52	0·41	2·8	18·0

rationalise completely. While one might expect some decrease due to the lesser availability of abstractable hydrogens compared to cyclohexanone the decrease is more than can be explained on a statistical basis. Indeed, with 3-methylcyclohexanone only the alkenal (21) is obtained (Scheme 1). Steric factors could be invoked as an explanation of this problem.

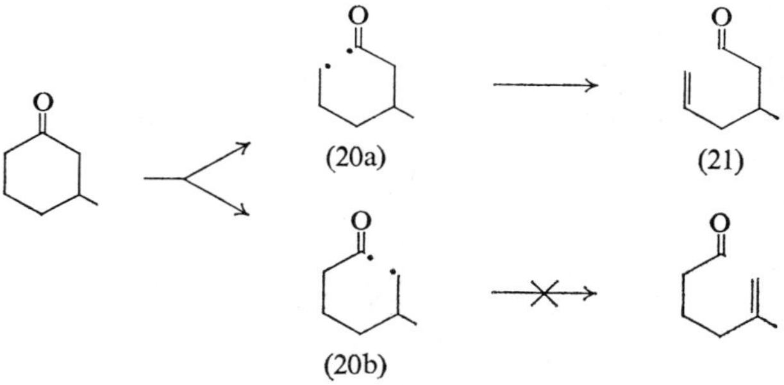

Scheme 1

The effect of spin multiplicity on the Norrish Type I cleavage of cycloalkanones has shown [25] that there is considerable difference between the $n\pi^*$ singlet and triplet states in their reactivities towards the Type I process. The results are shown in Table 5. From the quenching studies it can be

Table 5 *Photoisomerisation of cycloalkanones to unsaturated aldehydes in benzene solution* [25]

Ketone	Aldehyde	Φ	$k_q \tau_T$	τ_T (ns)	τ_s (ns)
(22a)	(24a)	0·11	4·7	9·5	1·9
(22b)	(24b)	0·26	1·4	0·28	2·7
(22c)	(24c)	0·61	≤0·1	≤0·02	8·7
(23d)	(25d)	0·28	0·3	≤0·02	5·2
(23e)	(25e)	0·25	≤0·1	≤0·02	4·2
(23f)	(25f)	0·26	≤0·1	≤0·02	3·8
(23g)	(26g)	0·12	≤0·1	≤0·02	4·4

[25] J. C. Dalton, D. M. Pond, D. S. Weiss, F. D. Lewis, and N. J. Turro, *J. Amer. Chem. Soc.*, 1970, **92**, 2564.

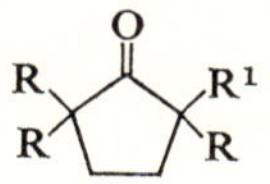

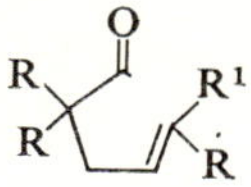

(22)
(a) $R^1 = R = H$
(b) $R^1 = Me, R = H$
(c) $R^1 = R = Me$

(24)

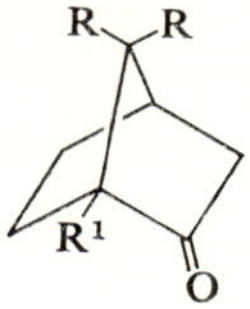

(23)
(d) $R^1 = R = H$
(e) $R^1 = Me, R = H$
(f) $R^1 = Pr^n, R = H$
(g) $R = R^1 = Me$

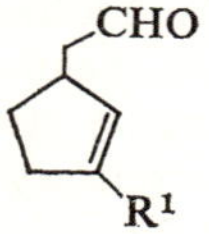

(25)
(d) $R^1 = H$
(e) $R^1 = Me$
(f) $R^1 = Pr^n$

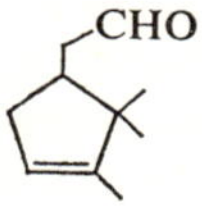

(26g)

seen that only for the first two cases in the Table is there efficient suppression of the aldehyde formation. Indeed, for the remaining entries in the Table no quenching of aldehyde formation was detected. It is argued that although this at first sight might imply the intermediacy of the singlet state, this is only the case if the rate of cleavage (k_r^s) is greatly in excess of the intersystem crossing rate constant (k_{ST}). If $k_r^s \gg k_{ST}$ then the singlet lifetimes (τ_s) will be determined by k_r^s. However, experiments show that this is not the case and that τ_s is determined by k_{ST}. Thus the α-cleavage in cases 3—7 (Table 5) arises at least in part from the triplet state. The authors [25] conclude by calculating that there is a difference of at least two orders of magnitude in the reactivity of the $n\pi^*$ singlet and $n\pi^*$ triplet states towards α-cleavage.

Apart from the mechanistic investigations of this reaction, synthetic studies have also appeared. A convenient way to effect ring-expansion of a cyclohexanone derivative by three carbon units has been described.[26] The process involves the photoreaction of 2-cyclopropylcyclohexanone (27). Irradiation ($\lambda = 300$ nm) into the $n\pi^*$ band (294 nm, $\varepsilon = 25$) of the ketone brings about bond-cleavage to yield the more stable diradical (28: bond a fission). Subsequent ring-opening of the cyclopropylcarbinyl radical to produce the homo-allyl diradical (29), followed by rebonding, accounts for the major products, namely the *cis*- and the *trans*-cyclonon-4-enone (30) and (31), which are formed in 29% and 44% yields respectively. These products are readily inter-converted by further photolysis, but at low

[26] R. G. Carlson and E. L. Biersmith, *Chem. Comm.*, 1969, 1049.

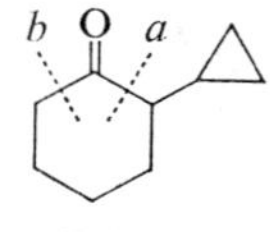

(27) 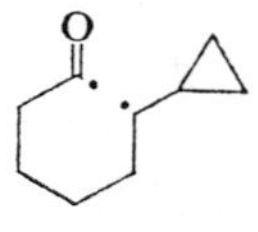 (28) 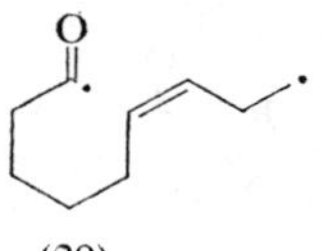 (29)

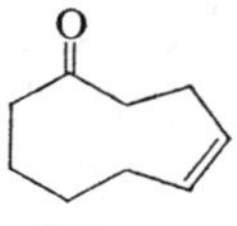

(30) 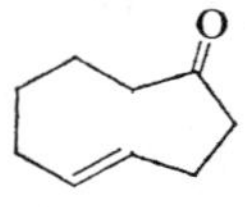 (31)

conversions the *trans*-isomer is favoured. An additional isolated product, 1-cyclopropylhex-1-en-6-al (21%), results from cyclohexanone ring-opening in the other mode (bond *b* fission). Preliminary studies suggest that the reactions arise from a singlet excited state contrary to conclusions from most other studies: compare for example ref. 22.

The photolysis of friedelin (32) has been further studied.[27] In reagent-grade chloroform the seco-ester (33) is isolated as the major product. However, in ether decarbonylation is the main reaction path, leading to a complex mixture of hydrocarbons from which 5-ethyl-10β-vinyldes-A-friedelin (34) was isolated. Norfriedelin (35), from the recombination of diradical (36) rather than hydrogen loss, was also isolated. Aldehydic

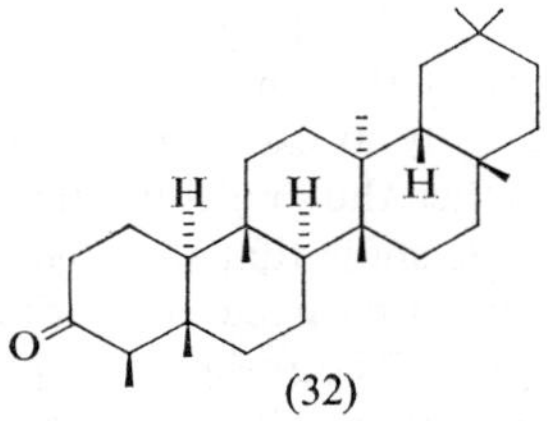

(32) 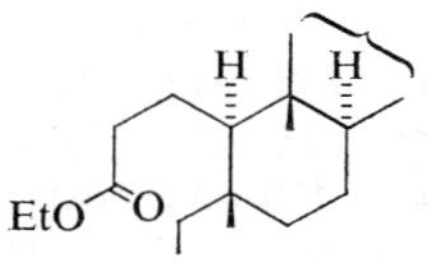 (33) 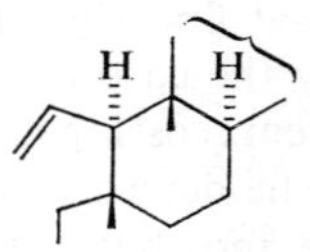 (34)

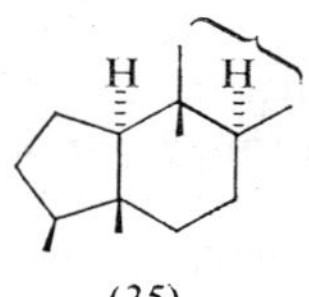

(35) 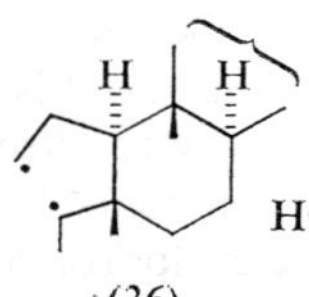 (36) 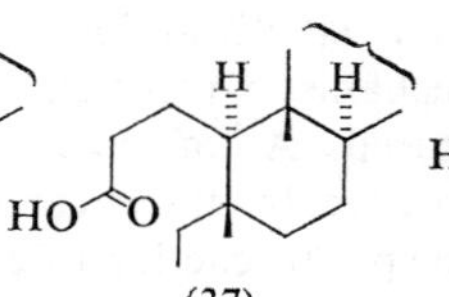 (37) 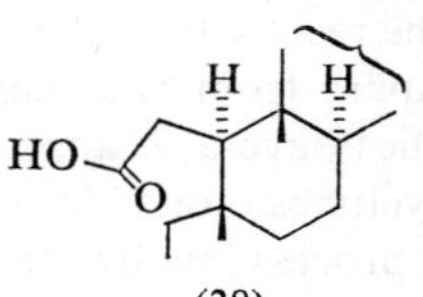 (38)

material was also isolated but could not be separated. Other workers[28] have found that if the photolysis of friedelin (32) is carried out in n-hexane the expected seco-acid (37) can be isolated. In addition to this an unexpected nor-acid (38) was isolated. Although no route to this compound was

[27] F. Kohen, A. S. Samson, and R. Stevenson, *J. Org. Chem.*, 1969, **34**, 1355.
[28] M. Takai, R. Aoyagi, S. Yamada, T. Tsuyuki, and T. Takahashi, *Bull. Chem. Soc. Japan*, 1970, **43**, 972.

suggested, one may reasonably infer the involvement of atmospheric oxygen.

Photolysis [29] of camphor (39a) gave the substituted acetaldehyde (40).[30] The corresponding oxime (39b) forms a mixture of camphor (39a), the acetonitriles (41a) and (41b), and the reduced nitrile (42).[29] The results are

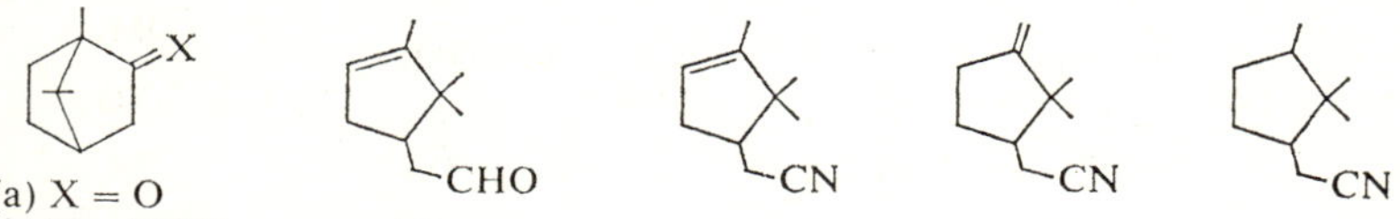

(a) X = O
(b) X = NOH
(39)

(40) CHO

(41a) CN

(41b) CN

(42) CN

rationalised by the proposals that fission of the N–O bond takes place on photolysis and that the resultant $>$C=N· moiety behaves similarly to the excited carbonyl function.

The synthesis of the homocubanone (43) has been accomplished [31] by the photolysis of ketone (44) in methylene chloride. The necessary intermediate (45) for the cycloaddition must arise by the rearrangement of the initial ketone, as shown in Scheme 2. This is an example of a type of

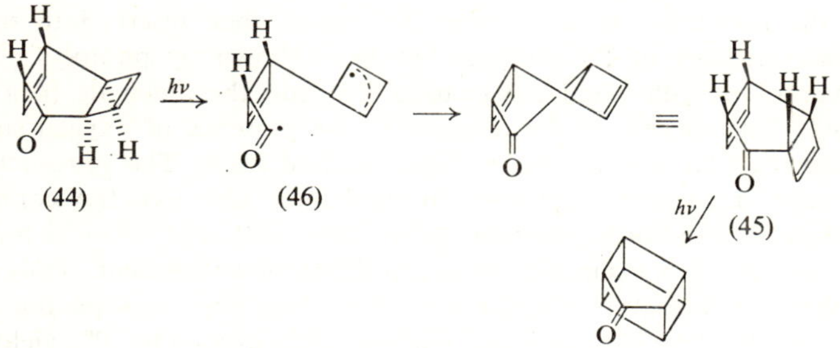

(44)   (46)   (45)

(43)

Scheme 2

rearrangement discussed last year [32] involving α-fission to form a diradical (46) capable of rebonding to form (45).

Two reports concerning the study of spiro[2,4]heptanones have appeared. The work of both groups [33, 34] has shown that although the

[29] T. Sato, *Asaki Garasu Kogyo Gijutsu Shorei-kai Kenkyu Hokoku*, 1968, **14**, 425 (*Chem. Abs.*, 1970, **72**, 21,785m).
[30] See also work cited by W. M. Horspool, in 'Photochemistry,' ed. D. Bryce-Smith, (Specialist Periodical Reports), The Chemical Society, London, 1970, vol. I, pp. 152, 153.
[31] R. L. Cargill and T. Y. King, *Tetrahedron Letters*, 1970, 409.
[32] E. J. Forbes, J. Griffiths, and R. A. Ripley, cited by W. M. Horspool, in 'Photochemistry,' ed. D. Bryce-Smith, (Specialist Periodical Reports), The Chemical Society, London, 1970, vol. I, p. 216.
[33] J. K. Crandall and R. J. Seidewand, *J. Org. Chem.*, 1970, **35**, 697.
[34] A. Sonoda, I. Moritani, J. Miki, and T. Tsuji, *Tetrahedron Letters*, 1969, 3187.

principal reaction is Norrish Type I cleavage, no involvement of the spiro three-membered ring occurs. The photolysis of the spiro-heptanone (47a) in methanol affords three products (48a), (49a), and (50) which are the result of Norrish Type I cleavage. The cleavage of the ring is noteworthy in that it occurred solely in the direction which yields a primary radical

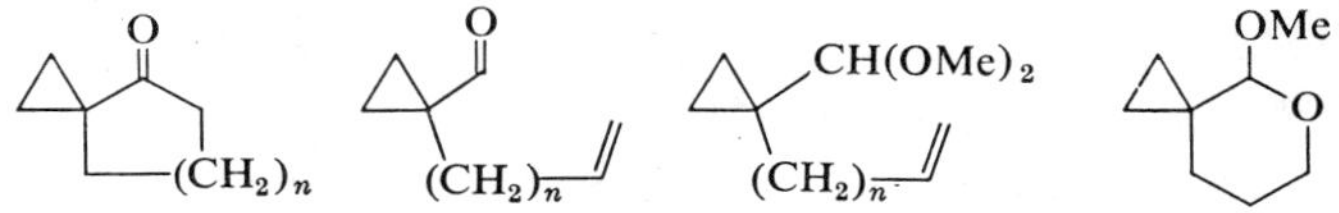

(a) $n = 1$
(b) $n = 2$
(47)

(a) 32%
(b) 7%
(48)

(a) 6%
(b) 49%
(49)

(50)

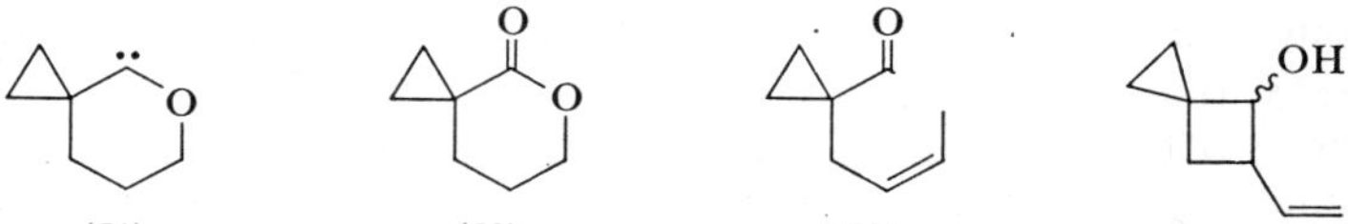

(51) (52) (53) (54)

rather than a cyclopropyl radical. The formation of the ketal (50) results from the formation of a carbene (51) which then inserts into solvent. The intermediacy of the carbene has been shown by photolysis in the presence of oxygen when the lactone (52) and the aldehyde (48a) were isolated. The success of the reaction in the presence of oxygen suggests that the reaction may involve a singlet excited state. The spiro-octanone (47b) also undergoes photolysis in methanol, and like the compound described above affords products (48b), (49b), (53), and (54) which do not arise from direct participation of the cyclopropyl substituent. Only in the photolysis of the larger ring ketone (47b) does any ring-opening of the cyclopropane occur [34] to produce 1-ethylcyclohexanone in 10% yield. The lack of involvement of the cyclopropyl substituent has yet to be successfully explained. The isomeric spiroketones (55a and b) undergo a Norrish Type I process to give the cyclopropylcarbinyl radical (56).[34] Ring-opening and rebonding can account for the formation of the observed ring-expanded

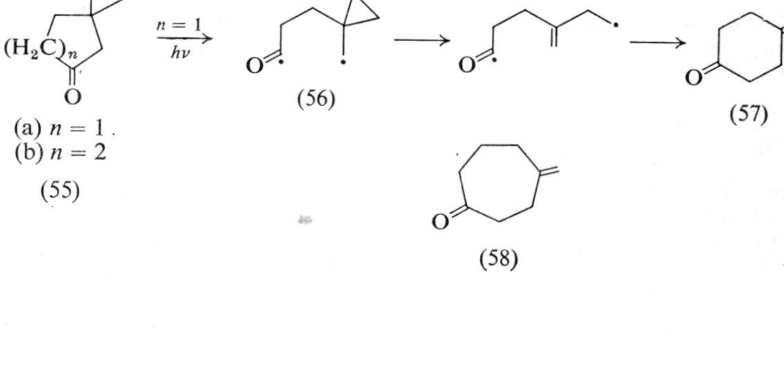

(a) $n = 1$
(b) $n = 2$
(55)

$\xrightarrow{n=1, h\nu}$

(56)

(57)

(58)

ketones [(57) and (58) respectively]. The difference between this and earlier reports of the photolysis of ketone (55) where ring cleavage is the major pathway [35] is attributed to conformational interactions. A similar lack of involvement of a cyclopropyl substituted in a bicyclic ketone has been studied by Paquette and Eizember.[36] They found that irradiation of *cis*-bicyclononanone (59) in t-butanol affords two aldehydes (60) and (61). Only aldehyde (60) is the result of initial photolysis and is formed as shown in Scheme 3 by a Norrish Type I fragmentation. The occurrence of aldehyde (61) can be explained by a second photoprocess involving a Norrish Type II reaction of the initial aldehyde (60) (Scheme 4). The *trans*-isomer (62)

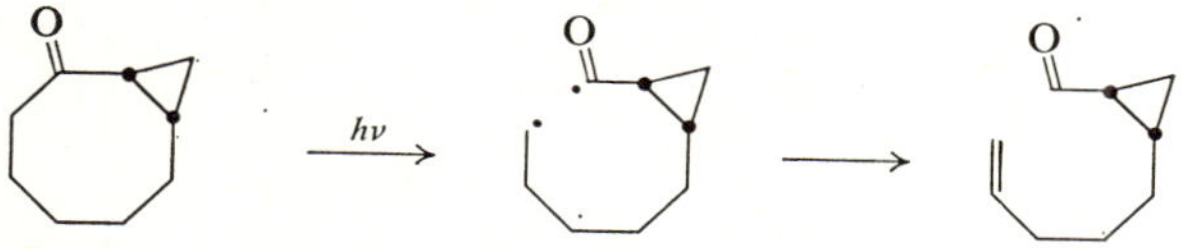

Scheme 3

Scheme 4

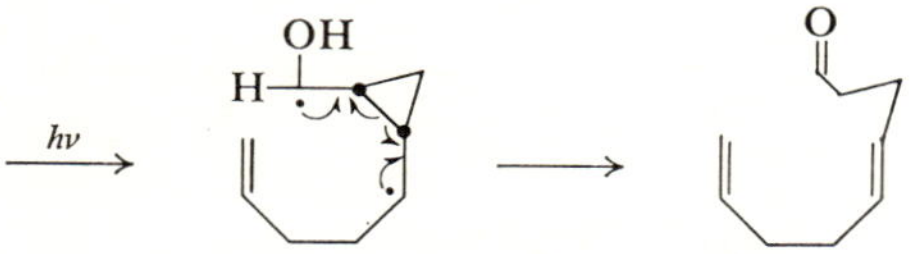

which is also isolated from the photolysis described above undergoes ring-opening in t-butanol or t-butanol(OD) on photolysis [36] to yield the ester (63) as shown in Scheme 5.

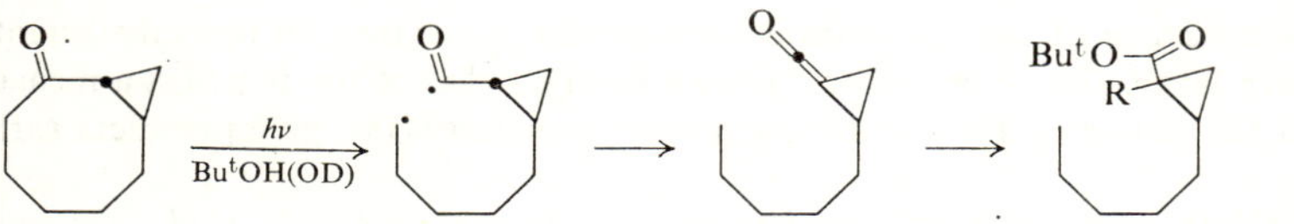

Scheme 5

(63) R = H or D

3 Norrish Type II Processes (Hydrogen-abstraction Reactions: Cyclobutanol Formation)

Considerable interest has been shown in the nature of the diradical intermediate generated by γ-hydrogen transfer in the Norrish Type II

---

[35] W. G. Dauben and G. W. Shaffer, *Tetrahedron Letters*, 1967, 4415.
[36] L. Paquette and R. F. Eizember, *J. Amer. Chem. Soc.*, 1969, **91**, 7108.

process. In a recent study [37] caution is advocated in the interpretation of results from experiments on the Norrish Type II elimination reaction of alkyl aryl ketones. A wrong impression may be obtained if the only results available are those from a hydrocarbon and a neat alcohol solvent. Wagner and Schott [37] have found that the quantum yield for the reaction of substituted valerophenones is dependent on the amount of alcohol present in the reaction medium: see Figure 1. The extreme values of the quantum

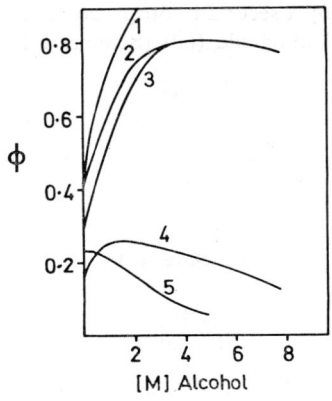

1. Valerophenone.
2. *p*-Methylvalerophenone.
3. *p*-Chlorovalerophenone.
4. *p*-Methoxyvalerophenone.
5. *p*-Methoxyvalerophenone + methanol.

Figure 1

yield of product formation in neat hydrocarbon solution and the quantum yield for alcohol addition are listed in Table 6. It can be seen that the dependence of the quantum efficiency of the reaction on the substituent is very large. The effect is best shown by the value of Φ_P which is a measure of the tendency for the diradical intermediate (64) to give products rather

Table 6 *Quantum efficiencies for substituted n-butyl aryl ketones* [37] *(0·1M soln. of ketone, 313·0 nm, < 5% conversion)*

Aryl group	Φ (benzene)	Φ (alcohol)	Φ_P
p-Anisyl	0·18	0·26	0·69
p-Tolyl	0·39	0·80	0·50
Phenyl	0·40	1·00	0·40
p-Chlorophenyl	0·30	0·80	0·37
3-Pyridyl	0·29	1·00	0·29
2-Pyridyl	0·18	1·00	0·18

[37] P. J. Wagner and H. N. Schott, *J. Amer. Chem. Soc.*, 1969, **91**, 5383.

than to disproportionate to the starting material. Thus electron-withdrawing substituents favour disproportionation of the biradical while electron-donating groups favour product formation. This influence is interpreted as

(64)

(65)

a dependence of disproportionation on the acidity of the hydroxyl function. Thus the transition state for the disproportionation is stabilised by charge separation (65). Additional problems arise from the mixing of $\pi\pi^*$ and $n\pi^*$ states, and with chloro-, methyl-, and methoxy-substituted valerophenones, which show long-lived phosphorescence, high concentrations of alcohol begin to decrease the quantum yields. This is not the case with valerophenone and the pyridyl ketones which have $n\pi^*$ lowest triplet states.

Yang and Elliot [38] have pointed out that the photoprocesses from simple alkanones having γ-hydrogens have quantum yields which fall below unity, in complete contrast with the case of arylalkyl ketones. The results obtained from the study of three ketones are recorded in Table 7. It is

Table 7 Details for photofragmentation of alkyl ketones [38]

	Pentan-2-one	Hexan-2-one	5-Methylhexan-2-one
$K_q \tau_s$ (l mol^{-1})	20·2	7·3	4·1
τ_s	2·02 × 10^{-9}	7·3 × 10^{-10}	4·1 × 10^{-10}
Φ_{ST}	0·63	0·27	0·11
k_{ST} (s^{-1})	3·1 × 10^8	3·7 × 10^8	2·7 × 10^8
Φ_s	0·025	0·10	0·10
k_s (s^{-1})	1·2 × 10^7	1·4 × 10^8	2·4 × 10^8
Φ_{-s}	0·35	0·63	0·79
k_{-s} (s^{-1})	1·7 × 10^8	8·5 × 10^8	1·9 × 10^9
Φ_t	0·36	0·23	0·09

particularly important to notice that the intersystem crossing efficiency (Φ_{ST}) decreases from pentan-2-one through hexan-2-one to 5-methylhexan-2-one while the rates for intersystem crossing remain virtually the same. The important difference in the compounds considered is the marked difference in the environment of the γ-hydrogen and it is proposed that the bond strength of this hydrogen plays an important part in the k_s

[38] N. C. Yang and S. P. Elliot, *J. Amer. Chem. Soc.*, 1969, **91**, 7550.

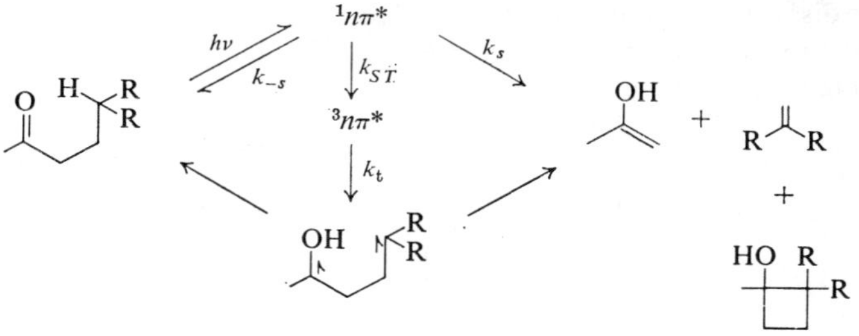

Scheme 6

and k_{-s} processes (Scheme 6). Indeed, a chemical process is indicated, perhaps one involving hydrogen transfer to give a singlet diradical (66) which can revert to starting material or else give products. Thus the low efficiency of the reactions is attributed to an increase in k_{-s} causing decrease

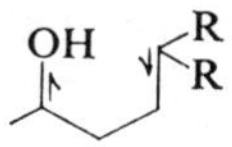

(66)

in Φ_{ST} and Φ_t. A detailed study of the photochemistry of (S)-(+)-5-methylheptan-2-one at $\lambda = 313$ nm has also shown that two processes take place, one from the triplet and one from the singlet excited state.[39a] The intersystem crossing efficiency has been estimated as 0·11, which is considerably lower than with other carbonyl compounds. The results show that the singlet and the triplet reactions have different pathways. The efficiencies of the various processes are shown in Table 8. It should be noticed that

Table 8 Details for the photolysis of (S)-(+)-5-methylheptan-2-one [39a]

Solvent	Φ	Φ (Type II)	Φ (cyclobutanols)	Φ (rac.)	Φ (photored)
Hexane	0·158	0·128	0·025	0·08	0·004
t-Butyl alcohol	0·18	0·157	0·029	0·04	
Hexane +2·5M cis-piperylene	0·073	0·071	0·006	0·002	
t-Butyl alcohol +2·5M cis-piperylene	0·07	0·058	0·004		

racemisation occurs in both solvents but only in the absence of the quencher diene. This implies that the triplet state is the intermediate which brings about racemisation; but see also ref. 10. The results concerning the increased efficiency in t-butanol are in agreement with Wagner's observations.[39b] Thus products are formed from the triplet state by the collapse of the diradical intermediate, as shown in Scheme 7, to form cyclobutanols,

[39a] N. C. Yang, S. P. Elliot, and B. Kim, *J. Amer. Chem. Soc.*, 1969, **91**, 7551.
[39b] P. J. Wagner, *J. Amer. Chem. Soc.*, 1967, **89**, 5898.

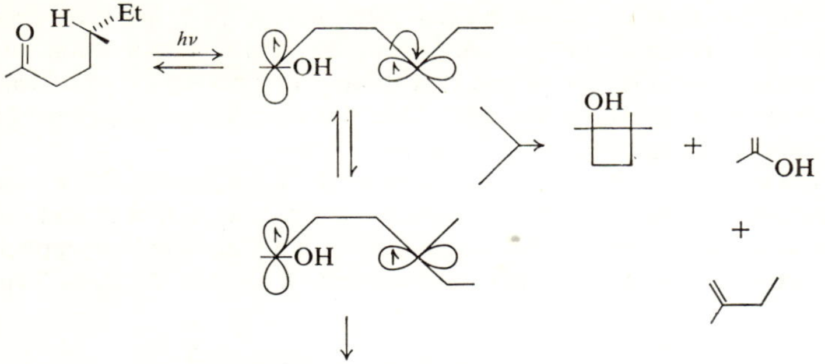

(R)-(−)-5-methylheptan-2-one
Scheme 7

racemic starting material, 2-methylbut-1-ene, and acetone. The combined quantum yield for the reaction in t-butanol, and hence the reactions from the triplet state, is ca. 0·2, and it is calculated that in hexane the triplet processes have a total quantum yield of 0·14. Thus there is a large energy loss ($\Phi = 0.78$): this is attributed to radiationless decay from the singlet state.

Other approaches to the determination of the nature of the diradical intermediate have used either optically active esters [40] or butyrophenone derivatives [41, 42] with which some idea of the freedom of rotation in the diradical can be inferred. In the Norrish Type II photodecomposition [40] of threo- and erythro-1,2-dimethylbutyl acetate (67a and b, respectively), the quantitative results (Table 9) show that both esters give a mixture of cis- and trans-olefins (68a and b respectively). Hence there seems to be con-

Table 9 Photolysis of esters (67a and b) [40]

	%cis-olefin	Φ_{cis} (× 10²)	Φ_{trans} (× 10²)
threo-1,2-Dimethylbutyl acetate (a)	25	3·4	10·7
	23	3·2	10·7
erythro-1,2-Dimethylbutyl acetate (b)	55	4·7	3·9
	53	4·9	4·3

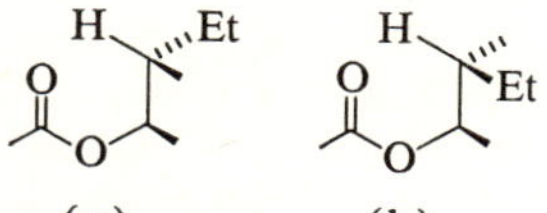

(67) (a) (b)   (68) (a) (b)   (69)

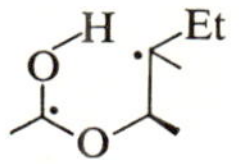

[40] J. E. Gano, Tetrahedron Letters, 1969, 2549.
[41] R. A. Caldwell and P. M. Fink, Tetrahedron Letters, 1969, 2987.
[42] P. J. Wagner and P. A. Kelso, Tetrahedron Letters, 1969, 4151.

siderable rotation* in the diradical intermediate (69) prior to product formation. There is no evidence from an n.m.r. examination of the unchanged ester after photolysis that there is any reversal of the biradical intermediate to the starting esters. This, however, is in disagreement with the results of Yang et al.[39]

A less elegant result which also illustrates the same point, i.e. free rotation in the diradical intermediate (71), has been reported by Caldwell and Fink.[41] These authors have shown that the photolysis of 3,4-diphenylbutyrophenone (70) produces both cis- and trans-stilbene from the Norrish Type II

PhCOCH$_2$CHPhCH$_2$Ph
(70)

(71)

fragmentation induced by irradiation at 366 nm in benzene solution. The critical analysis of the results shows that the initial stilbene ratio is 40 : 1 trans-, but as the irradiation progresses, sensitized irradiation of the stilbene leads to a cis : trans ratio of 0·59 : 0·41 which is in agreement with the decay ratio of the stilbene triplet.[43] Thus the initial ratio of cis- and trans- stilbenes suggests that there is at most only 3—4% participation of stilbene triplets in the fragmentation.

Similar conclusions have been reached by Wagner and Kelso.[42] They have established that the photodecomposition of 3,4-diphenylbutyrophenone (70) is a triplet process and can be quenched by dienes producing linear Stern–Volmer plots: with 2,5-dimethyl-2,4-hexadiene the slope is 2·41 mol^{-1}. The quantum yield for the Norrish Type II process in benzene is 0·19. The quenching results give a value for the triplet lifetime of $0·5 \times 10^{-9}$ s.

Both results suggest that spin correlation may be so weak that distinctions between singlet and triplet are practically non-existent.[41, 42]

Alkylcyclohexanones have not previously been examined for the involvement of Norrish Type II reactions. However, Barltrop and Coyle[44] have examined the photolysis (λ ca. 350 nm) of a series of 2-alkylcycloalkanones (72). The irradiations in methanol or benzene gave products arising from the more normal mode of reaction of such cycloalkanones, the Norrish Type I process forming aldehydes (73) and ketens (74) which were trapped

(72)   (73)   (74)

[43] G. S. Hammond et al., J. Amer. Chem. Soc., 1964, **86**, 3197.

* The author[40] suggests that the yield of cis-olefin would be 39% if the diradical is equilibrated. Thence it can be calculated that 39% of the threo-ester and 25% of the erythro-ester decomposes by a concerted route.

as esters in the methanol experiments. The Norrish Type II reaction was found to occur in several of the ketones examined. The results are shown in Table 10, and it should be noticed that there is a marked difference for

Table 10 *Percentage of Type II elimination in 2-alkanones* [44]

R	Cyclopentanone (n = 1)	Cyclohexanone (n = 2)
2-Et	<1	<1
2-Et-4-Bu^t	—	<1
2-Pr^n	<1	65
2-Pr^i	<1	14
2-Pr^i-5-Me	—	17
2-Bu^i	20	77
2-Bu^t	—	49

the process when the environment of the abstractable hydrogen changes from primary to secondary.[44] However, while this effect is expected to be noticeable (in line with other experiments [38]), the marked effect is thought to reflect some conformational problems already recognised by other workers.[45]

Lewis and Turro have extended [46] the results reported last year [47] from the photolysis of arylalkoxyketones (75).

$$ArCOCH_2CHR^1OR^2$$

(75)

The synthetic utility of intramolecular hydrogen abstraction processes in ketones for the formation of cyclobutane rings continues to be exploited. An interesting example of such cyclobutanol formation following γ-hydrogen abstraction from a proximate methyl group in a Norrish Type II reaction has been reported in the photolysis of the adduct (76) in dioxan.[48] The isomeric cyclobutanols (77a) and (77b), and a de-acetylated product (78) (the result of a Norrish Type I cleavage) were isolated. Highly strained ring

(76) (77a) 35% (77b) 20% (78)

[44] J. A. Barltrop and J. D. Coyle, *Chem. Comm.*, 1970, 390.
[45] N. J. Turro and D. S. Weiss, *J. Amer. Chem. Soc.*, 1968, **90**, 2185.
[46] F. D. Lewis and N. J. Turro, *J. Amer. Chem. Soc.*, 1970, **92**, 311.
[47] N. J. Turro and F. D. Lewis, cited by W. M. Horspool, in 'Photochemistry,' ed. D. Bryce-Smith, (Specialist Periodical Reports), The Chemical Society, London, 1970, Vol. I, p. 133.
[48] P. Sunder-Plassman, P. H. Nelson, P. H. Boyle, A. Cruz, J. Iriarte, P. Crabbe, J. A. Zderic, J. A. Edwards, and J. H. Fried, *J. Org. Chem.*, 1969, **34**, 3779.

systems can also be synthesised by this route, as has been shown previously by Padwa et al.[49] An extension of this approach has led to the synthesis of the tricyclic alcohol (79). This was achieved by the photolysis (Pyrex filter) of a benzene solution of the ketone (80).[50] Two products were isolated, the ketone (81) and the expected alcohol (79). The quantum yield for the disappearance of the starting material was established as 0·06, which is considerably lower than the figure obtained with related phenyl alkyl ketones. The quantum yield was not increased by the addition of t-butyl alcohol. However, the reaction was quenched by the addition of piperylene. The inefficiency of the reaction cannot arise from the disproportionation of the 1,4-diradical (82) formed by hydrogen abstraction (Scheme 8), and

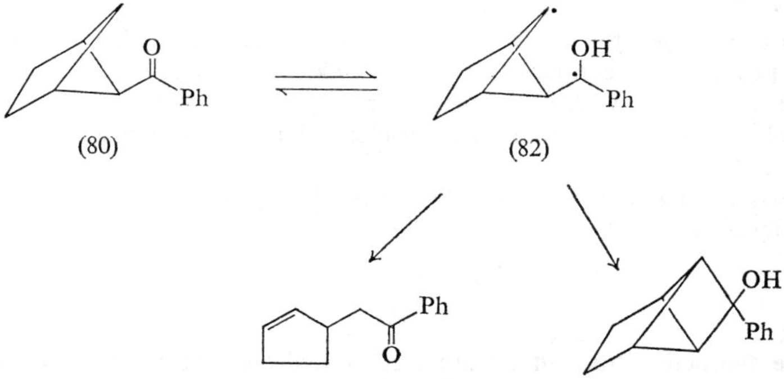

Scheme 8

it is thought that unfavourable geometry in the transition state for the hydrogen abstraction process is the major cause of the inefficiency.[50]

Further examples of hydrogen abstractions of alkyl aryl ketones (83a—c) in alcoholic solution have been reported.[51] In common with others of this type,[22] the hydrogen abstraction is predominantly intramolecular and gives rise to the enol form (84). In the present examples the products from the

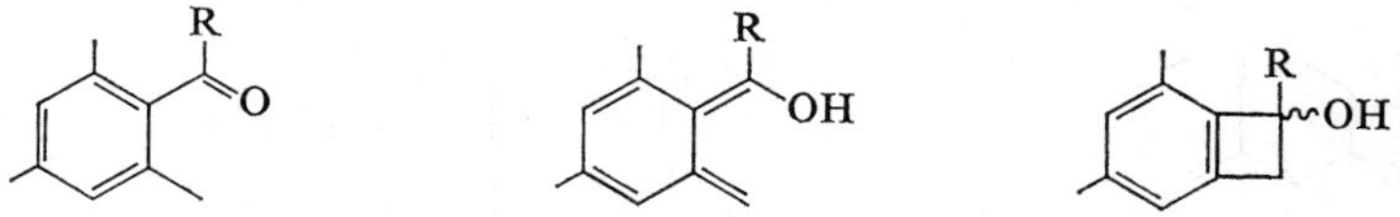

(a) R = Me
(b) R = Et
(c) R = Pr^i

(83)

---

[49] A. Padwa, E. Alexander, and M. Niemcyzk, cited by W. M. Horspool, in 'Photochemistry,' ed. D. Bryce-Smith, (Specialist Periodical Reports), The Chemical Society, London, 1970, vol. I, p. 142.
[50] A. Padwa and W. Eisenberg, J. Amer. Chem. Soc., 1970, **92**, 2590.
[51] T. Matsuura and Y. Kitaura, Tetrahedron, 1969, **25**, 4487.

enol form are cyclobutenols (85a—c). The formation of the photo-enol is proposed to take place from the $n\pi^*$ triplet state, and indeed the reaction can be quenched by the addition of known triplet quenchers. It is not yet known whether cyclisation of the photo-enol to the cyclobutenol is a thermal or a photochemical process. Consideration of the steric requirements of the photo-enol suggests that a thermal conrotatory cyclisation might be favoured.

An intramolecular hydrogen abstraction is also involved[52] in the photolysis of the aryl ketone (86). Bonding between the diradical centres (87) forms the furan skeleton (88), and final dehydration yields the

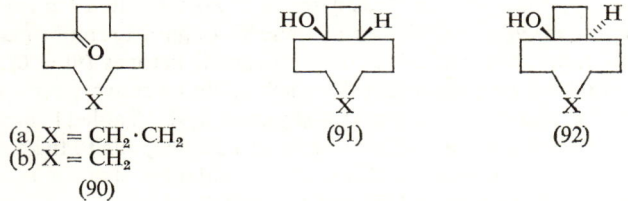

product (89). The cyclisation appears to have certain similarities to others which have been reported, e.g. the photocyclisation of 4,6-di-t-butyl-2-methoxybenzophenone to 5-t-butyl-7-methoxy-3,3-dimethyl-1-phenyl-1-indanol.[53] This report is not novel and indeed the authors[52] omit an extremely pertinent reference to the earlier work of Pappas and Blackwell[54] on the similar cyclisation of 2-benzyloxybenzaldehyde to 3-hydroxy-2-phenyl-2,3-dihydrobenzofuran.

Intramolecular hydrogen transfer is also important in the photolysis of large-ring cycloalkanones (90). Cyclobutanes (91) and (92) are the major products. The products mostly arise from the excited singlet state.[55] In Norrish Type II reactions, hydrogen abstraction usually occurs from the

[52] G. R. Lappin and J. S. Zannucci, *Chem. Comm.*, 1969, 1113.
[53] E. J. O'Connell, cited by W. M. Horspool, in 'Photochemistry,' ed. D. Bryce-Smith, (Specialist Periodical Reports), The Chemical Society, London, 1970, vol. I, p. 139.
[54] S. P. Pappas and J. E. Blackwell, *Tetrahedron Letters*, 1966, 1171.
[55] T. Mori, K. Matsui, and H. Nozaki, *Tetrahedron Letters*, 1970, 1175.

γ-position relative to the excited carbonyl group. Such abstraction allows either cyclobutanol formation or fragmentation to the olefin and ketone. In certain cases hydrogen abstraction can take place from the δ-carbon when this is conformationally possible, as in previously discussed examples of furan formation.[52] Yates and Pal have now studied the influence of radical stability on δ-hydrogen abstraction.[56] Thus the photolysis of the β-alkoxyketones (93 a—d) in pentane gave several products. From the cyclisations of ketones (93a and b), the furanols (94a and b) were isolated, rather than the expected cyclobutanols. However, ketones (93c and d) gave

PhCOCH$_2$C(OCH$_2$R^2)(CH$_2$R^1)—CH$_2$R^1

(a) R^1 = R^2 = H
(b) R^1 = H, R^2 = Me
(c) R^1 = Me, R^2 = H
(d) R^1 = R^2 = Me

(93)

(94)

(95)

(96)

(97)

both the furanols (94c and d) and the cyclobutanols (95c and d). The preference shown for the δ-hydrogen is thought to reflect the stability of the oxalkyl radical (96) over the alkyl radical (97) formed by γ-H-abstraction.

Hydrogen transfer can apparently take place over even greater distances than from γ- or δ-carbon atoms.

Breslow and Winnik report the use of a triplet excited carbonyl group as a means of functionalising a remote methylene.[57] This is claimed as the first chemical application of a principle often found in enzymic systems, namely the ability of an enzyme to convert say stearic to oleic acid. The system used by these workers was a series of p-substituted benzophenone carboxylic esters (98). Irradiation of these through a uranium glass filter in CCl$_4$ solution gave rapid loss of the original ketones (Φ = 0·2). The products obtained were alcohols (99) which could be converted into olefins by dehydration and into diketones (100) by ozonolysis. Table 11 summarises the results. Functionalisation does not occur below C$_9$ due to the geometric restrictions of the molecule. The sites of oxidation show a reasonable amount of selectivity, particularly for C$_{14}$ and C$_{16}$ ester side-chains where oxidation takes place at C$_{12}$ and C$_{14}$ respectively. It is obvious that intramolecular hydrogen transfer can take place over fairly large distances.

[56] P. Yates and J. M. Pal, *Chem. Comm.*, 1970, 553.
[57] R. Breslow and M. A. Winnik, *J. Amer. Chem. Soc.*, 1969, **91**, 3083.

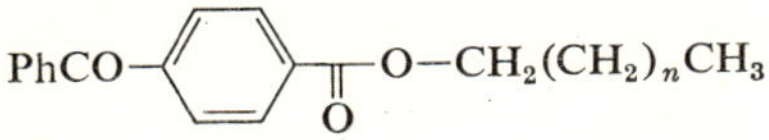

PhCO—⟨C₆H₄⟩—C(=O)—O—CH₂(CH₂)ₙCH₃

(98)

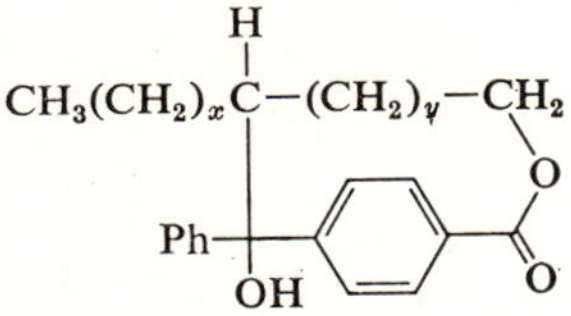

CH₃(CH₂)ₓ—C(H)—(CH₂)ᵧ—CH₂
        | \
        Ph—⟨C₆H₄⟩—C(=O)—O
        |
        OH

(99)

CH₃(CH₂)ₓ—C(=O)—(CH₂)ᵧ—CH₂
                              \
        Ph—C(=O)—⟨C₆H₄⟩—C(=O)—O

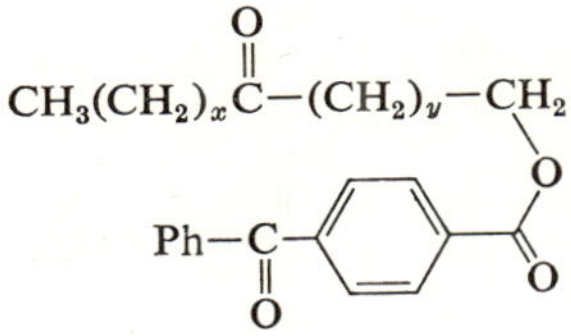

(100)

Extensions of the remote oxidative techniques to steroid molecules have been reported by two groups.[58, 59] Breslow and Baldwin [58] have found that the photolysis of the steroid derivative (101a) in CCl_4 affords after a five-step work-up, the keto-steroid (102) in 16% yield. The low yield is attributed

Table 11 *Details of remote oxidation of alkyl side-chains* [57] *(% yields)*

Oxidation site	Number of carbon atoms in side-chain			
	C_{14}	C_{16}	C_{18}	C_{20}
C-9	1·4	1·1	0·1	2
C-10	3	7·8	8	5
C-11	11	12	17	15
C-12	49	13	21	20
C-13	22	10	18	19
C-14		56	12	19
C-15		7	5	13
C-16			13	8
C-17			6	0·7

in part to the fact that only one-third of the diradicals formed by hydrogen abstraction form a new C–C bond while the rest abstract chlorine from the solvent. A different result is obtained from the photolysis in acetonitrile. The cleavage of the products from the photolysis affords the two olefins (103a) and (104) in 27 and 8% respectively. The yield of the major product is increased to 20% in acetone as solvent and to 35% in benzene. In acetonitrile the primary photoproduct is assumed to be the alcohol (105a).

Baldwin, Bhatnagar, and Harper [59] have studied the irradiation of the steroid derivative (101b) through a Pyrex filter in benzene solution. This led to the isolation of the olefin (103b) (44%), olefin (106) (9%), and (105b) and (107) in a combined yield of 25%. All the products isolated can be accounted for utilising radical abstraction and recombination processes.

---

[58] R. Breslow and S. W. Baldwin, *J. Amer. Chem. Soc.*, 1970, **92**, 732.
[59] J. E. Baldwin, A. K. Bhatnagar, and R. W. Harper, *Chem. Comm.*, 1970, 659.

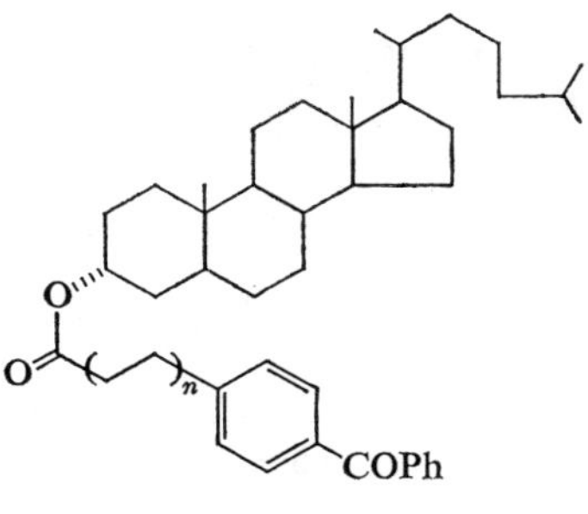

(101) (a) n = 1 (b) n = 2

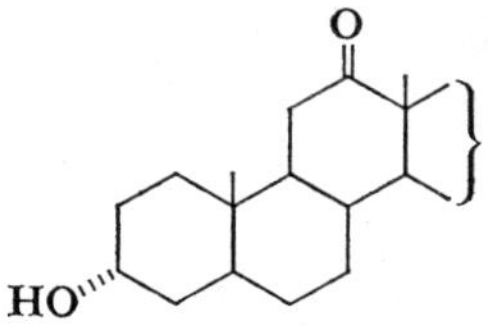

(102)

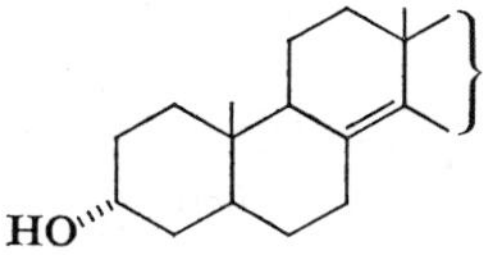

(103) (a) R = H (b) R = COR²

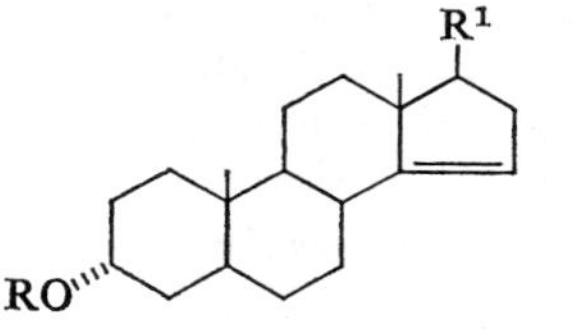

$R^2 = (CH_2)_4C_6H_4\overset{H}{\underset{OH}{C}}Ph$

(104)

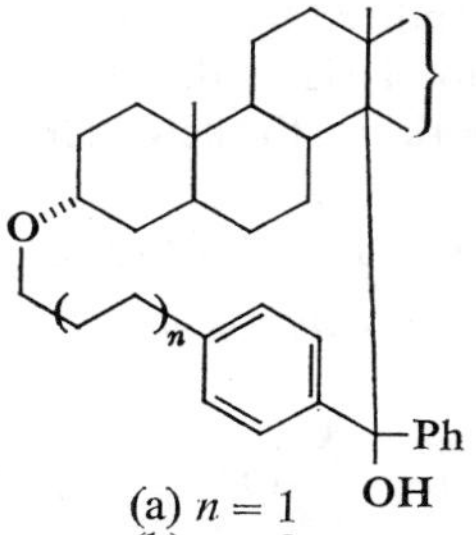

(105) (a) n = 1 (b) n = 2

There are slight differences between this report[59] and that of Breslow and Baldwin[58] in that the former describes a greater preponderance of olefinic material. There is also the difference in position of attack (cf. products 106 and 107), but this could be attributable to the greater

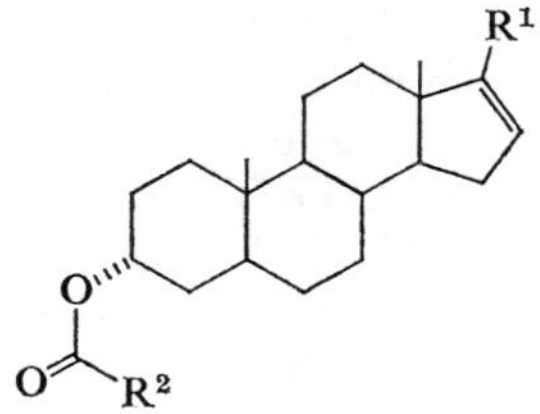

(106)

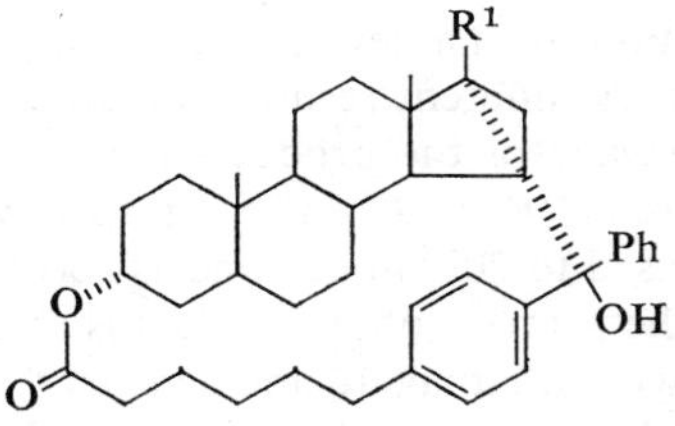

(107)

reach imparted to the excited carbonyl group by the additional two methylenes in the side-chain.

These examples of the functionalisation of remote positions seem likely to make a dramatic impact on steroid chemistry. Since the processes involved are free-radical in type, it would seem profitable to look for non-photochemical analogues.

The results of the photolysis of aroylazetidines (108) reported [60] last year have been elaborated.[61, 62]

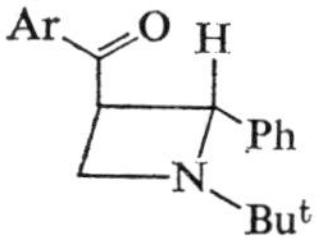

(108)

Interest in the photolysis of cyclopropyl ketones has been maintained.[63, 64] Photolysis of the bicyclononanone (109) in ether affords the reduction products (110, 111) by a singlet process which involves ring-opening of the cyclopropyl moiety.[64]

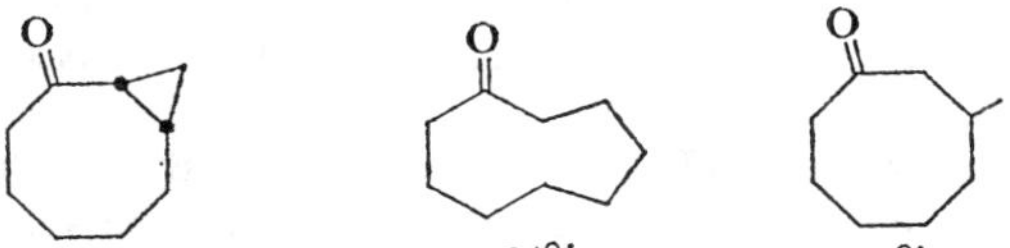

(109)   (110) 21%   (111) 10%

Remarkably, the internal cyclopropane bond breaks in preference to an external one. This is in contrast to the results of Dauben et al.[64] who have shown that a cyclopropylcarbinyl radical (112) produced by hydrogen abstraction from the solvent (Pr^{i}OH) is involved in the photolysis of the ketones (113). The reactions of the intermediate radical can be controlled

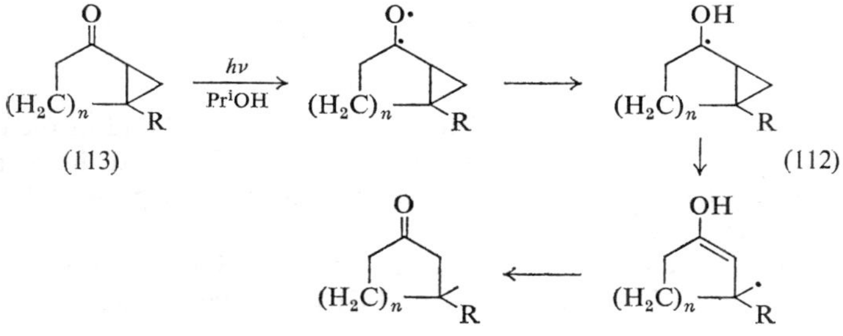

(113) → (hv, Pr^{i}OH) → → (112)
↓
(114) ←

[60] L. R. Hamilton, and A. Padwa and R. Gruber, cited by W. M. Horspool, in 'Photochemistry,' ed. D. Bryce-Smith, (Specialist Periodical Reports), The Chemical Society, London, vol. 1, p. 141.
[61] A. Padwa and R. Gruber, J. Amer. Chem. Soc., 1970, 92, 100.
[62] A. Padwa and R. Gruber, J. Amer. Chem. Soc., 1970, 92, 107.
[63] L. A. Paquette and R. F. Eizember, J. Amer. Chem. Soc., 1969, 91, 7108.
[64] W. G. Dauben, L. Schutte, R. E. Wolf, and E. J. Deviny, J. Org. Chem., 1969, 34, 2512.

by steric rather than thermodynamic features. The intermediacy of a triplet state in the reaction has been demonstrated by quenching and sensitization studies. The results for a series of ketones are shown in Table 12. The

Table 12 *Products from irradiation of bicyclic ketones* (113) [64]

Ketone	Irrad. time (hr)	Recovered starting material	Product (%)
$n = 2$; R = H	4	30	60
$n = 2$; R = Me	0·5	63	32
$n = 1$; R = H	2	39	28
$n = 1$; R = Me	2·5	38	51

overall reaction scheme involves ring-opening of the intermediate radical and subsequent hydrogen abstraction to yield the product, (114). These reactions are stereospecific (at room temperature), with rupture of an outside bond of the cyclopropane, *i.e.* the bond which best overlaps with the new radical centre.

A series of experiments (Table 13) showed that the products of photo-reduction were dependent on the temperature of the reaction. As the

Table 13 *Effect of temperature on the photolysis of ketone* (113; $n = 1$, R = Me) [64]

Temperature (°C)	Hours	Products (%) (114)+(115)	Ratio (115):(114)
25	0·5	12	1:9
	1	23	1:9
80	0·5	15	4:6
	0·8	18	3·5:6·5
135	0·5	12	5:5
	2	7	5:5

temperature is increased the tendency for the 'internal' bond of the cyclopropane to break is increased, to yield (115). This effect is attributed to the excess of energy possessed by the intermediate radical produced at the higher temperatures.

Correlations between mass spectral fragmentation patterns and the photochemistry of a series of bicyclo[4,1,0]heptan-2-ones have been

(113; $n = 1$, R = Me) ⟶ [cyclopentane-cyclopropane with OH] ⟶ [cyclohexenol with OH] ⟶ [cyclohexanone] (115)

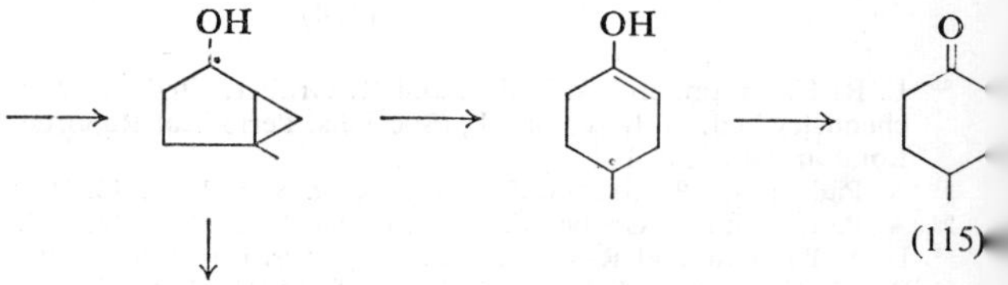

↓

(114; $n = 1$, R = Me)

reported.[65a] In the study [65b] of the photolysis of a simpler cyclopropyl ketone (116), ring-opening of the three-membered ring is brought about by Norrish Type II hydrogen abstraction. This intramolecular process

appears to be preferred over intermolecular hydrogen abstraction from the solvent. The product from the reaction is hex-1-en-5-one.

The photolysis of (+)-cis-caran-5-one (117) in ethereal solution gives three major products (118), (119), and (120), all of which result from ring-opening of the cyclopropane moiety.[66] The results are considerably different

from those reported for the photolysis of caran-2-one (121) whence dienal (122) was obtained.[67] No evidence was found for similar ring-cleavage in the former case.[66] No mechanistic examination of the reaction was put forward.

Applications of cyclopropane ring-opening reactions to uridines and thymines have also been reported.[68] Thus the photolysis of nucleosides (123) gave rise to the diazepine nucleosides (124).

Ring-opening of the aziridine (125) by photolysis in dioxan affords the 1,3-dipolar reagent (126) which was trapped by dimethyl acetylenedicarboxylate.[69]

[65a] C. Fenselau, G. W. Shaffer, and W. G. Dauben, *Org. Mass Spectrometry*, 1970, **3**, 1.
[65b] W. G. Dauben, L. Schutte, and R. E. Wolf, *J. Org. Chem.*, 1969, **34**, 1849.
[66] R. S. Carson, W. Cocker, S. M. Evans, and P. V. R. Shannon, *Chem. Comm.*, 1969, 726.
[67] W. G. Dauben and G. W. Shaffer, *Tetrahedron Letters*, 1967, 4415.
[68] T. Kunieda and B. Witkop, *J. Amer. Chem. Soc.*, 1969, **91**, 7751.
[69] S. Oida and E. Ohki, *Chem. Pharm. Bull. Tokyo*, 1969, **17**, 2461.

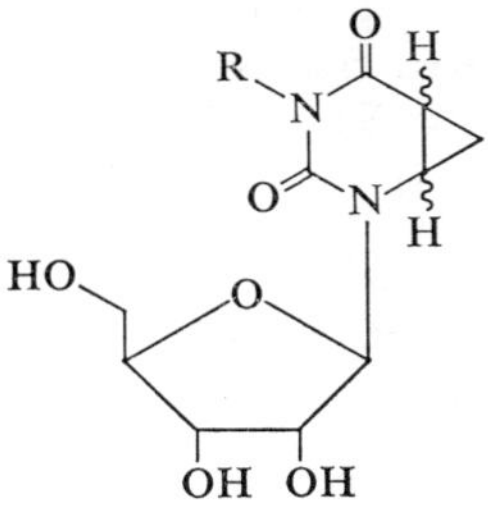

R = Me (5,6 β)
R = Me (5,6 α)
R = H (5,6 β)

(123)

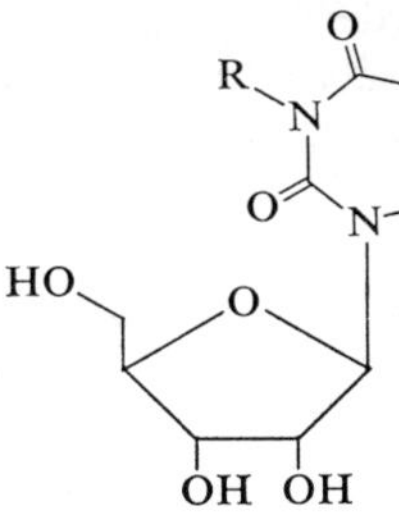

(124)

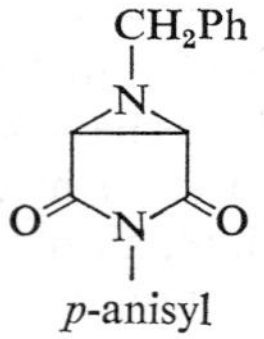

(125) p-anisyl

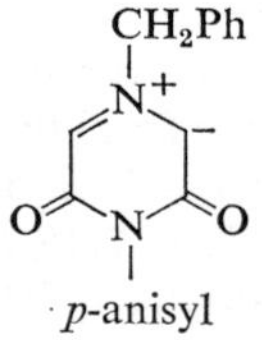

(126) p-anisyl

This section concludes by reference to a study of phosphonate photolysis.[70,71] The aroylphosphonates (127) have been shown to have enhanced $n \to \pi^*$ absorptions which are probably due to interaction with d-orbitals of the phosphorus. Irradiation of the dialkyl phosphonate (127a) was the only one which was atypical in that the trimeric product (128) was formed in a variety of solvents. The other alkyl phosphonates (127b, c) and the p-substituted derivatives (127d—f) all formed pinacols (129) by pinacolisation of the arylcarbonyl function from the triplet state. Quantum yield measurements showed that the influence of the p-aryl substituent had a marked effect on the reaction, and this could be correlated with the Hammett σ coefficient showing that the pinacolisation was dependent on the electron availability at the carbonyl function (Table 14). The nature of the solvent also appeared to have an effect probably relatable to the hydrogen-donating properties of the solvent (Table 15). In chlorohydrocarbon solvents, hydrogen abstraction must take place from another molecule of phosphonate.

Table 14 *Influence of a* p-*substituent on the photoreaction of aroylphosphonates* [70]

Compound	Φ	σ
(127f)	2·100	−0·268
(127e)	2·000	−0·197
(127c)	1·850	0
(127a)	1·792	+0·227

[70] K. Terauchi and H. Sakurai, *Bull. Chem. Soc. Japan*, 1970, **43**, 883.
[71] Y. Ogata and H. Tomioka, *J. Org. Chem.*, 1970, **35**, 596.

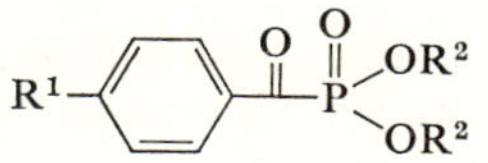

Photolysis of Carbonyl Compounds

(127)

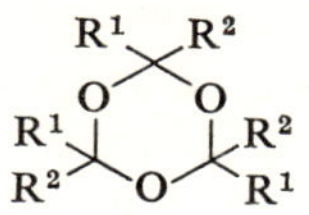

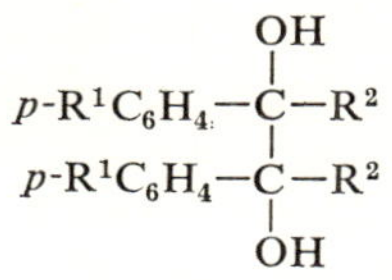

(129)

[128; R^1 = Ph, R^2 = $\overset{O}{\underset{\|}{P}}(OEt)_2$]

(a) R^1 = H, R^2 = Et
(b) R^1 = H, R^2 = Bu^n
(c) R^1 = H, R^2 = Pr^i
(d) R^1 = Cl, R^2 = Et, Pr^i, or Bu^n
(e) R^1 = Bu^t, R^2 = Et, Pr^i, or Bu^n
(f) R^1 = MeO, R^2 = Et, Pr^i, or Bu^n

(a) R^1 = H, R^2 = $PO(OBu^n)$
(b) R^1 = H, R^2 = $PO(OPr^i)$
(c) R^1 = Cl, R^2 = $PO(OR^3)$
(d) R^1 = Bu^t, R^2 = $PO(OR^3)$
(e) R^1 = OMe, R^2 = $PO(OR^3)$

Table 15 *Influence of solvents on photolysis of aroylphosphonates* [70]

Solvent	Φ	conc. (mol l^{-1})
Cyclohexane	0·745	1·0 × 10^{-4}
Ether	0·782	1·1 × 10^{-4}
Acetonitrile	0·660	1·0 × 10^{-4}
Carbon tetrachloride	0·621	1·0 × 10^{-4}

Contrary to these observations, the aroylphosphonates (130) are not reported [71] to show *d*-orbital participation in the excited state. The irradiations were carried out in benzene solution using either a Pyrex or a quartz reaction vessel. The reaction is presumed to involve an $n\pi^*$ triplet. The basic reaction is shown here, and is similar in the initial stages to a Norrish Type II elimination, in that the first step is thought to be the abstraction of a hydrogen atom by the excited carbonyl group (130a). Subsequently the molecule is thought to rearrange *via* the epoxide (131). This part of the process is thought to be thermal and involves a 1,2-acyl migration concomitant with epoxide ring-opening.

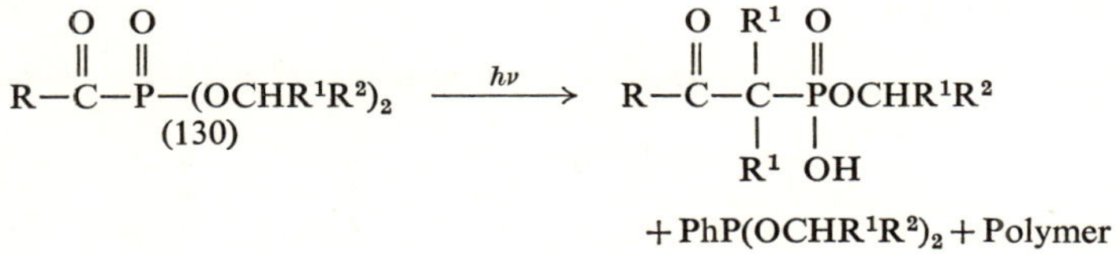

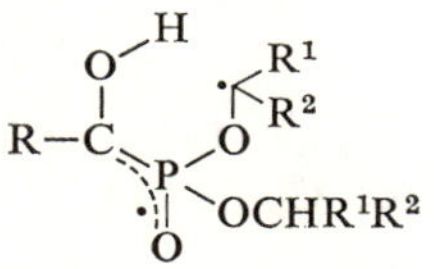

$$R-\overset{O}{\underset{\|}{C}}-\overset{O}{\underset{\|}{P}}-(OCHR^1R^2)_2 \xrightarrow{h\nu} R-\overset{O}{\underset{\|}{C}}-\overset{R^1}{\underset{R^1}{\underset{|}{C}}}-\overset{O}{\underset{\|}{P}}OCHR^1R^2$$
(130) |
                                                      OH

+ PhP(OCHR^{1}R^2)$_2$ + Polymer

(130a) 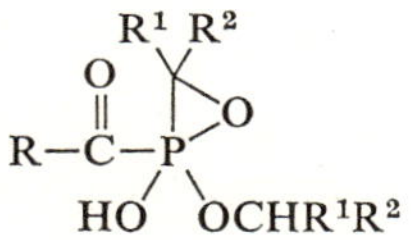   (131)

4 Rearrangement Reactions

Cyclopropane Rearrangements.—Cyclopropyl ketones can still undergo an isomerisation process in the absence of hydrogen abstraction. Thus photolysis of the ketone (132a) leads to isomerisation to the *cis*-isomer (132b) *via* the diradical (132c).[66] A more detailed study of this isomerisation

(132a)   (132b)   (132c)

process has been reported by Zimmerman *et al.*,[72] who have shown that the photolysis of the *trans*-diphenylcyclopropane (133a) in benzene affords two isomeric compounds (133b) and (133c) in a kinetically-controlled ratio

(133a)   (133b)   (133c)

of 3·5 : 1. The direct irradiation proceeds with a quantum yield of 1·06 for disappearance of starting material, an observation which is interpreted to show that once the ring is opened in the excited state it does not reclose in its original conformation, possibly because of an unfavourable eclipsed conformation of the opened species. The excited state has been reported to be a triplet (which stereoisomerises at 2×10^{11} s^{-1}). The same product distribution was obtained from both the direct irradiation and sensitized experiments (Table 16). This observation does not in itself establish a triplet pathway for the unsensitized process, although it points in that direction.

Table 16 *Product distribution from the photolysis of cyclopropyl ketone (133)* [72]

Additive	% Conversion	Quantum yield Φ (133)			Ratio b : c
		a	b	c	
—	6·1 (in Bu^{t}OH)	1·03	0·62	0·15	4·13
—	7 (in C$_6$H$_6$)	1·06	0·64	0·20	3·2
Propiophenone	10·4 (in C$_6$H$_6$)	0·29	0·18	0·044	4·09
Acetophenone	7·2 (in C$_6$H$_6$)	0·32	0·21	0·062	3·38
Piperylene	5·1 (in C$_6$H$_6$)	0·87	0·53	0·18	2·95

Experiments with the optically active ketone (133a) indicate that the C_1–C_2 bond is broken in preference to the C_1–C_3 bond (in a ratio of

[72] H. E. Zimmerman, S. S. Hixson, and E. F. McBride, *J. Amer. Chem. Soc.*, 1970, **92**, 2000.

8 : 1). The products from these experiments are (−) ketone (133b) and (+) ketone (133c).

Further results concerned with the above problem have been reported from the photolyses of the methoxy-substituted ketone (134a) which affords the two isomeric ketones (134b) and (134c) in a ratio of 3·4 : 1.[73]

(134a)

(134b) $\Phi = 0\cdot 548$

(134c) $\Phi = 0\cdot 130$

A quantum yield study showed that the direct reaction was considerably more efficient than the sensitized one, but that the product ratio was the same from both approaches. It was considered that the lowest triplet state ($k_r = 1\cdot 5 \times 10^{10}$ s^{-1}) was produced in the acetophenone sensitized runs and was also responsible for the direct irradiation isomerisation; but such reasoning is not watertight. Quenching studies showed that as the concentration of quencher was increased there was a small but real increase in the specificity of the reaction, *i.e.* the yield of the major product, ketone (134b), increased. This effect was thought to be due to the formation of an exciplex between the excited state ketone and the quencher. These results, taken in their entirety, imply that there is no charge separation in the excited states of the cyclopropylketones studied, and that the bond-opened species is presumably a diradical.[74]

Cyclobutane and Cyclopentane Rearrangements.—In last year's account, one of the more interesting problems associated with phenanthraquinone photochemistry was the genesis of the dioxoles (135) formed by the photo-addition of certain olefins to phenanthrenequinone.[75] It appeared fairly obvious at that time that the route to these compounds involved ring-opening of the oxo-oxetans (136) also found among the reaction products. However, the actual mode of rearrangement had not been verified.

A further investigation into this problem has proved that the dioxoles do indeed arise from subsequent photolysis of the oxo-oxetans (136). Farid and Scholz interpret the reaction as involving cleavage of the oxetan-ring into a diradical.[76] This process could be sensitized by low-energy sensitizers, and gave rise to the same products (Scheme 9) as were obtained

---
[73] H. E. Zimmerman and C. M. Moore, *J. Amer. Chem. Soc.*, 1970, **92**, 2023.
[74] K. G. Hancock *et al.*, cited by W. M. Horspool, in 'Photochemistry,' ed. D. Bryce-Smith, (Specialist Periodical Reports), The Chemical Society, London, 1970, vol. I, p. 153.
[75] S. Farid, D. Hess, G. Pfundt, K. H. Scholz, and G. Steffan, cited by W. M. Horspool, in 'Photochemistry,' ed. D. Bryce-Smith, (Specialist Periodical Reports), The Chemical Society, London, 1970, vol. I, p. 222.
[76] S. Farid and K. H. Scholz, *Chem. Comm.*, 1969, 572.

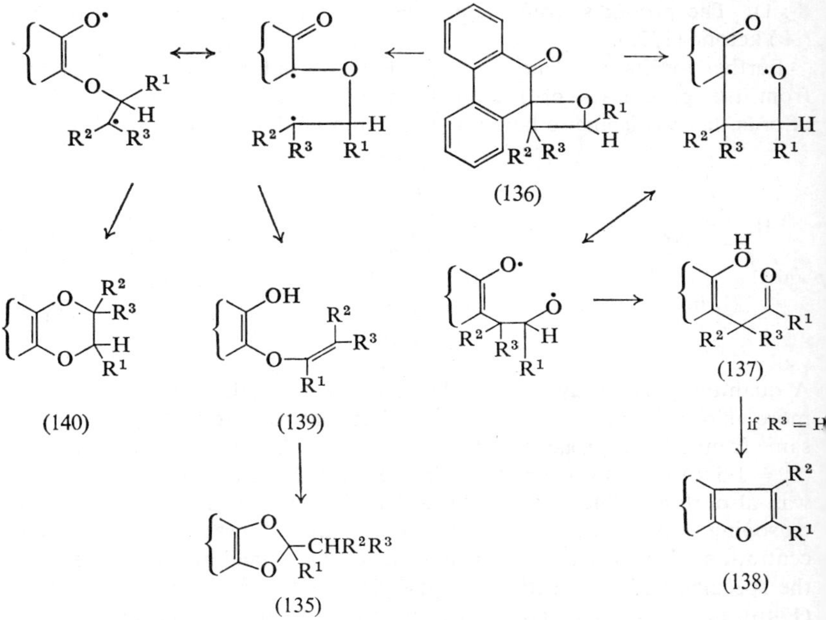

Scheme 9

from the direct irradiations. Thus cleavage of the O–C bond allows for formation of the ketone (137), the precursor of the furan (138), while C–C bond rupture gives the phenanthrol ether (139) which can be cyclised by traces of acid to the dioxole (135). The formation of the dioxen (140) lacks stereospecificity, especially as the temperature is increased. The temperature effect is reasonably attributed to increased rotation in the intermediate diradical.

These experiments do not appear to eliminate an attractive alternative route to the dioxoles involving intramolecular hydrogen abstraction as a primary step, as shown in Scheme 10.

The inefficient decarbonylation ($\Phi = 0.02$ in benzene) of 4,5,6,7-tetrahydroindan-2-one (141) to form 1,2-dimethylenecyclohexane has been shown to be a triplet process by the effect of triplet quenchers on the

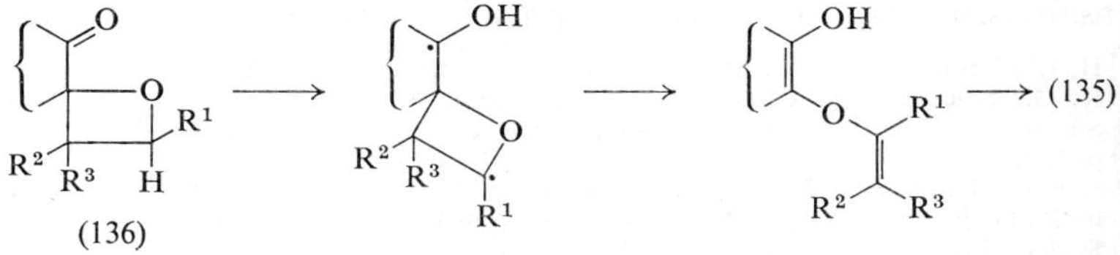

Scheme 10

evolution of carbon monoxide.[77] In iso-octane, ethanol, isopropanol, and methylene dichloride, two other products have been identified as the cis-

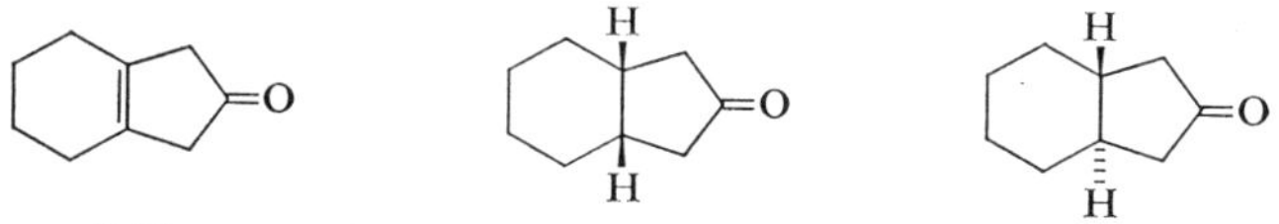

(141)   (142)   (143)

and the *trans*-hexahydroindan-2-ones, (142, 9%) and (143, 3%) respectively. Maximum yields of the products were obtained (26%) by prolonged irradiation, although under these conditions several other products were also detected. Although this was not mentioned, it seems possible that this photoreduction is an example of intramolecular energy transfer.

Cyclohexane Rearrangements.—A cyclopropanone intermediate (144) is proposed[78] to account for the photochemical rearrangement of the chromanone (145). The authors suggest that a second photon is necessary for the ring-opening of the cyclopropanone to form the product (146); however, intramolecular ketalisation and thermal ring-opening seem equally plausible (Scheme 11).

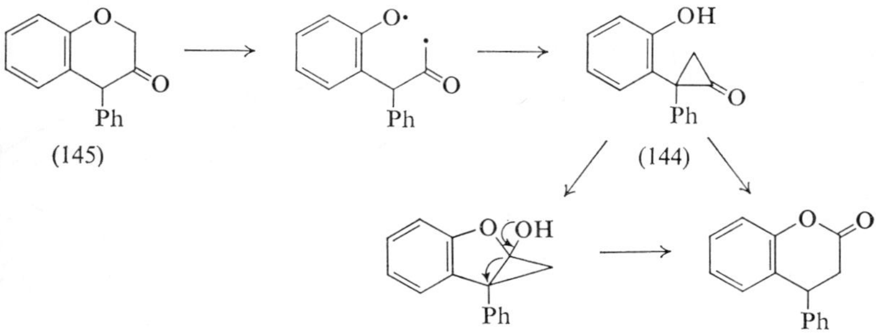

Scheme 11

The involvement of sulphur has also been studied in the photolysis of isothiochromanones (147): these produce the isomeric thiochroman-3-ones (148).[79] Photolysis of the substituted derivatives (147b, c) helps to show that the sulphur atom becomes bonded to the aryl carbon atom to which the carbonyl group was originally bound. This suggested the intermediacy of the spirocyclobutanone (149) and this was verified by the photolysis of the simpler thiopyranone (150) which afforded the thiacyclobutanone (151).

[77] P. S. Engel and H. Ziffer, *Tetrahedron Letters*, 1969, 5181.
[78] P. K. Grover and N. Anand, *Chem. Comm.*, 1969, 982.
[79] W. C. Lumma and G. A. Berchtold, *J. Org. Chem.*, 1969, **34**, 1566.

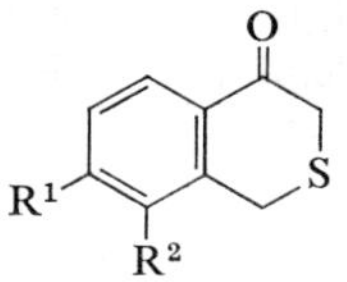

(a) $R^1 = R^2 = H$
(b) $R^1 = MeO$, $R^2 = H$
(c) $R^1 = H$, $R^2 = Me$
(147)

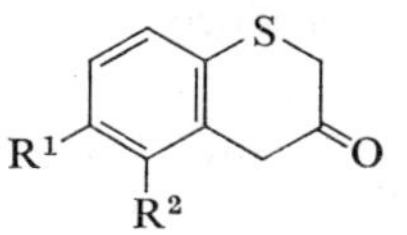

(148) 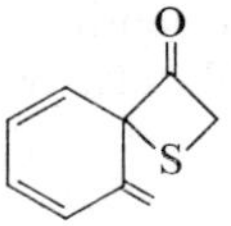 (149) 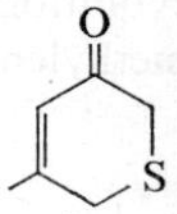 (150)

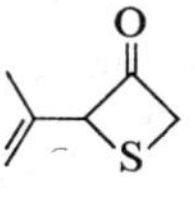

(151) 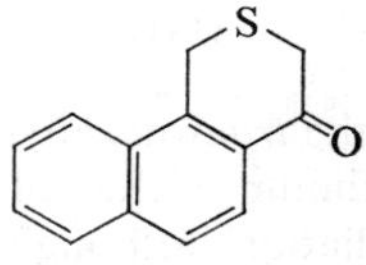 (152)

In molecules where $\pi \to \pi^*$ excitation is dominant, *e.g.* derivative (152), no rearrangement was observed and it is suggested that this is reasonable evidence for the necessary intermediacy of an $n\pi^*$ state.

Pete and Villaume [80] have studied the photoreactions of the steroidal epoxyketones (153a—d) to give the identical diketone (154) (which exists in the monoenolic form shown). However, minor differences in the reaction products can be related to the stereochemical relationship of the initially formed diradical centres. Thus, the photolysis of ketone (153a) produced

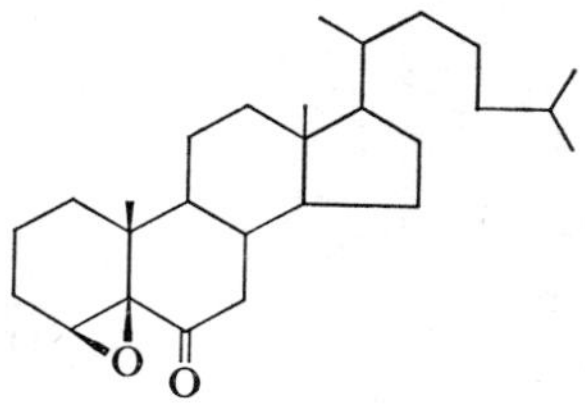

(153a) 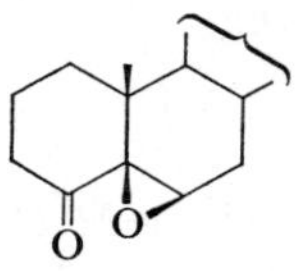 (153b) 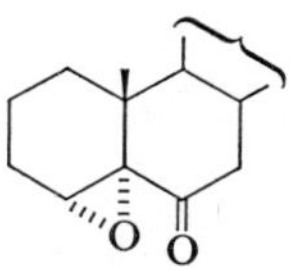 (153c) 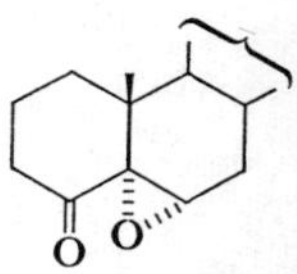 (153d)

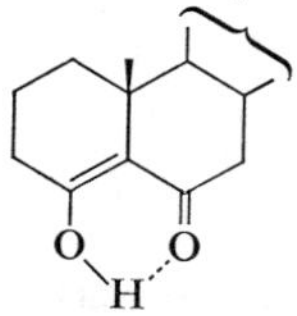

(154) 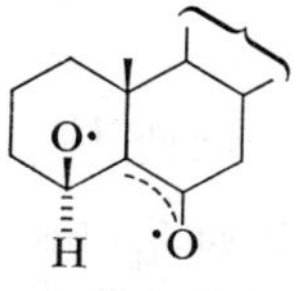 (155) 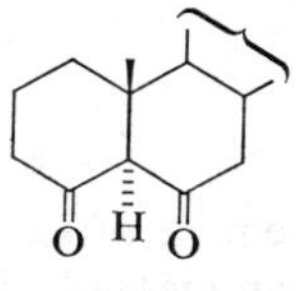 (156)

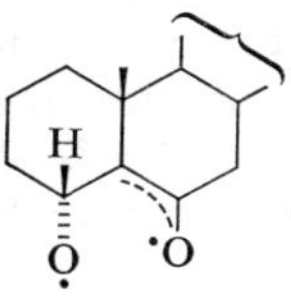

(157) 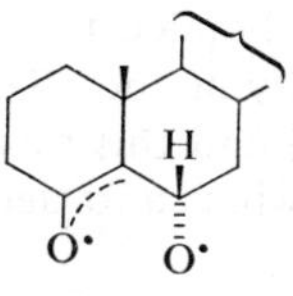 (158) 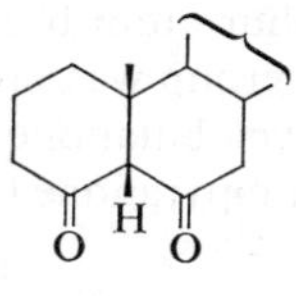 (159)

[80] J. P. Pete and M. L. Villaume, *Tetrahedron Letters*, 1969, 3753.

intermediate (155) which yields a *trans*-fused ring junction (156a), whereas ketones (153c and d) give intermediates (157 and 158) which produce *cis*-fused rings (159).

Bicyclic and Spiro Systems.—The formation of ketals from the photolysis of fenchone (160) has previously been reported.[81, 82] A new report has helped to elucidate the route by which the ketals are formed.[83] Irradiation of fenchone (160), in methanol containing a trace of toluene-*p*-sulphonic acid gives, after 2—3 weeks, four products. The major products (161, 162) have been identified as ketals of two aldehydes formed by Norrish Type I cleavage of fenchone. The two minor products (163, 164) are cyclic ketals.

The ketal (164) arises from a carbene mechanism resulting from fission of bond *b* in fenchone followed by rebonding to yield the intermediate carbene (165) which then inserts into the solvent to give an epimeric mixture of ketals (164). The divergence of the pathway for the fate of the diradical formed by fission of bond *b* in fenchone is not fully understood. It is indeed surprising that no products of carbene insertion arising from bond *a* fission have been isolated, although the authors claim that hydrogen abstraction is more facile in the case of the diradical formed by fission of bond *a*.

A carbene intermediate (166) also explains the ketals (167) and (168) formed by photolysis of the thiospiroketone (169).[84] With light of $\lambda > 280$ nm, and methanol solutions, an $n \rightarrow \pi^*$ transition in the carbonyl group brings about a Norrish Type I cleavage to give a product (170, 22%), resulting from elimination of carbon monoxide, in addition to the ketals. With 253·7 nm radiation, the major product (171, 56%) is formed by elimination of sulphur. The transition involved in the reaction bringing about elimination of sulphur is proposed to be of $n \rightarrow \sigma^*$ type. From these

---
[81] P. Yates and A. G. Fallis, *Tetrahedron Letters*, 1967, 4621.
[82] A. G. Fallis, *Diss. Abs.*, 1969, **29B**, 2348.
[83] P. Yates and G. Hagens, *Tetrahedron Letters*, 1969, 3623.
[84] J. G. Pacifici and C. Diebert, *J. Amer. Chem. Soc.*, 1969, **91**, 4595.

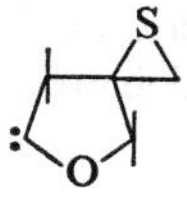

(166) 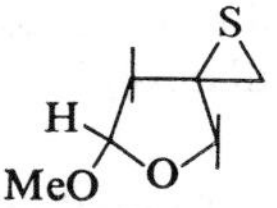 (167) 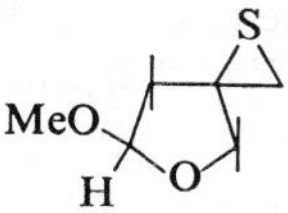 (168)

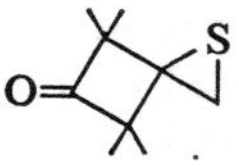

(169)  (170) 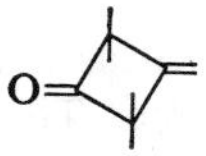 (171)

results it is thought that there is little intramolecular energy transfer between the $n\sigma^*$ and $n\pi^*$ states.

5 Oxetan Formation

A quantitative study of the addition of acetone to *cis*- and *trans*-1-methoxybut-1-ene has been reported.[85] The qualitative aspects of the work reported briefly last year showed that four oxetans (172—175) resulted from the addition.[86] The latest work [85] indicates that two intermediates are essential

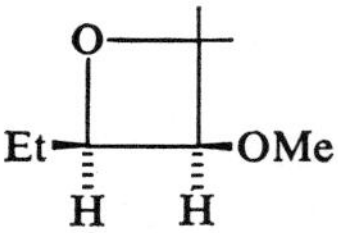

(172) 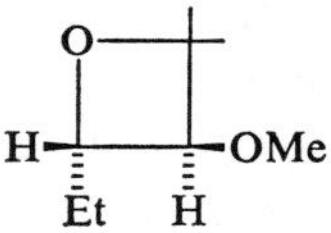 (173) 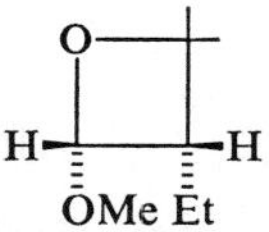 (174) 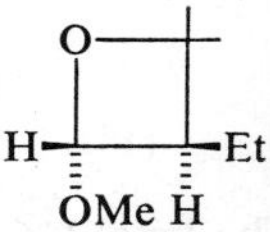 (175)

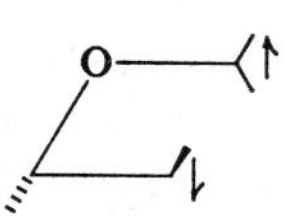

(176) 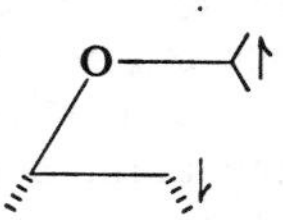 (177) 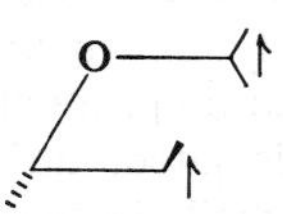 (178) 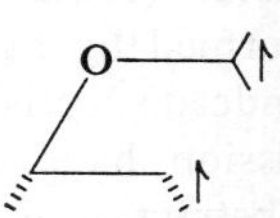 (179)

in order to explain the results shown in Table 17. The authors state that the best rationale for the reactions involves four biradical intermediates (176—179). The first pair are the result of the addition of singlet acetone to the olefin and the second pair arise from the addition of triplet acetone. This treatment appears to explain all the problems encountered in the analysis of the reaction.

Other points worthy of mention are the observations that the additions of singlet and triplet acetone to the olefin proceed *via* chemically similar

[85] N. J. Turro and P. A. Wriede, *J. Amer. Chem. Soc.*, 1970, **92**, 320.
[86] N. J. Turro and P. A. Wriede, cited by W. M. Horspool, in 'Photochemistry,' ed. D. Bryce-Smith, (Specialist Periodical Reports), The Chemical Society, London, 1970, vol. I, p. 156.

Photolysis of Carbonyl Compounds

transition states from which the reaction pathway diverges according to the multiplicity. Thus, if acetone is in its singlet state there are three ways in which the diradical intermediate can react: (*i*) fragmentation to acetone and olefin, (*ii*) rotation about all single bonds, and (*iii*) formation of oxetan. If acetone is in its triplet state only two chemical effects are reasonable: (*i*) intersystem crossing to the singlet state or (*ii*) rotation about single bonds.

Table 17 Ratio of cis- and trans-oxetans from 1-methoxybut-1-ene under various conditions [85]

trans-*olefin* 172 : 173	cis-*olefin* 172 : 173	cis-*olefin* 174 : 175	Conditions
1·20	1·35	0·88	Acetone solvent
2·70	4·50	2·20	0·4 M penta-1,3-diene–acetone
1·04	0·97	0·65	Extrapolated to zero [methoxybutene]

Thus, owing to the expected greater lifetime of the triplet-derived diradical relative to that of the singlet-derived one,* there should be greater loss of configuration at the carbon atom bearing the unpaired electron.

Aliphatic and alicyclic ketones do not normally add readily to simple olefins owing to their high triplet energies and tendency to abstract hydrogen from the substrate. However, vinyl ethers have been found to undergo efficient photoaddition of acetone, propionaldehyde, and cyclohexanone to give mixtures of 2- and 3-alkoxyoxetans (180) in a *ca.* 1 : 3 ratio which was almost completely independent of the solvents used.[87] The major product in each case arises from the attack of the triplet carbonyl group on the vinyl ether to produce the more stable diradical. Additions

(a) $R^1 = R^2 = Me$, $R = Et$, Bu^n, Bu^i
(b) $R = R^1 = Et$, $R^2 = H$
(c) R^1—$R^2 = (CH_2)_5$, $R = Et$

(180)

$R^1 = R^2 = Me$
$R^1 = H$, $R^2 = Et$
R^1—$R^2 = (CH_2)_5$
$R^1 = R^2 = Ph$

(181)

(182)

to keten diethylacetal seem to be more stereospecific and only 3,3-dialkoxyoxetans (181) could be isolated.[87] The increased specificity is presumably due to the comparatively higher stability of the diradical (182).

A singlet complex is reported [88] to account for the formation of oxetans (183a and b) from the photolysis ($\lambda > 310$ nm) of propionaldehyde in the

[87] S. H. Schroeter and C. M. Orlando, *J. Org. Chem.*, 1969, **34**, 1181.
[88] T. Kubota, K. Shima, S. Toki, and H. Sakurai, *Chem. Comm.*, 1969, 1462.

* Use of terms such as 'singlet diradical' is too loose to be recommended.

presence of cyclohexa-1,3-diene. This result arises from the fact that a different mechanism appears to be in operation for oxetan formation than for cyclohexadiene dimerisation, a known triplet process. The oxetans are presumably formed by way of the more stable diradical (184); but it is strange that no trace of the other isomers was found.

(a) $R^1 = H$, $R^2 = Et$
(b) $R^1 = Et$, $R^2 = H$

(183)   (184)

The results from the photoaddition of aldehydes and ketones to a variety of furans to form oxetans have been described.[89] The aldehydes used are reported to be more reactive than ketones in the formation of oxetans. No difference in the mechanistic interpretation of the results from that currently accepted for oxetan formation is put forward, unlike a previous report where oxetan formation by aldehydes was thought to arise from a singlet complex.[88] Biacetyl has also been successfully added photochemically to furan and several other substrates.[90] It was found that biacetyl is less reactive than acetone in oxetan formation.

Acetone and perdeuterioacetone have been shown[91] to add photochemically (at 75 °C) to phospholen (185). The initial adduct is thought to be the oxetan (185a) but this rearranges to the isolated product (185b).

(185)   (185a)   (185b)

Twelve products were detected from the direct irradiation of pure acetone through a quartz filter.[92] Some of these were diacetone alcohol (12%) acetonylacetone (25%), pinacol (12%), and a new dimer, 2,2,3-trimethyl-3-hydroxyoxetan (186, 50%). Conjecture concerning the mechanism for the formation of this dimer suggests that it is formed by cyclo-addition of acetone to its enol, *i.e.* the addition of acetone to an electron-rich olefin.

Considerable interest has been shown in the reactions of ketones with electron-deficient olefins where a different mechanism from that described for electron-rich olefin addition is in operation.

[89] H. Sakurai, K. Shima, and S. Toki, *J. Chem. Soc. Japan*, 1968, **89**, 537.
[90] H.-S. Ryang, K. Shima, and H. Sakurai, *Tetrahedron Letters*, 1970, 1091.
[91] W. G. Bentrude and K. R. Darnall, *Chem. Comm.*, 1969, 862.
[92] J. T. Przybytek, S. P. Singh, and J. Kagan, *Chem. Comm.*, 1969, 1224.

(186)

Direct irradiation of acetone in acetonitrile solution in the presence of cis- or trans-1,2-dicyanoethylene afforded the corresponding cis- (187a) and trans-oxetan (187b) as the sole product in each case.[93] The reaction is

(187a) (187b) (187c)

therefore remarkably stereospecific and it has been shown that, unlike the addition of acetone to electron-rich olefins where triplet states are involved, a singlet mechanism obtains in the present example. It is argued that the addition of acetone excited singlets to either of the olefins cannot proceed in the normal manner, *i.e.* the electrophilic addition of the half-filled n-orbital of the excited acetone to the olefin, because loss of the original geometry of the olefin might be expected to occur. Thus, a mechanism involving the electron-rich three-electron system above and below the plane of the carbonyl group in the acetone singlets has been invoked. This system can act nucleophilically towards dicyanoethylene (Figure 2) in a

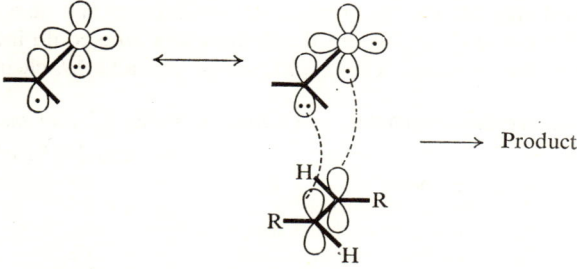

Figure 2

concerted or 'quasi-concerted' fashion. Such a mechanism involving the nucleophilic attack of a carbonyl excited state also explains the observations of Barltrop and Carless,[94] reported last year, where addition of acetone to methacrylonitrile gave only the 2-cyano-oxetan (187c). One should note, however, that stereospecificity in cycloadditions does not guarantee their concertedness, especially in reactions between donors and acceptors where

[93] J. C. Dalton, P. A. Wriede, and N. J. Turro, *J. Amer. Chem. Soc.*, 1970, **92**, 1318.
[94] J. A. Barltrop and H. A. J. Carless, cited by W. M. Horspool, in 'Photochemistry,' ed. D. Bryce-Smith, (Specialist Periodical Reports), The Chemical Society, London, 1970, vol. I, p. 158.

polar effects can promote interactions between reactive sites: see Vol. I, p. 223.

The addition of acetone to dicyanoethylene is rather inefficient (Table 18), possibly owing to the formation of an exciplex which decays to the ground

Table 18 Quantum yields for the formation of trans-oxetan and for olefin isomerisation [93]

[trans-dicyanoethylene] (mol l^{-1})	Φ^a	Φ^b
0·104	0·026	0·078
0·202	0·038	0·065
0·303	0·047	0·056
0·40	0·051	0·050
0·50	0·054	0·042

^{a} Quantum yield for the formation of trans-oxetan (187b).
^{b} Quantum yield for the formation of cis-olefin from the trans-isomer.

state without giving a photochemical product. In addition to the singlet processes described above, there is also triplet acetone sensitized isomerisation of the olefin. This is more efficient at low olefin concentrations (Table 18).

Further details concerning the proposed complex formation in such oxetan-forming reactions has come from a study of the quenching of fluorescence of cyclohexanones by 1,2-dicyanoethylene.[95] This has indicated that there are steric effects which influence the rate of formation of the complex necessary for quenching or oxetan formation. It can be seen that with a more bulky ketone the presence of substituents and also the preference for a boat or a chair form in the cyclohexane ring could influence the interaction in the complex. The results, shown in Table 19, clearly demon-

Table 19 Fluorescence quenching of cyclohexanone by 1,2-dicyanoethylene [95]

Ketone	k_c (l mol^{-1} s^{-1} 10^{-9})
1. Cyclohexanone	3·8
2. 4-t-Butylcyclohexanone	3·6
3. cis-3,5-Dimethylcyclohexanone	3·3
4. 3,3,5-Trimethylcyclohexanone	1·3
5. 2,2,5,5,-Tetramethylcyclohexanone	1·2
6. 2-Methylcyclohexanone	1·8
7. cis-2,6-Dimethylcyclohexanone	0·9
8. trans-2,6-Dimethylcyclohexanone	2·5
9. 2,2-Dimethylcyclohexanone	1·7
10. Adamantanone	5·0

strate this. The first three entries in the Table show that in comparison with cyclohexanone the introduction of equatorial methyl groups at C-3 and C-5 or the addition of an equatorial butyl group at C-4 does not seriously affect the rate of formation of the complex. However, axial

[95] J. C. Dalton, D. M. Pond, and N. J. Turro, *J. Amer. Chem. Soc.*, 1970, **92**, 2173.

methyl substituents at C-3 and C-5 (entries 4 and 5) inhibit the rate of formation. From these results and others reported in the Table it is argued that equatorial attack on chair cyclohexanone is preferred to axial attack. This hypothesis is being further tested.

The photoaddition of several alkyl ketones to 1,2-dicyanoethylene to produce oxetans (188) has been described.[96] The formation of the oxetans is specific with the *trans*-isomer being favoured over the *cis* in the ratio 3 : 1. The specificity of this reaction is explained on the basis of efficient addition of the ketone singlets to the substrate. The involvement of triplet states is known to isomerise the olefin, while singlets add to the olefin but do not bring about isomerisation.

(188)

Alkyl substituent	Total yield (%) based on trans-*olefin*
$R^1 = R^2 = Me$	53·7
$R^1R^2 = (CH_2)_5$	70·2
$R^1R^2 = (CH=CH)_2$	53·0
$R^1 = Me, R^2 = (CH_2)_2Me$	40·0
$R^1 = Me, R^2 = (CH_2)_3Me$	28·7

Additions of ketones to maleic anhydride have been complicated by dimerisation and polymerisation of the olefin, and only the adduct (189) between acetone and maleic anhydride has been isolated.[96]

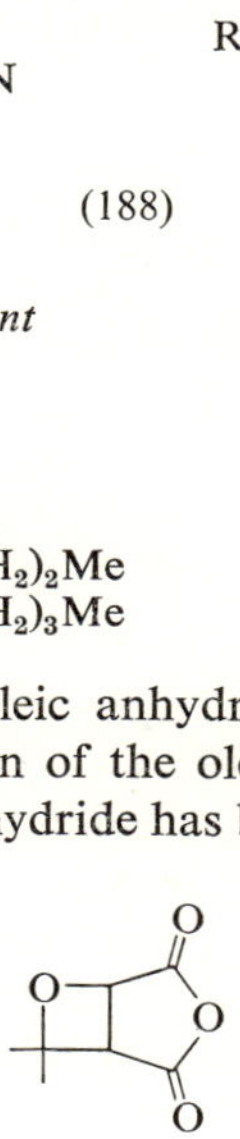

(189)

The addition of carbonates [97] and aromatic esters [98] to diphenylethylene has also afforded oxetans.

The formation of an oxeten (191) as an intermediate has been invoked to explain the formation of enones (192) in the photochemical additions of carboxylate esters (190) to diphenylacetylene;[99] an analogous mechanism

---
[96] N. J. Turro and P. A. Wriede, *J. Org. Chem.*, 1969, **34**, 3562.
[97] T. Tominaga and S. Tsutsumi, *Tetrahedron Letters*, 1969, 3175.
[98] Y. Shigemitsu, H. Nakai, and Y. Odaira, *Tetrahedron*, 1969, **25**, 3039.
[99] T. Miyamoto, Y. Shigemitsu, and Y. Odaira, *Chem. Comm.*, 1969, 1410.

has previously been advanced for the reaction of *p*-benzoquinone to the same acetylene.[100]

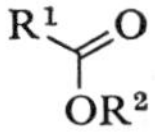

(190)

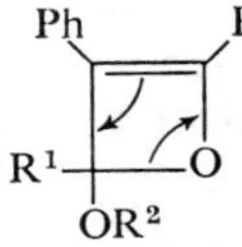

(191)

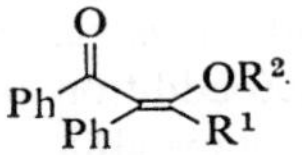

(192)

(a) R^1 = PhCO$_2$Me, R^2 = Me
(b) R^1 = *p*-cyanophenyl, R^2 = Me
(c) R^1 = ethoxycarbonyl, R^2 = Et

(a) 32% *trans*, 32% *cis*
(b) 80%
(c) 25%

Further reports [101, 102] of intramolecular cyclisation of acylbicyclo[2,2,2]-octenes (193) to the oxetans (194) have appeared, and of the norbornenyl derivatives (195) to oxetans (196) and (197).

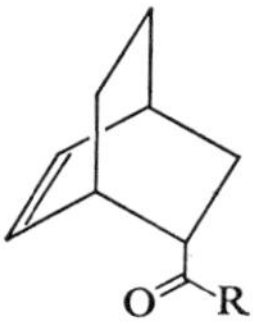

(a) R = H
(b) R = Ph
(c) R = Me
(193)

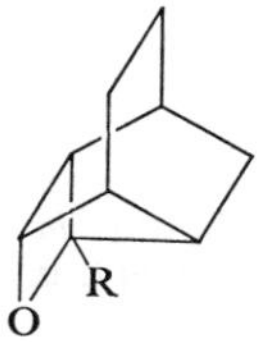

(194)

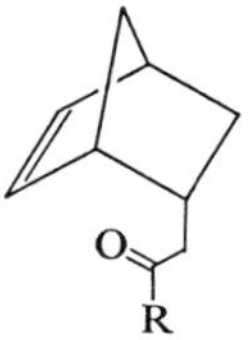

(a) R = Me
(b) R = H
(195)

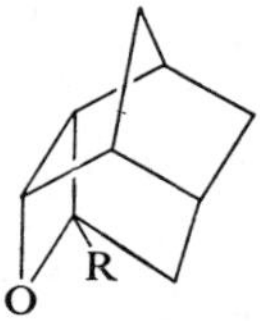

(196)

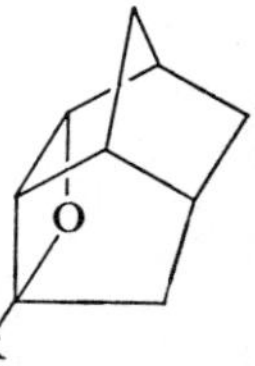

(197)

6 Alkylation Reactions

Reactions of aromatic carbonyl compounds having S_1 states of $\pi\pi^*$ character with alkyl substituted aromatic substrates has been reported.[103] The reactions were simple hydrogen abstractions by the carbonyl function and subsequent radical combination to produce pinacols (198), hydroxy-diaryl ethanes (199), and substituted bibenzyls. Benzophenone has also

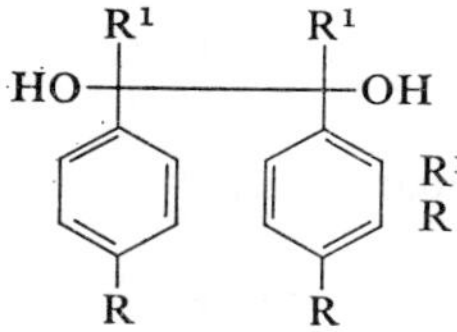

(198)

R^1 = Me
R = Me, MeO, or H

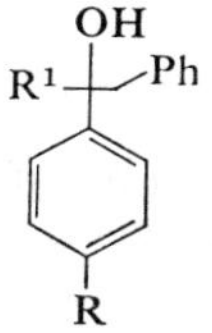

(199)

[100] H. E. Zimmerman and L. Craft, *Tetrahedron Letters*, 1964, 2131; D. Bryce-Smith, G. I. Fray, and A. Gilbert, *Tetrahedron Letters*, 1964, 2137.
[101] R. R. Sauers and J. A. Whittle, *J. Org. Chem.*, 1969, **34**, 3579.
[102] R. R. Sauers and K. W. Kelly, *J. Org. Chem.*, 1970, **35**, 498.
[103] D. Bellus and K. Schaffner, *Helv. Chim. Acta*, 1969, **52**, 1010.

been shown [104] to pinacolise in the presence of triphenylsilane, giving 56% yields of silyl ether (200). This is contrary to an earlier report.[105]

$$Ph_3SiOCPh_2CPh_2OH$$
(200)

Irradiation of ethyl acetoacetate in methanol solution through a quartz filter gave a product (201a, 55%), the result of addition of the solvent to the carbonyl group of the acetyl function in a mixed pinacolisation.[106] An analogous product (201b) was obtained from the irradiation in propan-2-ol. The formation of product (201a) was shown to be time dependent and after 4 hr (the initial irradiation had been carried out for 2 hr) (201a) was no longer present in the irradiation mixture but had been replaced by the

(201) (a) R = H (b) R = Me   (202)

lactone (202, 60%), which had also been isolated from the initial experiment in trace amounts. Little evidence has been reported concerning the mechanism of the reaction apart from the fact that (201a) was converted photochemically into (202). It was assumed that an $n \to \pi^*$ excitation was essential for the process, and in agreement with this no reaction was found to take place in benzene or t-butanol, a result consistent with the need for a hydrogen abstraction process.

7 Ketone Fragmentation Reactions

The photoreaction of triphenylphosphine with benzophenone has been shown to involve the $n\pi^*$ state of benzophenone.[107] The intermediacy of the $n\pi^*$ state was inferred from the failure of 2-hydroxy-4-methoxybenzophenone, a compound known to have a low-lying $\pi \to \pi^*$ state, to undergo the same reaction. The outcome of the reaction is the deoxygenation of the benzophenone to form, on reaction with an additional mole of the phosphine, diphenylmethylenetriphenylphosphorane. The formation of the ylide is thought to involve the formation of diphenylmethylene, although no direct evidence was obtained for the intermediacy of carbenes. Deoxygenation presumably results from the interaction of the triphenyl phosphine with the excited state benzophenone forming a 1 : 1 complex (203). How-

$$Ph_2\bar{C}-\overset{..}{\underset{..}{O}}-\overset{+}{P}Ph_3$$
(203)

---

[104] H. D. Becker, *J. Org. Chem.*, 1969, **34**, 2469.
[105] C. Eaborn, 'Organosilicon Compounds,' Butterworth, 1960, p. 213.
[106] S. P. Singh and J. Kagan, *Chem. Comm.*, 1969, 1121.
[107] L. D. Westcott, H. Sellers, and P. Poh, *Chem. Comm.*, 1970, 586.

ever, whether the deoxygenation involves a three-membered transition state or the intervention of another mole of triphenyl phosphine has yet to be established.

The most common photoelimination reactions involve the loss of nitrogen. In the main these will be discussed in a later Chapter, but a few of them involving carbonyl compounds will be included here.

The irradiation of the bis-diazoketone (204) in methanol–tetrahydrofuran has been shown to afford a mixture of products.[108] These were separated

$$\text{Ph}\underset{\underset{O}{\|}}{\overset{N_2}{\overset{\|}{C}}}\underset{}{\overset{N_2}{\overset{\|}{C}}}\text{Ph}$$

(204)

by t.l.c. and identified as the methyl esters of *erythro-* and *threo-*2,3-diphenyl-3-methoxypropionic acids (41 and 13% respectively), methyl α-phenylcinnamate (4%), and diphenylacetylene (24%). In toluene, the irradiation gave only diphenylacetylene (62%). Irradiation (λ = 436 nm) of the ketone (204) at −40 °C in an unspecified solvent showed that diphenylcyclopropenone was formed. Thus, it seems reasonable to assume that the diphenylcyclopropenone is the initial product in toluene and efficient decarbonylation at the higher temperature affords diphenylacetylene. The esters presumably arise from a Wolff rearrangement of the carbene intermediate.

The photo-production of carbenes by the elimination of nitrogen from ketones (205) has also been used to form the sulphonium ylides (206) by trapping the carbene with dialkyl sulphides.[109] The ylide (207a) and thiiran (207b) have been investigated [110] as potential intermediates in the formation of γ-butyrolactone from the photolysis of diethyl diazomalonate in the

$N_2C(COR)_2$ $R^1{}_2\overset{+}{S}-\overset{-}{C}(COR)_2$ $Ph_2C=\overset{+}{S}-\overset{-}{C}(CO_2Et)_2$

(a) R = OMe
(b) R = OEt   (206)   (207a)
(c) R = Me

(205) $Ph_2\underset{S}{\triangle}(CO_2Et)_2$

(207b)

(208)

presence of thiobenzophenone. Among the other products isolated is ethyl 2,2-diphenylethylene-1,1-dicarboxylate, presumably formed by extrusion of sulphur from thiiran (207b). Loss of sulphur is also reported [111] in

---

[108] P. J. Whitman and B. M. Trost, *J. Amer. Chem. Soc.*, 1969, **91**, 7534.
[109] W. Ando, T. Yagihara, S. Tozune, and T. Migita, *J. Amer. Chem. Soc.*, 1969, **91**, 2786.
[110] J. A. Kauffman and S. J. Weininger, *Chem. Comm.*, 1969, 593.
[111] N. Ishibe, M. Odani, and K. Teramura, *Chem. Comm.*, 1970, 371.

the formation of the desulphurised dimer (208) from 2,6-diphenyl-4H-pyran-4-thione in dioxan. This dimerisation is thought to occur from a $\pi\pi^*$ triplet since the reaction could be quenched or sensitized (anthraquinone, E_T = 203·6 kJ mol^{-1}, 48·7 kcal mol^{-1}).

Decarbonylations.—Decarbonylations have continued to be a source of interest. Several years ago decarbonylation was mainly a gas-phase phenomenon but in recent times solution- and even solid-phase examples have come to light.

The cyclopentenealdehydes (209a—d) undergo facile decarbonylations on photolysis (high-pressure Hg arc, Pyrex filter) in benzene.[112] The products formed are the cyclopentenes (210) which undergo further photolysis by cyclisation of the aryl rings of the stilbene moiety to form the phenanthrocyclopentenes (211), the major products of the reactions.

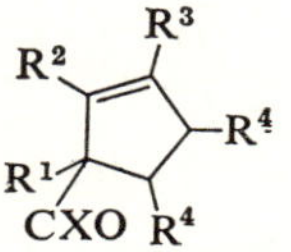

(209) 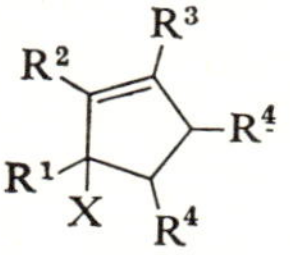   (210) 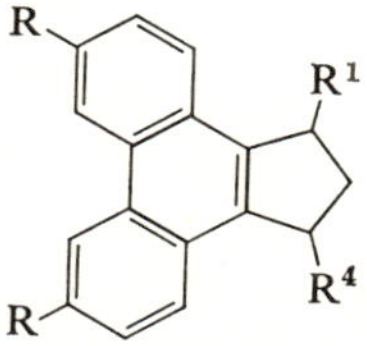   (211)

(a) $R^1 = R^2 = R^3 = R^4$ = Ph, X = H
(b) $R^1 = R^4$ = Ph, $R^2 = R^3$ = p-anisyl, X = H
(c) $R^1 = R^4$ = Ph, $R^2 = R^3$ = p-chlorophenyl, X = H
(d) $R^1 = R^2 = R^3 = R^4$ = Ph, X = D

(a) $R^1 = R^4$ = Ph, R = H
(b) $R^1 = R^4$ = Ph, R = OMe
(c) $R^1 = R^4$ = Ph, R = Cl

Dowd[113] reports the efficient decarbonylation of the ketone (212) by u.v. irradiation at −196 °C to afford a species whose e.s.r. spectrum is consistent with formulation as perdeuteriotetramethylene-ethane (213).

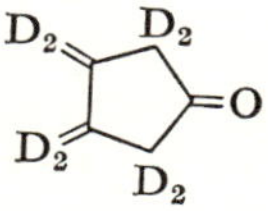

(212) 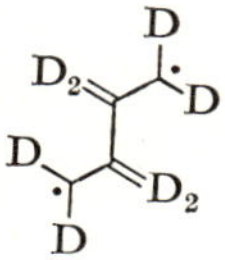   (213) 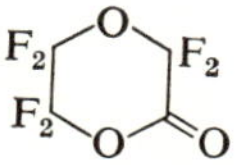   (214)

The solution-phase photolysis of perfluorovalerolactone (214) is reported to produce good yields (70%) of perfluoro-1,3-dioxolan by decarbonylation.[114]

Inevitably the application of photochemical decarbonylation techniques have been applied to carbohydrate chemistry to afford ring-contracted

---
[112] H. Dürr, P. Heitkämper, and P. Herbst, *Tetrahedron Letters*, 1970, 1599.
[113] P. Dowd, *J. Amer. Chem. Soc.*, 1970, **92**, 1066.
[114] J. R. Throckmorton, *J. Org. Chem.*, 1969, **34**, 3438.

products. The di-isopropylidene diulose (215) has been shown[115] to undergo a ready photo-decarbonylation by irradiation with or without the use of a Pyrex filter. Two products (216a and b) were isolated. The route to the products is thought to involve a Norrish Type I ring-cleavage followed

by equilibration of the radical pair prior to rebonding, as shown. The method is of interest since the major product (216b) is not the isomer produced by the acid-catalysed isopropylidenation of D-ribulose.

The photolysis ($\lambda = 253.7$ nm) of N-aryloxindoles (217) has been described.[116] The isolated products have been identified as the spiro-compounds (218a and b). The quinone-imine (219) was thought to be the most likely intermediate, but quantum yield studies on the conversion of the oxindole (217a) to the final product showed this to be inefficient (Table 20), with a maximum value of 0·1 at 1·9% conversion. T.l.c.

Table 20 *Percentage conversion of oxindoles and influence on* Φ [116]

% Conversion	Φ
0·15	0·063
0·31	0·073
0·67	0·079
0·90	0·100
2·54	0·094
3·30	0·089

examination of the product of irradiation at low conversions showed the presence of the azetidine (220), and the authors argue that there is an equilibrium between the open and closed forms in the intermediate with an efficient conversion ($\Phi = 0.48$) from the azetidine (220) to the final dimeric product. The low quantum efficiency for the overall reaction is presumably a reflection of the two photochemical steps required for the overall conversion (Scheme 12).

[115] P. M. Collins and P. Gupta, *Chem. Comm.*, 1969, 1288.
[116] M. Fischer and F. Wagner, *Chem. Ber.*, 1969, **102**, 3486.

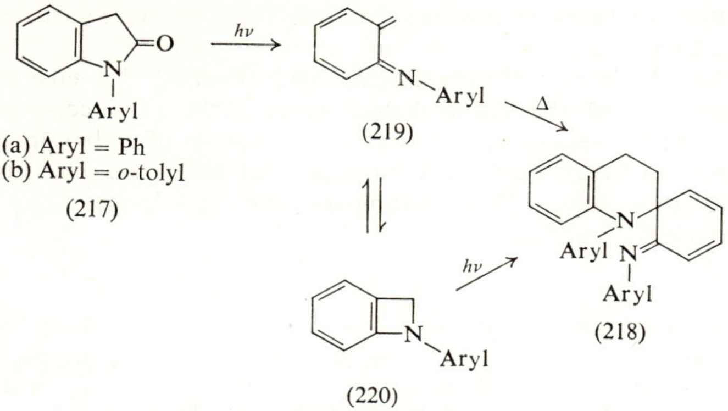

Photolysis of Carbonyl Compounds

Scheme 12

(217) (a) Aryl = Ph
(b) Aryl = o-tolyl

The intermediate quinone-imine (219a) or azetidine (220a) can be trapped [117] by nucleophiles to afford the adducts (221).

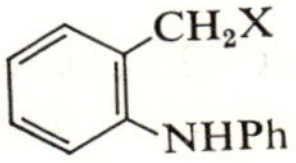

(221)

X	%		X	%
(a) OEt	61		(e) NMe_2	92
(b) OAc	75		(f) N-piperidinyl	97
(c) NHMe	65		(g) NHPh	94
(d) $NHBu^n$	92		(h) N-(2-pyridonyl)	98

The diazabicycloheptanones (222a and b) gave the pyrazole (223, 17%) as the major product from photolysis in benzene (light filtered by a Pyrex sleeve), via decarbonylation and rearrangement (Scheme 13).[118] The

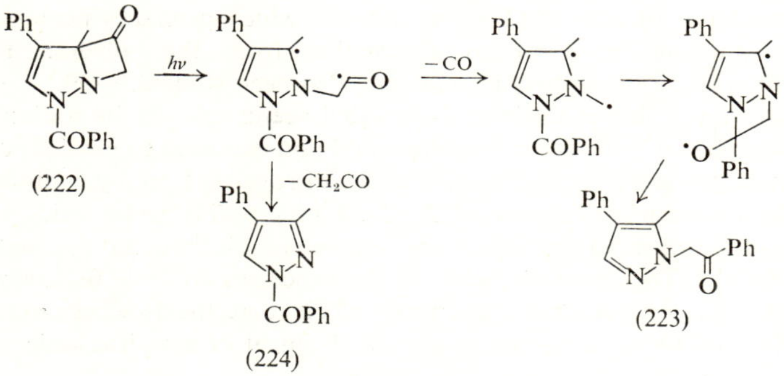

Scheme 13

[17] M. Fischer, *Chem. Ber.*, 1969, **102**, 3495.
[18] E. J. Voelker and J. A. Moore, *J. Org. Chem.*, 1969, **34**, 3639.

elimination of keten to give the pyrazole (224) occurs concurrently—a remarkable process.

Efficient elimination of carbon monoxide is reported [119] to take place on the photolysis of the dithio-oxalate esters (225). The corresponding sulphides (226) are formed by loss of two molecules of carbon monoxide, together with lower yields of dithiocarbonates (227), the products from monodecarbonylation. Trithiocarbonates yield the sulphides (226) by the photoelimination of CS.[119]

$$\begin{array}{ccc}
\text{O O} & \text{X} & \\
\| \ \| & \| & \\
\text{R-S-C-C-S-R} & \text{R-S-C-S-R} & \text{R-S-S-R} \\
\text{(a) R = Ph} & \text{(a) R = Ph, X = O} & \text{(a), (b), and (c)} \\
\text{(b) R = } p\text{-tolyl} & \text{(b) R = } p\text{-tolyl, X = O} & \text{(227)} \\
\text{(c) R = } p\text{-Cl-phenyl} & \text{(c) R = } p\text{-Cl-phenyl, X = O} & \\
\text{(225)} & \text{(d) R = Ph, X = S} & \\
& \text{(e) R = } p\text{-tolyl, X = S} & \\
& \text{(f) R = } p\text{-Cl-phenyl, X = S} & \\
& \text{(226)} &
\end{array}$$

Decarboxylations (see also Part III, Chapter 7, Section 5).—A diradical pair is involved in the photochemical decarboxylation of derivatives of pentaphenylcyclopenta-2,4-dien-1-ol (228a and b).[120] The products from

(a) R = Me
(b) R = Et

(228)   (229)

irradiation in benzene are the hydrocarbons (229a and b) formed by fission to the radical pair, decarboxylation of the acyloxyl radical, and recombination. Confirmatory evidence for the involvement of radicals comes from the photolysis of acetate (228a) in cumene which produces pentaphenylcyclopentadiene by hydrogen abstraction from the solvent. Photodecarboxylation of carboxylic acids has also been studied.[121, 122]

Under anaerobic conditions, 1-naphthyl acetic acid (as its sodium salt) has been shown [121] to decarboxylate readily on exposure to sunlight or u.v. The main product from the reaction under nitrogen is 1-methylnaphthalene.

Efficient decarboxylations of 2-, 3-, and 4-pyridylacetic acids to the corresponding methylpyridines in aqueous solution have also been described.[122] The quantum yields for the processes are 0·49, 0·46, and 0·20 respectively. The reactions are most efficient at the isoelectronic point (pH 4·0—4·2) and this suggests the involvement of a zwitterionic species.

[119] H. G. Heine and W. Metzner, *Annalen*, 1969, **724**, 223.
[120] J. J. Basselier and J. C. Cherton, *Compt. rend.*, 1969, **269**, 1412.
[121] D. G. Crosby and C. S. Tang, *J. Agric. Food Chem.*, 1969, **17**, 1291.
[122] F. R. Stermitz and W. H. Huang, *J. Amer. Chem. Soc.*, 1970, **92**, 1446.

such as (230) as an essential intermediate in the process prior to decarboxylation to (231). The reaction has been extended to the conversion of

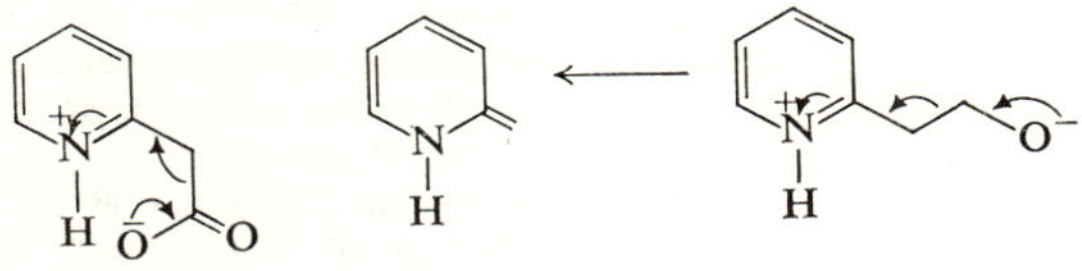

(230)   (231)

2-(2-pyridyl)-1-ethanol to 2-methylpyridine.[122] A mechanism involving loss of formaldehyde is proposed, which is similar to that for decarboxylation of the acids.

Irradiation of cyclic carbonates and sulphites (232) in methanol gives benzophenone and benzhydryl methyl ether,[123] as do other carbene

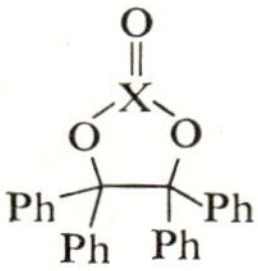

(232)   X = C or S

generators such as oxirans, dioxoles, and phospholans. No evidence was cited for the intermediacy of tetraphenyloxiran although cyclic carbonates are known to ring-contract thermally to epoxides. Phenylcarbene, formed by photolysis of a variety of carbene precursors, inserts into pentane with little selectivity between secondary and primary hydrogens (Table 21).[124]

Table 21 *Difference of reactivity of carbenes from different sources towards insertion into pentane* [124]

Carbene precursor	Insertion ratio (*sec.* : *prim.*)
Stilbene oxide	8·33
dl-Hydrobenzoin carbonate	8·27
meso-Hydrobenzoin carbonate	8·47
dl-Hydrobenzoin sulphite	8·00
meso-Hydrobenzoin sulphite	8·48
Phenyldiazomethane	{ 7·14 / 8·38 }

Formally analogous to the decarbonylation of cyclic ketones is the work described by Krull and Arnold [125] on the gas-phase photodecarboxylation and photodecarbonylation of cyclic anhydrides. The results summarised in Table 22 can be accommodated by a reaction (Scheme 14) where the

---

[123] R. L. Smith, A. Manmade, and G. W. Griffin, *J. Heterocyclic Chem.*, 1969, **6**, 443.
[124] R. L. Smith, A. Manmade, and G. W. Griffin, *Tetrahedron Letters*, 1970, 663.
[125] I. S. Krull and D. R. Arnold, *Tetrahedron Letters*, 1969, 4349.

Table 22 *Products from anhydride decomposition*[125]

Anhydride	% Conversion	Products (%)
Succinic	33	ethylene (15·4), acetylene (38·2)
Glutaric	44	ethylene (4·7), acetylene (23·5), cyclopropane (53·8), propene (0·4), keten (trace)
Adipic	30	ethylene (23·3), acetylene (54·3), cyclopropane (trace), propene (1·9), cyclobutane (5·0), keten (trace)
Maleic	20	acetylene (83·1)
Dimethylmaleic	26	but-2-yne (33), ethylene (1·7), acetylene (1·1), propyne (0·7), buta-1,3-diene (0·6), but-1-yne (0·6), buta-1,2-diene (0·6)
cis-Cyclobutane-dicarboxylic	20	ethylene (7·5), acetylene (30), buta-1,3-diene (3·8), cyclobutene (7·5)

production of the diradical (233) is the major step. Secondary reactions can occur from the vibrationally-excited intermediate or else by secondary photolysis. Unexpectedly, glutaric and adipic anhydrides give keten.

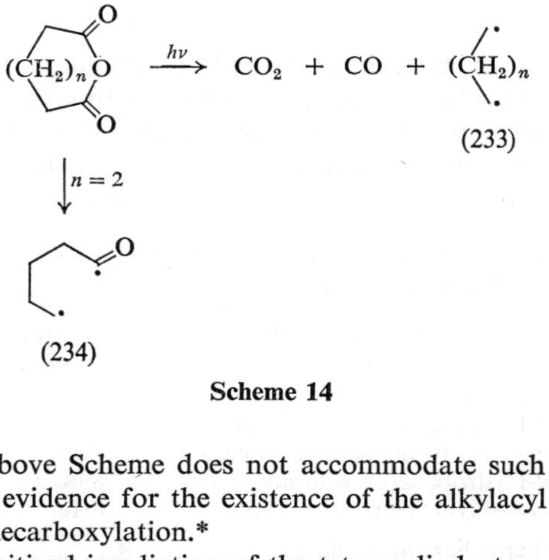

$$(\dot{C}H_2)_n\begin{matrix}O\\O\\O\end{matrix} \xrightarrow{h\nu} CO_2 + CO + (\dot{C}H_2)_n$$

(233)

$$\downarrow n=2$$

(234)

Scheme 14

Although the above Scheme does not accommodate such a product, the result could be evidence for the existence of the alkylacyl diradical (234) resulting from decarboxylation.*

Direct or sensitized irradiation of the tetracyclic lactone (235) has been shown [126] to produce, in addition to CO_2, the ketone (236) ($\Phi = 0.5$). In addition to this major product, naphthalene, keten, and benzobicyclo[2,2,2]-octadienone (237) are also produced. It appears that the most likely route

---

[126] R. S. Givens and W. F. Oettle, *Chem. Comm.*, 1969, 1164.

* The authors [125] do, however, point out that there is no evidence for the formation of keten in the decarbonylation of cyclopentanone. R. Srinivasan, *Adv. Photochem.*, 1963, **1**, 83.

to the formation of the benzotricyclo-octanone (236) is by the photolysis of the dienone (237), irradiation of which, labelled with deuterium at the sites shown, indicates that the di-π-methane route shown in Scheme 15 is

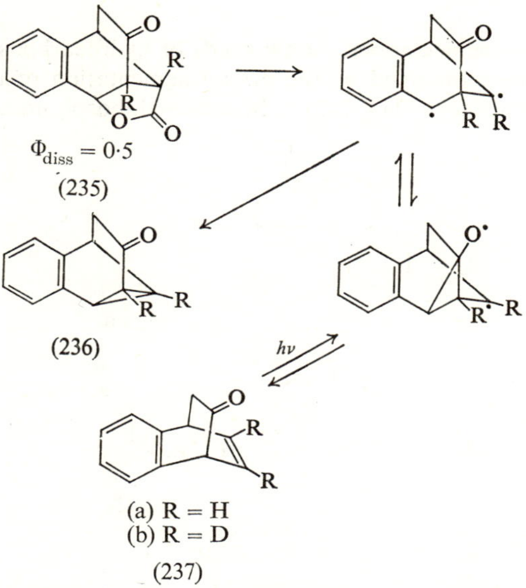

$\Phi_{diss} = 0.5$
(235)

(236)

(a) R = H
(b) R = D
(237)

Scheme 15

preferred. While it is reasonable that the ketone (236) may arise solely from the benzobicyclo-octadienone (237) in the sensitized runs, the high conversion by direct photolysis may involve the diradical intermediate.

A further account of the photochemistry of lactones in solution has been published.[127a]

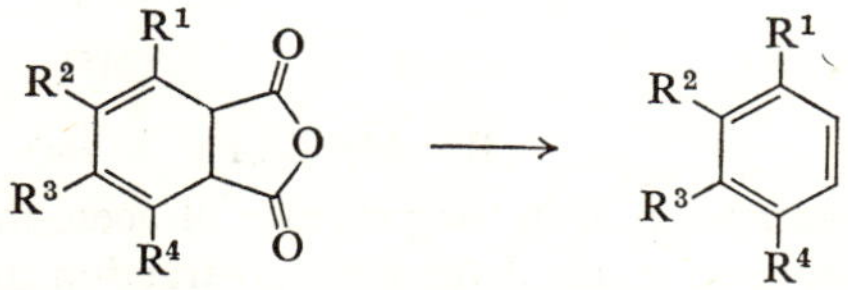

(238) (a) $R^1 = R^2 = R^3 = R^4 = Me$ 94%
(b) $R^1 = R^4 = Ph; R^2 = R^3 = H$ 70%
(c) $R^1 = R^4 = Me; R^2 = R^3 = Ph$ 82%
(d) $R^1-R^2 = (CH=CH)_2; R^3 = R^4 = H$ 97%
(e) $R^1-R^2 = (CH=CH)_2 = R^3-R^4$ 69%
(f) $R^3 = R^4 = Me; R^1-R^2 = 2,2'$-biphenyl   42%

Scheme 16

[127a] R. Simonaitis, *Diss. Abs.*, 1969, **29B**, 2834.

A number of anhydrides have been shown to undergo facile elimination of CO_2 and CO on irradiation in ether.[127b] The driving force for the reaction could well be the aromatisation of the cyclohexadienyl moiety (238) present in all the examples studied. The successful examples are shown in Scheme 16.

Miscellaneous Fragmentations.—Low yields of hydrocarbons (239 and 240 respectively) were obtained by the photofragmentation of the cholestane acetates and benzoates (241, 242) in hexamethylphosphoramide.[128]

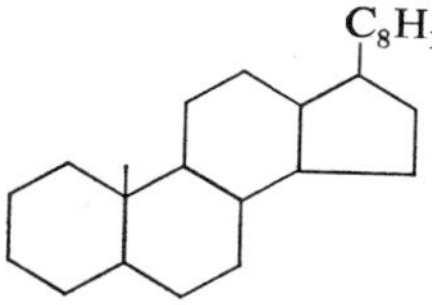

(239) 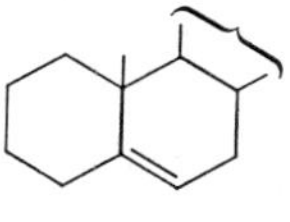   (240)

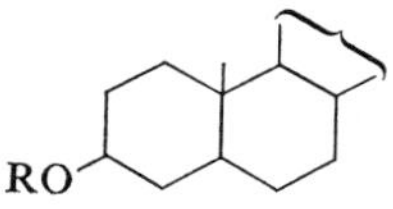

(241) 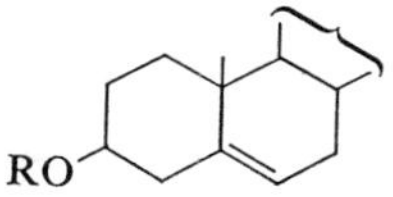   (242)

(a) R = Ac
(b) R = COPh

Other fragmentations involving radical hydrogen abstraction have been described. Of particular interest is the synthesis [129] of a β-lactone (243) from the oxetans (244). This reaction is suggested to involve hydrogen abstraction by photoexcited acetone followed by breakdown of the resultant oxetanyl radical (245). 2-Phenoxy-1-phenylethanol undergoes

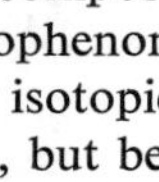

(243) 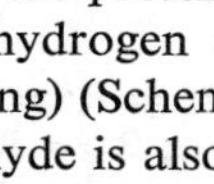   (244) 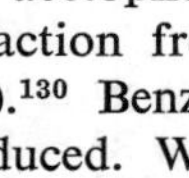   (245)

R = Me, Et, Pr^n, or Bu^n

photochemical decomposition in the presence of acetophenone to produce phenol and acetophenone *via* hydrogen abstraction from the benzylic centre (proved by isotopic labelling) (Scheme 17).[130] Benzophenone shows the same reaction, but benzaldehyde is also produced. With quinones the reaction yielded only benzaldehyde and phenol. Acetonaphthone and fluorenone were ineffective.[130]

[127b] R. Kitzing and H. Prinzbach, *Helv. Chim. Acta*, 1970, **53**, 158.
[128] R. Beugelmans, M. T. Le-Goff, and H. C. De-Marchville, *Compt. rend.*, 1969, **269**, 1309.
[129] S. H. Schroeter, *Tetrahedron Letters*, 1969, 1591.
[130] Y. Saburi, T. Yoshimoto, and K. Minami, *J. Chem. Soc. Japan*, 1969, **90**, 587.

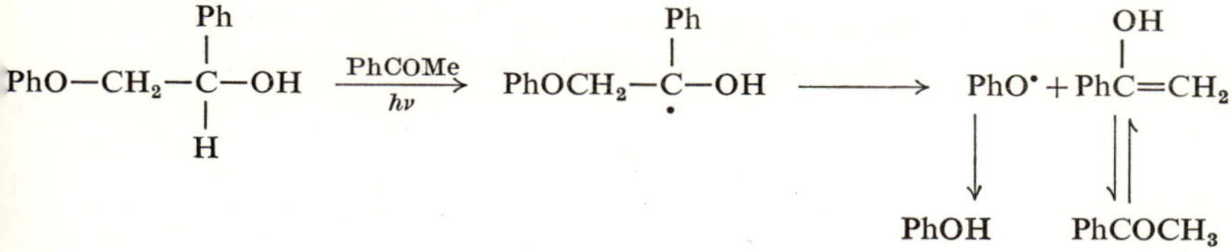

Scheme 17

PhO—CH$_2$—C(Ph)(H)—OH $\xrightarrow[h\nu]{\text{PhCOMe}}$ PhOCH$_2$—Ċ(Ph)—OH ⟶ PhO• + PhC(OH)=CH$_2$

PhO• → PhOH

PhC(OH)=CH$_2$ → PhCOCH$_3$

Pinacols have also been cleaved by photolysis in the presence of quinone.[131]

A remarkable photoconversion of the insecticide carbamate (246a) to the oxazinedione (247) in 85% yield in methanolic solution has been

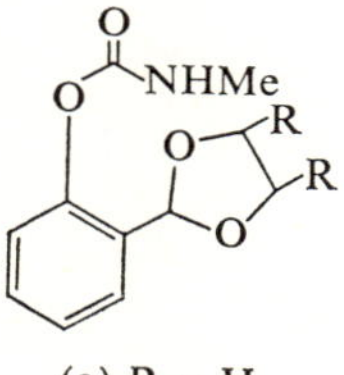

(a) R = H
(b) R = Me
(246)

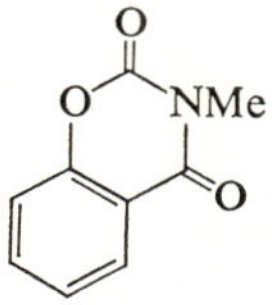

(247)

described.[132] This conversion appears to be restricted to the use of an unsubstituted dioxolan ring since experiments with (246b) afforded no new material. The mechanism for the formation of the observed product is not understood but the involvement of aerial oxidation is not precluded.

Interest in the photolysis of ylides has continued. The two ylides (248a and b) fragment on photolysis to afford products resulting from the insertion of carbene into the solvent.[133] The photochemistry of the

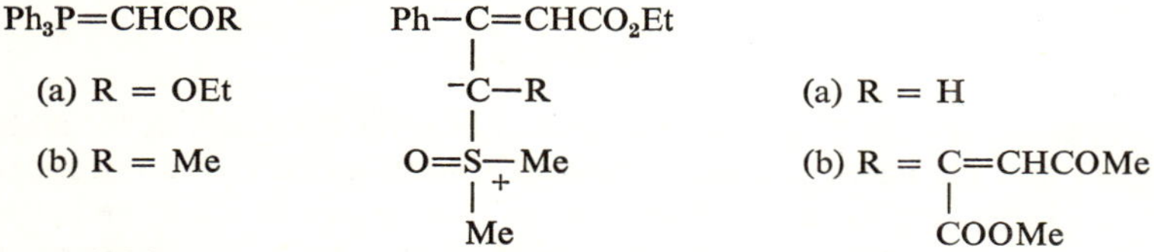

Ph$_3$P=CHCOR

(a) R = OEt
(b) R = Me

(248)

Ph—C(=CHCO$_2$Et)—C(R)(O=S$^+$(Me)Me)

(249)

(a) R = H
(b) R = C=CHCOMe | COOMe

sulphoxonium ylides (249a and b) has also been studied.[134] These, however, yield many products and a clear understanding of the reaction has not been obtained. The phenacylsulphonium salt (250) has been shown to

---

[131] R. S. Davidson, F. A. Younis, and R. Wilson, *Chem. Comm.*, 1969, 826.
[132] B. E. Pape, M. F. Para, and M. J. Zabik, *J. Agric. Food Chem.*, 1970, **18**, 490.
[133] Y. Nagao, K. Shima, and H. Sakurai, *Kogyo Kagaku Zasshi*, 1969, **72**, 236 (*Chem. Abs.*, 1969, **70**, 114,372).
[134] Y. Kishida, T. Hiraoka, and J. Ide, *Chem. Pharm. Bull. Tokyo*, 1969, **17**, 1591.

undergo cleavage of a C–S bond on photolysis.[135, 136] The products isolated can be accounted for by the intermediacy of a phenacyl radical.

$$PhCOCH_2\overset{+}{S}Me_2Br^-$$
(250)

The u.v. spectra of certain cyclic keto-sulphides show the presence of both charge transfer and $n \to \pi^*$ bands.[137] It appears that irradiation in the charge transfer band leads to products resulting from ring contraction. This process is illustrated below for the case of the simple keto-sulphide (251a). It is suggested that the reaction involves electron transfer* as shown in Scheme 18.

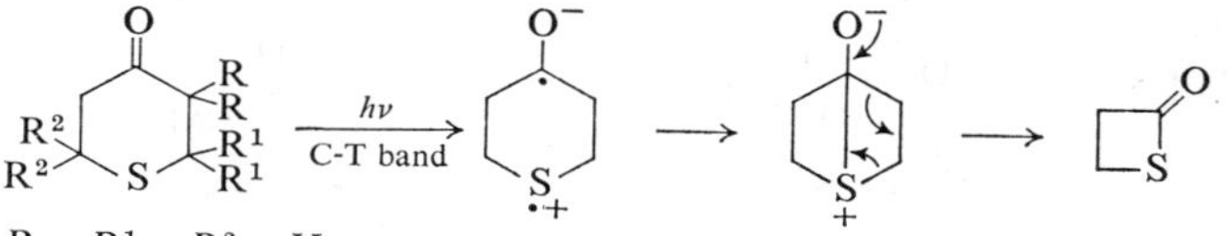

(a) $R = R^1 = R^2 = H$
λ_{max} 237 nm($\varepsilon = 435$)
287 nm ($\varepsilon = 22$)

(251)

Scheme 18

On the other hand, excitation in the $n \to \pi^*$ band yields products arising from Norrish Type I cleavage (Scheme 19). Selected results are recorded in Table 23.

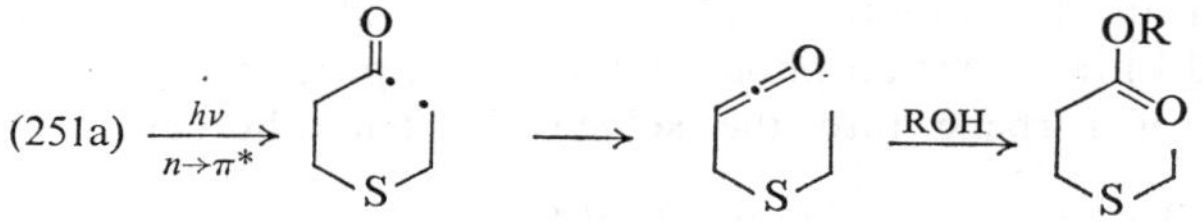

Scheme 19

A reinvestigation [138] of the photolysis of benzyl desyl sulphide (252a) has shown that the major product is benzo[b]thiophen (253a) and not bidesyl as originally reported.[139] Photolysis of the corresponding methylated derivatives (252b and c) has shown that the benzoyl phenyl group ends up

[135] T. Laird and H. Williams, *Chem. Comm.*, 1969, 560.
[136] T. Laird and H. Williams, *Chem. Comm.*, 1969, 561.
[137] P. Y. Johnson and G. A. Berchtold, *J. Org. Chem.*, 1970, **35**, 584.
[138] J. R. Collier and J. Hill, *Chem. Comm.*, 1969, 640.
[139] A. Schonberg, A. K. Fateen, and S. M. A. R. Omran, *J. Amer. Chem. Soc.*, 1956, **78**, 1224.

* A mechanism involving electron transfer was proposed by Cohen to account for the quenching of ketone fluorescence by sulphides. See W. M. Horspool, in 'Photochemistry,' ed. D. Bryce-Smith (Specialist Periodical Reports), The Chemical Society, London, 1970, vol. I, p. 165.

Photolysis of Carbonyl Compounds

Table 23 *Products from the photolysis of some cyclic keto-sulphides* [137]

Ketosulphide (251)	Solvent	Products (%)	
R = Me, R¹ = R² = H	Bu^{t}OH	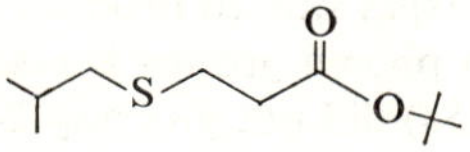 iBu-S-CH₂CH₂-CO-O-CMe₃ (9)	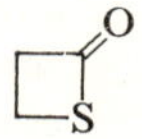 β-thiolactone (51)
R¹ = Me, R = R² = H	Bu^{t}OH	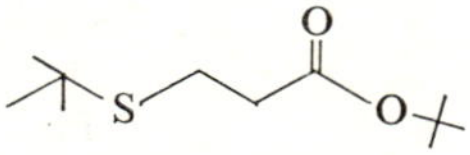 tBu-S-CH₂CH₂-CO-O-CMe₃ (2)	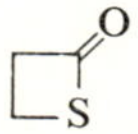 β-thiolactone (11)
		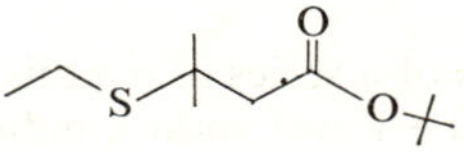 Et-S-CMe-CH₂-CO-O-CMe₃ (1)	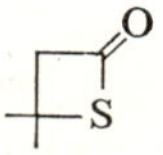 Me-substituted β-thiolactone (11)
		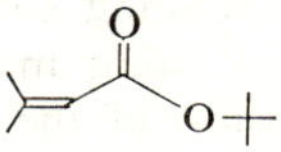 CH₂=CMe-CO-O-CMe₃ (10)	
R¹ = R² = Me, R = H	Bu^{t}OH	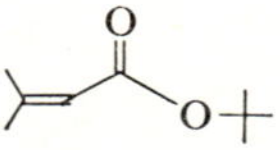 CH₂=CMe-CO-O-CMe₃ (3)	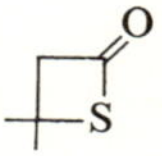 Me-substituted β-thiolactone (11)

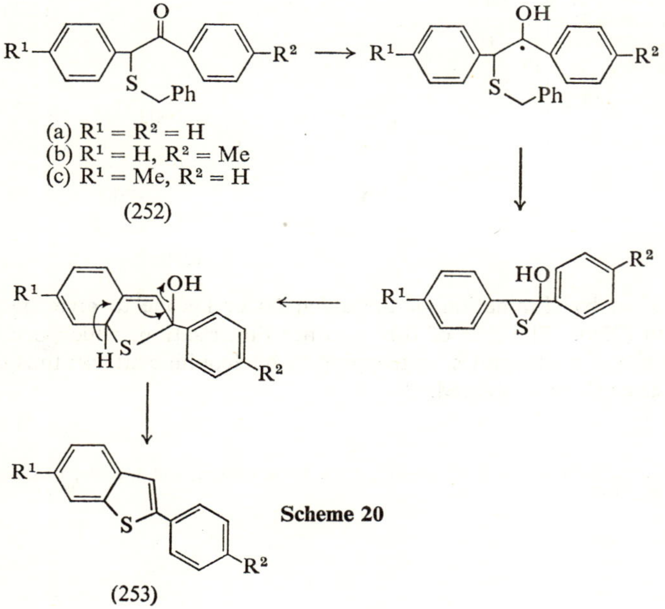

Scheme 20

(a) R¹ = R² = H
(b) R¹ = H, R² = Me
(c) R¹ = Me, R² = H

(252) → (253)

as the 2-aryl group in the product thiophens (253b and c). The proposed mechanism, shown in Scheme 20, involves hydroxythiiran formation and subsequent ring expansion to produce the products. It is proposed that the presence of both phenyl groups is essential for the success of the reaction since sulphide (254) did not give any 2-methylbenzthiophen.

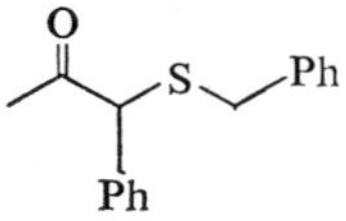

(254)

The photolysis of a series of oxathianones (255) at 350 nm in chloroform solution yields the novel dithiocandione (256) with a quantum yield of 0·025 [when R^1 = H and R = Ph in (255)].[140] The other product from the reaction is the corresponding carbonyl compound which results from fragmentation of the starting material. The route to the products is thought to involve fission of the S–C bond to give the diradical (257)

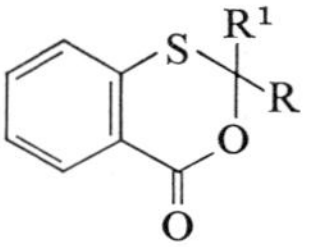

(a) R^1 = H, R = Ph
(b) R^1 = R = H
(c) R^1 = Me, R = H
(d) R^1 = R = Ph

(255)

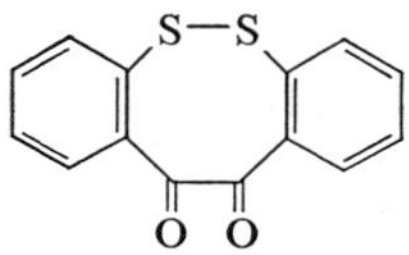

(256)

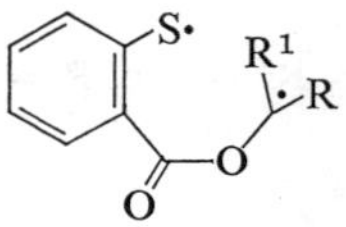

(257)

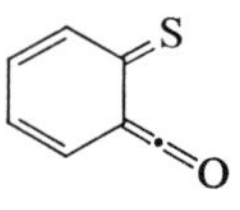

(258)

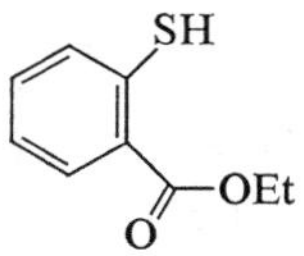

(259)

followed by elimination of the aldehyde or ketone to give the thioketo-keten (258). The fate of this is either dimerisation, which occurs in the absence of nucleophiles, or trapping by nucleophilic attack: thus in ethanol the ester (259) is isolated.[140]

W. M. H.

[140] A. O. Pederson, S. O. Lawesson, P. D. Klemmensen, and J. Kolc, *Tetrahedron*, 1970, **26**, 1157.

2
Enone Rearrangements and Cycloadditions: Photoreactions of Cyclohexadienones, Tropones, Quinones, *etc.*

1 Enone Rearrangements

αβ-**Unsaturated Carbonyl Compounds.**—CNDO calculations on the triplet states of acraldehyde have shown that the lowest triplet state is of 3A″ symmetry and $n\pi^*$ in nature, while the next triplet state is of 3A′ symmetry and $\pi\pi^*$ in character.[1] Thus it is obvious when considering the possible intervention of either of these states in photochemical reactions that the planarity of the $n\pi^*$ state and the non-planarity of the $\pi\pi^*$ state must be taken into account.

One of the most exciting results reported[2] is the formation of the stable oxeten (1). Attempts to isolate this oxeten from the photolysis of enone (2) through a Vycor filter yielded an oil having an n.m.r. spectrum in accord with the proposed structure (1), hydrogenation of which gave the oxetan (3). Attempts to prove the stereochemistry of the reduced product by the direct synthesis of the oxetan structure by the photolysis of acetaldehyde and 2-methylbut-2-ene (Scheme 1) gave the required isomer only as a trace. It is interesting that in this addition the rule that the more stable diradical is formed in the initial mode of addition is followed, but the specificity

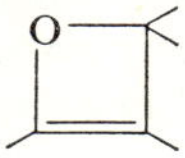

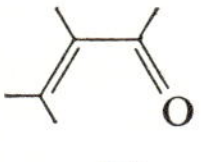

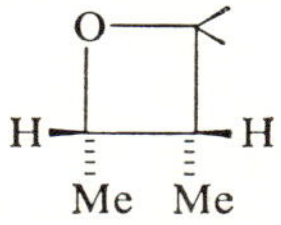

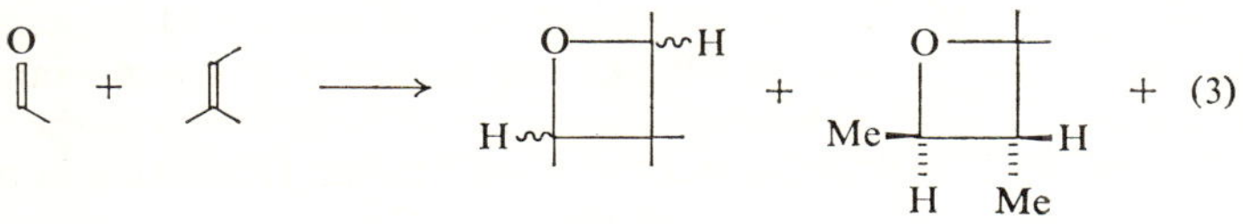

Scheme 1

[1] J. J. McCullough, H. Ohorodnyk, and D. P. Santry, *Chem. Comm.*, 1969, 570.
[2] L. E. Freidrich and G. B. Schuster, *J. Amer. Chem. Soc.*, 1969, **91**, 7204.

(9 : 1) is much greater than that reported by Turro and Wriede for the addition of acetone to 1-methoxybut-1-ene. The authors[2] claim that the rapid formation of the oxeten (1) is due to the preferred *cisoid* configuration of the enone. They also claim that $\pi \to \pi^*$ excitation is responsible even though in intermolecular oxetan formation only $n \to \pi^*$ states are reported to be involved. The non-conjugated dienone (4), although closely similar to enone (2), behaves differently. The singlet state reaction of the dienone (4a) (the reaction cannot be sensitized or quenched) has been shown[3] to produce the dihydropyran (5) as the result of a 2+4 addition. This is the major product (60%) from a 30-hr irradiation in petroleum ether. It is,

(4a)　　(4b)　　(5)　　(6)

(7)　　(8)

however, accompanied by two minor products, the cyclopentenyl ketone (6, R = H; 15%) and the cyclopropane (7; 25%). The formation of (7) is negligible in benzene but assumes major importance in methanol where hydrogen abstraction is important. The rationale for the formation of the dihydropyran has already been reported.[4] The formation of the cyclopentenyl ketone (6, R = H) has been shown to involve the intermediacy of a remarkable 9-membered transition state (8). The excited carbonyl group abstracts hydrogen from one of the vinyl methyl substituents and rebonding forms the new product. Evidence confirming this postulate is obtained from an irradiation in D_2O when incorporation of one deuterium (6, R = D) is observed.

Short irradiation times (0·5 hr) result in a facile *trans–cis* isomerisation of the starting material (4a → 4b).[3] Indeed, this is essential to obtain the correct alignment of the groups for the subsequent reactions.

A full report[5] of the photochemical rearrangements of a series of vinylcyclopropanecarboxylates (9) has appeared.[6] The esters examined all have intense $\pi \to \pi^*$ absorption maxima at 217—250 nm and the weak $n \to \pi^*$ transition is occluded under the $\pi \to \pi^*$ band. Irradiation into this

[3] J. Meinwald and J. W. Kobzina, *J. Amer. Chem. Soc.*, 1969, **91**, 5177.
[4] R. A. Scheider and J. Meinwald, *J. Amer. Chem. Soc.*, 1967, **89**, 2023.
[5] M. J. Jorgenson and C. H. Heathcock, *J. Amer. Chem. Soc.*, 1965, **87**, 5264; M. J. Jorgenson, *J. Amer. Chem. Soc.*, 1966, **88**, 3463.
[6] M. J. Jorgenson, *J. Amer. Chem. Soc.*, 1969, **91**, 6432.

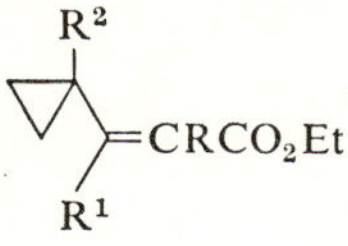

Enone Rearrangements and Cycloadditions

(9) 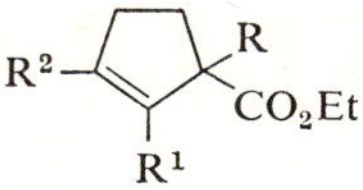 (10) 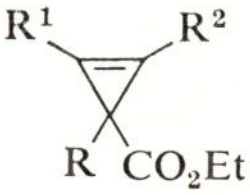 (11)

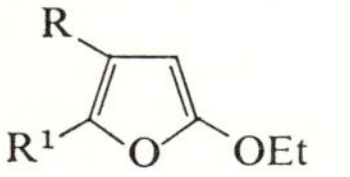

(12) 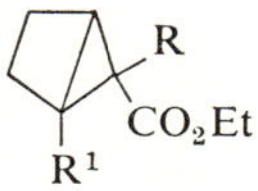 (13)

intense band gives rise to the products shown in Table 1. In the main, the products from the reactions come from several primary photoprocesses. The furan (12) and the cyclopropene (11) products can be accounted for on the basis of a fragmentation (Scheme 2). The puzzle in this particular

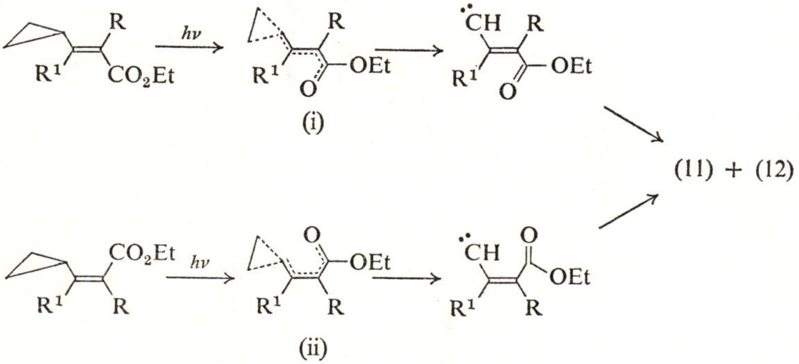

Scheme 2

scheme involves the preference for cyclopropene formation, but the author suggests [6] that the substitution pattern on the double bond may be important (esters 9c, d, and e would give a crowded furan), and that steric influences therefore control the formation of cyclopropene from the excited state represented by (ii). The cyclopentenes (10) are proposed to arise from ring-opening of the esters in the *s-cis*-conformation (reaction 1). Attempts to establish the concertedness of the conversion of the cyclopropane to the cyclopentene proved futile, although sensitization studies on the esters (9c and d) produced only the cyclopentenes (10c and d), perhaps pointing to a diradical rather than a concerted mechanism. Probably the most important result as far as the chemistry of vinylcyclopropanes is concerned is the isomerisation to produce the bicyclopentanes (13), the path to which (reaction 2)

Table 1 Products from the irradiation of vinylcyclopropyl esters [6]

Ester	Products (%)			
	(10)	(11)	(12)	(13)	(Others)
(9a) R = R^1 = R^2 = H	R = R^1 = R^2 = H (50)	R = R^2 = H, R^1 = Me (8)	R = R^1 = H (20)		(10)
(9b) R = R^2 = H, R^1 = Me	R = R^2 = H, R^1 = Me (70)	R = R^1 = Me, R^2 = H (24)	R = Me, R^1 = H (3)		(20)
(9c) R = R^1 = Me, R^2 = H	R = R^1 = Me, R^2 = H (10)	R = Me, R^1 = R^2 = H (25)		R = Me, R^1 = H (48)	(10)
(9d) R^1 = R^2 = H, R = Me	R^1 = R^2 = H, R = Me (6)	R = H, R^1 = R^2 = Me (40)		R = R^1 = Me (14)	(53)
(9e) R = H, R^1 = R^2 = Me	R = H, R^1 = R^2 = Me (60)				
(9f) R = R^2 = H, R^1 = cyclopropyl	R^1 = cyclopropyl (88) R = R^2 = H, R^1 = propenyl (12)				

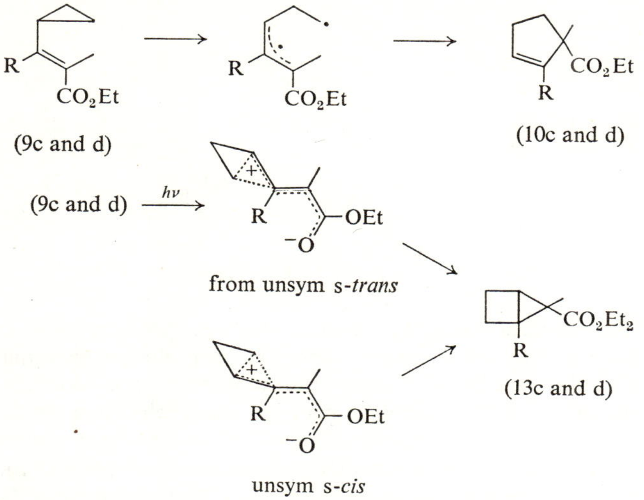

(9c and d) → (9c and d) → (10c and d) (1)

(9c and d) —hv→ [cyclopropyl cation intermediate with R, OEt, O⁻]

from unsym s-*trans*

→ (13c and d) (2)

[cyclopropyl cation intermediate with R, OEt, O⁻]

unsym s-*cis*

is very selective, and only the esters (9c and d) were effective. The evidence available does not allow a decision on whether or not the reaction is concerted. The photochemical decomposition of di-t-butylcyclopropenone by $n \to \pi^*$ excitation (253·7 nm irradiation) gave a good yield of di-t-butylacetylene,[7] a compound otherwise only accessible by a long sequence of thermal reactions.

The irradiation of cyclobutanones to give tetrahydrofurans *via* carbene intermediates has been known for a number of years.[8] In a new report,[9] evidence is cited for the first example of intramolecular trapping of the postulated carbene intermediate (14) by a hydroxy-group, *viz*. from the

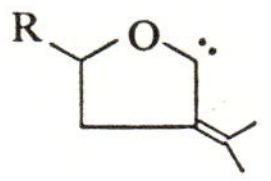

(14)

photolysis of cyclobutanone (15a) in benzene, to yield the ketal (16). This adduct can be converted into the methyl acetal (17a), the product from the photolysis of cyclobutanone (15a) in methanol. The irradiation of the simpler cyclobutanone (15b) in methanol also gives a methyl ketal (17b). Irradiation in pentane with methanol, however, produces an unstable compound (18) which reacts thermally to give the ketal (17b) or on pyrolysis

---

[7] J. Ciabattoni and E. C. Nathan, tert., *J. Amer. Chem. Soc.*, 1969, **91**, 4766.
[8] N. J. Turro and R. M. Southam, *Tetrahedron Letters*, 1967, 545.
[9] N. J. Turro, E. Lee-Ruff, D. R. Moton, and J. M. Conia, *Tetrahedron Letters*, 1969, 2991.

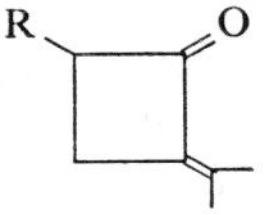

(15) (a) R = HOCMe$_2$ (b) R = H

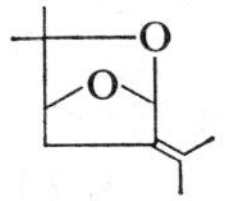

(16)

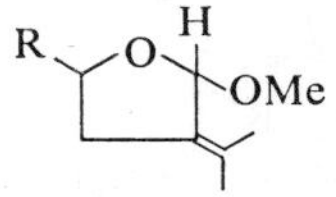

(17) (a) R = HOCMe$_2$ (b) R = H

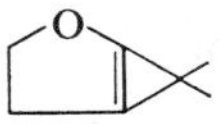

(18)

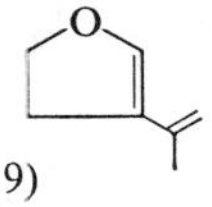

(19)

to yield the dihydrofuran (19). Product (18) also provides good evidence for the intermediacy of a carbene since its formation is presumably the result of an allylcarbene–cyclopropene reorganisation. Jorgenson has shown examples of the rearrangement of vinylcyclopropanes to cyclopentenes.[6] The reverse of this rearrangement has also been reported [10] in a study of the photolysis of substituted cyclopentenones (20a and b). These photolyses in acetonitrile–water (in the case of 20a) or methanol (for 20b) produce the acid (21a) and the ester (21b), respectively. Spectroscopic studies on the rearrangement of cyclopentenone (20b) have shown the

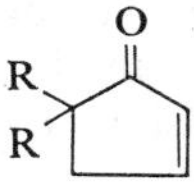

(20) (a) R = Ph (b) R = Me

(21) (a) R = Ph, R^1 = H (b) R = Me = R^1

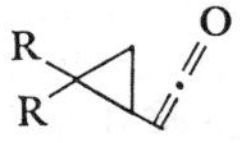

(22)

existence of a cyclopropylketen (22); a keten was also detected (i.r. spectrum) in the irradiation product of cyclopentenone. The multiplicity of the reactive state was not established. It is possible, however, in view of the concertedness of the reverse reaction, that the reaction is an example of a photo-allowed concerted reaction occurring from the singlet state. Such a rearrangement also appears possible on first examination for the reorganisation encountered in the photolysis of the pyrazolinone (23). However, the product imidazolone (24) is best accounted for by the intermediacy of the α-lactam (25).[11] The imidazolone is formed under all the conditions used, *i.e.* methanol, ethanol, and acetone as solvents. The reactions in methanol and ethanol also give the products (26)—(28). In acetone, the major product is the keto-aldehyde (29). Some doubt may be cast on the intermediacy of the α-lactam, since α-lactams undergo

---

[10] W. C. Agosta, A. B. Smith, A. S. Kende, R. G. Eilerman, and J. Benham, *Tetrahedron Letters*, 1969, 4517.
[11] S. N. Ege, *J. Chem. Soc.* (*C*), 1969, 2624.

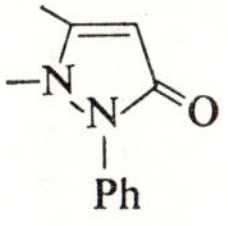

Enone Rearrangements and Cycloadditions

(23) [structure]

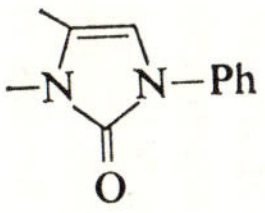

(24) [structure]

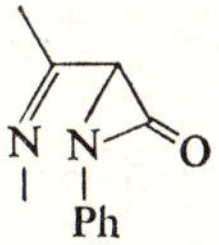

(25) [structure]

(26) (a) R = Me
(b) R = Et

(27) [structure]

(28) [structure]

(29) [structure]

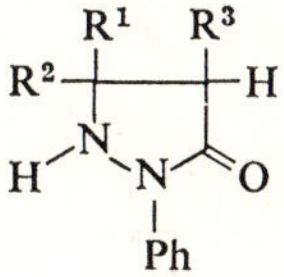

(30) [structure]

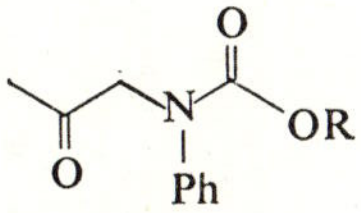

(31) [structure]

(a) R^1 = Me, R^2 = R^3 = H
(b) R^1 = R^2 = H, R^3 = Me
(c) R^1 = R^2 = Me, R^3 = H

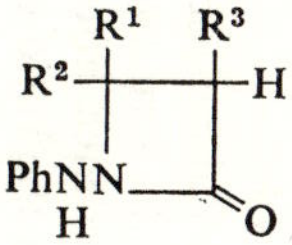

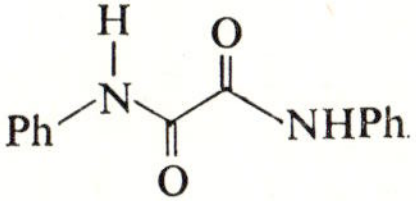

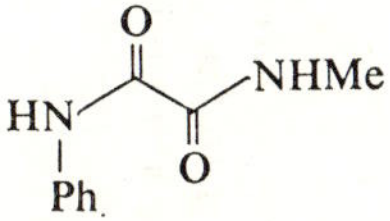

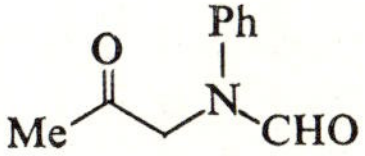

efficient decarbonylation.[12] Photolysis of the pyrazolidinones (30) produced 1-amino-azetidinones (31) as the major products.[11]

The photolysis of the unsaturated lactone (32a) in hexane using 253·7 nm radiation gave the geometric isomer (32b) and the structurally rearranged dione (33).[13] The photostationary state was reached after 2 hr and consisted of the starting olefin and the products in the ratio of 1 : 1 : 2. The

(a) $R^1 = H, R^2 = CN$
(b) $R^1 = CN, R^2 = H$
(32)

(33)

(34)

(35)

conjugated ketone (34) has been reported [14] to be exceptionally stereospecific in its photoisomerisation to the cyclopropane (35). The fact that the optically active forms retain their optical activity during the rearrangement implies that, even though the reaction proceeds by a triplet state, the diradical intermediate must also be optically active. A flash spectrophotometric study has shown that two species are present.[15] The longer-lived one (*ca.* 340 nm) is a ketyl radical formed by hydrogen abstraction from the solvent. The shorter-lived species (*ca.* 430 nm) was not detected directly, but by an indirect method involving sensitization of the naphthalene $\pi\pi^*$ triplet. This intermediate is not the precursor of the rearranged product, and neither of the excited states was optically active.

In the chemistry of simpler cyclohexenones, a mechanistic examination of the photochemistry of 4,4-dimethylcyclohexenone has been reported.[16]

[12] J. C. Sheehan and M. M. Nafassi-V., and E. R. Talaty, A. E. Dupuy, and T. H. Golson, cited by W. M. Horspool in 'Photochemistry,' ed. D. Bryce-Smith, (Specialist Periodical Reports), The Chemical Society, London, 1970, Vol. 1, p. 172.
[13] H. W. Moore, H. R. Shelden, D. W. Deters, and R. J. Wikholm, *J. Amer. Chem. Soc.*, 1970, **92** 1675.
[14] H. E. Zimmerman, R. G. Lewis, J. J. McCullough, A. Padwa, and M. Semmelhack, *J. Amer. Chem. Soc.*, 1966, **88**, 159; O. L. Chapman, J. B. Sieja, and W. J. Welstead, *J. Amer. Chem. Soc.*, 1966, **88**, 162.
[15] G. Rämme, R. L. Strong, and H. H. Richtol, *J. Amer. Chem. Soc.*, 1969, **91**, 5711.
[16] T. H. Koch, *Diss. Abs.*, 1969, **29B**, 3263.

Enone Rearrangements and Cycloadditions

An examination of the photochemistry of a dihydrosantonin (36) has shown fragmentation to be remarkably stereospecific, the cyclopropane (37) being the sole product.[17] The reaction is interpreted as proceeding by excitation of the enone chromophore followed by rupture of the lactone C–O bond to produce the intermediate (36a). Decarboxylation of this diradical and rebonding gives the isolated product. The isomeric dihydrosantonin (38) is surprisingly photochemically inert.

(36)   (36a)   (37)

(38)

Further investigations of the reactions of steroidal enones (39) have been reported. During the past few years two major types of reaction have been uncovered for this type of molecule, *viz*. the molecular rearrangement to bicyclic ketones, and the double bond migration out of conjugation to give the $\beta\gamma$-ketone (40). Yet another process has now been uncovered, that of addition of the substrate to the double bond.[18] All the above processes occur from the triplet state and can be quenched or sensitized. The question whether the excited state is $n\pi^*$ or $\pi\pi^*$ in type has been discussed. It is proposed [18] that the rearrangement and the addition processes occur solely from the $\pi\pi^*$ states and the deconjugation reaction from the $n\pi^*$ state. An explanation for the observed effects of solvents in changing the type of reaction is based on the consideration that in alcohols the $\pi\pi^*$ state is lower in energy, but in less polar solvents, *e.g.* benzene, the $n\pi^*$ state is lower. A typical example of the reactions encountered is provided by isomerisation of the testosterone (39) to the $\beta\gamma$-isomer (40) on irradiation at 327 nm in benzene solution. Similar irradiation conditions using toluene as solvent produce an additional product (41) arising from addition of the solvent to the enone by a radical process. Irradiation of the steroidal ketone (39) in t-butanol at 327 nm produces the bicyclic ketone (42) and the ring-cleaved product (43). Separate irradiation of (42) shows that the formation of the bicyclic ketone is reversible, producing the starting ketone

[17] G. W. Perold and G. Ourisson, *Tetrahedron Letters*, 1969, 3871.
[18] D. Bellus, D. R. Kearns, and K. Schaffner, *Helv. Chim. Acta*, 1969, **52**, 971.

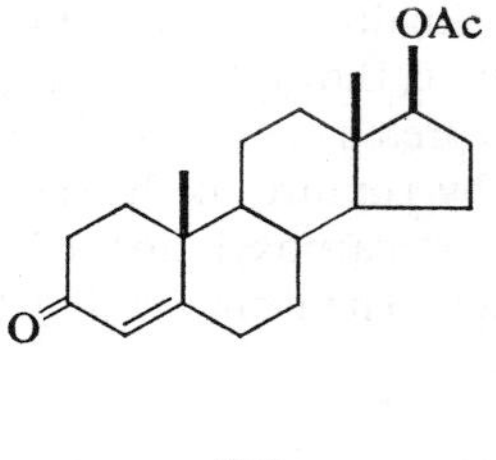

(39) 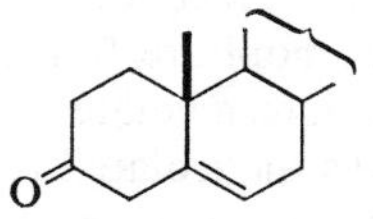 (40) 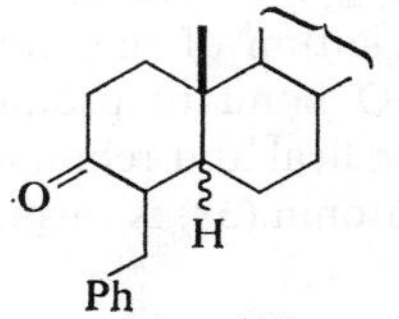 (41)

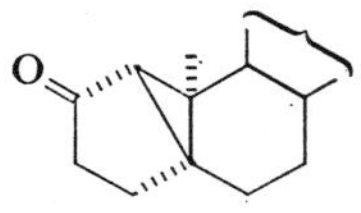

(42) 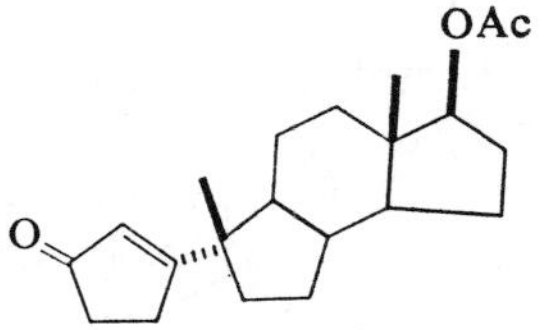 (43)

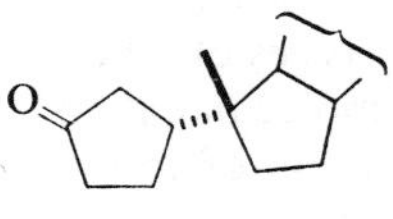

(44) 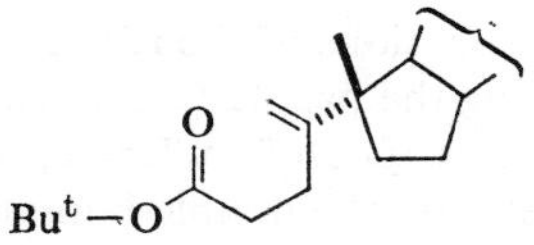 (45)

and the cleaved product. It is argued from these and earlier results that this process could be concerted in both directions. Irradiation of the ring-cleaved product at > 327 nm in t-butanol gives the reduced compound (44), while at > 280 nm further ring-cleavage takes place to give the ester (45).

The photo-rearrangements of the epoxyenones (46) have been interpreted.[19] The photoreactions are stereospecific owing in the most part to the geometric restrictions of the molecules in question. Thus the epoxyenones (46a and b) on irradiation ($\lambda = 253$ nm) in Bu^{t}OH undergo ring expansion of ring C to form the product (47) as shown in Scheme 3. The isomeric 9β,10β-epoxyenone (46c) also rearranges in a like manner. Two pathways for rearrangement are open to the intermediate (i, Scheme 4), and two products (48 and 49) are obtained in the ratio of 3 : 1. As with all epoxyketone photoreactions, the initial excitation brings about bond cleavage in the epoxide to produce the more stable diradical. These reactions differ remarkably from those of the 6,7-epoxy-enones described

[19] M. Debono, R. M. Molloy, D. Bauer, T. Iizuka, K. Schaffner, and O. Jeger, *J. Amer. Chem. Soc.*, 1970, **92**, 420.

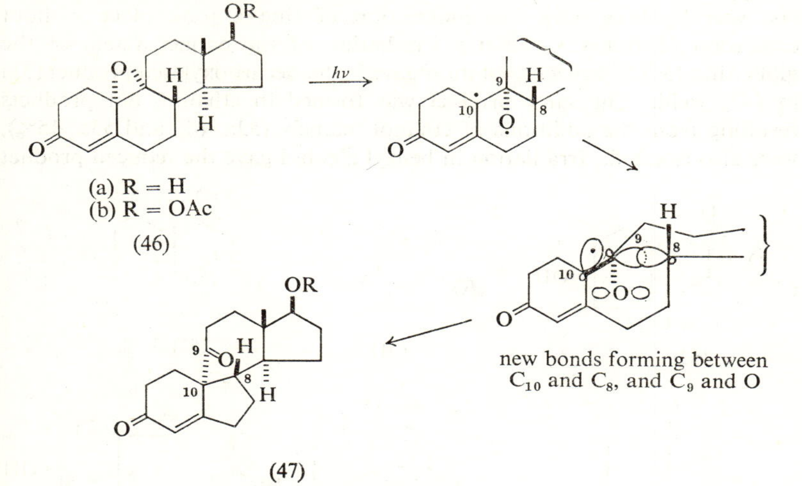

Enone Rearrangements and Cycloadditions

(a) R = H
(b) R = OAc
(46)

new bonds forming between C_{10} and C_8, and C_9 and O

(47)

Scheme 3

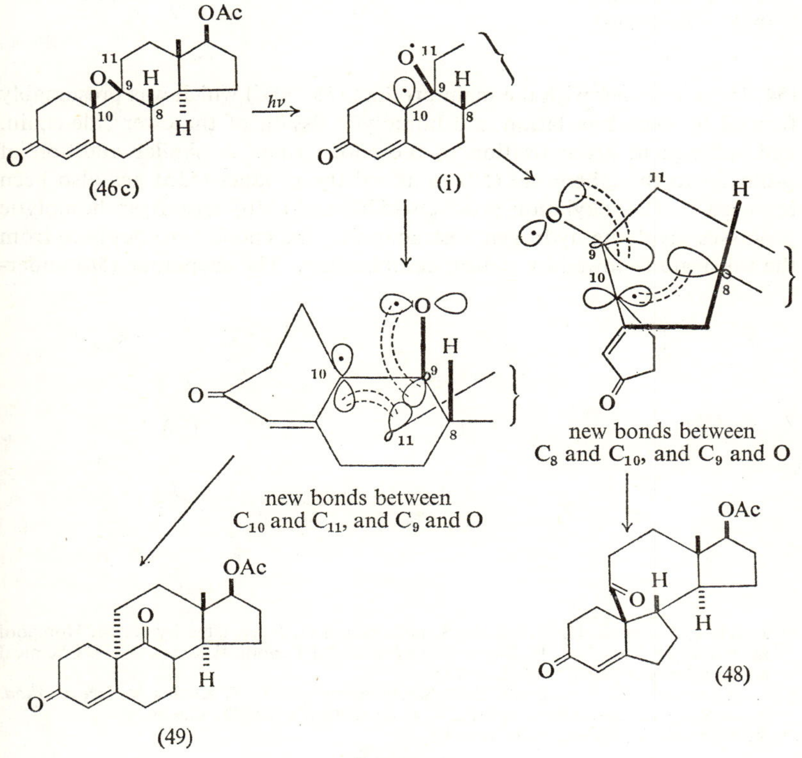

(46c)

(i)

new bonds between C_{10} and C_{11}, and C_9 and O

new bonds between C_8 and C_{10}, and C_9 and O

(49)

(48)

Scheme 4

last year [20] where only ring-contraction of ring B took place without expansion of ring C. The u.v. irradiation of the enone system of the gibberillin (50) in benzene solution gave [21] the decarboxylated product (51) in 14% yield. The same product was formed in ethanol, but products resulting from the additions of ethanol, namely (52a, 6% and 53a, 15%), were also isolated. Irradiation in benzyl alcohol gave the reduced product

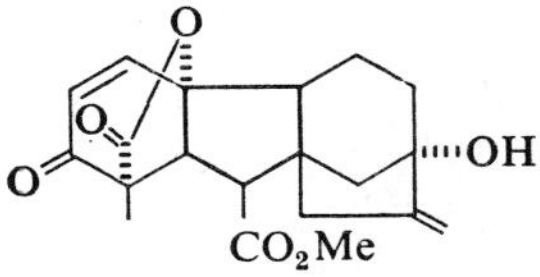

(50)

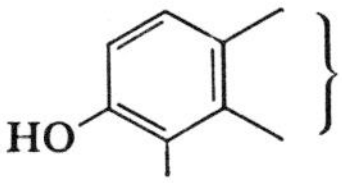

(51)

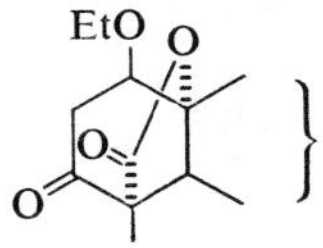

(52)

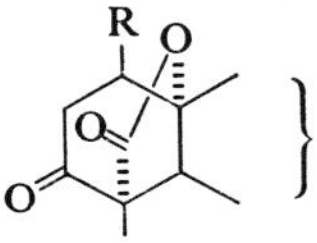

(a) R = MeCHOH
(b) R = 2-dioxanyl
(53)

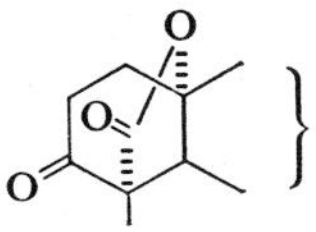

(54)

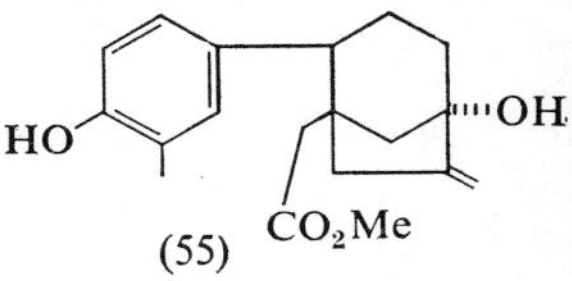

(55)

(54, 15%), together with the main product (55, 60%) which was presumably formed by decarboxylation and homolytic fission of the ester side-chain, and subsequent aromatisation of the enone ring. A similar addition of p-dioxan to the gibberillin (50) to afford the product (53b) has also been reported.[22] The alkylation products (53a) and (53b) arise from homolytic processes involving hydrogen abstraction by the enone chromophore from the substrate followed by radical combination. The azepinone (56) under-

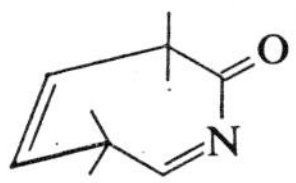

(56)

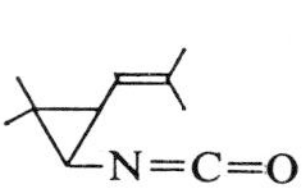

(57)

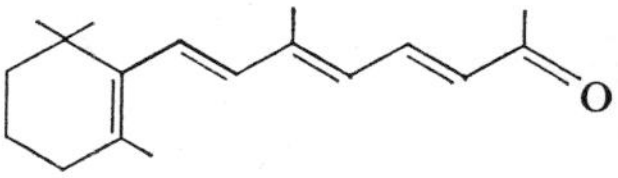

(58)

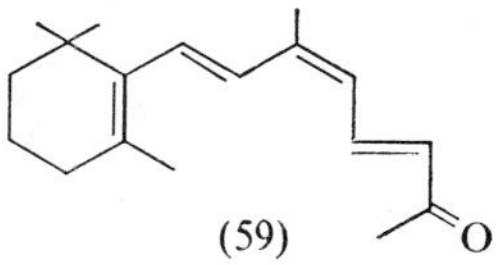

(59)

[20] J. Saboz, T. Iizuka, H. Wehrli, K. Schaffner, and O. Jeger, cited by W. M. Horspool in 'Photochemistry,' ed. D. Bryce-Smith (Specialist Periodical Reports) The Chemical Society, London, 1970, Vol. 1, p. 185.
[21] I. A. Gurvich, N. S. Kobrina, E. P. Serebtyakov, and V. F. Kucherov, *Izvest. Akad. Nauk S.S.S.R., Ser. Khim.*, 1969, 2342 (*Chem. Abs.*, 1970, **72**, 42,916v.)
[22] R. Tümmler, *Z. Chem.*, 1970, **10**, 140.

goes a slow photochemical isomerisation to yield the cyclopropyl isocyanate (57, 20%).[23]

Finally in this section, a study of retinal is reported.[24] From the effects of various triplet energy donors of known lifetimes and energies, the triplet energy of all-*trans*-retinal is placed at 158·8 kJ mol^{-1} (38 kcal mol^{-1}). This result was obtained from the changes in the decay time of the observed phosphorescence when the retinal (acceptor) concentration was altered. Correlation of this with the emission spectra of the donors and the absorption spectrum of the acceptor permitted calculation of the triplet energy. Brief irradiation (sunlight + iodine, or 125 watt medium-pressure Hg arc through Pyrex) of an ethanolic solution of the ionylidene-ethylidene acetone (58) gave a mixture of the starting material and a new isomer (59).[25] The percentage of the new isomer was dependent on the length of the irradiation, and best results (12% yield) were obtained after 10—12 min. Irradiation for longer times brought about extensive degradation.

$\beta\gamma$-**Unsaturated Carbonyl Compounds.**—The practical applications of photochemical ring-expansion of a cyclohexanone to a larger ring ketone have already been discussed (Chapter 1). This route eliminates unfavourable entropy terms by the utilisation of the ring-opening of a three-membered ring. One must remember that the ring-expansion of 2-vinylcyclohexanone to cyclo-oct-3-en-1-one is extremely inefficient. However, nothing daunted, Carlson and Henton[26] report the photochemical ring-expansion *via* the singlet state of 2-ethynylcycloheptanone to the unstable allene, cyclonona-2,3-dien-1-one and the unsaturated aldehyde non-6-en-7-yn-1-al. The formation of both compounds arises from the same intermediate (60). Hydrogen abstraction produces the aldehyde, and radical recombination produces the allene. Evidence for

(60)   (61)

formation of the allene was obtained from spectroscopic analysis of the irradiation product and from hydrogenation to give cyclononanone, and from Diels–Alder reactions, *e.g.* with cyclopentadiene to yield adduct (61). The involvement of an 'oxa-di-π-methane' rearrangement, *i.e.* an example of a $\pi_a^2 + \sigma_a^2$ photoallowed rearrangement, has been described.[27] This is

[23] T. Sasaki, S. Eguchi, and M. Ohno, *J. Amer. Chem. Soc.*, 1970, **92**, 3192.
[24] A. V. Guzzo and G. L. Pool, *J. Phys. Chem.*, 1969, **73**, 2512.
[25] M. Mousseron-Canet and J. L. Olive, *Bull. Soc. chim. France*, 1969, 3242.
[26] R. G. Carlson and D. E. Henton, *Chem. Comm.*, 1969, 674.
[27] W. G. Dauben, M. S. Kellogg, J. I. Seeman, and W. A. Spitzer, *J. Amer. Chem. Soc.*, 1970, **92**, 1786.

involved in the sensitized isomerisation of the enone (62) to the cyclopropylketone (63, 93%). In contrast, direct irradiation affords only the product (64) of 1,3-acyl migration. No evidence for either of these processes

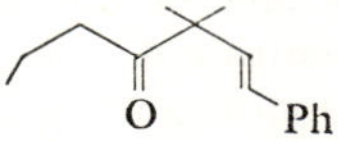

(62) 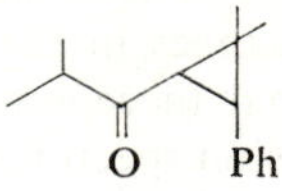   (63) 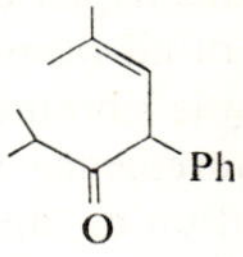   (64)

was found in a study [28] of the photochemistry of 2,2,4-trimethylpent-3-en-1-al. Only decarbonylation products were isolated. Acyl migration has been reported to occur in the photolysis [29] of the optically-active ketone (65). This provides an example of photochemical 1,3-sigmatropic migration. Evidence for this comes from an experiment when irradiation was stopped after 13% conversion, and it was observed that the optical rotation of the starting material had dropped from $[\alpha]_{436}^{23} + 1212°$ to $+1071°$. The photoracemisation (reaction 3) accounts for over 32% of the photoreaction of

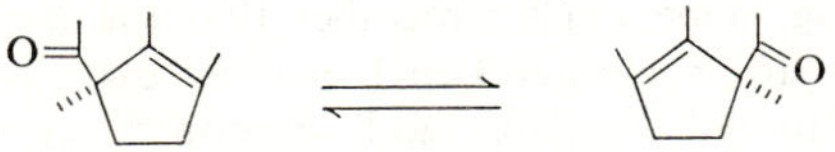

(65a) ⇌ (65b)   (3)

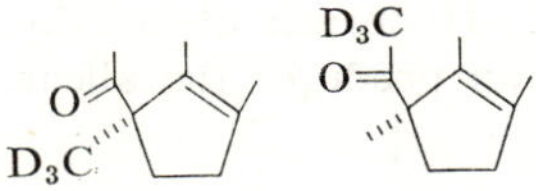

(66) 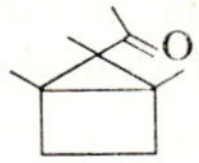   (67)

the starting ketone. The intramolecularity of the reaction, and its dependence on acetyl migration, were established by the photolysis (>280 nm) of a mixture of trideuteriated ketones (66) whence only trideuterioketones were formed. This reaction could not be affected by triplet quenchers. However, when the irradiation was repeated at 253·7 nm a new photoisomer (67) was obtained. Schaffner et al.[29] propose that both of the reactions described above are examples of $\sigma_a^2 + \pi_a^2$ rearrangements. Taking these results in conjunction with those of Dauben et al.,[27] it is apparent that both processes represent further examples of different reaction paths from different excited states. Both the acyclic 1,3-migration (62 → 64) and the alicyclic processes (65a → 65b) arise from the singlet state. There is

[28] R. G. Tonkyn and R. J. Cotter, *J. Polymer Sci. Part A-1, Polymer Chem.*, 1969, **7**, 2744.
[29] E. Baggiolini, K. Schaffner, and O. Jeger, *Chem. Comm.*, 1969, 1103.

apparently no 'oxa-di-π-methane' rearrangement of (65a) corresponding to the conversion of (62) to (63). However, if one can differentiate between a di-π-methane and $\pi_a^2 + \sigma_a^2$ process (and the difference is not clearly apparent: see Chapter 3) then product (67) could be the result of such an oxa-di-π-methane process rather than a discrete 1,2-migration of the acetyl function (see reaction 4).

$$(65a) \xrightarrow{h\nu} \quad\longrightarrow\quad (67) \qquad (4)$$

Differences in reaction path have also been reported for the rearrangements of the steroidal enone (68).[30] Sensitized isomerisation of the $\beta\gamma$-unsaturated ketones (68a and b) has been shown to give the ring-contracted products (69).[31] The authors suggest the intermediacy of a diradical, but a concerted process would be equally plausible.[32] Direct

(a) R = H
(b) R = Me
(68)

(69)

(70)

irradiation, however, gave rise to different products which were identified as the oxetans (70a and b).[30] It is suggested by the authors that a different mechanism obtains in this example, viz. Norrish type I cleavage followed by hydrogen transfer to yield an aldehydo-diene (reaction 5) which undergoes secondary photolysis to yield the oxetans. The photorearrangement of the diketone (70) to the isomers (71a) and (71b) does not appear to be either wavelength- or solvent-dependent.[33] Two interpretations of the reaction are possible, viz. either the stepwise Norrish type I process shown in the Scheme, or else a concerted $\sigma_a^2 + \pi_a^2$ photoallowed process. Both the carbonyl groups appear to be essential for the process, since irradiation of the acetal (72) in dioxan, in which the diketone rearranges, forms only a diastereoisomeric mixture of the dioxan adduct (73).

[30] K. Kojima, K. Sakai, and K. Tanabe, Tetrahedron Letters, 1969, 3399.
[31] K. Kojima, K. Sakai, and K. Tanabe, Tetrahedron Letters, 1969, 1925.
[32] J. R. Williams and H. Ziffer, cited by W. M. Horspool in 'Photochemistry,' ed. D. Bryce-Smith, (Specialist Periodical Reports), The Chemical Society, London, 1970, Vol. 1, p. 192.
[33] S. Domb, G. Bozzato, J. A. Saboz, and K. Schaffner, Helv. Chim. Acta, 1969, 52, 2436.

(68) 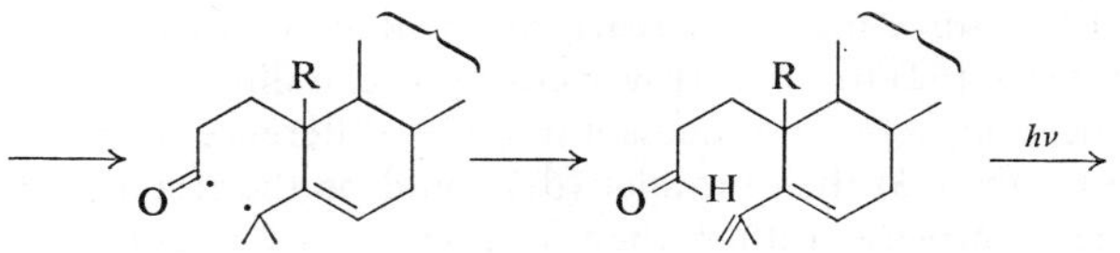 (70) (5)

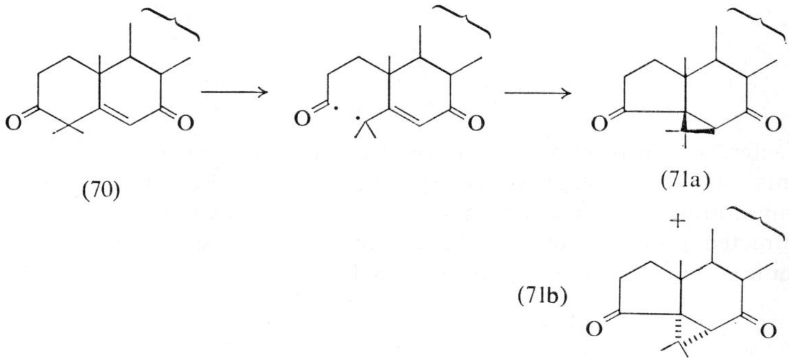

(70) (71a) + (71b)

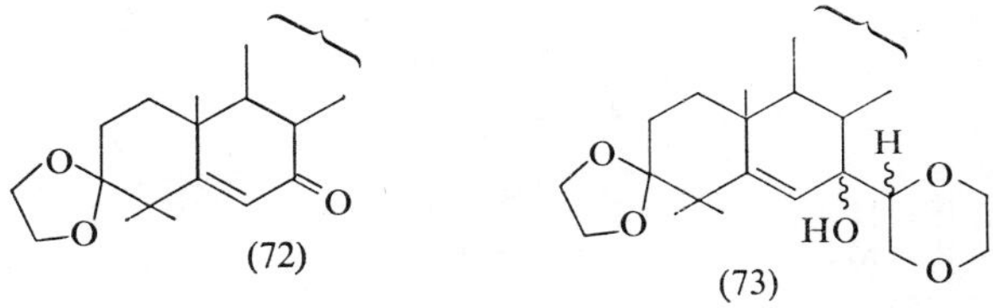

(72) (73)

γδ-Unsaturated Carbonyl Compounds.—The irradiation of the ketols (74) in ethereal solution led to the oxetan (75), in addition to *cis–trans*-isomerisation of the starting material. The formation of oxetans is one of the most

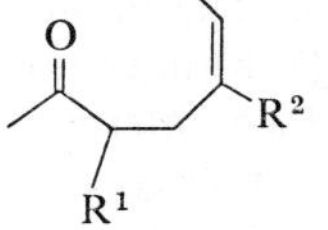

(a) $R^1 = R^2 = H$
(b) $R^1 = H$, $R^2 = Me$
(c) $R^1 = Me$, $R^2 = H$

(74)

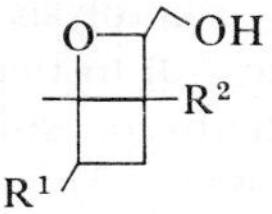

(75)

common processes in inter- and intra-molecular photoreactions between ketones and olefins. Aldehydes form oxetans less readily than ketones unless an electron-rich olefin is employed as the addend. This possibly explains why the sensitized photochemical reaction of hex-5-en-1-al gives 2-methylcyclopentanone (4%) and cyclohexanone (2%), together with

[84] Y. Baherel, G. Descotes, and F. Pautel, *Compt. rend.*, 1970, **270** *C*, 1528.

various other products.[34] A triplet state is inferred from the fact that direct photolysis of the aldehyde in pentane affords no evidence for the formation of cyclic products. It is suggested that the olefinic triplet is accessible by energy transfer from the excited acetone solvent. Hydrogen transfer and cyclisation can account for the observed products. This work is put forward [35] as an explanation for the isolation of 2-methylcyclopentanone from the photolysis of cyclohexanone.[36] Irradiation of hepta-5,6-dien-2-one also yields an oxetan, the methylene oxetan (76a), and the divinyl ether (76b).[37] In methanol, providing the irradiation time was kept short, only the oxetan was isolated, but more prolonged irradiation led to a complex mixture of as yet unidentified secondary products. In methanol–water only the oxetan was isolated. The irradiation by Pyrex-filtered u.v. light is thought to produce $n\pi^*$ triplets followed by nucleophilic attack of their oxygen on the allene. The intermediate diradical (i) can then be invoked to explain the formation of both the oxetan and the divinyl ether (Scheme 5).

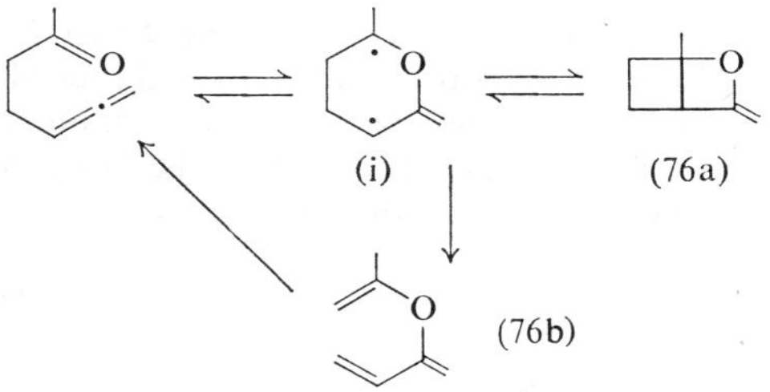

Scheme 5

Also of interest and pertinent to the re-formation of the starting material is the thermal rearrangement of the oxetan to the diradical intermediate which can ring-open to the divinyl ether. Under mild conditions this can undergo a Cope rearrangement to produce the keto-allene.[37] Interposition of a cyclopropane ring between the double bond and the carbonyl function brings about different reactions. cis–trans Mixtures of three cyclopropyl carbonyl compounds (77a, b, and c) have been investigated.[38] An $n\pi^*$ excited state is proposed to be the result of initial excitation bringing about ring-cleavage of the cyclopropane. The diradical intermediate is preferred over concerted pathways to the products since such an intermediate can also be used to explain the observed cis–trans-isomerisation of the original compounds (see also Chapter 1). The ratio of cis : trans-isomers at equilibrium could not be determined owing to the rapid conversion of the compounds to the products, dihydrofuran (78), and ester (79) being the major products in the case of (76a and b). The esters presumably arise

[35] W. C. Agosta, D. K. Herron, and W. W. Lowrance, Tetrahedron Letters, 1969, 4521.
[36] R. Srinivasan and S. E. Cremer, J. Amer. Chem. Soc., 1965, 87, 1647.
[37] J. K. Crandall and C. F. Mayer, J. Org. Chem., 1969, 34, 2814.
[38] W. G. Dauben and G. W. Shaffer, J. Org. Chem., 1969, 34, 2301.

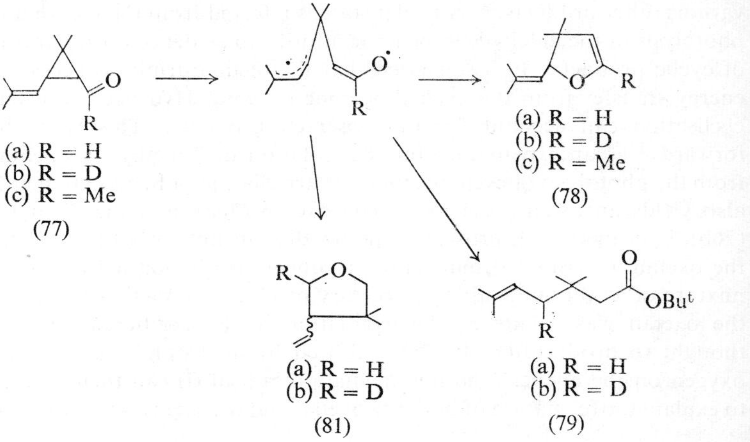

(a) R = H
(b) R = D
(c) R = Me
(77)

(a) R = H
(b) R = D
(c) R = Me
(78)

(a) R = H
(b) R = D
(81)

(a) R = H
(b) R = D
(79)

from addition of the solvent t-butanol to the keten intermediate (80) formed by a 1,3-hydrogen migration either by a diradical or a concerted process. The third product isolated from the photoreactions of the aldehydes (76a and b) is tentatively assigned the structure (81a and b). The ketone (76c) gave only the dihydrofuran (78c).

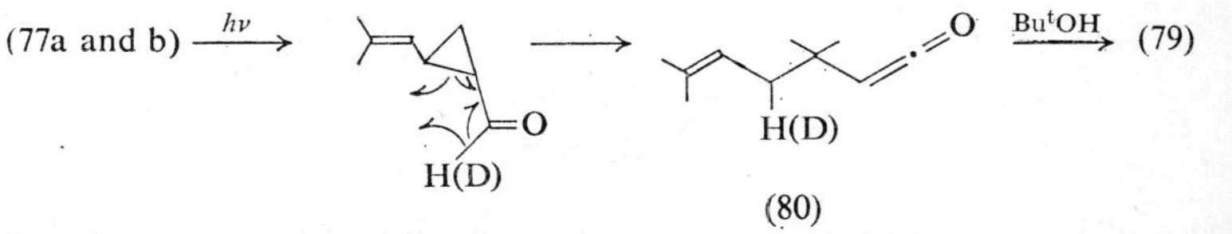

(77a and b) $\xrightarrow{h\nu}$ ⟶ ⟶ $\xrightarrow{Bu^tOH}$ (79)

(80)

Bridged Unsaturated Carbonyl Compounds.—Irradiation of the benzonorbornenone (82) affords the indenone (83) via the diradical (equation 6).[39]

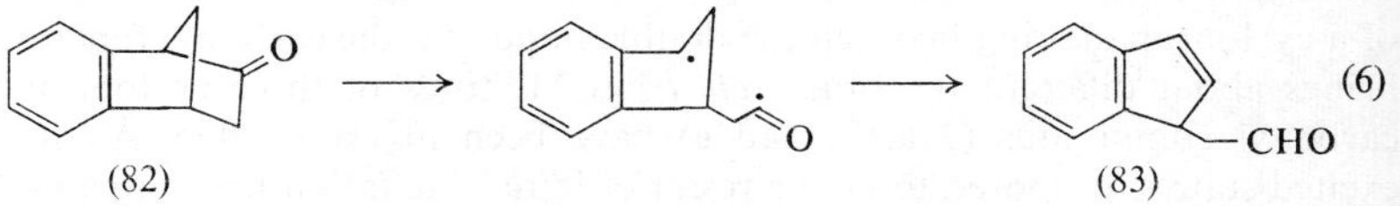

(82) ⟶ ⟶ (83) (6)

No rearrangement similar to that of the norbornenone (84) to (85) was encountered since this would have meant disruption of the aromaticity of the benzo-substituent. A comparison of the photochemistry of small ring compounds (86) and their mass spectral fragmentation patterns has been reported.[40] A 1,3-sigmatropic rearrangement is invoked to explain the low-temperature photolytic rearrangement of bicyclic ketone (87) to the cyclopropane (88).[41] Evidence for the existence of this intermediate is

[39] J. Ipaktschi, *Tetrahedron Letters*, 1969, 2153.
[40] H. Dürr and P. Heitkaemper, *Z. Naturforsch.*, 1969, **24b**, 779.
[41] L. L. Barber, O. L. Chapman, and J. D. Lassila, *J. Amer. Chem. Soc.*, 1969, **91**, 3664.

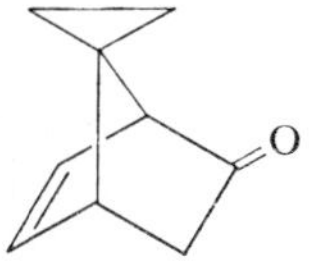

(84) 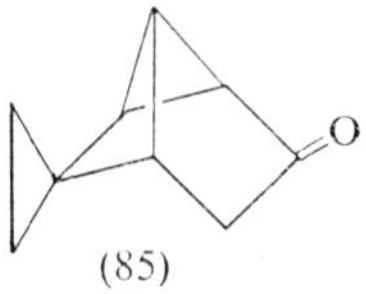 (85)

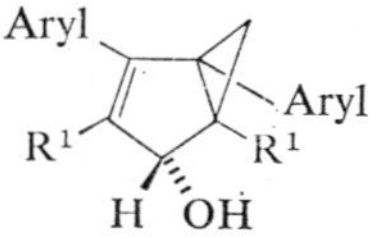

(86a) 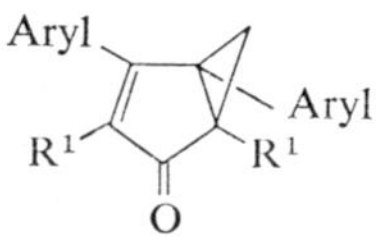 (86b)

obtained from the i.r. spectrum of the irradiation product which shows an absorption at 1812 cm^{-1}, typical of cyclopropanones. Further evidence

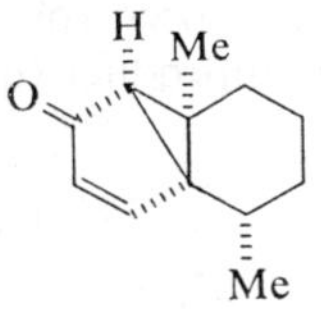

(87) 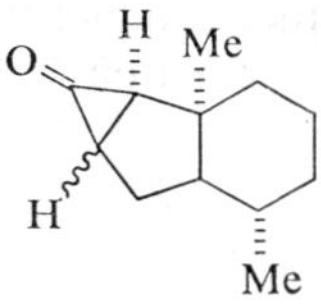 (88) 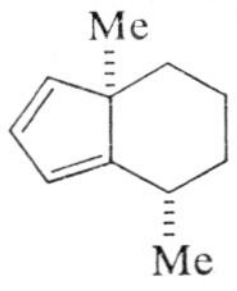 (89)

is obtained from the thermal reaction with furan which yields an adduct from cycloaddition to the cyclopropanone. The wavelength used for the process is surprisingly critical, for 253·7 nm or > 360·0 nm radiation gives the cyclopropane in good yield, whereas 300—360 nm radiation decarbonylates the product to give CO and a diene (89).

Several reports have appeared concerning the photochemical rearrangements of the bicyclo[3,2,2]dienones (90). Low-temperature studies [42] have shown the presence of ketens which can be trapped as the esters (91). In the absence of nucleophiles, the photoproduced ketens revert to starting material. Two groups [42, 43] have studied dienone (90c). From the room-temperature photolysis in methanol Chapman et al.[42] isolated two isomeric esters (92a) and (93a) while Kende et al.[43] obtained the corresponding acids (92b) and (93b). At lower temperatures ($-80\,°C$), only one keten (94) could be observed in the i.r. spectrum, and gave rise to the *endo*-ester (93a).[42] This specificity was also found by the other workers,[43] who found that the *endo*-acid (93b) thermally inverts to the *exo*-isomer (92b). This rearrangement is an example of a [3,3] photorearrangement arising from the singlet

[42] O. L. Chapman, M. Kane, J. D. Lassila, R. L. Loeschen, and H. E. Wright, *J. Amer. Chem. Soc.*, 1969, **91**, 6856.
[43] A. S. Kende, Z. Goldschmidt, and P. T. Izzo, *J. Amer. Chem. Soc.*, 1969, **91**, 6858.

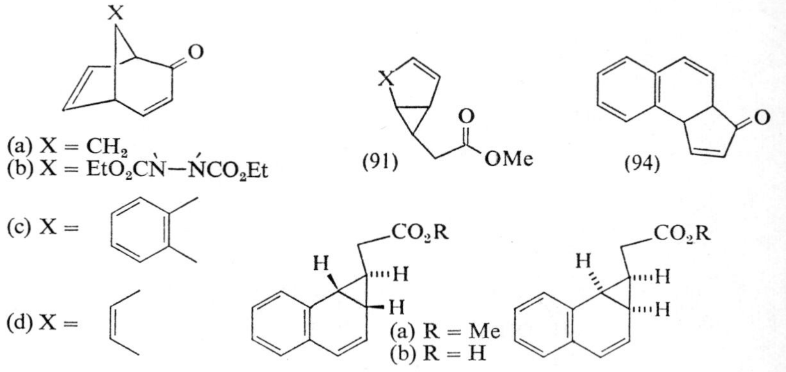

(a) X = CH$_2$
(b) X = EtO$_2$CN—NCO$_2$Et

(c) X =

(d) X =

(90)

(91)

(92) (a) R = Me (b) R = H

(93)

(94)

state. In addition to this process, a slower photoreaction afforded the product (94) of 1,3-acyl migration.[43] Sensitization (benzophenone) of ketone (90c) affords yet another product (95). The proposed route to the observed product is shown in reaction 7.[44] The photoproduct (quantum

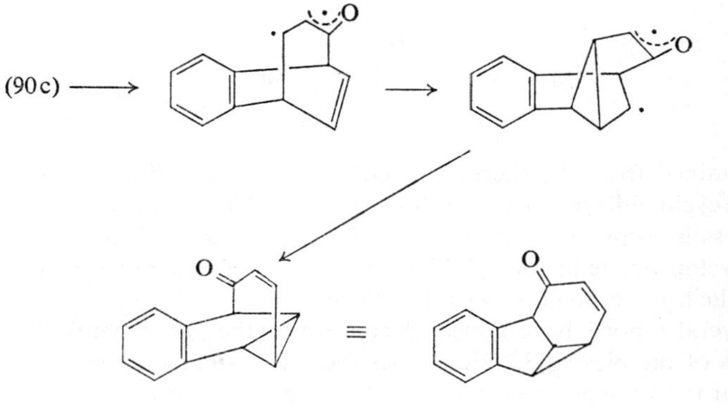

(90c) ⟶ ⟶

(95)

(7)

yield of 0·16) from the photolysis (366 nm) of 9-thiabicyclo[3,3,1]non-6-en-2-one in pentane has been unambiguously identified by X-ray analysis as 2-thiabicyclo[6,1,0]non-6-en-3-one (96).[45] The formation of the rearrangement product could not be quenched by naphthalene or by piperylene, and it is suggested that the reaction takes place by way of a singlet excited state. A possible route to the product (96) is shown in Scheme 6, although at the present time no differentiation between the charge-transfer route and the $n\pi^*$ route can be made.

[44] A. S. Kende and Z. Goldschmidt, *Tetrahedron Letters*, 1970, 783.
[45] A. Padwa, A. Battisti, and E. Shefter, *J. Amer. Chem. Soc.*, 1969, **91**, 4000.

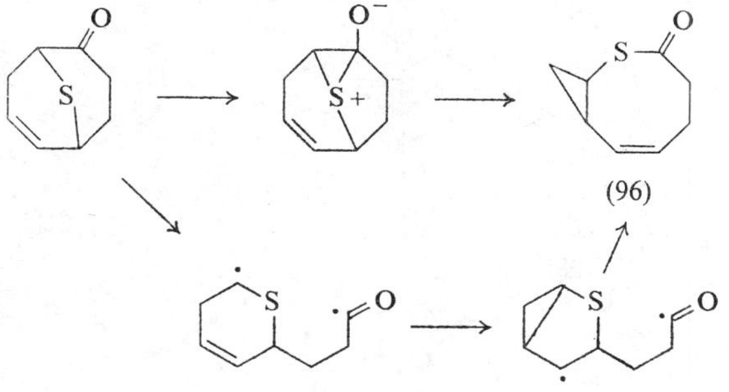

Scheme 6

Unsaturated Nitriles, Acids, and Esters.—Direct irradiation of 2-cyanobutadiene in ether solution affords the two products 1-cyanocyclobutene ($\Phi = 0.009$) and the 1-cyanobicyclobutane ($\Phi = 0.029$).[46] Attempts to sensitize the reaction were futile, and it is implied that the observed products arise by a singlet process. The formation of the bicyclobutane is thought to be due to the intermediacy of a biradical (97), as proposed in similar work.[47]

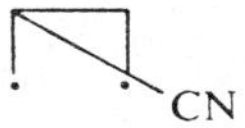

(97)

Acrylonitrile photodimerises to the two head–head dimers, the *cis*- and *trans*-dicyano-compounds.[46] The ratio of the two dimers (*cis* : *trans* = 0.8) is dependent on the medium in which the reaction is carried out, but not on the triplet sensitizer, thus implying the intermediacy of a triplet state. Only sensitizers with $E_T > 69$ kcal mol^{-1} were effective in the dimerisation. On the other hand, crotonaldehyde did not dimerise but on triplet sensitization underwent *cis–trans*-isomerisation.

Further reports on the *trans–cis* isomerisation of acyl-[48] and aroyl-[49] acrylic acids have been published. A lactonisation process accompanies the *trans–cis*-isomerisation of the acylacrylic acids.[48] This becomes possible since in the *cis*-form of the acid the carbonyl group is adjacent to the carboxyl function. Similar lactonisation has been reported[50] for the androstane derivative (98), again following *trans–cis* isomerisation of the $\alpha\beta$-unsaturated acid (reaction 8). An involvement of a suitably placed hydroxy-function with respect to a carbonyl function is reported in the

---

[46] D. M. Dale, *J. Org. Chem.*, 1970, **35**, 970.
[47] R. Srinivasan, *J. Amer. Chem. Soc.*, 1968, **90**, 4498.
[48] N. Sugiyama, H. Kataoka, C. Kashima, and K. Yamada, *Bull. Chem. Soc. Japan*, 1969, **42**, 1098.
[49] N. Sugiyama, H. Kataoka, C. Kashima, and K. Yamada, *Bull. Chem. Soc. Japan*, 1969, **42**, 1353.
[50] M. Debono and R. M. Molloy, *J. Org. Chem.*, 1970, **35**, 483.

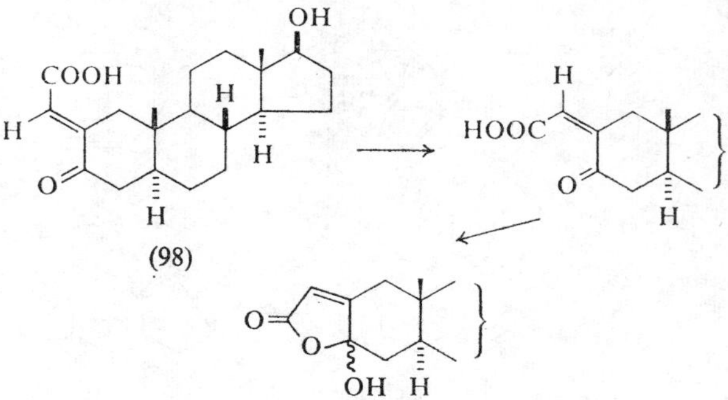

photoconversion of a hydroxy-chalcone (99) to the flavene (100) on irradiation in methanol or ethanol.[51] No evidence was cited for the identity of the chalcone as purely the *cis*-isomer shown above. It is obviously possible that the *trans*-isomer could well be isomerised to the *cis*-form. It is

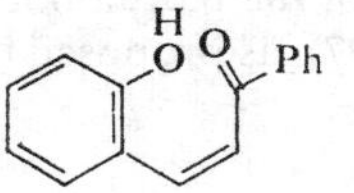

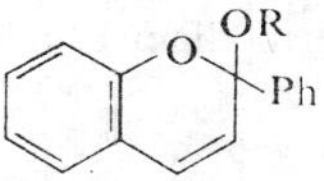

(a) R = Et
(b) R = Me

suggested that there is a possible biosynthetic significance in this photochemical route to flavenes.

cis- and *trans*-Cinnamic acids have been shown to undergo photoisomerisation.[52] The photostationary states approached from both the *trans*- and the *cis*-isomers of cinnamic acid have been measured in methanol solution. The results (Table 2) are somewhat lower than those reported

Table 2 *Photostationary state of cinnamic acid in methanol* [52]

Irradiation time (hr)	% cis acid present	
	from cis	from trans
0	99.0	0
24	54.0	52.0
48	57.7	58.3
72	60.2	59.0

previously,[53] but this could be due to the higher concentrations (0.27 mol l^{-1}) used in this work than in the earlier studies (0.002 and 0.003 mol l^{-1}). The

[51] D. Dewar and R. G. Sutherland, *Chem. Comm.*, 1970, 272.
[52] M. B. Hocking, *Canad. J. Chem.*, 1969, **47**, 4567.
[53] B. K. Vaidya (*Proc. Roy. Soc.*, 1930, **129**, 299) reported 72 and 75% *cis*-isomer at the photostationary state.

influence of concentration on the photostationary state has been noted by other workers.[54]

Deconjugation reactions have also been studied further during the year. Debono and Molloy[50] have observed some steric requirements for the reaction in some androstane derivatives. Photosensitization was required to effect the deconjugation of the acid (101) to acid (102) and they propose

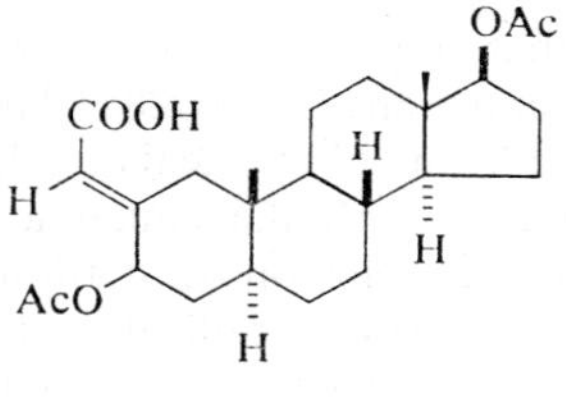

(101) 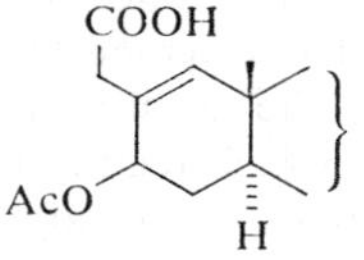   (102)

that the side-chain and the C-1 equatorial proton must be coplanar to allow formation of the 6-membered transition state necessary for deconjugation. Further observations by Jorgenson and Patumtevapibal[55] indicate that the deconjugation reactions of certain esters could well be governed by rules of orbital symmetry and involve 1,5-migrations of hydrogen atoms, following $\pi \to \pi^*$ excitation. This process is seen to advantage in the deconjugation of esters (103a and b) as a 20 : 80 mixture by irradiation through a Vycor filter to afford the deconjugated ester (104). It is proposed

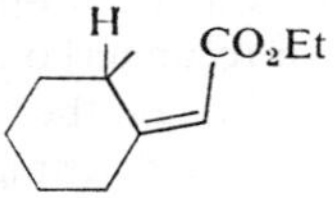

(103a) 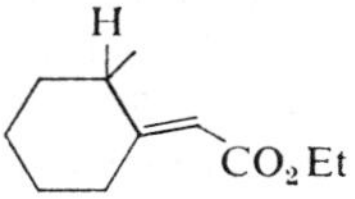   (103b) 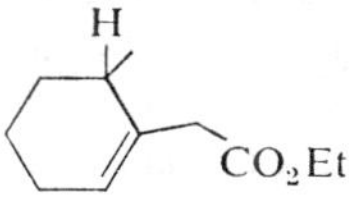   (104)

that conformational effects influence the direction and the unusual preference for primary over secondary hydrogen abstraction. Other esters have also been examined and the results show that deconjugation into a 6-membered ring is preferred to that into 7-, 5-, and 4-membered rings, where only polymerisation took place. The rates for deconjugation into the 5- and 7-membered rings were approximately 0·1, and 0·05 that for for the 6-membered ring. The styrylamides (105a and b, and 106) have been shown to undergo photochemical *cis–trans*-isomerisation.[56] With both amides (105a and b) unfiltered light was used and in case of (105a) the same photostationary state could be attained by photolysis of both the *trans*- and the *cis*-isomers of the amide. By the use of filtered light (>300 nm), the isomerisation of amide (106) has been shown to be a

---

[54] A. A. Zimmerman, C. M. Orlando, and M. H. Gianni, *J. Org. Chem.*, 1969, **34**, 73.
[55] M. J. Jorgenson and S. Patumtevapibal, *Tetrahedron Letters*, 1970, 489.
[56] R. W. Hoffmann and K. R. Eicken, *Chem. Ber.*, 1969, **102**, 2987.

(a) $n = 1$, (b) $n = 2$ $R = $ Ph or Me
(105)   (106)

triplet process and it is thought that the triplet energy of the amide is ca. 50 kcal mol^{-1}, since both benzophenone [E_T = 284 kJ mol^{-1} (68 kcal mol^{-1})] and fluorenone [E_T = 221·5 kJ mol^{-1} (53 kcal mol^{-1})] are equally effective in the sensitized isomerisation. Unlike cinnamic acid and its derivatives, which readily forms truxillic acids on irradiation, atropic acid (107a) is photochemically inert.[57] However, radical additions can be brought about by the photolysis of atropic acid ester (107b) in the presence of dioxan or tetrahydrofuran and benzophenone. Benzophenone acts as a hydrogen-abstracting agent. The compounds isolated from these reactions

(a) $R = H$ (a) $R = $ 2-dioxanyl
(b) $R = $ Et   (b) $R = $ 2-tetrahydrofuranyl
(107)   (108)

are the dioxanyl ether (108a) and the tetrahydrofuranyl ether (108b). In the presence of a good hydrogen donor (alcohol–water), $\alpha\beta$-unsaturated carboxylate anions of cinnamates have been shown to undergo abstraction on photolysis either at 253·7 nm or > 280 nm, depending on the substitution pattern of the cinnamate.[58] The carboxylates are not as efficient in hydrogen abstraction as ketones, and do not abstract hydrogen from methanol. The intermediate is thought to be a hitherto unknown class of carboxylate radical anions (109). Although direct evidence concerning the structure of the intermediate is lacking, the ability of the carboxylate oxygen to abstract hydrogen has been demonstrated by a photolysis in D_2O–MeOD when incorporation of deuterium in the α-methylene group of the product was found to take place. The products from the reaction and the wavelength dependence of the processes are shown in Tables 3 and 4. As can

(109)   (a) *erythro*
                      (b) *threo*
                      (113)

[57] W. Droste, H. D. Scharf, and F. Korte, *Annalen*, 1969, **724**, 71.
[58] E. F. Ullman, E. Babad, and M. T. Sung, *J. Amer. Chem. Soc.*, 1969, **91**, 5792.

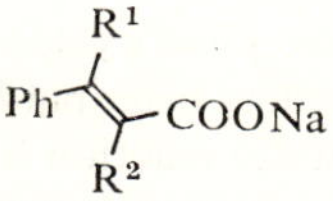

Enone Rearrangements and Cycloadditions

(a) $R^1 = R^2 = H$
(b) $R^1 = H, R^2 = Me$
(110)

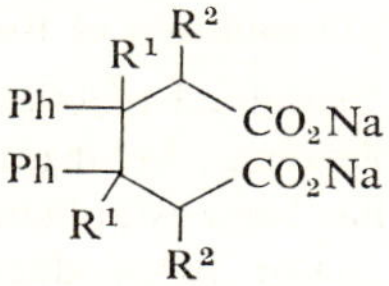

(c) $R^1 = Me, R^2 = H$
(d) $R^1 = Ph, R^2 = H$
(111)

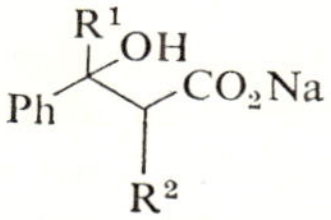

(e) $R^1 = H, R^2 = Ph$
(112)

Table 3 Products from the photolysis of substituted sodium cinnamates (110a—c) [58]

Reactant	Solvent	λ nm	Products (%)		
			(110)[a]	(111a–c)	(112)
(110a)	Pr^{i}OH	253·7	26	7	15
	Pr^{i}OH	280	6	36	
	MeOH	253·7			38
	MeOH	280	99		1
(110b)	Pr^{i}OH	253·7			24
	Pr^{i}OH	280	9	8	
	MeOH	253·7			
	MeOH	280	27		29
(110c)	Pr^{i}OH	253·7			
	Pr^{i}OH	280	32	29	
	MeOH	253·7	5		3
	MeOH	280	99		

[a] A photostationary mixture of *cis*- and *trans*-(110).

Table 4 Products from the photolysis of substituted cinnamates (110d and e) [58]

Reactant	Solvent	λ nm	Products (%)				
			Na 3,3-diphenyl propionate	Na benzoate	1,2-dihydroxy-1,2-diphenyl ethane	(113a)	(113b)
(110d)	Pr^{i}OH	253·7	66				
	Pr^{i}OH	280	83				
(110e)	MeOH	253·7		26	31		
	MeOH	280		56	50		
	Pr^{i}OH	253·7		6	6	18	14
	Pr^{i}OH	280		3	3	16	45

be seen, the hydroxylation to produce β-hydroxy-propionates is dependent to a great extent on the pattern of the substituents and their nature, and also on the wavelength of the exciting radiation. In general, however, the presence of phenyl substituents appears to make the photolysis independent of exciting wavelength, for 253·7 nm and >280 nm radiation were equally effective. The nature of the excited state was not determined.

2 Addition and Cycloaddition of Enones

One of the most extensively examined reactions is the photocycloaddition of dimethyl maleate to cyclohexene. Most attention has been paid to the product cyclobutanes and it has been established that the reaction is, in the main, non-concerted. The major by-product, the ene-reaction product (114), has been shown by the use of 1,2-dideuteriocyclohexene to arise by a partly non-concerted process which probably involves cyclohexenyl radicals. Thus the two ene-products (114a and b) were isolated.[59]

(a) $R^1 = D$, $R^2 = H$
(b) $R^1 = H$, $R^2 = D$ (115)

(114)

Ahlgren and Akermark [59] point out that the photo-ene reaction is expected to be non-concerted since only the thermal process is symmetry allowed. There are two criteria which must be fulfilled before the photoreaction for $n\pi^*$ maleate is allowed: (i) the interaction between the highest occupied orbital of the maleate and the lowest unoccupied orbital of the cyclohexene must be bonding, and (ii) the complementary interaction between the lowest unoccupied orbital of the maleate and the highest occupied orbital of the cyclohexene must also be bonding.[59] In practice this is not the case and (ii) is disallowed. Thus the reaction is considered to be 'half-allowed'. This analysis emphasises the undesirability of applying the Woodward–Hoffmann rules to photoreactions involving $n\pi^*$ states. Observations on the mass spectral fragmentation patterns of *cis,cis*-muconic acid, *cis,trans*-muconic acid, and *trans,trans*-muconic acid and the three possible cyclobutene 3,4-dicarboxylic acids show that there are certain similarities.[60] Thus the pattern from *cis,cis*-muconic acid is closely similar to those from *cis,cis*-cyclobutene (115) and the *trans,trans*-muconic acid. The Woodward–Hoffmann rules imply that these three compounds should be interconvertible *via* an electronic excited state. On these grounds it has been suggested that there is a connection between the photochemical process and the mass spectral fragmentation. It is pointed out that it is not safe to generalise; but there are now several apparent examples of such correlations.

Cyclopentenone-type Cycloadditions.—A study of thermal effects on the cycloadditions of cyclopentenone has been reported.[61] The quantum

---
[59] C. Ahlgren and B. Akermark, *Tetrahedron Letters*, 1970, 1885.
[60] M. K. Hoffmann, M. M. Bursey, and R. E. K. Winter, *J. Amer. Chem. Soc.*, 1970, **92**, 727.
[61] R. O. Loutfy, P. de Mayo, and M. F. Tchir, *J. Amer. Chem. Soc.*, 1969, **91**, 3984.

yields for the cycloadditions with four olefins at three temperatures are shown in Table 5. Previous studies on the inefficiency of this particular cycloaddition have shown that either a complex or a diradical intermediate is initially formed, and that this can either form the product or revert to the

Table 5 *Temperature dependence of quantum yield of cycloadditon of olefins to cyclopentenone* [61]

Olefin	ϕ_{27}	ϕ_{-10}	ϕ_{-71}
Cyclohexene	0·46	0·51	0·62
Cyclopentene	0·23	0·31	0·61
cis-Dichloroethylene	0·23		0·34
trans-3-Hexene	0·22		0·19

starting materials.[62] From the kinetic analysis of the reaction represented in Scheme 7, the quantum yield ϕ is given by Equation (*1*). The change in the

$$\Phi = \Phi_{ic} K(k_r[\text{olefin}]/k_r[\text{olefin}] + k_d) \quad (1)$$
when $K = (k_4/k_3 + k_4)(k_2/k_1 + k_2)$.

quantum efficiency has been shown to be due to a change in the partition function (K) by experiments (Table 6) which showed the changes in k_d and k_r, and that the value of the bracketed term in Equation (*1*) did not alter appreciably. However, it was not possible to differentiate between the two

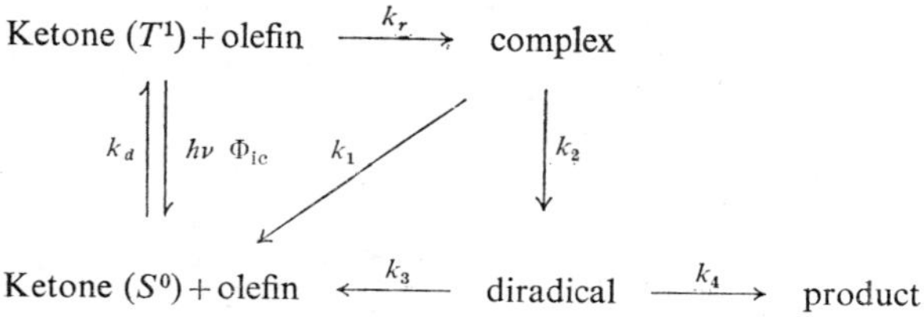

Ketone (T^1) + olefin $\xrightarrow{k_r}$ complex

$k_d \updownarrow h\nu \, \Phi_{ic} \quad k_1 \quad \downarrow k_2$

Ketone (S^0) + olefin $\xleftarrow{k_3}$ diradical $\xrightarrow{k_4}$ product

Scheme 7

possible intermediates in the reaction, although the differences in activation energy for the competing processes can be calculated to be $-0·9$ and $-2·1$ kcal mol^{-1} for cyclohexene and cyclopentene respectively.

Independently, Dilling *et al.*[63] have proposed that in the cycloaddition of *cis*- or *trans*-1,2-dichloroethylene to cyclopentenone two steps are involved with initial addition at C-3 of the enone. Thus a diradical is produced

Table 6 *Temperature dependence of* k_d *and* k_r *in olefin cycloaddition to cyclopentenone* [61]

Temperature °C	$k_{\text{diff}} \times 10^{-1}$ l mol^{-1} s^{-1}	$k_d \times 10^{-8}$ s^{-1}	$k_r \times 10^{-8}$ l mol^{-1} s^{-1}
20	2·0	1·5	10·3
−71	0·36	1·3	3·2

[62] See 'Photochemistry,' ed. D. Bryce-Smith, (Specialist Periodical Reports), The Chemical Society, London, 1970, Vol. 1, pp. 203, 204, for the relevant work.

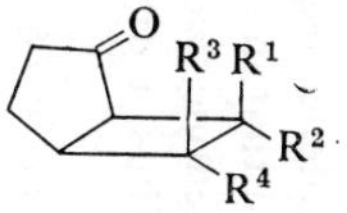

(a) $R^1 = R^3 = H$, $R^2 = R^4 = Cl$
(b) $R^1 = R^4 = Cl$, $R^2 = R^3 = H$
(c) $R^2 = R^3 = Cl$, $R^1 = R^4 = H$

(116)

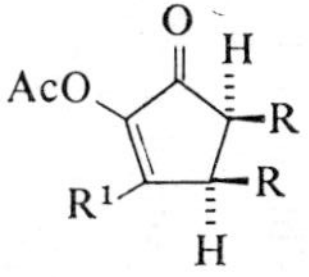

(a) $R = R^1 = H$
(b) $R^1 = Me$, $R-R = CH_2(CMe_2)CH_2$
(c) $R = H$, $R^1 = Me$

(117)

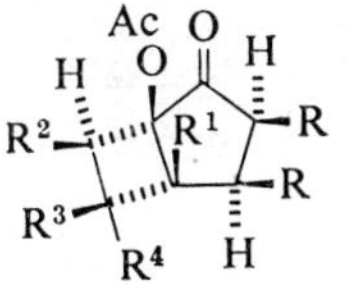

(a) $R = R^1 = R^4 = H$, $R^2-R^3 =$ H$_2$C—CH$_2$ (cyclopropane)
(b) $R^1 = Me$, $R-R = CH_2(CMe_2)CH_2$, $R^2 = H$, $R^3 = R^4 = OEt$
(c) $R = R^2 = H$, $R^1 = Me$, $R^3 = R^4 = OEt$
(d) $R = R^1 = R^2 = H$, $R^3 = R^4 = OEt$

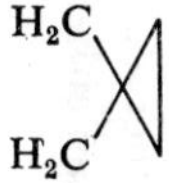

(118)

which can thermally equilibrate prior to formation of the second bond. The three products (116a), (116b), and (116c) were obtained (Table 7). Considerable interest has been shown in the use of 2-acetoxycyclopentenones as intermediates in the synthesis of natural products. Thus the

Table 7 *Product distributions from the photoaddition of cyclopentenone to cis- and trans-1,2-dichloroethylene extrapolated to zero conversion* [63]

Starting olefin	Products (%)		
	(116a)	(116b)	(116c)
cis-1,2-Dichloroethylene	28·6	19·1	52·4
trans-1,2-Dichloroethylene	47·8	30·0	22·2

photoaddition of spiro[2,4]hept-4-ene to 2-acetoxycyclopentenone (117a) affords adduct (118a), a precursor for the synthesis of isomarasmate.[64] Similarly, enone (117b) can be used to form adduct (118b) for the synthesis of protoilludane.[65] 1,1-Diethoxyethylene has also been added photochemically to enones (117c) and (117a) to afford adducts (118c) and (118d) respectively.[66] 2-Acetoxy-3-methylcyclohexenone is surprisingly unreactive under the conditions used for the cyclopentenone additions.[66] Examples of the synthetic value of photocycloaddition of olefins to 5-membered ring ketones have been published. The addition of ethylenes, acetylenes, and allenes to 3-β-hydroxpregna-5,16-diene-20-one acetate gives cyclobutanes, cyclobutenes, and isomeric methylenecyclobutanes.[67] The photochemical addition of 1,2-dichloroethylene and but-2-yne to the bicyclic enone (119) has been reported [68] to give the tricyclic ketones (120a) and (120b) respectively.

(119)   (120a)   (120b)

Cyclohexenone-type Cycloadditions.—A mechanistic study of the photoreactions of cyclohexenone has been carried out.[69] From these studies the unimolecular decay constant has been calculated as $1·7 \times 10^8$ s^{-1} and the

---

[63] W. L. Dilling, T. E. Tabor, F. P. Boer, and P. P. North, *J. Amer. Chem. Soc.*, 1970, **92**, 1399.
[64] D. Helmlinger, P. de Mayo, M. Nye, L. Westfelt, and R. B. Yeats, *Tetrahedron Letters*, 1970, 349.
[65] S. Kagawa, S. Matsumoto, S. Nishida, S. Yu, J. Morita, A. Ichihara, H. Shirahama, and T. Matsumoto, *Tetrahedron Letters*, 1969, 3913.
[66] T. Matsumoto, H. Shirahama, and A. Ichihara, *Tetrahedron Letters*, 1969, 4103.
[67] P. Sunder-Plassman, P. H. Nelson, P. H. Boyle, A. Cruz, J. Iriarte, P. Crabbe, J. A. Zderic, J. A. Edwards, and J. H. Fried, *J. Org. Chem.*, 1969, **34**, 3779.
[68] R. L. Cargill and J. W. Crawford, *J. Org. Chem.*, 1970, **35**, 356.
[69] P. de Mayo, A. A. Nicholson, and M. F. Tchir, *Canad. J. Chem.*, 1970, **48**, 225.

Table 8 *Influence of sensitizer triplet energy on cycloadditions reactions of Cyclohexenone* [69]

Sensitizer	E_T (sens) kJ mol^{-1} (kcal mol^{-1})	$\Phi_{add_1}{}^a$	$\Phi_{add_2}{}^b$	$\Phi_{dim}{}^c$	Φ_S/Φ_{add_1}	Φ_S/Φ_{add_2}	λ nm
—	—	0·48	0·22	0·096	—	—	366
Acetophenone	307·6 (73·6)	0·33	0·22	—	0·69	1·0	313
p-Trifluoromethyl acetophenone	295·9 (70·8)	0·10	—	—	0·208	—	313
Benzophenone	287·6 (68·8)	0·03	0·077	0·086	0·064	0·34	366
4,4-Dichlorobenzophenone	284·2 (68·0)	—	0·03	0·095	—	0·14	366
4-Cyanobenzophenone	277·6 (66·4)	—	0·002	0·04	—	0·005	366

Φ_{add} = cycloaddition, Φ_{dim} = dimerisation, Φ_S = sensitization.
[a] In neat cyclohexene Φ_{add_1} at 313 = 0·48. Cyclohexenone concentration = 0·25 mol l^{-1}
[b] 1·96M Cyclohexene in benzene; cyclohexenone concentration = 0·45 mol l^{-1}
[c] Cyclohexenone concentration = 0·8 mol l^{-1}

bimolecular decay constant as $2.5 \times 10^7 \, l \, mol^{-1} \, s^{-1}$. It has also been established that the reactive and the lowest triplet states of cyclohexenone must be very close together (if indeed they are not identical) and have an energy of 66—68 kcal mol^{-1}. Other results obtained from the sensitized reactions are shown in Table 8.

The synthetic utility of the cycloaddition of olefins to cyclohexenones continues to be exploited. One of the first examples of cycloaddition of an olefin to a cyclohexenone substituted in position three with an electron-withdrawing group has been reported.[70] Irradiation through Pyrex of a solution of ethylene and the cyclohexenones (121a—c) in benzene gave good yields of the cyclobutane adduct (122). There is no mention of any

(a) R^1 = H, R = COOH
(b) R^1 = H, R = COOEt
(c) R = CN, R^1 = H
(d) R = R^1 = H
(e) R = H, R^1 = Me
(f) R = Me, R^1 = H

(121)

R = COOH, COOEt, or CN
(122)

adverse effect of electron-withdrawing substituents on the synthetic utility of the reaction. A full report[71] on the additions of cyclohexenones (121d—f) to norbornadiene has been published.[72] The irradiations were carried out using a Pyrex-filtered 450 watt mercury arc. The reactions all gave complex mixtures of products, the most important of which were formed by rearrangement of the norbornadiene skeleton. Details of the products are shown in Scheme 8. The cyclobutanes were the major components of the irradiation products, and in the case of cyclohexenone addition the percentages were: cyclobutanes, ca. 55%; α-substituted cyclohexenones, ca. 15%; and β-substituted cyclohexenones, ca. 30%. The reaction was virtually unaffected by solvent changes or by added naphthalene, although in the presence of naphthalene a marked decrease in the rate was noticed. It is inferred that these arise from the triplet state, in common with other cyclohexenone reactions, although no definite assignment could be made about the nature of the triplet state, *i.e.* whether it is $n\pi^*$ or $\pi\pi^*$. The products (123) and (124) are formed by *endo* attack of the cyclohexenone triplet on the norbornadiene, followed by bridging and 1,4-hydrogen migration (reaction 9). The norbornene products (125) and

---

[70] W. C. Agosta and W. W. Lowrance, *Tetrahedron Letters*, 1969, 3053.
[71] J. J. McCullough and J. M. Kelly, *J. Amer. Chem. Soc.*, 1966, **88**, 5935; J. J. McCullough and P. W. W. Rasmussen, *Chem. Comm.*, 1969, 387.
[72] J. J. McCullough, J. M. Kelly, and P. W. W. Rasmussen, *J. Org. Chem.*, 1969, **34**, 2933.

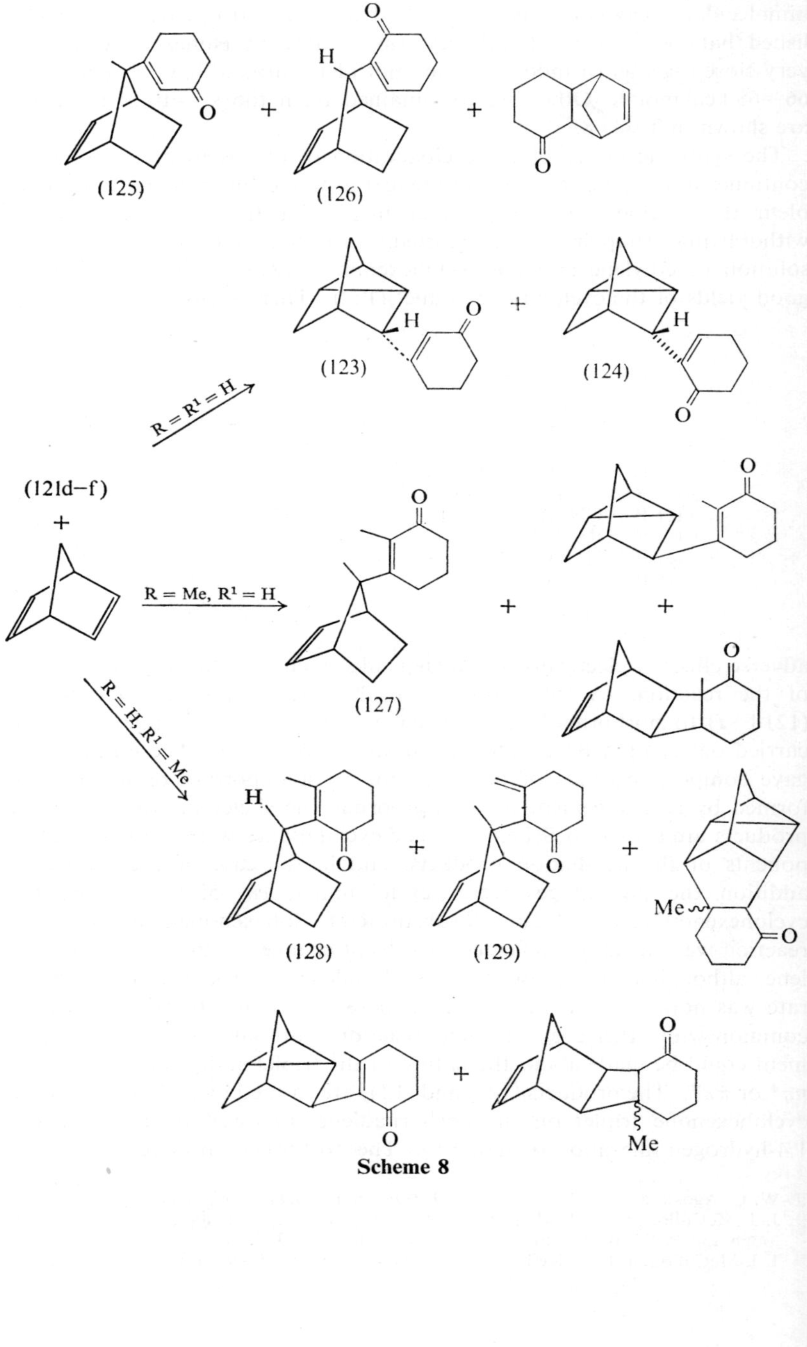

Scheme 8

(126) are the result of *exo* attack, bridging, rearrangement, and 1,4-hydrogen migration. The authenticity of this route was checked by the photolysis carried out on 3-deuteriocyclohex-2-enone, and the position of the deuterium atom was shown to be solely *exo* as postulated by the reaction scheme (reaction 10). Products (127), (128), and (129) are formed in an

analogous manner (reaction 10), both (127) and (128) requiring a 1,4-migration while the formation of (129) requires a 1,6-migration (reaction 11). The reaction paths which give the tricyclene and norbornene products almost certainly involve radical processes and are novel in the photochemistry of cyclohexenone additions. A fuller report [73] of the work

[73] J. W. Hanifin and E. Cohen, *J. Amer. Chem. Soc.*, 1969, **91**, 4494.

originally reported as notes [74] has been published concerning the cycloadditions of chromone (130). Emission spectra of chromone show strong phosphorescence and allow the assignment of the first triplet state as of $n\pi^*$ type with a lifetime of 38 ns and a triplet energy of 313·9 kJ mol^{-1} (75·1 kcal mol^{-1}) corresponding to an 0–0 band at 381 nm. The cycloadditions of various olefins to chromone give rise to mixtures of products. With tetramethylethylene, four products are isolated, *viz.* a cyclobutane (131a), an oxetan (132a), and two products resulting from hydrogen abstraction. The cyclobutane rings in all the compounds are *cis*-fused to the chromone. Cyclopentene and but-2-yne also give 4-membered ring compounds as well as a product from hydrogen abstraction and subsequent

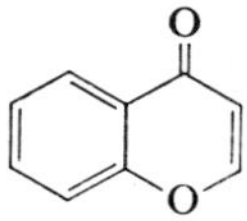

(130)

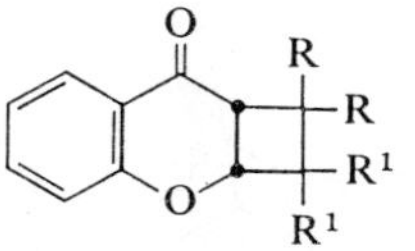

(131)
(a) R = R^1 = H
(b) R = H, R^1 = OMe

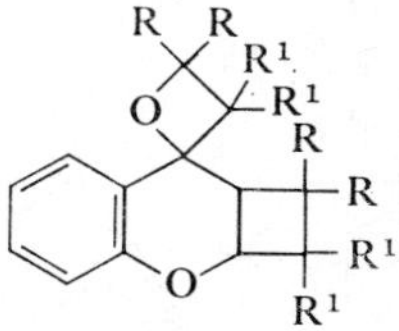

(132)

addition. The experimental evidence cited points towards initial attack and bond formation at the α-carbon, contrary to the results obtained by Dilling *et al.*[63] for some additions to cyclopentenone, which suggested the formation of a diradical intermediate. Thus, the addition of 1,1-dimethoxyethylene yields the cyclobutane (131b) as well as the oxetan (132b). Evidence for the intermediacy of a π-complex is still being sought.[75] The synthesis of tricyclo[4,3,2,0^{1,6}]undec-10-en-2-one (133a) has been achieved by the photocycloaddition of *cis*- and *trans*-1,2-dichloroethylene to the enone (134), followed by removal of the chlorine atoms by sodium in ammonia.[68] The synthesis of tricyclic ketone (133b) was accomplished by the photoaddition of but-2-yne to the same enone (134). 1,2-Dichloroethylene was also successfully added to 3-methylcyclohexenone by photolysis through a Corex filter to give the adduct (135), which was subsequently converted to the bicyclic ketone (136).[68]

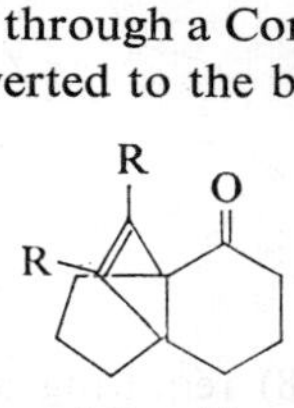

(133)
(a) R = H
(b) R = Me

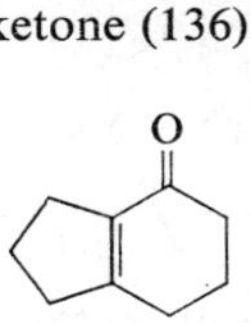

(134)

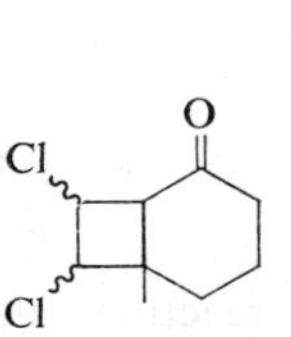

(135)

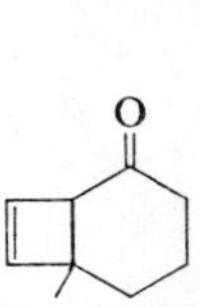

(136)

[74] J. W. Hanifin and E. Cohen, *Tetrahedron Letters*, 1966, 1419, 5421.
[75] See ref. 62, p. 200, for other examples involving π-complexes.

3 Dimerisation, Intramolecular Cycloaddition, and some Addition Reactions of Enones

The photoreaction of pseudo-ionone (137) in hexane solution has been reinvestigated.[76a] The original workers[76b] could find only polymeric materials, but the present group were able to isolate two cyclobutane dimers (138, gross structure shown) as well as what was presumed to be polymeric material. Direct irradiation (>280 nm) of *trans*-cinnamimide in 1,2-dimethoxyethane solution give rise to α-truxinimide (139, 48%), as well as

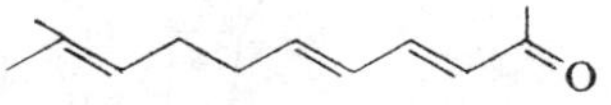

cis- and *trans*-cinnamamide.[77] Lalonde and Davis[77] suggest that the *cis–trans*-isomerisation takes place once the amide is formed and not at the imide stage, since no evidence for the presence of other truxinimide isomers was obtained. Thus, cyclisation to the cyclobutane is more rapid than *cis–trans*-isomerisation of the double bonds in the imide. A complex mixture of products results from the photolysis (253·7 and 350 nm) of an acetonitrile solution of the divinyl ester [140, λ_{max} 258 nm (ε = 31,800) and 292 nm (ε = 25,800)].[78] The major product (140b) (90% yield) was formed by a facile *trans–cis*-isomerisation of one of the double bonds. The minor products were a cyclobutane dimer (141), a bis-cyclobutane dimer (142), and a dihydronaphthalene (143). The authors propose that this last product could be formed by an 'electrocyclic' process although they keep their options open and do not discount the involvement of radical intermediates and a 1,2-hydrogen migration.[78] The solid-phase photodimerisation of 2-benzyl-5-benzylidene-cyclopentanone (144a and b) has been observed to give one product, the antiparallel dimer (145).[79a] This work

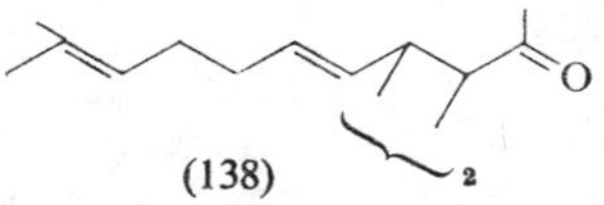

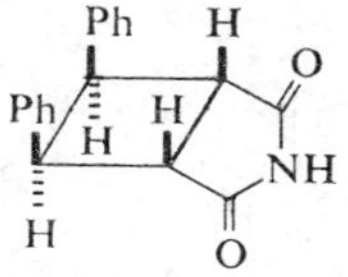

[76a] N. Sugiyama, Y. Sato, M. Yoshioka, K. Yamada, and H. Kataoka, *Bull. Chem. Soc. Japan*, 1969, **42**, 1153.
[76b] G. Büchi and N. C. Yang, *J. Amer. Chem. Soc.*, 1957, **79**, 2318.
[77] R. T. Lalonde and C. B. Davis, *Canad. J. Chem.*, 1969, **47**, 3250.
[78] D. F. Tavares and W. H. Ploder, *Tetrahedron Letters*, 1970, 1567.
[79a] G. C. Forward and D. A. Whitind, *J. Chem. Soc. (C)*, 1969, 1868.

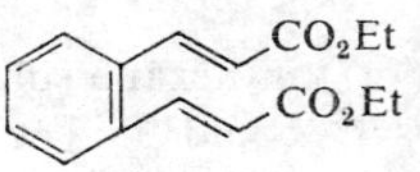

(140a)

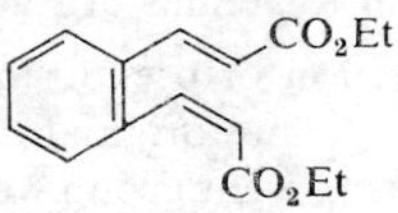

(140b)

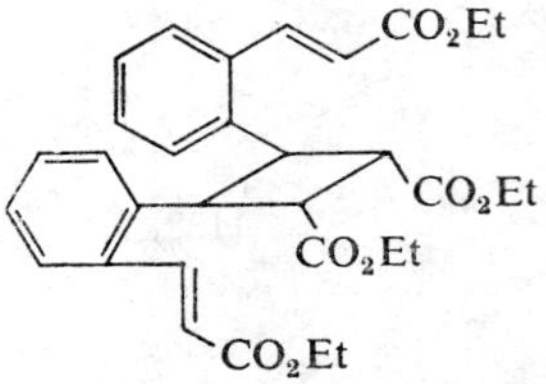

(141)

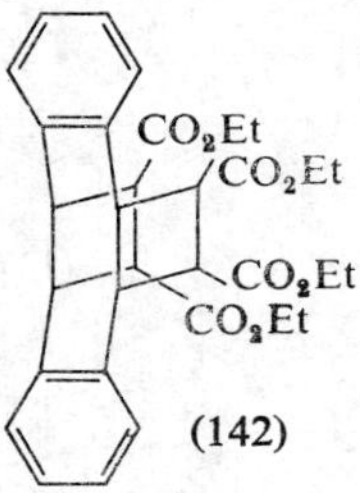

(142)

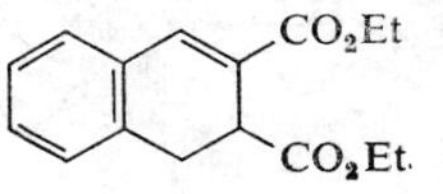

(143)

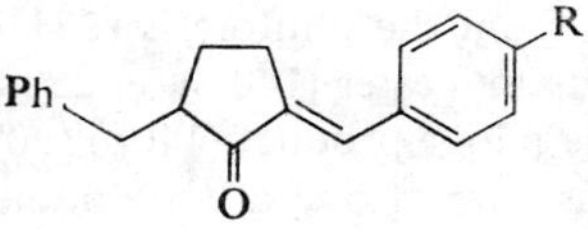

(144) (a) R = H
(b) R = Br

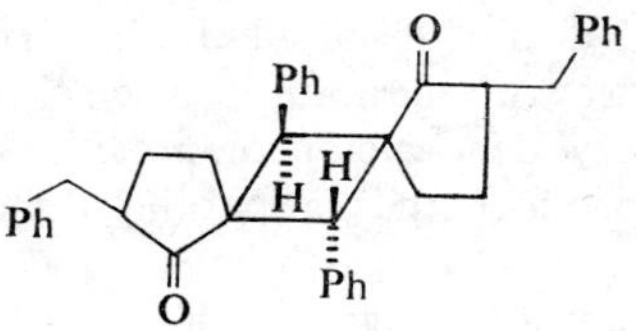

(145)

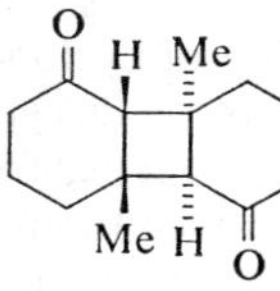

(146a) 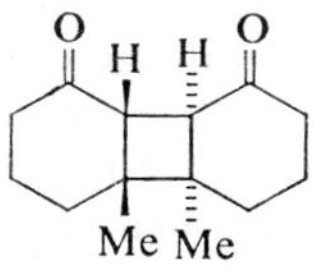 (146b)

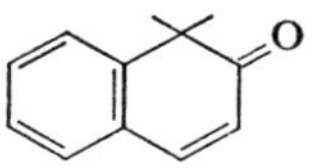

(147)

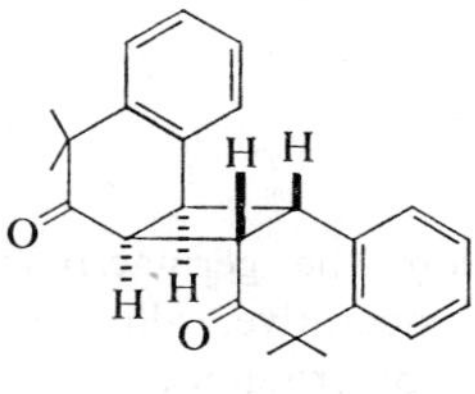

(148a) 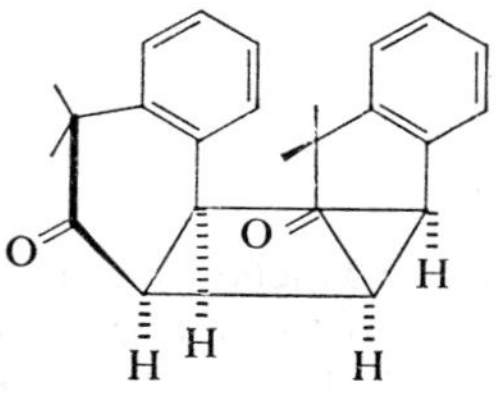 (148b)

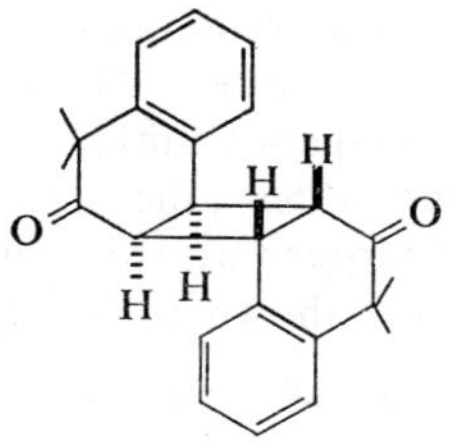

(148c)

has helped to correct the older literature where it was reported that the cyclopentanone (144a) gave a trimer on photolysis.[79b]

Interest is still shown in the dimerisation of cyclohexenones. Ziffer and Mattews [80] have shown that the presence of a 3-alkyl substituent does not affect the dimerisation of 3-methylcyclohexenone, for both the head–tail and head–head dimers (146a and b, respectively) were obtained. The more complex naphthalenone (147) has been reported to form only one dimer (148a),[81] but other workers [82] have isolated two additional dimers (148b and c).

The two possible 'straight' photoaddition products (150a) and (150b) have been isolated [83] from the photolysis of isogermacrone (149): no evidence for the presence of cross-bonded compounds was found. The

(149)   (150a)   (150b)

(151)

fact that two-products were isolated from the germacrone photolysis suggests that a two-step mechanism is operative, allowing rotational equilibration of the diradical (151). Isopiperitenone (152a) and the 4-acetoxy-derivative (152b) have been demonstrated to yield two products on photolysis;[84] the tricyclic ketones (153a and b) were obtained in cyclohexane. There is evidence for the intermediacy of a keten, and in methanol the esters (154a and b) are formed. A novel mode of ring-opening has been proposed. This involves cleavage of the 1,6-bond to give an acyl and an allyl radical (155), followed by rebonding to give the keten which reacts with methanol to produce the ester. The photolysis [85a] of 1-acetylcyclohexene in the presence of ethanol initially gave rise to cis- and trans-1-acetyl-2-ethoxycyclohexane (156, the cis-isomer predominates). No evidence was found for hydrogen abstraction. The quantum yield for the disappearance of enone was shown to be higher for the $n \to \pi^*$ (313 nm)

[79b] R. Cornubert, M. de Remo, R. Joby, P. Louis, R. Pobinet, and A. Strebel, *Bull. Soc. chim. France*, 1938, **5**, 513.
[80] H. Ziffer and B. W. Mattews, *Chem. Comm.*, 1970, 294.
[81] J. Carnduff, J. Iball, D. G. Leppard, and J. N. Low, *Chem. Comm.*, 1969, 1218.
[82] T. Mukai, T. Oine, and H. Sukawa, *Chem. Comm.*, 1970, 271.
[83] J. R. Scheffer and B. A. Boire, *Tetrahedron Letters*, 1969, 4005.
[84] W. F. Erman and T. W. Gibson, *Tetrahedron*, 1969, **25**, 2493.
[85a] M. B. Rubin, *Israel J. Chem.*, 1969, **7**, 49.

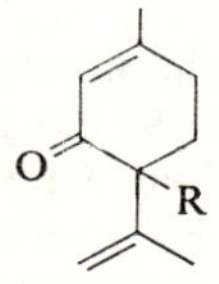

(a) R = H
(b) R = OAc
(152)

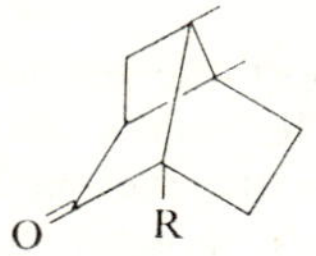

(153)

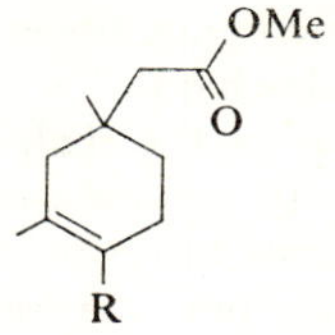

(154)

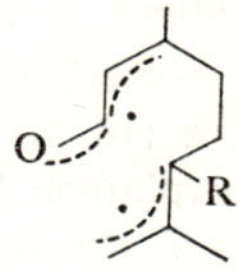

(155)

excitation than the $\pi \rightarrow \pi^*$ (253·7 nm) excitation. This appears to rule out the possibility that $\pi \rightarrow \pi^*$ excitation is essential, as had been proposed

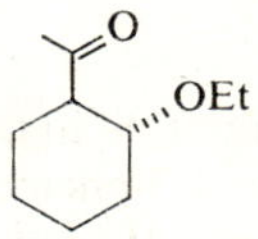

(156)

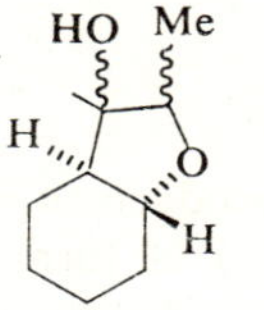

(157a)

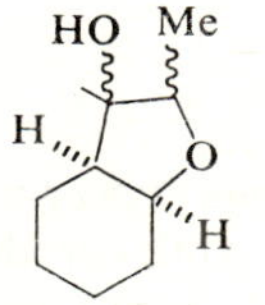

(157b)

originally [85b] to account for the addition of t-butanol to acetylcyclohexene. Indeed, it had been suggested that prior *cis–trans*-isomerisation was necessary, followed by a thermal addition of the alcohol to the reactive *trans* double bond. However, in this report there is no evidence that such an intermediate is formed. Subsequent photolysis of the adducts (156) converted them into tetrahydrofurans (157a) and (157b). The formation

Table 9 *Photochemically induced addition of trialkylboranes to some enones* [86]

Organoborane R in R_3B	Carbonyl compound	Product	(%)
Et	Ethylidene acetone	4-Me-2-hexanone	(85)
Cyclohexyl	Ethylidene acetone	4-cyclohexyl-pentan-2-one	(96)
Et	Crotonaldehyde	3-methylpentanal	(60)
Cyclohexyl	Crotonaldehyde	3-cyclohexylbutanal	(100)
Et	2-Cyclohexenone	3-ethylcyclohexanone	(95)
Cyclohexyl	2-Cyclohexenone	3-cyclohexylcyclohexanone	(100)

[85b] B. J. Ramey and P. D. Gardner, *J. Amer. Chem. Soc.*, 1967, **89**, 3949; P. E. Eaton, *Accounts Chem. Res.*, 1968, **1**, 50.
[86] H. C. Brown and G. W. Kabalka, *J. Amer. Chem. Soc.*, 1970, **92**, 712.

of these compounds is readily explained by hydrogen abstraction from the ether side-chain.

Brown and Kabalka[86] report the efficient photoaddition of trialkyl-boranes to certain enones. The results obtained (Table 9) are generally better than those produced in the thermal reaction initiated by diacetyl peroxide; but the reaction is thought to proceed by a free-radical mechanism.

4 Photoreactions of Thymines, *etc.*

Irradiation of the pyridone (158) in methanol using a Vycor filter and a low-pressure mercury arc lamp gave rise to three products (159a—c).[87] The first product (159a) is readily explained by a retro-Michael reaction. The other two products (159b and c) are more difficult to explain, but

(158) (159a) (159b) (159c)

formally they appear to be the products of group migration in the pyridone. However, while product (159b) is formed by a 1,3-phenyl migration (conceivably a photoallowed reaction) the rearrangement required to form product (159c) is not instantly obvious and requires further study.

One of the most widely investigated photoreactions in the thymine system is that of dimerisation. Closely associated with this is the description[88] of the dimerisation (unsensitized or sensitized) of the caprolactam derivative (160) to afford two head-tail dimers (161a and b). Flash excitation of thymine (162a) has given a value of 14 μs for the lifetime of the triplet state[89a] which is to be compared with the lower value of 4·3 μs obtained indirectly by Wagner and Bucheck.[89b] The authors[89a] of this

(160) (161a, major) (161b, minor)

[87] C. Kashima, M. Yamamoto, Y. Sato, and N. Sugiyama, *Bull. Chem. Soc. Japan*, 1969, **42**, 3596.
[88] E. Cavalieri and S. Horoupian, *Canad. J. Chem.*, 1969, **47**, 2781.
[89a] A. G. Szabo, W. D. Riddell, and R. W. Yip, *Canad. J. Chem.*, 1970, **48**, 694.
[89b] P. J. Wagner and D. J. Bucheck, *J. Amer. Chem. Soc.*, 1968, **90**, 6530; see also ref. 62, p. 205.

Structure (162)

(a) $R^1 = R^2 = R^3 = H$, $R = Me$
(b) $R^1 = R^2 = R = Me$, $R^3 = H$
(c) $R^1 = R^2 = R = H$, $R^3 = CO_2H$
(d) $R^1 = R^2 = R = H$, $R^3 = CO_2Me$
(e) $R^1 = R^2 = R^3 = R = H$

(162)

article claim that the error in the earlier results arises from the use of the wrong value of k_q for penta-1,3-diene. They suggest that the correct value should be 3.4×10^9 l mol^{-1} s^{-1} rather than 1.1×10^{10} l mol^{-1} s^{-1}. The triplet state of thymine could be quenched by hexa-2,4-dien-1-ol with a rate constant of 8.1×10^9 l mol^{-1} s^{-1}. The rate of dimerisation was found to be 5.3×10^8 l mol^{-1} s^{-1}. This value is in good agreement with that reported by Wagner and Bucheck.[89b] Wagner and Bucheck have elaborated their results.[90] Their kinetic data concerning the photodimerisation of thymine and uracil in acetonitrile solution show that the triplet state ($\pi\pi^*$) dimerisation of uracil is almost an order of magnitude greater than that of thymine (Table 10). It has been shown that the primary triplet state processes have little

Table 10 *Kinetic data for photodimerisation of thymine and uracil* [90]

	Thymine	Uracil
k_i, 10^5 s^{-1}	2.2	1.6
k_a, 10^9 s^{-1} l mol^{-1}	0.7	2.0
Φ_{dim}	0.0025	0.019
Φ_{isc}	0.18	0.40
Φ_P	0.02	0.06

to do with the overall quantum efficiencies, and indeed the radiationless decay rates of both triplets are the same. The difference, as reported originally,[89] appears to be due to the intermediacy of a metastable dimeric intermediate (163). The 1,4-diradical form of this intermediate is preferred although this proposal does not exclude the involvement of an excimer prior to the diradical formation. The diradical can conceivably either form products (*e.g.* 164) which are known to have *cis*-geometry at the bridgeheads, or else revert to starting material. The latter possibility could account for the low values of Φ_p. Thus, the authors conclude that only 1.4% of thymine triplets and 4.9% of uracil triplets undergo dimerisation. Other workers [91] have shown that the photosensitized dimerisation of

---

[90] P. J. Wagner and D. J. Bucheck, *J. Amer. Chem. Soc.*, 1970, **92**, 181.
[91] B. H. Jennings, S. C. Pastra, and J. L. Wellington, *Photochem. Photobiol.*, 1970, **11**, 215.

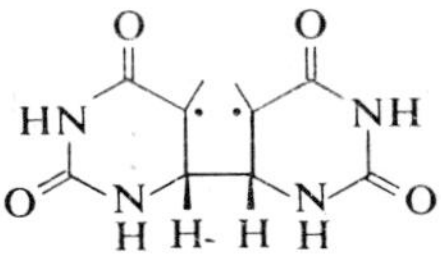

(163) 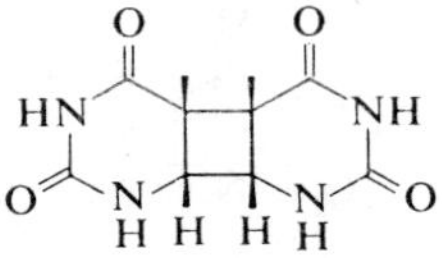 (164a) 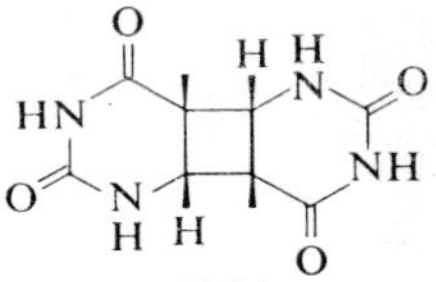 (164b)

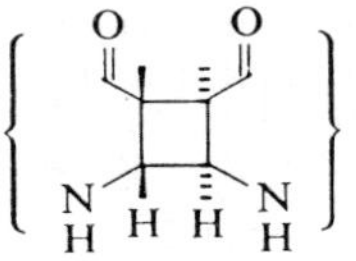

(164c) 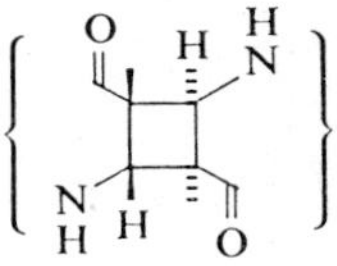 (164d)

thymine using acetone and propiophenone as triplet sensitizers and light of $\lambda > 290$ nm gives rise to all four dimers (164a), (164b), (164c), and (164d). The relative amounts of each appear to be largely independent of sensitizer concentration (Table 11), but not wholly so since at 83% acetone concen-

Table 11 *Influence of sensitizer concentration on the formation of thymine dimer* [91]

Sensitizer	(vol. %)	(mol l^{-1})	% reaction	% isomer formed			
				(164a)	(164b)	(164c)	(164d)
Acetone	0·1	0·014	0				
	2	0·27	43	22	21	33	23
	5	0·68	39	22	21	34	23
	10	1·4	79	22	22	33	24
	83	11	100	20	22	18	40
Propio- phenone, sat		0·014	30	24	20	33	23

tration the head–tail dimer (164c), normally the predominant one, falls to 20% yield, a significant drop in efficiency. Attempts to rationalise the preference for specific dimer formation on the grounds of the dipole moments of the reactant (triplet thymine) and the products, appear to be unsatisfactory. X-Ray crystallographic data have confirmed [92] the structure of photodimer A of 1,3-dimethylthymine (162b) as the *cis-syn*-5,5:6,6 dimer (165). Although other dimers are obtained from the solution phase dimerisation of thymine, only the above is formed in the photolysis of DNA.

Photodimerisation of 1′-(2′-desoxy-β-D-ribofuranosyl)-[4-^3H]-5-ethylura-cil (166) has been shown [93] to take to an extent of 48%, and thymidine also dimerises (49%) under identical conditions. The extent to which the compound had dimerised was shown by paper chromatography, but the structure of the dimeric material was not established. One of the main

---

[92] N. Camerman and A. Camerman, *J. Amer. Chem. Soc.*, 1970, **92**, 2523.
[93] K. K. Gauri, K. W. Pflughaupt, and R. Mueller, *Z. Naturforsch.*, 1969, **24b**, 833.

(165)   (166)

interests in pyrimidine dimerisation stems from the fact that DNA and RNA can be deactivated by photo- or gamma-irradiation, thereby upsetting the replication mechanism of the nucleic acids. Thus it is of interest to find ways for the reversal of this effect. Ben-Hur and Rosenthal [94] have found that *cis–syn* and *trans–syn* cyclobutane dimers of pyrimidine and the *cis–syn* thymine : uracil adduct can be monomerised by the use of sensitization techniques. The sensitizers effective in this are 1-anthraquinone sulphonic acid (Na salt) (A), 2-anthraquinone sulphonic acid (Na salt) (B), and the 2,7-anthraquinone sulphonic acid (di Na salt) (C): the magnitude of the effect can be seen in Table 12. It has also been shown that the

Table 12 *Photosensitized monomerisation of thymine dimers* [94]

Dimer	Quinone sensitizer					
	A		B		C	
	I	II	I	II	I	II
(164a) *cis–syn* Thymine	76	69	82	75	96	89
(164b) *trans–syn* Thymine	67	63	93	82	96	92
cis–syn Thymine-uracil	78	77	85	83	96	92

I = % dimer disappearing after 20 min irradiation.
II = % thymine produced.

rate of splitting of the dimers is dependent on the initial geometrical arrangement; thus the *trans–syn* dimer cleaves three times as rapidly as the *cis–syn* dimer.

A triplet energy of 251 kJ mol^{-1} (60 kcal mol^{-1}) has been deduced from the results of flash photolytic studies on orotic acid (162c) and its methyl ester (162d). The excited state was initially identified as the triplet by sensitization and quenching studies (Table 13). From these results, the rate constant for dimerisation at 77 K was established as $2\cdot20 \times 10^{-9}$ l mol^{-1} s^{-1} for the methyl ester, and $1\cdot91 \times 10^{-9}$ l mol^{-1} s^{-1} for the acid (at pH 0·6). The dimerisation of orotic acid (162c) in aqueous solution has been demonstrated [96] to occur *via* the triplet state, and can be sensitized by benzophenone. The dimerisation process is dependent on the concentration of the

[94] E. Ben-Hur and I. Rosenthal, *Photochem. Photobiol.*, 1970, **11**, 163.
[95] R. W. Yip, W. D. Riddell, and A. G. Szabo, *Canad. J. Chem.*, 1970, **48**, 987.
[96] M. Charlier, C. Helene, and M. Duorlent, *J. Chim. Phys.*, 1969, **66**, 700.

Table 13 *Quenching data for methyl orotate* (162b) *(aqueous solutions at pH 5)* [95]

Sensitizer conc. (mol l^{-1})	E_T (kcal mol^{-1})	T_0 (μs)	$K_q \times 10^{-9}$ (l mol^{-1} s^{-1})
Xanthen-9-one ($1 \cdot 14 \times 10^{-5}$)	74	37·7	2·78 ±0·24
Naphthalene ($1 \cdot 72 \times 10^{-5}$)	60·9	246	2·35 ±0·16
2-Acetonaphthone (1×10^{-5})	59·3	244	0·71 ±0·09
1-Acetonaphthone (1×10^{-5})	56·4	277	0·82 ±0·04
9-Fluorenone ($3 \cdot 7 \times 10^{-5}$)	53·3	110	0·026±0·01

^{a} T_0 is the lifetime in μs of the unquenched sensitizer.

sensitizer and the pH. (Table 14 shows the rate of the reaction at various concentrations of orotic acid.) The reaction can also be quenched by the introduction of known triplet quenchers (piperylene), but these results are less meaningful since the piperylene adds to the acid to give low yields of

Table 14 *Effect of concentration on rate of dimerisation of orotic acid* (162c) [96]

Concentration of acid	Rate i $\times 10^8$	Rate ii $\times 10^8$
10^{-4} mol l^{-1}	8·3	2·1
2×10^{-4} mol l^{-1}	26	7·4
4×10^{-4} mol l^{-1}	62·6	16·1
$6 \cdot 6 \times 10^{-4}$ mol l^{-1}	111	30·5

Rate i = rate of dimerisation in the absence of quencher.
Rate ii = rate in presence of piperylene.
Rates are measured in mol l^{-1} s^{-1}.

unidentified adducts. The dimer is thought to be a cyclobutane, and only one isomer has been obtained (chromatography). The unidentified adducts from piperylene and orotic acid can be degraded by direct irradiation at 253 nm in mild base. Orotic acid also produces polymers from irradiation in the presence of acrylonitrile.[97] It is proposed that the polymerisation is induced by radicals (167) formed by hydrogen abstraction. Irradiation of the same solution at >320 nm affords no polymerisation, but gives an adduct (168a). Thymine (162a) also affords such a cyclobutane adduct

(167)

(168)
(a) $R^1 = H$, $R^2 = CO_2H$
(b) $R^1 = Me$, $R^2 = H$

[97] C. Helene and F. Brun, *Photochem. Photobiol.*, 1970, **11**, 77.

(168b).⁹⁷ The triplet state of orotic acid has been shown to be reactive in the formation of the cycloadducts. The photolysis at 257 nm of tritiated samples (for ease of analysis) of thymine (162a) has been studied at $-196\,°C$. A thymine dimer is formed in a maximum yield of 3·2% and is characterised by the appearance of u.v. absorption at 315 nm. The authors [98] suggest that the dimer is in fact an oxetan (169) which is stable at low temperatures but rearranges at higher temperatures ($-80\,°C$) to the isolated product (170). A similar adduct (169b) is reported [99] from the photolysis of uracil (162e). However, the isolated product in this instance is the dehydrated dimer (171).

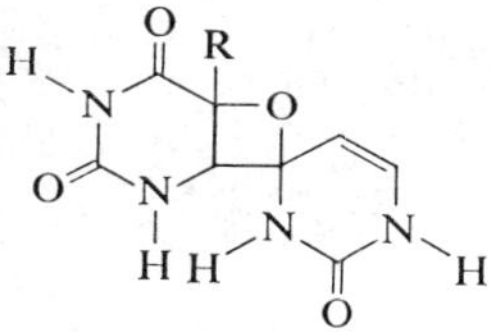

(a) R = Me
(b) R = H
(169)

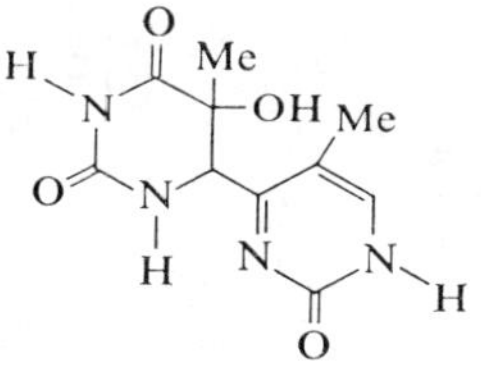

(170)

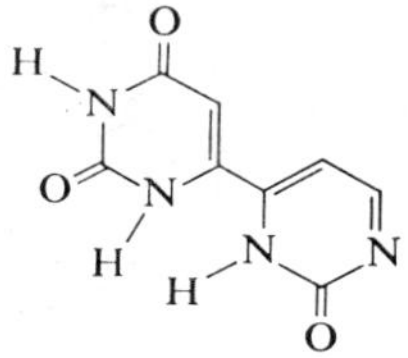

(171)

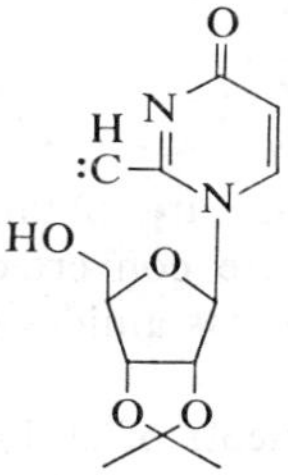

(172)

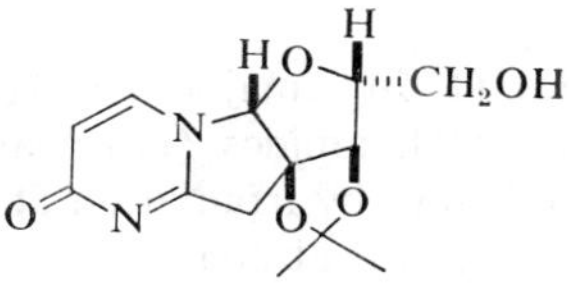

(173)

Several studies concerning the photohydration of pyrimidines have appeared.[100-102] Intramolecular insertion of a carbene (172) formed by the photolysis of the corresponding sulphoxonium ylide yielded the novel nucleoside (173).[103]

5 Photochemistry of Dienones

Linearly-conjugated Dienones.—A u.v. study of the photoreaction of cyclohexadienone (174a) in water or cyclohexylamine has shown that the process is a 'simple reaction' (a series of straight lines pass through the

---

[98] R. O. Rahn and J. L. Hosszu, *Photochem. Photobiol.*, 1969, **10**, 131.
[99] M. N. Khattack and S. Y. Wang, *Science*, 1969, **163**, 1341.
[100] J. C. Nnadi, *Diss. Abs.*, 1969, **29B**, 2792.
[101] L. Kittler and G. Lober, *Photochem. and Photobiol.*, 1969, **10**, 35.
[102] P. V. Hariharan, *Diss. Abs.*, 1969, **29B**, 3239.
[103] T. Kunieda and B. Witkop, *J. Amer. Chem. Soc.*, 1969, **91**, 7752.

origin of the co-ordinates of an extinction difference diagram).[104] The initial products of the reactions of dienones (174a and b) are ketens for which u.v. and n.m.r. structural evidence has been obtained.[105, 106] Reaction of the ketens with the substrate gave the corresponding acids and amides. The reaction of dienone (174b) with cyclohexylamine is wavelength dependent. At 365 nm, amides (175a and b) are isolated, but at 313 nm the

(174) (a) R = Me (b) R = Ph

(175) (a) R = Ph, R¹ = Me (b) R = Me, R¹ = Ph

(176) (a) R = Ph, R¹ = Me (b) R = Me, R¹ = Ph

amides (175b) (major) and (175a) (minor) are obtained. In separate experiments, amides (175a) and (176a) were converted to the isomeric amide (176b) by irradiation at 313 nm, whereas amides (175a and b) were not isomerised at 365 nm.

Interest in the photochemical rearrangements of cyclohexa-2,4-dienones to bicyclic ketones has continued. Further examples provided by the rearrangement of cyclohexadienones (177a and b) have shown that the initial photolysis gives a keten which can be trapped efficiently by cyclohexylamine, and which thermally rearranges in the absence of nucleophiles by a $(\pi_a^4 + \pi_a^2)$-allowed process to a bicyclic ketone (178).[107] † The authors [107] claim that the pronounced solvent effect reflects a highly asymmetric transition state, but the observations of Hart et al.[108] that solvents can

[104] G. Quinkert, M. Hintzmann, P. Michaelis, and P. Juerges, Angew. Chem., 1970, **82**, 219 (Internat. Edn., 1970, **9**, 238).
[105] G. Quinkert, B. Bronstert, P. Michaelis, and U. Krüger, Angew. Chem. Internat. Edn 1970, **9**, 240.
[106] H. H. Perkampus, G. Prescher, B. Bronstert, and G. Quinkert, Angew. Chem., 1970 **82**, 222.
[107] M. R. Morris and A. G. Waring, Chem. Comm., 1969, 526.
[108] H. Hart et al., cited by W. M. Horspool in 'Photochemistry,' ed. D. Bryce-Smith (Specialist Periodical Report), The Chemical Society, London, 1970, Vol. 1, p. 215.

† The results and proposals reported here are in direct conflict with the accepted mechanism for rearrangement of cyclohexadienones to bicyclic ketones. There is an increasing weight of evidence to suggest that the formation of ketens as intermediates i inessential: see refs. 108 and 110. Yet ketens can certainly be formed in some cases.[109]

Enone Rearrangements and Cycloadditions

(a) $R^1 = OAc, R^2 = R^3 = R^4 = Me$
(b) $R^1 = OAc, R^2 = R^4 = CH_2Cl, R^3 = Me$
(c) $R^1 = R^2 = Me, R^3\text{---}R^4 = CH_2\text{---}CH_2$

(177)

(178)

(179)

influence the nature of the lowest excited state ($n\pi^*$ or $\pi\pi^*$) provide the basis of an apparently more attractive suggestion. Photolysis of the strained cyclohexadienone (177c) leads to the keten (179) which gives only the methyl ester in methanol, and not the bicyclic ketone.[109] Extension of this study to the photolysis of the naphthalenones (180) has shown that two rearrangements take place.[110] Thus, initially the naphthalenones rearrange to the tricyclic ketones (181) which in the case of compounds (181a and b) could be isolated. Further photolysis of these ketones brings about another rearrangement to form an isomeric naphthalenone (182). The selection of the compounds studied in this paper was made so as to differentiate between the possible routes for rearrangement. It has been established that there is indeed a similarity between the photolysis of these compounds and those studied previously by the same authors.[109] Thus, in the rearrangement to the tricyclic ketones there was no evidence for the intermediacy of ketens since these intermediates could not be trapped by conventional techniques, e.g. the addition of nucleophiles to the irradiation product. The preferred mechanism therefore involves a 'bond-switching' mechanism, probably via the $\pi\pi^*$ singlet state. The second rearrangement has been shown to involve alkyl group migration; e.g. in the rearrangement of ketone (181c) the naphthalenone (182c) is isolated. The rationale for this process involves ring-opening to a zwitterionic intermediate (i) followed by methyl migration (Scheme 9). Further study has been made of the photochemistry of eucarvone (183; E_T ca. 255 kJ mol^{-1}, 61 kcal mol^{-1}). On photolysis (313 nm) in a variety of solvents it undergoes[111] an intramolecular cycloaddition to yield the bicyclic ketone (183, $\Phi = 2\cdot5 \times 10^{-3}$). The nature of the solvent affected the quantum yield for the disappearance

---

[109] R. J. Bastiani, D. J. Hart, and H. Hart, *Tetrahedron Letters*, 1969, 4841.
[110] H. Hart and R. K. Murray, *J. Org. Chem.*, 1970, 35, 1535.
[111] D. I. Schuster and D. H. Sussman, *Tetrahedron Letters*, 1970, 1657.

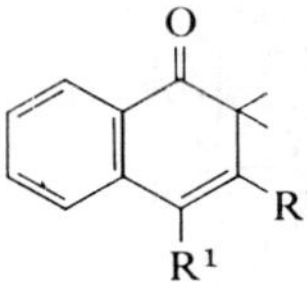

Scheme 9

(180), (181) (a) $R^1 = R^2 = Me$; (b) $R^1 = Me, R^2 = H$; (c) $R^1 = Et, R^2 = H$; (182)

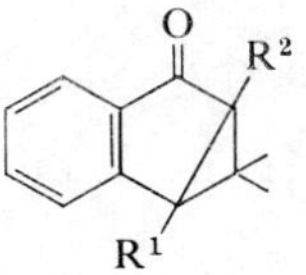

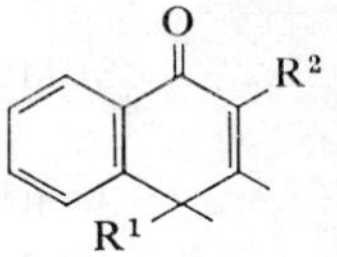

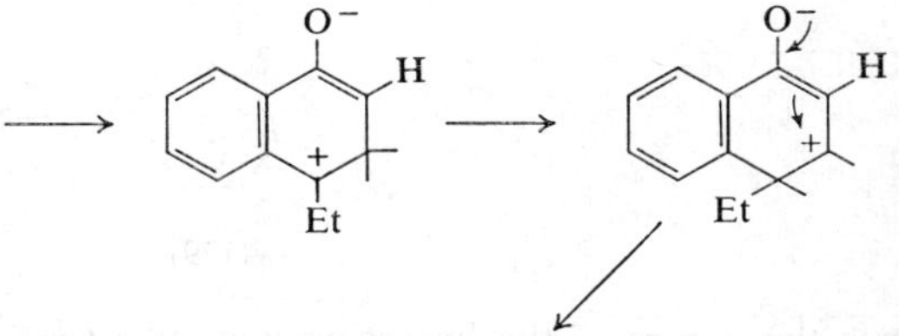

(181c) → intermediate → (182c)

of the starting material, the efficiency for the disappearance being greater in more polar solvents. The conversion of (183) into (184a) could be effected by triplet sensitization ($E_T > 259.2$ kJ mol^{-1}, 62 kcal mol^{-1}). Results obtained from quenching studies suggest that two excited states are involved in the reaction, only one of which is quenchable. Thus,

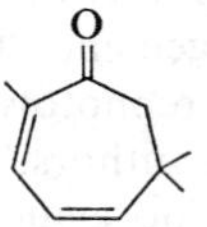

(183) 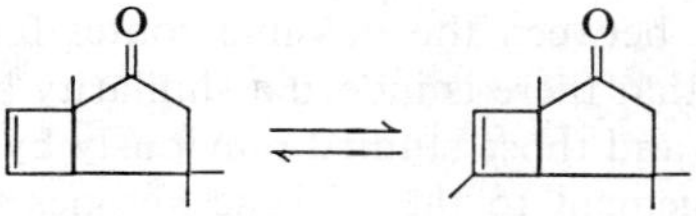   (184a) ⇌ (184b)

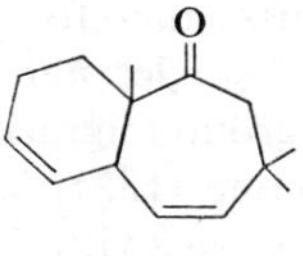

(185)

values of the triplet decay rate (1.3×10^7 s^{-1}) and rate of triplet isomerisation (1.4×10^4 s^{-1}) were calculated. The use of cyclohexa-1,3-diene as the quencher gave dimers of the cyclohexadiene as well as two additional products, the major one being identified as the 1 : 1 adduct (185) which subsequent experiments have shown, could arise from an upper triplet state of eucarvone. The bicyclic ketone (184a) undergoes conversion ($\phi = 0.041$) into the isomeric ketone (184b) on photolysis (313 nm) in benzene solution.[112] The reaction could also be photosensitized ($E_T > 271 \cdot 7$ kJ mol^{-1}, 65 kcal mol^{-1}), and quenching studies providing curved Stern–Volmer plots were interpreted as evidence for the isomerisation occurring from the singlet state (74%). This is to be contrasted with the results of Ipaktschi[113] and Baggiolini et al.[29] who have reported triplet rearrangements of ketones.

An endeavour has been made to unravel the problems surrounding the difference in behaviour of variously substituted tropones. Thus, 2-aminotropone (186a) gave no isolable products on irradiation in benzene or ethanol through a Pyrex filter.[114] However, the acetylated and benzoylated compounds (186b and c) undergo conversion to the bicyclic dienones (187b and c) and the isomeric compounds (188b and c). The observation was also made that the process was more rapid in aprotic than in protic solvents. The same effect was noticed for the irradiation of the anilino-derivative (186d) which afforded the rearranged material (187d) in aprotic solvents but not in protic ones. The explanation[114] for this effect is found

(186)   (187)   (188)

(a) R = H
(b) R = COMe
(c) R = COPh
(d) R = Ph

n an examination of the u.v. spectrum which shows that in aprotic solvents the substituted tropones (186b—d) exhibit a blue shift, while the tropone (186a) shows a red shift. This is interpreted as being evidence that the 2-aminotropone (186a) has a low-lying charge-transfer band which does not lead to products, while the other tropones (186b—d) have an $n\pi^*$ excited state which gives rise to the rearranged products. The introduction of a second electron-donating substituent into the tropone nucleus alters the photo-induced reactions. Thus, 5-aminotropolone (189) undergoes a photo-induced ring-contraction to the aminocyclopentenone (190) by irradiation in aqueous hydrochloric acid. The mechanism proposed by the authors[115] required protonation and valence bond isomerisation to the bicyclo[3,2,0]heptadiene (i, Scheme 10), followed by a series of thermal steps.

Continued interest in the photochemistry of α-pyrones has resulted in several publications on this topic. The low-temperature study of 2-pyrone

---

[112] D. I. Schuster and D. H. Sussman, *Tetrahedron Letters*, 1970, 1661.
[113] J. Ipaktshi, *Tetrahedron Letters*, 1969, 2153.
[114] T. Mukai and M. Kimura, *Tetrahedron Letters*, 1970, 717.
[115] S. Seto, H. Sugiyama, S. Takenaka, and H. Wanatabe, *J. Chem. Soc. (C)*, 1969, 1625.

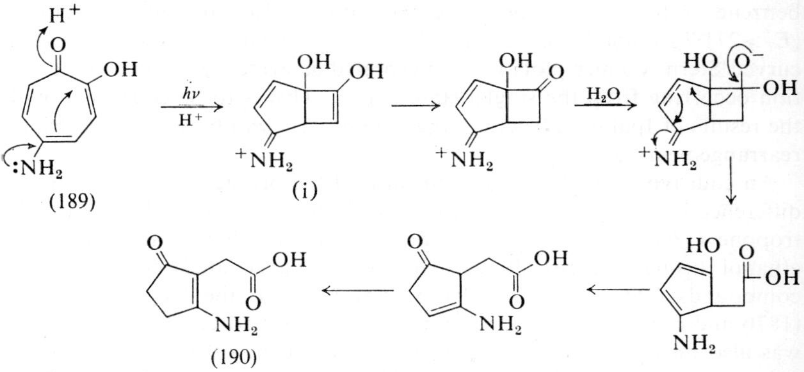

(189) (i) (190)

Scheme 10

(191a) showed i.r. absorptions for both the keten (192a) produced only by irradiation through quartz and the β-lactone (193a) formed by irradiation through Pyrex.[116] In contrast with this, irradiation of the pyrone (191b) through quartz or Pyrex led only to absorptions at 1807 and 1600 cm^{-1} assignable to the lactone (193b). Thus the irradiation of this pyrone is considered to proceed mainly through the lactone intermediate. The β-lactone (193c) is also the essential intermediate in the photoreaction of

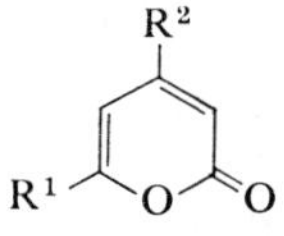

(191)

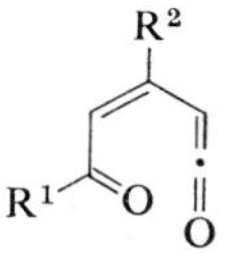

(a) $R^1 = R^2 = H$
(b) $R^1 = Me$, $R^2 = MeO$
(c) $R^1 = Me$, $R^2 = OH$

(192)

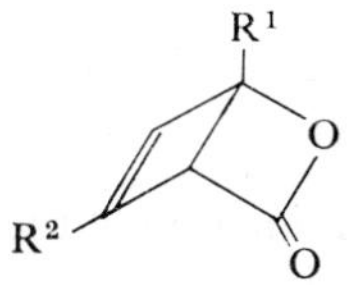

(193)

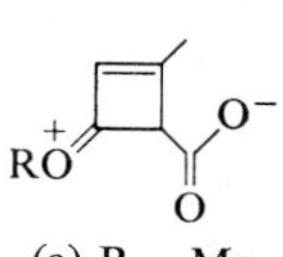

(a) R = Me
(b) R = H

(194)

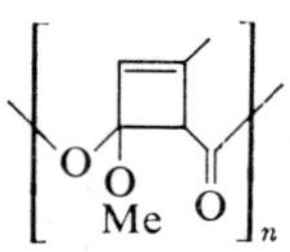

(195)

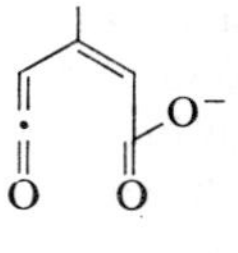

(196)

triacetic acid lactone (191c).[117] The presence of the electron-donating substituent allows for either ring-opening of the lactone to afford the zwitterion (194), which forms an oligomer (195) if R = Me,[116] or else complete ring-opening of R = H to form the keten (196).[117] This keten

[116] J. P. Guthrie, C. L. McIntosh, and P. de Mayo, *Canad. J. Chem.*, 1970, **48**, 237.
[117] C. T. Bedford and T. Money, *Chem. Comm.*, 1969, 685.

(196) cyclises to form β-methylglutaconic anhydride, the sole photolysis product of pyrone (191c) in benzene or ether. The 'normal' reactions of α-pyrones on photolysis are exemplified[118] by the photolysis of the bufadienolides (197) in alcohol: β-lactone formation is followed by ring-opening to the keten, trapping by alcohol, intramolecular ketalisation by the 14-β-hydroxy-substituent, and finally dehydration to afford the products

(a) R = R¹ = R² = H
(b) R = Ac, R¹ = R² = H

R^3 = Et or Me
(198)

(198). In the absence of the 14-β-hydroxy-substituent, no intramolecular ketalisation takes place.[119]

Cross-conjugated Cyclohexadienones.—The question whether a zwitterionic intermediate or a diradical is involved in the photorearrangement of 4,4-diphenylcyclohexa-2,5-dienone is still of interest. Yet another non-photochemical approach to the rearrangement of this molecule to 6,6-diphenylbicyclo[3,1,0]hex-3-en-2-one has now been reported. This approach combines an electrochemical technique[120] with the postulate that the photochemical excitation gives rise to a 5-electron system, not to the zwitterion. Electron transfer gave the radical anion (199), and thence only the pinacol, not the bicyclic ketone. The cyclohexadienone (200) has proved to be almost completely photostable in benzene or t-butanol.[121,122]

[118] Y. Kamano and M. Komatsu, *Chem. Pharm. Bull. Tokyo*, 1969, **17**, 1698.
[119] Y. Kamano, Y. Tanaka, and M. Komatsu, *Chem. Pharm. Bull. Tokyo*, 1969, **17**, 1706.
[120] A. Mazzenga, D. Lomnitz, and J. Villegas, *Tetrahedron Letters*, 1969, 1665.
[121] H. E. Zimmerman and G. Jones, *J. Amer. Chem. Soc.*, 1969, **91**, 5678.
[122] H. E. Zimmerman and G. Jones, *J. Amer. Chem. Soc.*, 1970, **92**, 2753.

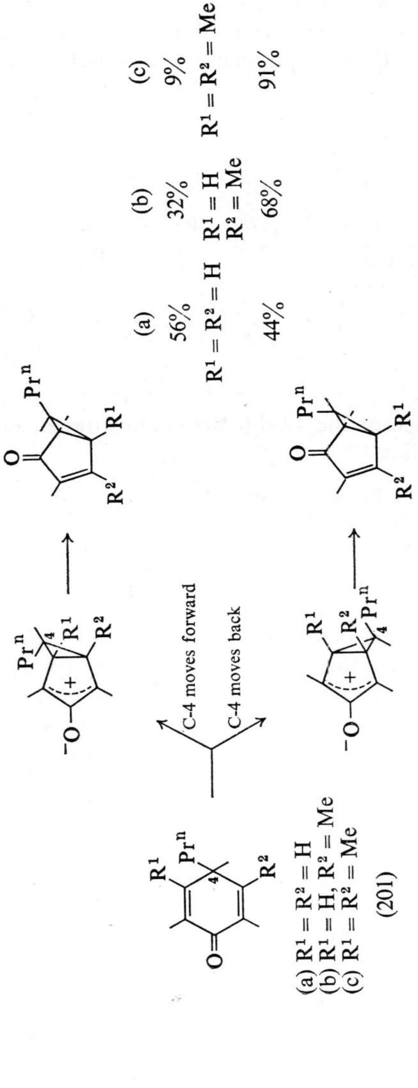

	(a)	(b)	(c)
	$R^1 = R^2 = H$	$R^1 = H$, $R^2 = Me$	$R^1 = R^2 = Me$
(202)	56%	32%	9%
	44%	68%	91%

(201)
(a) $R^1 = R^2 = H$
(b) $R^1 = H$, $R^2 = Me$
(c) $R^1 = R^2 = Me$

(203)

C-4 moves forward
C-4 moves back

In solvents which are good hydrogen donors, no evidence for protonation of possible zwitterionic intermediates was obtained, and only reactions resulting from hydrogen abstraction were found, as would be consistent with the photoreactivity expected for $n\pi^*$ triplets. (Stern–Volmer plots using 2,5-dimethylhexa-2,4-diene as quencher gave a hydrogen-abstraction rate of $k = 3.6 \times 10^4 \text{ l mol}^{-1} \text{ s}^{-1}$). A similar relationship is to be seen in the decay rate of the dienone triplet state compared with that of benzophenone ($k_d = 4.2 \times 10^6 \text{ s}^{-1}$ and $k_d = 1 \times 10^5 \text{ s}^{-1}$ respectively). The decay rate with this highly constrained dienone is 10^{-3} times that of 4-methyl-4-phenyl-cyclohexa-2,5-dienone ($k_d = 4 \times 10^9 \text{ s}^{-1}$). From these observations, Zimmerman and Jones [121, 122] propose that a rapid decay rate of a dienone triplet is essential for a 'type A' rearrangement, or else an orthogonal triplet state [which would be impossible for the rigid dienone (200)].

Further mechanistic details concerning the stereospecificity of the photoisomerisation of 4,4-dialkylcyclohexa-2,5-dienones (201) have been provided by Rodgers and Hart in a study of the influence of steric effects on the rearrangement to bicyclo[3,1,0]hex-3-en-2-ones (202).[123] The rationale behind the study concerns the formation of the bicyclic zwitterions (203) by moving C-4 either forward or back. These authors reason that the bulkier substituent on C-4 will tend to move C-4 in the direction of less steric impedance (in this case C-4 moves forward) and so the results indicate that the n-propyl group prefers to be in the *endo*-position as one increases the substitution at C-3 and C-5.

Two products, 3,6-di-t-butyl-2-methoxy-phenol and the cyclobutane (204) are isolated from the photolysis of 2,4,6-tri-t-butyl-3-methoxycyclohexa-2,5-dienone in benzene solution utilising pyrex-filtered u.v. irradiation.[124] The formation of these products is both wavelength- and sensitizer-dependent (Table 15). From these results it is reasoned that an $n\pi^*$ state

Table 15 *Influence of sensitizer and wavelength on product distribution in photolysis of 2,4,6-tri-t-butyl-3-methoxycyclohexa-2,5-dienone* [124]

Additive	λ nm	Recovered dienone %	Phenol %	Cyclo-butane (204)%	Phenol: (204)	Irrad. time (hr)
None	>318	81	11	8	1.37	19
None	>280	75	5	20	0.25	19
Acetophenone	>280	35	2	11	0.18	9

is responsible for the formation of the phenol while a $\pi\pi^*$ state produces the cyclobutane (204). The route proposed for the formation of the phenol is shown in Scheme 11, and involves the isomerisation of the starting cyclohexadienone *via* a three-photon process. The cyclobutane is formed by what appears to be the first example of an intramolecular hydrogen abstraction in a cyclohexadienone.

---

[123] T. R. Rodgers and H. Hart, *Tetrahedron Letters*, 1969, 4845.
[124] K. Ogura and T. Matsuura, *Tetrahedron*, 1970, **26**, 445.

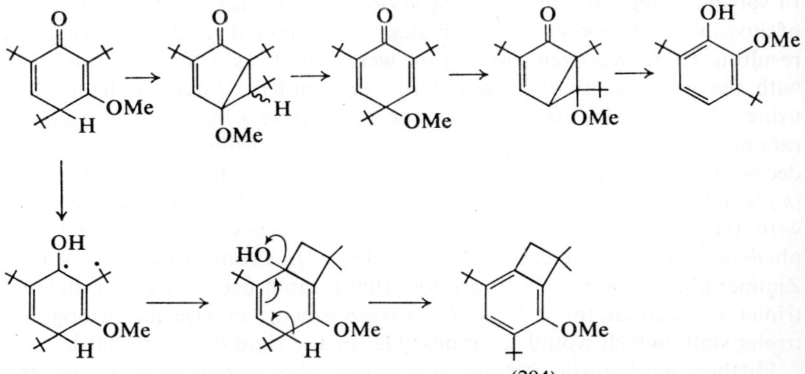

Scheme 11

An unusual vinyl-cyclobutenone (205) has resulted from further work on the photochemical reactions of 2,4,6-tri-t-butyl-4-hydroxycyclohexa-2,5-dienone.[125] Matsuura and Ogura had previously demonstrated that compounds (206) to (208) were produced.[126] The latest paper [125] reports the isolation of the cyclobutenone (205) in 33% yield (benzene solution, Pyrex filter) together with small amounts of the previously reported products. No mechanistic explanation was advanced, but the report of a preliminary experiment has ruled out the intermediacy of the 1,3-dione (206), the major product reported by Matsuura and Ogura.[126] The discrepancy is puzzling.

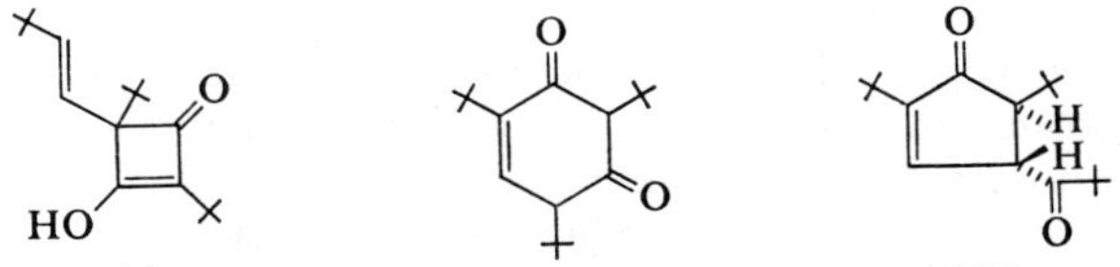

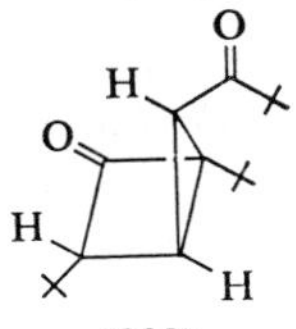

In a previous report [127] of the photolysis of 4-methyl-4-trichloromethyl-cyclohexa-2,5-dienone, an additional product, apart from the readily identifiable cyclopentenone, was obtained from the reaction in acidified

---

[125] D. A. Plank, J. C. Floyd, and W. H. Starnes, *Chem. Comm.*, 1969, 1003.
[126] T. Matsuura and K. Ogura, cited in ref. 62, p. 209.
[127] D. J. Patel and D. I. Schuster, *J. Amer. Chem. Soc.*, 1968, **90**, 5153, 5145; see also ref. 62, pp. 208, 209.

methanol. This adduct had been tentatively assigned structure (209), and has now been isolated [128] and its identity confirmed, although the question whether the trichloromethyl group has an *endo* or an *exo* configuration has not yet been resolved. Finally, the dimerisation of a thiacyclohexadienone

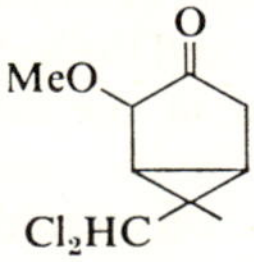

(209)

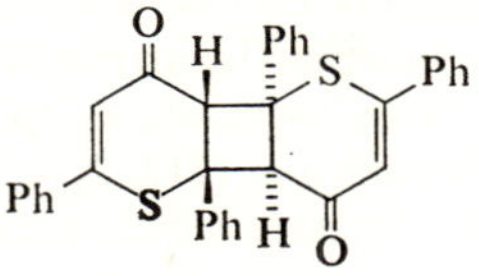

(210)

has been achieved by the direct irradiation of 2,6-diphenyl-4*H*-thiopyran-4-one in benzene solution using a high-pressure mercury lamp and an unspecified filter.[129] This produced a low (25%) yield of the *cis-anti-cis* head–tail dimer (210).

Santonin-type Rearrangements.—Fisch [130] has added to the mystery which surrounds the dilemma concerning the intermediacy of zwitterions (211) in the condensed-phase photochemistry of cross-conjugated dienones. Last year we reported on a gas-phase study on such molecules which pointed to

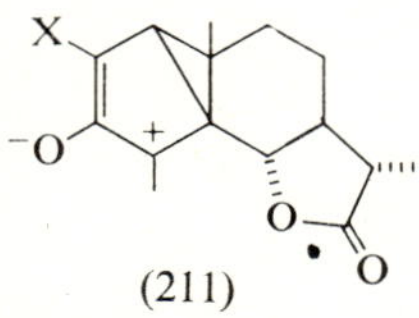

(211) 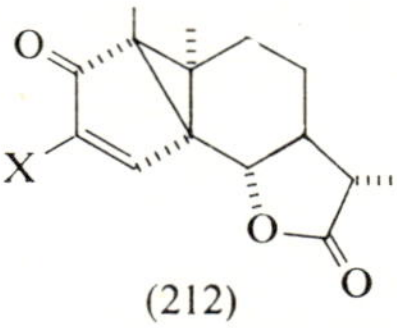   (212) 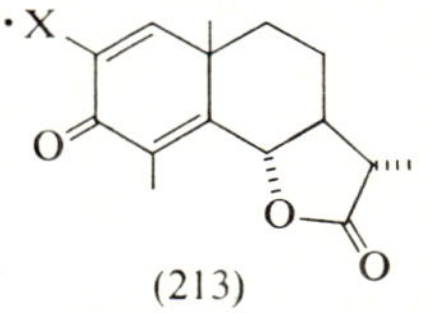   (213)

X = H, Cl or Br

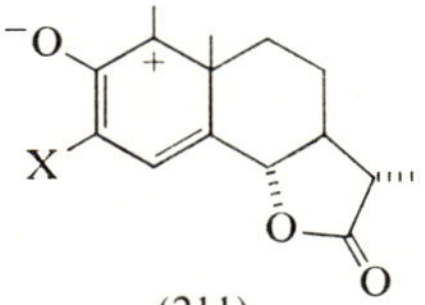

(211)

the intermediacy of radicals rather than zwitterions in the formation of the lumiketone.[131] The latest study reports further on the observation of highly coloured species produced on low-temperature photolysis of lumisantonin (212) and the santonins (213). The halogeno-substituents had no effect on the absorption spectrum of the trapped intermediate (λ_{max} = 412 nm). The i.r. spectra excluded the presence of cyclopropanone intermediates. Thus, Fisch [130] concludes that the intermediates are the zwitter-

[128] D. I. Schuster and V. Y. Abraitys, *Chem. Comm.*, 1969, 419.
[129] N. Sugiyama, Y. Sato, H. Kataoka, C. Kashima, and K. Yamada, *Bull. Chem. Soc. Japan*, 1969, **42**, 3005.
[130] M. H. Fisch, *Chem. Comm.*, 1969, 1472.
[131] J. S. Swenton, E. Samborn, R. Srinivasan, and F. I. Sontag, cited in ref. 62, p. 207.

ionic species (211) and that solvent viscosity has no effect on their stability. The low-temperature stability must therefore be associated with an energy barrier to further reaction. Phosphorescence excitation spectroscopy of α-santonin, 6-episantonin, and 2-bromosantonin has been carried out.[132] The results obtained with α-santonin complement the observations of Fisch [130] that two reactive triplet states, the $\pi\pi^*$ and the $n\pi^*$, are involved. However, for both these states to be involved in the isomerisation of santonin to lumisantonin (212, X = H), intersystem crossing must proceed largely *via* the $n\pi^*$ state which must then react very rapidly in order to compete with internal conversion to the lower lying $\pi\pi^*$ state, and both states must rearrange to the same intermediate. A study of the photochemistry of α-santonin (213, X = H) in protic solvents has been published.[133] The results show that, in alcoholic solution, lumisantonin (212, X = H) is the predominant product, but compounds arising from its photorearrangement were also isolated. Derivatives of the lactone (214) and dehydro-lactone (215) were isolated, and it was established (Table 16)

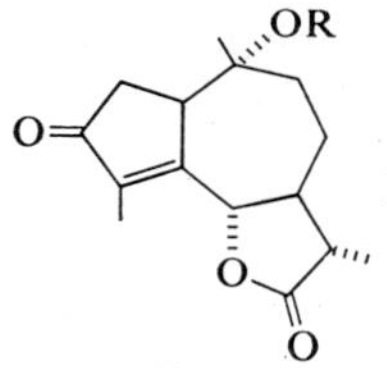

R = Me, Et, Pr^i, Bu^t

(214)   (215)

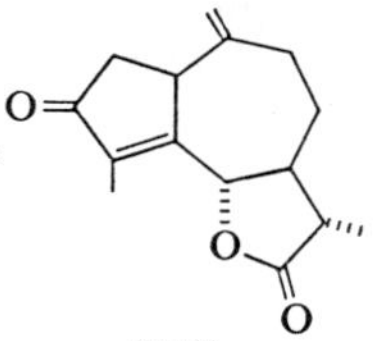

Table 16 *Irradiation of α-santonin in dioxan solution in the presence of acetic and trichloroacetic acids* [133]

Acid	Conc. (%)	Product yield (%)		
		(212) (X = H)	(214)	(215)
AcOH	25	30	26	10
AcOH	50	39	29	6
AcOH	100	64	37	2
CCl$_3$COOH	6	22	31	26
CCl$_3$COOH	10	35	28	14
CCl$_3$COOH	20	45	34	7

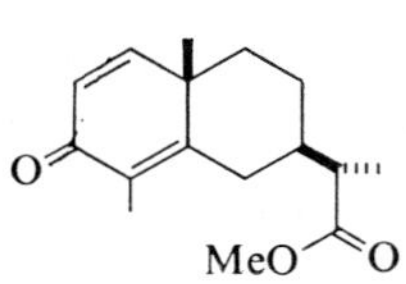

(216) 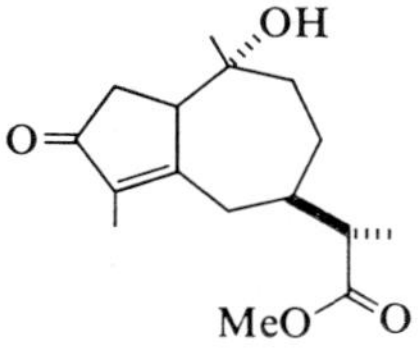   (217) 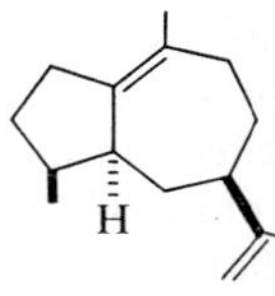   (218)

[112] G. Marsh, D. R. Kearns, and M. Fisch, *J. Amer. Chem. Soc.*, 1970, **92**, 2252.
[133] K. Schaffner-Sabba, *Helv. Chim. Acta*, 1969, **52**, 1237.

that the formation of these products was, to a certain extent, dependent on the acidity of the medium in which the photolysis was being carried out. The santonin-type rearrangement of cross-conjugated dienone (216) to the hydroazulene (217) which occurs on photolysis in aqueous acetic acid has been used as the basis for the synthesis of α-bulnesene (218).[134]

Quinone Methides.—As an extension of Becker's report [135] that dilute solutions of quinone methide (219) could be converted into the phenol (220a), by direct or acetophenone-sensitized irradiation in isopropanol, Matsuura and Ogura [136] have obtained alcohol addition products, together

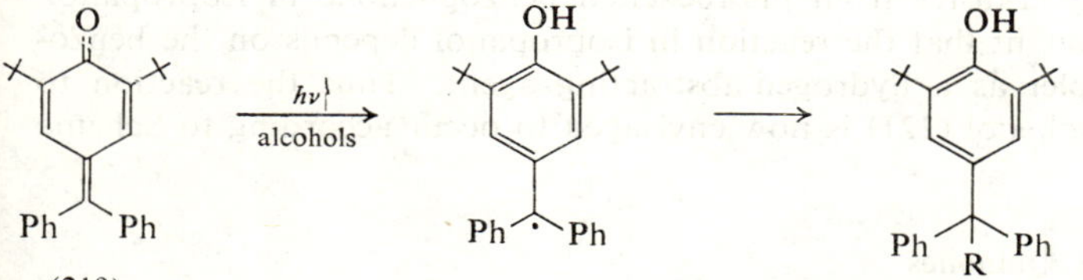

(a) R = H, (b) R = CH$_2$OH, (c) R = CMe$_2$OH
(220)

Scheme 12

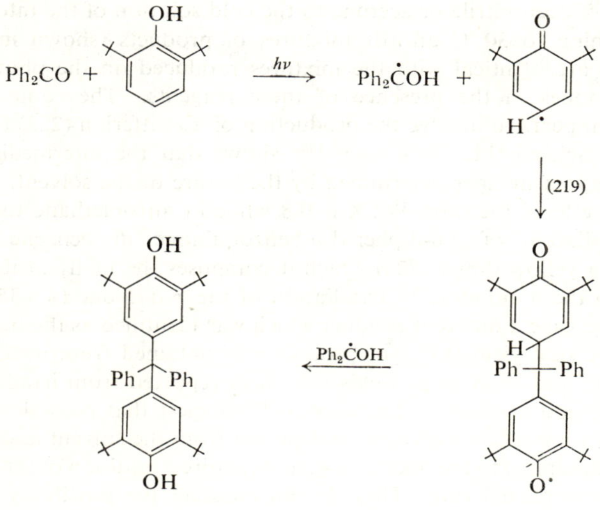

Scheme 13

[134] E. Piers and K. F. Cheng, *Chem. Comm.*, 1969, 562.
[135] H. D. Becker, *J. Org. Chem.*, 1967, **32**, 2115.
[136] T. Matsuura and K. Ogura, *Bull. Chem. Soc. Japan*, 1969, **42**, 2970.

with the phenol. In methanol, using acetophenone as sensitizer, the addition product (220b, 74%) and the phenol (220a, 7%) were isolated. In isopropanol, a better hydrogen donor, the phenol (220a, 54%) is the major product, but the addition product (220b, 11%) is also formed. Experiments suggest that the addition products arise from hydrogen abstraction by the sensitized quinone methide followed by radical addition (Scheme 12). A reassessment of the photoreactions of certain quinone methides (219) with 2,6-disubstituted phenols has been published.[137] Originally it was proposed that the reaction of quinone methide (219) with 2,6-di-t-butylphenol to give the bisphenol (221) was a process involving triplet energy transfer from photoexcited benzophenone in isopropanol. It is now thought that the reaction in isopropanol depends on the benzophenone triplet as a hydrogen-abstracting agent. Thus the reaction to form the bisphenol (221) is now envisaged to occur according to Scheme 13.

6 Quinones

p-Quinones.—A reinvestigation of the photolysis of t-butyl-*p*-quinones has shown the presence of a thermally unstable intermediate.[138a] This intermediate, formed at $-80\,°C$ in 1,2-dimethoxyethane, is presumed to be the spirocyclopropyldienone (222) first proposed by Orlando *et al*.[138b] Addition of acetonitrile or acetone to the cold solution of the intermediate and warming to 30 °C affords mixtures of products (shown in Scheme 14) which are identical with the mixtures produced on the photolysis of the *p*-quinones in the presence of these reagents. The route to these products appears to involve the production of a zwitterion (223) formed as shown in Scheme 14. This study [138a] shows that the intermediate (222) rearranges in a manner determined by the nature of the solvent. Thus, in dimethoxyethane the ratio W : X is 0·8 while in nitromethane the ratio is 5·5. Irradiation of 2,6-diphenyl-*p*-benzoquinone in benzene solution produces a yellow dimer (224) which decomposes thermally at its melting point into the *p*-quinone.[139] Irradiation of the *p*-quinone ($\lambda > 350$ nm) in acetonitrile gave a different product which was identified as the benzofuran (225). The best yield (81%) of product was obtained from irradiation in acetonitrile, although good yields were also reported from irradiations in acetic acid and methanol. The authors [139] suggest that possible routes to the product are (*i*) via hydrogen abstraction from the solvent and addition of the semiquinone to the phenyl ring, and (*ii*) direct addition of the carbonyl group to the phenyl ring. They do not consider the possibility of direct hydrogen abstraction from the aryl ring followed by the formation of a

---

[137] H. D. Becker, *J. Org. Chem.*, 1969, **34**, 2472.
[138a] S. Farid, *Chem. Comm.*, 1970, 303.
[138b] C. M. Orlando, H. Mark, A. K. Bose, and M. S. Manhas, *Tetrahedron Letters*, 1966, 3003; also cited in ref. 62, p. 218.
[139] H. J. Hageman and W. G. B. Huysman, *Chem. Comm.*, 1969, 837.

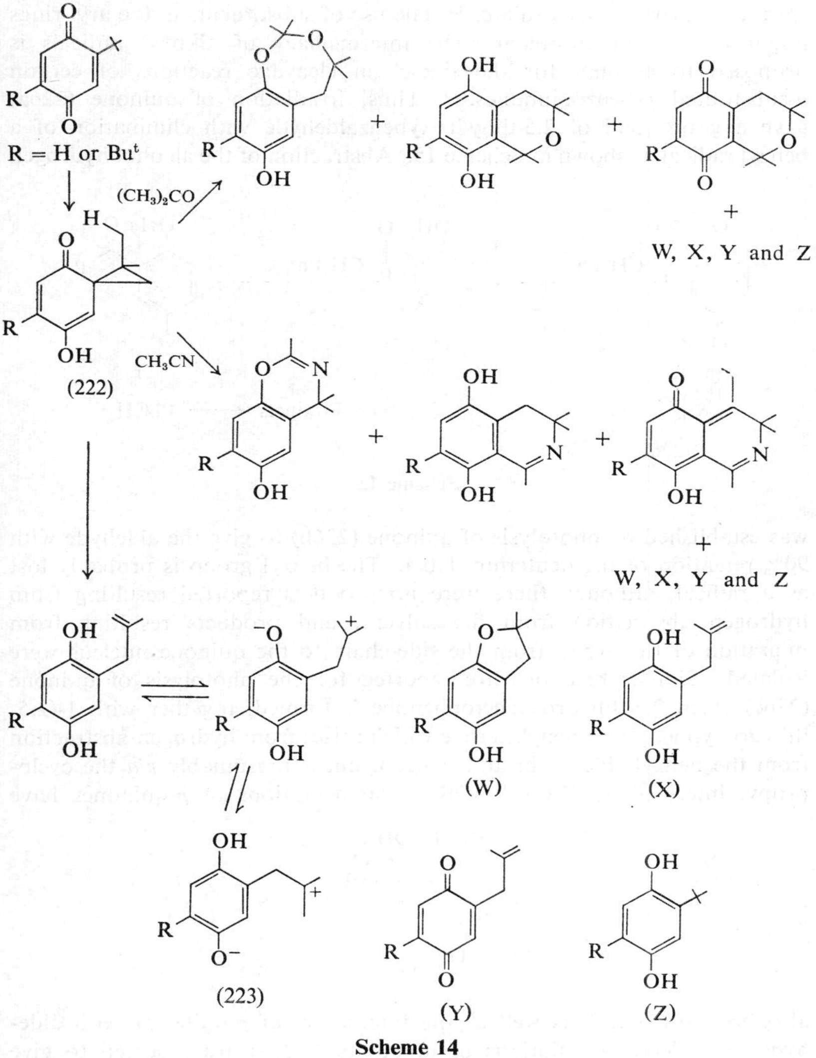

Scheme 14

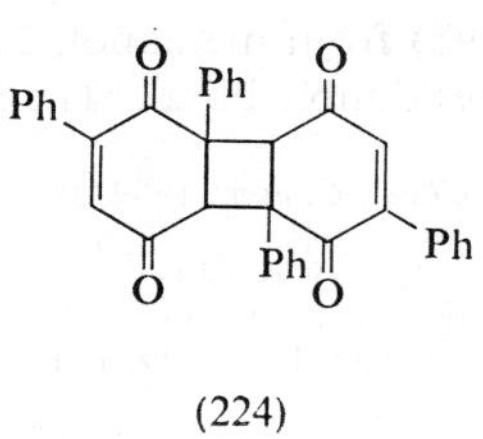

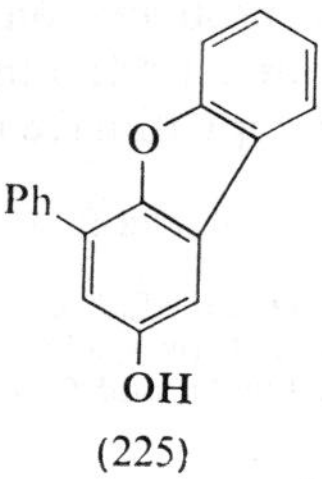

spiro-cyclopropyl intermediate.[138] The use of substituents in the aryl rings might resolve this problem. The intermediacy of alkoxyl radicals is proposed to account for the side-chain cleavage reactions of certain *o*-substituted *p*-benzoquinones.[140] Thus, irradiation of quinone (226a) gave a good yield of 2,5-dihydroxybenzaldehyde with elimination of a benzyl radical as shown in Scheme 15. Abstraction of the alcohol hydrogen

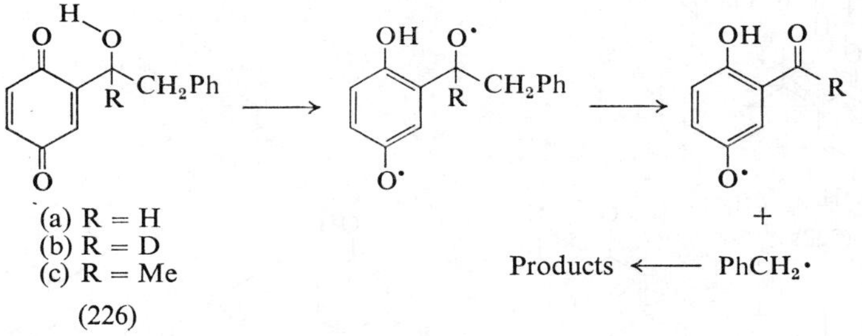

(a) R = H
(b) R = D
(c) R = Me
(226)

Products ⟵ PhCH$_2\cdot$

Scheme 15

was established by photolysis of quinone (226b) to give the aldehyde with 90% retention of the deuterium label. The benzyl group is probably lost as a radical, although there were no products reported resulting from hydrogen abstraction from the solvent, and products resulting from migration of the group from the side-chain to the quinone nucleus were isolated. Similar reactions are reported for the photolysis of quinone (226c) when 2,5-dihydroxyacetophenone is formed, together with 1-(2,5-dihydroxyphenyl)-1-phenylacetone which arises from hydrogen abstraction from the benzyl side-chain and rearrangement presumably *via* the cyclopropyl intermediate (227).[141] Other fragmentations of *p*-quinones have

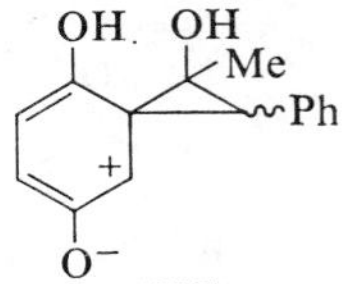

(227)

also been studied,[142] as well as the interaction of *p*-quinones with aldehydes.[143] Alcoholic solutions of ubiquinone (228) are reported to give alcohol addition products on exposure to sunlight.[144] The products [(229a) from ethanol, 19%; and (229b) from methanol, 24%] were shown by n.m.r. spectroscopy to have a *trans* double bond. The formation of the

[140] J. M. Bruce, D. Creed, and K. Dawes, *Chem. Comm.*, 1969, 594.
[141] Ref. 62, p. 218.
[142] J. M. Bruce and D. Creed, *J. Chem. Soc. (C)*, 1970, 649.
[143] J. M. Bruce and K. Dawes, *J. Chem. Soc. (C)*, 1970, 645.
[144] H. Morimoto, I. Imada, and G. Goto, *Annalen*, 1969, **729**, 184.

Enone Rearrangements and Cycloadditions

products is rationalised by invoking $n\pi^*$ excitation followed by hydrogen transfer (Scheme 16). This allows for the formation of the vinylogous o-quinone methide which adds solvent (alcohol) to give, after oxidation, the observed products.

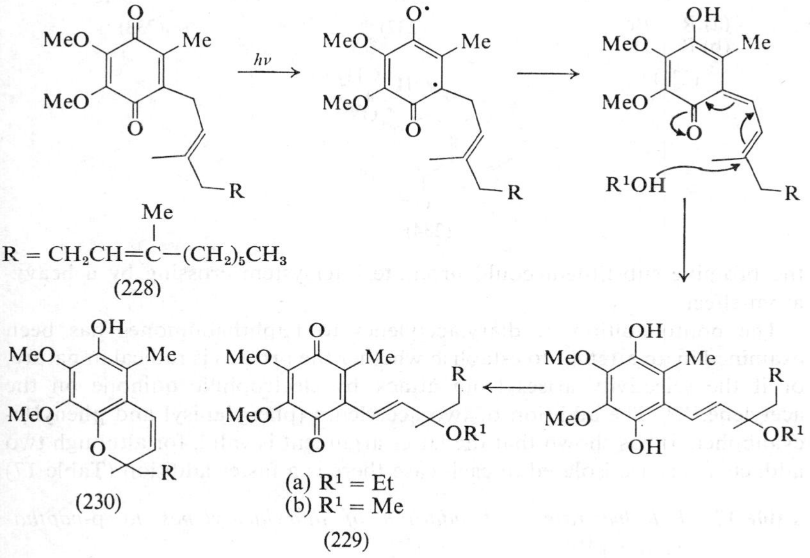

$R = CH_2CH=\overset{\underset{\displaystyle Me}{|}}{C}-(CH_2)_5CH_3$

(228)

(a) $R^1 = Et$
(b) $R^1 = Me$

(229)

(230)

Scheme 16

Contrary to these observations is the report [145] of the photolysis of ubiquinone (228) in alcoholic solution in the presence of oxygen (air) when ubichromenol (230), amongst other products, was isolated. It is interesting that demethylated products were also isolated. The involvement of an o-methoxy-substituent has also been reported in the photolysis of the naphthaquinones (231).[146] The photolysis of a naphthaquinone (231a) was undertaken, the function of the bromo-substituent being to compress the methoxy-group into the correct configuration for hydrogen abstraction. Two products were isolated from the irradiation in acetic anhydride, viz. a triacetate (232) and the desired methylenedioxy-ether (233). The quinone (231b), lacking the bromo-substituent, gave only a product from hydrogen abstraction from the solvent and photo-Fries rearrangement. Mechanistically, it is proposed that after hydrogen abstraction an oxonium ylide (234) is produced by electron demotion, and that addition of the adjacent hydroxy-group to this intermediate produces the cyclic ether. The nature of the excited state was not determined, nor was it apparently realised that

---
[145] H. M. Cheng and J. E. Casida, *J. Labelled Compounds*, 1970, **6**, 66.
[146] Jack E. Baldwin and J. E. Brown, *Chem. Comm.*, 1969, 167.

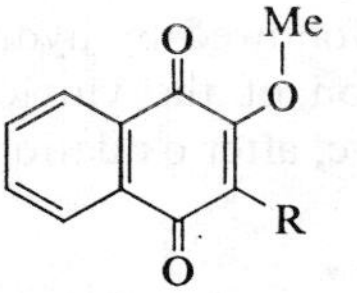

(231) (a) R = Br (b) R = H

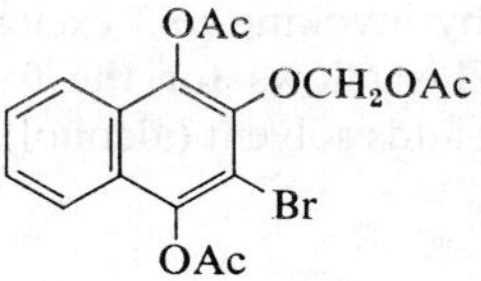

(232)

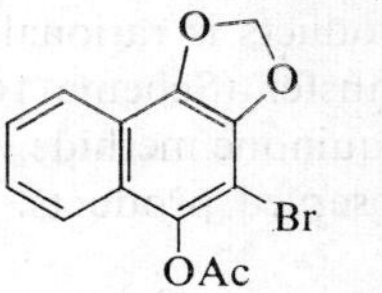

(233)

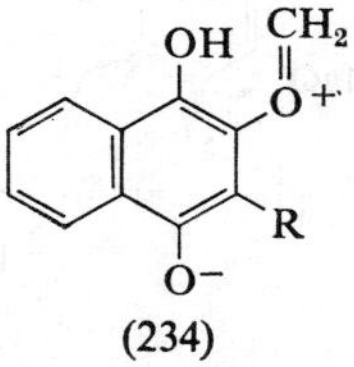

(234)

the bromine substituent could promote intersystem crossing by a heavy-atom effect.

The photoaddition of diarylacetylenes to naphthaquinones has been examined in an attempt to establish whether the process is radical in nature, or if the selectivity arises from attack by electrophilic quinone on the acetylene.[147a] The addition of two acetylenes (phenylanisyl and phenyl-p-cyanophenyl) has shown that the latter argument is valid, for although two adducts (235) are isolated in each case there is a faster addition (Table 17)

Table 17 *Relative rates† of addition of diarylacetylenes to p-napthaquinones* [147a]

Adduct	Benzene	Acetonitrile
(235a)	5·1 ± 0·3	3·1 ± 0·3
(235b)	2·3 ± 0·3	1·9 ± 0·3
(235c)	0·9 ± 0·2	1·0 ± 0·2
(235d)	0·7 ± 0·2	0·8 ± 0·2

† Relative to diphenyl acetylene in each solvent.

for formation of the adduct which best stabilises the dipolar form of the transition state represented in diagram (236). Another study has extended

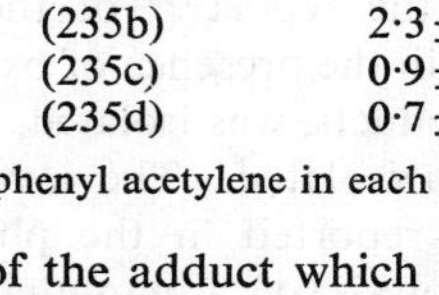

(a) R^1 = Ph, R^2 = p-anisyl
(b) R^1 = p-anisyl, R^2 = Ph
(c) R^1 = Ph, R^2 = p-CN-phenyl
(d) R^1 = p-CN-phenyl, R^2 = Ph

(235)

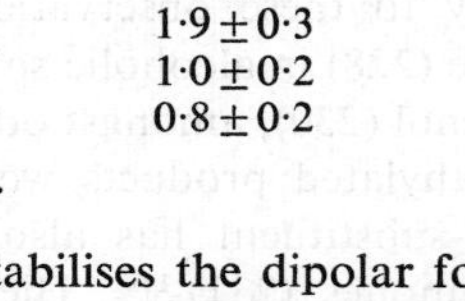

(236)

[147a] S. P. Pappas and N. A. Portnoy, *Chem. Comm.*, 1969, 597.

the scope of this addition.[147b] Photochemical studies of piperidino-1,4-naphthaquinones,[148] piperidinoanthraquinones,[149,150] and anthraquinone-2-sulphonate [151] have been reported.

o-Quinones.—Further photoreactions of phenanthrenequinone with olefins have been described. Farid and Hess [152] have reported the addition of the quinone to di-, tri-, and tetra-chloroethylene. The major photoreaction with these olefins is to produce dioxens (237) and keto-oxetans (238). There are also minor products which result from the elimination of HCl

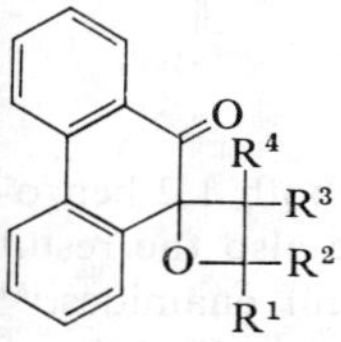

(238)

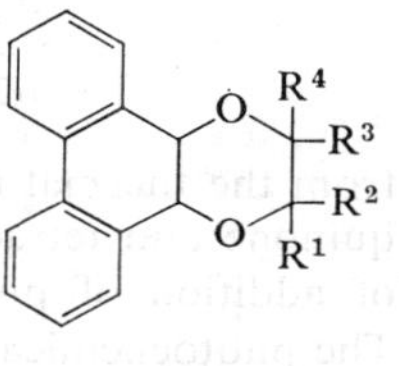

(237)

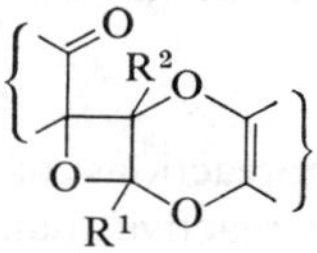

(a) $R^1 = H$, $R^2 = Cl$
(b) $R^1(R^2) = H$, $R^2(R^1) = Cl$
(239)

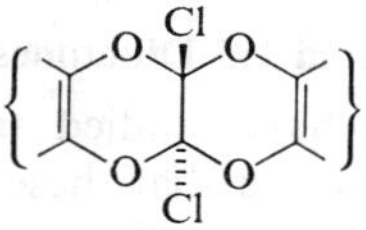

(240)

from some of the oxetan adducts (3, 4, and 5, Table 18) and addition of phenanthrenequinone to the resulting double bond produces adducts (239a, 239b, and 239a respectively). In the case of the trichlorodioxen (Table 18; 12), elimination of HCl and addition of the quinone gives the

Table 18 *Formation of oxetans and dioxens from the photoaddition of phenanthrenequinone to chloroethylenes*

Oxetan (238) (%)	R^1	R^2	R^3	R^4	Dioxen (237) (%)
3 (15)	H	H	Cl	Cl	9 (31)
4 (6)	H	Cl	H	Cl	cis 10 (4)
5 (7)	H	Cl	H	Cl	trans 11 (7)
6 (63)	H	Cl	Cl	Cl	12 (30)
7 (44)	Cl	Cl	Cl	Cl	

adduct (240). The addition of phenanthrenequinone to trichloroethylene has been shown to be slightly temperature dependent, with the keto-oxetan being favoured at lower temperature (-23 °C) and the dioxen at 70 °C.

[147b] W. Kothe, *Tetrahedron Letters*, 1969, 5201.
[148] E. P. Fokin and A. M. Detsina, *Izvest. sibirsk Otdel, Akad. Nauk, S.S.S.R., Ser. Khim. Nauk*, 1969, 95 (*Chem. Abs.*, 1970, **72**, 12,500n).
[149] G. O. Phillips, A. K. Davies, and J. F. McKellar, *Chem. Comm.*, 1970, 519.
[150] G. O. Phillips, A. K. Davies, J. F. McKellar, and D. Price, *Chem. Comm.*, 1969, 1097.
[151] K. P. Quinlan, *J. Phys. Chem.*, 1969, **73**, 2058.
[152] S. Farid and D. Hess, *Chem. Ber.*, 1969, **102**, 3747.

Phenanthrenequinone has also been reported to add photochemically to isobenzofurans, giving a dioxole (241) and a dioxocin derivative (242).[153a] It is interesting to note that similar dioxole derivatives were reported [153b]

(241)   (242)

from the thermal reaction of diphenylisobenzofuran with both 1,2-benzoquinone and tetrachloro-1,2-benzoquinone. Dioxens were also the result of addition of phenanthrenequinone and benzil to certain enamides.[154] The photochemical reaction of benzil in cumene has been described.[155]

7 1,2- and 1,3-Diketones

Turro and Lee [156] have studied the photoreactions of aliphatic 1,2-diketones and have found that these are less reactive than aliphatic monoketones in hydrogen abstractions. The study involved the photoconversion (435 nm) of diketones (243) in benzene solution into the corresponding cyclobutanols (244). The reactions involved the triplet state and could be

(a) $R^1 = R^2 = H$
(b) $R^1 = Me, R^2 = H$
(c) $R^1 = R^2 = Me$

(243)   (244)

quenched by the use of pyrene [$E_T = 200.6$ kJ mol^{-1} (48 kcal mol^{-1})]. The results obtained from Stern–Volmer plots in benzene, acetonitrile, and t-butanol (Table 19) demonstrate environmental effects on the hydrogen abstraction. The rate constants for the abstraction of hydrogen are much lower than those measured for the corresponding mono-ketones ($k_r > 10^8$ s^{-1}). The reactions are remarkably solvent independent although a degree of reversibility in the hydrogen-transfer step is suggested by the slight change in Φ_{cy} (quantum yield for cyclobutanol formation) when changing solvent from benzene to acetonitrile. It is suggested that the overall low reactivity of the 1,2-diketones could be due to the low triplet energy [$E_T = 221.5$ kJ mol^{-1} (53 kcal mol^{-1})] making the formation of the

---

[153a] W. Friedrichsen, *Tetrahedron Letters*, 1969, 1219.
[153b] W. M. Horspool, J. M. Tedder, and Z. U. Din, *J. Chem. Soc. (C)*, 1969, 1964.
[154] K. R. Eicken, *Annalen*, 1969, **724**, 66.
[155] D. L. Bunbury and T. C. Chuang, *Canad. J. Chem.*, 1969, **47**, 2045.
[156] N. J. Turro and T. J. Lee, *J. Amer. Chem. Soc.*, 1969, **91**, 5651.

Enone Rearrangements and Cycloadditions

intermediate diradical more endothermic than formation of such intermediates by the monoketones. The unusual formation of cyclopentenones (245a and b) from the irradiation of αβ-unsaturated diketones (246a and b) has been reported.[157, 158] The authors propose that the enol (247) is not a

Table 19 *Details for photolysis of α-diketones in various solvents* [156]

α-Diketone	Solvent	k_qT	$1/T$	k_r	Φ_{cy}
(243a)	Benzene	1.07×10^5	4.7×10^4	2.5×10^3	0.054
(243a)	Benzene	0.81×10^{5a}	6.2×10^4		
(243a)	Acetonitrile	1.6×10^5	7.0×10^4	4.4×10^3	0.062
(243a)	t-Butanol	0.43×10^5	5.4×10^4	3.7×10^3	0.069
(243b)	Benzene	1.8×10^4	2.7×10^5	1.3×10^5	0.50
(243c)	Benzene	3.2×10^3	1.5×10^6	8.5×10^5	0.57
(243c)	Acetonitrile	8.5×10^3	1.3×10^6	8.5×10^5	0.66
(243c)	t-Butanol	2.9×10^3	0.8×10^6	5.0×10^5	0.62

a Obtained from Stern–Volmer quenching of phosphorence of (234a), error = ±10%.

likely precursor of the final product, although photoenolisation usually occurs with *o*-alkyl-benzophenones. The representation of the photoenol as the *o*-quinonoid form (247) is perhaps misleading since the triplet state of benzophenone would be likely to undergo hydrogen transfer to produce

(245) (a) R = H (b) R = Me   (246)   (247)   (248)

the intermediate (248) which would be significantly different in reactivity and lifetime from the enol (247). Subsequent ring-closure and hydrogen transfer could account for the products formed in this reaction,[158] and also for the absence of deuterium in the product following irradiation in MeOD.

Oddly enough, the diketone (249) also undergoes photocyclisation, but hydrogen abstraction does not occur in this case and cyclisation is thought to proceed *via* formation of the zwitterion (250a) and hydride migration to give the product (250).[158] Another unexpected product has been reported from the photolysis of the acyclic α-diketone (251) when the bicyclic keto-oxetan (252) was isolated.[159] This presumably results from intramolecular addition of a carbonyl function to the ethylenic bond. However, the structure (252) assigned to this oxetan is not that which would have resulted from formation of the more stable diradical, a factor normally dominant in oxetan formation. The photochemical decomposition of the

---
[157] R. Bishop and N. K. Hamer, *Chem. Comm.*, 1969, 804.
[158] T. L. Burkoth and E. F. Ullman, *Tetrahedron Letters*, 1970, 145.
[159] R. Bishop and N. K. Hamer, *Chem. Comm.*, 1969, 804.

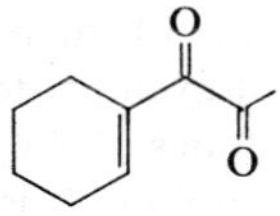

(249)

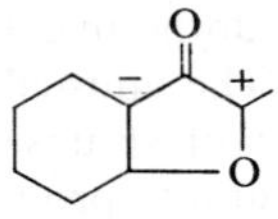

(250a)

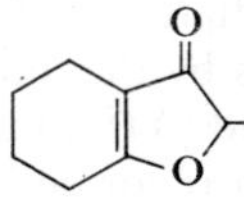

(250)

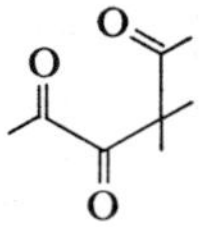

(251)

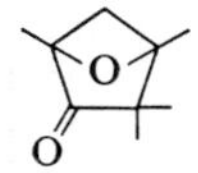

(252)

pink 3,4-di-t-butyl-cyclobutane-1,2-dione produces the comparatively stable 2,3-di-t-butylcyclopropanone by decarbonylation.[160] Turro and Cole[161] have studied the photolysis of 2,2,4,4-tetramethyl-cyclobutane-1,3-dione and the isomeric β-enol. There is evidence that both compounds give rise to a common intermediate (i, Scheme 17). The ratio of the

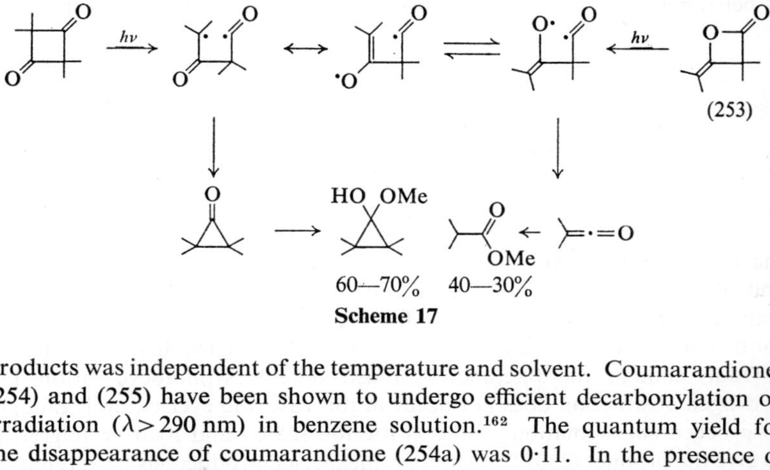

(253)

60—70% 40—30%

Scheme 17

products was independent of the temperature and solvent. Coumarandiones (254) and (255) have been shown to undergo efficient decarbonylation on irradiation ($\lambda > 290$ nm) in benzene solution.[162] The quantum yield for the disappearance of coumarandione (254a) was 0·11. In the presence of nucleophiles the intermediate (256) of the decarbonylation could be efficiently trapped to afford the products (257). The excited state was thought to be $n\pi^*$ in nature since an oxetan (258) was obtained following the irradiation of coumarandione (254a) in a benzene solution of *cis*-1,2-dichloroethylene. Irradiation of the ginkgolide equilibration mixture (259a and c) in methanol leads to photoreduction to the corresponding

160 A. De Groot, D. Oudman, and H. Wynberg, *Tetrahedron Letters*, 1969, 1529.
161 N. J. Turro and T. Cole, *Tetrahedron Letters*, 1969, 3451.
162 W. M. Horspool and G. D. Khandelwal, *Chem. Comm.*, 1970, 257.

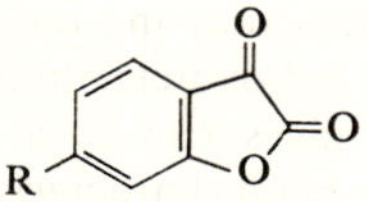

(a) R = H
(b) R = Me
(254)

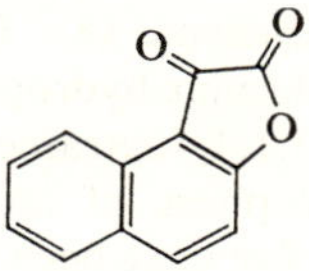

(255)

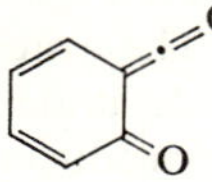

(256)

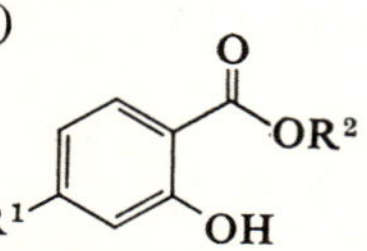

(a) $R^1 = R^2 = H$
(b) $R^1 = Me, R^2 = H$
(c) $R^1 = H, R^2 = Ph$
(d) $R^1 = H, R^2 = COPh$
(257)

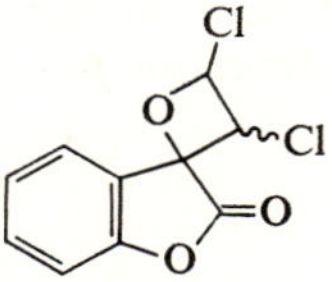

(258)

alcohol and pinacols (Scheme 18).[163] Photolysis of the acetal (259b) in methanol or methanol[$^2H_4$] led to the isolation of a methanol adduct (260) together with the corresponding alcohol (261). It is concluded that two mechanisms are operative. Thus the product (261) may be formed *via* intramolecular hydrogen abstraction which leads to a diradical and thence a

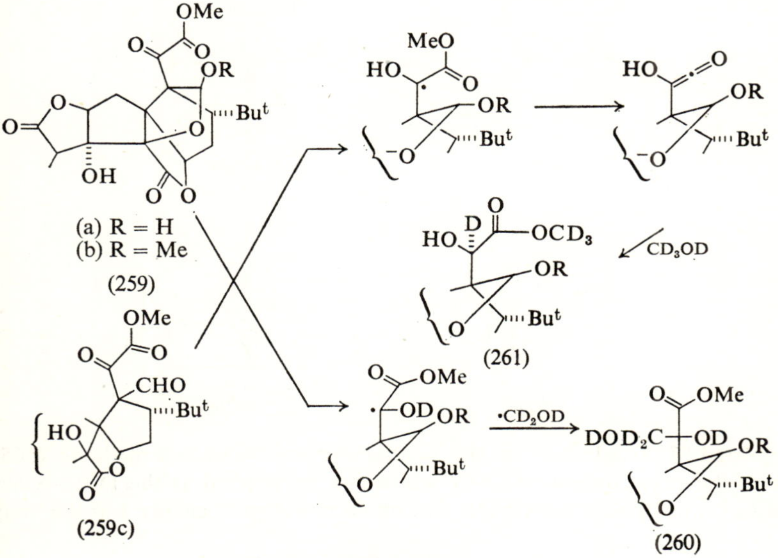

(a) R = H
(b) R = Me
(259)

(259c)

(261)

(260)

Scheme 18

[163] Y. Nakadaira, Y. Hirota, and K. Nakanishi, *Chem. Comm.*, 1969, 1469.

keten which adds methanol (Scheme 18). On the other hand, the pinacols appear to arise from intermolecular hydrogen abstraction and radical combination. The irradiation of 1,1,4,4-tetramethyl-2,3-dioxotetralin in carbon tetrachloride under an atmosphere of nitrogen gives rise to a complex mixture of products of which five have been isolated and characterised. Surprisingly, decarbonylation contributed little to the overall reaction, for indanone (262) was formed in only 0·5% yield.[164] The photolysis of α-oximinocyclodecanone has been described.[165]

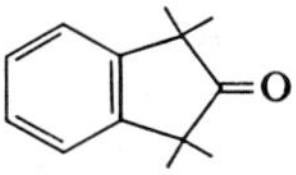

(262)

8 1,4- and Higher Diketones

3-Caren-2,5-dione has been reported to form three cyclobutane-type dimers on photolysis in acetone.[166] All the dimers were *cis*-fused and most were head–head. The intramolecular photocycloaddition (253·7 nm, solid phase) of the adduct (263) to form the cage compound (264) has been

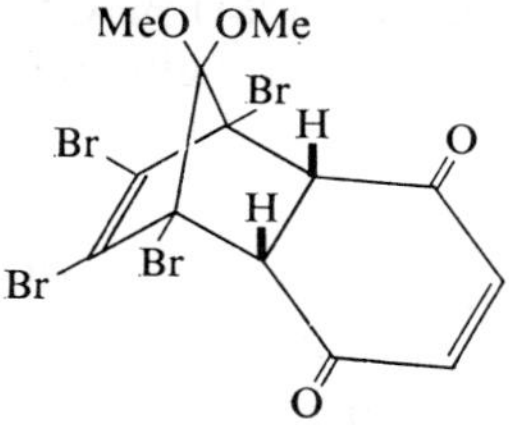

(263) 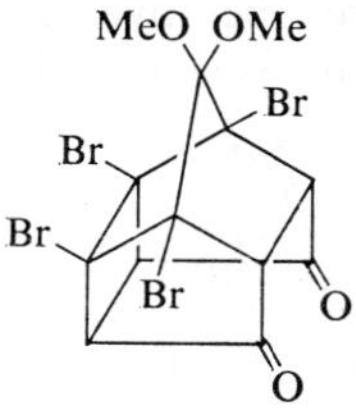   (264)

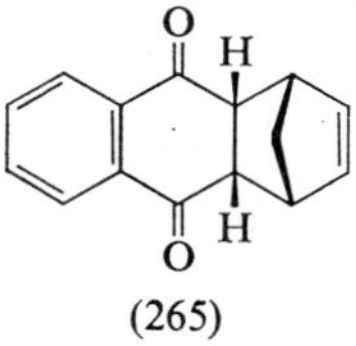

(265)    (266)

reported.[167] Irradiation of the methanohexahydroanthracenedione (265) gives the cage compound (266), which can be trapped as the Diels–Alder adduct with maleic anhydride: intramolecular triplet energy transfer may

[164] G. E. Gream and J. C. Paice, *Austral. J. Chem.*, 1969, **22**, 1249.
[165] A. Stojilkovic and R. Tasovac, *Tetrahedron Letters*, 1970, 1405.
[166] I. W. J. Still, C. J. MacDonald, and Y. N. Oh, *Canad. J. Chem.*, 1970, **48**, 1526.
[167] R. G. Pews, C. W. Roberts, and C. R. Hand, *Tetrahedron*, 1970, **26**, 1711.

play a major part.[168] Cholest-4-ene-3,6-dione (267) in cyclohexane (medium-pressure lamp, Pyrex filter) gave two products, *viz.* cholestane-3,6-dione (268), which was also produced by irradiation in benzene, and the

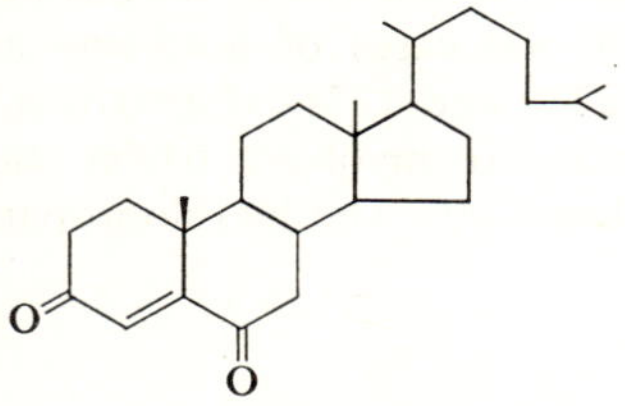

(267)

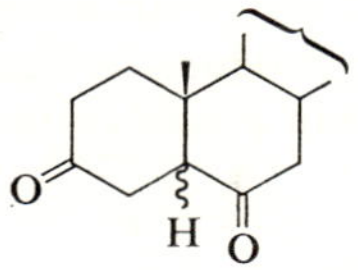

(268)

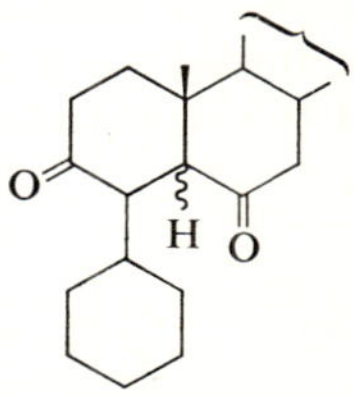

(269)

product from radical combination, 4-cyclohexylcholestane-3,6-dione (269).[169] The synthesis and some reactions of thiomaleic anhydride have been described by Verbeek, Scharf, and Korte.[170] The compound undergoes photoaddition to ethylene to give the cyclobutanedicarboxylic

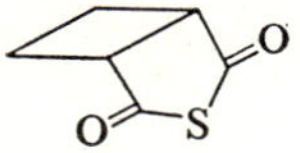

(270)

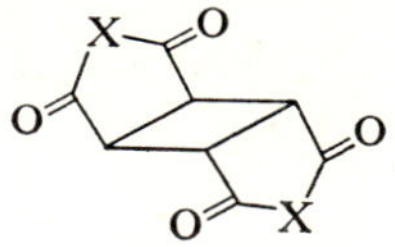

(a) X = S
(b) X = O

(271)

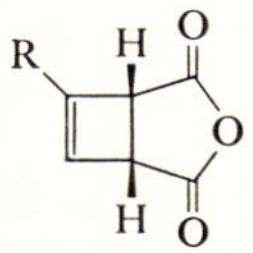

(a) R = H (36%)
(b) R = Me (55)
(c) R = Et (75)
(d) R = Pr^n (79)
(e) R = Bu^n (77)
(f) R = Bu^t (86)

(272)

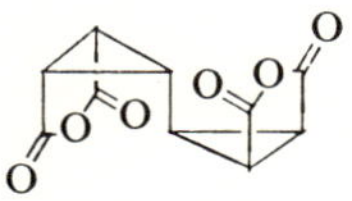

(273a)

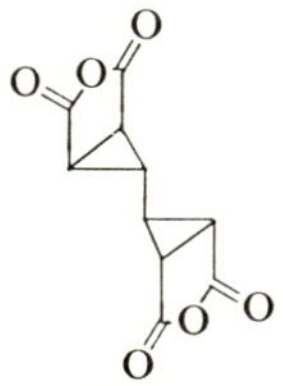

(273b)

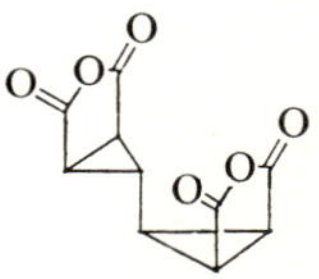

(273c)

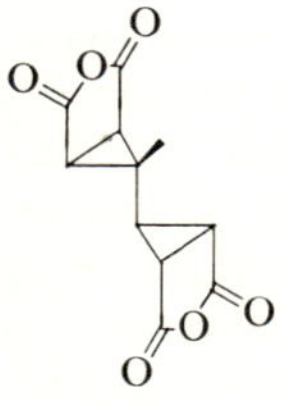

(274a)

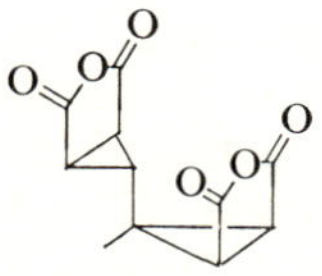

(274b)

[168] N. Filipescu and J. M. Menter, *J. Chem. Soc.* (*B*), 1969, 616.
[169] H. Hikino and H. Takeshita, *J. Pharm. Soc. Japan*, 1968, **88**, 98.
[170] M. Verbeek, H. D. Scharf, and F. Korte, *Chem. Ber.*, 1969, **102**, 2471.

anhydride (270), together with the dimer (271a) of the anhydride. In contrast, maleic anhydride gives the dimer (271b) and a copolymer of ethylene and maleic anhydride under similar conditions.[171] The acetone-photosensitized addition of maleic anhydride to terminal alkynes has been shown to yield cyclobutenes (272) and, in the cases of acetylene and propyne, a mixture of bicyclopropanes (273) and (274) respectively.[172] The formation of these products is interpreted to involve a triplet state, and the proposed reaction is shown in Scheme 19. The bicyclopropanes

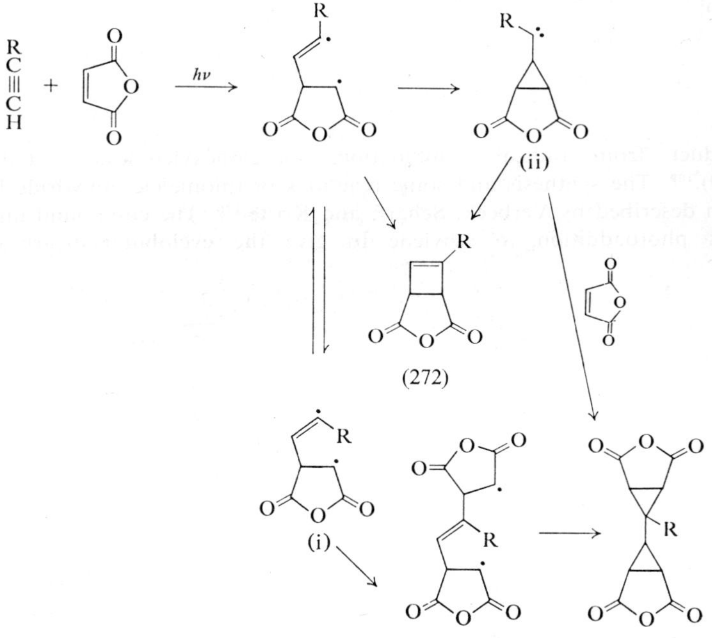

Scheme 19

are also formed by this route. Thus the intermediate (i) is common to both products, although in bicyclopropane formation the diradical reacts further with maleic anhydride to produce the final product (274). The formation of (272a) and (273a) has been shown to be temperature- and concentration-dependent (Table 20). As might be expected, the formation of the bicyclopropane is favoured at high concentrations of maleic anhydride, while at lower concentrations the cyclobutene is the predominant product. There is also a slight enhancement of the yield of the cyclobutene on

---

[171] H. D. Scharf and F. Korte, *Chem. Ber.*, 1965, **98**, 3672.
[172] W. Hartmann, *Chem. Ber.*, 1969, **102**, 3974.

Table 20 *Temperature dependence of the photoaddition of acetylene to maleic anhydride*[172]

Maleic anhydride (mol l^{-1})	Temp. (°C)	(272a) (mol %)	(273a–c) (mol %)
1·0	+7	59	41
1·0	−18	62	38
1·0	−36	64	36
1·0	−58	67	33
1·0	−68	72	28
0·8	+7	65	35
1·0	+7	59	41
1·2	+7	45	55
1·5	+7	31	69
1·7	+7	27	73
2·0	+7	24	76

lowering the temperature. This of course could be due to the increased viscosity of the solution preventing the diffusion of the diradical (i) to another molecule of maleic anhydride. No study of this viscosity effect was made. An alternative scheme for product formation was also considered. This involved the formation of a carbene intermediate (ii, Scheme 19). The photochemical addition of dichloromaleimide to the phospholen

(275) affords the isomeric 1 : 1 adducts (276a) and (276b).[173] The photochemistry of diformylpyridines has been described.[174] Photolysis of the diendione (277a) in benzene through a Pyrex filter gave the triasteranedione (278a) in 40% yield.[175] This compound arises by a 1,2-acyl migration (Scheme 20). A similar rearrangement is reported for the diendione (277b), which gives the triasteranedione (278b). Additional products were also isolated, *viz.* an unidentified dihydrocoumarin from diendione (277a) formed by a 1,5-diacyl migration, and two enones (279a) and (279b) produced by 1,3-diacyl migrations from diendione (277b).[175]

A further communication[176] on the photolysis of *cis-cis*-cyclodeca-3,8-diene-1,6-dione (280a) has been published.[177] The authors[176] report

---

[173] G. Märkl and H. Schubert, *Tetrahedron Letters*, 1970, 1273.
[174] G. Queguiner and A. Godard, *Compt. rend.*, 1969, **269***C*, 1646.
[175] P. A. Knott and J. M. Mellor, *Tetrahedron Letters*, 1970, 1829.
[176] J. W. Stankorb and K. Conrow, *Tetrahedron Letters*, 1969, 2395.

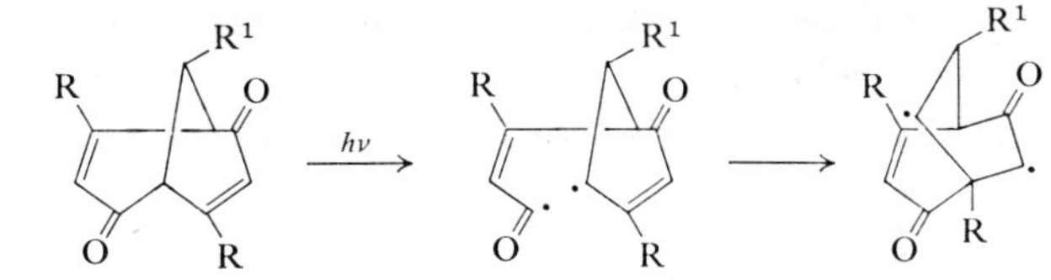

(a) R = R¹ = H
(b) R = H, R¹ = Ph
(277)

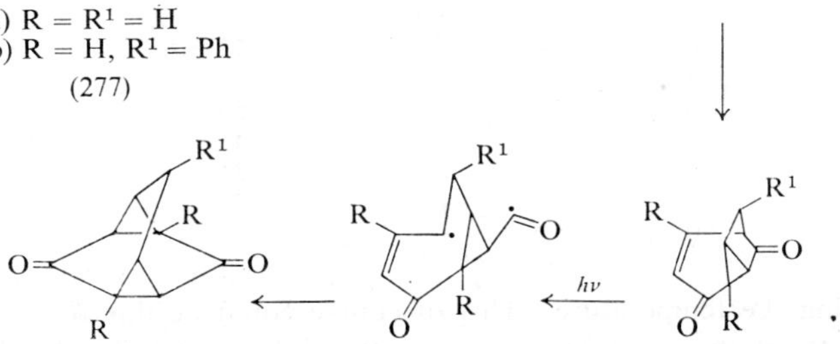

(278)

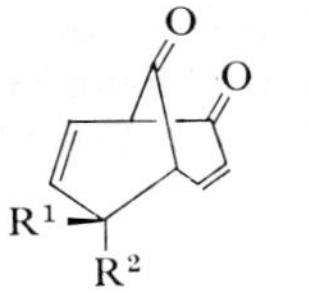

(a) R¹ = Ph, R² = H
(b) R¹ = H, R² = Ph
(279)

Scheme 20

that the use of solvent acetone as sensitizer leads to a complex mixture of products from which the *cis-trans-cis*-tricyclodione (281) was isolated. In

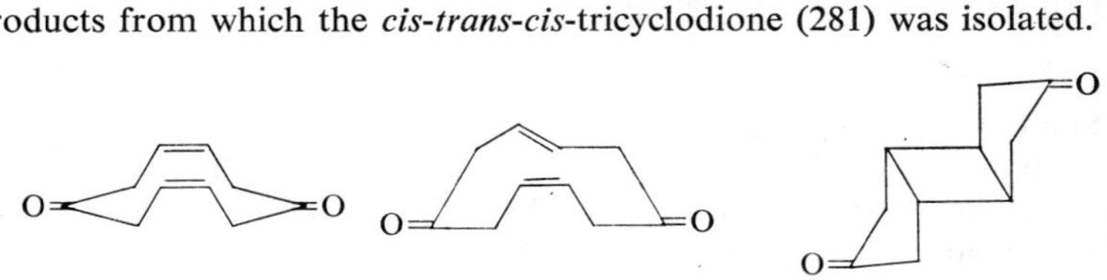

(280a) (280b) (281)

agreement with Scheffer and Lungle,[177] they propose that initial *cis–trans* isomerisation of the starting dione takes place prior to intramolecular cycloadditions. Indeed, Stankorb and Conrow[176] were able to isolate *cis-trans*-cyclodecadione (180b). The initial excitation of the molecule is proposed to be $n \to \pi^*$ in character, and is followed by intersystem crossing to produce a triplet state. Intramolecular energy transfer to one of the double bonds can account for the isomerisation and subsequent cycloadditions.

[177] See ref. 62, p. 202 for previous work in this area.

3
Photochemistry of Olefins, Acetylenes, and Related Compounds

1 Reactions of Alkenes

cis–trans-Isomerisation.—A review of the literature on the photochemical isomerisation of olefins has given examples consistent and inconsistent with the Schenck mechanism for *cis–trans*-isomerisation.[1] Caldwell has reported inefficiencies in the Schenck mechanism for the isomerisation of simple olefins and the quenching of ketonic triplets.[2] The results from a study of the hydrogen-isotope effect on the benzophenone-sensitized isomerisation of but-2-ene are interpreted as involving the production of an intermediate prior to the formation of the normal biradical. The compounds analysed were olefins (1a and b) and (1c and d). The isotopic ratio k_a/k_b for the isomerisation of (1a) and (1b) was found to be 1·02.[3] This study has examined the same effect in the formation of oxetans, and has shown that there is little evidence for isotopic selectivity

MeCR¹=CR²Me
(a) $R^1 = R^2 = H$
(b) $R^1 = R^2 = D$
(c) $R^1 = D, R^2 = H$
(d) $R^1 = H, R^2 = D$
(1)

$Ar_2\dot{C}-O-\underset{\underset{Me}{|}}{\overset{\overset{R^1}{|}}{C}}-\underset{\underset{Me}{|}}{\overset{\overset{R^2}{|}}{\dot{C}}}$

(2)

$Ar_2\overset{+}{C}-\overset{..}{\underset{..}{O}}\overset{-}{:}$

(3)

$\underset{R^1\ \ R^2}{\overset{Me\ \ \ Me}{\text{+>}\!\!-\!\!\!<\!\text{·}}}$

($k_a/k_b = 1·03$), in the partitioning of the diradical (2). The olefin (1c) gives a ratio $k_c/k_d = 1·10$ for oxetan formation, showing that there is a slight preference for the carbon bearing a deuterium atom to end up next to oxygen. Further evidence is cited concerning the effect of substituents in the benzophenone on the quenching of the ketones by *cis*-but-2-ene (Table 1). All this is claimed as evidence for the existence of a complex resulting from the donation of an electron from the olefin to the ketone during excitation (3). Saltiel *et al.*[4] have also investigated the mechanism for the isomerisation of simple olefins (pent-2-enes) but do not report

---
[1] N. J. Turro, *Photochem. and Photobiol.*, 1969, **9**, 555.
[2] R. A. Caldwell, *J. Amer. Chem. Soc.*, 1970, **92**, 1439.
[3] R. A. Caldwell and S. P. James, *J. Amer. Chem. Soc.*, 1969, **91**, 5184.
[4] J. Saltiel, K. R. Neuberger, and M. Wrighton, *J. Amer. Chem. Soc.*, 1969, **91**, 3658.

Table 1 Quenching of benzophenones by cis-*but*-2-ene [2]

Compound	$\Phi_{cis-trans}$	k_0	E_T (kJ mol^{-1})	E_T (kcal mol^{-1})
4,4-Dimethoxybenzophenone	0·35	4×10^6	290·1	69·4
Benzophenone	0·3	7×10^7	286·7	68·6
4-Trifluoromethylbenzophenone	0·34	$2·2 \times 10^8$	282·6	67·6
4-Benzoylpyridine	0·34	$3·3 \times 10^8$	280·5	67·1

evidence for electron transfer in the Schenck mechanism. The reaction scheme proposed by them for the process is shown below, where X* is a common intermediate (Scheme 1) from which isomerisation takes place.

$$S^0 \to S^1 \to S^3 \xrightarrow{k_4} S^0$$

$$S^3 + t^0 \xrightarrow{k_5} X^*$$

$$S^3 + c^0 \xrightarrow{k_6} X^*$$

$$X^* \xrightarrow{k_7} \alpha t^0 + (1+\alpha)c^0$$

$t^0 =$ *trans*-pent-2-ene in ground state $c^0 =$ *cis*-pent-2-ene

Scheme 1

The important part of the analysis of the results relies on the measurement of the excitation ratio k_6/k_5, and the decay ratio $\alpha/(1-\alpha)$. However, for the isomerisation of the pent-2-enes studied, the results show that with different sensitizers a single decay ratio does not account for the results obtained (Table 2). This variation in the decay ratio implies that different

Table 2 Effect of sensitizer on the decay and excitation ratios for pent-2-enes [4]

Sensitizer	E_T (kJ mol^{-1})	E_T (kcal mol^{-1})	$\alpha/(1-\alpha)$	k_6/k_5
Benzene	351·1	84	1·00	0·92
Acetone	334·4	80	1·17	1·30
Acetophenone	309·3	74	1·90	2·85

intermediates are produced from different sensitizers. When the triplet energy of the sensitizer is close to 342·8 kJ mol^{-1} (82 kcal mol^{-1}), a triplet mechanism is obviously important, but as the triplet energy decreases below this value there is a concomitant rise in the value of the decay ratio which is reasonably interpreted as an increase in the involvement of the Schenck mechanism for isomerisation. Thus, with acetone, and more so with acetophenone, an adduct diradical (4), where rotation competes favourably with bond rupture, is important.

A method of synthetic value for the preparation of *trans*-cyclo-octene (5) by the xylene-photosensitised isomerisation of commercial *cis*-cyclo-octene has been described.[5] The method, utilising a recyclisation of the product of irradiation, produced 97% pure *trans*-isomer after four cycles. Cyclo-octa-1,5-diene was also detected but was thought to have been merely an impurity in the cyclo-octene, not a photoproduct. The Reporters comment that this diene can be removed from *cis*-cyclo-octene by heating with maleic anhydride in the presence of oxygen or a peroxide. The irradiation of *cis,trans*-cyclodeca-1,5-diene (6a) has been shown [6] to yield the *cis,cis*-isomer (6b) on irradiation at 253·7 nm: the photostationary state contained (6a) and (6b) in the ratio 1 : 4. No traces of either 'straight' or 'crossed' cycloaddition products were detected.

The possibility of both inter- and intra-molecular energy transfer in the photolysis of 1-phenylbut-2-ene has been investigated.[7] It has been found that four energy processes are operative, *viz.* singlet–singlet and triplet–triplet intermolecular, and singlet–singlet and triplet–triplet intramolecular. However, at low concentrations the preferred energy transfer process seems to be intramolecular. The rates for the intramolecular processes have been calculated as 1×10^6 s^{-1} for the singlet–singlet process and 2×10^6 s^{-1} for the triplet–triplet transfer. Intersystem crossing efficiency in the phenyl moiety is $\leq 65\%$. The irradiations were carried out utilising light of $\lambda = 253$ nm and solutions of the olefin in cyclohexane. The quantum yields for the processes are recorded in Table 3.

Table 3 *Quantum yield values for* cis–trans-*isomerisation of* 1-*phenylbut-2-ene* [7]

Starting conc. (mol l^{-1})				
cis	trans	Φ_{cis}	Φ_{trans}	Φ_c/Φ_t
$2·54 \times 10^{-3}$	$2·53 \times 10^{-3}$	0·09	0·08	1·12
$3·25 \times 10^{-3}$	$3·25 \times 10^{-3}$	0·09	0·08	1·12
$1·97 \times 10^{-2}$	$1·97 \times 10^{-2}$	0·20	0·18	1·11

An unexpected di-π-methane rearrangement has been discovered in the gas-phase photolysis of 1-phenylbut-2-ene at various wavelengths (266,

---
[5] J. S. Swenton, *J. Org. Chem.*, 1969, **34**, 3217.
[6] J. G. Traynham and H. H. Hsieh, *Tetrahedron Letters*, 1969, 3905.
[7] C. S. Nakagawa and P. Sigal, *J. Chem. Phys.*, 1970, **52**, 3277.

261, 252, 247 nm) to give the *cis*-isomer of the starting olefin and the rearrangement product, 1-methyl-2-phenylcyclopropane. The results are shown in Table 4. It should be noticed that the quantum yields decrease at

Table 4 *Effect of exciting wavelength on the photolysis of* trans-1-*phenyl-but-2-ene* [8]

Wavelength (nm)	$\Phi_{trans \to cis}$	$\Phi_{cyclopropane}$
247	0·09	0·088
252	0·27	0·24
261	0·57	0·46
266	0·66	0·50

the shorter wavelengths, and it is suggested that this is due to the isomerisation of the phenyl moiety at the expense of the isomerisation of the alkenyl group. The introduction of an inert gas (n-butane) reduces the efficiency of the cyclopropane formation and increases that of the *trans–cis*-isomerisation process.

The rearrangement to the cyclopropane does not seem to occur in solution. This observation, coupled with the quenching effect of n-butane, suggests that vibrationally 'hot' molecules may be involved in formation of the cyclopropane, and that the balance between this process and isomerisation of the phenyl moiety may be very sensitive to such energy factors.

Further examples of intramolecular energy transfer have been reported for compounds in the series (7a—d).[9] When $n = 1$ (7a) there is strong evidence for interaction between the two chromophores, and the $n \to \pi^*$ transition is intensified by a factor of 6. There is, however, very little

$$Ph\overset{O}{\underset{}{\|}}(CH_2)_n\diagdown Ph$$

(a) $n = 1$
(b) $n = 2$
(c) $n = 3$
(d) $n = 4$

(7)

interaction in compound (7b), and none at all in (7c) and (7d). Selective excitation of the carbonyl chromophore in these compounds brought about efficient isomerisation of the olefinic moiety (Table 5). Quenching experiments established that intermolecular energy transfer cannot compete with the intramolecular process, and led to the rate values shown in Table 5. The involvement of a Schenck mechanism for the isomerisation was discounted, and it is proposed, from a study of the influence of change of solvent, that the energy transfer mechanism is the same as for bimolecular quenching and requires a collision between the donor and the acceptor. The quantum yield (Φ_c) for the photochemical conversion of *cis*-stilbenes

---

[8] M. Comtet, *J. Amer. Chem. Soc.*, 1969, **91**, 7761.
[9] D. O. Cowan and A. A. Baum, *J. Amer. Chem. Soc.*, 1970, **92**, 2153.

bearing *meta*-substituents is reported to decrease with increasing heavy-atom effect (Table 6).[10] At the same time, the mole fraction (x_t) of the *trans*-isomer in the photostationary state is also diminished. The Table also illustrates the fact that Φ_t, the quantum yield for *trans–cis*-isomerisa-

Table 5 *Olefinic isomerisation data and rate constants for energy transfer* [9]

Compound	$\Phi_{trans \to cis}$	k_q/k_{et}	$\Phi_{et} \times \Phi_{ISC}$ [a]	k_{et} (s^{-1})
$n = 1$	0·53	—	1·01	—
$n = 2$	0·52	0·069	0·99	7.2×10^{10}
$n = 3$	0·52	0·55	0·99	1.0×10^{10}
$n = 4$	0·51	1·5	0·98	3.3×10^{9}

[a] Φ_{et} is quantum yield of energy transfer. Φ_{ISC} is quantum yield of intersystem crossing.

tion, is practically unaffected by the presence of heavy atoms. The heavy-atom effect resulting from the use of methyl iodide as a solvent also enhances intersystem crossing in stilbene following direct irradiation at 436 nm.[11] Both *trans* → *cis* and *cis* → *trans* isomerisation were observed in this study, and the photostationary state contained 70—80% of *cis*-stilbene. The quantum yields for the two processes were estimated as 0·75 ± 0·25 and 0·95 ± 0·25 respectively. The uncertainty of the values of the quantum yields is due mainly to the absence of accurate values for the extinction coefficients of the two isomers in methyl iodide.

Table 6 *Effect of* m-*substitution on stilbene cis–trans-isomerisation* [10]

Stilbene	Φ_{cis}	Φ_{trans}	x_t
Stilbene	0·35	0·50	0·07
3-Chlorostilbene	0·54	0·44	0·11
3-Bromostilbene	0·18	0·46	0·06
3-Iodostilbene	0·13	0·40	0·05
3,3-Dichlorostilbene	0·25	0·51	0·06
3,3-Dibromostilbene	0·05	0·53	0

The photoisomerisation of *cis*-styrylferrocene (8) in benzene solution by direct or sensitized irradiation has been described.[12] In both cases, a triplet state is thought to be the reactive intermediate with, in the direct irradiation experiments, assistance from the heavy atom (iron) in intersystem crossing.

(8)

(9)

(a) X = $\overset{+}{\text{N}}$Me I$^-$, Y = H, Φ = 0·41
(b) Y = $\overset{+}{\text{N}}$Me I$^-$, X = H, Φ = 0·04
(c) X = $\overset{+}{\text{N}}$Me I$^-$, Y = H

---

[10] K. Krüger and E. Lippert, *Z. phys. Chem. (Frankfurt)*, 1969, **66**, 293.
[11] G. Fisher, K. A. Muszkat, and E. Fischer, *Israel J. Chem.*, 1969, **6**, 965.
[12] J. H. Richards and N. Pisker-Trifunac, *J. Paint Technol.*, 1969, **41**, 363.

The product from irradiation of the *cis*-isomer was solely the *trans*-isomer. The quantum yields for the processes are shown in Table 7. *cis–trans*-Isomerisation does not occur in the photolysis of *trans*-1,2-bis(4-pyridyl)-ethylene and *trans*-1,2-bis(3-pyridyl)ethylene as their bis-methiodide salts

Table 7 *Quantum yield values for the photoisomerisation of* cis-*styrylferrocene* [12]

Exciting wavelength (nm)	% conversion cis *to* trans	$\Phi \times 10^3$
313 direct irrad.	12	5·4
366 sensitized (Ph$_2$CO)	12	5·0
366 direct irrad.	15	1·3

in the presence of nucleophiles; efficient conversion into ethers or alcohols (9) takes place instead.[13] Charge-transfer mechanisms involving the transfer of an electron from the iodide ion to the pyridinium salt were rejected since the dichloride salt also behaves equally efficiently in its photohydration, and chloride ion is a weaker donor than iodide. But this argument does not seem strong enough to justify complete exclusion of a charge-transfer component. The authors prefer a singlet mechanism since the reaction could not be quenched, and fluorescence of the pyridylethylenes was quenched by the addition of water to acetonitrile solutions.[13] A study of the quenching of singlet excited states of 1,2-bis(4-pyridyl)ethylene, methylacridinium fluoroborate, and riboflavin by nucleophiles has been reported.[14]

Rearrangement Reactions.—The gas-phase mercury-sensitized decomposition of *cis*-3,4-dimethylcyclobutene has been shown to be a complex reaction giving rise to hexa-1,3-diene, *trans,trans*-hexa-2,4-diene, and *cis,trans*-hexa-2,4-diene.[15] The reaction is pressure dependent, and at higher pressures the *cis,trans*-isomer is favoured. This product arises from the excited ground state of the cyclobutene, and is the consequence of a conrotatory ring-opening. The presence of inert gas (ether) also brings about a preferential formation of the *cis,trans*-hexa-2,4-diene. The *trans,trans*-hexa-2,4-diene must arise from an excited state since it is the result of a disrotatory ring-opening and the triplet state is favoured. The assignment of the states from which these two products are formed is based on the Woodward–Hoffmann Rules for the conservation of orbital symmetry. The formation of the hexa-1,3-diene is thought to take place in the initially produced vibrationally-excited triplet state. Possible complications resulting from the isomerisation of the *trans,trans*-diene during the photolysis to the *cis,trans*-isomer were discounted. The more complex cyclobutenes (10) and (11) undergo stereospecific disrotatory ring-opening in solution to yield

[13] M. T. McCall and D. G. Whitten, *J. Amer. Chem. Soc.*, 1969, **91**, 5681.
[14] D. C. Whitten, J. W. Happ, G. L. B. Carlson, and M. T. McCall, *J. Amer. Chem. Soc.*, 1970, **92**, 3499.
[15] R. Srinivasan, *J. Amer. Chem. Soc.*, 1969, **91**, 7557.

(10)   (11)   (12)

(13)   (14)

cis- and *trans*-cyclododecenynes (12) and (13), respectively.[16] In the former, example, an additional product, *cis,cis*-1,1-bicyclohexenyl was isolated from direct photolysis, but in the presence of triphenylene only the *cis*-cyclododecenyne (12) was obtained. Direct irradiation of the *trans*-cyclobutene (11) did not give the bicyclohexenyl (14).[16] The authors attributed this non-event to the high strain energy expected for the *trans*-cyclohexene ring in such a product. The reactions described in this work are examples of $2\pi-2\sigma$ concerted rearrangements. If the products are formed in electronic excited states (as application of the Woodward–Hoffmann Rules to singlet processes would suggest), the fact that no isomerisation of the olefinic moiety is observed suggests that the excitation might be associated with the acetylenic system. On the other hand, the authors propose that the products are more likely to be produced in their electronic ground state from the S_1 cyclobutenes.[16] This sort of problem is tied up with a growing realisation among photochemists that potential energy surfaces for even apparently simple photochemical processes may be in fact more complex than is implied by the more unsophisticated orbital symmetry considerations.

A review dealing with photochromic substances has been published.[17] Also of interest in this area is the report of photochromism in thin films of microcrystalline 1,3,3-trimethylspiro(indoline-2,2′-benzopyrans).[18]

Table 8 *Absorption characteristics of the spiropyrans* (15) [19]

Spiropyran (15)	Absorption maxima (nm)				
(a)	324(sh),	312·5(sh)	296	265	243·5
(b)	324,	310	305		
(c)	372·5(sh),	340	300	269	
(d)		370(sh)	310	268	254·5
(e)		370[a]			
Absorbing chromophore:	chromene		indoline	chromene	indoline

[a] Mixture of both indoline and chromene absorptions.

[16] J. Saltiel and L. N. Lim, *J. Amer. Chem. Soc.*, 1969, **91**, 5405.
[17] G. Wettermark, *Kem. Tidskr.*, 1970, **82**, 48.
[18] H. Kobayashi, I. Shimizu, M. Nakazawa, H. Kokado, and E. Inoue, *Bull. Chem. Soc. Japan*, 1969, **42**, 2735.
[19] N. W. Tyer and R. S. Becker, *J. Amer. Chem. Soc.*, 1970, **92**, 1289.

A spectroscopic study of several photochromic spiropyrans (15) has shown that the chromophores are orthogonal and that the absorption transitions are mostly localised on particular parts of the molecule (Table 8). Substituent and temperature effects showed that the indoline absorptions, which show a marked red shift, are $\pi \to \pi^*$ in nature. The chromene absorptions in (15a) are assigned as $\pi \to \pi^*$ transitions, but the introduction

(a) $R^1 = R^2 = H$
(b) $R^1 = Cl, R^2 = H$
(c) $R^1 = H, R^2 = NO_2$
(d) $R^1 = Cl, R^2 = NO_2$
(e) $R^1 = R^2 = NO_2$

(15)

(a) $R = CHO$
(b) $R = H$

(16)

of the 6'-nitro-substituent introduces a shoulder at 370 nm which appears to be an $n \to \pi^*$ absorption localised on the chromene. These observations are backed up by emission studies.[20] The simpler pyran (16a) also underwent photolysis.[21] Light absorption in this case is by the $n\pi^*$ state of the aldehydic function, and leads to products from decarbonylation and ring-contraction. Dihydropyran (16b) gave carbon monoxide, ethylene, cyclobutane, acrolein, and also the ring-contracted product cyclobutane carboxaldehyde following mercury-sensitized photodecomposition in the gas-phase. This last product presumably arises by O—C bond rupture, followed by rebonding (Equation 1).

(16b) $\xrightarrow{\text{Hg}\\ h\nu}$ [structure] $\longleftrightarrow$ [structure] $\longrightarrow$ [structure] CHO (1)

Interest has continued to be shown in photochemically induced isomerisations of octalin.[22] The reactions involve the intermediacy of carbonium ions, and these entities have also been evoked to explain the photochemical fragmentation in benzene–ether of the steroidal allylic alcohol (17).[23] This forms the aldehydic olefin (17a), which was not isolated, and the related oxetan (17b) (Scheme 2).

Addition Reactions: Alkenes and Alkynes.—Different results from the above were encountered for the photolysis of the steroidal alcohol (17) by changing the medium to a mixture of t-butyl alcohol, o-xylene (as sensitizer), and water.[24] The oxetan (17b) is common to both accounts, but in the latter [24] the diol (18) was also reported. This is good evidence for the existence of the cation (i, Scheme 2). Cholesterol forms the same photoproducts on photolysis under the same conditions.

---

[20] N. W. Tyer and R. S. Becker, *J. Amer. Chem. Soc.*, 1970, **92**, 1295.
[21] R. Srinivasan, *J. Org. Chem.*, 1970, **35**, 786.
[22] A. R. Hochstetler, *Diss. Abs.*, 1969, **29**, B, 3678.
[23] D. Guénard and R. Beugelmans, *Tetrahedron Letters*, 1970, 1705.
[24] J. A. Waters and B. Witkop, *J. Org. Chem.*, 1969, **34**, 3774.

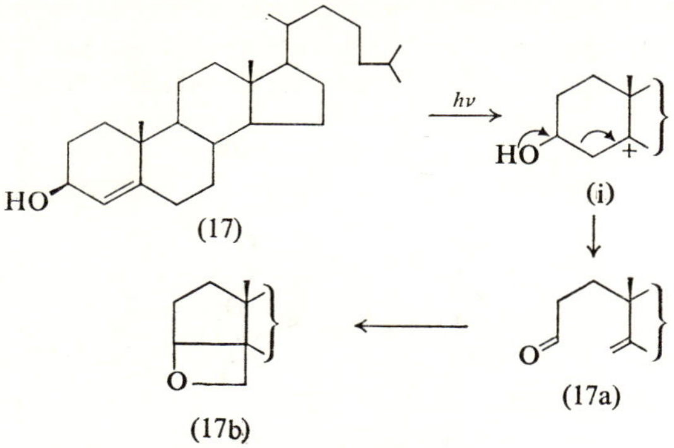

Scheme 2

The direct irradiation of cholest-4-ene and cholest-5-ene in methanol–cyclohexane gave products from addition of methanol (19a and b) as well as the reduced hydrocarbon cholestane.[25] When benzene was used as the photosensitizer, a sole product of methanol addition (19a) was obtained.

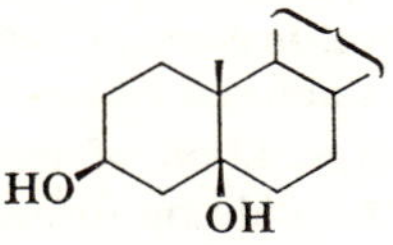

The authors propose that the non-stereospecific addition of methanol observed from the direct irradiations is the result of direct excitation of the end absorption of the double bond.[25] The use of a Vycor filter (substantially opaque to wavelengths shorter than 200 nm) did not yield products of addition. It is suggested that a $\pi \to \sigma^*$ absorption plays a part in the excitation and may induce a $\pi \to \pi^*$ transition.[25]

The addition of alcohols (methanol and isopropanol) to the double bond of isopropylidene phthalide (20) has been reported.[26] The products from the reaction were the adducts (21a) and (21b), as well as a rearrangement

---

[25] H. Compaignon de-Marcheville and R. Beugelmans, *Tetrahedron Letters*, 1969, 1901.
[26] S. F. Nelson and P. J. Hintz, *J. Amer. Chem. Soc.*, 1969, **91**, 6190.

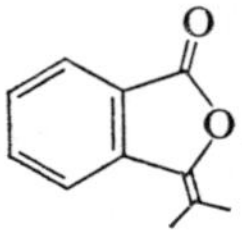

(20)

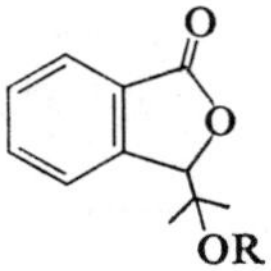

(21) (a) R = Me
(b) R = Pr^i

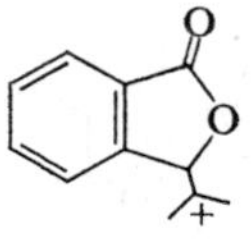

(22)

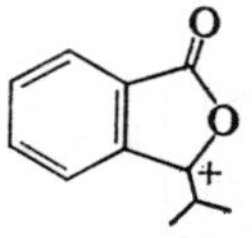

(22a)

product and a small amount of dimeric material. No evidence for the nature of the excited state was put forward, although it was suggested that protonation of the carbonyl function in the excited state might influence the polarity of the double bond. Another alternative not suggested by the authors could simply involve intramolecular energy transfer followed by protonation (resulting from enhanced nucleophilicity of the double bond) and addition of alcohol to the carbonium ion (22). On the other hand, one might reason that protonation of the excited olefin would be more likely to afford the more stabilised cation (22a).

Earlier studies [27] of the addition of tertiary amines to $\alpha\beta$-unsaturated ketones had shown that the addition forms a new C—C bond between the β-carbon of the unsaturated system and the α-carbon of the amine. In a new report from the same school, conditions are described for the related

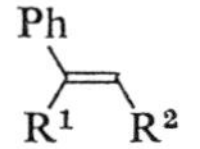

(a) $R^1 = R^2 = H$
(b) $R^1-R^2 = C_4H_8$
(c) $R^1-R^2 = C_3H_6$

(23)

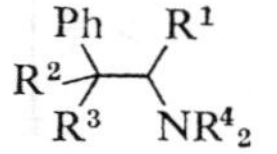

(a) $R^1 = R^2 = Me$, $R^4 = Et$, $R^3 = H$
(b) $R^1 = H$, $R^4 = Me$, $R^2-R^3 = C_5H_{10}$
(c) $R^1 = Me$, $R^4 = Et$, $R^2-R^3 = C_5H_{10}$
(d) $R^1 = Me$, $R^4 = Et$, $R^2-R^3 = C_4H_8$

(24)

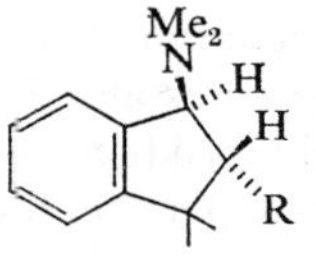

(24e) R = H or D

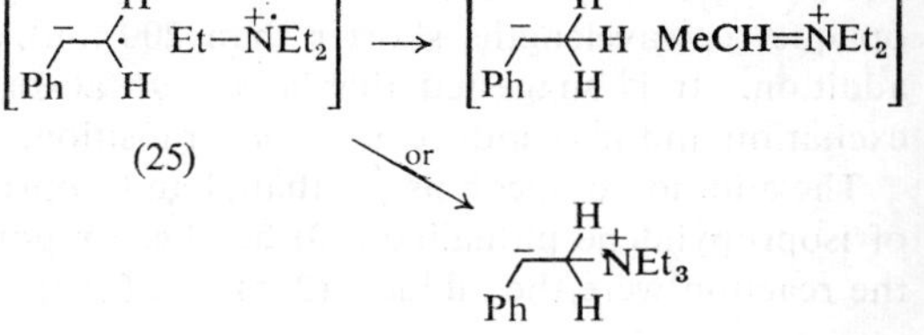

(25)

(25a)

[27] R. C. Cookson, J. Hudec, and N. A. Mizra, *Chem. Comm.*, 1968, 180.

addition of tertiary amines (triethyl- and trimethyl-amine) to styrenes (23) to form adducts (24).[28] In cases where it was possible to observe *cis* and *trans* addition, n.m.r. analysis showed that the process is stereospecifically *cis*, and produces (24e). Although no mechanistic details of the reaction were reported, it is possible by comparison with investigations of related charge-transfer systems to suggest that the essential step in the addition is electron transfer from the amine to the styrene to form the complex (25), and/or a corresponding zwitterion (25a). The observed stereospecificity suggests a degree of concertedness between hydrogen (proton ?) transfer and formation of the new C—C bond. The zwitterion (25a) could, in principle, be transformed into the isolated product *via* intramolecular proton shift and Stevens-type rearrangement of the resulting ylide.

The acid-catalysed photochemical addition of alcohols to olefins has been previously demonstrated to proceed in an entirely predictable manner based on the stability of the more stable carbonium ion possible from the protonation of a given olefin.[29] One new result has thrown doubt on this general thesis.[30] The photochemical addition of methanol to the furocoumarone (26) gave two 'normal' adducts (27a) and (27b) in 24 and 30% yields respectively, together with 38% of the positionally different adduct (27c).

(a) $R^1 = R^3 = H, R^2 = MeO$
(b) $R^1 = MeO, R^2 = R^3 = H$
(c) $R^1 = R^2 = H, R^3 = MeO$

No mechanistic study of the reaction was reported. However, it was proposed that the use of a Pyrex filter and the absence of a high-energy sensitizer implies that the reaction must proceed by enone excitation (singlet or triplet) and intramolecular energy transfer. It is pointed out that

[28] R. C. Cookson, S. M. de B. Costa, and J. Hudec, *Chem. Comm.*, 1969, 753.
[29] 'Photochemistry,' ed. D Bryce-Smith, Specialist Periodical Reports, The Chemical Society, London, 1970, vol. 1, p. 260, for lists.
[30] A. C. Waiss and M. Wiley, *Chem. Comm.*, 1969, 512.

although additions by this ionic mechanism to cyclopentenes were thought to be impossible, the influence of the dihydrofuran ring oxygen is an unknown factor, and indeed a completely different mechanism may be operative in this system.[31]

The photolysis of an $\alpha\beta$-unsaturated oxime (28) in benzene solution through a quartz filter has led to the synthesis of a hetero-ring E (29) by the addition of the oxime hydroxy-group to the $\alpha\beta$-double bond.[32] The stereochemistry of the new ring-junction has not been fully elucidated. Irradiation of the oxime in tetrahydrofuran brings about isomerisation of the oxime.

The photochemical addition of protic solvents to double bonds (of which the above is an intramolecular example) has continued to be a subject of great interest. The xylene-photosensitized addition of methanol and several carboxylic acids to 2-cyclohexenyl acetate gives the adducts (30).[33] The presence of methyl groups, known to stabilise ground-state carbonium ions, appeared to have an adverse effect in these reactions: thus 1,5,5-trimethyl-3-acetoxycyclohexene gave only a 30% yield of the diacetate (31) accompanied

(a) R = CH$_3$CO,   70%
(b) R = CH$_3$CH$_2$CO,   53%
(c) R = Pr^{n}CO,   61%
(d) R = Me,   51%
(30)

(31)

(a) $n = 1$, R = Me   54%
(b) $n = 1$, R = Ac   31%
(c) $n = 2$, R = Me   64%
(d) $n = 2$, R = Ac   41%
(32)

by 1-methylene-3,3-dimethyl-5-acetoxycyclohexane. The presence of the phenyl group in cyclic styrylolefins no doubt enhances the stability of a cation adjacent to the phenyl group during certain photochemical additions. Thus, 1-phenylcyclohexene and 1-phenylcycloheptene undergo photosensitized [34] or non-sensitized [36] additions of methanol to give (32a) and (32c) and acetic acid to yield (32b) and (32d).[35] Together with the acetoxy-derivatives, products of reduction and methylation were formed in the reactions with acetic acid, presumably as a result of concurrent radical reactions.

The addition reactions could involve the *cis–trans*-isomerisation of the olefin to produce a highly reactive olefin which would add methanol thermally. Such high-energy intermediates would be difficult or impossible

[31] P. J. Kropp, *J. Amer. Chem. Soc.*, 1967, **89**, 3650; P. J. Kropp and H. J. Krauss, *ibid.*, p. 5199.
[32] R. P. Ghandi and V. K. Chadha, *Indian J. Chem.*, 1969, **7**, 633.
[33] T. Okada, K. Shibata, M. Kawanisi, and H. Nozaki, *Tetrahedron Letters*, 1970, 859.
[34] M. Tada and H. Shinozaki, *Bull. Chem. Soc. Japan*, 1970, **43**, 1270.
[35] S. Fujita, T. Nomi, and H. Nozaki, *Tetrahedron Letters*, 1969, 3557.
[36] P. J. Kropp, *J. Amer. Chem. Soc.*, 1969, **91**, 5783.

to form from cyclic olefins of small ring-size, and in the photolysis of 1-phenylcyclopentene no products of addition were indeed isolated.[34] A detailed study of the effect of ring-size has been reported by Kropp.[36]

In the case of the unsubstituted cycloalkenes (33, $n = 2$ or 3), dimerisation is the predominant reaction on photolysis in xylene–methanol. The

$(CH_2)_n$

(a) $n = 2$
(b) $n = 3$
(c) $n = 4$

(33)

H OR
$(CH_2)_n$

(a) $n = 2$, R = Me
(b) $n = 3$, R = Me
(c) $n = 4$, R = Ac
(d) $n = 4$, R = CHO
(e) $n = 4$, R = Ph
(f) $n = 4$, R = Me

(34)

addition of 1% sulphuric acid changes the course of the reactions, and ethers (34) are isolated at the expense of the dimeric products (Table 9). At a larger ring-size (33, $n = 4$) decomposition occurs but no clearly defined products have been obtained.

Table 9 *Product distribution from photolysis of cycloalkenes* [36]

Olefin (33)	Solvent	Sensitizer	Time (hr)	Product (%) (33)	(34)	Dimers
(a)	MeOH	xylene	8	(a) 17		72
	MeOH, H⁺ (1%)	xylene	4		(a) 62	
	MeOH, H⁺ (1%)	toluene	4	(a) 12	(a) 61	
	Xylene	xylene	8	(a) 3		88
(b)	MeOH	xylene	8	(b) 65		20
	MeOH, H⁺ (1%)	xylene	4	(b) 13	(b) 51	
	Xylene	xylene	8	(b) 33		
(c)	MeOH	*m*-xylene	8	(c) 50		
	MeOH, H⁺ (1%)	*m*-xylene	4	(c) 59		

Norbornene is substantially more reactive than cyclopentene in these reactions, possibly since it is more strained, and gave the products shown in Scheme 3, together with dimers. No evidence for the formation of norbornyl methyl ethers was obtained. The formation of norbornane also involves transfer of hydrogen from solvent, but does not involve the transfer of the hydroxyl hydrogen since irradiations in MeOD led to only 6% inclusion of deuterium in the product.

Contrary to Kropp's observations,[36] protic solvents can be added photochemically to *cis*-cyclo-octene under sensitization by benzene.[37] Thus, the acetate (34c, 18%) and formate (34d, 11%) were prepared by the photolysis

[37] H. Kato and M. Kawanisi, *Tetrahedron Letters*, 1970, 865.

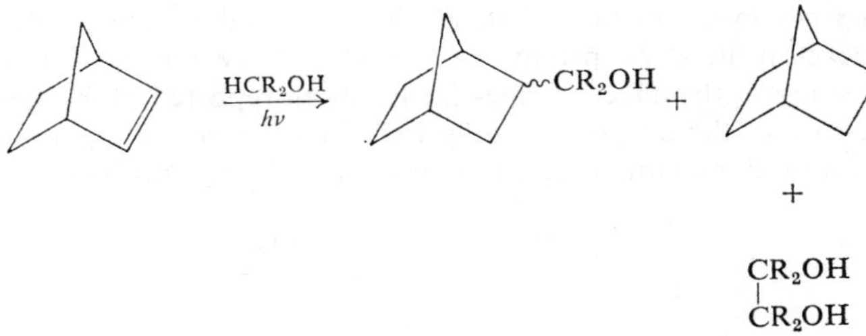

Scheme 3

of the olefin in the presence of the acid and benzene. Ethers (34e, 12%) and (34f, 10%) were also prepared. A slight difference was noticed in the synthesis of these ethers in that phenol addition requires no acid to effect the addition, while methanol required the addition of a few drops of sulphuric acid. This presumably reflects the increased acidity of phenol. It should not be necessary to sensitize this particular reaction, since phenols have been reported to add to olefinic double bonds on direct irradiation.[38]

The intramolecular addition of the carboxylate group of the acid (35) to form the bicyclic lactone (36) has also been reported.[37]

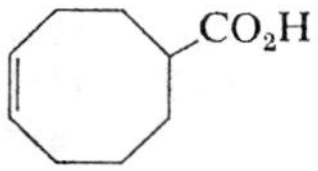

Benzene-sensitized photolysis of the bicyclic olefin (37a) in the presence of acetic acid (1%) gives rise to the addition products (38a) and (39) in 22 and 4% yields, respectively.[39] The reaction was not as simple as it first appeared, and photolysis of the phenyl derivative (37b) gave adducts (40) and (38b) in 34 and 28% yields, respectively. This latter product must arise by a transannular hydride migration to the carbonium ion formed on protonation of the excited olefin. It is presumed that the same hydride migration takes place in the formation of the unsubstituted acetate (39).

Intramolecular addition of alcohols is also of interest, and has been studied by Kropp and Krauss.[40] Thus, the photolysis of the *endo*-alcohols (41) and (42) gave rise to the cyclic ethers (43) and (44), respectively.

The alcohol (45) undergoes a different reaction to give products of fragmentation [(46) and acetaldehyde], presumably *via* the intermediate

---

[38] W. M. Horspool and P. L. Pauson, *Chem. Comm.*, 1967, 195.
[39] M. Kawanisi and H. Kato, *Tetrahedron Letters*, 1970, 721.
[40] P. J. Kropp and H. J. Krauss, *J. Amer. Chem. Soc.*, 1969, **91**, 7466.

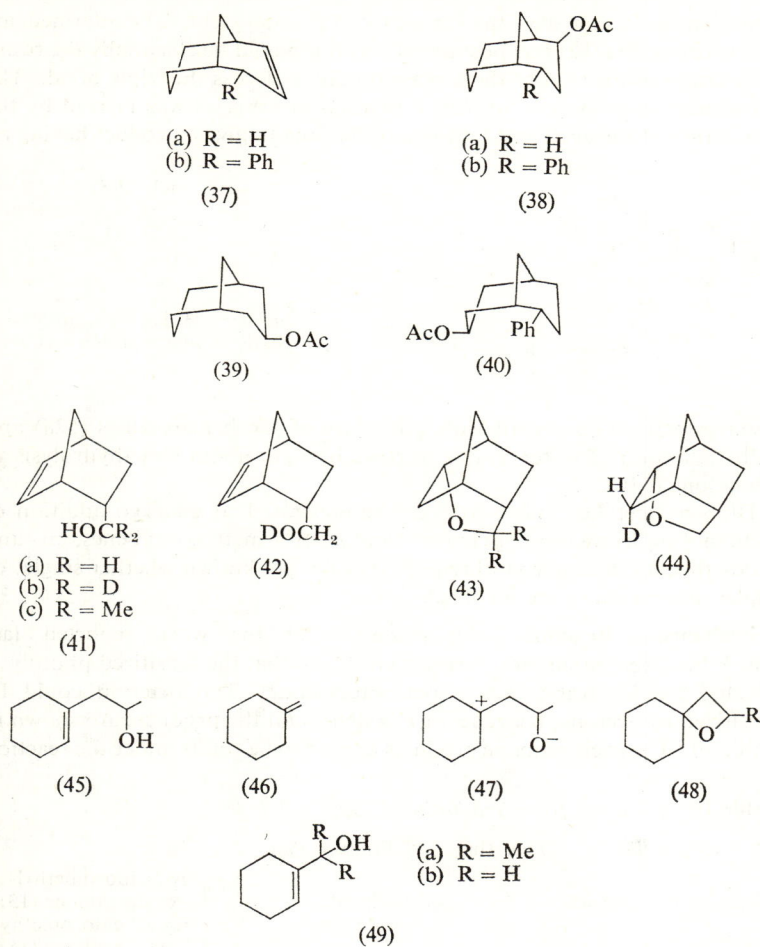

(47). The involvement of the oxetan (48) was discounted on the grounds that no spectral evidence could be obtained for its existence.

The photosensitized reaction of the cyclohexenylcarbinol (49) with water has been studied.[41]

A review article on the photochemistry of additions to multiple-bonded systems has been published.[42]

The participation of an *ortho*-acetamido-group in the hydration of a diarylacetylene has been described.[43] Thus, the acetylene (50a) on irradia-

---

[41] J. A. Marshall and A. E. Greene, *Tetrahedron*, 1969, **25**, 4183.
[42] D. Elad, *Org. Photochem.*, 1969, **2**, 168.
[43] T. D. Roberts, H. Schechter, and L. Ardemagni, *J. Amer. Chem. Soc.*, 1969, **91**, 6185.

tion in moist hexane gave the ketone (51a) in good yield. The intermediacy of a benzoxazine (52) was postulated and this would be essentially the result of addition of the enol of the acetamido-group across the triple bond. The subsequent hydrolysis is, in fact, a thermal process, as was proved by the irradiation of the acetylene (50b) in dry hexane to give a product having an

(50)
(a) Ar = Ph
(b) Ar = *m*-anisyl

(51)

(52)
(a) R^1 = H, R^2 = *m*-anisyl
(b) R^1 = *m*-anisyl, R^2 = H

n.m.r. spectrum consistent with a mixture of the benzoxazines (52a) and (52b): addition of water to this mixture brought about slow hydrolysis to the ketone (51b).

Diphenylacetylene has also been demonstrated to undergo addition of methanol on irradiation, giving *cis*- and *trans*-α-methoxystilbenes, *cis*- and *trans*-stilbenes, and phenanthrene.[43] It is not yet known whether singlet or triplet intermediates are involved.

Miscellaneous Reactions.—An extension of the work reported last year[44] has been published. It has been shown that the sensitized photolysis of allyl halides brings about rearrangement.[45] The reaction could be sensitized by acetone, benzene, or *m*-xylene, and the products are shown in Table 10. The cyclopropane products appear to be stable under the reaction

Table 10 *Products from irradiation of allyl halides* [45]

Allyl halide	Time (hr)	Products (%)	
γ-Methylallyl chloride	56	α-methylallyl chloride (16)	*cis*-2-chloromethyl-cyclopropane (13) *trans*-2-chloromethyl-cyclopropane (15)
Allyl chloride	24	chlorocyclopropane (19)	
Allyl bromide	16	bromocyclopropane (11)	

conditions. The authors[45] suggest that the mechanism for the reaction involves triplet diradical intermediates (Scheme 4) with 1,2- and 1,3-chlorine migrations as essential steps. Alternative processes cannot be excluded, however, and the involvement of carbonium ions has not been rigorously ruled out. An independent study[46] of the photo-rearrangement of com-

---

[44] 'Photochemistry,' ed. D. Bryce Smith, Specialist Periodical Reports, The Chemical Society, London, 1970, vol. 1, p. 242.
[45] S. J. Cristol and G. A. Lee, *J. Amer. Chem. Soc.*, 1969, **91**, 7554.
[46] B. B. Jarvis and R. O. Fitch, *Chem. Comm.*, 1970, 408.

pound (53) in bromotrichloromethane gave the bromo-derivative (54b), conceivably *via* the allyl radical (55). Photolysis of the compound (53) in acetone afforded the isomer (54a).[47] Cyclopropanes (56) have also been

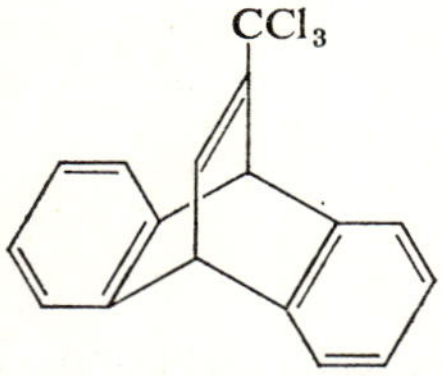

(53)

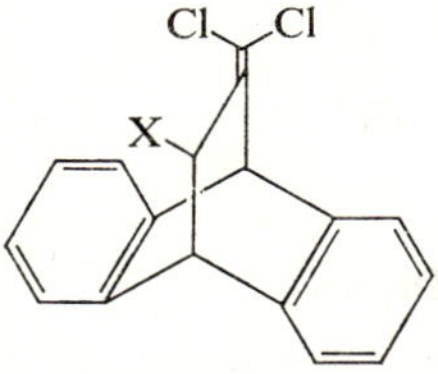

(54) (a) X = Cl
(b) X = Br

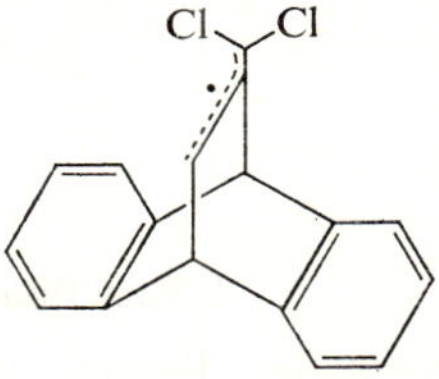

(55)

obtained [48] from the photolysis of certain allyl ethers (57) by the migration of a phenyl group, presumably *via* a radical mechanism.

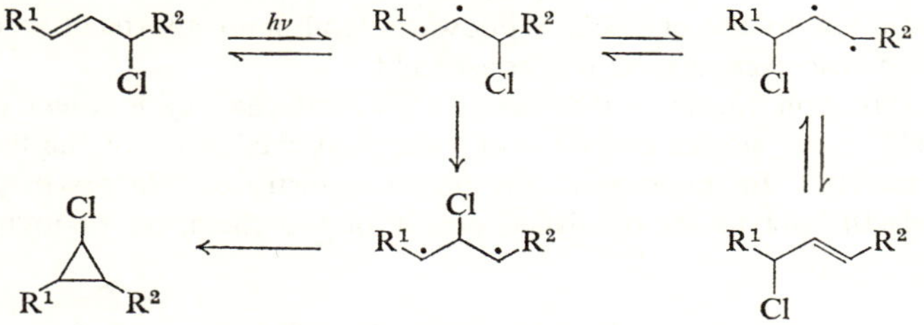

Scheme 4

Further photochemical reactions of some tetracyclanes have been reported.[49] Methyl sterculate has been synthesised by the addition of methylene to methyl stearolate.[50] The photolysis of vitamin K_1 (58) in oxygen-free propan-2-ol afforded a variety of products, details of which will be found in the original paper.[51]

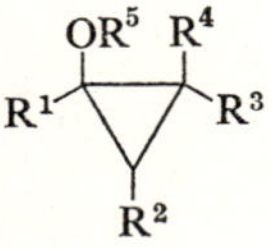

(56)
(a) $R^1 = R^2 = R^3 = R^4 = Ph$, $R^5 = Me$
(b) $R^1 = R^2 = R^3 = Ph$, $R^4 = H$, $R^5 = Me$
(c) $R^1 = R^2 = R^3 = Ph$, $R^4 = H$, $R^5 = CH_2Ph$
(d) $R^1 = H$, $R^2 = R^3 = R^4 = Ph$, $R^5 = Me$

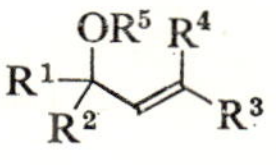

(57)

[47] B. B. Jarvis, *J. Org. Chem.*, 1969, **33**, 4075.
[48] J. J. Brophy and G. W. Griffin, *Tetrahedron Letters*, 1970, 493.
[49] G. Kaupp and H. Prinzbach, *Annalen*, 1969, **725**, 52.
[50] M. M. Schlosser, A. J. Longo, J. W. Berry, and R. J. Deutschman, *J. Amer. Oil Chemists' Soc.*, 1969, **46**, 171.
[51] M. Ohmae and G. Katsui, *Bitamin*, 1970, **41**, 178.

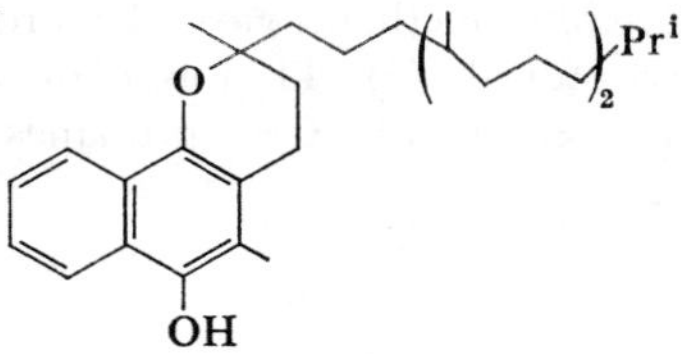

(58)

2 Reactions of Dienes

Conjugated Dienes.—A full account of the theoretical conclusions of Van der Lugt and Oosterhoff on the butadiene–cyclobutene transformation has now appeared,[52] following a previous preliminary account.[53] The approach allowed the establishment of potential energy surfaces from which reaction paths could be predicted. The improbability of the thermal disrotatory cyclisation is due to the need for a large activation energy [418 kJ mol^{-1} (100 kcal mol^{-1})]. The photochemical disrotatory cyclisation is considered to derive from a favourable symmetric excited state, in direct contradiction to the classical Woodward–Hoffmann approach where an antisymmetric excited state was considered.

The direct photolysis of β-farnesene (59) produces a cyclobutene derivative (60) by the singlet excited state concerted cyclisation of the terminal diene moiety.[54] In this respect, the photochemistry of β-farnesene closely parallels the previously published details of the chemistry of myrcene.[55]

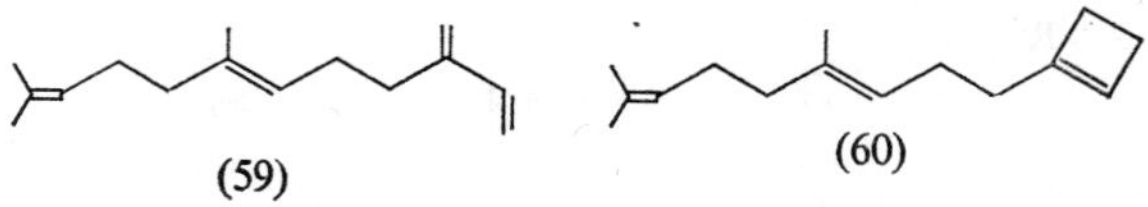

(59)   (60)

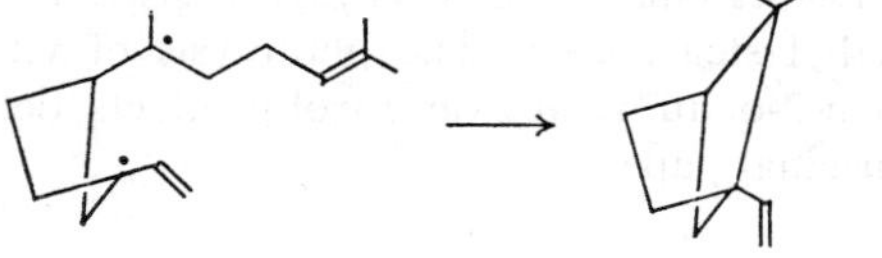

(61)

(a) R^1 = Me, R^2 = (CH$_2$CH$_2$CH=CMe$_2$)
(b) R^1 = (CH$_2$CH$_2$CH=CMe$_2$), R^2 = Me

Two other products (61a) and (61b) which were obtained in small amounts have been shown to be the major products arising from the naphthalene-

---

[52] W. T. A. M. Van der Lugt, and L. J. Oosterhoff, *J. Amer. Chem. Soc.*, 1969, **91**, 6042.
[53] W. T. A. M. Van der Lugt and L. J. Oosterhoff, *Chem. Comm.*, 1968, 1235.
[54] J. D. White and D. N. Gupta, *Tetrahedron*, 1969, **25**, 3331.
[55] K. J. Crowley, *Proc. Chem. Soc.*, 1962, 334; *Tetrahedron*, 1965, **21**, 1001.

sensitized (triplet) process, and White and Gupta[54] propose that this is evidence for the inefficiency of intersystem crossing in dienes. That the bicyclohydrocarbons (61a) and (61b) are formed from the triplet state of the molecule is evidenced by the fact that they are stereoisomeric, and arise by initial bond formation to produce an intermediate which can undergo bond rotation to produce the isomeric compounds. Benzophenone was also effective as a sensitizer, although in this case it was partially consumed, presumably by oxetan formation (although this was not rigorously verified). The mechanism suggested for the reaction involves a complex between the excited diene and an ethylenic group in the ground-state diene, which collapses to the more stable diradical.

Caldwell has studied the influence of added isoprene on the benzophenone-sensitized isomerisation of stilbene.[56] It had been observed that the photostationary state, predicted to be 59% *cis*-stilbene, actually contained 79·6% of the *cis*-isomer in 2—3M isoprene and 83·7% in neat isoprene. The influence of the isoprene was determined quantitatively, and the results are shown in Table 11. The only reasonable explanation is

Table 11 *Influence of isoprene concentration on sensitized cis–trans-isomerisation of stilbene* [56]

trans-*Stilbene* (mol l^{-1})	*Isoprene* (mol l^{-1})	$\Phi_{t \to c}$ (*observed*)	Φ (*predicted*)
0·056	0	0·55	0·56
0·045	0·108	0·231	0·173
0·045	0·52	0·106	0·0475
0·045	1·05	0·084	0·0242
0·045	5·2	0·0454	0·0050
0·022	3·0	0·0258	0·0043

that the isoprene triplets must be capable of transferring energy to the stilbene. The magnitude of the effect is such that 12% of the isoprene triplets transfer energy to *trans*-stilbene and about 3% transfer energy to *cis*-stilbene, owing to the difference of the triplet energies of the *trans*- and the *cis*-stilbenes. *trans*-Stilbene appears to quench isoprene triplets at a rate two orders of magnitude less than diffusion-controlled rates. This is explained by the assumption that isoprene triplets are deactivated by a non-vertical transition.

Two investigations of the photolysis of the hexa-2,4-dienes have been reported.[57, 58] Originally, Hammond et al.[57] proposed that a chain-transfer mechanism was operative for the observed isomerisation and they measured $\Sigma\Phi$ for high concentrations of diene as $\geqslant 2$. Saltiel, Metts, and Wrighton[58] have now re-examined the process, and have found the initial report[57] to be in error. They found no change in $\Sigma\Phi$ for a ten-fold increase in diene

---

[56] R. A. Caldwell, *J. Amer. Chem. Soc.*, 1970, **92**, 3229.
[57] H. L. Hyndman, B. M. Monroe, and G. S. Hammond, *J. Amer. Chem. Soc.*, 1969, **91**, 2852.
[58] J. Saltiel, L. Metts, and M. Wrighton, *J. Amer. Chem. Soc.*, 1969, **91**, 5684.

Table 12 *Total quantum yields and composition of photostationary states for photosensitized isomerisation of hexa-2,4-diene isomers* [58]

Diene (mol l^{-1})	$\Phi_{tt \to ct}$	$\Phi_{cc \to ct}$	$\Phi_{ct \to tt}$	$\Phi_{cc \to tt}$	$\Phi_{ct \to cc}$	$\Phi_{tt \to cc}$	$\Sigma\Phi$
0.09	0.45	0.50	0.31	0.28	0.15	0.16	1.85
	0.48	0.51	—	0.30	—	0.20	—
0.8	0.52	0.48	0.36	0.29	0.16	0.19	2.00
	0.48	0.50	0.33	0.29	0.16	0.18	1.94

Diene (mol l^{-1})	% tt	% ct	% cc
tt 0.09	31.3	50.2	18.5
cc 0.09	31.3	50.0	18.7

concentration (Table 12). The results show that the absorption of one quantum of light isomerises both double bonds in the diene. The mechanism suggested for the observed isomerisation involved the production of a common triplet state having the geometry of a 1,4-biradical (62).[57, 58]

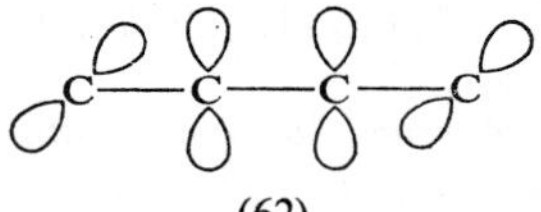

(62)

The direct photoisomerisation of hexa-2,4-dienes takes a different course from the sensitized process.[59] Quantitative measurements on the quantum yields for the various processes are recorded in Table 13. The

Table 13 *Direct isomerisation of hexa-2,4-dienes* [59]

$\Phi_{tt \to ct}$	$\Phi_{cc \to ct}$	$\Phi_{ct \to tt}$	$\Phi_{cc \to tt}$	$\Phi_{ct \to cc}$	$\Phi_{tt \to cc}$	$\Sigma\Phi$
0.37	0.41	0.17	<0.03	0.29	<0.03	1.31

small values for $\Phi_{cc \to tt}$ and $\Phi_{tt \to cc}$ indicate that more than 90% of the excited molecules bypass the triplet manifold, thus making the quantum yield for intersystem crossing <0.1. Thus intersystem crossing is inefficient. The more attractive of the mechanisms proposed for the isomerisation is

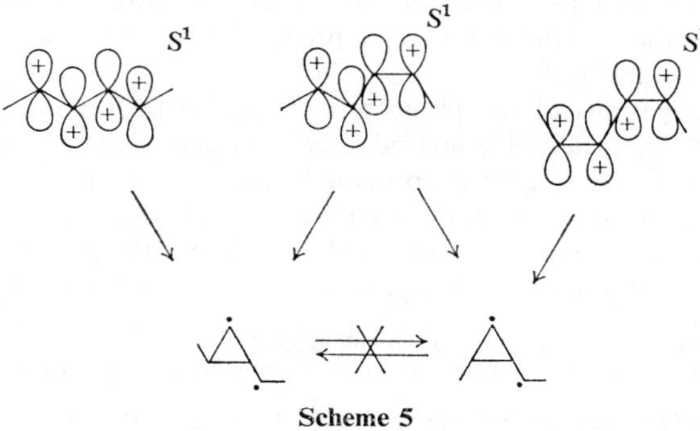

Scheme 5

shown in Scheme 5, and involves the formation of a cyclopropyl diradical. However, the authors [59] point out that either formation of the intermediate must not be so totally efficient that it overwhelms the competing decay to the starting dienes, or else, if formation of the intermediates is completely efficient, the lifetimes involved are not long enough to allow for complete randomisation.

The penta-1,3-dienes [59, 60] undergo relatively inefficient *cis–trans*-isomerisation ($\Phi_{c \to t} = 0.091$, $\Phi_{t \to c} = 0.115$). The isomerisation to 1,3-dimethylcyclopropene arises from either of the two intermediate diradicals (63a) and (63b) shown in Scheme 6. The final step in the formation of the

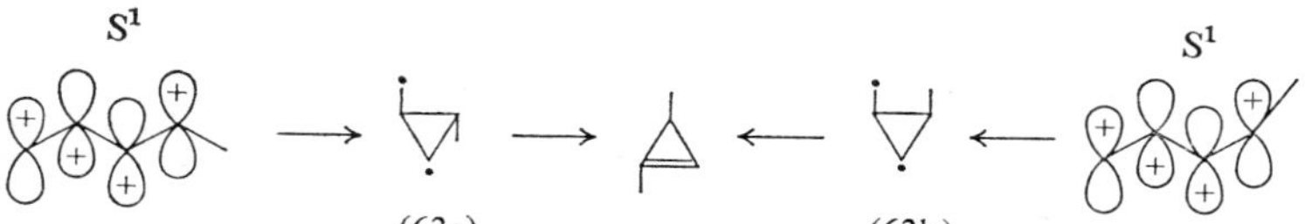

Scheme 6

cyclopropene is a 1,2-hydrogen migration [a 1,3-hydrogen migration is also possible in the case of intermediate (63a)]. Valence isomerisation to 3-methylcyclobutene also occurs. In agreement with Saltiel, Metts, and Wrighton,[59] *cis* → *trans* and *trans* → *cis* isomerisation constitute less than 10% of the total processes.[60] These results suggest that the excited singlet states of the *cis*- and the *trans*-olefins are chemically distinct species and do not interconvert in the manner observed for the photosensitized process.[60]

The photochemical reactions of acyclic dienes have therefore been rationalised as involving a diradical intermediate (*e.g.* 63a) which could form a cyclopropene by hydrogen migration. Srinivasan and Boue [61] have developed a mercury-sensitized gas-phase reaction of dienes which likewise exemplifies this mode of reaction. The results from four dienes are shown in Table 14.

Table 14 *Results from gas-phase mercury-sensitized irradiation of linear dienes* [61]

Diene	Cyclopropene	Conversion (%)	Yield (%)
Butadiene	3-methylcyclopropene	8	30
Penta-1,3-diene	1,3-dimethylcyclopropene	8	20
Hexa-1,3-diene	1-ethyl-3-methyl- and 1-methyl-3-ethyl-cyclopropenes	65	5
Isoprene	3,3-dimethyl- and 2,3-dimethyl-cyclopropenes	10	20

[59] J. Saltiel, L. Metts, and M. Wrighton, *J. Amer. Chem. Soc.*, 1970, **92**, 3227.
[60] S. Boue and R. Srinivasan, *J. Amer. Chem. Soc.*, 1970, **92**, 3226.
[61] R. Srinivasan and S. Boue, *Tetrahedron Letters*, 1970, 203.

Rigid (s)-*trans*-dienes also react photochemically, but several reports in past years have shown that bicyclobutanes are the usual products. The irradiation of the geometric isomeric olefins (64a) and (64b) is no exception,

(64a)   (64b)

(65a)   (66)

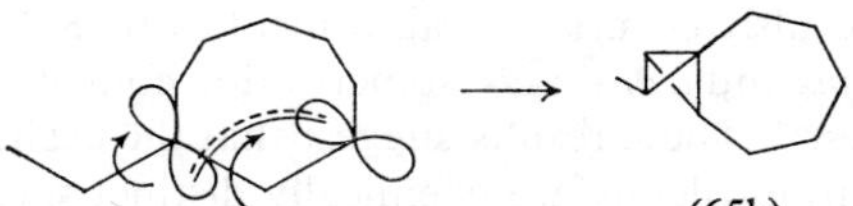

(65b)

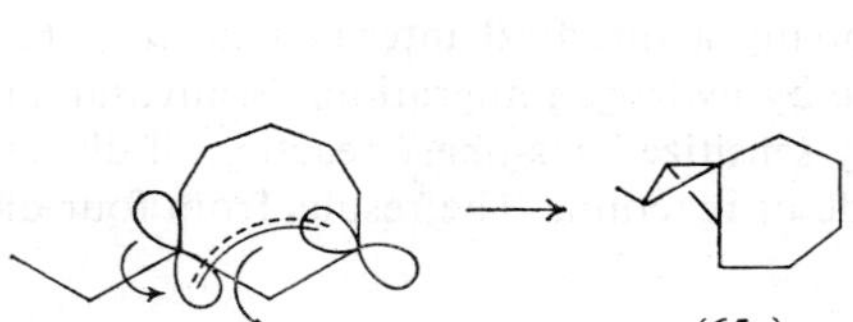

(65c)

Scheme 7

and affords the bicyclobutane (65a) from (64a), and (65b) and (65c) from (64b).[62] The authors consider the formation of products from the concept of orbital symmetry, and propose that the reaction proceeds from a vibrationally relaxed singlet state having a configuration closely similar to that of the zwitterion (66).[62] Orbital symmetry controls the cyclisation of the allylic system. It was observed that the endocyclic double bond of the initial diene undergoes ready isomerisation to the strained *trans* configuration, and that the cyclisation occurs from this twisted state with concomitant retention of stereochemistry. Thus, in Scheme 7, clockwise rotation

[62] W. G. Dauben and J. S. Ritschers, *J. Amer. Chem. Soc.*, 1970, **92**, 2925.

gives the *cis*-bicyclobutane (65b) and anticlockwise rotation gives the *trans*-bicyclobutane (65c). In the case under study, the strain induced by the eight-membered ring assures preferential formation of the *cis*-bicyclobutane (65b). The (*s*)-*trans*-diene in methyl neoabietate (67) does not undergo a similar reaction on irradiation in methanol with light from a low-pressure mercury lamp,[63] but yields the ether (68) instead. The usual mechanisms for etherification do not apply in this example. It is possible that an alternative to $\pi \rightarrow \pi^*$ electronic excitation is operative, possibly the formation of a Rydberg excited state.

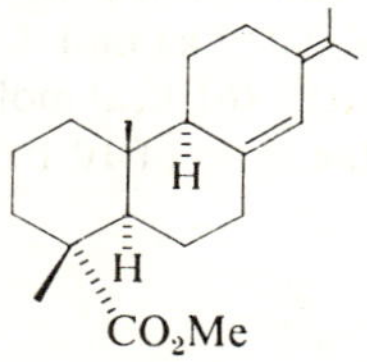

(67)

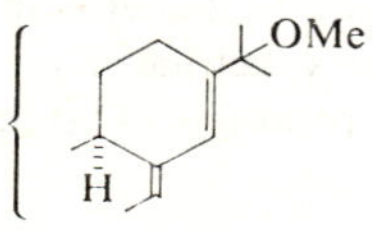

(68)

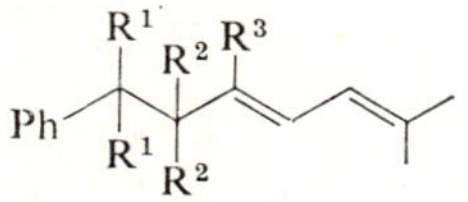

(a) R^1 = Me, $R^2 = R^3$ = H
(b) $R^1 = R^3$ = H, R^2 = Me
(c) $R^1 = R^3$ = Me, R^2 = H

(69)

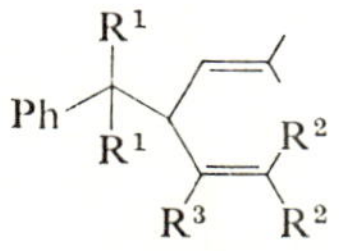

(70)

(71)

Singlet excited states are involved in the rearrangement of the non-conjugated aryl dienes (69) to the aryl divinylmethanes (70) and vinylcyclopropanes (71).[64] The divinylmethanes (70) formally arise by a 1,3-migration of the benzyl group and the vinylcyclopropanes (71) by a 1,2-migration. The mechanism by which the reaction proceeds is not clear. However, since the molecules (69) do not exhibit the fluorescence which would be expected from an ethylbenzene, it appears that the diene and the aryl chromophores are coupled and the corresponding excited systems are mixed. The authors [64] discounted the possibility of mechanisms formally analogous to 'di-π-methane' rearrangements. The formation of the vinylcyclopropanes (71) by the 1,2-migration of a benzyl group is formally

---

[63] J. C. Sircar and G. S. Fisher, *Chem. and Ind.*, 1970, 26.
[64] E. C. Sanford and G. S. Hammond, *J. Amer. Chem. Soc.*, 1970, **92**, 3497.

analogous to the conversion of allyl ethers [48] and allyl chlorides [45] into cyclopropanes.

Further observations on 'di-π-methane' rearrangements have been reported during the year and are discussed in this and the following section dealing with non-conjugated dienes. It has been generally assumed that acyclic dienes utilise an excited singlet state in 'di-π-methane' rearrangements, whereas dienes with steric constraint require a triplet state reaction for efficient rearrangement. Two independent reports have examined this concept.[65,66] Thus, irradiation of 5,5-diphenylcyclohexa-1,3-diene in benzene under sensitization conditions [fluorenone $E_T = 174.7$ kJ mol^{-1} (53.3 kcal mol^{-1}), 2-acetonaphthone $E_T = 249.9$ kJ mol^{-1} (59.6 kcal mol^{-1}), and Michler's ketone $E_T = 255.0$ kJ mol^{-1} (61 kcal mol^{-1})] affords two monomeric products (72a) and (72b) in the ratio of 10.1 : 1.[65] Zimmerman

and Epling obtained the same products (72a and b) in a ratio of 13.5 : 1 from a benzophenone-sensitized reaction.[66] The difference in results is attributed[66] to the 33.4 kJ mol^{-1} (8 kcal mol^{-1}) difference in energy between benzophenone and Michler's ketone. Direct irradiation of the starting diene gave only 1,1-diphenylhexa-1,3,5-triene.[66] The diene (73) also underwent a 'di-π-methane' rearrangement to the isomeric bicyclohexanes (74a) and (74b), but only on direct irradiation.[67] The triplet state of diene (73) was unreactive. The *trans*-isomer (74a) predominated due to minimisation of phenyl–phenyl interaction. Secondary photolysis of the bicyclohexanes brings about their interconversion.

The gas-phase photolysis of 2,5-dimethylfuran afforded the cyclopentenone (75) as the major product.[68] The route to this is envisaged to involve isomerisation of the excited furan to a keto-carbene (76), followed by insertion to form the product (75). The gas-phase photodecomposition of tetrahydrofuran has been reported.[69]

[65] J. S. Swenton, A. E. Crumrine, and T. J. Walker, *J. Amer. Chem. Soc.*, 1970, **92**, 1406.
[66] H. E. Zimmerman and G. A. Epling, *J. Amer. Chem. Soc.*, 1970, **92**, 1411.
[67] H. E. Zimmerman and G. E. Samuelson, *J. Amer. Chem. Soc.*, 1969, **91**, 5307.
[68] S. Boue and R. Srinivasan, *J. Amer. Chem. Soc.*, 1970, **92**, 1824.
[69] B. C. Roquitte, *J. Amer. Chem. Soc.*, 1969, **91**, 7664.

(75) (76)

(77)

(a) $R^1 = R^2 = H$
(b) $R^1 = D, R^2 = H$
(c) $R^1 = H, R^2 = D$

(78)   (2)

Direct irradiation of 1,2-dihydronaphthalene (77a) has been shown to produce benzobicyclo[3,1,0]hexene (78a).[70] The absence of hydrogen migration during the reaction was established by a study of the deuteriated compounds (77b) and (77c), which gave rise to the bicyclohexenes (78b) and (78c) respectively. The authors suggest that the simplest mechanism to account for the rearrangement is one involving ring-opening and ring-closing of the hexatriene intermediate (Equation 2).[70] An alternative is possible, for the reaction could be an example of an allowed $2\pi + 2\sigma$ process. Evidence for concertedness, or lack of it, in the reaction is no doubt being sought.

Considerable interest has been shown in the photoreactions of cyclohexa-1,3-dienes. Typical of this is the photolysis of the enol ether (79) to form the *cis*-fused isomer (80).[71] Attempts to isolate the intermediate triene were unsuccessful; but its proposed involvement is very plausible since conrotatory ring-opening of (79) would lead specifically to triene (81) and thermal disrotatory ring-closure of this would afford the isolated product (80).

Trienes do result from the photolysis of less-complex systems. α-Phellandrene (82) has been shown to photochemically ring-open to a mixture of trienes (83) from the $\pi\pi^*$ singlet state.[72] It has been established that only trienes (83a) and (83c) are genuine primary photo-products, and that trienes (83b) and (83d) arise by secondary processes. The photo-ring-opening is temperature-dependent (Table 15), a fact which is believed to reflect the conformational equilibria in α-phellandrene. Baldwin and Krueger reasoned that the ratio of trienes (83c) : (83a) should be a measure of the equilibrium constant for the conversion of one α-phellandrene conformer into the other.[72] The reaction is highly stereospecific, conformer (82a) giving triene (83a) by a conrotatory ring-opening, and conformer (82b) giving triene (83c). At low temperatures, the pseudo-equatorial conformer (82b) is favoured (see Table 15), and it is proposed that this

[70] R. C. Cookson, S. M. de B. Costa, and J. Hudec, *Chem. Comm.*, 1969, 1272.
[71] M. Miyashita, H. Uda, and A. Yoshikoshi, *Chem. Comm.*, 1969, 1396.
[72] J. E. Baldwin and S. M. Krueger, *J. Amer. Chem. Soc.*, 1969, **91**, 6444.

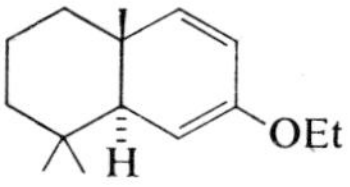

(79) 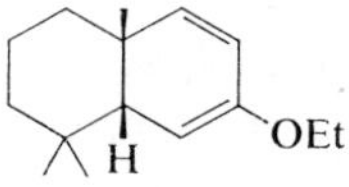 (80) 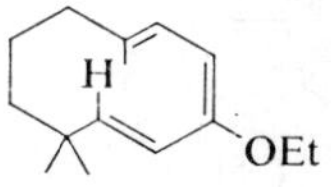 (81)

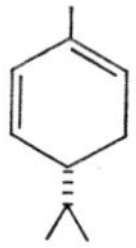

(82) 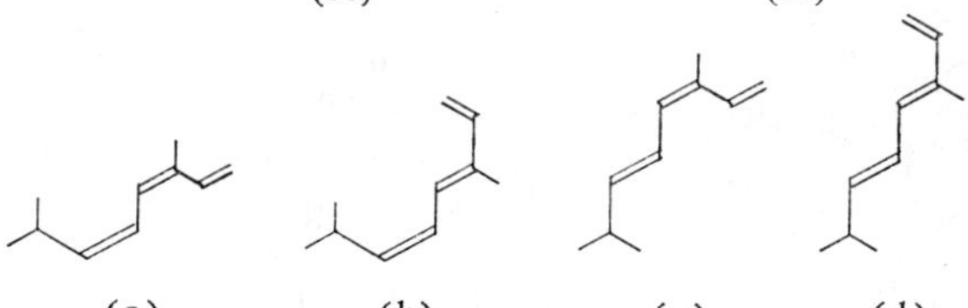 (a) (b) (c) (d)
      (83)

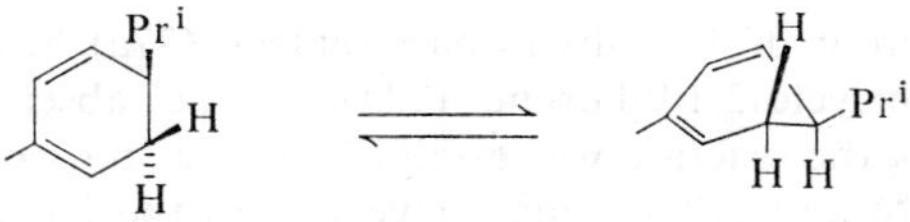

(82a) ⇌ (82b)

study represents a further case where the mode of a photochemical reaction is determined by ground-state geometry.

Table 15 *Temperature dependence of triene product ratios from α-phellandrene* [72]

Temp. (°C)	Ratio (83c) : (83a)
−196	29 ± 1
−160	9.7 ± 1.6
−110	5.2 ± 0.4
0	3.7 ± 0.2
30	3.0 ± 0.1
100	2.7 ± 0.1

Photochemical ring-opening of the cyclohexadienes (84a) and (85a) afforded the triene (86).[73] This arises from the *trans*-isomer (84a) by conrotatory ring-opening to the all-*cis*-triene (87a), which then undergoes a 1,7-hydrogen migration to afford the product (86) which may be isolated. The *cis*-isomer must have given the *cis,cis,trans*-triene (87b). The *cis*-diester (85b) also afforded a triene (87c) as the initial product of photolysis.[74] However, the isomeric *trans*-ester (84b) did not afford an isolable intermediate triene on photolysis.[74] Evidence has been reported [75] for photo-equilibration between the norcaradiene (88) and the cycloheptatriene (89).

---

[73] P. Courtot and R. Rumin, *Tetrahedron Letters*, 1970, 1849.
[74] P. Courtot and R. Rumin, *Bull. Soc. chim. France*, 1969, 3665.
[75] T. Toda, M. Nitta, and T. Mukai, *Tetrahedron Letters*, 1969, 4401.

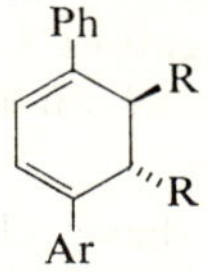

(a) Ar = Ph, R = Me
(b) Ar = p-tolyl or p-Cl·C$_6$H$_4$
R = CO$_2$Me

(84)

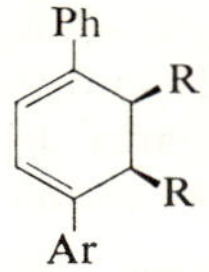

(a) R = Me, Ar = Ph
(b) R = CO$_2$Me, Ar = Ph

(85)

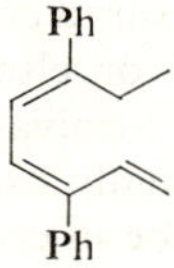

(86)

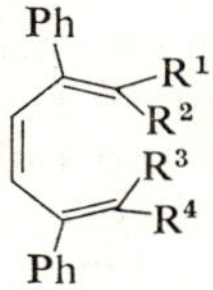

(a) R^2 = R^3 = Me, R^1 = R^4 = H
(b) R^1 = R^3 = Me, R^2 = R^4 = H
(c) R^1 = R^3 = CO$_2$Me, R^4 = R^2 = H

(87)

Owing to the constraint on the molecule, thermal rather than photochemical ring-opening of the norcaradiene (88) affords an intermediate cycloheptatriene which undergoes an allowed photo-1,7-hydrogen-migration to yield the observed product (89). Other thermal- and photo-reactions occur concurrently.

The syntheses and reactions of several photochromic cyclohexadienes (90) have been described.[76] The efficient photochemical conversion of ketofurans (91) to ketovinylfurans (92) has been reported.[77] Low-temperature irradiation of 1,2,3,4,5,6-hexamethylbenzenonium cation gave the

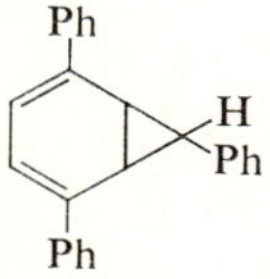

(88) 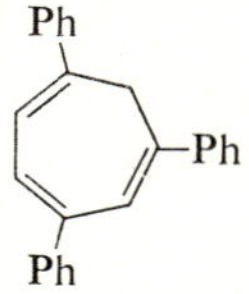 (89) 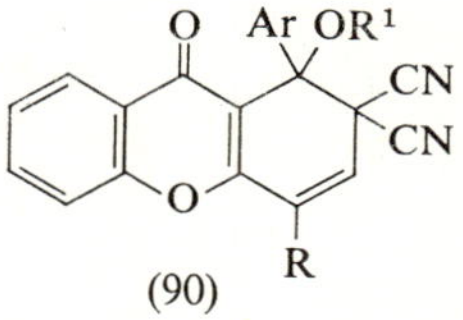 (90)

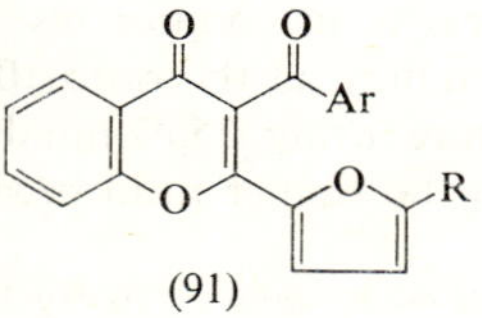

(91) 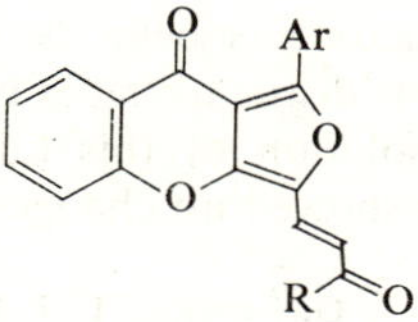 (92) 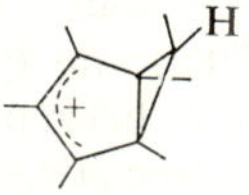 (93)  (94)

[76] K. R. Huffman, M. Burger, W. A. Henderson jun., M. Loy, and E. F. Ullman, *J. Org. Chem.*, 1969, **34**, 2407.
[77] K. R. Huffman, C. E. Kuhn, and A. Zweig, *J. Amer. Chem. Soc.*, 1970, **92**, 599.

isomer cation (93),[78] while tropylium fluoroborate was converted on photolysis in FSO_3H into the norbornadien-7-yl cation.[79] The norbornadienyl cation must arise by thermal conversion of an initially formed bicyclo[3,2,0]heptenyl cation (94).

Non-conjugated Dienes.—Ward and Karafiath have elaborated their previous observations on the photochemical reactions of the allenes, hexa-1,2-diene, and cyclonona-1,2-diene.[80] The mercury-sensitized gas-phase decarbonylation of 3,3-dideuterionorcamphor afforded the isomeric pair of dienes 1,1-dideuteriohexa-1,5-diene and 3,3-dideuteriohexa-1,5-diene, from which the major products of the reaction are formed by secondary photolysis.[81] These arise from cyclopentyl diradicals (95), which can cyclise to the two isolated bicyclohexanes (96) or else undergo rearrangement to allylcyclopropanes (97) (Scheme 8). Srinivasan originally claimed that the cyclopropanes arose from vibrationally excited bicyclohexanes, but the new work shows this conclusion to be in error.[82]

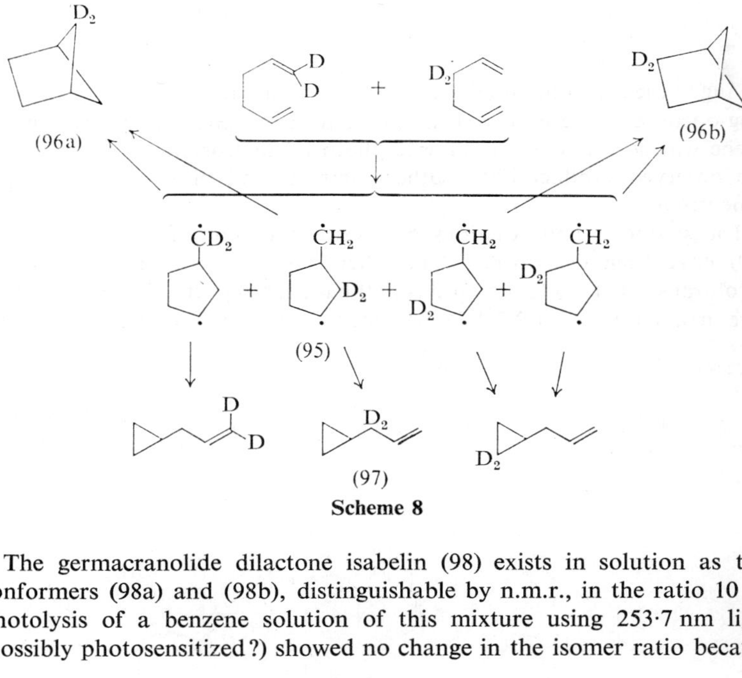

Scheme 8

The germacranolide dilactone isabelin (98) exists in solution as two conformers (98a) and (98b), distinguishable by n.m.r., in the ratio 10 : 7. Photolysis of a benzene solution of this mixture using 253·7 nm light (possibly photosensitized?) showed no change in the isomer ratio because

[78] I. S. Isaev, V. I. Mamatyuk, T. G. Egorova, L. I. Kuzoubova, and V. A. Koptyug, *Izvest. Akad. Nauk S.S.S.R., Ser. khim.*, 1969, 2089.
[79] R. F. Childs and V. Taguchi, *Chem. Comm.*, 1970, 695.
[80] H. R. Ward and E. Karafiath, *J. Amer. Chem. Soc.*, 1969, **91**, 7475.
[81] J. E. Baldwin and J. E. Gano, *J. Org. Chem.*, 1969, **34**, 612.
[82] R. Srinivasan, *J. Amer. Chem. Soc.*, 1961, **83**, 4923.

of rapid equilibration between the conformers.[83] The product was the pentacyclic dilactone (99a) formed by a (2+2) cycloaddition within the conformer (98b). Photolysis of the dihydroisabelin, which exists as a single

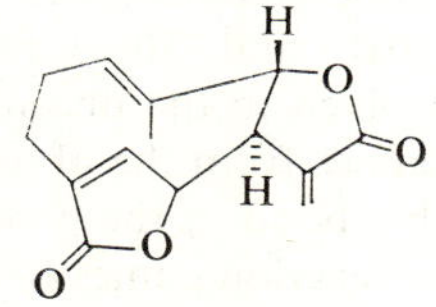

(98a)

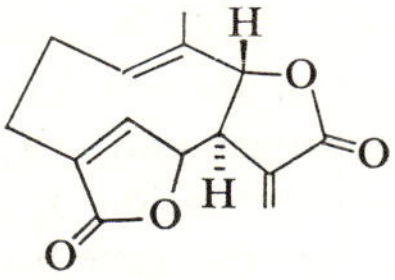

(98b)

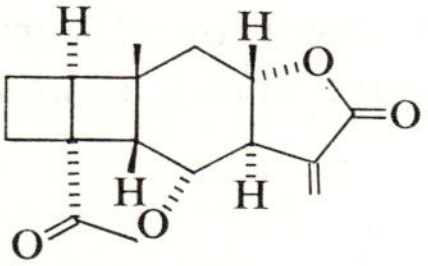

(99a)

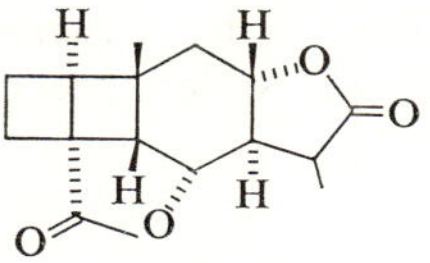

(99b)

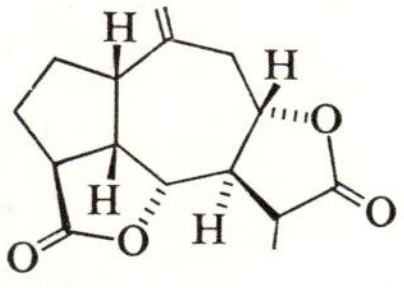

(100)

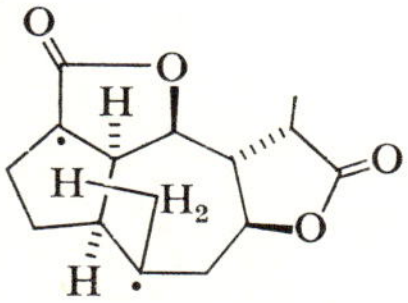

(100a)

conformer in solution, gave the pentacyclic lactone (99b) as well as a new product assigned structure (100): formation of the latter provides an example of a photochemically induced 'ene-reaction' involving an intramolecular transfer of hydrogen, as depicted by structure (100a). Both the results from the isabelin and its dihydro-derivative are presumably controlled by the conformations of the species present in solution.

'Di-π-methane' rearrangements of non-conjugated dienes still excite much interest, and examples have been reported by many groups. Thus, Zimmerman and Pratt have reported that direct irradiation of the diene (101a) gives only the cyclopropane (102a) ($\Phi = 0.097$).[84] The corresponding

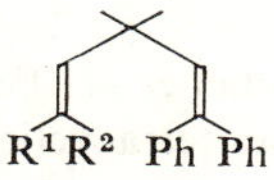

(101) (a) $R^1 = R^2 = Me$ (102)
       (b) $R^1 = H, R^2 = Me$
       (c) $R^1 = Me, R^2 = H$

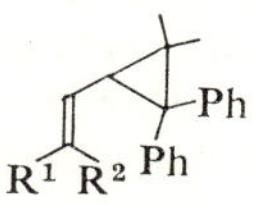

---

[83] H. Yoshioka, T. J. Mabry, and A. Higo, *J. Amer. Chem. Soc.*, 1970, **92**, 923.
[84] H. E. Zimmerman and A. C. Pratt, *J. Amer. Chem. Soc.*, 1970, **92**, 1407.

benzophenone-sensitized triplet process is much less efficient ($\Phi = 0.008$). It is suggested that the preference for the mode of rearrangement is influenced by the development of appreciable free valencies, and the intermediate diradical formed, assuming non-concertedness, will localise the diradical least of all. A study of dienes (101b) and (101c), has suggested that only one of the ethylenic bonds is involved in the unsensitized di-π-methane rearrangement. This conclusion comes from the observation that there is no isomerisation of the other double bond in the products (102a) and (102b). The evidence for a singlet-state reaction comes from sensitisation studies with benzophenone, when only *cis–trans*-isomerisation of olefins (101a) or (101b) took place. Zimmerman and Pratt [85] explain the lack of isomerisation in the unsensitized rearrangement on the grounds of a Möbius system which prevents free rotation about a diradical centre (Scheme 9).†

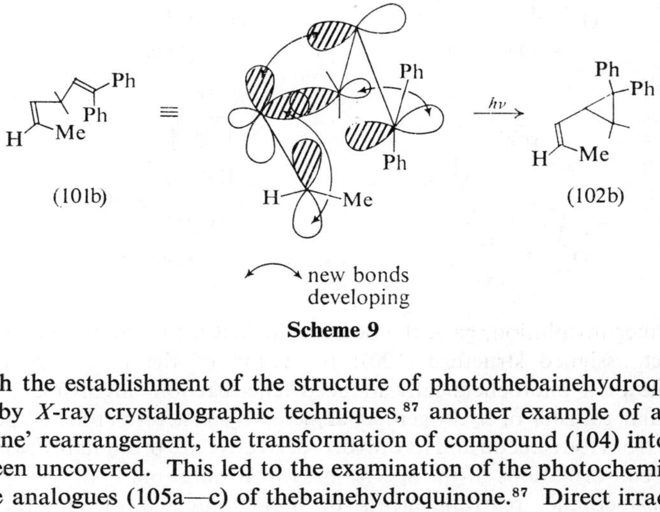

Scheme 9

With the establishment of the structure of photothebainehydroquinone (103) by X-ray crystallographic techniques,[87] another example of a 'di-π-methane' rearrangement, the transformation of compound (104) into (103) has been uncovered. This led to the examination of the photochemistry of simple analogues (105a—c) of thebainehydroquinone.[87] Direct irradiation of these bicyclic hydroquinone [87] systems (105a—c) in ether by a 450 W medium-pressure mercury arc lamp leads to the formation of the rearranged products (106a—c). The structures are deduced from the spectral properties of the molecules and also by the results obtained from the photolysis of the deuterium-labelled products.

Erdman has shown that benzonorbornadienes (105d—g) undergo sensitized but not direct photolysis into the tetacyclic hydrocarbons

[85] H. E. Zimmerman and A. C. Pratt, *J. Amer. Chem. Soc.*, 1970, **92**, 1409.
[86] R. B. Woodward and R. Hoffmann, *Angew. Chem. Internat. Edn.*, 1969, **8**, 797.
[87] M. G. Waite, G. A. Sim, C. R. Olander, J. Warnet, and D. M. S. Wheeler, *J. Amer. Chem. Soc.*, 1969, **91**, 7763.

† Systems of this type have been described by Woodward and Hoffmann [86] as typical $\sigma_a{}^2 + \pi_a{}^2$ processes.

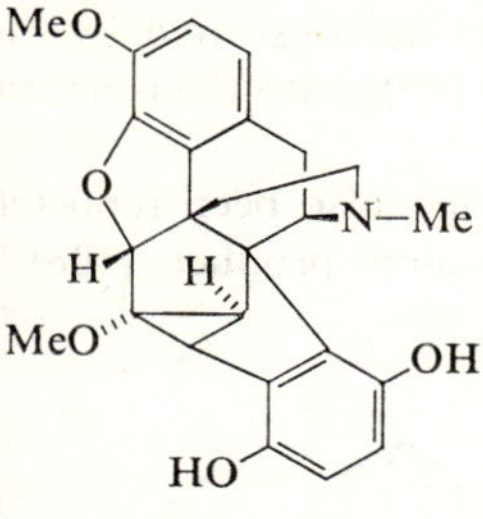

(103)

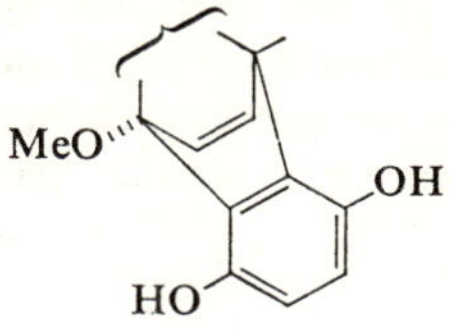

(104)

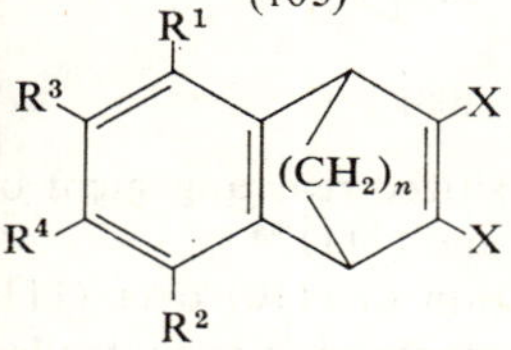

(a) $R^1 = R^2 = OH$, $R^3 = R^4 = X = H$, $n = 1$
(b) $R^1 = R^2 = OH$, $R^3 = R^4 = X = H$, $n = 2$
(c) $R^1 = R^2 = OH$, $R^3 = R^4 = H$, $X = D$, $n = 1$
(d) $R^1 = R^2 = R^3 = R^4 = X = H$, $n = 1$
(e) $R^1 = R^2 = R^3 = R^4 = H$, $X = D$, $n = 1$
(f) $R^1 = Me$, $R^2 = R^3 = R^4 = X = H$, $n = 1$
(g) $R^1 = R^3 = R^4 = X = H$, $R^2 = Me$, $n = 1$

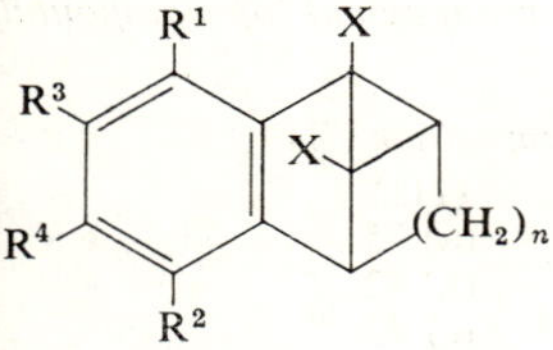

(105)

(106)

(f) { $R^1 = Me$, $R^2 = R^3 = R^4 = X = H$ 70%
    { $R^1 = R^3 = R^4 = X = H$, $R^2 = Me$ 30%
(g) { $R^1 = R^2 = R^4 = X = H$, $R^3 = Me$ 50%
    { $R^1 = R^2 = R^3 = X = H$, $R^4 = Me$ 50%

(106d—g).[88] The triplet energy of the norbornadiene (105d) was established as 292·6 kJ mol^{-1} (70 kcal mol^{-1}). The results in Table 16 illustrate that lower-energy sensitizers with reasonably accessible T_2 states are also effective in bringing about the reaction. Benzo–vinyl bridging is the preferred

Table 16 *Effect of sensitizer triplet energy on product formation from benzonorbornadiene* [88]

Sensitizer	T_1		Product
	kJ mol^{-1}	kcal mol^{-1}	
None	—	—	none
Xanthone	310·2	74·2	+
Acetophenone	307·6	73·6	+
Benzophenone	286·3	68·5	+
Triphenylene	278·4	66·6	+
Fluorenone	213·2	51	none
Anthracene	178·1	42·6	+
	[T_2 = 311·0	74·4]	
9,10-Dichloroanthracene	168·0	40·2	+
	[T_2 = ?]		
9,10-Dibromoanthracene	168·0	40·2	+
	[T_2 = ?]		

[88] J. R. Edman, *J. Amer. Chem. Soc.*, 1969, **91**, 7103.

mechanistic route to products, and this was further demonstrated by the rearrangement of the methyl derivatives (105f) and (105g) into the isomeric pairs of products (106f) and (106g).[88]

The spiro-substituted benzonorbornadiene (107) has also been reported to undergo a similar triplet rearrangement, yielding product (108).[89]

(107)   (108)

Triplet states are also involved in the acetone-sensitized rearrangement of the lactams (109) to the azasemibullvalones (110) and (111).[90]

There is a three-fold preference for the formation of (110) over (111) (Table 17). It is believed that the course of the photoreaction is controlled

Table 17 *Product distribution in sensitized rearrangement of compounds (109)* [90]

Starting material (109)	Product composition (%)	
	(110)	(111)
(a)	(a) 25	(a) 75
(b)	(b) 18	(b) 82
(c)	(c) 18	(c) 82
(d)	(d) 9	(d) 91
(e)	(e) 41	(e) 59

by localisation of free valence, and of the two intermediates (i) and (ii) formed by benzo–vinyl bridging (Scheme 10), (ii) is preferred and the energy balance is not shifted even by alkyl substituents. The preference for intermediate (ii) is thought to arise from the stabilisation of the radical sites by the cyclopropylcarbonyl moiety.

The foregoing results contrast with the photorearrangements of benzobicyclo-octadienones, where acyl migration is the sole route to products.[91]

Irradiation of bicyclo[3,2,0]octadiene (112) leads to the rapid formation of a 4 : 1 mixture of isomeric tetracyclic hydrocarbons, the major isomer of which has been shown to be the tetracyclic hydrocarbon (113), derived from the starting olefin by rearrangement to the norbornadiene (114) followed by intramolecular cycloaddition.[92] Three mechanisms (Scheme 11) are suggested to account for the formation of the norbornadiene: (1) a bond rupture to an unlikely vinyl radical, followed by recyclisation; (2) a $\pi + \pi$ addition; and (3) a concerted 1,3-migration. The nature of the excited

[89] B. M. Trost, *J. Org. Chem.*, 1969, **34**, 3644.
[90] L. A. Paquette and R. H. Meisinger, *Tetrahedron Letters*, 1970, 1479.
[91] H. Hart and R. K. Murray, *Tetrahedron Letters*, 1969, 379.
[92] A. A. Gorman and J. B. Sheridan, *Tetrahedron Letters*, 1969, 2569.

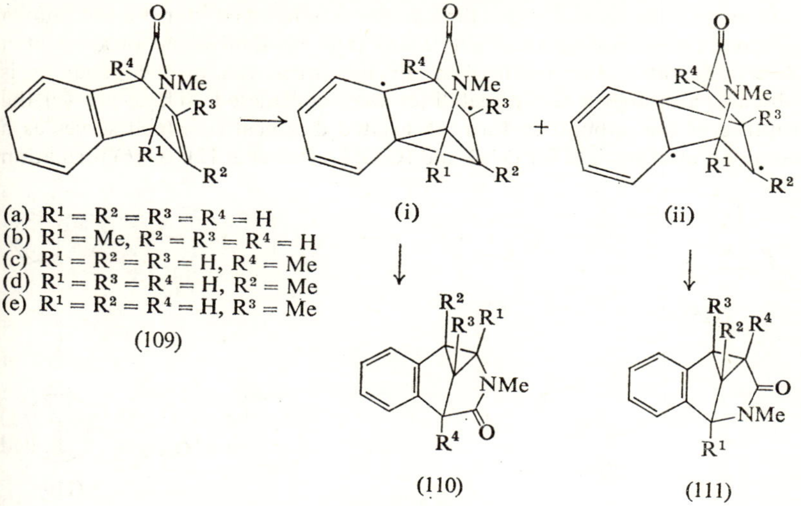
O
R4
NMe
R3
R1
R2
(a) R1 = R2 = R3 = R4 = H
(b) R1 = Me, R2 = R3 = R4 = H
(c) R1 = R2 = R3 = H, R4 = Me
(d) R1 = R3 = R4 = H, R2 = Me
(e) R1 = R2 = R4 = H, R3 = Me
(109)
(i)
(ii)
(110)
(111)

Scheme 10

state was not determined. However, the use of benzene as a solvent suggests the possibility of a triplet sensitization process, in which case a concerted process would be unlikely.

The recently prepared tricyclo[3,3,0,0^{2,6}]octa-3,7-diene (115) has been observed to have a remarkably intense long-wavelength absorption in the u.v. (300 nm, ε = 190).[93] The photochemistry of this compound has been investigated, and at −60 °C, using a Vycor filter, the diene (115) is converted into a mixture of semibullvalene (116) and cyclo-octatetraene (C.O.T.). It was established that semibullvalene was not converted photochemically

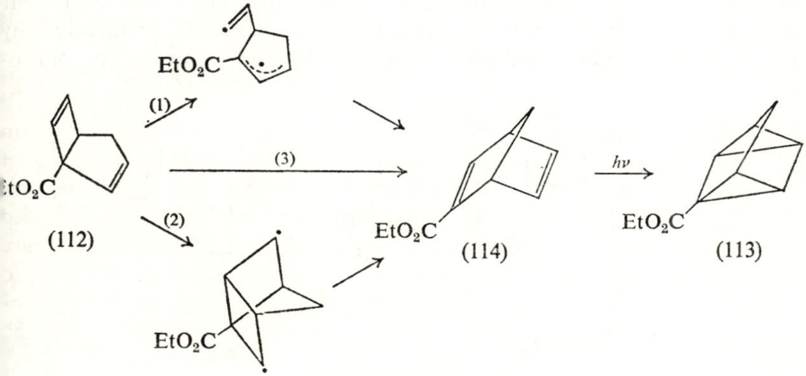
EtO2C
(1)
(3)
(2)
hν
(112)
(114)
(113)

Scheme 11

[93] J. Meinwald and H. Tsuruta, *J. Amer. Chem. Soc.*, 1970, **92**, 2579.

into cyclo-octatetraene under the conditions employed, nor did the known thermal conversion of the diene (116) into semibullvalene proceed at a detectable rate. The mechanism for the formation of the products is obscure, although a few possibilities were considered. These are (a) the rupture of one σ-bond to form an excited diradical (i) which serves as a common intermediate for (116) and (C.O.T.) (Scheme 12), or (b) formation

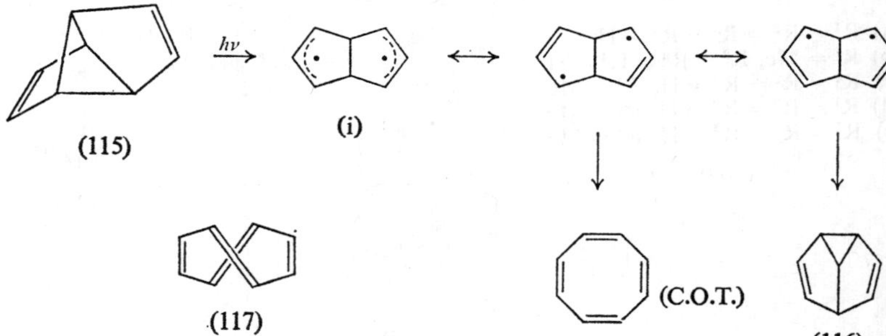

Scheme 12

of the diradical (i) and thence only semibullvalene, cyclo-octatetraene being formed by a retro(2+2)π reaction *via* a *cis,trans,cis,trans*-C.O.T. (117) which is rapidly isomerised under the conditions employed, and (c) formation of (116) from the diene (115) by a concerted suprafacial 1,3-migration.[93] However, the exact nature of the intermediate(s) still requires to be established.

3 Reactions of Trienes and Higher Polyenes

Irradiation ($\lambda = 370$ nm) of the silyl ethers of tachysterol (118a) and previtamin D_3 (118b) in outgassed benzene solution has shown that their interconversion by *cis–trans*-isomerisation is capable of sensitization by benzil, anthraquinone, and most effectively by fluorenone.[94] The photo-

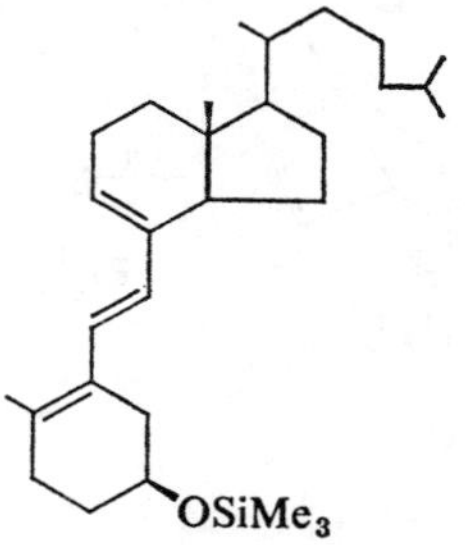

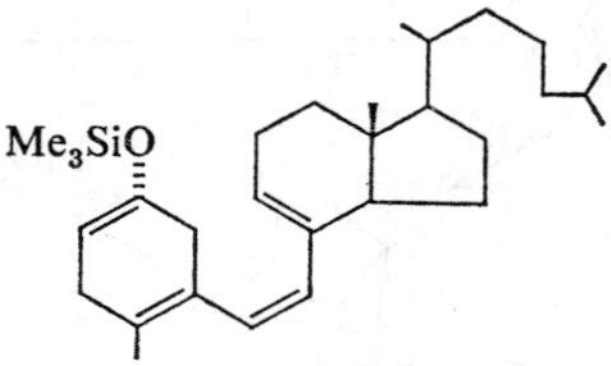

---

[94] A. E. C. Snoeren, M. R. Daha, J. Lugtenburg, and E. Havinga, *Rec. Trav. chim.*, 1970, **89**, 261.

stationary state reached using this last sensitizer contains 80% of (118). The quantum yields of the processes have been established as $\Phi_{(a)\to(b)} = 0{\cdot}58$ and $\Phi_{(b)\to(a)} = 0{\cdot}15$. Anthracene, benzpyrene, or azulene were ineffective as sensitizers. The results obtained suggest that, in addition to the above isomerisation taking place from the triplet state, the cyclisation of previtamin D_3 to 7-dehydrocholesterol and lumisterol takes place from the singlet state.[94]

Accounts of the isomerisation of homofulvenes to spirocyclopentadienes have been given.[95, 96] The homofulvenes (119) for one of the studies were synthesized by the photolysis ($\lambda = 253$ nm) of the triene (120).[95] The initial

(119a)   (119b)   (119c)

(120)   (121)

ratio of the two products was 62:38, but continued photolysis produced three other products. It was shown that homofulvene (119) was converted into the triene (119c) and the spirocyclopentadiene (121).

A similar reaction has been reported[96] for the rearrangement of the bicyclohexene (122) to the isomeric mixture of spirodienes (123).[96] This particular reaction arises from a $\pi \to \pi^*$ singlet ($\Phi = 0{\cdot}039$). The *anti*-product (123b) predominates at low conversions. Both groups[95, 96] interpret the reaction as a 'slither' mechanism where, as depicted in Scheme 13, the one-carbon unit moves around the five-membered ring.

The 7-arylcycloheptatrienes (124) have been shown to undergo photochemical (253·7 nm) 1,7-hydrogen migrations to form isomeric cycloheptatrienes (125a), which on further photolysis (300 nm) undergo a further 1,7-hydrogen migration to give cycloheptatriene (125b).[97] No evidence for 1,3-suprafacial hydrogen migration was found. Similar 1,7-hydrogen migrations account for the isomerisation of the cycloheptatriene products formed from the photochemical addition of ethyl diazoacetate to benzene.[98] A 1,7-methyl migration accounts for the major product (126) from the direct photolysis of 2,3,7,7-tetramethylcycloheptatriene.[99]

[95] T. Tabata and H. Hart, *Tetrahedron Letters*, 1969, 4929.
[96] H. E. Zimmerman, D. F. Juers, J. M. McCall, and B. Schroeder, *J. Amer. Chem. Soc.*, 1970, **92**, 3474.
[97] T. Tezuka, M. Kimura, A. Sato, and T. Mukai, *Bull. Chem. Soc. Japan*, 1970, **43**, 1120.
[98] G. Linstumelle, *Tetrahedron Letters*, 1970, 85.
[99] L. B. Jones and V. K. Jones, *J. Org. Chem.*, 1969, **34**, 1298.

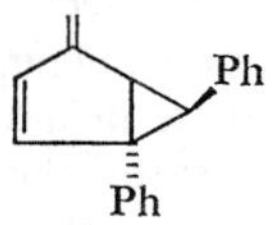

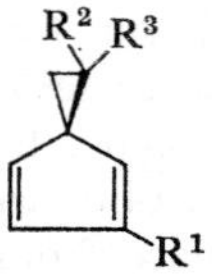

(a) $R^1 = R^3 = Ph$, $R^2 = H$
(b) $R^1 = R^2 = Ph$, $R^3 = H$

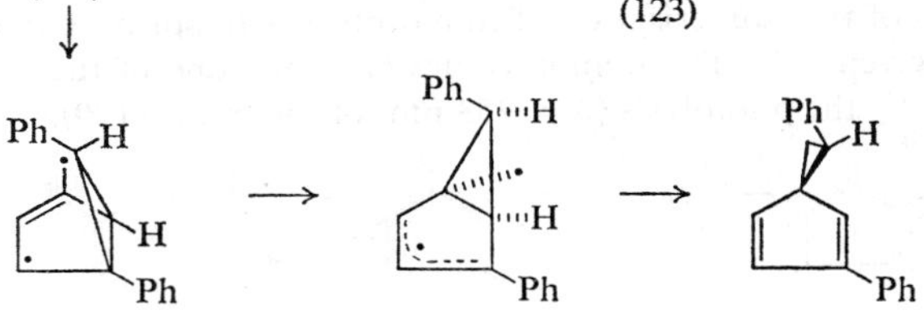

Scheme 13

Cyclononatetraenide shows enhanced basicity on photolysis in the presence of a proton donor (hex-1-yne). The products formed in the reaction are the result of series of thermal- and photo-reactions of the initial product, cyclononatetraene.[100a] This compound has also been obtained by the photolysis of *cis*-bicyclo[6,1,0]nona-2,4,6-triene.[100b]

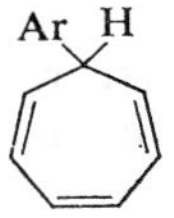

Photochemically synthesized annulenes have also been reported during the year.[101, 102] [10]Annulene has been reported as a product in the low-temperature photolysis of *cis*-9,10-dihydronaphthalene, in contrast with an earlier report by van Tamelen and Burkoth.[103] Masamune and Seidner[101] interpret the n.m.r. spectrum of the product of irradiation as consistent with the [10]annulenes (127a) and (127b). Similar photolysis of bicyclo[6,2,0]deca-2,4,6,9-tetraene leads to the same spectrum, although *trans*-9,10-dihydronaphthalene was unproductive. The highest yield (7%) of the annulene (127b) was obtained from photolysis at $-60\,°C$. The other products of the reaction were identified as the *cis,trans*-isomer (127a) (15%), bullvalene (23%), and tetracyclo[4,4,0,0^{2,10}, 0^{5,7}]deca-3,8-diene (40%). The n.m.r. evidence for the existence of [10]annulenes rests on the appearance of a temperature-dependent signal at 4·16τ and a temperature-

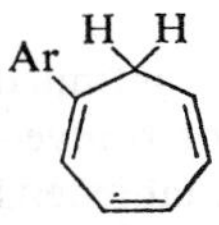

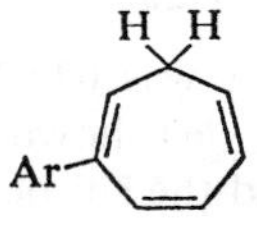

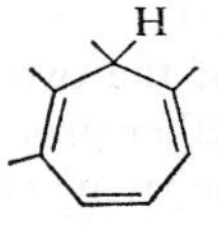

---

[100] (a) J. Schwartz, *Chem. Comm.*, 1969, 833; (b) A. G. Anastassiou, V. Orfanos, and J. H. Gebrian, *Tetrahedron Letters*, 1969, 4491.
[101] S. Masamune and R. T. Seidner, *Chem. Comm.*, 1969, 542.
[102] H. Röttele, W. Martin, J. F. M. Oth, and G. Schröder, *Chem. Ber.*, 1969, **102**, 3985.
[103] E. E. van Tamelen and T. L. Burkoth, *J. Amer. Chem. Soc.*, 1967, **89**, 151.

independent signal at 4·34τ, attributed to (127a) and (127b) respectively. Thus the reaction is summed up as a specific conrotatory process to yield the *cis,trans*-annulene followed by a *trans* → *cis* isomerisation to the all-*cis*-annulene (127b), photochemical ring-closure of which produces the

(127a)   (127b)   (3)

trans-9,10-dihydronaphthalene (Equation 3). Isomers of [12]annulene (128) may be formed from the photolysis of tricyclo[8,2,0,0^{2,9}]dodeca-3,5,7,11-tetraene.[102] A photochemical disrotatory ring-opening of one of the four-membered rings gave bicyclo[8,2,0]dodeca-2,4,6,8,11-pentaene, which either thermally isomerises to the [12]annulene (128a) or photochemically ring-opens to the [12]annulene (128b). A subsequent series of thermal and

(128a)   (128b)

photochemical events produces the isolated products. The unsensitized photolysis of *cis,trans,cis*-cyclododeca-1,5,9-triene gave the *cis,cis,cis*-cyclododecatriene isomer as well as a mixture of isomeric 1,2,4-trivinylcyclohexenes.[104]

Last year, a preliminary account of the photolysis of cyclo-octatetraene showed that semibullvalene (129a) was formed and that the reaction was a good route to this complex molecule. A new paper reports a study of the mechanism by which semibullvalene is formed, and discounts the intermediacy of bicyclo[4,2,0]octa-2,4,7-triene since the photolysis of this compound only affords benzene.[105] The authors argue that in the absence of the suspected di-π-methane rearrangement, a concerted 1,5- and 4,6-bonding would be the most likely mode of cyclisation. Indeed, the evidence cited by White *et al.*[106] for the existence of a *trans,cis,cis,cis*-cyclo-octatetraene has added weight to the idea that the isomerisation of cyclo-octatetraene to semibullvalene is a two-photon process. The first photon brings about *cis–trans* isomerisation of one of the bonds and the second effects the conversion to (129a). The need for *cis–trans* isomerisation is

[104] C. J. Attidge and S. J. Baker, *Tetrahedron Letters*, 1970, 387.
[105] H. E. Zimmerman and H. Iwamura, *J. Amer. Chem. Soc.*, 1970, **92**, 2015.
[106] 'Photochemistry,' ed. D. Bryce-Smith, Specialist Periodical Reports, The Chemical Society, London, 1970, Vol. 1, p. 258.

suggested further by the fact that 1,3,5,7-tetramethylcyclo-octatetraene does not afford the corresponding semibullvalene (129b): *cis* → *trans* isomerisation here would introduce severe methyl–methyl interactions. The semibullvalene (129b) does, however, revert to the cyclo-octatetraene on photolysis. Semibullvalene (129a) is also formed by the acetone-sensitized photolysis of barrelene (130a).[107]

The direct photolysis of barrelene (bicyclo[2,2,2]octa-2,5,7-triene) is reported to produce cyclo-octatetraene rather than semibullvalene.[107] The molecular contortions of barrelene to produce semibullvalene have been classified as di-π-methane rearrangements. Irradiation of dideuteriobarrelene (130b) shows that only two of the three double-bonds are primarily involved in the reaction. Scheme 14 shows the triplet reaction taking place by vinyl–vinyl bridging to produce a 1:1 mixture of two labelled barrelenes, and is consistent with the absence of a secondary

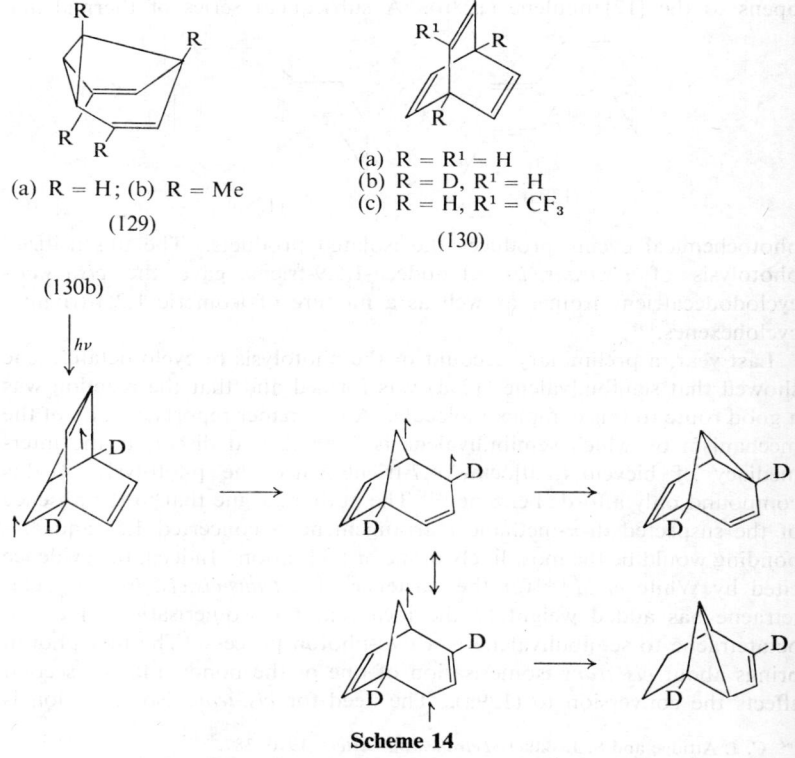

(a) R = H; (b) R = Me
(129)

(a) R = R^1 = H
(b) R = D, R^1 = H
(c) R = H, R^1 = CF$_3$
(130)

(130b)

Scheme 14

[107] H. E. Zimmerman, R. W. Binkley, R. S. Givens, G. L. Grunwald, and M. A. Sherwin *J. Amer. Chem. Soc.*, 1969, **91**, 3316.

isotope effect. The di-π-methane rearrangement of barrelene at 13 °C can be brought about by sensitization ($E_T > 60$ kcal mol^{-1}).[108]

Benzbarrelenes (131) [108, 109] and the tetrafluorodihydro-analogue (132) [110] also undergo di-π-methane rearrangements. Interestingly, the benzbarrelenes required sensitization but the dihydro-analogue reacted on direct irradiation. The authors of ref. 110 suggest that the dihydro-analogue (132) rearranges *via* a discrete quadricyclane intermediate, but evidence from other

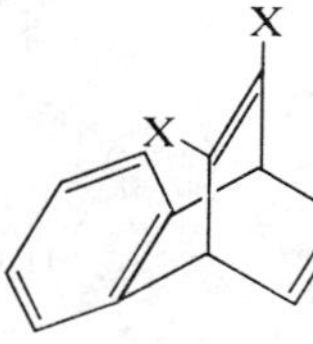

(a) X = CF$_3$
(b) X = H
(131)

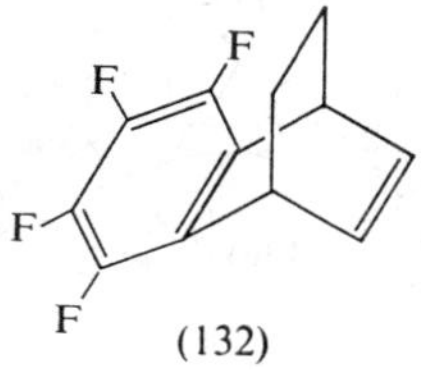

(132)

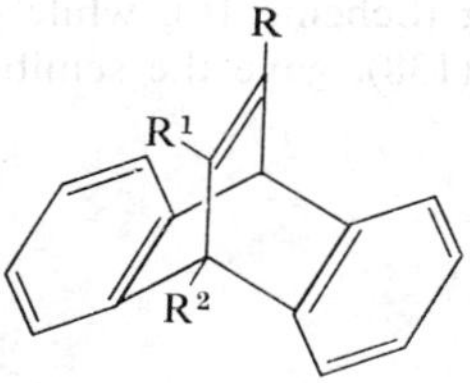

(133)

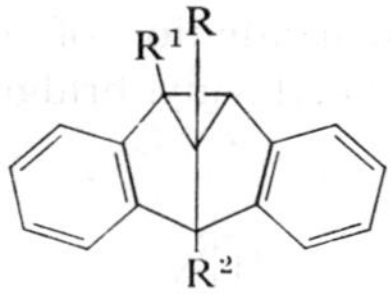

(a) R^1 = R^2 = H, R = Me
(b) R^1 = Me, R = R^2 = H

(134)

workers points towards a diradical route to the semibullvalene.[110] Dibenzbarrelenes (133) also photoisomerise to the corresponding dibenzsemibullvalenes (134).[111] A closely related report concerns rearrangement of the methylenebicyclo-octadiene (135) both by direct and sensitized irradiation to form traces of the dibenzsemibullvalenes (134a and b) as well as the major product (136).[112] This could be interpreted as another example of a di-π-methane process (Scheme 15).

In most of the examples so far reported, the di-π-methane rearrangement involves benzo-vinyl bridging. Extension of the reactions to naphthobarrelenes (137, 138) has shown that vinyl-vinyl bridging can take place even though an aryl group is available for the usual interaction.[113] Thus, the deuterium-labelled 1,2-naphthalene (137) and 2,3-naphthalene (138) derivatives undergo benzophenone-sensitized photolysis (H = remaining protons

[108] R. S. H. Liu and C. G. Krespan, *J. Org. Chem.*, 1969, **34**, 1271.
[109] R. M. Pagni, *Diss. Abs.*, 1969, **29**, B, 3267.
[110] I. F. Eckhard, H. Heaney, and B. A. Marples, *Tetrahedron Letters*, 1969, 3273.
[111] E. Ciganek, U.S. Pat. 3,489,791 (*Chem. Abs.*, 1970, **72**, 66,707d).
[112] S. J. Cristol and G. O. Mayo, *J. Org. Chem.*, 1969, **34**, 2363.
[113] H. E. Zimmerman and C. O. Bender, *J. Amer. Chem. Soc.*, 1969, **91**, 7516.

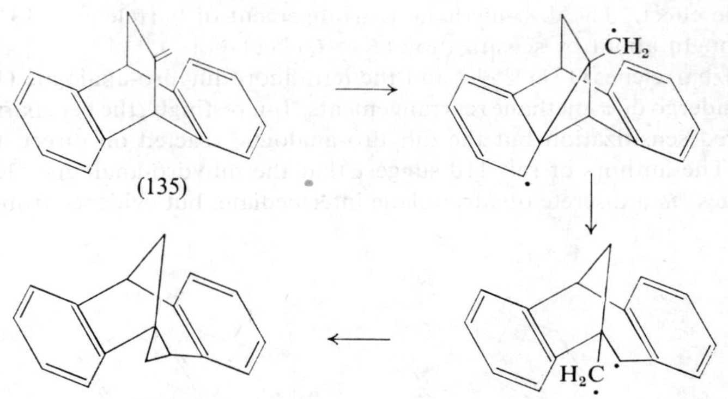

(135)

(136)

Scheme 15

in polydeuteriated molecules). Isomer (137) gave the semibullvalenes (139a) and (139b) by benzo–vinyl bridging (Scheme 16), while direct or sensitized irradiation of the other isomer, (138), gave the semibullvalene (140) by vinyl–vinyl bridging (Scheme 17).[113]

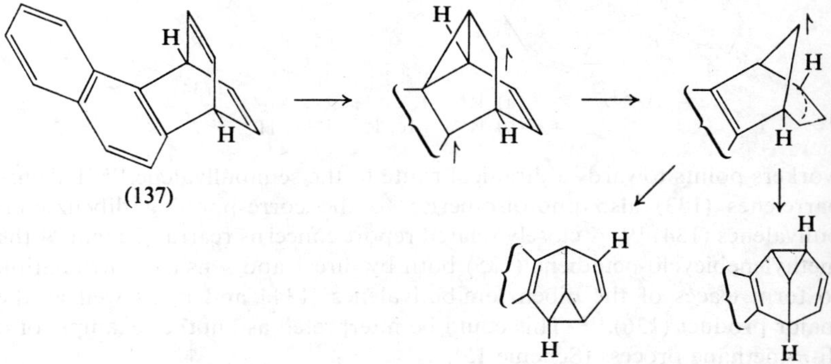

(137)

(139b) (139a)

Scheme 16

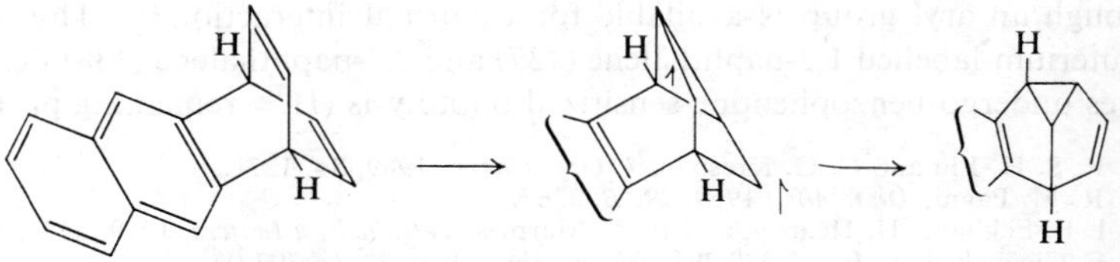

(138) (140)

Scheme 17

Although the triplet energies of the naphthalene moieties were established as 255 kJ mol^{-1} (61 kcal mol^{-1}), it is not at first clear why there should be a difference in the reaction pathways to the formation of the semibullvalenes. However, it seems that in the first example, (137), the bridging to the aromatic ring is *via* the more reactive α-naphthyl position, whereas in the second example, (138), bridging would involve the less reactive β-naphthyl position. Exclusive vinyl–vinyl bridging is used to explain the results from the sensitized or direct irradiation of the tricycloimine ether (141) to give

(141)   (142)

(143)   (144)

the benzazabullvalene (142).[114] But, as with all di-π-methane reactions, there is no need to invoke vinyl–vinyl bonding at all, for the same result can be achieved by invoking a photo-allowed $\sigma_a^2 + \pi_a^2$ rearrangement. The lactam (143) affords the rearranged product (144) on acetone-sensitized photolysis.

The course of photolysis of the bridged phosphine (145) and its oxide depends on whether or not the reaction is sensitized.[115] Thus, the photolysis of a benzene solution of the phosphine (145) yields the tetracyclic compound (146) as the major product in a singlet di-π-methane rearrangement (Scheme 18). The acetone-sensitized reaction of the bridged phosphine oxide affords the homocubane (147) *via* the intermediate valence tautomer (148), which can itself be isolated following direct or brief sensitized irradiation of the phosphine oxide.

Continued interest in the photochemical synthesis of oxonin (149a) from *cis*-cyclo-octatetraene oxide has produced corrections [116–118] of assignments

[114] L. A. Paquette, J. R. Malpass, and J. R. Krow, *J. Amer. Chem. Soc.*, 1970, **92**, 1980.
[115] T. J. Katz, J. C. Carnahan, G. M. Clarke, and N. Acton, *J. Amer. Chem. Soc.*, 1970, **92**, 734.
[116] S. Masumune, S. Takada, and R. T. Seidner, *J. Amer. Chem. Soc.*, 1969, **91**, 7769.
[117] J. M. Holovka, R. R. Grabbe, P. D. Gardner, C. B. Strow, M. L. Hill, and T. V. VanAuken, *Chem. Comm.*, 1969, 1522.
[118] A. G. Anastassiou and R. P. Cellura, *Chem. Comm.*, 1969, 1521.

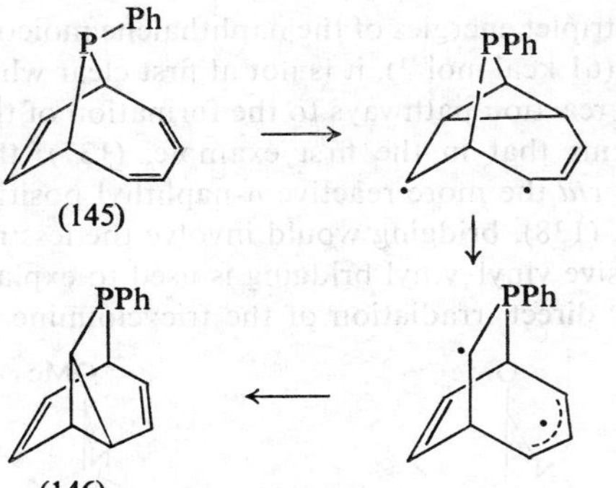

(145) → (146)

Scheme 18

made last year.[119] Masamune et al.[116] have shown that oxonin (149a) undergoes a thermal cyclisation to afford a *cis*-dihydrobenzofuran (150) and not a *trans*-isomer. Corrections have also been made by the original authors.[117, 118]

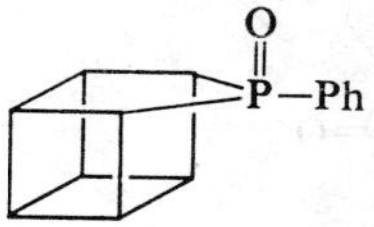

(147) 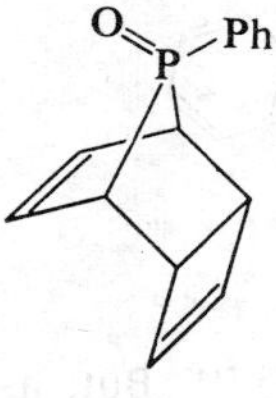   (148)   (149) (a) X = O
                      (b) X = NCO$_2$Et

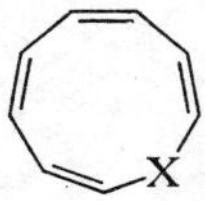

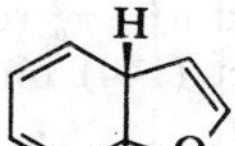

(150) 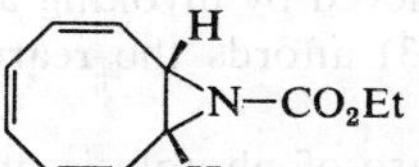   (151)

N-Carbethoxyazonin (149b) forms the azabicyclic olefin (151) on direct photolysis.[120]

In an attempt to rationalise the synthesis of the tricyclic triene (152) from a variety of precursors, Katz and Cheung[121] propose that the photorearrangements proceed *via* the intermediate (153), which is readily accessible from all the precursors (154a—d) reported so far.[122]

[119] 'Photochemistry,' ed. D. Bryce-Smith, Specialist Periodical Reports, The Chemical Society, London, 1970, Vol. 1, p. 254; A. G. Anastassiou and R. P. Cellura, *Chem. Comm.*, 1969, 903.
[120] A. G. Anastassiou and J. H. Gebrian, *Tetrahedron Letters*, 1969, 5239.
[121] T. J. Katz and J. J. Cheung, *J. Amer. Chem. Soc.*, 1969, **91**, 7772.
[122] (a) M. Jones, *J. Amer. Chem. Soc.*, 1967, **89**, 4236; (b) S. Masamune, C. G. Chin, K. Hojo, and R. T. Seidner, *ibid.*, 1967, **89**, 4804; (c) S. Masamune, H. Zenda, M. Wiesel, N. Nakatsuka, and G. Bigam, *ibid.*, 1968, **90**, 2727.

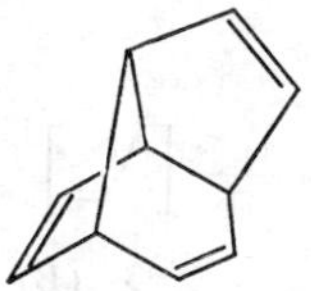

(152)    (153)

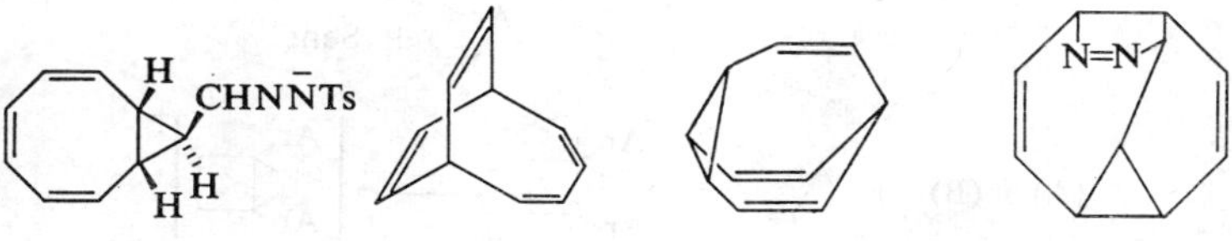

(a)   (b)   (c)   (d)

(154)

4 Dimerisations and Intermolecular Cycloadditions

Dimerisations.—A review article on photodimerisation reactions has been published.[123] The unsensitized [124] and sensitized [125] dimerisations of N-vinylcarbazole have been reported.[124, 125] Quantum yields for sensitization by p-chloranil or fluorenone are listed in Table 18. The values are in

Table 18 *Quantum yields (Φ_d) for photosensitized dimerisation of N-vinylcarbazole* [125]

Sensitizer	Conc.(M)	Solvent	Atmosphere	Φ_d
p-Chloranil	3×10^{-3}	Me_2CO	air	8·5
p-Chloranil	3×10^{-3}	Me_2CO	N_2	4·3
p-Chloranil	10^{-3}	MeOH	air	2·9
p-Chloranil	10^{-3}	MeOH	N_2	1·1
Fluorenone	2×10^{-4}—10^{-2}	Me_2CO	air	3·6
Fluorenone	10^{-2}	Me_2CO	N_2	2·1

excess of unity, thereby indicating a chain mechanism. The fact that the reactions proceed both in the presence and absence of air requires a modification of the original mechanism reported last year (Scheme 19).[126] However, a radical-cation process is still favoured. The sensitizers presumably function through their well-known electron-accepting properties.

One of the least understood reactions of stilbene is its dimerisation to cyclobutanes. A series of stilbenes (155) has now been examined.[127] In one

---

[123] D. J. Trecker, *Org. Photochem.*, 1969, **2**, 63.
[124] J. W. Breitenbach, F. Sommer, and G. Unger, *Montash.*, 1970, **101**, 32.
[125] R. A. Crellin, M. C. Lambert, and A. Ledwith, *Chem. Comm.*, 1970, 682.
[126] 'Photochemistry,' ed. D. Bryce-Smith, Specialist Periodical Reports, The Chemical Society, London, 1970, Vol. 1, p. 226.
[127] H. Ulrich, D. V. Rao, F. A. Stuber, and A. A. R. Sayigh, *J. Org. Chem.*, 1970, **35**, 1121.

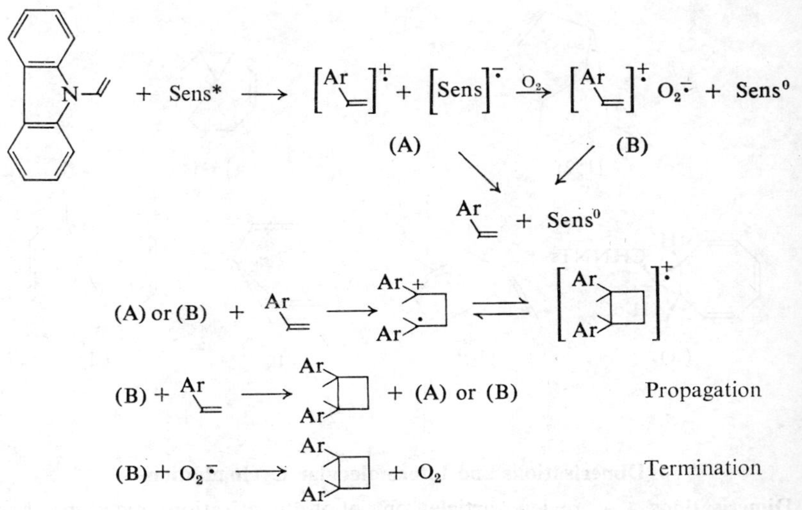

Scheme 19

$$\text{Carbazole-N-CH=CH}_2 + \text{Sens}^* \longrightarrow \begin{bmatrix} \text{Ar} \\ \diagdown \end{bmatrix}^{+\cdot} + [\text{Sens}]^{-\cdot} \xrightarrow{O_2} \begin{bmatrix} \text{Ar} \\ \diagdown \end{bmatrix}^{+\cdot} O_2^{-\cdot} + \text{Sens}^0$$

(A)   (B)

$$\text{(A) or (B)} \longrightarrow \text{Ar}\diagdown + \text{Sens}^0$$

(A) or (B) + Ar⟍ ⟶ [Ar–Ar cyclobutane radical cation] ⇌ [Ar/Ar cyclobutane]⁺·

(B) + Ar⟍ ⟶ Ar/Ar cyclobutane + (A) or (B)   Propagation

(B) + $O_2^{-\cdot}$ ⟶ Ar/Ar cyclobutane + O_2 Termination

example, the dimerisation of (155e) has been shown to have a quantum yield of 0·06 and to exhibit first-order kinetics from 5—70% conversion. Electronic, steric, and solvent effects are important, and the presence of electron-donating groups (OMe, and $NHCO_2Me$) enhances

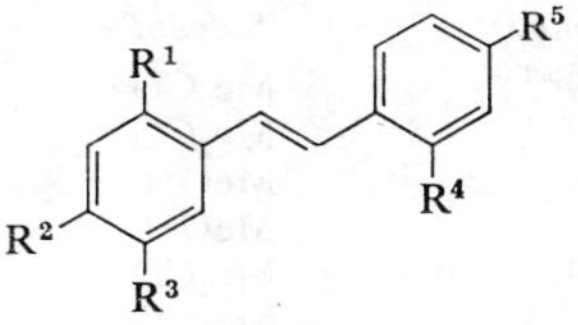

(155) stilbene with R^1, R^2, R^3 on one ring and R^4, R^5 on the other

	R^1	R^2	R^3	R^4	R^5	Dimer (%)
(a)	H	H	H	H	H	11
(b)	H	H	H	H	$NHCO_2Me$	27
(c)	MeO	H	H	H	$NHCO_2Me$	51
(d)	H	MeO	H	H	$NHCO_2Me$	66
(e)	MeO	H	MeO	H	$NHCO_2Me$	74[a]
						68[b]
						91[c]
(f)	H	H	H	$NHCO_2Me$	$NHCO_2Me$	10
(g)	H	$NHCO_2Me$	H	H	$NHCO_2Me$	45[c]
(h)	H	MeO	H	$NHCO_2Me$	$NHCO_2Me$	35
(i)	MeO	H	MeO	$NHCO_2Me$	$NHCO_2Me$	22

[a] the same yield of dimer is obtained in benzene solution.
[b] dimerisation in dimethylformamide.
[c] dimerisation in tetrahydrofuran.

dimerisation. Dimerisation occurred least readily in *NN*-dimethylformamide, but was unaffected by oxygen (therefore singlet process ?). Photolysis of the compound (156) in cyclohexane afforded a mixture of the dimers (157a) and (157b) in low (13%) yield.[128]

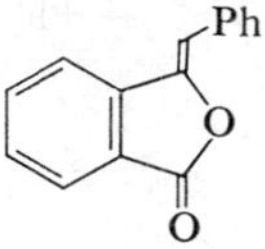

(156)

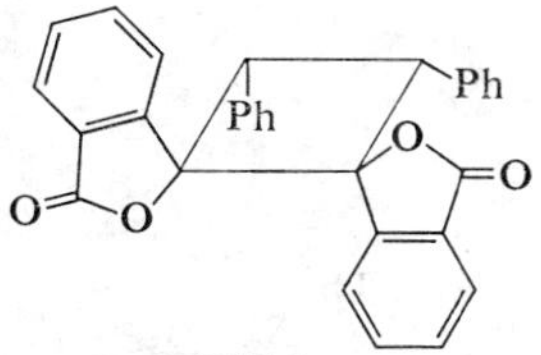

(157a)

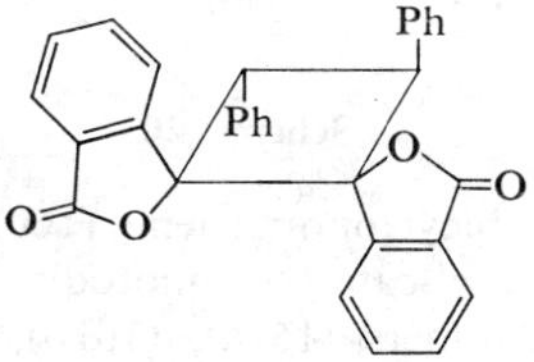

(157b)

A more extensive study of the photodimerisation of 1,2,3-triphenylcyclopropene has been reported.[129] The direct and photosensitized reactions produce two dimers (158a) and (158b), and 1,2-diphenylindene. From the sensitized reactions, it was concluded that the triplet energy of the cyclopropene is 229·5 kJ mol^{-1} (54·9 kcal mol^{-1}). From the sensitized reactions, the *cis,anti,cis*-dimer (158a) and the indene were obtained as the initial products, and an independent experiment established that direct irradiation of the *anti*-dimer (158a) can give the *cis,syn,cis*-dimer (158b). The proposed mechanism involves triplet addition to ground-state cyclopropene (Scheme 20). Similar results were obtained when the cyclopropene was generated *in situ* by decomposition of the potassium salt of 1-*p*-tolylsulphonylhydrazino-1,2,3-triphenylcycloprop-2-ene.[129]

Irradiation of indene in the presence of benzophenone as sensitizer gives rise to the four possible dimers (159a—d) in the ratio 84 : 8 : 3 : 5.[130] The dimerisation process was not markedly temperature-dependent over the range − 50 to + 50 °C. The unsensitized dimerisation led to a corresponding ratio of 69 : 7 : 18 : 6, and probably involves both triplet and singlet excited states (cyclohexa-1,3-diene as quencher). With increasing diene concentration, and therefore decreasing triplet contribution, the proportions of dimers (159a) and (159d) decrease, while that of the head–head *cis,syn,cis*-dimer (159c) increases. Photosensitized dimerisations of 1,2-dimethylene-

[128] E. V. Blackburn and C. J. Timmons, *J. Chem. Soc.* (*C*), 1970, 172.
[129] H. Dürr, *Annalen*, 1969, **723**, 102.
[130] W. Metzner and D. Wendisch, *Annalen*, 1969, **730**, 111.

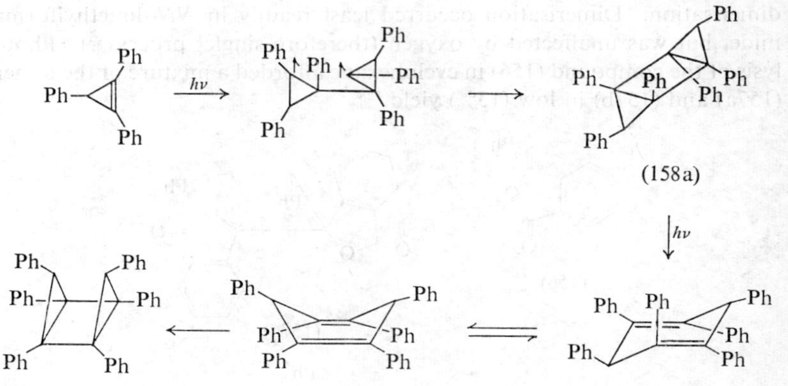

(158a)

(158b)

Scheme 20

cyclobutane[131] and the bicyclononatriene (160)[132] have been reported. The chromium-hexacarbonyl-catalysed photodimerisation of norbornadiene has afforded the three dimers (161a), (161b), and (161c) in the ratio 1·8 : 1·0 : 1·4, and in an overall yield of 40%.[133] This is the first metal-carbonyl-catalysed photodimerisation which has afforded all the three known cyclobutane dimers from norbornadiene. The reaction is thought to

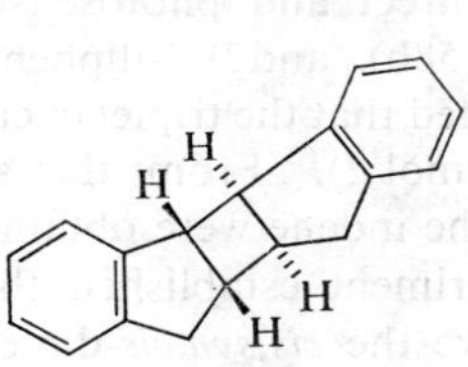

(159a)

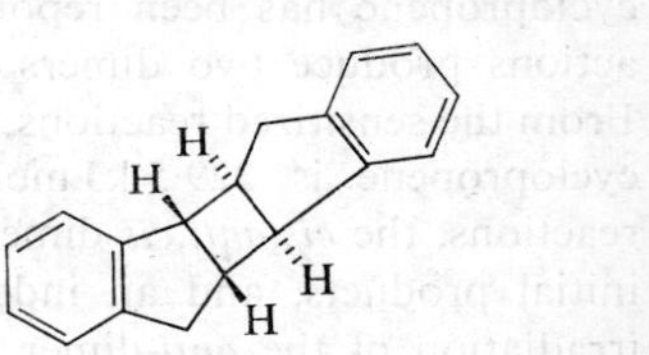

(159b)

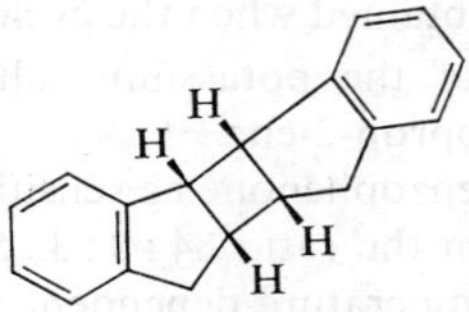

(159c)

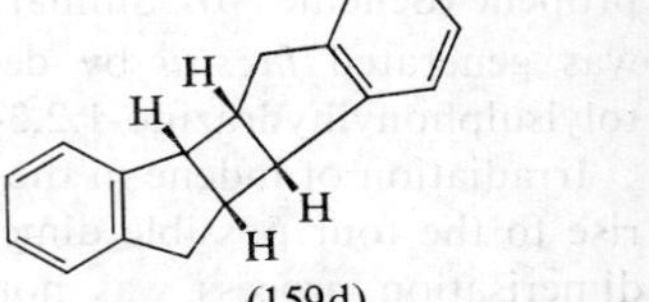

(159d)

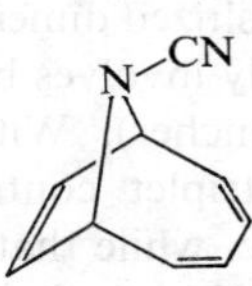

(160)

[131] W. T. Borden, L. Sharpe, and I. V. Reich, *Chem. Comm.*, 1970, 461.
[132] A. G. Anastassiou and R. M. Lazarus, *Chem. Comm.*, 1970, 373.
[133] W. Jennings and B. Hill, *J. Amer. Chem. Soc.*, 1970, **92**, 3199.

(161a)   (161b)   (161c)

proceed *via* the intermediacy of norbornadienylchromium tetracarbonyl (50%) formed by a photochemical reaction.

Intermolecular Cycloadditions.—Considerable interest in the photochemical additions to electron-rich olefins has been shown in the past year.

Direct irradiation ($\lambda = 253 \cdot 7$ nm) or photosensitized irradiation (triphenylene, $\lambda = 350$ nm) of tolan in 2-methyl-4,5-dihydrofuran affords a good yield (82%) of the bicyclic compound (162).[134] From the effectiveness of the sensitization, it is reasoned that the reaction involves the T_1 state of tolan ($E_T = 214$ kJ mol^{-1}, 51 kcal mol^{-1}). But a singlet state was thought to be involved in addition of the same acetylene to dihydropyran.

(162)   (163)   (164a)   (164b)

1,1-Diphenylethylene has been shown to undergo benzophenone-sensitized ($E_T = 289$ kJ mol^{-1}, 69 kcal mol^{-1}) addition to dihydropyran.[135] The most abundant product (55%) was identified as the 2+2 cycloadduct (163). The direction of the addition appears to be different from that found in the addition of ketonic triplets to other electron-rich olefins, the difference being speculatively attributed to the involvement of bond-polarisation. Two further unusual products (164a) and (164b) were isolated in 20 and 25% yields respectively. They apparently arise by hydrogen abstraction from the dihydropyran to produce an allylic radical followed by recombination of the radical pair. Hydrogen abstraction by triplet olefins has not hitherto been reported.

The position of the cyano-group in the products (165a) and (165b) from photoaddition of acrylonitrile to indene has been established by chemical degradation,[136] in confirmation of previous work.[137]

Cyanoacetylene has been shown to undergo sensitized addition to indene to yield the adduct (166).[136]

[134] M. P. Serve and H. M. Rosenberg, *J. Org. Chem.*, 1970, **35**, 1237.
[135] P. Serve, H. M. Rosenberg, and R. Rondeau, *Canad. J. Chem.*, 1969, **47**, 4295.
[136] R. M. Bowman, J. J. McCullough, and J. S. Swenton, *Canad. J. Chem.*, 1969, **47**, 4503.
[137] J. J. McCullough and C. W. Huang, *Chem. Comm.*, 1967, 815; J. J. McCullough and C. W. Huang, *Canad. J. Chem.*, 1969, **47**, 757.

Photoaddition of acrylonitrile to acenaphthylene has been shown to afford two adducts (167a) and (167b), the *syn*- and *anti*-isomers.[138] Table 19

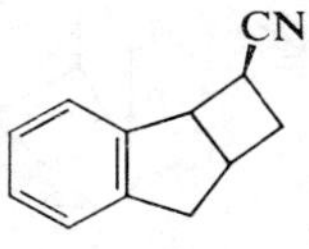

(165a) 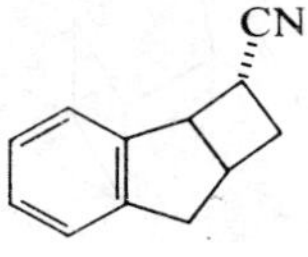 (165b) 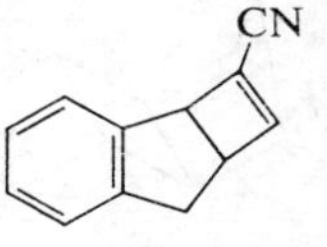 (166)

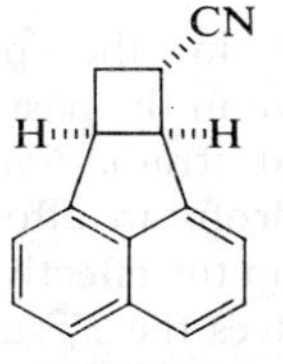

(167a) 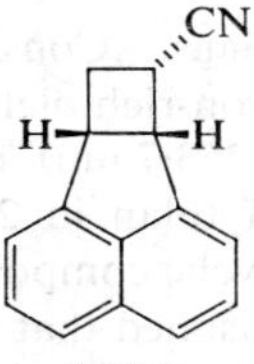 (167b)

shows the yields of products obtained from the reaction, and the effects of heavy-atom solvents on the rates and isomer ratios. Results reported recently on the effect of heavy-atom solvents on the dimerisation of acenaphthylene suggest that intersystem crossing is favoured.[139] Triplet acenaphthylene gave product ratios which were different from those found in the singlet-state reaction. Thus, the effects shown in Table 19 may reflect enhanced intersystem crossing.

Table 19 *Influence of solvent on rates of additions to acenaphthylene*[138]

Solvent	Reaction time (min)	Overall yield (%)	Isomer ratio syn : anti
Dibromomethane	36	43	3·8
Bromoethane	58	40	2·9
1-Bromopropane	99	52	3·4
1-Bromobutane	95	40	3·0
Acetonitrile	240	31	2·5
Cyclohexane	360	42	4·3

Cycloadditions of tolan[140] and dimethyl acetylenedicarboxylate[141] to benzo[*b*]thiophen (168) can be effected by direct irradiation. With tolan, two adducts (169a) and (169b) were isolated, but only low yields of 1 : 1 adducts (169c—e) were isolated from the irradiations in the presence of dimethyl acetylenedicarboxylate. The conversion of adduct (169a) into (169b) can be induced photochemically, and while this process is formally analogous to the rearrangement of the cyclopentenone–but-2-yne adduct,

[138] B. F. Plummer and R. A. Hall, *Chem. Comm.*, 1970, 44.
[139] D. O. Cowan and R. L. Driska, *J. Amer. Chem. Soc.*, 1967, **89**, 1255, 3068.
[140] W. H. F. Sasse, P. J. Collin, and D. B. Roberts, *Tetrahedron Letters*, 1969, 4791.
[141] D. C. Neckers, J. H. Dopper, and H. Wynberg, *Tetrahedron Letters*, 1969, 2913.

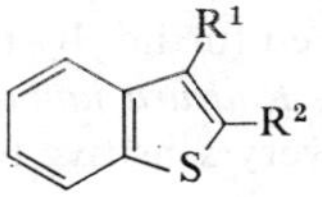

(168)
(a) $R^1 = R^2 = H$
(b) $R^1 = H, R^2 = Me$
(c) $R^1 = Me, R^2 = H$
(d) $R^1 = R^2 = Me$

(169)
(a) $R^1 = R^3 = H, R^2 = R^4 = Ph$
(b) $R^1 = R^2 = H, R^3 = R^4 = Ph$
(c) $R^1 = H, R^2 = Me, R^3 = R^4 = CO_2Me$
(d) $R^1 = Me, R^2 = H, R^3 = R^4 = CO_2Me$
(e) $R^1 = R^2 = Me, R^3 = R^4 = CO_2Me$

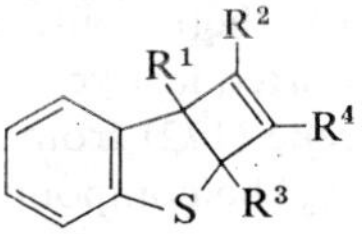

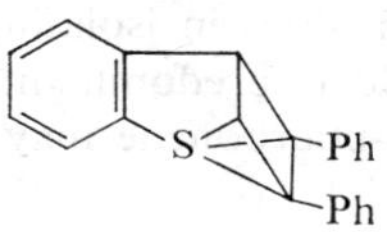

(170)

which involves diradicals,[141] the other authors[140] prefer a bridged intermediate (170). The corresponding additions of halogeno-olefins to benzo[b]thiophens required photosensitization.[142] Diradical intermediates are proposed. Energy transfer from excited benzothiophen also results in the dimerisation of the halogeno-olefins.[142]

An elaboration of the results of cross-addition of isoprene and cyclopentadiene to α-acetoxyacrylonitrile reported last year[143] has now been published[144] and full details have appeared of earlier work on the photoaddition of butadiene to α-acetoxyacrylonitrile.[144] The photosensitized (acetophenone) cycloaddition of butadiene to acrylonitrile has been reported to afford the mixed adducts (171a) and (171b) together with the expected

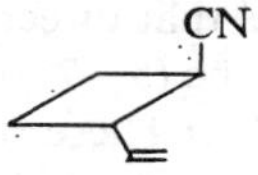

(171a)    (171b)

dimers of butadiene and acrylonitrile. The cross-addition is thought to take place from the triplet state of the diene.[145]

5 Intramolecular Cycloadditions of Dienes *etc.*

(2+2) Cycloaddition: Formation of Cage Compounds.—A review concerning the cycloaddition reactions of non-aromatic conjugated dienes and polyenes has been published.[146]

---

[142] D. C. Neckers, H. J. Dopper, and H. Wynberg, *J. Org. Chem.*, 1970, **35**, 1582.
[143] 'Photochemistry,' ed. D. Bryce-Smith, Specialist Periodical Reports, The Chemical Society, London, 1970, Vol. 1, p. 227.
[144] W. L. Dilling, R. D. Kroening, and J. C. Little, *J. Amer. Chem. Soc.*, 1970, **92**, 928.
[145] W. L. Dilling and R. D. Kroening, *Tetrahedron Letters*, 1970, 695.
[146] W. L. Dilling, *Chem. Rev.*, 1969, **69**, 845.

A possible biosynthetic route to bourbonene, the photochemical synthesis of which was reported last year,[147] has been elucidated further by the isolation of Germacrene (172) from *Pittosporum Tobira, Kadsura japonica*, and *Piper Kadsura*.[148] The compound was found to be very sensitive to acid, a

(172)   (173)

fact which led to great difficulties in isolating the pure material. Photochemical isomerisation yielded predominantly (−)-β-bourbonene (173). The authors suggest that β-bourbonene may arise solely from this intermediate in the plant.

Details of the isomerisation of *cis,cis*-cyclo-octa-1,5-diene in the presence of CuCl have been given.[149] The major product is tricyclo[3,3,0,0^{2,6}]octane, as reported by other workers,[150] and *cis,trans*- and *trans,trans*-cyclo-octa-1,5-diene were also isolated as their complexes with cuprous chloride. The appearance of the tricyclo-octane is slow at the start, but becomes more rapid as the conversion proceeds. This rate effect may result from the intermediate formation of *cis,trans*-cyclo-octa-1,5-diene.

A kinetic study of the cyclisation of norbornadiene to quadricyclane has been published.[151] Prinzbach *et al.* have shown that the $2\pi + 2\sigma$ additions which take place in the tetracyclic compounds (174a) and (175b) involve addition of the ethylenic bond to the *exo*-cyclopropane and not to the *endo*-oxiran or cyclopropane.[152] Two reports [153, 154] on the photochemistry of the tricyclic triene (176) differ remarkably. One group [153] have found the photolysis to be temperature-independent, and that direct irradiation gives dimethyl phthalate (37—40%), cyclo-octatetraene (5—7%), benzene (trace), and the *syn*-dimer of cyclobutadiene (16—18%). These results suggest the intermediate formation of butadiene. The reaction could be partly inhibited by the addition of known triplet quenching dienes, a result which provides some evidence for the intermediacy of a triplet excited state. However, other workers report that cycloaddition product (177) is the only product from direct irradiation (313 nm) at −50 °C.[154] A more extended investigation seems to be needed. Other intramolecular cycloadditions have dealt

---

[147] 'Photochemistry,' ed. D. Bryce-Smith, Specialist Periodical Reports, The Chemical Society, London, 1970, Vol. 1, p. 198.
[148] K. Yoshihara, Y. Ohta, T. Sakai, and Y. Hirose, *Tetrahedron Letters*, 1969, 2263.
[149] G. M. Whitsides, G. L. Goe, and A. C. Cope, *J. Amer. Chem. Soc.*, 1969, **91**, 2608.
[150] R. Srinivasan, *J. Amer. Chem. Soc.*, 1964, **88**, 5084.
[151] G. Kaupp and H. Prinzbach, *Helv. Chim. Acta*, 1969, **52**, 956.
[152] H. Prinzbach, M. Klaus, and W. Mayer, *Angew. Chem.*, 1969, **81**, 902 (*Angew. Chem. Internat. Edn*, 1969, **8**, 883).
[153] R. D. Miller and E. Hedaya, *J. Amer. Chem. Soc.*, 1969, **91**, 5401.
[154] S. F. Nelson and J. P Gillespie, *Tetrahedron Letters*, 1969, 5059.

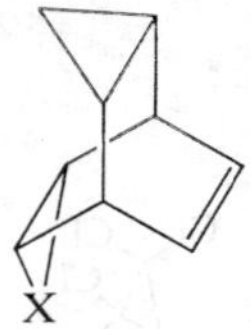

(a) X = O
(b) X = CHCO$_2$Me
(174)

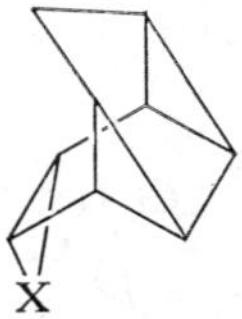

(175)

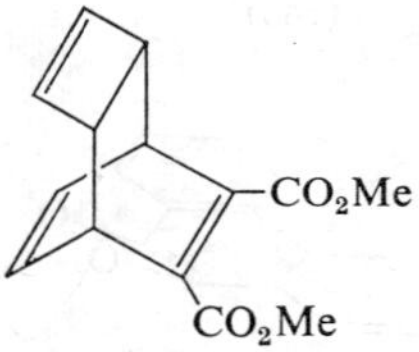

(176)

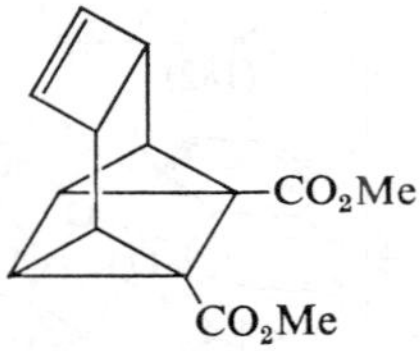

(177)

with the synthesis of compound (178) from (179),[155] and the bishomocubane (180) from (181).[156]

Studies have continued on the sensitized and direct photolysis of cyclodienic insecticides.[157–159] Of special interest is the isolation of a solvent

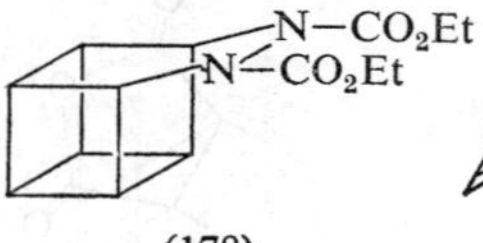

(178)

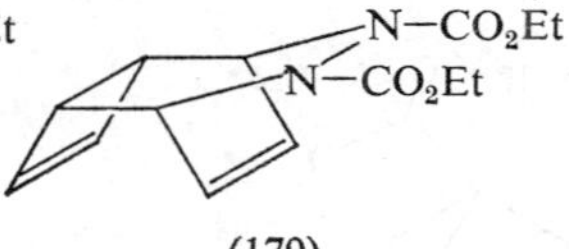

(179)

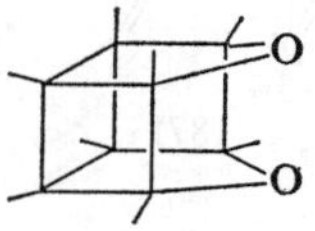

(180)

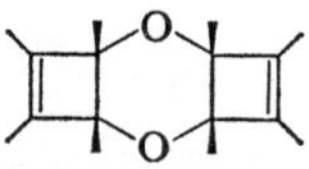

(181)

addition product (182) from the acetone-sensitized photolysis of 'Heptachlor' (183). This seems to be good evidence for the intermediacy of radical reactions in the sensitized photolysis of this class of compound.[159] Two intriguing examples of cycloaddition reactions have been reported in the isolation of adducts (184a) and (184b) from the direct irradiation of (185a)

[155] R. Askani, *Chem. Ber.*, 1969, **102**, 3304.
[156] R. Criegee and R. Rucktäschel, *Chem. Ber.*, 1970, **103**, 50.
[157] M. M. Fischler and F. Korte, *Tetrahedron Letters*, 1969, 2793.
[158] L. Vollner, W. Klein, and F. Korte, *Tetrahedron Letters*, 1969, 2967.
[159] R. R. McGuire, M. J. Zabik, and R. D. Schuetz, *J. Agric. Food. Chem.*, 1970, **18**, 319.
[160] S. Farid and D. Hess, *Chem. Ber.*, 1969, **102**, 3747.

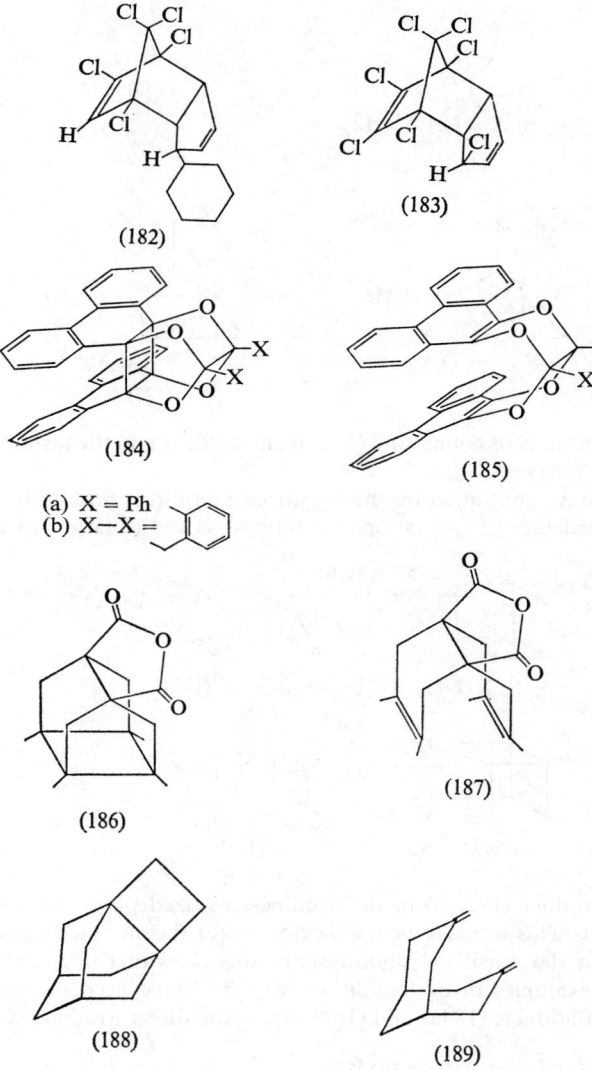

(a) X = Ph
(b) X—X =

and (185b),[160] and compound (186) from sensitized irradiation of (187).[161] The tetracyclic hydrocarbon (188) has been prepared from the bicyclic diene (189).[162]

Valence-bond Isomerisation.—Irradiation of 7,7-dimethoxycycloheptatriene has given the isomeric 4,4-dimethoxybicyclo[3,2,0]hepta-2,6-diene.[163] Similar valence-bond isomerisation to form 2-oxabicyclo[3,2,0]hepta-3,6-diene is reported for the direct irradiation of oxepin.[164]

The photolysis of both 1,3,5-cyclo-octatriene and bicyclo[4,2,0]octa-2,4-diene in an argon matrix at 20 K (240—270 nm) has been shown to yield the same mixture of cis,cis-1,3,5,7-octatetraene, benzene, and ethylene, with little evidence for the formation of more-strained polycyclic isomers.[165] But in the case of the cyclo-octatriene, an isomer was observed which reverted to the starting material on warming the product of irradiation to room-temperature. Direct evidence for the identity of this isomer is tenuous, but the authors suggest that it may be a cyclic isomer having a *trans* double bond in the ring.[165]

Irradiation of the 7,8-dimethylenecyclo-octa-1,3,5-trienes (190a) and (190b) leads mainly to polymerisation reactions, but (190a) gave 10% of bicyclo[6,2,0]deca-1,3,5,7-tetraene (191).[166] The same product was also isolated from photolysis of the bromo-derivative (190b), together with a trace of a mixture of isomers (192a) and (192b). The less labile chloro-derivative (190c) gave much better yields (11%) of the analogous isomers

(190)
(a) R = H
(b) R = Br
(c) R = Cl

(191)

(192)
(a) R^1 = H, R^2 = Br
(b) R^1 = Br, R^2 = H
(c) R^1 = H, R^2 = Cl
(d) R^1 = Cl, R^2 = H

(193)
(a) R^1 = H, R^2 = Cl
(b) R^1 = Cl, R^2 = H

(192c) and (192d) as well as a mixture of hydrocarbons (193a) and (193b). It should be noticed that the chlorine substituent apparently diminishes the reactivity of the *exo*-methylene-diene system, while relatively increasing the reactivity of the ring-triene system. There is a precedent for such rearrangements of cyclic trienes of this type.[167]

[161] W. B. Avila and R. A. Silva, *Chem. Comm.*, 1970, 94.
[162] A. G. Lyurchenko, A. T. Voroshchenko, and F. N. Stepanov, *Zhur. org. Khim.*, 1970, **6**, 189 (*Chem. Abs.*, 1970, **72**, 89,871S).
[163] R. W. Hoffmann, K. R. Eicken, H. J. Luthardt, and B. Dittrich, *Chem. Ber.*, 1970, **103**, 1547.
[164] J. M. Holovka, *Diss. Abs.*, 1969, **29**, B, 3260.
[165] P. Datta, T. D. Goldfarb, and R. S. Boikess, *J. Amer. Chem. Soc.*, 1969, **91**, 5429.
[166] J. A. Elix, M. V. Sargent, and F. Sondheimer, *J. Amer. Chem. Soc.*, 1970, **92**, 969.
[167] See J. Meinwald and P. H. Mazzocchi, *J. Amer. Chem. Soc.*, 1967, **89**, 1755.

6 Fragmentation Reactions

Apparent violations of the Woodward–Hoffmann Rules for the conservation of orbital symmetry occur when the bicyclic diazo-compounds (194a—c) are photolysed at −70 °C through a Pyrex filter.[168] The products from these reactions are the dienes (195a), (195b), and (195c) respectively; and they

(194a)   (194b)   (194c)   (194d)

(195a)   (195b)   (195c)

are also formed by the *thermal* elimination of nitrogen. Berson and Olin [168] suggest that the presence of a strong '*extrasymmetric*' factor opposes the *trans*-photoelimination of nitrogen. This is shown in structure (194d) for the *cis,anti*-isomer and represents a conrotatory twisting of the carbon atoms which bear methyl groups, and would, for a *trans*-elimination, require considerable disruption of the non-reacting N–N π-bond. However, the correlation diagrams which are promised for a later date are said to suggest that the answer to the problem is more complex.

Over the past few years Dürr has published a number of reports relating to the elimination of nitrogen from diazacyclopentadienes and pyrazoles.[169] Further reports on this work have appeared and deal with the addition of cyclopentadienylidene to olefins [170] and acetylenes.[171] A report about a new synthesis of benzocyclopropene by the elimination of nitrogen from a spiropyrazole has been published.[172] The sensitized or direct photolysis of the cyclopropylpyrazoline (196) gives rise to two methylenecyclopropanes (197a) and (197b).[173] These products result from the production of a diradical (197c) which can recombine in two possible ways. There was no evidence for rearrangement of this diradical.

Photolysis of the tricyclic azo-compound (198) in ether solution through a Pyrex filter gave a quantitative conversion to (199).[174] Subsequent pyrolysis of this material gave the norbornadiene (200a), which could be photochemically cyclised to the quadricyclane (200b).

[168] J. A. Berson and S. S. Olin, *J. Amer. Chem. Soc.*, 1970, **92**, 1086.
[169] See 'Photochemistry,' ed. D. Bryce-Smith, Specialist Periodical Reports, The Chemical Society, London, 1970, Vol. 1, pp. 256, 407.
[170] H. Dürr and L. Schrader, *Chem. Ber.*, 1969, **102**, 2026.
[171] H. Dürr and L. Schrader, *Angew. Chem.*, 1969, **81**, 426.
[172] H. Dürr and L. Schrader, *Angew. Chem.*, 1969, **81**, 426 (*Angew. Chem. Internat. Edn*, 1969, **8**, 446).
[173] T. Sanjiki, M. Ohta, and H. Kato, *Chem. Comm.*, 1969, 638.
[174] L. A. Paquette and L. M. Leichter, *J. Amer. Chem. Soc.*, 1970, **92**, 1765.

Binkley and Binkley have reported further on photochemical deiodination of the iodo-sugar (201).[175] The photochemical hydrolysis of the phosphate (202) is best carried out in neutral or acidic solutions. The corresponding benzyl alcohol is obtained in 60—80% yield, and is considered to arise *via* a benzyl carbonium formed by fission of the α-C—O bond.[176]

Caldwell has reported that the sensitized decomposition of 7,8-diacetoxy-bicyclo[4,2,0]octa-2,4-diene gives *cis-* and *trans-*1,2-diacetoxyethylene; but direct irradiation does not give these ethylenes.[177] The triplet nature of the reaction has been inferred from sensitization studies, and the quantum yield has been measured. The formation of both olefins requires that there should be a step where the steric identity of the molecule is lost. Two possibilities have been considered, one involving a step-wise formation of diradical intermediate where rotation about the central bond can take place, and the other where the loss of triplet olefin takes place followed by isomerisation of the excited olefin prior to loss of excitation energy (Scheme 21).[178] Attempts to measure the decay ratio in order to determine the stage at which electronic energy is lost were not successful, so the question remains open. The possibility of an intermediate complex with the sensitizer will also need to be considered in any future discussions of the mechanism.

---

[175] W. W. Binkley and R. W. Binkley, *Carbohydrate Res.*, 1969, **11**, 1.
[176] V. M. Clark, J. B. Hobbs, and D. W. Hutchinson, *Chem. Comm.*, 1970, 339.
[177] R. A. Caldwell, *J. Org. Chem.*, 1969, **34**, 1886.
[178] A. P. Bindra, J. A. Elix, and M. V. Sargent, *Austral. J. Chem.*, 1969, **22**, 1449.

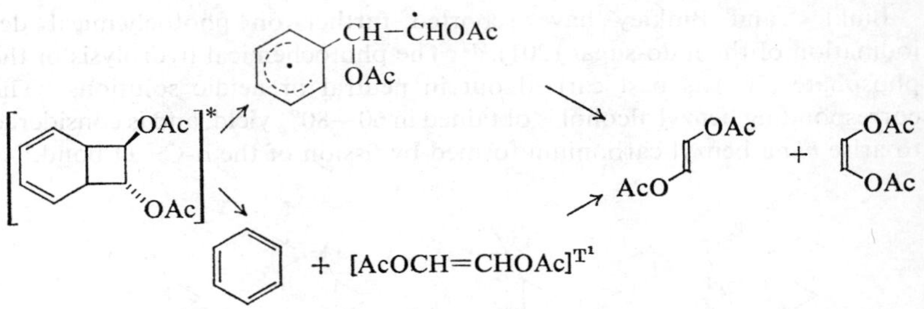

Scheme 21

Irradiation of di-*trans*-1,2,3,4-dibenzo[7,8-*c*]furo[10]annulene (203) in ethanol gave phenanthrene and phenol (Equation 4).[178] Note that hydrogen-transfer, presumably from the solvent, must be involved.

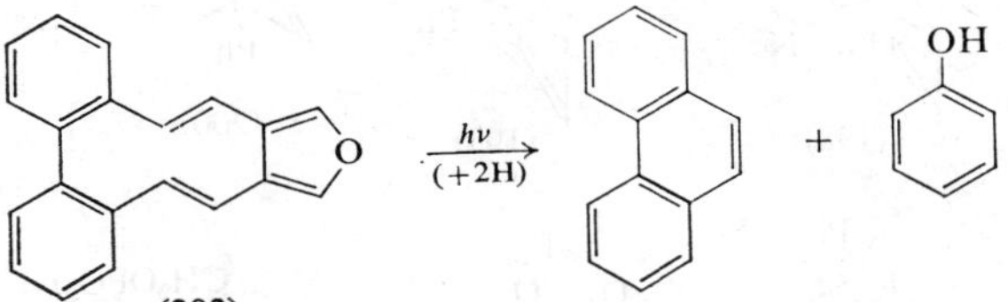

Photolysis of the ozonide (204) at $-20\,°C$ in methylene chloride is reported as a useful method for the synthesis of the Dewar-benzene derivative (205).[179] 2,3-Dimethylnaphthalene is also formed, presumably by a subsequent thermal isomerisation.

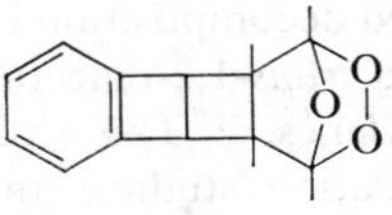

7 Cyclopropyl Compounds

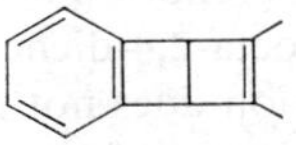

Photolysis of the cyclopropyl-steroid (206) in dry dioxan using a Corex-filtered light source afforded one product (207) in 35% yield.[180] The proposed route to this product involves decarbonylation, a Norrish type I process, and ring-opening of the resultant cyclopropylcarbinyl radical (Scheme 22). The reaction sequence is completed by hydrogen abstraction from an unidentified source (solvent ?). This behaviour is typical of the reactions of cyclopropylcarbinyl radicals, and adds little to our knowledge of such processes.

[179] D. T. Carty, *Tetrahedron Letters*, 1969, 4753.
[180] K. Kojima, R. Hayashi, and K. Tanabe, *Chem. and Pharm. Bull. (Japan)*, 1970, **18**, 88.

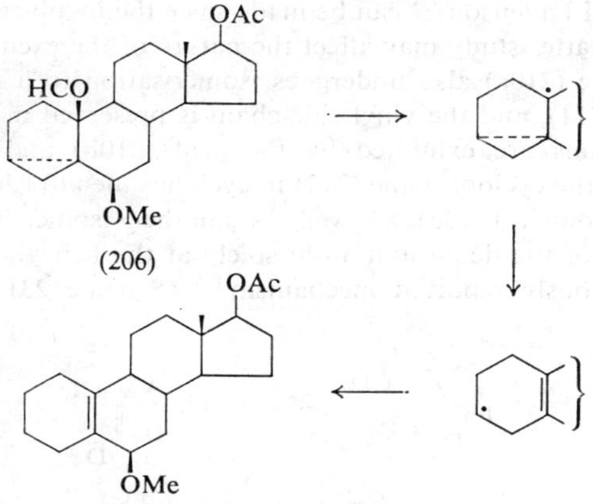

Scheme 22

A study of simple vinylcyclopropanes (208a—d) has shown that both conformational and electronic effects play a large part in the nature of the product formed on prolonged photolysis in hexane.[181] The final products are cyclopentenes (209). Prior to the formation of these, there is a rapid cis–trans-isomerisation of the starting olefin, and it is therefore possible to start from either isomer, or from a mixture of the two. It should be noticed

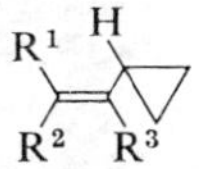

(208)
(a) $R^1 = R^3 = H$, $R^2 = Ph$
(b) $R^1 = Ph$, $R^2 = R^3 = H$
(c) $R^1 = H$, $R^2 = Ph$, $R^3 = Me$
(d) $R^1 = Ph$, $R^2 = H$, $R^3 = Me$
(e) $R^1 = Pr^i$, $R^2 = Ph$, $R^3 = H$
(f) $R^1 = Ph$, $R^2 = Pr^i$, $R^3 = H$
(g) $R^1 = R^2 = H$, $R^3 = Me$

(209)
(a) $R = H$
(b) $R = Me$

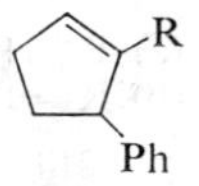

(210)
(a) $R = CO_2Et$
(b) $R = CO_2H$
(c) $R = CONH_2$

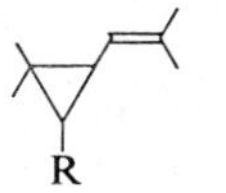

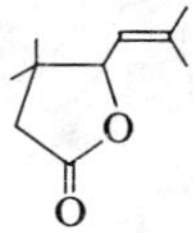

(211)

that the olefins which undergo this reaction bear only one substituent on C-1 of the double bond. Indeed, when C-1 is disubstituted (208e, f), no reaction other than isomerisation and polymerisation takes place. The analogous rearrangement of vinylcyclopropane (208g) to 1-methylcyclopentene has been shown by Cooke to involve a singlet excited state.[182] The formation of the cyclopentene showed zero-order kinetics ($1 \cdot 1 \times 10^{-3}$ mol l^{-1} hr^{-1}). No comparison between this study [182] and that of

[181] P. H. Mazzocchi and R. C. Ladenson, *Chem. Comm.*, 1970, 469.
[182] R. S. Cooke, *Chem. Comm.*, 1970, 454.

Mazzocchi and Ladenson [181] can be made since the involvement of the aryl groups in the latter study may affect the nature of the excited state. Ethyl chrysanthemate (210a) also undergoes isomerisation to a five-membered-ring lactone (211), and the vinyl side-chain is preserved in the product.[183] Similar reactions are exhibited by the acid (210b) and amide (210c). Irradiation of the cyclopropane (212) in cyclohexane afforded the expected rearranged product (213a) as well as another isomer (213b).[184] The incorporation of the deuterium atom solely at the benzylic carbon shows that the previously reported mechanism [185] (Scheme 23) involving ring

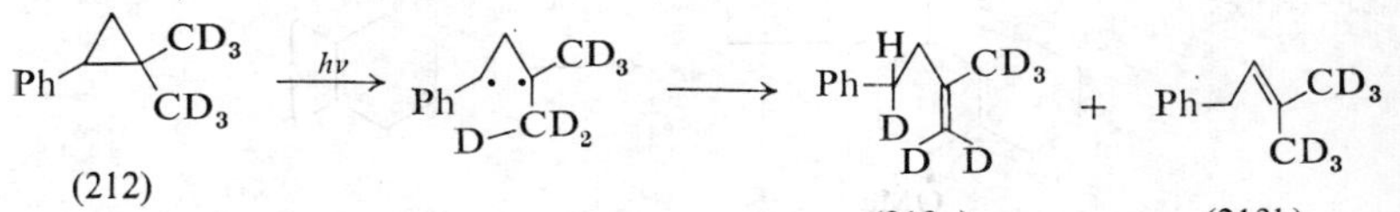

$$Ph\underset{(212)}{\triangle}\genfrac{}{}{0pt}{}{CD_3}{CD_3} \xrightarrow{h\nu} Ph\underset{D-CD_2}{\triangle}CD_3 \longrightarrow \underset{(213a)}{Ph-\overset{H}{\underset{D}{C}}=\underset{D\;D}{C}-CD_3} + \underset{(213b)}{Ph-\overset{}{C}=\underset{CD_3}{C}-CD_3}$$

Scheme 23

fission and proton or deuteron transfer, to afford the olefin, has some chemical justification. It has not yet been ascertained if the reaction is concerted. Benzonorcaradiene (214a) is reported to rearrange on photolysis

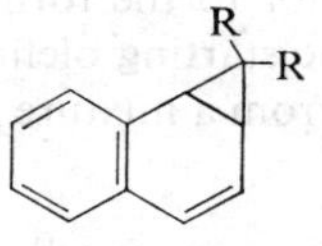

(a) R = H
(b) R = D
(214)

to afford naphthalene (46%), 1-methylnaphthalene (4%), benzobicyclo-[3,2,0]hepta-2,6-diene (2%), 1,2-benzotropilidene (2%), and 3,4-benzotropilidene (2%).[186] The photolysis of the deuteriated material (214b) has elucidated the route by which the transformation comes about. Thus the benzotropilidene products arise by a suprafacial (1,5) carbon migration in much the same way as reported by Zimmerman et al.[96] and Tabata and Hart.[95]

The photolysis of azabicyclohexene (215a) in benzene solution using a Pyrex filter has been shown to be temperature dependent.[187] Thus at 15 °C, the thermally labile triene (216a) was detected, and was converted

---

[183] T. Sasaki, S. Eguchi, and M. Ohno, *J. Org. Chem.*, 1970, **35**, 790.
[184] P. H. Mazzocchi, R. S. Lustig, and G. W. Graig, *J. Amer. Chem. Soc.*, 1970, **92**, 2169.
[185] H. Kristinsson and G. W. Griffin, *J. Amer. Chem. Soc.*, 1966, **88**, 378; *Tetrahedron Letters*, 1966, 3259.
[186] G. W. Gruber and M. Pomerantz, *J. Amer. Chem. Soc.*, 1969, **91**, 4004.
[187] A. Padwa, S. Clough, and E. Glazer, *J. Amer. Chem. Soc.*, 1970, **92**, 1778.

(215)
(a) R^1 = Ph, R^2 = H
(b) R^1 = H, R^2 = Ph

(216)
(a) R^1 = R^4 = Ph, R^2 = R^3 = H
(b) R^1 = R^3 = H, R^2 = R^4 = Ph

(217a)

(217b)

into the cis-pyrazine (217a) on warming the solution to 50 °C. This same product was isolated when the photolysis was carried out at 50 °C. Irradiation of the isomeric bicyclohexene (215b) at 50 °C also gave the cis-pyrazine (217a) exclusively. This was an unexpected result since the formation of the cis-pyrazine from the isomer (215a) could be explained on the basis of the Woodward–Hoffmann Rules and it was expected that the other isomer would be isolated from the photolysis of (215b) The result is explained on the grounds that there are two isomeric forms of the triene (216a and b), and after initial formation of the isomer (216b) from bicyclohexene (215b) there is a rapid conversion to the alternative form (216a), which thermally cyclises to the pyrazine (217a). Photolysis of the trans-pyrazine (217b) gives the triene (216), which thermally cyclises to the cis-isomer, while the photolysis of the cis-pyrazine (217a) affords a triene (216b) which regenerates starting material on warming.[187]

An elaboration of the results [188] on the photolysis of several oxirans published in note form last year has appeared.[189] The spectroscopic results published in the new paper [188] show that the oxirans (218a—e) exhibit highly structured u.v. absorptions at 271 nm at −196 °C. A more intense structureless band appears at 230 nm. Irradiation into the 0—0 band causes each of the oxirans to fluoresce (λ_{max}. 305—310 nm). Tetraphenyloxiran (218e) shows phosphorescence at 404 nm. Apart from phenyloxiran (218a), all the oxirans show the formation of a coloured intermediate (Table 20) on photolysis. This coloured intermediate on further photolysis either reverts to the oxiran, or else fragments to the carbene and ketonic products which have become associated with this type of fragmentation. The coloured intermediate was shown conclusively to be involved in the reaction pathway

[188] R. S. Becker, R. O. Bost, J. Kolc, N. R. Bertoniere, R. L. Smith, and G. W. Griffin, J. Amer. Chem. Soc., 1970, 92, 1302.
[189] 'Photochemistry,' ed. D. Bryce-Smith, Specialist Periodical Reports, The Chemical Society, London, 1970, Vol. 1, pp. 266, 267.

Table 20 U.v. data for coloured intermediates formed by photolysis of oxirans [188]

Oxiran (218)	U.v. absorptions of intermediate (nm)	Colour
(a) $R^1 = Ph; R^2 = R^3 = R^4 = H$	—	—
(b) $R^1 = R^3 = Ph; R^2 = R^4 = H$	506·5	orange-red
(c) $R^1 = R^4 = Ph; R^2 = R^3 = H$	501	orange-red
(d) $R^1 = R^2 = R^3 = Ph; R^4 = H$	398, 547	pink
(e) $R^1 = R^2 = R^3 = R^4 = Ph$	320, 465, 435, 605	blue

by the observation that irradiation into the absorption bands of this intermediate brought about product formation. The products from these oxiran fragmentations usually contain olefinic material, and it was thought that olefins could be formed by the dimerisation of the carbene intermediates (Equation 5). However, it has been shown by product analysis,

$$\underset{R^2\quad R^4}{\overset{R^1\quad O\quad R^3}{\diagdown\!\diagup}}\quad (218)$$

$$(218) \longrightarrow \underset{R^2\quad\quad R^4}{\overset{R^1\quad O^{\cdot}\quad R^3}{\diagdown\!\diagup}}\overset{\cdot}{C}-\overset{\cdot}{C} \longrightarrow \underset{R^2}{\overset{R^1}{\diagdown}}C=O + \underset{R^4}{\overset{R^3}{\diagup}}C: \longrightarrow \underset{R^4\quad R^4}{\overset{R^3\quad R^3}{\diagdown\!\diagup}} \quad (5)$$

$$(218) \longrightarrow \underset{\underset{(i)}{R^2\quad\quad R^4}}{\overset{R^1\quad O^{\cdot}\quad R^3}{\diagdown\!\diagup}}\overset{\cdot}{C}-\overset{\cdot}{C} \longrightarrow \underset{R^2\quad R^4}{\overset{R^1\quad R^3}{\diagdown\!\diagup}} + [O] \quad (6)$$

viscosity effects, and by a comparative study of *cis-* and *trans-*diphenyloxiran (218b) and (218c), that the olefins are formed by a hetero-atom extrusion process (Equation 6). The results are best explained by the implication of a common intermediate (i) for the formation of all the products detected.

Further interest in the coloured intermediates produced in the low-temperature photolysis of aryl-oxirans in rigid media has been shown in the last year.[190, 191] It has been shown that the ring-opening and ring-closing of stilbene epoxides following photolysis are stereospecific. It is suggested [190, 191] that the intermediate is a carbonyl ylide (219) formed by disrotatory ring-opening. Thus, the intermediate (219a) (λ_{max} 490 nm, $\varepsilon > 10^4$) is formed from the *cis-*stilbene epoxide by disrotatory opening outwards, thereby avoiding serious steric interaction between the phenyl

---

[190] T. Do-Mihn, A. M. Trozzolo, and G. W. Griffin, *J. Amer. Chem. Soc.*, 1970, **92**, 1402.

[191] D. R. Arnold and L. A. Karnischky, *J. Amer. Chem. Soc.*, 1970, **92**, 1404.

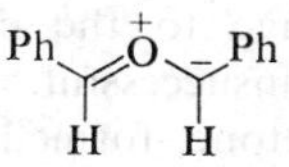

(219a) 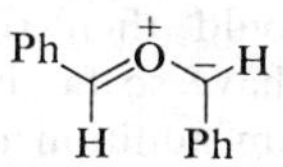 (219b) 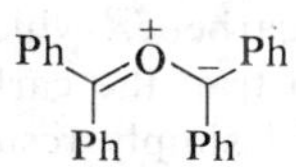 (219c)

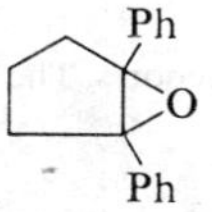

(220a) 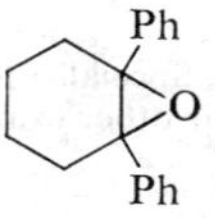 (220b) 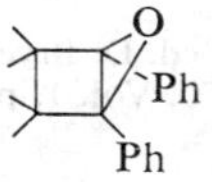 (220c) 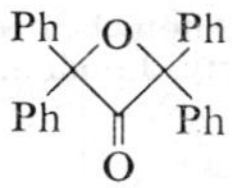 (221)

groups. In the case of the *trans*-epoxide, the intermediate (219b) (λ_{max} 510 nm, $\varepsilon > 10^4$) has a bathochromic shift due to the phenyl interaction. Similar highly-coloured intermediates result from the photolysis of epoxides (220a), (220b),[190] and (220c),[191] and have absorptions at 525, 538, and 550 nm respectively. A blue intermediate (219c), (λ_{max} 318, 420, and 585 nm) results from the photolysis of the oxetanone (221).

Benzene-sensitized irradiation of oxaspiropentane (222) gives rise to three volatile products, *viz.* a cumulene (223), an allene alcohol (224), and an allene-oxetan (225).[192] The oxetan (225) was proposed as a secondary product, since g.l.c. monitoring of the reaction showed only the cumulene and the alcohol as the initial products. Scheme 24 shows possible routes

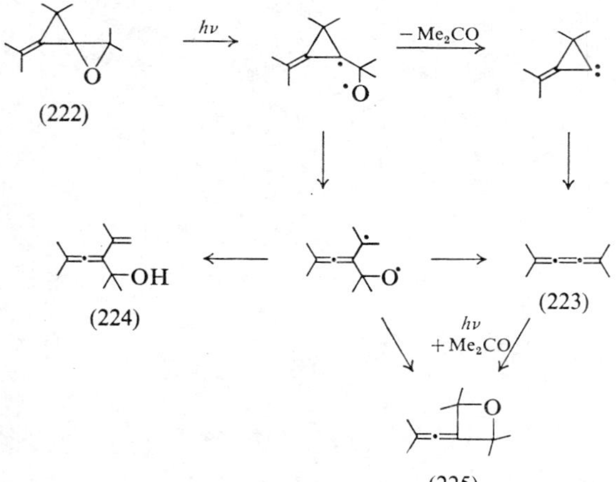

Scheme 24

to the products. The cumulene could result from two routes, (*a*) by the two-step elimination of acetone from the biradical intermediate, the precursor of the alcohol, or (*b*) by fragmentation of the oxiran to yield the

[192] J. K. Crandall and D. R. Paulson, *Tetrahedron Letters*, 1969, 2751.

cyclopropylcarbene, which would then rearrange to the cumulene.[193] Attempts to trap the carbene have so far been unsuccessful. The allene-oxetan could simply result from addition of acetone, formed during the reaction, to the cumulene.

W. M. H.

[193] See 'Photochemistry,' ed. D. Bryce-Smith, Specialist Periodical Reports, The Chemical Society, London, 1970, Vol. 1, p. 267, for other examples.

4
Photochemistry of Aromatic Compounds

Again this year there have been many publications concerned with the varied aspects of the photochemistry of aromatic systems. Apart from energy-transfer phenomena, the fields of aromatic substitution and cycloaddition have received particular attention.

1 Energy transfer: Isomerization Reactions

Hexyl azide is reported to be an efficient quencher of the fluorescence of naphthalene, phenanthrene, triphenylene, pyrene, and 1,2-benzanthracene.[1] In accord with an endothermic collisional energy transfer mechanism, the efficiency of fluorescence quenching is related to the singlet energy of the donor. The observed quenching is consistent with the reaction scheme outlined in equations (1—6.)

$$S \xrightarrow{h\nu} {}^1S \qquad S \equiv \text{Aromatic compound} \qquad (1)$$
$$^1S \longrightarrow S + h\nu \qquad (2)$$
$$^1S \longrightarrow {}^3S \qquad (3)$$
$$^1S + A \longrightarrow S + {}^1A \qquad A \equiv \text{Azide} \qquad (4)$$
$$^1A \longrightarrow \text{Products} \qquad (5)$$
$$^1A \longrightarrow A \qquad (6)$$

Energy transfer by donors with low singlet energies was found to be more efficient than expected for classical endothermic energy transfer, as was previously observed for triplet sensitization of the decomposition of alkyl azides.[2] The results for singlet and triplet sensitization of the azides are best explained by vertical energy transfer to a bent ground-state azide thereby generating a bent excited-state.[3] Singlet sensitization leads to azide decomposition with an efficiency similar to that for the direct photolysis.

The fluorescence yields and triplet yields of toluene vapour as a function of exciting wavelengths and pressures have been determined.[4] The technique based on sensitized emission of biacetyl was used to determine

---

[1] F. D. Lewis and J. C. Dalton, *J. Amer. Chem. Soc.*, 1969, **91**, 5260.
[2] F. D. Lewis and W. H. Saunders, *J. Amer. Chem. Soc.*, 1968, **90**, 7033.
[3] F. D. Lewis and W. H. Saunders, *J. Amer. Chem. Soc.*, 1968, **90**, 7031.
[4] S. L. Lem, G. P. Semeluk, and I. Unger, *Canad. J. Chem.*, 1969, **47**, 4711.

the triplet yields. With 266, 260, and 254 nm exciting radiation, the sum of the two yields is near to unity. The reported data confirm recently published results of Burton and Noyes with toluene, when the *cis* → *trans* isomerization of olefins (Cundall's technique) was used as the triplet probe.[5] Cundall and Tippett have also determined triplet yields of toluene and benzene using the *cis* → *trans* isomerization of but-2-ene.[6] At 25 °C, the respective quantum yields of benzene and toluene triplets were 0·57 and 0·45 with internal conversion efficiencies of *ca.* 0·41 and 0·48. Addition of xenon to the systems increases the triplet yield to unity.

Unger and co-workers have continued their investigations of the photochemistry of fluorinated benzenes and have determined the singlet and triplet quantum yields of monofluorobenzene with radiation in the 235—271 nm region.[7] The triplet yield was based on the sensitized emission of biacetyl. The sums of the quantum yields with exciting wavelengths of 247, 255, 259, and 267 nm were 0·67, 0·75, 0·93, and 0·82 respectively. It is also suggested that the values require some correction in the light of previous work.[8,9] The effects of carbon dioxide, chloroform, and an external heavy atom on fluorobenzene photochemistry have also been examined.[10,11] With exciting wavelengths of 247, 259, and 267 nm, the yield of triplet fluorobenzene in the presence of both carbon dioxide and chloroform is increased by over 40%, *ca.* 30%, and *ca.* 20% respectively.[10] In the latter measurements, chloroform was found to be slightly more effective than carbon dioxide. The quantum yield for fluorescence is unaffected by either of these two additives. On the other hand, experiments with benzene with 254-nm radiation show that the biacetyl fluorescence yield decreases with increasing chloroform pressure. This confirms previously reported work,[12] but the present results suggest that the effect is not as great as originally described. Carbon dioxide does not affect the benzene triplet yield, and again the fluorescence is unaffected. The external heavy atom for this study was provided by xenon, and as found previously with benzene and toluene (ref. 6), xenon evidently enhances the intersystem crossing in both fluorobenzene and biacetyl.[11] The key processes are outlined in equations (7) and (8).

$$^1C_6H_5F + Xe \longrightarrow {}^3C_6H_5F + Xe \qquad (7)$$

$$^3(MeCO)_2 + Xe \longrightarrow (MeCO)_2 + Xe \qquad (8)$$

The spectrofluorometric studies with *o*-difluorobenzene have been reexamined with narrower exciting band widths at 249, 254, 258, 266, and

---

[5] C. S. Burton and W. A. Noyes, *J. Chem. Phys.*, 1968, **49**, 1705.
[6] R. B. Cundall and W. Tippett, *Trans. Faraday Soc.*, 1970, **66**, 350.
[7] M. E. McBeath, G. P. Semeluk, and I. Unger, *J. Phys. Chem.*, 1969, **73**, 995.
[8] G. M. Almy and P. R. Gillette, *J. Chem. Phys.*, 1943, **11**, 188.
[9] D. Phillips, *J. Phys. Chem.*, 1967, **71**, 1839.
[10] M. E. McBeath and I. Unger, *Canad. J. Chem.*, 1970, **48**, 1607.
[11] A. Cook, G. P. Semeluk, and I. Unger, *Canad. J. Chem.*, 1969, **47**, 4527.
[12] S. H. Ng, G. P. Semeluk, and I. Unger, *Canad. J. Chem.*, 1968, **46**, 2461.

270 nm.[13] The data indicate that the fluorescence yields are nearly the same as those reported previously.[14] The effect of vibrational quenchers on such data has also been investigated at these wavelengths. Carbon dioxide and chloroform do not affect the fluorescence yield at wavelengths greater than 249 nm, but at 249 nm addition of both of these compounds increases the fluorescence yield, and in each case the triplet yield, as with fluorobenzene, increases at every exciting wavelength examined.

In contrast with previous years, there have been few reports in the year under review concerned with the photoisomerization of aromatic compounds. But one most interesting account again originates from the Argonne Laboratories and describes the first valence isomer of pyridine, 2-azabicyclo[2,2,0]hexa-2,5-diene (1), 'Dewar' pyridine.[15] The irradiation is carried out in the liquid phase at 254 nm, and the half-life of (1) is reported to be 2 min at 25 °C and 36 min at 0 °C. In the presence of

$$H_2N-CH=CH-CH=CH-CHO$$
(3)

aqueous sodium borohydride, photo-reduction occurs to yield 2-azabicyclo-[2,2,0]hex-5-ene (2). The 'Dewar' pyridine has been shown to be the intermediate in the formation of (2) as well as in the photohydration of pyridine to give 5-aminopenta-2,4,-dienal (3). These results coupled with preliminary studies of the photolysis of monoazoles and diazoles in aqueous sodium borohydride suggest that the formation of transient non-aromatic isomers upon irradiation may be a general phenomenon among nitrogen heterocycles.

The photo-interconversion of the xylenes has been known for several years, and the reaction has been shown to be intramolecular,[16] probably involving mainly benzvalene (4) and (less probably) prismane (5) intermediates. The conversion of *m*-xylene to the *o*- and *p*-isomers in n-hexane, E.P.A., and perfluorohexane solutions has been further studied with 248 and 275 nm radiation.[17] The u.v. spectra of *m*-xylene in the three solvents are identical, so the media are suggested to act only as vibrational energy

---

[13] S. H. Ng, G. P. Semeluk, and I. Unger, *Ber. Bunsengesellschaft Phys. Chem.*, 1970, **74**, 29.
[14] J. L. Durham, G. P. Semeluk, and I. Unger, *Canad. J. Chem.*, 1968, **46**, 3177.
[15] K. E. Wilzbach and D. J. Rausch, *J. Amer. Chem. Soc.*, 1970, **92**, 2178.
[16] L. Kaplan, K. E. Wilzbach, W. G. Brown, and S. S. Yang, *J. Amer. Chem. Soc.*, 1965, **87**, 675.
[17] D. Anderson, *J. Phys. Chem.*, 1970, **74**, 1686.

sinks. The quantum yields for formation of the isomers were found to be the same at these two wavelengths and with the three solvents, but they increased with reaction temperature. The reaction was studied at temperatures between -10 and $+80\,°C$. For isomerization of m-xylene, the quantum yields for formation of the o- and p-isomers were 0·00075 and 0·0032 respectively. Since both yields have an activation energy of $ca.$ 4·7 kcal mol^{-1}, the $o:p$ ratio is independent of the reaction parameters studied. From the data it is not possible to decide whether intermediates such as (4) and (5) are involved in the reaction or not. The isomerization reaction accounts for only a fraction of the parent compound which disappears on photolysis, and no other products than xylenes were identified. Specifically, no hydrogen, methane, or ethyltoluene was observed. The loss of xylene may be due to polymer formation.

The formation of the 'Dewar' isomer (6) from the irradiation of hexafluorobenzene was reported in 1966.[18] The extent of this photoisomerization with substituted pentafluorobenzenes has now been investigated.[19] Isomers of types (7) and (8) are produced with H, CF_3, CH_3, and OCH_3 substituents, but no reaction is observed when Cl or $CH=CH_2$ groups are present. Reasons for this are not discussed. Most chemists would consider that 'Dewar' isomers are inherently more thermally unstable than the parent benzenoid compounds, and there is ample documented evidence for this. However, in contrast with this belief, Haszeldine and co-workers report that perfluorohexamethylbenzene (9) yields the *para*-bonded valence isomer (10), in a flow *pyrolysis* apparatus at 400 °C with a contact time of 1 s, thereby affording a ready quantitative route to this structural system.[20] Compound (10) is converted back to the aromatic form by heating at 140 °C for 6 h. The thermal formation of isomer (10) by a concerted process would be forbidden on orbital symmetry grounds, and the only obvious driving force is the relief of overcrowding in the aromatic isomer (9). This is one of the most remarkable reactions to have been described during the year.

[18] G. Camaggi, F. Gozzo, and C. Cevidalli, *Chem. Comm.*, 1966, 313.
[19] E. Ratajczak, *Roczniki Chem.*, 1970, **44**, 447.
[20] E. D. Clifton, W. T. Flowers, and R. N. Haszeldine, *Chem. Comm.*, 1969, 1216.

R = H, CF₃, Me, OMe

The mechanism of the photoisomerization of 1,2-dihydronaphthalene has been studied by separate deuterium labelling of the olefin (11) and a pair of methylene hydrogen atoms (12).[21] By this approach, it has been

shown that the photoisomerization to benzobicyclo[3,1,0]hexene proceeds by opening and reclosure of the cyclohexadiene ring rather than by hydrogen migration.

2 Addition Reactions

Since the first formulation of the Woodward–Hoffmann Rules, many workers have applied the principle of the conservation of orbital symmetry to both inter- and intra-molecular photochemical and thermal reactions of aliphatic compounds. In 1966, Bryce-Smith and Longuet-Higgins reported their analysis of the photoisomerization of, and addition to, the benzene ring in terms of the orbital and state symmetries of the excited singlet and

[21] R. C. Cookson, S. M. de B. Costa, and J. Hudec, *Chem. Comm.*, 1969, 1272.

triplet states of benzene.[22] This treatment has now been extended, and numerous known and unknown concerted cycloadditions to the benzene ring have been termed as 'forbidden' or 'allowed' following orbital symmetry analysis of each system.[23] Reactions considered include 1,2-, 1,3-, and 1,4-cycloadditions of olefins and dienes to yield compounds (13), (14),

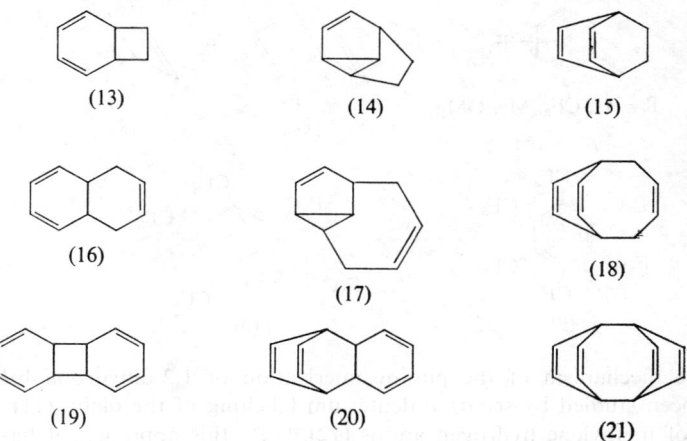

and (15), and (16), (17), and (18) respectively. Dimerizations of benzene to (19), (20), and (21) have also been treated in the same manner. The results of the orbital symmetry analysis have been tabulated for the following cases: (a) excitation of the aliphatic addends, (b) excitation to the $B_{2u}(S_1)$ and $B_{1u}(S_2, T_1)$ states of benzene, (c) excitation of charge-transfer from donor benzene and to acceptor benzene, and (d) thermal initiation. The analysis indicates that the Woodward–Hoffmann Rules apply to the thermal additions,† but not to the light-induced reactions. For example, ss and as additions may be equally allowed, in marked contrast with the aliphatic systems analysed classically by Woodward and Hoffmann.‡ The conclusions are in accord with observed results, and provide various interesting predictions. The author makes the observation that although the

[22] D. Bryce-Smith and H. C. Longuet-Higgins, *Chem. Comm.*, 1966, 593.

† There is an obvious ambiguity in applying the Rules to cases of 1,4-addition to benzene.
‡ The Woodward–Hoffmann Rules were derived on the basis of the orbital and state symmetry properties of non-aromatic hydrocarbon systems and their ($\pi\pi^*$) excited states. The inapplicability of the Rules to excited states of benzene may be understood to result from the orbital degeneracy associated with this D_{6h} system which leads to two orbital components in each of the $S_1(^1B_{2u})$ and $S_2(^1B_{1u})$ states. The two components will normally have different symmetry properties with respect to the symmetry elements preserved in the transition state of a cycloaddition process, but since each is equally part of the excited state, either may serve to establish a valid correlation with orbitals in a low-lying state of the product. Although this duality causes the Woodward–Hoffmann Rules as such to be inapplicable in benzene photochemistry, the principles underlying the Rules are equally valid for aromatic and non-aromatic systems.

concerted pathway for reaction normally provides a kinetic advantage, the mode and speed of the process will be governed by kinetic, steric, and thermodynamic factors as usual. Thus where 1,2- and 1,4-cycloaddition processes are 'allowed', the former may be favoured since its transition state is likely to involve less distortion of the planar benzene ring. Although the relationships do not necessarily apply to condensed aromatic hydrocarbons, they should retain some force for cases of substituted benzenes and addends of different molecular symmetry where the substituents do not greatly perturb the electronic systems of the parent molecules.[23]

The 1,3-inter- and -intra-molecular cycloadditions of olefins to benzene to yield such products as (22) and (23) have been investigated by several groups of workers.[24-27] The kinetics of the intermolecular process of cycloaddition of *cis*- and *trans*-but-2-ene to excited benzene in the gas phase have been studied.[28] The adducts (24) and (25) are formed stereospecifically from the *cis*- and *trans*-olefins respectively. These results, together with a study of the effects of oxygen and biacetyl on the system, indicate the involvement of singlet rather than triplet states of benzene. This is in accord with prediction.[22, 23] The cadmium-photosensitized reaction of benzene with *cis*-but-2-ene has also been reported,[29] but in this investigation the workers were more concerned with the lifetime of the benzene triplet, which was estimated to be of the order of 10^{-5} s, than with the chemistry.

[23] D. Bryce-Smith, *Chem. Comm.*, 1969, 806.
[24] K. E. Wilzbach and L. Kaplan, *J. Amer. Chem. Soc.*, 1966, **88**, 2066.
[25] D. Bryce-Smith, A. Gilbert, and B. H. Orger, *Chem. Comm.*, 1966, 512.
[26] H. Morrison and W. I. Feree, *Chem. Comm.*, 1969, 268.
[27] K. Kraft, Ph.D. Thesis, University of Bonn, 1968.
[28] A. Morikawa, S. Brownstein, and R. J. Cvetanovic, *J. Amer. Chem. Soc.*, 1970, **92**, 1471.
[29] S. Tsunashima, S. Saton, and S. Sato, *Bull. Chem. Soc. Japan*, 1969, **42**, 1531.

The liquid-phase irradiation of hexafluorobenzene in the presence of cis-cyclo-octene is reported to yield seven 1 : 1 adducts.[30] The spectroscopic properties of the separated adducts were consistent with the structures (26), (27), (28), and (29). Although the other three components of the

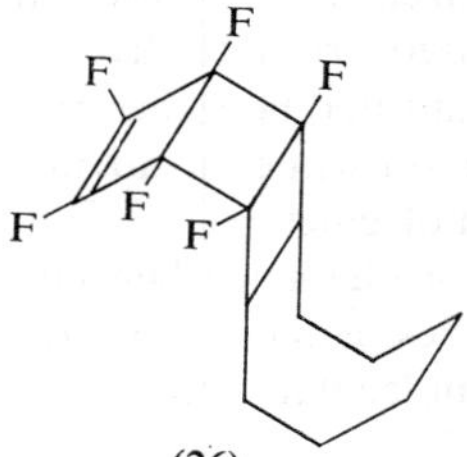

(26) 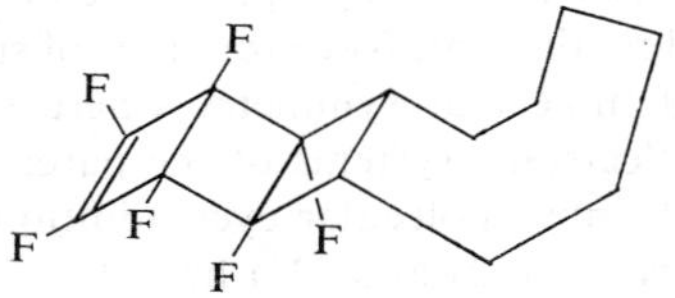   (27) 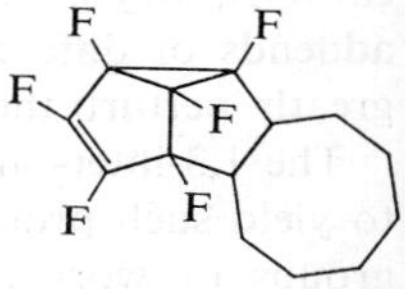   (28)

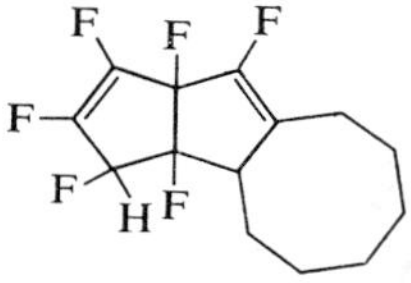

(29) 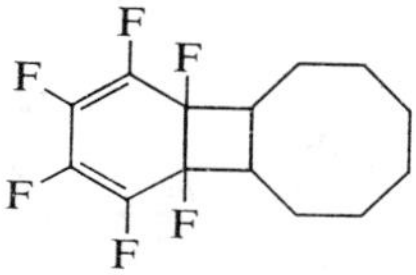   (30) 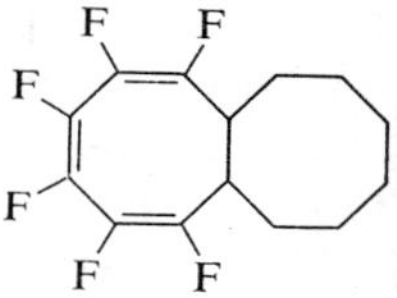   (31)

mixture were not obtained pure, the two major components were believed to have structures (30) and (31). These assignments are consistent with the observed thermal rearrangement of adducts (26) and (27) to this three-component mixture. The relative proportions of the adducts (26)—(29) remained constant throughout the irradiation, thereby indicating that they are all primary photochemical products. This is a surprising conclusion in the case of adduct (29), and since this is also formed thermally from adduct (28) by a suprafacial 1,5-hydrogen shift, the attractive genesis for (29) in the irradiation is from the vibrationally-excited S_0 adduct (28) rather than photoisomerization of (28) by the sterically unattractive antarafacial pathway. The well-known 'Dewar' isomer of hexafluorobenzene was shown not to be the direct precursor of adducts (26) and (27). From orbital symmetry considerations, it appears that formation of adducts (26), (27), and (30) should involve species of B_{1u} symmetry, whereas adduct (28) should be formed via B_{2u} symmetry species. In the present unfiltered radiation from a medium-pressure mercury arc, both symmetry species could well have been generated since population of the first three excited singlet states is possible. Formation of the 1,3-adduct (28) provides the only similarity yet found between the photochemistry of benzene and hexafluorobenzene: cf. ref. 25. Hexafluorobenzene is evidently less prone than benzene to undergo meta-bonding. Prior to this work, the only known photoaddition reactions of hexafluorobenzene were the

[30] D. Bryce-Smith, A. Gilbert, and B. H. Orger, Chem. Comm., 1969, 800.

production of an uncharacterized 2:2 adduct with buta-1,3-diene,[31] and the formation of solvent-derived products from cycloalkane solutions.[32]

The 2:1 photoaddition of maleic anhydride and maleimides to simple benzenoid compounds is well known.[33] The synthesis of maleic thioanhydride (32) has now been reported, and irradiation of its benzene solution with a high-pressure mercury lamp yields the adduct (33).[34] The *exo–endo*-stereochemistry of (33) was established by its conversion to the maleic anhydride–benzene adduct, the stereochemistry of which is known.[35] In the 2:1 addition of maleic anhydride to benzene, the inability of powerful dienophiles such as tetracyanoethylene to intercept the originally suggested intermediate 1:1 adduct (34a) and the fact that the radiation fruitful of reaction is absorbed by a charge-transfer transition in a complex between the reactants, has led to the suggestion that the intermediate may have the zwitterionic form (35), which reacts with another maleic anhydride molecule before closure of the cyclobutane ring.[33] The effect of acid on this reaction has now provided further support for the zwitterion mechanism.[36] Formation of the 2:1 adduct is completely suppressed in

---

[31] G. Koltsenburg and K. Kraft, *Tetrahedron Letters*, 1966, 389.
[32] D. Bryce-Smith, B. E. Connett, A. Gilbert, and T. E. Kendrick, *Chem. and Ind.*, 1966, 855; I. Haller, *J. Chem. Phys.*, 1967, **47**, 1117.
[33] D. Bryce-Smith, *Pure Appl. Chem.*, 1968, **16**, 47.
[34] M. Verbeek, H. D. Scharf, and F. Kote, *Chem. Ber.*, 1969, **102**, 2471.
[35] D. Bryce-Smith, G. I. Fray, and B. Vickery, *J. Chem. Soc. (C)*, 1967, 390.
[36] D. Bryce-Smith, R. Deshpande, A. Gilbert, and J. Grzonka, *Chem. Comm.*, 1970, 561.

the presence of trifluoroacetic acid, and phenylsuccinic anhydride (36) is obtained instead. These results can be accounted for by protonation of zwitterion (35), followed by deprotonation. The superficially analogous 2 : 1 additions of maleimides to benzene occur by a different mechanism which involves excitation of the addend. The orbital symmetry analysis referred to earlier in this Section indicates that a concerted 1,2-addition of an S_1 ethylene to S_0 benzene would be allowed.[23] So it is particularly satisfying to find that the presence of acid has little or no effect on the addition of maleimides to benzene, and that the intermediate 1 : 1 adduct (34b) may be trapped by tetracyanoethylene.[36] In the same communication, the effect of acid on the irradiation of other dienophile–benzene systems is reported. Thus the irradiation of *p*-benzoquinone in benzene in the presence of trifluoroacetic acid leads to 4-phenoxyphenol, presumably *via* the zwitterionic intermediate (37): no photoadduct is formed from this

charge-transfer system in the absence of acid. The effect of acid on the irradiation of acetylenedicarboxylic ester–benzene solutions is more complex, for while dimethyl phenylfumarate and phenylmaleate are formed at the expense of the cyclo-octatetraene (38), dimethyl phthalate is one of the major products. However, it has been shown that irradiation of (38) yields dimethyl phthalate, and it is interesting that this light-induced massive fragmentation is acid catalysed. From the formation of the fumarate and maleate at the expense of (38), it is tempting to suggest that a zwitterionic species such as (39) is involved in the photoaddition of acetylenedicarboxylic ester to benzene. If this is so, intramolecular ring closure to form the diene (40) occurs before interaction with another acetylenic molecule, since (40) may be trapped with tetracyanoethylene as the 1 : 1 : 1 adduct (41): contrast the case of maleic anhydride–benzene described above. Evidently the isomerization of (40) to (38) is faster than its addition to acetylenedicarboxylic ester, for no 2 : 1 adducts of type (42) have been isolated from the reactions.[36, 37] The excited species in the acetylene additions to benzene, previously unidentified, has now been

[27] D. Bryce-Smith, A. Gilbert, and J. Grzonka, *Chem. Comm.*, 1970, 498.

shown to be the acetylene rather than the benzene, at least in the formation of the cyclo-octatetraene (43) from benzene solutions of methyl phenyl-propiolate.[37] The reaction proceeds with light of wavelength longer than 290 nm where only the acetylene absorbs to a significant degree. This experimental observation is consistent with the proposal from orbital symmetry considerations that concerted 1,2-addition of S_1 ethylene or acetylene to S_0 benzene is allowed, but not the corresponding addition of S_1 benzene to S_0 ethylene or acetylene.[23] The photoaddition of acetylene itself to benzene has hitherto been reported only tentatively.[38] Irradiation of acetylene-saturated benzene is now confirmed to yield cyclo-octatetraene as the only C_8 product, and 1,2,3,4,5,6-hexadeuteriocyclo-octatetraene is formed from a similar reaction with hexadeuteriobenzene.[37] The quantum yield of the process is estimated to be less than 0.001. It appears that non-conjugated acetylenes add very inefficiently to benzene, forming cyclo-octatetraenes rather than the semibullvalenes (44) which might have been expected by analogy with the 1,3-photoaddition of non-conjugated olefins to benzene. In contrast to this, perfluorobut-2-yne is reported to undergo

[38] D. Bryce-Smith and J. E. Lodge, *Proc. Chem. Soc.*, 1961, 333; *J. Chem. Soc.*, 1963, 695.

photoaddition to benzene in the vapour phase with 253·7 nm radiation to yield adducts (45), (46), and (47) in the relative proportions 25 : 12 : 5, together with the cyclo-octatetraene (48).[39] Adduct (45) could in principle

arise directly by an analogue of the 1,3-addition process known for olefins, but adducts (46) and (47) could not be formed in this way. It is significant that the known thermal adduct (49) of perfluorobut-2-yne undergoes photosensitized isomerization to form (45), (46), and (47) in the relative proportions 4 : 2 : 1. These proportions are so similar to those from the photoaddition reaction that the thermal adduct (49) is strongly implicated as the precursor of compounds (45), (46), and (47). Following private correspondence between the Reporters and the authors of ref. 39, it is now agreed that the contrary conclusion reached in this paper was not adequately supported by the experimental evidence. The only argument against the intermediacy of (49) in the photoaddition lies in the failure to detect it by direct methods: further checking of this point would be desirable. Prolonged irradiation of solutions of perfluorobut-2-yne in benzene yields the 2 : 1 adduct (50) (70%) with minor amounts of (46) (9%) and (48) (14%).[39]

Again, the thermal adduct (49) seems a likely precursor, although the photoadduct (48) might be involved *via* its [4,2,0] valence-isomer.

The photoaddition of cyanoacetylene to indene is reported to yield one adduct, 7-cyano-2,3-benzobicyclo[3,2,0]hepta-2,6-diene (51), and the structures of the two cyclobutane adducts from the photosensitized addition of acrylonitrile to indene have been unambiguously established as (52) and (53).[40]

[39] R. S. H. Liu and C. G. Krespan, *J. Org. Chem.*, 1969, **34**, 1271.
[40] R. M. Bowman, J. J. McCullough, and J. S. Swenton, *Canad. J. Chem.*, 1969, **47**, 4503.

Relatively few photoadditions to the naphthalene nucleus have hitherto been reported, so the recent communication by Arnold and co-workers is of particular interest.[41] Their reaction involves the cycloaddition of methyl cinnamate to 2-acetylnaphthalene, and yields adduct (54). The 1:1 *cis:trans* photo-stationary state of the addend is rapidly attained, but the reaction, which requires one month of irradiation under the specified conditions, is stereospecific. The mechanism as yet is unknown, but the initial excitation is reported to involve the $n \rightarrow \pi^*$ transition in 2-acetylnaphthalene. In view of the inefficiency of the reaction, however, direct excitation of the cinnamate (weak absorption at $\lambda > 300$ nm) cannot be ruled out. There was no spectroscopic evidence for direct charge-transfer excitation. No dimers of the reactants were found, and only the *trans*-isomer (54) of the adduct was apparently produced, despite the presence of *cis*- and *trans*-forms of the addend. The authors' conclusion that the adduct (54) is derived from the *cis*-isomer of the addend is based on simple application of the Woodward–Hoffmann Rules to the case of naphthalene; but in view of the inapplicability of the Rules to benzene,[23] and the uncertainty in their application to $n\pi^*$ states, the conclusion should probably be regarded with reserve.

Sasse has carried out a mechanistic study of his earlier work [42] on the photoaddition of diphenylacetylene to naphthalene to yield adduct (55).[43] An exciplex derived from singlet naphthalene and diphenylacetylene is suggested to be involved. The reaction is retarded by oxygen and completely inhibited by benzophenone. The mechanistic proposal is outlined in equations (9—18).

$$N + h\nu \longrightarrow {}^1N^* \qquad N \equiv \text{Naphthalene} \qquad (9)$$

$$ {}^1N^* \longrightarrow {}^3N^* \qquad (10)$$

$$ {}^3N^* \longrightarrow N \qquad (11)$$

$$ {}^1N^* \longrightarrow N + h\nu \qquad (12)$$

$$ {}^1N^* \longrightarrow N \qquad (13)$$

$$ {}^1N^* + D \longrightarrow (ND)^* \qquad D \equiv \text{Diphenylacetylene} \qquad (14)$$

---

[41] D. R. Arnold, L. B. Gillis, and E. B. Whipple, *Chem. Comm.*, 1969, 918.
[42] W. H. F. Sasse, P. J. Collin, and G. Sagowdz, *Tetrahedron Letters*, 1965, 3373.
[43] W. H. F. Sasse, *Austral. J. Chem.*, 1969, **22**, 1257.

$$(ND)^* \longrightarrow {}^1N^* + D \qquad (15)$$
$$(ND)^* \longrightarrow N + D \qquad (16)$$
$$(ND)^* \longrightarrow (56) \qquad (17)$$
$$(56) \xrightarrow{h\nu} (55) \qquad (18)$$

(56)   (55)

Acetylenedicarboxylic ester has been reported to undergo photoaddition to naphthalene in the molten state (90 °C) to yield compounds (57), (58), (59), and (60).[44] Adduct (60) was shown to arise by photochemical cyclization of (57), and it was the major 1 : 1 adduct formed when methanol was used as a solvent. The reaction in molten naphthalene is complicated by the formation of the thermal 1 : 1 adduct (61), which on brief irradiation in methanol yields mainly the cyclo-octatetraene (62), and on prolonged irradiation, or by irradiation in acetone solution, gives 'isomeric products' one of which, (58), is of a type known to be formed by photoisomerization of benzbarrelenes.[45] It might be argued that (58) is not a true 1,3-photoadduct, but originates by photoisomerization of the thermal adduct (61). This could well be the case in molten solutions which yield 0·3% of (61) and 3·6% of (58), and in the presence of benzophenone, a trace of (61) and 1% of (58) is formed. [Irradiation of (61) in molten naphthalene in the presence and absence of benzophenone might have led to clarification of the point, but the experiment was not described]. However, irradiation of methanol and benzene solutions of the reactants leads to quite good yields of compound (58), whereas compound (61) evidently shows little tendency to isomerize to (58) under the conditions employed. Thus the formation of adduct (58) under these conditions may provide the first authentic example of the direct 1,3-cycloaddition of an acetylene to an aromatic ring: cf. the foregoing remarks concerning ref. 39. The suggested mechanism is outlined in Scheme 1. The composition of the product mixture shows an interesting solvent dependency. In molten naphthalene the ratio of [(57)+(60)] to (58) is 1·4. It drops to approximately half this value in cold benzene, but in hot or cold methanol it is ca. 12, and is only slightly less in cold cyclohexane. The relative yield of (59)/(58) follows a similar trend with solvent composition. The Reporters note that these trends in the case of methanol are qualitatively what one would expect if adducts (57), (59), [and hence (60)] were being formed by a proton-transfer process of the type

[44] E. Grovenstein, T. C. Campbell, and T. Shibata, *J. Org. Chem.*, 1969, **34**, 2418.
[45] H. E. Zimmerman, R. S. Givens, and R. M. Pagni, *J. Amer. Chem. Soc.*, 1968, **90**, 4191, and references therein.

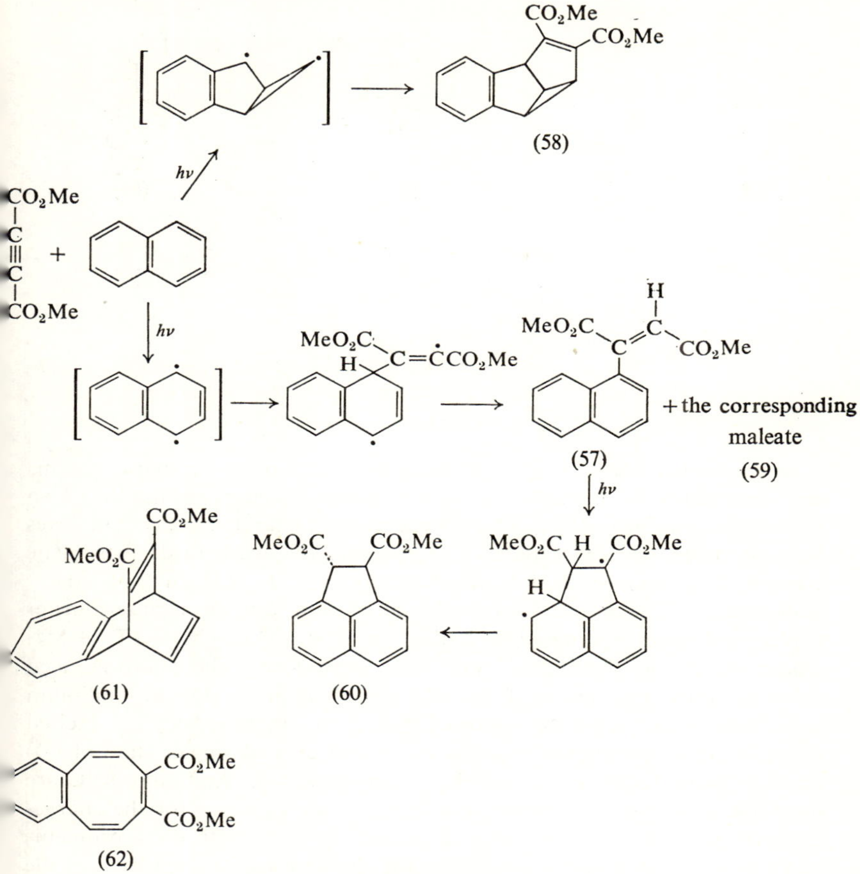

Scheme 1

demonstrated in the case of benzene plus dimethyl acetylenedicarboxylate.[36] The effect of cyclohexane is surprising, and requires a different explanation.

The 1,2-addition of maleic anhydride to the 9,10-positions of phenanthrene has been previously reported,[45] and this year the analogous addition of diphenylacetylene to three phenanthrenes has also been described.[46] None of the adducts (63), (64), and (65) showed any tendency to isomerize to dibenzocyclo-octatetraenes (see ref. 37). One of the phenanthrenes used, (4H-cyclopenta[d,e,f]phenanthrene) (66) also dimerizes photochemically to yield the cyclobutane derivative (67). This is an example of the dimerization of a phenanthrene which is unsubstituted in the 9-position.[47]

[46] G. Sugowdz, P. J. Collins, and W. H. F. Sasse, *Tetrahedron Letters*, 1969, 3843.
[47] M. V. Sargent and C. J. Timmons, *J. Chem. Soc.*, 1964, 5544.

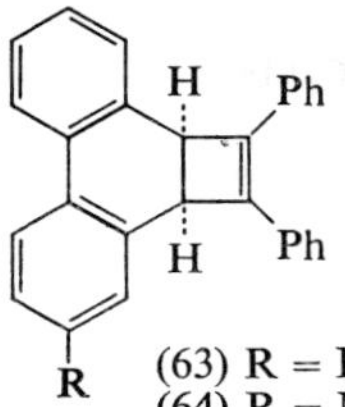

(63) R = H
(64) R = Me

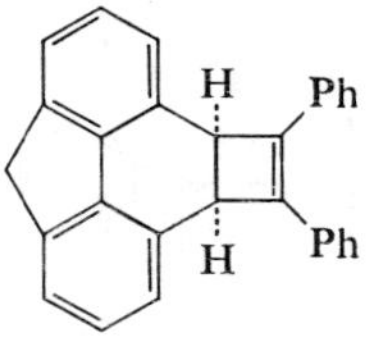

(65)

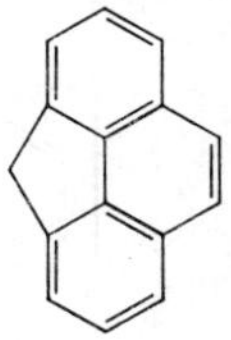

(66)

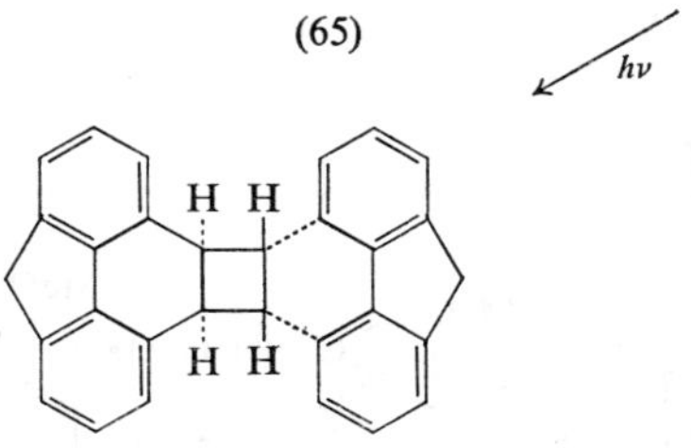

(67)

As mentioned earlier in this section, photoaddition to aromatic systems can sometimes be brought about by charge-transfer excitation. Two reports this year have dealt with the irradiation of charge-transfer complexes between aromatic compounds. The pyromellitic dianhydride–mesitylene complex has been studied by flash photolysis at 77 K in 2:1 ether–isopentane.[48] The results lead to the conclusion that the charge-transfer triplet undergoes spontaneous ionization thus: $^3(PD-M) \rightarrow PD^- + M^+$, where PD is the dianhydride and M is mesitylene. The photo-induced electron transfer reactions from NN-dimethylaniline (D) to anilinium salts have been previously reported,[49] and the intermediacy of excited charge-transfer complexes between the lowest excited singlet state of (D) and the ground state of the salt has been suggested. Anilinium salts are insoluble in non-polar solvents and thus definite evidence for the charge-transfer suggestion has not been obtained. To overcome these problems and to obtain unambiguous evidence for charge-transfer complexes in the reaction, halogenobenzenes have now been used as the acceptors.[50] Latowsi has carried out a similar investigation and again suggested the intermediacy of charge-transfer complexes, but without confirmatory evidence.[51] Ref. 50 refers to spectroscopic and kinetic evidence for the intermediacy of excited charge-transfer complexes in the dimethylaniline–halogenobenzene systems. Irradiation of methanol solutions of NN-dimethylaniline with chloro-, bromo-, or iodo-benzene in the presence of triethylamine leads to the quantitative formation of the triethylamine hydrohalide, together with benzene, biphenyl, N-methylaniline, and o- and p-NN-dimethylaminobiphenyls.[50]

[48] R. Postashnik, C. R. Goldschmidt, and M. Ottolenghi, *J. Phys. Chem.*, 1969, **73**, 3170.
[49] C. Pac and H. Sakurai, *Tetrahedron Letters*, 1968, 1865.
[50] T. Tosa, C. Pac, and H. Sakurai, *Tetrahedron Letters*, 1969, 3635.
[51] T. Latowski, *Z. Naturforsch*, 1968, **23a**, 1127.

There are few examples of intramolecular addition to aromatic rings in which the product is non-aromatic (but see ref. 51). The photolysis of N-chloroacetyl-3,4-dimethoxyphenethylamine (68) has been previously reported to yield four products.[52] One of these products has now been

identified unambiguously by X-ray analysis as 1,2,5α,7β-tetrahydro-5α,5β-dimethoxy-5βH-cyclobuta[1,4]cyclobuta[1,3,3-g,h]pyrrolidin-4(5H)-one (69) and is the product of a most unusual photocyclization reaction seemingly involving addition to the 'Dewar' isomer or its immediate precursor with the elimination of hydrogen chloride.[53] The photolysis of N-chloroacetyl derivatives of aromatic amino-acids [e.g. (70)] and amines

has been previously described to lead to a cyclization product (71) without change of the original chromophore.[54] The long-sought-after products from the photolysis of (72), N-chloroacetyl mescaline, the trimethoxy-analogue of (68), have now been isolated and the structures proved.[55] Compound (72) in ethanol–water gives a 10% yield of 7,8,9-trimethoxy-1,2,4,5-tetrahydro-3(3H)benzazepin-2-one (73), thereby providing the first example of such a ring closure in the mescalin series in the absence of o- or p-activation by a phenolic hydroxy-group. In addition to (73), 33% of the non-aromatic product (74) is also formed by addition to the *para*-position of the original aromatic ring rather than to the *ortho*-position as in (69). These two reports are extremely interesting and further reactions of similar systems readily spring to mind.

---

[52] O. Yonemitsu, Y. Okuno, Y. Kanaoka, I. L. Karle, and B. Witkop, *J. Amer. Chem. Soc.*, 1968, **90**, 6522.
[53] I. L. Karle, J. W. Gibson, and J. Karle, *Acta Cryst.*, 1969, **B25**, 2034.
[54] O. Yonemitsu, P. Cerutti, and B. Witkop, *J. Amer. Chem. Soc.*, 1966, **88**, 3941.
[55] O. Yonemitsu, H. Nakai, Y. Kanaoka, I. L. Karle, and B. Witkop, *J. Amer. Chem. Soc.*, 1969, **91**, 4591.

(72) → (73) + (74)

3 Substitution Reactions

Examples of photo-electrophilic aromatic substitution have been provided by the acid-catalysed photoreactions of acceptors such as maleic anhydride with benzene:[36] these are described in the previous section.

Most of the aromatic substitution studies reported during the year have been concerned with photonucleophilic substitution,[56] but examples involving attack on aromatic rings by photochemically-generated free-radicals continue to appear: these latter are not of direct concern to the photochemist.

An eighteen-page review has appeared of the 'Mechanism for Nucleophilic and Photo-nucleophilic Aromatic Substitution Reactions,' but only two pages are concerned with photoreactions.[57] This review provides some useful references.

Reports in this section are in presented the order monocyclic benzenoid compounds, naphthalene derivatives, and finally heterocyclic compounds.

Bowie and Musgrave have now published the full paper corresponding to their earlier communication [58a] on the photochemical reactions of boron halides with aromatic compounds.[58b] The irradiation and subsequent hydrolysis of boron tri-iodide or the tribromide with benzene, toluene, biphenyl, or naphthalene yields the corresponding arylboronic acids. The initial products of the reaction, the aryl boron dihalides, appear to be formed by the photolysis of the boron trihalide followed by reaction of the boron dihalide radical with the aromatic compound, as suggested earlier.[59] This conclusion is somewhat different from the proposal involving 1,2-addition to the ring.[58a, 60] Complexes may be involved.[58a, 58b, 60] Thus

[56] E. Havinga and M. E. Kronenberg, *Pure Appl. Chem.*, 1968, **16**, No. 1, 137.
[57] F. Pietra, *Quart. Rev.*, 1969, **23**, 504.
[58a] R. A. Bowie and O. C. Musgrave, *Proc. Chem. Soc.*, 1964, 15.
[58b] R. A. Bowie and O. C. Musgrave, *J. Chem. Soc. (C)*, 1970, 485.
[59] D. Bryce-Smith, *Ann. Reports*, 1964, **61**, 334.
[60] Y. Ogata, Y. Izawa, H. Tomioka, and T. Ukigai, *Tetrahedron*, 1969, **25**, 1817.

ref. 58b describes the irradiation of boron tribromide and benzene with visible light to give a low yield of phenylboronic acid and *para*-phenylenediboronic acid, and it is suggested that this reaction may arise by excitation of a loosely bonded complex between the reactants. Aromatic ethers have also been examined in the reaction, and diphenyl ether with boron tribromide yields, after hydrolysis, 10-hydroxy-10-bora-9-oxa-*aro*-anthracene (75). In preliminary work, phosphorus tribromide and benzene have been found to give a trace of phenylphosphorus dibromide.[58b].

$$Ph_2O + BBr_3 \xrightarrow[\text{(ii) hydrolysis}]{\text{(i) } h\nu}$$

(75)

Havinga and his co-workers are well-known for their work on photonucleophilic substitution processes: these usually have involved reactions between excited aromatic compounds and charged nucleophiles.[56] These workers have now investigated substitution reactions which involve uncharged species such as (liquid) ammonia as well as ammonia–methanol and ammonia–water systems.[61, 62] The results show that aromatic compounds may be divided into two classes according to their behaviour in these reactions. The first class contains *m*-nitroanisole and some substituted *m*-methoxynitrobenzenes which on irradiation in the above systems undergo substitution of the *m*-methoxy-group by the amino-group. It is suggested that the reactions may arise from the lowest $\pi\pi^*$ singlets of the aromatic compounds and are comparable with the photohydrolysis of these compounds. Nitrobenzene, dinitrobenzene, and chloro-nitrobenzenes form the second class of compounds, and by similar irradiation in liquid ammonia undergo substitution at the *o*- and *p*-positions but not at the *m*-positions. The reactions do not occur as well in ammonia–methanol and ammonia–water. The reactions with compounds of this second type are more complicated than those with the first type, and the case of nitrobenzene has been investigated in detail. Irradiation of nitrobenzene in liquid ammonia yields two unstable products, the nitrobenzene radical anion and an unidentified compound which has an absorption maximum at 335 nm. The radical anion is considered to be a possible intermediate in the reaction, but the second product does not seem to be a precursor of the nitroaniline. The radical anion of nitrobenzene has been generated by electrolytic means in a similar environment and led to low yields of *o*- and *p*-nitroaniline. If formation of the radical anion were the sole route in the

---

[61] A. Van-Vliet, M. E. Kronenberg, J. Corvelisse, and E. Havinga, *Tetrahedron*, 1970, **26**, 1061.
[62] A. Van-Vliet, J. Cornelisse, and E. Havinga, *Rec. Trav. chim.*, 1969, **88**, 1339.

photochemical reaction, oxygen would be expected to inhibit the amination process. The light-induced reaction in the presence of oxygen shows a remarkable wavelength dependence, and at wavelengths longer than 300 nm, no reaction is observed, while the use of unfiltered light from a mercury vapour lamp led to p-nitroaniline in 30% yield. This implies an alternative route for the reaction in which the radical anion is not involved. This second route may be analogous to the conversion of m-nitroanisole to m-nitroaniline. However, some of the effect of oxygen at wavelengths longer than 300 nm may be due to triplet quenching since the reaction is sensitized by benzophenone. It is suggested that at wavelengths shorter than 300 nm the reaction involves the S_1 state of the aromatic compound. Thus both singlet and triplet states are suggested as the starting levels for the amination process in reactions of the second class of compounds. With nitrobenzene, formation of the p-nitroaniline as the major product led to the conclusion that charge densities in excited states are not the only decisive factors for the $p:m:o$ isomer ratios, particularly in the case of reaction with uncharged nucleophiles.

Nitroanisoles seem to be popular compounds for the study of photonucleophilic substitution reactions, and have been the subject of three reports by Letsinger and his co-workers this year. These workers suggest that the excited state for substitution in nitroaromatics is an open question and that previous evidence [63] for singlet states of nitroaromatics is not compelling.[64] In an attempt to obtain further evidence concerning this aspect, an investigation into the possibility of sensitizing a nucleophilic substitution reaction has been carried out. The system chosen was the irradiation of p-nitroanisole with light of wavelength longer than 290 nm in the presence of hydroxyl ion and benzophenone. The reaction gave 80% of p-methoxyphenol and 20% of p-nitrophenol. It was found that benzophenone, absorbing at 252 nm, sensitized the reaction of p-nitroanisole with hydroxyl ion, leading to p-nitrophenol under nitrogen, but under oxygen the reaction was inhibited. Yet in the unsensitized reaction the product yield was slightly enhanced in the presence of oxygen (Table 1). The data clearly demonstrate that the nitrophenol formation in the sensitized reaction results from excitation of the benzophenone, and thus triplet

Table 1 *Sensitized and unsensitized reaction of* p-*nitroanisole with hydroxyl ion*

Benzophenone concentration (mol l^{-1})	Irradiation atmosphere	10^4 Rate constant (s^{-1})
0	N$_2$	0·62
1·0 × 10^{-4}	N$_2$	2·46
0	O$_2$	0·77
1·0 × 10^{-4}	O$_2$	0·86

[63] E. Havinga, R. O. de Jongh, and M. E. Kronenberg, *Helv. Chim. Acta*, 1967, **50**, 2550.
[64] R. L. Letsinger and K. E. Steller, *Tetrahedron Letters*, 1969, 1401.

p-nitroanisole is involved in the sensitized process. The fact that the same products in the same ratios are formed in the sensitized and unsensitized reactions strongly suggests that the triplet state of the nitroaromatic may also be involved in the unsensitized reaction: compare refs. 61 and 62.

It is reported that substituent groups and solvents may have striking effects on rates of bimolecular reactions involving nucleophiles and photoexcited nitroaromatic compounds.[65] These effects have emerged from studies of the reactions of cyanide anion with p-nitroanisole, 1-nitronaphthalene, and 4-methoxy-1-nitronaphthalene. Thus aqueous cyanide ion and p-nitroanisole give 2-cyano-4-nitroanisole, and since nitrobenzene itself is inert in this reaction, the methoxy substituent appears to be acting (most uncharacteristically) as an activating group for the nucleophilic attack. In contrast, the methoxy-group may also serve as a deactivating substituent, as evidenced by the results from photoreactions of aqueous cyanide ion with naphthalenes: 4-methoxy-1-nitronaphthalene is inert whereas 1-nitronaphthalene gives 1-cyanonaphthalene. Displacement of the nitro- by the cyano-group has not previously been reported, although displacements by pyridine, piperidine, and hydroxyl ion are known. The solvent effects are dramatic. The quantum yield for reaction of cyanide ion with p-nitroanisole in 95% acetonitrile–5% water is one-fiftieth of that in 90% water–10% acetonitrile, and yields 2-cyano-4-nitroanisole (9%), p-cyanoanisole (6%), p-nitrosoanisole (13%), and a substance not identified, plus 32% of the starting material. On the other hand, in 95% acetonitrile–5% water, 4-methoxy-1-nitronaphthalene becomes reactive to yield 70% of 4-methoxy-1-cyanonaphthalene while the 1-nitronaphthalene reaction is little affected by the solvent change. These solvent effects are suggested to be very similar to those observed for the reduction of aromatic ketones, *i.e.* the lowest reactive level ($n\pi^*$, $\pi\pi^*$, charge-transfer) varies in differing solvents. In the present case, the lowest state necessary for reaction is $\pi\pi^*$. The light-induced reaction of the three isomers of nitroanisole with potassium cyanide under air has also been investigated (Scheme 2).[66] o-Nitroanisole yields a mixture of 4-cyano-2-nitroanisole and 6-cyano-2-nitroanisole, m-nitroanisole forms only m-nitrobenzonitrile, while p-nitroanisole (as previously reported) gives 2-cyano-4-nitroanisole. These results clearly demonstrate the m-activation by the nitro-group in each instance. The rates and quantum yields of the reaction of the p-isomer under oxygen have been studied as a function of cyanide ion concentration. It is concluded from these results that cyanide ion intercepts the photoexcited aromatic compound with high efficiency, forming a short-lived species which is oxidized by molecular oxygen to 2-cyano-4-nitroanisole (Scheme 3). Competition experiments utilizing cyanide ion with pyridine, 2,4,6-trimethylpyridine, and iodide ion provide evidence that: (*a*) pyridine and cyanide ion compete for the same excited species of the anisole,

---

[65] R. L. Letsinger and R. R. Hautala, *Tetrahedron Letters*, 1969, 4205.
[66] R. L. Letsinger and J. H. McCain, *J. Amer. Chem. Soc.*, 1969, **91**, 6425.

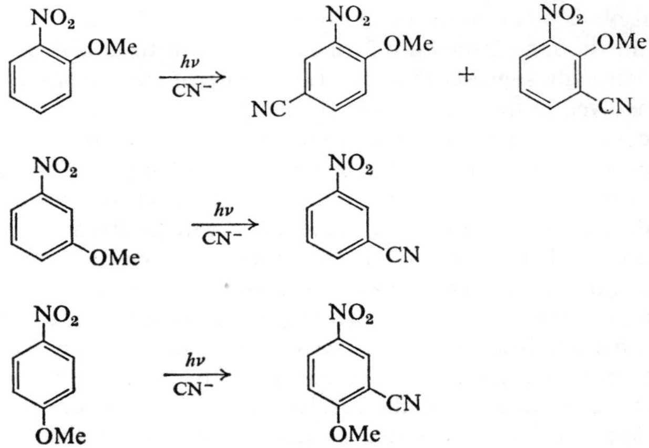

Scheme 2

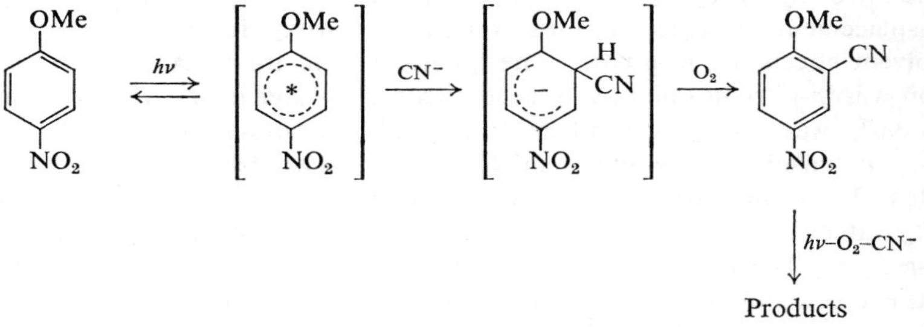

Scheme 3

(b) pyridine converts the excited species to ground state p-nitroanisole as well as undergoing the substitution reaction, (c) the quenching reaction is subject to steric hindrance, and (d) that iodide quenches the excited state of the anisole. These and the previous data are combined in Table 2, which shows the relative reactivities at 3 °C of several nucleophiles towards p-nitroanisole (which is generally inert to common nucleophilic reagents in the absence of light). The types of products observed from light-induced substitution of p-nitroanisole (Scheme 4) pose the intriguing question why pyridine displaces the nitro-group whereas cyanide ion adds to the carbon atom *meta* to the nitro-group. An explanation of this is that reactive nucleophiles in general tend to attack at the same position in a given compound. The positions of attack are determined by electronic structures of the excited compounds, and steric effects are important. The course of the reaction is then dependent upon the stability of the resulting intermediate and the relative leaving tendencies of the nucleophile and the group which is being displaced. As shown by Scheme 2, the position *meta* to the nitro-group is activated. Cyanide ion forms a carbon–carbon

Table 2 *Light-induced reaction rate of several nucleophiles with* p-*nitroanisole*

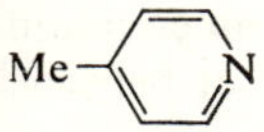

Nucleophile	Relative rate of photoreaction compared with that of pyridine
CN⁻	70
OH⁻	4·5
(A)	4·7
(B)	1
(C)	~0·04
H₂O	0
I⁻	0

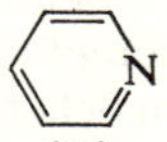

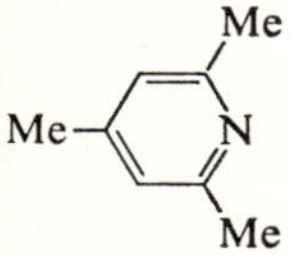

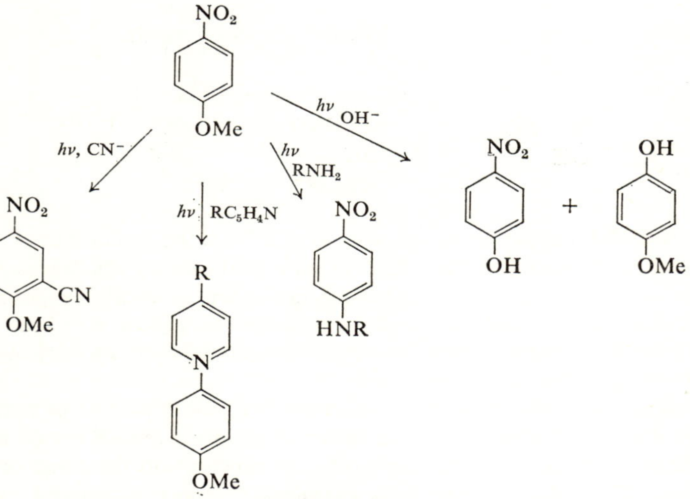

(A) 4-methylpyridine
(B) pyridine
(C) 2,4,6-trimethylpyridine

Scheme 4

bond here, and the intermediate has a sufficient lifetime to interact with oxygen and disproportionate (Scheme 3). Since pyridine is a good leaving group, however, the intermediate is unstable, and the pyridine thus 'backs away' and the anisole returns to the ground state. Positions of secondary reactivity have nitro and methoxy substituents, and replacement of the nitro-group by pyridine is the more favourable since nitrite is a better leaving group than methoxide. Thus the pyridinium salt is formed, but with a low quantum yield because of the unproductive reaction at the *meta*-carbon atom. With *m*-nitroanisole, the cyanide ion displaces the methoxy-group but pyridine fails to react. Yet the reaction of *p*-nitroanisole with pyridine is complete under the same conditions.

In view of all the interest in the photonucleophilic substitution of *p*-nitroanisole (76), it is interesting to note that light-induced isomerization also occurs with (76).[67] Thus irradiation of (76) in acetonitrile or benzene forms 2-nitro-4-methoxyphenol and 4-nitrosoanisole, each with a quantum yield of $2 \cdot 3 \times 10^{-3}$. Trace amounts of nitric oxide and nitrous oxide are detected in the reaction but there is no evidence for the formation of either nitrogen dioxide or oxygen. The reaction sequence is outlined in Scheme 5.

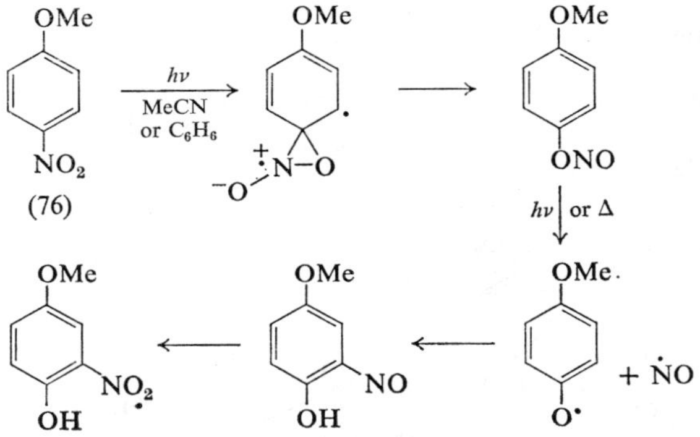

Scheme 5

2-Nitroso-4-methoxyphenol was not isolated from the reaction nor could it be synthesized owing to the further rapid oxidation step which is activated by the *m*-methoxy-group. From sensitization and quenching experiments, *p*-nitroanisole was deduced to have a triplet energy of 60·8 kcal mol^{-1} (255 kJ mol^{-1}). Significantly, the *o*- and *m*-isomers of nitroanisole were inert in the isomerization process.

Photonucleophilic substitution reactions are well known to be markedly affected by ring substituents. In order to estimate the effects of distant substituents on this process, light-induced substitution reactions of compounds (77) and (78) have been examined.[68] Electron-donating substituents (X) were found markedly to retard nucleophilic attack on compound (77); and similarly in (78) for X = OMe, reaction on the nitroaromatic ring was also slowed. Emission spectral data and experiments with model compounds suggest that the observed deactivation results from interaction of the photoexcited nitroaromatic ring with the ethylenedioxy-group in (78) to yield a transient species which is inactive or of low reactivity in the nucleophilic substitution process. The reaction medium used in this investigation was aqueous pyridine. Two types of processes were observed with compound (77) (Scheme 6), *viz.*, (a) displacement of nitrite by pyridine,

[67] L. B. Jones, J. C. Kudrna, and J. P. Foster, *Tetrahedron Letters*, 1969, 3263.
[68] K. E. Steller and R. L. Letsinger, *J. Org. Chem.*, 1970, **35**, 308.

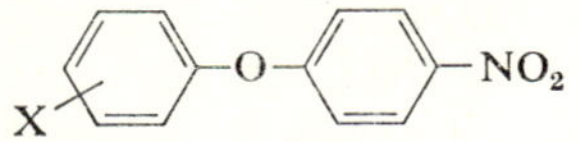

Photochemistry of Aromatic Compounds

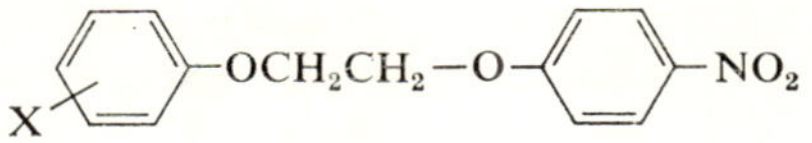

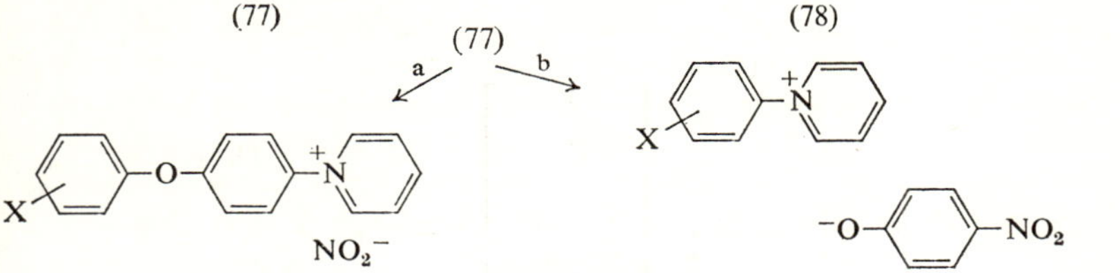

Scheme 6

and (b) displacement of 4-nitrophenoxide by pyridine. Path (b) is unusual in that the nucleophile attacks a benzene ring which does not bear a nitro substituent. With compounds (78), only type (a) reaction was observed. In (78), the methylenes severely limit the transmission of inductive and resonance effects, but in (77) the effects of substituents are clearly transmitted from one ring to the other. Three conclusions have been drawn from this work, *viz*. (*a*) that electron-withdrawing groups in the 3- and 4-positions have little effect, (*b*) that strong electron-withdrawing substituents (*e.g.* nitro- and cyano-groups) favour path (a) relative to path (b), and (*c*) that electron-donating substituents (*e.g.* hydroxy-, methoxy-, and amino-groups) have a major quenching or retarding effect. A preliminary investigation has been made into intermolecular effects on the photosubstitution process. Thus the reaction of *p*-nitroanisole with hydroxyl ion is reported to be inhibited by 1,4-dimethoxybenzene.[68] This interesting phenomenon merits further investigation.

There are many examples in the literature concerned with the light sensitivity of aromatic nitro compounds having a methine group in an *ortho* position. Such reactions include the classical Ciamician–Silber rearrangement of *o*-nitrobenzaldehyde (79) to *o*-nitrosobenzoic acid.[69] The *m*- and *p*-isomers of (79) have been reported to be relatively light-stable, failing to isomerize in methanol, ether, toluene, benzene, and in the solid state. The irradiation of *p*-nitrobenzaldehyde with 290—400 nm light in aqueous media is now reported to give *p*-nitrosobenzoic acid (80).[70] Neither oxygen nor iodide ion have any appreciable effect on the process, and in agreement with earlier work, the *m*-isomer is found to be essentially inert. The *p*-isomer reacts slowly in methanol or ether, seemingly to follow a different reaction path from that in water. In hexane, the *p*-isomer is inactive. The quantum yield for the formation of (80) in water is 0·037, and the reaction is thus considerably less efficient than that with the *o*-isomer, for which a quantum yield of 0·5 in a variety of solvents has been reported. Intermediate (81) is suggested in the rearrangement of the

---
[69] G. Ciamician and P. Silber, *Ber.*, 1901, **34**, 2040.
[70] G. W. Wubbels, R. R. Hautala, and R. L. Letsinger, *Tetrahedron Letters*, 1970, 1689.

p-isomer, and forms p-nitrosobenzoic acid by addition and elimination of water. The intermediacy of such a species as (81) in the rearrangement would account for the stability of the m-isomer. The gas-phase photolysis of aromatic aldehydes such as benzaldehyde and pentafluorobenzaldehyde has been observed to involve a Norrish Type I elimination of carbon monoxide with the formation of benzene and pentafluorobenzene respectively; but the major product is 'a yellowish polymer'.[71]

During an investigation into the light-induced cyanation of p-iodophenol to yield p-cyanophenol, the isolation of a little p-hydroxybenzaldehyde prompted an investigation into the photo-lability of the major product.[72] Indeed, irradiation of p-cyanophenol in aqueous alkali has been found to form p-hydroxybenzaldehyde in yields up to 80% in the presence of potassium iodide. In the absence of the iodide the yield fell to 30%. The only very minor by-product of the reaction is p-hydroxybenzoic acid. Previous investigations concerned with aromatic nitriles have not described transformation of the nitrile group itself, and indeed the direct reduction of a nitrile to an aldehyde photochemically or thermally has not previously been observed. Indeed, virtually no photoreactions of the nitrile group have hitherto been reported.

Aromatic halogen compounds are well known to be photo-labile and give rise to a variety of reaction products, dependent upon the media and their molecular environment, by nucleophilic substitution and radical arylation reactions. This year there have been several reports concerned with such photolyses which lead, sometimes in the loosest possible sense, to substitution of the original aromatic ring as one of the reaction paths. U.v. irradiation of benzene solutions of chlorobenzene is now reported to yield biphenyl and hydrogen chloride in a ratio of 1 : 1·04.[73] The phenylation reaction has been studied with a variety of substituted chlorobenzenes, and the biphenyls formed account for 20—81% of the loss of the starting

[71] J. R. Majer, S-Ama Naman, and J. C. Robb, *Trans. Faraday Soc.*, 1969, **65**, 1846.
[72] K. Omura and T. Matsuura, *Chem. Comm.*, 1969, 1516.
[73] G. E. Robinson and J. M. Vernon, *Chem. Comm.*, 1969, 977.

compounds. 4-Chlorobiphenyl is known to be relatively photo-stable but the 2-chloro isomer is reported to yield a mixture of biphenyl, o-terphenyl, and hydrogen chloride, but no chlorine. Similarly, photolysis of 1-o-chlorophenylnaphthalene in benzene with 313 nm light results in dechlorination and phenylation, as well as intramolecular cyclization.[74] It was observed in this work that 4,4′-dichlorobibenzyl only suffered replacement of one of the chlorines and that the product (82) was photo-stable. To account for this, it is suggested that the excitation in (82) is quenched intramolecularly in the biphenyl group and thus this moiety acts as a protecting group

for the C—Cl bond. Such intramolecular energy transfer is known to be very efficient.[75] This explanation is also offered to account for the light-stability of 1-α-naphthyl-2-p-chlorophenylethane (83). In the course of studies on the photoreduction of aromatic halogen compounds, it was found that irradiation of o- and p-chlorophenols in isopropyl alcohol led to the formation of phenol.[76] In an attempt to extend the reaction to m-chlorophenol, it was found that phenol was only a minor product, plus traces of resorcinol: the major product was m-isopropoxyphenol (84).[77] Thus m-chlorophenol undergoes both C—Cl bond homolysis and nucleophilic substitution on irradiation (Scheme 7). Photolysis in ethanol rather than isopropyl alcohol led to a relatively higher proportion of substitution (giving m-ethoxyphenol) than reduction. The reactivity of m-chlorophenol towards nucleophiles is further demonstrated by photolysis in dioxan–water when, although phenol was the major product, 2-(β-chloroethoxy)-ethyl m-hydroxyphenyl ether (85) and small amounts of resorcinol were also formed. m-Bromophenol in the present reaction showed a very marked increase in formation of the reduction product with corresponding decrease in the yield of the substitution reaction. The increase in the phenol formation here is consistent with the decreasing strength of the C—X bond. The yield of anisole from m-chloroanisole in isopropyl alcohol is lower than that from the o- and p-isomers, but is still greater than that ($<10\%$) of the substitution product. However, in methanol, significant amounts of 1,3-dimethoxybenzene were formed from m-chloroanisole. p-Chloroanisole in methanol likewise gives 1,4-dimethoxybenzene in low yield, but

---

[74] W. A. Henderson and A. Zweig, *J. Amer. Chem. Soc.*, 1967, **89**, 6778.
[75] O. Schnepp and M. Levy, *J. Amer. Chem. Soc.*, 1962, **84**, 172; A. A. Lamola, P. A. Leermakers, G. W. Byers, and G. S. Hammond, *ibid.*, 1965, **87**, 2322; R. D. Raun, T. R. Evans, and P. A. Leermakers, *ibid.*, 1968, **90**, 6897.
[76] J. T. Pinhey and R. D. G. Rigby, *Tetrahedron Letters*, 1969, 1267.
[77] J. T. Pinhey and R. D. G. Rigby, *Tetrahedron Letters*, 1969, 1271.

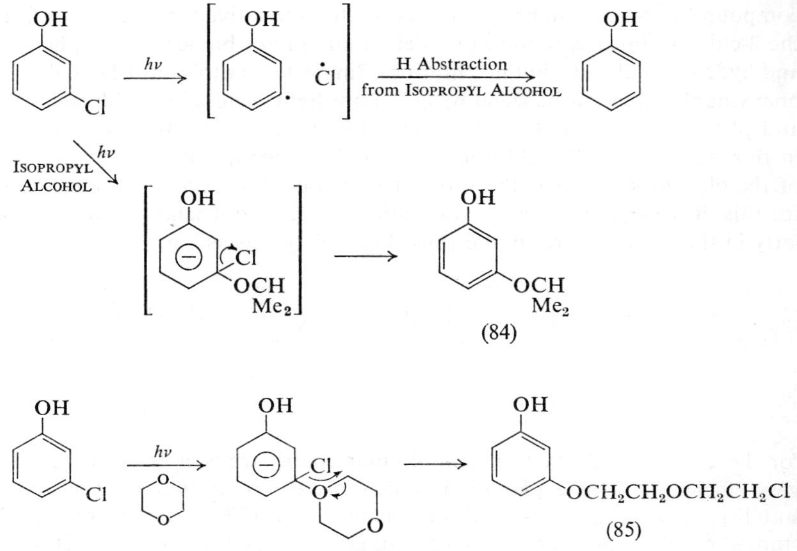

[Scheme 7]

Scheme 7

there is no substitution product from the *o*-isomer. The photolysis of *p*-halogenophenols in aqueous alkali solution also yields both substitution and reduction products (quinol and phenol respectively) as well as 2,4′-dihydroxybiphenyl (86), the major reaction product.[78] The yield of

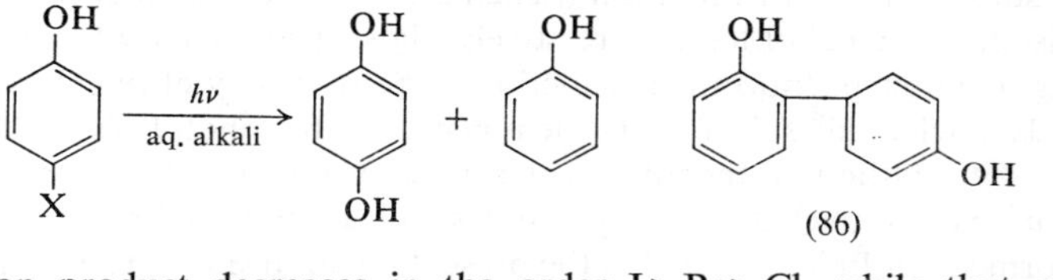

reduction product decreases in the order I > Br > Cl, while that of the quinol decreases in the reverse order. Photo-cyanation of the *p*-halogenophenols gave *p*-cyanophenol in good yield accompanied by small amounts of quinol and (86). Although *m*-chlorophenol also gave the substitution product in the cyanation process in small amount, the major product was resorcinol in high yield, and no reduction was reported. Since all these reactions in aqueous media gave no detectable amount of di- or polysubstituted halogenophenols, it is considered that the halogen is removed from the parent molecule as halide ion rather than atomic halogen. The photoreaction of pentachlorophenol in aqueous solution provides a complex mixture resulting from the above processes, as might be expected.[79]

[78] K. Omura and T. Matsuura, *Chem. Comm.*, 1969, 1394.
[79] M. Kuwahara, N. Shindo, N. Kato, and K. Munakata, *Agric. and Biol. Chem.* (*Japan*), 1969, **33**, 892.

Three of the products have been previously identified, and a fourth has now been shown to be the C_{18} compound (87).

Irradiation of *p*-chlorophenoxyacetic acid (88) in 95% ethanol gives the same products as are obtained from phenoxyacetic acid, *viz.* 2-coumaranone

(87)

(89) (29%) and phenol (8%).[76] In particular, neither *p*-chlorophenol nor 5-chloro-2-coumaranone was detected, but phenoxyacetic acid was formed in low yield. Thus loss of chlorine does occur, at least in part, prior to homolysis of the O—CH$_2$ bond. *p*-Chlorophenol rapidly produces phenol under the same conditions, and 2-hydroxy-5-chlorophenylacetic acid (90) yields 2-coumaranone. It is thus possible that the aforementioned two compounds are formed in the photolysis of (88), perhaps *via* O—CH$_2$ bond homolysis and lateral-nuclear rearrangement followed by reduction. The reduction process was found to proceed more rapidly in isopropyl alcohol, and even chlorobenzene underwent appreciable reduction in this solvent. The chloro-, bromo-, and iodo-aromatic phenols examined in this study were reduced at comparable rates, and the yields were good in all cases. This photochemical reduction appears so convenient and occurs with such good yields that it may be a useful alternative to Raney-nickel–alkali and magnesium–isopropyl alcohol reduction of halogenoaromatics in some cases. The formation of pinacol from isopropyl alcohol in greater than 50% yield indicates that, after the absorption of light, the aromatic undergoes C—X bond homolysis to give radicals which abstract hydrogen from the solvent (Scheme 8). The observed production of hydrogen halide is in keeping with this mechanism.

The photo-substitution and -reduction of chlorine on other aromatic compounds than phenols or their derivatives has also been observed. Thus irradiation of the three isomeric chlorobenzoic acids or their sodium salts in aqueous solution leads to the corresponding hydroxy-acids and benzoic acid.[80] In the case of the 4-isomer, benzoic acid is formed in 90% yield together with a mixture of terephthalic and 4-acetylbenzoic acid: despite an intensive investigation, the source of the acetyl group is still a mystery. Photolysis of the three isomeric sodium monochlorophenylacetates, on the other hand, does not proceed by such a simple reaction pathway, and two

[80] D. G. Crosby and E. Leitis, *J. Agric. Food Chem.*, 1969, **17**, 1033.

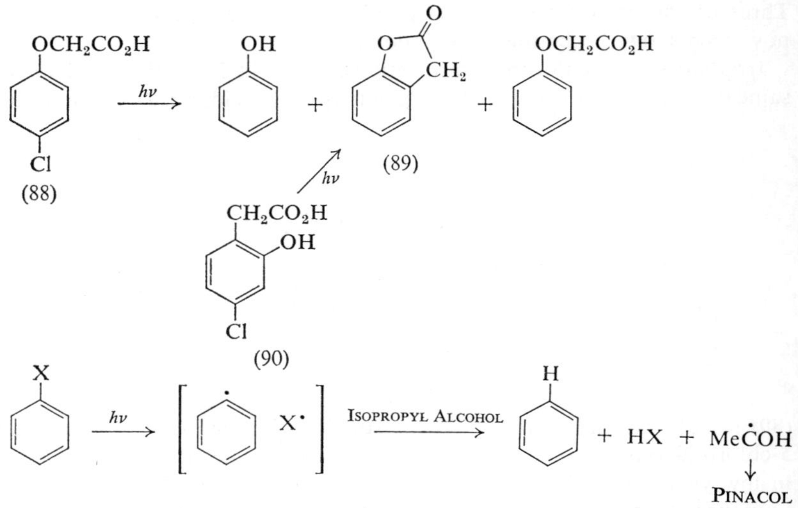

Scheme 8

major processes appear to be operative.[81] In one sequence of reactions, phenylacetic acid, benzyl alcohol, benzaldehyde, and benzoic acid are formed, while the corresponding hydroxyphenylacetic acids, hydroxybenzyl alcohols, and humic acid appear to be produced in a simultaneous sequence of events.

Photolysis of m-chloronitrobenzene in aqueous methanolic sodium nitrite is reported to lead to dechlorination, denitration, and substitution, with the formation of benzene, chlorobenzene, nitrobenzene, anisole, m-dimethoxybenzene, and m-nitrophenol.[82] The reaction pathways are obviously complex, and a more detailed description of the processes and factors affecting the reaction is awaited.

Aromatic bromo-compounds are well known to be more photo-labile than their chloro-analogues. The photochemical reaction of tribenzyl borate in bromobenzene, followed by hydrolysis, yields dibenzyl ether (27·5%), benzyl bromide (29·5%), benzaldehyde (25%), and benzene.[83] Similarly, the irradiation of tris-1-phenylethylborate (91a) in bromobenzene leads to the formation of bis-1-phenylethyl ether (91b), 1-phenylethyl bromide, acetophenone, p-dibromobenzene, and benzene. The reaction was shown to be intermolecular since 'mixed' products were formed from irradiations with 'mixed' borates. The primary process involves excitation of the bromobenzene rather than the borate. Naturally, the bromobenzene yields phenyl radicals and bromine atoms which abstract benzylic hydrogen

[81] D. G. Crosby and E. Leitis, *J. Agric. Food Chem.*, 1969, **17**, 1036.
[82] A. N. Frolov and A. V. El-Tsov, *Zhur. org. Khim.*, 1970, **6**, 637.
[83] Y. Ogata and T. Ukigai, *J. Chem. Soc. (C)*, 1969, 2413.

from the borate compound to yield benzene and hydrogen bromide respectively. This process starts the reaction under way, and the suggested involvement of benzyl alcohol, which is always present from hydrolysis of the borate, enables a mechanism to be postulated (equations 19—26).

$$PhBr \xrightleftharpoons{h\nu} Ph\cdot + Br\cdot \qquad (19)$$

$$Br\cdot + (PhCH_2O)_3B \longrightarrow HBr + (PhCH_2O)_2BO\dot{C}HPh \quad (20)$$

$$3HBr + (PhCH_2O)_3B \longrightarrow 3PhCH_2Br + H_3BO_3 \qquad (21)$$

$$PhCH_2Br + PhCH_2OH \rightleftharpoons PhCH_2OCH_2Ph + HBr \qquad (22)$$

$$PhCH_2OH + HBr \rightleftharpoons PhCH_2Br + H_2O \qquad (23)$$

$$(PhCH_2O)_3B + 3H_2O \longrightarrow 3PhCH_2OH + H_3BO_3 \qquad (24)$$

$$Ph\cdot + H\cdot \text{ (from various sources)} \longrightarrow PhH \qquad (25)$$

$$PhCH_2OH + Ph\cdot \longrightarrow Ph\dot{C}HOH \longrightarrow PhCHO \qquad (26)$$

$$\begin{pmatrix} Me \\ | \\ Ph-CHO \end{pmatrix}_3 B + PhBr \xrightarrow{h\nu}$$
(91a)

$$\begin{pmatrix} Me \\ | \\ Ph-CH \end{pmatrix}_2 O + Ph-\overset{Me}{\underset{|}{C}}HBr + PhCOMe + p\text{-}Br_2C_6H_4 + PhH$$
(91b)

From previous knowledge of the photolysis of aromatic bromo-compounds in benzene, it would be expected that the light-induced reaction of (92) in such a medium would yield the biphenyl (93) and the reduction product (94) as major products. Both (93) and (94) are indeed formed, but their respective yields are only 7·9 and 24·2%. The major product (43·2%) from the reaction is reported to be 3,3,5,5-tetra-t-butyldiphenoquinone (95).[84] It is interesting to note that (95) is also the major product from the photoreaction of iodobenzene and 2,6-di-t-butylphenol, and from 2,6-di-t-butylphenol in benzene in the presence of iodine. It is thus inferred that the formation of (95) requires the intermediacy of a phenoxy radical (96), which is stabilized by steric hindrance as well as electron delocalization. The process is outlined in Scheme 9.

Irradiation (300 nm) of dilute solutions of various aromatic iodo-compounds, including anisoles and iodobenzoic acid, in carbon tetrachloride, gives good yields of the corresponding chloro-compounds.[85] Previous to this report the only substitution reaction of iodide by chloride in carbon tetrachloride solution involved formation of 4-chlorobiphenyl: other substitution processes required the use of iodine monochloride.[86] The

---

[84] G. R. Lappin and J. S. Zannucci, *Tetrahedron Letters*, 1969, 5085.
[85] F. Kienzle and E. C. Taylor, *J. Org. Chem.*, 1970, **35**, 528.
[86] B. Milligan, R. L. Bradow, J. E. Rose, H. E. Hubbert, and A. Roe, *J. Amer. Chem. Soc.*, 1962, **84**, 158.

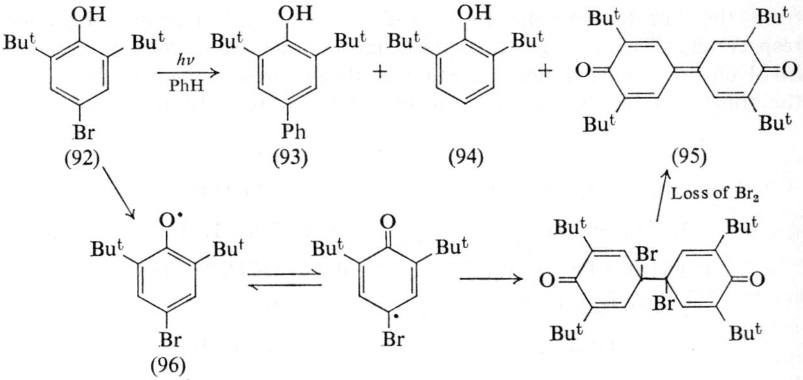

Scheme 9

suggested mechanism is outlined in equations 27—30, and also accounts for the formation of hexachloroethane. As may be expected, aromatic compounds bearing methyl groups undergo secondary processes leading to benzylic products.

$$ArI \xrightarrow{h\nu} Ar\cdot + I\cdot \qquad (27)$$

$$Ar\cdot + CCl_4 \longrightarrow ArCl + \cdot CCl_3 \qquad (28)$$

$$2I\cdot \longrightarrow I_2 \qquad (29)$$

$$2\cdot CCl_3 \longrightarrow C_2Cl_6 \qquad (30)$$

Irradiation of ethyl chloroacetate in excess of benzene leads as expected to homolysis of the C—Cl bond. Traces of ethyl phenylacetate, diethyl succinate, and biphenyl were formed.[87] But when the ester was used in excess, a 78·4% yield of the ethyl phenylacetate was obtained, and an increased amount of biphenyl was formed. The effects of changes in various parameters were investigated. At 65 °C the formation of the phenylacetate was increased slightly, but irradiation at 254 nm gave the same result as was obtained with unfiltered light from a medium-pressure mercury arc lamp. Irradiation of 0·1 M-solutions of each reactant in either acetone or methanol yielded no product. The reaction with excess of benzene was also investigated in the presence of aluminium chloride, when ethyl phenylacetate was the sole product. The intermediacy of a 1 : 1 complex (97) between the chloroacetate and aluminium chloride is suggested, and indeed an optimum yield of the phenylacetate was produced at 1 : 1 molar ratios of aluminium chloride and the chloro-compound.

$$AlCl_3 + ClCH_2CO_2Et \rightarrow [\overset{\delta-}{AlCl_3} \ldots \overset{\delta+}{ClCH_2CO_2Et}]$$

$$(97)$$

[87] Y. Ogata, T. Itoh, and Y. Izawa, *Bull. Chem. Soc. Japan*, 1969, **42**, 794.

Solutions of aluminium chloride in ethyl chloroacetate show additional absorption maxima at 275 and 365 nm consistent with the presence of a complex. Intermolecular reactions of the Friedel–Crafts type have been reported from irradiations (254 nm) of phenol and anisole in the presence of chloroacetamide.[88] Aqueous solutions of anisole and the chlorocompound yield o-, m-, and p-methoxyphenylacetamides in the proportions 20·8, 4·34, and 6·08. The product distribution suggests a mechanism which involves an amidomethyl radical ($\cdot CH_2CONH_2$) formed by homolysis of the C—Cl bond. In acetonitrile, dioxan, or acetone, the yields of the Friedel–Crafts products are greatly decreased. Irradiation of chloroacetamide with phenol yields (as well as a mixture of hydroxyphenyl-acetamides) phenoxyacetamide, which on photolysis yields some of the observed intramolecular products in a photo-Fries type process.[88]

The hydroxylation of phenols by photodecomposition of hydrogen peroxide has been suggested earlier as a synthetic procedure.[89] The same workers now report that the best solvent found for the process is acetonitrile.[90] With 254 nm radiation the o- and p-dihydroxy-compounds are the main products. The apparent reactivity of the phenols was found to decrease in the order p-phenyl > p-acetyl, p-methyl, p-chloro > p-CO_2H, o-nitro, p-cyano, p-t-butyl > m-CO_2H, p-nitro > 2,4-$(CO_2Me)_2$ > 2,4-dinitro. Substituent effects on the decomposition rate of hydrogen peroxide were also found to decrease in the same order, but no distinct cause for this has been found.

The mechanism proposed for the formation of aryl benzoates from irradiation of benzoyl peroxide in aromatic solvents is outlined in Scheme 10.

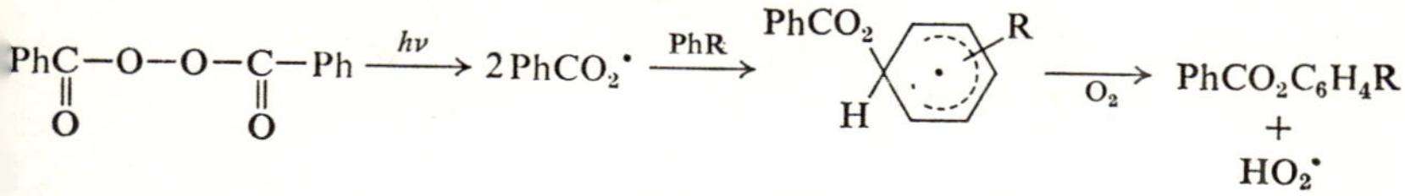

$$PhC(=O)-O-O-C(=O)-Ph \xrightarrow{h\nu} 2\, PhCO_2 \cdot \xrightarrow{PhR} \underset{H}{\overset{PhCO_2\quad R}{\bigcirc\!\cdot}} \xrightarrow{O_2} PhCO_2C_6H_4R + HO_2\cdot$$

Scheme 10

Other processes, such as reaction of excited benzene with the peroxide (Scheme 11), are readily envisaged. By means of ^{18}O-labelling of a carbonyl group of the peroxide, the former mechanism has now been confirmed.[91]

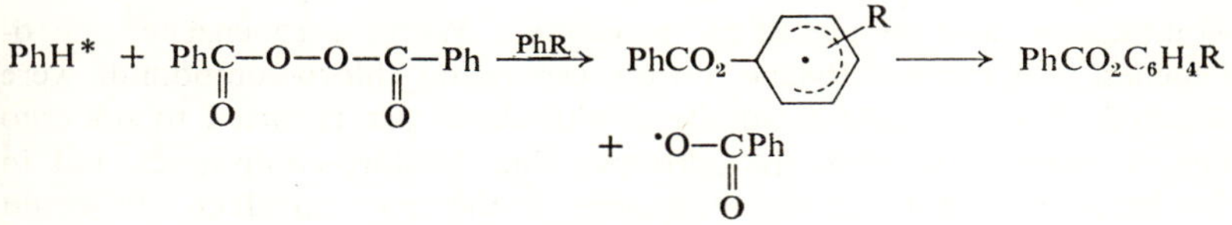

$$PhH^* + PhC(=O)-O-O-C(=O)-Ph \xrightarrow{PhR} PhCO_2-\underset{}{\overset{R}{\bigcirc\!\cdot}} \longrightarrow PhCO_2C_6H_4R$$
$$+\ \cdot O-C(=O)Ph$$

Scheme 11

---

[88] O. Yonemitsu and S. Naruto, *Tetrahedron Letters*, 1969, 2387.
[89] K. Omura and T. Matsuura, *Tetrahedron*, 1968, **24**, 3475.
[90] K. Omura and T. Matsuura, *Tetrahedron*, 1970, **26**, 255.
[91] M. Kobayashi, H. Minato, and Y. Ogi, *Bull. Chem. Soc. Japan*, 1969, **42**, 2737.

Solvolysis reactions were among the first examples to be reported in the field of heterolytic photosubstitution processes. Similar hydrolysis is now reported for 3,5-dimethoxybenzyl phosphate (98) to yield (99).[92] The

$$\text{MeO-C}_6\text{H}_3(\text{OMe})\text{-CH}_2\text{OPO}_3\text{H}_2 \xrightarrow[\text{H}_2\text{O}]{h\nu} \text{MeO-C}_6\text{H}_3(\text{OMe})\text{-CH}_2\text{OH}$$
(98) → (99)

reaction proceeds most readily in neutral or acidic solutions. The probable formation of carbonium ions in the process has led the workers to consider a photo-initiated phosphoryl transfer reaction. Thus irradiation of 3,5-dimethoxybenzyl acetate and monophenyl phosphate in dry acetonitrile gives H_3PO_4 (5%), unchanged monophenyl phosphate (71%), and P^1P^2-diphenylpyrophosphate (24%). The reaction is suggested to arise by attack of a carbonium ion on the acetonitrile to give a nitrilium salt which reacts with the phenylphosphate to form an imidoyl phosphate and then the pyrophosphate.

A new type of light-induced aromatic substitution reaction in the naphthalene series has been reported by Frater and Havinga.[93,94] 1-Nitronaphthalenes (100) irradiated in alkyl chlorides or HCl–CHCl$_3$, HCl–CCl$_4$, HCl–acetic acid, or HCl–hexane mixtures give good yields (80—90% in

$$\text{1-NO}_2\text{-2-R}^1\text{-naphthalene} + R^2Cl \xrightarrow[\lambda > 300 \text{ nm}]{h\nu} \text{1-Cl-2-R}^1\text{-naphthalene}$$

R^1 = H, Me, OMe
R^2 = H, alkyl
(100) → (101)

some cases) of the corresponding chloro-compounds (101). The quantum yield for formation of (101) from alkyl halides is ca. 0·01, while in hydrochloric acid-containing media, values of 0·1—0·5 have been obtained: the overall product yields suggest reactions of synthetic utility. In contrast, 2-nitronaphthalene reacts slowly to give relatively low yields of 2-chloronaphthalene and unidentified products. With nitrobenzene, nitrotoluenes, and nitromesitylene, no corresponding chloro-compounds were detected. The 1,5- and 1,8-dinitronaphthalenes are reported to undergo smooth reaction to yield the corresponding dichloro-compounds, but in the latter case, a trichloronaphthalene is the main product. It would appear that the lowest triplet states of the naphthalene are involved, and a new type of substitution in media of relatively low dielectric constant is

[92] V. M. Clark, J. B. Hobbs, and D. W. Hutchinson, *Chem. Comm.*, 1970, 339.
[93] G. Frater and E. Havinga, *Tetrahedron Letters*, 1969, 4603.
[94] G. Frater and E. Havinga, *Rec. Trav. chim.*, 1970, **89**, 273.

suggested. The displacement of nitrite by chloride is probably not involved, nor is a free-radical mechanism considered to be attractive. The authors propose a type of 'concerted' reaction mechanism involving the photo-excited nitro compound and alkyl chloride which leads to the chloro-naphthalene in its ground state (Scheme 12).[93, 94]

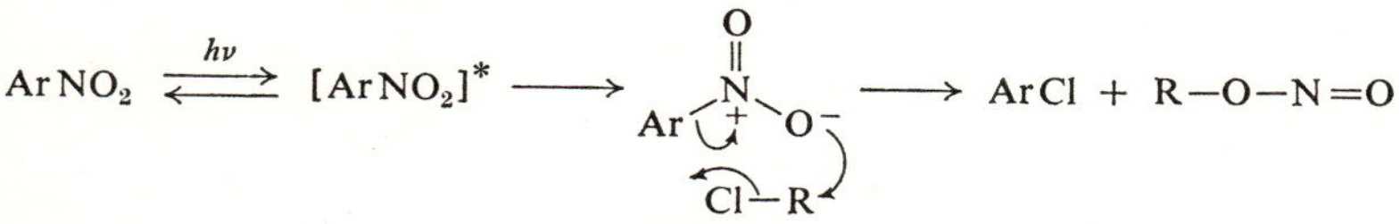

$$ArNO_2 \xrightleftharpoons{h\nu} [ArNO_2]^* \longrightarrow \underset{\underset{Cl-R}{Ar \overset{+}{\underset{}{\bigcirc}} O^-}}{\overset{\overset{O}{\underset{\|}{N}}}{}} \longrightarrow ArCl + R-O-N=O$$

Scheme 12

The irradiation of 2,3-dimethoxy-5-nitronaphthalene in alkaline solution yields either 2-hydroxy-3-methoxy-5-nitronaphthalene (102), or 2-methoxy-3-hydroxy-5-nitronaphthalene (103), but the exact structure was not

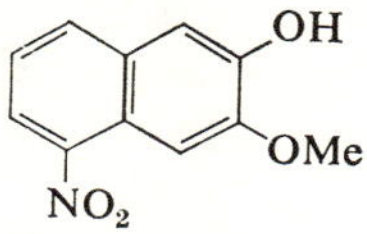

(102)

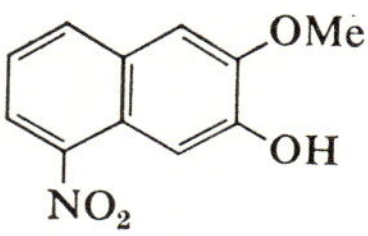

(103)

previously known.[56] The nuclear Overhauser effect has now been used to establish that the product has structure (102).[95]

The naphthalene radical-anion prepared from lithium and naphthalene in tetrahydrofuran has been subjected to irradiation from a high-pressure mercury lamp in the same medium.[96] Reductive substitution occurs with the formation of 30% of the 1,4-adduct (104) and 5% of the 1,2-adduct

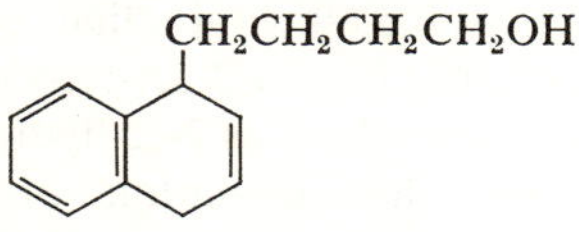

(104)

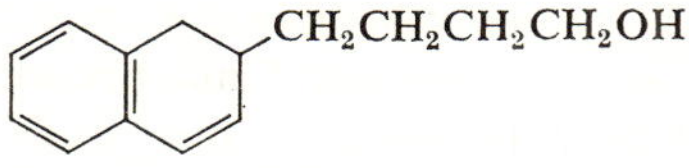

(105)

(105). Analogous reaction also occurs with anthracene, phenanthrene, and pyrene when the 9- (65%), 10- (54%), and 10- (40%) derivatives are formed respectively.

Attempts to photo-alkylate pyridine have in the past been unsuccessful.[97] It has now been shown that irradiation (254 nm) of pyridine in methanol in the presence of hydrochloric acid leads to the formation of 2- and 4-methylpyridines, 1-(2-pyridyl)-2-(4-pyridyl)ethane (106) and 1,2-di-(4-pyridyl)ethane (107).[98] The intermediates for the reaction are suggested to be (108) and (109), and a pathway similar to that in the related reaction

[95] J. Lugtenburg and E. Havinga, *Tetrahedron Letters*, 1969, 1505.
[96] K. Suga, S. Watanabe, and T. Fujita, *Chem. and Ind.*, 1970, 402.
[97] H. Nozaki, M. Kato, R. Noyori, and M. Kawanisi, *Tetrahedron Letters*, 1967, 4259.
[98] E. F. Travecedo and V. I. Stenberg, *Chem. Comm.*, 1970, 609.

of quinoline is suggested.[99] Dehydration of such species as (108) and (109) is favoured in strong acid. An intermediate similar to these has been isolated from the photolysis of 3,5-dialkoxycarbonyl-substituted pyridines.[100] The irradiation of quinoline and 8-methylquinoline in acidic ethanol likewise yields 2- and 4-ethylquinolines.[99,101] In 95% ethanol, however, no dehydration of the intermediates to the alkylquinolines occurs, and 2α-hydroxyethylquinoline and the corresponding 1,2,3,4-tetrahydroquinolines are formed. Similar reaction in t-butanol yields 2-(2-hydroxy-2-methylpropyl)quinoline, but there are no photoalkylation products formed in isopropyl alcohol, only a low yield of the quinoline dimer. In the same report, 2-substituted quinolines bearing γ-hydrogens are demonstrated to undergo photoelimination (see Chapter 7, ref. 142) corresponding to a Type II process. The decomposition probably arises from the $n\pi^*$ singlet, and has a quantum yield between 0·014 and 0·29, depending on structure, and only those compounds which undergo this elimination show a McLafferty rearrangement ion as the base peak in the mass spectrum.

Diethylamine is reported to accelerate the rate of photolysis of 5-bromopyrimidines and of oxindole through a photo-induced electron transfer reaction.[102] Irradiation of 5-bromo-2-methylpyrimidine (110) in methanol

[99] F. R. Stermitz, C. C. Wei, and W. H. Huang, *Chem. Comm.*, 1968, 482.
[100] R. M. Kellog, T. J. van Bergen, and H. Wynberg, *Tetrahedron Letters*, 1969, 5211.
[101] F. R. Stermitz, C. C. Wei, and C. M. O-Donnell, *J. Amer. Chem. Soc.*, 1970, **92**, 2745.
[102] J. Nasielski, A. Kirsch-Demesmae, P. Kirsch, and R. Nasielski-Hinke, *Chem. Comm.*, 1970, 302.

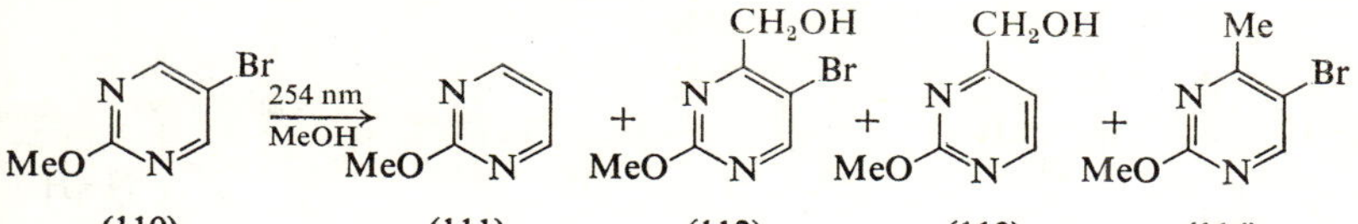

(110) → (111) + (112) + (113) + (114) 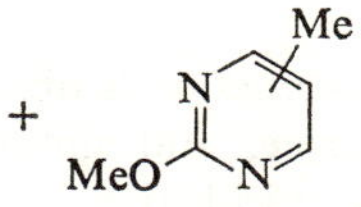 + (115)

254 nm, MeOH

at 253·7 nm gives low yields of compounds (111)—(115): thus both reduction and substitution processes occur. In the presence of diethylamine, the dehalogenated pyrimidine (111) is the major product. The influence of the amine increases with concentration, but no amine-derived products are observed. The results are rationalized by the amine acting as an electron donor towards the excited state of the pyrimidine (equation 31). Dehalogenation of the radical-anion of the pyrimidine occurs, as is typical in

$$(P)^* + Et_2NH \longrightarrow (P)^{\bar{\cdot}} + Et_2NH^{+\cdot} \qquad (31)$$

Birch-type reductions. Oxindole (116) is fairly light-stable in benzene, cyclohexane, methanol, and acetonitrile. In 0·13 M-diethylamine solution, however, (116) yields o-toluidine, and the amine adduct (117); and in the presence of isopropyl alcohol as solvent, (118) is also formed. Again the

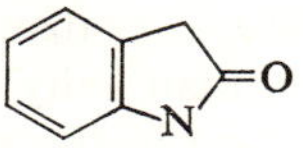

(116) 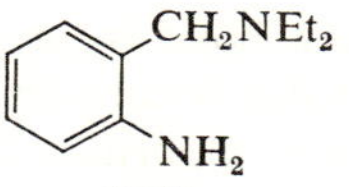   (117) 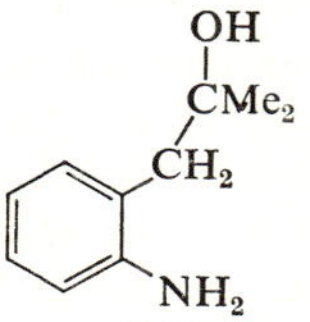   (118)

electron transfer process offers an attractive mechanism: the radical-anion of oxindole then undergoes typical fragmentations leading to a decarbonylated intermediate which reacts with the medium to give the observed products.

The chloro-substituent in atrazine, [2-chloro-4-ethylamino-6-isopropyl-amino-s-triazine (119)] and in related compounds propazine and simazine, undergoes substitution in methanol or water solutions under 254 nm irradiation to yield the corresponding 2-methoxy and 2-hydroxy compounds in yields of the order of 85—95%.[103] In ethanol and n-butanol, the 2-methoxy- and 2-butoxy-derivatives are obtained. At wavelengths longer

[103] B. E. Pape and M. J. Zabik, *J. Agric. Food Chem.*, 1970, **18**, 202.

than 300 nm, no reaction is observed. The 2-methylthio derivatives (120) of atrazine, propazine, and simazine, by similar irradiation in any of the above solvents, yield the unsubstituted 2-compounds (121) (see also Chapter 7, ref. 106).

4 Cyclization Reactions

This aspect of the photochemistry of aromatic compounds has been the subject of many reports in the year. The study of the cyclization of polynuclear aromatic hydrocarbon-substituted ethylenes has attracted the interest of several groups. Reviews on the photocyclization of stilbene analogues, and photo-aryl coupling, have appeared.[104, 105] The first review by Blackburn and Timmons is most comprehensive and is a useful source of references for the subject. Unfortunately for the Reporters, the second review is in Japanese, so appreciation has been limited to the numerous structural examples.

The photo-oxidative cyclization of stilbenes to yield phenanthrenes *via* dihydrophenanthrenes has been known for many years, but much useful and interesting work is still published in this area. Some stilbene analogues do not undergo the photocyclization process, and two groups of workers have commented on this.[106, 107] For the reaction to occur, it is reported that the sum of the free valence indices (ΣF^*) for the first excited state at atoms between which the new bond is formed, must be greater than unity. Compounds in which this value is calculated to be less than unity do not cyclize. But other factors are important, and substitution in the stilbene by groups (*e.g.* acetyl, nitro) which promote intersystem crossing also prevents the cyclization.[106] Compounds which do not show a 'cyclization mass number' in their mass spectra also fail to photocyclize. The alternative process to the cyclization is cyclobutane formation, which is observed in the case of *trans*-benzylidenephthalide (122).

Timmons and co-workers have examined the photolysis of sterically hindered styrylnaphthalenes in an attempt to form the dihydro-intermediates in the general photocyclo-dehydrogenation process.[108] The

---

[104] E. V. Blackburn and C. J. Timmons, *Quart. Rev.*, 1969, **23**, 482.
[105] T. Sato, *J. Soc. Org. Synth. Chem. Japan*, 1969, **27**, 715.
[106] E. V. Blackburn and C. J. Timmons, *J. Chem. Soc.* (*C*), 1970, 172.
[107] W. H. Laarhoven, T. J. H. M. Cuppen, and R. J. F. Nivard, *Tetrahedron*, 1970, **26**, 1069.
[108] E. V. Blackburn, C. E. Loader, and C. J. Timmons, *J. Chem. Soc.* (*C*), 1970, 163.

(122)

photoreactions of the styrylnaphthalenes were carried out in hexane solution with u.v. light filtered through a cobalt sulphate–nickel sulphate solution, and a study was made of the kinetics of the thermal ring-opening of the resulting yellow dihydro-intermediates. First-order kinetics were obeyed. 1,2-Di-(2-naphthyl)ethylene (123) was found to yield the most stable intermediate (124), which is hardly surprising since the expected product of the cyclo-dehydrogenation reaction, dibenzo[c,g]phenanthrene

(27) $R^1 = R^2 = R^3 = Me$
(28) $R^1 = Me, R^2 = R^3 = H$
(29) $R^1 = H, R^2 = R^3 = Me$
(30) $R^1 = R^3 = H, R^2 = Cl$

(131) $R^1 = Br, R^2 = H$
(132) $R^1 = H, R^2 = Me$

(133) $R^2 = H$
(134) $R^2 = Me$

(125), shows considerable steric overcrowding and loss in delocalization energy due to distortion of the ring system from planarity. Formation of (125) from (124) is slow, but (125) is somewhat unstable and undergoes rapid further cyclization and dehydrogenation to benzo[g,h,i]perylene

(126). 2-Methyl-1-(2,4,6-trimethylstyryl)naphthalene (127), 2-methyl-1-styrylnaphthalene (128), 1-(2,4,6-trimethylstyryl)naphthalene (129), and 1-(2,6-dichlorostyryl)naphthalene (130) all gave the expected chrysene derivatives. Thus in no case was migration observed, and all reactions involved elimination. The yield of 1,3-dimethylchrysene from (127) was very low, but the chrysene itself decomposed under the reaction conditions, a fact which may account for the observed low yield. Cyclization has not previously been reported for compounds in which all possible positions of cyclization were substituted. Cyclization of 1-bromo-2-styrylnaphthalene (131) and 2-(2,4,6-trimethylstyryl)naphthalene (132) yielded benzo[c]-phenanthrene (133) and the 2,4-dimethyl derivative (134) respectively. The effects of electron impact on the styrylnaphthalenes correlate with their photochemical behaviour.

Photocyclization of 1,2-di-(1-naphthyl)ethylene has been reported by two groups of workers to yield the dihydro-derivative (135) which is

oxidized to picene (136).[109, 110] The other two 1,2-dinaphthylethylenes have also been investigated by one group who observe that ring closure takes place in a stereospecific way with respect to the possible rotational conformers.[110]

A photo-oxidative cyclization has been used in the synthesis of 4,11-diphenylbisanthene (137), one of the few aromatic compounds for which fluorescence has been detected in the i.r. region.[111] The previous thermal synthesis involves three stages with an overall yield of 20% whereas the present reaction involves two stages with a yield of 59%. The photochemical step involves the cyclization of (138) to (137). The intermediate

[109] R. N. Nirmukhametov and G. I. Grishina, *Zhur. fiz. Khim.*, 1969, **43**, 2925.
[110] C. Goedicke and H. Stegemeyer, *Ber. Bunsengesellschaft Phys. Chem.*, 1969, **73**, 782
[111] D. R. Maulding, *J. Org. Chem.*, 1970, **35**, 1221.

in the process is the diol (139) since, in the presence of iodine, (139) yields 79% of (137). The photo-oxide (140) is a by-product of the reaction.

The structures of polynuclear aromatic hydrocarbons produced from photocyclizations involving 1,2-diarylethylenes are generally deduced by u.v., n.m.r., and mass spectra, and from circular dichroism and optical rotatory dispersion curves. Martin and Schurter have examined the possibility of obtaining simpler and more definite structural proof by using partially deuteriated starting materials.[112] Their approach has been tested in the case of hexahelicene (141) for which the structure has been fully established. The monodeuterio aryl-ethylene (142) can photocyclize either with the loss of two hydrogens to give [7-^{2}H]hexahelicene (143), or one hydrogen and one deuterium forming benzo[*a*]naphtho[1,2-*h*]anthracene (144). Distinction between (143) and (144) can then be made by mass spectroscopy. Compound (143) was formed in 87% yield from (142), thus showing the validity of the method. This approach has been used to assist in the structural determination of a more complex cyclization. Tridecahelicene (145), the first benzologue of the multilayer helicene, has been synthesized in 52% yield by the double cyclization of (146) in benzene solution.[113] The reaction could yield twelve structural isomers, but by using the deuteriation technique, nine of these have been rejected. Of the three remaining possibilities ($\alpha\beta'/\alpha\beta'$, $\alpha\beta'/\alpha\alpha'$, and $\alpha\alpha'/\alpha\alpha'$), the one resulting from a double cyclization in the β' positions of the phenanthrene moiety is excluded by examining the mono-cyclized product formed at

[112] R. H. Martin and J. J. Schurter, *Tetrahedron Letters*, 1969, 3679.
[113] R. H. Martin, G. Morren, and J. J. Schurter, *Tetrahedron Letters*, 1969, 3683.

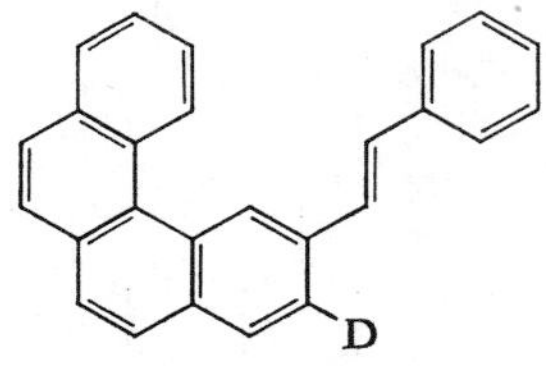

(142)

(141) R = H
(143) R = D

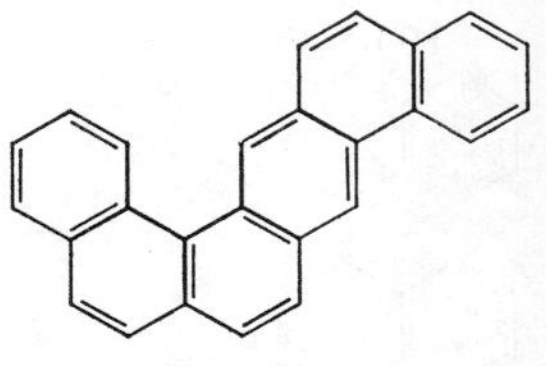

(144)

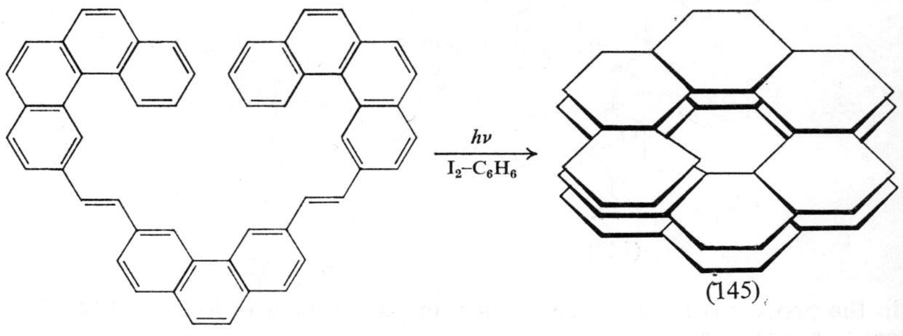

(146) $\xrightarrow{h\nu}{I_2-C_6H_6}$ (145)

short retention time, and only the αα′/αα′ isomer (145) is fully compatible with the spectral data.

The photocyclization reactions of distyrylbenzenes may yield several isomers. Published information on this topic has been somewhat unsatisfactory as well as incomplete. In a most comprehensive paper, Laarhoven and co-workers now describe the results of a thorough investigation of the photolysis of the distyrylbenzene isomers and point out again (see also ref. 106) that the cyclization only occurs when the sum of the free valence number of the excited state (ΣF^*) of atoms involved is greater than unity.[107] The benzo[c,d,e]perylene (148) which is formed on irradiation of p-distyrylbenzene (147) has been shown to arise *via* 3-styrylphenanthrene (149) as an intermediate. However, (149) only arises from (147) by an indirect route involving dimerization of (147) (see Scheme 13a). The cyclization of (149) to (150) does not occur owing to the low ΣF^* value of this reaction. In accord with predictions, no dibenzo[a,j]anthracene (151) was formed from the *m*-isomer (152) (see Scheme 13b). Besides benzochrysene (153), (70%), minor amounts of dimers such as (154) and 4-phenylpyrene (155) were formed in this reaction. These results implicate 4-styrylphenanthrene (156) as an intermediate since (156) is the precursor of (155). A similar approach has been successfully applied to the photocyclization of *o*-distyrylbenzene (157), and the formation of picene (158) in this reaction is analogous to the conversion of (147) to (148) *via* dimeric compounds. The formation

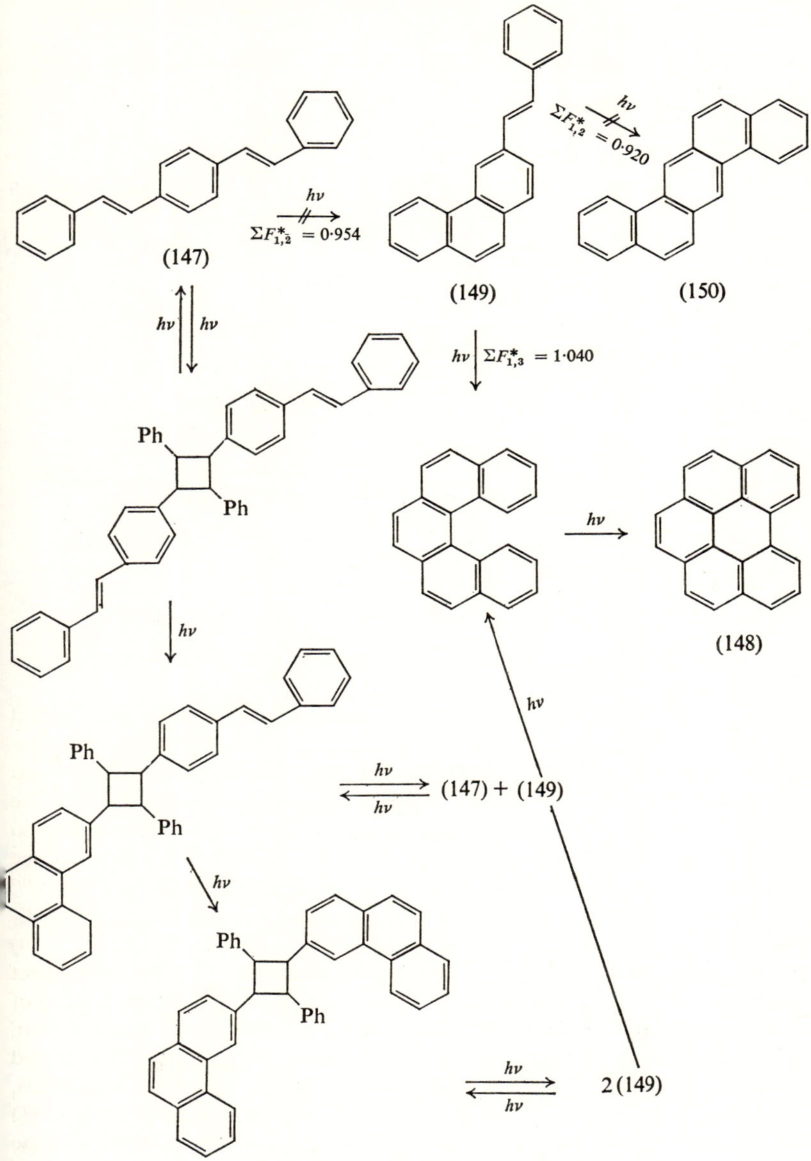

Scheme 13a

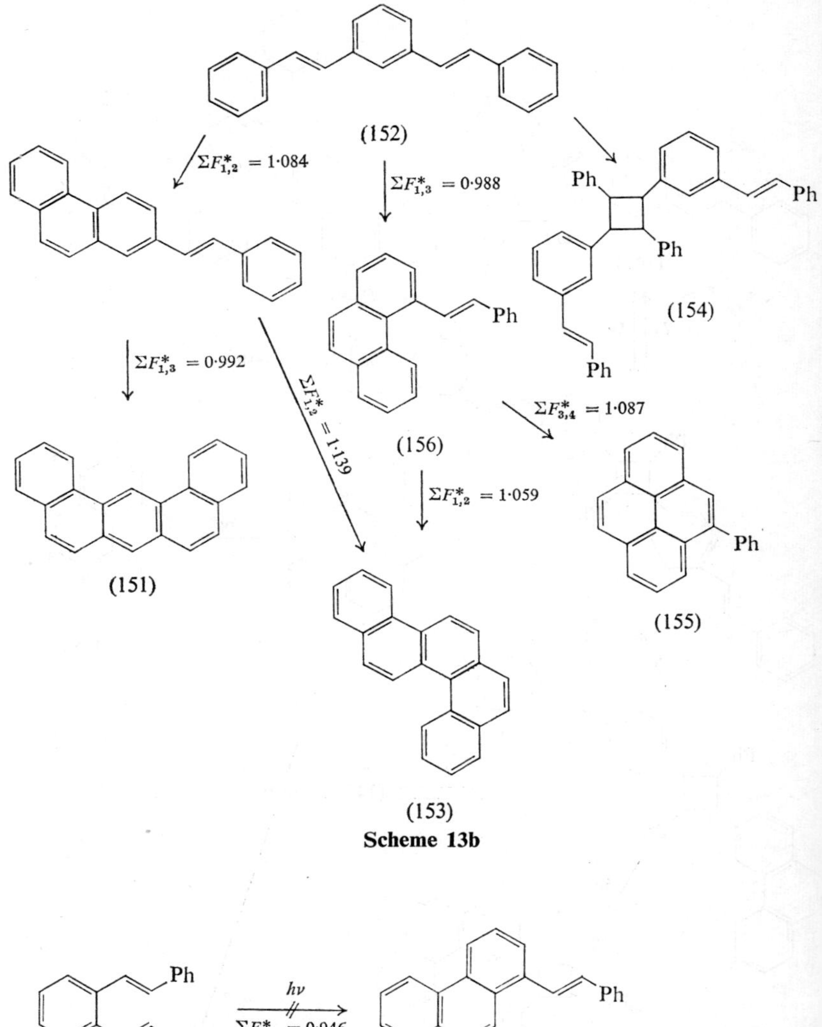

(152)
$\Sigma F^*_{1,2} = 1{\cdot}084$
$\Sigma F^*_{1,3} = 0{\cdot}988$
Ph
Ph
Ph
Ph
(154)
Ph
$\Sigma F^*_{1,3} = 0{\cdot}992$
$\Sigma F^*_{1,2} = 1{\cdot}139$
(156)
$\Sigma F^*_{3,4} = 1{\cdot}087$
$\Sigma F^*_{1,2} = 1{\cdot}059$
Ph
(151)
(155)
(153)

Scheme 13b

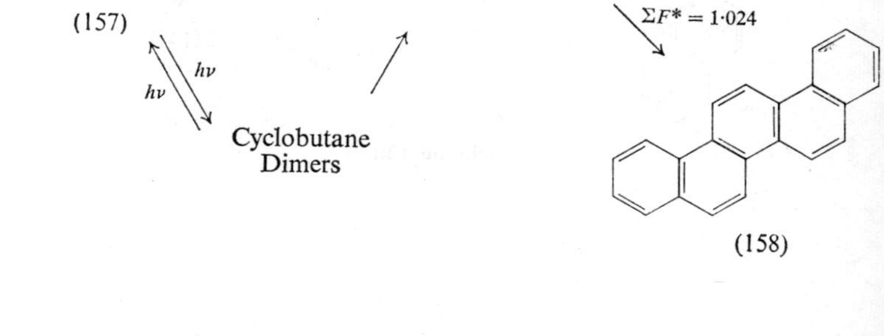

Ph
Ph
(157)
$h\nu$
$\Sigma F^*_{1,2} = 0{\cdot}946$
Ph
$\Sigma F^* = 1{\cdot}024$
$h\nu$
$h\nu$
Cyclobutane
Dimers
(158)

of stilbene and phenanthrene from the photolysis of (157), and the discrepancies between these results and Müller's earlier work [114] are discussed.

Both 2- and 4-styrylpyridines undergo the photocyclization process, in contradiction of an earlier report.[115] 2-Stilbazole derivatives (159a—f) in general undergo oxidative cyclization to yield benzo[f]quinolines (160a—f).[116] The effects of solvent and additives on the process have been investigated. In two cases (159a) and (159b), the additional photo-products (161)

(a) $X = H, R^1 = R^2 = H$
(b) $X = CN, R^1 = R^2 = H$
(c) $X = CN, R^1 = H, R^2 = OMe$
(d) $X = CO_2Me, R^1 = R^2 = H$
(e) $X = H, R^1 = H, R^2 = NHCOMe$
(f) $X = H, R^1 = H, R^2 = NO_2$

(159)   (160)   (161)   (162)

and (162) respectively are formed from irradiation of the cyclohexane solutions in the presence of oxygen. However, (160b) was shown to yield (162) under the same photolysis conditions. For most of the starting 2-stilbazoles, the yields of (160) are synthetically useful, and although in polar solvents the rate of disappearance of starting material is much faster, the yields of cyclized materials are generally less than in non-polar solvents. Irradiation of (159f) resulted in disappearance of the starting material, but no (160f) was formed. The photochemical behaviour of α-substituted 4-styrylpyridines has been studied.[117] With phenyl substituents, pyridylphenanthrenes (163) are formed together with the *trans*-isomer (164), which yields its own dehydrocyclization product (165).

The stilbene-phenanthrene oxidative cyclization has been used as a key step in the new synthetic route to the aporphine alkaloids nuciferine (166) and glaucine (167).[118] Thus 1-benzylidene-2-carboethoxy-1,2,3,4-tetrahydro-6,7-dimethoxyisoquinoline (168) and the 1-veratrylidene derivative

---

[114] E. Mueller, M. Sauerbier, and J. Heiss, *Tetrahedron Letters*, 1966, 2473; *Trans. New York Acad. Sci.*, 1966, **28**, 845.
[115] C. S. Wood and F. B. Mallory, *J. Org. Chem.*, 1964, **29**, 3373.
[116] P. L. Kumler and R. A. Dybas, *J. Org. Chem.*, 1970, **35**, 125.
[117] G. Galiazzo, P. Bortolus, G. Cauzzo, and U. Mazzucato, *J. Heterocyclic Chem.*, 1969, **6**, 465.
[118] M. P. Cava, M. J. Mitchell, S. C. Havlicek, A. Lindert, and R. J. Spangler, *J. Org. Chem.*, 1970, **35**, 175.

(169) yield respectively *N*-carboethoxy-6a,7-dehydronornuciferine (170) and *N*-carboethoxy-6a,7-dehydronorglaucine (171). The conversion of (170) and (171) into the desired products involves a two-step reduction. A cleaner formation of (170) and (171) is realized by non-oxidative photocyclization of the 2^1-chloro derivative of (166) and the 6^1-bromo derivative of (167) respectively. The latter transformations represent the first examples of photochemical synthesis of phenanthrenes by the loss of hydrogen chloride or hydrogen bromide from a simple stilbene system. Loss of hydrogen iodide has also been observed. Thus 12-fluoro-7-methyl- and 7-fluoro-benz[*a*]anthracene, (172) and (173), are formed by photocyclization of 1-(1-fluoro-2-naphthyl)-2-(*o*-iodophenyl)ethylene (174) and the 1-(4-fluoro-1-methyl-2-naphthyl) derivative (175).[119] It is reported that (172) and (173) were only formed with light sources of relatively low intensity.

[119] J. Blum, F. Grauer, and E. D. Bergmann, *Tetrahedron*, 1969, **25**, 3501.

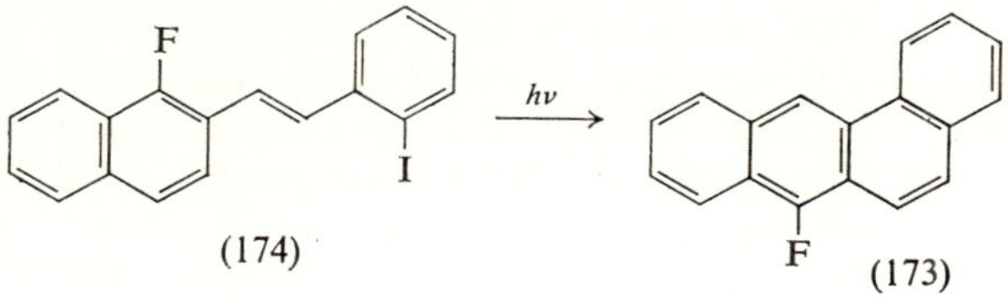

(174) → (173)

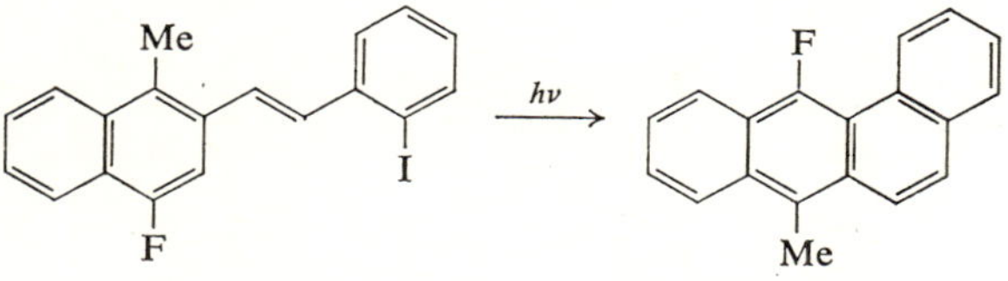

(175) → (172)

Stilbene-type cyclizations have also been reported for compounds which have one or two hetero-atoms in the 'ethylenic' bond. Acetophenone anils (176) undergo the reaction to form (177),[120] and boron compounds of type (178) lose hydrogen iodide to yield borazarophenanthrene (179), boroxarophenanthrene (180), and borathiarophenanthrene (181).[121]

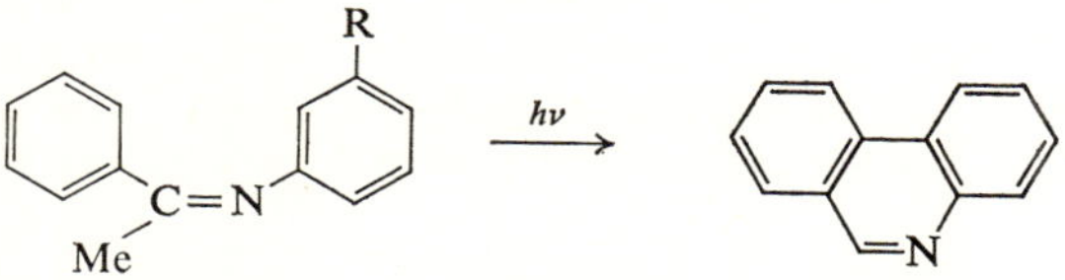

(176) R = H or Me → (177)

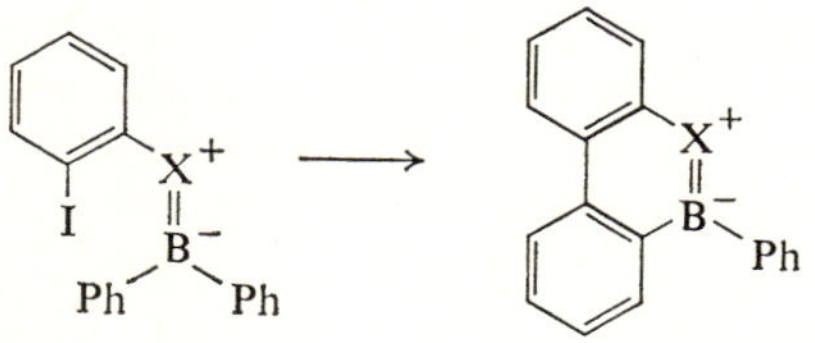

(178) → (179) X = NH
(180) X = O
(181) X = S

The formation of all compounds reported so far has involved the cyclization of one aromatic ring to another with an ethylenic bond in conjugation with each. The remaining reports in this section are concerned with cyclization of the aromatic ring onto an olefin, attack on another aromatic ring (*e.g.* triphenylene synthesis), and miscellaneous cyclizations.

Heller and his group have carried out further studies on the photochemistry of overcrowded molecules. *trans*-2-Benzylidene-1-diphenyl-methyleneindane (182) has been shown to undergo a disrotatory thermal

[120] A. V. El-Tsov, O. P. Studzinskii, and N. V. Ogol-Tsova, *Zhur. org. Khim.*, 1970, 6, 405.
[121] P. J. Grisdale and J. L. R. Williams, *J. Org. Chem.*, 1969, 34, 1675.

ring-closure to (183) which by a suprafacial 1,5-hydrogen shift yields cis-10,10a-dihydro-5,10-diphenyl-11H-benzo[b]fluorene (184).[122] With 366 nm radiation, (182) undergoes a conrotatory closure and forms the trans-isomer (185) of (184). The 3-methyl derivative of (182) undergoes

analogous rearrangement to produce only one dihydrobenzofluorene in each of the thermal and photochemical reactions, indicating that steric effects determine which of the two possible disrotatory and conrotatory processes occur. Similarly, photochemical ring-closure of 1,2-bisdiphenyl-

[122] H. G. Heller and K. Salisbury, *J. Chem. Soc.* (C), 1970, 399.

methylene-3-methylindane (186) occurs by only one of the possible conrotatory modes, followed by the 1,5-hydrogen shift to yield compound (187) which has the methyl group *anti* to the 4b hydrogen.[123]

By analogy with the formation of 9,10-dicyanodihydrophenanthrene (188) from the dicyanodiphenylethylene,[124] it was hoped to synthesize compound (189) from 1-(α-indolyl)-2-(β-pyridyl)acrylonitrile (190).[125] However, irradiation of (190) in the presence of ferric chloride or iodine yielded the dehydro derivatives (191) and (192) in 5 and 30% yields respectively. When the reaction was performed under nitrogen, a complex mixture of products was formed from which 1-cyano-3-(β-formylvinyl)-carbazole (193) (30%) was isolated.[125]

It is known that photolysis of *N*-acyl enamines of acyclic and aromatic systems leads to products of acyl-migration.[126] The results of a study of similar reactions of *N*-acylenamines of cyclic ketones such as β-tetralone and cyclohexanone have now been reported.[127] Analogous migration occurs in the case of *N*-acetylenamines (194), but with *N*-benzoylenamines

[123] H. G. Heller and K. Salisbury, *J. Chem. Soc.* (*C*), 1970, 873.
[124] M. V. Sargent and C. J. Timmons, *J. Amer. Chem. Soc.*, 1963, **85**, 2186; *J. Chem. Soc.*, 1964, 5544.
[125] H. P. Husson, C. Thal, P. Potier, and E. Wenkert, *J. Org. Chem.*, 1970, **35**, 442.
[126] R. W. Hoffman and K. R. Eicken, *Tetrahedron Letters*, 1968, 1759, and references therein.
[127] I. Ninomiya, T. Naito, and T. Mori, *Tetrahedron Letters*, 1969, 2259.

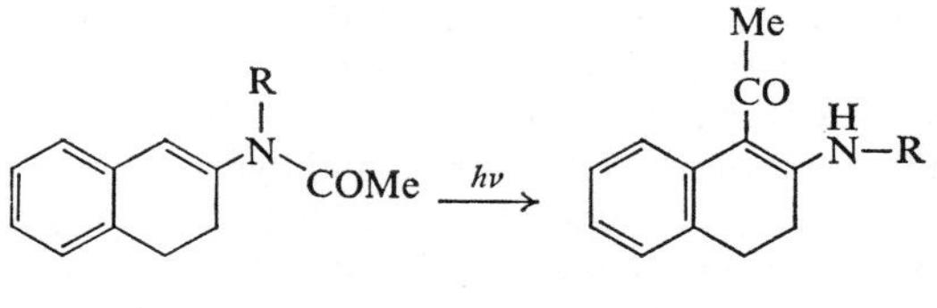

(194) 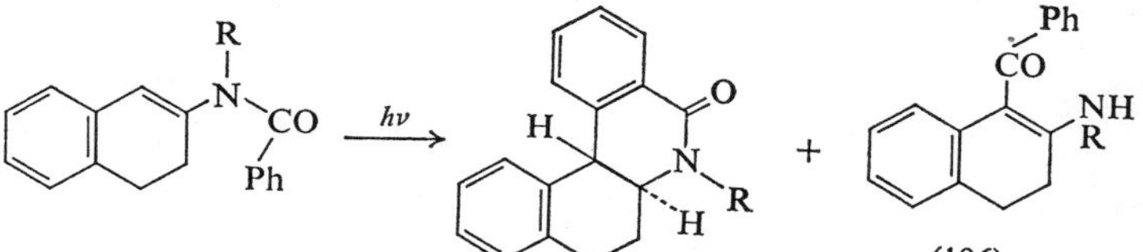 → (195) + (196)

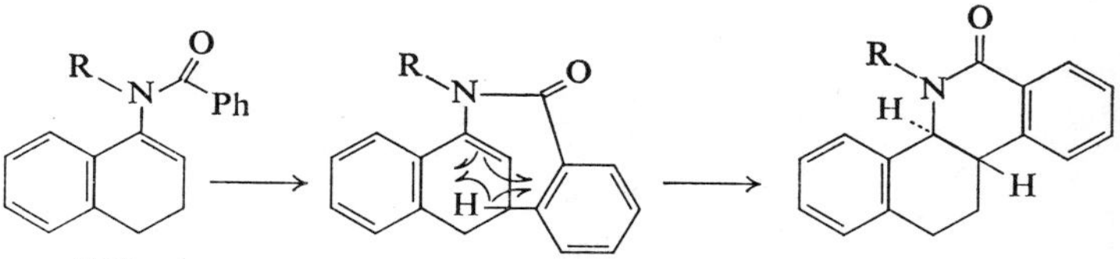

(197) → (199) → (198)

a hitherto unknown photocyclization occurred to form benzo[*a*]phenanthridines (195) with the migration product (196) representing only a minor proportion of the total reaction products. This reaction is a useful synthesis for ring systems such as that in (195). The process has been extended to the α-tetralone derivatives (197), when benzo[*c*]phenanthridines (198) are formed.[128] The stereospecific cyclization to the *trans*-fused ring systems could be considered to proceed through such an intermediate as (199). These cyclizations provide a synthetic route to a group of alkaloids such as chelidonine and make a promising approach to the total synthesis of some benzo[*c*]phenanthridines.

The photochemical formation of triphenylenes (200) from *o*-terphenyls in the presence of iodine is well known.[129] The electronic effects of substituents on this reaction have been studied by reference to the three systems (201), (202), and (203).[130] Electron-withdrawing substituents (*e.g.* NO_2) hinder the cyclization but with methoxy, fluoro, chloro, and bromo substituents the corresponding triphenylene derivatives are formed. With iodo- and bromo-compounds, a complex reaction mixture is formed due to radical reactions resulting from C—X bond cleavage; and with the iodo-compounds, it is unnecessary to add iodine to the reaction since this is

---

[128] I. Ninomiya, T. Naito, and T. Mori, *Tetrahedron Letters*, 1969, 3643.
[129] T. Sato, Y. Goto, and K. Hata, *Bull. Chem. Soc. Japan*, 1967, **40**, 1994.
[130] T. Sato, S. Shimada, and K. Hata, *Bull. Chem. Soc. Japan*, 1969, **42**, 766.

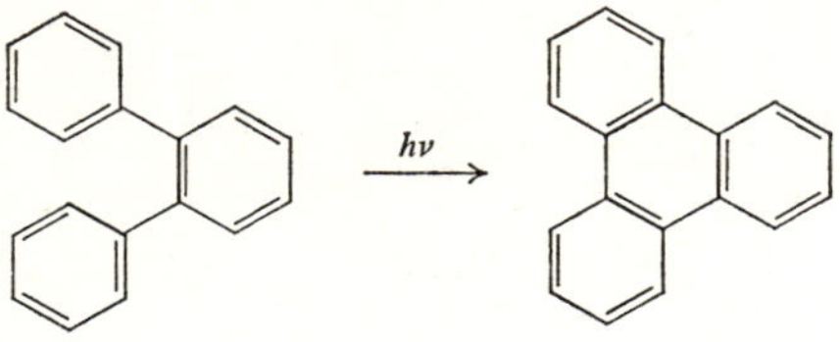

(200)

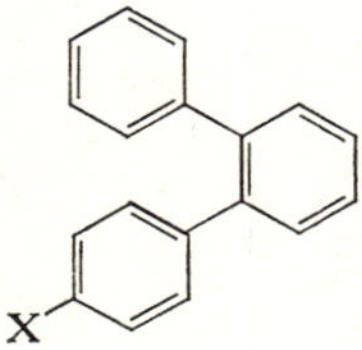

(201) 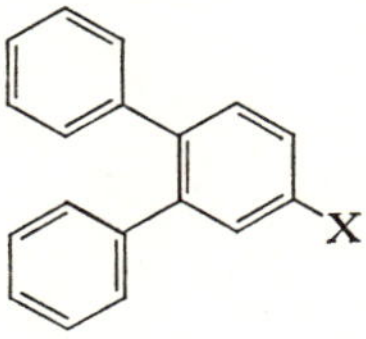   (202) 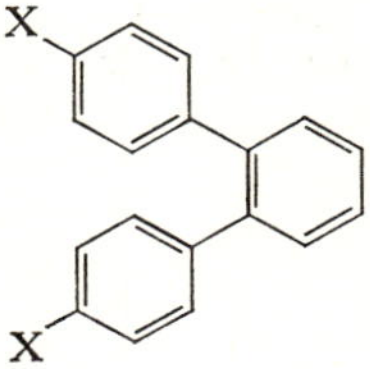   (203)

formed anyway. Among the halogeno-compounds, except iodo-, cyclization occurred more readily with (202) than (201), and (203) reacted with difficulty. Benzene was found to be a better reaction medium than cyclohexane and led the workers to suggest that the aromatic solvent may act as a sensitizer. Light-filter experiments would readily settle this point.

1,1-Dichloro-2,2-bis(p-chlorophenyl)ethylene (D.D.E.) is produced by metabolism of D.D.T. in insects, birds, and mammals. Its photolysis in methanol (see also Chapter 7, ref. 120) in the presence of oxygen is reported to yield 3,6-dichlorofluorenone.[131] Many other products are formed including those from loss of chlorine from the aromatic rings or the ethylene group of D.D.E. by a radical mechanism. Hydrogen abstraction from the solvent played a significant part in subsequent reactions of the radicals, and reductively dechlorinated products were identified. The reaction mixture also contained difluorobiphenyl and pp'-dichlorobenzophenone; but the latter was shown not to be the precursor of the fluorenone, the formation of which may proceed by way of a C_1 hydroperoxide.

In an attempt to understand the requirements for photolytic generation of 5-membered rings in polycyclic aromatic systems, the photochemistry of o-halogenophenylnaphthalenes has been studied.[132] The u.v. spectra of these compounds are very similar to that of phenylnaphthalene; thus the halogen has little effect on the energy or radiative lifetime of the singlet states. All have triplet energies of the order of 60 kcal mol^{-1}, but great differences in the (quantum yields of phosphorescence) : (quantum yield of fluorescence) ratio show that intersystem crossing increases markedly with increase in atomic weight of the halogen, as one would expect. Irradiation of the 1-chloro-derivative (204) in benzene gives a high yield of fluoranthene

[131] J. R. Plimmer and U. I. Klingebiel, *Chem. Comm.*, 1969, 648.
[132] W. A. Henderson, R. Lopresti, and A. Zweig, *J. Amer. Chem. Soc.*, 1969, **91**, 6049.

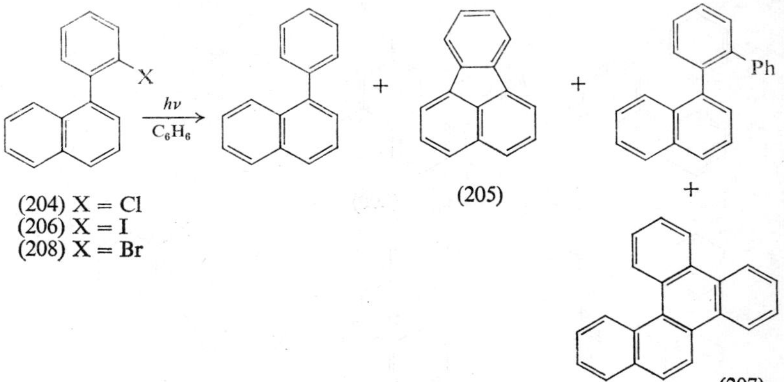

(204) X = Cl
(206) X = I
(208) X = Br

(205), whereas the *o*-iodo-compound (206) yields a variety of rearrangement and solvent-derived products, *e.g.* benztriphenylene (207), but virtually no fluoranthene (see Table 3). Intermediate behaviour is demon-

Table 3 *Photolysis products from* o-*halogenophenylnaphthalenes*

Halogen	% Conversion	% Phenyl-naphthalene	% (206)	% Arylation products
Cl	13	17	59	24
	5	14	72	14
Br	30	58	5	37
	72	46	8	46
I	80	18	0	82
	45	15	1	84

strated by the *o*-bromo-derivative (208). Formation of (205) from the chloro-compounds occurs with a better quantum yield in polar solvents. A 5-methoxy-group in (204) produces only a slightly higher ratio of the (205) derivative to other products than that found for the parent compound. Results are consistent with C—X bond homolysis to give an aryl radical and a halogen atom. The radical may undergo several reactions but the product distribution is controlled by the reactivity of the halogen atom. From sensitization experiments, it would appear that (204) and (208) react only from their excited singlet states, whereas with (206) the triplet state is the major, if not the sole, source of the products. Compound (204) in bromobenzene does not react photochemically, owing to the external heavy-atom effect. In this investigation it was also found that (204), (206), and (208) were interconvertible with 1-halogeno-8-phenylnaphthalene, and that treatment of (206) with thermally-generated phenyl radicals gave (205).

An interesting photocyclization has been reported for the sodium enolate of diethyl *o*-biphenylyl malonate (209), which yields 70% of ethyl-9-phenanthrol-10-carboxylate (210).[133] The enolates of phenylethylidene

[133] N. C. Yang, L. C. Lin, A. Shani, and S. S. Yang, *J. Org. Chem.*, 1969, **34**, 1845.

malonate and phenylethylidene cyanoacetate systems form the corresponding naphthol derivatives on irradiation.

Irradiation of 2,6-diphenyl *p*-benzoquinone in benzene leads to a dimeric product, the precise structure of which is not known.[134] In acetonitrile, the reaction path changes drastically and the quinone forms 2-hydroxy-4-phenyldibenzofuran (211).[134] Formation of (211) could occur by two

pathways, (*a*) hydrogen abstraction from the solvent and reaction of the semiquinone radical with the phenyl substituent, and (*b*) addition of the carbonyl group to a phenyl ring and reorganization of the bonds as previously suggested for isopropenyl-*p*-benzoquinone.[135] Monoaryl-substituted *p*-benzoquinones have been said to react similarly, but details have not yet been published.

5 Dimerization Reactions

There are few reported photodimerizations of aromatic compounds other than those which occur with anthracene and its derivatives. Until recently, the only naphthalene compounds which were known to dimerize were the β-alkoxy-derivatives, and the reported structures for the dimers (212) were analogous to that of the anthracene dimer.[136] Evidence has been described recently which strongly suggests that the dimerization is of a 1,2-/1,2-type not observed in the anthracene series.[137] Photo-'dimerization' does occur with the naphthalene paracyclophane (213),[138] and this year a similar

---

[134] H. J. Hageman and W. G. B. Huysmans, *Chem. Comm.*, 1969, 837.
[135] J. M. Bruce and P. Knowles, *J. Chem. Soc.* (*C*), 1966, 1627.
[136] J. S. Bradshaw and G. S. Hammond, *J. Amer. Chem. Soc.*, 1963, **85**, 3953.
[137] J. Christie and B. K. Selinger, *Photochem. and Photobiol.*, 1969, **9**, 471.
[138] H. H. Wasserman and P. M. Keehn, *J. Amer. Chem. Soc.*, 1969, **91**, 2374; *ibid.*, 1967, **89**, 2270.

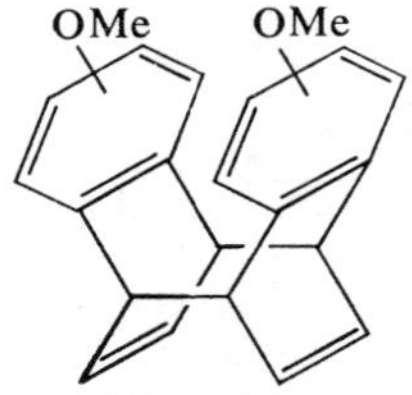

(212)

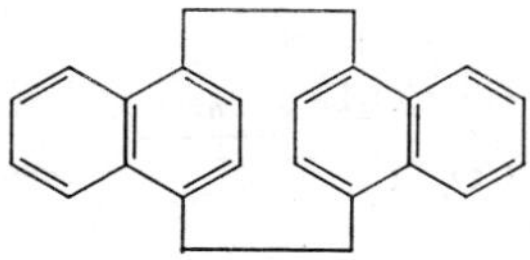

(213)

reaction has been reported in the conversion of 1,3-bis(α-naphthyl)propane (214) into the intramolecular 'dimer' (215).[139] The photo-'dimer' is

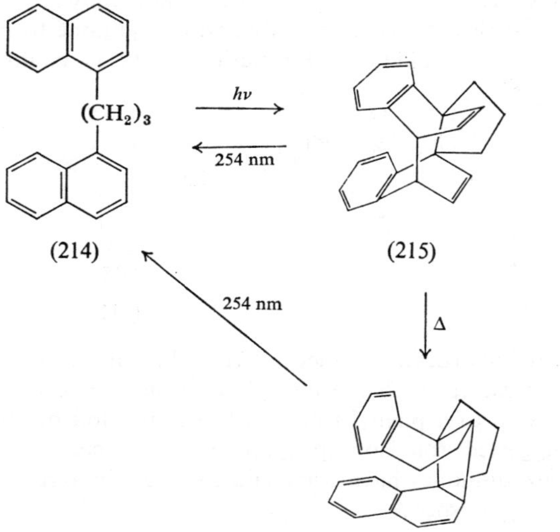

(214) ⇌ (hν / 254 nm) (215) →Δ (216); (216) →254 nm (214)

thermally unstable and yields the cyclobutane isomer (216) when the crude photo-product is boiled in chloroform. Both (215) and (216) are converted back into (214) on irradiation at 254 nm. Of the various dinaphthylalkanes examined in this study, only the symmetrical 1,3-dinaphthylpropanes exhibit strong excimer interaction, since they can form overlapping sandwich pairs. However, the β-naphthyl derivative is unchanged by irradiation. Compounds such as 1,2-di-α-naphthylethylene and 1,4-di-α-naphthylbutane do not exhibit excimer fluorescence at 25 °C and do not photodimerize, nor indeed does α-methylnaphthalene. The differences in the behaviour of the various compounds studied appear to be due to stabilization of the intramolecular excimer of (214) by the cyclopropane chain. This stabilization is attributable to the decrease in the entropy of

[139] E. A. Chandross and C. J. Dempster, *J. Amer. Chem. Soc.*, 1970, **92**, 703.

association. 1,3-Diphenylpropane did not photodimerize but 'some fulvene derivative apparently was formed'. The same workers in the following paper describe their work on the excimer fluorescence and dimer phosphorescence from a naphthalene sandwich pair.[140]

The sensitized and unsensitized photodimerization of indene yields the four possible cyclobutane isomers.[141] The unsensitized reaction is reported to proceed *via* singlet and triplet states, and the reaction from the singlet state is said to lead predominantly to the *syn* head-to-head dimer (217).

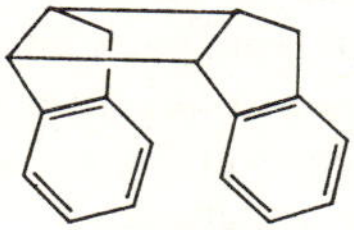

(217)

Factors affecting the photodimerization of anthracene and its derivatives have been previously published.[142] It is often difficult to determine the exact structures of such dimers, *i.e.* whether head-to-head (218) or head-to-tail (219). Dipole moment measurements of the photo-products have

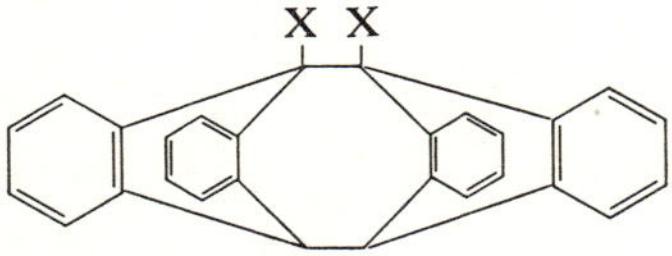

(218)

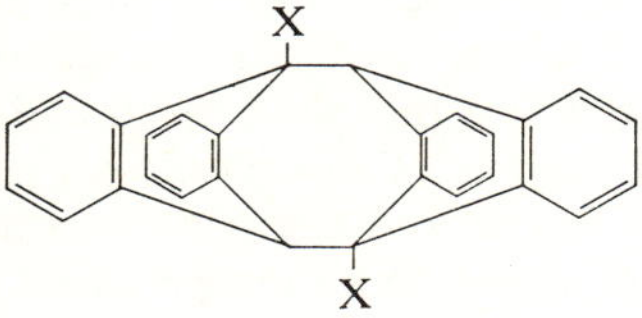

(219)

been successfully used in the past, but a different approach to the problem of structural determination of such products has now been reported.[143] The present approach is based on the ^{13}C sideband technique which was first reported by Anet.[144] From measurement of the magnitude of the

---

[140] E. A. Chandross and C. J. Dempster, *J. Amer. Chem. Soc.*, 1970, **92**, 704.
[141] W. Metzner and D. Wendisch, *Annalen*, 1969, **730**, 111.
[142] L. Burnelle, J. Lahiri, and R. Detrano, *Tetrahedron*, 1968, **24**, 3517.
[143] O. L. Chapman and K. Lee, *J. Org. Chem.*, 1969, **34**, 4166.
[144] R. Anet, *Tetrahedron Letters*, 1965, 3713.

coupling constant of the bridgehead protons in (218) and (219), a distinction between the two may be made, since in (218) the rigidly eclipsed bridgehead protons would be expected to have a coupling constant of 6—10 Hz whereas in (219) the value should approach zero. The authors point out that it is not possible to obtain the value of $J_{AA'}$ from an A_2 spectrum, but the 1·1% natural abundance of ^{13}C allows measurements to be made. In dimers with ^{13}C at the bridgehead, the protons here will constitute an AA'X system and the coupling constant between the protons ($J_{AA'}$) may be obtained by analysis of the ^{13}C sidebands of the bridgehead protons. The $J_{AA'}$ value ($\simeq 0$) from such analysis of the dimer of 9-aminoanthracene shows that the head-to-tail structure is formed in this case. The validity of the method was checked by converting the 9-aminoanthracene dimer into a dimer of known structure.

The stereochemistry of the solid-state dimerization of 9-cyanoanthracene [145] has initiated a study of the complementary roles of topochemical preformation theory [146] and dislocation theory in the interpretation of reactions of organic solids.[147] 9-Cyanoanthracene on irradiation yields the head-to-tail dimer (220), but in the monomer lattice, the molecules

(220)

have a head-to-head arrangement. These data imply that displacements along the c-axis must occur either before, or during, irradiation in order to bring the molecules into the required configuration. It is difficult to envisage monomer rotation within the crystal, and thus it is suggested that reaction is centred at regions of disorder, for example, at lattice defects such as dislocations. The dislocations emergent at bc and ac faces have been characterized and shown to be the preferred sites of photodimerization. In particular, partial dislocations on (211) planes give rise to a *trans* arrangement of molecules across contiguous planes and thus facilitate formation of the head-to-tail dimer.

Several groups of workers have studied the 'mixed' dimerization of different anthracenes. This process has now been further investigated by photolysis of a mixture of an electron donor, 9,10-dimethylanthracene

---

[145] G. M. J. Schmidt, '13th Solvay Conference, Brussels,' Interscience, 1966; D. P. Craig and P. Sarti Fantoni, *Chem. Comm.*, 1966, 742.
[146] M. D. Cohen and G. M. J. Schmidt, *J. Chem. Soc.*, 1964, 1966.
[147] M. D. Cohen, Z. Ludmer, J. M. Thomas, and J. O. Williams, *Chem. Comm.*, 1969, 1172.

(221), and an electron acceptor, 9-cyanoanthracene (222), when formation of the mixed dimer (223) was greatly facilitated.[148] There was no spectroscopic evidence for formation of a complex in the ground state, but both monomers are known to yield excimers,[149] and studies on the fluorescent exciplexes between (222) and several donors such as 2,3-dimethylnaphthalene,[150] suggest the formation of an exciplex between the two anthracenes. Compound (221) does not photodimerize, so only compound (223)

and the 'pure' dimer (224) are to be expected from the irradiation. The 0–0 singlet levels of the two monomers are 72·6 and 72 kcal mol^{-1} respectively, and it would be expected that a statistical ratio of (223) to (224) would be 1·4 : 1. However, the ratio of (223) to (224) after half-hour irradiation was 99·5 : 0·5, and the combined yield was 21%. The 'mixed' dimer (223) is not thermally stable, and reverts to starting materials. Thus irradiation at 80 °C only yields the 'pure' dimer (224), and it is concluded that the mixed dimer is the kinetically controlled product. When the dielectric constant of the solvent is increased, the yield of (223) decreases while that of (224) increases. These observations are consistent with Weller's[151] results concerning the effect of solvent polarity on the formation of exciplexes and solvated ion-pairs from electron-donor and -acceptor compounds. Mixed dimers (225) have also been formed by the irradiation of ether solutions of benz[*a*]anthracene with anthracene, 9,10-dimethylanthracene, and 9,10-dimethoxyanthracene.[152] The adducts regenerate the starting materials on heating.

---

[148] H. Bouas-Laurent and R. Lapouyade, *Chem. Comm.*, 1969, 817.
[149] H. Bouas-Laurent and R. Lapouyade, *Compt. rend.*, 1967, **264**, *C*, 1061; R. L. Barnes and J. B. Birks, *Proc. Roy. Soc.*, 1966, **A291**, 570.
[150] E. A. Chandross and J. Ferguson, *J. Chem. Phys.*, 1967, **47**, 2557.
[151] A. Weller, *Pure Appl. Chem.*, 1968, **16**, No. 1, 115.
[152] R. Lapouyade, A. Castellan, and H. Bouas-Laurent, *Tetrahedron Letters*, 1969, 3537.

(225)

Further details of the oxidative dimerization of anthracene in carbon disulphide have been published.[153] The products of the reaction are anthraquinone and bianthrone (226). The intermediate in the reaction is the well-known anthracene peroxide, and further oxidation of (226) with singlet oxygen yields anthraquinone *via* the intermediate dioxetan (227).

The irradiation of anthracene or acenaphthylene in benzene–t-amine solvents is reported to yield the corresponding hydrocarbon dimers.[154] In acetonitrile–t-amine solvents, the hydrocarbons undergo photoreduction, the products being, in the case of anthracene, 9,10-dihydroanthracene

[153] N. Sugiyama, M. Iwata, M. Yoshioka, K. Yamada, and H. Aoyama, *Bull. Chem. Soc. Japan*, 1969, **42**, 1377.
[154] R. S. Davidson, *Chem. Comm.*, 1969, 1450.

Photochemistry of Aromatic Compounds

(228), tetrahydrobianthryl (229), and aminoanthracenes. The results are interpreted in terms of Scheme 14 and involve charge-transfer and formation of radical ions. n-Butylamine does not photoreduce anthracene in acetonitrile solution, but the photodimer is formed, possibly owing to the

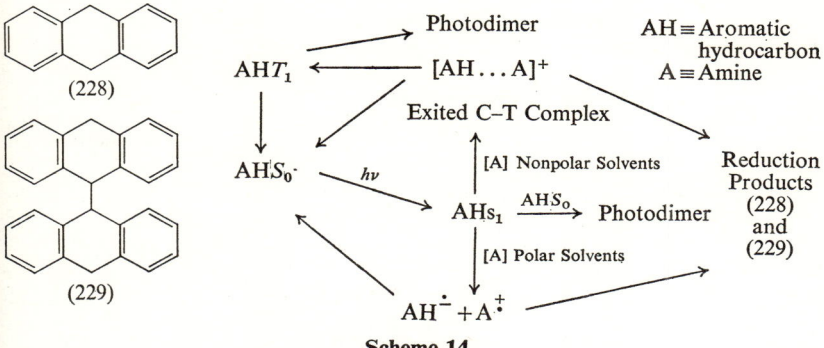

Scheme 14

lack of interaction between the amine and the excited singlet state of anthracene. Furthermore, the fact that no reduction by n-butylamine is noted in either benzene or acetonitrile indicates that triplet anthracene does not abstract hydrogen. Another group of workers have also reported the photoreactions of anthracene in acetonitrile solutions of another tertiary amine, NN-dimethylaniline.[155] In this case the products are the anthracene dimer, 5—10% of (228), 10—20% of (229), and 60—65% of 9-(p-dimethylaminophenyl)-9,10-dihydroanthracene (230). But the use of benzene in

(230)

place of acetonitrile again led only to the anthracene dimer. A mechanism has been tentatively proposed, and is outlined in Scheme 15. The formation of the excited charge-transfer complex between anthracene and the amine may compete with that of the photodimer of anthracene, and thus the formation of the dimer in benzene solution may be understood by assuming exclusive decay of the excited charge-transfer complex to ground state molecules. In acetonitrile, however, the excited charge-transfer complex

[155] C. Pac and H. Sakurai, *Tetrahedron Letters*, 1969, 3829.

leads to a solvent shared ion-pair which may transform into (230) or dissociate into free ion-radicals, since a highly polar solvent facilitates electron transfer. The photoadditions of aniline and NN-dimethyl-m-toluidine to anthracene are also reported. The corresponding *ortho*-isomer

D ≡ Dimethylaniline

A ≡ Anthracene

$$D + A \xrightarrow{h\nu} (D^+ - A^-)$$

$$(D^+ - A^-) \longrightarrow (D^+ \ldots A^-) \xrightarrow[D_2O]{H_2O} (230) \text{ or deuterio derivative}$$

$$\downarrow$$

$$D^+ + A^-$$

$$2A^- \dashrightarrow {}^-A-A^- \xrightarrow{H_2O} (229)$$

$$A^- \xrightarrow[D_2O]{H_2O} AH + AD$$

$$AH \text{ or } AD + A^- \longrightarrow A + AH^- \text{ or } AD^- \longrightarrow (228) \text{ or deuterio derivative}$$

Scheme 15

gave only low yields of adducts, but the anthracene dimer was formed in 70% yield. On the other hand, the major products from the photoreaction of NN-dimethyl-p-toluidine with anthracene were (228) and (229), formed in 50 and 30% yields respectively.

6 Lateral-nuclear Rearrangements

The most extensively studied reaction in this class is the photo-Fries rearrangement. The reaction has been known for a number of years, and yet fundamental aspects of the process still require further investigation. Trecker and co-workers recently reported that the reaction with p-tolyl acetate involved singlet states or very short-lived triplet states.[156] This fact was deduced from the inability to sensitize or quench the process with common triplet 'additives'. The multiplicity of the process has been further studied with phenyl benzoate.[157] No sensitization of the reaction was observed by acetophenone, xanthone, or benzophenone, but it was felt that the sensitizer energy in all cases could be insufficient for efficient energy-transfer to the benzoate. The reduction of the ketones was unaffected by phenyl benzoate (or phenyl acetate) and thus the triplet energy of the benzoate is considered to be greater than 74 kcal mol^{-1} (xanthone). Sensitization of the photo-Fries reaction of the benzoate can, however, be achieved by triphenylmethane, which has a triplet energy of 81 kcal mol^{-1}, and since the same product distribution is obtained in the

---

[156] M. R. Sander, E. Hedaya, and D. J. Trecker, *J. Amer. Chem. Soc.*, 1968, **90**, 7249.
[157] D. A. Plank, *Tetrahedron Letters*, 1969, 4365.

absence and presence of sensitizer, it is strongly suggested that the unsensitized reaction involved triplet intermediates. The lack of quenching of the photo-Fries rearrangement by naphthalene which was observed by Trecker, is here explained by the formation of the products in an intramolecular process the rate of which is sufficiently close to the rate of diffusion to preclude quenching at normal quencher concentrations. Indeed, preliminary experiments with *p*-tolyl benzoate and concentrations of naphthalene greater than 0·3 mol l^{-1} show that the ratio of rearranged product : cresol increases, thus indicating that at least a portion of the reaction may be quenched with naphthalene.

The photo-Fries reaction of phenyl acetate in cyclohexane yields *o*-hydroxyacetophenone, *p*-hydroxyacetophenone, and phenol with respective quantum yields of 0·17, 0·15, and 0·06.[158] Two basic mechanisms have been outlined for the process: (*a*) radical reaction within a solvent cage and phenol production by escape of the radicals from the cage, and (*b*) a concerted rearrangement process, and formation of phenol in a separate reaction. The mechanistic conclusions previously described have been based on studies of the effects of solvent viscosity and polarity on the quantum yield.[159] In an attempt to distinguish clearly between the two mechanistic possibilities, Meyer and Hammond have removed the solvent parameter, and examined the photo-Fries reaction of phenyl acetate in the vapour phase where solvent cages are not possible, and any ketone formation must involve a concerted mechanism.[160] Irradiation of phenyl acetate vapour at 254 nm in a static system yielded 25 products, as evidenced by gas chromatography. Similar irradiation in a flow system formed six products, the major (65%) of which was phenol, but no *o*- and *p*-hydroxyacetophenones were found, although these are the main products formed in solution. Thus no concerted mechanism was operating in the vapour phase. Likewise, the absence of toluene and anisole also demonstrates that the decarboxylation and decarbonylation processes do not occur. The results are consistent with formation of acyl and phenoxyl radicals, which drift apart in the absence of a solvent cage. The rearrangement of phenyl acetate and *p*-substituted phenyl acetates in cyclohexane solution has been investigated by Shizuka and co-workers.[158] The quantum yield for rearrangement of phenyl acetate is independent of irradiation time, concentration of the acetate ($<2\times10^{-2}$ mol l^{-1}), temperature over the range 20—70 °C, and the presence of piperylene or oxygen. From a comparison of the quantum yields for rearrangement products and phenol formation from phenyl acetate and the *p*-methoxy-derivative, the quantum yields for predissociation of the O—CO bond, the back reaction, and the recombination ratio were estimated to be 0·61, 0·23, and 0·9 respectively

---

[158] H. Shizuka, T. Morita, Y. Mori, and I. Tanaka, *Bull. Chem. Soc. Japan*, 1969, **42**, 1831.
[159] M. R. Sander and D. J. Trecker, *J. Amer. Chem. Soc.*, 1967, **89**, 5725; J. S. Bradshaw, E. L. Loveridge, and L. White, *J. Org. Chem.*, 1968, **33**, 4127.
[160] J. W. Meyer and G. S. Hammond, *J. Amer. Chem. Soc.*, 1970, **92**, 2187.

in cyclohexane. The results obtained in this study demonstrate that the reaction sequence for phenyl acetate in the liquid phase with 254 nm radiation is similar to that previously reported for acetanilide.[161]

Hageman has continued his work into factors affecting the photo-Fries process *versus* photo-decarboxylation, and has irradiated a number of substituted-phenyl acetates.[162] In contrast with the conclusions in ref. 157, all the reactions were considered to arise from the first excited singlet state. Methoxy-substituents at the *o*- and *p*-positions are displaced by the migrating acyl moiety, and the decarboxylation process is considerably enhanced by substitution at the *o*- and/or *m*-positions. In isopropyl alcohol and cyclohexane, little of the decarboxylation process is observed, but in diethyl ether the process is most pronounced. The substituents and solvents studied are suggested to affect the energy difference between the

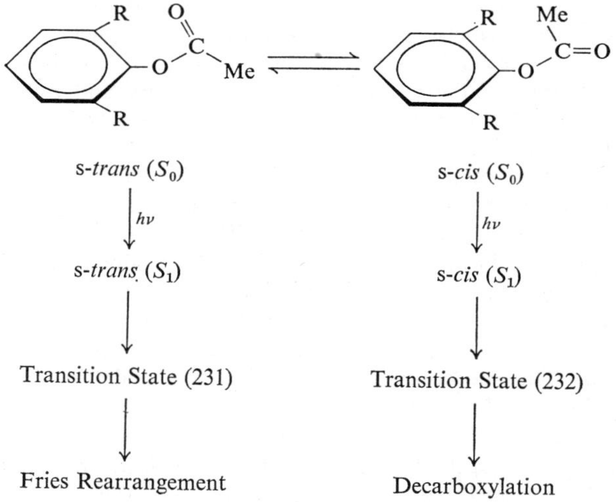

Scheme 16

two transition states (231) and (232) (see Scheme 16); for example, ethers decrease this difference, particularly for *o*- and/or *m*-substituted esters.

The photo-rearrangements of acetanilides are generally related to the work of Shizuka, who carried out useful fundamental work on the system.[161] The photochemistry of acetanilide in the vapour phase and in solid matrices has been reported from the same research school, and in the absence of the solvent cage (vapour, 120 °C, 254 nm radiation) no rearrangement is observed (*cf.* ref. 160).[163] In this case only gross decomposition occurs,

[161] H. Shizuka, *Bull. Chem. Soc. Japan*, 1969, **42**, 52, 57.
[162] H. J. Hageman, *Tetrahedron*, 1969, **25**, 6015.
[163] H. Shizuka and I. Tanaka, *Bull. Chem. Soc. Japan*, 1969, **42**, 909.

giving carbon monoxide and ethane as the major products in the ratio of ca. 2:1. Small amounts of methane and traces of hydrogen and benzene were also found. The quantum yield for the dissociation is ca. 0·28 and the scheme to account for the reaction is given in equations (32—36). In a rigid matrix at −196 °C, acetanilide does not rearrange (254 nm), and only a back reaction resulting in the distortional structure of acetanilide is observed. The lack of reaction (equation 36) is attributed to the stiffness of the solvent cage since in P.V.C. sheet at room temperature, low quantum yields for the formation of o- and p-aminoacetophenones are recorded. This report again stresses the importance of cage effects in the rearrangement process.

$$S_1(\pi\pi^*) \longrightarrow PhNH\cdot + \cdot COMe \tag{32}$$

$$\cdot COMe \longrightarrow CO + \cdot Me \tag{33}$$

$$\cdot Me \longrightarrow C_2H_6 \tag{34}$$

$$\cdot Me + RH \longrightarrow CH_4 + R\cdot \tag{35}$$

$$S_1(\pi\pi^*) \longrightarrow (Ph\dot{N}H + \cdot COMe) \quad \text{Stiff solvent cage} \rightarrow S_0 \tag{36}$$

As noted previously,[162] methoxy-groups may be displaced by migrating groups in the photo-Fries rearrangement. 2,4-Dimethoxyacetanilide has been examined to see if the displacement also occurs in the N-aryl amide system.[164] The major process (63%) involves migration of the COMe group to the unoccupied o-position, as may have been expected. The total amount of the rearrangement–displacement reaction is 16%, while 8% of 2,4-dimethoxyaniline is also observed in addition to five minor unidentified products. Excited singlet states are again implicated, and the mechanism is considered to be the same as that previously reported.[161] It is not, however, thought that the methoxy displacement occurs by predissociation of C—OMe, but rather by a concerted process.

Both N-acetyl diphenylamine (233) and N-acetylcarbazole (234) undergo photo-Fries rearrangement with 254 nm radiation to yield the o- and p-isomers and the corresponding amines.[165] With (233) no cyclization to the carbazole (234) was observed, although diphenylamine is well known

[164] J. S. Bradshawm, R. D. Knudsen, and E. L. Loveridge, J. Org. Chem., 1970, 35, 1219.
[165] H. Shizuka, M. Kato, T. Ochiai, K. Matsui, and T. Morita, Bull. Chem. Soc. Japan, 1970, 43, 67.

to yield carbazole on irradiation.[166] Excited singlet states are involved, and the results of this work are consistent with the acetanilide mechanism involving predissociation by intersystem crossing $[S_1(\pi\pi^*) \rightarrow T_1(\sigma\pi^*)_{C-N}]$ forming a pair of radicals in a solvent cage. The reaction mechanisms are also discussed on the basis of a relation between rate constant of rearrangement and the odd π-electron densities obtained from simple Hückel molecular orbital calculations.

The photo-Fries reaction has also been investigated for heteroaromatic compounds. Thus the three isomeric pyridinium compounds (235), (236), and (237) have been irradiated, and while both (235) and (236) yield rearrangement products, and 3-hydroxypyridine in the case of (236), (237) undergoes no Fries reaction; but 20% of benzoic acid is isolated.[167]

Light-induced lateral-nuclear migration occurs with aromatic ethers to yield substituted phenols. The reaction has been extended [168] to a study of photo-rearrangements of aryloxyacetic acids (238) and allyl aryl ethers (photo-Claisen rearrangement).[169] In both classes of compounds studied, the o- and p-substituted phenols and phenol were formed by an intramolecular process involving a solvent cage.

The u.v. irradiation of N-phenylbenzylamine (239) and N-α-phenethylaniline (240) in various solvents is reported to yield o- and p-alkylated anilines (5—42%), the parent aniline, and coupling products (241) of the

---

[166] C. A. Parker and W. J. Barnes, *Analyst*, 1957, **82**, 606; H. Stegemeyer, *Naturwiss.*, 1966, **53**, 582.
[167] M. T. Le-Goff and M. R. Beugelmans, *Tetrahedron Letters*, 1970, 1355.
[168] D. P. Kelly, J. T. Pinhey, and R. D. G. Rigby, *Austral. J. Chem.*, 1969, **22**, 977.
[169] G. Koga, N. Kikuchi, and N. Koga, *Bull. Chem. Soc. Japan*, 1968, **41**, 745.

alkyl radicals.[170] In a typical reaction of 0·1M solutions of (239) in isopropanol–t-butanol, 28·9% of o-benzylaniline, 13·6% of p-benzylaniline, 5·8% of aniline, and 3% of bibenzyl were formed. The quantum yield for disappearance of (239) was 0·48. The ortho:para product isomer ratio is

2—3, but rises to ca. 20 in the presence of radical scavengers. There is no photo ortho ↔ para rearrangement, and no 'crossed' products are observed in mixed irradiations of (239) and (240). Irradiation of optically active (240) gives o- and p- rearrangement products in which an appreciable retention of activity is observed. As with the other migration reactions, the process here is considered to involve combination of radicals in a solvent cage for the formation of the o- and p-alkylated anilines, and out-of-cage reactions for formation of the p-isomer and aniline.

The Smiles rearrangement involves the conversion of (242) into (243), and has been extensively studied in the ground state. The corresponding light-induced reaction (254 nm) has been investigated, and while some aromatic systems [e.g. (244)] give the Smiles rearrangement product and some side reactions, other compounds [e.g. (245)] are inert.[171] Unfortunately, some irradiations were carried out in benzene and others in ethanol, so mechanistic conclusions must await the further results which are promised for 'the near future'.

---

[170] Y. Ogata and K. Takagi, *J. Org. Chem.*, 1970, **35**, 1642.
[171] K. Matsui, N. Maeno, S. Suzuki, H. Shizuka, and T. Morita, *Tetrahedron Letters*, 1970, 1467.

(242) → (243)

X = O, S, CO₂, SO₂
Y = O, NH, CONH
   NAryl, S
A ≡ Substituted phenyl
    or other aromatic ring
B ≡ Substituted phenyl
    or N-heteroaromatic
    ring

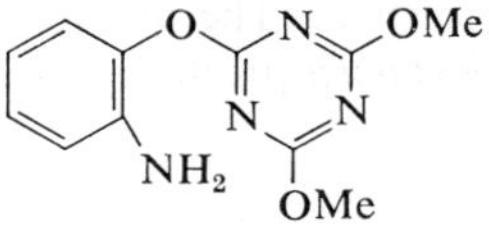

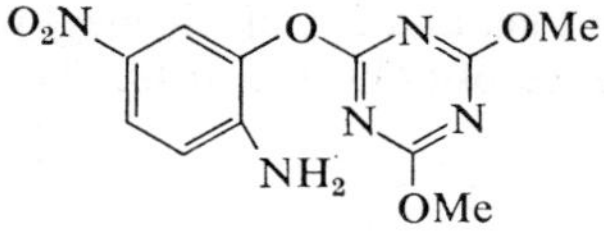

(244)   (245)

7 Photochemistry of Furan, Thiophen, and other Heterocyclopentadienes

Last year there were several significant papers published concerned with the photolysis of furan, methylfurans, and t-butylfurans. The intermediacy of cyclopropenecarboxaldehyde in these photolyses has been well-established.[172, 173] 2-Methyl-, 3-methyl-, and 2,4-dimethyl-furans have been subjected to both direct and mercury-sensitized vapour-phase photolyses.[174] From the sensitized reaction, 3-methyl-, 1-methyl-, and 1,3-dimethyl-cyclopropenes are the respective main products, and 3-methyl-furan also yields the 2-methyl isomer. The major reaction products from the direct irradiation of the 2- and 3-methylfurans are ring-opened isomers of the methylcyclopropene, although small quantities of the cyclic isomers were also formed. The 3-methyl- ↔ 2-methyl-furan light-induced interconversion occurs by direct irradiation, but in low yield. In the sensitized process, the mechanism (Scheme 17) again involves the cyclopropenecarboxaldehyde, which undergoes decarbonylation to yield the cyclopropenes. Selective ring contraction to one of the two possible isomers (246) and (247) can be explained by assuming that it is directed by initial breakage of the weakest bond in the lowest excited state. Similar sensitized vapour-phase photolysis has been carried out with 2,5-dimethyl-furan, when eleven products were formed in characterizable amounts.[175]

---

[172] R. Srinivasan, *Pure Appl. Chem.*, 1968, **16**, No. 1, 65.
[173] E. E. van Tamelen and T. H. Whitesides, *J. Amer. Chem. Soc.*, 1968, **90**, 3894.
[174] H. Hiraoka, *J. Phys. Chem.*, 1970, **74**, 574.
[175] S. Boue and R. Srinivasan, *J. Amer. Chem. Soc.*, 1970, **92**, 1824.

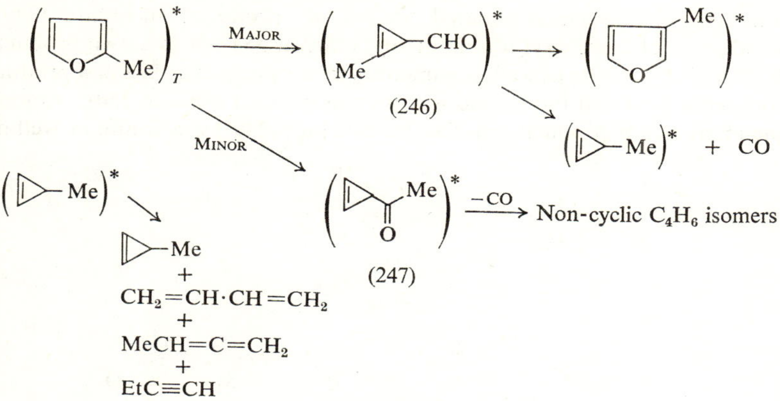

Scheme 17

The products in decreasing order of abundance were 4-methylcyclopent-2-enone, cis- and trans-penta-1,3-dienes, carbon monoxide, 1-methyl-3-acetylcyclopropene, hexa-3,4-dien-2-one, 2-ethyl-5-methylfuran, isoprene, 1,3-dimethylcyclopropene, pent-2-yne, and propylene. Only 2-ethyl-5-methylfuran was totally eliminated when the reaction was performed in the presence of oxygen and thus it is believed to be the only product of free-radical origin. Secondary photolysis of penta-1,3-diene accounts for the formation of the three minor C_5H_8 products. The intermediate which yields 4-methylcyclopent-2-enone and penta-1,3-diene and carbon monoxide is considered to be the carbene (248), while the intermediate for the other two isomeric ketones is possibly a 1,3-diradical.

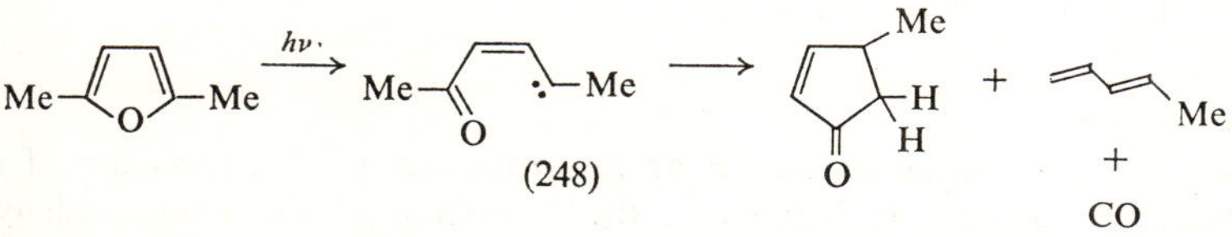

The primary process in the above reactions is thus C—O cleavage, and the behaviour is thus similar to that of a vinyl ether. The mercury-sensitized and unsensitized reactions of 2-methoxy- and 2-acetoxy-furans have been studied.[176] In these compounds there is an exocyclic ether linkage which is in conjugation with the ethylene in the ring, and thus it is of interest to examine which (exo- or endo-)cyclic ether linkage undergoes the cleavage reaction on irradiation. The reactions from the sensitized process were essentially the same as those by direct irradiation in the vapour phase or in diethyl ether or cyclohexane solution. 2-Methoxyfuran gave numerous products, but the major one was the isomer (249). Small

---

[176] R. Srinivasan and H. Hiraoka, *Tetrahedron Letters*, 1969, 2767.

amounts of oxygen eliminated almost all products, including (249). Irradiation of 2-acetoxyfuran in diethyl ether at 254 nm gave two products in 10 and 15% yields as well as numerous minor products. The first product was acetic acid and the second was the unsaturated lactone (250). Among the many minor products, another lactone [not (249)] was found as well as

furan. The results suggest that the major cleavage occurs at the first C—O bond in the side chain. The oxyfuran radical (251) can react in its lactone form with a methyl radical to yield the isolated lactones. Irradiation of isopropyl alcohol solutions of 2-nitrofuran is reported to cause photoreduction of the nitro-group to form (252) with concomitant oxidation of the alcohol to acetone.[177] There is little in the report concerning product

identification, but since there is an absorption in the i.r. spectrum of the irradiated solution at 1650 cm^{-1}, the formation of the oxime–aldehyde (253) in the reaction is suggested.

The mercury-sensitized decomposition of 2,5-dihydrofuran in the gas phase at 20 °C has been studied.[178] The major products are carbon monoxide, propene, and hydrogen, while diallyl, allene, methylacetylene, furan, 2,3-dihydrofuran, tetrahydrofuran, and small amounts of unidentified products are also formed. The quantum yields of all the products are decreased by increasing substrate pressure, indicating collisional deactivation of an excited state precursor which at low pressure decomposes according to Scheme 18. It is not known what 'P' is, but apparently it is not furan.

---

[177] W. Kemula and J. Zawadowska, *Bull. Acad. polon. Sci. Sér. Sci. Chim.*, 1969, **17**, 599.
[178] B. Francis and A. G. Sherwood, *Canad. J. Chem.*, 1970, **48**, 25.

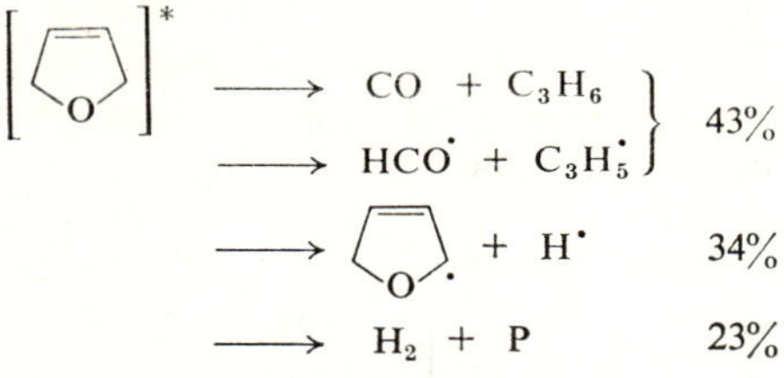

$$[\text{furan}]^* \longrightarrow CO + C_3H_6$$
$$\phantom{[\text{furan}]^*} \longrightarrow HCO^\bullet + C_3H_5^\bullet \Big\} \; 43\%$$
$$\phantom{[\text{furan}]^*} \longrightarrow \text{(oxiranyl)}^\bullet + H^\bullet \quad 34\%$$
$$\phantom{[\text{furan}]^*} \longrightarrow H_2 + P \quad 23\%$$

Scheme 18

Two reports have appeared within the year in which the photosynthesis of benzofurans is described. Hill and co-workers have published the full paper concerned with their work on the photolysis of α-aryloxy ketones, (254) in methanol.[179] Fission of the C—OAr bond occurs to yield phenol, a ketone ($R^1CH_2COR^2$), and a product (255) of *o*-rearrangement which

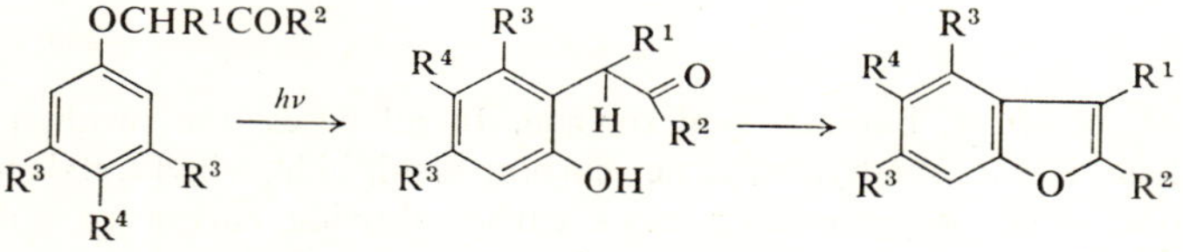

(254) OCHR¹COR², R³, R³, R³, R⁴ — $h\nu$ → (255) with R³, R⁴, R³, R¹, R², H, OH → (256) benzofuran with R³, R⁴, R³, R¹, R²

cyclizes during the work-up procedure to form a substituted benzofuran (256). In a few cases, the aryloxy ketone yields the corresponding dimethyl acetals [$ArO \cdot CHR^1C(OMe)_2R^2$] on photolysis.

The photochemical cyclization of 4,6-di-t-butyl-2-methoxybenzophenone (257) to 5-t-butyl-7-methoxy-3,3-dimethyl-1-phenylindan-1-ol (258) is known.[180] A reaction which has apparent mechanistic similarity to this has now been reported.[181] Irradiation (310 nm) of 2-benzyloxy-4-methoxybenzophenone (259) yields 2,3-diphenyl-6-methoxybenzofuran (260) (50%), presumably *via* internal hydrogen transfer followed by cyclization of the resulting biradical. 2-Hydroxy-4-methoxybenzophenone was also formed in 8% yield. The reaction provides a simple synthesis of a variety of substituted 2,3-diarylbenzofurans.

The photolysis of thiophen vapour has received a very detailed study.[182] Thiophen vapour has a weak absorption band starting at *ca.* 260 nm, but at 240 nm and below the $\pi\pi^*$ band is very intense. Thus light of wavelength 213·9 and 228·8 nm has been utilized as well as mercury sensitization to cause the decompositions. The reaction has been studied at various temperatures, pressures, and light intensities, and in the presence of carbon

---

[179] J. R. Collier, M. K. M. Dirania, and J. Hill, *J. Chem. Soc.* (*C*), 1970, 155.
[180] E. J. O'Connell, *J. Amer. Chem. Soc.*, 1968, **90**, 6550.
[181] G. R. Lappin and J. S. Zannucci, *Chem. Comm.*, 1969, 1113.
[182] H. A. Wiebe and J. Heicklen, *Canad. J. Chem.*, 1969, **47**, 2965.

dioxide, ethylene, butenes, and oxygen. In all cases, the products were acetylene, allene, methylacetylene, carbon disulphide, vinylacetylene, and polymer. When oxygen was present, carbon dioxide, carbon oxysulphide, sulphur dioxide, and carbon monoxide were also formed. The quantum yield of product formation decreased on increase of pressure. The hydrocarbons are all produced in primary processes with additional vinylacetylene and acetylene coming respectively from the postulated intermediates ·CH:CHCH:CHS· and C_2H_2S·. Along with the C_3 fragments, excited CS species are produced which may either react with the radical intermediates to yield carbon disulphide, be deactivated to give polymer, or form oxidation products in the presence of oxygen.

The photolysis (254 nm) of thiophens in the presence of primary amines as solvents is reported to yield pyrroles.[183] Hydrogen sulphide is given off during the reaction. A mechanistic sequence (Scheme 19) is suggested by analogy with the furan photolyses, and with the rearrangement of 3,5-diarylisoxazoles to 2,5-diaryloxazoles when the 3-aroyl-2-aryl-1-azirine intermediate has been isolated.[184] Wynberg's intermediates (261) could also explain the observed reaction.[185] Formation of the two pyrroles from 2-methylthiophen can be explained by either mechanism. However, 3-methylthiophen only leads to 3-methylpyrrole, and this result is inexplicable at present.

The mechanism outlined for the photolysis of methylfurans (Scheme 17) has also been suggested (Scheme 20) to account for the irreversible

---

[183] A. Couture and A. Lablache-Combie, *Chem. Comm.*, 1969, 524.
[184] E. F. Ullman and B. Singh, *J. Amer. Chem. Soc.*, 1967, **89**, 6911.
[185] H. Wynberg, R. M. Kellogg, H. van Driel, and G. E. Beekhuis, *J. Amer. Chem. Soc.*, 1967, **89**, 3501.

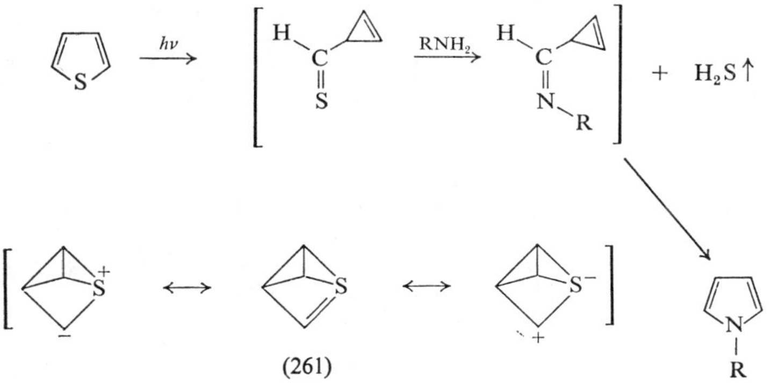

Photochemistry of Aromatic Compounds

Scheme 19

rearrangement of 2-phenylthiophen to 3-phenylthiophen, and seems to be the only one to date which accounts for this specific rearrangement.[174]

Two groups of workers have been concerned with the photoaddition of olefins and acetylenes to benzothiophens. *cis*- and *trans*-Dichloroethylenes undergo photosensitized addition to benzothiophen to yield 1 : 1 adducts

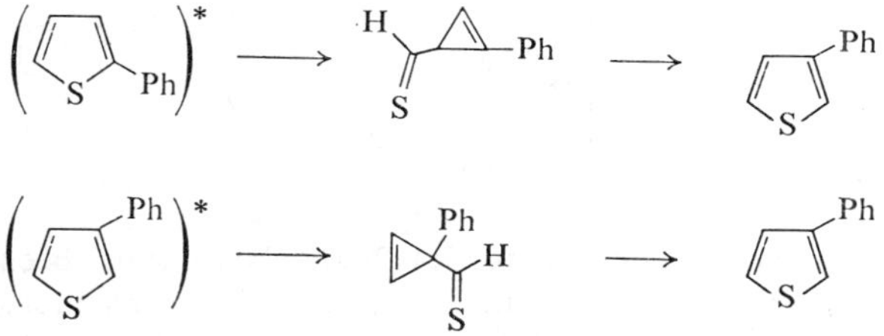

Scheme 20

(262).[186] The reaction would appear to be general for olefin addition, and with 2,3-dimethylbenzothiophen and the dichloroethylenes, all four possible isomers are formed. The sensitizers employed were benzophenone, acetophenone, and 2-acetonaphthone. The same workers report that irradiation of acetylenedicarboxylic ester in the presence of benzofuran in benzene solution surprisingly leads to the formation of (263), and they suggest that this product is formed by photo-rearrangement of the initial 1,2-adduct (264).[187] Compound (263) is thermally unstable and forms naphthalene derivatives (265) *via* the ring-opened adduct (266). Other workers have been able to demonstrate that (264) is indeed the first-formed adduct which on further irradiation yields the rearranged adduct (263); but they suggest (267) as the intermediate in the conversion of (264) into (263) in preference to the diradical (268) proposed by the first group.[188]

[186] D. C. Neckers, J. H. Dopper, and H. Wynberg, *J. Org. Chem.*, 1970, 35, 1582.
[187] D. C. Neckers, J. H. Dopper, and H. Wynberg, *Tetrahedron Letters*, 1969, 2913.
[188] W. H. F. Sasse, P. J. Collin, and D. B. Roberts, *Tetrahedron Letters*, 1969, 4791.

α ≡ CO₂Me or Ph

The irradiation of aryl desyl sulphides (269) in benzene has been reported to yield bidesyl (270) and thiophenol.[189] From a similar reaction of benzyl desyl sulphide (271) in methanol, however, only a small amount of bidesyl was formed, the major product being 2-phenylbenzo[b]thiophen (272).[190] In order to ascertain which of the benzene rings becomes the phenyl substituent of the thiophen, the methyl analogues (273) and (274) were irradiated. Formation of the benzothiophens (275) and (276) respectively demonstrates that direct cyclization of the homolysis product (277) to the benzothiophen precursor does not occur (see Scheme 21). Differentiation between paths (a) and (b) in the reaction sequence leading to the episulphide (278) cannot be made at present, and ionic character in path (b) is also not excluded.

It is known that the formation of 2- and 3-substituted pyrroles from N-substituted derivatives occurs thermally with 75—80% retention of configuration.[191] The corresponding photoreaction with 254 nm radiation has been investigated for the case of N-benzylpyrrole, when 12·6% of

[189] A. Schönberg, A. K. Fateen, and S. M. A. R. Omran, *J. Amer. Chem. Soc.*, 1956, **78**, 1224.
[190] J. R. Collier and J. Hill, *Chem. Comm.*, 1969, 640.
[191] J. M. Patterson, L. T. Burka, and M. R. Boyd, *J. Org. Chem.*, 1968, **33**, 4033.

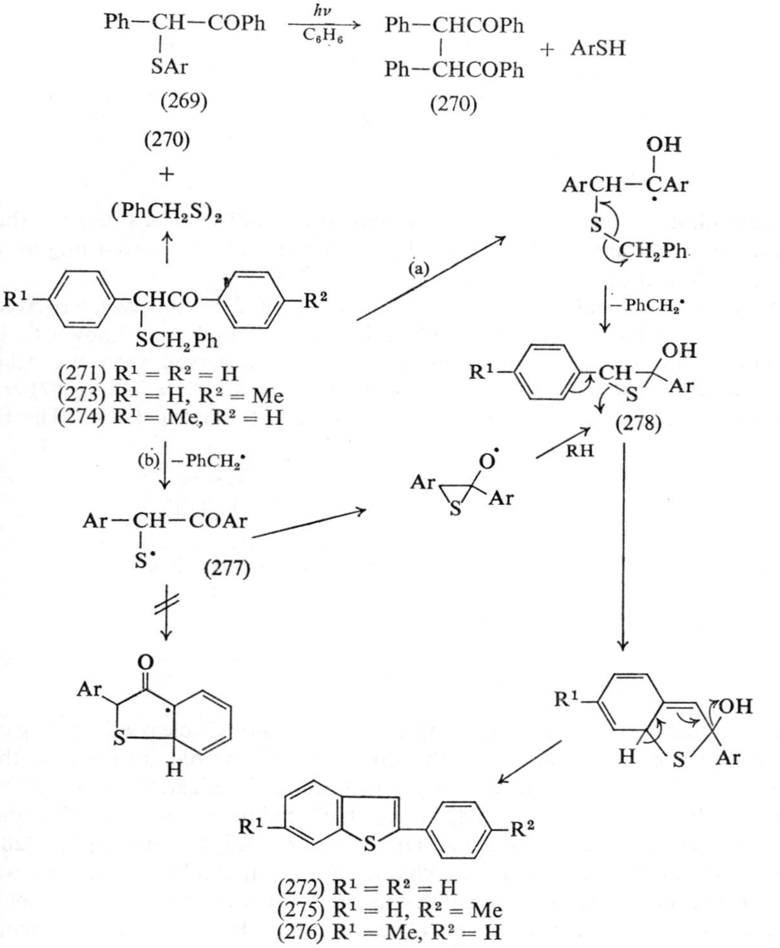

Scheme 21

(Scheme 21 depicts photochemical reactions:)

Ph—CH(SAr)—COPh (269) →[hν, C₆H₆]→ Ph—CH(COPh)—CH(COPh)—Ph (270) + ArSH

(270) + (PhCH₂S)₂ ← [from] R¹—C₆H₄—CH(SCH₂Ph)—CO—C₆H₄—R²

(271) R¹ = R² = H
(273) R¹ = H, R² = Me
(274) R¹ = Me, R² = H

Pathway (a): leads to ArCH(S···)—C(OH)(Ar)(CH₂Ph) → (278) R¹—C₆H₄—CH(S)—C(OH)(Ar) → R¹—C₆H₄—CH(S)—C(OH)(Ar)(H) → benzothiophene products

Pathway (b): −PhCH₂• gives Ar—CH(S•)—COAr (277) → thietane-type intermediate, ≠ to cyclized Ar—CH(S)—indanone

Products:

(272) R¹ = R² = H
(275) R¹ = H, R² = Me
(276) R¹ = Me, R² = H

Scheme 21

2-, and 2·5% of 3-benzylpyrroles were formed.[192] In methanol, the yields are increased, and since the 2- and 3-isomers are photo-stable, the 2-isomer is not a precursor of the 3-isomer, as with the thermal isomerizations. With optically active groups, migration occurs with 54% retention of configuration. To estimate the effect of blocking groups, neat (+)-N-(1-phenylethyl)-2,5-dimethylpyrrole was irradiated at 254 nm. This process gave 3-(1-phenylethyl)-2,5-dimethylpyrrole (279) (5·4%), and 2-(1-phenylethyl)-2,5-dimethyl-2H-pyrrole (280) (11·9%). Since (280) does not yield

[192] J. M. Patterson and L. T. Burka, *Tetrahedron Letters*, 1969, 2215.

(279) photochemically (although it does so thermally), it is suggested that the 3-isomers are produced directly, and do not involve initial migration to the 2-position.

The photochemical behaviour of a series of 2,5-diphenylheterocyclopentadienes has been reported.[193] 2,5-Diphenylthiophen is known to be photochemically inert but some reaction was observed with the other diphenyl compounds studied, except for the pyrrole (281). Thus (282) and (283) both gave crystalline cyclobutane dimers (284) and (285). The tin compound (286), not surprisingly, showed Sn—C cleavage and gave polymeric compounds. Since the thiophen and pyrrole are inactive, this suggests that no aromatic stability is lost in the dimerization in the present cases; but the phosphole (287) (for which claims of aromatic character have been made) also dimerizes. On the other hand, the tellurophen (288) (also claimed to have aromatic character) is light stable. No success was achieved in trapping intermediates in the dimerization process by olefins and acetylenes: the dimers were always formed. However, 'mixed dimers' were obtained from the irradiation of (282) in the presence of (287).

A. G.

[193] T. J. Barton and A. J. Nelson, *Tetrahedron Letters*, 1969, 5037.

5
Photo-oxidation and -reduction Reactions

1 Conversion of C=O to C—OH

Again this year, factors affecting the photoreduction of aryl ketones have been the subject of several publications. Much of the work has been concerned with the reactions of benzophenones, fluorenone, and acetophenones, in the presence of amines and alcohols.

The photoreduction of aryl ketones by amines may show high quantum yields (0·6—1·4) in comparison with the theoretical value of 2·0, and low sensitivity to diffusion-controlled physical quenchers and to amine concentration. These observations have led Cohen and his co-workers to propose a stepwise reaction proceeding *via* rapid charge-transfer interaction of the triplet ketone with amine with a high rate constant, k_{ir}.[1] This is followed by either charge destruction or by hydrogen transfer and formation of radicals (see Scheme 1). The relative importance of these two processes

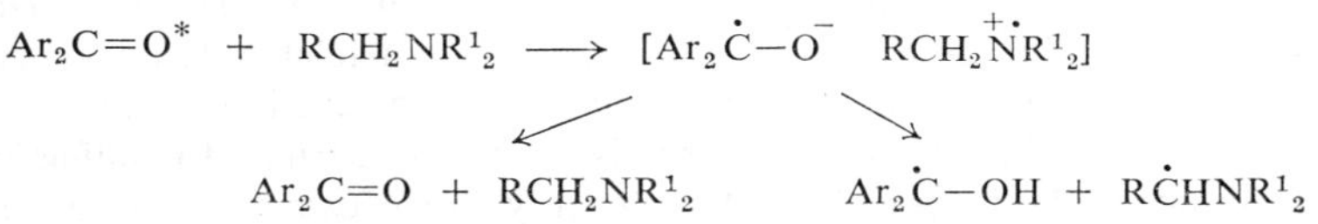

$$Ar_2C=O^* + RCH_2NR^1{}_2 \longrightarrow [Ar_2\dot{C}-O^- \quad RCH_2\overset{+\cdot}{N}R^1{}_2]$$

$$Ar_2C=O + RCH_2NR^1{}_2 \qquad Ar_2\dot{C}-OH + R\dot{C}HNR^1{}_2$$

Scheme 1

largely determines the quantum yields. The k_{ir} values for benzophenone in benzene in a variety of amines and alcohols have now been reported.[2] The least squares values of k_{ir} were calculated from emission intensities and the Stern–Volmer equation (1). These k_{ir} values are discussed along

$$I_0/I = 1 + \tau_0 k_{ir}[Q] \qquad (1)$$

with those obtained by other methods.

The intermediacy of the α-hydroxydiarylmethyl radical in the photoreduction of benzophenone by alkyl amines[3] and *N*-alkylaryl amines[4] (in particular triethylamine and *NN*-dimethylaniline) has been verified by

---

[1] S. G. Cohen and J. I. Cohen, *J. Amer. Chem. Soc.*, 1967, **89**, 164; *J. Phys. Chem.*, 1968, **72**, 3782.
[2] S. G. Cohen and A. D. Litt, *Tetrahedron Letters*, 1970, 837.
[3] S. G. Cohen and H. M. Chao, *J. Amer. Chem. Soc.*, 1968, **90**, 165.
[4] R. S. Davidson and P. F. Lambeth, *Chem. Comm.*, 1967, 1265; 1968, 511.

e.s.r. spectroscopy and flash photolysis studies respectively.[5] Tertiary amines with no abstractable hydrogens can quench the reaction of photo-excited benzophenone. To account for this observation, it was suggested that electron transfer from the amine to the excited ketone occurred. More compelling evidence for this postulate has now come from a flash photolysis study of the photoreaction of benzophenone with tri-p-tolyl amine in acetonitrile. In both air- and nitrogen-saturated solutions, a transient is formed with an absorption maximum at 670 nm. In the absence of benzophenone the absorption of the transient is much weaker. This absorption of the transient is very similar to that previously recorded for the amine radical-cation.[6] The position and intensity of the absorption of the radical-cation have made it possible to tell that the benzophenone radical-anion had also been produced via the electron-transfer reaction shown in equation (2).

$$Ph_2C=O + (p\text{-}MeC_6H_4)_3N \longrightarrow Ph_2\dot{C}-O^- + (p\text{-}MeC_6H_4)_3N^{+\cdot} \quad (2)$$

Benzophenone is reported not to be reduced in aniline, and its reduction by isopropyl alcohol is also suppressed in the presence of this amine.[7] Further, the aniline is photo-oxidised to azobenzene if oxygen is not excluded from these reactions. This suppression of photoreduction is apparently due to triplet energy transfer from benzophenone to aniline in a diffusion-controlled process. There is no evidence for complex formation from the u.v. spectrum, and the participation of exiplexes is considered to be only very nominal under the conditions employed. The sensitized oxidation of aniline to azobenzene is quenched by triplet state quenchers, and thus the intermediacy of triplet aniline in the reaction is suggested. The energy-transfer rate constant from benzophenone triplet to aniline has been shown to be independent of the energy gap between the donor and acceptor levels: the favourable geometry of the donor–acceptor pair evidently controls the energy-transfer process.

The reactions of tervalent phosphorus compounds with excited carbonyl compounds have been further investigated since excited charge-transfer complex radical-ion formation may well occur, as in the photoreaction of amines with carbonyl compounds (cf. Scheme 1). Thus, the effect of triphenylphosphine on the photoreduction of benzophenone by diphenylmethanol,[8] and its fluorescence quenching of biacetyl ($n\pi^*$ excited singlet state) and anthracene ($\pi\pi^*$ excited singlet state) have been studied. The conclusions are that triphenylphosphine and triphenylamine can quench $n\pi^*$ excited singlet and triplet, and $\pi\pi^*$ excited singlet states. The amine is more efficient, and its efficiency is increased by the use of polar solvents

---

[5] R. S. Davidson, P. F. Lambeth, J. F. McKellar, P. H. Turner, and R. Wilson, *Chem. Comm.*, 1969, 732.
[6] S. Granick and L. Michaelis, *J. Amer. Chem. Soc.*, 1940, **62**, 2241.
[7] M. Santhanam and V. Ramakrishnan, *Chem. Comm.*, 1970, 344.
[8] R. S. Davidson and P. F. Lambeth, *Chem. Comm.*, 1969, 1098.

(*cf.* reactions of fluorenone and *p*-aminobenzophenone in amines and solvents of various polarities, refs. 10 and 12). Evidence for radical-ion formation in the quenching by triphenylphosphine was not obtained; the quenching is believed to involve formation of an excited charge-transfer complex. Radical-ion formation probably only occurs in those systems in which the electron donor has a low ionisation potential and in which solvation of the radical-ion is efficient.[8]

$$R\cdot + Bu^tNO \longrightarrow R-\overset{+\cdot}{N}-O^- \qquad Bu^t-\overset{+\cdot}{N}-O^- \qquad (3)$$
$$\underset{Bu^t}{|} \qquad\qquad \underset{H}{|}$$
$$(1)$$

The radical intermediates from the photoreduction of benzophenone by alcohols, amines, phenols, sulphides, thiols, ethers, hydrocarbons, and amides have been trapped by t-nitrosobutane.[9] This procedure results in the formation of a nitroxide radical which is characterized *in situ* by e.s.r. spectroscopy (equation 3). From benzhydrol, the trapped radical is (1) and from isopropyl alcohol, triethylamine, and thiophenol in benzene, $Me_2\dot{C}OH$, $EtN\dot{C}HMe$, and $PhS\cdot$ radicals are trapped.

In the photoreduction of *p*-aminobenzophenone (PAB), the effects of solvent on the intersystem crossing have been studied.[10] It is known that PAB is photoreduced very inefficiently or not at all by alcohols and primary amines, but is reduced by triethylamine. The efficiency increases on dilution with cyclohexane but falls with dilution by polar solvents such as acetonitrile. These results are attributed to the formation of an unreactive charge-transfer state or $\pi\pi^*$ triplet of PAB in more polar media, and formation of the reactive $n\pi^*$ triplet in non-polar solvents.[11] It is now reported that the failure of PAB to be reduced by, or in, polar solvents results essentially from complete absence of intersystem crossing as indicated by a study of the isomerization of *trans*-stilbene.[10] *trans*-Stilbene is an effective quencher of triplet PAB, and in amine solutions PAB causes the *trans–cis* stilbene isomerization. In isopropyl alcohol solutions of PAB and the stilbene, however, there is no isomerization, and thus no intersystem crossing. Hence, the lack of light-induced reduction of PAB does not result from unreactive charge-transfer triplet but from the absence of triplets (at least triplets which can cause the *trans–cis* isomerization of stilbene). It may be that in cyclohexane the singlet and triplet $n\pi^*$ states have energies below that of the charge-transfer triplet, so that rapid intersystem crossing can occur. In isopropyl alcohol, on the other hand, the singlet and triplet charge-transfer levels may lie below the $n\pi^*$ states, and intersystem crossing between them may be slow. It is also conceivable that a triplet is formed but is quenched by the alcohol. In PAB–acetonitrile solutions, an intermediate degree of isomerization of the stilbene is

[9] I. H. Leaver and G. C. Ramsay, *Tetrahedron*, 1969, **25**, 5669.
[10] S. G. Cohen, M. D. Saltzman, and J. B. Guttenplan, *Tetrahedron Letters*, 1969, 4321.
[11] S. G. Cohen and J. I. Cohen, *J. Phys. Chem.*, 1968, **72**, 3782.

observed; and with *m*-aminobenzophenone there is no isomerization in isopropyl alcohol, but the reaction does proceed in hydrocarbons, and to an intermediate extent in acetonitrile. *o*-Aminobenzophenone induces no stilbene isomerization in isopropyl alcohol or benzene, presumably because of intramolecular quenching of the excited carbonyl group by the *ortho*-amino-group, a process which could in principle involve transfer of either an electron or a hydrogen atom.

Davidson and co-workers have studied the photoreduction of aromatic ketones by amines and alcohols, and report that in general the photoreactions lead to formation of the pinacol as the major product.[12] This contrasts with an earlier report for 4-benzoylbenzoic acid in aqueous triethylamine, when the secondary alcohol was the exclusive product.[13] In Davidson's work, solutions of benzophenone, 4-phenylbenzophenone, and 2-acetylnaphthalene in ethanolic triethylamine all yielded the corresponding pinacols. The products from fluorenone included 9-hydroxyfluorene and 9,9′-bisfluorenyl-9,9′-diol (the former product is also formed by a thermal reaction of fluorene with sodium isopropoxide in isopropyl alcohol *via* the radical-anion as an intermediate. The radical anions from photolysis of fluorene in ethanolic solutions of triethylamine, and of benzophenone in aqueous and ethanolic solutions of isobutylamine and t-butylamine were characterized by e.s.r. spectroscopy in each case.

The photoreduction characteristics of fluorenone are similar to those of *p*-aminobenzophenone in that reduction occurs in triethylamine but not in isopropyl alcohol or cyclohexane solutions.[14] The efficiency of reduction by triethylamine is increased by dilution with cyclohexane from a quantum yield of 0·049 for pure triethylamine to 0·37 in 1M- and 0·92 in 0·2M-triethylamine solutions.[15] A study of the quenching of the fluorenone fluorescence by triethylamine has shown that the lack of reduction at high amine concentrations is due to quenching of the ketone singlets by triethylamine.[16] There is also a remarkable solvent effect on the singlet lifetimes of the ketone in that the intersystem crossing rate decreases markedly with increasing solvent polarity. The singlet lifetimes measured are orders of magnitude larger than those estimated for typical aromatic ketones. Intersystem crossing of fluorenone is known to occur with greater than 90% efficiency in cyclohexane [17] and benzene.[18] Thus, to a good approximation, the lifetimes represent the inverse of the rate constants for intersystem

---

[12] R. S. Davidson, P. F. Lambeth, F. A. Younis, and R. Wilson, *J. Chem. Soc.* (C), 1969, 2203.
[13] S. G. Cohen, N. Stein, and H. M. Chao, *J. Amer. Chem. Soc.*, 1968, **90**, 521.
[14] J. N. Pitts, H. W. Johnson, and T. Kuwana, *J. Phys. Chem.*, 1962, **66**, 2456; K. Yoshihara and D. R. Kearns, *J. Chem. Phys.*, 1966, **45**, 1991; W. E. Bachmann, *J. Amer. Chem. Soc.*, 1933, **55**, 391.
[15] G. A. Davis, J. D. Gresser, P. A. Carapellucci, and K. Szoc, *J. Amer. Chem. Soc.*, 1969, **91**, 2264.
[16] R. A. Caldwell, *Tetrahedron Letters*, 1969, 2121.
[17] S. G. Cohen and J. B. Guttenplan, *Tetrahedron Letters*, 1968, 5353.
[18] A. A. Lamola and G. S. Hammond, *J. Chem. Phys.*, 1965, **43**, 2129.

crossing. Fluorenone is reported to be essentially unreactive in 1M triethylamine in isopropyl alcohol or 0·1M triethylamine in acetonitrile. As with p-aminobenzophenone, the photoreaction of fluorenone has now been studied in the presence of stilbene, and again the results show most conclusively that the inefficient reduction in polar media is due to decreased rates of intersystem crossing to triplets.[19]

The photoreduction of fluorenone has also been investigated by kinetic methods and flash photolysis.[20] By these methods, the previously noted lack of reactivity of fluorenone with isopropyl alcohol and negligible activity with toluene and methylcyclohexane was confirmed. Reduction does occur, not surprisingly, in the presence of tri-n-butylstannane, giving the pinacol. The reduction proceeds at least in the major part from the triplet state. Primary, secondary, and tertiary amines, having α-hydrogens, all photoreduce the triplet fluorenone in both benzene and acetonitrile. Again observation is made that the fluorescence of the ketone is quenched by amines. The efficiency in the quenching is inversely proportional to the ionization potential of the amines, and is greater in acetonitrile than in benzene. The data support an electron-transfer mechanism for the fluorescence quenching, although flash photolysis failed to provide conclusive evidence for this. These workers also found [20] that the quantum yield of reduction was decreased at high amine concentrations by amines which were effective at quenching the fluorescence. With triethylamine in benzene, kinetic evidence is presented for the possibility of some reaction from singlet fluorenone. The only isolated products from this study are fluorenone pinacol (2) and 9-hydroxyfluorene (3) which arises by thermal

cleavage of (2) in the presence of amines. Cohen isolated a cross-coupling product between the 9-hydroxyl-9-fluorenyl and 1-diethylamino-1-ethyl radicals with triethylamine.[21]

The kinetics for the photoreduction of aromatic ketones by primary amines were reported in 1968.[22] Such studies with secondary and tertiary amines are complicated by the production of light-absorbing intermediates. This difficulty has, however, been overcome by carrying out the reaction in

---

[19] J. B. Guttenplan and S. G. Cohen, *Tetrahedron Letters*, 1969, 2125.
[20] G. A. Davis, P. A. Carapellucci, K. Szoc, and J. D. Gresser, *J. Amer. Chem. Soc.*, 1969, **91**, 2264.
[21] S. G. Cohen and J. B. Guttenplan, *Tetrahedron Letters*, 1968, 5353.
[22] S. G. Cohen and H. M. Chao, *J. Amer. Chem. Soc.*, 1968, **90**, 165.

aqueous solutions in which the secondary and tertiary amines are very efficient reducers. The kinetic constants for the photoreduction of 4-benzoylbenzoate anion by the three classes of amines in 1:1 water: pyridine have been determined.[23] The value of k for the interaction of triplet carboxyketone with isobutylamine is $6.3 \times 10^7 \, l \, mol^{-1} \, s^{-1}$. This value of k rises in passing from primary to secondary to tertiary amines. This order of increasing reactivity is that of decreasing ionization potential of the amines, and is consistent with earlier proposals that the reaction proceeds *via* a rapid charge-transfer interaction which leads to partial quenching and partial reduction (see Scheme 1). The reaction with the tertiary amine is very fast, $k = 8 \times 10^8 \, l \, mol^{-1} \, s^{-1}$, and interaction of the ketone triplet with the non-bonding electrons may be diffusion controlled, particularly if a small steric probability factor lowers the value slightly. The rates have also been determined as a function of concentration of a quencher, naphthalene. The product of the reduction (quantum yield 0·7) of the carboxyketone by amines in aqueous alkaline media is 4-carboxybenzhydrol formed by disproportionation of the ketyl radicals. A second reducing moiety is transferred from the amine-derived radical to the ground state ketone (Scheme 2) to yield either an enamine, which has been characterized in the photoreduction of benzophenone by triethylamine in benzene,[24] or to a charged species (4), both of which may lead in the aqueous system to the carbonyl compound and secondary amine.

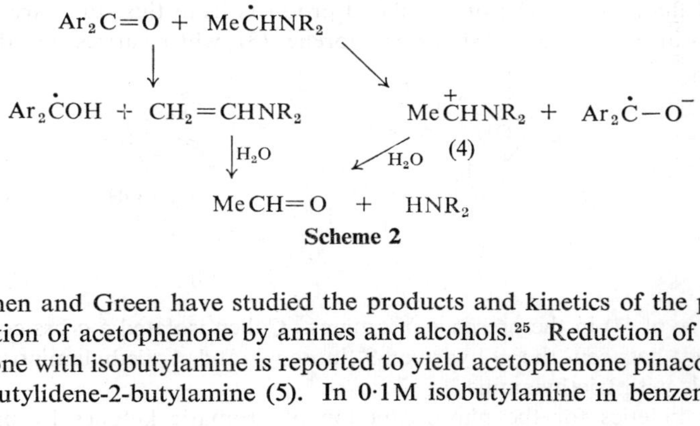

$$Ar_2C=O + Me\dot{C}HNR_2$$
$$\downarrow \qquad \searrow$$
$$Ar_2\dot{C}OH + CH_2=CHNR_2 \qquad Me\overset{+}{C}HNR_2 + Ar_2\dot{C}-O^-$$
$$\downarrow H_2O \qquad \swarrow H_2O \quad (4)$$
$$MeCH=O + HNR_2$$

Scheme 2

Cohen and Green have studied the products and kinetics of the photoreduction of acetophenone by amines and alcohols.[25] Reduction of acetophenone with isobutylamine is reported to yield acetophenone pinacol, and *N*-2-butylidene-2-butylamine (5). In 0·1M isobutylamine in benzene, the

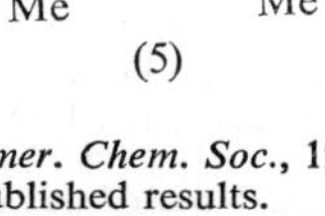

$$\underset{Me}{\overset{Et}{\diagdown}}C=N-\underset{Me}{\overset{Et}{\diagup}}CH$$
(5)

[23] S. G. Cohen and N. Stein, *J. Amer. Chem. Soc.*, 1969, **91**, 3690.
[24] N. Stein and S. G. Cohen, unpublished results.
[25] S. G. Cohen and B. Green, *J. Amer. Chem. Soc.*, 1969, **91**, 6824.

quantum yield for this reaction is approximately 1·1 and is 1·4 times greater than that in neat butylamine. A light-absorbing transient is not formed in this amine, and the low efficiency in neat amine may be due to $\pi\pi^*$ character of the ketone triplet in this medium, or to deactivating solvation of ketone and triplet by the amine. Acetophenone is not reduced by α-methylbenzyl acetate, but it is reduced with low efficiency by N-acetyl-α-methylbenzylamine. The quantum yield for the reduction of acetophenone by 0·5M α-methylbenzylamine in benzene is 0·49, by 1·0M α-methylbenzyl alcohol is 0·37, and by 0·5M isopropyl alcohol is 0·75. Light-absorbing transients are formed to a greater extent and photoreduction is less efficient at high concentrations of each of these reducing agents. The efficiency of the photoreduction first increases with dilution with benzene, and then decreases at high dilution: extrapolation of the dilute solution values of the quantum yield leads to hypothetical limiting values of 0·54 for α-methylbenzyl alcohol and 1·2 for isopropyl alcohol in benzene. The photoreduction of acetophenone and p-methylacetophenone by α-methylbenzylamine yields three products of coupling of the radicals formed by initial hydrogen abstraction: these are pinacols (6), amine alcohols (7), and diamines (8). Thus aryl alkyl ketone- and amine-derived radicals are of similar stability and both survive uncoupled. The radicals derived from purely aliphatic alcohols and amines reduce a second molecule of the ketone.

$$p\text{-MeC}_6\text{H}_4\overset{*}{\text{C}}\text{OMe} + \text{PhCH}(\text{NH}_2)\text{Me}$$

$$\downarrow$$

$$p\text{-MeC}_6\text{H}_4\dot{\text{C}}(\text{OH})\text{Me} + \text{Ph}\dot{\text{C}}(\text{NH}_2)\text{Me}$$

$$\downarrow$$

$$\left[\begin{array}{c}\text{Me}\\|\\p\text{-MeC}_6\text{H}_4-\text{C}-\\|\\\text{OH}\end{array}\right]_2 + p\text{-MeC}_6\text{H}_4-\overset{\text{Me}}{\underset{\text{OH}}{\overset{|}{\underset{|}{\text{C}}}}}-\overset{\text{Me}}{\underset{\text{NH}_2}{\overset{|}{\underset{|}{\text{C}}}}}\text{Ph} + \left[\begin{array}{c}\text{Me}\\|\\\text{C}_6\text{H}_4-\text{C}-\\|\\\text{NH}_2\end{array}\right]_2$$

(6)   (7)   (8)

In accord with 4-hydroxybenzophenone (and 4-aminobenzophenone, see ref. 10), 3,5-di-t-butyl-4-hydroxyacetophenone (9) and 3,5-di-t-butyl-4-hydroxybenzophenone (10) are found to be unreactive in isopropyl alcohol but in triethylamine they yield the pinacol (11) and 2,6-di-t-butyl-4-hydroxybenzhydrol (12) respectively.[26] The photolysis of (9) and (10) in cyclohexane, however, resulted in de-t-butylation to yield 3-t-butyl-4-hydroxyacetophenone (13) and 3-t-butyl-4-hydroxybenzophenone (14)

[26] T. Matsuura and Y. Kitaura, *Tetrahedron*, 1969, **25**, 4501.

respectively, in addition to isobutene. The debutylation was shown by quenching experiments to arise from the triplet state of the ketone. Photo-oxygenation of (9) and (10) in cyclohexane resulted in oxidative cleavage to yield acetic acid and benzoic acid respectively, in addition to 2,6-di-t-butylbenzoquinone (15). 2-Hydroxy-5-methylbenzophenone and 3,5-t-butyl-2-hydroxybenzophenone are both photo-unreactive under various conditions while 3,5-di-t-butyl-4-methoxybenzophenone is readily reduced. Mechanisms involving triplet states and phototautomerization are discussed.

The steric effects in the photoreduction of aryl alkyl ketones have been studied by Lewis by reference to α-substituted acetophenones (16—19), which until this report had not been investigated to any extent.[27] The reductions were carried out in isopropyl alcohol and a decrease in the rate constant for hydrogen abstraction was observed as the size of the alkyl group increased, until with pivalophenone (19) no photoreduction products were observed, but instead a low conversion to benzaldehyde, and presumably isobutene. The lower rate constants are thought to be simply due to steric hindrance of the abstraction process, rather than a change in triplet energy or an increase in the $\pi\pi^*$ character of the excited state.

Ph—CO—R

(16) R = Me   (18) R = CHMe$_2$
(17) R = Et   (19) R = CMe$_3$

Steric effects have also been noted for the hydrogen donor. Thus, the quantum yield for ketone formation from 2,4-di-methyl-3-heptanol and the rate constants for hydrogen abstraction from this substrate show a substantial decrease in comparison with those of isopropyl alcohol.

The photolysis of pivalophenone (19) has also been reported by Heine, who again finds α-cleavage to yield benzaldehyde, isobutane, and isobutene:[28] 4-t-butyl and 2,4,6-trimethylpivalophenones (20) and (21) react analogously. One slightly odd feature of this report is that the α-cleavage of (19) and (20) was found to *compete* with photoreduction, giving

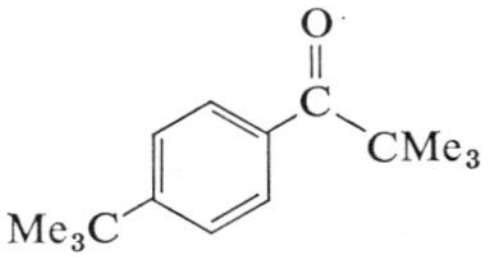

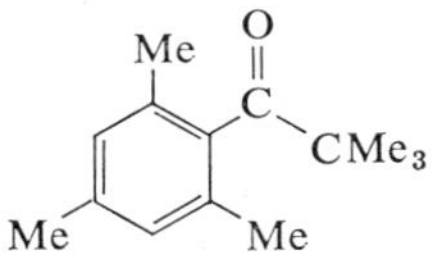

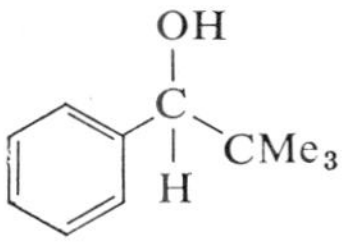

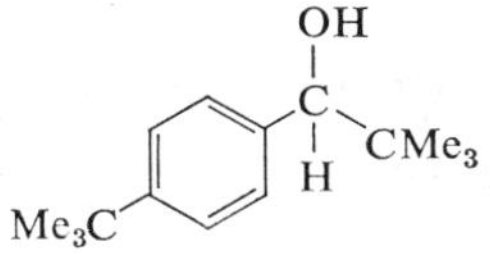

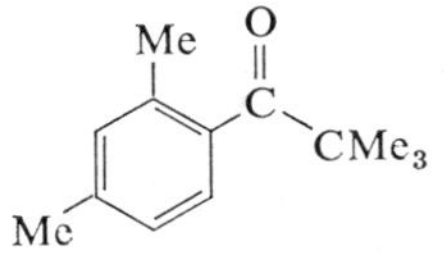

the corresponding benzyl alcohols (22) and (23) instead of the expected pinacols. No pinacol products were reported in ref. 27. 2,4-Dimethylpivalophenone (24) reacts anomalously in comparison with (19), (20), and (21), giving its isomeric benzocyclobutenol derivative (25). All the reactions involve triplet states. 1-Pivalonaphthone is reported to be stable under these conditions.

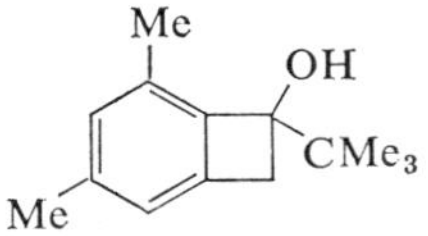

---

[27] F. D. Lewis, *Tetrahedron Letters*, 1970, 1373.
[28] H. G. Heine, *Annalen*, 1970, **732**, 165.

New and previously published data [29] for the photochemical and electrochemical bimolecular reduction of aldehydes and unsymmetrical ketones have been tabulated, and the stereochemical results from the two techniques have been compared.[30] In all cases, both methods of pinacolization give essentially the same ratios of *dl*- to *meso*-pinacols in acid solution with corresponding changes of ratios in basic media. Mechanisms previously proposed for the two systems are shown to be mutually compatible and several examples of contrasting behaviour (*i.e.* initiation of reaction by only one of the techniques) are reported and discussed. Only the final step in the pinacolization is under consideration: the two techniques have different mechanisms prior to this step and other reactions of intermediates may occur which are unique to a particular process. The lack of photopinacolization of benzaldehyde and 2-acetylpyridine in strong base is due to as yet undetermined light-accelerated alternative reactions, and phenylcyclohexyl ketone reacts so slowly that alternative photoreactions also occur.[31] Failure of trifluoroacetophenone to pinacolize electrochemically is due to complete carbon–fluorine fission at a potential lower than that required for pinacolization.[32] Electrochemical pinacolization has also been reported for a well-known photorearrangement system.[33] Thus, it is reported that cross-conjugated 4,4′-disubstituted-2,5-cyclohexadienones (26) do not undergo bond crossing reactions when converted electro-

chemically into a 7-π-electron system, (*cf.* photochemical case), but yield the pinacols (27) in acids or bases.

Three groups of workers report on the photolysis of aromatic ketones in toluene. The formation of benzopinacol, bibenzyl, and the radical coupling

[29] J. H. Stocker and D. H. Kern, *J. Org. Chem.*, 1968, **33**, 1271; J. H. Stocker and R. M. Jenevein, *J. Org. Chem.*, 1969, **34**, 2807, and refs. therein.
[30] J. H. Stocker, R. M. Jenevein, and D. H. Kern, *J. Org. Chem.*, 1969, **34**, 2810.
[31] J. H. Stocker and D. H. Kern, *Chem. Comm.*, 1969, 204.
[32] J. H. Stocker and R. M. Jenevein, *Chem. Comm.*, 1968, 934.
[33] A. Mazzenga, D. Lomnitz, J. Villegag, and C. J. Polowczyk, *Tetrahedron Letters*, 1969, 1665.

product (28) from benzophenone in toluene is well known.[34] This reaction has been used to demonstrate the usefulness of carrying out photolyses in an n.m.r. spectrometer cavity and observing the nuclear spin polarization.[35] Benzophenone in toluene under these conditions gives rise to a single absorption line (apart from the solvent resonances) at 76 Hz downfield from the toluene methyl protons. This signal disappears and reappears when the light source is switched off and on, so is a true photo-signal. It arises from the methylene protons in (28). The steady-state concentration of polarized (28) is given by $[(28)^*]_S = nk_L T_1$ where n is the yield of (28) in the reaction, k_L is the zero order rate constant for the disappearance of benzophenone, and T_1 is the spin lattice relaxation time of the methylene protons of (28). Independent determination of n, k_L, and T_1 give a value of $2.0 \times 10^{-4}\,\mathrm{mol\,l^{-1}}$ for (28), and comparison of the intensity of the photo-signal with that of a standard solution of (28) gives a polarization of 250 ± 25. Ethylbenzene gave the quartet of the methine proton in the radical coupled product [*cf.* (28)] as a strong photo-signal. The data presented in this report clearly demonstrate that the product (28) is formed from paramagnetic precursors, in agreement with the chemical studies: they also point to the potential value of dynamic polarization in mechanistic photochemistry.

Bellus and Schaffner have examined the photolysis of a series of aromatic ketones in toluene.[36] The triplet state reactivity of the ketones examined is known to be mostly $\pi\pi^*$. Their reasons for this study arise from their observation that novel primary photoprocesses of triplet ($\pi\pi^*$) excited $\alpha\beta$-unsaturated cycloalkanones (29) occur in toluene in that abstraction of benzylic hydrogen yields olefin reduction products (30) and not carbonyl reduction.[37] With the aromatic ketones examined, however, no analogous reaction (*i.e.* formation of benzylcyclohexa-1,3- and -1,4-dienes) was observed, and only in the one case of 4-methoxyacetophenone (31) was any reaction observed at all. This reaction was the slow formation of the tertiary carbinol (32), the pinacol (33), and bibenzyl. These products are typically those formed from other acetophenones (34—36), benzophenone, and cyclopropyl phenyl ketone which exhibit $n\pi^*$ triplet reactivity.

The photoreaction of camphorquinone (37) in isopropyl alcohol was reported last year,[38] and the reaction of benzophenone in toluene to yield (38), (39), and bibenzyl is known.[39] The latter reaction has been reinvestigated for both camphorquinone and benzophenone.[40] The adducts (39) and (39a and b) could be formed by two pathways: (*a*) immediate coupling of the initially formed geminate pair of radicals, or (*b*) separation of the

---

[34] C. Walling and M. J. Gibian, *J. Amer. Chem. Soc.*, 1965, **87**, 3361, and refs. therein.
[35] G. L. Closs and L. E. Closs, *J. Amer. Chem. Soc.*, 1969, **91**, 4550.
[36] D. Bellus and K. Schaffner, *Helv. Chim. Acta*, 1969, **52**, 1010.
[37] D. Bellus, D. R. Kearns, and K. Schaffner, *Helv. Chim. Acta*, 1969, **52**, 971.
[38] B. M. Monroe and S. A. Weiner, *J. Amer. Chem. Soc.*, 1969, **91**, 450.
[39] M. B. Rubin and R. G. La Barge, *J. Org. Chem.*, 1966, **31**, 3283.
[40] M. B. Rubin, *Tetrahedron Letters*, 1969, 3931.

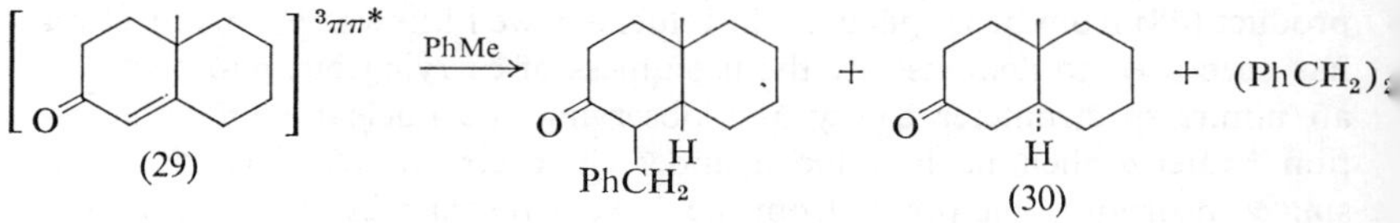

$$[\text{(29)}]^{3\pi\pi^*} \xrightarrow{\text{PhMe}} \text{(30)} + \cdots + (\text{PhCH}_2)_2$$

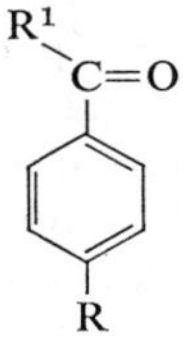

(31) R = OMe, R¹ = Me
(34) R = Me, R¹ = Me
(35) R = H, R¹ = Me
(36) R = R¹ = H

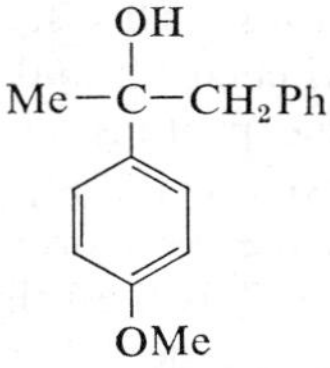

(32)

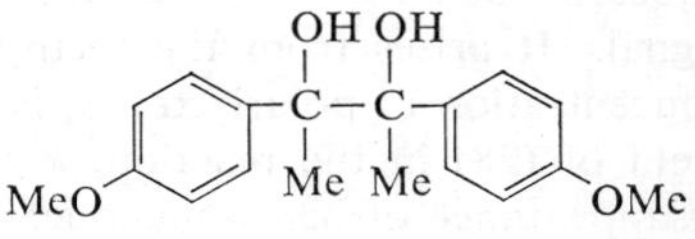

(33)

radicals and coupling during subsequent encounters. Data from sensitization experiments and composition of product mixtures now indicate that the path (*a*) occurs for (37), but path (*b*) is the major if not the exclusive route for benzophenone. The reason for this difference in reaction path is certainly not obvious, but the author points out that a possible factor may be that hydrogen abstraction by the triplet state of one ketone generates a pair of radicals with parallel spins of their unpaired electrons, and in order for such geminate radicals to couple, spin inversion must occur. Benzophenone has also been photolysed in the presence of triphenylsilane to give

$$\text{Ph}_2\text{CO} + \text{ArMe} \xrightarrow{h\nu} (\text{ArCH}_2)_2 + (\text{Ph}_2\text{COH})_2 + \text{Ph}_2\overset{\text{OH}}{\underset{|}{\text{C}}}-\text{CH}_2\text{Ar}$$

(38) (39)

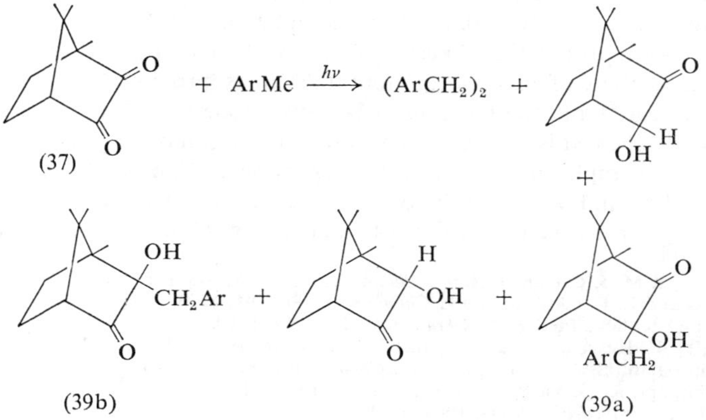

(37) + ArMe $\xrightarrow{h\nu}$ (ArCH$_2$)$_2$ + ...

(39b)   (39a)

Photo-oxidation and -reduction Reactions

the monotriphenylsilyl ether of benzpinacol (40) formed in 56% yield.[41] The mechanism is outlined in Scheme 3. A previous statement [42] that this reaction does not occur results from an error in translation: the reaction had not previously been investigated.

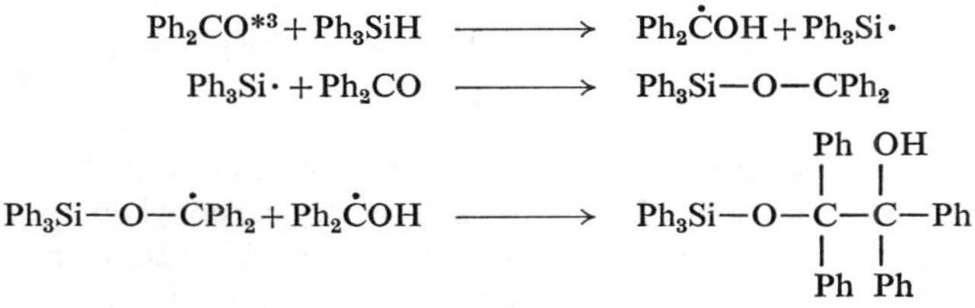

$$Ph_2CO^{*3} + Ph_3SiH \longrightarrow Ph_2\dot{C}OH + Ph_3Si\cdot$$

$$Ph_3Si\cdot + Ph_2CO \longrightarrow Ph_3Si-O-\dot{C}Ph_2$$

$$Ph_3Si-O-\dot{C}Ph_2 + Ph_2\dot{C}OH \longrightarrow Ph_3Si-O-\underset{\underset{Ph}{|}}{\overset{\overset{Ph}{|}}{C}}-\underset{\underset{Ph}{|}}{\overset{\overset{OH}{|}}{C}}-Ph$$

(40)

Scheme 3

Until the present year, there have been no reports on the photoreduction of aromatic esters. Fukui and Odaira have now reported that irradiation of dilute solutions of dimethyl terephthalate (41) in hydrocarbons with

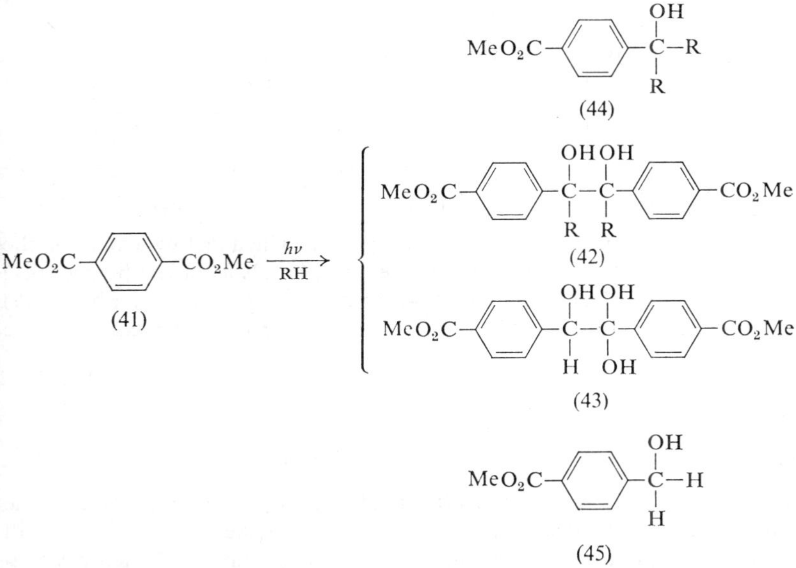

labile hydrogens yields the pinacols (42) and (43), the tertiary carbinols (44) and (45), dimers of the aromatic hydrocarbons, and appreciable amounts of methyl alcohol.[43] It is noteworthy that there are no products

---
[41] H. D. Becker, *J. Org. Chem.*, 1969, **34**, 2469.
[42] C. Eaborn, 'Organosilicon Compounds,' Butterworth and Co. Ltd., London, 1960, p. 213.
[43] K. Fukui and Y. Odaira, *Tetrahedron Letters*, 1969, 5255.

derived from direct coupling of the radical (46). The basic mechanism is discussed but refinements of this are promised shortly.

$$\text{MeO}_2\text{C}-\underset{(46)}{\text{C}_6\text{H}_4}-\underset{\underset{\text{OMe}}{|}}{\overset{\overset{\text{OH}}{|}}{\text{C}\cdot}}$$

The photolysis of pinacols in the presence of ketones or quinones has been known for a number of years to yield the arylcarbonyl compound and reduction products of the ketone or quinone.[44] Davidson and his co-workers rightly point out that a knowledge of the mechanism of this reaction is necessary to ascertain the part it plays (equations 4 and 5) in the well known reduction of aryl ketones.[45] The formation of α-hydroxyl-arylmethyl radicals from the photolysis of arylpinacols was unequivocally established by the irradiation of xanthopinacol (47) and fluoropinacol (48) in a variety of solvents. The 9-hydroxyxanthyl and 9-hydroxyfluorenyl

(47) (48) (49)

radicals were identified by their e.s.r. spectra. Chemical evidence for the formation of such radicals was obtained by photolysis of the pinacol (47) in the presence of acridine when 9-(9'-acridaryl)-9-hydroxyxanthene (49) was obtained. From sensitization and quenching data, it seems that the

$$\text{Ar}_2\text{C}=\text{O} + \text{R}_2\text{CHOH} \rightarrow \text{Ar}_2\dot{\text{C}}\text{OH} + \text{R}_2\dot{\text{C}}\text{OH} \qquad (4)$$

$$\text{Ar}_2\dot{\text{C}}\text{OH} \rightleftharpoons \text{Ar}_2-\underset{\underset{\text{OH}}{|}}{\text{C}}-\underset{\underset{\text{OH}}{|}}{\text{C}}\text{Ar}_2 \qquad (5)$$

cleavage does not occur efficiently, if at all, from the triplet state. The conclusion of this work is that cleavage of arylpinacols (equation 4) will not occur in photoreductions unless the exciting light is of such a wavelength that it is absorbed directly by the pinacol.

There have been several reports within this year of the photoreduction of aliphatic ketones. Radiation from a sunlamp of solutions of benzhydrol in acetone causes the formation of isopropanol, benzpinacol, and some

---

[44] A. Schönberg and A. Mustafa, *J. Chem. Soc.*, 1944, 67; D. C. Neckers and D. P. Calenbrandez, *Tetrahedron Letters*, 1968, 5045.
[45] R. S. Davidson, F. A. Younis, and R. Wilson, *Chem. Comm.*, 1969, 826.

benzophenone, apparently from the free radicals (50) and (51).[46] In the presence of mesityl mercaptan, isopropyl alcohol is formed more rapidly

$$Me_2\dot{C}OH + RSH \longrightarrow Me_2CHOH + RS\cdot \quad (6)$$
(50)

$$Ph_2\dot{C}OH + RS\cdot \longrightarrow Ph_2C=O + RSH \quad (7)$$
(51)

and benzophenone persists longer, and the yields are larger. This is because radical (50) is reduced by the mercaptan (equation 6) while radical (51) is oxidised by the thiyl radical (equation 7). These processes inhibit the photoreduction of benzophenone by isopropyl alcohol by converting the radicals back to the starting material, thereby altering the course of the reaction and converting the benzhydrol to benzophenone and the acetone to isopropyl alcohol. The irradiation of acetone yields isopropyl alcohol, and this is formed more readily in the presence of the mercaptan, but with benzophenone in acetone there is no apparent reaction. The acetone–benzhydrol system is more complicated than the benzophenone–isopropyl alcohol system because of the persistence and reactions of both initially formed radicals and the photoreduction of acetone by acetone.

Dauben and his co-workers have made a comparative study of the photoreduction of a series of bicyclo[4,1,0]- and [3,1,0]-alkan-2-ones (52) in isopropyl alcohol, and the ground-state radical rearrangement of the corresponding bicyclo[4,1,0]- and [3,1,0]-alkan-2-ols (53) produced by di-t-butylperoxide as initiator.[47] The product ratios in either the photoreductive process or the radical rearrangement change markedly with temperature. The similarity of product distributions from the two processes at different temperatures indicates that a common intermediate, the α-hydroxycyclopropylcarbinyl radical (54), is involved. Overlap between the p-orbital of the radical and the outside cyclopropane bond leads to selective cleavage (55) of this bond at room temperature to yield (56). At higher temperatures, the products (58) of inside bond cleavage (57) are also observed. It is suggested that the similarity between specific ring-opening of the outside bond in photo-isomerization [(59) → (60)],[48, 49] and the photoreduction arises from the energy features of the ring system rather than a common intermediate. The authors emphasize that the photoreductions only occur in efficient hydrogen donors; in poor hydrogen donors the triplet of the ketones falls back to the ground state, or other reactions may occur.

Both α-diketones and quinones are labile in photoreductive systems. In the case of α-diketones (61), (62), and (63), having primary, secondary,

---

[46] S. G. Cohen, S. Aktipis, and H. Rubinstein, *Photochem. and Photobiol.*, 1969, **10**, 45.
[47] W. G. Dauben, L. Schutte, R. E. Wolf, and E. J. Deviny, *J. Org. Chem.*, 1969, **34**, 2512.
[48] L. D. Hess and J. N. Pitts, *J. Amer. Chem. Soc.*, 1967, **89**, 1973.
[49] W. G. Dauben and G. W. Shaffer, *Tetrahedron Letters*, 1967, 4415.

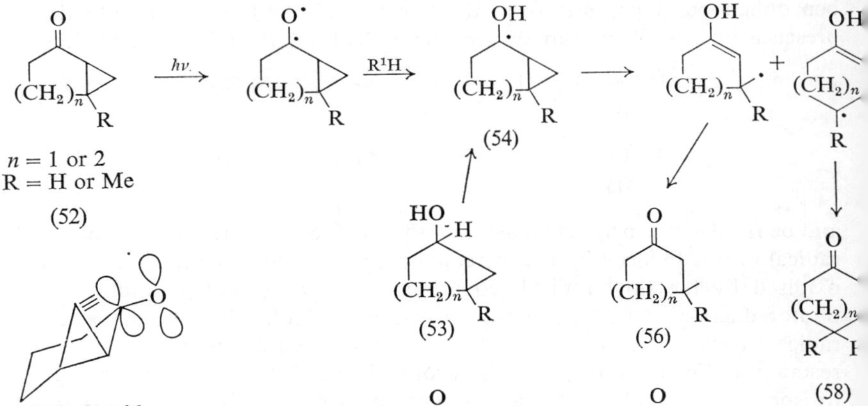

$n = 1$ or 2
R = H or Me

(52) (53) (54) (56) (58)

(55) Outside Cleavage

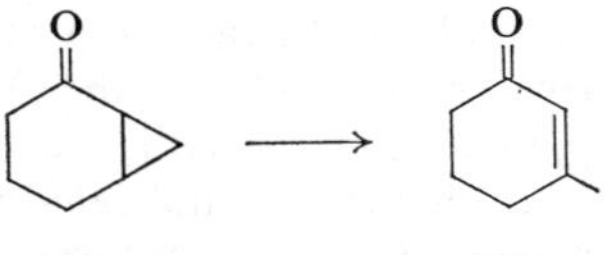

(59) (60)

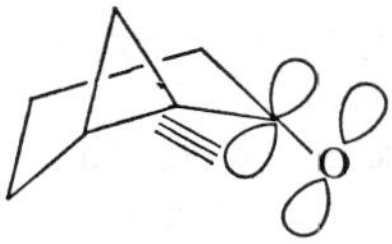

(57) Inside Cleavage

and tertiary hydrogen respectively, the conversion of carbonyl to C–OH occurs intramolecularly with the formation of cyclobutane derivatives

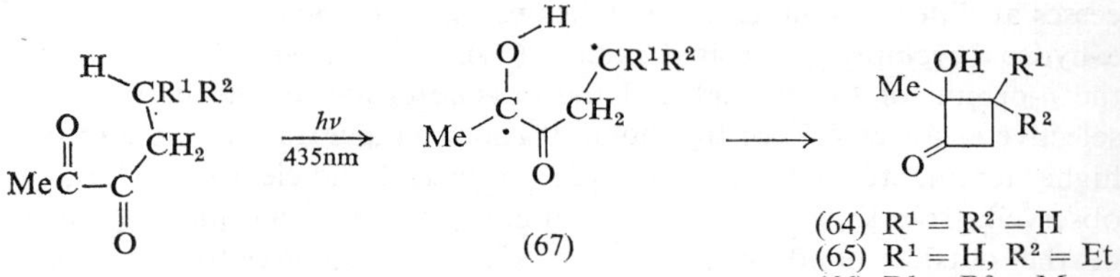

(61) $R^1 = R^2 = H$
(62) $R^1 = H$, $R^2 = Et$
(63) $R^1 = R^2 = Me$

(67)

(64) $R^1 = R^2 = H$
(65) $R^1 = H$, $R^2 = Et$
(66) $R^1 = R^2 = Me$

(64), (65), and (66).[50] The reactions are induced by 435 nm radiation in 0·15M benzene solution. Each reaction is essentially quantitative and is quenched by piperylene, a diffusion-controlled quencher of triplets. (*N.B.* Piperylene can quench S_1 states of carbonyl compounds at high diene concentrations.) The main features of the reactions are (*a*) the order of magnitude in increase in the quantum yield in going from primary (0·054) to secondary (0·50) or tertiary (0·57) hydrogen abstraction, (*b*) very

[50] N. J. Turro and T. J. Lee, *J. Amer. Chem. Soc.*, 1969, **91**, 5651.

low rate constants compared to monoketones for the intramolecular abstraction, (c) the insensitivity of rate of reaction and quantum yield to solvent polarity and hydrogen bonding, and (d) the increased reactivity in passing from (61) to (62) to (63). The low quantum yield for (61) could result from a low rate constant for the formation of (67) relative to the decay of the triplet of (61), or reversal of the hydrogen transfer [*i.e.* (67) → (61)]. The low reactivity of α-diketones compared with monoketones is accounted for by the fact that the low energy of the starting triplet may make biradical formation more endothermic than is the case for analogous triplet monoketones.

The irradiation of benzil in isopropyl alcohol or cumene is reported to form benzoin (68), 1,2,3,4-tetraphenyl-2,3-dihydroxybutane-1,4-dione (69)

$$PhCOCOPh + Me_2CHOH \longrightarrow Ph\dot{C}OHCOPh + Me\dot{C}OHMe$$

$$PhCOCOPh + Me\dot{C}OHMe \longrightarrow Ph\dot{C}OHCOPh + Me_2CO$$

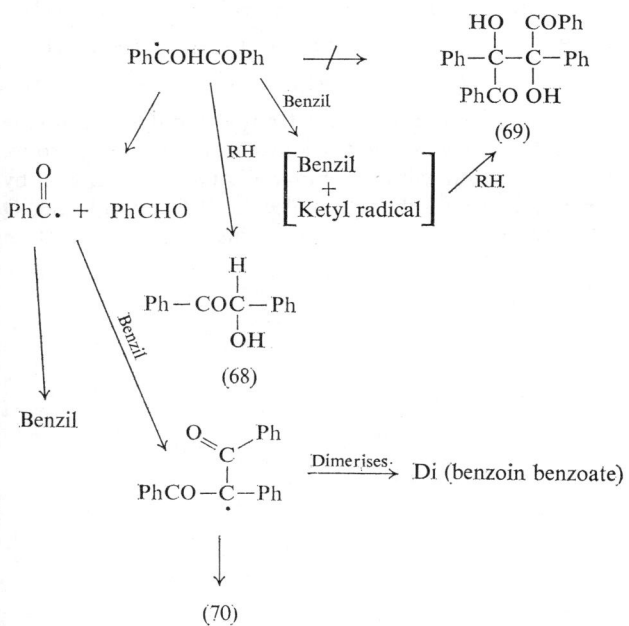

(benzilpinacol), benzoin benzoate (70), benzaldehyde, benzoic acid, the dibenzoate of (69), and some unidentified high-molecular-weight compounds.[51] The yield of (69) is strongly dependent on the initial benzil concentration, temperature, and solvent. The quantum yield for benzil

[51] D. L. Bunbury and T. T. Chuang, *Canad. J. Chem.*, 1969, **47**, 2045.

disappearance depends on the initial benzil concentration to a lower limit which varies with solvent. Unlike the formation of benzpinacol from benzophenone, the formation of (69) does not seem to involve coupling of ketyl radicals. Instead, the formation of a complex between a benzil molecule and the ketyl radical is suggested. The principal path for benzaldehyde and benzoyl radicals is dissociation of the ketyl radical. Excited benzil is reported here to abstract hydrogen from isopropyl alcohol at least ten times faster than it dissociates into two benzoyl radicals.

The photopinacolization of benzoylphosphonates (71) is also considered to arise from the ketyl radical.[52] The intermediacy of triplet states is again suggested since the reaction is effectively quenched by penta-1,3-diene or oxygen. The quantum yields for the disappearance of several p-substituted aryl phosphonates are highly sensitive to the electron-inductive abilities of the substituents. Thus, the quantum yield falls from 2·10 for the p-methoxy-derivative to 1·79 for the p-chloro-derivative at room temperature. It is

$$P \longrightarrow P^* \quad (8)$$
$$P^* + Q \longrightarrow P + Q^* \quad (12)$$
$$P^* + RH \longrightarrow PH + R\cdot \quad (9)$$
$$P^{1\cdot} + RH \longrightarrow P + R\cdot \quad (13)$$
$$P^* \longrightarrow P \quad (10)$$
$$R\cdot + P \longrightarrow PH + R^1 \quad (14)$$
$$P^* + P \longrightarrow PH + P^{1\cdot} \quad (11)$$
$$2PH \longrightarrow P-P \quad (15)$$

suggested that this substituent effect results from the difference in the conformation between the carbonyl group and the phosphonyl group. The mechanism of the reaction with 320 nm radiation is described by equations (8—15), where P is phosphonate molecule, PH ketyl radical, Q quencher, and RH solvent. Equation (11) explains why the quantum

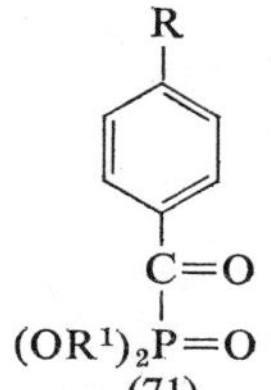

(71)

(72) R = H, R^1 = Et

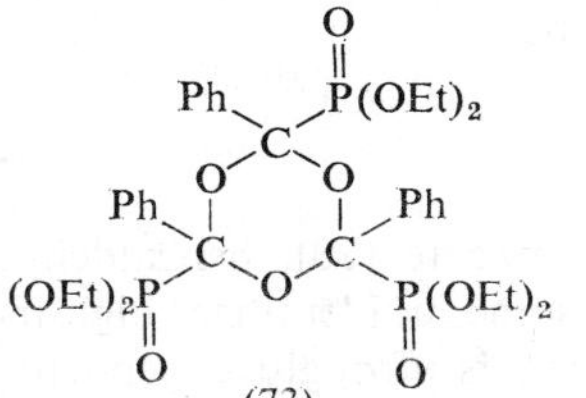

(73)

[52] K. Terauchi and H. Sakurai, *Bull. Chem. Soc. Japan*, 1970, **43**, 883.

yield is observed to be *ca.* 2. The photopinacolization was observed with a number of phosphonates, but diethylbenzoyl phosphate (72) behaved anomalously. Thus, (72) in a variety of solvents yielded 2,4,6-triphenyl-2,4,6-tris[diethylphosphonyl]-1,3,5-trioxan (73) almost quantitatively. The mechanistic reasons for this deviation in reaction mode should prove interesting reading when they appear in print.

Over the years there have been a number of reports concerning the photochemistry of *ortho*-quinones and, in particular, phenanthrene-quinone (74). Its photoreduction in ethyl alcohol to give 9,10-phenanthrenediol (75) and acetaldehyde has been reported to have a quantum yield in excess of unity for the disappearance of the quinone.[53] In isopropyl alcohol, phenanthrenequinone is now reported to be cleanly reduced to the diol (75) and acetone, and this reaction has been studied in some detail in an attempt to elucidate the mechanism.[54] The quantum yield for disappearance of the quinone is independent of the incident light intensity (435 nm), and increases with alcohol concentration to a maximum of 1·6 in pure isopropyl alcohol. Flash photolysis experiments demonstrate the presence of two transients in the reaction. The transient which absorbs in the 340—500 nm region decays with second-order kinetics, and appears in all the solvents examined: it is assigned as the 10-hydroxy-9-phenanthrenyloxyl radical (76). The other transient absorbs in the 500—760 nm region, and decays with first-order kinetics. The latter transient is only formed in benzene and glacial acetic acid, and is tentatively identified as the first excited triplet state of the quinone. The quantitative results of this study are interpreted in terms of the initial formation of internally-excited semiquinone radicals by hydrogen abstraction from the solvent.

[53] P. Walker, *J. Chem. Soc.*, 1963, 5545.
[54] P. A. Carapellucci, H. P. Wolf, and K. Weiss, *J. Amer. Chem. Soc.*, 1969, **91**, 4635.

A simplified mechanism is proposed (equations 16—24) in which the

$$Q \longrightarrow Q^1 \quad (16) \qquad Q^3 + RH_2 \longrightarrow QH^* + RH\cdot \quad (20)$$
$$Q^1 \longrightarrow Q^3 \quad (17) \qquad QH^* + RH_2 \longrightarrow QH_2 + RH\cdot \quad (21)$$
$$Q^3 \longrightarrow Q \quad (18) \qquad QH^* \longrightarrow QH\cdot \quad (22)$$
$$Q^1 \longrightarrow Q \quad (19) \qquad RH\cdot + Q \longrightarrow QH\cdot + R(Me_2CO) \quad (23)$$
$$2QH\cdot \longrightarrow QH_2 + Q \quad (24)$$

vibrationally-excited semiquinone radical (QH*) plays an important part, and accounts for the independence of quantum yield on light intensity. Some reduction of the quinone in benzene to give the phenyl ether (77) has previously been observed.[55, 56] In another report this year, the great ability of photo-activated phenanthrenequinone (74) to abstract hydrogen from donors is commented upon, and (74) is stated to be *easily* reduced when irradiated in benzene; but these workers did not mention the formation of (77) in this 'facile' process.[57] The rate of abstraction is reported to be strongly inhibited in the presence of oxygen, from which triplet intermediates are deduced. Also, the rate of disappearance of the quinone, in large excess of suitable substrates, follows first-order kinetics. Tertiary hydrogens are abstracted faster than secondary hydrogens which in turn are abstracted faster than primary ones, but the differences are really rather small.

The photobleaching of yellow solutions of (74) to yield colourless (75) solutions has been studied in the presence of magnesium, calcium, and zinc divalent ions.[58] In such reactions the solution becomes blue-green, and a long-lived intermediate is formed which may be reduced to the diol (75). The system has now been examined by e.s.r. spectroscopy and the blue-green intermediate is deduced to be a cation-semiquinone complex.[59] The blue-green colour is also formed in air oxidation of (75), and the solutions are paramagnetic. In the oxidation of catechol in the presence of calcium, structure (78) was proposed for the coloured intermediate.[60]

It has been suggested that the simultaneous reduction of two quinones offers special opportunities for determining the mechanism of the photo-reduction of quinones in hydrogen donors,[61] because of the similarity of such a procedure to a titration in which first one quinone and then the other is reduced. The point at which both quinones absorb the light equally resembles a 'mid-point.' The present theoretical paper presents a method for determining the rate constants of a 24-step mechanism which describes

[55] J. Bohning and K. Weiss, *J. Amer. Chem. Soc.*, 1966, **88**, 2895.
[56] M. B. Rubin and Z. Neuwirth-Weiss, *Chem. Comm.*, 1968, 1607.
[57] K. Maruyama, K. Ono, and J. Osugi, *Bull. Chem. Soc. Japan*, 1969, **42**, 3357.
[58] L. Michaelis and S. Granwick, *J. Amer. Chem. Soc.*, 1948, **70**, 624.
[59] J. Rennert, M. Mayer, J. Levy, and J. Kaplan, *Photochem. and Photobiol.*, 1969, **10**, 267.
[60] D. R. Eaton, *Inorg. Chem.*, 1964, **3**, 1268.
[61] J. Beutel, R. J. Ruszkay, and J. F. Brennan, *J. Phys. Chem.*, 1969, **73**, 3240.

the process involving the two quinones. It is very satisfying to note that two of the same workers have, in a subsequent paper, reported an experimental analysis of such a two quinone reduction system, *viz.* the simultaneous reduction with 436 nm radiation of 9,10-phenanthrenequinone and 2-t-butyl-9,10-anthraquinone in ethanol.[62] The rate constants were determined by flash photolysis and by fitting optical density–time data into the mathematical model given in the previous paper.[61] Equations

$$PQ \xrightarrow{h\nu} {}^1PQ \quad (25)$$

$$AQ \xrightarrow{h\nu} {}^1AQ \quad (26)$$

$$^1PQ \longrightarrow PQ \quad (27)$$

$$^1AQ \longrightarrow AQ \quad (28)$$

$$^1PQ \longrightarrow {}^3PQ \quad (29)$$

$$^1AQ \longrightarrow {}^3AQ \quad (30)$$

$$^3PQ \longrightarrow PQ \quad (31)$$

$$^3AQ \longrightarrow AQ \quad (32)$$

$$^3PQ + PQ \longrightarrow 2PQ \quad (33)$$

$$^3PQ + RH_2 \longrightarrow P\dot{Q}H + \dot{R}H \quad (34)$$

$$^3AQ + RH_2 \longrightarrow A\dot{Q}H + \dot{R}H \quad (35)$$

$$P\dot{Q}H + AQ \longrightarrow A\dot{Q}H + PQ \quad (36)$$

$$A\dot{Q}H + PQ \longrightarrow P\dot{Q}H + AQ \quad (37)$$

$$PQ + \dot{R}H \longrightarrow P\dot{Q}H + R \quad (38)$$

$$AQ + \dot{R}H \longrightarrow A\dot{Q}H + R \quad (39)$$

$$2P\dot{Q}H \longrightarrow PQ + PQH_2 \quad (40)$$

$$2A\dot{Q}H \longrightarrow AQ + AQH_2 \quad (41)$$

$$2\dot{R}H \longrightarrow R + RH_2 \quad (42)$$

$$PQH_2 + AQ \longrightarrow AQH_2 + PQ \quad (43)$$

$$AQH_2 + PQ \longrightarrow PQH_2 + AQ \quad (44)$$

$$A\dot{Q}H + P\dot{Q}H \longrightarrow PQH_2 + AQ \quad (45)$$

$P\dot{Q}H$ and $A\dot{Q}H \equiv$ semiquinone radicals

$\dot{R}H \equiv Me\dot{C}HOH$

$R \equiv MeCHO$

[62] J. F. Brennan and J. Beutel, *J. Phys. Chem.*, 1969, **73**, 3245.

25—45 summarize the mechanism. The work is thought to be easily applicable to any pair of quinones as long as there is a redox potential between the respective hydroquinones and quinones.

The photoreduction of the quinone methide (79) and the bis-spirodienone (80) has been previously reported to involve triplet energy trans-

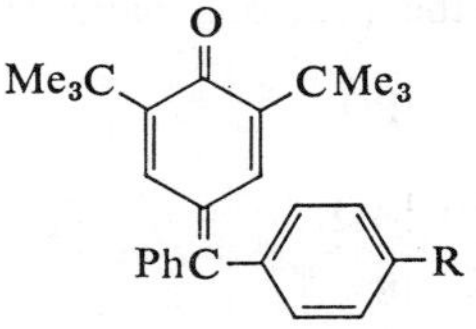

(79)

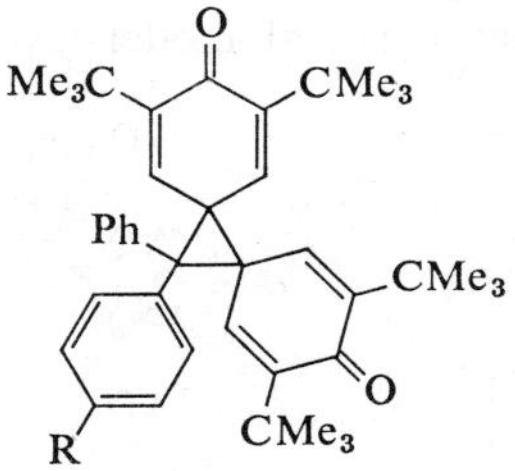

(80)

fer.[63,64] The observation that quinonoid compounds Q are easily reduced to QH_2[65] by diphenylhydroxymethyl radicals generated non-photochemically[66] (equations 46 and 47) has now led to a modification of the mechanism, and a proposal that the reaction proceeds *via* diphenylhydroxymethyl radicals and dimethylhydroxymethyl radicals.[65] The earlier

$$Ph_2C(OH)-C(OH)Ph_2 \xrightarrow{\Delta} 2\,Ph_2C(OH)\cdot \qquad (46)$$

$$Q + 2\,Ph_2C(OH)\cdot \longrightarrow QH_2 + 2\,Ph_2C=O \qquad (47)$$

mechanism[63,67] of energy transfer for the photosensitized addition of phenols to quinone methides now also seems doubtful. A more likely pathway is that the sensitizer (benzophenone or acetophenone) in the addition reaction acts as a hydrogen carrier, as outlined in Scheme 4.

One of the major problems in flavonoid biosynthesis is the role of sunlight, since exposure to light is essential for the formation of anthocyanins. Chalcones have been shown to be the precursors for many classes of flavonoid compounds[68] and hence their photochemistry is of importance. Irradiation of alcoholic solutions of 2-hydroxychalcone (81) with 350 nm radiation is reported to promote an intramolecular rearrangement, and conversion of the carbonyl group to C–OH with the formation

---

[63] H. D. Becker, *J. Org. Chem.*, 1967, **32**, 2115.
[64] H. D. Becker, *J. Org. Chem.*, 1967, **32**, 2136.
[65] H. D. Becker, *J. Org. Chem.*, 1969, **34**, 2472.
[66] D. C. Neckers and A. P. Schaap, *J. Org. Chem.*, 1967, **32**, 22.
[67] H. D. Becker, *J. Org. Chem.*, 1967, **34**, 2131.
[68] H. Grisebach, 'Biosynthetic Patterns in Microorganisms and Higher Plants,' Wiley, New York, 1967.

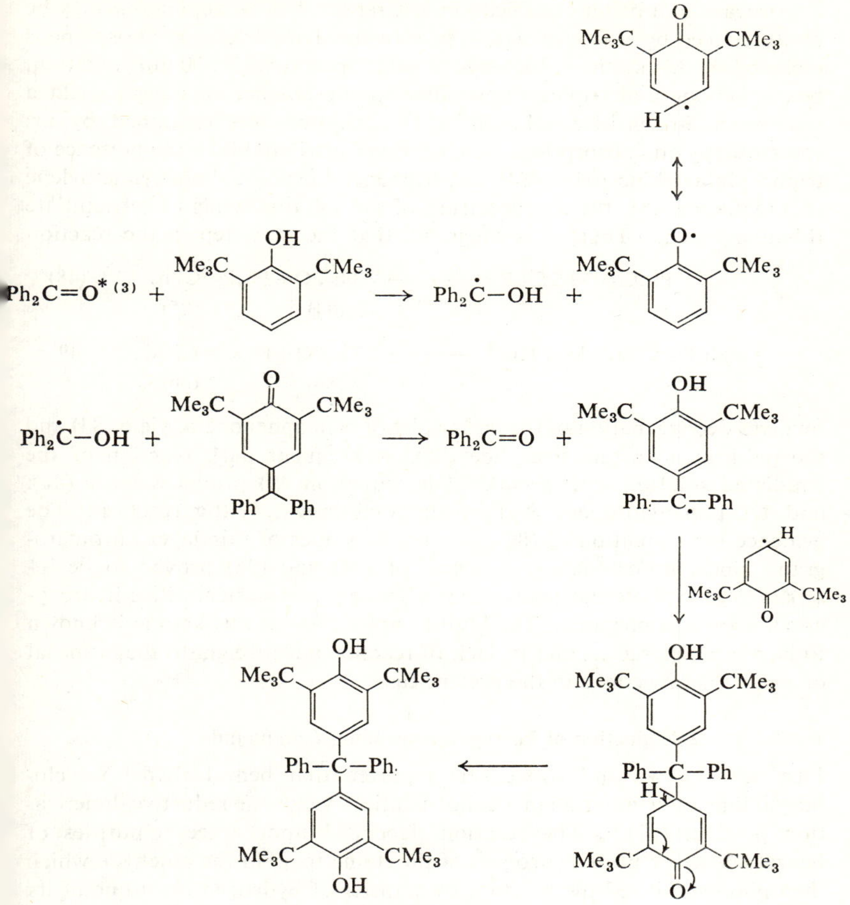

Scheme 4

of 2-alkoxy-flav-3-ene (82).[69] Three products are formed in the reaction; (82) is the major one, with flavone and an unidentified compound making 2% of the total reaction mixture.

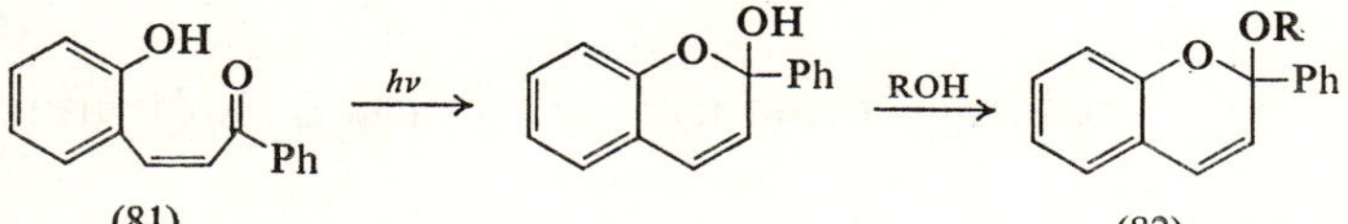

[69] D. Dewar and R. G. Sutherland, *Chem. Comm.*, 1970, 272.

For want of a better classification, the report that benzophenone may be photoreduced by a deoxygenation process involving triphenyl phosphine is included in this section. In benzene solution and with 350 nm radiation, benzophenone and triphenylphosphine in the absence of oxygen yield a red-brown air-sensitive solution.[70] The solution was examined by i.r. spectroscopy and absorptions were observed attributable to the presence of triphenylphosphine oxide (84) and triphenylphosphine diphenylmethylene (85) (equation 48): the u.v. spectrum of the solution tends to substantiate this assignment. Thus, it is suggested that the first step in the reaction

$$Ph_2C=O + 2Ph_3P \longrightarrow Ph_3PO + Ph_3P=CPh_2 \quad (48)$$
$$ (84) \quad\ \ (85)$$

$$Ph_3P=CPh_2 + MeCHO \longrightarrow Ph_3PO + Ph_2C=CHMe \quad (49)$$
$$ (84) \quad\ \ (86)$$

involves deoxygenation of the $n\pi^*$ triplet of benzophenone to yield (84) and the red-brown air-sensitive ylide (85). Subsequent dark reaction of the irradiated solution with acetaldehyde (equation 49) produces more (84), and 1,1-diphenylpropene (86) by the well-known Wittig reaction. The evidence for formation of (86) rests on R_F values of thin-layer chromatograms since, unfortunately, mixtures of (84) and (86) proved to be inseparable. No photoreaction is observed in the present system with 2-hydroxy-4-methoxybenzophenone. The lowest triplet state of this ketone is known to have $\pi\pi^*$ character, and its lack of reaction adds weight to the proposal of $n\pi^*$ triplet reactions in the present case.

2 Reduction of Nitrogen-containing Compounds

Last year Padwa and co-workers reported that benzaldehyde N-cyclohexylimine, on irradiation in alcohol solution, yields the reductive dimerization product (87).[71] The reaction depended upon trace impurities of benzaldehyde or prior hydrolysis of the imine to yield the aldehyde which then photosensitized the reaction by transfer of hydrogen from one of its own reduction intermediates (*e.g.* PhĊHOH) to the imine. A full report of this work and associated reactions has now appeared.[72] Irradiation of a series of benzaldehyde N-alkyl imines in 95% ethanol affords dihydrodimers, whereas the benzophenone N-alkyl imines (88) undergo simple reduction to yield benzhydrylamines (89) rather than the dihydro-dimer

$$Ph_2CO^* + Me_2CHOH \longrightarrow Ph_2\dot{C}OH + Me_2\dot{C}OH \quad (50)$$

$$Ph_2\dot{C}OH + ArCH=NR \longrightarrow Ph_2CO + Ar\dot{C}HNHR \quad (51)$$

---

[70] L. D. Wescott, H. Sellers, and P. Poh, *Chem. Comm.*, 1970, 586.
[71] A. Padwa, W. Bergmark, and D. Pashayan, *J. Amer. Chem. Soc.*, 1968, **90**, 4458.
[72] A. Padwa, W. Bergmark, and D. Pashayan, *J. Amer. Chem. Soc.*, 1969, **91**, 2653.

$$Me_2\dot{C}OH + ArCH=NR \longrightarrow Me_2CO + Ar\dot{C}HNHR \quad (52)$$

$$2Ar\dot{C}HNHR \longrightarrow \underset{\underset{ArCH-NHR}{|}}{ArCH-NHR} \quad (53)$$

$$Ph_2C(OH)-C(OH)Ph_2 + PhCH=NR \xrightarrow{\Delta} Ph_2CO + \underset{\underset{PhCHNHR}{|}}{PhCHNHR} \quad (54)$$

$$Bu^tO-OBu^t + Me_2CHOH + Ph_2C=NR \xrightarrow{\Delta}$$
$$\underset{\underset{Ph}{|}}{PhCHNHR} + Bu^tOH + Me_2CO \quad (55)$$

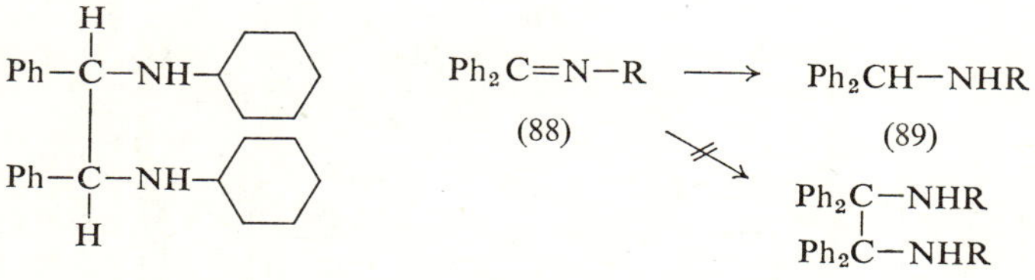

$$\underset{(87)}{\begin{array}{c} H \\ Ph-\overset{|}{C}-NH-\bigcirc \\ Ph-\overset{|}{C}-NH-\bigcirc \\ H \end{array}} \qquad \underset{(88)}{Ph_2C=N-R} \longrightarrow \underset{(89)}{Ph_2CH-NHR}$$

$$\underset{(90)}{\begin{array}{c} Ph_2\overset{|}{C}-NHR \\ Ph_2\overset{|}{C}-NHR \end{array}}$$

(90). This latter observation confirms an earlier literature report.[73] The mechanism for the aryl N-alkyl imine reaction may be summarized by equations 50—53. The tendency of the imino-radicals to dimerize or disproportionate (Scheme 5) is shown in the reduction of the imines by

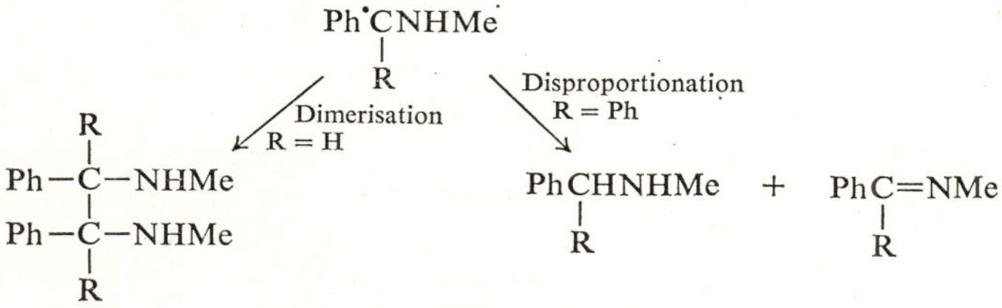

$$\underset{R=H}{\overset{\text{Dimerisation}}{\swarrow}} \underset{R}{\overset{Ph\dot{C}NHMe}{|}} \underset{R=Ph}{\overset{\text{Disproportionation}}{\searrow}}$$

$$\underset{\underset{R}{|}}{Ph-\overset{R}{\underset{|}{C}}-NHMe} \qquad \underset{\underset{R}{|}}{PhCHNHMe} + \underset{\underset{R}{|}}{PhC=NMe}$$

Scheme 5

ketyl radicals generated by chemical methods.[66] Benzophenone ketyl radical with the imine yields imino-radicals with the same tendency to dimerize or disproportionate (equation 54). The thermal decomposition of di-t-butyl peroxide in isopropyl alcohol solutions of the imine causes formation of dihydro-dimers from benzaldehyde N-alkyl imines and amines from benzophenone N-alkyl imines (equation 55). When the

[73] M. Fischer, *Tetrahedron Letters*, 1966, 5273; *Chem. Ber.*, 1967, **100**, 3599.

starting materials are rigorously dried and purified, photoreduction still occurs, and this is attributed to a prior photo-oxidative generation of a carbonyl compound followed by the sensitization process. The lack of reactivity of the imine excited states themselves is suggested to be due to very efficient non-radiative decay to the ground state. It had previously been suggested that irradiation of benzaldehyde *N*-methyl imine in methyl alcohol yields *dl*- and *meso*-1,3-dimethyl-4-diphenylimidazolidines (91) and (92) as primary products.[74] The present investigation indicates that the dihydro-dimer is initially formed, and that this process is followed by condensation of the dimer with formaldehyde formed from the methanol.

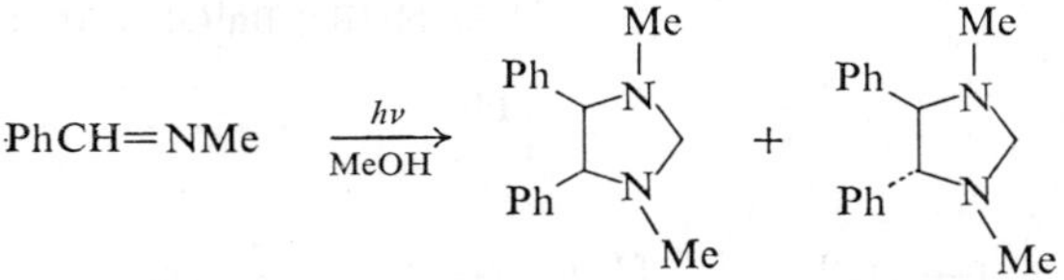

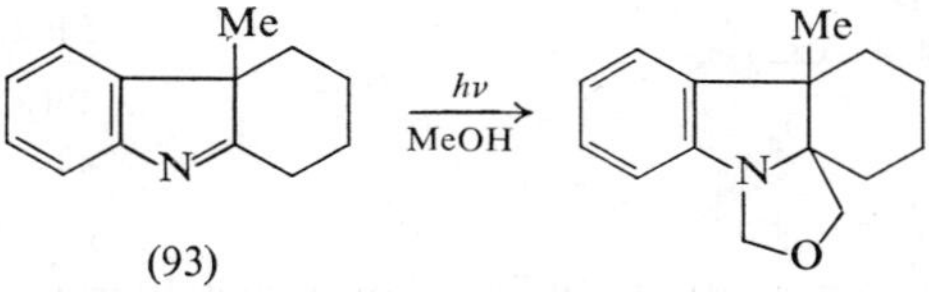

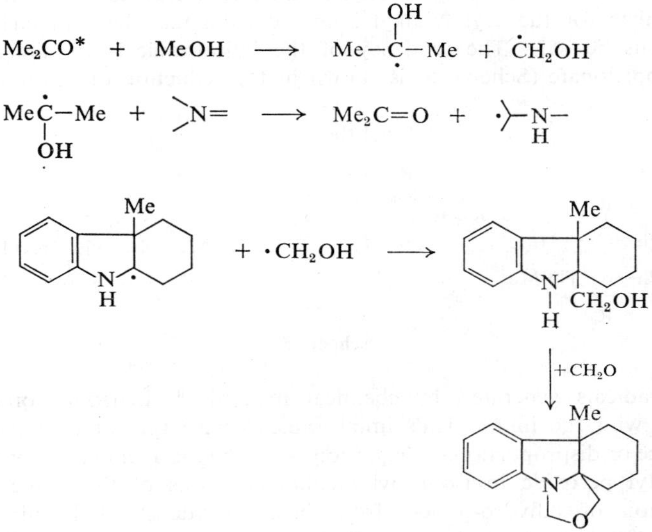

Scheme 6

[74] P. Cerutti and H. Schmid, *Helv. Chim. Acta*, 1962, **45**, 1992; 1964, **47**, 203.

The evidence for this is provided by a dark reaction of the dihydro-dimer with formaldehyde to give the observed products (91) and (92). From this it is suggested that other literature examples of photo-alkylation of imines [e.g. of indolenes (93); see Scheme 6] is best described by this kind of mechanism. The photochemical reactions of imines have also been investigated by reference to the benzophenone-sensitized reaction of benzophenone N-benzyl imine (94) in alcohol solution.[75] With 200 W light intensity, there is a predominant formation of the unsymmetrical dihydro-dimer (95); the dihydro-monomer (96) (i.e. the simple reduction product) is also formed, but significantly the presence of the symmetrical dihydro-dimers (97) and (98) was not detected. When the light intensity

$$Ph_2C=N-CH_2Ph \xrightarrow[Ph_2C=O]{350\,nm, ROH} Ph_2CHNHCH_2Ph$$
(94)   (96)

Ph$_2$CHNHCHPh
|
Ph$_2$CHNHCHPh
(97)

Ph$_2$CHNHCHPh
|
PhCH$_2$NHCPh$_2$
(95)

Ph$_2$C—NHCH$_2$Ph
|
Ph$_2$C—NHCH$_2$Ph
(98)

was increased to 450 W, only (96) and no dimers were formed. The formation of (95) is suggested not to occur entirely by secondary reactions of (96), since irradiation of (96) under similar conditions yields both the symmetric dihydro-dimer (97), as well as (95), in a 1 : 2 ratio. Another report concerns benzophenone photosensitized reactions of N-alkyl imines, viz. benzaldehyde N-methyl imine (99) and benzaldehyde N-cyclohexyl imine (100).[76] Irradiation of these imines in isopropyl alcohol with radiation of wavelength longer than 300 nm yields the dihydro-dimers (101), and (102) respectively,[76] as one would expect from consideration of ref. 71.

Ph—CH=N—Me   Ph—CH=N—C$_6$H$_{11}$ R—NH—CH—CH—NHR
(99)   (100) | |
                                                                 Ph   Ph

(101) R = Me

(102) R = C$_6$H$_{11}$

Witkop has extended his earlier work[77] on the photoreduction of the ethylenic bond in uridine by sodium borohydride to a similar system not containing the sugar residue.[78] Thus with 253·7 nm radiation, 1,3-dimethyluracil (103) in the presence of sodium borohydride yields 5,6-dihydro-1,3-

---

[75] B. Fraser-Reid and E. W. Usherwood, Canad. J. Chem., 1969, 47, 4511.
[76] G. Balogh and F. C. De-Schryner, Tetrahedron Letters, 1969, 1371.
[77] B. Witkop, Photochem. and Photobiol., 1968, 7, 813.
[78] Y. Kondo and B. Witkop, J. Amer. Chem. Soc., 1969, 91, 5264.

dimethyluridine (104) and a 'dimeric' product (not strictly a dimer) which is assigned structure (105), 1,3-dimethyl-5,4'[1',3'-dimethyltetrahydropyrimidone-2']uracil, on the basis of spectroscopic evidence.[78] The reduction of (103) to (104) requires five-fold longer radiation times than similar conversions with uridine and thymidine. Compound (105) is derived from (104) with sodium borohydride in a novel type of light-independent condensation. A by-product of this thermal reaction is γ-(NN'-dimethylureido)propenal-1 (106). 4-Thiouridine (107) is known to undergo reduction with sodium borohydride in the dark to yield 4-ribosyl-2-oxohexahydropyrimidine (108).[79] The reduction is now reported to proceed more rapidly in the presence of light.[80] By a similar procedure the light-induced reduction and desulphuration of 4-thiouracil (109) gives 2-oxohexahydropyrimidine (110), while 2-thiouracil (111) under the same conditions yields 2-thioureidopropanol (112).

[79] P. Cerutti, K. Ikeda, and B. Witkop, *J. Amer. Chem. Soc.*, 1965, **87**, 2505; P. Cerutti, J. W. Holt, and N. Miller, *J. Mol. Biol.*, 1968, **34**, 505.
[80] E. Sato and Y. Kanaoka, *Tetrahedron Letters*, 1969, 3547.

Acridine,[81, 82] quinoline,[83] and purine[84] undergo light-induced addition and reduction (which appear to involve radical intermediates), under conditions where the corresponding hydrocarbons are unreactive. One plausible explanation for this is the involvement of $n\pi^*$ states. However, the absorption and fluorescence spectra show no sign of $n \to \pi^*$ transitions, and are almost identical with the spectra of the corresponding hydrocarbons. A novel photoreduction of, and addition to, the ethylenic bond of 1,2-bis(4-pyridyl)ethylene (113) is now reported.[85] From (113), compounds

$$\text{Py-CH=CH-Py} \quad (113) \xrightarrow[\text{or } C_7H_{14}]{h\nu, \text{Me}_2\text{CHOH}} \quad \text{Py-CH}_2\text{-CH}_2\text{-Py} \quad (114)$$

$$+$$

$$\text{Py-CH}_2\text{-CHR-Py} \quad R = \text{CMe}_2\text{OH or } C_7H_{13} \quad (115)$$

$$\text{Py-CH=CH-Py} \xrightarrow{h\nu, RH} \text{HN=}\langle\rangle\text{=CH-ĊH-Py} \quad (56)$$

$$\text{=CH-ĊH-Py} \xrightarrow{RH \text{ or } R\cdot} \text{HN=}\langle\rangle\text{=CH-CHR(H)-Py} \longrightarrow (114) \text{ and } (115) \quad (57)$$

(114) and (115) are formed: stilbene does not, of course, undergo this reaction. These results suggest the presence and reaction of $n\pi^*$ states, inaccessible by direct absorption and which have lower excitation energies than the initially formed fluorescent $\pi\pi^*$ singlets. It is suggested that the reduction and addition occur via hydrogen abstraction by nitrogen followed by rearrangement (equations 56 and 57). The $n\pi^*$ singlet is responsible for these reactions, and isomerization arises from the triplet states.

The photoreduction of 4-diethylamino-4'-nitroazobenzene (116) has been studied in detail. The reaction has been investigated by direct

[81] H. Göth, P. Cerutti, and H. Schmid, *Helv. Chim. Acta*, 1965, **48**, 1395.
[82] A. Kira, S. Kato, and M. Koizumi, *Bull. Chem. Soc. Japan*, 1966, **39**, 1221.
[83] F. R. Stermitz, C. C. Wei, and W. H. Huang, *Chem. Comm.*, 1968, 482.
[84] H. Linschitz and J. S. Connolly, *J. Amer. Chem. Soc.*, 1968, **90**, 2979.
[85] D. G. Whitten and Y. J. Lee, *J. Amer. Chem. Soc.*, 1970, **92**, 415.

$O_2N-\langle C_6H_4 \rangle-N=N-\langle C_6H_4 \rangle-NEt_2$

(116)

$O_2N-\langle C_6H_4 \rangle-NH-NH-\langle C_6H_4 \rangle-NEt_2$

(117)

$H_2N-\langle C_6H_4 \rangle-N=N-\langle C_6H_4 \rangle-NEt_2$

(118)

irradiation at 254, 310, and 366 nm in alcoholic solutions, when 4-diethylamino-4′-nitrohydrazobenzene (117) is formed with a low quantum yield.[86] There is no evidence for the formation of the nitro-reduction product (118) in the present system. Both naphthalene and benzene are inefficient as sensitizers, and there is no isomerization in the direct or attempted sensitized process. The effect of ketonic sensitizers on the conversion of (116) to (117) is also reported, and both acetone and benzophenone are found to be efficient in the presence of hydrogen donors such as benzhydrol or isopropyl alcohol.[87] On the other hand, fluorenone is an inefficient sensitizer for the reaction. Two possible mechanisms for the ketonic sensitization

$$B \xrightarrow{h\nu} B^* \longrightarrow {}^3B \quad (58)$$

$${}^3B \longrightarrow B \quad (59)$$

$${}^3B + BH_2 \longrightarrow 2BH\cdot \quad (60)$$

$$BH\cdot + A \longrightarrow AH\cdot + B \quad (61)$$

$$AH\cdot + BH\cdot \longrightarrow AH_2 + B \quad (62)$$

$$AH\cdot + BH\cdot \longrightarrow A + BH_2 \quad (63)$$

are suggested: (a) energy-transfer from the triplet state of the ketone to the reactive triplet state of (116), and (b) transfer of a hydrogen atom from the ketyl radical (formed by hydrogen abstraction from the solvent by triplet ketone) to ground state (116). The latter process is known to be important in other systems.[88,89] Path (b) is suggested to be the major process and the reaction mechanism may be represented by equations 58—63, where B is benzophenone, A the azo-compound, and BH_2 benzhydrol. Chain-transfer of the hydrazyl radical (AH·) with benzhydrol and termination by a combination of two ketyl radicals, are precluded by the first-order dependence of the reaction on light intensity. The mechanism accounts for the inability of benzene and naphthalene to sensitize the process, and since fluorenone is known to be inefficient at hydrogen abstraction, these

[86] G. Irick and J. G. Pacifici, *Tetrahedron Letters*, 1969, 1303.
[87] J. G. Pacifici and G. Irick, *Tetrahedron Letters*, 1969, 2207.
[88] A. Padwa, W. Bergmark, and D. Pashayan, *J. Amer. Chem. Soc.*, 1968, **90**, 4458.
[89] E. Vander-Donckt and G. Porter, *J. Chem. Phys.*, 1967, **46**, 1173.

data give further evidence against a triplet sensitization process. The products and kinetics of the photoreduction of (116) have been found to be markedly influenced by the nature of the solvent, and the results in alcohols and butylamine allow a distinction to be made between a mechanism involving hydrogen abstraction by the excited state of (116), and an electron transfer to the excited state of (116).[90] The reduction in butylamine yields none of (117), but a product is formed which has its u.v. absorption maximum at 420 nm with a shoulder at 460 nm. There is no evidence for the formation of such a compound in the photolysis of (116) in alcohols. Admission of air to the irradiated butylamine solution converted the product to a species absorbing at 507 nm. This second product is identified as 4,4'-bis{[p-(diethylamino)phenyl]azo}azoxybenzene (119), so the first formed compound is suggested to be the hydroxylamine

$$Et_2N-\langle\rangle-N=N-\langle\rangle-\overset{O}{\overset{\uparrow}{N}}=N-\langle\rangle-N=N-\langle\rangle-NEt_2$$

(119)

$$\underset{H}{HON}-\langle\rangle-N=N-\langle\rangle-NEt_2$$

(120)

(120). Thus, in butylamine the reaction takes a completely different pathway from that in alcohols, and photoreduction of the nitro-group occurs.[91] This selective nitro-group reduction is rationalised by an electron-transfer process.[92] It is interesting to note that both benzophenone and fluorenone sensitize this reaction, indicating that both of the ketyl radical anions formed in this medium transfer an electron rapidly to (116) before proton transfer from the amine radical-cation occurs. Products resulting from interaction of the ketyl radical with (116) would be formed if proton transfer to the ketyl radical-anion competed significantly with electron transfer to (116).

The photoreduction of nitrobenzene in the presence of hydrochloric acid is known to yield aniline, p-aminophenol, and p-chloroaniline.[93] Similar reaction with 1-nitronaphthalene has now been reported when the product is 4-chloro-1-naphthylamine.[94] The reactions are carried out with 366 nm radiation in 50% isopropyl alcohol–water mixtures. There is no detectable reduction in pure isopropyl alcohol and the highest quantum

[90] J. G. Pacifici, G. Irick, and C. G. Anderson, *J. Amer. Chem. Soc.*, 1969, **91**, 5654.
[91] R. Hurley and A. C. Testa, *J. Amer. Chem. Soc.*, 1966, **88**, 4330.
[92] S. G. Cohen and J. I. Cohen, *J. Phys. Chem.*, 1968, **72**, 3782.
[93] J. A. Barltrop and N. J. Bunce, *J. Chem. Soc.* (C), 1968, 1467.
[94] W. Trotter and A. C. Testa, *J. Phys. Chem.*, 1970, **74**, 845.

yield for disappearance of the nitro-group is 1.28×10^{-2} for 6M hydrochloric acid in the solvent mixture. The results of the study are interpreted in terms of protonation of nitronaphthalene triplet (equations 64—68). Although protonation of both singlet and triplet excited states may account for the current data, the former seems less likely because of the very short lifetime of this state of the naphthalene, and the usual involvement of the triplet state in hydrogen abstraction. The photoreduction of 1-nitronaphthalene and nitrobenzene by tri-n-butylstannane was reported last year.[95]

$$N + h\nu \longrightarrow {}^1N^* \rightsquigarrow {}^3N^* \quad (64)$$

$$^3N^* + H^+ \rightleftharpoons {}^3NH^{+*} \quad (65)$$

$$^3N^* + Me_2CHOH \longrightarrow\!\!/\!\!/\!\!\longrightarrow \quad (66)$$

$$^3N^* \rightsquigarrow N \quad (67)$$

$$^3NH^{+*} + MeCHOH \longrightarrow \dot{N}H + Me_2\dot{C}OH + H^+ \quad (68)$$
$$\searrow \text{Product}$$

$$N \equiv \text{Nitronaphthalene}$$

There have been several reports over the years concerned with the photoreduction of acridine, and intermediates for the process have been discussed. The reduction product has now been examined by flash photolysis in benzene and methanol solutions, and a transient, characterized by its 350 and 560 nm absorption, has been reported.[96] This transient species is different from the acridine semiquinone reported already, and is identified as another acridine semiquinone radical since it is converted to acridine, and its low reactivity to oxygen denies the possibility of triplet acridan. Thus, it is concluded that there are two isomeric forms, the C-radical (121)

(121)   (122)

and the other transient which is most reasonably assigned the N-radical structure (122). The transient radical which does not show the present 560 nm absorption is concluded to be the C-radical since it is formed from biacridan which could only yield this radical.

The lowest-lying excited state of unprotonated 1- and 2-piperidinoanthroquinones has been shown to be of the intramolecular charge-transfer type, and with protonation of the 1-derivative, at least, the band corresponds to an $n \to \pi^*$ transition.[97] Thus, there are marked differences in the photoreactivity of unprotonated and protonated forms of the quinone.

[95] W. Trotter and A. C. Testa, *J. Amer. Chem. Soc.*, 1968, **90**, 7044.
[96] A. Kira and M. Koizami, *Bull. Chem. Soc. Japan*, 1969, **42**, 625.
[97] G. O. Phillips, A. K. Davies, J. F. McKellar, and D. Price, *Chem. Comm.*, 1969, 1097.

Photo-oxidation and -reduction Reactions

Photolysis of neutral anaerobic solutions of 2-piperidino-anthroquinone yields the anthrahydroquinone (AH_2).[98] At pH values greater than 11, similar irradiation forms the radical anion ($A \cdot ^-$), which is stable for periods in excess of twelve hours, and has been characterized by u.v. and e.s.r. spectroscopy.[99] It is estimated that the ratio of the quantum yield for $A \cdot ^-$ formation at pH 12 and AH_2 at pH 7 is greater than 100, and this magnitude indicates two distinct mechanisms for the photoreduction at the two pHs. At the higher pH it is postulated that formation of $A \cdot ^-$ from the 2-derivative proceeds mainly via an electron-transfer process from hydroxyl or alkoxyl ion to the charge-transfer excited state responsible for the reduction (equation 69). On the other hand, the 1-derivative did not give a similar reaction at high pH, but formation of the dianion A^{2-} occurred and appeared to proceed via the well established process shown in equation 70.[100] This is then followed by $AH \cdot \rightarrow A \cdot ^- + H^+$ and $2A \cdot ^- \rightarrow A^{2-} + A$. Recent work has shown that 2-piperidino-anthroquinone is much more efficient than

$$A^* + AH^- \longrightarrow A \cdot ^- + OH \cdot \qquad (69)$$

$$A^* + RCH_2OH \longrightarrow AH \cdot + R\dot{C}HOH \qquad (70)$$

the 1-derivative at sensitizing the photodegradation of nylon fibre.[101] This situation can be explained by electron transfer from the polymer to the photoexcited quinone which, like the electron-transfer process in the reduction, proceeds more readily with the 2-derivative.

The irradiation of 1-methyl-6,7-dimethoxy-1,2,3,4-tetrahydroisoquinoline-N-tosylates (123) in neutral or basic (Na_2CO_3) ethanol solution is reported to lead to detosylation, and the formation of 3,5-dihydro-compounds (124).[102] The photoreaction of (123) and similar compounds (125) with sodium borohydride has been studied in the same media, and

(123) R = Me
(125) R = Ph, CH_2Ph

---

[98] G. O. Phillips, A. K. Davies, and J. F. McKellar, *Chem. Comm.*, 1970, 519.
[99] J. H. Sharp, T. Kuwana, A. Osborne, and J. N. Pitts, *Chem. and Ind.*, 1962, 508; P. J. Baugh, G. O. Phillips, and J. C. Arthur, *J. Phys. Chem.*, 1966, **70**, 3061.
[100] N. K. Bridge and G. Porter, *Proc. Roy. Soc.*, 1958, A, **244**, 259, 276; C. F. Wells, *Trans. Faraday Soc.*, 1961, **57**, 1703, 1719.
[101] G. S. Egerton, N. E. N. Assaad, and N. D. Uffindell, *J. Soc. Dyers and Colourists*, 1967, **83**, 409.
[102] B. Umezawa, O. Hoshino, and S. Sawaki, *Chem. Pharm. Bull. Tokyo*, 1969, **17**, 1115.

gives the amines (126) in good yields.[103] Photoreductive cleavage is also observed with the acyclic toluene-*p*-sulphonamides (127) and (128) to yield benzylamine and aniline respectively.[104]

The reinvestigation of early reports that potential anti-malarial compounds are formed in the u.v. irradiation of aqueous acid solutions of quinine (129)[105] has resulted in the characterization of a type of reduction which seemingly does not conform with previously known processes.[106] The reaction was investigated in 2N hydrochloric acid–isopropyl alcohol solution and gave up to a 32% yield of deoxyquinine (130). Quinidine undergoes an analogous reaction to yield deoxyquinidine (131). The proposed mechanism for the transformation is shown in Scheme 7. The

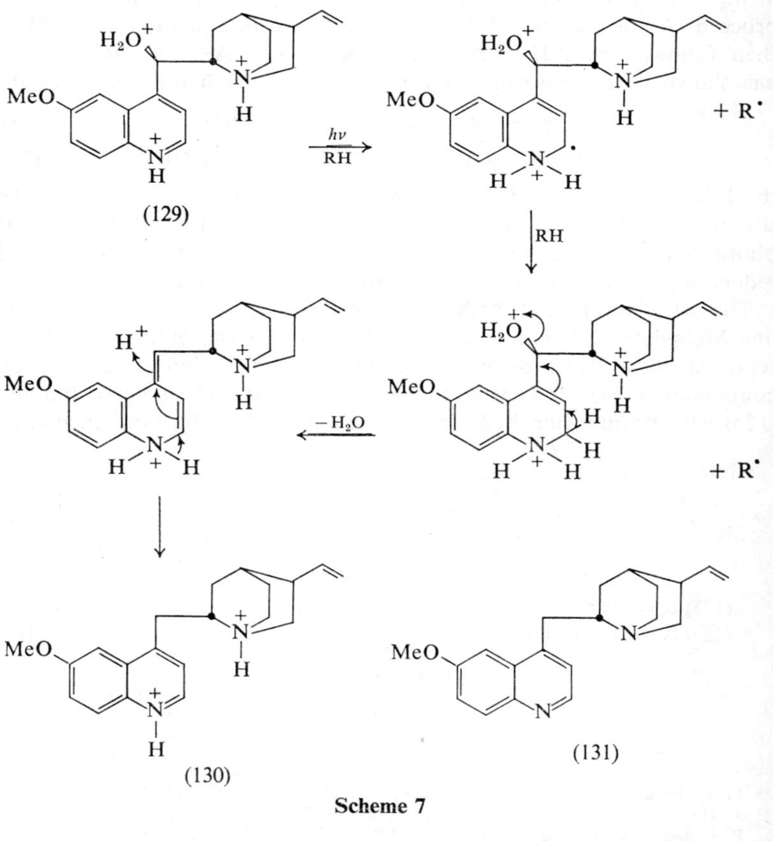

Scheme 7

[103] B. Umezawa, O. Hoshino, and S. Sawaki, *Chem. Pharm. Bull. Tokyo*, 1969, **17**, 1120.
[104] O. Hoshino, S. Sawaki, and B. Umezawa, *Chem. Pharm. Bull. Tokyo*, 1970, **18**, 182.
[105] G. C. Kyker, M. M. McEwen, and W. E. Cornatzer, *Arch. Biochem.*, 1947, **12**, 191, and refs. cited therein.
[106] V. I. Stenberg, E. F. Travecedo, and W. E. Musa, *Tetrahedron Letters*, 1969, 2031.

presence of a hydrogen donor enhances the yield of (130), which was greatest in relatively concentrated acid solutions. In dilute acid solutions (129) undergoes a rapid photoreaction to yield mainly polymeric products. The remaining asymmetric centres of deoxyquinine and deoxyquinidine survived the irradiation procedure. Testing of the anti-malarial properties of these photoproducts is reported to be in hand.

1,3-Disubstituted thioparabanates (132) are all yellow and have an $n \rightarrow \pi^*$ transition at 400—440 nm and a $\pi\pi^*$ transition at approximately 300 nm. Irradiation of their ethanol solutions is now reported to yield 2-mercaptoimidazolidine-4,5-dione (133), *i.e.* reduction of the thioketone in preference to carbonyl reduction.[107] The alcohol photoaddition product, 2-(1-hydroxyethyl)-2-mercaptoimidazolidine-4,5-dione (134), is also formed. The reaction is reported to be initiated by excitation of the $n \rightarrow \pi^*$ transition of the lone pair in the 2-thiocarbonyl group. In the same report the addition of amylene (2-methylbut-2-ene) to 1,3-dimethylthioparabanate to give the thio-oxetan (135) is also described. Other thioparabanates do not apparently undergo the cycloaddition process, but no reasons have been given.

The role of the hydrated electron in the photoreduction of cysteine in the presence of indole has been investigated by flash photolysis.[108] The (indole)$^+$ band is observed, together with a new band at 410 nm which is

---

[107] T. Yonezawa, M. Matsumoto, Y. Matsumura, and H. Kato, *Bull. Chem. Soc. Japan*, 1969, **42**, 2323.
[108] L. I. Grossweiner and Y. Usui, *Photochem. and Photobiol.*, 1970, **11**, 53.

attributed to the cysteine radical anion (equation 71). It would seem that any aromatic compound which can provide photochemically hydrated electrons can reduce aqueous solutions of cysteine.

$$e_{aq} + RSSR \longrightarrow RSSR \cdot^- \qquad (71)$$

3 Oxidation of Aromatic Compounds

Many of the oxygenations described in this and other sections of this chapter involve the dye-sensitized formation of singlet oxygen and its subsequent reaction with the organic substrate. The theoretical and experimental aspects of the role of singlet oxygen in the sensitized photo-oxygenation have been discussed in a paper by Kearns and Khan.[109] Various factors which control the generation and reaction of singlet oxygen are considered in detail. A relatively simple theoretical procedure is developed to predict the relative reactivities of $^1\Sigma$, $^1\Delta$, and $^3\Sigma$ oxygen towards various organic acceptors, and is used to discuss the chemical and phtotchemical properties of some of the oxygenated products. This publication also considers the properties of dioxetans in connection with the role which they may play in chemi- and bio-luminescence. In dye-sensitized oxygenations, it is useful to know the triplet energy of the dye sensitizers in order to ensure that other sensitization processes are unlikely to occur. Acridines, isoalloxazines, and xanthenes are typical sensitizers for oxygenation. Measurement of the triplet energies is technically difficult since the dyes often form dimers and higher aggregates, and are difficult to purify. A study of the phosphorescence of such dyes at 77 K in ethanol–methanol, 9 : 1, has been reported.[110] The data suggest that an approximate value of the triplet energy may be obtained by subtracting given amounts from the energies of the lowest excited singlet states in each class of dyes. In practice, with reactions involving sensitizers, the solutions are so concentrated that dimers are probably the sensitizing species. For calculation of the location of dimer triplet states, preliminary studies have indicated that a given additional amount should be abstracted from the monomer energies. Literature discrepancies with current triplet energy data are indicated.

The photo-oxidation of air-saturated neat liquid toluene has been studied with 253·7 nm and 313 nm radiation, and it is claimed that the results indicate the course of the reaction to be strongly influenced by the presence of a charge-transfer complex between toluene and oxygen.[111] The reaction does not proceed *via* a chain process, and attack by singlet excited oxygen does not contribute materially to the oxidation. Benzaldehyde, benzyl alcohol, and benzoyl hydroperoxide are the major products from 253·7 nm radiation, and contrary to previous reports [112] the formation

---

[109] D. R. Kearns and A. U. Khan, *Photochem. and Photobiol.*, 1969, **10**, 193.
[110] R. W. Chambers and D. R. Kearns, *Photochem. and Photobiol.*, 1969, **10**, 215.
[111] K. S. Wei and A. H. Adelman, *Tetrahedron Letters*, 1969, 3297.
[112] E. J. Bowen and A. H. Williams, *Trans. Faraday Soc.*, 1939, **35**, 765; A. J. Krasnovskii, *J. Gen. Chem.* (*USSR*), 1940, **10**, 1094.

of benzoic acid was not observed in the present reaction, nor was there any evidence for the formation of a transannular peroxide of toluene. The quantum yield for the combined processes was 0.08 ± 0.01 at room temperature. The individual quantum yields of benzaldehyde and benzyl alcohol are little affected by increase in temperature but the quantum yield for hydroperoxide formation increases from approximately 0·01 at 15 °C to approximately 0·05 at 90 °C. The value of 0·08 at room temperature is increased to 0·12 when the oxygen concentration is increased from 1.8×10^{-3} mol l^{-1} to 9.0×10^{-3} mol l^{-1}. Pure toluene does not absorb radiation of wavelengths longer than ca. 280 nm, but in toluene saturated with oxygen there appears an additional absorption extending to 350 nm which the authors attribute to contact charge-transfer between toluene and oxygen.* It has recently been reported that irradiation in this latter region causes a slow photoreaction with the consumption of oxygen.[113] The reaction has now been reinvestigated using 313 nm radiation, and benzaldehyde, benzyl alcohol, and benzoyl hydroperoxide were again formed in ratios nearly identical with those from the 253·7 nm radiation. Thus, the authors contend that at both wavelengths the oxygen–toluene 'complex' is involved in the reaction. At 253·7 nm the toluene absorbs the radiation, and the reaction is said to proceed by the mechanism outlined in equations 72—76. The

$$\text{Ph—Me} \xrightarrow{h\nu} \text{PhCH}_3{}^*, \tag{72}$$

$$\text{PhCH}_3{}^* + (\text{PhCH}_3.\text{O}_2) \longrightarrow \text{PhCH}_3 + (\text{PhCH}_3.\text{O}_2)^* \tag{73}$$

$$(\text{PhCH}_3.\text{O}_2)^* \longrightarrow \text{PhCHO} + \text{H}_2\text{O} \tag{74}$$

$$\longrightarrow \text{PhCH}_2\text{O}_2\text{H} \tag{75}$$

$$\longrightarrow \text{PhCH}_2\text{O}\cdot + \text{OH}\cdot \tag{76}$$

benzoyl hydroperoxide dissociates photochemically to benzaldehyde and benzyl alcohol, although it is thermally stable. Although the concentration of the 'complex' is said to decrease with increase in temperature, there is no reduction in the photo-oxidation rate: this might be accounted for by the marked compensating increase in the rate of the propagation steps with temperature.[111] In the Reporters' opinion, more work will be needed before the reality of the proposed 'complex' and its necessary involvement in these processes can be accepted without reserve. The possible intermediacy of free benzyl radicals (which are certainly formed by gas-phase photolysis of toluene) will need to be seriously considered.

It has previously been shown that cleavage of 3-hydroxyflavones by photosensitized oxygenation, possibly by singlet oxygen, provides a model for the enzymatic degradation of quercetin by the action of

[113] J. C. W. Chien, *J. Phys. Chem.*, 1966, **69**, 4317.

* Note that oxygen-promoted $S_0 \to T_1$ absorption of toluene would also appear in this region. One must also beware of referring to a 'charge-transfer complex' when contact charge-transfer is involved.

dioxygenase.[114] The same workers have now reported additional examples of the reaction which provide models for the enzymatic cleavage of aromatic rings.[115] Thus, irradiation of 4,6-di-t-butylresorcinol (136) in methanol in the presence of oxygen and the dyestuff Rose Bengal yields the keto-ester (137) (16%) and the lactone (138) (6%). Two moles of oxygen are consumed in the process. In the absence of sensitizer, approximately 1·5 moles of oxygen were slowly absorbed after a long induction period to yield a complex mixture which contained no (137) or (138). The reaction, which represents a possible model for enzymatic cleavage of homogentisic acid,[116] is outlined in Scheme 8. The photo-oxidative cleavage of (136) is

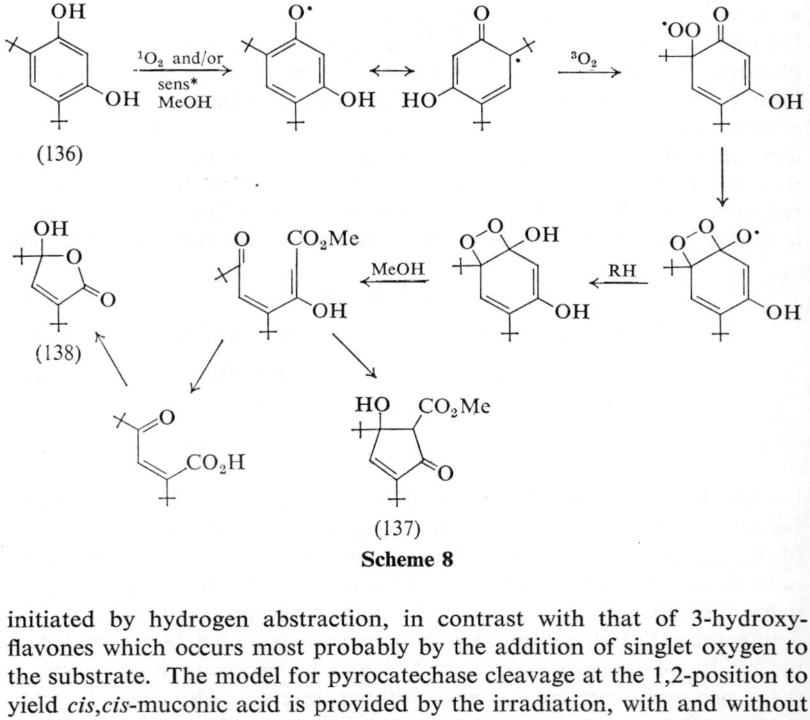

Scheme 8

initiated by hydrogen abstraction, in contrast with that of 3-hydroxy-flavones which occurs most probably by the addition of singlet oxygen to the substrate. The model for pyrocatechase cleavage at the 1,2-position to yield *cis,cis*-muconic acid is provided by the irradiation, with and without sensitizer, of 3,5-di-t-butylpyrocatechol (139), when the acid (140) and the ester (141) are formed. An examination of the photo-oxygenation of 2,5-di-t-butylhydroquinone, again in the presence and absence of Rose Bengal, showed that products (142) and (143) were formed in both cases.

[114] T. Matsuura, H. Matsushima, and H. Sakamoto, *J. Amer. Chem. Soc.*, 1967, **89**, 6370.
[115] T. Matsuura, A. Nishinaga, N. Yoshimura, T. Arai, K. Omura, H. Matsushima, S. Kato, and I. Saito, *Tetrahedron Letters*, 1969, 1673.
[116] D. I. Crandall, R. C. Krueger, F. Anan, K. Yasunobu, and H. S. Mason, *J. Biol. Chem.*, 1960, **235**, 3011.

The reactions (Scheme 9) were shown to involve the intermediacy of 2,5-di-t-butyl-*p*-benzoquinone (144).

The first example of singlet oxygen addition to monocyclic rings has been observed by the Rose Bengal-sensitized oxygenation of the monomethyl ether of the resorcinol (136) in methanol solution.[117] The hydro-

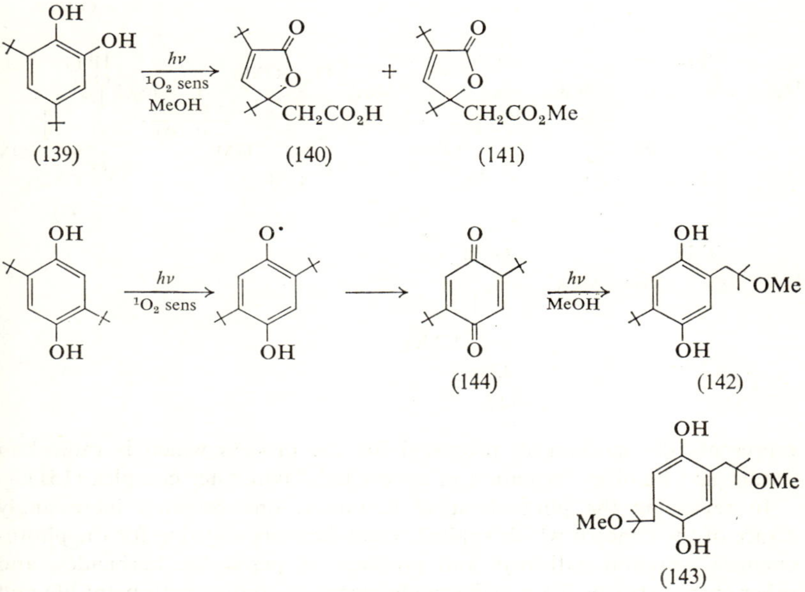

Scheme 9

peroxide (145) is formed in 48% yield. The dimethyl ether (146) under similar conditions yields the epoxyketone (147) (70%) and both ethers are inert in the absence of sensitizer. Since the ethers give similar results from hydrogen peroxide and hypochlorite, the photoreaction is suggested to involve singlet oxygen. The corresponding diethers (148) and (149) of the catechol and quinol are resistant to oxidation, but the monomethyl ether (150) of the quinol undergoes photo-oxidation to form di-t-butyl-*p*-benzoquinone and formaldehyde.

Photo-oxidative decarboxylation of dihydrophthalates is reported to be sensitized by both riboflavin and 3-methyl-lumiflavin.[118] The products of the reaction of the aromatic methyl ester are benzoic acid and methyl benzoate. The reaction was found to be favoured by ionisation of one of the carboxy-groups, but hindered by ionisation of the flavin. Scheme 10

[117] I. Saito, S. Kato, and T. Matsuura, *Tetrahedron Letters*, 1970, 239.
[118] G. D. Weatherby and D. O. Carr, *Biochemistry*, 1970, **9**, 344.

represents the mechanism proposed for the process which is radical in nature and involves formation of an excited flavin–ester complex (151).

In reviewing the photochemical literature, one becomes increasingly aware of the concern which applied researchers are showing for the photochemical reaction pathways and products of pesticides, herbicides, and other chemicals which are deliberately placed in contact with plant life and hence sunlight. The photodecomposition of [1,1,1-trichloro-2,2-bis(*p*-chlorophenyl)ethane] (DDT) has been widely discussed.[119] The photo-

[119] A. R. Mosier, W. D. Guenzi, and L. L. Miller, *Science*, 1969, **164**, 1083.

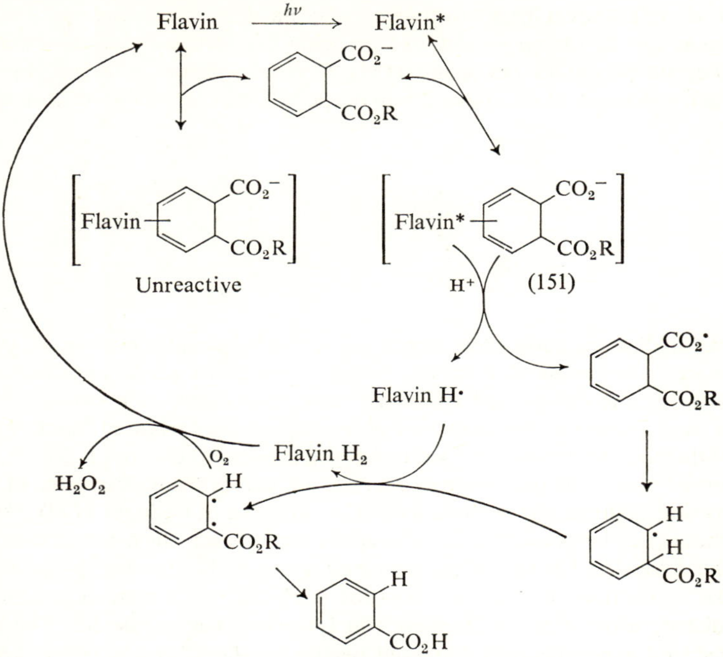

Photo-oxidation and -reduction Reactions

Scheme 10

oxidations of DDT and its metabolite DDE in methanol solution have now been examined.[120] The process involves the photogeneration of free radicals which may abstract hydrogen from the solvent, react with oxygen, or abstract hydrogen from unreacted substrate. Secondary decompositions also occur. The oxidation products are benzoic acid, aromatic ketones, and chlorinated phenols. DDE also undergoes photocyclization to yield the dichlorofluorenone (152).

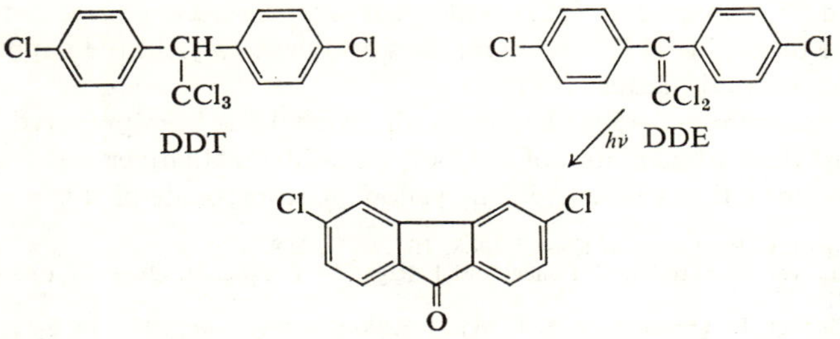

[120] J. R. Plimmer, U. I. Klingebiel, and B. E. Hummer, *Science*, 1970, **167**, 67.

It is well known that anthracene and its derivatives undergo photo-oxygenation involving singlet oxygen with the formation of 1,4- and/or 9,10-cyclic peroxides (153) and (154).[121] The position of attack by the oxygen is dependent upon the nature of the substituents X and Y. Attempts

(153)   (154)

to rationalize the substituent effect, using the Wheland model or an intermediate complex similar to the initial state, have given unsatisfactory results.[122] The delocalized transition state model described earlier [123] has now been applied to the formation of photoperoxides from anthracene and naphthalene derivatives.[124] In the present system the term ΔE is the difference between the electron energies of the delocalized model of the transition state of the 9,10-adduct (154) and the 1,4-adduct (153). With naphthalene compounds, ΔE is the difference between the electron energies of the delocalized model of the transition state of the 5,8- and 1,4-adducts. If ΔE is positive, the 9,10- (or 5,8- for naphthalene derivatives) adduct is predicted, while if ΔE is negative the 1,4-product is predicted. The predicted and experimentally observed results tabulated in this report are in complete agreement.

Stevens and Algar have studied the photoperoxidation of 9,10-dimethylanthracene and 9,10-dimethyl-1,2-benzanthracene in benzene sensitized by azulene, anthanthrene, and perylene, as a function of dissolved oxygen concentration.[125] The conclusion from this work is that the singlet oxygen utilized in the reaction is produced solely by energy-transfer from the sensitizer triplet state even when transfer from the excited singlet state is spin-allowed and exothermic.

The 1,4-peroxide (155) of 1,4-dibenzyloxyanthracene (156) has been reported.[126] Compound (155) undergoes light-induced rearrangement to yield the diepoxide (157) and eliminates acetylene to yield the naphthalene dicarboxylic acid diester (158).

Pure helianthrene (159) (dibenzo[a,o]perylene) has been prepared in 50% yield by dehydrogenation of 7,16-dihydro-dibenzo[a,o]perylene, and its photo-oxygenation studied.[127] The yellow photoperoxide of this polycyclic

[121] S. Rigaudy, *Pure Applied Chem.*, 1968, **16**, No. 1, 169.
[122] O. Chalvet, R. Daudel, C. Ponce, and J. Rigaudy, *Internat. J. Quantum Chem.*, 1968, **2**, 521.
[123] O. Chalvet, R. Daudel, and T. F. W. McKillop, *Tetrahedron*, 1970, **26**, 349.
[124] O. Chalvet, R. Daudel, G. H. Schmid, and J. Rigaudy, *Tetrahedron*, 1970, **26**, 365.
[125] B. Stevens and B. E. Algar, *J. Phys. Chem.*, 1969, **73**, 1711.
[126] J. Rigaudy, R. Dupont, and K. C. Nguyen, *Compt. rend.*, 1969, **269**, C, 416.
[127] H. Brockmann and F. Dicke, *Chem. Ber.*, 1970, **103**, 7.

aromatic compound has been assigned structure (160) and consequently the other yellow photoperoxides of the derivatives of (159) [*i.e.* (159a—e)] are assigned structures (160a—e).

A study has been made of the methylene-blue-sensitized photo-oxygenation of aryl ethylene compounds (161) to carbonyl compounds.[128] The similar photoreaction of bifluorenylidene (162) has been the subject of two reports this year. Richardson and Hodge have carried out competitive oxidation of (162) with 2-methylbut-2-ene in both the thermal oxidation with hydrogen peroxide and hypochlorite, and the photochemical reaction using methylene blue as the sensitizer.[129] The chemical reaction gave 44% of fluorene (163) from (162) and 41% of reaction of the aliphatic olefin while in the photoprocess the yields were 96% and 66% respectively. Compound (162) is reported to be 0·034 times as reactive as the most reactive olefin to singlet oxygen, and 720 times as reactive as the least reactive olefin. Thus, allylic hydrogens are not a prerequisite for facile reaction of singlet oxygen with olefins. The present reaction is suggested to proceed by the dioxetan (164), which is not isolated. The intermediacy of

---

[128] G. Rio and J. Berthelot, *Bull. Soc. chim. France*, 1969, 3609.
[129] W. H. Richardson and V. Hodge, *J. Org. Chem.*, 1970, **35**, 1216.

$$\underset{R}{\overset{Ph}{>}}C=C\underset{R^2}{\overset{R^1}{<}} \xrightarrow[{}^1O_2]{h\nu} \underset{R}{\overset{Ph}{>}}C=O \ + \ O=C\underset{R^2}{\overset{R^1}{<}}$$

(161) $\quad\quad\quad\quad\quad\quad$ R, R^1, R^2 = H, Ph

(162) $\xrightarrow{{}^1O_2}$ [(164)] $\searrow$ (163)

(164) in the photo-oxidation of (162) is not suggested in the second report although the reaction is again reported to yield fluorenone.[130] The reaction proceeds with wavelengths shorter than 300 nm in air. In the absence of oxygen, cyclo-dehydrogenation occurs to yield fluoreno(1,9a,9-a,b)fluoranthrene (165) and the reduced product 9,9'-bifluorene (165) in very low yield. In another report, the unusual photoreduction behaviour of (162) to form (166) is described.[131] The quantum yield for photoreduction of aromatic compounds in solution is usually wavelength independent owing to fast internal conversion and vibrational relaxation in excited states. But 9,9'-bifluorenylidene (162) in hydrogen donor solvents differs in this respect,[130] and a mechanism has now been proposed for this reaction which is dependent on the exciting wavelength. Although hydrocarbon (162) has a strong absorption in the visible spectrum (λ_{max} = 455 nm), in the absence of sensitizer, the reduction only occurs at wavelengths shorter than 315 nm. The unsensitized reaction is autocatalytic and it appears that (165) (AH_2) is able to sensitize the reaction. This situation might be rationalized by either Scheme 11 or 12, but since the use of 9,9'-bifluorene deuteriated in the 9,9'-position (*i.e.* AD_2) does not lead to any of the mixed product (ADH),

$A^* + AH_2 \longrightarrow 2AH\cdot$ $\quad\quad$ $2AH_2^* + A \longrightarrow AH_2 + A^*$

$AH\cdot \xrightarrow{Solvent} AH_2$ $\quad\quad\quad\quad$ $A^* \xrightarrow{Solvent} AH_2$

$\quad\quad$ Scheme 11 $\quad\quad\quad\quad\quad\quad\quad\quad$ Scheme 12

[130] G. P. de-Gunst, *Rec. Trav. chim.*, 1969, **88**, 801.
[131] J. Nasielski, M. Jauquet, E. Vander-Donckt, and A. Van-Sinoy, *Tetrahedron Letters*, 1969, 4859.

Scheme 11 may be rejected. The reduction occurs in methanol, ethanol, and isopropyl alcohol, but not in benzene. The products of the reaction are 9,9′-bifluorene (165) (AH$_2$), and the pinacol when the concentration of

(162) is less than 10^{-4} mol l^{-1}. At concentrations of *ca.* 10^{-3} mol l^{-1} of (162) in isopropyl alcohol, small amounts of (167) and (168) were isolated but no (164).[130] At higher concentrations no reaction was observed. Triplet intermediates are suggested since the singlet lifetimes are considered to be insufficient for reactions to occur at these concentrations.

The sensitized photo-oxygenation of furan and its derivatives to yield cyclic perioxides (169) is very well known,[121] and the similar reaction

[2,2](2,5)furanophane (170) in methanol has been reported to produce an oxygenated intermediate which undergoes an intramolecular Diels–Alder reaction to give (171).[132] It has now been demonstrated that the course of the photo-oxygenation process of (170) is solvent-dependent, and that in dichloromethane a new product is formed in 75% yield.[133] The new product is assigned structure (172), and is suggested to be formed by rearrangement of the *endo*-peroxide (173). An *X*-ray analysis by Fratini has established the configuration of (172) as (174).[134] It has also been found that with the furanocyclophane (175), the product in dichloromethane differs from those [(176), (177), and (178)] in methanol, and is tentatively assigned

[132] H. H. Wasserman and A. R. Doumaux, *J. Amer. Chem. Soc.*, 1962, **84**, 4611.
[133] H. H. Wasserman and R. Kitzing, *Tetrahedron Letters*, 1969, 5315.
[134] A. Frantini, to be published (quoted in ref. 133).

structure (179). A dependence of product ratio on solvent is also reported in the reaction of photogenerated singlet oxygen with 2,5-dimethylthiophen (180).[135] Thus, in methanol solution with 520 nm radiation and in the presence of methylene blue, the thiophen (180) forms the *cis*-sulphine (181) in 70% and the *trans*-diketone (182) in 2% yield, whereas in chloroform solution the yields of (181) and (182) are 56 and 28% respectively. It is considered that (181) and (182) arise from (180) by two different pathways. Compound (182) is a result of Diels–Alder addition of singlet oxygen to the thiophen to yield the thio-ozonide (183) which undergoes desulphuration to give the *cis*-isomer (184) of (182). It is suggested that intermediate (183)

[135] C. N. Skold and R. H. Schlessinger, *Tetrahedron Letters*, 1970, 791.

may be involved in the formation of (181) by conversion to the oxathiairane (185). An alternative route to (181) involves the sulphoxonium ion (186) and the ylide dioxetan (187). The isolation of (181) from (180) has also been reported by other workers who, in addition, have studied the methylene-blue-sensitized oxygenation of 2,3-dimethyl-4,5,6,7-tetrahydrothianaphthene (188) to give the cyclic sulphine (189).[136] The suggested intermediates in this process are similar to those suggested for one route from (180) to (181).[135]

4 Oxidation of Aliphatic Unsaturated Systems

Simple olefins and polyenes are susceptible to photo-oxidation, and give a variety of products. As mentioned in Section 3, many of the oxygenations

[136] H. H. Wasserman and W. Strehlow, *Tetrahedron Letters*, 1970, 795.

proceed by molecular oxygen in its low-lying singlet states. States $O_2(^1\Sigma_g^+)$ and $O_2(^1\Delta_g)$ have been implicated in many of these processes. The rate constants for the deactivation or reaction of ($^1\Delta_g$) oxygen with some simple olefins have been determined directly from measurements of the intensity of the oxygen ($^1\Delta_g$) to oxygen ($^3\Sigma_g^-$) emission at 126·9 nm in a discharge system.[137] The olefins examined were ethylene, propylene, but-1-ene, pent-1-ene, and tetramethylethylene; the reaction with methyl chloride was also studied. For terminal olefins, the rate constant is reported to increase with molecular weight. The fact that methyl chloride is a relatively polar molecule apparently has little effect on its ability to deactivate ($^1\Delta_g$) oxygen. With tetramethylethylene, the rate constant is more than two orders of magnitude larger than with terminal olefins.

The photo-oxidation of propylene with nitrogen dioxide has been investigated with 366 nm radiation, and quantum yields for nitrogen dioxide consumption as a function of propylene pressure, nitrogen dioxide pressure, total pressure (N_2), and temperature have been determined.[138] The quantum yield increases as a function of propylene concentration, and approaches a value of 6. Photolysis of the nitrogen dioxide yields oxygen atoms which add to the propylene, forming intermediates which are either collisionally stabilized or dissociate into free radicals. Nitrogen dioxide reacts rapidly with the free radicals to yield stable compounds. A 29-step mechanism is proposed (!), and the products of the reaction are shown in Scheme 13.

$$\text{MeCH=CH}_2 + \text{NO}_2 \xrightarrow{h\nu} \text{CO}_2, \text{MeCHO}, \text{C}_2\text{H}_4\text{O}, \text{MeCH}_2\text{CHO}, \text{C}_3\text{H}_6\text{O},$$
$$\text{Me}_2\text{CO}, \text{MeNO}_2, \text{MeONO}_2, \text{MeCH}_2\text{NO}_2,$$
$$\text{MeCH}_2\text{ONO}_2$$

Scheme 13

Two groups of workers report the formation of stable dioxetans by the sensitized photo-oxygenation of olefins. Bartlett and Schaap have studied the reaction of *cis*- and *trans*-diethoxyethylene at −78 °C in trichloro-fluoromethane with tetraphenylporphin as sensitizer.[139] It is extremely interesting to note that the stereochemistry of the olefins is retained in the products (190) and (191). From orbital symmetry considerations, formation of the dioxetans should not proceed by suprafacial 1,2-addition of the singlet oxygen: thus the process either involves antarafacial ($2s + 2a$) mode of reaction (192) or occurs in a stepwise manner. A biradical stepwise process would result in loss of stereospecifity, but cycloadditions involving dipolar ions [*e.g.* (193)] would show more retention of configuration, and such zwitterionic species have previously been postulated for the essentially stereospecific photochemical cycloadditions of stilbenes to

---

[137] R. A. Ackerman, J. N. Pitts, and R. P. Steer, *J. Chem. Phys.*, 1970, **52**, 1603.
[138] S. Jaffe and R. C. S. Grant, *J. Chem. Phys.*, 1969, **50**, 3477.
[139] P. D. Bartlett and A. P. Schaap, *J. Amer. Chem. Soc.*, 1970, **192**, 3223.

o-chloranil.[140] Each dioxetan decomposed quantitatively to ethyl formate on warming. The second report is concerned with the addition of singlet oxygen to tetramethoxyethylene.[141] Again the reaction is carried out at low temperature (-70 °C), and the sensitizer in this case is zinc tetraphenylporphin or dinaphthalenethiophen. The reaction consumed 1 mol of oxygen per mol of olefin, and yielded 94% of a pale yellow liquid which contained approximately 10% of dimethyl carbonate. The pure product (194) had a m.p. of -8 to -9 °C and was found to be remarkably stable with $t_{\frac{1}{2}} = 102$ min at 56 °C. Both groups of workers were aware of the similarity of their products to the suggested intermediates in many chemiluminescence reactions.[142, 143] Indeed, solutions of dioxetan (194) in boiling benzene containing various fluorescent hydrocarbons have been observed to emit luminescence, the colour of which corresponds to the hydrocarbon fluorescence.[141] These reports represent the first examples of 1,2-dioxetan formation by photo-oxygenation.

A novel photo-oxidative cyclization involving the cyclic olefin and hydroxyl moieties of cannabidiolic acid (195) has provided a synthesis of cannabielsoic acid A (196), a compound extracted from hashish.[144] The irradiation of (195) in cyclohexane in the presence of oxygen, followed by sodium bisulphite reduction, yields two isomers, (196) (17%) and 1-iso-(196) (16%). With no bisulphite reduction the corresponding hydroperoxides are formed which, on irradiation with wavelengths longer than 290 nm, yield the above products. The stereochemistry at C-2–C-3 is

---

[140] D. Bryce-Smith and A. Gilbert, *Chem. Comm.*, 1968, 1701.
[141] S. Mazur and C. S. Foote, *J. Amer. Chem. Soc.*, 1970, **92**, 3225.
[142] M. M. Rauhut, *Accounts Chem. Res.*, 1969, **2**, 80.
[143] J. P. Paris, *Photochem. and Photobiol.*, 1965, **4**, 1059.
[144] A. Shani and R. Mechoulam, *Chem. Comm.*, 1970, 273.

deduced to be *cis* from consideration of the fact that radical processes which form related tricyclic dihydrobenzofuran systems give the more stable *cis*-compounds.[145] This reaction apparently represents the first reported example of a new type of photo-oxidative cyclization involving attack on an olefin by a phenoxy-group and molecular oxygen. Since the total synthesis of (195) is reported, the present reaction completes the total synthesis of cannabielsoic acid (196). In the absence of the *o*-phenolic hydroxy-group such systems as (195) do not undergo cyclization, and heamatoporphyrin-sensitized photo-oxygenation of aryl dihydropyrans (197) in dry pyridine leads to ring-fission and the formation of the ester–aldehyde (198).[146] The dioxetan (199) is the suggested intermediate, rather than the allylic hydroperoxide. A similar reaction is involved in the photo-sensitized oxygenation of 3-hydroxyflavones (200).[147] This particular work was carried out to investigate further the similarity between photo-sensitized oxygenation and dioxygenase-catalysed enzymatic oxidation. Compounds (200) are reported to undergo Rose Bengal-sensitized oxygenation to yield the corresponding depsides (201) with the evolution of carbon monoxide and carbon dioxide. The formation of the depsides and carbon monoxide represents a model for the enzymatic degradation of quercetin (200a) by quercetinase to yield the depside (201a). The mechanism of the process is considered to involve formation of the hydroperoxide (202) by

[145] D. H. R. Barton, A. M. Deflorin, and O. E. Edwards, *J. Chem. Soc.*, 1956, 530.
[146] R. S. Atkinson, *Chem. Comm.*, 1970, 177.
[147] T. Matsuura, H. Matsushima, and R. Nakashima, *Tetrahedron*, 1970, **26**, 435.

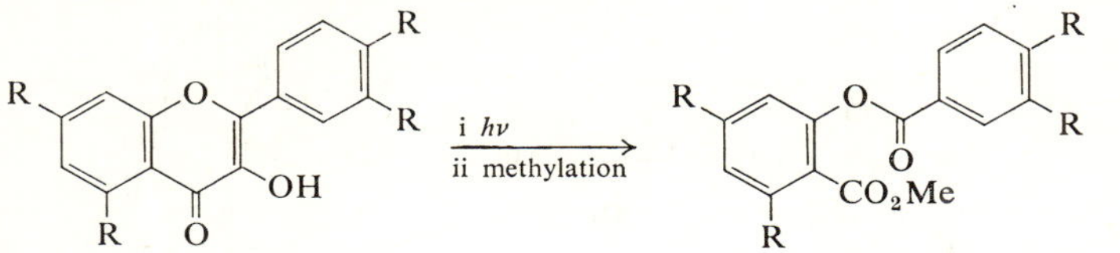

(200) (a) R = H
(b) R = OMe
(c) R = H

(201)

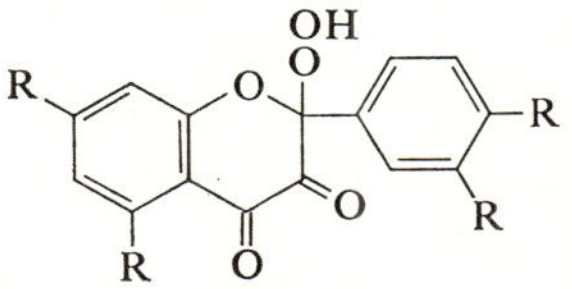

(202)

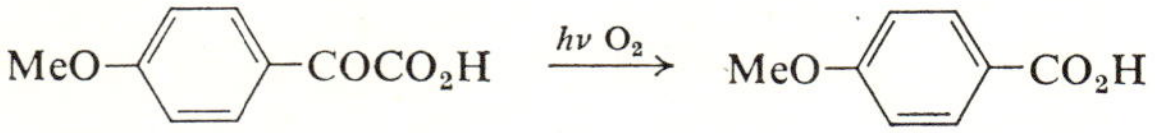

(204) MeO-C6H4-COCO2H $\xrightarrow{h\nu\ O_2}$ MeO-C6H4-CO2H

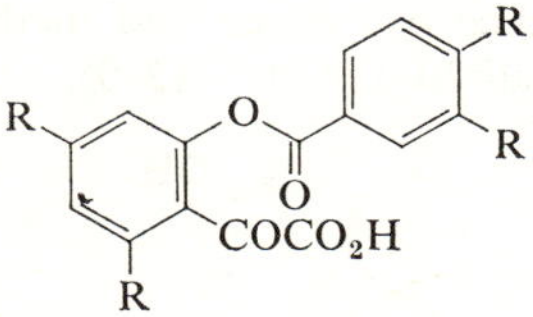

(203)

attack of singlet oxygen. It is concluded that since the photo-oxygenation of *p*-anisylglyoxylic acid (204) yields *p*-anisic acid and carbon dioxide, the carbon dioxide formed from (200) is formed by photo-oxidative decarboxylation of the keto-acid intermediate (203). In the reaction of (204), it seems that direct excitation may be advantageous in the oxidative decarboxylation.

2,3-Dimethyl-5-(3-methylbut-2-enyl)-1,4-benzoquinone (205) (plastoquinone) is a model compound for the electron transport quinones in plants. Its photolysis is thus of interest. Photo-oxidation of (205) with 370 nm radiation in benzene or isopropyl alcohol yields 20—45% of the novel tricyclic peroxide 4,5-dihydro-3,3,8,9-tetramethyl-4,9a-epoxy-9a*H*-1,2-benzodioxepin-7(3*H*)-one (206).[148] The formation of (206) from (205) contrasts with the recently reported formation of hydroperoxides from irradiation of menaquinones (207), from which side-chain degradation followed.[149]

The dye-sensitized photo-oxygenation of thujopsene (208), a tricyclic sesquiterpene with an α-cyclopropyl olefinic system, has been studied and

[148] D. Creed, H. Werbin, and E. T. Strom, *Chem. Comm.*, 1970, 47.
[149] C. D. Snyder and H. Rapoport, *J. Amer. Chem. Soc.*, 1969, **91**, 731.

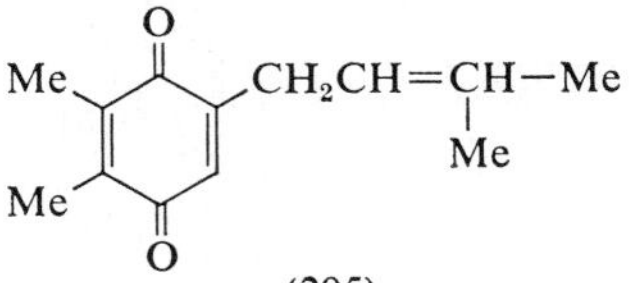

(205)

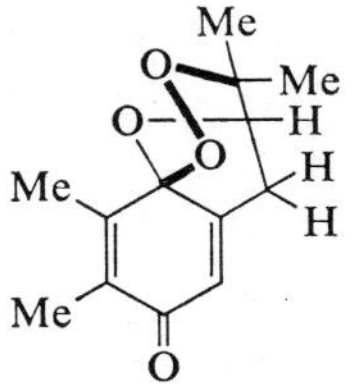

(206)

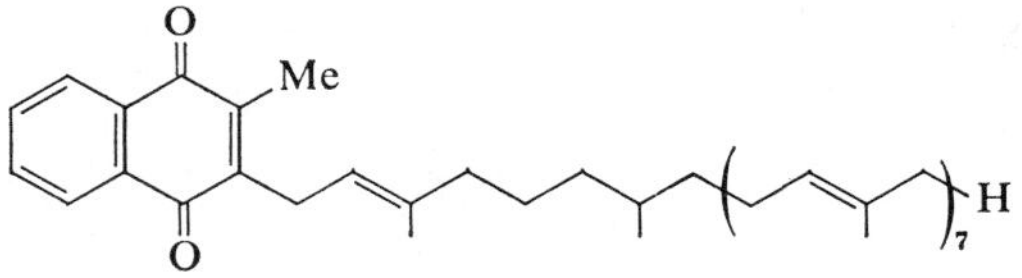

(207)

reported by two groups. The first report describes results which determine the conformation of thujopsene and outline the formal total synthesis of thujopsadiene (209), a natural thujopsane sequiterpene.[150] The reaction of (208) in methanol in the presence of oxygen and methylene blue yields, after reduction by sodium bisulphite, (209), (210), (211), and (212) in

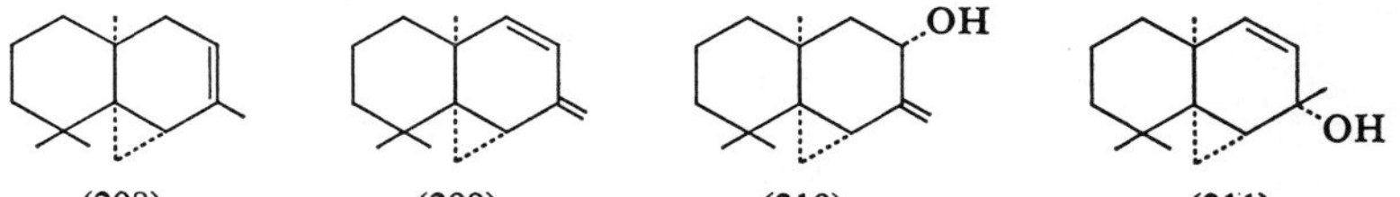

(208) (209) (210) (211)

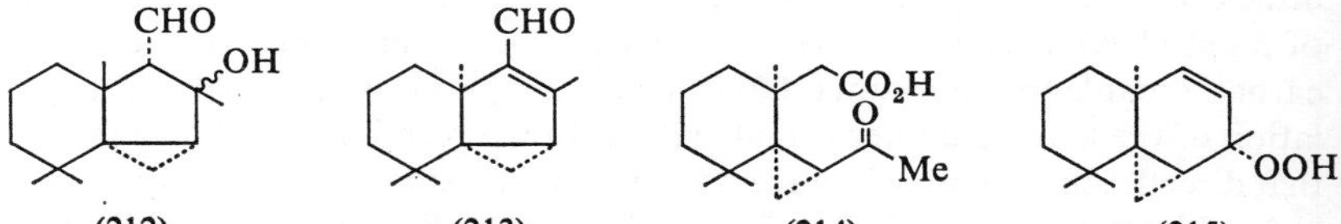

(212) (213) (214) (215)

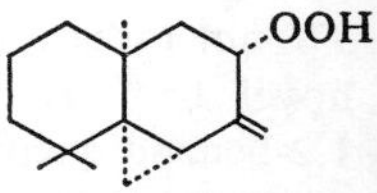

(216)

addition to the known compounds (213) and (214), which were obtained previously by chemical degradation of (208). All the reaction products are shown to originate from (215) with the exception of (210) which is formed from another hydroperoxide (216). A cyclic mechanism is proposed for

[150] S. Ito, H. Takeshita, T. Muroi, M. Ito, and K. Abe, *Tetrahedron Letters*, 1969, 3091.

this type of reaction and preferred abstraction of an axial allylic hydrogen over methyl hydrogens by singlet oxygen is clearly demonstrated. It is concluded from the results that the α-side of thujopsene is less hindered in spite of the cyclopropane ring, and the 9α-hydrogen is axially orientated, defining the preferred conformation as (217). This is in agreement with the conformation inferred from circular dichroism measurements.[151] The second report of the photo-oxygenation of thujopsene only describes two of the foregoing products [(210) and (211)].[152] These products are used to demonstrate the complete stereospecificity of the addition of ($^1\Delta_g$) oxygen to the acceptor system, and in this way (208) is shown to exist in only one conformation in solution [*i.e.* (−)-thujopsene (217)]. After reductive decomposition of the hydroperoxide formed from the photo-oxygenation of thujopenol (218), five products have been isolated, characterized, and

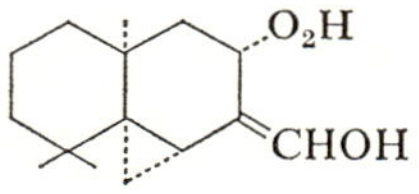

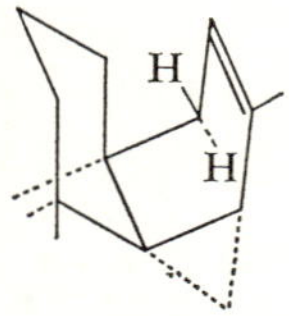

(217)

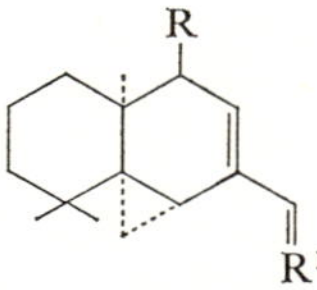

(218) R = H, R¹ = H, OH
(220) R = H, R¹ = O
(221) R = OMe, R¹ = O

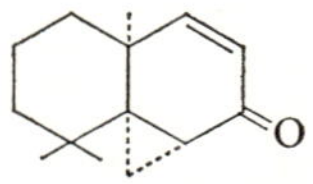

(224)

(219)

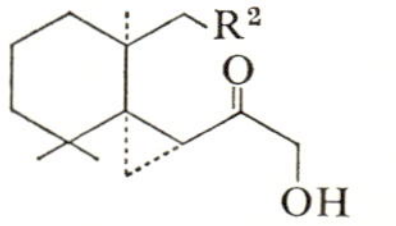

(222) R² = CH(OMe)₂
(223) R² = CO₂H

assigned structures (219), (220), (221), (222), and (223).[153] Compound (219) is mayurone, a norsesquiterpenic ketone isolated from several conifers. The main features of this process are explained by a mechanism similar to that proposed for thujopsene, but in this case the secondary allylic hydroperoxide (224), formed by abstraction of one of the hydrogens of the hydroxymethylene, is a β-hydroperoxylaldehyde which eliminates hydrogen peroxide to give (220).

Photo-oxygenation of acyclic and cyclic dienes yielding cyclic peroxides is very well authenticated.[121] The sensitized reaction of *trans,trans*-diphenyl-1,4-butadiene is reported to form *cis*-diphenyl-3-*cis*-dihydro-3,6-dioxine-1,2

---

[151] K. Kuriyama, S. Ito, T. Norin, and W. Klyne, to be published.
[152] G. Ohloff, H. Strickler, B. Willhalm, C. Borer, and M. Himder, *Helv. Chim. Acta*, 1970, **53**, 623.
[153] H. Takeshita, T. Sato, T. Muroi, and S. Ito, *Tetrahedron Letters*, 1969, 3095.

(225),[154] and the 1,2-diphenylcyclopentadienes (226) and (227) yield (228) and (229) under typical photo-oxygenation conditions.[155] The diene acid (230), however, under similar reaction conditions, does not yield (231) as expected, and the intermediacy of (231) is suggested to account for the observed products (232), (233), and (234).[155] Such cyclic peroxide formation

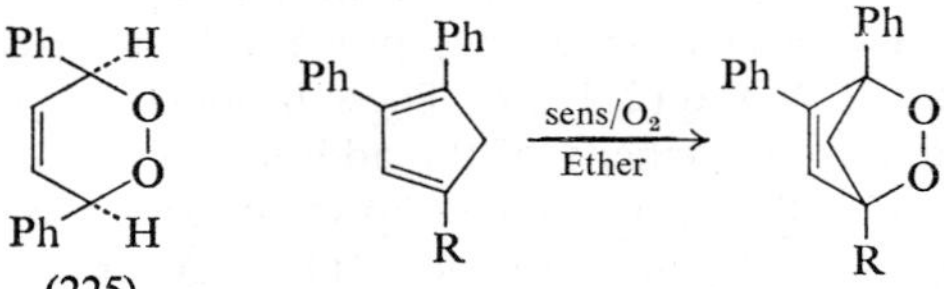

(225)

(226) R = H
(227) R = Me
(230) R = CO_2H

sens/O_2
Ether

(228)
(229)
(231)

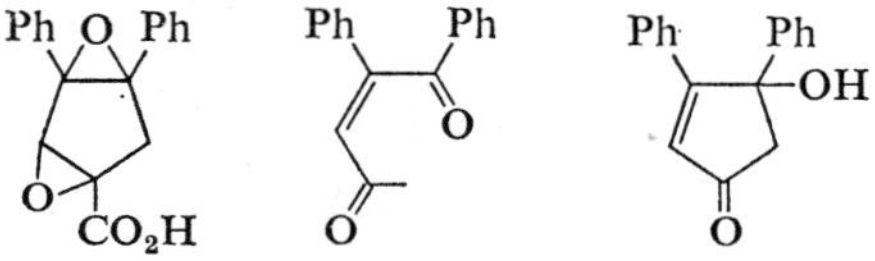

(232)
(233)
(234)

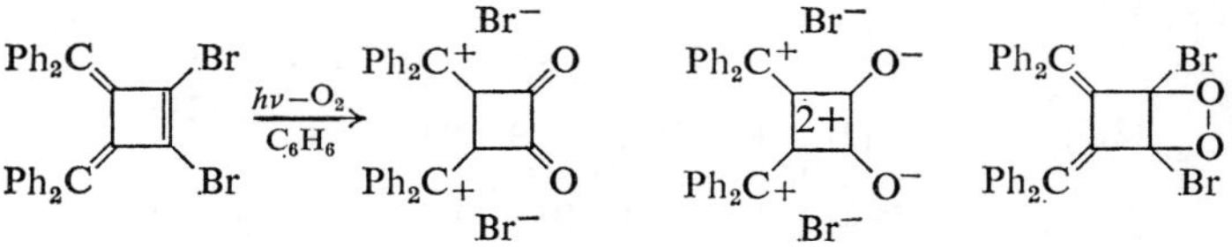

(235)

$h\nu - O_2$
C_6H_6

(236)
(237)
(238)

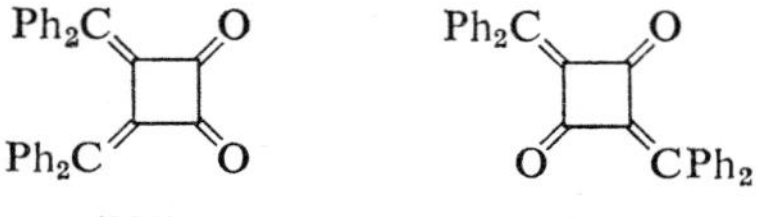

(239)
(240)

is not found in the direct irradiation of the triene (235) under oxygen in benzene solution.[156] A novel photo-oxygenation does, however, occur with the formation of 51% of the dibromocyclobutadienequinone (236). The λ_{max} absorption of (236) (crimson) at 460 nm is supposed to be due to contribution of the cyclobutadiene di-cation structure (237). The proposed intermediate for the formation of (236) is the dioxetan (238). In alcoholic solution, (236) reacts with silver fluoroborate to form 3,4-bis(diphenylmethylene)cyclobuteno-1,2-dione (239), which is the first reported example

[154] G. Rio and J. Berthelot, *Bull. Soc. chim. France*, 1969, 1664.
[155] G. Frio and M. Charifi, *Compt. rend.*, 1969, **268**, C, 1960.
[156] F. Toda, H. Ishihara, and K. Akagi, *Tetrahedron Letters*, 1969, 2531.

of this system, although the isomeric 1,3-dione (240) has recently been prepared.[157]

The sunlight-induced oxygenation of the fixed *trans*-diene, cholesta-3,5-dieno-[3,4-*b*]-1,4-oxathian (241) has been reported to yield the 3,4-seco-3,4-dioxo-compound (242) (37%) in addition to a trace of (243).[158] There is no reaction under nitrogen and the oxygenation is not induced thermally by peroxy-acid or hydrogen peroxide. When (241) was irradiated in wet ether, the formation of (243) (40%) was favoured in comparison with that of (242) (7%). Again, reaction is suggested to proceed by formation of the dioxetan (244), a process doubtless favoured by the mesomeric effect of the sulphur and oxygen functions in the conformationally rather rigid oxathiano-3-ene ring system.

The photo-oxygenation of tropolone methyl ether to yield the epidioxide (245) has been previously reported.[159] Tropone itself is now reported to form the analogous epidioxide which on treatment with triethylamine in ethanol at room temperature undergoes facile cleavage of the epidioxide linkage to yield 5-hydroxy-tropolone (246) quantitatively.[160] [This is probably the most convenient procedure for the formation of (246) reported to date.[161]] The same workers report that the photo-oxygenation of 8-cyanoheptafulvene in acetone gives the isomers (247) and (248).[162]

[157] G. A. Taylor, *Chem. Comm.*, 1968, 1314.
[158] A. Miyake and M. Tomoeda, *Chem. Comm.*, 1970, 240.
[159] E. J. Forbes and J. Griffiths, *J. Chem. Soc.* (*B*), 1968, 575.
[160] M. Oda and Y. Kitahara, *Tetrahedron Letters*, 1969, 3295.
[161] T. Nozoe, S. Seto, S. Ito, and M. Sato, *Proc. Japan Acad.*, 1951, **27**, 426.
[162] M. Oda and Y. Kitahara, *Angew. Chem.*, 1969, **81**, 702.

(245)   (246)   Et₃N / EtOH →

(246)

(247)   (248)

For many years there has been considerable interest in the photo-oxygenation reactions of ionones and carotenoids. Carotenoids and vitamin A may react photochemically as sensitizers, substrates, quenchers, and scavengers in such processes. β-Carotene has been shown to act as a sensitizer in type 2 photo-oxygenation of dienes and olefins, generating singlet oxygen by the Kautsky–Schenck mechanism *via* excited carotene–oxygen complexes.[163] Strong quenching of singlet oxygen by β-carotene has been found, especially in solvents containing benzene, but not in methanol or isopropyl alcohol. The quenching of singlet oxygen seems to be associated with oxygenation of β-carotene and has been interpreted as a catalytic action proceeding by complexing of singlet oxygen with β-carotene. The difference in behaviour of the excited sensitizer–oxygen complex and of the quencher/acceptor–oxygen complex is explained by the different states of aggregation of the β-carotene in various solvents. Association seemingly favours sensitization, and occurs in alcohols. On the other hand, the stronger solvation by benzene favours acceptor and associated catalytic or quenching activities. all-*trans* β-Carotene (249) has been photo-oxidised in the presence of chlorophyll (contrast the reaction of β-carotene in the presence of Rose Bengal [164]), to yield the new C₄₀ carotenoids (250), (251), (252), and (253), as well as the *cis*-isomers, the 5,6-monoepoxide, and possibly the 5,8-furanoid oxide of (249).[165] This is the first case in which photo-oxygenation has yielded both 6,7- and 7,8-dehydro-derivatives of C₄₀ carotenoids *in vitro*. Mousseron-Canet has continued her exacting studies in the field of photo-oxygenation of polyenes of biological importance, and reports the formation of the endocyclic peroxides (254) and the exocyclic peroxide (255) from compounds (256), (257), (258), and (259).[166]

[163] G. O. Schenck and G. Schade, *Chimia*, 1970, **24**, 13.
[164] S. Isoe, S. B. Hyeon, and T. Sakan, *Tetrahedron Letters*, 1969, 279.
[165] K. Tsukida, S. Cho, and M. Yokota, *Chem. Pharm. Bull. Tokyo*, 1969, **17**, 1755.
[166] J. L. Olive and M. Mousseron-Canet, *Bull. Soc. chim. France*, 1969, 3252.

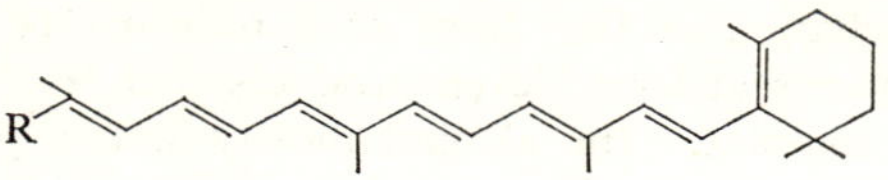

(249) R =

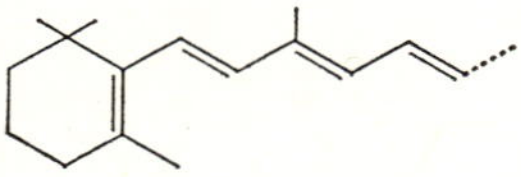

(250) R =

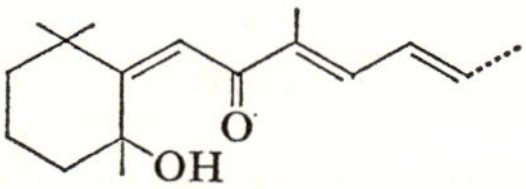

(251) R =

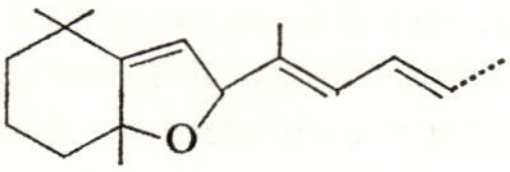

(252) R = 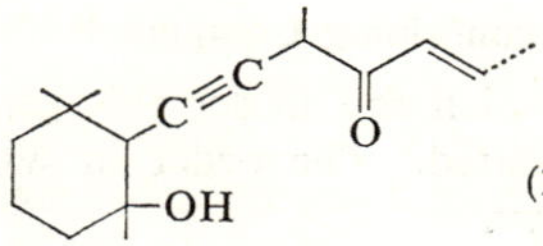   (253) R =

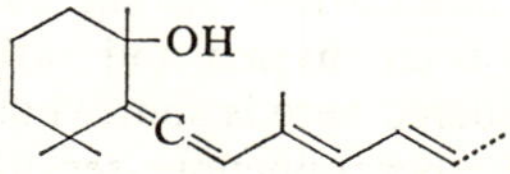

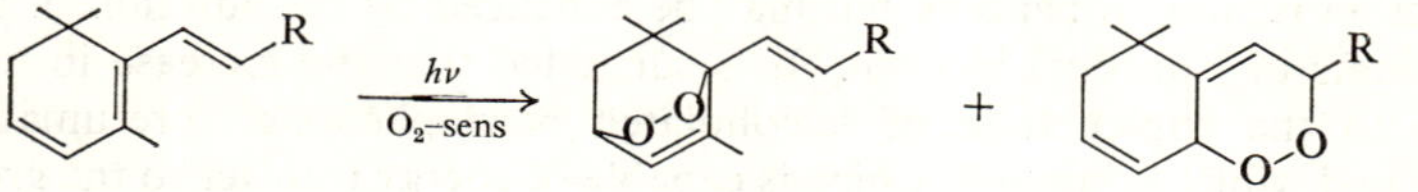

(254)   (255)

(256) R = 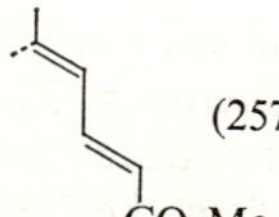   (257) R =

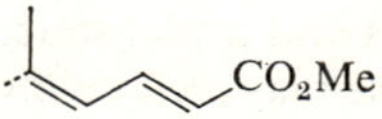

(258) R =

(259) R =

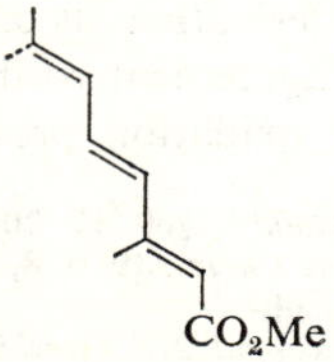

The sunlight irradiation of thin films of pyrethrin (260) and related compounds in air yields at least eleven products each.[167] The processes are evidently very complex. The alcohol moiety suffers photochemical attack but the reactions involved are not known. However, the changes in the acid moiety of each of the pyrethroids studied are the same, and

(260)

involve (*a*) stepwise oxidation of the *trans*-methyl group of the isobutenyl moiety to the respective alcohol, aldehyde, and carboxylic derivative, (*b*) oxidation of the isobutenyl double bond to a keto-derivative, (*c*) fission of this double bond to form esters of *trans*-caronic acid, and (*d*) other reactions resulting in at least six different modifications of the acid moiety.

5 Oxidation of Nitrogen-containing Compounds

The publications within the year which fall under this general heading have been diverse and rather unrelated. The order in which they are considered here is somewhat arbitrary.

The benzophenone-sensitized oxidation of diphenylamine in benzene has been reinvestigated,[168] and the product has been shown to be diphenyl nitroxide, in agreement with a multitude of earlier reports. The rate of reaction is slow in benzene but may be enhanced by the addition of small amounts of isopropyl alcohol. It is suggested that the increase in rate is due to the triplet state of 'alcoholated benzophenone' (presumably a solvated triplet is implied) which is capable of energy transfer to the ground state amine several times faster than the triplet state of benzophenone. The reader may usefully compare this result with those given in ref. 8 where polar solvents are reported to promote the quenching by triphenylamine of $n\pi^*$ singlet and triplet states of aryl ketones. The reported energy transfer to the amine may well involve charge-transfer *from* the amine.

The photolysis of organic nitrites involved in the Barton reaction is a well-known and synthetically useful process.[169] The photo-oxidation of such compounds in benzene solution has now been described.[170] Photolysis of 1-pentyl nitrite and menthyl nitrite (261) in the presence of oxygen yields the 4-nitratopentyl and 10-nitratomenthyl derivatives as oxidised rearrangement products. These findings are not in agreement with an earlier report which had implied that the bimolecular oxidation processes leading to

---

[167] Y. L. Chen and J. E. Casida, *J. Agric. Food Chem.*, 1969, **17**, 208.
[168] M. Santhanam and V. Ramakrishnan, *Indian J. Chem.*, 1970, **8**, 374.
[169] D. H. R. Barton, *Helv. Chim. Acta*, 1959, **42**, 2604.
[170] Y. L. Chow, T. Hayasaka, and J. N. S. Tam, *Canad. J. Chem.*, 1970, **48**, 508.

(261)

1-octylnitrate from 1-octylnitrite were far faster than the intramolecular hydrogen–nitroso exchange which is a major process in the absence of oxygen.[171] In contrast, the new examples indicate that the intramolecular hydrogen–nitroso exchange reaction can be much faster than the oxidation.

The photochemical conversion of nitrones to oxazirans is already known.[172] It has now been observed that u.v. irradiation of nitrones in solution at room temperature yields long-lived (days) radicals which are considered to arise by a photo-oxidation.[173] The radicals have been identified as N-benzoylnitroxides (262). The nitrone solutions were de-aerated with argon before irradiation, and it is considered likely that oxygen was formed during the irradiation, since liberation of oxygen from nitrones has

$$R^1-CH=\overset{\overset{O}{\uparrow}}{N}-R^2 \xrightarrow{h\nu} R^1-HC\overset{O}{\underset{}{\diagdown}}N-R^2 \xrightarrow[[O]]{h\nu} R^1CHO + R^2N=O$$
(263)

$$R^1CHO + R^1HC\overset{O}{\underset{}{\diagdown}}N-R^2 \xrightarrow{h\nu} R^1C=O \xrightarrow{R_2N=O} R^1-\overset{\overset{O}{\|}}{C}-\overset{\overset{O}{\uparrow}}{N}-R^2$$
(262)

been reported.[174] It was found that, after a brief irradiation period, the radical signal continued to grow in the dark, suggesting a chain reaction or a thermal process. The oxaziran (263) may be an intermediate.

It has previously been reported that the photolysis of nitrosamides leads to homolytic cleavage with the formation of nitric oxide and amido-radicals.[175] This process is followed by a series of reactions involving exchange of the nitroso-group and a hydrogen atom to yield 4-nitroso-alkylamides (intramolecular exchange), the parent amide (intermolecular exchange), and other oxidation products. The photochemistry of N-hexyl-N-nitrosoacetamide (264) in benzene solution is now reported.[176] Under nitrogen, and with light of wavelength longer than 400 nm, (264) yields hexylacetamide (265) and the anti-dimer of N-(4-nitrosohexyl)acetamide

---

[171] P. Kabasakalian and E. R. Townley, J. Amer. Chem. Soc., 1962, **84**, 2711.
[172] K. Shinazawa and I. Tanaka, J. Phys. Chem., 1964, **68**, 1205.
[173] A. L. Bluhm and J. Weinstein, J. Amer. Chem. Soc., 1970, **92**, 1444.
[174] N. Hata and I. Tanaka, J. Chem. Phys., 1962, **36**, 2072; A. Alkaitis and M. Calvin, Chem. Comm., 1968, 292.
[175] L. P. Kunn, G. G. Kleinspehn, and A. C. Duckworth, J. Amer. Chem. Soc., 1967, **89**, 3858.
[176] Y. L. Chow and J. N. S. Tam, Chem. Comm., 1969, 747.

(266). With radiation of wavelengths longer than 280 nm, still under nitrogen, two further products (267) and (268) were detected, but these arose from (266) by secondary processes. In the presence of oxygen there is a dramatic increase in the proportion of the oxidation products (267) and (268) at the expense of (266), thus demonstrating the vital role of oxygen

$$\text{Me(CH}_2)_5\text{NAcNO} \xrightarrow[>400\text{ nm}]{h\nu} \text{Me(CH}_2)_5\text{NAcH} + [\text{EtCHNO(CH}_2)_3\text{NHAc}]_2$$
$$(264) \qquad\qquad (265) \qquad\qquad (266)$$

$$\Big\downarrow >280\text{ nm}$$

$$\text{Me(CH}_2)_5\dot{\text{N}}\text{Ac} \xrightarrow{+\text{NO},\frac{1}{2}\text{O}_2} \text{Me(CH}_2)_5\text{NAcNO}_2$$
$$(267)$$

$$\Big\downarrow$$

$$\text{Et}\dot{\text{C}}\text{H(CH}_2)_3\text{NHAc} \xrightarrow{+\text{NO},\frac{1}{2}\text{O}_2} \text{EtCHONO}_2(\text{CH}_2)_3\text{NHAc}$$
$$(268)$$

in the oxygenation. Addition of nitric oxide to the irradiation does not assist in the oxidation. It is assumed that nitric oxide from photodissociation of (264) or the C-nitroso dimer (266) is quickly oxidised to ·NO$_3$ which attacks the C- or N-radicals to yield (267) and (268). The present results suggest that excited nitric oxides yield oxygen, and this oxygen is the oxidation carrier in experiments performed under nitrogen. Participation of singlet oxygen is ruled out by the inability of Rose Bengal to sensitize the process.

Four reports have appeared during the year under review on the photochemical conversion of a methylene group to a carbonyl group. In each case the methylene was either situated on nitrogen, α to it, or in a conjugated system involving nitrogen. The first example is provided by the oxidation of the N-methyl groups in pseudopelletieriene (269) and tropinone (270) to N-formyl groups by singlet oxygen.[177] Singlet oxygen is here

$$\overset{\diagup(\text{CH}_2)_n\diagdown}{\diagdown\qquad\diagup}\text{NMe}\!\!=\!\!\text{O} \xrightarrow[\text{O}_2]{h\nu} \overset{\diagup(\text{CH}_2)_n\diagdown}{\diagdown\qquad\diagup}\text{N}\!-\!\text{CHO}\!\!=\!\!\text{O}$$

(269) $n = 1$
(270) $n = 0$

behaving in a way different from any previously described, but is definitely implicated in the reaction since the process is sensitized by Rose Bengal, naphthalene, or triphenylene, but quenched by the known singlet oxygen

---

[177] M. H. Fisch, J. C. Gramain, and J. A. Oleson, *Chem. Comm.*, 1970, 13.

quencher, 1,4-diazabicyclo-octane.[178] The results are rationalized by a mechanism in which the carbonyl group acts as a sensitizer in the direct irradiation and presumably its triplet state is quenched by dissolved oxygen. Hence, singlet oxygen is generated adjacent to the N-methyl group. The reaction of structurally related compounds in which the reactant groups are farther apart would be illuminating. Land and Frasca earlier in the year reported that a methyl group α to nitrogen on a pyrazole ring was oxidised photochemically to a formyl group.[179] The work has been extended to other compounds, and the oxidation of a methyl group on the benzene ring in indazoles is reported.[180] Thus, irradiation of glacial acetic acid solutions of the indazoles (271), (272), and (273) yielded two aldehydic products each. In one set of products (274), (275), and (276), the methyl

(271) R^1 = Me, R^2 = R^3 = H
(272) R^1 = R^3 = H, R^2 = Me
(273) R^1 = H, R^2 = R^3 = Me

(274) R^1 = Me, R^2 = R^3 = H
(275) R^1 = R^3 = H, R^2 = Me
(276) R^1 = H, R^2 = R^3 = Me

+

(277) R^1 = CHO, R^2 = R^3 = H
(278) R^1 = R^3 = H, R^2 = CHO
(279) R^1 = H, R^2 = CHO, R^3 = Me

group on the nitrogen 5-ring is oxidised, in the other set (277), (278), and (279), a methyl group on the benzene ring has been oxidised. Facile autoxidation of the ethyl group of 2-ethyl-3,3-disubstituted-3H-indoles (280) to an acetyl group is reported.[181] In this case there is no effect of sensitizers

[178] C. Ouannès and T. Wilson, *J. Amer. Chem. Soc.*, 1968, **90**, 6527.
[179] H. B. Land and A. R. Frasca, *Chem. and Ind.*, 1969, 1594.
[180] H. B. Land and A. R. Frasca, *Chem. and Ind.*, 1970, 500.
[181] Y. Kanaoka, K. Miyashita, and O. Yonemitsh, *Chem. Pharm. Bull. Tokyo*, 1970, **18**, 634.

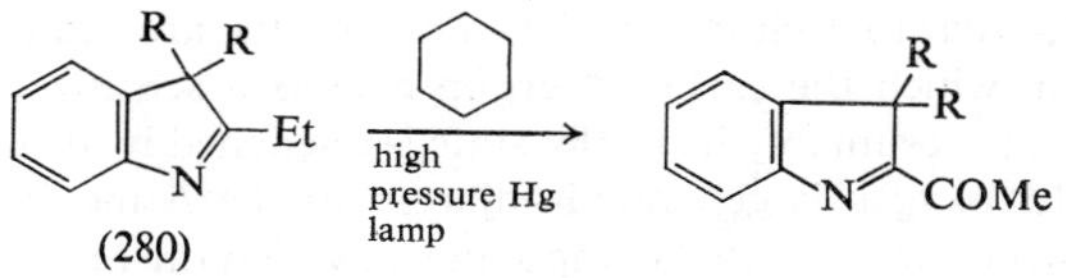

(280) [structure] → high pressure Hg lamp → [structure with COMe]

such as methylene blue and benzophenone, so it is suggested that singlet oxygen is not involved. A mechanism involving direct photo-oxidation with triplet oxygen [182] is proposed.

The photo-oxidation of imines to yield ketones and amides is reported, and seemingly also does not involve singlet oxygen.[183] Irradiation of 1,1-diphenylmethyleneimine in isopropyl alcohol with unfiltered radiation from a high pressure mercury lamp leads to an almost quantitative formation of benzophenone. Benzylidene aniline yields benzoic acid and, unexpectedly, benzanilide. From a comparison of the photo-oxidation and -reduction data of imines, it is suggested that the same reaction intermediate is involved. The suggested mechanism is outlined in Scheme 14.

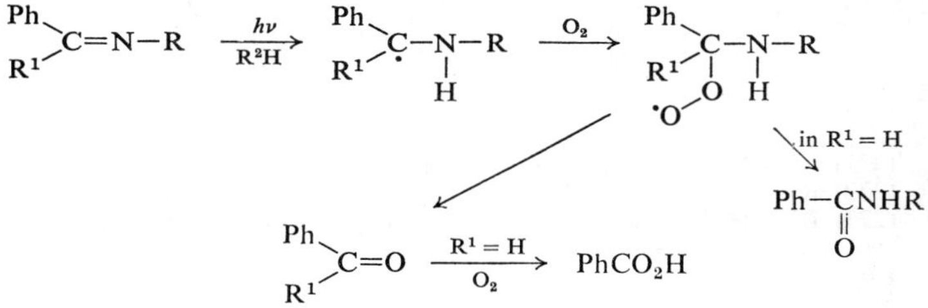

$$\underset{R^1}{\overset{Ph}{>}}C=N-R \xrightarrow{h\nu, R^2H} \underset{R^1}{\overset{Ph}{>}}\overset{\cdot}{C}-N(H)-R \xrightarrow{O_2} \underset{R^1}{\overset{Ph}{>}}C(OO\cdot)-N(H)-R$$

$$\underset{R^1}{\overset{Ph}{>}}C=O \xrightarrow[O_2]{R^1=H} PhCO_2H$$

in $R^1 = H$

$$Ph-\underset{\overset{\|}{O}}{C}NHR$$

Scheme 14

Phenylhydrazones on photo-oxidation do not apparently undergo such a fission process since the product from such reaction with the benzaldehyde derivative is reported to have structure (281).[184]

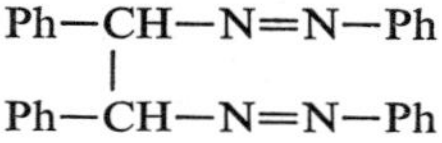

$$\begin{array}{c} Ph-CH-N=N-Ph \\ | \\ Ph-CH-N=N-Ph \end{array}$$
(281)

Oxidation by loss of hydrogen from partially hydrogenated nitrogen heterocyclic compounds has been previously reported,[185,186] and another example has appeared this year with the synthesis of pyrrole nucleosides [e.g. (282)] from Δ^3-pyrroline derivatives (283).[187] Several examples of dehydrogenation of nucleoside pyrrolidines are given, as well as the formation of

[182] K. Gollnichk, *Adv. Photochem.*, 1968, **6**, 1.
[183] N. Toshima and H. Hirai, *Tetrahedron Letters*, 1970, 433.
[184] J.-C. Bloch, *Tetrahedron Letters*, 1969, 4041.
[185] Y. Kanaoka, E. Sato, and O. Yonemitsh, *Chem. and Ind.*, 1968, 1250.
[186] M. Calvin and G. D. Dorough, *J. Amer. Chem. Soc.*, 1948, **70**, 699.
[187] M. Kawana and S. Emoto, *Bull. Chem. Soc. Japan*, 1969, **42**, 3539.

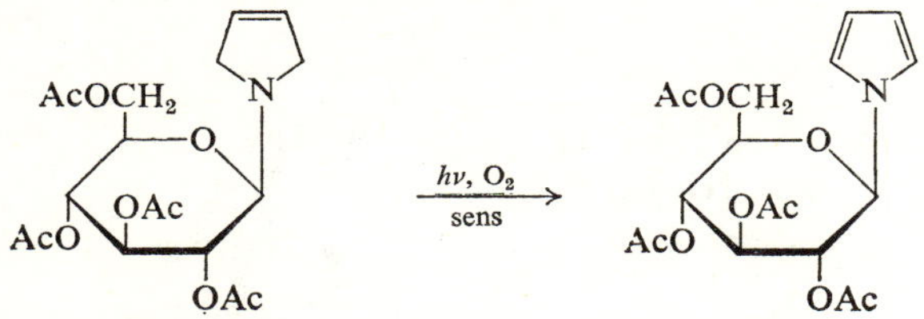

Photo-oxidation and -reduction Reactions

(283) → hv, O₂, sens → (282)

pyrrole from pyrroline. The reaction is carried out in the presence of oxygen and benzophenone: the fate of the latter is not seemingly considered. Thus, does the benzophenone act simply as a triplet sensitizer, or does it abstract hydrogen and end up as benzhydrol and benzpinacol? Pyrroles are also photolabile in oxidative systems. 2,5-Diphenyl- and 2,3,4,5-tetraphenyl-pyrroles in various solvents are reported to react with singlet oxygen to form the hydroperoxides (284).[188]

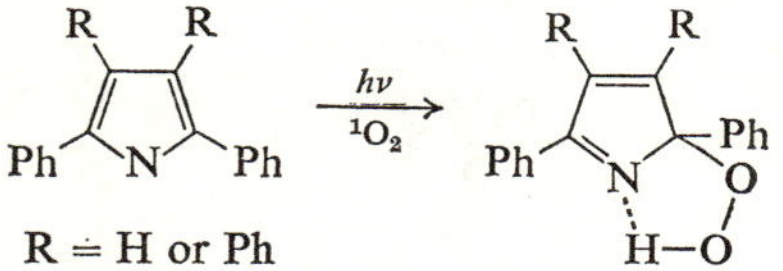

R = H or Ph

(284)

Still in connection with the photodynamic action of biological systems, Matsuura has studied the photosensitized oxygenation of 2,4,5-triphenylthiazole (285).[189] In methanol and in the presence of Rose Bengal, the thiazole (285) yields benzil (11%) and benzamide (18%), and 0·45 mol of oxygen is consumed. On the other hand, photo-oxidation sensitized by methylene blue in chloroform produces NN-dibenzylthiobenzimide (286) in good yield with the consumption of 0·9 mol of oxygen. Mechanisms involving an endo-peroxide (287) or a zwitterionic peroxide (288) are discussed.

The kinetics of the photosensitized decomposition of deoxyguanosine with methylene blue and oxygen have been studied.[190] The reaction rate increases with increase in concentration of either the nucleotide or oxygen. Evidence is presented for specific attack on the base part (289) of the nucleotide in the photoreaction. The deoxyribose residue is, not surprisingly, photostable under these conditions.

Further work on the photo-oxidation of histidine (290) and its N-benzoyl derivative has been reported.[191] From histidine, aspartic acid and urea have been identified, but the N-benzoyl derivative yields a 17-component

[188] G. Rio, A. Ranjon, O. Pouchot, and M. J. Scholl, *Bull. Soc. chim. France*, 1969, 1667.
[189] T. Matsuura and I. Saito, *Bull. Chem. Soc. Japan*, 1969, **42**, 2973.
[190] H. Fujita and H. Yamazaki, *Bull. Chem. Soc. Japan*, 1970, **43**, 1177.
[191] M. Tomita, M. Irie, and T. Ukita, *Biochemistry*, 1969, **8**, 5149.

mixture. It is considered that histidine (290) is first converted into a cyclic peroxide by the 1,4-addition of singlet oxygen to the imidazole ring. The cyclic peroxide then undergoes fission to yield such hydroxylic compounds as (291) and (292).

The photo-oxidation of N-butylcaprolactam (293) is, seemingly, also a complex process.[192] The reaction has been studied with radiation mainly in the 250—270 nm band, and yields N-butyrylcaproamide, caproamide,

[192] B. Lanska and J. Sebenda, *Coll. Czech. Chem. Comm.*, 1969, **34**, 1911.

(293) [structure: N-butyl caprolactam]

butyramide, water, carboxylic acids, aldehydes, carbon dioxide, carbon monoxide, hydrogen, hydrocarbons, and compounds containing active oxygen. It is hardly surprising that 2·55 mol of oxygen was absorbed during the reaction.

The 'photochemical' oxidation product, 5-phenylazo-3-phenyl-1,3,4-thiadiazole-2-one (294), of 1,5-diphenyl-3-mercaptoformazan (295) in chloroform would appear to involve nothing more than the light-induced

$$CHCl_3 \xrightarrow[{[O]}]{h\nu} COCl_2 + HCl$$

$$COCl_2 + \underset{(295)}{\begin{array}{c}Ph\\ \diagdown\\ N-N\\ H \quad \diagdown\\ \quad C-N=N-Ph\\ H-S \diagup\end{array}} \xrightarrow{-2HCl} \underset{(294)}{\begin{array}{c}Ph\\ \diagdown\\ N-N\\ \diagdown\\ O= \quad C-N=N-Ph\\ S \diagup\end{array}}$$

formation of phosgene from chloroform, and its subsequent reaction with (295), since the thermal reaction of (295) with phosgene certainly yields (294).[193]

6 Miscellaneous Oxidations

The photo-oxidation of *dl*- and *meso*-2,4-dichloropentanes has been studied as a model for the similar light-induced reaction of polyvinyl chloride (PVC).[194] The reactions were carried out with 253·7 nm radiation with PVC in thin films and the dichloropentanes at 0—10 °C. The structural unit —CHCl—CH_2—COCH_2— was identified, and the initial rates of reaction of the model compounds were comparable with that of PVC. The *dl*-dichloropentane was oxidised 1·5 times faster than the *meso*-compound and thus it seems likely that the syndiotactic sequences in PVC might be more easily oxidised than the isotactic sequences.

Various inorganic ethers have been prepared photochemically by the irradiation of silicon hydrides and tin hydrides in the vapour phase in the presence of nitric oxide.[195] Thus, the oxidation of silane or trimethylstannane (Me_3SnH) under these conditions yielded disiloxane, $(SiH_3)_2O$, and hexamethyl distannoxane, $(Me_3Sn)_2O$, respectively.

---

[193] H. M. N. H. Irving and D. C. Rupainwar, *Analyt. Chim. Acta*, 1969, **48**, 187.
[194] K.-P. S. Kwei, *J. Polymer Sci. Part A*, 1969, **7**, 237.
[195] R. Varma, A. K. Ray, and B. K. Sahay, *Inorg. Nuclear Chem. Letters*, 1969, **5**, 497.

In the presence of titanium dioxide, the u.v. irradiation of paraffins and olefins at room temperature leads selectively to ketones.[196] By such a procedure, n-butane has been converted into butanone and propane into acetone; the yields were about 30%.

The photochemistry of biacetyl has been the subject of numerous publications. The photo-oxidation in the liquid phase, using 350—500 nm radiation at 20 °C has now been reported.[197] In n-decane solutions of biacetyl, acetic anhydride is formed, while in carbon tetrachloride solution both acetic anhydride and possibly acetyl chloride have been detected by i.r. spectroscopy.

The rates of oxygen uptake in the liquid phase photo-oxidation of ethers, correlate with the orders of basicity of acyclic and cyclic ethers,[198] in further substantiation of the earlier proposed role of charge-transfer complexes in such reactions.[199] Tetrahydrofuran yields butyrolactone, α-hydroxy-tetrahydrofuran, and water, while diethyl ether leads to formation of ethyl acetate, ethyl formate, ethanol, and acetaldehyde. Thus, two

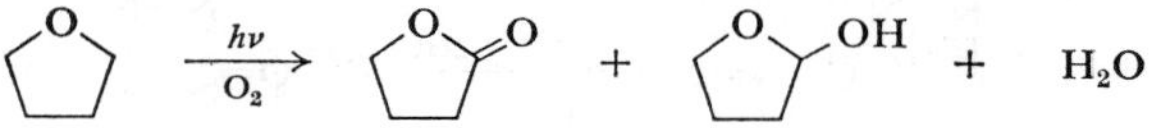

Scheme 15

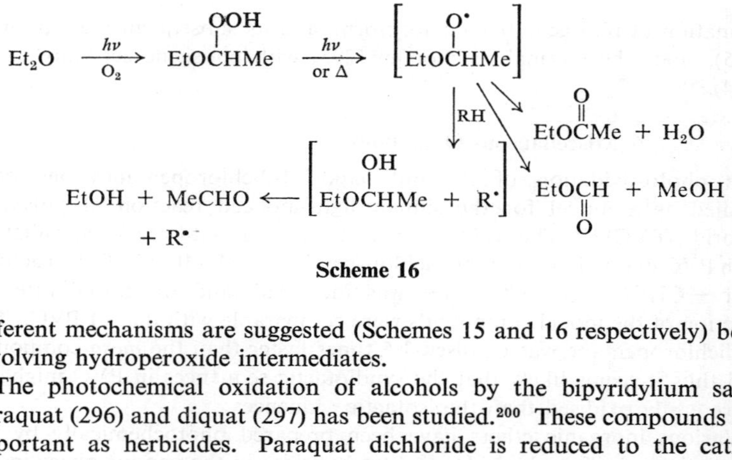

Scheme 16

different mechanisms are suggested (Schemes 15 and 16 respectively) both involving hydroperoxide intermediates.

The photochemical oxidation of alcohols by the bipyridylium salts, paraquat (296) and diquat (297) has been studied.[200] These compounds are important as herbicides. Paraquat dichloride is reduced to the cation radical $PQ^{+\cdot}$ by irradiation in aqueous solutions containing primary and secondary alcohols, but no reaction occurred in pure water or when

[196] M. Formenti, F. Juillet, and S. J. Teichner, *Compt. rend.*, 1970, **270**, C, 138.
[197] A. Cicolella, X. De Glise, M. Bouchy, J. C. Andre, J. Lemaire, and M. Niclause, *Compt. rend.*, 1969, **268**, C, 1929.
[198] N. Kulevsky, C. T. Wang, and V. I. Stenberg, *J. Org. Chem.*, 1969, **34**, 1345.
[199] J. C. W. Chien, *J. Phys. Chem.*, 1965, **69**, 4317.
[200] A. S. Hopkins, A. Ledwith, and M. F. Stam, *Chem. Comm.*, 1970, 494.

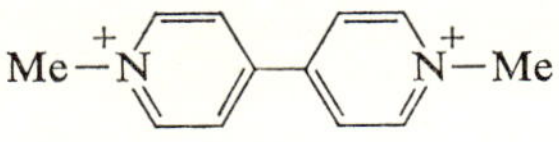

$$\text{Me}-\overset{+}{\text{N}}\underset{}{\diagup\!\!\diagdown}-\underset{}{\diagup\!\!\diagdown}\overset{+}{\text{N}}-\text{Me}$$
(296)

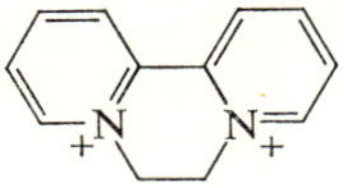

(297)

t-butanol was added. The alcohols are oxidised to carbonyl compounds according to the stoichiometry shown in equation 77. Both 313 and 334 nm

$$2PQ^{2+} + R^1R^2CHOH \xrightarrow{h\nu} 2PQ^{+\cdot} + R^1R^2C{=}O + 2H^+ \qquad (77)$$

radiation are effective in the reaction, but that of 366 nm is not. The rate of reaction is decreased as the reaction proceeds owing to quenching of the excited states or intermediates by the radical cation. However, when the irradiation vessels are open to the atmosphere, the radical cation is oxidised back to paraquat and high yields of carbonyl compound are obtained. Paraquat does not luminesce at room temperature, but diquat fluoresces (quantum yield = 0.04 ± 0.02) in water but not in alcohol. The deuterium kinetic isotope effect (MeOH/MeOD, $k_H/k_D = 1.4$) suggests that hydrogen abstraction is not the rate-determining step, and fluorescence quenching results have indicated that electron-transfer reactions are rate determining.

The formation of methanol in the photo-oxidation of methane has been reported,[201] and methanol also has been shown to be photolabile under 147 nm irradiation.[202] In the latter case, aqueous solutions of methanol were irradiated under argon, carbon monoxide, and oxygen atmospheres to give hydrogen, formaldehyde, glycolaldehyde, glycol, and formic acid in each case; but under oxygen, a peroxide was also detected among these products.

A. G.

---

[201] M. D. Museridze, A. A. Mantashyan, V. I. Kokochashvili, and A. A. Nalbandyan, *Armyan. khim. Zhur.*, 1969, **22**, 547.
[202] W. Zich and N. Getoff, *Monatsh.*, 1969, **100**, 1745.

6
Photoreactions of Compounds Containing Heteroatoms other than Oxygen

One review article has appeared during the last year which is relevant to this chapter. The subject matter is concerned with the photochemical reactions of azoxy-compounds, nitrones, and aromatic amine N-oxides, and the first three chapters deal in a most comprehensive way with these compounds.[1] There are two other chapters, one of which deals with mechanistic aspects of aromatic amine N-oxide photochemistry, and the other considers the mass spectrometric aspects of such compounds. The review is most complete, and will be of use to all workers in the field as well as those who wish to commence research in this area.

1 Rearrangement Reactions

N-Oxide Rearrangements.—The photorearrangement of aromatic amine N-oxides is well known to yield ring-expansion and -contraction products, and the parent amine; in some instances oxidation of the substrate also occurs.[1] The isomerization processes usually involve an oxaziridine intermediate (1). Pyridine N-oxides have been examined previously, but further interesting reports of work with various substituted derivatives continue to appear. 2,5-Dimethylpyridine N-oxide (2) on u.v. irradiation displays two types of reaction.[2] These are (a) transfer of oxygen to the substrate, and (b) isomerization to yield the acetylpyrrole (3) and the β-hydroxypyridine (4). The reaction is considered to arise via singlet states, and oxaziridine (5), oxazepine (6), and pyrrolenine (7) intermediates are suggested. The conversion of (6) to (7) involves a 1,3-sigmatropic displacement process. Such oxidation of aromatic molecules has been observed by other workers who were investigating pyridine N-oxide photolysis as a mechanistic model for enzymatic oxidation.[3] Thus irradiation at 254 nm of methylene chloride solutions of naphthalene and pyridine N-oxide leads to a 1% yield of α-naphthol and the epoxide (8). Formation of (8) constitutes the first example of chemical epoxidation of an aromatic bond. (Analogous aromatic epoxides are probably involved as inter-

---

[1] G. G. Spence, E. C. Taylor, and O. Buchardt, *Chem. Rev.*, 1970, **70**, 231.
[2] J. Streith and C. Sigwalt, *Bull. Soc. chim. France*, 1970, 1157.
[3] D. M. Jerina, D. R. Boyd, and J. W. Daly, *Tetrahedron Letters*, 1970, 457.

mediates in the biological oxidative degradation of aromatic rings.) It would seem that this photolysis of pyridine-*N*-oxides resembles many oxidation reactions typical of mixed function oxidases.

Last year saw the first report of photolysis of an *N*-oxide which had an internal azo function [pyridazine *N*-oxide (9)].[4] The reaction was carried out in benzene solution and gave phenol. Further work on such systems has now been published, and the reaction in the presence of olefins is reported.[5] Irradiation of 6-methylpyridazine *N*-oxide (10) in the presence of cyclohexene gives a 20—30% yield of a 5 : 1 mixture of cyclohexene oxide and cyclohexanone. A small amount of cyclohexane-1,2-diol is also formed. Many olefins were studied in the reaction, and the epoxide was generally the major product. Styrene gave styrene oxide rather than a phenol. The mechanism of such transfer oxidation reactions is seemingly not known, but phenol may arise by rearrangement of an initially produced benzene oxide (11).

Irradiation of 3,6-diphenylpyridazine *N*-oxide (12) gave the ring-contraction product, 3-benzoyl-5-phenylpyrazole (13), possibly *via* the

---

[4] H. Igeta, T. Tsuchiya, M. Yamada, and H. Arai, *Chem. and Pharm. Bull. (Japan)*, 1968, **16**, 767.
[5] T. Tsuchiya, H. Arai, and H. Igeta, *Tetrahedron Letters*, 1969, 2747.

diazo-compound (14).[6] The reaction has now been studied with saturated azine *N*-oxides in the hope of finding a route to the unexplored β-, γ-, and δ-diazoketones.[7] Irradiation of (15) gave two products. The first was shown to be 1,5-diphenylpent-4-en-1-one (16) (44%), which most likely arises by decomposition of the diazoketone (17). All the spectroscopic and chemical properties of the second product are consistent with structure (18). This formation of a four-membered-ring azoxy-compound represents the first authentic example of an electrocyclic ring-forming reaction in a diazabutadiene system. Compound (18) shows remarkable stability, with a half-life of 12 h in refluxing 'glyme'. The inability of (18) to revert thermally to (15), and its stability, are consistent with qualitative Woodward–Hoffmann predictions for such systems. Irradiation of (18) with 260 nm light yields the ketone (16), probably *via* the azine *N*-oxide (15).

[6] P. L. Kumler and O. Buchardt, *J. Amer. Chem. Soc.*, 1968, **90**, 5640.
[7] W. R. Dolbier and W. M. Williams, *J. Amer. Chem. Soc.*, 1969, **91**, 2818.

Irradiation of the pyrimidine *N*-oxide (19) in benzene solution gives the parent 5-methylpyrimidine and phenol, together with a photoisomer, probably (20).[8] The genesis of (20) is considered to involve a 1,3-cyclization of (19) to form an oxaziridine, followed by valence tautomerism to give 1-oxa-2,6-diazepine (21). Rupture of the N—O bond in (21) occurs concomitant with a shift of the C-3 proton to the oxygen, and keto–enol tautomerism finally leads to the formamide function in (20).

Quinoline- and isoquinoline-*N*-oxides have been popular compounds to study since Buchardt really started things moving in this area of photochemistry. Both reports concerned with such compounds this year have originated from the Danish research school. The composition of the products of the photolysis of phenylquinoline *N*-oxides in solution is found to be solvent-dependent.[9] Thus, while 3-phenylquinoline *N*-oxide (22) in

[8] J. Streith and P. Martz, *Tetrahedron Letters*, 1969, 4899.
[9] O. Buchardt and P. L. Kumler, *Acta Chem. Scand.*, 1969, **23**, 2149.

96% ethanol yields 3-phenylcarbostyril (23), in acetone the major path leads to 4-phenylbenz[d][1,3]oxazepine (24), and (23) is a minor product. 4-Phenylquinoline N-oxide (25) yields 3-phenyl-2-indolecarboxaldehyde (26), and although such a process is well known with pyridine N-oxides, production of (26) in the quinoline series is novel and is here explained by the intermediacy of the nitrene (27). The observed solvent effects may be partly accounted for by the mechanism previously proposed for such rearrangements,[10] but in the new report it is suggested that the oxaziridine (28) undergoes heterolysis to the zwitterionic intermediates (29a) and

(29b), which react by pathway *a* or *b* to yield the carbostyril (30). The authors point out that the competing rearrangement of the intermediate (28) to (31) can be formulated as a thermally allowed suprafacial [1,5]- or [1,9]-sigmatropic shift. The conversion of (31) into (32) has many precedents in the literature. From X-ray data, the main product from photolysis of 1-cyano-4-bromoisoquinoline N-oxide (33a) in acetone, is deduced to be 2-cyano-5-bromobenz[f][1,3]oxazepine (34).[11] Comparison of the i.r. and u.v. spectra of (34) with those of products formed by photolysis of other isoquinoline N-oxides (33b—e) provides strong evidence for the benz[f][1,3]oxazepine structure in these compounds also. Again, the rearrangement of the oxaziridine intermediate to yield the oxazepines (34)

[10] O. Buchardt, P. L. Kumler, and C. Lohse, *Acta Chem. Scand.*, 1969, **23**, 159.
[11] O. Simonsen, C. Lohse, and O. Buchardt, *Acta Chem. Scand.*, 1970, **24**, 268.

can be formulated as a [1,5]-sigmatropic shift to (35) followed by ring expansion.

Until this year, the photochemistry of N-alkoxy-derivatives of aromatic amine N-oxides had not been studied, but Hamana and Noda now report on the light-induced reactions of N-ethoxyquinolinium perchlorate (36).[12] In methanol solution, (36) yields 2- and 4-hydroxymethylquinolines [(37) and (38) respectively], and the diol (39). Compound (39) is shown to be formed by a secondary reaction of (37) with oxygen, and (37) is isolated as the perchlorate salt (40). The authors have not yet discussed the mechanisms in any detail.

A further example [13] of N-oxide rearrangement in the quinazoline series has been reported, and involves the photorearrangement of (41) in benzene solution to the benz[f][1,3,5]oxadiazepine (42).[14] The chloromethylene compound (43) undergoes photo-cleavage and eventually gives the benzoxazole (44).

---

[12] M. Hamana and H. Noda, *Chem. and Pharm. Bull. (Japan)*, 1969, **17**, 2633.
[13] G. F. Field and L. H. Sternbach, *J. Org. Chem.*, 1968, **33**, 4438.
[14] K. H. Wuensch and H. Bajdala, *Z. Chem.*, 1970, **10**, 144.

The photo-conversions of tetrahydroacridine N-oxides (45) into the bridged oxazepine (46) and the ketone (47) have previously been described.[15] A report of further work on this system has now appeared, and both (46) and (47) are suggested to arise from the oxaziridine, with (48) being a second intermediate for (47).[16] The reactions are carried out in benzene solution. In methanol or ethanol solutions, irradiation (> 300 nm) of the N-oxides (45) yields the spiro-compounds (49) (35—70%) and small amounts of the keto-indole derivatives (50) and (47). Earlier attempts to obtain the oxazepine (51) from acridine N-oxide (52) had produced good yields of the unsaturated ketone (53), but the intermediacy of (52) had been

[15] C. Kaneko, S. Yamada, I. Yokoe, and M. Ishikawa, *Tetrahedron Letters*, 1967, 1873; M. Ishikawa, C. Kaneko, and S. Yamada, *ibid.*, 1968, 4519.
[16] C. Kaneko, I. Yokoe, S. Yamada, and M. Ishikawa, *Chem. and Pharm. Bull.* (*Japan*), 1969, **17**, 1290.

suggested.[15] This reaction has been further studied with 9-cyanoacridine-10-oxides (54).[17] Five products are formed from the photolysis of (54), viz. the sought-after 2-aza-1,6-oxido[10]annulenes (55), azepinoindoles (56), oxepino[2,3-b]quinolines (57), oxabicyclo[3,2,0]heptadienes (58), and the parent amine, the latter in low yield. The formation of (58) was shown to arise by light-induced ring-closure of (57), and again oxaziridine intermediates are suggested in the process. A different mode of reaction from those outlined above has been observed for the photolysis of quinoline N-oxide in methanol or ethanol.[18] Here a rearrangement of the amine oxygen occurs together with attachment of a solvent molecule. The photoreduction products are thus substituted dihydrodibenzoxazepines (59).

Reactions of Pyridinium Ylides.—The interest in the photochemistry of pyridinium ylides is only recent, but already the topic has received considerable attention. The first report on this topic concerned compounds of

---

[17] C. Kaneko, S. Yamada, and M. Ishikawa, *Chem. and Pharm. Bull. (Japan)*, 1969, **17**, 1294.
[18] H. Mantsch, V. Zanker, W. Seffert, and G. Prell, *Annalen*, 1969, **723**, 95.

type (60),[19] but later publications have described iminopyridinium ylide (61) rearrangements,[20, 24] when processes of ring expansion to 1H-1,2-diazepines (62), rearrangement to 2-aminopyridines (63), and photo-Curtius rearrangements have been observed.[21] The first type of reaction appears to be general for such compounds as (61), and provides access to new diazepines. The originators of this area of research report that the dicyano-compound (64) in benzene undergoes light-induced rearrangement to yield the vinylpyrrole (65) and dicyanonorcaradiene (66).[22] The scope of the rearrangements of (61)-type compounds has been explored in detail by both Canadian and Japanese groups. Photolysis of the acetylpyridinium ylides (67) in methylene chloride at 350 nm again leads to ring expansion and formation of diazepines (68), but in this reaction fragmentation also occurs to yield the pyridines (69) and methyl isocyanate, which is the result of a Curtius rearrangement.[21] On the other hand, the ylides (70) and (71) give high yields of aniline and the parent pyridines.[23] The reaction is suggested to involve heterolytic cleavage of the N—N bond to yield the pyridine and phenylnitrene. Singlet phenylnitrene undergoes C—H bond insertion, and

---

[19] J. Streith and J.-M. Cassal, *Compt. rend.*, 1967, **264**, *C*, 1307; J. Streith, B. Danner, and C. Sigwalt, *Chem. Comm.*, 1967, 979.
[20] J. Streith and J.-M. Cassal, *Angew. Chem. Internat. Edn.*, 1968, **7**, 129; J. Streith and J.-M. Cassal, *Bull. Soc. chim. France*, 1969, 2175.
[21] V. Snieckus, *Chem. Comm.*, 1969, 831.
[22] J. Streith, A. Blind, J.-M. Cassal, and O. Sigwalt, *Bull. Soc. chim. France*, 1969, 948.
[23] V. Snieckus and G. Kan, *Chem. Comm.*, 1970, 172,

in the case of (70), compound (72) is formed but undergoes hydrogen abstraction and hydrolysis of the resulting Schiff's base during chromatography so that (73) is the isolated compound. Further evidence for nitrene

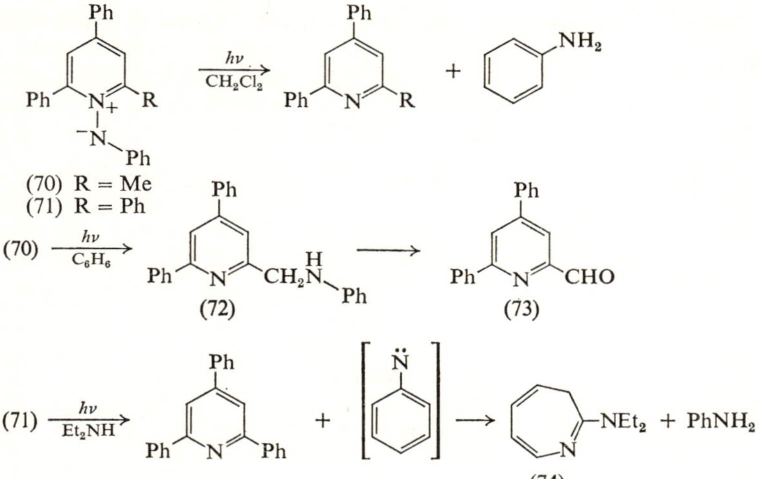

(70) R = Me
(71) R = Ph

intermediates is provided by the photolysis of (71) in diethylamine, when, in addition to the pyridine, minor amounts of 2-diethylamino-3H-azepine (74) are detected.

The irradiation of 1-ethoxycarbonyliminopyridinium ylides is the subject of three reports. A series (75) of such compounds has been studied, and in

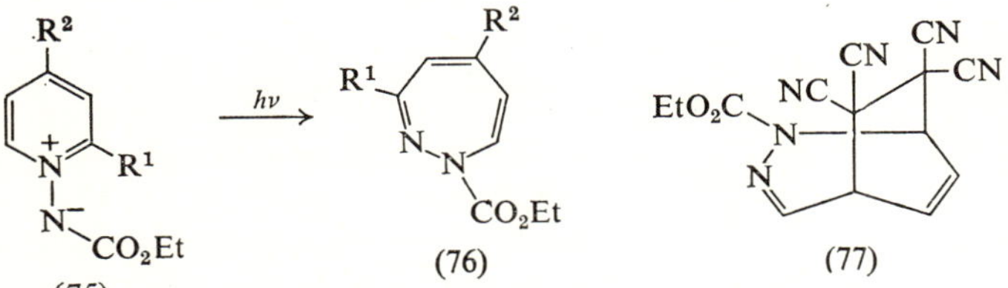

(a) R^1 = Me, R^2 = H
(b) R^1 = H, R^2 = Me
(c) R^1 = R^2 = Me
(d) R^1 = R^2 = H

each case the 1-ethoxycarbonyl-1H-1,2-diazepines (76) were formed.[24] Compounds (76) were found to be surprisingly inert to dienophiles such as maleic anhydride and acetylenedicarboxylic ester, but with tetracyanoethylene, the 1 : 1 adducts (77) were formed. The other two reports describe

[24] T. Sasaki, K. Kanematsu, and A. Kakehi, *Chem. Comm.*, 1969, 432.

photoreactions of many examples of this system in various solvents.[25, 26] The mechanism of formation of (76) is suggested to involve the intermediate diazabicyclo[4,1,0]heptadiene, and results with the ylide (78) suggest that the intramolecular cyclization occurs on the less-hindered α-carbon (Scheme 1).[25] In the case of 2,6-lutidine, and the 2,4,6-collidine derivatives

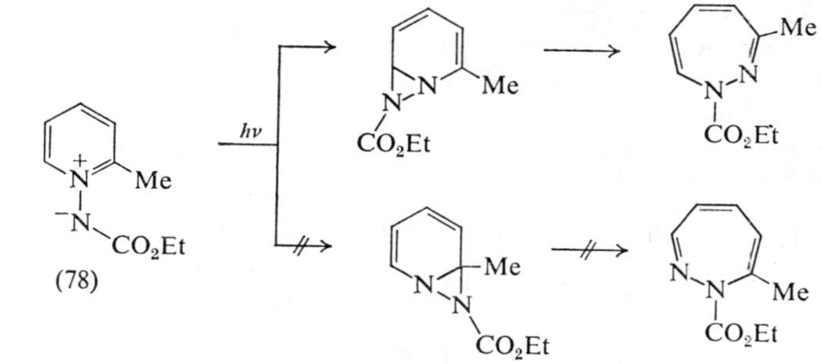

Scheme 1

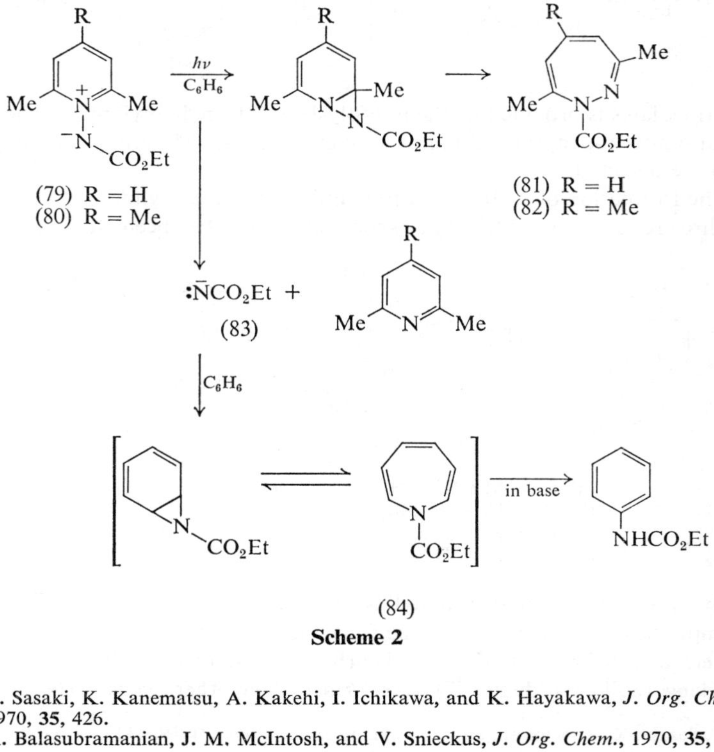

Scheme 2

[25] T. Sasaki, K. Kanematsu, A. Kakehi, I. Ichikawa, and K. Hayakawa, *J. Org. Chem.*, 1970, **35**, 426.
[26] A. Balasubramanian, J. M. McIntosh, and V. Snieckus, *J. Org. Chem.*, 1970, **35**, 433.

(79) and (80), in which the α-positions are occupied, the photoreaction yields the 1H-1,2-diazepine compounds (81) and (82), plus phenylurethane, in respective yields of 47, 76, and 2—5%. This suggests that cleavage of the ylides yields ethoxycarbonylnitrene (83) (Scheme 2), which reacts with the solvent benzene to give N-ethoxycarbonylazepine (84). Compound (84) rearranges to phenylurethane, a process which is known to occur under basic conditions.[27] Compounds (85a—f) and (86a, b) all demonstrate the photorearrangement with the formation of (87a—f) and (88a, b) respectively.[26] However, whereas (85f) undergoes the present rearrangement,

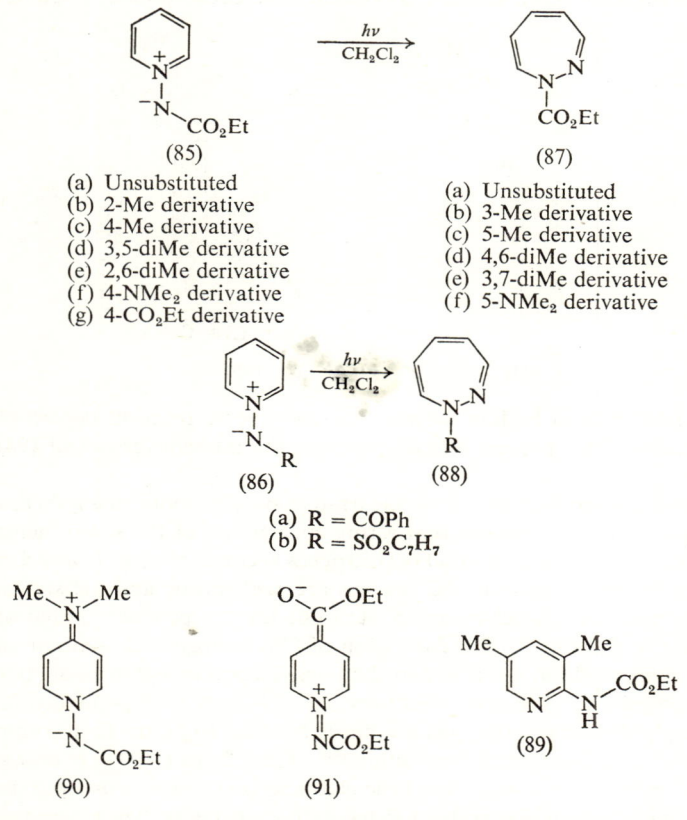

(85g) was found to be light-stable. The reason for this is suggested to lie in the large contributions of forms (90) and (91) to the respective excited states of the two ylides. The second major product from (85d) is shown to be (89) by independent synthesis.

[27] K. Hafner, D. Zinser, and K.-L. Moritz, *Tetrahedron Letters*, 1964, 1733; W. Lwowski, *Angew. Chem. Internat. Edn.*, 1967, **6**, 897.

Ring Contraction and Intramolecular Cyclization of Cyclic Nitrogen-containing Compounds.—Photochemical ring contraction and intramolecular cyclization are well-known processes with unsaturated hydrocarbons. Many of the mono- and di-nitrogen analogues of such systems have been synthesized and photolysed. Reports of such reactions, together with publications describing reactions confined to nitrogen heterocycles, are considered here.

The thermal- and photo-[3,3]-sigmatropic rearrangements in the *cis*-2,2-dimethyl-3-isobutenylcyclopropyl isocyanate (92)–3,6-dihydro-3,3,6,6-tetramethyl-2*H*-azepin-2-one (93) system have been studied.[28] Compound

(92) yields (93) in boiling xylene, and the reverse reaction occurs photochemically. This process exactly parallels the interconversion of (94) and (95).[29]

One of the products (96) from the irradiation of tropone phenylhydrazone (97) not only demonstrates that a ring contraction of the seven-membered ring has occurred, but also that the nitrogens become involved in a cyclization process.[30] By-products of the process are azobenzene and the substituted azobenzene (98). Mechanisms to account for the product formation are outlined in Scheme 3. The formation of (98) is suggested to occur *via* the phenylazotropylium ion (99) and the phenylazocycloheptatrienol (100).

Photolysis of benzene solutions of 1,1-dibenzyl-1,2-dihydro-2,4-diphenylphthalazine (101) yields what is believed to be the first example of a stable, photogenerated azomethine.[31] Thus from (101), an orange-red compound is formed which has been characterized by spectroscopic analysis and from its 1,3-dipolar adducts to have structure (102). The rearrangement is suggested to arise from the first excited singlet state, and intermediates such as (103) and (104) cannot yet be discounted.

---
[28] T. Sasaki, S. Eguchi, and M. Ohno, *J. Amer. Chem. Soc.*, 1970, **92**, 3192.
[29] O. L. Chapman, M. Kane, J. D. Lassila, R. L. Loeschen, and H. E. Wright, *J. Amer. Chem. Soc.*, 1969, **91**, 6856, and references therein.
[30] T. Tezuka, A. Yanagi, and T. Mukai, *Tetrahedron Letters*, 1970, 637.
[31] B. Singh, *J. Amer. Chem. Soc.*, 1969, **91**, 3670.

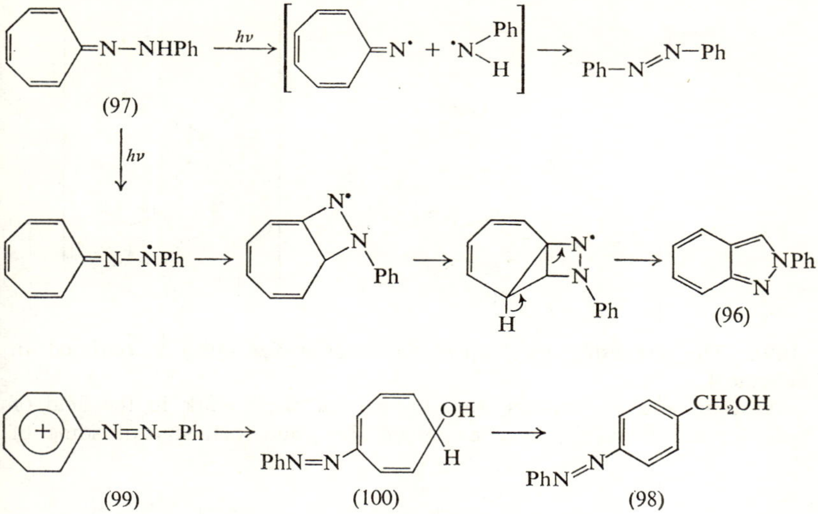

Scheme 3

From a knowledge of the photochemical reactions of benzocyclobutadienoquinone (105) in methanol, it was expected that benzo[3,4]cyclobuta[1,2-b]quinoxaline (106) might undergo a similar reaction with the formation of (107), via (108).[32] However, irradiation of (106) in methanol through Pyrex led to an 80% yield of 11H-isoindolo[2,1-a]benzimidazole

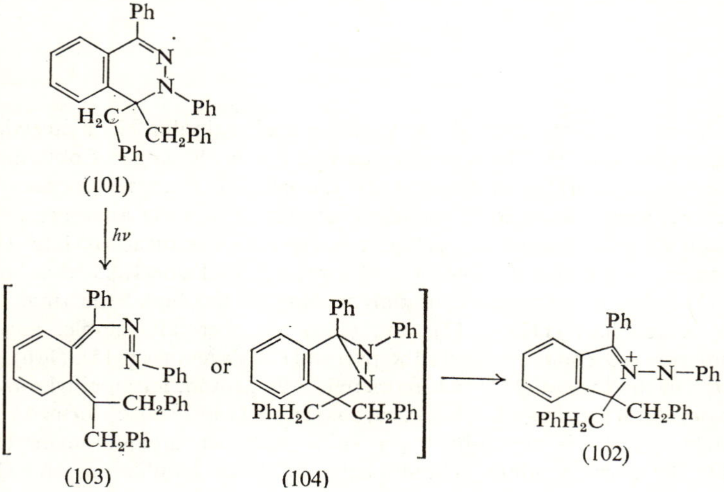

[32] J. I. Sarkisian and R. W. Binkley, *J. Org.Chem.*, 1970, **35**, 1228

(109). The suggested mechanism to account for (109) is outlined in Scheme 4.

Acheson and co-workers, well known for their work in the field of heterocyclic chemistry, have examined the photochemistry of some of the known adducts formed by pyridine and quinoline with acetylenedicarboxylic ester.[33] The work was carried out in the hope of obtaining a better route to 4*H*-quinolizines than the existing thermal process. The reactions were examined in methanol solution in quartz apparatus. The photolysis of tetramethyl 4*H*-quinolizine-1,2,3,4-tetracarboxylate (110) gave tarry material and a low yield of trimethylindolizine-1,2,3-tricarboxylate (111) by, it is suggested, initial opening of the butadiene ring. The 9a*H*-quinolizines (112), (113), and (114) are thermally stable, and on irradiation were found to yield the pyrrolo[1,2-*a*]azepines (115), (116), and (117). In the case of (112), where there is a 9a-proton, tetramethyl 6-*trans*-styryl-4*H*-quinolizine-1,2,3,4-tetracarboxylate (118) is also formed, presumably by a 1,3- or 1,5-hydrogen shift. 4a*H*-Benzo[*c*]quinolizine (119) yields the corresponding 1*H*-quinolizine (120) on irradiation, and (121)

---

[33] R. M. Acheson and J. K. Stubbs, *J. Chem. Soc.* (*C*), 1969, 2316.

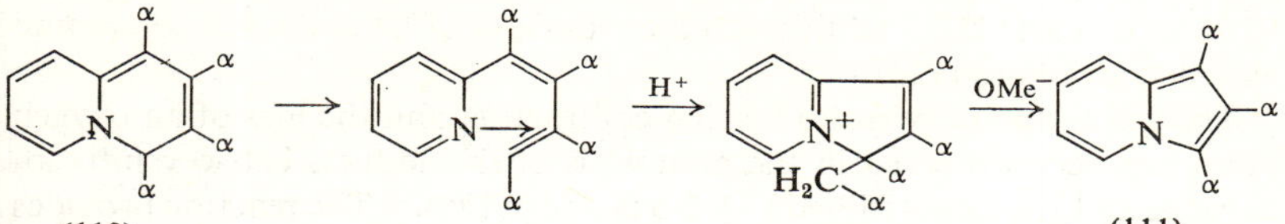

Photoreactions of Compounds with Heteroatoms other than Oxygen

(110) α≡CO₂Me

(111)

(112) 9aH-6-trans-styryl
(113) 6,9a-dimethyl
(114) 9a-vinyl

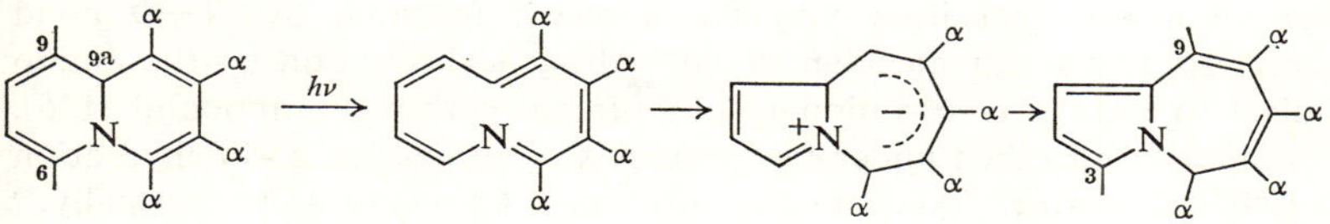

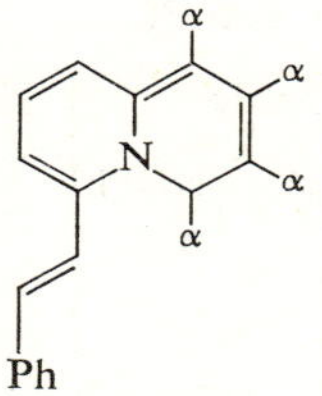

(115) 3-trans-styryl
(116) 3,9-dimethyl

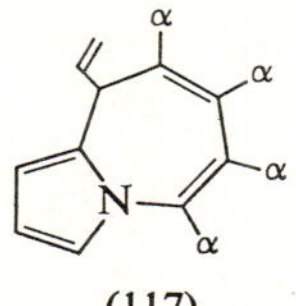

(117)

(118)

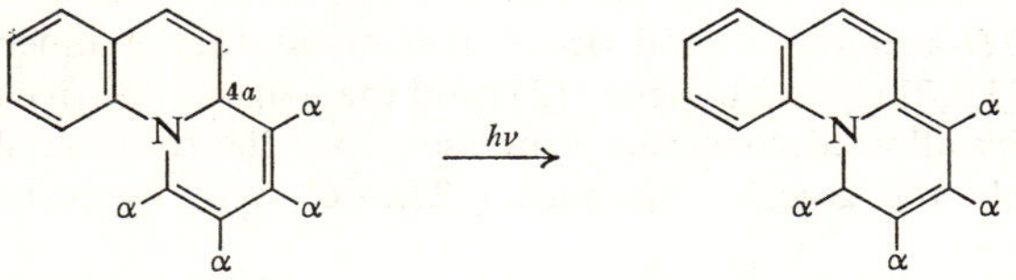

(119) → (120)

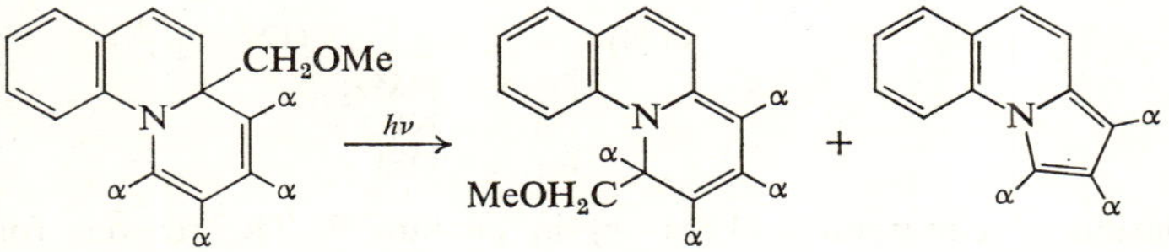

(121) → (122) + (123)

+

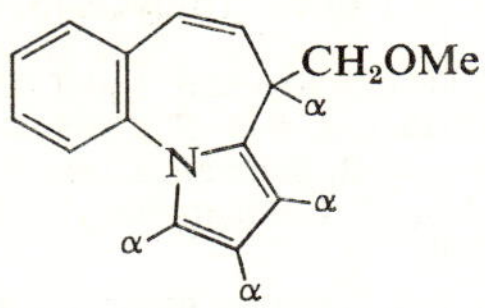

(124)

likewise forms (122) in addition to the indolizine (123) and a red compound assigned structure (124).†

The formation of pyrroles by ring contraction and the loss of an oxygen atom has been observed in the photolysis of the adducts between nitrosobenzene and 1,3-dienes (dihydro-1,2-oxazines) (125).[34] The reaction provides a convenient synthesis for phenylpyrroles. Neither oxygen nor naphthalene was found to be an efficient quencher of the reaction. The results suggest that an $n \to \sigma^*$ transition may be involved, followed by N—O bond homolysis and intramolecular allylic hydrogen abstraction by the amino radical to yield a cis-γ-amino-$\alpha\beta$-unsaturated carbonyl compound (126). Compound (126) then undergoes processes of cyclization and dehydration to yield the pyrrole. Cyclization of this type is known to occur thermally.[35] In support of the mechanism and the intermediacy of (126), irradiation of (125) at $-180\,°C$ led to the appearance of an absorption at 1695 cm^{-1}, which disappeared on warming to $-46\,°C$.

Because of the constraints of a seven-membered ring, the intramolecular cycloaddition of 3H-azepines should occur in a disrotatory manner and both 2-azabicyclo[3,2,0]hepta-2,6-diene (127) and the 6-aza-derivative (128) could be formed by allowed concerted processes. But the reaction shows selectivity and only the 2-aza-compounds (127a—c) were formed from

(a) X = NMe$_2$
(b) X = NH$_2$
(c) X = OEt

irradiation of compounds (129a—c) in pentane.[36] The reasons for the specificity of the reaction are obscure, since in direct contrast to the present observations, 1-ethoxycyclohepta-1,3,5-triene [an analogue of (127a)] is known to yield 1-ethoxybicyclo[3,2,0]hepta-3,6-diene.[37] The authors point

[34] P. Scheiner, O. L. Chapman, and J. D. Lassila, *J. Org. Chem.*, 1969, **34**, 813.
[35] R. M. Rodebaugh and N. H. Cromwell, *Tetrahedron Letters*, 1967, 2859.
[36] R. A. Odum and B. Schmall, *Chem. Comm.*, 1969, 1299.
[37] G. W. Borden, O. L. Chapman, R. Swindell, and T. Tezuka, *J. Amer. Chem. Soc.*, 1967, **89**, 2979.

† It is instructive to compare these results with those in ref. 150 described later in this chapter.

out that in (128), the isomer which is not formed, there is a 1-azetidine moiety, and it is known that such systems are found in the products of intramolecular electrocyclizations.[38] In such reactions, however, there was no alternative mode of reaction, and formation of the 1-azetidine did not require concomitant loss of the resonance energy of amidine or imidate ester. This may be a factor which determines the reaction path.

The thermal reaction of methoxycarbonylnitrene with cycloheptatriene yields 2,3- and 4,5-homo-1H-azepines (130a) and (130b). The photolysis of both these products has been examined, and while (130b) appears light-stable, (130a) yields the cyclized product 5-carbomethoxy-5-aza-1α, 2α, 4α, 6α-tricyclo[4,2,0,0^{2,4}]oct-7-ene (131), and 2-carbomethoxy-2-azabicyclo[3,3,0]octa-3,6-diene (132).[39] The reaction of (130a) is presumed to involve the $\pi\pi^*$ singlet state. The stereochemistry has not been definitely established, but (131) is thought to arise from a disrotatory closure (133) of (130a) rather than formation of the 1β, 2α, 4α, 6β-isomer from (134), since in this latter case there are adverse steric interactions which develop

between the two terminal vinyl and the *endo* cyclopropyl hydrogens in the transition state. Photolysis of 3H-azepine systems such as (135) also leads to an intramolecular cyclization product (136) as well as other rearrangement products (137) and (138).[40]

---

[38] C. Lohse, *Tetrahedron Letters*, 1968, 5625.
[39] L. A. Paquette and R. J. Haluska, *J. Org. Chem.*, 1970, **35**, 132.
[40] M. Ogata, H. Matsumoto, and H. Kano, *Tetrahedron*, 1969, **25**, 5217.

1-Azatwistanes have been synthesized photochemically.[41] Thus, irradiation of the N-chloro-compound (139) in trifluoroacetic acid gives an 80% yield of the two isomeric chloroazatwistanes (140) and (141) in the ratio 4:3. Separation of (140) and (141) is realized in benzene solution by addition of sufficient toluene-p-sulphonic acid to neutralize the more basic (140).

There are many examples in the literature of di-π-methane photorearrangements of olefinic hydrocarbons, and the similar aza-compounds have now been studied. Irradiation of 5,6-benzo-2-azabicyclo[2,2,2]octa-5,7-dien-3-ones (142) in acetone yields 92% of a 1:3 mixture of compounds (143) and (144).[42] Compound (145) undergoes similar rearrangement to (146), either by direct irradiation, or when photosensitized by acetone.[43]

The light-induced elimination of keten from such systems as (147), and the sensitized 1,2-acyl migration to yield a semibullvalene, are both known processes.[44,45] The system (148), in which there is a σ-bond joining the two atoms (X) bearing n-electrons, has now been examined.[46] Direct irradiation of (149) in ether has been found to give two products, the relative

[41] K. Heusley, *Tetrahedron Letters*, 1970, 97.
[42] L. A. Paquette and R. H. Meisinger, *Tetrahedron Letters*, 1970, 1479.
[43] L. A. Paquette, J. R. Malpass, and G. R. Krow, *J. Amer. Chem. Soc.*, 1970, **92**, 1980.
[44] R. K. Murray and H. Hart, *Tetrahedron Letters*, 1968, 4995.
[45] H. Hart and R. K. Murray, *Tetrahedron Letters*, 1969, 379.
[46] H. Hart, R. K. Murray, and G. D. Appleyard, *Tetrahedron Letters*, 1969, 4785.

yields of which are dependent upon irradiation time. Thus (150) and (151) are formed, and (150) loses carbon monoxide to yield (151). Sensitized photolysis of (149) by irradiation in acetone produces minor amounts of (150) and (151), but the major product (90% yield) is a new crystalline

compound (152). The isomerization of (149) to (150) is suggested to arise from the excited singlet state, while the formation of (152) is evidently a triplet-state reaction.

Reactions of Five-membered Heterocyclic Compounds.—In the past, photorearrangements of five-membered ring heteroaromatic compounds have been studied in some detail. Such rearrangements involve substituent-group migration (cf. photoisomerization in the xylenes) and apparent 1,2-migrations of heteroatoms as common modes of reaction. Examples of the latter are provided in the isomerizations of indazole to benzimidazole, pyrazole to imidazole, 3,5-diphenylisoxazole to 2,5-diphenyloxazole, and benzisoxazole to benzoxazole. Rearrangement of 1,3- to 1,2-systems had not been described until two reports appeared this year. Thus irradiation of 2,5-diphenyloxazole (153) in ethanol yields 3,5-diphenylisoxazole (154) (3%).[47] Other products from the reaction are 4,5-diphenyloxazole (155) (20%), phenanthra[9,10]oxazole (156) (1·5%), and benzoic acid (1%). Compound (155) was found to be the precursor of (156), as expected. In benzene solution, (153) yields (154) (7%), benzoic acid (2%), dibenzamide (0·4%), and a new product, 2,4-diphenyloxazole (157) (4·5%). The conver-

[47] M. Kojima and M. Maeda, *Tetrahedron Letters*, 1969, 2379.

sion of (154) into (153) is known to involve the intermediacy of (158),[48] but this species could not be isolated in the latest work.[47] The reaction mechanism for the formation of (155) in ethanol solutions is probably analogous to that given for the sulphur compounds in the upper line of Scheme 5. The second report of a 1,3- to 1,2-conversion in such systems was from the same research group and concerned the photorearrangements of 2,5- and 2,4-diphenylthiazole.[49] The latter compound yields 3,5-diphenyl-isothiazole (159), while the 2,5-diphenyl isomer is converted to products (160), (161), and (162). The mechanism cannot be accounted for by a ring contraction–expansion sequence, and involvement of the bicyclic intermediates (163), (164), (165), and (166) is suggested (Scheme 5). On the other hand, irradiation of isothiazole in propylamine is reported to yield thiazole and at least five other products.[50] In ether as solvent, the conversion is very low, and no matter what the solvent, the reverse reaction of the thiazole to isothiazole (cf. refs. 47 and 49) does not occur.

The molecular environment can dictate the course of photorearrangement, and in 1,2-heteroaromatics such as the anthranil (167), ring expansion occurs on irradiation to yield (168) or (169).[51] The scheme proposed for the rearrangement is similar to that earlier suggested for ring expansion of phenyl azide,[52] and involves conversion of the anthranil to a nitrene which then forms an azirene. The azirene to azepine step has been shown to

[48] E. F. Ullman and B. Singh, *J. Amer. Chem. Soc.*, 1966, **88**, 1844.
[49] M. Kojima and M. Maeda, *Chem. Comm.*, 1970, 386.
[50] J. P. Catteau, A. Lablache-Combie, and A. Pollet, *Chem. Comm.*, 1969, 1018.
[51] M. Ogata, H. Matsumoto, and H. Kano, *Tetrahedron*, 1969, **25**, 5205.
[52] W. von E. Doering and R. A. Odum, *Tetrahedron*, 1966, **22**, 81; R. Huisgen, D. Voissius, and M. Appl, *Chem. Ber.*, 1958, **91**, 1.

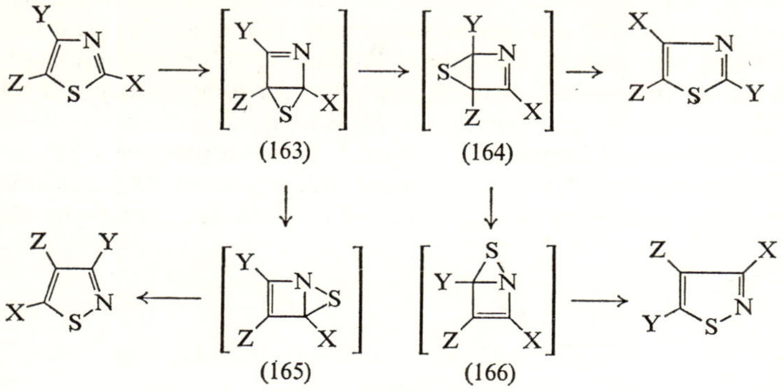

Photoreactions of Compounds with Heteroatoms other than Oxygen

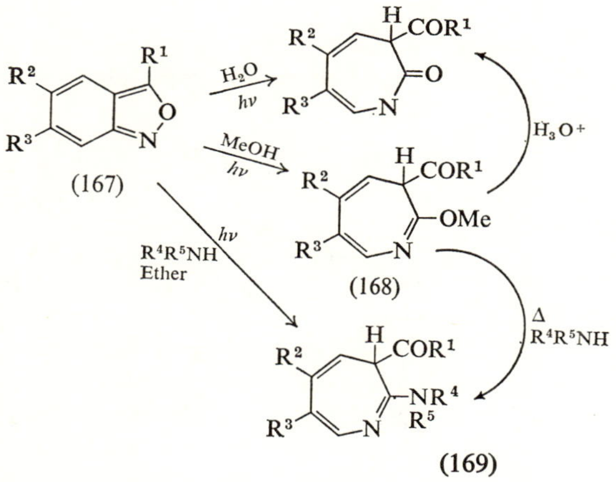

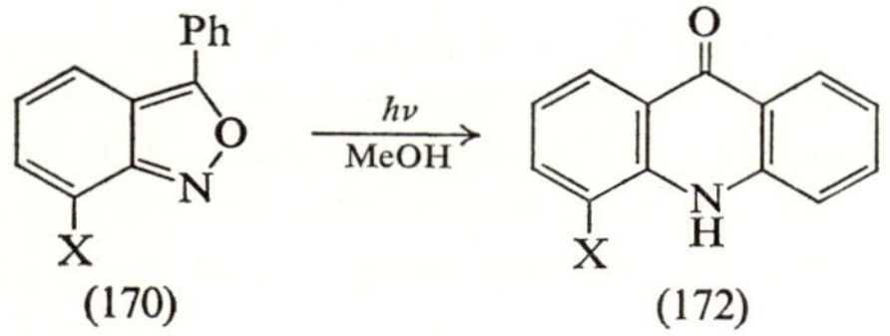

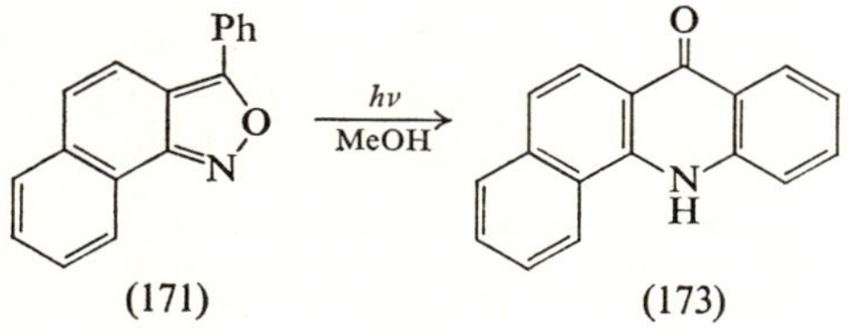

Scheme 5

involve a thermal reaction of the intermediate with protic solvents. On the other hand, 7-substituted-3-phenylanthranils (170) and (171) yield the corresponding 9-acridanone derivatives (172) and (173) on irradiation.

Methyl group migration has been observed for some N-methyldiazoles.[53] 1,4-Dimethylimidazole (174) in t-butanol is found to give a 40% conversion to the 1,2-dimethyl isomer (175). Under the same photolysis conditions, the back-reaction [(175) → (174)] occurs with a yield of 50% and in three-quarters the irradiation time. In ethanol solution, the trimethylimidazole

(176) is converted into the trimethyl isomer (177), and 1,3,5-trimethylpyrazole yields (178) and (179). Bicyclic intermediates (180) and (181) are again suggested for the interconversion of the imidazoles (174) and (175), and for rearrangement of (176) to its isomer (177). In contrast with the previously observed sensitized photorearrangements of some pyrazoles to imidazoles,[54] the present reactions were not sensitized by acetophenone or benzophenone.

General transposition of substituent groups is thus well known in aromatic heterocycles, but not so well authenticated for non-aromatic systems. However, irradiation of 2,3-dimethyl-1-phenyl-3-pyrazolin-5-one

---

[53] P. Beak and W. Messer, *Tetrahedron*, 1969, **25**, 3287.
[54] H. Tiefenthaler, W. Dorschelen, H. Göth, and H. Schmid, *Helv. Chim. Acta*, 1967, **50**, 2244, and references therein; S. N. Ege, *Chem. Comm.*, 1967, 488.

(182) in alcohols has now been reported to lead to transposition of the ring atoms to yield 1,5-dimethyl-3-phenyl-4-imidazolin-2-one (183), plus the appropriate ring-fragmented product, *N*-acetonyl phenyl carbamate (184).[55] Minor products (185), (186), and (187) of the reaction are also formed by ring opening and fragmentation. Irradiation of (182) in acetone yields (188), together with (183), (186), and (187). The effects of further saturation on the photo-lability of such systems were examined by photolysis of a series of 2-phenylpyrazolidin-3-ones (189a—c). The major products of the reaction in methanol were 1-anilinoazetidin-2-ones (190a—c) in all cases: minor products were the *cis*- and *trans*-isomers of phenylazoalkanoates (191) and (192). The *trans*-isomer was found to be partly converted to the *cis*-isomer by sunlight. Compounds (190a—c) are of interest since they represent the only known 1-aminoazetidin-2-ones. Previous reports of such compounds have involved the production of inner salts.[56] An imidazolin-2-one analogous to (183) is also formed from the irradiation of 5-imino-2,3,4-triphenyl-Δ^3-isoxazoline (193).[57] This reaction is carried

---

[55] S. N. Ege, *J. Chem. Soc.* (*C*), 1969, 2624.
[56] N. P. Zapevalova and T. A. Sokolova, *Bull. Acad. Sci., U.S.S.R.*, 1965, 1398.
[57] H. G. Aurich, *Annalen*, 1970, **732**, 195.

out in ethanol solution, and 3-[(α-phenylamino)benzylidene]indolin-2-one (194) is also formed. The aziridinone (195) is believed to be the intermediate in the formation of the imidazolin-2-one.

Continuing with their work on the photochemical transformations of small-ring heterocyclic compounds, Padwa and co-workers have reported on the light-induced isomerization of triphenyl-1,3-diazabicyclo[3,1,0]hex-3-ene (196).[58] In benzene solution at 50 °C, (196) gives a 94% yield of cis-2,3-dihydro-2,3,5-triphenylpyrazine (197). When the irradiation is carried out at 15 °C, virtually no (197) is formed, but a compound is produced which at 50 °C in the dark gives (197). From this thermal instability, and the behaviour on hydrolysis, the thermally labile photoproduct is assigned the ene-di-imine structure (198). The concerted transformation of (196) to (197) would be allowed by Woodward–Hoffmann

[58] A. Padwa, S. Cloush, and E. Glazer, *J. Amer. Chem. Soc.*, 1970, **92**, 1778.

Rules. The isomer (199) of (196) should, by similar argument, yield the *trans*-isomer (200) of (197), but irradiation at 50 °C led again to the *cis*-isomer (197). The offered explanation for the formation of the *cis* product is that the ene-di-imine exists as an equilibrium mixture of the two isomers (198) and (201), the more prevalent being (198). In support of this, irradiation of the *trans*-isomer (200) at 15 °C gave a labile ene-di-imine which thermally produced the *cis*-isomer (197), and irradiation of (197) at 15 °C gave an ene-di-imine which re-formed (197) on warming.

The diverse photo-processes of *N*-oxides have been reported earlier in this chapter. The *N*-oxide function was always found to be lost on subsequent rearrangement, but a system in which this does not occur has now been reported.[59] Thus irradiation of the 4*H*-pyrazole *N*-oxides (202a) and (202b) gives 70 and 29% yields of (203a) and (203b) respectively. Formation of the 1-oxide in preference to the 2-oxide is evidently favoured, and a

a) $R^1 = Ph, R^2 = Me$
b) $R^1 = Me, R^2 = Et$

reasonable mechanism to account for this involves the bicyclic intermediates (204) and (205). The conversion of (204) into (205) is analogous to the rearrangements in 4,4-disubstituted cyclohexadienone systems,[60] and the final step is similar to that reported from studies of the valence-bond isomerization of 3*H*-pyrazoles.[61]

A new diaziridine synthesis is reported in the photocyclization of azomethinimines (206) in dioxan.[62] In methanol, ring-opening occurs to yield (207). Re-formation of (206) from the diaziridine (208a) occurs either thermally or by acid catalysis. Photochemical synthesis has also been employed in the preparation of *N*-unsubstituted-1,2,4-triazoles (208b).[63] Compounds (208b) and sulphur are formed from any of the three precursors (209), (210), or (211). The yields are nearly quantitative and the compounds thus prepared are all found to be light-stable, apart from the 3-phenyl derivative.

Substituted five-membered tetrazoles are photo-labile. Irradiation of 2-methyl-5-phenyltetrazole (212) in dioxan gives rise to the triazole (213)

---

[59] W. R. Dolbier and W. M. Williams, *Chem. Comm.*, 1970, 289.
[60] H. E. Zimmerman and J. S. Swenton, *J. Amer. Chem. Soc.*, 1967 **89**, 906.
[61] G. L. Closs, W. A. Boll, H. Heyn, and V. Dev. *J. Amer. Chem. Soc.*, 1968, **90**, 173.
[62] M. Schultz and G. West, *J. prakt. Chem.*, 1970, **312**, 161.
[63] A. J. Blackman, *Austral. J. Chem.*, 1970, **23**, 631.

(22%) and a compound which is tentatively assigned the structure 1,2-di(methylazo)-1,2-diphenylethylene (214) (6%), the product of ring-opening.[64] The only other product formed in the reaction is the triazole (215), and this in less than 4% yield. These results are in contrast with previously reported studies on 2,5-diphenyltetrazole, when the nitrilimine was formed.[65] In the present work, the formation of the products is rationalized to occurring by an intermediate photodimer (216a). The triazole (215) might have arisen from a nitrilimine or a head-to-tail dimer. The possibility cannot be ruled out that a nitrilimine is first formed and adds to a molecule of the tetrazole. Loss of methyl azide from such an adduct would yield the triazole (213). However, attempts to trap the possible nitrilimine in excess of benzonitrile failed, and the precursor of (213) was shown not to be (214) since this was shown to be stable under the reaction conditions. On the other hand, the action of 253·7 nm radiation on the tetrazole (216b) leads to loss of nitrogen, and the product formed is solvent dependent.[66] Irradiation in isopropyl alcohol led to a 30% yield of the pseudo-urea derivative (217), whereas the use of acetonitrile led to 25% of benzimidazole (218). The intermediate for both reactions is suggested to be the azomethine biradical/nitrene (219).

[64] R. R. Fraser and K. E. Haque, *J. Org. Chem.*, 1969, **34**, 4118.
[65] J. S. Clovis, A. Eckell, R. Huisgen, and R. Sustmann, *Chem. Ber.*, 1967, **100**, 60.
[66] F. L. Bach, J. Karliner, and G. E. van Lear, *Chem. Comm.*, 1969, 1110.

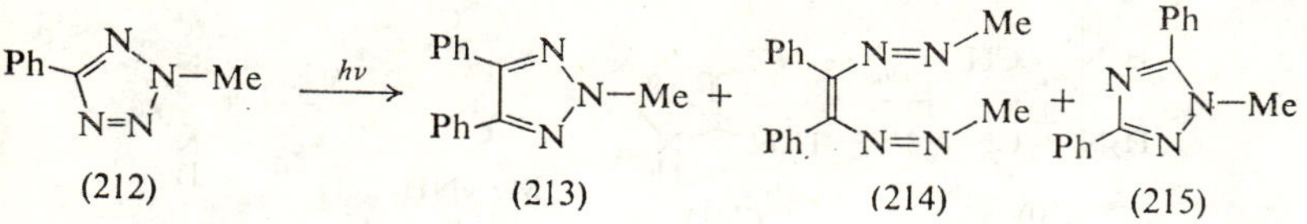

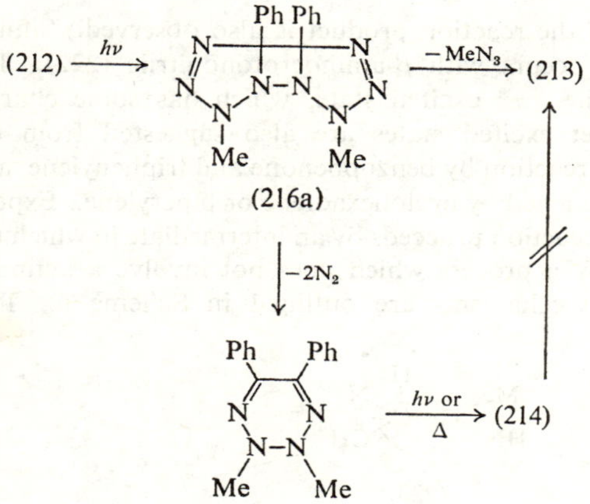

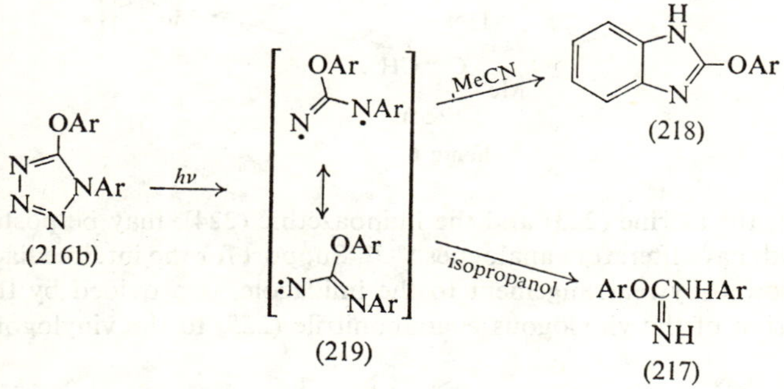

The remainder of the publications considered in this section do not involve reactions of five-membered nitrogen heterocycles, but are included here since they are concerned with the formation of such compounds by photorearrangement of other nitrogen-containing systems.

A critical step in the proposed prebiotic synthesis of purines is the conversion of the tetramer of hydrogen cyanide, diaminomaleonitrile (220), into 2-aminoimidazole-5-carbonitrile (221). The reaction is now reported to occur in high yield in the absence of oxygen in a variety of solvents; in fact the only solvent found to be unsuitable was chloroform.[67, 68]

[67] J. P. Ferris, J. E. Kuder, and A. W. Catalano, *Science*, 1969, **166**, 765.
[68] J. P. Ferris and J. E. Kuder, *J. Amer. Chem. Soc.*, 1970, **92**, 2527.

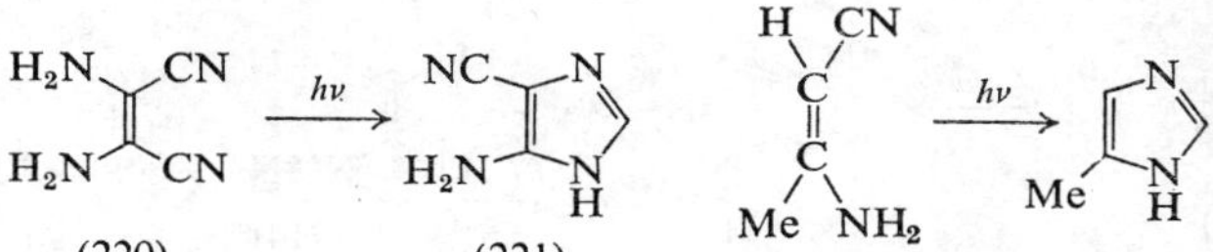

Some photodestruction of the reaction product is also observed. Similar photorearrangement also occurs with β-aminocrotononitrile (222). The reaction proceeds from the $\pi\pi^*$ excited state, which has some charge-transfer character. Singlet excited states are also suggested from the lack of sensitization of the reaction by benzophenone and triphenylene, and that the reaction is not quenched by cyclohexadiene or piperylene. Experimental data show that the reaction proceeds by an intermediate in which the C—CN bond is broken by a process which does not involve a ketimine intermediate. Suggested mechanisms are outlined in Scheme 6. Two

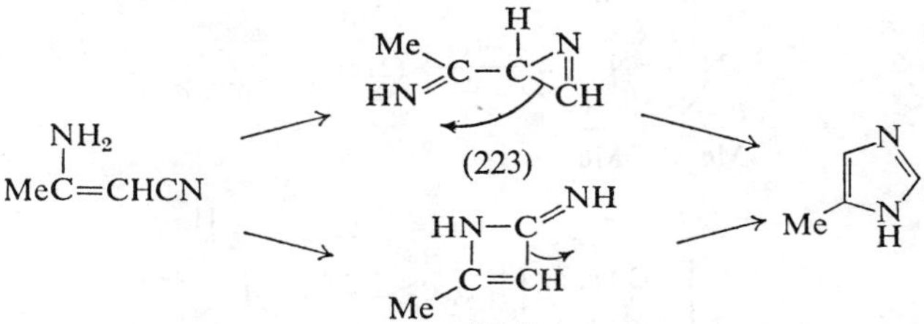

Scheme 6

intermediates, the azirine (223) and the iminoazetine (224), may be postulated, and both have literature analogues.[69, 70] Support for the intermediacy of (224), followed by rearrangement to the imidazole, is provided by the photoconversion of the vinylogous enaminonitrile (225) to the vinylogous

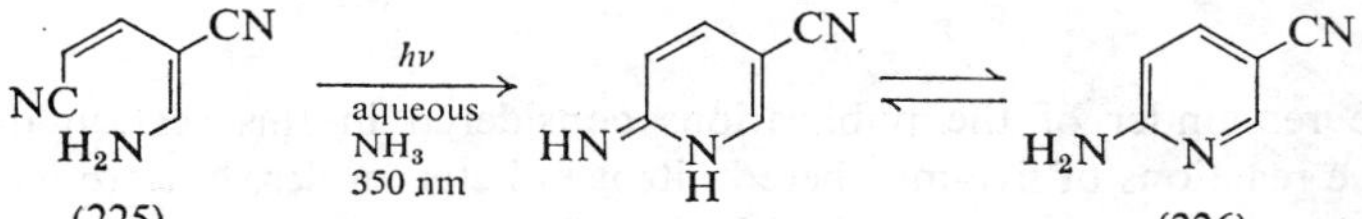

azetidine, 6-aminonicotinonitrile (226): here the intermediate is 'trapped.' This latter reaction also affords an efficient route to a nicotinamide derivative from cyanoacetylene.

The formation of pyrroles from photoinduced ring-expansion of aroylazetidines has been previously reported by Padwa and Gruber,[71] and the

---

[69] B. Singh and E. F. Ullman, *J. Amer. Chem. Soc.*, 1967, **89**, 6911; and ref. 54.
[70] K. Wierzchowski and D. Shugar, *Photochem. and Photobiol.*, 1963, **2**, 377.
[71] A. Padwa and R. Gruber, *J. Amer. Chem. Soc.*, 1968, **90**, 4456.

same workers this year have described further work on the reaction.[72] With a *cis*-azetidine (227), only one primary product (228) is formed, but

Ph_/OPh   Ph   Ph_/O   Ph
 [azetidine] hv→ [pyrrole]   [azetidine] hv→   [pyrrole] + (228)
   | | | |
  Bu^t Bu^t Bu^t Bu^t
 (227)   (228)   (229)

from the *trans*-isomer (229), both 2,3- and 2,4-diphenyl-*N*-t-butylpyrroles are produced. It is deduced from sensitization and emission studies that the reaction occurs from the lowest triplet state. The lack of quenching observed implies that the reaction of (227) is too rapid for diffusion of the excited-state molecule to quencher molecule to occur. It is suggested that a mechanism involving transfer of an electron from nitrogen to the excited triplet-state ketone, followed by proton transfer and electron reorganization, accounts for the observed rearrangement patterns. The quantum yield is 0·046, and is increased slightly by deuteriation on the ring. These findings are rationalized by assuming that the hydrogen transfer step is reversible, and this assumption can also account for the low quantum efficiency.

NN-Dialkyl-*o*-nitroanilines (230) are known to photocyclize under a variety of conditions to yield benzimidazoles (231).[73] Suschitzky and co-workers have postulated the intermediate formation of the benzimidazole-*N*-oxide (232) in the reaction, and now report their efforts to terminate the

R^{1}H$_2$C_N_/CH$_2$R^1 R^{1}H$_2$C_N_$\overset{R^1}{\diagup}$ R^{1}H$_2$C_N_$\overset{R^1}{\diagup}$
   [benzene with NO$_2$]   [benzimidazole]   N→O
        | | |
       R^2 R^2 R^2
      (230)   (231)   (232)

reaction at this latter stage. The photo-cyclization of such compounds as (230) has also been investigated in the presence of acid, when it is found that *either* (231) *or* (232) is produced.[74] The cyclization proceeds slowly in the absence of acid and the direction of the reactions is apparently dependent upon a combination of steric and electronic factors. A mechanism (Scheme 7) is proposed to account for the acid-catalysed photolysis, and evidence for it is presented. A similar reaction has also been reported for *N*-2,4-

---

[72] A. Padwa and R. Gruber, *J. Amer. Chem. Soc.*, 1970, **92**, 100, 107.
[73] *E.g.* H. Suschitzky and M. E. Sutton, *Tetrahedron Letters*, 1967, 3933; W. M. Lauer, M. M. Spring, and C. M. Langkammerer, *J. Amer. Chem. Soc.*, 1936, **58**, 225.
[74] R. Fielden, O. Meth-Cohn, and H. Suschitzky, *Tetrahedron Letters*, 1970, 1229.

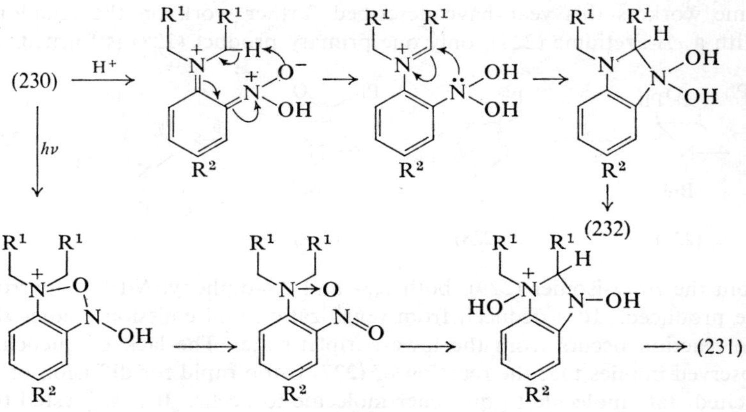

Scheme 7

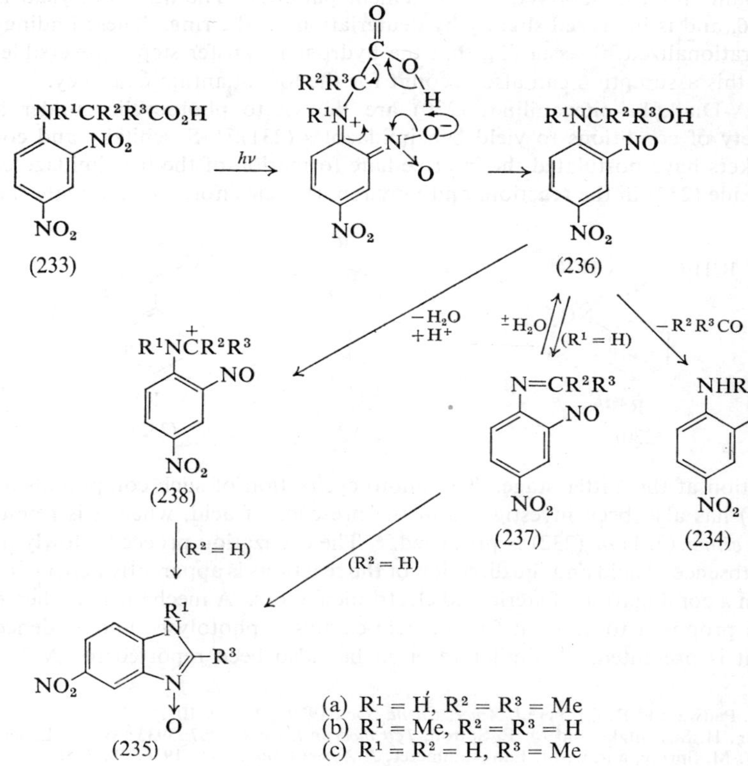

Scheme 8

(a) $R^1 = H$, $R^2 = R^3 = Me$
(b) $R^1 = Me$, $R^2 = R^3 = H$
(c) $R^1 = R^2 = H$, $R^3 = Me$

dinitrophenyl-amino-acids (233).[75] Thus irradiation of (233) in aqueous solution may proceed by two routes, *viz.* (*a*) decarboxylation to yield the 4-nitro-2-nitrosoaniline (234) and the aldehyde with one carbon atom less than in the parent amino-acid,[76] and (*b*) formation of substituted 6-nitrobenzimidazole 1-oxides (235).[77] The relative importance of each route depends upon the pH of the solution,[78] and it is now reported that there are structural requirements for the formation of (235).[75] Optimum yields of (235) from (233) are realized at very low pH (-3) and at pH approximately 3. The reaction at pH 3 is found to require a hydrogen atom on the amino-group, whereas at lower pH this is not so. The suggested mechanism is outlined in Scheme 8. The intermediate (236) may yield the anil (237) by reversible dehydration where $R^1 = H$, and where $R^2 = H$ the anil could cyclize to form the 6-nitrobenzimidazole 1-oxide (235). The alternative route *via* the carbonium ion (238) is probably the path which operates at very low pH.

Photochemistry of Oximes.—The flash photolysis of formaldoxime and acetaldoxime has been studied under isothermal and adiabatic conditions.[79] The predominant primary photoreaction produces $R^1R^2CN\cdot$ and $OH\cdot$ free radicals, and the isothermal reactions show that the former radicals are not very reactive, whereas hydrogen abstraction (equation 1) by hydroxyl radical is fast. Under adiabatic conditions, the oximes decompose by a short chain mechanism, with hydroxyl as the chain-propagating radical. Nitric oxide accelerates the decomposition of the oximes by sensitizing the production of the hydroxyl radicals.

$$\cdot OH + CH_2NOH \longrightarrow \dot{C}HNOH \qquad (1)$$

The photo-Beckmann rearrangement was reported in 1963 by Amin and de Mayo.[80] Other processes from the photolysis of oximes involve the formation of nitriles, and *syn–anti* isomerization. The intermediacy of an oxaziridine in the rearrangement seems to be generally accepted, particularly after the work of Oine and Mukai last year,[81] and singlet-state intermediates are suggested by all workers in this field.

The photolysis of camphoroxime has been reported to yield the parent ketone and various nitriles,[82] and is thus at variance with the 'normal' photoreaction of oximes to form amides. In order to ascertain whether this situation was common for all bicyclo[2,2,1]heptanone oximes, the photolysis of norcamphoroxime (239) has been studied.[83] Irradiation of

---

[75] D. J. Neadle and R. J. Pollitt, *J. Chem. Soc.* (*C*), 1969, 2127.
[76] D. W. Russell, *J. Chem. Soc.*, 1964, 2829; *ibid.*, 1963, 894.
[77] R. J. Pollitt, *Chem. Comm.*, 1965, 262.
[78] D. J. Neadle and R. J. Pollitt, *J. Chem. Soc.* (*C*), 1967, 1764.
[79] D. G. Horne and R. G. W. Norrish, *Proc. Roy. Soc.*, 1970, **A315**, 287.
[80] J. H. Amin and P. de Mayo, *Tetrahedron Letters*, 1963, 1585.
[81] T. Oine and T. Mukai, *Tetrahedron Letters*, 1969, 157.
[82] T. Sato and H. Obase, *Tetrahedron Letters*, 1967, 1633; R. T. Taylor, N. Douek, and G. Just, *Tetrahedron Letters*, 1966, 4143.
[83] B. L. Fox and H. M. Rosenberg, *Chem. Comm.*, 1969, 1115.

the oxime (239) in anhydrous methanol solution (253·7 nm) gave four principal products (240), (241), (242), and (243) in the yields shown. The formation of (240) and (241) thus shows that, at least in the case of (239), some bicyclic ketoximes photoreact similarly to aldoximes and acyclic and

cyclic ketoximes. The lack of detection of nitriles from this reaction is noteworthy. The *cis* stereochemistry of (242) was assigned on the basis of the mode of formation. It has been suggested earlier that protonation may be involved in some step of the rearrangement, or the oxaziridine formation may proceed by a polarized excited state.[81] The latest work supports this proposal, since there is a reduction in the rate of oxime consumption and lactone formation when base (sodium methoxide) is present in the reaction solution.[83] Further, determination of the oxaziridine concentration during the course of the irradiation showed that it was significantly lower in basic than in neutral solutions. These results add further weight to the proposals in ref. 81. Other workers have also examined the mechanism of the photo-rearrangement of oximes, and have studied the light-induced reactions of cyclohexane solutions of the steroid oximes (244) and (245), and also the acetates of the oximes of benzophenone and acetophenone.[84] In each case, the parent oxime and ketone were formed, probably *via* oxaziridines. On the other hand, aromatic oxime benzoates (246) yield the corresponding azines (247) as well as the ketones and ammonium benzoate.[85] Benzoic acid and biphenyl are formed in all cases. The reaction course is suggested to involve N—O bond homolysis as the initial step, and sensitization experiments establish that the triplet state of the oxime benzoate is involved. The photochemistry of cyclic oximes with an α-keto-group is reported this year for the first time.[86] α-Oximinocyclododecanone (248) yields four products on extended exposure from unfiltered radiation of a high-pressure mercury

[84] R. Beugelmans and J. P. Vermes, *Bull. Soc. chim. France*, 1970, 342.
[85] T. Okada, M. Kawanisi, and H. Nozaki, *Bull. Chem. Soc. Japan*, 1969, **42**, 2981.
[86] A. Stojilkovic and R. Tasovac, *Tetrahedron Letters*, 1970, 1405.

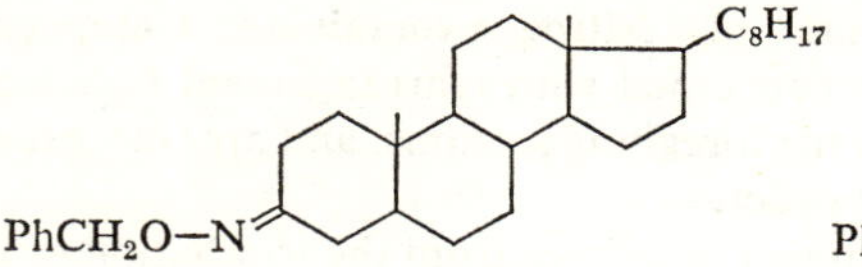

Photoreactions of Compounds with Heteroatoms other than Oxygen 663

(244) PhCH$_2$O—N= (steroid with C$_8$H$_{17}$ side chain)

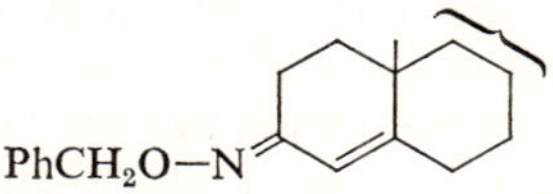

(245) PhCH$_2$O—N= (decalin derivative)

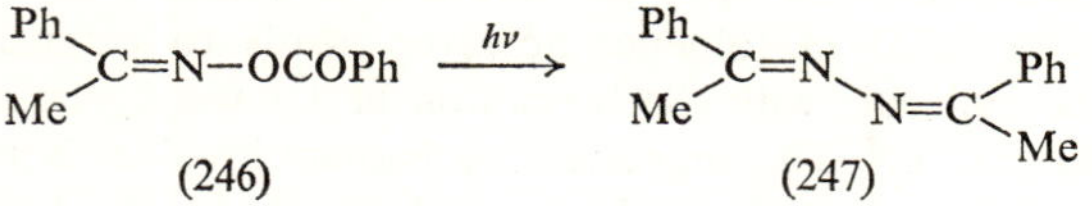

(246) Ph(Me)C=N—OCOPh $\xrightarrow{h\nu}$ (247) Ph(Me)C=N—N=C(Ph)Me (49%)

+

PhCOMe + PhCO$_2$H + Ph—Ph
(30%)   (9%)   (11%)

lamp. The reaction seems to be somewhat similar to that reported for camphoroxime since the second-order Beckmann product, the nitrile (249), is formed, in addition to products from ring cleavage (250), photocyclization (251), and geometrical isomerization (252). The irradiation of

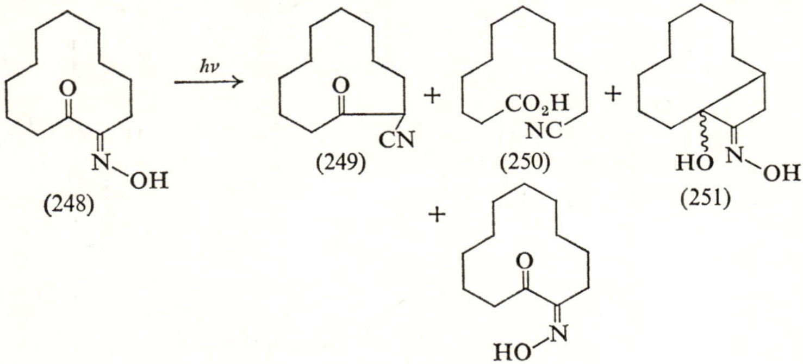

(248) cyclic ketoxime $\xrightarrow{h\nu}$ (249) + (250) + (251) + (252)

anti-16-dehydropregnenolone oxime-3-acetate (253) does not yield products of the forementioned type, although the syn-oxime isomer is formed,[87] but rearrangement to an isoxazoline (254) occurs instead.[87]

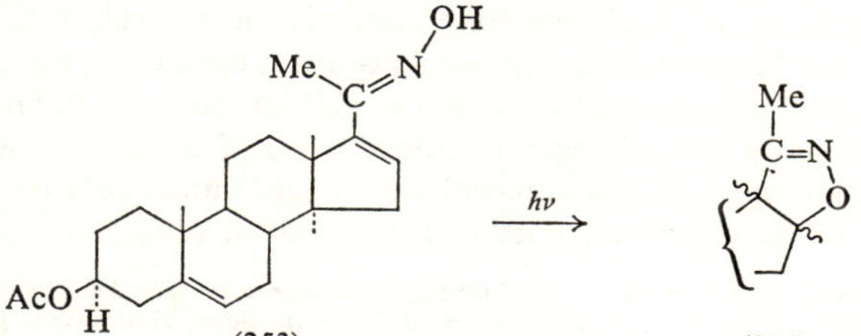

(253) → (254)

[87] R. P. Gandi and V. K. Chadha, *Indian J. Chem.*, 1969, **7**, 633.

Miscellaneous Rearrangements of Nitrogen-containing Compounds.—The remainder of the reports concerned with rearrangement reactions are difficult to classify into any of the foregoing sections, and are thus presented here in a somewhat arbitrary order.

The formation of nitroso-dimers or oximes from the irradiation of nitrite esters is known as the Barton reaction, and has found wide usage in the activation of normally inactive sites in steroids. Irradiation of a fused cyclopentanol nitrite (255) in toluene, however, leads to extensive reorganization of the molecule with the formation of a novel cyclic nitrone (256) in good yield (59%).[88] The suggested mechanism involves homolysis of the nitrite followed by homolytic cleavage of the (C-11)–(C-12) single bond in (255). The *C*-nitrosoaldehyde (257) is then formed, which undergoes isomerization to an oximinoaldehyde (258). Photocyclization of (258) yields the observed product (256).

The irradiation of *N*-nitroso-compounds is the subject of three reports this year. The stabilizing interaction between a cyclopropyl group and an adjacent developing free-radical centre is well established. In an attempt to generate the unknown nitrogen analogue (259) of the carbon and oxygen radicals (260) and (261), the photolyses of (262) and (263) have been examined.[89] In fluorotrichloromethane at −60 °C, (262), however, gave a

---

[88] H. Suginome, N. Sato, and T. Masamune, *Tetrahedron Letters*, 1969, 3353.
[89] E. E. J. Dekker, J. B. F. N. Engberts, and Th. J. de Boer, *Tetrahedron Letters*, 1969, 2651.

▷—ṄR   ▷—ĊR^{1}R^2 ▷—Ö   ▷—N(NO)—C(=O)—⟨C$_6$H$_4$⟩—Me
(259)   (260)   (261)   (263)

p-MeC$_6$H$_4$SO$_2$—N—▷   —CFCl$_3$/−60 °C, hν→   (p-MeC$_6$H$_4$SO$_2$N=CHCH$_2$CH$_2$NO)$_2$
(262)   (264)

hν, MeOH, −60 °C ↘   ↙ MeOH

(p-MeC$_6$H$_4$SO$_2$NHCHCH$_2$CH$_2$NO)$_2$
                    |
                   OMe
(265)

90% yield of the rather unstable *trans*-nitroso-dimer (264). Under similar conditions in methanol, (265) was formed both from (262) and the dimer (264). Analogous reactions were observed for (263). Thus the products arise by homolysis of the N—N bond and ring opening to yield the RN=CHCH$_2$CH$_2$· radical, combination of which with nitric oxide gives the observed products. Thermolysis of such *N*-nitroso-compounds leads to denitrosation to the parent compounds, *e.g.* *N*-cyclopropylsulphonamides, rather than rearrangement. Chow and co-workers have also worked on the irradiation of *N*-nitroso-compounds, but with the objective of studying the amido-radicals.[90, 91] With the compounds (266) and (267), the initial step was again found to be N—N bond homolysis to form nitric oxide and the amido-radicals, which abstracted hydrogen atoms intermolecularly from the solvent as well as intramolecularly from the δ-carbon atom of the alkyl chain.[90] These two pathways are outlined in Schemes 9 and 10. By such photolysis, the functionalization of the non-activated δ-carbon site in the alkyl chain was achieved in 55% yield. Compounds (268a—d) have been photolysed in order to investigate the photochemistry of such systems when hydrogen–nitroso-group exchange cannot occur.[91] The reactions were studied in a variety of solvents with various donating powers. In cyclohexane, cyclohexyl nitrate and the dimer of nitrosocyclohexane were again formed, while the products in mesitylene included 3,5-dimethylbenzaldehyde and its oxime; and in two reactions, 3,5-dimethylbenzyl alcohol and di(3,5-dimethylbenzyl) ether were also formed. Nitrobenzene and *o*- and *p*-nitrophenols were isolated from some experiments in benzene, and were detected in others. The products derived from the amide part of the nitrosamide included the parent amides, and *gem*-bis-acetamido-aralkyl compounds (269); other compounds formed are benzaldehyde from (268b), and *N*-formylacetamide from (268a,c,d). In

[90] Y. L. Chow, J. N. S. Tam, and A. C. H. Lee, *Canad. J. Chem.*, 1969, **47**, 2441.
[91] Y. L. Chow and J. N. S. Tam, *J. Chem. Soc.* (*C*), 1970, 1138.

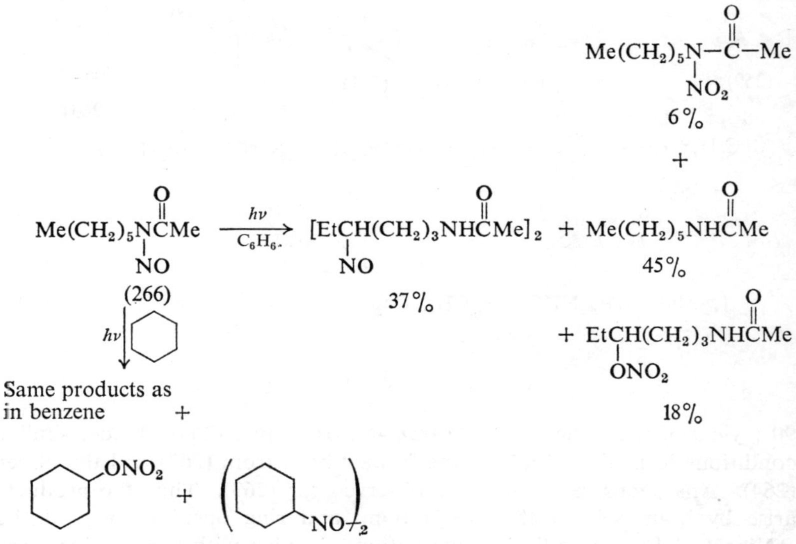

Scheme 9

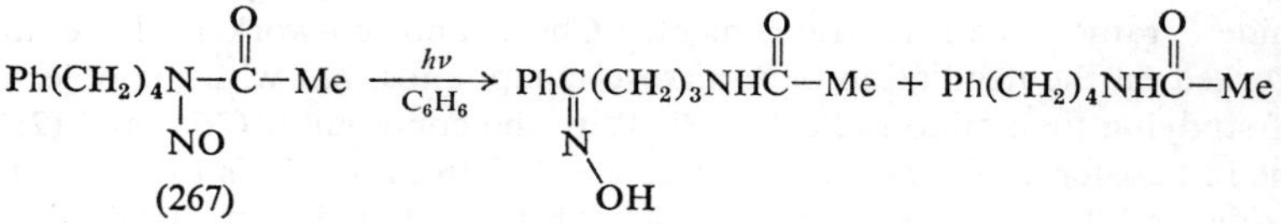

$$Me(CH_2)_5\underset{NO}{\underset{|}{N}}\overset{O}{\overset{\|}{C}}Me \xrightarrow[C_6H_6]{h\nu} [EtCH(CH_2)_3NH\overset{O}{\overset{\|}{C}}Me]_2 + Me(CH_2)_5NH\overset{O}{\overset{\|}{C}}Me$$

(266)

$$Me(CH_2)_5\underset{NO_2}{\underset{|}{N}}\overset{O}{\overset{\|}{C}}Me \quad 6\%$$

$$+$$

$$[EtCH(CH_2)_3NH\overset{O}{\overset{\|}{C}}Me]_2 \quad 37\% \qquad Me(CH_2)_5NH\overset{O}{\overset{\|}{C}}Me \quad 45\%$$

$$+ EtCH(CH_2)_3NH\overset{O}{\overset{\|}{C}}Me$$
$$\underset{ONO_2}{|}\quad 18\%$$

$h\nu$ ↓ (benzene)

Same products as in benzene +

cyclohexyl-ONO$_2$ + (cyclohexyl-NO)$_2$

Scheme 9

Scheme 10

$$Ph(CH_2)_4\underset{NO}{\underset{|}{N}}\overset{O}{\overset{\|}{C}}Me \xrightarrow[C_6H_6]{h\nu} Ph\underset{\|}{\overset{N-OH}{C}}(CH_2)_3NH\overset{O}{\overset{\|}{C}}Me + Ph(CH_2)_4NH\overset{O}{\overset{\|}{C}}Me$$

(267)

Scheme 10

addition, photolysis of (268a) yields N-ethoxymethylacetamide (270) in benzene, and the N-acetyl-N-methylhydrazone of 3,5-dimethylbenzaldehyde (270a) in mesitylene. Although the amido-radicals generated here show β-scission of the C—H and C—C bonds and reactions similar to those of alkoxy-radicals, β-scission of the C—Ar bond was not detected. The presence of oxygen does not quench the primary photochemical process or

$$\underset{NO}{\underset{|}{RCH_2-NCOMe}} \qquad RCH(NHCOMe)_2 \qquad EtOCH_2NHCOMe$$

(268) (a) R = H
       (b) R = Ph
       (c) R = CH$_2$Ph
       (d) R = CH$_2$CH$_2$Ph

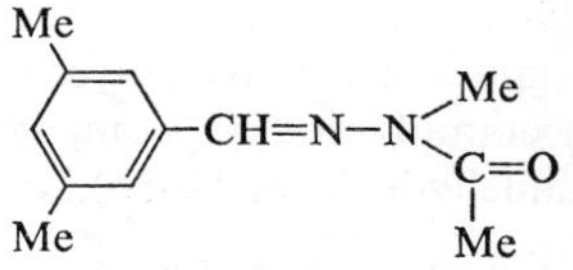

(269)   (270)

3,5-Me$_2$C$_6$H$_3$—CH=N—N(Me)—C(=O)Me

(270a)

interfere with the reaction of the amido-radicals except to increase the yields of the oxidation products.

The thermal- and photo-isomerizations of nitro-olefins have been studied as a function of the nature, number, and position of the substituents in the chain.[92] Irradiation of diethyl ether or hexane solutions of the nitro-olefins (271) may cause both isomerization to the *cis*-isomer and rearrangement to the non-conjugated nitro-olefin (272). The differences in reaction mode can be large: thus while (273) gives 49% *trans*, 36% *cis*, and 15% of (272), compound (274) gives only traces of the non-conjugated olefin and a 60:40 mixture of *trans*- and *cis*-isomers. The nitro-olefin (275) also demonstrates ethylenic bond migration by the formation of (276) in 100% yield. The thermal and photochemical condensations of conjugated nitro-olefins (271) and (272) with unsaturated aldehydes have also been studied.[93] The proportions of the Michael-type condensation products (277) and (278) vary considerably, but the formation of (277) generally is favoured.

[92] G. Pingeon and R. Rostaing, *Bull. Soc. chim. France*, 1970, 290.
[93] G. Pingeon and R. Rostaing, *Bull. Soc. chim. France*, 1970, 295.

The photorearrangement of o-nitrobenzylideneaniline (279) to o-nitrosobenzanilide (280) was reported in 1902.[94] In 1914, the reaction was examined in the solid state, when the crystals underwent a photochromic change into a permanent deeper coloured 'dimorphic' variety having a melting point 2 °C lower than the starting compound.[95] Work has now been described which shows that (279) still gives (280) in the crystalline state, and which

$$(279) \xrightarrow{h\nu} (280)$$

clarifies the nature and stability of the so-called dimorphic form.[96] Irradiation of the yellow crystals of (279) with light from a high-pressure mercury lamp having a filter which cuts off light of wavelengths shorter than 320 nm, causes formation of an orange-red compound, the colour of which persists for many hours. Further irradiation of the system yields a deep yellow compound. It is suggested that these changes resemble a phase transformation. Consistent with this but contrary to the early work,[95] unphotolysed (279), the orange-red, and the deep yellow forms of (279), all yield compounds after melting which have similar i.r. spectra. These spectra are different from the i.r. of (279) prior to irradiation and melting.

Many aromatic nitro-compounds and other nitrogen-containing compounds exhibit photochromism. There is no intention here to present the literature concerned with photochromism, but rather to illustrate, by examples from the current literature, the types of compounds which undergo light-induced rearrangement to coloured, or more deeply coloured (transient), isomers. The primary photochemical processes in the 366 nm irradiation of 5,7-dichloro-6-nitro-(1,3,3-trimethylindolino)benzopyrylospiran (281) have been studied in acetonitrile solution.[97] Both direct and sensitized irradiations have been investigated, and while benzophenone and 2-acetonaphthone sensitized the conversion of (281) into (282), benzil has little effect as a sensitizer or quencher; but 9-fluorenone quenches the reaction in the direct irradiation. From sensitization experiments, the triplet energy of (281) is deduced to be 53—57 kcal mol^{-1}, which is in reasonable agreement with the value of 58 kcal mol^{-1} from the 0—0 band in the phosphorescence spectrum of (281). The results suggest that in the direct photolysis both excited singlet and triplet states participate in the rearrangement. The quantum yield data for the system are summarized in Scheme 11. Thin layers of a trimethylindolinobenzopyrylospiran are also reported to display

---

[94] F. Sachs and R. Kempf, *Ber. deut. chem. Gesellschaft*. 1902, **35**, 2704.
[95] A. Senier and R. Clarke, *J. Chem. Soc.*, 1914, **105**, 1917.
[96] E. Hadjoudis and E. Hayon, *J. Phys. Chem.*, 1970, **74**, 2224.
[97] H. Bach and J. G. Calvert, *J. Amer. Chem. Soc.*, 1970, **92**, 2608.

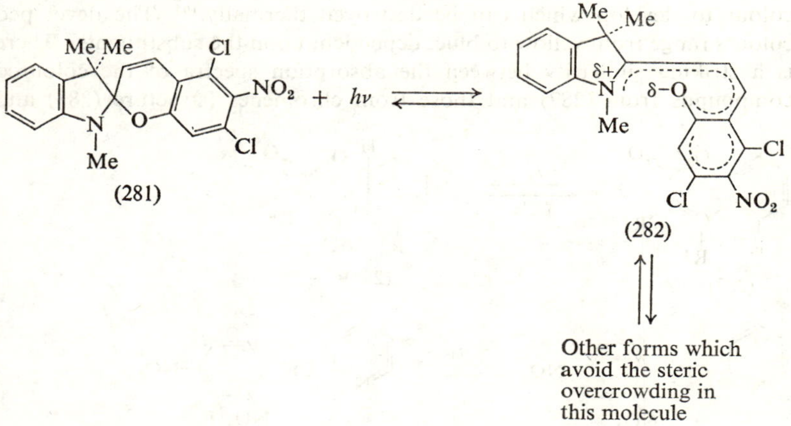

Photoreactions of Compounds with Heteroatoms other than Oxygen 669

(281)

(282)

⇅

Other forms which avoid the steric overcrowding in this molecule

photochromism,[98] and while such compounds are known to be photochromic in solution, on adsorbents, and in polymer films, this report is the first to describe the reaction of microcrystalline spiropyrans. Many nitrogen-containing aromatic systems more simple than (281) exhibit

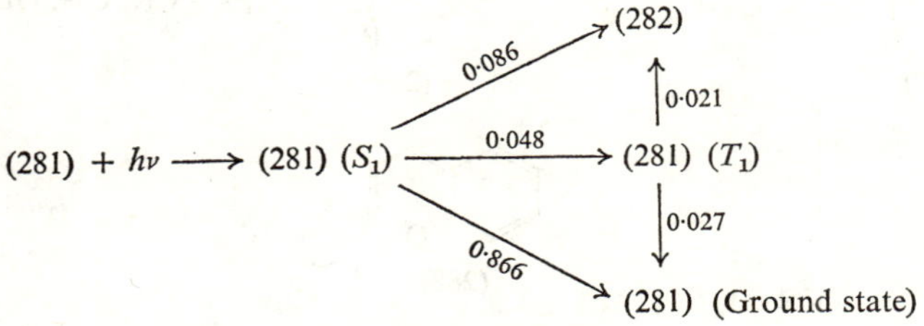

$$(281) + h\nu \longrightarrow (281)\,(S_1) \xrightarrow{0\cdot 086} (282)$$
$$(281)\,(S_1) \xrightarrow{0\cdot 048} (281)\,(T_1) \xrightarrow{0\cdot 021} (282)$$
$$(281)\,(S_1) \xrightarrow{0\cdot 866} (281)\,(\text{Ground state}) \xleftarrow{0\cdot 027} (281)\,(T_1)$$

Scheme 11

photochromism, and in the coloured (transient) isomer, bond location occurs in the aromatic ring to form conjugated cyclohexa-1,3-diene systems. Thus irradiation of 1-aryl-2-nitroalkenes (283) yields coloured transients which absorb in the 410—450 nm region.[99] The structure of these transients is considered to be (284), and in this reaction only compounds with a *cis* relationship of the nitro and phenyl groups are photochromic. The kinetics of the reversible and irreversible photoreactions of photochromic 2-(2,4-dinitrobenzyl)pyridine (285) have been reported.[100] The transient here is (286). Irradiation of colourless solutions of 1,2-dihydroquinolines (287) at −196 °C in E.P.A. with 250—390 nm light causes a

---

[98] H. Kokado and E. Inoue, *Bull. Chem. Soc. Japan*, 1969, **42**, 2735.
[99] J. A. Sousa, J. Weinstein, and A. L. Bluhm, *J. Org. Chem.*, 1969, **34**, 3320.
[100] A. A. Parshutkin, V. P. Bazov, and V. A. Krongauz, *Khim. vysok. Energii*, 1970, **4**, 131.

colour to develop which can be destroyed thermally.[101] The developed colours range from yellow to blue, dependent upon the substituents. There is a striking similarity between the absorption spectra of the coloured compounds from (287) and those from chromenes [structure (288) and

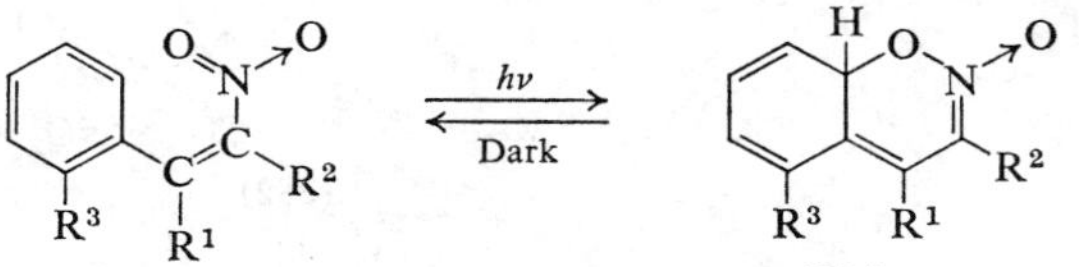

(283) ⇌ (hv/Dark) (284)

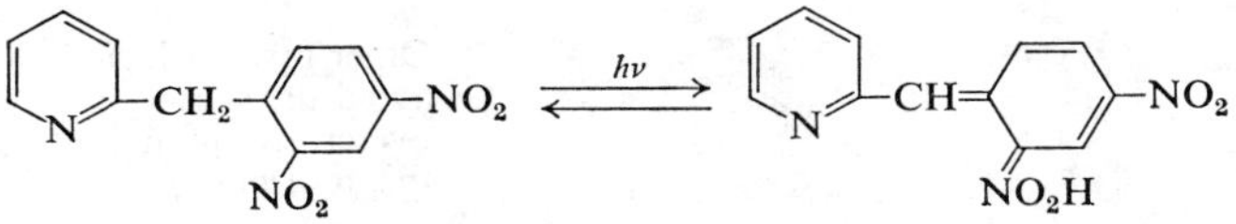

(285) ⇌ hv (286)

(287) —hv, −196 °C→ (289)

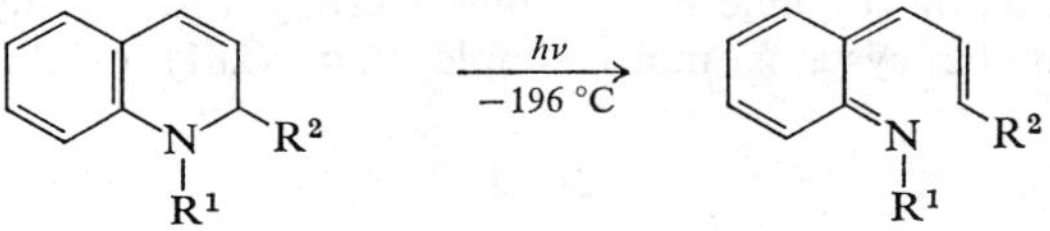

R^1 = COPh, CO_2Ph, CN, SO_2Ph—Me
R^2 = CN, OH, OEt, H

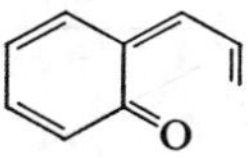

(288)

derivatives]. Thus the transient formed by irradiation of (287) is most reasonably suggested to have structure (289). The spectra and photochemistry of the different isomers of the photochromic anils of salicylaldehyde and hydroxynaphthaldehyde [e.g. (290) and (291)] have been investigated.[102] From flash-photolysis experiments, it is shown that both the intramolecularly hydrogen-bonded enol (292) and the cis-keto-compound (293) yield the trans-keto-isomer (294) as a common photoproduct. In rigid paraffin glasses, the photoisomerization of (292) to (294) may be induced, but not that of (293) to (294). The relaxation of (294) to (292) in the dark occurs in low-polarity systems via dimer intermediates.

Evidence for similar non-aromatic intermediates to those reported for the above photochromic systems has been obtained in the photodecarbonylation (see Part III, Chapter 1, Section 7) of N-phenyloxindole (295).[103]

[101] J. Kolc and R. S. Becker, J. Amer. Chem. Soc., 1969, 91, 6513.
[102] R. Potashnik and M. Ottolenghi, J. Chem. Phys., 1969, 51, 3671.
[103] M. Fischer and F. Wagner, Chem. Ber., 1969, 102, 3486.

Compound (295) is shown to yield the decarbonylated product (296) *via* formation of the cyclohexa-1,3-diene (297), since the spiro-dimer (298) is obtained from the reaction, and photolysis of (295) in the presence of *N*-phenylmaleimide yields the adduct (299). Further evidence for the intermediacy of (297) is obtained by the isolation of such compounds as (300) from photochemical reaction of (295) with nucleophilic reagents HX.[104]

X = OEt, OCOMe, NHMe, NHBu^n

Photo-transposition of atoms and groups in five-membered nitrogen heterocycles has been previously considered in this section. A report this year describes the occurrence of a similar process in the six-membered nitrogen heterocycle 1,2-diphenyl-6-methyl-2,3-dihydro-4-pyridone (301). Irradiation in methanol or benzene solution yields the migration products

[104] M. Fischer, *Chem. Ber.*, 1969, **102**, 3495.

(302) and (303), as well as the ring-opened isomer (304).[105] The reaction does not occur with light of wavelength longer than 290 nm, but is induced with 254 nm radiation. The paper is disappointing in that there is no discussion of mechanistic aspects or proposals for this interesting rearrangement. One would judge by the absence of mono- and di-substituted

derivatives in the product mixture, that the reaction does not involve homolysis and combination of free radicals. However, it is difficult to conceive a mechanism to account for the formation of (303) where there is migration of phenyl from nitrogen to carbon and apparent displacement of the 6-methyl group in (301) to the 3-position in (303). In contrast to this, the photochemistry of 5,6-dihydro-4,6,6-trimethyl-2(1H)-pyridone (305) shows no migration reactions but the reaction paths differ for the direct and sensitized irradiations.[106] In the absence of sensitizer, (305) yields almost exclusively the cleavage products (306) and (307), whereas a single cyclobutane dimer is formed in the presence of acetophenone (350 nm radiation). However, both direct and sensitized irradiation of the homo-derivative

(308) yield approximately the same ratio of the two cyclobutane compounds (309) and (310). It is suggested that dimer formation involves triplet states, whereas the cleavage reaction most probably arises from excited singlet states.

[105] C. Kashima, M. Yamamoto, Y. Sato and N. Sugiyama, *Bull. Chem. Soc. Japan*, 1969, **42**, 3596.
[106] E. Cavalieri and S. Horoupian, *Canad. J. Chem.*, 1969, **47**, 2781.

It is proposed that the rearrangement observed on photolysis of enamides (311) occurs *via* homolysis of the N—CO bond, a 1,3-acyl shift, and the intermediate formation of (312).[107] Compound (312) then isomerizes to the isolated compound (313). The irradiations are carried out with cyclohexane, benzene, or methanol solutions, and the rearrangements also occur with cyclic analogues (314) to give compounds such as (315).

The photochemical and thermal isomerizations of 2,3- and 3,4-diformylpyridines and 2,3-diformylquinoline have been studied.[108] For example, photoisomerization (360 nm) of the 2,3-pyridine isomer (316) in benzene at 20 °C leads to the lactones (317) and (318).

N-Amino-2-ethanethiols (319) are reported to undergo photoisomerization with the formation of thiazolidines.[109] The substituent groups (R *etc.*) examined were hydrogen, methyl, and butyl, and the proportions of the reaction products (320), (321), and (322) were found to be very dependent

---

[107] R. W. Hoffmann and K. R. Eicken, *Chem. Ber.*, 1969, **102**, 2987.
[108] G. Queguiner and A. Godard, *Compt. rend.*, 1969, **269**, C, 1646.
[109] J. M. Surzur and M. P. Crozet, *Compt. rend.*, 1969, **268**, C, 2109.

upon the degree of substitution. A saturated analogue of (319) was also found to undergo photocyclization on photolysis; thus NN-dibutylamino-2-ethanethiol (323) gave the n-propylthiazolidine (324).[109]

A photocleavage of the three-membered ring of 5,6-cyclopropyluridines (325) has been reported to yield the diazepine nucleosides (326).[110]

The photorearrangement of azoxybenzene to 2-hydroxyazobenzene is well known.[111] The rearrangement and isomerization of *trans*-azoxybenzene are considered to arise from the excited singlet state, and the

[110] T. Kunieda and B. Witkop, *J. Amer. Chem. Soc.*, 1969, **91**, 7751.
[111] See, for example, R. Tanikagar, *Bull. Chem. Soc. Japan*, 1968, **41**, 1664, 2151.

reduction to yield azobenzene is thought to be a triplet process since it only occurs in the presence of benzophenone in ethanol solution. Further work on the unsensitized photoreaction of azoxybenzene (327) has been carried out, with results different from those reported previously; and a product (328) formerly 'overlooked' has been obtained.[112] The irradiation of (327) in chloroform with a high-pressure mercury lamp yields at least seven coloured products, which have been separated by column chromatography and their structures determined by high-resolution mass spectrometry. In contrast with the results obtained for ethanol solutions,

Ph—N(→O)=N—Ph $\xrightarrow{h\nu}$ Ph—N=N—C$_6$H$_4$—OH

(327)   (328)

+

Ph—N=N—(2-hydroxyphenyl)

+ cis and trans isomers
of azobenzene

reduction occurred in the absence of sensitizer, and *cis*- and *trans*-azobenzenes were isolated. 2-Hydroxyazobenzene was formed as expected, but 4-hydroxyazobenzene (328) was also isolated and had not been previously reported as a product from the irradiation of (327). The other three photoproducts all contained impurities, and the use of mass spectrometry for the structural determination was not feasible. The workers suggest that the formation of the *cis*- and *trans*-isomers of azobenzene and the 4-hydroxyazobenzene may have been overlooked by earlier researchers since the yields of these products from the reaction are very low, but in the present large-scale photolysis, sufficient quantities were obtained for detection, isolation, and structural determination.[112]

2 Synthesis, Substitution, and Addition Reactions of Nitrogen-containing Compounds

There have been many scattered reports this year which may be loosely classified under this heading. A two-part review on the photochemical preparation of heterocycles has appeared within the year, and although in Japanese, it can be followed quite well through the numerous structural examples.[113, 114]

[112] M. Iwata and S. Emoto, *Bull. Chem. Soc. Japan*, 1970, **43**, 946.
[113] T. Sasaki, *J. Synthetic Org. Chem., Japan*, 1969, **27**, 879.
[114] T. Sasaki, *J. Synthetic Org. Chem., Japan*, 1969, **27**, 1000.

The determination of quantum yields for the photonitrosation of aliphatic hydrocarbons is complicated by the insolubility of the oxime hydrochloride in the hydrocarbon media. Attempts have been made to overcome this difficulty by using an integrating sphere to remove the effect of light-scattering by the precipitated product.[115] From such procedures, a quantum yield of 0·72 has been recorded which is independent of wavelength within an estimated error of ±20%. The measurements of quantum yield for the process have now been carried out in a homogeneous system containing chloroform, in which the oxime hydrochloride formed is soluble.[115a] The values obtained with 365, 436, and 578 nm exciting radiation are 0·80 ± 0·06, 0·79 ± 0·06, and 0·97 ± 0·13 respectively. These workers agree with Müller's proposal that different mechanisms may operate at different wavelengths. The two suggested mechanisms are shown in Schemes 12 and 13. Both should give quantum yields of unity unless side-reactions occur,

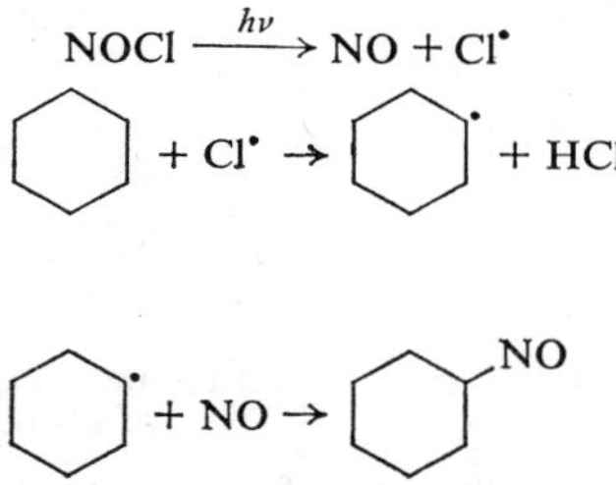

$$NOCl \xrightarrow{h\nu} NO + Cl^{\bullet}$$

$$C_6H_{12} + Cl^{\bullet} \rightarrow C_6H_{11}^{\bullet} + HCl$$

$$C_6H_{11}^{\bullet} + NO \rightarrow C_6H_{11}NO$$

Scheme 12

$$NOCl \xrightarrow{h\nu} NOCl^{*}$$

$$C_6H_{12} + NOCl^{*} \rightarrow \text{[intermediate]} \rightarrow C_6H_{11}NO + HCl$$

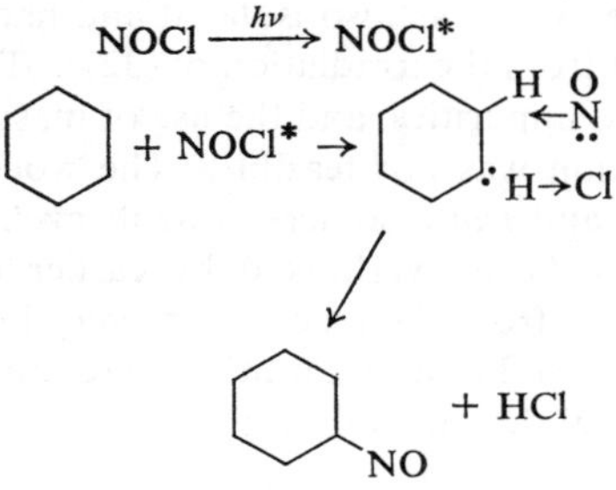

Scheme 13

such as the formation of cyclohexyl chloride. The photochemical reaction of nitrosyl chloride with hexanoic, heptanoic, and octanoic acids has been investigated.[116] After hydrolysis of the reaction mixture, keto-acids and di-acids with the same number of carbon atoms are obtained. Light-induced nitrosation of cyclohexyl chloride has also been investigated and

---

[115] H. Miyama, N. Harumiya, Y. Ito, and S. Wakamatsu, *J. Phys. Chem.*, 1968, **72**, 4700.
[115a] H. Miyama, K. Fukuzawa, N. Harumiya, Y. Ito, and S. Wakamatsu, *J. Phys. Chem.*, 1969, **73**, 4345.
[116] E. Barale and A. Guillemonat, *Compt. rend.*, 1969, **268**, C, 1201.

the products described.[117] The photonitrosation of hydrocarbons is an important industrial process since it is claimed that caprolactam (and hence Nylon-type polymers) can be made at lower cost *via* photonitrosation of cyclohexane than by conventional thermal routes. The nitrosyl chloride which is used in the large-scale reaction is very corrosive, and research workers have therefore investigated other potential nitrosation agents. Thus the photonitrosation of cyclohexane by t-butyl nitrite has been studied as a model system for reactions between alkyl nitrites and hydrocarbons.[118] The reaction products are the *trans*-dimer (329) and minor amounts of

cyclohexanone oxime (330): the total yield is 81%. Conversion of (329) into (330) may be achieved by various catalysts.[119] Most of (329) is formed from the *cis*-dimer, which can be isolated at low temperature. Optimum conditions for the reaction involve the use of a t-alkyl nitrite in low concentrations, wavelengths around 400 nm, and temperatures slightly below ambient. A free-radical mechanism is suggested for the reaction, and has been supported by e.s.r. studies. The reaction has also been reported for other cycloalkanes and alkylbenzenes having abstractable primary or secondary hydrogen atoms.[120] Again the *trans*-azodioxy-compounds and corresponding oximes are obtained in total yields ranging from 55—80%. With alkylbenzenes, attack occurs preferentially at the α-carbon since this yields a resonance-stabilized radical. Monomeric nitroso-compounds are formed from the light-induced reaction of t-butyl nitrite with some branched-chain hydrocarbons which have tertiary hydrogens.[121] Yields are again good (70%), and it is reported that 400 nm radiation is essential for the reaction. The monomeric nitroso-compounds are very sensitive to light, heat, and free-radicals, and this may explain why Müller[122] did not observe tertiary nitrosation in his system in the presence of hydrogen chloride (which is known to destroy monomeric nitroso-compounds). Heating such dimers as (329) yields the monomeric primary and secondary nitrosoalkanes, which on irradiation yield oximes (330) or, in the presence of oxygen, nitroalkanes and small amounts of alkyl nitrates.[123] With cyclo-

[117] G. N. Semina, L. G. Zelenskaya, L. A. Levashova, K. E. Kuznetsova, and A. A. Strel'Tsova, *Neftekhimiya*, 1970, **10**, 103.
[118] A. Mackor, J. U. Veenland, and Th. J. de Boer, *Rec. Trav. chim.*, 1969, **88**, 1249.
[119] M. Pape, *Fortschr. Chem. Forsch.*, 1967, **7**, 559.
[120] A. Makor and Th. J. de Boer, *Rec. Trav. chim.*, 1970, **89**, 151.
[121] A. Makor and Th. J. de Boer, *Rec. Trav. chim.*, 1970, **89**, 159.
[122] E. Müller and M. Salamon, *Chem. Ber.*, 1965, **98**, 3501.
[123] A. Makor and Th. J. de Boer, *Rec. Trav. chim.*, 1970, **89**, 164.

hexane as solvent in the latter reaction, considerable amounts of cyclohexanol are formed, presumably *via* a nitroxide-type intermediate. The same report describes the photochemical oxidation of t-butyl nitrite to t-butyl nitrate in 90% yield by the action of intermediate NO and NO_2.

The photo-difluoramination of alkanes and alkenes has been studied by Bumgardner and co-workers.[124] The NF_2 radicals from the $N_2F_4 \rightleftharpoons 2NF_2$ equilibrium absorb radiation in the 260 nm region, and irradiation at 254 nm of N_2F_4 in alkanes has led to the substitution of hydrogen by NF_2. The reaction with alkenes and alkynes results in the addition of the elements of NF_3 to the multiple bond. Formation of these products has been rationalized by the postulate that NF_2 is photolysed to fluoronitrene (NF) and atomic fluorine. This hypothesis has been tested, and the same workers have now ascertained selectivity patterns for photo-difluoramination of n-butane, n-butyl fluoride, ethyl chloride, and ethyl difluoramine by analysing the isomeric mixture of alkyl difluoramines produced by photolysis of N_2F_4 in these substrates.[125] From a study of the photolysis of N_2F_4 with methane, kinetic and stoicheiometric information was obtained concerning the photo-difluoramination process. Orientational preferences in the addition reaction were investigated by photo-difluoramination of propene and isobutene, when not only substitution of hydrogen by NF_2 was observed, but also the addition of fluorine and NF_2 to the double bond: the fluorine was found mainly on the terminal carbon atom. The experimental results are rationalized by schemes involving $NF_2 \rightarrow NF + F$ as a common step.

Further studies on the photoaddition of formamide to acetylenes have been reported.[126] Terminal acetylenes give 2:2 adducts (331) as the major products, while non-terminal acetylenes yield 2:1 adducts (332). The

$$\begin{array}{cc} \text{CONH}_2 & \text{H}_2\text{NOC} \quad \text{CONH}_2 \\ | & | \quad\quad\quad | \\ \text{RCH}-\text{CHCONH}_2 & \text{Me(CH}_2)_2\text{CH}-\text{CH(CH}_2)_2\text{Me} \\ | & \\ \text{HC}=\text{CHR} & \\ (331) & (332) \end{array}$$

process is sensitized by acetone or benzophenone, and a free-radical mechanism is proposed.

A full report of Elad and Sterling's studies on the photochemical modification of glycine dipeptides has appeared.[127] In irradiation of glycylglycine, glycylalanine, and glycyl-leucine methyl esters with isobutene, but-1-ene, and toluene, preferential conversion of the glycine residue occurs to give yields up to 60% of leucine, norleucine, and phenylalanine respectively. A free-radical chain mechanism (Scheme 14) is proposed for this process.

---

[124] C. L. Bumgardner, *Tetrahedron Letters*, 1964, 3683; C. L. Bumgardner and K. G. McDaniel, *J. Amer. Chem. Soc.*, 1969, **91**, 1032.
[125] H. Carmichael, *J Amer. Chem. Soc.*, 1970, **92**, 1311.
[126] D. Elad and G. Friedman, *J. Chem. Soc. (C)*, 1970, 893.
[127] D. Elad and J. Sperling, *J. Chem. Soc. (C)*, 1969, 1579.

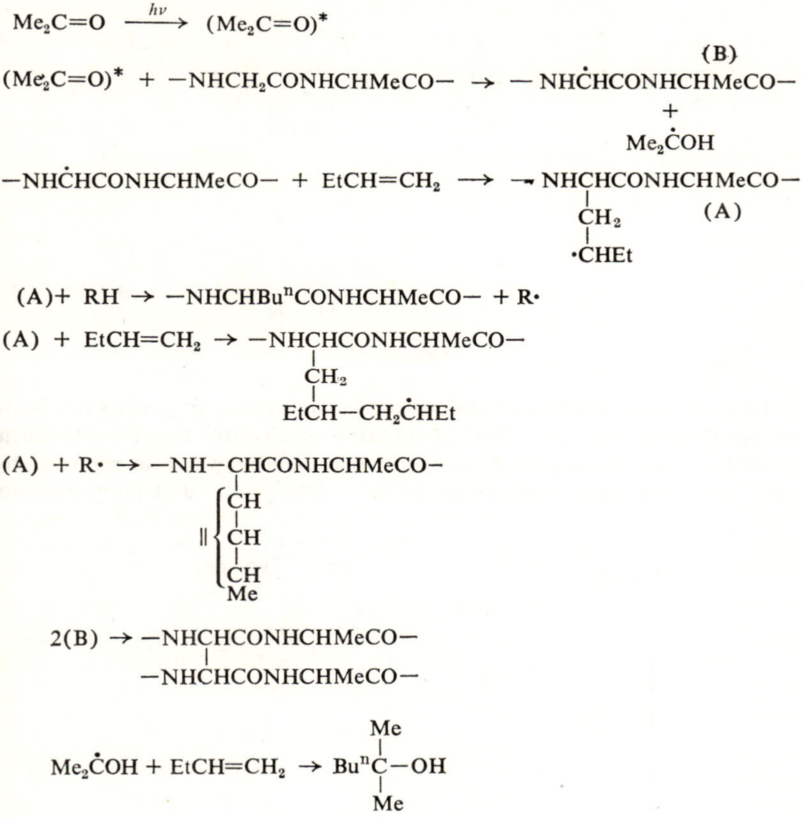

$Me_2C=O \xrightarrow{h\nu} (Me_2C=O)^*$

$(Me_2C=O)^* + -NHCH_2CONHCHMeCO- \rightarrow -NH\dot{C}HCONHCHMeCO-$ (B)
                                                            $+ Me_2\dot{C}OH$

$-NH\dot{C}HCONHCHMeCO- + EtCH=CH_2 \rightarrow -NHCHCONHCHMeCO-$
                                                                       $|$
                                                                       CH_2 (A)
                                                                       $|$
                                                                       $\cdot CHEt$

(A) + RH $\rightarrow$ $-NHCHBu^nCONHCHMeCO- + R\cdot$

(A) + $EtCH=CH_2 \rightarrow$ $-NHCHCONHCHMeCO-$
                                          $|$
                                          CH_2
                                          $|$
                                          $EtCH-CH_2\dot{C}HEt$

(A) + R· $\rightarrow$ $-NH-CHCONHCHMeCO-$
                                    $|$
                              $\begin{cases} CH \\ CH \\ CH \end{cases}$
                                     Me

2(B) $\rightarrow$ $-NHCHCONHCHMeCO-$
             $|$
        $-NHCHCONHCHMeCO-$

$Me_2\dot{C}OH + EtCH=CH_2 \rightarrow Bu^n\underset{Me}{\overset{Me}{\underset{|}{\overset{|}{C}}}}-OH$

Scheme 14

Light-induced peptide synthesis and peptide bond-cleavage have been reported, although the latter process occurs to the lesser degree.[128] Thus when glycine and diglycine are subjected to either 242 or 300 nm radiation on pre-coated thin-layer chromatography sheets, diglycine is formed from glycine, and glycine from diglycine.

The mechanisms for the photoaddition of cysteine to uracil (333) and the simultaneous formation of dihydrouracil have been investigated.[129] The reaction is shown to proceed from the triplet state of the uracil, which can abstract hydrogen from cysteine to give dihydrouracil. The thiyl radicals thus produced add to the ground-state uracil (see Scheme 15). The triplet state of uracil involved in the reaction is the same as that responsible for its dimerisation. Scheme 16 summarizes the mechanistic proposals.

[128] M. M. Cosgrove, M. A. Collins, and R. A. Grant, *Photochem and Photobiol.*, 1969, **10**, 141.
[129] T. Jellinek and R. B. Johns, *Photochem. and Photobiol.*, 1970, **11**, 349.

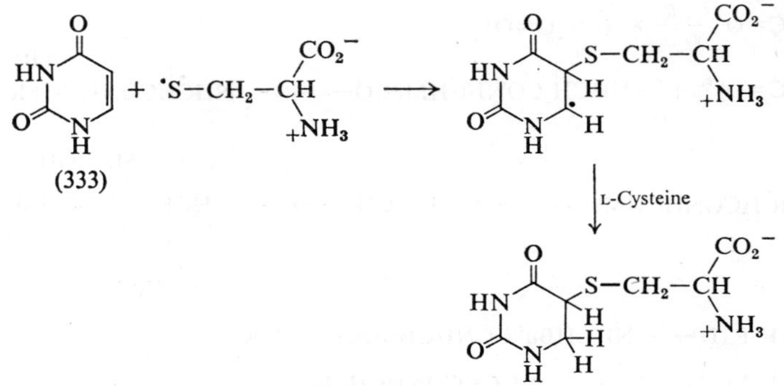

[Scheme 15 structural diagram showing uracil (333) reacting with ·S–CH₂–CH(CO₂⁻)(⁺NH₃) to form intermediates with L-Cysteine]

Scheme 15

Light-induced addition of various organic molecules to purines has been reported again this year. The 1,6-addition of alcohols to purine to form (334) has been previously described,[130] and caffeine was also reported to yield the C-8 substitution product (335).[131] The photo- and γ-ray-induced

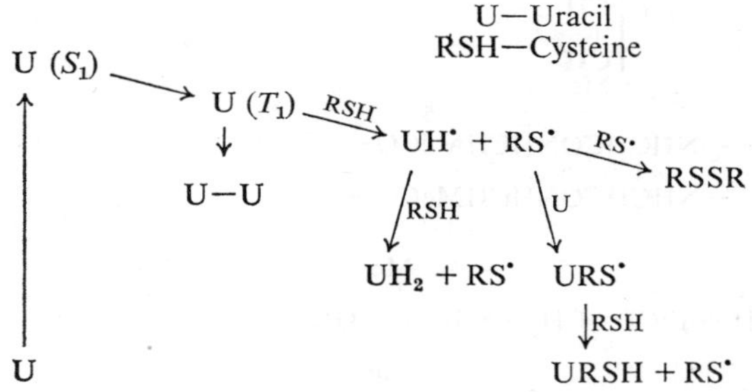

[Scheme 16 mechanism diagram]

U—Uracil
RSH—Cysteine

$$U(S_1) \rightarrow U(T_1) \xrightarrow{RSH} UH^{\cdot} + RS^{\cdot} \xrightarrow{RS^{\cdot}} RSSR$$

with branches: U–U; UH₂ + RS·; URS· → URSH + RS·

Scheme 16

reactions of purines and purine nucleosides (336a—d) with isopropyl alcohol all lead to substitution at the C-8 position of the purine ring system:[132] contrast the formation of (334) from purine itself. The present photoreactions occur by direct irradiation or acetone sensitization, and quantum yields are generally low (*e.g.* ca. 0·0015) in the direct irradiation. It would seem that the sugar moiety does not effect the site of attack in the purine ring, since in purine riboside evidence for 1,6-addition of the isopropyl alcohol was obtained from the u.v. spectral changes which occur on irradiation. Substitution of the 8-position in caffeine is also observed when its solutions are irradiated with amino-acids and ethers. The former reaction yields 8-alkyl derivatives (337), and since it is known that glycine

---

[130] J. S. Connolly and H. Linschitz, *J. Amer. Chem. Soc.*, 1968, **90**, 2979.
[131] D. Elad, I. Rosenthal, and H. Steinmaus, *Chem. Comm.*, 1969, 305.
[132] H. Steinmaus, I. Rosenthal, and D. Elad, *J. Amer. Chem. Soc.*, 1969, **91**, 4921.

Me-C(Me)(OH)-CH(H)- attached to purine
(334)

(335) R = Me₂C—OH
      R = MeCHOH

(336)
(a) X = H, Y = NH₂, R = H (Adenine)
(b) X = H, Y = NH₂, R = ribose (Adenosine)
(c) X = NH₂, Y = OH, R = ribose (Guanosine)
(d) X = H, Y = OH, R = H (Hypoxanthine)

caffeine + RCHCO₂H(NH₂) —hν, λ > 260 nm→ (337) R = alkyl

undergoes fragmentation with the formation of free-radicals derived from the amino-acid, in the presence of photo-activated adenosine, it is suggested that a similar mechanism may be operative.[133] Caffeine in this case serves as the light-absorbing system which induces the amino-acid fragmentation. The reaction of caffeine with ethers occurs in aqueous solutions and is photosensitized by benzophenone or acetophenone.[134] It seems very probable that the sensitization results from hydrogen abstraction by the ketone rather than energy transfer, since the ketone is converted into the pinacol.

Although addition of alcohols to the azomethine (C=N) has been observed previously [*e.g.* formation of (334)], the corresponding addition to an amidine C=N bond was not reported until this year.[135] Thus irradiation (350 nm) of 5-phenyl-7-acetamidofurazano[3,4-*d*]pyrimidine (338) in anhydrous ethyl alcohol leads to both the alcohol addition product (339) and the C=N reduction product (340). Similar reaction occurs in the presence of isopropyl alcohol, but with t-butanol (no α-hydrogen), the 1:1 adduct (341), in which a methyl group has added across the 6,7-amidine C=N bond, is formed in 35% yield.

The technique of spin trapping has been recently developed and involves addition of reactive free radicals to phenyl t-butyl nitrone ('spin-trap') to yield relatively stable t-butyl (α-substituted benzyl) nitroxides ('spin-

[133] D. Elad and I. Rosenthal, *Chem. Comm.*, 1969, 905.
[134] S. Jerumanis and A. Martel, *Canad. J. Chem.*, 1970, **48**, 1716.
[135] E. C. Taylor, Y. Maki, and B. E. Evans, *J. Amer. Chem. Soc.*, 1969, **91**, 5181.

adducts'). The nitroxides are then examined by e.s.r. spectroscopy and the structure of the α-substituent is deduced from a comparison of the N and β-H coupling constants for the spin-adducts with values obtained from authentic samples of the nitroxide. The technique has now been applied to the photolysis of nitrobenzene in tetrahydrofuran,[136] which from earlier work was suggested to yield the radicals shown in equation 2.[137] The new

work shows that the suggested alternative (342) is in fact formed. Many m- and p-substituted nitrobenzenes were examined in this reaction, and all with electron-withdrawing groups gave corresponding radicals (342) in tetrahydrofuran. With electron-donating groups, only 4-phenyl-, 3- and 4-methyl-, 4-ethyl-, 4-isopropyl-, and 4-phenoxy-nitrobenzenes gave the (342) radicals.

N-Nitroso-compounds undergo photoaddition to olefins and dienes as well as the rearrangement reactions which were considered in the previous section. The former reactions have been studied in particular with N-nitrosopiperidine (343).[138] Irradiation of acidic methanol solutions of

[136] E. G. Janzen and J. L. Gerlock, *J. Amer. Chem. Soc.,* 1969, **91**, 3108.
[137] R. L. Ward, *J. Chem. Phys.,* 1963, **38**, 2588.
[138] Y. L. Chow, S. C. Chen, and D. W. L. Chang, *Canad. J. Chem.,* 1970, **48**, 157.

(343) and 3,3-dimethylbut-1-ene yields 3,3-dimethyl-1-piperidino-2-(*N*-nitrosohydroxylamino)butane (344) as the major product. Different results were obtained with hex-1-ene and other olefins, the reactions for which are summarized in Scheme 17. In addition to (344), minor amounts

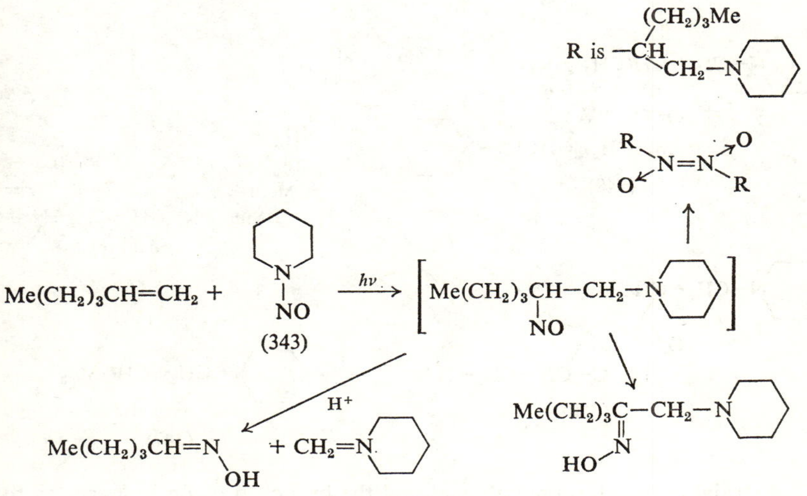

Scheme 17

of 3,3-dimethyl-1-piperidino-2-(*N*-hydroxyformamido)butane (345) and 3,3-dimethyl-1-piperidino-2-butanone oxime (346) were formed under certain conditions. Other products from this reaction are outlined in Scheme 18. The proposed reaction path for the butane derivatives (344)

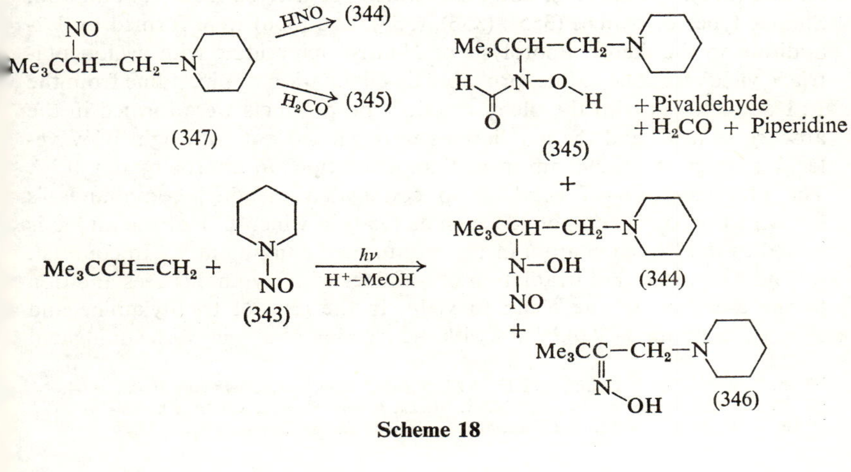

Scheme 18

and (345) involves reaction of the C-nitroso-monomer (347) with the HNO and H_2CO respectively which are photo-generated from (343) and methanol. In agreement with this, photolysis of a mixture of the *trans*-dimer of 1-piperidino-2-nitrosohexane (348) and (343) yields 1-piperidino-2-(*N*-nitrosohydroxylamino)hexane (349). The difference in behaviour between

the straight-chain terminal alkenes and the branched olefin is suggested to be due to the effect of the bulky t-butyl group in slowing down the tautomerism of (347) and accelerating the conversion of (350) into (347). In the presence of acid, (343) undergoes photoaddition to conjugated dienes to give good yields of 1 : 1 adducts.[139] The dienes examined were butadiene, cyclopentadiene, cyclohexa-1,3-diene, and *cis,cis*-cyclo-octa-1,3-diene. The major products were the *syn*- and *anti*-isomers of piperidino $\alpha\beta$-unsaturated oximes (351), (352), (353), and (354) which are derived from 1,4-addition. Similar types of oximes (355), (356), (357), and (358) were formed by 1,2-addition to the diene. Photolysis of *N*-nitrosopiperidine with cycloheptatriene yields tropone oxime, probably by elimination of piperidine from the 1 : 1 adduct. As with the olefin reaction, no products were formed in the absence of acid, and the irradiations were carried out with light of wavelengths longer than 290 nm in methanol solutions at approximately 0 °C. The photoaddition of t-amines to conjugated carbonyl compounds is known to produce compounds such as (359) in which C-1 of the amine is joined to the β-carbon atom of the unsaturated compound.[140] In contrast, it is now reported that irradiation of t-amines with styrenes causes addition to the α-carbon of the olefin to yield, in the case of triethylamine and styrene, compound (360).[141] As with the reaction of amines with conjugated

---

[139] Y. L. Chow, C. J. Colon, and D. W. L. Chang, *Canad. J. Chem.*, 1970, **48**, 1664.
[140] R. C. Cookson, J. Hudec, and N. A. Mirza, *Chem. Comm.*, 1968, 180.
[141] R. C. Cookson, S. M. de Costa, and J. Hudec, *Chem. Comm.*, 1969, 753.

MAJOR PRODUCTS — (351), (352), Pip is N(piperidine)

MINOR PRODUCTS — (355), (356), (353), (354), (357), (358)

carbonyl compounds, the adducts are accompanied by formation of the dihydro-derivatives of the olefins, and an increase in the reaction temperature promotes formation of the dihydro-compound. The photoaddition was shown to be stereospecifically *cis* by an n.m.r. analysis of the adducts (361)

(359), (360), (361) R = H, (362) R = D

and (362) from trimethylamine and 3,3-dimethylindene and its 2-deuterio-analogue. Addition of t-amines to olefins to form quinoline derivatives has been initiated by both u.v.- and γ-radiation.[142] Thus *NN*-dimethylaniline and *N*-phenylmaleimide yield 1,2,3,4-tetrahydro-1-methylquinoline-3,4-*N*-phenyldicarboximide (363) by a radical process. The reaction has been reported to occur with other *NN*-dialkylanilines, *N*-phenylpyrrolidine, and *N*-phenylpiperidine in the presence of such olefinic compounds as *N*-phenylmaleimide, diethyl maleate, cyclohexene, and cyclopentene. In the addition of azodicarbonyl compounds to mono-olefins to yield both 1,2- (364) and 1,4-adducts (365), the photo-*trans* → *cis* isomerization of the azo-compound appears to be a much more important process than the addition of any

[142] J. M. Fayadh and G. A. Swan, *J. Chem. Soc.* (*C*), 1969, 1781.

PhNMe$_2$ + [N-phenylmaleimide] $\xrightarrow{h\nu}$ (363)

electronically excited species to the unsaturated substrate.[143] Thus the acceleration of the process by light is due to formation of the *cis*-isomer, which then adds thermally more rapidly than the *trans*-isomer.

(364) (365)

Light-induced 1,3-dipolar cycloaddition reactions have been observed with the aziridine dicarboximide (366).[144] In the presence of dimethyl acetylenedicarboxylate, (366) yields three 1 : 1 adducts which are assigned structures (367), (368), and (369), and also a 2 : 1 adduct (370). Adduct (369) is very interesting since the compound results from an internal 1,2-addition of the dimethyl maleate moiety to the aromatic ring of the benzyl group. The photochemical nature of its genesis from (367) has been verified, and it is also of interest that such a 1,2-aromatic addition product is found to be photo-labile, regenerating (367). Compound (368) is also found to arise from (367) on photolysis. The 2 : 1 adduct (370) is formed thermally from (369) by Diels–Alder addition of the acetylene, and other 1,4-addition products [*e.g.* (371)] with dienophiles have also been obtained.

The 253·7 nm irradiation of aqueous solutions of pyridine yields a product with aldehydic properties and which has an absorption at 365 nm. This compound is suggested to be 5-aminopenta-2,4-dienal (372), which by further hydrolysis yields glutonic aldehyde and ammonia.[145] In this work, the hydroxyl enamine (373) was thought to be the intermediate, but it could not be isolated. In their work on the photoisomerization of pyridine to the 'Dewar' isomer (374), Wilzbach and Rausch have found that the aldehydic product is indeed (372) by its reduction to 5-aminopentan-1-ol, and further have shown that the intermediate in the hydration is (374).[146] Evidence implicating (374) in the formation of (372) was obtained by

---

[143] E. K. von Gustorf, D. V. White, B. Kim, D. Hess, and J. Leitich, *J. Org. Chem.*, 1970, **35**, 1155.
[144] S. Oida and E. Ohki, *Chem. and Pharm. Bull. (Japan)*, 1969, **17**, 2461.
[145] J. Joussot-Dubien and J. Houdard-Pereyre, *Bull. Soc. chim. France*, 1969, 2619.
[146] K. E. Wilzbach and D. J. Rausch, *J. Amer. Chem. Soc.*, 1970, **92**, 2178.

irradiating pyridine in acetonitrile and adding water to the product in the dark. The quantum yield of 0·06 for the disappearance of pyridine in water is essentially the same as those determined in inert solvents and in aqueous sodium borohydride solution. This work is further discussed in Chapter 4.

The light-induced reduction of 4-nitropyridine in isopropyl alcohol has been studied.[147] Irradiation of the system in the presence of hydrogen chloride leads to a 29% yield of 4,4′-azopyridine. It is suggested, however, that the primary photoproduct of the reaction is 4-hydroxyaminopyridine, and that this is subsequently converted to the azo-compound by base during the work-up procedure. Irradiation of 4-nitropyridine in neutral solutions gives an unidentified brown compound.

It has been observed that 3,5-dicarboalkoxy-substituted pyridines undergo a variety of photoreactions. In ethanol solution, 3,5-dicarboethoxy-2,4,6-trimethylpyridine (375) yields 3-carboethoxy-2,4,5-trimethylpyrrole (376) and the 1,4-dihydropyridine (377).[148] On the other hand, the

---

[147] S. Hashimoto, K. Kano, and K. Ueda, *Tetrahedron Letters*, 1969, 2733.
[148] R. M. Kellogg, T. J. van-Bergen, and H. Wynberg, *Tetrahedron Letters*, 1969, 5211.

only isolable product (55%) from the 2,6-dimethyl derivative (378) is the dihydropyridine (379), and no pyrrole could be detected. However, a change of solvent from ethanol to methanol in the latter reaction resulted in the formation of only 9% of (379) and the production of the previously unknown 1,2- and 1,4-adducts (380) and (381) with alcohol in 61 and 19% yields respectively. The pyridine reduction products formed by irradiation in —OD alcohols were found to have greater than 90% of the deuterium in the 4-position. The authors speculate that the α-hydroxyalkyl radical formed by hydrogen-abstraction, transfers the O—H hydrogen (deuterium) to the 4-position to yield the 1,4-dihydropyridine or to the 3-position, when the intermediate (382) is formed by bond switching. Decomposition of (382) then yields the pyrrole. The alcohol adducts are considered to arise by coupling between the pyridyl and α-hydroxyalkyl radicals. The 1,4-dihydropyridine (377) is reported to be stable to 254 nm radiation, and is not the pyrrole precursor. Other workers have observed that 1,4-dihydropyridines are photo-labile, but that the presence of 2,6-substituents inhibits reaction except in the case of (383), when disproportionation occurs.[149] Three types of light-induced reaction have been found for 1,4-dihydropyridines, viz. (a) disproportionation, (b) isomerization, and (c) dimerization. The photochemistry of such systems is remarkably sensitive to the

---

[149] U. Eisner, J. R. Williams, B. W. Matthews, and H. Ziffer, *Tetrahedron*, 1970, **26**, 899.

substituents present, and has been investigated in detail for butanol solutions of the diester (384) and the diketone (385). The corresponding 1,2-isomers and photodimers are formed. Compound (384) yields minor amounts of the unstable 1,2-isomer (386) and the dimers (387) and (388). The former dimer cyclizes to form the 'cage' compound (389) on further irradiation. The diketone (385) likewise yields the 1,2-dihydro-compound and a photodimer, the detailed structure and stereochemistry of which are still in doubt.

Tetramethyl 11bH-benzo[a]quinolizine-1,2,3,4-tetracarboxylate (390) and its 4H-isomer (391) both yield the same photodimer (392) on irradiation in benzene solution [150] (*cf.* the isomerization of such systems described in ref. 33). The reaction only occurred in the presence of 1-bromonaphthalene as sensitizer. One wonders whether a heavy-atom effect may be involved. In contrast, the 6-methyl derivative (393) is converted into the 4H-isomer (394) on photolysis, and 9aH-dimethylquinolizine (395) undergoes ring contraction to the indolizine (396). Two features of these results [150] are at

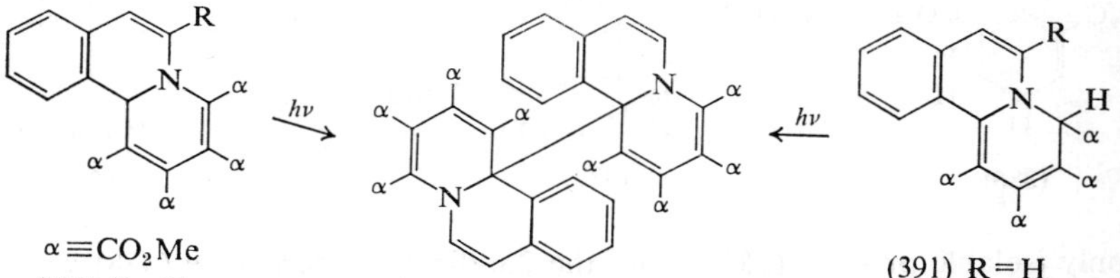

$\alpha \equiv CO_2Me$
(390) R = H
(393) R = Me

(392)

(391) R = H
(394) R = Me

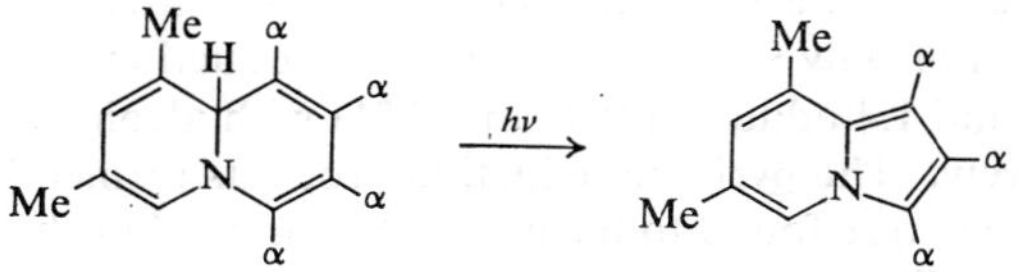

(395)

(396)

variance with the studies reported in ref. 33, *viz.* (*a*) the present reactions do not occur in the absence of 'sensitizer' whereas the previously reported isomerization processes were initiated by direct irradiation, and (*b*) the 9aH-dimethylquinolizine here yields an indolizine, whereas a similar system was found to form pyrrolo[1,2-a]azepines by the other workers. The latter difference may be explained if (395) initially undergoes photoisomerization to the 4H-quinolizine, which rearranges to the observed product (396); analogy for this idea is found in refs. 33 and 150.

The photoinduced reactions of aqueous solutions of monuron, 3-(p-chlorophenyl)-1,1-dimethylurea (397), lead to several products.[151] The

[150] A. O. Plunkett, *Chem. Comm.*, 1969, 1044.
[151] D. G. Crosby and C. S. Tang, *J. Agric. Food Chem.*, 1969, **17**, 1041.

principal pathways involve stepwise oxidation and demethylation of the
N-methyl group, hydroxylation of the aromatic nucleus, and polymerization.

Photolysis of chlorine isocyanate (398) has been used as a preparative
route to chlorocarbonyl isocyanate.[152, 153] Nitrogen, carbon monoxide, and
phosgene are also formed.

$$8\text{Cl-NCO} \xrightarrow{h\nu} 2\text{Cl-}\overset{\text{O}}{\underset{\|}{\text{C}}}\text{-NCO} + 3\text{Cl-}\overset{\text{O}}{\underset{\|}{\text{C}}}\text{-Cl} + 3\text{N}_2 + \text{CO}$$
(398)

3 Reactions of Sulphur-containing Compounds

The report has appeared of a mechanistic study of the photocycloaddition
of thiocarbonyl compounds, in particular thiobenzophenone, to olefins,[154]
and this extends the work described in preliminary publications.[155, 156]
Electron-rich (*i.e.* nucleophilic) olefins are found to react with $n\pi^*$ triplet
states of thiobenzophenone to yield either 1,4-dithians (399) or thietans
(400), and the reaction path is dependent upon steric factors. Electron-
deficient (*i.e.* electrophilic) olefins react with the $\pi\pi^*$ singlet state of the

[152] W. Gottardi and D. Henn, *Monatsh.*, 1969, **100**, 1860.
[153] W. Gottardi and D. Henn, *Monatsh.*, 1970, **101**, 11.
[154] A. Ohno, Y. Ohnishi, and G. Tsuchihashi, *J. Amer. Chem. Soc.*, 1969, **91**, 5038.
[155] A. Ohno, M. Fukuyama, and G. Tsuchihashi, *J. Amer. Chem. Soc.*, 1968, **90**, 7038.
[156] A. Ohno, Y. Ohnishi, and G. Tsuchihashi, *Tetrahedron Letters*, 1969, 283.

thioketone to yield thietans also. From kinetic studies and product analysis, it is deduced that the former type of olefin reacts *via* a radical mechanism whereas the second proceeds by nucleophilic attack of excited thiobenzophenone on the olefin, or *via* complexes between the olefins and the $\pi\pi^*$ singlet state of the thioketone. When derivatives of 1,4-dithian are formed, *trans*-addition of the two molecules of thiobenzophenone has been verified.

Thiacyclohexan-3-one has been previously reported to form reduction and addition products on irradiation in alcohols, whereas in inert solvents it gives ring-expansion and -contraction products.[157] The same research group have examined the reaction with isothiochroman-4-one (401) and find that isomerization occurs to yield thiochroman-3-one (402).[158] That the 7-methoxy and the 8-methyl derivatives of (401) were converted into the 6-methoxy and 7-methyl derivatives of (402) respectively, establishes that the sulphur atom in the product is attached to the aryl carbon atom which held the carbonyl group in (401). From other work with substituted compounds, C-3 in the starting material was found to end up as C-2 in the product. From these results, it may be suggested that (403) is the intermediate in the photolysis, and evidence for this is obtained from irradiation of (401) for short periods, when an i.r. absorption at 1770 cm^{-1} appears. The formation of (404) from 5-methyl-2,3-dihydro-2H,6H-thiopyran-3-one

(405) adds weight to the idea that (403) is an intermediate in the reaction, and the stability of (404) suggests that the conversion of (403) to (402) proceeds by excitation of the triene moiety in (403). Further evidence for this reaction path is provided by irradiation of (401) at 330 nm, when the

---

[157] K. K. Maheshwari and G. A. Berchtold, *Chem. Comm.*, 1969, 13.
[158] W. C. Lumma and G. A. Berchtold, *J. Org. Chem.*, 1969, **34**, 1566.

1770 cm^{-1} absorption in the i.r. spectrum is again observed, and only polymeric materials and not (402) are formed, *i.e.* there is insufficient energy in the photon at this wavelength to excite the triene system. Irradiation of (406), which may be expected to have a lower energy $\pi\pi^*$ triplet than $n\pi^*$ triplet, did not yield any (407). It therefore seems likely that a $n\pi^*$ configuration is necessary for the rearrangement process.

β-Keto-sulphides have been well studied, and excited-state interaction between the two chromophores is known.[157, 159] The photolysis of the 6-hydroxy bridge-compound (408) has been previously reported,[159] but further work on the primary processes and structure of the products has now been described.[160] Two primary processes appear to occur, and the efficiencies of each depend markedly on the nature of the excitation, *viz.* whether it is charge-transfer or $n \rightarrow \pi^*$ in type. In methanol, charge-transfer gives almost exclusively (409) *via* C$_\alpha$—S fission. Excitation to the $n\pi^*$ state yields nearly equal amounts of (409), (410), and (411), which are the result of α-cleavage. A previously unidentified product has now been

---

[159] C. Ganter and J. F. Moser, *Helv. Chim. Acta*, 1968, **51**, 300.
[160] C. Ganter and J. F. Moser, *Helv. Chim. Acta*, 1969, **52**, 725.

assigned structure (412).[161] Irradiation of the unsaturated β-ketosulphide (413) at 366 or 253·7 nm yields the cyclopropyl compound (414).[162] The reaction is suggested to arise *via* a zwitterionic species such as (415) or a Norrish Type I cleavage involving intermediates (416) and (417).

Cyclic γ-ketosulphides and some cyclic δ-ketosulphides show an excited-state interaction which is probably similar to that observed in β-ketosulphides.[163] The photolysis of the γ-ketosulphide (418) in t-butanol is reported to yield thiacyclobutan-2-one (27%) and t-butyl 4-thiahexanoate (18%).[164] Derivatives of (418) were found to undergo similar reactions.

$$\underset{(418)}{\text{[cyclic ketosulphide]}} \xrightarrow[\text{Bu}^t\text{OH}]{h\nu} \text{[thiacyclobutanone]} + \text{MeCH}_2\text{SCH}_2\text{CH}_2\overset{\text{O}}{\underset{\|}{\text{C}}}-\text{O}-\text{CMe}_3$$

$$\underset{(419)}{\text{[Ph-S-Ph ketosulphide]}} \xrightarrow[\text{C}_6\text{H}_6]{h\nu} \underset{(420)}{\text{[trans-dimer]}}$$

The unsaturated γ-ketosulphide (419) in benzene solution yields the *trans*-dimer (420).[165] Irradiation in diethyl ether, methylene dichloride, or ethanol gives a very low yield of (420).

The major product from irradiation of 2-methyldihydrobenzo[*b*]thiophen (421) is reported to be thiochroman.[166] Minor amounts of 2-methylbenzo[*b*]thiophen are formed if the reaction is performed in hydrocarbon solvents, but the yield of this product, which doubtless arises by dispro-

---

[161] C. Ganter and J. F. Moser, *Helv. Chim. Acta.*, 1969, **52**, 967.
[162] A. Padwa, A. Battisti, and E. Shefter, *J. Amer. Chem. Soc.*, 1969, **91**, 4000.
[163] *E.g.* G. Bergson, G. Claeson, and L. Schotte, *Acta Chem. Scand.*, 1962, **16**, 1159; N. J. Leonard, T. W. Milligan, and T. L. Brown, *J. Amer. Chem. Soc.*, 1960, **82**, 4075.
[164] P. Y. Johnson and G. A. Berchtold, *J. Org. Chem.*, 1970, **35**, 584.
[165] N. Sugiyama, Y. Sato, H. Kataoka, C. Kashima, and K. Yamada, *Bull. Chem. Soc. Japan*, 1969, **42**, 3005.
[166] D. C. Neckers and J. Dezwaan, *Chem. Comm.*, 1969, 813.

portionation of the radical at the 2-position, increases in chlorinated solvents. The mechanism for the formation of the thiochroman is suggested to involve S—(C-2) cleavage followed by abstraction of hydrogen from the methyl group, and ring closure of the olefin on to the thiophenol. Dihydrobenzo[*b*]thiophen itself yields benzo[*b*]thiophen on irradiation, even in hydrocarbon solvents, and again the yields are increased in solvents which themselves could produce radicals (cumene, chloroform, carbon tetrachloride).

The naphthalene-sensitized photo-racemization of several alkyl aryl sulphoxides (422) has been shown to depend on transfer of energy from the

R = Me, Br, Cl, OH, H, or CMe$_3$

(422)

singlet excited state of the sensitizer molecule.[167] The proposed reaction mechanism requires exciplex formation and subsequent radiationless decay. Substitution in the sulphoxide shows that steric or electronic effects on the rate of fluorescence quenching or racemization are small.

The photolysis of aqueous solutions of cysteine (423) in the presence of benzyl chloride has been studied.[168] Evidence for the presence of CyS· and CySS· radicals has been obtained by the formation of *S*-benzylcysteine and *S*-benzylthiocysteine. Significantly, no CyBz or CyCy were obtained.

HO$_2$CCH—CH$_2$—S—S—CH$_2$—CH—CO$_2$H
   | |
   NH$_2$ NH$_2$
            (423)

The addition of sulphur trioxide to fluoro-olefins is known to give cyclic adducts, mainly β-sultones.[169] The reaction of perfluoroethylene and perfluoroisoprene with sulphur dioxide has now been observed, and irradiation in the condensed phase produces 1 : 1 adducts in good yield.[170] These adducts are α-(fluorosulphinyl)acyl fluorides (424). The cyclic

$$CFX=CF_2 + SO_2 \xrightarrow{h\nu} \begin{bmatrix} CFX-CF_2 \\ | \quad\quad | \\ S\text{———}O \\ \| \\ O \end{bmatrix} \longrightarrow \begin{matrix} CFX-COF \\ | \\ SOF \end{matrix}$$

(425)   (424)

X = F or CF$_3$

---

[167] R. S. Cooke and G. S. Hammond, *J. Amer. Chem. Soc.*, 1970, **92**, 2739.
[168] C. J. Dixon and D. W. Grant, *J. Phys. Chem.*, 1970, **74**, 941.
[169] D. C. England, M. A. Dietrich, and R. V. Lindsey, *J. Amer. Chem. Soc.*, 1960, **82**, 6181.
[170] D. Sianesi, G. C. Bernardi, and G. Moggi, *Tetrahedron Letters*, 1970, 1313.

compound (425) is suggested as an intermediate. Photoreaction of N-chloroperfluoromethylethylamine (426) with sulphur chloride pentafluoride yields several products, as outlined in equation 3.[171]

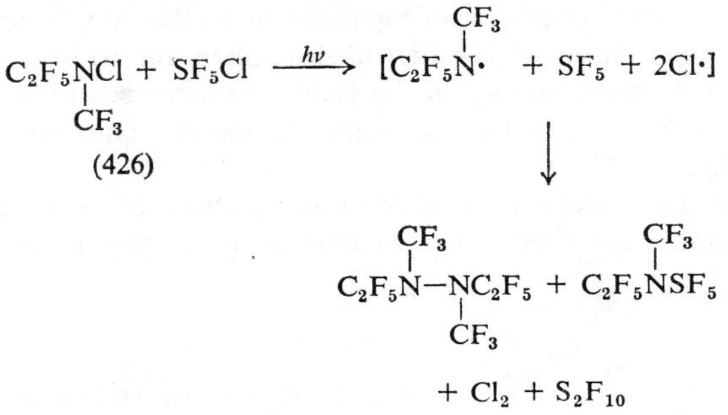

$$C_2F_5\underset{\underset{CF_3}{|}}{N}Cl + SF_5Cl \xrightarrow{h\nu} [C_2F_5\underset{\underset{}{|}}{\overset{\overset{CF_3}{|}}{N}}\cdot + SF_5 + 2Cl\cdot]$$

(426)

$$\downarrow$$

$$C_2F_5\underset{\underset{CF_3}{|}}{\overset{\overset{CF_3}{|}}{N}}-NC_2F_5 + C_2F_5\overset{\overset{CF_3}{|}}{N}SF_5$$

$$+ Cl_2 + S_2F_{10}$$

(3)

A. G.

[171] D. D. Moldavskii and V. G. Temchenko, *Zhur. org. Khim.*, 1970, **6**, 185.

7
Photoelimination Reactions

This chapter is concerned with the many diverse processes of light-induced fragmentation of organic molecules. Typically these are the loss of small molecules (*e.g.* nitrogen, carbon dioxide, hydrogen chloride) or decomposition into two or more sizeable fragments. Where it is considered of general chemical interest, the subsequent reactions of the various fragments are reviewed, but publications in which the photolyses have been simply used for the generation of free radicals, and where the subject matter is concerned mainly with radical studies, are not reported here. Norrish types I and II reactions of carbonyl compounds are reviewed in Part III, Chapter 1.

1 Elimination of Nitrogen from Azides

The photochemical loss of nitrogen from azides is a well-documented process and leads to the formation of electron-deficient species known as nitrenes (also less commonly termed azylenes, azacarbenes, imine radicals, and imido-intermediates) which react inter- or intra-molecularly by insertion, hydrogen abstraction, and coupling processes.

The direct and photosensitized decomposition of alkyl azides has been studied in some detail by reference to the photolysis products of n-butyl-, n-amyl-, 4-heptyl-, 2-phenylethyl-, 3-phenylpropyl-, 4-phenylbutyl-, phenylmethyl-, 1-phenylbutyl-, 1,1-diphenylethyl-, and cyclohexyl azides in benzene solution.[1] The primary products are nitrogen, and imines which are derived from hydrogen, alkyl, or acyl migration to the nitrogen. The rate of nitrogen evolution was found to be the same for all azides and the migratory order was hydrogen : alkyl, 5 : 1; phenyl : hydrogen, 1 : 1; phenyl : alkyl, 1 : 1. For cyclohexyl azide only α-hydrogen migration occurred. Chrysene is a sensitizer for the reaction and yields the same product distribution as in the direct photolysis (which is also unaffected by the presence of oxygen). The results are explained in terms of very short-lived singlet and triplet electronically excited azides. Evidence is presented which favours closely synchronous migration of the α-substituent with elimination of nitrogen from the excited azide. The authors propose that the migratory ratios may be determined by conformational factors which are operative in the ground state of the alkyl azides, but point out that gas-phase

[1] R. M. Moriarty and R. C. Reardon, *Tetrahedron*, 1970, **26**, 1379.

photolysis and solution thermolysis of alkyl azides, which are the most direct ways of testing this conformation basis for selectivity, are complicated by the instability of the products.

The photolysis products of polyfluorinated alkyl azides have been examined both from reactions of the neat compound and in the presence of cyclohexane, methylcyclohexane, and cyclohexene.[2] Nitrogen and 3H-hexafluoro-2-azabut-1-ene (1) are observed from 2H-hexafluoropropylazide (2). Compound (1) undergoes hydrolysis to trifluoroacetaldehyde and also

$$CF_3CHFCF_2N_3 \xrightarrow[-N_2]{h\nu} CF_3CHFN=CF_2 \xrightarrow[SiO_2]{Quartz} CF_3CHFNCO + SiF_4$$
$$(2) \qquad\qquad\qquad (1) \qquad\qquad\qquad (4)$$

$$\downarrow H_2O$$

$$CF_3CHFNCO + HF \qquad \nearrow CF_3CHFNHCF_3 \; (3)$$

suffers fluoride-initiated isomerization to 3H-hexafluoro-2-azabut-2-ene. I.r. spectrometric analysis of the photolysis mixture indicates that traces of a secondary amine (3) and isocyanate (4) are also formed in the reaction. In the presence of the above hydrocarbon diluents, the photolysis of (2) yields, after hydrolysis, N-cyclohexyl-, N-(1-methylcyclohexyl)-, and N-(cyclohex-2-enyl)-$\alpha\beta\beta\beta$-tetrafluoropropionamide, respectively. Similar reactions are reported for 2-chloro-1,1,2-trifluoroethyl azide.

Swenton has continued his work on 2-azidobiphenyl systems.[3] Previously it had been concluded from work on the photolysis of triarylmethyl azides that the same intermediate was formed from the singlet and triplet states of the azide.[4] The ketone-sensitized decomposition of 2-azidobiphenyl (5), however, is now reported to give remarkably different results from those obtained with aromatic hydrocarbon sensitization, and both results differ from those of direct photolysis. While the direct photolysis of (5) in diethyl ether, benzene, or isopropyl alcohol yields the carbazole (6) (68—71%) and 2-azobiphenyl (7) (8—11%), sensitization of the decomposition in benzene by acetophenone, m-methoxyacetophenone, or benzophenone produces

[2] R. E. Banks, D. Berry, M. J. McGlinchey, and G. J. Moore, *J. Chem. Soc.* (*C*), 1970, 1017.
[3] J. S. Swenton, T. J. Ikeler, and B. H. Williams, *Chem. Comm.*, 1969, 1263.
[4] F. D. Lewis and W. H. Saunders, *J. Amer. Chem. Soc.*, 1967, **87**, 645.

Photoelimination Reactions

(7), and none of the carbazole (6) is formed. In contrast, triphenylene, naphthalene, or pyrene sensitize the decomposition of azide (5) to form carbazole (6) as the major product. In the unsensitized reaction, the formation of the azo-compound (7) is quenched by piperylene or oxygen. It is thus concluded that hydrocarbon sensitization yields singlet azides and thence carbazole (6), whereas ketones transfer triplet energy to the azide, and azo-compound (7) is ultimately formed. The authors quite rightly add the cautionary note that data on hydrocarbon sensitization experiments must be treated carefully, as hydrocarbon singlet states have recently been implicated where triplet states had previously been supposed.[5,6] A report of the direct unsensitized photolysis of a series of 2-azidobiphenyls by the same workers gives further evidence for the differing reactions from the two excited states, and indicates in this reaction the uniqueness of pyrene which quenches the triplet azide but sensitizes the singlet decomposition.[7]

The photolysis of acyl azides has come under the scrutiny of two Japanese workers who have used the procedure as a 'synthesis' of 5,5-dimethyl-9-azabicyclo[6,2,0]decane-3,7,10-trione cyclic-3,7-bis(ethyleneacetyl) (8) and 7-azabicyclo[4,2,1]nonan-8-one (9) from precursors (10) and (11) respectively.[8] It is really not too clear why the photolysis of such a complicated example as (10) was undertaken but may be a reflection of the fact that more

simple acyl azides such as cyclohexyl and cyclopentyl carbonyl azides tend to yield complex mixtures.

Last year the photolysis of the carbohydrate azide 2,3,4-tri-*O*-acetyl-6-azido-6-deoxy-α-D-glucopyranoside (12) in cyclohexane was reported to yield the aldehyde (13) *via* the imine (14).[9] Other workers have now exam-

---
[5] R. S. Cole and G. S. Hammond, *J. Amer. Chem. Soc.*, 1968, **90**, 2958.
[6] P. D. Bartlett and P. S. Engel, *J. Amer. Chem. Soc.*, 1968, **90**, 2961.
[7] J. S. Swenton, T. J. Ikeler, and B. H. Williams, *J. Amer. Chem. Soc.*, 1970, **92**, 3103.
[8] K. Kawashima and I. Agata, *J. Pharm. Soc. Japan*, 1969, **89**, 1426.
[9] D. Horton, A. E. Luetzow, and J. C. Wease, *Carbohydrate Res.*, 1968, **8**, 366.

ined the photolysis of (12) in cyclohexene and although no reaction with the olefin is reported, the former reaction intermediate (14) has now been isolated along with another compound which is assigned the dimeric structure (15).[10] The imine (14) is stable as a solid but in chloroform

solution yields (15). Photolysis of azide (12) in dry ethanol, followed by acetylation, is deduced to yield (16) on the basis of spectroscopic data.

Inevitably, diazido-compounds are being photolysed; thus Ege and Jooss report on the reaction of diphenyldiazidomethane in acetone with radiation from a high pressure mercury lamp.[11] The major product (60% yield) is 2-phenylbenzimidazole (17) with benzophenone azine and benzophenone benzhydrylimine (18) formed in 4 and 8% yields respectively. The mechanism for the formation of the major product would seem to involve mononitrene formation, and a phenyl shift from carbon to nitrogen to give (19): this is followed by a further nitrene formation and an insertion.

The photolysis of azide ion (sodium azide) in aqueous solution has been investigated at 228·8 and 213·9 nm,[12] and is discussed here because of its relevance to the behaviour of organic azides. The onset of the charge-transfer to solvent band (CTTS) of azide ion is at approximately 235 nm,

---

[10] R. L. Whistler and A. K. M. Anisuzzaman, *J. Org. Chem.*, 1969, **34**, 3823.
[11] G. Ege and G. Jooss, *Chem.-Ztg.*, 1970, **94**, 215.
[12] I. Burak, D. Shapira, and A. Treinin, *J. Phys. Chem.*, 1970, **74**, 568.

Photoelimination Reactions

$$\underset{Ph}{\overset{Ph}{>}}C\underset{N_3}{\overset{N_3}{<}} \xrightarrow{h\nu} \underset{\underset{H}{N}}{\overset{N}{\bigg\langle}}\!\!\!\!-Ph \;+\; \underset{Ph}{\overset{Ph}{>}}C=N-N=C\underset{Ph}{\overset{Ph}{<}}$$

(17)

$$\underset{N_3}{\overset{Ph}{>}}C=N\!-\!Ph \qquad \underset{Ph}{\overset{Ph}{>}}C=N-CHPh_2$$

(19)   (18)

but at this wavelength tends to be overshadowed by the $n \rightarrow \pi^*$ transition. Thus, photolyses of azide ion at 228·8 and 253·7 nm hardly differ, and both probably involve the $n \rightarrow \pi^*$ excitation. Radiation at 213·9 nm is well within the CTTS band, and new features of the photolysis of the azide ion are apparent. Approximately 10% of the excited ions undergo ionization and about 30% react to form hydroxylamine, probably by being first converted to the $n\pi^*$ state. It would seem from this work that the azide radical abstracts a hydrogen atom from alcohols and, more readily, from acetone and hydroxylamide. The photolysis of azide ion at the two wavelengths is summarized by reactions (1)—(9).

$$N_3^- + 2H_2O \xrightarrow{h\nu} NH_2OH + N_2 + OH^- \quad \text{at 228·8 and 254 nm} \quad (1)$$

$$N_3^- \xrightarrow{h\nu} N_3^{-*} \qquad (2)$$
$$N_3^{-*} + H_2O \longrightarrow NH + OH^- + N_2 \qquad (3)$$
$$N_3^{-*} \longrightarrow N_3 + e_{aq}^- \qquad (4)$$
$$NH + H_2O \longrightarrow NH_2OH \qquad (5)$$
$$e_{aq}^- + NH_2OH \longrightarrow NH_2 + OH^- \qquad (6)$$
$$NH_2 + NH_2OH \longrightarrow NH_3 + NHOH \qquad (7)$$
$$N_3 + NH_2OH \longrightarrow N_3^- + H^+ + NHOH \qquad (8)$$
$$2NHOH \longrightarrow N_2 + 2H_2O \qquad (9)$$

at 213·9 nm

2 Decompositions of Diazo-compounds

It is always pleasing to read accounts in which researchers have used photochemical methods to overcome difficult synthetic problems. While non-photochemical synthetic routes for the formation of the tryptamine system are adequate, the isotryptamine system (20) [2-(2-aminoalkyl)indole], has hitherto been less accessible. This year, however, a facile route has been described for the synthesis of the latter system which involves a photochemical Wolff rearrangement of the diazo-compound (21) as the

---

[13] V. Snieckus and K. S. Bhandari, *Tetrahedron Letters*, 1969, 3375.

(21) → hν, R¹R²NH → [indole-2-CH2-CO-NR¹R²]

THF | LiAlH₄ ↓

(20) [indole-2-CH2CH2-NR¹R²]

key step.[13] The present synthesis makes compounds of type (20) readily available for the first time in preparative quantities.

Simons and his co-workers have employed the photolysis of diazomethane in *cis*-but-2-ene, propane, and isobutane to study the chemically activated hydrocarbon species thus produced.[14, 15, 16] The wavelengths employed in the studies were 366 and 435·8 nm and in all cases the reactions involved singlet methylene. With *cis*-but-2-ene, data were obtained on the geometric and structural isomerization rates of the chemically activated *cis*-1,2-dimethylcyclopropane and the decomposition rate of pent-2-ene, compounds formed respectively by addition and insertion of the singlet methylene into the butene. Insertion of singlet methylene into the secondary and tertiary C—H bonds of propane and isobutane yields isobutane and neopentane respectively. The decomposition rates for energised isobutane are 1·9 and $3·6 \times 10^7$ s^{-1} at 435·8 and 366 nm and for energised neopentane are 4·4 and $6·8 \times 10^6$ s^{-1} at 435·8 and 366 nm. From the same research school the wavelength dependency of the relative proportions of singlet/triplet methylenes from photolysis of diazomethane is reported.[17] These investigations involve the photolysis of mixtures of diazomethane with *cis*-but-2-ene and ethylene at the same two wavelengths as employed previously, and the data indicate that the proportion of ground triplet state methylene radicals produced in the primary photolysis is independent of the wavelength of the exciting radiation.

The photolysis of phenyldiazocyclopentadienes (22) in the presence of alkynes has been used as a route for the synthesis of benzoindenes (23): a two-photon process is indicated, proceeding *via* the spiro-compound (24).[18] Irradiation of compounds (25), analogous to (22) but with an unsubstituted 4-position, in the presence of benzene and its simple methyl derivatives (26), leads to bicyclo[6,3,0]undecapentaenes (27), the cyclopentadiene (28), and

[14] J. W. Simons and G. W. Taylor, *J. Phys. Chem.*, 1969, **73**, 1274.
[15] G. W. Taylor and J. W. Simons, *J. Phys. Chem.*, 1970, **74**, 464.
[16] R. L. Johnson, W. L. Hase, and J. W. Simons, *J. Chem. Phys.*, 1970, **52**, 3911.
[17] G. W. Taylor and J. W. Simons, *Canad. J. Chem.*, 1970, **48**, 1016.
[18] V. Durr and L. Schrader, *Angew. Chem.*, 1969, **81**, 426.

the insertion product (29).[19] A sequence for the formation of (27) *via* spiro-norcaradienes (30) and tropylidienes (31) is discussed. On the other hand, irradiation of 1,4-disubstituted diazocyclopentadienes (32) in benzene yields 7*H*- and 5*H*-benzocycloheptenes (33) and (34) respectively, again *via* carbenacyclopentadienes [$(4n+2)\pi$-electrons] as intermediates. The side-reactions give an insertion product (35) and the carbene dimer (36). The mechanism of the process is outlined in Scheme 1, and photoinitiated 1,3- and/or 1,7-hydrogen shifts are proposed.[20] Photolysis of (33) and (34) produces fragmentation to the naphthalene derivative (37). It is interesting to note that similar photolysis of 3-diazo-2,4,5-triphenyl pyrrole in benzene gave only the insertion product (38).[21]

[18] H. Duerr and G. Scheppers, *Chem. Ber.*, 1970, **103**, 380.
[20] H. Duerr and G. Scheppers, *Annalen*, 1970, **734**, 141.
[21] R. F. Bartholomew and J. M. Tedder, *J. Chem. Soc.* (*C*), 1968, 1601.

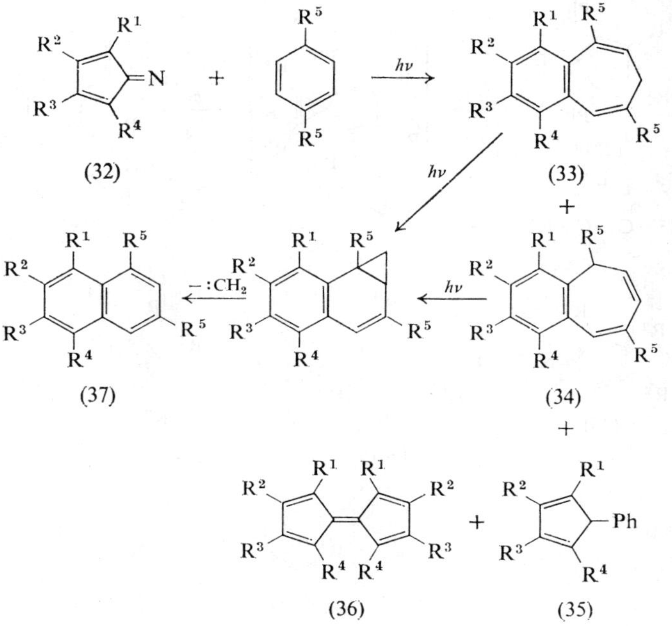

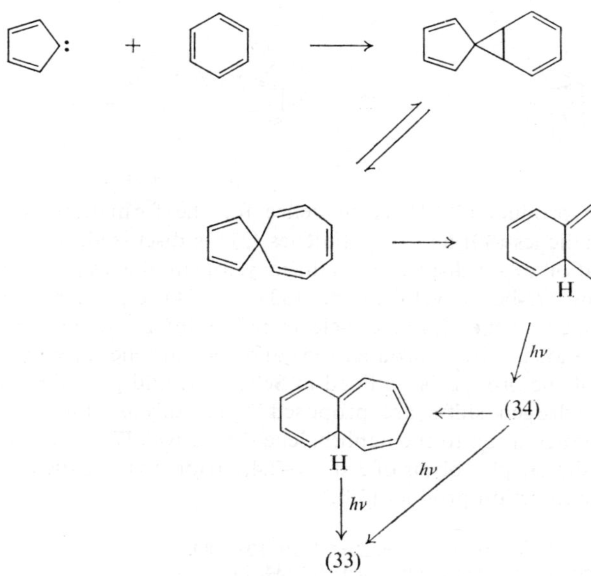

Scheme 1

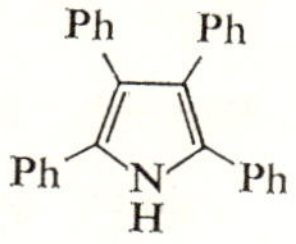

$$\underset{\text{(38)}}{\underset{H}{\text{Ph}}\underset{}{\overset{\text{Ph Ph}}{\diagup\!\!\!\diagdown}}\text{Ph}}$$

The carbene (39) formed from the photolysis of trimethylsilyl diazoethylacetate (40) undergoes 95% *cis* stereospecific addition to olefins, as well as yielding the usual insertion products (41).[22] In contrast, ethoxycarbonyl-t-butylcarbene (42), analogous to (39), formed from the photolysis of ethyl 2-diazo-3,3-dimethylbutanoate (43) reacts almost exclusively in either

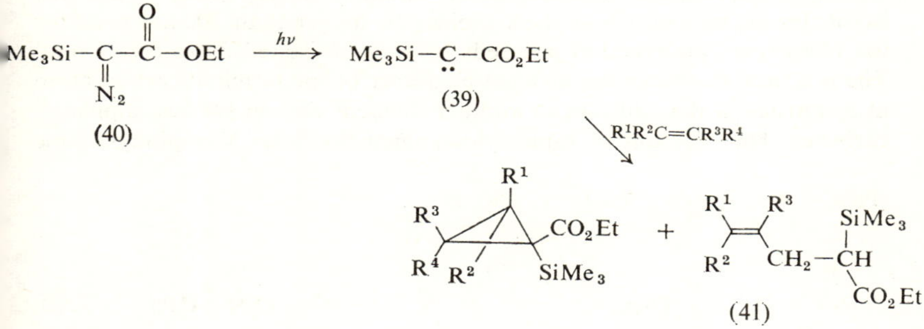

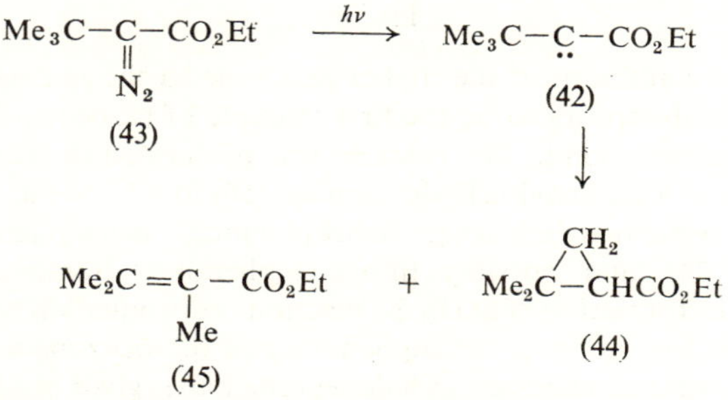

$$\underset{(40)}{Me_3Si-\underset{\underset{N_2}{\|}}{C}-\overset{\overset{O}{\|}}{C}-OEt} \xrightarrow{h\nu} \underset{(39)}{Me_3Si-\underset{..}{C}-CO_2Et}$$

$\searrow R^1R^2C=CR^3R^4$

(41): cyclopropane with $R^1, R^2, R^3, R^4, CO_2Et, SiMe_3$ substituents + $\underset{R^2}{\overset{R^1}{>}}=\underset{CH_2-\underset{CO_2Et}{\overset{SiMe_3}{|}}CH}{\overset{R^3}{<}}$

$$\underset{(43)}{Me_3C-\underset{\|}{\underset{N_2}{C}}-CO_2Et} \xrightarrow{h\nu} \underset{(42)}{Me_3C-\underset{..}{C}-CO_2Et}$$

$\downarrow$

$\underset{(45)}{Me_2C=\underset{\underset{Me}{|}}{C}-CO_2Et}$ + $\underset{(44)}{Me_2\overset{CH_2}{\overset{\diagup\!\!\diagdown}{C}}-CHCO_2Et}$

cyclopentane or isobutene by an intramolecular path to yield ethyl 2,2-dimethylcyclopropanecarboxylate (44) and ethyl trimethylacrylate (45).

The thermal and photolytic decompositions of diaryldiazomethanes (46) and diarylmethylenetriphenylphosphazines (47) appear in the same report.[23] The products of both reactions of each series of compounds indicate that the major pathway is one in which nitrogen is eliminated with the formation of carbenes and, in the case of (47), triphenylphosphine. The carbenes from both precursors undergo reactions which are dependent upon their inherent stability and the medium in which they are generated. Both

[22] U. Schoellkopf, D. Hoppe, N. Rieber, and V. Jacobi, *Annalen*, 1969, **730**, 1.
[23] D. R. Dalton and S. A. Liebman, *Tetrahedron*, 1969, **25**, 3321.

Ph_2CN_2 (46)

$Ph_2C=N-N=PPh_3$ (47)

$\xrightarrow[\text{or }\Delta]{hv}$

$Ph_2C=CPh_2$, $Ph_2CHCHPh_2$, $Ph_2C=O$,

$Ph_2C=N-N=CPh_2$, $Ph_2CH-O-CHPh_2$,

+ [in the case of (47)], Ph_3P and Ph_3PO

thermal and photolytic decompositions of (47) are apparently stepwise: firstly, the diazo-compound and triphenylphosphine are formed, followed by decomposition of the diazo-compound to the carbene.

Diphenyldiazomethane has been photolysed in the presence of isonitriles [24] and N-sulphinylaniline.[25] The former reaction gives up to 50% yields of ketenimines (48), and the major side-product is tetraphenylethylene: notably the formation of benzophenone azine is not reported.[26] Since the isonitriles do not absorb at the wavelengths longer than 300 nm used for the photolysis, the reaction proceeds by way of excited diazo-compound. The reaction illustrates the divalent character of the terminal carbon atom in isonitriles in the addition to another divalent carbon species, diphenylcarbene. The reaction of diphenyldiazomethane with N-sulphinylaniline

$Ph_2C=N_2 \xrightarrow[RNC]{hv} Ph_2C=C=N-R$
(48)

$PhN=S=O$ + Ph_2C: $\longrightarrow$
(49)

$\begin{array}{c}Ph\quad\ O\\ \diagdown\ //\\ N-S\\ \diagup\\ C\\ \diagup\diagdown\\ Ph\quad Ph\end{array}$

$\longrightarrow$ $PhN=CPh_2$ + SO
(50)

(49) proceeds by addition of the diphenylcarbene to the nitrogen–sulphur bond of (49), and appears to be the first example of carbene addition to a non-carbon multiple bond. The reaction was performed under nitrogen in hexane and gave N-diphenylmethyleneaniline (50) in 67% yield, with minor amounts of tetraphenylethylene, benzophenone, benzophenone azine (contrast ref. 24), and possibly thiobenzophenone. Some evidence is presented for the formation of SO in the reaction: its production would nicely account for the formation of benzophenone and thiobenzophenone.

The reaction of diazomethane in halogenomethanes gives products which have been rationalized in terms of a chain reaction.[27] In this case the quantum yields are high. With methyl diazoacetate decomposition in halogenomethanes, the quantum yields of products are low: this feature seems to make a chain mechanism somewhat unlikely.[27] On the other hand, some nuclear polarization studies of this system have provided evidence for the possibility of a chain mechanism.[28] Such n.m.r. studies are possible

[24] J. A. Green and L. A. Singer, *Tetrahedron Letters*, 1969, 5093.
[25] J. O. Stoffer and H. R. Musser, *Chem. Comm.*, 1970, 481.
[26] T. Oncesch, D. Bogdan, M. Contneanu, and G. Balaceanu, *Ber. Bunsengesellschaft Phys. Chem.*, 1968, **72**, 274.
[27] W. H. Urry, J. R. Eiszner, and J. W. Wilt, *J. Amer. Chem. Soc.*, 1957, **79**, 918.
[28] M. Cocivera and H. D. Roth, *J. Amer. Chem. Soc.*, 1970, **92**, 2573.

when free-radical intermediates are involved since the unpaired electrons can cause nuclear spin polarization. The polarization is retained in the diamagnetic product and is observed as enhanced n.m.r. absorption and/or emission. In the reaction of methyl diazoacetate and chloromethanes, a product was formed which gave n.m.r. emission lines which decayed at rates slower than the nuclear spin relaxation rates, and these slower rates are thought to imply a radical chain mechanism. It is at present rather difficult to reconcile all the reported features of these systems.

Two groups of workers report on the photochemical reactions of dialkyl diazomalonates in the presence of sulphur compounds. The first report concerns irradiation at 253·7 nm of cyclohexane solutions of diethyl diazomalonate and thiobenzophenone.[29] The major product is γ-butyrolactone (51), which represents only the second recorded example of intramolecular insertion of an alkoxycarbonylcarbene: in the first example the lactone was a minor reaction product.[30] While the presence of α-ethoxycarbonyl-γ-butyrolactone (52), the expected insertion product, has not been confirmed, photolysis of this ester lactone does yield (51). The major side-products in the diazomalonate–thiobenzophenone reaction are diethyl cyclohexylmalonate (53), diethyl 2,2-diphenylethylene-1,1-dicarboxylate, and benzophenone: the latter is probably formed in the work-up

procedure. Compound (53) is the major product in the absence of thiobenzophenone: reduction in the formation of the malonate (53) from this reaction has also been observed in the presence of benzophenone.[31] The possible intermediacy of the ylide (54) and/or the thiiran (55) in these transformations is being investigated. The formation of stable sulphonium ylides by the irradiation of dimethyl diazomalonate solutions in dimethyl

[29] J. A. Kaufmann and S. J. Weininger, *Chem. Comm.*, 1969, 593.
[30] W. Kirmse, H. Dietrich, and H. W. Bücking, *Tetrahedron Letters*, 1967, 1833.
[31] M. Jones, W. Ando, and A. Kulczycki, *Tetrahedron Letters*, 1967, 1391.

sulphide forms the subject of the second report.[32] The reaction is carried out in borosilicate glass and yields 88% of dimethylsulphonium bismethoxycarbonylmethylide (56). Since the reaction does not appear to be sensitized

$$Me_2S + N_2C(CO_2Me)_2 \xrightarrow{h\nu} \underset{Me}{\overset{Me}{>}}\overset{+}{S}-\overset{-}{C}\underset{CO_2Me}{\overset{CO_2Me}{<}} \quad (56)$$

$$\downarrow \begin{array}{c} h\nu \\ MeOH \end{array}$$

$$MeOCH(CO_2Me)_2 + H_2C(CO_2Me)_2$$

by benzophenone, the involvement of a singlet carbene is suggested: this carbene is shown to react four times faster with dimethyl sulphide than with cyclohexene. The reaction of biscarbonylcarbenes with carbon–oxygen bonds is well known,[31] but this report is one of the few concerned with reactions involving carbon–sulphur bonds. Further photolysis at 253·7 nm of the ylide (56) in methanol yields dimethyl methoxymalonate and dimethyl malonate.

Previous investigations into the photolysis[33,34] and silver-oxide-catalysed[35] decomposition of $\alpha\alpha'$-bisdiazocyclohexanones have resulted in the isolation of the corresponding ring-contracted $\alpha\beta$-unsaturated ester or acid. The products formed by both procedures were rationalized as having been formed by a Wolff rearrangement followed by hydrogen migration. The possibility of a cyclopropenone intermediate seems not to have been considered. In order to study the intermediacy of such a reactive species in the processes, Whitman and Trost have examined the reactions of 1,3-bisdiazo-1,3-diphenylpropan-2-one (57) in some detail.[36] The photolysis of compound (57) in methanol or tetrahydrofuran solutions at $-40\,°C$

[Reaction scheme: (57) Ph-C(=N$_2$)-C(=O)-C(=N$_2$)-Ph → hv/MeOH → Ph-CH(OMe)-CH(CO$_2$Me)-Ph (58)(59) + Ph$_2$C=CH-CO$_2$Me (60); (57) → Ag$_2$O-MeOH → (60) + (62) + Ph-C(=O)-CH(Ph)-CO$_2$Me (63); (61) PhC≡CPh ← hv/-CO ← cyclopropenone (62)]

[32] W. Ando, T. Yagihara, S. Tozune, and T. Migita, *J. Amer. Chem. Soc.*, 1969, **91**, 2786.
[33] W. Kirmse, *Angew. Chem.*, 1959, **71**, 539.
[34] R. Tasovac, M. Stefanovic, and A. Stojiljkovic, *Tetrahedron Letters*, 1967, 2731.
[35] R. F. Borch and D. Fields, *J. Org. Chem.*, 1969, **34**, 1481.
[36] P. J. Whitman and B. M. Trost, *J. Amer. Chem. Soc.*, 1969, **91**, 7534.

through borosilicate glass leads to the formation of the *erythro-* (58) and the *threo-* (59) forms of methyl 2,3-diphenyl-3-methoxypropionate, methyl α-phenylcinnamate (60), and diphenylacetylene (61). In toluene solution (57) yields (61) as the major product. The ester (60) is found to be photostable under these reaction conditions but diphenylcyclopropenone (62) decarboxylates quantitatively to (61). Under conditions where (62) is stable, it is formed from the diazo-compound (57) with no trace of (61). Thus the cyclopropenone (62) is the precursor of the acetylene (61). In the silver-oxide-catalysed decomposition of (57) in methanol, the major product is (60), with 11% of (62) and 8% of methyl benzylphenylacetate (63). These results suggest that the cyclopropenone (62) is the sole precursor of (61) in the photolysis and of (60) in the silver-oxide-catalysed reaction, and now require that cyclopropenones are considered as the intermediates in the decomposition of bisdiazoketones in general and bisdiazocyclohexanone in particular. In the latter case, straining the cyclopropenone with a short carbon chain would increase the carbonyl reactivity towards water and alcohol addition, thus leading to the observed products (*i.e.* αβ-unsaturated acids and ketones).

3 Decomposition of Azo-compounds

The photochemical loss of nitrogen from azo-compounds is well known. Aryl azo-compounds also undergo *trans-cis* isomerization, and it may be expected by analogy with the photochemistry of these compounds and of olefins, that azoalkanes would also undergo the isomerization in addition to the usual elimination of nitrogen. Only with azomethane and azoisopropane, however, has the *trans-cis* conversion been unequivocally demonstrated.[37] The isomerization and decomposition aspects of azoalkane photochemistry have now been investigated further by reference to the photolysis of a variety of *trans*-azoalkanes in fluorotrichloromethane, acetone, methanol, or pentane solutions below −50 °C.[38] After several hours irradiation the solutions became intensely yellow but only a trace of nitrogen was detected. On warming the solutions to 25 °C the colour disappeared and nitrogen was freely evolved. This intriguing state of affairs has been examined in detail with azoisobutane in CH_3OD solution at −70 °C by n.m.r. spectroscopy. After irradiation, the solution showed two n.m.r. singlets at τ 8·88 due to the starting compound and τ 8·55 due to a photoisomer: small amounts of isobutane and isobutene (approximately 2%) were also observed. When the solution was warmed to −23 °C, the hydrocarbon proportion increased at the expense of the τ 8·55 singlet, and at 0 °C further decomposition of the photochemical product was evident. At −80 °C in the presence of acid the photoisomer, which is normally stable at these temperatures, was rapidly and quantitatively

---

[37] I. I. Abram, G. S. Milne, B. S. Soloman, and C. Steel, *J. Amer. Chem. Soc.*, 1969, **91**, 7221.

[38] T. Mill and R. S. Stringham, *Tetrahedron Letters*, 1969, 1853.

converted to *trans*-azobutane. These data are wholly consistent with the photoisomer being the *cis*-form of azoisobutane. The rapid thermal decomposition with nitrogen evolution below 0 °C of the *cis*-azoalkanes thus formed is in striking contrast to the *trans*-isomers, and suggests that the two pairs of non-bonded electrons in *cis*-azoisobutane produce an electrostatic interaction which could account for the difference in steric strain between the azo-compound and the comparable model, *cis*-di-t-butylethylene. Since the decomposition is only very minor at low temperatures, it is suggested that at room temperature the major path for formation of the t-alkyl radicals and elimination of nitrogen is *via* photoisomerization of the *trans* to the *cis* isomer and rapid thermolysis of the latter. Such a procedure explains why in previous work no differences have been observed in the ratios of recombination to disproportionation for t-alkyl radicals generated thermally or photochemically from corresponding azoalkanes. Both processes yield spin-correlated radical pairs. Other workers have also examined the thermal and photochemical decomposition of one of the compounds, azobisisobutyronitrile, studied in the previous reference.[39] The fraction of kinetically free radical products has been measured in benzene and cyclohexane. The photochemical process is somewhat more efficient than the thermal one in both solvents, but both processes are less efficient in cyclohexane than in benzene.

Secondary processes in the thermal and photolytic decomposition of azoethane have been reported.[40] In the thermal process at 115 °C the important secondary products are ethyl-2-butyldi-imide and acetaldehyde diethylhydrazone. Photolysis at 366 nm at 100 °C formed tetraethylhydrazine in substantial amounts, in addition to the 'thermal' products. Reactions (10)—(20) account for the formation of the observed products.

$$\text{EtN=NEt} \xrightarrow{\Delta \text{ or } h\nu} 2\text{Et}\cdot + \text{N}_2 \quad (10)$$

$$2\text{Et}\cdot \longrightarrow \text{C}_4\text{H}_{10} \quad (11)$$

$$2\text{Et}\cdot \longrightarrow \text{C}_2\text{H}_4 + \text{C}_2\text{H}_6 \quad (12)$$

$$\text{Et}\cdot + \text{EtN=NEt} \longrightarrow \text{C}_2\text{H}_6 + \text{Me}\dot{\text{C}}\text{H}-\text{N}-\text{N}-\text{Et} \quad (13)$$

$$\text{Et}\cdot + \text{Me}\dot{\text{C}}\text{H}-\text{N}-\text{N}-\text{Et} \longrightarrow \text{MeCHEtN=NEt} \quad (14)$$

$$\text{Et}\cdot + \text{Me}\dot{\text{C}}\text{H}-\text{N}-\text{N}-\text{Et} \longrightarrow \text{MeCH=N}-\text{NEt}_2 \quad (15)$$

$$\text{RH} + \text{Me}\dot{\text{C}}\text{H}-\text{N}-\text{N}-\text{Et} \longrightarrow \text{R}\cdot + \text{MeCH=N}-\text{NHEt} \quad (16)$$

$$\text{EtN=NEt} \xrightarrow{\text{Wall}} \text{MeCH=N}-\text{NHEt} \quad (17)$$

$$\text{RH} + \text{Me}\dot{\text{C}}\text{H}-\text{N}-\text{N}-\text{Et} \longrightarrow \text{R}\cdot + \text{EtN=NEt} \quad (18)$$

$$\text{Et}\cdot + \text{EtN=NEt} \longrightarrow \text{Et}_2\text{N}-\dot{\text{N}}\text{Et} \quad (19)$$

$$\text{Et}\cdot + \text{Et}_2\text{N}-\dot{\text{N}}\text{Et} \longrightarrow \text{Et}_2\text{N}-\text{NEt}_2 \quad (20)$$

[39] R. D. Burkhart and J. C. Merrill, *J. Phys. Chem.*, 1969, **73**, 2699.
[40] O. P. Strausz, R. E. Berkley, and H. E. Gunning, *Canad. J. Chem.*, 1969, **47**, 3470.

Photoelimination Reactions

Photolysis of azoethane in solution (and of hexafluoroazomethane) has also been used to generate ethyl (and trifluoromethyl) radicals.[41]

The loss of nitrogen from cyclic azo-compounds has in the past been used extensively for the generation and study of biradicals and small ring compounds, and this year there have been several accounts of this general reaction. 1-Pyrazolines are of interest in this area of photochemistry. The parent compound (64) has been reported earlier to eliminate nitrogen with the formation of 'hot' cyclopropane.[42] The reaction has been subsequently examined in detail in the gas phase at 313 nm, when among the products are cyclopropane, propylene, and ethylene.[43] The propylene : cyclopropane ratio was found to be pressure-dependent and the ethylene formation only slightly so. Propylene is formed from at least two sources, only one of which is pressure-dependent. The photolysis is interpreted in terms of a singlet 'biradical' trimethylene [44] which forms 'hot' cyclopropane: this can then isomerize to propylene or be deactivated to cyclopropane. The 'hot' cyclopropane is not monoenergetic but has energy distributed around 75 kcal mol^{-1} (314 kJ mol^{-1}). The mechanism is outlined in reactions (21)—(30). Formation of the pressure-independent propylene is shown in

$$\underset{(64)}{\underset{N=N}{\triangle}} \xrightarrow[\text{Gas Phase}]{h\nu} \triangleright \ + \ \diagup\!\!\!\diagdown\text{Me} \ + \ H_2C=CH_2$$

(64) + $h\nu$ ⟶ $^1(64)^*$	(21)
$^1(64)^*$ ⟶ $^3(64)^*$	(22)
$^3(64)^*$ ⟶ N_2 + $\diagup\!\!\!\diagdown$Me	(23)
$^3(64)^*$ ⟶ $^3?$ + O_2 ⟶ ?	(24)
$^3(64)^*$ ⟶ $H_2C=CH_2$ + CH_2N_2	(25)
$^3(64)^*$ + O_2 ⟶ (64)	(26)
$^1(64)^*$ ⟶ (64)†	(27)
$^1(64)$ ⟶ $H_2C=CH_2$ + CH_2N_2 ?	(28)
(64)† ⟶ $^1\triangle$ + N_2	(29)
$^1\triangle$ ⟶ $\triangle$†	(30)

this mechanism to originate through reactions (22) and (23) from triplet pyrazoline. Pressure-dependent formation of propylene and cyclopropane probably arises from a vibrationally-excited ground state. The ethylene is formed from both singlet and triplet pyrazoline. Oxygen probably acts as a physical rather than a chemical quencher.

Dorer has examined the photolysis of both 3- and 4-methyl-1-pyrazolines in order to study the product energy distribution in photochemical re-

---

[41] A. P. Stefani, G. F. Thrower, and C. F. Jordan, *J. Phys. Chem.*, 1969, **73**, 1257.
[42] B. H. Al-Sader and R. J. Crawford, *Canad. J. Chem.*, 1968, **46**, 3301.
[43] P. Cadman, H. M. Meunier, and A. F. Trotman-Dickenson, *J. Amer. Chem. Soc.*, 1969, **91**, 7640.
[44] See A. Gilbert, in 'Photochemistry,' ed. D. Bryce-Smith (Specialist Periodical Report), The Chemical Society, London, 1970, Vol. 1, 404.

actions.[45, 46] The wavelength of light used in these studies is again 313 nm. The methylcyclopropane from the 4-methyl isomer (65) has sufficient energy to undergo isomerization to butene isomers (*cf.* ref. 43). From a consideration of the structural changes that occur on the loss of nitrogen from (65) with the formation of the trimethylene intermediate, a qualitative explanation is offered of the partitioning of the energy between the two products of the reaction. The partitioning of the excess of energy between the cyclic hydrocarbon and the nitrogen products in the photolysis of 1-pyrazoline is

$$Me-\underset{(65)}{\underset{N}{\overset{N}{\bigtriangleup}}} \xrightarrow{h\nu} Me-\triangleleft^{*} \longrightarrow \begin{array}{l} MeCH_2CH=CH_2 \\ MeCH=CHMe \\ Me_2C=CH_2 \\ Me-\triangleleft \end{array} \quad \underset{(66)}{\underset{N=N}{\bigtriangleup}}-Me$$

non-random. The internal degrees of freedom of the hydrocarbon fragment receive only about 62% of the available energy. All examples of photolysis of 1-pyrazolines reported until the present have involved symmetrical compounds, in the sense that rupture of the carbon–nitrogen bonds is equivalent. Thus the previous studies of energy partitioning between methylcyclopropane and nitrogen have been repeated with the use of an 'unsymmetrical' compound, 3-methyl-1-pyrazoline (66), to ascertain what differences may be observed by introducing a small difference in the strength of the carbon–nitrogen bonds.[45] The products of the reaction at 313 nm in propane were as expected. The conclusion from this study is that substitution in the 3-position, as against the 4-position, alters the relative amount of olefin formation in the primary photolysis process, but has little effect on the partitioning of energy between the methylcyclopropane and nitrogen fragments. The author points out that greater structural changes are probably necessary in order to alter the potential energy surface sufficiently to affect the energy distribution in the reaction products to a measurable degree.

Although cyclopropyl carbonium ions have been well studied, little appears to be known about the cyclopropylcarbinyl radicals. An investigation of the photolysis and thermolysis of the 3-cyclopropyl-1-pyrazoline derivatives (67) and (68) was undertaken to ascertain if structural differences in the biradical would affect the fate of the cyclopropylcarbinyl radical.[47] Compound (67) yielded the same proportions of (69), (70), (71), and (72), both photochemically and thermally. Similarly, (68) photochemically, sensitized or unsensitized, and thermally at 250 °C, gave (73) and (74) in the same proportions. Thus in no case was rearrangement of the cyclopropylcarbinyl radical observed.

[45] F. H. Dorer, *J. Phys. Chem.*, 1970, **74**, 1142.
[46] F. H. Dorer, *J. Phys. Chem.*, 1969, **73**, 3109.
[47] T. Sanjiki, M. Ohta, and H. Kato, *Chem. Comm.*, 1969, 638.

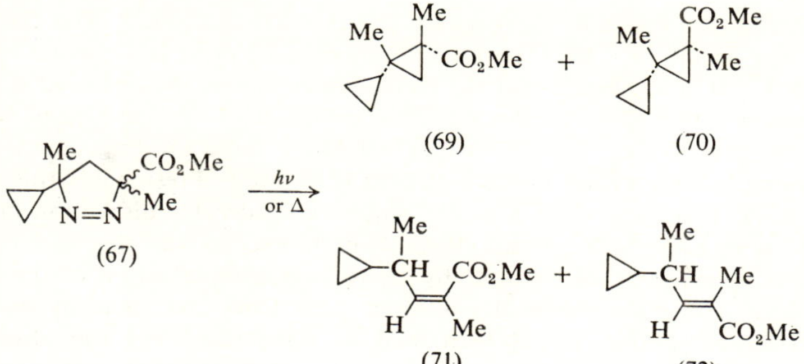

Photoelimination Reactions

(67) —hv or Δ→ (69) + (70)

(71) + (72)

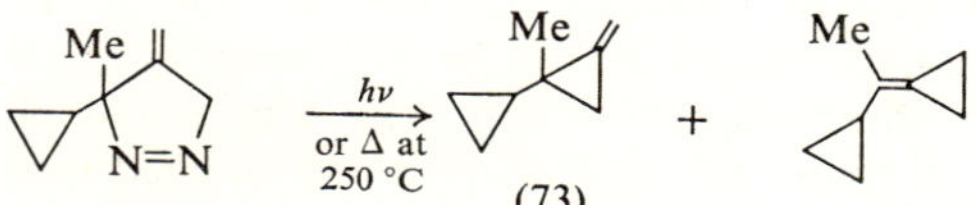

(68) —hv or Δ at 250 °C→ (73) + (74)

There are many reports in the literature which confirm the expected change in stereochemistry arising from the orbital symmetry prediction that the selection rule for an allowed thermal process will be reversed in the photochemical reaction. Berson and Olin, however, report a clear violation of this in the photochemical retro-homo-Diels–Alder reaction of azo-compounds (75), (76), and (77).[48] A disrotation of the methyl-bearing carbon atoms gives the same stereochemistry in the product as that reported

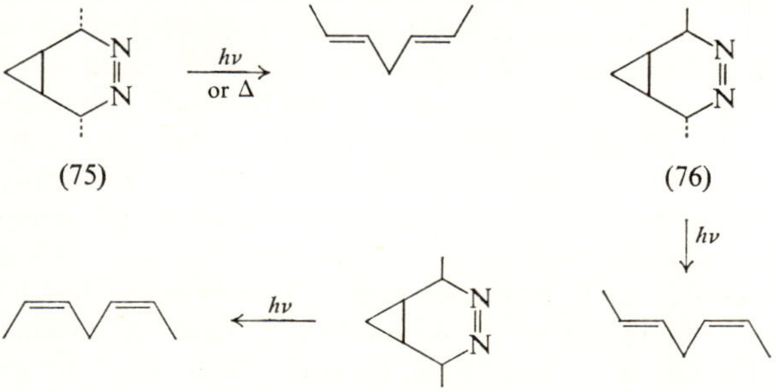

(75) —hv or Δ→ ... (76)

(77) ←hv— ... ↓hv

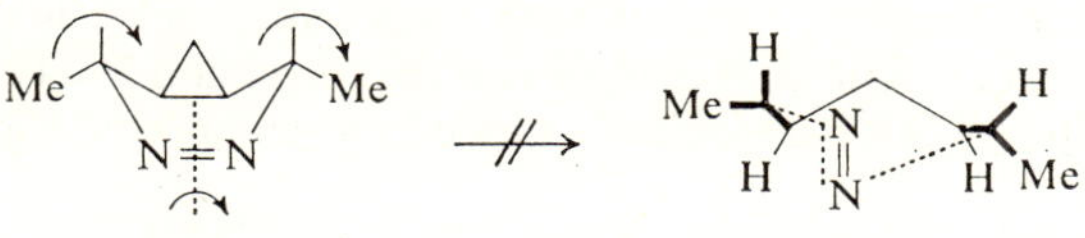

(78) —//→ (79)

[48] J. A. Berson and S. S. Olin, *J. Amer. Chem. Soc.*, 1970, **92**, 1086.

previously by the same workers for the thermal reaction.[49] The thermal reaction occurs at -10 °C and the photochemical process is carried out at -70 °C. It is suggested that the factor which causes the observed stereochemistry is that which opposes the *trans* expulsion of nitrogen: the pathway for the process [*i.e.* from (78) to (79)] has a large twisting component resulting in conrotation which is thermally allowed. Presumably this does not occur thermally since it would require considerable disruption of the non-reacting azo N–N π-bond. Possibly in the excited state this prohibition may be overcome, for example by *cis* to *trans* isomerization of the azo-linkage. The photochemically allowed path from electronically excited *trans*-azo-compound would then lead to disrotation and the observed product. The same products would be formed by internal conversion of electronically excited states to vibrationally excited ground *cis*-azo-compound followed by 'thermal' reaction of the latter. The detailed correlation diagrams, which are to be published later, are reported to suggest that the observations here may have a more complex origin. The authors emphasize that theoretical analysis of concerted reactions involving hetero-atoms should be treated with caution.

The stereochemistry of the products from nitrogen elimination from *exo*- and *endo*-5-methoxy-2,3-diazabicyclo[2,2,1]hept-2-ene systems (80) has also been examined very carefully.[50] The direct photolysis in the liquid phase yields a different *cis* : *trans* ratio for each epimer. An excess of *trans*-isomer is formed from the *exo*-azo-compound while an excess of *cis*-isomer is formed from the *endo*-azo-compound (but see ref. 50a for a reaction in which the configuration is apparently retained). Thus there is a net cross-over in configuration. Benzophenone sensitization of the reaction gives the same product ratio for each epimer, whereas photolysis of the crystalline compounds forms products with high retention of configuration in each case. All these observations are interpreted on the basis of a scheme involving short-lived interconverting pyramidal diradical-like intermediates. To account for the product cross-over it is proposed that structurally inverted 'diradicals' arise directly upon nitrogen elimination and undergo ring-closure before they can equilibrate. The inverted diradical is thought to be formed as a consequence of recoil from carbon–nitrogen bond breaking. The significant differences in the product ratios from the thermal, benzophenone-sensitized, and unsensitized photochemical reactions are said to provide evidence that ground-state singlet diradicals, triplet diradicals, and electronically excited singlet diradicals are involved respectively. Sensitization with triphenylene involves both singlet and triplet intermediates.[51] In the solid-state reactions, it is suggested that inversion from carbon–nitrogen bond rupture of an electronically excited azo-compound should be severely

---

[49] J. A. Berson and S. S. Olin, *J. Amer. Chem. Soc.*, 1969, **91**, 777.
[50] E. L. Allred and R. L. Smith, *J. Amer. Chem. Soc.*, 1969, **91**, 6766.
[50a] L. A. Paquette and L. M. Leichter, *J. Amer. Chem. Soc.*, 1970, **92**, 1765.
[51] P. D. Bartlett and P. S. Engel, *J. Amer. Chem. Soc.*, 1968, **90**, 2960.

restricted because of the fixed position of the nearest neighbour. As the medium becomes less rigid it may thus be expected that there is a decrease in retention of configuration, and indeed this is observed in going from crystals to glasses to liquid phases.

The photolysis of 2,3-diazabicyclo[2,2,1]hept-2-ene (81) has also been studied in hydrocarbon solution at 20 °C.[52] Aromatic hydrocarbons whose first excited singlet states lie higher than that of the azo-compound cause singlet sensitized decomposition. Evidence is presented that (81) does not cross to the triplet state before decomposing, but the triplet (59·3 to 61 kcal mol^{-1}) is also photolabile.

It has recently been reported that *anti*-6,7-diazatricyclo[3,2,2,0^{2,4}]non-6-ene (82) eliminates nitrogen to form cyclonona-1,4-diene (83) at room temperature or photochemically at -60 °C.[53] The reaction has now been reinvestigated by a group of Japanese workers whose past experience with similar decompositions had led them to believe that the products from the two processes differ.[54] Indeed, while confirming that the diene is formed thermally they found that photolysis of (82) in diethyl ether at -20 °C gave (84) as well as (83). Again, by the production of (84) from (82) the elimination shows an inversion of stereochemistry. Similar photolysis with *anti*-7,8-diazatricyclo[4,2,2,0^{2,5}]dec-7-ene (85) leads again to the inversion product (86), as well as to *cis*-cyclo-octa-1,5-diene, the cyclobutane derivative (87), and an unidentified compound.[55] Thermal treatment of (85) yields the cyclo-octadiene only. These two reports are the first of stereospecific inversion in ring formation in the photolytic decomposition of polycyclic tetrahydropyridazines.

The photoelimination of nitrogen from cyclopentadiene-spiropyrazoles (88) in benzene solution has furnished a new and novel synthesis of benzcyclopropenes (89).[18, 56] Formation of (89) from (88) involves $n \to \pi^*$

---

[52] P. S. Engel, *J. Amer. Chem. Soc.*, 1969, **91**, 6903.
[53] M. Martin and W. R. Roth, *Chem. Ber.*, 1969, **102**, 811.
[54] H. Tanida, S. Teratake, Y. Hata, and M. Watanabe, *Tetrahedron Letters*, 1969, 5345.
[55] H. Tanida, S. Teratake, Y. Hata, and M. Watanabe, *Tetrahedron Letters*, 1969, 5341.
[56] H. Duerr and L. Schrader, *Chem. Ber.*, 1970, **103**, 1334.

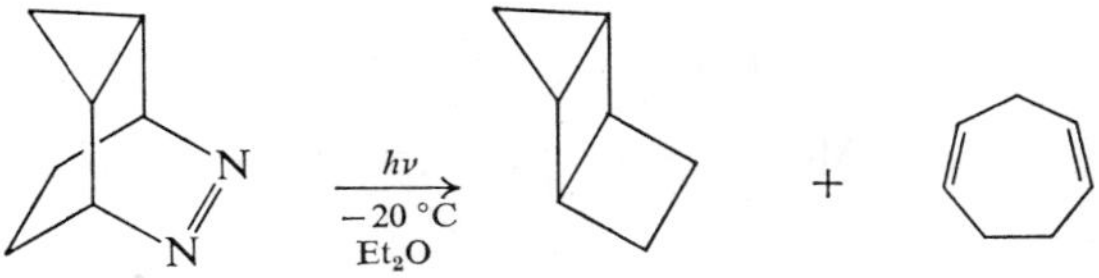

(82) →[hν, −20 °C, Et₂O] (84) 20% + (83) 60%

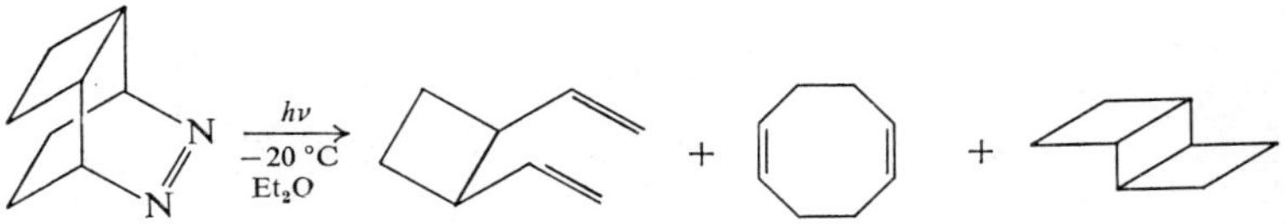

(85) →[hν, −20 °C, Et₂O] (87) 5% + 7% + (86) 18%

excitation followed by ring-expansion to (90), then light-induced loss of nitrogen in the now familiar manner. Triplet excited states of (89) and (90) are involved. Compounds (89) rearranged thermally to benzfurans (91).

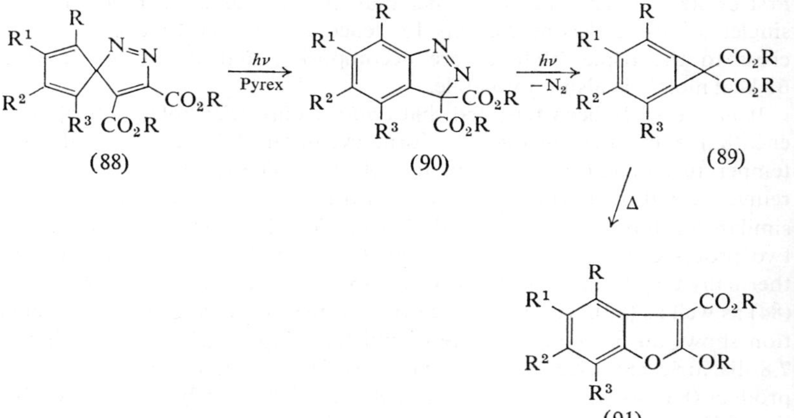

(88) →[hν, Pyrex] (90) →[hν, −N₂] (89) →[Δ] (91)

Hexafluorobenzene is known to form its *para*-bonded isomer (92) photochemically.[57] The thermal phenyl azide adducts (93) and (94) of (92) have been photolysed in acetone solution and yield the aziridine (95) and diaziridine (96), respectively.[58] The photoelimination of nitrogen from other triazoles has been reported within the year. Thus benztriazole (97), subjected to 300 nm radiation in methanol solution, yields aniline and *o*-anisidine, while its *N*-methyl derivative (98) similarly forms *N*-methyl-

[57] G. Camaggi and F. Gozzo, *J. Chem. Soc.* (*C*), 1969, 489.
[58] M. G. Barlow, R. N. Haszeldine, and W. D. Morton, *Chem. Comm.*, 1969, 931.

Photoelimination Reactions

aniline and N-methyl-o-anisidine.[59] Carbazole is the only product from N-phenylbenztriazole (99), and while linear naphthotriazole (100) forms β-naphthylamine but none of the methoxy-derivative, angular naphthotriazole (101) yields no identifiable products. In terms of the suggested mechanism this last result is difficult to understand. The aliphatic triazole, triazolo[1,5-a]pyridine (102) forms α-picoline, α-picolyl methyl ether, and α-picolylmethanol in low yield. The products from the red 1-methyl-naphtho[1,8-de]triazine (103) suggest the intermediacy of a biradical formed by loss of nitrogen from the triazine.[60] Examples of olefin addition, [e.g. (104) from photolysis in the presence of α-methylstyrene], hydrogen

abstraction, and aromatic substitution have all been observed with this system. In particular, with vinyl bromide, (103) yields 1-methyl-1-azaphenalene (105), apparently by a two-step process involving cycloaddition and loss of hydrogen bromide. By the use of (106) as the triazine in this reaction with vinyl bromide, the new aromatic heterocycle 1-methyl-acenaphtho[5,6-bc]pyridine (107) has been synthesized.

The photolysis of biacetyl di-p-tosylhydrazone dianion (108) yields but-2-yne, and similar reaction with the benzyl di-p-tosylhydrazone dianion in methanol gives cis- and trans-α-methoxy-stilbenes (ratio 38 : 62). It has now been shown, by carrying out such irradiations through suitable

[59] J. H. Boyer and R. Selvarajan, J. Heterocyclic Chem., 1969, 6, 503.
[60] P. Flowerday and M. J. Perkins, J. Chem. Soc. (C), 1970, 298.

filters, that the intermediates for the two reactions are the *p*-sulphonamido-1,2,3-triazoles (109) and (110) respectively.[61] The alternative mechanism for formation of the butyne and solvent-derived benzyl products by way of a bis-elimination process involving the loss of two toluene-*p*-sulphinate anions and two molecules of nitrogen, is thus discounted.

5-Phenyl-tetrazolide anion (111) is known to undergo photodecomposition with the formation of phenylcarbene.[62] In this work the nitrogen evolution was noted to be inhibited by oxygen. Further work on this system now demonstrates that (111) can act as a 'self-indicating' sensitizer in that when sensitization occurs no nitrogen evolution is observed, but when there is no sensitization, (111) decomposes and nitrogen is evolved.[63] Thus, while conjugated dienes inhibit the decomposition of the anion (111), the dimers of the dienes thereby formed from sensitization are identical with those formed from the photolysis of the dienes in the presence of benzophenone. Triplet decomposition of (111) and sensitized dimerization are thus suggested. Similarly, *cis-trans* isomerization sensitized by the anion has been observed, as has the norbornadiene–quadricyclene conversion. From emission studies it would appear that the triplet energy of the anion (111) is greater than 75 kcal mol^{-1} (314 kJ mol^{-1}) and that the 0–0 band is at 79 kcal mol^{-1} (331 kJ mol^{-1})

4 Decomposition of Other Compounds with N–N Bonds

Other classes of N–N compounds which have been shown to undergo photodecomposition are hydrazones, tosylhydrazones, azines, and hydrazines: examples of such photolyses are considered here in this order.

The hydrazones (112), (113), and (114) have all beeen photolysed in methanol solution.[64] Product analysis of the systems shows that two

---

[61] P. K. Freeman and R. C. Johnson, *J. Org. Chem.*, 1969, **34**, 1746.
[62] P. Scheiner, *J. Org. Chem.*, 1969, **34**, 199.
[63] P. Scheiner, *Tetrahedron Letters*, 1969, 4863.
[64] R. W. Binkley, *Tetrahedron Letters*, 1969, 1893.

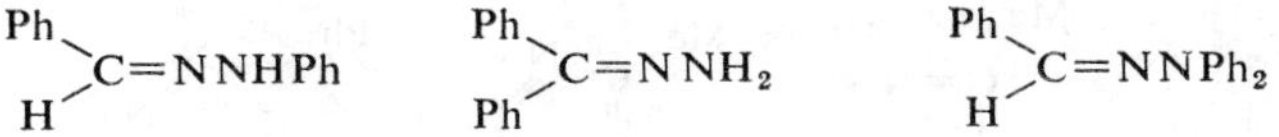

Ph\C=NNHPh   Ph\C=NNH$_2$ Ph\C=NNPh$_2$
H/   Ph/   H/
(112)   (113)   (114)

processes occur. These are (i) N–N bond cleavage, and (ii) hydrogen migration from nitrogen to carbon. In the case of (112), the latter process results in a reaction analogous to the thermal Wolff–Kishner reaction. The photoprocess may occur by one or both of the mechanisms given in Scheme 2. The present evidence favours hydrogen migration.

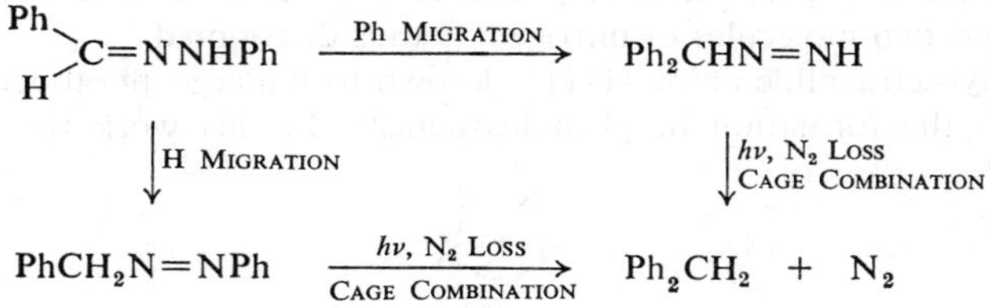

Ph\C=NNHPh —Ph Migration→ Ph$_2$CHN=NH
H/

↓ H Migration   ↓ $h\nu$, N$_2$ Loss
                          Cage Combination

PhCH$_2$N=NPh —$h\nu$, N$_2$ Loss→ Ph$_2$CH$_2$ + N$_2$
              Cage Combination

Scheme 2

Photolysis of the tosylhydrazones (115a—f) of $\alpha\beta$-unsaturated carbonyl compounds in ethereal solvents (tetrahydrofuran, dioxan) yields alkyl and aryl cyclopropenes (116), probably following $n \to \pi^*$ excitation.[65] The cyclopropenes (116a, b, d, and e) have been isolated, but (116c and f) undergo further photolysis to form 2,3-diphenyl-1-(1,2-diphenyl-2-cyclopropenyl)

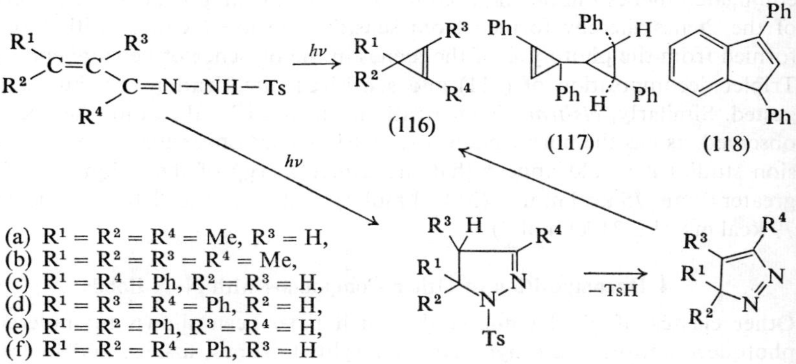

(a) $R^1 = R^2 = R^4$ = Me, R^3 = H,
(b) $R^1 = R^2 = R^3 = R^4$ = Me,
(c) $R^1 = R^4$ = Ph, $R^2 = R^3$ = H,
(d) $R^1 = R^3 = R^4$ = Ph, R^2 = H,
(e) $R^1 = R^2$ = Ph, $R^3 = R^4$ = H,
(f) $R^1 = R^2 = R^4$ = Ph, R^3 = H

(115)

(116) (117) (118) (119)

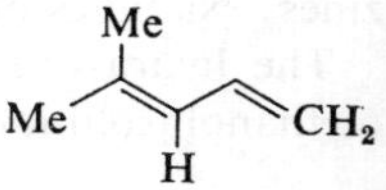

[65] H. Duerr, *Chem. Ber.*, 1970, **103**, 369.

cyclopropane (117) and 1,3-diphenylindene (118) respectively. In the case of (115a), the diene (119) is formed together with (116a).

The disodium salt of the di-*p*-tosylhydrazone (120) of tetramethylcyclobutane-1,3-dione is not able to form a triazole intermediate on photolysis (see ref. 61), and nitrogen is quantitatively eliminated by irradiation of its methanol solutions, with the concomitant formation of 2,2,4-trimethylpent-3-enal dimethylacetal (121) and 2,2,4-trimethylpent-3-enal (122).[66] It is found that despite the spectral data supporting transannular

conjugation and the favourable geometry of the system which should allow a bis-elimination process, there is no evidence from this work which suggests or even requires a bicyclobutane-type intermediate such as (123). Instead, the data are rationalized most simply in terms of a stepwise carbonium ion process.

Although *N*-tosyldiphenylcyclopropenimine (124) contains no N–N moiety, the report of its photolysis is included here with the other tosyl compounds. Since cyclopropenes dimerize and cyclopropenones lose carbon monoxide photochemically, the choice of photochemical reaction path in the case of (124) is of interest. The reactions were carried out in benzene solution with a high-pressure mercury lamp and produced a mixture of the six compounds (125), (126), (127), (128), (129), and (130).[67] As expected, (127) was formed from the irradiation of (126). In methanol solution, photolysis of (124) yielded 9-cyanophenanthrene and α-carbomethoxystilbenes in addition to all the previous six compounds.

This year there have been four reports by Binkley concerning the photolysis of azines. The mechanism for the formation of stilbenes from benzylazines involves an intermolecular process initiated by direct interaction between two azine systems.[68] Thus the azine (131) yields *cis*- and *trans*-stilbenes, 4-methoxystilbenes, 4,4'-dimethoxystilbenes, and small

[66] P. K. Freeman and R. C. Johnson, *J. Org. Chem.*, 1969, **34**, 1751.
[67] N. Obata, A. Hamada, and T. Takizawa, *Tetrahedron Letters*, 1969, 3917.
[68] R. W. Binkley, *J. Org. Chem.*, 1969, **34**, 931.

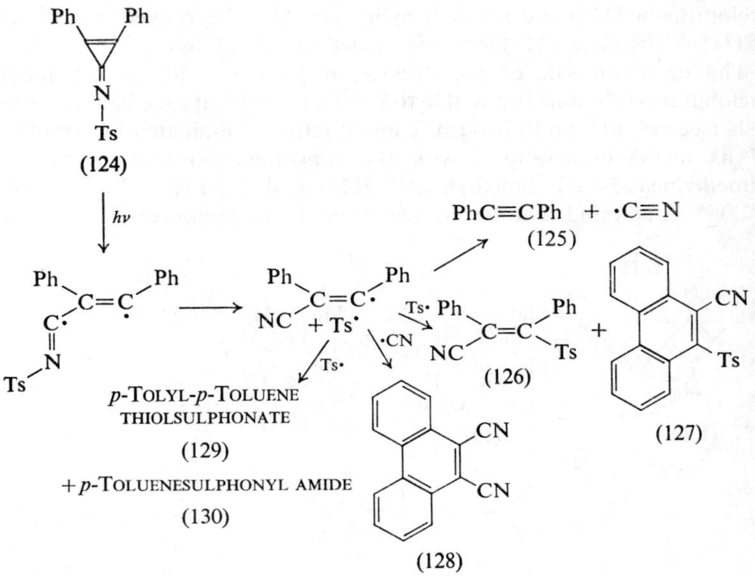

Scheme 3

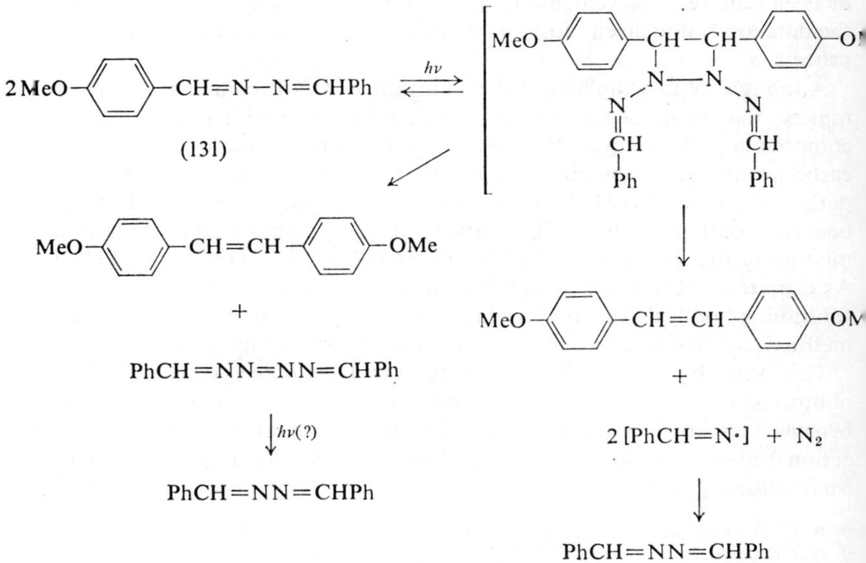

amounts of benzalazine and anisalazine. Unlike the thermal process, the light-induced reaction does not involve phenyldiazomethane, and phenylcarbene is also not importantly involved. The most likely mechanism is given in Scheme 3. The mechanism for the formation of benzonitrile and benzyldi-imine during the photolysis of benzalazine has also been examined.[69] The reaction was carried out in the presence of hydrogen donors in order to ascertain whether the transformations are inter- or intramolecular (Scheme 4). From the data, there are indications of both types

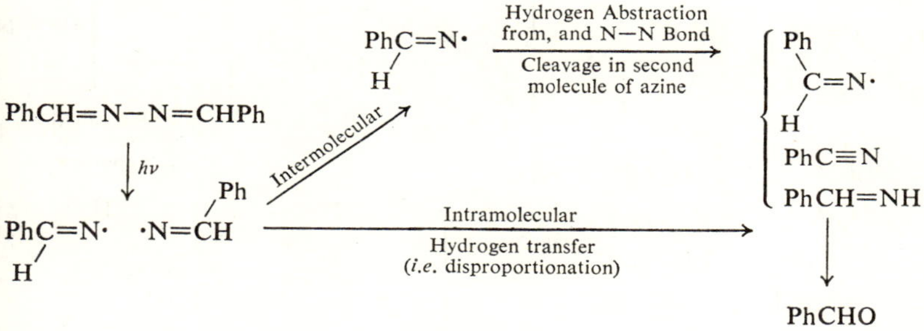

$$
\begin{array}{c}
\text{PhCH}=\text{N}-\text{N}=\text{CHPh} \\
\downarrow h\nu \\
\text{PhC}=\text{N}\cdot \quad \cdot\text{N}=\text{CH} \\
H \text{Ph}
\end{array}
$$

Intermolecular → PhC=N·/H —Hydrogen Abstraction from, and N—N Bond Cleavage in second molecule of azine→ Ph\C=N·/H, PhC≡N, PhCH=NH ↓ PhCHO

Intramolecular Hydrogen transfer (*i.e.* disproportionation) →

Scheme 4

of molecular process. Addition of benzhydrol and decyl mercaptan cause a definite decrease in the product yield; but a limiting value for the yield is reached beyond which further addition of trapping agents has no effect. Previous to this work, it had been reported that benzalazine undergoes benzophenonesensitized decomposition to benzonitrile *via* a mechanism outlined in Scheme 5.[70] Since Binkley found that benzonitrile is

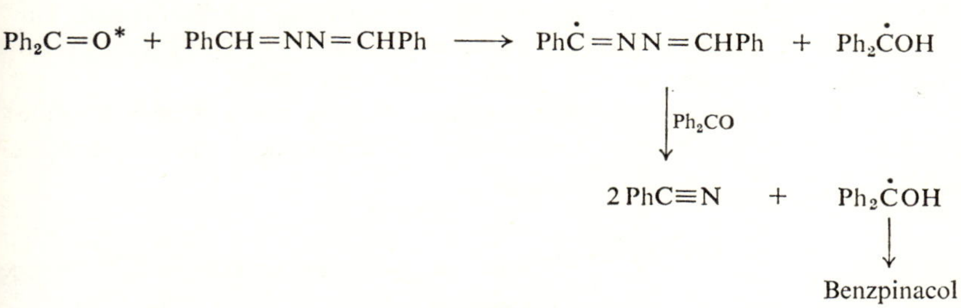

$$\text{Ph}_2\text{C}=\text{O}^* + \text{PhCH}=\text{NN}=\text{CHPh} \longrightarrow \text{Ph}\dot{\text{C}}=\text{NN}=\text{CHPh} + \text{Ph}_2\dot{\text{C}}\text{OH}$$

$$\downarrow \text{Ph}_2\text{CO}$$

$$2\,\text{PhC}\equiv\text{N} + \text{Ph}_2\dot{\text{C}}\text{OH}$$
$$\downarrow$$
$$\text{Benzpinacol}$$

Scheme 5

formed in the absence of benzophenone, he has re-examined the effect of this sensitizer on the reaction.[71] The new results demonstrate that the vital factor in the formation of benzonitrile is the presence of oxygen, not

[69] R. W. Binkley, *J. Org. Chem.*, 1969, **34**, 2072.
[70] J. E. Hodgkins and J. A. King, *J. Amer. Chem. Soc.*, 1963, **85**, 2680.
[71] R. W. Binkley, *J. Org. Chem.*, 1969, **34**, 3218.

benzophenone, and the mechanism now suggested is that for the direct photolysis in which there is homolytic N–N cleavage to yield PhCH=N· radicals which are subsequently oxidised to benzonitrile. Thus it is suggested that the earlier work was probably carried out in the presence of oxygen. Following this work, the photolysis of D-galactose azine (132) in methanol solution and under oxygen, has been investigated with the expectation of obtaining D-galactononitrile by analogy with the above

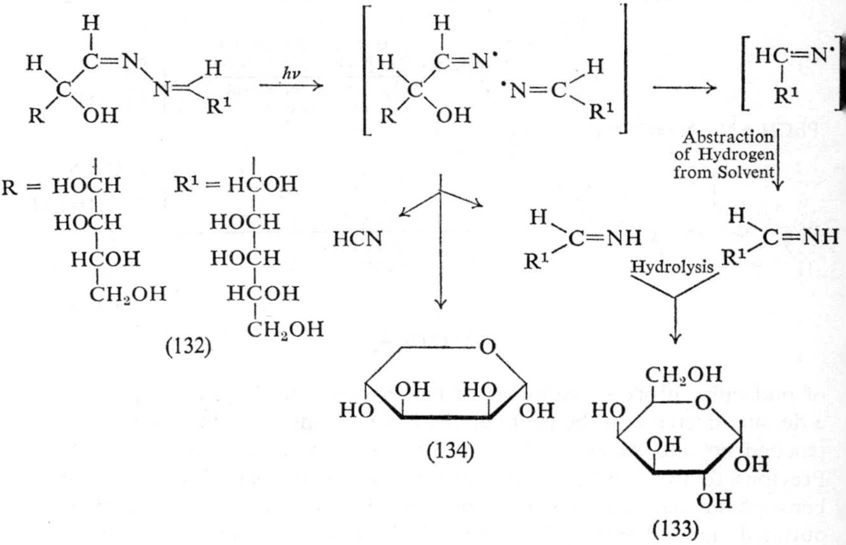

results.[72] Neither under oxygen nor nitrogen was the nitrile formed, but D-galactose (133) and D-lyxose (134) were isolated. The (133) : (134) ratio was higher than under nitrogen in an oxygen atmosphere.

The photolysis of tetraphenylhydrazine at 77 K in rigid matrices (Scheme 6) has been studied by e.s.r. spectroscopy.[73] The radical pair has a spin–

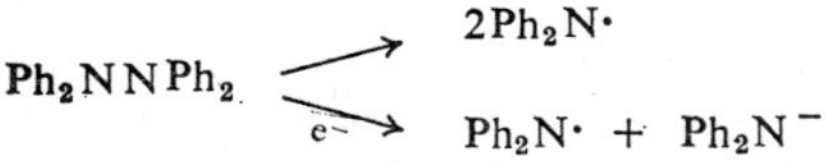

$$Ph_2NNPh_2 \rightleftarrows \begin{array}{c} 2Ph_2N\cdot \\ Ph_2N\cdot + Ph_2N^- \end{array}$$

Scheme 6

spin interaction manifested by e.s.r. spectra, but is not responsible for the electronic absorption band in the near-i.r. region, contrary to previous proposals.[74]

[72] R. W. Binkley and W. W. Binkley, *Carbohydrate Res.*, 1970, **13**, 163.
[73] T. Shida and A. Kira, *J. Phys. Chem.*, 1969, **73**, 4315.
[74] D. W. Wiersma and J. Kommandeur, *Mol. Phys.*, 1967, **13**, 241.

5 Loss of the Elements of Carbon Dioxide (see also Part III, Chapter 1, Section 7)

The decarboxylation of arylacetic acids or their anions is seemingly a general process and there have been several reports concerned with this reaction during the year.

The photolysis of *o*-nitrophenoxyacetic acid provides a new synthesis of *o*-nitrosophenols and an alternative to the Baudisch method.[75] Thus, α-(2,4-dinitrophenoxy)propionic acid (135) at pH 7 evolves both carbon dioxide and acetaldehyde, and after acidification of the reaction mixture yields 65% of 4-nitro-2-nitrosophenol. Other derivatives have been examined, and throughout the pH range 1—12 the nitrosophenols are formed exclusively: the reaction thus differs from the photolysis of 2,4-dinitrophenylamino-acids (136) in aqueous solution.[76] A mechanism to account

for the reactions of the latter compounds in aqueous solution has recently been proposed, and involves the intermediacy of (137), which is an analogue of the intermediate for related reactions of *o*-nitro-*NN*-dialkylanilines.[77]

While the nitrophenylacetic acids themselves are photostable, their anions readily decarboxylate.[78] Thus 3-nitrophenylacetate, 4-nitrophenylacetate, and 4-nitrohomophthalate anions decarboxylate with a quantum yield of approximately 0·6 at 367 nm. The decarboxylation of 2-nitrophenylacetate and 2,4-dinitrophenylacetate anions is much less efficient with a quantum yield of only 0·04. Aci-nitro-intermediates have been identified spectroscopically by flash photolysis from the *ortho*- and *para*-isomers but not from the *meta*-isomer. The results are consistent with

---

[75] P. H. McFarlane and D. W. Russell, *Chem. Comm.*, 1969, 475.
[76] D. J. Neadle and R. J. Pollitt, *J. Chem. Soc. (C)*, 1967, 1764; *ibid.*, 1969, 2127.
[77] O. Meth-Cohn, *Tetrahedron Letters*, 1970, 1235.
[78] J. D. Margerum and C. T. Petrusis, *J. Amer. Chem. Soc.*, 1969, **91**, 2467.

decomposition of the aryl acetate into carbon dioxide and a nitrobenzyl anion which is stabilized as the aci-nitro-anion (138) in the *ortho* and *para* structures. A kinetic analysis of the intermediate detected spectroscopically

from the 2-nitrophenylacetate anion indicates that two different species (139) and (140) are present with the major reaction product arising from (139), hence the low quantum yield in this case. Although 3-nitrophenylacetate ions decarboxylate with a high quantum yield, there is no spectral evidence for a metastable intermediate. This fact is consistent with the general mechanism, since the 3-nitrobenzyl anion would not be resonance stabilized. The results from this work are somewhat inconsistent with those of Zimmerman [79] and Havinga [80] for '*meta*-activation'. Also in contrast to this work, aqueous solutions of pyruvic acid photodecarboxylate quite readily whereas the pyruvate ion shows little or no reaction.[81]

Aqueous solutions of the 1-naphthylacetate ion undergo photodecomposition to yield 1-methylnaphthalene, 1-naphthyl carbinol, 1-naphthaldehyde, 1-naphthoic acid, and phthalic acid.[82] Judged by this product analysis, the decomposition processes in this case are quite diverse and obviously similar to those for the photolysis of 1-naphthylacetic acid in ethanol.[83]

[79] H. E. Zimmerman and S. Somasekhara, *J. Amer. Chem. Soc.*, 1963, **85**, 922.
[80] E. O. Havinga, R. O. de Jongh, and W. Dorst, *Rec. Trav. chim.*, 1956, **75**, 378.
[81] P. A. Leermakers and G. F. Vesley, *J. Amer. Chem. Soc.*, 1963, **85**, 3776.
[82] D. G. Crosby and C. S. Tang, *J. Agric. Food Chem.*, 1969, **17**, 1291.
[83] D. A. M. Waktins and D. Woodcock, *Chem. and Ind.*, 1968, 1522.

Photoelimination Reactions

The 2-, 3-, and 4-pyridylacetic acids are both thermally-[84] and photolabile.[85] With 253·7 nm radiation, 0·01 molar aqueous solutions of these acids give rise to 2-, 3-, and 4-methylpyridines with the evolution of carbon dioxide. The quantum yields for the formation of the 2-, 3-, and 4-methylpyridines are 0·49, 0·46, and 0·25 respectively (cf. those for nitrophenylacetate ions). The initial rate of decarboxylation of each acid is at a maximum at or near the iso-electric point (pH 4·0—4·2). At higher or lower pH values, the photoreaction is slowed considerably and eventually a reaction involving the pyridine ring, 'presumably hydration and cleavage', becomes faster than the decarboxylation process. The reaction mechanism is outlined in Scheme 7. Other 2-substituted pyridine derivatives also

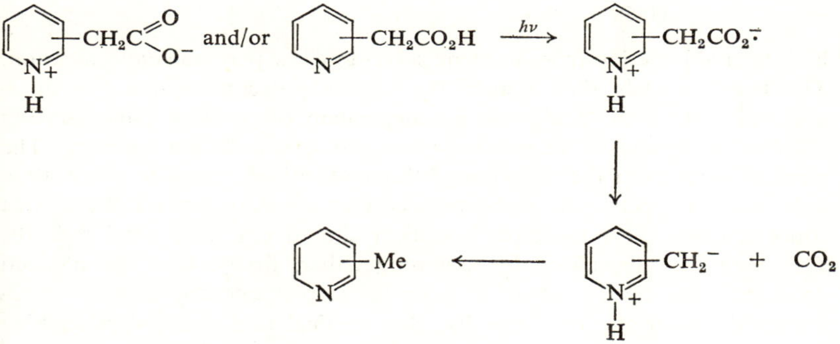

Scheme 7

undergo photocleavage to yield 2-methylpyridine. Significantly, the corresponding 3- and 4-isomers in these cases are photostable.

The decarboxylation process has also been observed for esters. Thus polyarylcyclopentadienyl esters [e.g. (144)] in benzene solution yield the corresponding hydrocarbons with the evolution of carbon dioxide.[86] The formation of the radical intermediate (145) is suggested since the photolysis of (144) in cumene produces (146).

In recent years there have been several publications concerned with the light-induced decomposition of peroxides and, in particular, benzoyl peroxide. The mechanism for the benzophenone- and 2-acetonaphthone-sensitized decomposition of benzoyl peroxide in benzene has now been investigated.[87] The main findings from the investigation are as follows. In the benzophenone-sensitized reaction, the yield of acid resulting from peroxide decomposition increases as the initial peroxide concentration decreases: this suggests that the benzophenone ketyl radicals are formed by

---

[84] W. von E. Doering and V. Z. Pasternak, *J. Amer. Chem. Soc.*, 1950, **72**, 143.
[85] F. R. Stermitz and W. H. Huang, *J. Amer. Chem. Soc.*, 1970, **92**, 1446.
[86] J. J. Basselier and J. C. Cherton, *Compt. rend.*, 1969, **269C**, 1412.
[87] W. F. Smith, *Tetrahedron*, 1969, **25**, 2071.

[Scheme showing compounds (144), (145), (146) with reaction:
(144) Ph₅C₅-OCOR →(hv, Cumene) [(145) Ph₅C₅• + •OCOR] → Ph₄C₅(R)Ph + CO₂
(146) Ph₄C₅(H)Ph]

hydrogen abstraction from benzene solvent at low peroxide concentration. The ketyl radicals then induce the peroxide decomposition. With 2-acetonaphthone, the acid yield is independent of peroxide concentration but does depend upon the sensitizer concentration and light intensity. The yield of acid from the 2-acetonaphthone-sensitized reaction is less than half that of the benzophenone-sensitized reaction in the region of peroxide concentration where the latter sensitizer should not form ketyl radicals. This may be rationalized by assuming either, firstly, that the induced decomposition of the peroxide is by benzophenone triplet but not 2-acetophenone triplet or, secondly, that vertical and non-vertical energy transfer with multi-bond cleavage occurs preferentially from the non-spectroscopic peroxide triplets or, thirdly, that triplet states are quenched be benzoyloxyl radicals.

The remaining reports in this section are concerned with the loss of the elements of carbon dioxide from cyclic compounds.

Mercury-sensitized vapour-phase photolysis of γ-lactones and simple butenolides causes elimination of carbon dioxide and the formation of olefins and acetylenes. γ-Butyrolactone (147), γ-crotonolactone (148), and β-methyl-$\Delta^{\alpha\beta}$-butenolide (149) have all been examined in this respect.[88] The liquid-phase photolysis of lactone (147) has been previously reported.[89] In contrast to the photoreactions of (147) and (148), the butenolide (149) yields a complex mixture of primary and secondary products. The but-1-yne and methyl allene could be secondary products of 1-methylcyclopropene or methylenecyclopropane, but either or both of these 'secondary' products may be formed via a 1,2-methyl migration in the proposed intermediate (150). The present processes might be useful in the generation of cyclopropene and cyclopropane derivatives, particularly if the undesirable secondary photolytic and/or thermal reactions of the primary products could be eliminated.

[88] I. S. Krull, and D. R. Arnold, *Tetrahedron Letters*, 1969, 1247.
[89] R. Simonaitis and J. N. Pitts, *J. Amer. Chem. Soc.*, 1968, **90**, 1389.

Photoelimination Reactions

(147) [β-propiolactone structure] $\xrightarrow[\text{Vapour}]{h\nu}_{\text{Hg}}$ CO_2 + [cyclopropene] + [methylcyclopropene Me] + $HC\equiv CH$
    64%   24·6%   7·0%

(148) [butenolide structure] $\xrightarrow[\text{Vapour}]{h\nu}_{\text{Hg}}$ $MeC\equiv CH$ + $CH_2=C=CH_2$ + [cyclopropene] + CO_2
    47%   2·1%   4·2%

(149) [butenolide structure] $\xrightarrow[\text{Vapour}]{h\nu}_{\text{Hg}}$ CO_2 + $\left[\rangle=\cdot \leftrightarrow \rangle-:\right]$ $\longrightarrow$ $CH_2=CH_2$ + $MeC\equiv CH$
                                            (150)   1·8%   1·0%

+ [propene] + $HC\equiv CEt$
  30·2%   3·8%

+ [Me-cyclopropene] 15% + $CH_2=C=CHMe$ 11·1%

+ $MeC\equiv CMe$ 35%

In the presence of acetylenedicarboxylic ester, 1,3,4-oxathiazole-2-ones (151) lose the elements of carbon dioxide thermally at 130 °C with the formation of adduct (152).[90] In contrast to this, the photolysis takes a different

PhCN + CO_2 + S $\xleftarrow[254\text{ nm}]{h\nu}$ [structure (151): Ph-C=N-S-C(=O)-O ring] $\xrightarrow[MeO_2CC\equiv CCO_2Me]{130\text{ °C}}$ [structure (152): Ph-C=N-S-C(CO_2Me)=C(CO_2Me) ring]

(151)   (152)

course in the presence of the acetylene, and in ethyl acetate solution with 254 nm radiation there is more extensive fragmentation with the formation of benzonitrile, sulphur, and carbon dioxide.

The elimination of carbon dioxide is also observed from 3-phenyl-Δ^2-oxadiazo-5-one (153) by irradiation of its methanol or dioxan solutions.[91, 92] The reactions of the resulting fragment(s) are diverse. Benzamide and the methyl ether of benzamidoxime (154) apparently result from hydrogen abstraction and insertion of an intermediate unrearranged azomethine nitrene (155). A migration of phenyl from carbon to nitrogen yields phenylcarbodi-imide (156) which tautomerizes to phenylcyanamide (157).

[90] J. E. Franz and L. L. Black, *Tetrahedron Letters*, 1970, 1381.
[91] J. H. Boyer and P. J. A. Frints, *J. Heterocyclic. Chem.*, 1970, **7**, 71.
[92] J. H. Boyer and P. J. A. Frints, *J. Heterocyclic. Chem.*, 1970, **7**, 59.

(153) ⇌ (154)   (154) Ph—C(NH₂)=NOMe   (155) Ph—C(N:)=NH

(156) PhN=C=NH   (157) PhNHCN   (158) 1,3-diphenyltriazole (N-N, H)   (159) 2,4,6-triphenyltriazine   (160) Ph-C(NH₂)=NCN

More extensive fragmentation yields benzonitrile, and the intermediacy of phenylcarbene could account for the formation of benzylmethyl ether. The production of 3,5-diphenyltriazole (158), triphenyltriazine (159), and *N*-cyanobenzamidine (160) remains unexplained.

Griffin and co-workers have examined the photofragmentation of cyclic carbonates (161) and sulphites (162) with 254 nm radiation in methanol solution.[93, 94] Benzophenone and diphenylcarbene, detected as benzhydryl

(161) Ph₂-cyclic carbonate   (162) Ph₂-cyclic sulphite   (163) Ph₂CHOMe   (164) hydrobenzoin carbonate   (165) hydrobenzoin sulphite

methyl ether (163), are formed from both precursors. This work has been extended to the *dl*- and *meso*-hydrobenzoincarbonates (164) and sulphites (165) in order to study the insertions of the phenylcarbenes thus generated. The insertion selectivity of the carbenes is compared with those obtained from accepted phenylcarbene precursors (such as *trans* stilbene oxide and phenyldiazomethane). The results obtained in all cases substantiate the original proposal that the photolysis of aryl cyclic carbonates and sulphites gives species similar to, if not identical with, those produced from the accepted carbene precursors.

6 Fragmentation Reactions of Organosulphur Compounds

There have been many reports within the year of the photolysis of a wide range of types of organosulphur compounds and few have any relation to

[93] R. L. Smith, A. Manmade, and G. W. Griffin, *J. Heterocyclic. Chem.*, 1969, **6**, 443.
[94] R. L. Smith, A. Manmade, and G. W. Griffin, *Tetrahedron Letters*, 1970, 663.

each other. Presentation of the year's publications in this area is thus inevitably on a somewhat *ad hoc* basis.

The transient spectra (λ_{max} = 420 nm) observed from the flash photolysis of 2-mercaptoethanol, thiophenol, and cysteine hydrochloride in aqueous solutions at various pH values have been identified as arising from the RSSR radical anion.[95] The formation of the radical anion from flash photolysis is proposed to involve a fast reaction of the primary RS· radical with RS⁻ anion. From the photolysis of ethanethiol in the presence of deuterium with 254 and 228·8 nm radiation, and using data obtained previously for the hydrogen bromide–deuterium system, the energy available for product excitation is shown to reside chiefly in the translational modes of motion rather than in the internal modes of the C_2H_5S· radical.[96]

The photolysis of 2-phenyl-3,1-benzoxathian-4-one (166) at 350 nm in chloroform solution causes overall fragmentation of CO–O and S–CPh bonds with resulting elimination of benzaldehyde and dimerization of the

residual sulphur fragment to yield the hitherto unknown compound dibenzo[*cg*][1,2]dithiocin-5,6-dione (167).[97] This is a new photofragmentation in the benzoxathianone system which also occurs with the 2*H*-, 2-methyl-, and 2,2-diphenyl derivatives of (166). The quantum yield for disappearance of (166) and benzaldehyde formation is 0·05 which, as may be expected, is twice that for the formation of the dione (167). The reaction has been examined at 77 K, a procedure which more and more photo-

[95] G. Caspari and A. Granzow, *J. Phys. Chem.*, 1970, **74**, 836.
[96] J. M. White, R. L. Johnson, and D. Bacon, *J. Chem. Phys.*, 1970, **52**, 5212.
[97] A. O. Pedersen, S. O. Lawesson, P. D. Klemmensen, and J. Kolc, *Tetrahedron*, 1970, **26**, 1157.

chemists are using to detect reaction intermediates which are suspected to be involved in processes carried out at much higher temperatures. In the present case it was found that the expulsion of benzaldehyde is a thermal process and that a fair proportion of light-induced bond fission is thermally reversed. Intermediates (168) and (169) are suggested, and some data are presented for a higher probability of (168) over (169). The proposed reaction sequence involves keten (170) as the immediate precursor of (167).

The formation of cage compounds from 4H-pyran-4-one and 4H-thiopyran-4-one derivatives is well known.[98, 99] On the other hand, however, 2,6-diphenyl-4H-pyran-thione (171) is now reported to undergo photo-desulphuration to 2,2′,6,6′-tetraphenyl-4,4′-bi(pyronylidine) (172).[100] The

$$(171) \xrightarrow[\text{Pyrex}]{h\nu} (172)$$

reactions were carried out in dioxan solution in borosilicate glassware with a medium-pressure mercury lamp. Similar irradiation of 4H-pyran-4-thione or 2,6-dimethyl-4H-pyran-4-thione only yielded small amounts (0—3%) of carbon–sulphur bond-cleavage products. It is suggested from sensitization and quenching experiments that the photoreaction involves triplet-state intermediates. The dimerization–desulphuration of (171) was also observed to occur thermally.

SS-Diaryl esters (173) of 1,2-dithio-oxalic acid, and the corresponding esters (174) of dithiocarbonic acid, are reported to undergo efficient photo-fragmentation with the elimination of carbon monoxide to yield diaryl

$$\text{Ar-S-CO-CO-S-Ar} \xrightarrow[-\text{CO}]{h\nu} \text{Ar-S-CO-S-Ar}$$
$$(173) \qquad\qquad (174)$$
$$\xrightarrow[-2\text{CO}]{h\nu} \qquad h\nu \downarrow -\text{CO}$$
$$\text{Ar-S-CS-S-Ar} \xrightarrow[-\text{CS}]{h\nu} \text{Ar-S-S-Ar}$$
$$(175)$$

sulphides.[101] The photodecarbonylation process is well known, but SS-diaryltrithiocarbonates (175) react in a similar way by the loss of carbon monosulphide, and this process is not so well documented.

[98] P. Yates and E. S. Hand, *J. Amer. Chem. Soc.*, 1969, **91**, 474.
[99] N. Sugiyama, Y. Sato, N. Kashima, and K. Yamada, *General Discussion on Radical Chemistry*, 1969, Osaka, p. 1.
[100] N. Ishibe, M. Odani, and K. Teramura, *Chem. Comm.*, 1970, 371.
[101] H. G. Heine and W. Metzner, *Annalen*, 1969, **724**, 223.

The wavelength effect in the photolysis of 4,4,6,6-tetramethyl-1-thiaspiro-(2,3)-hexan-5-one (176) is most marked.[102] Thus, in methyl alcohol solution with 235·7 nm radiation, (176) undergoes desulphuration to yield (177), while with wavelengths longer than 280 nm, the Norrish type I spiro-product (178) is formed, along with the solvent-derived products (179) and (180). Unlike the electronic spectra of other keto-sulphides, the spectrum of (176) shows no charge-transfer absorption resulting from transannular interaction of a lone pair of electrons on sulphur with the carbonyl group.[103] This is not surprising when it is realised that such interaction might give rise to a species such as (181). The band in the spectrum of (176) at 265 nm is considered to arise from the $n \to \sigma^*$ transition of the sulphur, and excitation within this band leads to the desulphuration.[104] Although the multiplicities of the excited states arising from the two chromophores are yet to be determined, the authors reasonably suggest that decarbonylation is from the $n\pi^*$ triplet state and the desulphuration from the $n\sigma^*$ singlet state. There is obviously no inter- or intra-molecular transfer of energy from the latter to the former states as evidenced by such a clear difference in the products at the two wavelengths.

Methylthio-bis(alkylamino)-S-triazine herbicides (182) are known to undergo desulphuration and be converted into 2-hydroxy-bis(alkylamino)-S-triazines (183) in plants by way of the intermediate sulphoxide (184) and sulphone (185).[105] This raises the intriguing possibility of photo-oxidation of (182) to (183). Oxidation of the methylthio-group did not occur, however, but an elimination with hydrogen transfer to form 2,4-bis(ethylamino)-S-triazine (186) was observed.[106] Compound (182) also demonstrates this latter conversion to (186) in a mass spectrometer.

---

[102] J. G. Pacifici and C. Diebert, *J. Amer. Chem. Soc.*, 1969, **91**, 4595.
[103] See, for example, P. Y. Johnson and G. A. Berchtold, *J. Amer. Chem. Soc.*, 1967, **89**, 2761.
[104] L. B. Clark and W. T. Simpson, *J. Chem. Phys.*, 1965, **43**, 3666.
[105] H. Gysin, *Chem. and Ind.*, 1962, 1393; P. W. Mueller and P. H. Payot, *Proc. Int. Atomic Energy Authority Sym.*, *Isotopes in Weed Research*, Vienna, 1966, 61.
[106] J. R. Plimmer, P. C. Kearney, and U. I. Klingebiel, *Tetrahedron Letters*, 1969, 3891.

β-Ketosulphonium ylides (187) undergo photocleavage of the dipolar sulphur–carbon bond to give products derived from ketocarbene and keten intermediates.[107] Somewhat similarly, dimethyl phenacylsulphonium bromide (188) undergoes photofragmentation in aqueous solution to give dimethylsulphide, and a decrease in pH of the solution.[108] From the reaction dibenzoylethane (189), p-bromodibenzoylethane (190), phenacyl bromide (191), acetophenone, and benzoin are isolated. These results are consistent

with a radical mechanism. In principle, the radicals may arise *via* three mechanisms, (i) homolysis of the C–S$^+$ bond to give the phenacyl radical and dimethylsulphide radical anion, (ii) nucleophilic substitution by bromide ion on excited phenacyl sulphonium ions to give excited phenacyl bromide which undergoes homolysis, and (iii) by the charge-transfer state analogous to ammonium and phosphonium salts.[109] Mechanism (i) was shown to be operative by carrying out the photolysis in the presence of a large excess of bromide anion, when no changes in rate or product yield

---

[107] E. J. Corey and M. Chaykovsky, *J. Amer. Chem. Soc.*, 1964, **86**, 1640; B. M. Trost, *J. Amer. Chem. Soc.*, 1966, **88**, 1587; *ibid.*, 1967, **89**, 138.
[108] T. Laird and H. Williams, *Chem. Comm.*, 1969, 561.
[109] J. W. Knapezyk and W. E. McEwen, *J. Amer. Chem. Soc.*, 1969, **91**, 145.

were observed: also dimethyl phenacyl sulphonium nitrate undergoes the same reaction as (188).

An investigation of the photolysis of 2-methylisothiochroman-4-one-3-ylide (192) in chloroform or methanol showed surprisingly that 1-indanone (193) is the only volatile product, *i.e.* no methyl ester or methyl ketone is obtained.[110] It was suggested that the photolysis of (192) leads to benzylic carbon–sulphur bond cleavage with the formation of 2-thiomethylindanone (194) since separate photolysis of (194) quantitatively yields (193). The acyclic analogue (195) of (192) gives nine products on photolysis, the major ones of which have been identified as (193), *o*-methylacetophenone (196), (197), (198), and (199). Further investigations of the processes are said to be in hand.

Stable sulphoxonium ylides have also been subjected to photolysis.[111] Dimethylsulphoxonium-3-ethoxycarbonyl-2-phenylallylide (200) in tetrahydrofuran gives low yields (0·95%) of diethyl-*p*-terphenyl-2,5-dicarboxylate (201). The first intermediate here is suggested to be a triplet carbene (202) which dimerizes and suffers dehydrogenation to give (201). On the other

---

[110] R. H. Fish, L. C. Chow, and M. C. Caserio, *Tetrahedron Letters*, 1969, 1259.
[111] Y. Kishida, T. Hiraoka, and J. Ide, *Chem. Pharm. Bull. Tokyo*, 1969, 1591.

hand, photolysis of dimethylsulphoxonium-1,2-dicarbomethoxy-5-carbethoxy-4-phenyl-1,4-pentadiene (203) gives the three products (204), (205), and (206) with (204) formed in 37·5% yield.

The analogous sulphoxonium ylide system in the pyrimidine nucleoside (207) again undergoes photocleavage of the sulphur–carbon bond with 254 nm radiation in aqueous solution.[112] Carbon–hydrogen insertion of the carbene thus produced leads to formation of the 2,2-methylenecyclouridine (208) in 40% yield. In the absence of the sugar moiety [*i.e.* (209)], the resulting carbene reacts with water to form 2-hydroxymethyl-4-hydroxypyrimidine (210).

Schlessinger and Schultz have recently observed that the photodesulphuration of diaryl sulphines (211) occurs exclusively *via* the singlet state.[112] This result has led them to re-examine their earlier work on the benzophenone-sensitized conversion of *cis*- and *trans*-(212) to (213).[113] They previously postulated the intermediacy of the short-lived oxathiairan (214), since no evidence was obtained for a two-photon process with the intermediacy of (215) and (216) which would be long lived. They now report that the desulphuration of (212) to (213) is in fact a multi-quantum process from the triplet state of (212) to the isolable sulphine intermediates (215) and (216). These intermediates then undergo singlet-state decomposition to (213). It is now realised that in the previous work the concentration of reactants was such that (215) and (216) would absorb half the incident

[112] T. Kunieda and B. Witkop, *J. Amer. Chem. Soc.*, 1969, **91**, 7752.
[113] A. G. Schultz and R. H. Schlessinger, *Chem. Comm.*, 1969, 1483.

light and decompose to (213): thus they were not observed. The same workers have also carried out further investigations into the photo-desulphuration of sulphines using the isomeric compounds (217) and (218).[114] The investigations have involved direct and sensitized photolyses. Compound (217) with Michler's ketone as the sensitizer yields (218) and none of the ketone (219). The same sensitized behaviour [*i.e.* conversion to (217)] is observed for (218): the quantum yield for both these isomerization processes is 0·3. On the other hand, direct irradiation of (217) or (218) in benzene or chloroform with 366 nm radiation causes not only the isomerization but also the decomposition of the sulphines. Gas chromatographic–mass spectrometric analysis of the decomposition mixture shows the presence of both the ketone (219) and the thioketone (220). Further experimental data indicate that the photo-desulphuration occurs by two different mechanisms which are concentration dependent. At high concentrations the sulphine forms a photo-dimer which subsequently undergoes light-induced decomposition to (219), whereas at low concentrations the quantum yield for (219) formation is equal to that for disappearance of the sulphine, and an oxathiairan intermediate (221) is suggested.

Abramovitch and Takaya have been studying a new photochemical source of aryl radicals.[115] Thus, the photolysis of *N*-arylsulphonyldimethyl-

[114] R. H. Schlessinger and A. G. Schultz, *Tetrahedron Letters*, 1969, 4513.
[115] R. A. Abramovitch and T. Takaya, *Chem. Comm.*, 1969, 1369.

sulphoximines [*e.g.* (222)] with 254 nm radiation in aromatic hydrocarbon solvents yields biphenyl. In a typical reaction in benzene solution, 74·5% of biphenyl, 7% of dimethylsulphoximine (223), sulphur dioxide, and traces of dimethyl sulphone and dimethyl sulphoxide were formed. From data

$$Ar_2C=S-O \xrightarrow{h\nu} Ar_2C=O + S$$
(211)

(212) $\xrightarrow{h\nu}_{CHCl_3}$ (215) + (216)

(212) → (214)

(215) $\xrightarrow{h\nu}$ (213)

(214) $\xrightarrow{\Delta}$ (213)

(217), (218), (219), (220), (221)

involving competition of phenylation of benzene and toluene, the phenyl radicals formed in the present photolytic study appear to be more selective (or electrophilic) than those generated thermally. The methylbiphenyl isomer ratio from photolysis of (222) in toluene is $o:m:p$, 11·3 : 48·5 : 40·5. These ratios appear inconsistent with attack by phenyl radicals, but it was found that under the photolysis conditions employed here, *o*- and *m*-methylbiphenyls isomerize to give the above ratios of isomers. It was shown that the dimethyl sulphone and sulphoxide are secondary light-induced products of (223). There is no decomposition of (222) in either nitrobenzene or pyridine, but this could be the trivial result of absorption of the radiation by the solvents, or possibly solvent quenching of the excited states of (222).

$$\text{PhSO}_2\text{N}{=}\overset{\overset{\text{O}}{\uparrow}}{\text{SMe}_2} + \text{PhH} \xrightarrow{h\nu} \text{Ph}{-}\text{Ph} + \text{SO}_2 + \text{Me}_2\overset{\overset{\text{O}}{\uparrow}}{\text{S}}{=}\text{NH} + \text{Me}_2\text{SO}$$
(222)   (223)   + Me$_2$SO$_2$

Earlier published work suggests the involvement of 3d-orbitals in the molecular orbitals of aromatic enethiol esters.[116] In an attempt to establish the influence of 3d-orbitals on the photolysis of enethiol esters, the light-induced reactions of p-tolyl thiolacetate (224) have been studied. The radiation with 254 nm light of cyclohexane solutions of (224) yields the disulphide (225) (77%) and the sulphide (226) (7%) via enethiol radicals.[117]

The behaviour is consistent with the fact that the spin density of the arylthio-radical is localized largely on the sulphur. Significantly there is no photo-Fries product (227). The Fries reaction of enolesters involves an intramolecular [1,3]-sigmatropic change[118] which demands that the sulphur–acyl bond be perpendicular to the plane of the aromatic ring. The 3d^2p hybrid orbitals of sulphur cause coplanarity of the aromatic ring and the thioacetoxy-group, thus preventing an intimate contact Fries process. The author points out that the absence of Fries product could also be due to poor overlap between the sulphur–acyl σ-bond and the 2p-orbitals of the aromatic ring. Sulphur–oxygen σ-bonds are significantly longer than carbon–oxygen σ-bonds.[119] With cyclohexen-1-yl thiolacetate (228), the thiophen (229) and cis- and $trans$-2-acetylcyclohexyl thiolacetate (230) are formed. In this case it is not possible to distinguish between dimerization at carbon or sulphur. α-Lipoic acid (231) (1,2-dithiolane-3-valeric acid) is an important biological compound which is thought to participate in

---

[116] G. Cilento, $Chem. Rev.$, 1960, **60**, 147; V. Baliah and R. Ganapathy, $J. Indian Chem. Soc.$, 1963, **40**, 1.
[117] J. R. Grunwell, $Chem. Comm.$, 1969, 1437.
[118] R. Sandner, E. Hedaya, and D. J. Trecker, $J. Amer. Chem. Soc.$, 1968, **90**, 7249.
[119] L. F. Fieser and M. Fieser, 'Advanced Organic Chemistry', Reinhold, New York, 1961, p. 1156.

photosynthesis. Its photolysis with lamps having emission peaks at 356 nm has been examined in a number of solvents.[120] The number and type of products formed in each solution are dependent upon the solvent, as are the amounts of products and their rates of formation. The mechanism postulated involves homolytic cleavage of the disulphide bond. The reaction then proceeds by chain transfer of hydrogen atoms from the solvent resulting in short chains, attack of the radicals on other lipoic acid molecules forming longer chains, or by hydrolysis of the thioketone (232) formed by intramolecular abstraction of tertiary hydrogens.

It was reported earlier that mesoionic anhydro-5-mercapto-2,3-diphenyl-1,3,4-thiadiazolium hydroxides (233) undergo photo-oxidative cyclization

[120] P. R. Brown and J. O. Edwards, *J. Org. Chem.*, 1969, **34**, 3131.

Photoelimination Reactions

to (234).[121] The same workers now report a new mode of fragmentation, as well as a wavelength dependency, for the photolysis of analogous mesoionic systems substituted with alkyl groups at C-2 or N-3 (235), (236), for which cyclization is most unlikely.[122] Thus, photolysis of anhydro-5-mercapto-3-methyl-2-phenylthiadiazolium hydroxide (236) with 253·7 nm

radiation in acetonitrile solution yields 21% of *N*-methylthiabenzamide (237) and 40% elemental sulphur. No fragmentation of (233) at wavelengths longer than 300 nm had been observed earlier, but with 253·7 nm radiation, (233) forms *N*-phenylthiabenzamide (238). In order to explain the fragmentations, an initial valence tautomerism to *N*-isothiacyanatothioamide (239) is proposed. Nitrogen–nitrogen cleavage of (239) then occurs. The thioamide is formed by hydrogen abstraction from the solvent which accounts for the increase in its yield in better hydrogen donor solvents. The photo-decomposition of the isothiocyanate group is known to proceed with the loss of elemental sulphur.[123] In support of the mechanism, photolysis of (236) in the presence of n-butylamine still yields (237), but no elemental sulphur is observed. This is interpreted as nucleophilic attack of n-butylamine on (239) to yield (240) which subsequently decomposes photochemically to (237). 5-Hydroxy-thiadiazolium analogues (241) behave similarly.

[121] R. M. Moriarty, J. M. Kliegman, and R. B. Desai, *Chem. Comm.*, 1967, 1255.
[122] R. M. Moriarty and R. Mukherjee, *Tetrahedron Letters*, 1969, 4627.
[123] U. Schmidt, I. Kabitzke, I. Boie, and C. Osterroht, *Chem. Ber.*, 1965, **98**, 3819.

(240) → (241) →[254 nm, MeCN] Ph-C(=S)-NH-Me 25%

Photochemical desulphuration has been observed with aromatic sulphonic acids to yield either the corresponding hydrocarbon (242) or amines (243), (244), and (245).[124] The formation of the amines under such mild conditions could be synthetically useful.

(243) ←[hν, NH$_4$OH]— anthraquinone-SO$_3$H —[hν, OH$^-$]→ Anthraquinone

naphthalene-SO$_3$H (X) —[hν, OH$^-$]→ naphthalene (X)   aniline-SO$_3$H —[hν, H$_2$O]→ aniline (245)

X = H (242)
X = NH$_2$ (244)

Elemental sulphur is also formed photochemically from the 254 nm irradiation of carbonyl sulphide in a variety of solvents.[125, 126] Mercaptans are formed also in the photolysis by the insertion of sulphur(^{1}D) atoms into the alkane solvents. In contrast with gas-phase reactions, the sulphur(^{1}D) atoms in solution insert selectively, and the rates of insertion increase in the order primary < secondary < tertiary C–H.

7 Miscellaneous Decomposition and Elimination Reactions

In the year there have been many publications concerning fragmentations which cannot be classified into any of the above sections and which have little relation to each other. These are loosely grouped together here in two broad sections: (i) decomposition into small fragments containing less than five atoms, and (ii) decomposition into fragments containing five or more atoms. The choice of the section in which to include these reports has been rather arbitrary in some instances.

[124] A. V. El-Tsov, O. P. Studzinskii, O. V. Kul-Bitskaya, N. V. Ogol-Tsova, and L. S. Efros, *Zhur. Org. Khim.*, 1970, **6**, 638.
[125] E. Leppin and K. Gollnick, *J. Amer. Chem. Soc.*, 1970, **92**, 2221.
[126] K. Gollnick and E. Leppin, *J. Amer. Chem. Soc.*, 1970, **92**, 2217.

Photoelimination Reactions

Decomposition into Small Fragments.—The direct and sensitized photolysis of neat 1,4-, 1,3-, 1,2-, and 2,3-dichlorobutanes has been studied with 253·7 nm radiation.[127] The principal reaction is reported to be the elimination of hydrogen chloride and the results are compared with those from the radiolysis reactions of the systems. There is also a significant amount of light-induced isomerization with 1,4-dichlorobutane, when the 1,3-isomer is formed. With other dichlorobutanes studied, this very interesting rearrangement is only a minor process. The ratio of quantum yields for isomerization and hydrogen chloride production for 1,4-dichlorobutane is approximately 2·6 : 1 for the photolysis reaction and approximately 2·8 : 1 for radiolysis. Sensitization of the reaction by acetone (313 nm radiation) yields the same major products as are formed by direct photolysis of the 1,4- and 1,3-isomers: the mechanism is considered to involve collisional deactivation of the excited states of acetone by the dichloride.

Using a helium–neon laser source (632·8 nm), the kinetics and mechanism of the photolysis of 2-chloro-2-nitrosobutane have been studied.[128] Such a process simplifies experimental techniques and reduces the reaction time, thereby minimising the effects of slow dark reactions. Concentration, oxygen, and solvent all effect the decomposition process. From this work, three types of primary process are postulated for *gem*-chloronitrosoalkanes (246) [reactions (31), (32), and (33)]. The competing paths (32) and (33)

$$\begin{array}{c} \text{Cl} \\ | \\ \text{R}-\text{C}-\text{CH}_2\text{R}^1 \\ | \\ \text{NO} \\ (246) \end{array} \longrightarrow \begin{array}{c} \text{HCl} + \text{RC}=\text{CHR}^1 \\ | \\ \text{NO} \end{array} \quad (31)$$

$$\longrightarrow \text{R}\dot{\text{C}}\text{CH}_2\text{R}^1 + \text{Cl}\cdot \quad (32)$$
$$\qquad\qquad\quad | $$
$$\qquad\qquad\;\; \text{NO}$$

$$\longrightarrow \text{R}\dot{\text{C}}\text{CH}_2\text{R}^1 + \text{NO}\cdot \quad (33)$$
$$\qquad\qquad\quad | $$
$$\qquad\qquad\;\; \text{Cl}$$

are important primary processes. Selectivity occurs in oxygen-saturated, but not oxygen-free solutions. In the former, the excited molecule reacts with oxygen to yield 2-chloro-2-nitrobutane and only traces of free-radical products; in the absence of oxygen the excited molecule either becomes collisionally deactivated or forms free radicals to give products which are dependent upon the nature of the radical and not its source.

The simple loss of ammonia is a comparatively rare photo-process but has been reported this year with 254 nm irradiation of tyrosine (247) and histidine (248) to yield acrylic acid derivatives.[129]

---

[127] M. A. Golub, *J. Amer. Chem. Soc.*, 1970, **92**, 2615.
[128] L. Creagh and I. Trachtenberg, *J. Org. Chem.*, 1969, **34**, 1307.
[129] B. Monties, *Compt. rend.*, 1969, **269C**, 1069.

$$R-CH_2-CH(NH_2)-CO_2H \xrightarrow[254\text{ nm}]{h\nu} R-CH=CH-CO_2H + NH_3$$

(247) R = Phenyl

(248) R = Imidazolyl

Formamide vapour undergoes complete decomposition with 206·2 nm radiation to yield carbon monoxide, hydrogen, and ammonia.[130] The reaction has been investigated at temperatures between 115 and 400 °C, and pressures of 8—50 Torr. The quantum yields for formation of carbon monoxide, hydrogen, and ammonia at 150 °C are respectively 0·8, 0·6, and 0·2. From experiments with added ethylene and decompositions of NH_2CDO and ND_2CHO, it is concluded that there are three major primary processes operative, as outlined in reactions (34), (35), and (36). These reactions are followed by secondary processes [reactions (37) and (38)]. The photolysis of formamides and acetamides has been studied by e.s.r.

$$NH_2CHO \longrightarrow NH_2\cdot + CO + H\cdot \quad \Phi = 0\cdot35 \quad (34)$$

$$\longrightarrow H\cdot + \cdot NHCHO \quad \Phi = 0\cdot22 \quad (35)$$

$$\longrightarrow NH_3 + CO \quad \Phi = 0\cdot45 \quad (36)$$

$$H\cdot + NH_2CHO \longrightarrow H_2 + NH_2\dot{C}O \quad (37)$$

$$NH_2\cdot + NH_2CHO \longrightarrow NH_3 + NH_2\dot{C}O \quad (38)$$

spectroscopy.[131] The most effective radiation in this study had wavelengths between 200 and 250 nm. With formamide the primary free-radical-forming step is shown to be carbon–hydrogen bond scission with the formation of H· and ·CONH, whereas acetamide undergoes carbon–carbon fission to yield Me· and ·$CONH_2$ radicals. Substitution of a methyl group on the nitrogen in these compounds does not change the primary step, in that carbon–hydrogen and carbon–carbon cleavage is observed for N-methylformamide and N-methylacetamide respectively. The radicals were identified at −196 °C by their spectra, and the authors wisely add the cautionary note that primary mechanisms may be different in the gas phase and in liquids at higher temperatures.

The gas-phase photolysis of tetramethylurea yields carbon monoxide, trimethylamine, dimethylamine, ethane, methane, and polymer.[132] The reaction has been studied at various temperatures, wavelengths of exciting radiation, and times of radiation. Reactions (39) to (47) can explain the variation in product yield with experimental parameters. A large fraction of the dimethylamino-radicals formed photochemically are not involved in the gaseous products, but disproportionate [reaction (43)] and thereby

[130] J. C. Boden and R. A. Back, *Trans. Faraday Soc.*, 1970, **66**, 175.
[131] S. R. Bosco, A. Cirillo, and R. B. Timmons, *J. Amer. Chem. Soc.*, 1969, **91**, 3140.
[132] J. R. Majer, Sama Naman, and J. C. Robb, *J. Chem. Soc.* (*B*), 1970, 93.

Photoelimination Reactions

$$Me_2N-CO-NMe_2 \xrightarrow{h\nu} Me_2NCO\cdot + Me_2N\cdot \quad (39)$$

$$Me_2NCO\cdot \longrightarrow Me_2N\cdot + CO \quad (40)$$

$$Me_2N\cdot \longrightarrow CH_2=NH \text{ (polymerised)} + Me\cdot \quad (41)$$

$$Me_2N\cdot + Me\cdot \longrightarrow Me_3N \quad (42)$$

$$Me_2N\cdot + Me_2N\cdot \longrightarrow Me_2NH + MeN=CH_2 \text{ (polymerised)} \quad (43)$$

$$Me_2N\cdot + RH \longrightarrow Me_2NH + R\cdot \quad (44)$$

$$Me\cdot + RH \longrightarrow CH_4 + R\cdot \quad (45)$$

$$Me\cdot + Me\cdot \longrightarrow C_2H_6 \quad (46)$$

$$Me_2H\cdot \longrightarrow Polymer \quad (47)$$

initiate polymerization which is responsible for about 70% of the total reaction. Both reactions (43) and (44) are important at higher temperatures, and the concentration of dimethylamine increases with time and increase in temperature. The results of this investigation indicate that the steps in the dissociation of tetramethylurea are similar to the familiar photolytic processes found with acetone.

Methylamine and trideuteriomethylamine have been subjected to 147 nm radiation (xenon resonance lamp) both in the presence and absence of free-radical scavengers.[133] The main process is formation of hydrogen atoms and in contrast with the longer wavelength studies, it is found that the major proportion of the hydrogen atoms arise from the methyl group: this was shown by photolysis of the deuterio-derivative. Direct molecular elimination of hydrogen definitely also occurs at 147 nm, but the total hydrogen formed by this route only represents a relatively minor percentage of the overall decomposition. Similarly, only a small fraction of the decomposition occurs *via* methyl radicals.

The kinetics of the photolysis of nitric acid vapour with 253·7 and 265 nm radiation, and at small conversions, has been studied under a variety of conditions. The direct photolysis yields hydroxyl radicals and NO_2 in the primary process, and the primary quantum yields at 265 and 253·7 nm are 0·1 and 0·3 respectively.[134] Experiments with added oxygen indicate that the contributions of the excited singlet and triplet states to the decomposition are comparable at 265 nm, but at 253·7 nm the decomposition originates almost exclusively from the excited singlet state. Nitric acid appears to decompose by a chain mechanism when irradiated at 265 nm in the presence of hydrogen and carbon monoxide.[135] The kinetics of the decomposition have also been investigated in the presence of hydrocarbons and with sensitization by nitrogen dioxide.[136]

---

[133] J. J. Magenheimer, R. E. Varnerin, and R. B. Timmons, *J. Phys. Chem.*, 1969, **73**, 3904.
[134] T. Berces and S Foergeteg, *Trans. Faraday Soc.*, 1970, **66**, 633.
[135] T. Berces, S. Foergeteg, and F. Marta, *Trans. Faraday Soc.*, 1970, **66**, 648.
[136] T. Berces and S. Foergeteg, *Trans. Faraday Soc.*, 1970, **66**, 640.

Decomposition into Fragments containing Five or more Atoms.—Isocyanic acid has been previously reported to undergo vapour-phase photolysis with the formation of carbon monoxide, nitrogen, hydrogen, and polymer.[137] Ethyl and acetyl isocyanates are now reported to undergo photolysis in the vapour phase and their reactions provide new examples of Norrish type II photodissociation.[138] The reactions are initiated by 228·8 nm radiation and occur at 25 °C at pressures of 2—30 Torr. Minor free-radical processes occur with both isocyanates and there is evidence for both R–NCO and RN–CO bond fission. The molecular dissociations are, however, of major importance, and lead to the formation of isocyanic acid and keten, and isocyanic acid and ethylene, from acetyl and ethyl isocyanates respectively. It is reasonably suggested that these products are formed *via* six-membered transition states (249) and (250), analogous to those involved in the type II process.

$$MeCONCO \xrightarrow{h\nu} \underset{(249)}{\begin{array}{c} O\!\!=\!\!C\!\!-\!\!N\!\!=\!\!C \\ | \quad\quad | \\ H_2C\!\!-\!\!H\cdots O \end{array}} \longrightarrow CH_2CO + HOCN$$

$$EtNCO \xrightarrow{h\nu} \underset{(250)}{\begin{array}{c} \quad\quad N\!\!=\!\!C \\ H_2C \quad | \\ | \quad\quad \\ H_2C\!\!-\!\!H\cdots O \end{array}} \longrightarrow C_2H_4 + HOCN$$

Various alcoholic side-chains attached to the N-9 in the isoalloxazine nucleus (251) have been photo-degraded under anaerobic and aerobic conditions.[139] Anaerobic photolysis in neutral aqueous solution yields two products. One product is alloxazine (252), and the other is an air-sensitive cyclic intermediate (253) which ultimately yields (256), the photo-fission product, *via* (254) and (255) intermediates.

The degradation products and processes of herbicides, insecticides, *etc.* are of interest not only chemically, but also because of their possible ecological importance. There have been several reports this year dealing with the photolysis of such compounds under a variety of conditions and some of this literature is considered in Chapter 5. Photochemical degradation of diquat (257) in dilute aqueous solution and on silica gel, both in natural sunlight and with filtered light from a mercury lamp, yields as many as nine breakdown products.[140] Some of these are transient and undergo further photo-degradation. Six of the compounds have been isolated and identified and thereby two pathways for the degradation are revealed. The minor route yields small amounts of fluorescent compounds with pyridone

---

[137] J. N. Bradley, J. R. Gilbert, and P. Svejda, *Trans. Faraday Soc.*, 1968, **64**, 911.
[138] N. J. Friswell and R. A. Black, *Canad. J. Chem.*, 1969, **47**, 4169.
[139] W. M. Moore and C. Baylor, *J. Amer. Chem. Soc.*, 1969, **91**, 7170.
[140] A. E. Smith and J. Grove, *J. Agric. Food Chem.*, 1969, **17**, 609.

structures (258) and (259), while the major pathway results in the formation of 1,2,3,4-tetrahydro-1-oxopyrido[1,2-a]-5-pyrazinium salt (260), picolinamide (261), and picolinic acid (262).

The photolysis products of the biologically active compound 2-(1,3-dioxolan-2-yl)-phenyl-N-methylcarbamate (263) are very dependent upon the solvent. In methanolic solution with 254 and 300 nm radiation, (263) undergoes scission of the dioxolan ring which leads to the formation of 3-methyl-2H-1,3-benzoxazine-2,4-(3H)dione (264) in yields in excess of 85%.[141] On the other hand, when the photolysis of (263) is carried out in aqueous solution with 300 and 360 nm radiation, two isolable products are

[141] B. E. Pape, M. F. Para, and M. J. Zabik, *J. Agric. Food Chem.*, 1970, **18**, 490.

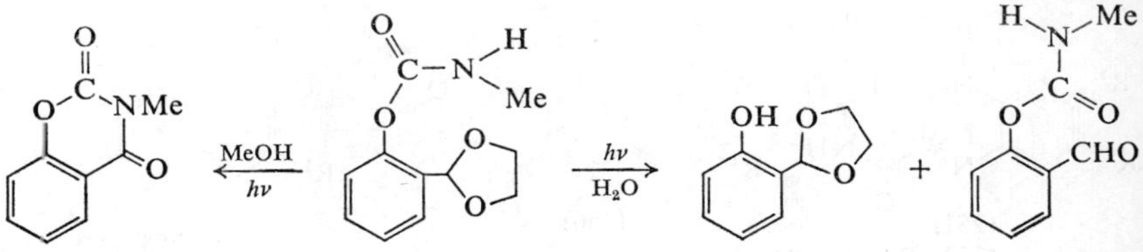

(264) (263) (265) (266)

formed. In one product, 2-(1,3-dioxolan-2-yl)-phenol (265), the dioxolan ring remains intact, while in the other product the dioxolan ring has been cleaved to yield 2-*N*-methylcarbonyl-benzaldehyde (266). Many aspects of these interesting degradations and effects have received no comment, and the field is thus wide open for a mechanistic and spectroscopic examination of these and related model systems.

The synthesis of α-methyl pyridines by photodecarboxylation of the corresponding pyridylacetic acids has been reported earlier in this chapter (ref. 85). α-Methyl-substituted quinolines have been formed by a similar photolytic procedure not involving the acid precursors. Thus both (267) and (268) in benzene solution (300 nm radiation) yield α-methylquinoline with the elimination of propylene and formaldehyde respectively.[142]

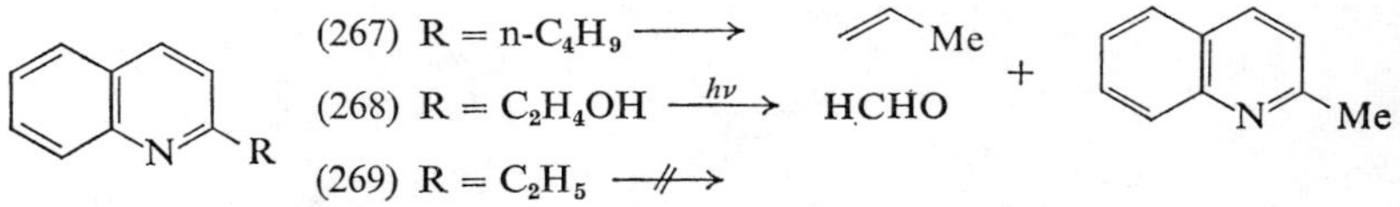

(267) R = n-C_4H_9 ⟶
(268) R = C_2H_4OH $\xrightarrow{h\nu}$ HCHO
(269) R = C_2H_5 ⟶̸

However, α-ethylquinoline (269) is photostable. The mechanism of the photolyses of (267) and (268) could be summarized by a process analogous to the Norrish type II involving intermediates (270) and (271). No such cyclic intermediate as (270) can be drawn for (269). These decomposition

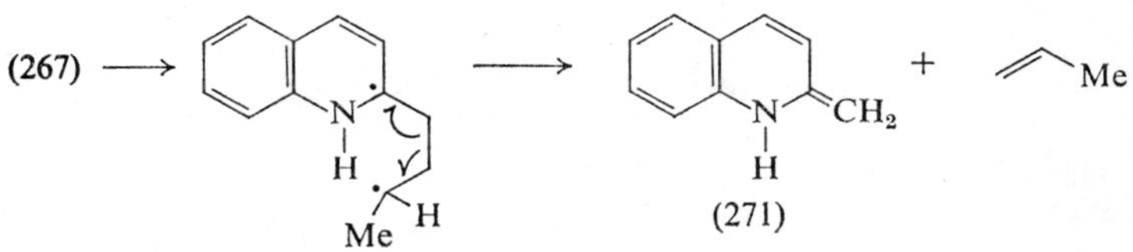

(267) ⟶ (270) ⟶ (271) +

(272)   (273)

[142] F. R. Stermitz and C. C. Wei, *J. Amer. Chem. Soc.*, 1969, **91**, 3103.

processes are further paralleled in the mass spectra of both (267) and (268) which have a base peak at $m/e = 143$, and this is interpreted as the ion (272). On the other hand, the base peak for (269) is at $m/e = 156$ ($M^+ - 1$) and is interpreted as the ion (273) for which there is no photochemical route Hence, mass spectrometry in this case is a diagnostic test for the photoelimination reaction. In the light-induced reaction of (268), the hydrogen abstraction in intermediate [*cf.* (270)] must occur from the hydroxy-group in a six-membered transition state rather than in a five-membered transition state, as indicated by the formation of α-monodeuteriomethylquinoline when the side-chain is CH_2CH_2OD. Such abstraction from a hydroxy-group is unusual, but has also been suggested for 9-substituted isoalloxazine reactions.[143]

The mercury-sensitized photochemical reaction of *cis-cis*-cyclo-octa-1,5-diene has been known for some time to yield two isomers (274) and (275)

together with products from allylic carbon–hydrogen fission,[144] but little attention has previously been paid to the degradation processes. Work has now been reported on the mercury-sensitized molecular elimination of ethylene from this diene.[145] The photolysis employs 253·7 nm radiation and yields isomers (274) and (275) as well as hydrogen, ethylene, acetylene, buta-1,3-diene, cyclohexa-1,3-diene, *cis-* and *trans*-hexa-1,3,5-triene, and benzene. The relative yields of the products are very dependent on the reaction time. Ethylene and the butadiene are primary products, and acetylene is a secondary product. The proportions of cyclic diene and acyclic triene also decrease with time, with concomitant increase in the benzene and hydrogen yields. The results from the photolysis in the presence of nitric oxide suggest an intramolecular pathway rather than a free-radical process. The suggested mechanism is outlined in Scheme 8.

Ethylene is eliminated photochemically from ethyl-lithium, as in the thermolysis (where it is the sole primary hydrocarbon product).[146] The

---
[143] W. M. Moore and C. Baylor, *J. Amer. Chem. Soc.*, 1966, **88**, 5677.
[144] I. Haller and R. Srinivasan, *J. Amer. Chem. Soc.*, 1964, **86**, 3318.
[145] S. Takamuku and H. Sakurai, *J. Phys. Chem.*, 1969, **73**, 1171.
[146] W. H. Glaze and T. L. Brewer, *J. Amer. Chem. Soc.*, 1969, **91**, 4490.

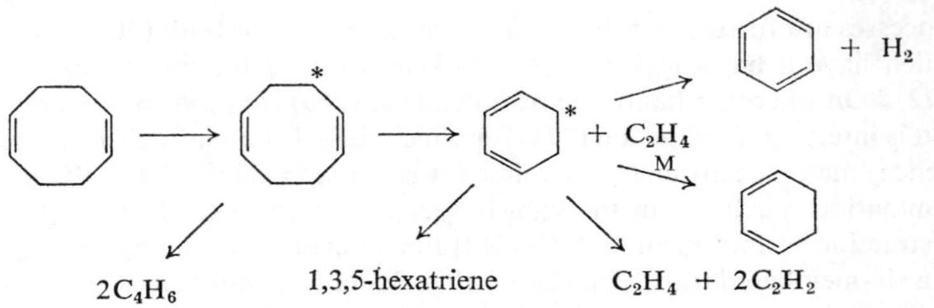

Scheme 8

reaction, in aliphatic hydrocarbons with 254 nm radiation, occurs by two competing photolytic mechanisms (reactions 48 and 49). It would appear,

$$\text{EtLi} \longrightarrow \text{LiH} + C_2H_4 \quad (48)$$

$$2\text{EtLi} \longrightarrow 2\text{Li} + C_2H_6 + C_2H_4 \quad (49)$$

(N.B. Ethyl-lithium is strongly associated in hydrocarbon solutions)

however, that the ethyl radicals implied in equation (49) do not behave as 'free' radicals (*i.e.* no butane is formed), and from the present data it is not possible to distinguish between a concerted process for the formation of lithium, ethylene, and ethane, and a mechanism involving caged or complexed radicals.

It is known that diazabicyclo[3,2,0]heptan-6-ones undergo several novel rearrangements under mild solvolytic and thermal conditions.[147] The photochemistry of the systems has now been reported and may be accounted for by a Norrish type I cleavage involving the elimination of the elements of keten.[148] Thus irradiation of (276) in benzene solution through borosilicate glass yields compounds (279), (280), and (281) in addition to benzil, by way of the intermediates (277) and (278). Similarly the Δ^2-diazabicycloheptenone (282) in benzene or methanol solutions yields 1,5-dimethyl-4-phenylpyrazol (283) while (284) yields (285) in methylene chloride solution and a mixture of both (283) and (285) in methanol. These solvent effects will doubtless be examined further.

A further example of the loss of the elements of keten is found in the thermolysis or photolysis of the reaction product (286) of tetrafluorobenzyne with methoxybenzenes.[149] In this case the aromatic compound (287) is formed.

Thermolysis of the adduct (288) of cyclo-octatetraene with acetylenedicarboxylic acid diester is reported to lead to rearrangement reactions and a minor fragmentation to buta-1,3-diene and dimethyl phthalate.[150] The

[147] J. M. Eby and J. A. Moore, *J. Org. Chem.*, 1967, **32**, 1346.
[148] E. J. Voelker and J. A. Moore, *J. Org. Chem.*, 1969, **34**, 3639.
[149] B. Hankinson and H. Heaney, *Tetrahedron Letters*, 1970, 1335.
[150] M. Avram, G. Mateescu, and C. D. Nenitzescu, *Ber.*, 1957, **90**, 1857.

Photoelimination Reactions

photolysis of adduct (288) in diethyl ether solution has now been reported.[151] The products of the reaction are dimethyl phthalate, cyclo-octatetraene, a small amount of benzene', unidentified higher-boiling material, and, (289) (16—18%) which might be ascribed to formation of a cyclobutadiene

intermediate. More convincing evidence for the transitory formation of cyclobutadiene is provided from the photolysis of (288) in the presence of piperylene and isoprene, when the adducts (290) and (291) respectively are formed at the expense of hydrocarbon (289).

Typically, bicyclo[4,2,0]octa-2,4-dienes [*e.g.* (292)] undergo '*b*'-type photolytic cleavage (known sometimes as the Barton reaction)[152] In

contrast, the photolysis of ester (292) sensitized by benzophenone, thioxanthone, or fluorenone at wavelengths longer than 300 nm leads to '*a*'-type cleavage with the formation of benzene and *cis-* and *trans*-diacetoxyethylenes as primary products.[153] Mechanisms to account for the formation of both stereoisomers of diacetoxyethylene are given in reactions (50) to (53). Reaction (52) is of theoretical significance since it postulates the elimination of a triplet excited molecule followed by the *cis-trans* isomerization process. This process may be favourable since the cleavage reaction is a retro-2+2 cycloaddition which is allowed photochemically: also the simple Woodward–Hoffmann correlations for 2+2 cycloadditions correlate excited-state reactants with excited-state products (although correlation with excited cyclobutane seems unattractive on energy grounds). The process should be exothermic and is thus energetically

[151] R. D. Miller and E. Hedaya, *J. Amer. Chem. Soc.*, 1969, **91**, 5401.
[152] D. H. R. Barton, *Helv. Chim. Atca.*, 1959, **42**, 2604.
[153] R. A. Caldwell, *J. Org. Chem.*, 1969, **34**, 1886.

$$\left[\begin{array}{c}\text{[bicyclic]}\diagup^{\text{OAc}}_{\diagdown\text{OAc}}\end{array}\right]^3 \longrightarrow \text{[cyclohexadienyl]}-\overset{\cdot}{\text{CH}}\text{CHOAc} \atop \text{OAc} \qquad (50)$$

$$\text{[cyclohexadienyl]}-\overset{\cdot}{\text{CH}}\text{CHOAc} \atop \text{OAc} \longrightarrow C_6H_6 + \text{either olefin isomer} \qquad (51)$$

$$\left[\begin{array}{c}\text{[bicyclic]}\diagup^{\text{OAc}}_{\diagdown\text{OAc}}\end{array}\right]^3 \longrightarrow C_6H_6 + [\text{AcOCH}=\text{CHOAc}]^3 \qquad (52)$$

$$[\text{AcOCH}=\text{CHOAc}]^3 \longrightarrow \text{Either olefin isomer} \qquad (53)$$

feasible. The experimental mechanistic work will obviously be difficult, but efforts to sort out the situation are reported to be in progress, and a further report is eagerly awaited.

The 1,2-photoaromatization of polycyclic systems, *e.g.* conversion of a dimer of cyclo-octatetrene into benzene and bullvalene,[154] has been further investigated with the other polycyclic systems (293), (294), and (295).[155] The presumed intermediates for these reactions [*e.g.* (296)] have apparently not been detected, and it seems most unclear, at least in the case of (295), whether decarbonylation is the first (route A) or second (route B) step in

[154] G. Schröder, *Chem. Ber.*, 1964, **97**, 3140.
[155] C. M. Anderson, J. B. Bremner, H. H. Westberg and R. N. Warrener, *Tetrahedron Letters*, 1969, 1585.

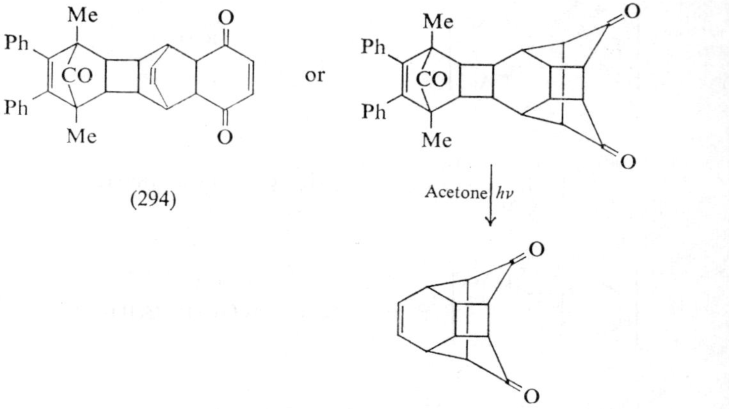

(294)

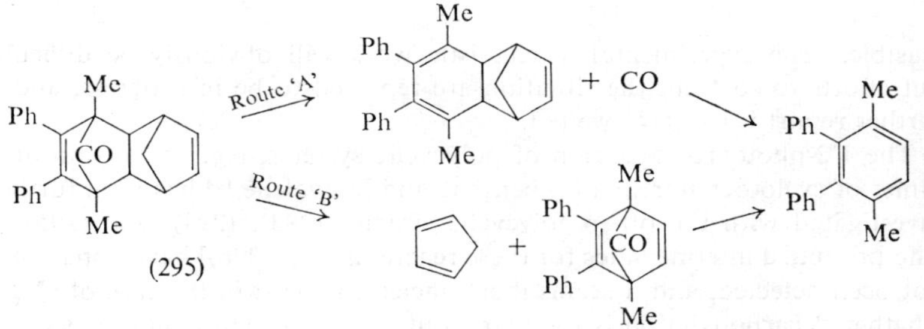

Acetone | $h\nu$

(295) Route 'N' / Route 'B'

the re-aromatization process. One notes that the concerted dissociation of an intermediate such as (296) should be symmetry-forbidden by the ground-state route,[156] although perturbation by substituents might well diminish the force of this prohibition, and aromatization could be a strong driving force.

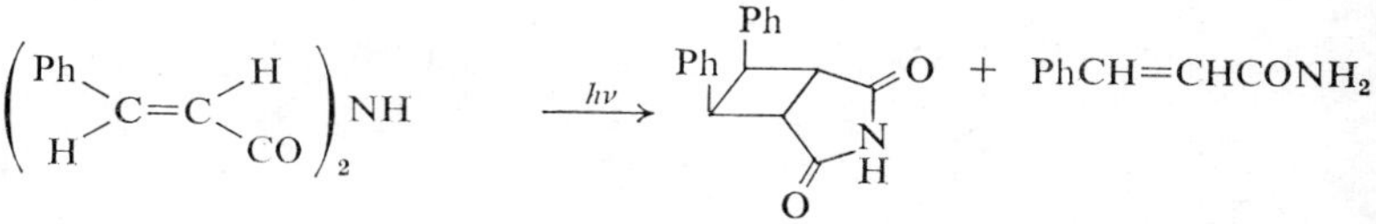

$$\left(\begin{array}{c}Ph\\H\end{array}C=C\begin{array}{c}H\\CO\end{array}\right)_2 NH \xrightarrow{h\nu}$$ (300) + PhCH=CHCONH$_2$

(299)

PhCH=CHSCH=CHPh $\xrightarrow{h\nu}$ (302) + (303)

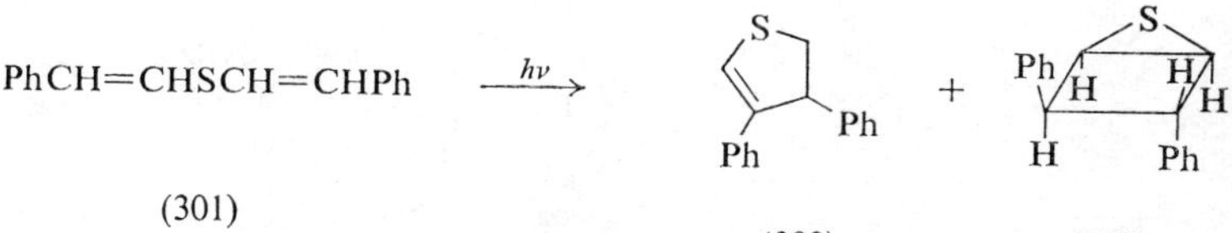

(301)

[156] D. Bryce-Smith, *Chem. Comm.*, 1969, 806.

The photosensitized monomerization of *cis-syn* and *trans-syn* cyclobutane-type thyamine dimers and the *cis-syn* thymine-uracil 'dimer' using phenanthraquinone derivatives as sensitizers has been described.[157]

trans-Cinnamide (299) undergoes light-induced massive fragmentation in 1,2-dimethoxyethane solution to yield *cis*- and *trans*-cinnamamide, together with the cyclobutane derivative, β-truxinimide (300).[158] The somewhat structurally similar sulphur compound, di-β-styryl sulphide (301) yields (302) and *trans*-2,3-diphenyl-5-thiabicyclo[2,1,0]pentane (303), unaccompanied by the decomposition products which might have been expected by analogy with the behaviour of other sulphur compounds.[159] Compound (303) is the first example of a cyclobutene episulphide.

A. G.

[157] E. Ben-Hur and I. Rosenthal, *Photochem. Photobiol.*, 1970, **11**, 163.
[158] R. T. Lalonde and C. B. Davis, *Canad. J. Chem.*, 1969, **47**, 3250.
[159] E. Block and E. J. Corey, *J. Org. Chem.*, 1969, **34**, 896.

Part IV

POLYMER PHOTOCHEMISTRY

As in Volume 1 of this series, the interaction between photochemistry and the chemistry of synthetic polymeric materials has been collated under four main headings. In the first section the use of additives which can initiate free-radical polymerizations in the presence of u.v. and visible radiation is discussed, and the short second section deals with the physical properties of polymeric materials under the influence of light. As before, however, the main interest of the polymer chemist in the interaction of light with polymers is structural changes within the polymer which can be brought about by u.v. and, to a lesser extent, visible radiation. These changes may be beneficial, leading to cross-linking and grafting, and a section is devoted to this aspect, but more often the effect of light upon the polymers under consideration leads to degradation, especially in the presence of atmospheric oxygen. The photoreactions leading to degradation and some of the ways of preventing such breakdown are the subject of the last main section.

Because of commercial interests, much of the literature pertaining to polymer photochemistry does not appear in the principal scientific journals, and for that reason no claim is made that the following review is complete. Much of the data on photostabilizers, for instance, is to be found in the patent literature, and the reader with interests in this direction is referred to this source. Thus, publications are reviewed below which are principally of scientific merit rather than of commercial interest. It is to be hoped that readers chiefly interested in polymer chemistry will refer to other parts of this Volume to gain an understanding of some of the basic principles of photochemistry before considering the specific applications to synthetic polymers.

1 Photopolymerization

Addition polymerisation may be initiated photochemically by any species which produce free radicals upon absorption of light. As an example, halogens and halogen-containing compounds are convenient sources of such chain-initiating species. Thus, methyl methacrylate may be polymerized readily in solution by the addition of molecular bromine,[1] by the following mechanism:

$$Br_2 + h\nu \longrightarrow 2Br\cdot \qquad (1)$$

$$Br\cdot + M \longrightarrow BrM\cdot \qquad (2)$$

$$BrM\cdot + Br_2 \longrightarrow BrMBr + Br\cdot \qquad (3)$$

$$BrM\cdot + M \longrightarrow BrMM\cdot \qquad (4)$$

$$BrMBr + h\nu \longrightarrow BrM\cdot + Br\cdot \qquad (5)$$

where M represents a methyl methacrylate molecule. Other vinyl monomers such as dibromomethyl methacrylate, dibromoacrylonitrile, dibromoethylacrylate, and dibromostyrene may also be photopolymerized by a similar

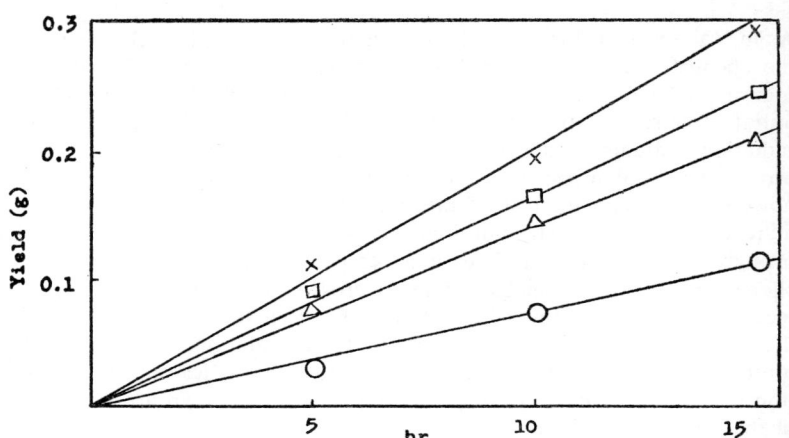

Figure 1 *Photopolymerization of* MMA *in the presence of* Me_3SiCl: (O) *no silane;* (△) $3\cdot12 \times 10^{-2}$ mol/l; (□) $6\cdot24 \times 10^{-2}$ mol l^{-1}; (×) $9\cdot36 \times 10^{-2}$ mol l^{-1} *Solvent, benzene;* MMA = $4\cdot68$ mol l^{-1}; 30 °C
(Reproduced by permission from *J. Polymer Sci., Part A*-1, 1969, **7**, 2837)

mechanism.[1] Methyl methacrylate has also been polymerized by the species produced by the photochemical breakdown of phosgene.[2] Chlorosilane compounds have also been shown to be efficient initiators of vinyl monomer polymerisation.[3] Of the silanes studied, the order of efficiency noted was as follows:

$$Me_4Si \simeq MeSiCl_3 > Me_3SiCl \simeq SiCl_4 > Me_2SiCl_2$$

An example of the enhancement of the rate of photopolymerization of methylmethacrylate is given in Figure 1, with Me_3SiCl as the photoinitiator. It was clear from an analysis that the polymer resulting from

---

[1] N. Sakota, T. Tanigaki, and K. Tabuchi, *J. Chem. Soc. Japan, Ind. Chem. Sect.*, 1969, **72**, 975.

[2] T. Tanigaki, A. Sakai, K. Inoue, and N. Sakota, *J. Chem. Sôc. Japan, Ind. Chem. Sect.*, 1969, **72**, 1580.

[3] Y. Minoura and H. Toshima, *J. Polymer. Sci., Part A*-1, 1969, **7**, 2837.

such initiation contained silyl groups, and the rate of polymerization (R_p) using this initiator obeyed the equation:

$$R_p = k(M)(Me_3SiCl)^{\frac{1}{2}} \qquad (6)$$

The same relationship was found to hold for the other silanes. Thus a mechanism concomitant with the above observation would be

$$Si + h\nu \longrightarrow Si^* \qquad (7)$$
$$Si^* \longrightarrow 2R\cdot \qquad (8)$$
$$R\cdot + M \longrightarrow RM\cdot \qquad (9)$$
$$RM\cdot + M \longrightarrow RM_2\cdot \qquad (10)$$
$$RM_i\cdot + RM_i\cdot \longrightarrow \text{dead polymer} \qquad (11)$$

where M refers to acrylic monomer, Si to silane initiator, and R to radicals.

However, the added initiator can also act as a chain-transfer agent, as in (12).

$$RM_i\cdot + Si \longrightarrow RM_iR + R\cdot \qquad (12)$$

Chain-transfer rate constants for the various silicon compounds compared with carbon tetrachloride are shown in Table 1.

Table 1 *Chain-transfer constants for various silicon compounds and carbon tetrachloride*

	$C_i \times 10^3$	
	Photopolymerization (30 °C)	Radical polymerization (50 °C)*
Me_4Si	0·5	0·13
$MeSiCl$	14·4	0·22
Me_2SiCl_2	17·5	0·245
$MeSiCl_3$	18·5	0·27
$SiCl_4$	25·6	0·30
CCl_4	2·0	0·50 (at 60 °C)

* Data from Y. Minoura and Y. Enomoto, *J. Polymer. Sci.*, Part A-1, 1968, **6**, 13.

Aqueous solutions of anthraquinone sulphonates may also be utilized in the polymerization of acrylic monomers.[4] It was found necessary to introduce a reducing agent, such as the chloride ion, in order to initiate polymerization. There are two possible mechanisms to explain the results. The first involves the excitation of a transient complex between the anthraquinone sulphonate sensitizer (AQ) sodium chloride, [reaction (13)], followed by radical formation [reaction (14)].

$$(AQ \ldots HCl) + h\nu \longrightarrow (AQ \ldots HCl)^* \qquad (13)$$
$$(AQ \ldots HCl)^* \longrightarrow AQH\cdot + Cl\cdot \qquad (14)$$

The chlorine atom may attack AQ to form AQ· (1).

[4] Q. Anwaruddin and M. Santappa, *J. Polymer. Sci.*, Part A-1, 1969, **7**, 1315.

(1)

$$Cl\cdot + AQ \longrightarrow HCl + AQ\cdot \qquad (15)$$

Either AQ· or Cl· may then initiate polymerization in the normal way. An alternative mechanism involves the intermediacy of the sensitizer triplet state, [reactions (16)—(18)].

$$AQ + h\nu \longrightarrow AQ^*(S) \qquad (16)$$

$$AQ^*(S) \longrightarrow AQ^*(T) \qquad (17)$$

$$AQ^*(T) + HCl \longrightarrow AQH\cdot + Cl\cdot \qquad (18)$$

Kinetic analysis shows that it is not possible to distinguish between the two mechanisms on kinetic grounds only. However, since there is no change in optical absorption of the sensitizer with increase in chloride ion concentration, it would appear that the second mechanism is more probably the correct one. Charge-transfer complexes can, however, be useful initiators. Thus, complexes are known to be formed between amides and maleic anhydride, and there is good spectroscopic evidence of such a complex between vinyl pyrrolidine and maleic anhydride,[5] with a 1:1 composition. This complex has been shown to be an efficient initiator of vinyl polymerization.[5] Other charge-transfer initiators have been described.[6]

Organic compounds which undergo scission to free radicals may induce photosensitized polymerization of unsaturated molecules. Thus, benzoin[7] and butylnitrates[8] have been used for this purpose. However, one of the requirements of an efficient sensitizer is that it can be an efficient absorber of the radiation provided, and for this reason dyestuffs have been much used as initiators. These also have the advantage that they can be incorporated into the polymer to give coloured materials. In general, these initiators have been utilized in aqueous solution, but a recent study has demonstrated that a variety of dyes may initiate polymerisation in non-aqueous media, especially in benzene solution.[9] Activators are necessary in this case, and generally long-chain aliphatic amines were used, such as cetyl dimethyl amine. For a typical system, eosin–cetyl dimethyl amine–methyl methacrylate in benzene, the rate of polymerization has a square-root

[5] H. Tamura, M. Tanaka, and N. Murata, *Bull. Chem. Soc. Japan*, 1969, **42**, 3041; *J. Chem. Soc. Japan, Ind. Chem. Sect.*, 1970, **73**, 800.
[6] S. Tazuke, *Fortschr. Hochpolym.-Forsch.*, 1969, **6**, 321.
[7] C. T. Chen and W. D. Huang, *J. Chinese Chem. Soc., Taipei*, 1969, **16**, 46.
[8] V. A. Sechkovskaya, C. V. Leplyanin, and C. P. Gladychev, *Izvest. Akad. Nauk. Kazakh. S.S.R. Ser. khim.*, 1970, **20**, 59.
[9] S. Maiti, M. K. Saha, and S. R. Patit, *Makromol. Chem.*, 1969, **127**, 224.

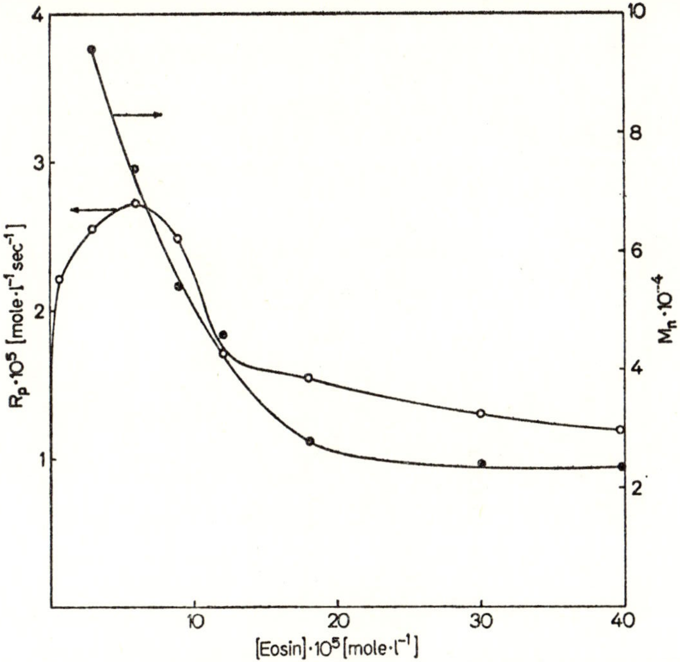

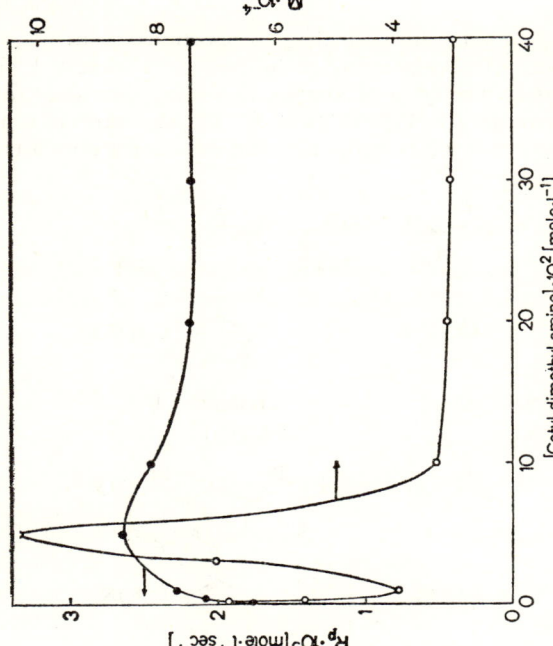

Figure 2 Effect of 'Eosin-Säure L neu' on the rate and the molecular weight of polymers for the dye-sensitized photopolymerization of methyl methacrylate in bulk at 40 °C. [Cetyl dimethylamine] = 5 × 10⁻² mol l⁻¹, [light intensity] = 4·347 × 10⁵ erg min⁻¹ cm⁻³ (Reproduced by permission from *Makromol. Chem*, 1969, **127**, 224)

Figure 3 Effect of cetyl dimethylamine on the rate and the molecular weight of polymers for the dye-sensitized photopolymerization of methyl methacrylate in bulk at 40 °C. [Eosin] = 6 × 10⁻⁵ mol l⁻¹, [light intensity] = 4·347 × 10⁵ erg min⁻¹ cm⁻³ (Reproduced by permission from *Makromol. Chem*, 1969, **127**, 224)

dependence upon light intensity and direct proportionality to monomer concentration, while with respect to eosin and cetyl dimethyl amine concentration it passes through a maximum, indicating their dual role as both initiator and retarder (see Figures 2 and 3). The structures of some of the dye-stuffs used are shown [(2)—(8)]. Electron-donating groups attached

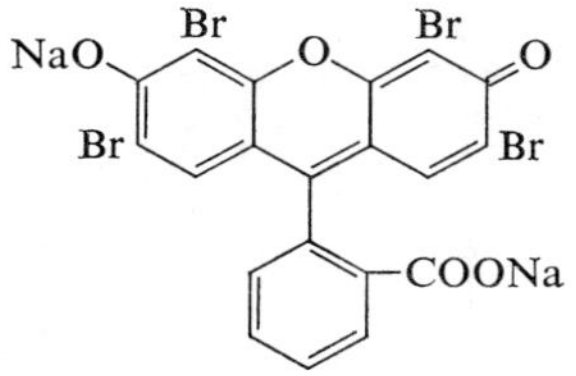

Eosin Säure L NEU
(2)

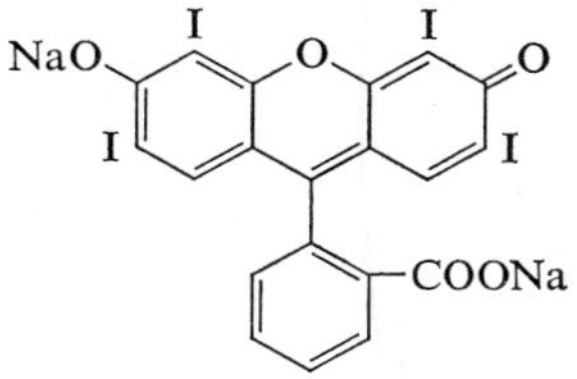

Erythrosin J
(3)

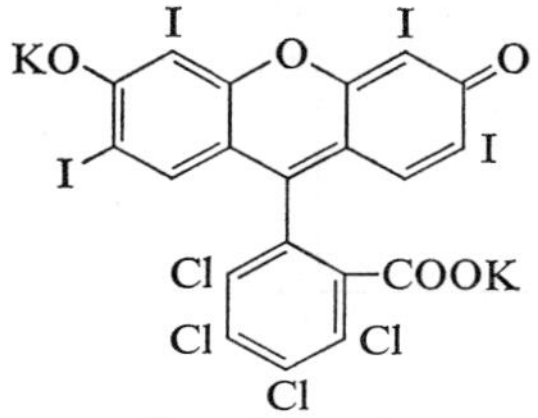

Rose Bengal
(4)

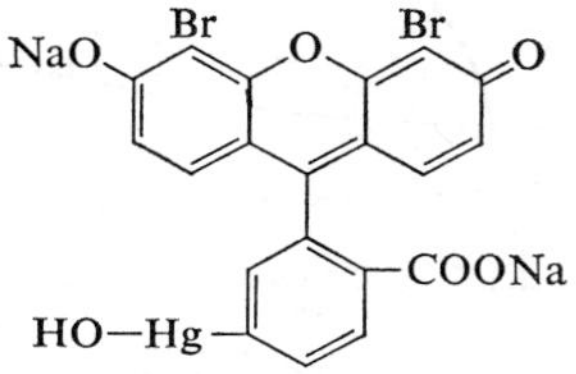

Mercurochrome
(5)

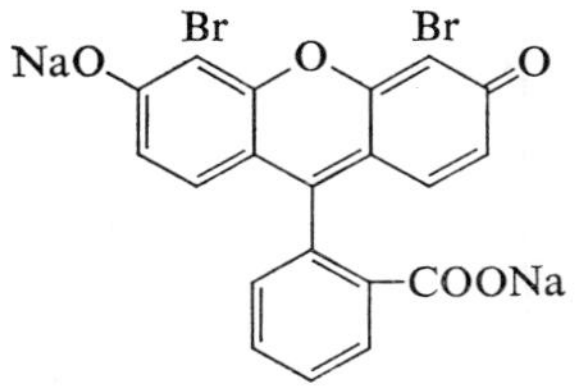

Eosin H8G
(6)

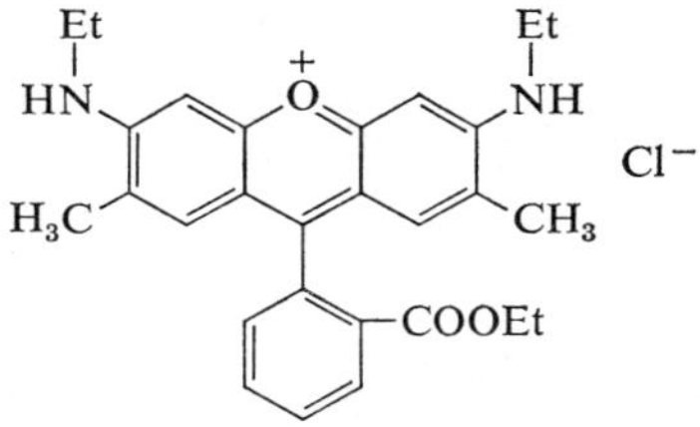

Rhodamine 6GEx
(7)

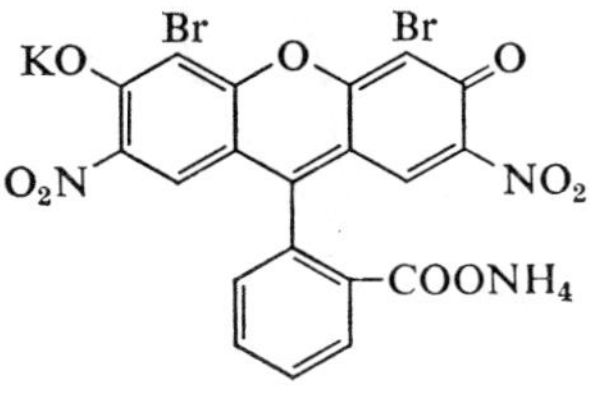

Eosin BNX
(8)

to the benzene nucleus in the halogenated fluorescein dyestuffs accentuate the sensitizing action of dyes, whereas electron-withdrawing groups deactivate it. The greater the chain length and amount of N-substitution in the aliphatic amine, the greater is the activating influence.

On the basis of the dependence of the observed rates of polymerization upon concentrations of additives, light intensity, *etc.*, the scheme shown in equations (19)—(30) has been proposed.

$$D + h\nu \longrightarrow D^*(S) \text{ (absorption)} \tag{19}$$

$$D^*(S) \longrightarrow D^*(T) \text{ (intersystem crossing)} \tag{20}$$

$$D^*(S) \longrightarrow D + h\nu_F \text{ (fluorescence)} \tag{21}$$

$$D^*(T) + RH \longrightarrow HD\cdot + R\cdot \text{ (reaction with reducing agent)} \tag{22}$$

$$D^*(T) + SH \longrightarrow (DSH)^* \longrightarrow DH\cdot + S\cdot \text{ (reaction with solvent)} \tag{23}$$

$$D^*(S) + RH \longrightarrow D + RH \text{ (fluorescence quenching)} \tag{24}$$

$$D^*(S) + O_2 \longrightarrow D + O_2 \text{ (oxygen quenching)} \tag{25}$$

$$D^*(S) + D \longrightarrow D + D \text{ (concentration quenching)} \tag{26}$$

$$HD\cdot + M \longrightarrow HDM\cdot$$

$$R\cdot + M \longrightarrow RM\cdot \text{ (initiation)} \tag{27}$$

$$S\cdot + M \longrightarrow SM\cdot$$

$$RM_i\cdot + M \longrightarrow RM_{i+1}\cdot \text{ etc. (propagation)} \tag{28}$$

$$RM_i\cdot + RM_j\cdot \longrightarrow \text{dead polymer (termination)} \tag{29}$$

$$\left.\begin{array}{l} RM_i\cdot + D \longrightarrow \text{dead polymer} + D\cdot \\ RM_i\cdot + RH \longrightarrow \text{dead polymer} + R\cdot \\ RM_i\cdot + SH \longrightarrow \text{dead polymer} + S\cdot \end{array}\right\} \text{transfer)} \tag{30}$$

In aqueous solution, the polymerization system consists of a vinyl monomer, with a photoreducible dye of the acridine, xanthene, or thiazine families in the presence of an electron donor. The light-excited dye is converted to the triplet state from which it abstracts an electron from the donor. The reduced dye usually then reacts with molecular oxygen to produce the initiating radical and regenerate the dye molecule. An alternative suggestion is that the reduced dye (semiquinone) and the oxidized electron donor serve as initiating species, but it has been shown that in the rigorous absence of oxygen, no initiation of polymerization occurs.[10] Thus, oxygen is necessary for the initiation process, and also will inhibit the chain polymerization. In such a situation there should exist an optimum concentration of oxygen such that the rate of polymerization is at a maximum. It is possible to maintain such an optimum concentration of

[10] N. L. Yang and G. Oster, *J. Phys. Chem.*, 1970, **74**, 856.

oxygen by utilizing a reversible oxygen carrier, which is a complex which combines with molecular oxygen in an oxygen-rich environment, but which releases it when the oxygen concentration is low. Cobalt(II) complexes fulfil this requirement,[10] and many of the ligands in such complexes are also electron donating, so that the complex may serve both as an electron donor and as a reversible oxygen carrier. In a model system, methylene

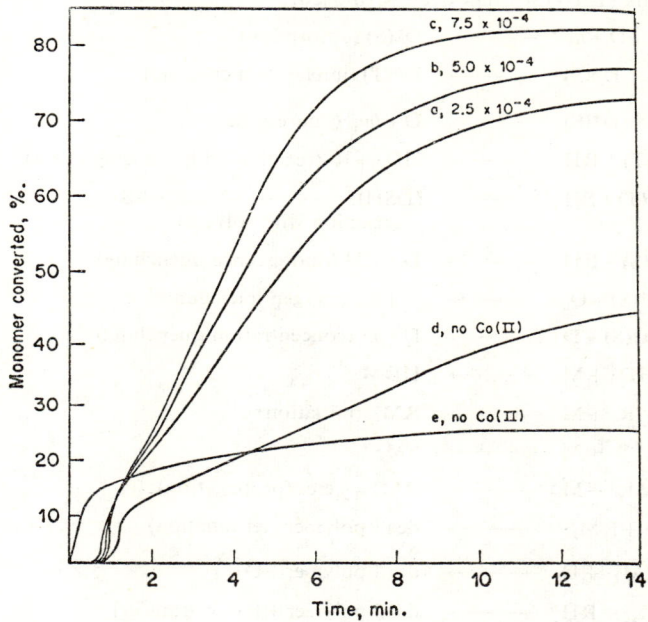

Figure 4 *Monomer conversion curve for aerobic photopolymerization of the acrylamide–methylene blue–triethylenetetramine system*
(Reproduced by permission from *J. Phys. Chem.*, 1970, **74**, 856)

blue was used as the sensitizer, acrylamide as the monomer, and triethylene tetramine (TETA) as both the electron donor and ligand for Co^{II}. Two molecules of this complex combine with one molecule of oxygen to form the oxygenated species $(TETA)Co^{II}O_2Co^{II}(TETA)$. With such a system, there is no need to purge oxygen in order to achieve high conversions of monomer, as Figure 4 demonstrates. The advantages and potential of systems like this are very great, and must be used increasingly in the future. Other systems utilizing methylene blue have been described,[11] and the photochemical reactions of vat dyes with polymer substrates also discussed.[12]

Metal salts are another class of compound which is used to initiate addition polymerization. Thus, vinyl monomers may be photopoly-

[11] N. Kamiya and M. Okawara, *J. Chem. Soc. Japan, Ind. Chem. Sect.*, 1969, **72**, 2639.
[12] C. S. Egerton and N. E. N. Assaad, *J. Soc. Dyers and Colourists*, 1970, **86**, 203.

merized by the addition of azidopentamminecobalt(III) chloride.[13] Oxidizing metal salts such as $UO_2(NO_3)_2$, $Ce(NH_4)_2(NO_3)_6$, $Hg(CH_3COO)_2$, and $AgNO_3$ were found to be good photosensitizers for 2-methyl-1-vinylimidazole (MVI) and 2-ethyl-1-vinylimidazole (EVI) polymerisation,[14] whereas non-oxidizing salts such as Zn^{II} did not act as photosensitizers (see Figure 5). The reaction was very sensitive to the nature of the monomer, since a

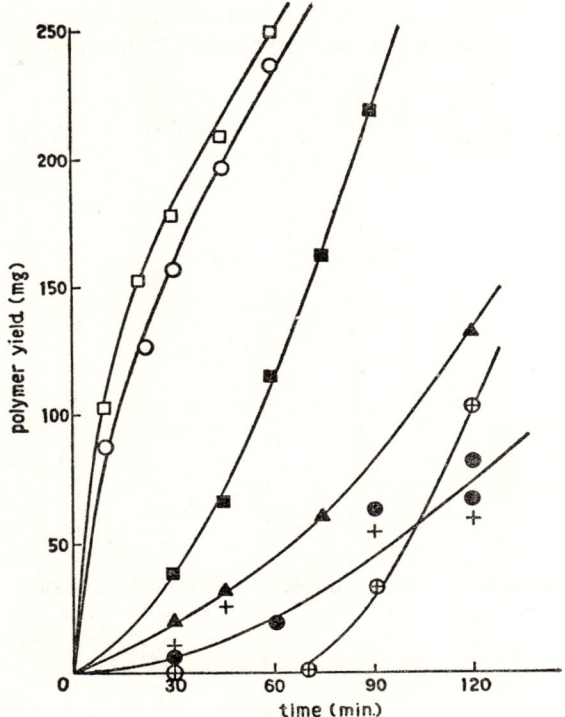

Figure 5 *Photopolymerization of EVI at 30 °C (EVI 1·5 ml; [metal salt] = 10^{-3} mol l^{-1}):* (●) *no metal salt;* (▲) $AgNO_3$; (■) $Hg(CH_3COO)_2,2H_2O$; (○) $Ce(NH_4)_2(NO_3)_6$; (□) $UO_2(NO_3)_2,6H_2O$; (+) $Zn(CH_3COO)_2,2H_2O$; (⊕) $NaAuCl_4,2H_2O$
(Reproduced by permission from *J. Polymer Sci. Part A*-1, 1969, **7**, 851)

non-complexing monomer (styrene) was not polymerized in the presence of these salts. It seems likely, therefore, that photosensitized electron transfer between monomer and metal salt *via* complex formation is the most probable initiation mechanism. Cupric acetate and sodium chloroaurate(III), which have been reported as efficient initiators for the poly-

[13] M. Santappa and L. V. Natarajan, *Proc. Indian Acad. Sci.*, 1969, **69A**, 284.
[14] S. Takuze and S. Okamura, *J. Polymer. Sci., Part A*-1, 1969, **7**, 851.

merization of vinylpyridine and *N*-vinylcarbazole respectively,[15] act as linear terminators in the above systems. The behaviour of Bunte salts in radical polymerization has also been discussed.[16] Very high molecular weight poly(ethylene oxide) can be prepared by u.v. irradiation using dimanganese decacarbonyl as a catalyst.[17]

One of the interesting aspects of photopolymerization is that it can be achieved in the solid state. Thus, 2,5-distyrylpyrazine (9) and related compounds undergo a four-centre polymerization in the solid state to give a linear highly crystalline product by the action of u.v. radiation [18] (31). A stepwise mechanism has been proposed to account for the observed nature of the polymer.

[15] S. Takuze, M. Asai and S. Okamura, *J. Chem. Soc. Japan, Ind. Chem. Sect.*, 1969, **72**, 1841.
[16] M. Tsunooka, M. Fujii, N. Ando, M. Tanaka, and N. Murata, *J. Chem. Soc. Japan, Ind. Chem. Sect.*, 1970, **73**, 805.
[17] W. Strohmeier and P. Hartmann, *Z. Naturforsch.*, 1969, **24b**, 939.
[18] H. Nakanishi, Y. Suzuki, and M. Hasegawa, *J. Polymer. Sci. Part A*-1, 1969, **7**, 743, 753.

$$(9) + (9) \longrightarrow (10) \tag{33}$$
$$(9) + (10) \longrightarrow (10) \tag{34}$$
$$(10) + (10) \longrightarrow (10) \tag{35}$$

It is clear that the product is typical of a condensation mechanism, and (9) and (10) must necessarily behave in a different manner because of the pronounced difference in the resonance stabilization of the olefinic double bonds. Since the crystalline polymer is produced from the crystalline

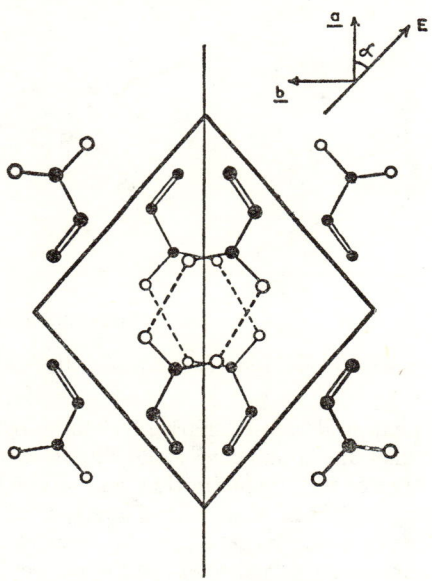

Figure 6 *The crystal structure of acrylic acid (projection on the* ab *plane) according to* Chatani et al.* (C *atoms,* ●; O *atoms,* ○), *showing also the shape of the monomer crystal and the definition of* α, *the angle between the* a-*axis and the plane of the electric vector* E *of the polarised radiation*
(Reproduced by permission from *Trans. Faraday Soc.*, 1969, **65**, 2497)

monomer and a simple relation exists between the crystal units, it is concluded that this type of photopolymerization belongs to a topochemical reaction. A similar polymeric type has been obtained with the phenylene diacrylic acid type of monomer.[19] The photopolymerization of acrylamide and methacrylamide in the solid state using chlorine as a catalyst has been achieved.[20] The solid-state polymerization of acrylic acid in the crystalline state has been investigated using plane polarized u.v. light.[21] Figure 6

---

[19] F. Suzuki, Y. Suzuki, H. Nakanishi, and M. Hasegawa, *J. Polymer. Sci. Part A*-1, 1969, **7**, 2319.
[20] T. Matsuda, T. Highashimura, and S. Okamura, *J. Macromol. Sci. Chem.*, 1970, **4**, 1.
[21] E. C. Eastmond, E. Haigh, and B. Taylor, *Trans. Faraday. Soc.*, 1969, **65**, 2497.

* Y. Chatani, Y. Sakata, and I. Nitta, *J. Polymer Sci., Part B*, 1963, **1**, 419.

shows the crystal structure of acrylic acid, and defines the angle α between the a-axis and the plane of the electric vector E of the polarized radiation. A correlation was obtained between the relative rate of polymerization observed and the magnitude of α, as seen in Figure 7. Thus, as α increased,

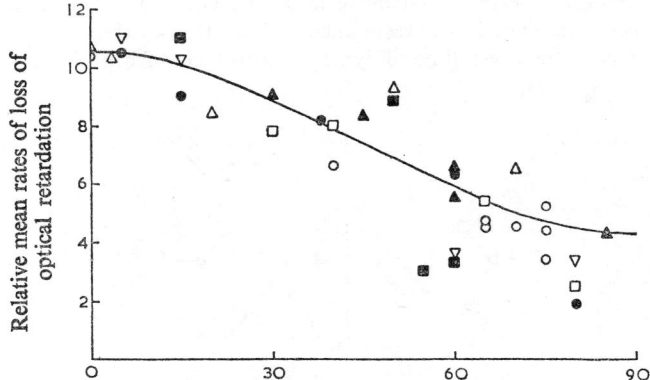

Figure 7 *Variation in mean relative rate of polymerization of acrylic acid with angle α*
(Reproduced by permission from *Trans. Faraday Soc.*, 1969, **65**, 2497)

the rate of polymerization decreased markedly. Although it might appear from these results that the monomer crystal lattice exerts a dominating influence on the course of the polymerization, with excited molecules in the perfect lattice reacting with their nearest neighbours, an alternative explanation is considered to be much more likely. While it is true that photons are absorbed predominantly by molecules in the perfect lattice, excitation of the monomer molecules will certainly produce mobile excitons, which will migrate through the lattice until ultimately becoming trapped in defects. Once trapped, the 'localized' exciton might initiate polymerization by non-radiative decay to produce free radicals, or conceivably the course of the photoreaction might be influenced by the formation of 'thermal spikes', with an enhancement of molecular mobility in the vicinity of the lattice defect caused by the non-radiative decay of the exciton. In either case, since absorption occurs primarily in the perfect lattice, a dependence of rate of polymerization upon incident angle of the polarised light, α, would be expected, as is observed.

2 Optical Properties of Photoexcited Polymers

Macromolecules exhibit the same optical characteristics as smaller molecules in that they have an absorption spectrum, may emit fluorescence and phosphorescence, and may demonstrate energy transfer, delayed fluorescence, *etc.*

The absorption and luminescence spectra of many polymers consist of broad structureless bands, which nevertheless are characteristic of the material under question. Thus, prolonged luminescence has been observed in several polymers,[22] and luminescence from aromatic polymers under high-energy electron excitation has also been recorded.[23] The fluorescence spectrum of polystyrene in solution consists of two main bands, one a structured band at short wavelengths and the other a broad featureless band at longer wavelengths.[24] This second band is entirely typical of an 'excimer' emission, but, as can be seen in Figure 8, the position of the

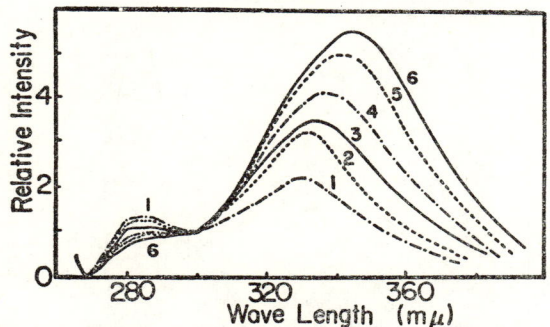

Figure 8 *The fluorescence spectra of polystyrene solution in ethyl acetate at various concentrations. The spectra are normalised at* 300 mμ. *Concentrations in* g/100 cm³: (1) 0·0282, (2) 0·480, (3) 1·016, (4) 2·236, (5) 4·575, (6) 6·28
(Reproduced by permission from *Makromol. Chem.*, 1969, **124**, 84)

maximum varies markedly with concentration of polystyrene. The shift in position of the band with change in concentration was also observed in other solvents, being most marked in dichloroethane. Reabsorption cannot account for the observed shifts, and it is proposed that in concentrated solutions of polystyrene there is a strong interaction between the benzene rings even in the ground state which accounts for the observed shifts, although this result is in contrast to effects seen in other excimer systems. However, in the paracyclophanes, the position of the fluorescence maxima shifts to the red as the benzene rings are sited nearer together and thus the explanation that in polystyrene increase in concentration causes the distance between benzene rings to decrease, resulting in a red shift, may be a valid one.

The phenomenon of energy transfer in polymers is one of great current interest in that it may be a possible means of stabilizing polymers against u.v. degradation. The energy transfer may be of two types, intermolecular and intramolecular, depending upon the nature of the polymer. An

---

[22] V. I. Gol'denberg and V. Ya Shlyapintokh, *Vysokomol. Soedineniya*, 1969, **11A**, 1958.
[23] D. H. Phillips and J. C. Schug, *J. Chem. Phys.*, 1969, **50**, 3297.
[24] T. Nishihara and M. Kaneko, *Makromol. Chem.*, 1969, **124**, 84.

example of the first has been discussed in which Nylon 66 is used as the donor and proflavine as the acceptor molecule.[25] The Nylon 66 excited at 295 nm phosphoresces with a maximum at around 400 nm, and increasing the concentration of proflavine from zero causes a reduction in the intensity of this peak, and the simultaneous production of two new peaks at 495 nm and 575 nm. These correspond, respectively, to delayed fluorescence and phosphorescence in the proflavine. In addition, lifetime measurements

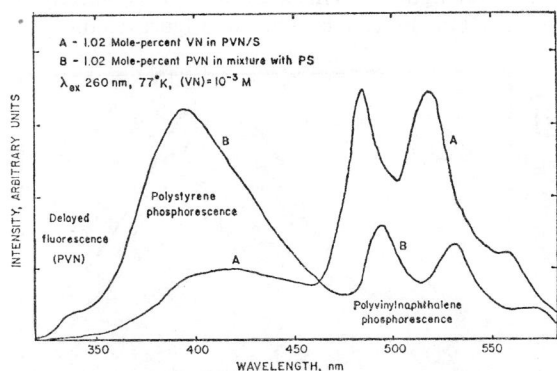

Figure 9 *Delayed emission spectra for a mixture of polystyrene* (PS) *and poly*(1-*vinylnaphthalene*) (PVN) *and the corresponding copolymer* (PVN/S) (Reproduced by permission from *Macromolecules*, 1969, **2**, 181)

clearly indicate that as the concentration of proflavine is increased, the lifetime of the Nylon 66 phosphorescence decreases. It is clear that energy transfer is occurring, and arguments are advanced to show that in fact the excited species of proflavine first formed by energy transfer is the lowest fluorescent singlet state. The phenomenon is thus an example of singlet–triplet energy transfer, and could occur *via* Forster long-range interaction, or by triplet exciton migration. Although the data do not favour either of these unequivocally, it is felt that the long-range interaction is the more probable mechanism. This study is one of the few in which a polymer itself has been used as the donor molecule. However, a recent investigation has been carried out in which the polymer acts as both donor and acceptor.[26] A styrene-1-vinylnaphthalene copolymer was prepared and it was shown that, relative to equivalent mixtures of polystyrene and poly(1-vinylnaphthalene), the copolymers excited by light absorbed by polystyrene show quenching of phosphorescence from the styrene-derived segments and sensitization of phosphorescence from the 1-vinylnaphthalene-derived segments (see Figure 9). It is concluded that triplet energy has been intramolecularly transferred from the styrene to the 1-vinylnaphthalene segments

[25] H. H. Dearman and F. T. Lang, *J. Polymer. Sci. Part A*-2, 1969, **7**, 497.
[26] R. B. Fox and R. F. Cozzens, *Macromolecules*, 1969, **2**, 181.

of the polymer chain. This result is extremely interesting, since the 1-vinylnaphthalene acts as an energy sink for the absorbed u.v. light, and may thus prevent photodegradation. It may thus be possible to stabilize certain polymers against u.v. degradation by the provision of such 'sinks' in the chain, and much work is in progress along these lines. Competition between triplet–triplet and triplet–singlet energy transfer in a polymer has been investigated,[27] and energy transfer in plastic scintillators described.[28] Triplet–triplet energy transfer has been observed in solid polyvinylbenzophenone–naphthalene systems.[29]

Photoconductivity in certain polymers is of considerable current interest. It has been known for some time that poly(N-vinylcarbazole), (PVK), (11),

(11)

which absorbs in the near u.v. region of the spectrum, exhibits transient photoconductivity, and that this can be increased by the addition of electron acceptors. A recent study examines the mechanism of photogeneration and subsequent transport of holes through thin films of PVK.[30] In this study, transient primary photoconductivity was measured using a light pulse of about 3 μs duration to create a sheet of charge carriers whose drift in an applied electrical field is then observed. The bulk effects, e.g. trapping, were separated from the surface effects (e.g. photogeneration) by conducting measurements on samples ranging from 1·44 to 30·5 μm thicknesses, and three exciting wavelengths were used (2540, 2800, and 3130 Å), at temperatures from 160 to 400 K. PVK films were coated on to aluminium substrates from toluene solution and, after drying, a semi-transparent aluminium electrode was evaporated on the top forming a sandwich cell. Measurements were then made on this cell using the experimental set-up as shown in Figure 10.

The voltage pulses generated by a light pulse of a few microseconds duration was exponential, and could be characterized by a time T_{eff} defined by the expression

$$\Delta V(t) = \Delta V_0[1 - \exp(-t/T_{\text{eff}})] \tag{36}$$

Although T_{eff} varied with the applied electric field and thickness of the film, the shape of the transient signal remained exponential in the entire

[27] V. Anisimov, K. Burshtein, E. Bogoyavlenskaya, and O. Karpukhin, *Optika Spektrosk.*, 1970, **28**, 814.
[28] F. Heisel and C. Laustriat, *J. Chem. Phys.*, 1969, **66**, 1895.
[29] C. David, W. Demarteau, and C. Geuskens, *European Polymer. J.*, 1970, **6**, 537.
[30] D. M. Pai, *J. Chem. Phys.*, 1970, **52**, 2285.

range (10^6—10^8 V m^{-1}) of the applied fields. Any successful model for the phenomenon would have to explain the following experimental observations:

(i) the transient signals are exponential in nature;
(ii) the signals can be characterized by an effective transit time T_{eff};
(iii) the transport is not characterized by a drift mobility, but rather by a T_{eff} that has a strong electric field and temperature dependence.

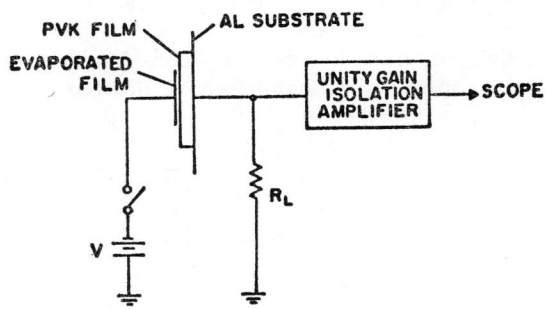

Figure 10 *Experimental arrangement for measurement of transient photoconductivity in* PVK *films*
(Reproduced by permission from *J. Chem. Phys.*, 1970, **52**, 2285)

The absence of a drift mobility precludes the existence of wide energy bands. An amorphous solid like PVK is likely to have only weak overlap of the wavefunctions of neighbouring molecules, which leads to localized energy states rather than the itinerant states observed in solids with strong overlap. A process which can explain all of the observed features of the transit pulse is one in which holes jump from one localized state to the next, with the release time from the localized state being much longer than the time of flight between the localized states. These release times are highly field dependent, depending on the way the potential barrier between the localized states is modified by the applied electric field. A simple development of this type of transport phenomenon can be given.

Consider a set of localized states numbered from 1 to $(m+1)$ equally spaced a distance d/m apart. Let n_0 carriers be generated by a flash of light in state 1 at time $t = 0$. Assuming a lifetime τ_0 for state 1, the carrier population in state 1 as a function of time is given by equation (37).

$$n_1 = n_0 \exp(-t/\tau_0) \qquad (37)$$

The lifetime τ_0 can also be expressed as an inverse probability per second of a carrier jumping out of a localized state. The kinetics for state 2 are given by equation (38), where the first term corresponds to the rate at

$$dn_2/dt = (n_0/\tau_0)\exp(-t/\tau_0) - (n_2/\tau_0) \qquad (38)$$

which the carriers jump from state 1 into state 2, and the second to the rate at which they leave state 2 for state 3.

Solving,

$$n_2 = n_0(t/\tau_0) \exp(-t/\tau_0) \tag{39}$$

Similar expressions can be derived for state 3 to state $(m+1)$. Assuming the time of flight between states is negligible compared to τ_0, the charge displacement ΔQ in a time interval t and $(t+\Delta t)$ is given by equation (40).

$$\Delta Q = \frac{e}{d} \cdot \frac{d}{m} \Delta t \left[\frac{n_0}{\tau_0} \exp\left(-\frac{t}{\tau_0}\right) + \frac{n_0 t}{\tau_0^2} \exp\left(-\frac{t}{\tau_0}\right) + \ldots \frac{n_0}{\tau_0^{m+1}} \cdot \frac{t^m}{m!} \exp\left(-\frac{t}{\tau_0}\right) \right] \tag{40}$$

Integrating, the expression for the transit is given by equation (41).

$$Q(t) = en_0 \left\{ 1 - \exp\left(-\frac{t}{\tau_0}\right) \left[1 + \frac{m}{m!} \frac{1}{1!} \cdot \frac{t}{\tau_0} + \frac{m-1}{m} \left(\frac{t}{\tau_0}\right)^2 \frac{1}{2!} + \ldots \left(\frac{t}{\tau_0}\right)^m \frac{1}{m} \cdot \frac{1}{m!} \right] \right\} \tag{41}$$

This expression can be used to give results which are in good agreement with experiment, and transit times of $m\tau_0$ can be obtained. It should be re-emphasized that τ_0 is a lifetime in a localised state before the carrier jumps to an adjoining state, and will have a strong electric field and temperature dependence which is characteristic of both the shape of the potential barrier between the states and the mode of transfer. Considering this, the theoretical curves derived from equation (41) can explain the observed features (ii) and (iii) outlined above. However, the approach is too idealised to explain feature (i). A more realistic picture of the model for the polymer will have a distribution of $m\tau_0$ values, but the exact mechanism of transfer from the localized states that explains the observed field and temperature dependence is not clear at present. The quantum efficiency of photogeneration of carriers is of the order of 0·1 at applied fields of 10^8 V m^{-1}. It is clear from this work that the concept of a trap may have to be broadened to include a localized state with very little interaction with its neighbouring states, especially in view of the evidence that PVK is not an isolated case, but the behaviour may be general for organic solids.

Photoconductivity may be observed in polystyrene when excited in the vacuum-u.v. region. The effective wavelengths are in the region from 1400 to 1050 Å. The behaviour of the photocurrent with applied voltage and light intensity is discussed.[31] In unsaturated polymers like PVK, a mechanism clearly exists for the transport of current *via* an electronic mechanism. In saturated polymers, such as polyethylene, no such mechanism should exist, but nevertheless quite high photocurrents can be registered in these polymers with light of wavelength as long as 365 nm. It has been shown that the origin of these photocurrents is not the polymer itself but the photoemission of electrons from the electrodes.[32] The charge may then be transported through the polymer, however, and it thus

---

[31] M. Ofran, N. Oron, and A. Weinreb, *J. Chem. Phys.*, 1969, **50**, 3131.
[32] A. E. Binks, A. C. Campbell, and A. Sharples, *J. Polymer. Sci., Part A*-2, 1970, **8**, 529.

appears that a route exists, even in simple saturated organic polymers, for the transport of charge by an electronic mechanism.

3 Photo-cross-linking and Grafting

As stated in the introduction to this chapter, absorption of u.v. and to a lesser extent visible radiation by polymer systems can give rise to structural changes. Often, especially in the presence of molecular oxygen, these changes are undesirable, leading to the degradation of the polymer and consequent loss of desirable mechanical properties. In some cases, disposable polymer is required and means are being sought to accelerate such photodegradations, but in the main the photodestruction of polymers is a serious problem, and ways are sought to prevent such effects. In certain cases, however, absorption of light causes desirable changes in mechanical properties *via* cross-linking, and associated with this are the phenomena of grafting and production of photosensitive polymers. The division of the discussion of the chemistry of photoexcited polymer species into two categories, cross-linking and degradation, is somewhat arbitrary since similar mechanisms may prevail, but in general degradation occurs in the presence of atmospheric oxygen and is thus considered separately from the other structural changes which may be brought about by u.v. and visible light.

The cross-linking of polymers may be achieved by the direct irradiation of the polymer itself, but is usually enhanced by the addition of photosensitizers. Thus poly(vinylcinnamate), upon absorption of light at wavelengths shorter than 340 nm, leads to effective cross-linking and subsequent reduction in solubility of the polymer. Addition of organic compounds which absorb at longer wavelengths can cause an enhancement of the rate at which cross-linking occurs, and it is thus believed that the triplet state of the cinnamoyl moiety is responsible for the cross-linking[33] (see Figure 11). For such a mechanism, an efficient sensitizer must have a high quantum yield of triplet state formation, and must also have a triplet energy level lying above that of the cinnamoyl moiety. An investigation has been made of the spectroscopic properties of several sensitizers and their relative efficiency in promoting cross-linking in this polymer.[33] Substitution in the sensitizer may cause an increase in efficiency by two mechanisms. If a heavy atom substituent is introduced, the quantum yield of triplet state formation will be enhanced, as shown in Table 2. Alternatively, a molecule which does not sensitize the cross-linking because its triplet energy level lies below that of the cinnamoyl group may be made to do so by substitution of groups which cause an increase in the triplet energy level to the point where transfer to the cinnamoyl becomes exothermic. This effect is also noted in the lower part of Table 2. It seems likely, therefore, on the basis of these results that the triplet mechanism is correct.

[33] M. Tsuda, *Bull. Chem. Soc. Japan*, 1969, **42**, 905.

Polyethylene may be effectively cross-linked by the addition of benzoyl chloride as a sensitizer, and it has been shown that the reaction proceeds *via* a two-quanta mechanism.[34] Oxalyl chloride has a similar effect, presumably *via* the mechanism shown in equations (42)—(45),[35] where (T) refers to a triplet state of the sensitizer and *(T) to a higher excited triplet. Phosphorous trichloride has also been utilized as an effective sensitizer of the

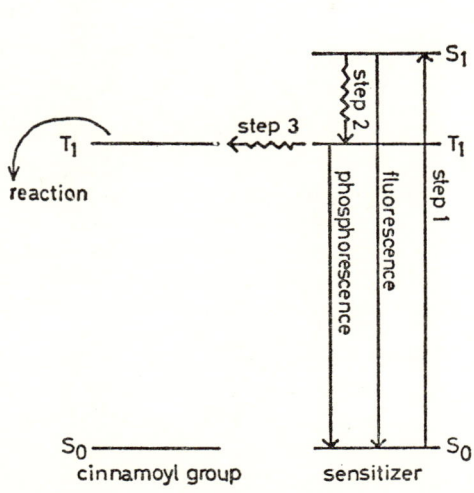

Figure 11 *Energy transfer diagram in spectral sensitization of poly(vinyl cinnamate)* (Reproduced by permission from *Bull. Chem. Soc. Japan*, 1969, **42**, 905)

$$C_2O_2Cl_2 + h\nu_1 \longrightarrow C_2O_2Cl_2(T) \quad (42)$$

$$C_2O_2Cl_2(T) + h\nu_2 \longrightarrow C_2O_2Cl_2^*(T) \quad (43)$$

$$C_2O_2Cl_2^*(T) \longrightarrow 2\dot{C}OCl \quad (44)$$

$$\downarrow + RH \text{ (polymer)}$$

$$C_2O_2Cl_2 + R\cdot + H\cdot \quad (45)$$

---

[34] D. A. Andrushchenko, A. A. Kachan, C. V. Chernyavskii, and V. A. Shrubovich, *Khim. vysok. Energii*, 1970, **4**, 169.

[35] D. A. Andrushchenko, A. A. Kachan, C. V. Chernyavskii, and V. A. Shrubovich, *Vysokomol. Soedineniya*, 1969, **11B**, 600.

Table 2 *Sensitization of the cross-linking of poly(vinyl cinnamate)*

Sensitizer	Relative efficiency	Sensitizer	Relative efficiency
NO_2–C$_6$H$_4$–NH_2 (para)	410	1-amino-anthraquinone	0
NO_2–C$_6$H$_4$–NH_2 (meta)	0	1,4-dihydroxy-anthraquinone (reduced)	0
NO_2–C$_6$H$_4$–NH_2 (ortho)	0	1,4-diaminonaphthalene	0
NO_2–C$_6$H$_4$–$N(Me)_2$	137	4-nitro-N-benzoyl-1-aminonaphthalene	769
2,6-dichloro-4-nitro-N,N-dimethylaniline	561	4-nitro-N-acetyl-1-aminonaphthalene	1100
2,6-dibromo-4-nitro-N,N-dimethylaniline	797		
anthraquinone	130	2-ethylanthraquinone	360

cross-linking of polyethylene [36] and polypropylene,[37] and the mechanical properties of the latter polymer have been described. Photocross-linking of stilbene modified polymers [38] and the photocyclization of 1,2-polybutadiene and 3,4-polyisoprene [39] have been reported.

Akin to photochemical cross-linking is the grafting of one polymer chain onto the backbone of a different synthetic or natural material, a process which naturally has considerable commercial significance. It has been demonstrated that poly(acrylic acid) and poly(vinylisocyanate) can be effectively grafted on to films of polyethylene using benzophenone as a sensitizer,[40] by a mechanism which is similar to that described for the cross-linking in the presence of benzoyl chloride in that it is a two-quanta process

$$B + h\nu_1 \longrightarrow {}^1B \quad (46)$$

$$^1B \longrightarrow {}^3B \quad (47)$$

$$^3B + h\nu_2 \longrightarrow {}^3B^* \quad (48)$$

$$^3B^* + RH \longrightarrow B + R\cdot + H\cdot \quad (49)$$

$$H\cdot + R'H \longrightarrow R'\cdot + H_2 \quad (50)$$

where B is benzophenone, and RH and R'H are the polymers which contain a methylene group. The subsequent formation of a bond between the $R\cdot$ and $R'\cdot$ species gives rise to the grafting. The grafting of some vinyl monomers on to collagen photosensitized by the triplet state of benzil,[41] and the photo-induced graft polymerization of styrene on to Nylon 66 [42] have been reported. Styrene has also been grafted on to the natural protein wheat gluten by photochemical means.[43] Poly(vinyl alcohol) provides a convenient base for the grafting of acrylonitrile [44] and other vinyl monomers [45] photochemically.

Also of considerable commercial interest is the production of photosensitive polymers which may be coated on metallic and other surfaces, and which then, under the influence of u.v. or visible light, harden into a

---

[36] A. Tkachenko, C. V. Chernyavskii, V. Shrubovich, and A. A. Kachan, *Khim. Volokna*, 1970, **70**, 30.

[37] V. I. Pavlov, V. A. Shrubovich, C. V. Chernyavskii, and A. A. Kachan, *Vysokomol. Soedineniya*, 1969, **11A**, 1760.

[38] F. A. Stuber, H. Ulrich, D. V. Rao, and A. A. R. Sayigh, *J. Appl. Polymer. Sci.*, 1969, **13**, 2247.

[39] M. A. Golub, *Macromolecules*, 1969, **2**, 550.

[40] A. A. Kachan, Yu. G. Lebo, and V. A. Shrubovich, *Vysokomol. Soedineniya*, 1970, **12A**, 214; Yu. C. Lebo, Z. A. Kostylova, A. A. Kachan, and V. A. Shrubovich, *Dopovidi Akad. Nauk Ukrain. R.S.R.*, Ser. B., 1969, 425; Z. A. Kostylova, V. A. Shrubovich, and A. A. Kachan, *Plast. Massy.*, 1970, **70**, 14.

[41] I. A. Kuduba, E. U. Chizhyunaite, and V. R. Yakubenaite, *Zhur. Vsesoyuz. Khim. Obshch. im. D. I. Mendeleeva*, 1969, **14**, 596.

[42] H. Ishibashi, *Chem. High Polymers (Japan)*, 1969, **26**, 331.

[43] K. Eskins and M. Friedman, *J. Macromol. Sci. Chem.*, 1970, **4**, 947.

[44] M. Tsunooka, M. Ishikawa, M. Tanaka, and N. Murata, *J. Chem. Soc. Japan, Ind. Chem. Sect.*, 1969, **72**, 1413.

[45] M. Tsunooka, M. Tanaka, and N. Murata, *J. Chem. Soc. Japan, Ind. Chem. Sect.* 1969, **72**, 1208.

protective film. There are several types of such polymers, some being based on a vinyl cinnamate structure.[46] However, a new group of such polymers has been discovered based upon a polyacetylene structure (12).[47]

$$\left[C\equiv C-CH_2O-\!\!\left\langle\!\!\!\begin{array}{c}\\\end{array}\!\!\!\right\rangle\!\!-\!\!\underset{\underset{Me}{|}}{\overset{\overset{Me}{|}}{C}}\!\!-\!\!\left\langle\!\!\!\begin{array}{c}\\\end{array}\!\!\!\right\rangle\!\!-OCH_2-C\equiv C \right]_n$$

(12)

When exposed to u.v. light, these polymers gave clear films by cross-linking. The efficiency of this cross-linking process was measured as a time of irradiation necessary, and values for different polymers are given in Table 3. It can be seen from Table 3 that addition of various additives

Table 3 *Efficiency of cross-linking of polyacetylenes*

Polymer	Cross-linking time [47] (sec)	Additive (10% of polymer)
(12) [—C≡C—CH₂O—C₆H₄—C(Me)₂—C₆H₄—OCH₂—C≡C—]ₙ	120	None
,,	60	Azobenzene
,,	50	Anthracene
,,	40	Maleic anhydride
,,	40	*p*-Benzoquinone
,,	30	*p*-Di-iodobenzene
,,	30	Acetophenone
,,	25	Rose Bengal
,,	25	*p*-Diethynyl-benzene
[—C≡C—CH₂O—C₆H₂(Ph)₂—C₆H₂(Ph)₂—OCH₂—C≡C—]ₙ	110	None
[—C≡C—CH₂O—C₆H₃(Ph)—C₆H₃(Ph)—OCH₂—C≡C—]ₙ	120	None
[—C≡C—CH₂O—C₆H₄—CPh₂—C₆H₄—OCH₂—C≡C—]ₙ	130	None

[46] M. Kato and M. Hasegawa, *J. Polymer Sci. Part B*, 1970, **8**, 263.
[47] A. S. Hay, D. A. Bolon, K. R. Leimer, and R. F. Clark, *J. Polymer Sci., Part B.*, 1970, **8**, 97; A. S. Hay, D. A. Bolon, and K. R. Leimer, *J. Polymer Sci., Part A*-1, 1970, **8**, 1022.

Table 3 (*cont.*)

Polymers	Cross-linking time [47] (sec)	Additive (10% of polymer)
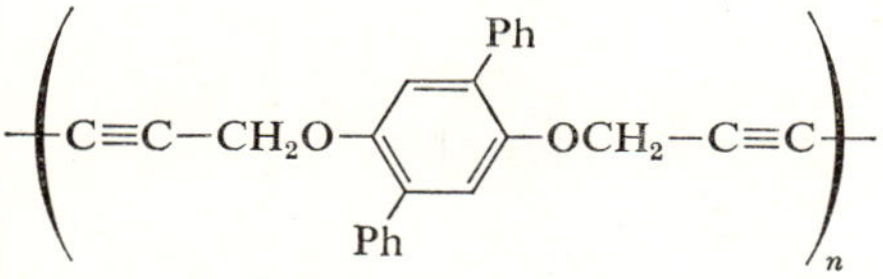 $+(C{\equiv}C{-}CH_2O{-}C_6H_3(Ph)_2{-}OCH_2{-}C{\equiv}C)_n+$	120	None
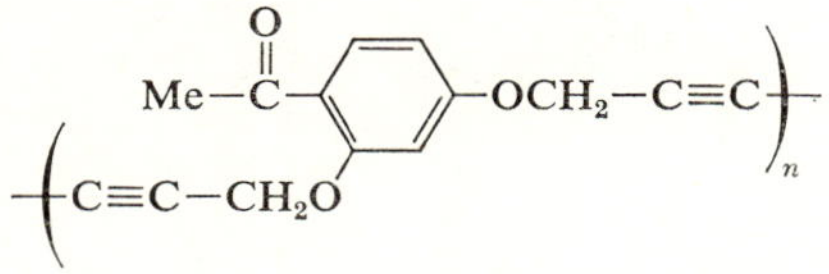 $+(C{\equiv}C{-}CH_2O{-}C_6H_3(COMe){-}OCH_2{-}C{\equiv}C)_n+$	35	None
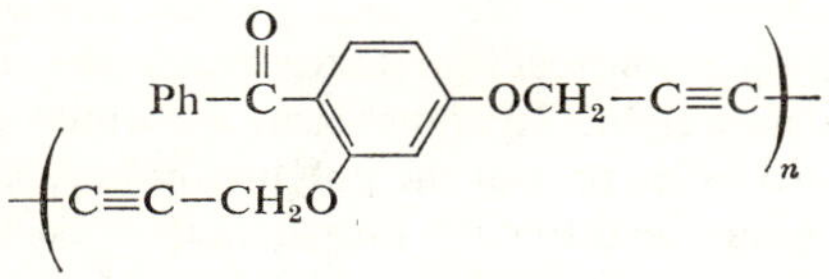 $+(C{\equiv}C{-}CH_2O{-}C_6H_3(COPh){-}OCH_2{-}C{\equiv}C)_n+$	40	None

to one of the polymers greatly enhanced the rate of the cross-linking process, and the advantages of introducing a carbonyl group into the polymer itself for increasing the efficiency of cross-linking are also apparent. Photosensitive polymers such as 1,2,3-thiadiazide polymers,[48] arylazido-polycondensates,[49] and many others [50–55] have been described, although in general the mechanisms of the cross-linking, film-forming reactions have not been investigated in detail. The applications of photolacquers to the electronics industry [56] and the paint industry [57] have been discussed.

It is known that 9,10-diphenyl anthracene (13) undergoes a reversible photo-oxidation, as shown in equation (51).

A polymer of *p*-vinyl-phenyl-9-phenyl-10-anthracene and its copolymer with styrene have been prepared which exhibit similar characteristics under the action of u.v. light and in the presence of oxygen.[58]

Other examples of light-sensitive polymers have been described.[59] In most of these reactions, which are of considerable technical interest

---

[48] C. A. Delzenne and U. Laridon, *Ind. chim. belge*, 1969, **34**, 395.
[49] C. A. Delzenne and U. Laridon, *J. Polymer Sci. Part C*, 1969, 1149.
[50] Y. Tomoda, *J. Soc. Org. Synth. Chem. Japan*, 1969, **27**, 1205.
[51] C. A. Delzenne, *European Polymer. J.*, 1969, **5**, 55.
[52] M. Okawara, *Kobunshi*, 1970, **19**, 94.
[53] M. Tsuda, *Kobunshi*, 1970, **19**, 109.
[54] M. Hasegawa, *Kobunshi*, 1970, **19**, 102.
[55] C. Nagamatu, *Kobunshi*, 1970, **19**, 120.
[56] K. Sato, *Kobunshi*, 1970, **19**, 136.
[57] M. Kinura, *Kobunshi*, 1970, **19**, 147.
[58] C. Meyer, *Bull. Soc. chim. France*, 1970, 702.
[59] R. C. Schulz, L. Rohe, and H. Adler, *European Polymer J.*, 1969, 309 (Bratislava supplement).

[Scheme (51): compound (13), 9,10-diphenylanthracene, reacts with $O_2 + h\nu$ / 180 °C to form endoperoxide]

because they are the basis of image-forming processes and the production of relief printing plates, the initiating photoreaction takes place on an initiator or sensitizer of low molecular weight. In some cases it would be desirable to have the light-sensitive groups as part of the polymer molecule, and several types of reaction which might lead to such polymers have been described.[59] These may be classified into groups.

(*i*) *Photochemical.* As an example, the light-induced rearrangement of polycarbonates may be given, which involves a partial degradation in addition to the formation of phenyl esters and benzophenone groups.

[Schemes (52) and (53): photo-Fries rearrangement of polycarbonate giving phenyl ester with ortho-OH, then benzophenone with ortho-OH]

Polymethacrylates with photochromic spiropyran groups can be prepared which gain colour on exposure to light:

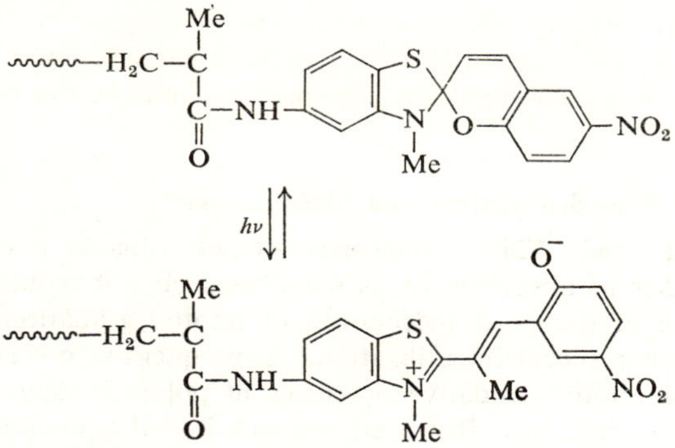

By the photolysis of poly(vinylbutyral), one could ideally photolytically produce a 1 : 1 copolymer of vinylbutyrate; however, this reaction proceeds only to an extent of 30% in practice (55):

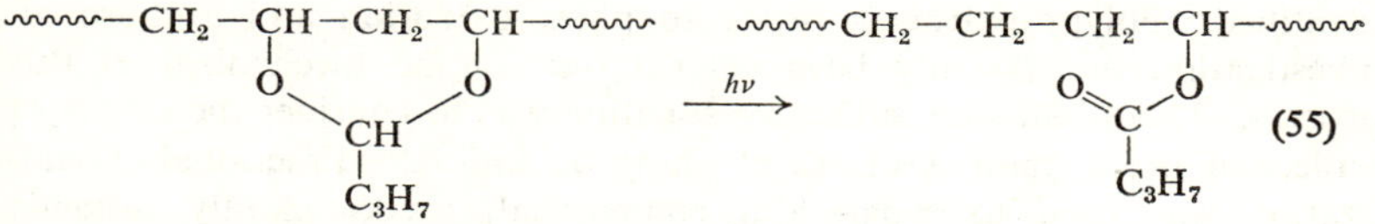

(ii) Photochemical Additions. The light-sensitive cinnamic acid group can be introduced to poly(vinyl alcohol) relatively easily, and the photosensitive reaction (56) is well established:

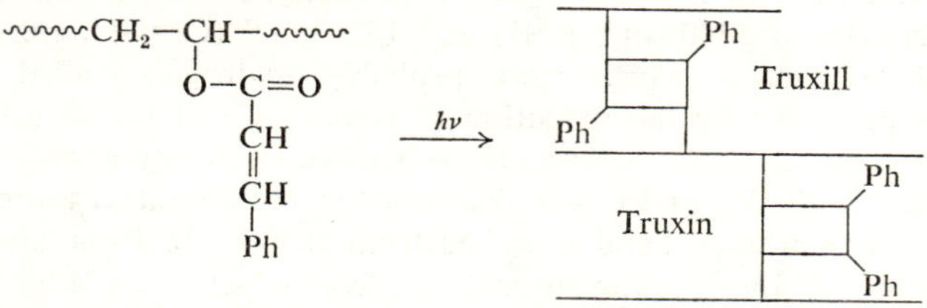

Furan may be made to undergo photoaddition to polyvinyl benzophenone, reaction (57), producing a polymer which may be naturally cross-linked if the second double bond of the furan also undergoes photoaddition.

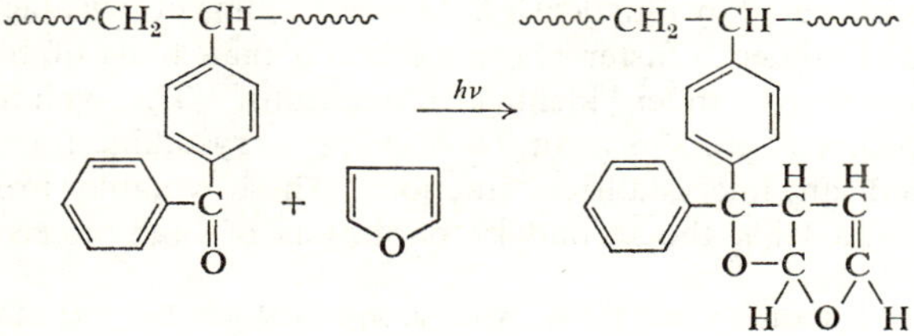

(*iii*) *Photoelimination of Nitrogen.* Reactions of polymer hydrazides with aldehydes can yield a very light-sensitive polymer with about 80% azidohydrazone groups. The light-sensitive nature is due to the photoinduced loss of nitrogen.

4 Photodegradation and Stabilization

Polypropylene and Vinyl Monomers.—Conventional photostabilizers operating either as absorbers or as scavengers do not require a detailed understanding of the exact mechanism of photodegradation in polymer systems for their successful utilization. Consequently, precise studies of the nature of photodegradative processes in polymers have only infrequently been carried out. However, research into the possibility of using electronic energy transfer as a means of preventing the photodestruction of polymeric materials is increasing, and for this type of stabilizer the details of the light-induced oxidation of particular systems must be understood and in particular the species responsible for the initial absorption of radiation identified. Polypropylene is prone to photo-oxidation and a thorough investigation has recently been carried out on the mechanism of this process. The initial step in the investigation was to consider the effects of molecular order upon the rates of photo-oxidation.[60] Fractional crystallization was used to separate a commercial, predominantly isotactic unstabilized polypropylene into fractions of relatively low and relatively high atacticity, but of similar molecular weight. I.r. spectroscopy was used to follow the photo-oxidations of films formed from these fractions during irradiation in air with light of wavelengths longer than 3000 Å. I.r. absorptions at 3340 cm^{-1} (OH) and 1716 cm^{-1} (C=O) were used to detect photoproducts which were probably hydrogen-bonded hydroperoxides produced by the oxidation of tertiary C—H bonds (OH) and ketonic species (C=O). No free OH was detected in any photo-oxidized sample, and both OH and C=O absorptions were absent in unirradiated polymer. The carbonyl band stretched from 1650 to 1850 cm^{-1}, and was separable into six overlapping absorptions, from which it could be deduced that the total concentration of ketonic species was proportional to three times the ketone absorption at 1716 cm^{-1}. The build-up of photo-oxidation products with time is illustrated in Figure 12. Films of constant tacticity but varying morphology were prepared by quenching from the melt. No distinct correlation was found between sample photostability and morphology, as indicated by film density. However, films of low atactic content were found to undergo faster photo-oxidation than films of high atactic content, irradiated under identical conditions. The volatile photo-oxidation products are shown in Table 4 for a crystalline fraction of the polymer and an uncrystallized fraction. These volatile products are probably formed *via* the secondary photolysis of high-molecular-weight

---

[60] Y. Kato, D. J. Carlsson and D. M. Wiles, *J. Appl. Polymer Sci.* 1969, **13**, 1347.

carbonyl compounds. The photo-oxidation of polypropylene (RH) is believed to propagate by a radical-chain mechanism:

$$R\cdot + O_2 \longrightarrow RO_2\cdot \qquad (58)$$

$$RO_2\cdot + RH \longrightarrow ROOH + R\cdot \qquad (59)$$

The differing photo-oxidative behaviour of atactic and isotactic polypropylenes must be caused by some conformational dependence of the

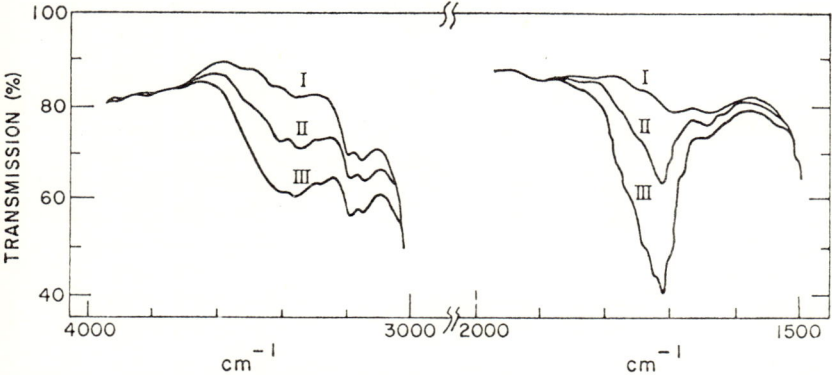

Figure 12 *Build-up of photo-oxidation products in polypropylene film.* OH $(3340\ cm^{-1})$ *and* C=O $(1716\ cm^{-1})$ *absorption after* (I) 0 h, (II) 65 h, *and* (III) 110 h *irradiation*
(Reproduced by permission from *J. Appl. Polymer Sci.*, 1969, **13**, 1347)

oxidation mechanism, although the rates of build-up of hydroperoxide are very similar in the two types of sample. Since carbonyl products mainly arise from the decomposition of hydroperoxides, the rate of formation and destruction of this species in the isotactic polymer must be higher

Table 4 *Volatile products of photo-oxidation of polypropylene*

Product	Crystallized polypropylene mol %	Uncrystallized polypropylene mol %
C_2 hydrocarbons	12	11
C_3 hydrocarbons	12	8
C_4 hydrocarbons	7	6
C_5 hydrocarbons	—	1
MeCHO	13	15
Me_2CO	41	48
Me_2CHOH	—	4
Me_2CHCHO	1	1
$MeCOC_2H_5$	1	3
MeCOOH	13	3
Total organics (μmol)	1·2	0·5

Films irradiated under identical conditions.

than that in the atactic polymer, in order to account for the overall observed higher rate of oxidation in the former. However, the differences in behaviour may arise in any of the oxidation steps, initiation, propagation, or termination, and a positive selection of the critical stereodependent step is not possible.

The production of the free radicals necessary for the chain oxidation in reactions (58) and (59) by photochemical means might arise by interaction of u.v. radiation in sunlight with carbonyl groups inadvertently introduced into commercial samples of polypropylene during processing, or by

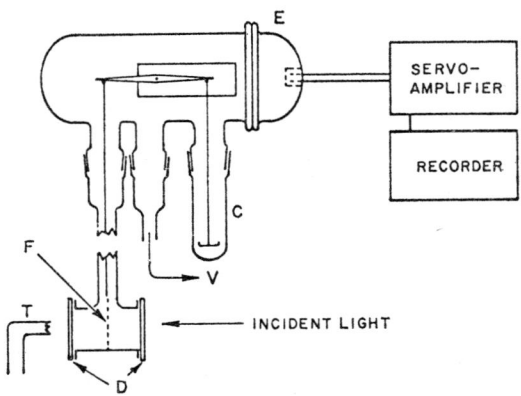

Figure 13 *Vacuum microbalance system used in study of photodegradation of polypropylene.* F = *film sample*, D = *removable quartz windows*, T = *calibrated thermopile for beam intensity measurements*, C = *balance*, E = *vacuum bottle*, V = *vacuum line*
(Reproduced by permission from *Macromolecules*, 1969, **2**, 587)

photolytic breakdown of hydroperoxides formed in reaction (59). The first possibility has been investigated thoroughly by producing polypropylene films from resin which had been extensively air-oxidized at 225 °C, and degrading these under vacuum by irradiating at various wavelengths in the region 2200 to 3800 Å.[61, 62] Changes in the film were followed by g.l.c., i.r. analysis as above, and by film weight loss using a vacuum microbalance shown in Figure 13. The i.r. spectra of the air-oxidized films and of model compounds related to polypropylene were compared, and from this consideration and the changes in i.r. spectra upon irradiation, the main oxidation products introduced during the processing of the films were shown to be most likely to be the polymeric ketones (13) and (14). The main products arising from the u.v. degradation of these oxidized films *in vacuo* are given in Table 5. Inspection of the

[61] D. J. Carlsson and D. M. Wiles, *Macromolecules*, 1969, **2**, 587.
[62] D. J. Carlsson and D. M. Wiles, *Macromolecules*, 1969, **2**, 597.

Table 5 *U.v. irradiation of oxidized polypropylene. Volatile products*

Oxidation product loss	Mol × 10^6	Mol % of total volatiles
$\left[\text{\textgreater}C=O\right]_{1726\ cm^{-1}}$	1·89	60
$\left[\text{\textgreater}C=O\right]_{1718\ cm^{-1}}$	1·36	43
Total loss	3·25	
Volatile products		
MeCOMe	1·54	49
CO	1·1	34
CO_2	0·03	0·9
MeCHO	0·21	6·6
$MeCH_2CHO$	0·01	0·4
MeCH(Me)OH	0·05	1·5
$MeCOCH_2Me$	0·03	0·9
MeCOOH	0·04	1·4
$MeCH_2COCH_2Me$	0·03	0·9
C_3H_6	0·05	1·5
C_2H_6	0·05	1·5
C_2H_4	0·02	0·6
CH_4	0·06	1·8
Total volatiles	3·17	

main products, and kinetic considerations, lead to the conclusions that the main mode of breakdown of (14) and (15) are as follows:

$$\begin{array}{c}\text{Me} \quad\ \ \ \text{O} \ \ \text{Me} \\ \sim\!\!\text{CH}_2-\underset{\underset{\text{H}}{|}}{\overset{|}{\text{C}}}-\text{CH}_2-\overset{\|}{\text{C}}-\underset{\underset{\text{H}}{|}}{\overset{|}{\text{C}}}\!\!\sim\!\!\sim + h\nu \\ (14)\end{array}$$

Scheme showing breakdown: ≤90% pathway (60) gives $\sim\!\!\text{C(Me)(H)}-\text{CH}_2^{\bullet} + \text{O}=\overset{\bullet}{\text{C}}-\text{CH}_2-\text{C(Me)(H)}\!\!\sim$, which yields $\text{CO} + \overset{\bullet}{\text{C}}\text{H}_2-\text{C(Me)(H)}\!\!\sim$.

≤10% pathway (61) gives $-\text{CH}=\text{C}(\text{Me})(\text{H})$ and $-\text{CH}_2-\text{C}(=\text{CH}_2)(\text{H})$ plus $\text{CH}_2=\text{C(OH)}-\text{CH}_2-\text{C(Me)(H)}\!\!\sim$, which yields $\text{Me}-\overset{\text{O}}{\overset{\|}{\text{C}}}-\text{CH}_2-\text{C(Me)(H)}\!\!\sim$.

$$\text{\large wwwCH}_2-\underset{\underset{H}{|}}{\overset{\overset{Me}{|}}{C}}-CH_2-C\underset{Me}{\overset{O}{\diagup}}$$
(15)

$$\xrightarrow{+h\nu}$$

(62) 15% → wwwCH$_2$–$\overset{\overset{Me}{|}}{\underset{\underset{H}{|}}{\dot{C}}}$–$\dot{C}H_2$ + Me$\dot{C}$O → MeCHO
                                          └→ ·Me + CO

(63) 85% → wwwCH=C$\overset{Me}{\underset{H}{\diagup}}$
                                         } + CH$_2$=$\overset{\overset{OH}{|}}{C}$–Me
            wwwCH$_2$–C$\overset{CH_2}{\underset{H}{\diagup}}$
                                                    ↓
                                                MeCOMe

Thus, (14) is photolysed principally *via* a Norrish type I reaction to give free radicals capable of initiating oxidation in the presence of molecular oxygen, whereas (15) dissociates primarily *via* a Norrish type II reaction in a molecular fashion. The quantum yields of dissociation of both (14) and (15) were determined as 0·08. Type II dissociation of (15) will cause the loss of a small molecule and generate unsaturation. Any type II dissociation of (14) will result in chain scission, although further photolysis of the resultant ketone (15) will not contribute to degradation. Hence, apart from the initial photolysis to give volatiles, and some chain scission, the type II reaction is unlikely to accelerate significantly the photodegradation of polypropylene under weathering conditions. Type I photolysis of (14) and (15) however, will generate radicals, and in the case of (14) cause chain scission. These alkyl radicals may combine or disproportionate and thus cause no further immediate damage, or in the presence of air may combine with dissolved oxygen to give peroxyl radicals. Radical combination of the two macroradicals formed in (60) will probably exceed $RO_2\cdot$ formation owing to the high internal viscosity of the polymer and the low dissolved O_2 concentration and O_2 permeability. However, radical combination after type I scission of (15) is unlikely, since the methyl and acetyl radicals formed can diffuse from the reaction site before recapture. Hydrogen abstraction by these species can then result in the formation of other isolated macroradicals. Hence, type I photolysis is likely to be an important contributor to polypropylene instability. The mechanism can be summarized as:

$$\text{wwwC}\underset{H}{\overset{Me}{|}}-CH_2-\underset{H}{\overset{Me}{|}}C\text{www} \xrightarrow[225\,°C]{O_2} \text{ketones (14)} \atop \text{and (15)} \xrightarrow{h\nu} (\!>\!C\!=\!O)^*$$

                                                                       ↓ ↘
                                                                           Norrish Type II
ROOH ←$_{RH}$— RO$_2^{\cdot}$ ←$_{O_2}$— R$^{\cdot}$ ←—— Norrish Type I ↓
↙$_{h\nu}$ molecular products
photo-oxidation   (64)

The mechanism above may dominate in the very early stages of photo-oxidation, but the build-up of hydroperoxides in the latter stages of (64) undoubtedly accelerates degradation owing to the photolysis of these species. The reaction has been investigated in a manner similar to that outlined for the ketonic species. The major primary step was shown to involve hydroperoxide cleavage into t-alkoxyl radicals and hydroxyl radicals, and the major volatile product was water formed from hydrogen abstraction by hydroxyl radicals. The t-alkoxyl radicals produced undergo β-scission (65) and (66) to produce ketonic products. One of these (66) leads to scission of the polymer backbone, and is probably responsible for the large drop in intrinsic viscosity which results from hydroperoxide photolysis. An experimental overall quantum yield of about 4 was found for hydroperoxide photolysis, although it was felt that the true value was probably closer to unity. Hydroperoxide photolysis by sunlight in the presence of air is believed to represent a major source of free radicals and backbone scission during the photodegradation of polypropylene. This photolysis is probably prevalent both in the very early and the advanced stages of photodeterioration of commercial polypropylene articles. Development

$$
\begin{array}{c}
 \\
\sim\!\!\sim\!\!\sim\underset{H}{\overset{Me}{C}}\!-\!CH_2\!-\!\underset{O^\bullet}{\overset{Me}{C}}\!-\!CH_2\!-\!\underset{H}{\overset{Me}{C}}\!\sim\!\!\sim\!\!\sim
\end{array}
\xrightarrow[]{\text{(65)}}
\sim\!\!\sim\!\!\sim\underset{H}{\overset{Me}{C}}\!-\!CH_2\!-\!\overset{O}{\underset{}{\overset{\|}{C}}}\!-\!CH_2\!-\!\underset{H}{\overset{Me}{C}}\!\sim\!\!\sim\!\!\sim\ +\ Me^\bullet
$$

$$
\xrightarrow{\text{(66)}}\ \sim\!\!\sim\!\!\sim\underset{H}{\overset{Me}{C}}\!-\!\overset{\bullet}{C}H_2\ +\ \underset{O}{\overset{Me}{\underset{\diagup\!\diagup}{C}}}\!-\!CH_2\!-\!\underset{H}{\overset{Me}{C}}\!\sim\!\!\sim
$$

of stabilizer systems which can completely prevent hydroperoxide photolysis will probably require the use of highly efficient energy transfer compounds capable of accepting energy from the excited hydroperoxide groups. The photo-oxidative degradation of polypropylene has also been discussed by other workers,[63] and the effect of fillers on the photo-oxidative stability of polyethylene reported.[64]

A study has been made of the photolysis of poly(methyl methacrylate) in the presence of photostabilizers of the 2-hydroxybenzophenone type, both added to a solution of the polymer and incorporated into the polymer chain.[65] The efficiency of stabilization was found to be independent of the way in which the stabilizer was added, and the additive was presumed to

[63] L. Balaban, J. Majer, and K. Vesely, *J. Polymer Sci.*, Part C, 1969, 1059.
[64] H. Schonhorn and J. P. Luongo, *Macromolecules*, 1969, 2, 364.
[65] T. Lukac, P. Hrdkovic, Z. Manasek, and D. Bellus, *European Polymer J.*, 1969, 523 (Bratislava supplement).

be acting as an u.v. absorber rather than an energy transfer agent. The photodegradation of copolymers of methyl methacrylate and methyl acrylate at elevated temperatures using 2537 Å radiation has been shown to proceed *via* a random chain scission process,[66] but the nature of the initial photochemical steps has not been investigated. Photorearrangement and photodegradation of poly(4-benzoyloxystyrene) and poly(*p*-cresylacrylate),[67] poly(*N*-dimethyl-β-aminoethyl methacrylate) in aqueous solutions,[68] and isotactic poly(methacrylic acid) copper complexes[69] have been the subject of recent reports.

The photodegradation of poly(vinyl chloride) (PVC) has been studied over the temperature range 30—150 °C, and the initiation with 2537 Å radiation correlated with the presence of minute amounts of ozone.[70] The oxidation gives rise to β-chloroketones, which decompose further by a Norrish type I process without loss of chlorine atoms. The gaseous products of PVC degradation at 30 °C were CO, CO_2, H_2, and methane. Attempts at stabilization using ferrocene and copper phthalocyanine caused acceleration of the degradation, whereas copper salicylate and commercial additives retarded the rate of oxidation. However, all additives accelerated the mechanical failure of the PVC. A stress relaxation technique has also been used to investigate the degradation of (PVC), and the following mechanism proposed to account for the chain scission and crosslinking:[71]

initiation
$$\sim\sim\sim CH_2-\underset{H}{\underset{|}{\overset{Cl}{\overset{|}{C}}}}-CH_2-\underset{H}{\underset{|}{\overset{Cl}{\overset{|}{C}}}}\sim\sim\sim + h\nu \longrightarrow \sim\sim\sim CH_2-\underset{H}{\underset{|}{\overset{\cdot}{C}}}-CH_2-\underset{H}{\underset{|}{\overset{Cl}{\overset{|}{C}}}}\sim\sim\sim + Cl^\cdot \quad (67)$$

oxidation propagation
$$\sim\sim\sim CH_2-\underset{H}{\underset{|}{\overset{\cdot}{C}}}-CH_2\sim\sim\sim + O_2 \longrightarrow \sim\sim\sim CH_2-\underset{H}{\underset{|}{\overset{O-O\cdot}{\overset{|}{C}}}}-CH_2\sim\sim\sim \quad (68)$$

scission
$$\sim\sim\sim CH_2-\underset{H}{\underset{|}{\overset{O-O\cdot}{\overset{|}{C}}}}-CH_2\sim\sim\sim + RH \longrightarrow \sim\sim\sim CH_2-\underset{H}{\underset{|}{\overset{O-OH}{\overset{|}{C}}}}-CH_2\sim\sim\sim + R^\cdot \quad (69)$$

[66] N. Grassie, B. J. D. Torrance, and J. B. Colford, *J. Polymer Sci.*, Part *A*-1, 1969, **7**, 1425.
[67] D. Bellus, P. Slama, P. Hrdkovic, Z. Manasek, and L. Durisinova, *J. Polymer Sci.*, Part C, 1969, 629.
[68] H. H. G. Jellinek and C. H. Chou, *J. Macromol. Sci. Chem.*, 1970, **4**, 255.
[69] H. H. G. Jellinek and S. N. Lipovac, *J. Polymer Sci.*, Part C, 1969, 621.
[70] K. P. S. Kwei, *J. Polymer Sci.*, Part *A*-1, 1969, **7**, 1075.
[71] W. C. Cox, D. J. Crawford, and P. L. D. Peill, *J. Appl. Polymer Sci.*, 1970, **14**, 611.

$$\text{wwCH}_2-\overset{\overset{\displaystyle OH}{|}}{\underset{\underset{\displaystyle H}{|}}{C}}-\text{CH}_2\text{ww} \longrightarrow \text{wwCH}_2-\overset{\overset{\displaystyle O^\bullet}{|}}{\underset{\underset{\displaystyle H}{|}}{C}}-\text{CH}_2\text{ww} + \text{OH}^\bullet \qquad (70)$$

$$\text{wwCH}_2-\overset{\overset{\displaystyle O^\bullet}{|}}{\underset{\underset{\displaystyle H}{|}}{C}}-\text{CH}_2-\overset{\overset{\displaystyle Cl}{|}}{\underset{\underset{\displaystyle H}{|}}{C}}-\text{CH}_2\text{ww} \longrightarrow \text{wwCH}_2-\text{CHO} + \overset{\bullet}{\text{C}}\text{H}_2-\overset{\overset{\displaystyle Cl}{|}}{\underset{\underset{\displaystyle H}{|}}{C}}-\text{CH}_2\text{ww} \qquad (71)$$

$$\begin{array}{c}\text{\{CH}_2 \\ | \\ \text{H}-\text{C}-\text{O}-\text{O}^\bullet \\ | \\ \text{CH}_2\text{\}}\end{array} + \text{R}^\bullet \longrightarrow \begin{array}{c}\text{\{CH}_2 \\ | \\ \text{H}-\text{C}-\text{O}-\text{O}-\text{R} \\ | \\ \text{CH}_2\text{\}}\end{array} \qquad (72)$$

cross-linking

The cross-linking step is favoured by increase in photon energy.

The effect of molecular weight on photocross-linking in the surface layer of a polystyrene film in a nitrogen atmosphere has been discussed.[72] The photolysis of phenoxyresin results in the evolution of hydrogen, methane, ethane, propene, propane, toluene, and the oxides of carbon. Photo-oxidation causes similar changes, but produces more oxidised chain fragments. The results can be explained by the assumption that the initial act of absorption produces random bond scission in the positions shown in reaction (73), although it seems more likely that the initial

$$(73)$$

scission will be more local. Photo-oxidation proceeds by the usual mechanism, peroxyl radical formation, followed by hydrogen abstraction.

[72] K. Kato, *J. Appl. Polymer Sci.*, 1969, **13**, 599.

Polyamides.—Titanium dioxide residues in polyamides such as Nylon 66 can cause a drastic increase in the rate of photodegradation. The mechanism of this action has recently been questioned, and proposals made.[73] The wavelength dependence of the photodegradation of TiO_2-free Nylon 66 (bright) and TiO_2-containing Nylon 66 is clearly different (Figure 14), indicating that the TiO_2 acts as a sensitizer. A possible mechanism would be absorption by the TiO_2 and subsequent crossover to the triplet state

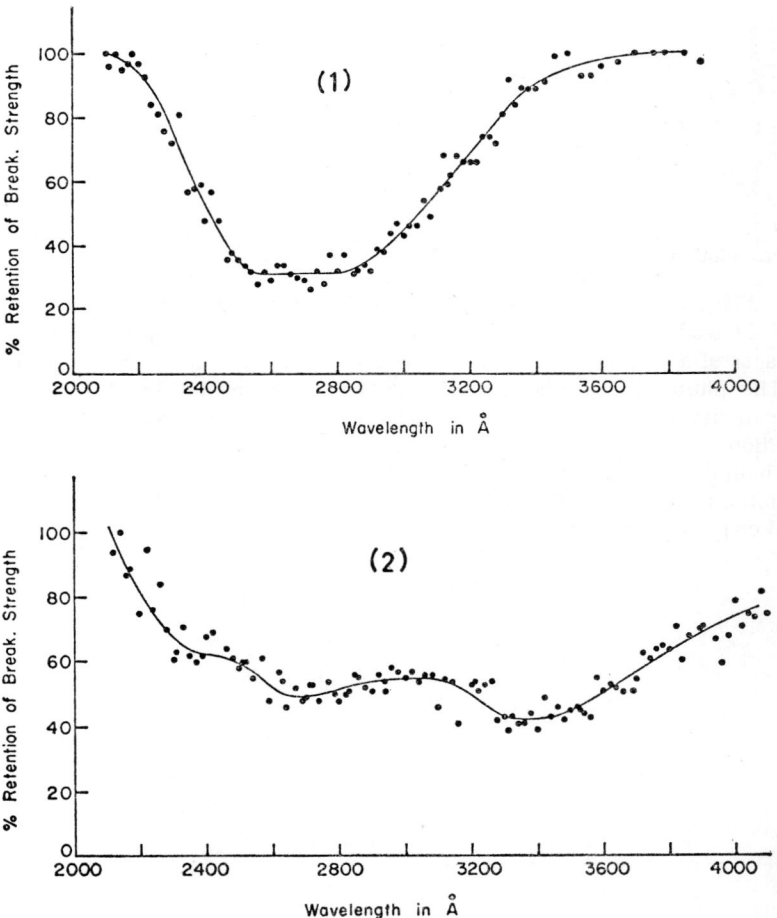

Figure 14 *Wavelength dependence of photodegradation of* (1) *bright Nylon 66* (TiO_2 *free) and* (2) *dull Nylon 66* (*containing* TiO_2 *residues*)
Reproduced by permission from *J. Appl. Polymer Sci.*, 1970, **14**, 141)

[73] P. G. Kelleher and B. D. Gesner, *J. Appl. Polymer Sci.*, 1969, **13**, 9.

followed by energy transfer to the triplet state of the polymer. Singlet transfer is not possible because the TiO_2 absorbs at longer wavelengths than the polymer. The phosphorescence spectra and excitation spectra shown in Figure 15 do not support this contention, however, since the new

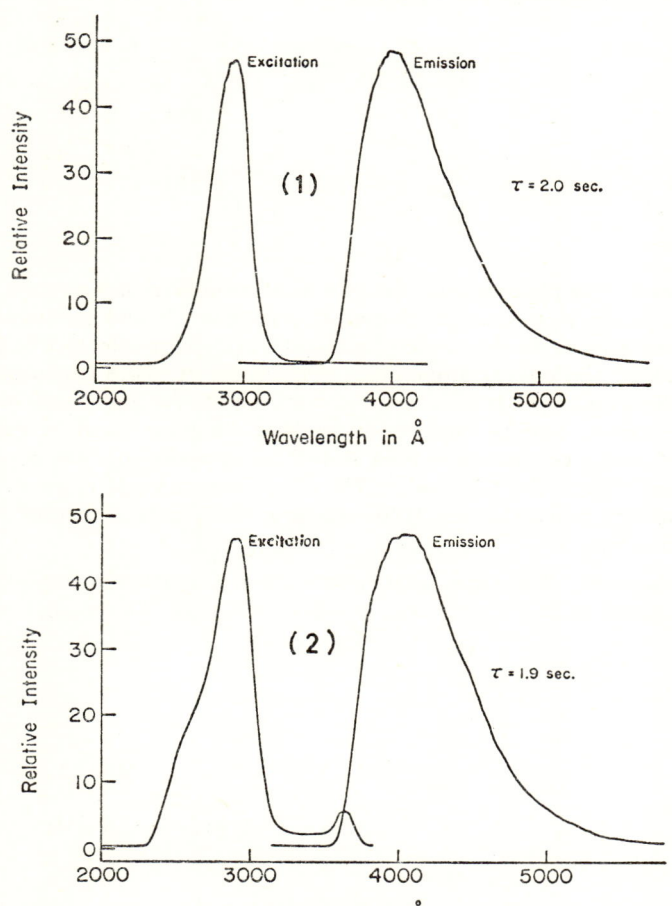

Figure 15 *Phosphorescence excitation and emission spectra of* (1) *bright Nylon* 66, (2) *dull Nylon* 66
(Reproduced by permission from *J. Appl. Polymer Sci.*, 1970, **14**, 141)

excitation band in the TiO_2-containing polymer at 2650 Å is too small to be a major contributor to the population of the polymer triplet state. The results therefore suggest that TiO_2 initiates Nylon 66 degradation by absorbing radiation and producing free radicals or peroxides which can chemically attack the polymer. Three polyamide polymers which have good thermal stability have been shown to be prone to ready photodegradation.[74] These are (16), (17), and a commercially available sample. All were degraded by light of much longer wavelengths than that which is harmful to other common polymers.

(16)

(17)

Cellulose.—The photodegradation of cotton cellulose in a pure state,[75] and in the presence of vat dyes,[76] optical brighteners,[77] and anthraquinone sulphonate sensitizer [78] has been discussed, chiefly with reference to loss of mechanical and optical properties. Mechanisms of photo-oxidation have been discussed, but these tend to be very speculative and based on little hard evidence. Surface effects have also been discussed,[79] and the degradation of acetyl cellulose has been shown to occur *via* a cleavage of the pyranose ring at the 1,2-bond.[80] The degradative process is shown to be inhibited by polyurethanes, these acting as both radical scavengers and u.v. absorbers.

Miscellaneous.—When irradiated in the vacuum u.v., poly(dimethylsiloxane) decomposed in two ways,[81] reactions (74) and (75). The methyl radicals

$$\left[\begin{array}{c}\text{Me}\\|\\-\text{Si}-\text{O}-\\|\\\text{Me}\end{array}\right] + h\nu \longrightarrow \sim\sim\sim\overset{\cdot}{\underset{|}{\text{C}H_2}}\text{Si}-\text{O}\sim\sim\sim + \text{H}^{\cdot} \quad (74)$$

$$\longrightarrow \sim\sim\sim\overset{\cdot}{\underset{|}{\text{Si}}}-\text{O}\sim\sim\sim + {}^{\cdot}\text{Me} \quad (75)$$

[74] H. A. Taylor, W. C. Tincher, and W. F. Hamner, *J. Appl. Polymer Sci.*, 1970, **14**, 141; L. D. Johnson, W. C. Tincher, and H. C. Bach, *ibid.*, 1969, **13**, 1825.
[75] A. H. Reine and J. C. Arthur, *Textile Res. J.*, 1970, **40**, 90.
[76] P. J. Baugh, G. O. Phillips, and N. W. Worthington, *J. Soc. Dyers and Colourists*, 1970, **86**, 19.
[77] W. B. Achwal and R. B. Chavan, *Indian J. Technol.*, 1970, **8**, 15.
[78] P. J. Baugh, G. O. Phillips, and N. W. Worthington, *J. Soc. Dyers and Colourists*, 1969, **85**, 241.
[79] R. L. Desai and J. A. Shields, *J. Colloid Interface Sci.*, 1969, **31**, 585.
[80] O. P. Kozmina, V. P. Dubyaga, V. K. Belyakov, and N. A. Zaichukova, *European Polymer J.*, 1969, 447 (Bratislava Supplement).
[81] S. Siegel and T. Stewart, *J. Phys. Chem.*, 1969, **73**, 823.

formed may combine with one another or abstract hydrogen from the parent polymer, so that the surface irradiation gave rise to ethane, methane, and molecular hydrogen as gaseous products. In the bulk only methane and hydrogen were formed. Solutions of a phenyl-containing siloxane gave reduced yields of volatile products because of radical addition to the phenyl group, as well as intermolecular energy transfer between the two polymers. There was no evidence of long-range transfer down a chain. Poly(methylsiloxane) irradiated between 3000 and 4000 Å gave volatile products similar to those above, and produced evidence of the formation of Si–CH_2–Si linkages.[82] With higher energy radiation from a mercury arc, Si–OH and Si–CH_2–CH_2–Si linkages were formed. An explanation is given which involves the excitation of molecular oxygen in the latter experiments.

Photostabilizers.—As has been pointed out in a recent review,[83] a single photon can cause considerable degradation of polymeric substrates, particularly when atmospheric oxygen is present. The species absorbing radiation may be situated in the polymer backbone or may be an adventitious impurity as has been described above. The prevention of degradation may be achieved in the following ways:

(*i*) Prevention of light absorption using u.v. absorbers.

(*ii*) Deactivation of initially absorbing species either by chemical quenching or electronic energy transfer.

(*iii*) Inhibition of the propagation of autoxidation. Generally reactive hydroperoxyl radicals are transformed into harmless radicals by 'antioxidants'.

(*iv*) Prevention of initiation of secondary autoxidation chains by hydroperoxides generally by the addition of synergists which ionically decompose them, or by suppressing the metal catalysis of the hydroperoxides by the addition of metal deactivators.

The first two types are of photochemical interest. The characteristics of an ideal u.v. absorber are as follows:

(*i*) The molecule should absorb all of the u.v. region of the spectrum, but none of the visible. In practice this is an impossible requirement and thus if absolute freedom of colour is required, a certain transparency in the u.v. is inevitable, or conversely, if good absorption in the near u.v. is necessary, as in Nylon 66 say, a yellow discolouration of the polymer must be accepted.

(*ii*) The u.v. absorber should not be photoactive and ideally would decay to its ground state *via* harmless radiationless transitions, converting the electronic energy to heat. For this reason, fluorescent materials

---

[82] A. D. Delman, M. Landy, and B. B. Simons, *J. Polymer Sci.*, Part A-1, 1969, 7, 3373.
[83] H. J. Heller, *European Polymer J.*, 1969, 105 (Bratislava Supplement).

such as optical brightening agents may not be good stabilizers, despite their having good characteristics as far as absorption profile is concerned.

(*iii*) The absorbers must be compatible with the polymer they are going to protect and must be in true solution. This condition is difficult to fulfil in highly crystalline polymers, where the additive tends to

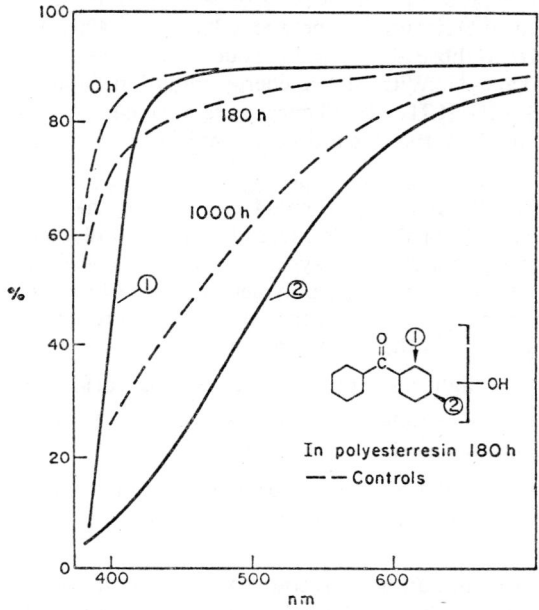

Figure 16 *Transmission of polyester resin plates containing hydroxybenzophenones* (0·25 per cent) *after* 180 h *Fadeometer irradiation.* o-*Hydroxybenzophenone* (1); p-*hydroxybenzophenone* (2); *controls after irradiation of indicated duration* (– –)
[Reproduced by permission from *European Polymer J.*, 1969, 105 (Bratislava Supplement)]

crystallize in the amorphous phase of the polymers, causing poor coverage.

(*iv*) A high permanence of absorption is required, but is difficult to meet, since the fastness properties are much more stringent than with normal textile dyestuffs.

(*v*) A variety of other specific conditions must be met depending on the polymer and its uses, such as thermal stability, resistance to sublimation, washing, dry-cleaning, *etc.*, and most important, the additive must not be toxic.

The chemical classes of absorbers which satisfy the conditions above to be in general use are as follows:

Carbonyl Oxygen Acceptors. Salicylates were first used,[83] and alkylated derivatives are still available, such as *p*-t-butylphenylsalicylate and *p*-t-octylphenylsalicylate. These are used infrequently because of low initial absorption. Polypropylene has recently been stabilized with salicylidene-2-amino-phenol.[84] Hydroxybenzophenones are still used widely, and a demonstration of the different protection given by the *o*- and *p*-isomers is given in Figure 16. The difference is striking, since the *ortho*-compound is a good stabilizer, whereas the *para*-compound actually acts as a photosensitizer for photodecomposition of the polymer. Such compounds may be of use in developing photodisposable polymeric items, for which there is a great current need. Successful stabilisers of the *o*-hydroxybenzophenone type are shown in Table 6.

Table 6 *U.v. absorbers of o-hydroxybenzophenone type*

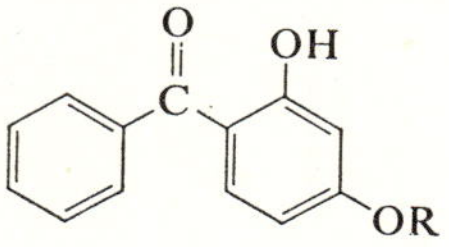

where R = H, Me, C_8H_{17}, $C_{10}H_{21}$, $C_{12}H_{25}$

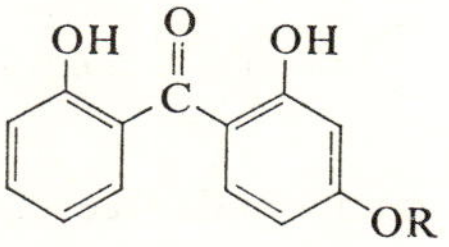

where R = Me, C_8H_{17}

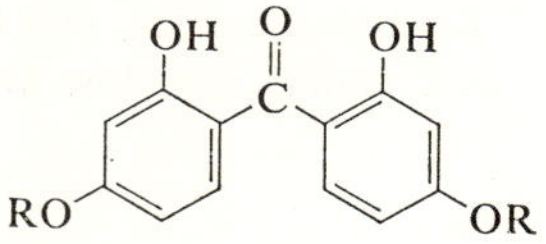

where R = H, Me

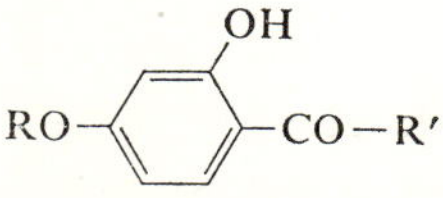

where R, R' are alkyl groups*

* From Yu. A. Gurvich, Yu. B. Zimin, L. S. Solodar, E. N. Mateeva, A. A. Kozodoi, P. I. Levin, and L. M. Lugova, *Plast. Massy.*, 1969, **69**, 30; U. B. Zimin, P. I. Levin, E. A. Mateeva, A. A. Kozodoi, and L. M. Sotnikova, *Plast. Massy.*, 1970, **70**, 20.

Absorbers with Nitrogen Groups.[83] Open-chain derivatives such as Schiff bases, hydrazones, and azines of salicylaldehyde have been proposed, but suffer from discolouring of the polymer host. Ring-compounds of the type (18) absorb in the correct region, but give little or no protection. Benzo-triazoles (19) form a large group which is being actively developed.[83]

---

[84] D. A. Akhmedzade, V. D. Yasnopol'skii, and Yu. I. Golovanova, *Azerb. khim. Zhur.*, 1969, 42.

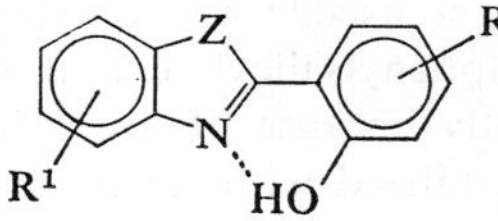

(18) where Z = O, S, NR³

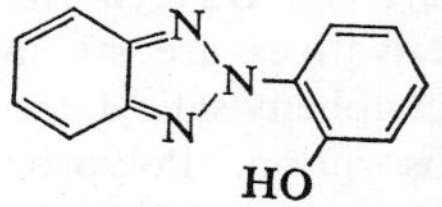

(19)

All *o*-hydroxyphenyl u.v. absorbers have their mode of action based on reaction (76). Unfortunately, this reaction is sensitive to alkali and heavy metal ions. In plastic applications this is not a severe limitation, but in

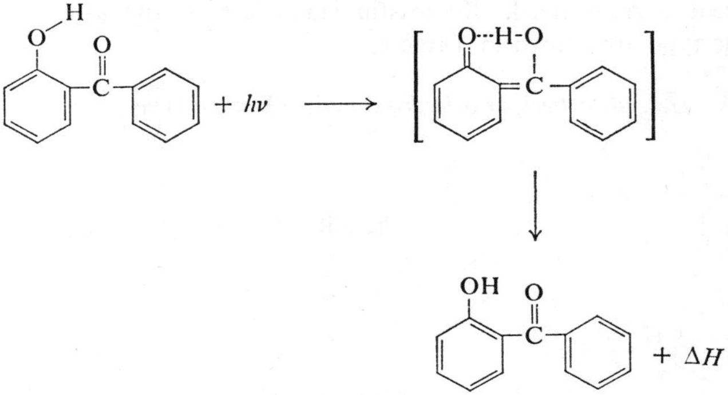

(76)

textiles it is, and research is under way for compounds free of phenolic groups for use in this area. Only one group has been discovered so far, which are the cinnamic acid derivatives (20),[83] discussed earlier, and their partially aliphatic analogues (21). However, these suffer from low absor-

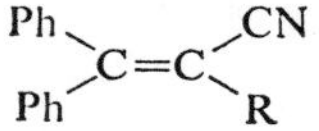

$$\underset{Ph}{\overset{Ph}{>}}C=C\underset{R}{\overset{CN}{<}} \quad \text{where R = CN, COOEt}$$
(20)

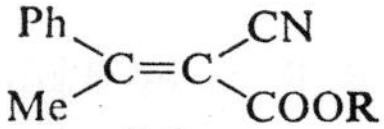

$$\underset{Me}{\overset{Ph}{>}}C=C\underset{COOR}{\overset{CN}{<}} \quad \text{where R = Me, Bu}$$
(21)

bancy and compounds have yet to be found which combine the alkali stability and high absorbance which will make them commercially desirable. Light stabilizers based on a quenching mechanism are much less common. The accidental discovery of the effect of nickel chelates in preventing photodegradation in polyolefins has led to their use in these polymers. Some of the structures of these additives are given in Table 7. The mode of action of these compounds may be short range. Recently there has been an upsurge of interest in the possibility of stabilization of polymers by an efficient long-range Forster-type of energy transfer, and several papers have been alluded to in this section concerning energy transfer in poly-

Table 7 *Nickel chelate photostabilisers*

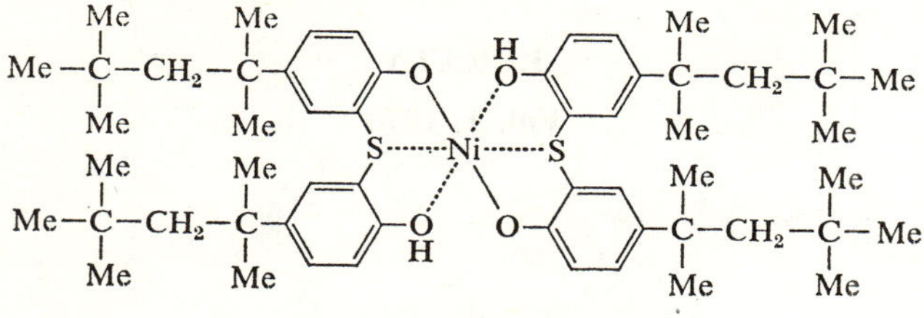

Ferro AM 101 (Ferro Corp.)

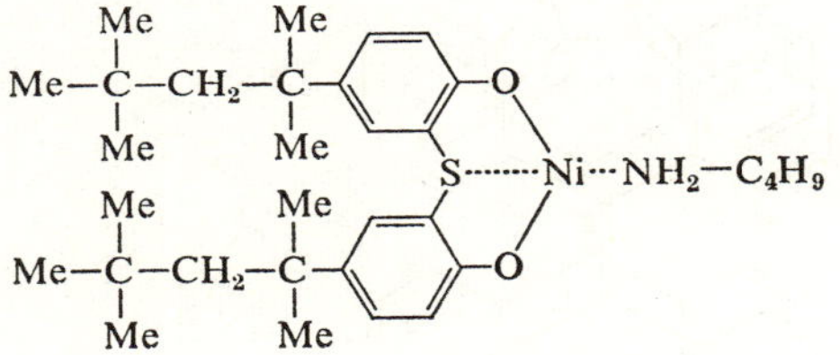

Cyasorb 1084 (Cyanamid Corp.)

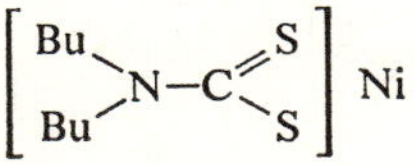

Nickeldibutyldithiocarbamate

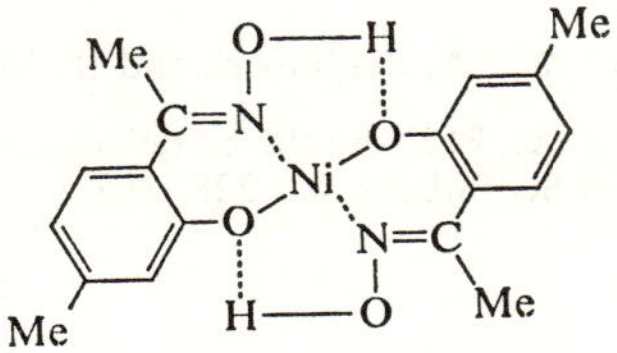

Negospex A (I.C.I.)

mers.[25, 26] Energy transfer in ethylene–carbon monoxide polymers,[85] and excitation energy transfer in alkane polymers [86] have been described, but as yet no efficient photostabilizer based on this principle has appeared commercially. This may in part be due to the fact that choice of energy transfer acceptor molecules depends upon a precise knowledge of the energy level and nature of the excited state in the polymer primarily responsible for degradation, and studies have not generally been carried out along these lines. Even in the few cases where a mechanism has been studied in detail, such as in polypropylene described earlier,[61, 62] it is evident that more than one donor can be important, and this choice of acceptor may be difficult. Nevertheless, because the transfer can be a long-range effect, such stabilizers could be very effective in small concentrations, and much work is continuing to enable this phenomenon to be usefully employed in stabilization in the future.

[85] M. Heskins and J. E. Guillet, *Macromolecules*, 1970, **3**, 224.
[86] R. H. Partridge, *J. Chem. Phys.*, 1970, **52**, 2501.

ERRATA
Vol. 1, 1970

Page 228. Formula (7) should have appeared as:

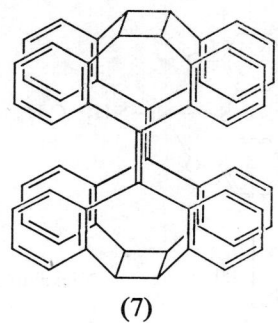

(7)

Page 315. Reference 25 should *read* A. Schönberg and R. Moubasher.

Page 467. The entry 'Schöberg, A., 315' should be deleted, and the last entry on the page should *read* Schönberg, A., 228, 315

Author Index

Abbott, S. R., 134
Abe, K., 614
Abell, P. I., 203
Abraitys, V. Y., 409
Abram, I. I., 709
Abramovitch, R. A., 737
Acheson, R. M., 644
Achwal, W. B., 794
Ackerman, J. P., 128
Ackerman, R. A., 223, 224, 610
Ackermann, F., 5
Acton, N., 467
Adam, K. R., 140
Adamson, A. W., 243, 248, 250, 256, 270
Addiss, R. R., 294
Adelman, A. H., 598
Adler, H., 781
Adler, S. E., 45
Adolf, P. K., 203
Adrian, F. J., 34
Agata, I., 699
Agosta, W. C., 141, 300, 360, 371, 385
Ahlgren, C., 380
Akagi, K., 616
Akermark, B., 380
Akhalkatsi, E. G., 145
Akhmedzade, D. A., 797
Aktipis, S., 577
Albrecht, A. C., 12, 118
Aleksandrov, P., 140
Aleshina, G. F., 203
Alexander, E., 318
Alfimov, M. V., 128
Algar, B. E., 134, 225, 604
Alkaitis, A., 621
Allen, E. R., 196, 224
Allen, G. C., 251, 265, 269
Allen, W. C., 241
Allnutt, L. A., 59
Allred, E. L., 714
Almy, G. M., 490
Al-Sader, B. H., 711
Amichai, O., 223, 240
Amin, J. H., 661
Anan, F., 600
Anand, N., 331
Anastassiou, A. G., 462, 467, 468, 472
Anderson, C. G., 593
Anderson, C. M., 753
Anderson, D., 141, 491
Anderson, E. M., 158
Anderson, L. G., 103, 190
Ando, N., 768
Ando, W., 342, 707, 708
Andre, J. C., 628

Andrushchenko, D. A., 777
Anet, R., 543
Anisimov, V., 773
Anisuzzaman, A. K. M., 700
Anwaruddin, Q., 761
Aoyagi, R., 308
Aoyama, H., 225, 546
Appl, M., 650
Appleyard, G. D., 648
Arai, H., 631
Arai, T., 600
Ardemagni, L., 441
Aristov, A. V., 98
Armstrong, D. A., 203
Arnold, D. R., 199, 347, 486, 501, 728
Artemev, E. V., 135
Arthur, J. C., 595, 794
Arvis, M., 223
Asai, M., 768
Ashpole, C. W., 170
Askani, R., 477
Asprey, L. B., 280
Assaad, N. E. N., 595, 766
Astier, R., 8, 118
Atkinson, R., 48
Atkinson, R. S., 612
Attidge, C. J., 463
Aurich, H. G., 653
Ausloos, P., 173, 179
Austel, V., 170
Avakian, P., 97
Avery, H. E., 190
Avigail, I., 108
Avila, W. B., 479
Avram, M., 750
Ayer, F. W., 168
Azumi, T., 120, 128

Babad, E., 378
Babenko, S. D., 98
Babko, A. K., 105
Bach, F. L., 656
Bach, H., 668
Bach, H. C., 794
Bachmann, W. E., 566
Back, R. A., 214, 744
Bacon, D., 731
Baddow, R. F., 153
Baessler, H., 128
Bagdasar'yan, K. S., 149
Bagger, S., 259
Baggiolini, E., 368
Baherel, Y., 370
Baird, N. C., 7

Baisley, V. C., 223
Bajdala, H., 635
Baker, R. T. K., 226
Baker, S. J., 463
Balaban, L., 789
Balaceanu, G., 706
Balasubramanian, A., 640
Balbo, S. A., 17
Baldwin, J. E., 321, 415, 451, 454
Baldwin, S. W., 321
Baliah, V., 739
Balling, L. C., 283
Balny, C., 106
Balogh, G., 589
Balzani, V., 262
Ban, G., 291
Banfield, D. L., 140
Banks, E., 290
Banks, R. E., 698
Barale, E., 676
Barat, F., 240
Barber, L. L., 372, 373
Barbucci, R., 266
Bard, A. J., 128
Barker, P. G., 146, 205
Barland, D. M., 33
Barlolo, R. D., 294
Barlow, M. G., 170, 171, 716
Barltrop, J. A., 305, 317, 337, 593
Barnes, R. L., 545
Barnes, W. J., 552
Bartholomew, R. F., 703
Bartlett, P. D., 610, 699, 714
Bartocci, C., 262, 264
Barton, D. H. R., 612, 620, 752
Barton, T. J., 562
Baruah, G. D., 108
Bass, A. M., 226, 287
Basselier, J. J., 346, 727
Bastiani, R. J., 401
Batekha, I. G., 128
Bathelt, W., 266
Battisti, A., 374, 694
Bauer, D., 364
Bauer, E., 230, 282
Bauer, S. H., 171
Baugh, P. J., 595, 794
Baum, A. A., 430
Baum, E. J., 14, 199
Baur, G., 289
Baybarz, R. D., 140
Bayes, K. D., 222
Bayliss, N. S., 17
Baylor, C., 746, 749

Bazov, V. P., 669
Beak, P., 652
Beardslee, R. A., 111
Becker, B. S., 146
Becker, H. D., 341, 411, 412, 575, 584
Becker, R. S., 145, 433, 434, 485, 670
Bedford, C. T., 404
Beekhuis, G. E., 558
Bell, S., 5
Bellus, D., 340, 363, 573, 789, 790
Belyakov, V. K., 794
Belyi, M. V., 276
Benasson, R., 154
Bender, C. O., 465
Benderskii, V. A., 98
Benham, J., 360
Ben-Hur, E., 397, 755
Benicke, D., 149
Bennett, R. J. M., 33
Bentrude, W. G., 336
Berces, T., 284, 745
Berchtold, G. A., 331, 352, 692, 694, 733
Berdnikov, V. M., 268
Berenfeld, V. M., 135
Beres, L. S., 249
Berg, R. A., 118
Bergmann, E. D., 534
Bergmark, W., 586, 592
Bergson, G., 694
Berkley, R. E., 209, 710
Berlman, I. B., 102
Berlmann, L. B., 155
Bernardi, G. C., 695
Bernstein, H. J., 203
Berry, D., 698
Berry, J. W., 443
Bersohn, R., 171
Berson, J. A., 480, 713, 714
Berthelot, J., 605, 616
Bertoniere, N. R., 485
Beugelmans, M. R., 552
Beugelmans, R., 350, 434, 435, 662
Beutel, J., 582, 583
Bew, M. J., 266
Beyer, K. D., 206
Bezrogova, E. V., 280
Bhandari, K. S., 701
Bhatnagar, A. K., 321
Bhawalker, D. R., 290
Bhujle, V. V., 120
Bibart, C. H., 131, 190
Biersmith, E. L., 307
Bigam, G., 468
Billing, D. E., 266
Bindra, A. P., 481
Binkley, R. W., 464, 481, 643, 719, 721, 723, 724
Binkley, W. W., 481, 724
Binks, A. E., 775
Birang, B., 294
Birks, J. B., 8, 545
Birss, F. W., 128
Bishop, R., 419
Bist, H. D., 10
Bixon, M., 20, 21, 26, 44

Black, C., 74
Black, E. D., 71
Black, G., 230
Black, K. D., 206
Black, L. L., 729
Black, R. A., 746
Blackburn, B. J., 286
Blackburn, E. V., 144, 471, 526
Blackman, A. J., 655
Blackwell, J. E., 319
Blasse, G., 291, 293, 294
Bleekrode, R., 231, 282
Blethen, M. L., 46
Bloch, J.-C., 624
Block, E., 755
Bluhm, A. L., 621, 669
Blum, J., 534
Blyumenfeld, L. A., 240
Boden, J. C., 214, 744
Boer, F. P., 383
Bogdan, D., 706
Bogoyavlenskaya, E., 773
Bohning, J., 582
Boie, I., 741
Boikess, R. S., 148, 479
Boire, B. A., 392
Bokobza, A., 118
Boll, W. A., 655
Bolon, D. A., 780
Bolotnikova, T. N., 126
Bonneau, R., 154
Borch, R. F., 708
Borden, G. W., 646
Borden, W. T., 472
Borer, C., 615
Borisevich, A. N., 154
Borrel, P., 179, 180
Bortner, M. H., 135
Bortolus, P., 533
Bosco, S. R., 78, 744
Bose, A. K., 412
Bost, R. O., 485
Bouas-Laurent, H., 545
Bouchy, M., 628
Boue, S., 229, 447, 450, 554
Boulon, G., 294
Bowen, E. J., 598
Bowers, P. G., 48, 208
Bowie, R. A., 506
Bowman, R. M., 473, 500
Boyd, D. R., 630
Boyd, M. R., 560
Boyd, R. K., 197
Boyer, J. H., 718, 729
Boyle, J. W., 238
Boyle, P. H., 317, 383
Bozzato, G., 369
Bradley, J. N., 746
Bradow, R. L., 519
Bradshaw, J. S., 541, 549, 551
Bragin, O. V., 154
Brandmueller, J., 60
Bratolyubov, A. S., 203
Brauer, G. M., 165
Braun, A. M., 131
Braun, W., 59, 226
Breckenridge, W. H., 70, 218

Breitenbach, J. W., 469
Bremner, J. B., 753
Brennan, J. F., 582, 583
Breslow, R., 320, 321
Breuer, G. M., 155
Brewer, T. L., 285, 749
Bridge, N. K., 595
Brie, A., 108
Briegleb, G., 106, 128, 132
Bril, A., 291, 293, 294
Brinen, J. S., 12, 128
Brith, M., 8, 9
Broadbent, A. D., 224
Brockmann, H., 604
Brody, S. S., 106
Broida, H. P., 67, 207, 283
Bronstert, B., 400
Brophy, J. J., 443
Brown, G. P., 150
Brown, H. C., 394
Brown, J. E., 415
Brown, P. R., 740
Brown, T. L., 694
Brown, W. G., 491
Brownstein, S., 161, 495
Bruce, J. M., 414
Brun, F., 398
Bruner, E. C., jun., 59
Brus, L. E., 282
Bryce, J. P., 541
Bryce-Smith, D., 51, 199, 285, 340, 494, 495, 496, 497, 498, 499, 506, 611, 754
Buchardt, O., 630, 632, 633, 634
Bucheck, D. J., 394, 395
Büchi, G., 389
Bücking, H. W., 707
Buller, E., 278
Bumgardner, C. L., 678
Bunbury, D. L., 418, 579
Bunce, N. J., 593
Burak, I., 241, 700
Burger, M., 453
Burich, I., 266
Burka, L. T., 560, 561
Burkhart, R. D., 710
Burkoth, T. L., 419, 462
Burland, D. M., 128
Burnelle, L., 543
Burnelle, L. A., 8
Burnett, J. L., 140
Burns, G., 203
Bursevitch, V. V., 276
Bursey, M. M., 380
Burshtein, K., 773
Burton, C. S., 38, 74, 155, 169, 199, 490
Busch, C. E., 45, 46, 76
Buschmann, H. W., 149
Butietynska, K., 279
Buxton, G., 239
Byers, G. W., 515
Bystitskii, C. I., 104

Cadle, R. D., 224
Cadman, P., 47, 209, 711
Cadogan, K. D., 118
Cala, F. R., 81

Caldwell, R. A., 108, 145, 315, 427, 445, 481, 566, 752
Calenbrandez, D. P., 576
Callear, A. B., 67, 69, 70, 205, 285
Callis, J. B., 61
Calvert, J. G., 219, 220, 221, 668
Calvin, M., 621, 624
Camaggi, G., 492, 716
Camassei, F. D., 252, 256
Camerman, A., 396
Camerman, N., 396
Campbell, A. C., 775
Campbell, J. M., 228, 229
Campbell, T. C., 502
Carapellucci, P. A., 566, 567, 581
Cardillo, M. J., 171
Cargill, R. L., 309, 383
Carless, H. A. J., 337
Carlson, G. L. B., 432
Carlson, K. D., 290
Carlson, R. G., 307, 367
Carlsson, D. J., 784, 786
Carmichael, H., 678
Carnahan, J. C., 467
Carnduff, J., 392
Carr, D. O., 601
Carr, R. W., jun., 195, 196
Carrington, T., 59
Carson, R. S., 325
Cartwright, C. J., 5
Carty, D. T., 482
Case, W. A., 114
Caserio, M. C., 735
Cashmore, P., 179, 180
Casida, J. E., 415, 620
Caspari, G., 218, 731
Cassal, J.-M., 638
Cassidy, H. G., 131
Castellan, A., 545
Castellano, E., 223
Castro, G., 113, 128
Catalano, A. W., 657
Catteau, J. P., 650
Cauzzo, G., 533
Cava, M. P., 533
Cavalieri, E., 394, 672
Cellura, R. P., 467, 468
Cerutti, P., 505, 588, 590, 591
Cesani, F. A., 207
Cevidalli, C., 492
Chachaty, C., 148
Chadha, V. K., 438, 663
Chakrabarti, S. K., 15, 110
Chalvet, O., 10, 604
Chambers, R. W., 132, 598
Chandross, E. A., 111, 542, 543, 545
Chang, D. W. L., 682, 684
Chang, H. W., 229
Chang, J., 111
Chao, H. M., 563, 566, 567
Chapman, O. L., 138, 301, 303, 362, 372, 373, 543, 642, 646
Charifi, M., 616

Charlier, M., 397
Charney, E., 120, 140
Chatalic, A., 283
Chaudhuri, M. K., 8
Chavan, R. B., 794
Chaykovsky, M., 734
Chen, C. T., 762
Chen, S. C., 682
Ch'en, S. Y., 283
Chen, Y. L., 620
Cheng, H. M., 415
Cheng, K. F., 411
Cherednichenko, A. E., 290
Chernyavskii, C. V., 777, 779
Cherton, J. C., 346, 727
Cheung, J. J., 468
Chiang, A., 250
Chibisov, A. K., 105, 239
Chien, J. C. W., 599, 628
Childs, R. F., 454
Chin, C. G., 468
Chizhikova, Z. A., 103
Chizhyunaite, E. U., 779
Cho, S., 618
Chong, S. L., 81
Chopin, C. M., 99
Chou, C. H., 790
Chow, L. C., 735
Chow, Y. L., 620, 621, 665, 682, 684
Christie, J., 112, 541
Chuang, T. C., 418
Chuang, T. T., 579
Chumaevskii, E. V., 135
Chutjian, A., 44, 203
Ciabattoni, J., 359
Cialdi, C., 266
Ciamician, G., 513
Cicolella, A., 628
Ciganek, E., 465
Cilento, G., 739
Cirillo, A., 78, 744
Claeson, G., 694
Clarc, M., 222
Clark, I. D., 223
Clark, L. B., 6, 733
Clark, R. F., 780
Clark, V. M., 481, 522
Clark, W. D. K., 126
Clarke, G. M., 467
Clarke, R., 668
Clarke, R. H., 118, 128, 132
Clifton, E. D., 492
Closs, G. L., 79, 80, 573, 655
Closs, L. E., 80, 573
Clough, S., 484, 654
Clovis, J. S., 656
Clyne, M. A. A., 230
Cocivera, M., 706
Cocker, W., 325
Cohen, E., 387, 388
Cohen, J. I., 563, 565, 593
Cohen, M. D., 544
Cohen, S. G., 563, 565, 566, 567, 568, 577, 593
Coillot-Demay, M. F., 290

Cole, R. S., 699
Cole, T., 420
Colford, J. B., 790
Colin, R., 5
Collier, J. R., 352, 557, 560
Collier, S. S., 219, 220
Collin, G. J., 173
Collin, P. J., 474, 501, 503, 559
Collins, M. A., 679
Collins, P. M., 344
Colon, C. J., 684
Comes, F. J., 171
Compaignon de-Marcheville, H., 435
Compton, L. E., 47, 48, 217
Comtet, M., 181, 430
Conia, J. M., 359
Connett, B. E., 497
Connolly, J. S., 591, 680
Conrow, K., 425
Conte, J. C., 140
Contneanu, M., 706
Conway, E. J., 290
Conway, J. G., 121
Cook, A., 490
Cooke, R. S., 483, 695
Cookson, R. C., 436, 437, 451, 493, 684
Coomber, J. W., 196
Coon, J. B., 207
Cope, A. C., 476
Corey, E. J., 734, 755
Cornatzer, W. E., 596
Cornelisse, J., 507
Cornelius, J. F., 76
Cornubert, R., 392
Cosgrove, M. M., 679
Costa, S. M. de B., 437, 451, 493, 684
Cotter, R. J., 368
Couch, D. A., 276
Courtot, P., 452
Couture, A., 558
Covicera, M., 80
Cowan, D. O., 430, 474
Cox, R. A., 193
Cox, W. C., 790
Coxon, J. A., 230
Coyle, J. D., 305, 317
Cozzens, R. F., 772
Crabbe, P., 317, 383
Craft, L., 340
Craig, D. P., 13, 544
Craig, G. W., 484
Crandall, D. I., 600
Crandall, J. K., 145, 309, 371, 487
Crawford, D. J., 790
Crawford, J. W., 383
Crawford, R. J., 711
Creagh, L., 743
Creed, D., 414, 613
Crein, F., 168
Crellin, R. A., 469
Cremer, S. E., 371
Criegee, R., 477
Cristol, S. J., 442, 465
Cromwell, N. H., 646

Crosby, D. G., 346, 517, 518, 690, 726
Cross, R. F., 287
Crowley, K. J., 444
Crozet, M. P., 673
Crumrine, A. E., 450
Cruz, A., 317, 383
Cuesten, H., 58
Cundall, R. B., 140, 141, 490
Cuppen, T. J. H. M., 526
Curtis, H. C., 113
Cvetanovic, R. J., 155, 161, 216, 222, 495
Czaja, J., 265
Czapskii, G., 240

Dacey, J. R., 178
Daha, M. R., 460
Dale, D. M., 375
Dalton, D. R., 705
Dalton, J. C., 97, 103, 300, 306, 337, 338, 489
Daly, J. W., 630
Damon, E., 220
Danielson, J. D. S., 61
Danner, B., 638
Darnall, K. R., 336
Darwent, B. de B., 200, 207
Datta, P., 148, 469
Dauben, W. G., 311, 323, 325, 367, 371, 448, 577
Dauchot, J. P., 140
Daudel, R., 604
Davenport, J. E., 178
David, C., 773
Davidson, R. S., 108, 351, 546, 563, 564, 566, 576
Davies, A. K., 417, 594, 595
Davis, C. B., 389, 755
Davis, D. D., 226
Davis, G. A., 566, 567
Dawes, K., 414
Dawson, W. R., 95, 96
Day, P., 259
Dearman, H. H., 722
Deb, S. K., 296
de Boer, T. J., 664, 677
Debono, M., 364, 375
De-Carlo, V. J., 230
Deckelmann, K., 259
Deflorin, A. M., 612
De Glise, X., 628
De Groot, A., 420
de-Gunst, G. P., 606
De Heer, J., 9
Dehler, J., 15
de Jongh, R. O., 508, 726
Dekker, E. E. J., 664
Delman, A. D., 795
Delosh, R. G., 139
Delzenne, C. A., 781
De-Marcheville, H. C., 350
Demare, G. R., 182
Demarteau, W., 773
de Mayo, P., 380, 383, 404, 661
DeMember, J. R., 138, 140
DeMeyer, D. E., 143, 181

De More, W. B., 223
Dempster, C. G., 111
Dempster, C. J., 542, 543
Denison, A. B., 117
de Remo, M., 392
Desai, R. B., 741
Desai, R. D., 108
Desai, R. L., 794
Desalos, J., 148
Deschamps, P., 283
De-Schryner, F. C., 589
Descotes, G., 370
Deshpande, R., 497
Deters, D. W., 362
Detrano, R., 543
Detsina, A. M., 417
Deutschman, R. J., 443
Dev, V., 655
Devadze, L. V., 145
Deviny, E. J., 323, 577
Dewar, D., 376, 585
Dewar, M. G. S., 132
Dezwaan, J., 694
Dheer, M. K., 154
Dhingra, R. C., 99
Dicke, F., 604
Diebert, C., 333, 733
Dieson, R. S., 45
Dietrich, H., 707
Dietrich, M. A., 695
Dilling, W. L., 383, 475
Din, Z. U., 418
Dirania, M. K. M., 557
Dittrich, B., 479
Dixon, C. J., 218, 695
Dixon, R. N., 241
Dmitriev, M. K., 239
Dodonova, N. Y., 102, 154
Doepker, R. D., 175, 182
Döpp, D., 302
Doering, W. von E., 650, 727
Dolbier, W. R., 632, 655
Domb, S., 369
Dombrowski, G. L., 146
Do-Minh, T., 486
Donovan, R. J., 218, 219, 231
Dopper, J. H., 474, 475, 559
Dorer, F. H., 47, 209, 712
Dorough, G. D., 624
Dorschelen, W., 652
Dorst, W., 726
Dothan, Y., 26
Douek, M., 661
Doumaux, A. R., 607
Dowd, P., 343
Doyle, L. C., 216
Dressler, K. P., 8
Driska, R. L, 474
Droste, W., 378
Drozdzynski, J., 279
Dubrin, J., 47, 217
Dubyaga, V. P., 794
Duckworth, A. C., 621
Dudley, R. J., 266
Dürr, H., 343, 372, 471, 480, 703, 715, 720
Dultz, W., 290

Duncan, A. B. F., 185
Duorlent, M., 397
Dupont, R., 604
Dupuy, A. E., 362
Durham, J. L., 168, 491
Durisinova, L., 790
Durocher, G., 10, 115
Durr, V., 702
Dwivedi, C. P. D., 15
Dybas, R. A., 533
Dym, S., 34
Dzantiev, B. G., 216
Dzaparidze, K. G., 145
Dzhagatspanyan, R. V., 135

Eaborn, C., 341, 575
Eastman, J. W., 90
Eastmond, E. C., 769
Eaton, D. R., 582
Eaton, P. E., 393
Eby, J. M., 750
Eckell, A., 656
Eckhard, I. F., 465
Edamura, T., 133
Eder, T. W., 195, 196
Edman, J. R., 457
Edwards, J. A., 317, 383
Edwards J. O., 740
Edwards, M. G., 230
Edwards, O. E., 612
Efros, L. S., 742
Egawa, A., 148
Ege, G., 213, 700
Ege, S. N., 360, 653
Egerton, G. S., 299, 595 766
Egorova, T. G., 454
Eguchi, S., 367, 484, 642
Eicken, K. R., 377, 418, 479, 537, 673
Eilerman, R. G., 360
Eisenberg, W., 318
Eisenthal, K. B., 136
Eisinger, J., 106
Eisner, U., 689
Eiszner, J. R., 706
Eizember, R. F., 311, 323
Elad, D., 441, 678, 680, 681
Elashvili, Z. M., 145
Elbanowski, M., 215
Elix, J. A., 479, 481
Elliot, S. P., 313, 314
El-Sayed, M. A., 115, 120, 122, 135
El-Sharkawy, G. A. M., 265
El-Tsov, A. V., 518, 535, 742
Emoto, S., 624, 675
Endicott, J. F., 248, 249
Engberts, J. B. F. N., 664
Engel, P. S., 331, 699, 714, 715
Engel, R., 103
England, D. C., 695
Englman, R., 29, 33
Epling, G. A., 450
Erman, W. F., 392

Ermolaev, V. L., 135, 140
Ershov, B. G., 148
Eskins, K., 779
Evans, B. E., 681
Evans, C. B., 140
Evans, S. M., 325
Evans, T. R., 515
Ewald, M., 113

Fallis, A. G., 333
Farid, S., 329, 412, 417, 477
Fateen, A. K., 352, 560
Faulkner, L. R., 128
Faure, E., 102
Favaro, G., 140
Fayadh, J. M., 685
Feitelson, G., 108
Felmeister, A., 63
Fenselau, C., 325
Fenical, W., 225
Feree, W. I., 495
Ferguson, J., 545
Ferris, J. P., 657
Field, G. F., 635
Field, R. J., 203
Fielden, R., 659
Fields, D., 708
Fieser, L. F., 739
Fieser, M., 739
Filipescu, N., 138, 140, 249, 251, 423
Filseth, S. V., 222, 229
Filss, P., 259
Findlay, F. D., 223
Fink, D. W., 260
Fink, P. M., 145, 315
Fiquet-Fayard, F., 235
Fisch, M., 115
Fisch, M. H., 409, 410, 622
Fischer, C., 13
Fischer, E., 144, 431
Fischer, E. O., 266
Fischer, H., 240
Fischer, M., 344, 345, 587, 670, 671
Fischer, P. H. H., 117
Fischler, M. H., 477
Fish, G. B., 5
Fish, R. H., 735
Fisher, E. R., 230, 282
Fisher, G., 144, 431
Fisher, G. S., 449
Fitch, R. O., 148, 442
Fleischauer, P., 243
Fleischauer, P. D., 243
Fletcher, A. N., 59
Flores, A. L., 207
Flouquet, F., 239
Flowerday, P., 718
Flowers, W. T., 492
Floyd, J. C., 408
Foergeteg, S., 284, 745
Förster, T., 109
Fokin, E. P., 417
Fomin, G. V., 240
Foote, C. S., 611
Forbes, E. J., 309, 617
Forchioni, A., 148

Ford, P. C., 269
Forest, H., 291
Formenti, M., 153, 628
Formosinho, S. J., 65, 170
Forster, L. S., 252
Fortin, C. J., 223
Forward, G. C., 389
Foster, J. P., 512
Fox, B. L., 661
Fox, R. B., 772
Fox, W. B., 286
Francis, B., 229, 556
Franz, J. E., 729
Frasca, A. R., 623
Fraser, R. R., 656
Fraser-Reid, B., 589
Frater, G., 522
Fratini, A., 607
Fray, G. I., 340, 497
Freed, K. F., 29, 30
Freeman, P. K., 719, 721
French, W. G., 148
Freund, R. S., 205
Frey, H. M., 191
Fricke, J., 230
Fried, J. H., 317, 383
Friedlander, S. K., 225
Friedman, G., 678
Friedman, M., 779
Friedrich, L. E., 355
Friedrichsen, W., 418
Frints, P. J. A., 729
Frio, G., 616
Friswell, N. J., 214, 746
Fritz, K., 15
Frolov, A. N., 518
Fronimos, N., 230
Frost, L. S., 283
Fuerxer, J., 294
Fuhr, H., 225
Fujii, M., 768
Fujimoto, T., 200
Fujisaki, Y., 113
Fujita, H., 625
Fujita, I., 257
Fujita, S., 438
Fujita, T., 523
Fukui, K., 575
Fukuyama, M., 691
Fukuzawa, K., 207, 676
Fulikov, O. F., 154

Gaily, T. D., 221
Galanin, M. D., 103
Galiazzo, G., 533
Gallivan, J. B., 12, 94
Ganapathy, R., 739
Gandi, R. P., 663
Ganguly, S. C., 8
Gann, R. G., 47, 217
Gano, J. E., 315, 454
Ganter, C., 693, 694
Gardner, P. D., 393, 467
Garg, C. L., 108
Garrett, B. B., 247
Garrison, R. L., 59
Gaume-Mahn, F., 294
Gauri, K. K., 396
Gausmann, H., 257

Gauthier, M., 223
Gayer, K. H., 280
Geactinov, N. E., 131
Gebhardt, W., 290
Gebrian, J. H., 462, 468
Geis, W., 256
Gelbart, W. M., 30, 37
Gelernt, B., 99
Genkin, V. N., 140
Gerkin, R. E., 128
Gerlock, J. L., 682
Gesner, B. D., 792
Getoff, N., 629
Geuskens, C., 773
Ghandi, R. P., 438
Ghormley, J. A., 238
Ghosh, A. K., 294
Gianni, M. H., 377
Gibian, M. J., 573
Gibson, J. W., 505
Gibson, T. W., 392
Gierke, T. D., 126
Gilbert, A., 340, 495, 496, 497, 498, 611, 711
Gilbert, J. R., 746
Gilles, L., 240
Gillespie, J. P., 476
Gillette, P. R., 490
Gillis, L. B., 501
Gilman, R. R., 9
Gilmore, F. R., 230, 282
Gindina, R. I., 276
Ginisty, J. C., 225
Givens, R. S., 348, 464, 502
Gladychev, C. P., 762
Glass, A. M., 294
Glasson, W. A., 225
Glaze, W. H., 285, 749
Glazer, E., 484, 654
Gleason, W. S., 224
Glikman, T. S., 265
Glockner, E., 94
Godard, A., 425, 673
Goe, G. L., 476
Goedicke, C., 528
Goedken, V. L., 247
Göth, H., 591, 652
Gol'danskii, V. I., 98
Gol'denberg, V. I., 771
Goldfarb, T. D., 148, 479
Goldfinger, P., 182
Goldman, A. G., 289
Goldschmidt, C. R., 111, 141, 504
Goldschmidt, Z., 373, 374
Gole, J. L., 47, 217
Gollnick, K., 218, 624, 742
Golovanova, Yu. I., 797
Golson, T. H., 362
Golub, M. A., 189, 743, 779
Gomes de Mesquita, A. H., 294
Gondo, Y., 128
Goode, D. A., 113
Gordeev, V. I., 128
Gordon, A. S., 189
Gorman, A. A., 458
Goto, G., 414
Goto, Y., 538

Gottardi, W., 691
Gouterman, M., 61, 106
Gozzo, F., 492, 716
Grabbe, R. R., 467
Graber, D. R., 132
Grabner, L., 294
Graduna, A. F., 106
Graham, D. M., 217
Gramain, J. C., 622
Granick, S., 564
Grant, D. W., 218, 695
Grant, R. A., 679
Grant, R. C. S., 208, 610
Grant, W. J. C., 139
Granwick, S., 582
Granzow, A., 218, 226, 731
Grasser. R., 289
Grassie, N., 790
Grauer, F., 534
Gray, H. B., 276
Gream, G. E., 422
Green, B., 568
Green, J. A., 706
Greene, A. E., 441
Greig, G., 229
Gresser, J. D., 566, 567
Griffin, G. W., 347, 443, 484, 485, 486, 730
Griffiths, J., 309, 617
Grigor'ev, R. V., 290
Grigoryan, G. L., 165
Grinev, M. L., 135
Grisdale, P. J., 535
Grisebach, H., 584
Grishina, G. I., 144, 528
Grivet, J. P., 117
Groff, R. P., 97
Groh, H. J., 187
Groncki, C. L., 146
Grossman, M., 169, 193
Grossweiner, L. I., 597
Groth, W., 149
Grove, J., 746
Grovenstein, E., 502
Grover, P. K., 331
Gruber, G. W., 484
Gruber, R., 323, 658, 659
Grunwald, G. L., 464
Grunwell, J. R., 739
Grzonka, J., 497, 498
Guénard, D., 434
Guenzi, W. D., 602
Güsten, G., 302
Gugava, M. T., 145
Guillemonat, A., 676
Guillet, J. E., 799
Gull, S. J., 132
Gunning, H. E., 190, 209, 226, 228, 229, 710
Gupta, D. N., 444
Gupta, P., 344
Gur'ev, M. V., 154
Gurov, F. I., 126
Gurskii, M. N., 108
Gurvich, I. A., 366
Guseva, L. N., 112
Guthrie, J. P., 404
Gutsche, C. D., 303

Guttenplan, J. B., 565, 566, 567
Guttridge, J., 67
Guzzo, A. V., 140, 367
Gysin, H., 733

Haake, P., 264
Haas, J., 230
Haas, R. M., 280
Haas, Y., 278
Hacker, H., 60
Hackmeyer, M., 15
Hadjoudis, E., 668
Hadley, S. C., 117
Hafner, K., 641
Hafner, M., 139
Hageman, H. J., 412, 541, 550
Hagens, G., 333
Haigh, E., 769
Hall, L., 115
Hall, R. A., 474
Haller, I., 497, 749
Halstead, M. P., 146, 205
Haluska, R. J., 647
Hamada, A., 721
Hamana, M., 635
Hamer, N. K., 419
Hamilton, L. R., 323
Hammer, W. F., 794
Hammond, C. S., 38
Hammond, G. S., 199, 300, 316, 445, 449, 515, 541, 549, 566, 695, 699
Hammond, H. A., 143, 181
Hammond, W. B., 131
Hancock, K. G., 329
Hand, C. R., 422
Hand, C. W., 223
Hand, E. S., 732
Hanifin, J. W., 387, 388
Haninger, G. A., jun., 161, 163
Hankinson, B., 750
Hanson, D. M., 118
Happ, J. W., 432
Haque, K. E., 656
Haranath, P. B. V., 283
Hariharan, P. V., 399
Harper, R. W., 321
Hart, D. J., 401
Hart, H., 400, 401, 407, 458, 461, 468
Harter, D. A., 155
Hartmann, H., 259, 294
Hartmann, P., 768
Hartmann, W., 424
Harumiya, N., 207, 676
Hase, W. L., 196, 213
Hasegawa, M., 768, 769, 780, 781
Hashimoto, S., 687
Hassell, C. L., 228
Haszledine, R. N., 170, 171, 492, 716
Hata, K., 259, 538
Hata, N., 621
Hata, Y., 715

Hatano, H., 148
Hathaway, B. J., 266
Hatton, E., 203
Haug, A., 132
Haugland, R. P., 133
Hautala, R. R., 508, 509, 513
Havinga, E., 460, 506, 507, 508, 522, 523
Havinga, E. O., 726
Havlicek, S. C., 533
Hay, A. S., 780
Hayakawa, K., 640
Hayasaka, T., 620
Hayashi, R., 482
Hayashi, T., 146
Hayes, D. M., 190
Hayon, E., 71, 118, 668
Heaney, H., 465, 750
Heathcock, C. H., 356
Hedaya, E., 476, 548, 739, 752
Hedges, R. E. M., 67, 69
Hefler, H., 240
Heicklen, J., 187, 218, 557
Heidt, L. J., 239, 279
Hein, D. E., 127
Heine, H. G., 304, 346, 571, 732
Heisel, F., 773
Heiss, J., 533
Heitkämper, P., 343, 372
Helene, A., 149
Helene, C., 149, 397, 398
Heller, H. G., 536, 537
Heller, H. J., 795
Helman, W. P., 85
Helmlinger, D., 383
Hempel, J. C., 253
Henderson, W. A., jun., 453, 515, 539
Henn, D., 691
Henneberg, D., 174
Henry, B. R., 18, 30, 34, 126
Henton, D. E., 365
Hentz, R. R., 177
Herberhold, M., 266
Herbert, C. R., 5
Herbst, P., 343
Hercules, D. M., 104, 111, 134
Herman, J., 179
Herman, K., 179
Herre, W., 106
Herron, D. K., 371
Heskins, M., 799
Hess, D., 329, 417, 477, 686
Hess, L. D., 199, 577
Heusley, K., 648
Hexter, R. M., 223
Heyn, H., 655
Hickel, B., 240
Highashimura, T., 769
Higo, A., 455
Hikino, H., 423
Hill, B., 472
Hill, K. L., 182

Author Index

Hill, J., 352, 557, 560
Hill, M. L., 467
Himder, M., 615
Hintz, P. J., 435
Hintzmann, M., 400
Hirai, H., 624
Hiraoka, H., 199, 554, 555
Hiraoka, T., 351, 735
Hirayama, F., 82, 86, 110, 178
Hirose, Y., 476
Hirota, Y., 421
Hixson, S. S., 328
Hobbs, J. B., 481, 522
Hochanadel, C. J., 238
Hochstetler, A. R., 434
Hochstrasser, R. M., 12 18, 34, 128, 132, 241, 286
Hocking, M. B., 376
Hodge, V., 605
Hodgkins, J. E., 723
Hoffmann, M. K., 380
Hoffman, M. Z., 226, 248, 249
Hoffmann, R., 14, 301, 456
Hoffmann, R. W., 377, 479, 537, 673
Hojo, K., 468
Holdy, K. E., 46
Hollinden, G. A., 217
Holmes, Q. A., 283
Holovka, J. M., 467, 479
Holroyd, R. A., 176
Holt, J. W., 590
Holuj, F., 294
Holzer, W., 203
Hoover, R. G., 118
Hopkins, A. S., 628
Hoppe, D., 705
Horne, D. G., 214, 661
Horoupian, S., 394, 672
Horrocks, D. L., 89, 90, 110, 111
Horsfield, A., 117
Horsley, J. A., 239
Horspool, W. M., 299, 309, 318, 319, 329, 334, 337, 362, 366, 369, 400, 418, 420, 440
Horton, D., 699
Hoshina, T., 292
Hoshino, O., 595, 596
Hosokawa, T., 286
Hosszu, J. L., 399
Houdard-Pereyre, J., 686
Houser, N., 132
Howard, B., 264
Hrdkovic, P., 789, 790
Hsieh, H. H., 429
Huang, C. W., 473
Huang, W. D., 762
Huang, W. H., 346, 524, 591, 727
Hubbard, R., 170
Hubbert, H. E., 519
Huber, J. R., 71
Huberman, F. P., 207
Hubrechts, G., 182
Hudec, J., 436, 437, 451, 493, 684

Hudson, A., 78
Huffman, K. R., 453
Huffman, R. E., 223
Hui Bon Hoa, G., 106
Huisgen, R., 650, 656
Hummer, B. E., 603
Hunter, T. F., 126
Hunziker, H. E., 74, 155, 170
Hurley, R., 132, 593
Husain, D., 140, 204, 218, 219, 231
Husson, H. P., 537
Hutchinson, D. W., 481, 522
Huyffer, P. S., 302
Huysmans, W. G. B., 412, 541
Hyeon, S. B., 618
Hyndman, H. L., 445

Iball, J., 392
Ibuki, T., 199
Ichihara, A., 383
Ichikawa, I., 640
Ide, J., 351, 735
Igeta, H., 631
Iizuka, T., 364, 366
Ikeda, K., 590
Ikeda, S., 276
Ikeler, T. J., 698, 699
Il'yikevich, L. A., 268
Imada, I., 414
Innes, K. K., 5, 17
Inoue, E., 146, 433
Inoue, S., 669, 760
Ip, J. K. K., 203
Ipaktschi, J., 372, 403
Iriarte, J., 317, 383
Irick, G., 592, 593
Irie, M., 625
Irving, H. M. N. H., 627
Isaev, I. S., 454
Ishibashi, H., 779
Ishibe, N., 342, 732
Ishihara, H., 616
Ishikawa, M., 636, 637, 779
Isoe, S., 618
Ito, M., 128, 614
Ito, S., 614, 615, 617
Ito, Y., 207, 676
Itoh, I., 520
Itoh, M., 7
Iwamura, H., 463
Iwanga, C., 225
Iwata, M., 546, 675
Izawa, Y., 506, 520
Izzo, P. T., 373

Jackson, P. T., 219
Jacobi, V., 705
Jacobsen, C. W., 223
Jacox, M. E., 149
Jaffe, S., 208, 610
Jakob, W., 265
Jakubowski, E., 190, 228
James, S. P., 427
James, T. C., 203
Janin, J., 102

Janzen, E. G., 682
Janzen, E. O., 286
Jarnagin, R. C., 132
Jarvis, B. B., 148, 442, 443
Jaseja, T. S., 154
Jauquet, M., 606
Jayswal, M. G., 120
Jeger, O., 364, 366, 368
Jellinek, H. H. G., 790
Jellinek, T., 679
Jenevin, R. M., 572
Jennings, B. H., 395
Jennings, K. R., 229
Jennings, W., 472
Jerina, D. M., 630
Jerumanis, S., 681
Jezowska - Trzebiatowska, B., 279
Joby, R., 392
John, G. A. St., 74, 206
Johns, R. B., 679
Johnson, D. A., 249
Johnson, H. W., 566
Johnson, L. D., 794
Johnson, P. Y., 352, 694, 733
Johnson, R. C., 719, 721
Johnson, R. L., 213, 702, 731
Johnson, S. E., 283
Johnston, H. S., 71, 201, 207
Jonas, E., 182
Jones, G., 405
Jones, I. T. N., 223
Jones, L. B., 461, 512
Jones, M., 468, 707
Jones, P. F., 95
Jones, V. K., 461
Jonsson, B. O., 10
Jooss, G., 213, 700
Jordan, C. F., 711
Jorgenson, M. J., 356, 377
Jortner, J., 8, 9, 12, 18, 20, 21, 26, 27, 29, 44
Joussot-Dubien, J., 154, 686
Joyce, T. A., 126
Jubert, A. H., 199
Judeikis, H. S., 128
Juerges, P., 400
Juers, D. F., 461
Juillet, F., 153, 628
Just, G., 661

Kabalka, G. W., 394
Kabasakalian, P., 621
Kabitzke, I., 741
Kachan, A. A., 777, 779
Kaczmar, U. B., 223
Kagan, J., 336, 341
Kagawa, S., 383
Kaiser, K., 290
Kakehi, A., 639, 640
Kalantar, A. H., 126
Kale, J. D., 190
Kalra, B. L., 229
Kamano, Y., 405
Kamiya, N., 766
Kan, G., 638

Kanaoka, Y., 505, 590, 623, 624
Kanda, Y., 120, 133
Kane, M., 373, 642
Kaneko, C., 636, 637
Kaneko, M., 771
Kanematsu, K., 639, 640
Kanev, S., 290
Kano, H., 647, 650
Kano, K., 687
Kantrowitz, E. R., 248
Kaplan, J., 582
Kaplan, L., 491, 495
Karafiath, E., 182, 454
Karle, I. L., 505
Karle, J., 505
Karliner, J., 656
Karnischky, L. A., 486
Karplus, S., 171
Karpukhin, O., 773
Karyakin, A. V., 105
Kasha, M., 18, 30, 118
Kashima, C., 375, 394, 409, 672, 694
Kashima, N., 732
Kaskan, W. E., 185
Kataev, D. I., 276
Kataoka, H., 375, 389, 409, 694
Kato, H., 439, 440, 480, 597, 712
Kato, K., 791
Kato, M., 523, 551, 780
Kato, N., 516
Kato, S., 225, 252, 257, 591, 600, 601
Kato, Y., 784
Katsui, G., 443
Katz, B., 8, 9
Katz, M., 225
Katz, T. J., 467, 468
Kaufman, J. A., 342, 707
Kaupp, G., 443, 476
Kawai, K., 98
Kawana, M., 624
Kawanisi, M., 438, 439, 440, 523, 662
Kawashima, K., 699
Kazachkov, V. C., 126
Kearney, P. C., 733
Kearns, D. R., 51, 114, 115, 132, 225, 363, 410, 566, 573, 598
Keehn, P. M., 541
Keenan, T. K., 280
Keene, J. P., 71
Kelleher, P. G., 792
Keller, O. L., 140
Keller, R. A., 117
Kelley, R. V., 177
Kellogg, M. S., 367
Kellogg, R. M., 524, 558, 687
Kelly, D. P., 552
Kelly, J. M., 385
Kelly, K. W., 340
Kelso, P. A., 315
Kemp, D. R., 118
Kempf, R., 668
Kemula, W., 556

Kende, A. S., 360, 373, 374
Kendrick, T. E., 497
Kennedy, T., 286
Kern, D. H., 572
Kerr, J. A., 226
Khachaturova, C. T., 104
Kh. Al-Ani, 166
Khan, A. U., 598
Khandelwal, G. D., 420
Khattack, M. N., 399
Kienzle, F., 519
Kikuchi, N., 552
Kim, B., 314, 686
Kim, J. J., 111
Kimel, S., 205
Kimura, K., 148
Kimura, M., 403, 461
King, J. A., 723
King, T. Y., 309
Kinura, M., 781
Kira, A., 591, 594, 724
Kirby, E. P., 102
Kirk, A. D., 257
Kirmse, W., 707, 708
Kirsanov, B. P., 103
Kirsch, L. J., 218, 231
Kirsch, P., 524
Kirsch-Demesmae, A., 524
Kiryushin, Yu. A., 201
Kishida, Y., 351, 735
Kistiakowsky, G. B., 158
Kitahara, Y., 617
Kitaura, Y., 318, 569
Kitroskii, N. K., 239
Kittler, L., 399
Kitzing, R., 350, 607
Klabunde, K. J., 78
Klasinc, L., 12
Klaus, M., 476
Klein, R., 222
Klein, W., 477
Kleinspehn, G. G., 621
Klemmensen, P. D., 354, 731
Kliegman, J. M., 741
Kliger, D. S., 12
Klimuskeva, C. V., 10
Klingebiel, U. I., 539, 603, 733
Klopffer, W., 140
Klotz, L. C., 46
Klyne, W., 615
Knapezyk, J. W., 734
Knight, A. R., 229
Knipe, R. H., 189
Knott, P. A., 425
Knowles, P., 541
Knudsen, R. D., 551
Kobayashi, H., 257, 433
Kobayashi, M., 521
Kobrina, N. S., 366
Kobzina, J. W., 356
Koch, T. H., 362
Kodama, S., 48, 225
Koenst, J. W., 195
Koga, G., 552
Koga, N., 552
Kohen, F., 308
Koizumi, M., 113, 139, 225, 591, 594

Kokado, H., 146, 433, 669
Kokochashvili, V. I., 629
Kokubun, H., 104, 113, 139
Kojima, K., 369, 482
Kojima, M., 649, 650
Kolc, J., 146, 354, 485, 670, 731
Kolobkov, V. P., 291
Koltsenburg, G., 497
Komatsu, M., 405
Kommandeur, J., 724
Kondo, M., 102
Kondo, Y., 589
Kondrat'ev, V. A., 149
Konstantatos, J., 189
Koob, R. D., 175
Kopashova, I. M., 280
Kopecky, K. R., 302
Koptyug, V. A., 454
Korolko, B. N., 289
Korte, F., 378, 423, 424, 477
Kostko, M. Ya., 215
Kostylova, Z. A., 779
Kostyshina, P., 105
Kote, F., 497
Kotel'nikov, V. A., 153
Kothe, W., 417
Kotov, E. I., 152
Kottis, P., 128
Kovaleva, I. V., 291
Koyonagi, M., 133
Kozlov, Yu. N., 268
Koz'menko, M. V., 154
Kozmina, O. P., 794
Kraft, K., 495, 497
Kramner, E. H. A., 139
Krasnovskii, A. J., 598
Krause, L., 230
Krauss, H. J., 438, 440
Krespan, C. G., 465, 500
Krishna, V. C., 12
Kristinsson, H., 484
Kroening, R. D., 475
Kronenberg, M. E., 506, 507, 508
Krongauz, V. A., 669
Kropp, J. L., 95, 96
Kropp, P. J., 438, 440
Krow, G. R., 467, 648
Krüger, K., 144, 431
Krueger, R. C., 600
Krueger, S. M., 451
Krüger, U., 400
Kruglyak, Yu. A., 8
Krull, I. S., 199, 347, 728
Kubota, T., 335
Kubota, Y., 113, 117
Kucherov, V. F., 366
Kuder, J. E., 657
Kudrna, J. C., 512
Kuduba, I. A., 779
Kuhn, C. E., 453
Kul-Bitskaya, O. V., 742
Kulczycki, A., 707
Kulevsky, N., 628
Kumler, P. L., 533, 632, 633, 634

Author Index

Kummler, R. H., 135
Kunieda, T., 325, 399, 674, 736
Kunn, L. P., 621
Kuriatnikov, U. I., 135
Kuriyama, K., 615
Kurylo, M. J., 217
Kutschke, K. O., 197
Kuvtin, W. E., 99
Kuwahara, M., 516
Kuwana, T., 566, 595
Kuwano, H., 102
Kuwata, K., 78
Kuzmin, M. G., 112
Kuz'min, V. A., 239
Kuznetsova, K. E., 207, 677
Kuzoubova, L. I., 454
Kwei, K.-P. S., 627, 790
Kyker, G. C., 596

Laarhoven, W. H., 526
La Barge, R. G., 573
Lablache-Combu, A., 558, 650
Ladenson, R. C., 483
Lahiri, J., 543
Laird, T., 352, 734
Lakova, M., 290
Lalonde, R. T., 389, 755
Lambert, M. C., 469
Lambeth, P. F., 108, 564, 566
Laming, F. P., 239
Lamola, A. A., 38, 140, 515, 566
Lance, E. J., 140
Land, E. J., 118
Land, H. B., 623
Landi, V. R., 239
Landy, M., 795
Lang, F. T., 772
Langford, C. H., 263
Langkammerer, C. M., 659
Lan Ngok Tkhiem, 104
Lanska, B., 626
Lantzke, I., 248
Lapouyade, R., 545
Lappin, G. R., 319, 519, 557
Lapworth, K. C., 59
Larach, S., 294
Laridon, U., 781
Larrabee, J. C., 223
Larson, C. W., 48, 183
Lashley, A., 290
Lassette, E. N., 226
Lassila, J. D., 372, 373, 642, 646
Latowski, T., 504
Lauer, W. M., 659
Laufer, A. H., 195
Laustriat, C., 773
Lavalette, D., 13, 118
Lavrushin, V. F., 104
Lawesson, S. O., 354, 731
Lawson, C. W., 86
Lazarus, R. M., 472
Leach, S., 10
Leaver, I. H., 565

Lebo, Yu. G., 779
Ledwith, A., 469, 628
Lee, A. C. H., 665
Lee, E. K. C., 155, 161, 170, 192, 218
Lee, G. A., 442
Lee, K., 543
Lee, P. S. T., 196
Lee, T. J., 418, 578
Lee, Y. J., 591
Leermakers, P. A., 515, 726
Lee-Ruff, E., 359
Le-Fahler, J. P., 128
Lefebure, T., 25
LeFevre, H. F., 190
Le-Goff, M. T., 350, 552
Leibovici, C., 10
Leichter, L. M., 480
Leimer, K. R., 780
Leitich, J., 686
Leitis, E., 517, 518
Lem, S. L., 169, 489
Lemaire, J., 38, 199, 628
Lemaistre, J. P., 128
Lemal, D. M., 170
Leonard, N. J., 694
Leplyanin, C. V., 762
Leppard, D. G., 392
Leppin, E., 218, 742
Lerner, D., 108
Le-Roux, Y., 225
Lesigne, B., 222
Letsinger, R. L., 508, 509, 513
Leubner, I. H., 126
Levashova, L. A., 207, 677
Levine, R. D., 139
Levshin, L. V., 104
Levy, J., 582
Levy, M., 515
Lewin, N., 299
Lewis, D., 204
Lewis, F. D., 97, 306, 317, 489, 571, 698
Lewis, J. W., 78
Lewis, R. G., 362
Leyland, L. M., 190
Lhoste, J. M., 117
Li, R., 120
Li, Y. H., 120
Lias, S. G., 173
Lichtin, N. N., 226
Liebman, S. A., 705
Lim, E. C., 120
Lim, L. N., 433
Lin, L. C., 540
Lindert, A., 533
Lindholm, E., 10, 172
Lindsey, R. V., 695
Linschitz, H., 591, 680
Linstumelle, G., 461
Lipovac, S. N., 790
Lippert, E., 141, 144, 431
Lipsky, S., 82, 86, 110, 178
Lispett, F. R., 113
Lister, D. H., 191
Litt, A. D., 126, 563
Little, J. C., 475
Littler, J. G. F., 178

Liu, R. S. H., 465, 500
Lloyd Braga, C., 125
Loader, C. E., 526
Lober, G., 399
Lodge, J. E., 499
Loeliger, H., 276
Loelkes, G., 128
Loeschen, R. L., 373, 642
Loewenthal, E., 111
Lohse, C., 634, 647
Lomnitz, D., 405, 572
Loners, J., 294
Longo, A. J., 443
Longuet-Higgins, H. C., 199, 494
Lopresti, R., 539
Lorquet, A. J., 42
Louis, P., 392
Loutfy, R. O., 380
Loveridge, E. L., 549, 551
Low, J. N., 392
Lowrance, W. W., 371, 385
Loy, M., 453
Lozykowskii, H., 294
Lubimova, A. K., 216
Luckey, G. W., 187
Ludmer, Z., 544
Ludwig, P. K., 155
Luetzow, A. E., 699
Lugtenburg, J., 460, 523
Lumb, M. D., 125
Lumma, W. C., 331, 692
Lustig, M., 286
Lustig, R. S., 484
Luthardt, H. J., 479
Luzanov, A. V., 10
Lwowski, W., 641
Lysenko, S. F., 289
Lythgoe, S., 63
Lyurchenko, A. G., 479

Mabry, T. J., 455
MacBeath, M. E., 165, 490
McBride, E. F., 328
McBurland, D., 113
McCain, J. H., 509
McCall, J. M., 461
McCall, M. T., 432
McCarville, M. E., 113
McClintock, M., 283
McCullough, J. J., 15, 355, 362, 385, 473, 500
McDaniel, K. G., 678
MacDonald, C. J., 422
McDonald, D. P., 269
McEwan, M. J., 231
McEwen, M. M., 596
McEwen, W. E., 734
McFarlane, P. H., 725
McGlinchey, M. J., 698
McGlynn, S. P., 113, 120, 123
McGuire, R. R., 477
McIntosh, C. L., 303, 404
McIntosh, J. M., 640
McKellar, J. F., 417, 564, 594, 595
Mackenzie, J. S., 286
McKillop, T. F. W., 604

Mackor, A., 677
McNesby, J. R., 177
McSwiney, H. D., jun., 17
McTique, P. T., 287
Mader, L., 289
Maeda, K., 146
Maeda, M., 649, 650
Maeno, N., 553
Märkl, G., 425
Magenheimer, J. J., 189, 213, 745
Mahbub'ul, A. S. M., 294
Maheshwari, K. K. 692
Mahoney, R. T., 45, 46, 76
Mains, G. J., 169, 204
Maisuradze, D. P., 145
Maiti, S., 762
Majer, J. R., 189, 190, 199, 215, 514, 744, 789
Makarov, I. E., 148
Maki, A. H., 128
Maki, Y., 681
Malhotra, B. R., 290
Malley, M. M., 70
Mallory, F. B., 38, 533
Malpass, J. R., 467, 648
Malpiece, C., 60, 299
Mal'tsev, A. A., 276
Malykharov, U. B., 10
Mamatyuk, V. I., 454
Manasek, Z., 789, 790
Manhas, M. S., 412
Manmade, A., 347, 730
Mansanet, J., 238
Mantashyan, A. A., 165, 629
Mantsch, H., 637
Marchetti, A. P., 12, 241, 286
Margerum, J. D., 725
Maria, H. G., 120, 123
Mark, H., 412
Marples, B. A., 465
Marsh, G., 115, 410
Marshall, J. A., 441
Marta, F., 745
Martel, A., 681
Martin, D. G., 127
Martin, J. E., 249, 256
Martin, M., 715
Martin, R. H., 529
Martin, R. M., 47, 48, 217
Martin, T. E., 126
Martin, W., 462
Martin-Brunetiere, F., 294
Martz, P., 633
Maruyama, K., 582
Marzzacco, C. J., 128
Masamune, S., 462, 467, 468
Masamune, T., 664
Masetti, F., 140
Mason, H. S., 600
Mastin, S. H., 264
Mataga, N., 106
Mateescu, G., 750
Matsen, F. A., 253
Matsuda, T., 769
Matsui, K., 319, 551, 553

Matsumoto, H., 102, 647, 650
Matsumoto, M., 597
Matsumoto, S., 383
Matsumoto, T., 383
Matsumura, Y., 597
Matsushima, H., 225, 600, 612
Matsuura, T., 225, 299, 318, 407, 408, 411, 514, 516, 521, 569, 600, 601, 612, 625
Mattee, H. D., 219
Matthews, B. W., 392, 689
Maudling, D. R., 528
Mayer, C. F., 371
Mayer, M., 582
Mayer, W., 476
Mayo, G. O., 465
Mazur, S., 611
Mazzenga, A., 405, 572
Mazzocchi, P. H., 479, 483, 484
Mazzucato, U., 140, 533
Mechoulam, R., 611
Meinwald, J., 356, 459, 479
Meisinger, R. H., 458, 648
Mele, A., 203, 214
Mellor, J. M., 425
Meluish, W. H., 118
Menter, J. M., 423
Merrifield, R. E., 97
Merrill, J. C., 710
Mersiner, R. P., 128
Messer, W., 652
Mestechkin, M. M., 10
Meth-Cohn, O., 659, 725
Mettee, H. D., 221
Metts, L., 113, 445, 447
Metzger, J. L., 121
Metzner, W., 346, 471, 543, 732
Meunier, H. M., 47, 209, 711
Meyer, B., 121
Meyer, C., 781
Meyer, J. A., 205
Meyer, J. W., 199, 549
Meyer, Y. H., 8, 118
Michaelis, L., 564, 582
Michaelis, P., 400
Middleton, F. A., 279
Miescher, E., 5
Migita, T., 342, 708
Miki, J., 309
Miley, G. W., 113
Mill, T., 213, 709
Miller, G. H., 178
Miller, L. L., 602
Miller, N., 590
Miller, R. D., 476, 752
Milligan, B., 519
Milligan, D. E., 149
Milligan, T. W., 694
Milne, G. S., 709
Minami, K., 350
Minato, H., 521
Minn, F. L., 138, 140
Minoura, Y., 760

Mirza, N. A., 436, 684
Misra, T. N., 128
Mitchell, M. J., 533
Mitra, R. P., 276
Mittsel', Yu. A., 104
Miura, M., 113, 117
Miyake, A., 617
Miyama, H., 207, 676
Miyamoto, T., 339
Miyashita, K., 623
Miyashita, M., 451
Mizutani, K., 252
Mobius, C. H., 257
Mode, V. A., 63
Modell, M., 153
Moesta, H., 153
Moggi, G., 695
Mohan, H., 276
Mok, C. Y., 191
Moldavskii, D. D., 696
Molloy, R. M., 364, 375
Money, T., 404
Monroe, B. M., 445, 573
Montague, D. C., 229
Monties, B., 743
Moomaw, W. R., 122
Moore, C. M., 329
Moore, G. J., 698
Moore, H. W., 362
Moore, J. A., 345, 750
Moore, W. M., 746, 749
Morantz, D. J., 127
Morgan, A. G., 299
Mori, T., 319, 537, 538
Mori, Y., 183, 549
Moriarty, R. M., 697, 741
Morikawa, A., 155, 161, 220, 495
Morimoto, H., 414
Morita, J., 383
Morita, T., 549, 551, 553
Moritani, I., 225, 286, 309
Moritz, K.-L., 641
Morren, G., 529
Morris, E. D., jun., 71, 201
Morris, E. R., 190
Morris, J. C., 59
Morris, M. R., 400
Morris, R. V., 229
Morrison, H., 495
Morse, R. I., 45, 46, 76
Morton, W. D., 171, 716
Moser, C., 290
Moser, J. F., 693, 694
Mosier, A. R., 602
Moton, D. R., 359
Mousseron-Canet, M., 367, 618
Muel, B., 60, 113, 299
Müller, E., 533, 677
Mueller, F. J., 259
Mueller, F. W. H., 287
Mueller, H. D., 152
Mueller, P. W., 733
Mueller, R., 396
Mukai, T., 392, 403, 452, 461, 642, 661

Author Index

Mukherjee, R., 741
Mulac, W. A., 155
Mullen, P. A., 15
Mulliken, R. S., 7
Mumford, C., 302
Munakata, K., 516
Murakami, Y., 133
Murata, N., 762, 768, 779
Murata, Y., 106
Muroi, T., 614, 615
Murphy, W. F., 203
Murray, R. K., 401, 458, 648
Murrell, J. N., 43
Murthy, A. A., 283
Murthy, N. S., 279
Musa, W. E., 596
Museridze, M. D., 629
Musgrave, O. C., 506
Musser, H. R., 706
Mustafa, A., 576
Muszkat, K. A., 144, 431

Nabandyan, A. A., 165, 629
Nadler, M. P., 286
Nafassi-V, M.-M., 362
Nagakura, S., 128
Nagamatu, C., 781
Nagao, Y., 351
Nair, K. P. R., 108
Naito, T., 537, 538
Nakadaira, Y., 421
Nakagawa, C. S., 144, 429
Nakahara, A., 133
Nakai, H., 339, 505
Nakanishi, H., 768, 769
Nakanishi, K., 421
Nakano, Y., 120
Nakashima, R., 612
Nakatsuka, N., 468
Nakazawa, M., 433
Naman, S.-A. M. A., 190, 199, 215, 514, 744
Narasimhamurthy, B., 279
Narula, S. P., 196
Naruto, S., 521
Nasielski, J., 524, 606
Nasielski-Hinke, R., 524
Natansohn, S., 291
Natarajan, L. V., 767
Nathan, E. C., tert., 359
Naumova, T. M., 126
Navan, G., 106
Navatte, J. C., 97, 148
Neadle, D. J., 661, 725
Neckers, D. C., 474, 475, 559, 576, 584, 694
Nelson, A. J., 562
Nelson, P. H., 317, 383
Nelson, S. F., 435, 476
Nemodruk, A. A., 280
Nemoto, N., 113, 139
Nenitzescu, C. D., 750
Ness, S., 134
Neuberger, K. R., 143, 180, 427
Neuwirth-Weiss, Z., 582

Ng, S. H., 166, 490, 491
Nguyen, K. C., 604
Nicholas, J. E., 203
Nicholls, P., 266
Nicholls, R. W., 5
Nicholson, A. A., 383
Niclause, M., 628
Nieman, G. C., 121
Niemcyzk, M., 318
Nieswandt, K., 174
Nikitchenko, V. M., 104
Nikolich, K., 266
Ninomiya, I., 537, 538
Nirmukhametov, R. N., 528
Nishi, N., 120
Nishida, S., 383
Nishihara, T., 771
Nishikawa, M., 155
Nishikida, K., 78
Nishinaga, A., 225, 600
Nitta, M., 452
Nivard, R. J. F., 526
Nizamov, N., 104
Nnadi, J. C., 399
Noda, H., 635
Nofre, C., 225
Nomi, S., 438
Norin, T., 615
Norrish, R. G. W., 203, 214, 661
North, P. P., 383
Novikov, B. V., 290
Noxon, O. F., 222
Noyes, W. A., jun., 38, 155, 169, 187, 199, 490, 491
Noyori, R., 523
Nozaki, H., 319, 438, 523, 662
Nozoe, T., 617
Nugent, I. J., 140
Nurmukhametov, R. N., 104, 144
Nye, M., 383

Obase, H., 661
Obata, N., 721
Ochiai, T., 551
O'Connell, E. J., 102, 319, 557
Oda, M., 617
Odaira, Y., 339, 575
Odani, M., 342, 732
O-Donnell, C. M., 524
Odum, R. A., 646, 650
O'Dwyer, M. F., 140
Oettle, W. F., 348
Offen, H. W., 17, 111, 127
Ofran, M., 775
Ogata, M., 647, 650
Ogata, Y., 326, 506, 518, 520, 553
Ogawa, M., 5
Ogdan, J., 240
Ogi, Y., 521
Ogilvie, P. F., 149
Ogol-Tsova, N. V., 535, 742

Ogura, K., 407, 408, 411
Oh, Y. N., 422
O'Hare, D. O., 5
Ohki, E., 325, 686
Ohloff, G., 615
Ohmae, M., 443
Ohnesorge, W. E., 260
Ohnishi, Y., 691
Ohno, A., 691
Ohno, M., 367, 484, 642
Ohno, S., 146, 259
Ohno, T., 252, 257
Ohno, Y., 691
Ohorodnyk, H., 15, 355
Ohta, M., 480, 712
Ohta, Y., 476
Oida, S., 325, 686
Oine, T., 392, 661
Okabe, H., 203, 214
Okada, T., 438, 662
Okamura, S., 767, 768, 769
Okano, T., 102
Okawara, M., 766, 781
Okuda, S., 221
Okuno, Y., 505
Olander, C. R., 456
Olavesen, C., 189, 190
Oldman, R. J., 46
Oleson, J. A., 622
Olin, S. S., 480, 713, 714
Olive, J. L., 367, 618
Omran, S. M. A. R., 352, 560
Omura, K., 514, 516, 521, 600
Oncesch, T., 706
O'Neal, H. E., 48, 183
Ono, K., 582
Oosterhoff, L. J., 51, 444
Orban, F., 287
Orfanos, V., 462
Orger, B. H., 495, 496
Orisheva, R. M., 249, 250
Orlando, C. M., 335, 377, 412
Orloff, M. K., 12, 15, 128
Oron, N., 775
Osborne, A., 595
Oster, C., 765
Osterroht, C., 741
Osugi, J., 582
O'Sullivan, M., 103
Oth, J. F. M., 462
Ottolenghi, M., 108, 111, 146, 240, 241, 504, 670
Ouannes, C., 623
Oude-Alink, B. A. M., 303
Oudman, D., 420
Ourisson, G., 363
Owens, D. V., 115, 120

Pac, C., 504, 547
Pacifici, J. G., 333, 592, 593, 733
Paden, J. D., 228
Padhye, M. R., 120
Padwa, A., 318, 323, 362, 374, 484, 586, 592, 654, 658, 659, 694
Page, T. F., 299

Pagni, R. M., 465, 502
Pai, D. M., 773
Paice, J. C., 422
Pal, J. M., 320
Pancir, J., 13
Pankratov, A. A., 152, 153
Pannetier, G., 283
Paoletti, P., 266
Pape, B. E., 351, 525, 747
Papee, H. M., 241
Pappas, S. P., 319, 416
Paquette, L., 311, 323, 458, 467, 480, 647, 648
Para, M. F., 351, 747
Paraskevopoulos, G., 222
Paris, J. P., 611
Parkanyi, C., 14
Parkash, V., 154
Parke, S., 276
Parker, C. A., 126, 552
Parker, G. B., 251
Parmenter, C. S., 103, 158, 159, 190, 191
Parshutkin, A. A., 669
Partridge, R. H., 84, 799
Pashayan, D., 586, 592
Pashchenko, G. S., 154
Fasternak, V. Z., 727
Fastra, S. C., 395
Faszyc, S., 268
Fatei, P. N., 280
Patel, D. J., 408
Patit, S. R., 762
Patterson, J. M., 560, 561
Patumtevapibal, S., 377
Paulson, D. R., 487
Pauson, P. L., 440
Pautel, F., 370
Pavlenishvili, I. Y., 145
Pavlopoulos, T. G., 118, 120
Pavlov, V. I., 779
Paxton, J., 117
Payne, W. A., 230
Payot, P. H., 733
Pearson, E. F., 5
Pedersen, A. D., 354, 731
Pedrini, C., 294
Peill, P. L. D., 790
Pellois, A., 96, 97, 148
Pen'kovskii, V. V., 8
Penneman, R. A., 280
Penzes, S., 226
Penzhorn, R. D., 200
Pereira, L. C., 125
Peretti, P., 105
Perkampus, H. H., 400
Perkins, M. J., 718
Perold, G. W. 363
Perrins, N. C., 150
Pete, J. P., 332
Peterson, D. B., 177
Petrak, P. Z., 251
Petriconi, G. L., 241
Petrovskii, G. T., 291
Petrusis, C. T., 725
Pews, R. G., 422
Pfughaupt, K. W. 396
Pfundt, G., 329

Phillips, D., 38, 166, 169, 199, 490
Phillips, D. H., 771
Phillips, D. J., 111
Phillips, G. O., 417, 594, 595, 794
Phillips, L. F., 190, 231
Phillips, R. W., 148
Philpott, M. R., 42, 128
Pierce, R. A., 118
Piers, E., 411
Pietra, F., 506
Pikulik, L. G., 215
Pingeon, G., 667
Pinhey, J. T., 515, 552
Pisker-Trifunac, N., 431
Pitts, J. N., jun., 14, 196, 199, 223, 224, 566, 577, 595, 610, 728
Plank, D. A., 304, 408, 548
Plimmer, J. R., 539, 603, 733
Ploder, W. H., 389
Plummer, B. F., 474
Plunkett, A. O., 690
Pobinet, R., 392
Poh, P., 341, 586
Poland, H. M., 191
Poleektov, V. A., 201
Pollet, A., 650
Pollitt, R. J., 661, 725
Pollock, T. L., 228
Polowczyk, C. J., 572
Pottorak, V. A., 104
Pomerantz, M., 484
Ponce, C., 604
Pond, D. M., 306, 338
Ponticelli, G., 266
Pool, G. L., 140, 367
Poole, G. A., 99
Pope, M., 131
Porter, G., 63, 65, 118, 170, 203, 592, 595
Porter, R. F., 286
Portnoy, N. A., 416
Potashnik, R., 146, 504, 670
Potier, P., 537
Pouchot, O., 625
Poznyak, A. L., 268
Pratt, A. C., 455, 456
Prell, G., 637
Prescher, G., 400
Preston, K. F., 193
Price, D., 417, 594, 595
Price, M. A., 225
Prinzbach, H., 350, 443, 476
Prochorow, J., 106
Procter, I. M., 266
Prudnikov, I. M., 152, 153
Przybytek, J. T., 336
Ptak, M., 149
Purmal, A. P., 268
Purnell, J. H., 146, 205

Quagliano, J. V., 247
Queguiner, G., 425, 673
Quinkert, G., 400

Quinlan, K. P., 417
Quinn, C. P., 189
Quinn, T. J., 59

Rabalais, J. W., 123
Rabani, J., 241
Rabinowitch, E., 247
Radlick, P., 225
Rämme, G., 362
Rahn, R. O., 399
Rai, D. K., 108
Ramakrishnan, V., 140, 564, 620
Ramey, B. J., 393
Ramme, G., 144
Ramsay, G. C., 565
Ranjon, A., 625
Ranson, P., 105
Rao, D. V., 469, 779
Rao, T. N., 219
Rapoport, H., 613
Rapoport, V. L., 152
Rasmussen, P. W. W., 385
Ratajczak, E., 170, 492
Rau, J. D., 162
Rauhut, M. M., 611
Raun, R. D., 515
Rausch, D. J., 491, 686
Ray, A. K., 627
Raz, B., 44
Reardon, R. C., 697
Rebbert, R. E., 173
Rebfeld, S. J., 90
Reich, I. V., 472
Reiffsteck, A., 222
Reine, A. H., 794
Reinhardt, G., 220
Reiser, A., 166
Renkes, G. D., 103, 300
Rennert, A. M., 200
Rennert, J., 582
Rentzepis, P. M., 70
Rest, A. J., 287
Rhodes, W., 30
Riccieri, P., 255
Rice, S. A., 18, 30, 37
Richards, J. H., 431
Richardson, W. H., 605
Richtol, H. H., 144, 146, 362
Riddell, W. D., 120, 132, 394, 397
Rieber, N., 705
Riehl, N., 289
Riess, I., 43
Rigaudy, J., 604
Rigby, L. J., 285
Rigby, R. D. G., 515, 552
Riley, J. F., 238
Rio, G., 605, 616, 625
Ripley, R. A., 309
Ripoche, J., 96, 97, 148
Ritchie, R. K., 5
Ritschers, J. S., 448
Robb, J. C., 189, 199, 215, 514, 744
Roberts, C. W., 422
Roberts, D. B., 474, 559
Roberts, T. D., 441

Author Index

Robinson, G. E., 514
Robinson, G. W., 33, 113, 161
Robinson, P. J., 63
Rockley, M. G., 103, 131, 190, 300
Rodebaugh, R. M., 646
Rodgers, T. R., 407
Roe, A., 519
Röttele, H., 462
Rohatgi, K. K., 279
Rohe, L., 781
Romkiewicz, Y., 141
Rondeau, R., 473
Roquitte, B. C., 172, 199, 450
Rose, J. E., 519
Rosenberg, H. M., 473, 661
Rosenthal, I., 397, 680, 681, 755
Rosiello, L. A., 264
Rossman, G. R., 276
Rostaing, R., 667
Roswell, D. F., 58, 301
Roth, H. D., 80, 706
Roth, W. R., 715
Rowland, F. S., 196, 229
Rubin, M. B., 392, 573, 582
Rubinstein, H., 577
Rublowsku, D., 265
Rucktäschel, R., 477
Rumin, R., 452
Rupainwar, D. C., 627
Russell, D. W., 661, 725
Russell, R. L., 196
Ruszkay, R. J., 582
Ryang, H.-S., 336
Rygalov, L. N., 105

Saboz, G. A., 369
Saboz, J., 366
Saburi, Y., 350
Sachs, F., 668
Sagowdz, G., 501
Saha, M. K., 762
Sahay, B. K., 627
Saito, I., 225, 600, 601, 625
Sakai, A., 760
Sakai, K., 369
Sakai, M., 226
Sakai, T., 476
Sakaki, T., 252
Sakamoto, H., 600
Sakan, T., 618
Sakota, N., 760
Sakurai, H., 181, 326, 335, 336, 351, 504, 547, 580, 749
Sakurai, K., 67, 207, 283
Salamon, M., 677
Salisbury, K., 536, 537
Saltiel, J., 113, 143, 180, 427, 433, 445, 447
Saltzman, M. D., 565
Salzman, W. R., 12
Samborn, E., 409
Sammes, P. G., 299

Samson, A. S., 308
Samuelson, G. E., 450
Sander, M. R., 548, 549
Sander, R. K., 46
Sanders, N., 259
Sandhu, H. S., 190, 226
Sandner, R., 739
Sanford, E. C., 449
Sanjiki, T., 480, 712
Santappa, M., 280, 761, 767
Santhanam, M., 140, 564, 620
Santos, P., 149
Santry, D. P., 15, 355
Sargent, M. V., 479, 481, 503, 537
Sarkisian, J. I., 643
Sarti Fantoni, P., 544
Sasaki, T., 367, 484, 639, 640, 642, 675
Sasse, W. H. F., 474, 501, 503, 559
Sastri, V. S., 263
Sato, A., 461
Sato, E., 590, 624
Sato, K., 781
Sato, M., 617
Sato, N., 664
Sato, S., 230, 495
Sato, T., 225, 259, 309, 526, 538, 615, 661
Sato, Y., 389, 394, 409, 672, 694, 732
Satoh, S., 230, 495
Sauerbier, M., 533
Sauers, R. R., 340
Saunders, W. H., 489, 698
Savolainen, J., 25
Sawaki, S., 595, 596
Sayigh, A. A. R., 469, 779
Scala, A. A., 222
Scandola, F., 262, 264
Schaap, A. P., 584, 610
Schade, G., 618
Schaffner, K., 340, 363, 364, 366, 368, 369, 573
Schaffner-Sabba, K., 410
Scharf, H. D., 378, 423, 424, 497
Scharmann, A., 289
Schaubman, R., 63
Schechter, H., 441
Scheer, M. D., 222
Scheffer, J. R., 392
Scheider, R. A., 356
Scheiner, P., 719
Scheiner, R., 646
Schenck, G. O., 618
Scheppers, G., 703
Schläfer, H. L., 252, 255, 256, 257
Schlag, E. W., 73
Schlessinger, R. H., 225, 608, 736, 737
Schlosser, D. W., 76
Schlosser, M. M., 443
Schmall, B., 646
Schmid, G. H., 604
Schmid, H., 588, 591, 652

Schmidt, G. M. J., 544
Schmidt, J., 128
Schmidt, M. W., 161, 218
Schmidt, U., 741
Schnepp, O., 515
Schoellkopf, U., 705
Schönberg, A., 560, 576
Scholl, M. J., 625
Scholz, K., 329
Schomburg, G., 174
Schonberg, A., 352
Schonhorn, H., 789
Schott, H. N., 312
Schotte, L., 694
Schrader, L., 480, 702, 715
Schroeder, B., 461
Schröder, G., 462, 753
Schroeter, S. H., 335, 350
Schroetter, H. W., 60
Schubert, H., 425
Schuetz, R. D., 577
Schug, J. C., 771
Schultz, A. G., 736, 737
Schultz, M., 655
Schulz, R. C., 781
Schumaker, H. J., 199, 223
Schurter, J. J., 529
Schuster, D. I., 401, 403, 408, 409
Schuster, G. B., 355
Schuster, H., 132
Schutte, L., 323, 325, 577
Schwartz, J., 462
Schwartz, R. W., 290
Schwartz, S. E., 71, 207
Schwoerer, M., 128
Scott, A. R., 144
Scott, P. M., 229
Sebenda, J., 626
Sechkovskaya, V. A., 762
Sedgwick, R. D., 63
Seeman, J. I., 367
Seffert, W., 637
Seidewand, R. J., 145, 309
Seidner, R. T., 462, 467, 468
Seinfeld, J. H., 225
Selinger, B., 67, 112, 541
Sellers, H., 341, 586
Selvarajan, R., 718
Semeluk, G. P., 165, 166, 168, 169, 193, 489, 490, 491
Semina, G. N., 207, 677
Semmelhack, M., 362
Sengupta, S. K., 279
Senier, A., 668
Senior, W. A., 238
Senkowski, T., 265
Serebtyakov, E. P., 366
Serve, M. P., 473
Seto, S., 403, 617
Setser, D. W., 205, 229, 230
Seybold, G., 106
Shaad, L. J., 8
Shaffer, G. W., 311, 325, 371, 577
Shagisultanova, G. A., 249, 250, 268

Shain, L. A., 128
Shamonovskaya, R. J., 280
Shani, A., 540, 611
Shannon, P. V. R., 325
Shapira, D., 241, 700
Sharf, B., 8, 9
Sharma, B. K., 276
Sharma, D., 290
Sharmann, A., 294
Sharnoff, M., 117
Sharp, J. H., 595
Sharpe, L., 472
Sharples, A., 775
Sheehan, J. C., 362
Shefter, E., 374, 694
Shekk, U. B., 128
Shelden, S. R., 362
Sheridan, J. B., 458
Sherwin, M. A., 464
Sherwood, A. G., 229, 556
Shetlar, M. D., 81
Shevel'kov, V. F., 276
Shibata, K., 438
Shibata, T., 502
Shida, T., 724
Shields, J. A., 794
Shigemitsu, Y., 339
Shigorin, D. N., 104
Shima, K., 335, 336, 351
Shimada, S., 259, 538
Shimizu, I., 433
Shimuda, R., 120
Shinazawa, K., 621
Shindo, N., 516
Shinozaki, H., 438
Shiokawa, J., 277
Shirahama, H., 383
Shirk, J. S., 287
Shishkin, L. P., 145
Shishkov, A. V., 216
Shizuka, H., 215, 549, 550, 551, 553
Shlyapintokh, V. Ya., 771
Sholina, S. I., 240
Shortridge, R. G., jun., 161, 192
Shrader, R. E., 294
Shrubovich, V. A., 777, 779
Shuali, U., 241
Shugar, D., 658
Sianesi, D., 695
Sicre, J. E., 199
Siebrand, W., 34, 126
Siegel, S., 95, 128, 794
Siegnoczynski, R., 106
Sieja, J. B., 362
Sigal, P., 144, 160, 429
Sigwalt, C., 630, 638
Sigwart, C., 269
Silaev, A. B., 104
Silber, P., 513
Silva, R. A., 479
Sim, G. A., 456
Simmons, J. D., 17, 226
Simonaitis, R., 349, 728
Simons, B. B., 795
Simons, J. P., 150
Simons, J. W., 195, 196, 213, 702
Simonsen, O., 634

Simpson, W. T., 733
Sinclair, R. S., 286
Singer, L. A., 706
Singh, A., 144, 290
Singh, A. N., 108
Singh, B., 558, 642, 650, 658
Singh, I. S., 108
Singh, O. N., 108
Singh, R. S., 108, 120
Singh, S. N., 120
Singh, S. P., 336, 341
Sinha, S. P., 278
Sircar, J. C., 449
Sixl, H., 128
Skeberle, A., 226
Skold, C. N., 225, 608
Skvortsov, B. V., 105
Slade, R. C., 266
Slama, P., 790
Slanger, T. G., 74, 206, 230
Slater, D. H., 220, 221
Small, C. J., 13
Smith, A. B., tert., 141, 300, 360
Smith, A. E., 746
Smith, B. E., 121
Smith, J. J., 121
Smith, J. M., 299
Smith, R. L., 347, 485, 714, 730
Smith, W. F., 727
Smith, W. H., 219
Snelling, D. R., 223
Snieckus, V., 638, 640, 701
Snoeren, A. E. C., 460
Snyder, C. D., 613
Snyder, P. A., 6
Sokolova, T. A., 653
Soloman, B. S., 709
Solomon, D. S., 97
Soltys, J. F., 217
Somasekhara, S., 726
Somerville, W. B., 33
Sommer, F., 469
Sommer, U., 12
Sondheimer, F., 479
Sone, K., 252
Song, P. S., 99
Sonoda, A., 309
Sontag, F. I., 409
Sopchyshyn, F., 144
Sousa, J. A., 669
Southam, R. M., 359
Sovers, O. J., 290
Spangler, R. J., 533
Speed, R., 67
Spees, S. T., 251
Speiser, S., 205
Spence, G. G., 630
Spence, J. T., 269
Spencer, H. E., 259
Spencer, R. D., 109
Sperling, H. P., 81
Sperling, J., 678
Spindler, R. J., jun., 4
Spitzer, W. A., 367
Spoerke, R. P., 127
Spoerke, R. W., 305
Spring, M. M., 659

Sprung, J. L., 224
Srinivasan, R., 181, 229, 247, 371, 375, 409, 432, 434, 447, 450, 454, 476, 554, 555, 749
Stacey, W. T., 117
Stam, M. F., 628
Stanislaus, J., 120
Stankorb, J. W., 425
Starnes, W. H., 408
Staros, J. V., 170
Stauff, J., 225
Stedman, D. H., 205, 230
Steel, C., 126, 709
Steer, R. P., 223, 224, 610
Stefani, A. P., 711
Stefanovic, M., 708
Steffan, G., 329
Stegemeyer, H., 528, 552
Stein, G., 278
Stein, N., 566, 568
Steinbach, F., 152
Steiner, R. F., 102
Steinmaus, H., 680
Steller, K. E., 512
Stenberg, V. I., 523, 596, 628
Stepanov, F. N., 479
Stephens, E. R., 225
Stephenson, L. M., 300
Stermitz, F. R., 346, 524, 591, 727, 748
Sternbach, L. H., 635
Stevens, B., 113, 134, 225, 604
Stevens, R. D. S., 168
Stevenson, C. D., 204, 219
Stevenson, R., 308
Stevenson, R. J., 204
Stewart, T., 794
Stickler, S. J., 126
Stief, L. J., 230
Still, I. W. J., 422
Stocker, J. H., 572
Stoffer, J. O., 706
Stoganov, V., 290
Stojilkovic, A., 422, 662, 708
Strachan, A. N., 197
Stranks, D. R., 273
Strausz, O. P., 190, 209, 226, 228, 229, 710
Strebel, A., 392
Strehlow, W., 609
Streit, P., 291
Streith, J., 630, 633, 638
Strel'tsova, A. A., 207, 677
Strickler, H., 615
Strickler, S. J., 158
Stringham, R. S., 213, 709
Strohmeier, W., 259, 768
Strom, E. T., 613
Strong, R. L., 144, 146, 362
Strow, C. B., 467
Stryer, L., 133
Stubbs, J. K., 644
Stuber, F. A., 469, 779
Studzinskii, O. P., 535, 742
Sturmer, D. H., 269
Stupavsky, M., 230

Sturm, G. P., jun., 47, 217
Sturm, J. E., 203
Suga, K., 523
Suginome, H., 664
Sugiyama, H., 403
Sugiyama, N., 225, 375, 389, 394, 409, 546, 672, 694, 732
Sugowdz, G., 503
Sukawa, H., 392
Sullivan, J. H., 63
Sundararajan, S., 266
Sunder-Plassman, P., 317, 383
Sundheim, B. R., 264
Sung, M. T., 378
Suppan, P., 10
Surzur, J. M., 673
Suschitzky, H., 659
Sussman, D. H., 401, 403
Sustmann, R., 656
Sutherland, R. G., 376, 585
Sutton, J., 240
Sutton, M. E., 659
Suzuki, S., 553
Suzuki, Y., 768, 769
Svejda, P., 214, 746
Sveshnikova, E. B., 135
Swan, G. A., 685
Swenberg, C. E., 117
Swenson, J. R., 14, 301
Swenton, J. S., 409, 429, 450, 473, 500, 655, 698, 699
Swindell, R., 646
Szabo, A. G., 120, 132, 394, 397
Szoc, K., 566, 567

Tabata, T., 461
Tabor, T. E., 383
Tabuchi, K., 760
Tachin, V. S., 140
Tada, M., 438
Taguchi, V., 454
Takada, S., 467
Takagi, K., 553
Takahashi, T., 308
Takai, M., 308
Takamuku, S., 181, 749
Takaya, T., 737
Takenaka, S., 403
Takeo, M., 283
Takeshita, H., 423, 614, 615
Takezaki, Y., 199
Takezawa, S., 171
Takita, S., 183
Takizawa, T., 721
Takuze, S., 767, 768
Talaty, E. R., 362
Tam, J. N. S., 620, 621, 665
Tamres, M., 106
Tamura, H., 762
Tanabe, K., 369, 482
Tanaka, I., 183, 215, 549, 550, 621
Tanaka, M., 277, 762, 768, 779
Tanaka, Y., 405

Tanei, T., 148
Tang, C. S., 346, 690, 726
Tanida, H., 715
Tanigaki, T., 760
Tanikagar, R., 674
Tanner, S. P., 140
Tanquary, R. L., 17
Tarrant, J. R., 140
Tasovac, R., 422, 662, 708
Taube, H., 218
Tavares, D. F., 389
Taylor, B., 769
Taylor, E. C., 519, 630, 681
Taylor, G. A., 617
Taylor, G. W., 195, 213, 702
Taylor, H. A., 794
Taylor, J. M., 43
Taylor, R. T., 661
Tazuke, S., 762
Tchir, M. F., 380, 383
Teague, M. W., 128
Tedder, J. M., 418, 703
Teichner, S. J., 153, 628
Teller, E., 34
Temchenko, V. G., 696
Teramura, K., 342, 732
Teratake, S., 715
Terauchi, K., 326, 580
ter Vrugt, J., 294
Testa, A. C., 103, 132, 593, 594
Tezuka, T., 299, 461, 642, 646
Thal, C., 537
Thomas, J. K., 64
Thomas, J. M., 544
Thomas, R. A., 63
Thomson, A. J., 106
Throckmorton, J. R., 343
Thrower, G. F., 711
Thrush, B. A., 48
Thynne, J. C. J., 190
Tiefenthaler, H., 652
Tilford, S. C., 17
Timm, D., 147
Timmons, C. J., 144, 471, 503, 526, 537
Timmons, R. B., 78, 189, 190, 213, 217, 744, 745
Tincher, W. C., 794
Tinti, D. S., 115, 122, 135
Tippett, W., 141, 490
Tkachenko, A., 779
T'Kint de Roodenbeke, A., 290
Toby, S., 81
Toda, F., 616
Toda, T., 452
Toki, S., 335, 336
Tolstorozhev, G. B., 154
Tomalesky, R. E., 203
Tominaga, T., 339
Tomioka, H., 326, 506
Tomita, M., 625
Tomkiewicz, Y., 96, 111
Tomlinson, A. A. G., 266
Tomoda, Y., 781
Tomoeda, M., 617

Tonkyn, R. G., 368
Topp, J. A., 60
Topp, M. R., 63
Torrance, B. J. D., 790
Tosa, T., 504
Toshima, H., 760
Toshima, N., 624
Toth, M., 182
Townley, E. R., 621
Tozune, S., 342, 708
Trachtenberg, I., 743
Trachtman, M., 169
Tramer, A., 132
Trappen, N., 153
Travecedo, E. F., 523, 596
Traynham, J. G., 429
Trecker, D. J., 469, 548, 549, 739
Tregay, G. W., 279
Treinin, A., 223, 240, 241, 700
Trencséni, J., 106
Trifunac, A. D., 79
Tripathi, B. N., 108
Troe, J., 208
Trost, B. M., 342, 458, 708, 734
Trotman-Dickenson, A. F., 47, 209, 711
Trotter, W., 593, 594
Trozzolo, A. M., 80, 486
Tsai, L., 120, 140
Tschuikow-Roux, E., 48, 225
Tsoy, A. N., 108
Tsubomura, H., 98, 148, 154
Tsuchihashi, G., 691
Tsuchiya, T., 631
Tsuda, M., 776, 781
Tsuji, T., 309
Tsukerman, S. V., 104
Tsukida, K., 618
Tsunashima, S., 230, 495
Tsunooka, M., 768, 779
Tsurikova, G. A., 291
Tsuruta, H., 459
Tsutsumi, S., 339
Tsuyuki, T., 308
Tuck, A. F., 46
Tümmler, R., 366
Tuesday, C. S., 225
Turetskaya, E. A., 280
Turner, J. J., 287
Turner, P. H., 564
Turro, N. J., 103, 144, 300, 306, 317, 334, 337, 338, 339, 359, 418, 420, 427, 578
Tyer, N. W., jun., 145, 433, 434
Tyerman, W. J. R., 196, 201

Uda, H., 451
Ueda, K., 687
Uehara, Y., 288
Uffindel, V. E., 132
Uffindell, N. D., 595
Ukigai, T., 506, 518

Ukita, T., 625
Ullman, E. F., 58, 302, 378, 419, 453, 558, 650, 658
Ulrich, H., 469, 779
Ulstrup, U., 259
Umezawa, B., 595, 596
Umreiko, D. S., 280
Unger, G., 469
Unger, I., 165, 166, 168, 169, 193, 489, 490, 491
Urry, W. H., 706
Usherwood, E. W., 589
Usui, Y., 225, 597

Vaidya, B. K., 376
Valadier, F., 102
Van-Auken, T. V., 467
Van Benthem, W., 231, 282
van Bergen, T. J., 524, 687
Van Cakenberghe, J., 140
Van-den-Berg, J. A., 251
Van-den-Bergh, H. E., 285
Van den Bogaerde, J., 71, 201
Vander Donckt, E., 99, 592, 606
Van der Lugt, W. Th. A. M., 51, 444
Van der Waals, J. H., 128
van Driel, H., 558
Van-Eldik, R., 251
van Lear, G. E., 656
Van-Sinoy, A., 606
van Tamelen, E. E., 462, 554
Van-Vliet, A., 507
Varga, L. P., 280
Varma, R., 627
Varnerin, R. E., 213, 745
Vasil'kova, L. I., 203
Vaubel, G., 128
Veenland, J. U., 677
Velashevich, K., 266
Venkaturao, K., 280
Verbeek, M., 423, 497
Verhaegen, J. P., 140
Verma, A. L., 10
Vermeil, C., 238
Vermes, J. P., 662
Vernon, J. M., 514
Verral, R. E., 238
Vesely, K., 789
Vesley, G. F., 726
Vickery, B., 497
Viktorova, E. N., 98
Villarejo, D., 4
Villaume, M. L., 332
Villegas, J., 405, 572
Vinogradov, A. P., 239
Vinogradov, I. P., 102, 154
Vinogradova, K. A., 104
Vitek, R., 286
Voelker, E. J., 345, 750
Vogel, F., 131
Vogler, A., 248, 250
Voissius, D., 650
Volkova, E. P., 140
Vollner, L., 477

Volman, D. H., 148, 214
Volob'ko, L. V., 280
Von-Bunau, G., 174
von Gustorf, E. K., 686
Voroshchenko, A. T., 479

Wachler, P., 291
Wagner, F., 344, 670
Wagner, P. J., 305, 312, 314, 315, 394, 395
Wagner, P. O., 127
Wahl, A. C., 290
Wahr, J. C., 45
Waiss, A. C., 437
Waite, M. G., 456
Wakamatsu, S., 207, 676
Wakim, F. G., 294
Walker, P., 581
Walker, T. J., 450
Walling, C., 573
Walsh, A. D., 5
Waltz, W. L., 270
Wampfler, G., 138, 301
Wanatabe, H., 403
Wang, C. T., 628
Wang, S. Y., 399
Wanmaker, W. L., 294
Ward, H. R., 182, 454
Ward, R. L., 682
Ware, W. R., 189
Waring, A. G., 400
Warnet, J., 456
Warren, K. D., 251, 265, 269
Warrener, R. N., 753
Warsop, P. A., 5
Wasgestian, H. F., 256
Wasserman, H. H., 541, 607, 609
Watanabe, M., 715
Watanabe, S., 523
Waters, J. A., 434
Watkins, D. A. M., 726
Watmann-Crajcar, L., 10
Watson, A. I., 276
Watts, R. J., 126, 158
Wauchop, T. S., 231
Way, H., 249, 251
Wayne, R. P., 223
Wease, J. C., 699
Weatherby, G. D., 601
Webber, S. E., 34
Weber, G., 109
Webman, I., 12
Wehrli, H., 366
Wehry, E. L., 266
Wei, C. C., 302, 524, 591, 748
Wei, K. S., 598
Weiner, S. A., 573
Weininger, S. J., 342, 707
Weinreb, A., 96, 111, 141, 775
Weiss, D. S., 306, 317
Weiss, K., 71, 581, 582
Weinstein, J., 621, 669
Welge, K. H., 206, 222
Weller, A., 97, 545
Wellington, J. L., 395

Wells, C. F., 595
Wells, C. H. J., 117
Welstead, W. J., 362
Wemer, H., 259
Wendisch, D., 471, 543
Wenkert, E., 537
Wenning, U., 171
Werbin, H., 613
Werner, G. K., 140
Werner, T. C., 104, 111
West, G., 655
West, M., 65
Westberg, H. H., 753
Westcott, L. D., 341, 586
Westfelt, L., 383
Wettack, F. S., 103, 131, 190, 300
Wettermark, G., 433
Wharton, J. H., 99
Wheeler, D. M. S., 456
Whipple, E. B., 501
Whistler, R. L., 700
White, A. H., 159
White, D. V., 686
White, E. H., 58, 301, 302
White, J. D., 444
White, J. M., 47, 199, 217, 731
White, L., 549
Whitesides, G. M., 476
Whitesides, T. H., 244, 554
Whitind, D. A., 389
Whitman, P. J., 342, 708
Whitten, D. G., 432, 591
Whitten, J. L., 15
Whittle, E., 224
Whittle, J. A., 340
Widman, R. P., 71
Wiebe, H. A., 557
Wiecko, J., 58, 301, 302
Wiersma, D. W., 724
Wierzchowski, K., 658
Wiesel, M., 468
Wijnen, M. H. J., 200
Wikholm, R. J., 362
Wiles, D. M., 784, 786
Wiley, M., 437
Wilkins, C. J., 276
Willard, J. E., 147, 148
Willets, F. W., 132
Willhalm, B., 615
Williams, A. H., 598
Williams, B. H., 698, 699
Williams, D. F., 115
Williams, H., 352, 734
Williams, J. L. R., 143, 181, 535
Williams, J. O., 544
Williams, J. R., 369, 689
Williams, W. M., 632, 655
Williamson, D. G., 222
Williamson, L. G., 121
Wilmarth, W. K., 239
Wilson, K. R., 45, 46
Wilson, K. W., 76
Wilson, R., 351, 564, 566, 576
Wilson, T., 133, 225, 623
Wilt, J. W., 706

Author Index

Wilzbach, K. E., 491, 495, 686
Windsor, M. W., 95
Winer, A. M., 128
Winnik, S. A., 320
Winter, R. E. K., 380
Winterle, J., 113
Witkop, B., 325, 399, 434, 505, 589, 590, 674, 736
Wolf, D., 128
Wolf, H. C., 94
Wolf, H. P., 581
Wolf, R. E., 323, 325, 577
Wong, K. T., 203
Wood, C. S., 533
Wood, G. O., 199
Wood, P. M., 205
Woodcock, D., 726
Woodward, R. B., 456
Woolley, G. R., 216
Worthington, N. W., 794
Wriede, P. A., 334, 337, 339
Wright, F. J., 203
Wright, H. E., 373, 642
Wright, M., 203
Wrighton, M., 113, 143, 180, 427, 445, 447
Wu, W.-T., 222
Wubbels, G. W., 513
Wuensch, K. H., 635
Wyatt, J. R., 14, 199
Wynberg, H., 420, 474, 475, 524, 558, 559, 687

Yagihara, T., 342, 708
Yagupol'skii, L. M., 10
Yakovenko, V. A., 215
Yakubenaite, V. R., 779

Yamada, K., 225, 375, 389, 409, 546, 694, 732
Yamada, M., 631
Yamada, S., 308, 636, 637
Yamaguchi, G., 277
Yamamoto, M., 394, 672
Yamamoto, N., 98
Yamanaka, C., 277
Yamawaki, K. R., 5
Yamazaki, H., 625
Yanagi, A., 642
Yandell, J. K., 273
Yang, K., 228
Yang, N. C., 313, 314, 389, 540
Yang, N. L., 765
Yang, S. S., 491, 540
Yaremko, R. V., 10
Yasnopol'skii, V. D., 797
Yasunobu, K., 600
Yates, P., 320, 333, 732
Yeats, R. B., 383
Yee, E. M., 135
Yelin, Z., 241
Yguerabide, J., 133
Yip, R. W., 120, 132, 394, 397
Yocem, P. N., 294
Yokoe, I., 636
Yokota, M., 618
Yokoyama, Y., 276
Yonemitsu, O., 505, 521, 623, 624
Yonezawa, T., 597
Yoshihara, K., 476, 566
Yoshikoshi, A., 451
Yoshimoto, T., 350
Yoshimura, N., 225, 600
Yoshioka, H., 455
Yoshioka, M., 225, 389, 546

Yoshioka, T., 290
Young, D. J., 287
Young, J. P., 264
Young, P. T., 229
Young, R. A., 74, 206
Younis, F. A., 351, 566, 576
Yu, S., 383
Yu, W. H. S., 200

Zabik, M. J., 351, 477, 525, 747
Zahradnik, R., 13
Zaichukova, N. A., 794
Zander, M., 102, 120
Zanker, V., 149, 637
Zannucci, J. S., 319, 519, 557
Zapevlova, N. P., 653
Zavgorodnyaya, L. N., 265
Zawadowska, J., 556
Zderic, J. A., 317, 383
Zelenskaya, L. G., 207, 677
Zenda, H., 468
Zich, W., 629
Ziegler, S. M., 120
Ziffer, H., 331, 369, 392, 689
Zimmerman, A. A., 377
Zimmerman, H. E., 302, 328, 329, 340, 362, 405, 450, 455, 456, 461, 463, 464, 465, 502, 655, 726
Zinato, E., 250
Zinser, D., 641
Zollweg, R. J., 283
Zugel, M., 139
Zuravich, R., 108
Zweig, A., 453, 515, 539

QD
601
A1
P46
v.2
1969/70

APR 17 1973